Reactions at Supports and Connections for a Three-Dime

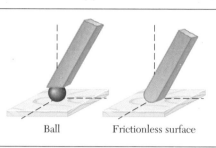

Ball   Frictionless surface   Force with known line of action (one unknown)

Cable

Force with known line of action (one unknown)

Roller on rough surface

Wheel on rail

Two force components

Rough surface

Ball and socket

Three force components

Universal joint   Three force components and one couple

Fixed support

Three force components and three couples

Hinge and bearing supporting radial load only

Two force components (and two couples; see page 191)

Pin and bracket   Hinge and bearing supporting axial thrust and radial load

Three force components (and two couples; see page 191)

242f

TENTH EDITION

VECTOR MECHANICS FOR ENGINEERS
Statics and Dynamics

Ferdinand P. Beer
Late of Lehigh University

E. Russell Johnston, Jr.
Late of University of Connecticut

David F. Mazurek
U.S. Coast Guard Academy

Phillip J. Cornwell
Rose-Hulman Institute of Technology

With the collaboration of
Brian P. Self
California Polytechnic State University—San Luis Obispo

McGraw Hill

Connect
Learn
Succeed™

The McGraw·Hill Companies

Mc Graw Hill — Connect Learn Succeed™

VECTOR MECHANICS FOR ENGINEERS: STATICS AND DYNAMICS, TENTH EDITION

Published by McGraw-Hill, a business unit of The McGraw-Hill Companies, Inc., 1221 Avenue of the Americas, New York, NY 10020. Copyright © 2013 by The McGraw-Hill Companies, Inc. All rights reserved. Printed in the United States of America. Previous editions © 2010, 2007, and 2004. No part of this publication may be reproduced or distributed in any form or by any means, or stored in a database or retrieval system, without the prior written consent of The McGraw-Hill Companies, Inc., including, but not limited to, in any network or other electronic storage or transmission, or broadcast for distance learning.

Some ancillaries, including electronic and print components, may not be available to customers outside the United States.

This book is printed on acid-free paper.

1 2 3 4 5 6 7 8 9 0 DOW/DOW 1 0 9 8 7 6 5 4 3 2

ISBN 978–0–07–339813–6
MHID 0–07–339813–6

Vice President, Editor-in-Chief: *Marty Lange*
Vice President, EDP: *Kimberly Meriwether David*
Senior Director of Development: *Kristine Tibbetts*
Global Publisher: *Raghothaman Srinivasan*
Editorial Director: *Michael Lange*
Executive Editor: *Bill Stenquist*
Developmental Editor: *Lora Neyens/Kathryn Neubauer*
Executive Marketing Manager: *Curt Reynolds*
Lead Project Manager: *Sheila M. Frank*
Senior Buyer: *Sherry L. Kane*
Senior Media Project Manager: *Tammy Juran*
Senior Designer: *Laurie B. Janssen*
Cover Designer: *Ron Bissell*
Cover Image: *©Andrea Capano/Alamy; ©Pictorial Press LTD/Alamy*
Lead Photo Research Coordinator: *Carrie K. Burger*
Photo Research: *Mary Reeg*
Compositor: *Aptara®, Inc.*
Typeface: *10.5/12 New Caledonia*
Printer: *R. R. Donnelley*

All credits appearing on page or at the end of the book are considered to be an extension of the copyright page.

Library of Congress Cataloging-in-Publication Data
Vector mechanics for engineers: statics and dynamics / Ferdinand Beer . . . [et al.]. — 10th ed.
   p. cm.
      Includes index.
         ISBN 978–0–07–339813–6 — ISBN 0–07–339813–6 (hard copy : alk. paper) 1. Mechanics, Applied.
      2. Vector analysis. 3. Statics. 4. Dynamics. I. Beer, Ferdinand P. (Ferdinand Pierre), 1915–2003.

TA350.V34   2013
620.1'05—dc23

2011034388

www.mhhe.com

About the Authors

As publishers of the books by Ferd Beer and Russ Johnston, we are often asked how they happened to write their books together with one of them at Lehigh and the other at the University of Connecticut.

The answer to this question is simple. Russ Johnston's first teaching appointment was in the Department of Civil Engineering and Mechanics at Lehigh University. There he met Ferd Beer, who had joined that department two years earlier and was in charge of the courses in mechanics.

Ferd was delighted to discover that the young man who had been hired chiefly to teach graduate structural engineering courses was not only willing but eager to help him reorganize the mechanics courses. Both believed that these courses should be taught from a few basic principles and that the various concepts involved would be best understood and remembered by the students if they were presented to them in a graphic way. Together they wrote lecture notes in statics and dynamics, to which they later added problems they felt would appeal to future engineers, and soon they produced the manuscript of the first edition of *Mechanics for Engineers* that was published in June 1956.

The second edition of *Mechanics for Engineers* and the first edition of *Vector Mechanics for Engineers* found Russ Johnston at Worcester Polytechnic Institute and the next editions at the University of Connecticut. In the meantime, both Ferd and Russ assumed administrative responsibilities in their departments, and both were involved in research, consulting, and supervising graduate students—Ferd in the area of stochastic processes and random vibrations and Russ in the area of elastic stability and structural analysis and design. However, their interest in improving the teaching of the basic mechanics courses had not subsided, and they both taught sections of these courses as they kept revising their texts and began writing the manuscript of the first edition of their *Mechanics of Materials* text.

Their collaboration spanned more than half a century and many successful revisions of all of their textbooks, and Ferd's and Russ's contributions to engineering education have earned them a number of honors and awards. They were presented with the Western Electric Fund Award for excellence in the instruction of engineering students by their respective regional sections of the American Society for Engineering Education, and they both received the Distinguished Educator Award from the Mechanics Division of the same society. Starting in 2001, the New Mechanics Educator Award of the Mechanics Division has been named in honor of the Beer and Johnston author team.

Ferdinand P. Beer. Born in France and educated in France and Switzerland, Ferd received an M.S. degree from the Sorbonne and an Sc.D. degree in theoretical mechanics from the University of Geneva. He came to the United States after serving in the French army during the early part of World War II and taught for four years at Williams College in the Williams-MIT joint arts and engineering program. Following his service at Williams College, Ferd joined the faculty of Lehigh University where he taught for thirty-seven years. He held several positions, including University Distinguished Professor and chairman of the Department of Mechanical Engineering and Mechanics, and in 1995 Ferd was awarded an honorary Doctor of Engineering degree by Lehigh University.

E. Russell Johnston, Jr. Born in Philadelphia, Russ holds a B.S. degree in civil engineering from the University of Delaware and an Sc.D. degree in the field of structural engineering from the Massachusetts Institute of Technology. He taught at Lehigh University and Worcester Polytechnic Institute before joining the faculty of the University of Connecticut where he held the position of chairman of the Department of Civil Engineering and taught for twenty-six years. In 1991 Russ received the Outstanding Civil Engineer Award from the Connecticut Section of the American Society of Civil Engineers.

David F. Mazurek. David holds a B.S. degree in ocean engineering and an M.S. degree in civil engineering from the Florida Institute of Technology and a Ph.D. degree in civil engineering from the University of Connecticut. He was employed by the Electric Boat Division of General Dynamics Corporation and taught at Lafayette College prior to joining the U.S. Coast Guard Academy, where he has been since 1990. He has served on the American Railway Engineering & Maintenance-of-Way Association's Committee 15—Steel Structures since 1991. Professional interests include bridge engineering, structural forensics, and blast-resistant design. He is a registered Professional Engineer in Connecticut and Pennsylvania.

Phillip J. Cornwell. Phil holds a B.S. degree in mechanical engineering from Texas Tech University and M.A. and Ph.D. degrees in mechanical and aerospace engineering from Princeton University. He is currently a professor of mechanical engineering and Vice President of Academic Affairs at Rose-Hulman Institute of Technology where he has taught since 1989. Phil received an SAE Ralph R. Teetor Educational Award in 1992, the Dean's Outstanding Teacher Award at Rose-Hulman in 2000, and the Board of Trustees' Outstanding Scholar Award at Rose-Hulman in 2001.

Brian P. Self. Brian obtained his B.S. and M.S. degrees in Engineering Mechanics from Virginia Tech, and his Ph.D. in Bioengineering from the University of Utah. He worked in the Air Force Research Laboratories before teaching at the U.S. Air Force Academy for seven years. Brian has taught in the Mechanical Engineering Department at Cal Poly, San Luis Obispo since 2006. He has been very active in the American Society of Engineering Education, serving on its Board from 2008–2010. With a team of five, Brian developed the Dynamics Concept Inventory to help assess student conceptual understanding. His professional interests include educational research, aviation physiology, and biomechanics.

Brief Contents

Contents

1 Introduction   1

2 Statics of Particles   14

3 Rigid Bodies: Equivalent Systems of Forces   74

4 Equilibrium of Rigid Bodies   158

5 Distributed Forces: Centroids and Centers of Gravity 218

6 Analysis of Structures 282

9 Distributed Forces: Moments of Inertia 468

10 Method of Virtual Work 556

13 Kinetics of Particles: Energy and Momentum Methods 762

14 Systems of Particles 866

15 Kinematics of Rigid Bodies   926

16 Plane Motion of Rigid Bodies: Forces and Accelerations   1040

17 Plane Motion of Rigid Bodies: Energy and Momentum Methods 1104

18 Kinetics of Rigid Bodies in Three Dimensions 1172

19 Mechanical Vibrations   1240

Preface

OBJECTIVES

The main objective of a first course in mechanics should be to develop in the engineering student the ability to analyze any problem in a simple and logical manner and to apply to its solution a few, well-understood, basic principles. It is hoped that this text, as well as the preceding volume, *Vector Mechanics for Engineers: Statics*, will help the instructor achieve this goal.†

GENERAL APPROACH

Vector algebra was introduced at the beginning of the first volume and is used in the presentation of the basic principles of statics, as well as in the solution of many problems, particularly three-dimensional problems. Similarly, the concept of vector differentiation will be introduced early in this volume, and vector analysis will be used throughout the presentation of dynamics. This approach leads to more concise derivations of the fundamental principles of mechanics. It also makes it possible to analyze many problems in kinematics and kinetics which could not be solved by scalar methods. The emphasis in this text, however, remains on the correct understanding of the principles of mechanics and on their application to the solution of engineering problems, and vector analysis is presented chiefly as a convenient tool.‡

Practical Applications Are Introduced Early.
One of the characteristics of the approach used in this book is that mechanics of *particles* is clearly separated from the mechanics of *rigid bodies*. This approach makes it possible to consider simple practical applications at an early stage and to postpone the introduction of the more difficult concepts. For example:

- In *Statics*, the statics of particles is treated first, and the principle of equilibrium of a particle was immediately applied to practical situations involving only concurrent forces. The statics of rigid bodies is considered later, at which time the vector and scalar products of two vectors were introduced and used to define the moment of a force about a point and about an axis.
- In *Dynamics*, the same division is observed. The basic concepts of force, mass, and acceleration, of work and energy, and of impulse and momentum are introduced and first applied to problems involving only particles. Thus, students can familiarize

---

---

†Both texts also are available in a single volume, *Vector Mechanics for Engineers: Statics and Dynamics*, tenth edition.

‡In a parallel text, *Mechanics for Engineers: Dynamics*, fifth edition, the use of vector algebra is limited to the addition and subtraction of vectors, and vector differentiation is omitted.

themselves with the three basic methods used in dynamics and learn their respective advantages before facing the difficulties associated with the motion of rigid bodies.

New Concepts Are Introduced in Simple Terms.   Since this text is designed for the first course in dynamics, new concepts are presented in simple terms and every step is explained in detail. On the other hand, by discussing the broader aspects of the problems considered, and by stressing methods of general applicability, a definite maturity of approach has been achieved. For example, the concept of potential energy is discussed in the general case of a conservative force. Also, the study of the plane motion of rigid bodies is designed to lead naturally to the study of their general motion in space. This is true in kinematics as well as in kinetics, where the principle of equivalence of external and effective forces is applied directly to the analysis of plane motion, thus facilitating the transition to the study of three-dimensional motion.

Fundamental Principles Are Placed in the Context of Simple Applications.   The fact that mechanics is essentially a *deductive* science based on a few fundamental principles is stressed. Derivations have been presented in their logical sequence and with all the rigor warranted at this level. However, the learning process being largely *inductive*, simple applications are considered first. For example:

- The kinematics of particles (Chap. 11) precedes the kinematics of rigid bodies (Chap. 15).
- The fundamental principles of the kinetics of rigid bodies are first applied to the solution of two-dimensional problems (Chaps. 16 and 17), which can be more easily visualized by the student, while three-dimensional problems are postponed until Chap. 18.

The Presentation of the Principles of Kinetics Is Unified.   The tenth edition of *Vector Mechanics for Engineers* retains the unified presentation of the principles of kinetics which characterized the previous nine editions. The concepts of linear and angular momentum are introduced in Chap. 12 so that Newton's second law of motion can be presented not only in its conventional form $\mathbf{F} = m\mathbf{a}$, but also as a law relating, respectively, the sum of the forces acting on a particle and the sum of their moments to the rates of change of the linear and angular momentum of the particle. This makes possible an earlier introduction of the principle of conservation of angular momentum and a more meaningful discussion of the motion of a particle under a central force (Sec. 12.9). More importantly, this approach can be readily extended to the study of the motion of a system of particles (Chap. 14) and leads to a more concise and unified treatment of the kinetics of rigid bodies in two and three dimensions (Chaps. 16 through 18).

Free-Body Diagrams Are Used Both to Solve Equilibrium Problems and to Express the Equivalence of Force Systems. Free-body diagrams were introduced early in statics, and their importance was emphasized throughout. They were used not only to solve equilibrium problems but also to express the equivalence of two

---

17.1   INTRODUCTION

In this chapter the method of work and energy and the method of impulse and momentum will be used to analyze the plane motion of rigid bodies and of systems of rigid bodies.

The method of work and energy will be considered first. In Secs. 17.2 through 17.5, the work of a force and of a couple will be defined, and an expression for the kinetic energy of a rigid body in plane motion will be obtained. The principle of work and energy will then be used to solve problems involving displacements and velocities. In Sec. 17.6, the principle of conservation of energy will be applied to the solution of a variety of engineering problems.

In the second part of the chapter, the principle of impulse and momentum will be applied to the solution of problems involving velocities and time (Secs. 17.8 and 17.9) and the concept of conservation of angular momentum will be introduced and discussed (Sec. 17.10).

In the last part of the chapter (Secs. 17.11 and 17.12), problems involving the eccentric impact of rigid bodies will be considered. As was done in Chap. 13, where we analyzed the impact of particles, the coefficient of restitution between the colliding bodies will be used together with the principle of impulse and momentum in the solution of impact problems. It will also be shown that the method used is applicable not only when the colliding bodies move freely after the impact but also when the bodies are partially constrained in their motion.

17.2   PRINCIPLE OF WORK AND ENERGY FOR A RIGID BODY

The principle of work and energy will now be used to analyze the plane motion of rigid bodies. As was pointed out in Chap. 13, the method of work and energy is particularly well adapted to the solution of problems involving velocities and displacements. Its main advantage resides in the fact that the work of forces and the kinetic energy of particles are scalar quantities.

In order to apply the principle of work and energy to the analysis of the motion of a rigid body, it will again be assumed that the rigid body is made of a large number n of particles of mass Δm_i. Recalling Eq. (14.30) of Sec. 14.8, we write

$$T_1 + U_{1\rightarrow2} = T_2 \qquad (17.1)$$

where T_1, T_2 = initial and final values of total kinetic energy of particles forming the rigid body

$U_{1\rightarrow2}$ = work of all forces acting on various particles of the body

The total kinetic energy

$$T = \frac{1}{2} \sum_{i=1}^{n} \Delta m_i v_i^2 \qquad (17.2)$$

is obtained by adding positive scalar quantities and is itself a positive scalar quantity. You will see later how T can be determined for various types of motion of a rigid body.

systems of forces or, more generally, of two systems of vectors. The advantage of this approach becomes apparent in the study of the dynamics of rigid bodies, where it is used to solve three-dimensional as well as two-dimensional problems. By placing the emphasis on "free-body-diagram equations" rather than on the standard algebraic equations of motion, a more intuitive and more complete understanding of the fundamental principles of dynamics can be achieved. This approach, which was first introduced in 1962 in the first edition of *Vector Mechanics for Engineers*, has now gained wide acceptance among mechanics teachers in this country. It is, therefore, used in preference to the method of dynamic equilibrium and to the equations of motion in the solution of all sample problems in this book.

A Careful Balance between SI and U.S. Customary Units Is Consistently Maintained.

Because of the current trend in the American government and industry to adopt the international system of units (SI metric units), the SI units most frequently used in mechanics are introduced in Chap. 1 and are used throughout the text. Approximately half of the sample problems and 60 percent of the homework problems are stated in these units, while the remainder are in U.S. customary units. The authors believe that this approach will best serve the need of the students, who, as engineers, will have to be conversant with both systems of units.

It also should be recognized that using both SI and U.S. customary units entails more than the use of conversion factors. Since the SI system of units is an absolute system based on the units of time, length, and mass, whereas the U.S. customary system is a gravitational system based on the units of time, length, and force, different approaches are required for the solution of many problems. For example, when SI units are used, a body is generally specified by its mass expressed in kilograms; in most problems of statics it will be necessary to determine the weight of the body in newtons, and an additional calculation will be required for this purpose. On the other hand, when U.S. customary units are used, a body is specified by its weight in pounds and, in dynamics problems, an additional calculation will be required to determine its mass in slugs (or lb · s²/ft). The authors, therefore, believe that problem assignments should include both systems of units.

The *Instructor's and Solutions Manual* provides six different lists of assignments so that an equal number of problems stated in SI units and in U.S. customary units can be selected. If so desired, two complete lists of assignments can also be selected with up to 75 percent of the problems stated in SI units.

Optional Sections Offer Advanced or Specialty Topics.

A large number of optional sections have been included. These sections are indicated by asterisks and thus are easily distinguished from those which form the core of the basic dynamics course. They can be omitted without prejudice to the understanding of the rest of the text.

The topics covered in the optional sections include graphical methods for the solution of rectilinear-motion problems, the trajectory

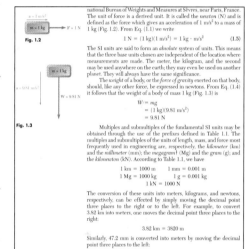

1.3 SYSTEMS OF UNITS

With the four fundamental concepts introduced in the preceding section are associated the so-called *kinetic units*, i.e., the units of *length, time, mass,* and *force*. These units cannot be chosen independently if Eq. (1.1) is to be satisfied. Three of the units may be defined arbitrarily; they are then referred to as *basic units*. The fourth unit, however, must be chosen in accordance with Eq. (1.1) and is referred to as a *derived unit*. Kinetic units selected in this way are said to form a *consistent system of units.*

International System of Units (SI Units†). In this system, which will be in universal use after the United States has completed its conversion to SI units, the base units are the units of length, mass, and time, and they are called, respectively, the *meter* (m), the *kilogram* (kg), and the *second* (s). All three are arbitrarily defined. The second,

† SI stands for *Système International d'Unités* (French).

national Bureau of Weights and Measures at Sèvres, near Paris, France. The unit of force is a derived unit. It is called the *newton* (N) and is defined as the force which gives an acceleration of 1 m/s² to a mass of 1 kg (Fig. 1.2). From Eq. (1.1) we write

$$1 \text{ N} = (1 \text{ kg})(1 \text{ m/s}^2) = 1 \text{ kg} \cdot \text{m/s}^2 \tag{1.5}$$

The SI units are said to form an *absolute* system of units. This means that the three base units chosen are independent of the location where measurements are made. The meter, the kilogram, and the second may be used anywhere on the earth; they may even be used on another planet. They will always have the same significance.

The *weight* of a body, or the *force of gravity* exerted on that body, should, like any other force, be expressed in newtons. From Eq. (1.4) it follows that the weight of a body of mass 1 kg (Fig. 1.3) is

$$W = mg = (1 \text{ kg})(9.81 \text{ m/s}^2) = 9.81 \text{ N}$$

Multiples and submultiples of the fundamental SI units may be obtained through the use of the prefixes defined in Table 1.1. The multiples and submultiples of the units of length, mass, and force most frequently used in engineering are, respectively, the *kilometer* (km) and the *millimeter* (mm); the *megagram†* (Mg) and the *gram* (g); and the *kilonewton* (kN). According to Table 1.1, we have

1 km = 1000 m	1 mm = 0.001 m
1 Mg = 1000 kg	1 g = 0.001 kg
1 kN = 1000 N	

The conversion of these units into meters, kilograms, and newtons, respectively, can be effected by moving the decimal point three places to the right or to the left. For example, to convert 3.82 km into meters, one moves the decimal point three places to the right:

$$3.82 \text{ km} = 3820 \text{ m}$$

Similarly, 47.2 mm is converted into meters by moving the decimal point three places to the left:

$$47.2 \text{ mm} = 0.0472 \text{ m}$$

Fig. 1.2

Fig. 1.3

of a particle under a central force, the deflection of fluid streams, problems involving jet and rocket propulsion, the kinematics and kinetics of rigid bodies in three dimensions, damped mechanical vibrations, and electrical analogues. These topics will be found of particular interest when dynamics is taught in the junior year.

The material presented in the text and most of the problems requires no previous mathematical knowledge beyond algebra, trigonometry, elementary calculus, and the elements of vector algebra presented in Chaps. 2 and 3 of the volume on statics.† However, special problems are included, which make use of a more advanced knowledge of calculus, and certain sections, such as Secs. 19.8 and 19.9 on damped vibrations, should be assigned only if students possess the proper mathematical background. In portions of the text using elementary calculus, a greater emphasis is placed on the correct understanding and application of the concepts of differentiation and integration, than on the nimble manipulation of mathematical formulas. In this connection, it should be mentioned that the determination of the centroids of composite areas precedes the calculation of centroids by integration, thus making it possible to establish the concept of moment of area firmly before introducing the use of integration.

---

†Some useful definitions and properties of vector algebra have been summarized in Appendix A at the end of this volume for the convenience of the reader. Also, Secs. 9.11 through 9.18 of the volume on statics, which deal with the moments of inertia of masses, have been reproduced in Appendix B.

Guided Tour

Chapter Introduction. Each chapter begins with an introductory section setting the purpose and goals of the chapter and describing in simple terms the material to be covered and its application to the solution of engineering problems. New chapter outlines provide students with a preview of chapter topics.

Chapter Lessons. The body of the text is divided into units, each consisting of one or several theory sections, one or several sample problems, and a large number of problems to be assigned. Each unit corresponds to a well-defined topic and generally can be covered in one lesson. In a number of cases, however, the instructor will find it desirable to devote more than one lesson to a given topic. *The Instructor's and Solutions Manual* contains suggestions on the coverage of each lesson.

Sample Problems. The sample problems are set up in much the same form that students will use when solving the assigned problems. They thus serve the double purpose of amplifying the text and demonstrating the type of neat, orderly work that students should cultivate in their own solutions.

Solving Problems on Your Own. A section entitled *Solving Problems on Your Own* is included for each lesson, between the sample problems and the problems to be assigned. The purpose of these sections is to help students organize in their own minds the preceding theory of the text and the solution methods of the sample problems so that they can more successfully solve the homework problems. Also included in these sections are specific suggestions and strategies that will enable the students to more efficiently attack any assigned problems.

Homework Problem Sets. Most of the problems are of a practical nature and should appeal to engineering students. They are primarily designed, however, to illustrate the material presented in the text and to help students understand the principles of mechanics. The problems are grouped according to the portions of material they illustrate and are arranged in order of increasing difficulty. Problems requiring special attention are indicated by asterisks. Answers to 70 percent of the problems are given at the end of the book. Problems for which the answers are given are set in straight type in the text, while problems for which no answer is given are set in italic.

Chapter Review and Summary. Each chapter ends with a review and summary of the material covered in that chapter. Marginal notes are used to help students organize their review work, and cross-references have been included to help them find the portions of material requiring their special attention.

Review Problems. A set of review problems is included at the end of each chapter. These problems provide students further opportunity to apply the most important concepts introduced in the chapter.

Computer Problems. Each chapter includes a set of problems designed to be solved with computational software. Many of these problems provide an introduction to the design process. For example, they may involve the determination of the motion of a particle under initial conditions, the kinematic or kinetic analysis of mechanisms in successive positions, or the numerical integration of various equations of motion. Developing the algorithm required to solve a given mechanics problem will benefit the students in two different ways: (1) It will help them gain a better understanding of the mechanics principles involved; (2) it will provide them with an opportunity to apply their computer skills to the solution of a meaningful engineering problem.

Concept Questions. Educational research has shown that students can often choose appropriate equations and solve algorithmic problems without having a strong conceptual understanding of mechanics principles.† To help assess and develop student conceptual understanding, we have included Concept Questions, which are multiple choice problems that require few, if any, calculations. Each possible incorrect answer typically represents a common misconception (e.g., students often think that a vehicle moving in a curved path at constant speed has zero acceleration). Students are encouraged to solve these problems using the principles and techniques discussed in the text and to use these principles to help them develop their intuition. Mastery and discussion of these Concept Questions will deepen students' conceptual understanding and help them to solve dynamics problems.

Free Body and Impulse-Momentum Practice Problems. Drawing diagrams correctly is a critical step in solving kinetics problems in dynamics. A new type of problem has been added to the text to emphasize the importance of drawing these diagrams. In Chaps. 12 and 16 the Free Body Practice Problems require students to draw a free-body diagram (FBD) showing the applied forces and an equivalent diagram called a "kinetic diagram" (KD) showing $m\mathbf{a}$ or its components and $\bar{\boldsymbol{\alpha}}$. These diagrams provide students with a pictorial representation of Newton's second law and are critical in helping students to correctly solve kinetic problems. In Chaps. 13 and 17 the Impulse-Momentum Practice Problems require students to draw diagrams showing the momenta of the bodies before impact, the impulses exerted on the body during impact, and the final momenta of the bodies. The answers to all of these questions are provided at **www.mhhe.com/beerjohnston.**

---

†Hestenes, D., Wells, M., and Swakhamer, G (1992). The force concept inventory. *The Physics Teacher*, 30: 141–158.

Streveler, R. A., Litzinger, T. A., Miller, R. L., and Steif, P. S. (2008). Learning conceptual knowledge in the engineering sciences: Overview and future research directions, *JEE*, 279–294.

What Resources Support This Textbook?

Instructor's and Solutions Manual. *The Instructor's and Solutions Manual* that accompanies the tenth edition features typeset, one-per-page solutions to the end of chapter problems. This Manual also features a number of tables designed to assist instructors in creating a schedule of assignments for their course. The various topics covered in the text have been listed in Table I and a suggested number of periods to be spent on each topic has been indicated. Table II prepares a brief description of all groups of problems and a classification of the problems in each group according to the units used. Sample lesson schedules are shown in Tables III, IV, and V, together with various alternative lists of assigned homework problems.

McGraw-Hill Connect Engineering McGraw-Hill Connect Engineering is a web-based assignment and assessment platform that gives students the means to better connect with their coursework, their instructors, and the important concepts that they will need to know for success now and in the future. With Connect Engineering, instructors can deliver assignments, quizzes, and tests easily online. Students can practice important skills at their own pace and on their own schedule.

Connect Engineering for *Vector Mechanics for Engineers* is available at **www.mhhe.com/beerjohnston** and includes algorithmic problems from the text, Lecture PowerPoints, an image bank, and animations.

Hands-on Mechanics. Hands-on Mechanics is a website designed for instructors who are interested in incorporating three-dimensional, hands-on teaching aids into their lectures. Developed through a partnership between the McGraw-Hill Engineering Team and the Department of Civil and Mechanical Engineering at the United States Military Academy at West Point, this website not only provides detailed instructions for how to build 3-D teaching tools using materials found in any lab or local hardware store, but also provides a community where educators can share ideas, trade best practices, and submit their own original demonstrations for posting on the site. Visit **www.handsonmechanics.com**.

CourseSmart. This text is offered through CourseSmart for both instructors and students. CourseSmart is an online browser where students can purchase access to this and other McGraw-Hill textbooks in a digital format. Through their browser, students can access the complete text online at almost half the cost of a traditional text. Purchasing the eTextbook also allows students to take advantage of CourseSmart's web tools for learning, which include full text search, notes and highlighting, and e-mail tools for sharing notes among classmates. To learn more about CourseSmart options, contact your sales representative or visit **www.coursesmart.com**.

Introduction

1.1 WHAT IS MECHANICS?

Mechanics can be defined as that science which describes and predicts the conditions of rest or motion of bodies under the action of forces. It is divided into three parts: mechanics of *rigid bodies*, mechanics of *deformable bodies*, and mechanics of *fluids*.

The mechanics of rigid bodies is subdivided into *statics* and *dynamics*, the former dealing with bodies at rest, the latter with bodies in motion. In this part of the study of mechanics, bodies are assumed to be perfectly rigid. Actual structures and machines, however, are never absolutely rigid and deform under the loads to which they are subjected. But these deformations are usually small and do not appreciably affect the conditions of equilibrium or motion of the structure under consideration. They are important, though, as far as the resistance of the structure to failure is concerned and are studied in mechanics of materials, which is a part of the mechanics of deformable bodies. The third division of mechanics, the mechanics of fluids, is subdivided into the study of *incompressible fluids* and of *compressible fluids*. An important subdivision of the study of incompressible fluids is *hydraulics*, which deals with problems involving water.

Mechanics is a physical science, since it deals with the study of physical phenomena. However, some associate mechanics with mathematics, while many consider it as an engineering subject. Both these views are justified in part. Mechanics is the foundation of most engineering sciences and is an indispensable prerequisite to their study. However, it does not have the *empiricism* found in some engineering sciences, i.e., it does not rely on experience or observation alone; by its rigor and the emphasis it places on deductive reasoning it resembles mathematics. But, again, it is not an *abstract* or even a *pure* science; mechanics is an *applied* science. The purpose of mechanics is to explain and predict physical phenomena and thus to lay the foundations for engineering applications.

1.2 FUNDAMENTAL CONCEPTS AND PRINCIPLES

Although the study of mechanics goes back to the time of Aristotle (384–322 B.C.) and Archimedes (287–212 B.C.), one has to wait until Newton (1642–1727) to find a satisfactory formulation of its fundamental principles. These principles were later expressed in a modified form by d'Alembert, Lagrange, and Hamilton. Their validity remained unchallenged, however, until Einstein formulated his *theory of relativity* (1905). While its limitations have now been recognized, *newtonian mechanics* still remains the basis of today's engineering sciences.

The basic concepts used in mechanics are *space, time, mass,* and *force*. These concepts cannot be truly defined; they should be accepted on the basis of our intuition and experience and used as a mental frame of reference for our study of mechanics.

The concept of *space* is associated with the notion of the position of a point P. The position of P can be defined by three lengths measured from a certain reference point, or *origin*, in three given directions. These lengths are known as the *coordinates* of P.

To define an event, it is not sufficient to indicate its position in space. The *time* of the event should also be given.

The concept of *mass* is used to characterize and compare bodies on the basis of certain fundamental mechanical experiments. Two bodies of the same mass, for example, will be attracted by the earth in the same manner; they will also offer the same resistance to a change in translational motion.

A *force* represents the action of one body on another. It can be exerted by actual contact or at a distance, as in the case of gravitational forces and magnetic forces. A force is characterized by its *point of application*, its *magnitude*, and its *direction*; a force is represented by a *vector* (Sec. 2.3).

In newtonian mechanics, space, time, and mass are absolute concepts, independent of each other. (This is not true in *relativistic mechanics*, where the time of an event depends upon its position, and where the mass of a body varies with its velocity.) On the other hand, the concept of force is not independent of the other three. Indeed, one of the fundamental principles of newtonian mechanics listed below indicates that the resultant force acting on a body is related to the mass of the body and to the manner in which its velocity varies with time.

You will study the conditions of rest or motion of particles and rigid bodies in terms of the four basic concepts we have introduced. By *particle* we mean a very small amount of matter which may be assumed to occupy a single point in space. A *rigid body* is a combination of a large number of particles occupying fixed positions with respect to each other. The study of the mechanics of particles is obviously a prerequisite to that of rigid bodies. Besides, the results obtained for a particle can be used directly in a large number of problems dealing with the conditions of rest or motion of actual bodies.

The study of elementary mechanics rests on six fundamental principles based on experimental evidence.

The Parallelogram Law for the Addition of Forces.   This states that two forces acting on a particle may be replaced by a single force, called their *resultant*, obtained by drawing the diagonal of the parallelogram which has sides equal to the given forces (Sec. 2.2).

The Principle of Transmissibility.   This states that the conditions of equilibrium or of motion of a rigid body will remain unchanged if a force acting at a given point of the rigid body is replaced by a force of the same magnitude and same direction, but acting at a different point, provided that the two forces have the same line of action (Sec. 3.3).

Newton's Three Fundamental Laws.   Formulated by Sir Isaac Newton in the latter part of the seventeenth century, these laws can be stated as follows:

FIRST LAW.   If the resultant force acting on a particle is zero, the particle will remain at rest (if originally at rest) or will move with constant speed in a straight line (if originally in motion) (Sec. 2.10).

SECOND LAW. If the resultant force acting on a particle is not zero, the particle will have an acceleration proportional to the magnitude of the resultant and in the direction of this resultant force.

As you will see in Sec. 12.2, this law can be stated as

$$\mathbf{F} = m\mathbf{a} \tag{1.1}$$

where $\mathbf{F}$, m, and $\mathbf{a}$ represent, respectively, the resultant force acting on the particle, the mass of the particle, and the acceleration of the particle, expressed in a consistent system of units.

THIRD LAW. The forces of action and reaction between bodies in contact have the same magnitude, same line of action, and opposite sense (Sec. 6.1).

Newton's Law of Gravitation. This states that two particles of mass M and m are mutually attracted with equal and opposite forces $\mathbf{F}$ and $-\mathbf{F}$ (Fig. 1.1) of magnitude F given by the formula

$$F = G\frac{Mm}{r^2} \tag{1.2}$$

where r = distance between the two particles

G = universal constant called the *constant of gravitation*

Newton's law of gravitation introduces the idea of an action exerted at a distance and extends the range of application of Newton's third law: the action $\mathbf{F}$ and the reaction $-\mathbf{F}$ in Fig. 1.1 are equal and opposite, and they have the same line of action.

A particular case of great importance is that of the attraction of the earth on a particle located on its surface. The force $\mathbf{F}$ exerted by the earth on the particle is then defined as the *weight* $\mathbf{W}$ of the particle. Taking M equal to the mass of the earth, m equal to the mass of the particle, and r equal to the radius R of the earth, and introducing the constant

$$g = \frac{GM}{R^2} \tag{1.3}$$

the magnitude W of the weight of a particle of mass m may be expressed as†

$$W = mg \tag{1.4}$$

The value of R in formula (1.3) depends upon the elevation of the point considered; it also depends upon its latitude, since the earth is not truly spherical. The value of g therefore varies with the position of the point considered. As long as the point actually remains on the surface of the earth, it is sufficiently accurate in most engineering computations to assume that g equals 9.81 m/s^2 or 32.2 ft/s^2.

Fig. 1.1

Photo 1.1 When in earth orbit, people and objects are said to be *weightless* even though the gravitational force acting is approximately 90% of that experienced on the surface of the earth. This apparent contradiction will be resolved in Chapter 12 when we apply Newton's second law to the motion of particles.

---

†A more accurate definition of the weight $\mathbf{W}$ should take into account the rotation of the earth.

The principles we have just listed will be introduced in the course of our study of mechanics as they are needed. The study of the statics of particles carried out in Chap. 2 will be based on the parallelogram law of addition and on Newton's first law alone. The principle of transmissibility will be introduced in Chap. 3 as we begin the study of the statics of rigid bodies, and Newton's third law in Chap. 6 as we analyze the forces exerted on each other by the various members forming a structure. In the study of dynamics, Newton's second law and Newton's law of gravitation will be introduced. It will then be shown that Newton's first law is a particular case of Newton's second law (Sec. 12.2) and that the principle of transmissibility could be derived from the other principles and thus eliminated (Sec. 16.5). In the meantime, however, Newton's first and third laws, the parallelogram law of addition, and the principle of transmissibility will provide us with the necessary and sufficient foundation for the entire study of the statics of particles, rigid bodies, and systems of rigid bodies.

As noted earlier, the six fundamental principles listed above are based on experimental evidence. Except for Newton's first law and the principle of transmissibility, they are independent principles which cannot be derived mathematically from each other or from any other elementary physical principle. On these principles rests most of the intricate structure of newtonian mechanics. For more than two centuries a tremendous number of problems dealing with the conditions of rest and motion of rigid bodies, deformable bodies, and fluids have been solved by applying these fundamental principles. Many of the solutions obtained could be checked experimentally, thus providing a further verification of the principles from which they were derived. It is only in the twentieth century that Newton's mechanics was found at fault, in the study of the motion of atoms and in the study of the motion of certain planets, where it must be supplemented by the theory of relativity. But on the human or engineering scale, where velocities are small compared with the speed of light, Newton's mechanics has yet to be disproved.

1.3 SYSTEMS OF UNITS

With the four fundamental concepts introduced in the preceding section are associated the so-called *kinetic units*, i.e., the units of *length, time, mass,* and *force*. These units cannot be chosen independently if Eq. (1.1) is to be satisfied. Three of the units may be defined arbitrarily; they are then referred to as *basic units*. The fourth unit, however, must be chosen in accordance with Eq. (1.1) and is referred to as a *derived unit*. Kinetic units selected in this way are said to form a *consistent system of units*.

International System of Units (SI Units†). In this system, which will be in universal use after the United States has completed its conversion to SI units, the base units are the units of length, mass, and time, and they are called, respectively, the *meter* (m), the *kilogram* (kg), and the *second* (s). All three are arbitrarily defined. The second,

†SI stands for *Système International d'Unités* (French).

which was originally chosen to represent 1/86 400 of the mean solar day, is now defined as the duration of 9 192 631 770 cycles of the radiation corresponding to the transition between two levels of the fundamental state of the cesium-133 atom. The meter, originally defined as one ten-millionth of the distance from the equator to either pole, is now defined as 1 650 763.73 wavelengths of the orange-red light corresponding to a certain transition in an atom of krypton-86. The kilogram, which is approximately equal to the mass of 0.001 m^3 of water, is defined as the mass of a platinum-iridium standard kept at the International Bureau of Weights and Measures at Sèvres, near Paris, France. The unit of force is a derived unit. It is called the *newton* (N) and is defined as the force which gives an acceleration of 1 m/s² to a mass of 1 kg (Fig. 1.2). From Eq. (1.1) we write

Fig. 1.2

$$1 \text{ N} = (1 \text{ kg})(1 \text{ m/s}^2) = 1 \text{ kg} \cdot \text{m/s}^2 \qquad (1.5)$$

The SI units are said to form an *absolute* system of units. This means that the three base units chosen are independent of the location where measurements are made. The meter, the kilogram, and the second may be used anywhere on the earth; they may even be used on another planet. They will always have the same significance.

The *weight* of a body, or the *force of gravity* exerted on that body, should, like any other force, be expressed in newtons. From Eq. (1.4) it follows that the weight of a body of mass 1 kg (Fig. 1.3) is

Fig. 1.3

$$W = mg$$
$$= (1 \text{ kg})(9.81 \text{ m/s}^2)$$
$$= 9.81 \text{ N}$$

Multiples and submultiples of the fundamental SI units may be obtained through the use of the prefixes defined in Table 1.1. The multiples and submultiples of the units of length, mass, and force most frequently used in engineering are, respectively, the *kilometer* (km) and the *millimeter* (mm); the *megagram*† (Mg) and the *gram* (g); and the *kilonewton* (kN). According to Table 1.1, we have

$$1 \text{ km} = 1000 \text{ m} \qquad 1 \text{ mm} = 0.001 \text{ m}$$
$$1 \text{ Mg} = 1000 \text{ kg} \qquad 1 \text{ g} = 0.001 \text{ kg}$$
$$1 \text{ kN} = 1000 \text{ N}$$

The conversion of these units into meters, kilograms, and newtons, respectively, can be effected by simply moving the decimal point three places to the right or to the left. For example, to convert 3.82 km into meters, one moves the decimal point three places to the right:

$$3.82 \text{ km} = 3820 \text{ m}$$

Similarly, 47.2 mm is converted into meters by moving the decimal point three places to the left:

$$47.2 \text{ mm} = 0.0472 \text{ m}$$

†Also known as a *metric ton*.

TABLE 1.1 **SI Prefixes** 1.3 Systems of Units **7**

Multiplication Factor	Prefix†	Symbol
$1\ 000\ 000\ 000\ 000 = 10^{12}$	tera	T
$1\ 000\ 000\ 000 = 10^{9}$	giga	G
$1\ 000\ 000 = 10^{6}$	mega	M
$1\ 000 = 10^{3}$	kilo	k
$100 = 10^{2}$	hecto‡	h
$10 = 10^{1}$	deka‡	da
$0.1 = 10^{-1}$	deci‡	d
$0.01 = 10^{-2}$	centi‡	c
$0.001 = 10^{-3}$	milli	m
$0.000\ 001 = 10^{-6}$	micro	μ
$0.000\ 000\ 001 = 10^{-9}$	nano	n
$0.000\ 000\ 000\ 001 = 10^{-12}$	pico	p
$0.000\ 000\ 000\ 000\ 001 = 10^{-15}$	femto	f
$0.000\ 000\ 000\ 000\ 000\ 001 = 10^{-18}$	atto	a

†The first syllable of every prefix is accented so that the prefix will retain its identity. Thus, the preferred pronunciation of kilometer places the accent on the first syllable, not the second.

‡The use of these prefixes should be avoided, except for the measurement of areas and volumes and for the nontechnical use of centimeter, as for body and clothing measurements.

Using scientific notation, one may also write

$$3.82 \text{ km} = 3.82 \times 10^{3} \text{ m}$$
$$47.2 \text{ mm} = 47.2 \times 10^{-3} \text{ m}$$

The multiples of the unit of time are the *minute* (min) and the *hour* (h). Since 1 min = 60 s and 1 h = 60 min = 3600 s, these multiples cannot be converted as readily as the others.

By using the appropriate multiple or submultiple of a given unit, one can avoid writing very large or very small numbers. For example, one usually writes 427.2 km rather than 427 200 m, and 2.16 mm rather than 0.002 16 m.†

Units of Area and Volume. The unit of area is the *square meter* (m^2), which represents the area of a square of side 1 m; the unit of volume is the *cubic meter* (m^3), equal to the volume of a cube of side 1 m. In order to avoid exceedingly small or large numerical values in the computation of areas and volumes, one uses systems of subunits obtained by respectively squaring and cubing not only the millimeter but also two intermediate submultiples of the meter, namely, the *decimeter* (dm) and the *centimeter* (cm). Since, by definition,

$$1 \text{ dm} = 0.1 \text{ m} = 10^{-1} \text{ m}$$
$$1 \text{ cm} = 0.01 \text{ m} = 10^{-2} \text{ m}$$
$$1 \text{ mm} = 0.001 \text{ m} = 10^{-3} \text{ m}$$

†It should be noted that when more than four digits are used on either side of the decimal point to express a quantity in SI units—as in 427 200 m or 0.002 16 m—spaces, never commas, should be used to separate the digits into groups of three. This is to avoid confusion with the comma used in place of a decimal point, which is the convention in many countries.

the submultiples of the unit of area are

$$1 \text{ dm}^2 = (1 \text{ dm})^2 = (10^{-1} \text{ m})^2 = 10^{-2} \text{ m}^2$$
$$1 \text{ cm}^2 = (1 \text{ cm})^2 = (10^{-2} \text{ m})^2 = 10^{-4} \text{ m}^2$$
$$1 \text{ mm}^2 = (1 \text{ mm})^2 = (10^{-3} \text{ m})^2 = 10^{-6} \text{ m}^2$$

and the submultiples of the unit of volume are

$$1 \text{ dm}^3 = (1 \text{ dm})^3 = (10^{-1} \text{ m})^3 = 10^{-3} \text{ m}^3$$
$$1 \text{ cm}^3 = (1 \text{ cm})^3 = (10^{-2} \text{ m})^3 = 10^{-6} \text{ m}^3$$
$$1 \text{ mm}^3 = (1 \text{ mm})^3 = (10^{-3} \text{ m})^3 = 10^{-9} \text{ m}^3$$

It should be noted that when the volume of a liquid is being measured, the cubic decimeter (dm^3) is usually referred to as a *liter* (L).

Other derived SI units used to measure the moment of a force, the work of a force, etc., are shown in Table 1.2. While these units will be introduced in later chapters as they are needed, we should note an important rule at this time: When a derived unit is obtained by dividing a base unit by another base unit, a prefix may be used in the numerator of the derived unit but not in its denominator. For example, the constant k of a spring which stretches 20 mm under a load of 100 N will be expressed as

$$k = \frac{100 \text{ N}}{20 \text{ mm}} = \frac{100 \text{ N}}{0.020 \text{ m}} = 5000 \text{ N/m} \quad \text{or} \quad k = 5 \text{ kN/m}$$

but never as $k = 5$ N/mm.

TABLE 1.2 Principal SI Units Used in Mechanics

Quantity	Unit	Symbol	Formula
Acceleration	Meter per second squared	. . .	m/s^2
Angle	Radian	rad	†
Angular acceleration	Radian per second squared	. . .	rad/s^2
Angular velocity	Radian per second	. . .	rad/s
Area	Square meter	. . .	m^2
Density	Kilogram per cubic meter	. . .	kg/m^3
Energy	Joule	J	$\text{N} \cdot \text{m}$
Force	Newton	N	$\text{kg} \cdot \text{m/s}^2$
Frequency	Hertz	Hz	s^{-1}
Impulse	Newton-second	. . .	$\text{kg} \cdot \text{m/s}$
Length	Meter	m	‡
Mass	Kilogram	kg	‡
Moment of a force	Newton-meter	. . .	$\text{N} \cdot \text{m}$
Power	Watt	W	J/s
Pressure	Pascal	Pa	N/m^2
Stress	Pascal	Pa	N/m^2
Time	Second	s	‡
Velocity	Meter per second	. . .	m/s
Volume			
Solids	Cubic meter	. . .	m^3
Liquids	Liter	L	10^{-3}m^3
Work	Joule	J	$\text{N} \cdot \text{m}$

†Supplementary unit (1 revolution = 2π rad = 360°).
‡Base unit.

U.S. Customary Units. Most practicing American engineers still commonly use a system in which the base units are the units of length, force, and time. These units are, respectively, the *foot* (ft), the *pound* (lb), and the *second* (s). The second is the same as the corresponding SI unit. The foot is defined as 0.3048 m. The pound is defined as the *weight* of a platinum standard, called the *standard pound*, which is kept at the National Institute of Standards and Technology outside Washington, the mass of which is 0.453 592 43 kg. Since the weight of a body depends upon the earth's gravitational attraction, which varies with location, it is specified that the standard pound should be placed at sea level and at a latitude of 45° to properly define a force of 1 lb. Clearly the U.S. customary units do not form an absolute system of units. Because of their dependence upon the gravitational attraction of the earth, they form a *gravitational* system of units.

While the standard pound also serves as the unit of mass in commercial transactions in the United States, it cannot be so used in engineering computations, since such a unit would not be consistent with the base units defined in the preceding paragraph. Indeed, when acted upon by a force of 1 lb, that is, when subjected to the force of gravity, the standard pound receives the acceleration of gravity, $g = 32.2$ ft/s² (Fig. 1.4), not the unit acceleration required by Eq. (1.1). The unit of mass consistent with the foot, the pound, and the second is the mass which receives an acceleration of 1 ft/s² when a force of 1 lb is applied to it (Fig. 1.5). This unit, sometimes called a *slug*, can be derived from the equation $F = ma$ after substituting 1 lb and 1 ft/s² for F and a, respectively. We write

$$F = ma \qquad 1 \text{ lb} = (1 \text{ slug})(1 \text{ ft/s}^2)$$

and obtain

$$1 \text{ slug} = \frac{1 \text{ lb}}{1 \text{ ft/s}^2} = 1 \text{ lb} \cdot \text{s}^2/\text{ft} \qquad (1.6)$$

Fig. 1.4

Fig. 1.5

Comparing Figs. 1.4 and 1.5, we conclude that the slug is a mass 32.2 times larger than the mass of the standard pound.

The fact that in the U.S. customary system of units bodies are characterized by their weight in pounds rather than by their mass in slugs will be a convenience in the study of statics, where one constantly deals with weights and other forces and only seldom with masses. However, in the study of dynamics, where forces, masses, and accelerations are involved, the mass m of a body will be expressed in slugs when its weight W is given in pounds. Recalling Eq. (1.4), we write

$$m = \frac{W}{g} \qquad (1.7)$$

where g is the acceleration of gravity ($g = 32.2$ ft/s²).

Other U.S. customary units frequently encountered in engineering problems are the *mile* (mi), equal to 5280 ft; the *inch* (in.), equal to $\frac{1}{12}$ ft; and the *kilopound* (kip), equal to a force of 1000 lb. The *ton* is often used to represent a mass of 2000 lb but, like the pound, must be converted into slugs in engineering computations.

The conversion into feet, pounds, and seconds of quantities expressed in other U.S. customary units is generally more involved and requires greater attention than the corresponding operation in SI units. If, for example, the magnitude of a velocity is given as $v = 30$ mi/h, we convert it to ft/s as follows. First we write

$$v = 30 \frac{\text{mi}}{\text{h}}$$

Since we want to get rid of the unit miles and introduce instead the unit feet, we should multiply the right-hand member of the equation by an expression containing miles in the denominator and feet in the numerator. But, since we do not want to change the value of the right-hand member, the expression used should have a value equal to unity. The quotient (5280 ft)/(1 mi) is such an expression. Operating in a similar way to transform the unit hour into seconds, we write

$$v = \left(30 \frac{\text{mi}}{\text{h}} \right) \left(\frac{5280 \text{ ft}}{1 \text{ mi}} \right) \left(\frac{1 \text{ h}}{3600 \text{ s}} \right)$$

Carrying out the numerical computations and canceling out units which appear in both the numerator and the denominator, we obtain

$$v = 44 \frac{\text{ft}}{\text{s}} = 44 \text{ ft/s}$$

1.4 CONVERSION FROM ONE SYSTEM OF UNITS TO ANOTHER

There are many instances when an engineer wishes to convert into SI units a numerical result obtained in U.S. customary units or vice versa. Because the unit of time is the same in both systems, only two kinetic base units need be converted. Thus, since all other kinetic units can be derived from these base units, only two conversion factors need be remembered.

Units of Length. By definition the U.S. customary unit of length is

$$1 \text{ ft} = 0.3048 \text{ m} \tag{1.8}$$

It follows that

$$1 \text{ mi} = 5280 \text{ ft} = 5280(0.3048 \text{ m}) = 1609 \text{ m}$$

or

$$1 \text{ mi} = 1.609 \text{ km} \tag{1.9}$$

Also

$$1 \text{ in.} = \tfrac{1}{12} \text{ ft} = \tfrac{1}{12}(0.3048 \text{ m}) = 0.0254 \text{ m}$$

or

$$1 \text{ in.} = 25.4 \text{ mm} \tag{1.10}$$

Units of Force. Recalling that the U.S. customary unit of force (pound) is defined as the weight of the standard pound (of mass 0.4536 kg) at sea level and at a latitude of 45° (where $g = 9.807$ m/s^2) and using Eq. (1.4), we write

$$W = mg$$
$$1\ \text{lb} = (0.4536\ \text{kg})(9.807\ \text{m/s}^2) = 4.448\ \text{kg} \cdot \text{m/s}^2$$

or, recalling Eq. (1.5),

$$1\ \text{lb} = 4.448\ \text{N} \qquad (1.11)$$

Units of Mass. The U.S. customary unit of mass (slug) is a derived unit. Thus, using Eqs. (1.6), (1.8), and (1.11), we write

$$1\ \text{slug} = 1\ \text{lb} \cdot \text{s}^2/\text{ft} = \frac{1\ \text{lb}}{1\ \text{ft/s}^2} = \frac{4.448\ \text{N}}{0.3048\ \text{m/s}^2} = 14.59\ \text{N} \cdot \text{s}^2/\text{m}$$

and, recalling Eq. (1.5),

$$1\ \text{slug} = 1\ \text{lb} \cdot \text{s}^2/\text{ft} = 14.59\ \text{kg} \qquad (1.12)$$

Although it cannot be used as a consistent unit of mass, we recall that the mass of the standard pound is, by definition,

$$1\ \text{pound mass} = 0.4536\ \text{kg} \qquad (1.13)$$

This constant may be used to determine the *mass* in SI units (kilograms) of a body which has been characterized by its *weight* in U.S. customary units (pounds).

To convert a derived U.S. customary unit into SI units, one simply multiplies or divides by the appropriate conversion factors. For example, to convert the moment of a force which was found to be $M = 47\ \text{lb} \cdot \text{in.}$ into SI units, we use formulas (1.10) and (1.11) and write

$$M = 47\ \text{lb} \cdot \text{in.} = 47(4.448\ \text{N})(25.4\ \text{mm})$$
$$= 5310\ \text{N} \cdot \text{mm} = 5.31\ \text{N} \cdot \text{m}$$

The conversion factors given in this section may also be used to convert a numerical result obtained in SI units into U.S. customary units. For example, if the moment of a force was found to be $M = 40\ \text{N} \cdot \text{m}$, we write, following the procedure used in the last paragraph of Sec. 1.3,

$$M = 40\ \text{N} \cdot \text{m} = (40\ \text{N} \cdot \text{m})\left(\frac{1\ \text{lb}}{4.448\ \text{N}}\right)\left(\frac{1\ \text{ft}}{0.3048\ \text{m}}\right)$$

Carrying out the numerical computations and canceling out units which appear in both the numerator and the denominator, we obtain

$$M = 29.5\ \text{lb} \cdot \text{ft}$$

The U.S. customary units most frequently used in mechanics are listed in Table 1.3 with their SI equivalents.

1.5 METHOD OF PROBLEM SOLUTION

You should approach a problem in mechanics as you would approach an actual engineering situation. By drawing on your own experience and intuition, you will find it easier to understand and formulate the problem. Once the problem has been clearly stated, however, there is

TABLE 1.3 U.S. Customary Units and Their SI Equivalents

Quantity	U.S. Customary Unit	SI Equivalent
Acceleration	ft/s^2	0.3048 m/s^2
	$in./s^2$	0.0254 m/s^2
Area	ft^2	0.0929 m^2
	in^2	645.2 mm^2
Energy	$ft \cdot lb$	1.356 J
Force	kip	4.448 kN
	lb	4.448 N
	oz	0.2780 N
Impulse	$lb \cdot s$	$4.448 \text{ N} \cdot \text{s}$
Length	ft	0.3048 m
	in.	25.40 mm
	mi	1.609 km
Mass	oz mass	28.35 g
	lb mass	0.4536 kg
	slug	14.59 kg
	ton	907.2 kg
Moment of a force	$lb \cdot ft$	$1.356 \text{ N} \cdot \text{m}$
	$lb \cdot in.$	$0.1130 \text{ N} \cdot \text{m}$
Moment of inertia		
Of an area	in^4	$0.4162 \times 10^6 \text{ mm}^4$
Of a mass	$lb \cdot ft \cdot s^2$	$1.356 \text{ kg} \cdot \text{m}^2$
Momentum	$lb \cdot s$	$4.448 \text{ kg} \cdot \text{m/s}$
Power	$ft \cdot lb/s$	1.356 W
	hp	745.7 W
Pressure or stress	lb/ft^2	47.88 Pa
	lb/in^2 (psi)	6.895 kPa
Velocity	ft/s	0.3048 m/s
	in./s	0.0254 m/s
	mi/h (mph)	0.4470 m/s
	mi/h (mph)	1.609 km/h
Volume	ft^3	0.02832 m^3
	in^3	16.39 cm^3
Liquids	gal	3.785 L
	qt	0.9464 L
Work	$ft \cdot lb$	1.356 J

no place in its solution for your particular fancy. *The solution must be based on the six fundamental principles stated in Sec. 1.2 or on theorems derived from them.* Every step taken must be justified on that basis. Strict rules must be followed, which lead to the solution in an almost automatic fashion, leaving no room for your intuition or "feeling." After an answer has been obtained, it should be checked. Here again, you may call upon your common sense and personal experience. If not completely satisfied with the result obtained, you should carefully check your formulation of the problem, the validity of the methods used for its solution, and the accuracy of your computations.

The *statement* of a problem should be clear and precise. It should contain the given data and indicate what information is required. A neat drawing showing all quantities involved should be included. Separate diagrams should be drawn for all bodies involved, indicating clearly the forces acting on each body. These diagrams are known as *free-body diagrams* and are described in detail in Secs. 2.11 and 4.2.

The *fundamental principles* of mechanics listed in Sec. 1.2 *will be used to write equations* expressing the conditions of rest or motion of the bodies considered. Each equation should be clearly related to one of the free-body diagrams. You will then proceed to solve the problem, observing strictly the usual rules of algebra and recording neatly the various steps taken.

After the answer has been obtained, it should be *carefully checked*. Mistakes in *reasoning* can often be detected by checking the units. For example, to determine the moment of a force of 50 N about a point 0.60 m from its line of action, we would have written (Sec. 3.12)

$$M = Fd = (50 \text{ N})(0.60 \text{ m}) = 30 \text{ N} \cdot \text{m}$$

The unit $N \cdot m$ obtained by multiplying newtons by meters is the correct unit for the moment of a force; if another unit had been obtained, we would have known that some mistake had been made.

Errors in *computation* will usually be found by substituting the numerical values obtained into an equation which has not yet been used and verifying that the equation is satisfied. The importance of correct computations in engineering cannot be overemphasized.

1.6 NUMERICAL ACCURACY

The accuracy of the solution of a problem depends upon two items: (1) the accuracy of the given data and (2) the accuracy of the computations performed.

The solution cannot be more accurate than the less accurate of these two items. For example, if the loading of a bridge is known to be 75,000 lb with a possible error of 100 lb either way, the relative error which measures the degree of accuracy of the data is

$$\frac{100 \text{ lb}}{75,000 \text{ lb}} = 0.0013 = 0.13 \text{ percent}$$

In computing the reaction at one of the bridge supports, it would then be meaningless to record it as 14,322 lb. The accuracy of the solution cannot be greater than 0.13 percent, no matter how accurate the computations are, and the possible error in the answer may be as large as (0.13/100)(14,322 lb) ≈ 20 lb. The answer should be properly recorded as 14,320 ± 20 lb.

In engineering problems, the data are seldom known with an accuracy greater than 0.2 percent. It is therefore seldom justified to write the answers to such problems with an accuracy greater than 0.2 percent. A practical rule is to use 4 figures to record numbers beginning with a "1" and 3 figures in all other cases. Unless otherwise indicated, the data given in a problem should be assumed known with a comparable degree of accuracy. A force of 40 lb, for example, should be read 40.0 lb, and a force of 15 lb should be read 15.00 lb.

Pocket electronic calculators are widely used by practicing engineers and engineering students. The speed and accuracy of these calculators facilitate the numerical computations in the solution of many problems. However, students should not record more significant figures than can be justified merely because they are easily obtained. As noted above, an accuracy greater than 0.2 percent is seldom necessary or meaningful in the solution of practical engineering problems.

Many engineering problems can be solved by considering the equilibrium of a "particle." In the case of this excavator, which is being loaded onto a ship, a relation between the tensions in the various cables involved can be obtained by considering the equilibrium of the hook to which the cables are attached.

CHAPTER

2

Statics of Particles

Chapter 2 Statics of Particles

2.1 INTRODUCTION

In this chapter you will study the effect of forces acting on particles. First you will learn how to replace two or more forces acting on a given particle by a single force having the same effect as the original forces. This single equivalent force is the *resultant* of the original forces acting on the particle. Later the relations which exist among the various forces acting on a particle in a state of *equilibrium* will be derived and used to determine some of the forces acting on the particle.

The use of the word "particle" does not imply that our study will be limited to that of small corpuscles. What it means is that the size and shape of the bodies under consideration will not significantly affect the solution of the problems treated in this chapter and that all the forces acting on a given body will be assumed to be applied at the same point. Since such an assumption is verified in many practical applications, you will be able to solve a number of engineering problems in this chapter.

The first part of the chapter is devoted to the study of forces contained in a single plane, and the second part to the analysis of forces in three-dimensional space.

FORCES IN A PLANE

2.2 FORCE ON A PARTICLE. RESULTANT OF TWO FORCES

A force represents the action of one body on another and is generally characterized by its *point of application,* its *magnitude,* and its *direction.* Forces acting on a given particle, however, have the same point of application. Each force considered in this chapter will thus be completely defined by its magnitude and direction.

The magnitude of a force is characterized by a certain number of units. As indicated in Chap. 1, the SI units used by engineers to measure the magnitude of a force are the newton (N) and its multiple the kilonewton (kN), equal to 1000 N, while the U.S. customary units used for the same purpose are the pound (lb) and its multiple the kilopound (kip), equal to 1000 lb. The direction of a force is defined by the *line of action* and the *sense* of the force. The line of action is the infinite straight line along which the force acts; it is characterized by the angle it forms with some fixed axis (Fig. 2.1). The force itself is represented by a segment of

Fig. 2.1
(a)
(b)

that line; through the use of an appropriate scale, the length of this segment may be chosen to represent the magnitude of the force. Finally, the sense of the force should be indicated by an arrowhead. It is important in defining a force to indicate its sense. Two forces having the same magnitude and the same line of action but different sense, such as the forces shown in Fig. 2.1*a* and *b*, will have directly opposite effects on a particle.

Experimental evidence shows that two forces **P** and **Q** acting on a particle *A* (Fig. 2.2*a*) can be replaced by a single force **R** which has the same effect on the particle (Fig. 2.2*c*). This force is called the *resultant* of the forces **P** and **Q** and can be obtained, as shown in Fig. 2.2*b*, by constructing a parallelogram, using **P** and **Q** as two adjacent sides of the parallelogram. *The diagonal that passes through A represents the resultant.* This method for finding the resultant is known as the *parallelogram law* for the addition of two forces. This law is based on experimental evidence; it cannot be proved or derived mathematically.

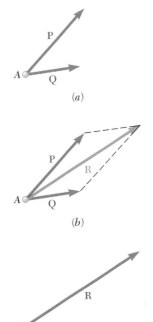

Fig. 2.2

2.3 VECTORS

It appears from the above that forces do not obey the rules of addition defined in ordinary arithmetic or algebra. For example, two forces acting at a right angle to each other, one of 4 lb and the other of 3 lb, add up to a force of 5 lb, *not* to a force of 7 lb. Forces are not the only quantities which follow the parallelogram law of addition. As you will see later, *displacements, velocities, accelerations,* and *momenta* are other examples of physical quantities possessing magnitude and direction that are added according to the parallelogram law. All these quantities can be represented mathematically by *vectors,* while those physical quantities which have magnitude but not direction, such as *volume, mass,* or *energy,* are represented by plain numbers or *scalars.*

Vectors are defined as *mathematical expressions possessing magnitude and direction, which add according to the parallelogram law.* Vectors are represented by arrows in the illustrations and will be distinguished from scalar quantities in this text through the use of boldface type (**P**). In longhand writing, a vector may be denoted by drawing a short arrow above the letter used to represent it ($\vec{P}$) or by underlining the letter (*P*). The last method may be preferred since underlining can also be used on a typewriter or computer. The magnitude of a vector defines the length of the arrow used to represent the vector. In this text, italic type will be used to denote the magnitude of a vector. Thus, the magnitude of the vector **P** will be denoted by *P*.

A vector used to represent a force acting on a given particle has a well-defined point of application, namely, the particle itself. Such a vector is said to be a *fixed,* or *bound,* vector and cannot be moved without modifying the conditions of the problem. Other physical quantities, however, such as couples (see Chap. 3), are represented by vectors which may be freely moved in space; these

Fig. 2.4

vectors are called *free* vectors. Still other physical quantities, such as forces acting on a rigid body (see Chap. 3), are represented by vectors which can be moved, or slid, along their lines of action; they are known as *sliding* vectors.†

Two vectors which have the same magnitude and the same direction are said to be *equal*, whether or not they also have the same point of application (Fig. 2.4); equal vectors may be denoted by the same letter.

The *negative vector* of a given vector **P** is defined as a vector having the same magnitude as **P** and a direction opposite to that of **P** (Fig. 2.5); the negative of the vector **P** is denoted by −**P**. The vectors **P** and −**P** are commonly referred to as *equal and opposite* vectors. Clearly, we have

$$\mathbf{P} + (-\mathbf{P}) = 0$$

Fig. 2.5

2.4 ADDITION OF VECTORS

We saw in the preceding section that, by definition, vectors add according to the parallelogram law. Thus, the sum of two vectors **P** and **Q** is obtained by attaching the two vectors to the same point *A* and constructing a parallelogram, using **P** and **Q** as two sides of the parallelogram (Fig. 2.6). The diagonal that passes through *A* represents the sum of the vectors **P** and **Q**, and this sum is denoted by **P** + **Q**. The fact that the sign + is used to denote both vector and scalar addition should not cause any confusion if vector and scalar quantities are always carefully distinguished. Thus, we should note that the magnitude of the vector **P** + **Q** is *not*, in general, equal to the sum *P* + *Q* of the magnitudes of the vectors **P** and **Q**.

Since the parallelogram constructed on the vectors **P** and **Q** does not depend upon the order in which **P** and **Q** are selected, we conclude that the addition of two vectors is *commutative*, and we write

$$\mathbf{P} + \mathbf{Q} = \mathbf{Q} + \mathbf{P} \tag{2.1}$$

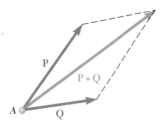

Fig. 2.6

†Some expressions have magnitude and direction, but do not add according to the parallelogram law. While these expressions may be represented by arrows, they *cannot* be considered as vectors.

A group of such expressions is the finite rotations of a rigid body. Place a closed book on a table in front of you, so that it lies in the usual fashion, with its front cover up and its binding to the left. Now rotate it through 180° about an axis parallel to the binding (Fig. 2.3*a*); this rotation may be represented by an arrow of length equal to 180 units and oriented as shown. Picking up the book as it lies in its new position, rotate

(a)   (b)

Fig. 2.3 Finite rotations of a rigid body

From the parallelogram law, we can derive an alternative method for determining the sum of two vectors. This method, known as the *triangle rule,* is derived as follows. Consider Fig. 2.6, where the sum of the vectors **P** and **Q** has been determined by the parallelogram law. Since the side of the parallelogram opposite **Q** is equal to **Q** in magnitude and direction, we could draw only half of the parallelogram (Fig. 2.7*a*). The sum of the two vectors can thus be found by *arranging* **P** *and* **Q** *in tip-to-tail fashion and then connecting the tail of* **P** *with the tip of* **Q**. In Fig. 2.7*b*, the other half of the parallelogram is considered, and the same result is obtained. This confirms the fact that vector addition is commutative.

The *subtraction* of a vector is defined as the addition of the corresponding negative vector. Thus, the vector **P** − **Q** representing the difference between the vectors **P** and **Q** is obtained by adding to **P** the negative vector −**Q** (Fig. 2.8). We write

$$\mathbf{P} - \mathbf{Q} = \mathbf{P} + (-\mathbf{Q}) \tag{2.2}$$

Here again we should observe that, while the same sign is used to denote both vector and scalar subtraction, confusion will be avoided if care is taken to distinguish between vector and scalar quantities.

We will now consider the *sum of three or more vectors.* The sum of three vectors **P**, **Q**, and **S** will, *by definition,* be obtained by first adding the vectors **P** and **Q** and then adding the vector **S** to the vector **P** + **Q**. We thus write

$$\mathbf{P} + \mathbf{Q} + \mathbf{S} = (\mathbf{P} + \mathbf{Q}) + \mathbf{S} \tag{2.3}$$

Similarly, the sum of four vectors will be obtained by adding the fourth vector to the sum of the first three. It follows that the sum of any number of vectors can be obtained by applying repeatedly the parallelogram law to successive pairs of vectors until all the given vectors are replaced by a single vector.

Fig. 2.7

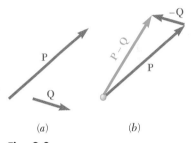

Fig. 2.8

it now through 180° about a horizontal axis perpendicular to the binding (Fig. 2.3*b*); this second rotation may be represented by an arrow 180 units long and oriented as shown. But the book could have been placed in this final position through a single 180° rotation about a vertical axis (Fig. 2.3*c*). We conclude that the sum of the two 180° rotations represented by arrows directed respectively along the *z* and *x* axes is a 180° rotation represented by an arrow directed along the *y* axis (Fig. 2.3*d*). Clearly, the finite rotations of a rigid body *do not* obey the parallelogram law of addition; therefore, they *cannot* be represented by vectors.

Fig. 2.9

If the given vectors are *coplanar,* i.e., if they are contained in the same plane, their sum can be easily obtained graphically. For this case, the repeated application of the triangle rule is preferred to the application of the parallelogram law. In Fig. 2.9 the sum of three vectors **P**, **Q**, and **S** was obtained in that manner. The triangle rule was first applied to obtain the sum **P** + **Q** of the vectors **P** and **Q**; it was applied again to obtain the sum of the vectors **P** + **Q** and **S**. The determination of the vector **P** + **Q**, however, could have been omitted and the sum of the three vectors could have been obtained directly, as shown in Fig. 2.10, by *arranging the given vectors in tip-to-tail fashion and connecting the tail of the first vector with the tip of the last one.* This is known as the *polygon rule* for the addition of vectors.

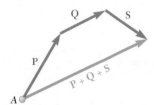

Fig. 2.10

We observe that the result obtained would have been unchanged if, as shown in Fig. 2.11, the vectors **Q** and **S** had been replaced by their sum **Q** + **S**. We may thus write

$$\mathbf{P} + \mathbf{Q} + \mathbf{S} = (\mathbf{P} + \mathbf{Q}) + \mathbf{S} = \mathbf{P} + (\mathbf{Q} + \mathbf{S}) \qquad (2.4)$$

which expresses the fact that vector addition is *associative.* Recalling that vector addition has also been shown, in the case of two vectors, to be commutative, we write

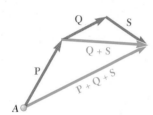

Fig. 2.11

$$\begin{aligned}\mathbf{P} + \mathbf{Q} + \mathbf{S} &= (\mathbf{P} + \mathbf{Q}) + \mathbf{S} = \mathbf{S} + (\mathbf{P} + \mathbf{Q}) \\ &= \mathbf{S} + (\mathbf{Q} + \mathbf{P}) = \mathbf{S} + \mathbf{Q} + \mathbf{P}\end{aligned} \qquad (2.5)$$

This expression, as well as others which may be obtained in the same way, shows that the order in which several vectors are added together is immaterial (Fig. 2.12).

Product of a Scalar and a Vector.   Since it is convenient to denote the sum **P** + **P** by 2**P**, the sum **P** + **P** + **P** by 3**P**, and, in general, the sum of n equal vectors **P** by the product $n\mathbf{P}$, we will define the product $n\mathbf{P}$ of a positive integer n and a vector **P** as a vector having the same direction as **P** and the magnitude nP. Extending this definition to include all scalars, and recalling the definition of a negative vector given in Sec. 2.3, we define the product $k\mathbf{P}$ of a scalar k and a vector **P** as a vector having the same direction as **P** (if k is positive), or a direction opposite to that of **P** (if k is negative), and a magnitude equal to the product of P and of the absolute value of k (Fig. 2.13).

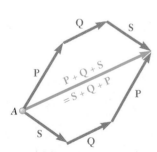

Fig. 2.12

2.5   RESULTANT OF SEVERAL CONCURRENT FORCES

Consider a particle A acted upon by several coplanar forces, i.e., by several forces contained in the same plane (Fig. 2.14*a*). Since the forces considered here all pass through A, they are also said to be *concurrent.* The vectors representing the forces acting on A may be added by the polygon rule (Fig. 2.14*b*). Since the use of the polygon rule is equivalent to the repeated application of the parallelogram law, the vector **R** thus obtained represents the resultant of the given concurrent forces, i.e., the single force which has the same effect on

Fig. 2.13

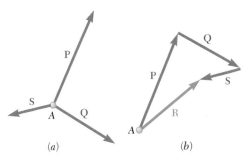

Fig. 2.14

the particle A as the given forces. As indicated in the previous section, the order in which the vectors **P**, **Q**, and **S** representing the given forces are added together is immaterial.

2.6 RESOLUTION OF A FORCE INTO COMPONENTS

We have seen that two or more forces acting on a particle may be replaced by a single force which has the same effect on the particle. Conversely, a single force **F** acting on a particle may be replaced by two or more forces which, together, have the same effect on the particle. These forces are called the *components* of the original force **F**, and the process of substituting them for **F** is called *resolving the force* **F** *into components.*

Clearly, for each force **F** there exist an infinite number of possible sets of components. Sets of *two components* **P** *and* **Q** are the most important as far as practical applications are concerned. But, even then, the number of ways in which a given force **F** may be resolved into two components is unlimited (Fig. 2.15). Two cases are of particular interest:

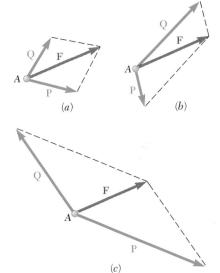

Fig. 2.15

1. *One of the Two Components,* **P**, *Is Known.* The second component, **Q**, is obtained by applying the triangle rule and joining the tip of **P** to the tip of **F** (Fig. 2.16); the magnitude and direction of **Q** are determined graphically or by trigonometry. Once **Q** has been determined, both components **P** and **Q** should be applied at A.

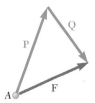

Fig. 2.16

2. *The Line of Action of Each Component Is Known.* The magnitude and sense of the components are obtained by applying the parallelogram law and drawing lines, through the tip of **F**, parallel to the given lines of action (Fig. 2.17). This process leads to two well-defined components, **P** and **Q**, which can be determined graphically or computed trigonometrically by applying the law of sines.

Many other cases can be encountered; for example, the direction of one component may be known, while the magnitude of the other component is to be as small as possible (see Sample Prob. 2.2). In all cases the appropriate triangle or parallelogram which satisfies the given conditions is drawn.

Fig. 2.17

SAMPLE PROBLEM 2.1

The two forces **P** and **Q** act on a bolt *A*. Determine their resultant.

SOLUTION

Graphical Solution. A parallelogram with sides equal to **P** and **Q** is drawn to scale. The magnitude and direction of the resultant are measured and found to be

$$R = 98 \text{ N} \qquad \alpha = 35° \qquad \mathbf{R} = 98 \text{ N} \angle 35° \blacktriangleleft$$

The triangle rule may also be used. Forces **P** and **Q** are drawn in tip-to-tail fashion. Again the magnitude and direction of the resultant are measured.

$$R = 98 \text{ N} \qquad \alpha = 35° \qquad \mathbf{R} = 98 \text{ N} \angle 35° \blacktriangleleft$$

Trigonometric Solution. The triangle rule is again used; two sides and the included angle are known. We apply the law of cosines.

$$R^2 = P^2 + Q^2 - 2PQ \cos B$$
$$R^2 = (40 \text{ N})^2 + (60 \text{ N})^2 - 2(40 \text{ N})(60 \text{ N}) \cos 155°$$
$$R = 97.73 \text{ N}$$

Now, applying the law of sines, we write

$$\frac{\sin A}{Q} = \frac{\sin B}{R} \qquad \frac{\sin A}{60 \text{ N}} = \frac{\sin 155°}{97.73 \text{ N}} \tag{1}$$

Solving Eq. (1) for sin *A*, we have

$$\sin A = \frac{(60 \text{ N}) \sin 155°}{97.73 \text{ N}}$$

Using a calculator, we first compute the quotient, then its arc sine, and obtain

$$A = 15.04° \qquad \alpha = 20° + A = 35.04°$$

We use 3 significant figures to record the answer (cf. Sec. 1.6):

$$\mathbf{R} = 97.7 \text{ N} \angle 35.0° \blacktriangleleft$$

Alternative Trigonometric Solution. We construct the right triangle *BCD* and compute

$$CD = (60 \text{ N}) \sin 25° = 25.36 \text{ N}$$
$$BD = (60 \text{ N}) \cos 25° = 54.38 \text{ N}$$

Then, using triangle *ACD*, we obtain

$$\tan A = \frac{25.36 \text{ N}}{94.38 \text{ N}} \qquad A = 15.04°$$

$$R = \frac{25.36}{\sin A} \qquad R = 97.73 \text{ N}$$

Again, $\alpha = 20° + A = 35.04°$ $\mathbf{R} = 97.7 \text{ N} \angle 35.0° \blacktriangleleft$

SAMPLE PROBLEM 2.2

A barge is pulled by two tugboats. If the resultant of the forces exerted by the tugboats is a 5000-lb force directed along the axis of the barge, determine (*a*) the tension in each of the ropes knowing that $\alpha = 45°$, (*b*) the value of α for which the tension in rope 2 is minimum.

SOLUTION

a. Tension for $\alpha = 45°$. *Graphical Solution.* The parallelogram law is used; the diagonal (resultant) is known to be equal to 5000 lb and to be directed to the right. The sides are drawn parallel to the ropes. If the drawing is done to scale, we measure

$$T_1 = 3700 \text{ lb} \qquad T_2 = 2600 \text{ lb} \quad \blacktriangleleft$$

Trigonometric Solution. The triangle rule can be used. We note that the triangle shown represents half of the parallelogram shown above. Using the law of sines, we write

$$\frac{T_1}{\sin 45°} = \frac{T_2}{\sin 30°} = \frac{5000 \text{ lb}}{\sin 105°}$$

With a calculator, we first compute and store the value of the last quotient. Multiplying this value successively by sin 45° and sin 30°, we obtain

$$T_1 = 3660 \text{ lb} \qquad T_2 = 2590 \text{ lb} \quad \blacktriangleleft$$

b. Value of α for Minimum T_2. To determine the value of α for which the tension in rope 2 is minimum, the triangle rule is again used. In the sketch shown, line *1-1'* is the known direction of $\mathbf{T}_1$. Several possible directions of $\mathbf{T}_2$ are shown by the lines *2-2'*. We note that the minimum value of T_2 occurs when $\mathbf{T}_1$ and $\mathbf{T}_2$ are perpendicular. The minimum value of T_2 is

$$T_2 = (5000 \text{ lb}) \sin 30° = 2500 \text{ lb}$$

Corresponding values of T_1 and α are

$$T_1 = (5000 \text{ lb}) \cos 30° = 4330 \text{ lb}$$
$$\alpha = 90° - 30° \qquad\qquad \alpha = 60° \quad \blacktriangleleft$$

SOLVING PROBLEMS ON YOUR OWN

The preceding sections were devoted to the *parallelogram law* for the addition of vectors and to its applications.

Two sample problems were presented. In Sample Prob. 2.1, the parallelogram law was used to determine the resultant of two forces of known magnitude and direction. In Sample Prob. 2.2, it was used to resolve a given force into two components of known direction.

You will now be asked to solve problems on your own. Some may resemble one of the sample problems; others may not. What all problems and sample problems in this section have in common is that they can be solved by the direct application of the parallelogram law.

Your solution of a given problem should consist of the following steps:

1. Identify which of the forces are the applied forces and which is the resultant. It is often helpful to write the vector equation which shows how the forces are related. For example, in Sample Prob. 2.1 we would have

$$\mathbf{R} = \mathbf{P} + \mathbf{Q}$$

You may want to keep that relation in mind as you formulate the next part of your solution.

2. Draw a parallelogram with the applied forces as two adjacent sides and the resultant as the included diagonal (Fig. 2.2). Alternatively, you can *use the triangle rule,* with the applied forces drawn in tip-to-tail fashion and the resultant extending from the tail of the first vector to the tip of the second (Fig. 2.7).

3. Indicate all dimensions. Using one of the triangles of the parallelogram, or the triangle constructed according to the triangle rule, indicate all dimensions—whether sides or angles—and determine the unknown dimensions either graphically or by trigonometry. If you use trigonometry, remember that the law of cosines should be applied first if two sides and the included angle are known [Sample Prob. 2.1], and the law of sines should be applied first if one side and all angles are known [Sample Prob. 2.2].

If you have had prior exposure to mechanics, you might be tempted to ignore the solution techniques of this lesson in favor of resolving the forces into rectangular components. While this latter method is important and will be considered in the next section, use of the parallelogram law simplifies the solution of many problems and should be mastered at this time.

PROBLEMS†

2.1 Two forces are applied at point B of beam AB. Determine graphically the magnitude and direction of their resultant using (a) the parallelogram law, (b) the triangle rule.

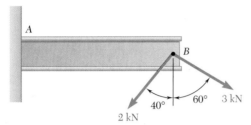

Fig. P2.1

2.2 The cable stays AB and AD help support pole AC. Knowing that the tension is 120 lb in AB and 40 lb in AD, determine graphically the magnitude and direction of the resultant of the forces exerted by the stays at A using (a) the parallelogram law, (b) the triangle rule.

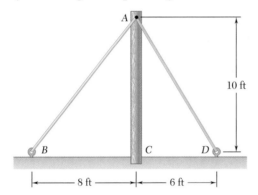

Fig. P2.2

2.3 Two structural members B and C are bolted to bracket A. Knowing that both members are in tension and that $P = 10$ kN and $Q = 15$ kN, determine graphically the magnitude and direction of the resultant force exerted on the bracket using (a) the parallelogram law, (b) the triangle rule.

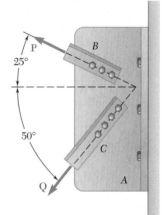

Fig. *P2.3* and *P2.4*

2.4 Two structural members B and C are bolted to bracket A. Knowing that both members are in tension and that $P = 6$ kips and $Q = 4$ kips, determine graphically the magnitude and direction of the resultant force exerted on the bracket using (a) the parallelogram law, (b) the triangle rule.

2.5 A stake is being pulled out of the ground by means of two ropes as shown. Knowing that $\alpha = 30°$, determine by trigonometry (a) the magnitude of the force **P** so that the resultant force exerted on the stake is vertical, (b) the corresponding magnitude of the resultant.

†Answers to all problems set in straight type (such as **2.1**) are given at the end of the book. Answers to problems with a number set in italic type (such as **2.3**) are not given.

Fig. P2.5

25

Fig. P2.6 and P2.7

Fig. P2.8 and *P2.9*

2.6 A trolley that moves along a horizontal beam is acted upon by two forces as shown. (*a*) Knowing that $\alpha = 25°$, determine by trigonometry the magnitude of the force **P** so that the resultant force exerted on the trolley is vertical. (*b*) What is the corresponding magnitude of the resultant?

2.7 A trolley that moves along a horizontal beam is acted upon by two forces as shown. Determine by trigonometry the magnitude and direction of the force **P** so that the resultant is a vertical force of 2500 N.

2.8 A telephone cable is clamped at *A* to the pole *AB*. Knowing that the tension in the left-hand portion of the cable is $T_1 = 800$ lb, determine by trigonometry (*a*) the required tension T_2 in the right-hand portion if the resultant **R** of the forces exerted by the cable at *A* is to be vertical, (*b*) the corresponding magnitude of **R**.

2.9 A telephone cable is clamped at *A* to the pole *AB*. Knowing that the tension in the right-hand portion of the cable is $T_2 = 1000$ lb, determine by trigonometry (*a*) the required tension T_1 in the left-hand portion if the resultant **R** of the forces exerted by the cable at *A* is to be vertical, (*b*) the corresponding magnitude of **R**.

2.10 Two forces are applied as shown to a hook support. Knowing that the magnitude of **P** is 35 N, determine by trigonometry (*a*) the required angle α if the resultant **R** of the two forces applied to the support is to be horizontal, (*b*) the corresponding magnitude of **R**.

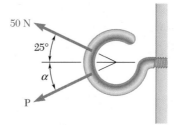

Fig. P2.10

2.11 A steel tank is to be positioned in an excavation. Knowing that $\alpha = 20°$, determine by trigonometry (*a*) the required magnitude of the force **P** if the resultant **R** of the two forces applied at *A* is to be vertical, (*b*) the corresponding magnitude of **R**.

2.12 A steel tank is to be positioned in an excavation. Knowing that the magnitude of **P** is 500 lb, determine by trigonometry (*a*) the required angle α if the resultant **R** of the two forces applied at *A* is to be vertical, (*b*) the corresponding magnitude of **R**.

2.13 A steel tank is to be positioned in an excavation. Determine by trigonometry (*a*) the magnitude and direction of the smallest force **P** for which the resultant **R** of the two forces applied at *A* is vertical, (*b*) the corresponding magnitude of **R**.

2.14 For the hook support of Prob. 2.10, determine by trigonometry (*a*) the magnitude and direction of the smallest force **P** for which the resultant **R** of the two forces applied to the support is horizontal, (*b*) the corresponding magnitude of **R**.

Fig. P2.11, *P2.12*, and P2.13

2.15 Solve Prob. 2.2 by trigonometry.

2.16 Solve Prob. 2.4 by trigonometry.

2.17 For the stake of Prob. 2.5, knowing that the tension in one rope is 120 N, determine by trigonometry the magnitude and direction of the force **P** so that the resultant is a vertical force of 160 N.

2.18 For the hook support of Prob. 2.10, knowing that $P = 75$ N and $\alpha = 50°$, determine by trigonometry the magnitude and direction of the resultant of the two forces applied to the support.

2.19 Two forces **P** and **Q** are applied to the lid of a storage bin as shown. Knowing that $P = 48$ N and $Q = 60$ N, determine by trigonometry the magnitude and direction of the resultant of the two forces.

2.20 Two forces **P** and **Q** are applied to the lid of a storage bin as shown. Knowing that $P = 60$ N and $Q = 48$ N, determine by trigonometry the magnitude and direction of the resultant of the two forces.

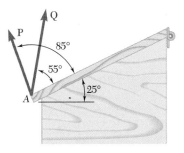

Fig. P2.19 and P2.20

2.7 RECTANGULAR COMPONENTS OF A FORCE. UNIT VECTORS†

In many problems it will be found desirable to resolve a force into two components which are perpendicular to each other. In Fig. 2.18, the force **F** has been resolved into a component $\mathbf{F}_x$ along the x axis and a component $\mathbf{F}_y$ along the y axis. The parallelogram drawn to obtain the two components is a *rectangle*, and $\mathbf{F}_x$ and $\mathbf{F}_y$ are called *rectangular components*.

Fig. 2.18

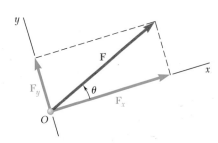

Fig. 2.19

The x and y axes are usually chosen horizontal and vertical, respectively, as in Fig. 2.18; they may, however, be chosen in any two perpendicular directions, as shown in Fig. 2.19. In determining the rectangular components of a force, the student should think of the construction lines shown in Figs. 2.18 and 2.19 as being *parallel* to the x and y axes, rather than *perpendicular* to these axes. This practice will help avoid mistakes in determining *oblique* components as in Sec. 2.6.

†The properties established in Secs. 2.7 and 2.8 may be readily extended to the rectangular components of any vector quantity.

Fig. 2.20

Fig. 2.21

(a)

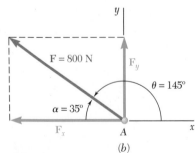

(b)

Fig. 2.22

Two vectors of unit magnitude, directed respectively along the positive x and y axes, will be introduced at this point. These vectors are called *unit vectors* and are denoted by **i** and **j**, respectively (Fig. 2.20). Recalling the definition of the product of a scalar and a vector given in Sec. 2.4, we note that the rectangular components $\mathbf{F}_x$ and $\mathbf{F}_y$ of a force **F** may be obtained by multiplying respectively the unit vectors **i** and **j** by appropriate scalars (Fig. 2.21). We write

$$\mathbf{F}_x = F_x\mathbf{i} \qquad \mathbf{F}_y = F_y\mathbf{j} \tag{2.6}$$

and

$$\mathbf{F} = F_x\mathbf{i} + F_y\mathbf{j} \tag{2.7}$$

While the scalars F_x and F_y may be positive or negative, depending upon the sense of $\mathbf{F}_x$ and of $\mathbf{F}_y$, their absolute values are respectively equal to the magnitudes of the component forces $\mathbf{F}_x$ and $\mathbf{F}_y$. The scalars F_x and F_y are called the *scalar components* of the force **F**, while the actual component forces $\mathbf{F}_x$ and $\mathbf{F}_y$ should be referred to as the *vector components* of **F**. However, when there exists no possibility of confusion, the vector as well as the scalar components of **F** may be referred to simply as the *components* of **F**. We note that the scalar component F_x is positive when the vector component $\mathbf{F}_x$ has the same sense as the unit vector **i** (i.e., the same sense as the positive x axis) and is negative when $\mathbf{F}_x$ has the opposite sense. A similar conclusion may be drawn regarding the sign of the scalar component F_y.

Denoting by F the magnitude of the force **F** and by θ the angle between **F** and the x axis, measured counterclockwise from the positive x axis (Fig. 2.21), we may express the scalar components of **F** as follows:

$$F_x = F \cos \theta \qquad F_y = F \sin \theta \tag{2.8}$$

We note that the relations obtained hold for any value of the angle θ from 0° to 360° and that they define the signs as well as the absolute values of the scalar components F_x and F_y.

EXAMPLE 1. A force of 800 N is exerted on a bolt A as shown in Fig. 2.22a. Determine the horizontal and vertical components of the force.

In order to obtain the correct sign for the scalar components F_x and F_y, the value 180° − 35° = 145° should be substituted for θ in Eqs. (2.8). However, it will be found more practical to determine by inspection the signs of F_x and F_y (Fig. 2.22b) and to use the trigonometric functions of the angle $\alpha = 35°$. We write, therefore,

$$F_x = -F \cos \alpha = -(800 \text{ N}) \cos 35° = -655 \text{ N}$$
$$F_y = +F \sin \alpha = +(800 \text{ N}) \sin 35° = +459 \text{ N}$$

The vector components of **F** are thus

$$\mathbf{F}_x = -(655 \text{ N})\mathbf{i} \qquad \mathbf{F}_y = +(459 \text{ N})\mathbf{j}$$

and we may write **F** in the form

$$\mathbf{F} = -(655 \text{ N})\mathbf{i} + (459 \text{ N})\mathbf{j} \quad \blacksquare$$

EXAMPLE 2. A man pulls with a force of 300 N on a rope attached to a building, as shown in Fig. 2.23a. What are the horizontal and vertical components of the force exerted by the rope at point A?

It is seen from Fig. 2.23b that

$$F_x = +(300 \text{ N}) \cos \alpha \qquad F_y = -(300 \text{ N}) \sin \alpha$$

Observing that $AB = 10$ m, we find from Fig. 2.23a

$$\cos \alpha = \frac{8 \text{ m}}{AB} = \frac{8 \text{ m}}{10 \text{ m}} = \frac{4}{5} \qquad \sin \alpha = \frac{6 \text{ m}}{AB} = \frac{6 \text{ m}}{10 \text{ m}} = \frac{3}{5}$$

We thus obtain

$$F_x = +(300 \text{ N})\tfrac{4}{5} = +240 \text{ N} \qquad F_y = -(300 \text{ N})\tfrac{3}{5} = -180 \text{ N}$$

and write

$$\mathbf{F} = (240 \text{ N})\mathbf{i} - (180 \text{ N})\mathbf{j} \quad \blacksquare$$

When a force **F** is defined by its rectangular components F_x and F_y (see Fig. 2.21), the angle θ defining its direction can be obtained by writing

$$\tan \theta = \frac{F_y}{F_x} \qquad (2.9)$$

The magnitude F of the force can be obtained by applying the Pythagorean theorem and writing

$$F = \sqrt{F_x^2 + F_y^2} \qquad (2.10)$$

or by solving for F one of the Eqs. (2.8).

EXAMPLE 3. A force $\mathbf{F} = (700 \text{ lb})\mathbf{i} + (1500 \text{ lb})\mathbf{j}$ is applied to a bolt A. Determine the magnitude of the force and the angle θ it forms with the horizontal.

First we draw a diagram showing the two rectangular components of the force and the angle θ (Fig. 2.24). From Eq. (2.9), we write

$$\tan \theta = \frac{F_y}{F_x} = \frac{1500 \text{ lb}}{700 \text{ lb}}$$

Using a calculator,† we enter 1500 lb and divide by 700 lb; computing the arc tangent of the quotient, we obtain $\theta = 65.0°$. Solving the second of Eqs. (2.8) for F, we have

$$F = \frac{F_y}{\sin \theta} = \frac{1500 \text{ lb}}{\sin 65.0°} = 1655 \text{ lb}$$

The last calculation is facilitated if the value of F_y is stored when originally entered; it may then be recalled to be divided by $\sin \theta$. $\blacksquare$

(a)

(b)

Fig. 2.23

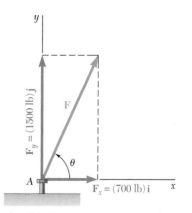

Fig. 2.24

†It is assumed that the calculator used has keys for the computation of trigonometric and inverse trigonometric functions. Some calculators also have keys for the direct conversion of rectangular coordinates into polar coordinates, and vice versa. Such calculators eliminate the need for the computation of trigonometric functions in Examples 1, 2, and 3 and in problems of the same type.

2.8   ADDITION OF FORCES BY SUMMING X AND Y COMPONENTS

(a)

(b)

(c)

(d)

Fig. 2.25

It was seen in Sec. 2.2 that forces should be added according to the parallelogram law. From this law, two other methods, more readily applicable to the *graphical* solution of problems, were derived in Secs. 2.4 and 2.5: the triangle rule for the addition of two forces and the polygon rule for the addition of three or more forces. It was also seen that the force triangle used to define the resultant of two forces could be used to obtain a *trigonometric* solution.

When three or more forces are to be added, no practical trigonometric solution can be obtained from the force polygon which defines the resultant of the forces. In this case, an *analytic* solution of the problem can be obtained by resolving each force into two rectangular components. Consider, for instance, three forces **P**, **Q**, and **S** acting on a particle A (Fig. 2.25a). Their resultant **R** is defined by the relation

$$\mathbf{R} = \mathbf{P} + \mathbf{Q} + \mathbf{S} \tag{2.11}$$

Resolving each force into its rectangular components, we write

$$
\begin{aligned}
R_x\mathbf{i} + R_y\mathbf{j} &= P_x\mathbf{i} + P_y\mathbf{j} + Q_x\mathbf{i} + Q_y\mathbf{j} + S_x\mathbf{i} + S_y\mathbf{j} \\
&= (P_x + Q_x + S_x)\mathbf{i} + (P_y + Q_y + S_y)\mathbf{j}
\end{aligned}
$$

from which it follows that

$$R_x = P_x + Q_x + S_x \qquad R_y = P_y + Q_y + S_y \tag{2.12}$$

or, for short,

$$R_x = \Sigma F_x \qquad R_y = \Sigma F_y \tag{2.13}$$

We thus conclude that *the scalar components R_x and R_y of the resultant **R** of several forces acting on a particle are obtained by adding algebraically the corresponding scalar components of the given forces.*†

In practice, the determination of the resultant **R** is carried out in three steps as illustrated in Fig. 2.25. First the given forces shown in Fig. 2.25a are resolved into their x and y components (Fig. 2.25b). Adding these components, we obtain the x and y components of **R** (Fig. 2.25c). Finally, the resultant $\mathbf{R} = R_x\mathbf{i} + R_y\mathbf{j}$ is determined by applying the parallelogram law (Fig. 2.25d). The procedure just described will be carried out most efficiently if the computations are arranged in a table. While it is the only practical analytic method for adding three or more forces, it is also often preferred to the trigonometric solution in the case of the addition of two forces.

---

†Clearly, this result also applies to the addition of other vector quantities, such as velocities, accelerations, or momenta.

SAMPLE PROBLEM 2.3

Four forces act on bolt A as shown. Determine the resultant of the forces on the bolt.

SOLUTION

The x and y components of each force are determined by trigonometry as shown and are entered in the table below. According to the convention adopted in Sec. 2.7, the scalar number representing a force component is positive if the force component has the same sense as the corresponding coordinate axis. Thus, x components acting to the right and y components acting upward are represented by positive numbers.

Force	Magnitude, N	x Component, N	y Component, N
F_1	150	+129.9	+75.0
F_2	80	−27.4	+75.2
F_3	110	0	−110.0
F_4	100	+96.6	−25.9
		$R_x = +199.1$	$R_y = +14.3$

Thus, the resultant **R** of the four forces is

$$\mathbf{R} = R_x\mathbf{i} + R_y\mathbf{j} \qquad \mathbf{R} = (199.1 \text{ N})\mathbf{i} + (14.3 \text{ N})\mathbf{j} \ \blacktriangleleft$$

The magnitude and direction of the resultant may now be determined. From the triangle shown, we have

$$\tan \alpha = \frac{R_y}{R_x} = \frac{14.3 \text{ N}}{199.1 \text{ N}} \qquad \alpha = 4.1°$$

$$R = \frac{14.3 \text{ N}}{\sin \alpha} = 199.6 \text{ N} \qquad \mathbf{R} = 199.6 \text{ N} \measuredangle 4.1° \ \blacktriangleleft$$

With a calculator, the last computation may be facilitated if the value of R_y is stored when originally entered; it may then be recalled to be divided by $\sin \alpha$. (Also see the footnote on p. 29.)

SOLVING PROBLEMS
ON YOUR OWN

You saw in the preceding lesson that the resultant of two forces may be determined either graphically or from the trigonometry of an oblique triangle.

A. When three or more forces are involved, the determination of their resultant **R** is best carried out by first resolving each force into *rectangular components.* Two cases may be encountered, depending upon the way in which each of the given forces is defined:

Case 1. The force F is defined by its magnitude F and the angle α it forms with the x axis. The x and y components of the force can be obtained by multiplying F by cos α and sin α, respectively [Example 1].

Case 2. The force F is defined by its magnitude F and the coordinates of two points A and B on its line of action (Fig. 2.23). The angle α that **F** forms with the x axis may first be determined by trigonometry. However, the components of **F** may also be obtained directly from proportions among the various dimensions involved, without actually determining α [Example 2].

B. Rectangular components of the resultant. The components R_x and R_y of the resultant can be obtained by adding algebraically the corresponding components of the given forces [Sample Prob. 2.3].

You can express the resultant in *vectorial form* using the unit vectors **i** and **j**, which are directed along the x and y axes, respectively:

$$\mathbf{R} = R_x\mathbf{i} + R_y\mathbf{j}$$

Alternatively, you can determine the *magnitude and direction* of the resultant by solving the right triangle of sides R_x and R_y for R and for the angle that **R** forms with the x axis.

PROBLEMS

2.21 and 2.22 Determine the x and y components of each of the forces shown.

Fig. P2.21

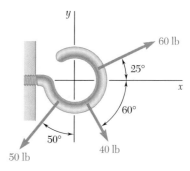

Fig. P2.22

2.23 and 2.24 Determine the x and y components of each of the forces shown.

Fig. P2.23

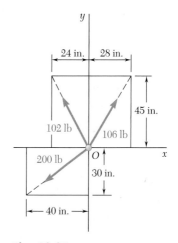

Fig. P2.24

2.25 The hydraulic cylinder *BD* exerts on member *ABC* a force **P** directed along line *BD*. Knowing that **P** must have a 750-N component perpendicular to member *ABC*, determine (*a*) the magnitude of the force **P**, (*b*) its component parallel to *ABC*.

Fig. P2.25

2.26 Cable *AC* exerts on beam *AB* a force **P** directed along line *AC*. Knowing that **P** must have a 350-lb vertical component, determine (*a*) the magnitude of the force **P**, (*b*) its horizontal component.

Fig. P2.26

2.27 Member *BC* exerts on member *AC* a force **P** directed along line *BC*. Knowing that **P** must have a 325-N horizontal component, determine (*a*) the magnitude of the force **P**, (*b*) its vertical component.

2.28 Member *BD* exerts on member *ABC* a force **P** directed along line *BD*. Knowing that **P** must have a 240-lb vertical component, determine (*a*) the magnitude of the force **P**, (*b*) its horizontal component.

Fig. P2.27

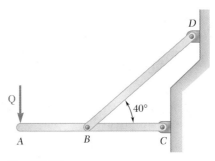

Fig. P2.28

2.29 The guy wire *BD* exerts on the telephone pole *AC* a force **P** directed along *BD*. Knowing that **P** must have a 720-N component perpendicular to the pole *AC*, determine (*a*) the magnitude of the force **P**, (*b*) its component along line *AC*.

2.30 The hydraulic cylinder *BC* exerts on member *AB* a force **P** directed along line *BC*. Knowing that **P** must have a 600-N component perpendicular to member *AB*, determine (*a*) the magnitude of the force **P**, (*b*) its component along line *AB*.

Fig. P2.29

2.31 Determine the resultant of the three forces of Prob. 2.23.

2.32 Determine the resultant of the three forces of Prob. 2.21.

2.33 Determine the resultant of the three forces of Prob. 2.22.

2.34 Determine the resultant of the three forces of Prob. 2.24.

Fig. P2.30

2.35 Knowing that $\alpha = 35°$, determine the resultant of the three forces shown.

Fig. P2.35

Fig. P2.36

2.36 Knowing that the tension in rope AC is 365 N, determine the resultant of the three forces exerted at point C of post BC.

2.37 Knowing that $\alpha = 40°$, determine the resultant of the three forces shown.

2.38 Knowing that $\alpha = 75°$, determine the resultant of the three forces shown.

2.39 For the collar of Prob. 2.35, determine (a) the required value of α if the resultant of the three forces shown is to be vertical, (b) the corresponding magnitude of the resultant.

2.40 For the post of Prob. 2.36, determine (a) the required tension in rope AC if the resultant of the three forces exerted at point C is to be horizontal, (b) the corresponding magnitude of the resultant.

2.41 A hoist trolley is subjected to the three forces shown. Knowing that $\alpha = 40°$, determine (a) the required magnitude of the force **P** if the resultant of the three forces is to be vertical, (b) the corresponding magnitude of the resultant.

2.42 A hoist trolley is subjected to the three forces shown. Knowing that $P = 250$ lb, determine (a) the required value of α if the resultant of the three forces is to be vertical, (b) the corresponding magnitude of the resultant.

Fig. P2.37 and P2.38

Fig. P2.41 and P2.42

2.9 EQUILIBRIUM OF A PARTICLE

In the preceding sections, we discussed the methods for determining the resultant of several forces acting on a particle. Although it has not occurred in any of the problems considered so far, it is quite possible for the resultant to be zero. In such a case, the net effect of the given forces is zero, and the particle is said to be in equilibrium. We thus have the following definition: *When the resultant of all the forces acting on a particle is zero, the particle is in equilibrium.*

A particle which is acted upon by two forces will be in equilibrium if the two forces have the same magnitude and the same line of action but opposite sense. The resultant of the two forces is then zero. Such a case is shown in Fig. 2.26.

Fig. 2.26

Fig. 2.27

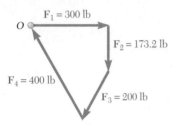

Fig. 2.28

Another case of equilibrium of a particle is represented in Fig. 2.27, where four forces are shown acting on A. In Fig. 2.28, the resultant of the given forces is determined by the polygon rule. Starting from point O with $\mathbf{F}_1$ and arranging the forces in tip-to-tail fashion, we find that the tip of $\mathbf{F}_4$ coincides with the starting point O. Thus the resultant $\mathbf{R}$ of the given system of forces is zero, and the particle is in equilibrium.

The closed polygon drawn in Fig. 2.28 provides a *graphical* expression of the equilibrium of A. To express *algebraically* the conditions for the equilibrium of a particle, we write

$$\mathbf{R} = \Sigma\mathbf{F} = 0 \qquad (2.14)$$

Resolving each force $\mathbf{F}$ into rectangular components, we have

$$\Sigma(F_x\mathbf{i} + F_y\mathbf{j}) = 0 \qquad \text{or} \qquad (\Sigma F_x)\mathbf{i} + (\Sigma F_y)\mathbf{j} = 0$$

We conclude that the necessary and sufficient conditions for the equilibrium of a particle are

$$\Sigma F_x = 0 \qquad \Sigma F_y = 0 \qquad (2.15)$$

Returning to the particle shown in Fig. 2.27, we check that the equilibrium conditions are satisfied. We write

$$\Sigma F_x = 300 \text{ lb} - (200 \text{ lb}) \sin 30° - (400 \text{ lb}) \sin 30°$$
$$= 300 \text{ lb} - 100 \text{ lb} - 200 \text{ lb} = 0$$
$$\Sigma F_y = -173.2 \text{ lb} - (200 \text{ lb}) \cos 30° + (400 \text{ lb}) \cos 30°$$
$$= -173.2 \text{ lb} - 173.2 \text{ lb} + 346.4 \text{ lb} = 0$$

2.10 NEWTON'S FIRST LAW OF MOTION

In the latter part of the seventeenth century, Sir Isaac Newton formulated three fundamental laws upon which the science of mechanics is based. The first of these laws can be stated as follows:

If the resultant force acting on a particle is zero, the particle will remain at rest (if originally at rest) or will move with constant speed in a straight line (if originally in motion).

From this law and from the definition of equilibrium given in Sec. 2.9, it is seen that a particle in equilibrium either is at rest or is moving in a straight line with constant speed. In the following section, various problems concerning the equilibrium of a particle will be considered.

2.11 PROBLEMS INVOLVING THE EQUILIBRIUM OF A PARTICLE. FREE-BODY DIAGRAMS

In practice, a problem in engineering mechanics is derived from an actual physical situation. A sketch showing the physical conditions of the problem is known as a *space diagram.*

The methods of analysis discussed in the preceding sections apply to a system of forces acting on a particle. A large number of problems involving actual structures, however, can be reduced to problems concerning the equilibrium of a particle. This is done by

choosing a significant particle and drawing a separate diagram showing this particle and all the forces acting on it. Such a diagram is
called a *free-body diagram.*

As an example, consider the 75-kg crate shown in the space
diagram of Fig. 2.29a. This crate was lying between two buildings,
and it is now being lifted onto a truck, which will remove it. The crate
is supported by a vertical cable, which is joined at A to two ropes
which pass over pulleys attached to the buildings at B and C. It is
desired to determine the tension in each of the ropes AB and AC.

In order to solve this problem, a free-body diagram showing a
particle in equilibrium must be drawn. Since we are interested in
the rope tensions, the free-body diagram should include at least one
of these tensions or, if possible, both tensions. Point A is seen to be
a good free body for this problem. The free-body diagram of point
A is shown in Fig. 2.29b. It shows point A and the forces exerted on
A by the vertical cable and the two ropes. The force exerted by the
cable is directed downward, and its magnitude is equal to the weight
W of the crate. Recalling Eq. (1.4), we write

$$W = mg = (75 \text{ kg})(9.81 \text{ m/s}^2) = 736 \text{ N}$$

and indicate this value in the free-body diagram. The forces exerted
by the two ropes are not known. Since they are respectively equal
in magnitude to the tensions in rope AB and rope AC, we denote
them by $\mathbf{T}_{AB}$ and $\mathbf{T}_{AC}$ and draw them away from A in the directions
shown in the space diagram. No other detail is included in the free-
body diagram.

Since point A is in equilibrium, the three forces acting on it
must form a closed triangle when drawn in tip-to-tail fashion. This
force triangle has been drawn in Fig. 2.29c. The values T_{AB} and T_{AC}
of the tension in the ropes may be found graphically if the triangle
is drawn to scale, or they may be found by trigonometry. If the latter
method of solution is chosen, we use the law of sines and write

$$\frac{T_{AB}}{\sin 60°} = \frac{T_{AC}}{\sin 40°} = \frac{736 \text{ N}}{\sin 80°}$$

$$T_{AB} = 647 \text{ N} \qquad T_{AC} = 480 \text{ N}$$

When a particle is in *equilibrium under three forces,* the problem
can be solved by drawing a force triangle. When a particle is in *equilibrium under more than three forces,* the problem can be solved graphically by drawing a force polygon. If an analytic solution is desired, the
equations of equilibrium given in Sec. 2.9 should be solved:

$$\Sigma F_x = 0 \qquad \Sigma F_y = 0 \qquad (2.15)$$

These equations can be solved for no more than *two unknowns;*
similarly, the force triangle used in the case of equilibrium under
three forces can be solved for two unknowns.

The more common types of problems are those in which the
two unknowns represent (1) the two components (or the magnitude
and direction) of a single force, (2) the magnitudes of two forces,
each of known direction. Problems involving the determination of
the maximum or minimum value of the magnitude of a force are also
encountered (see Probs. 2.57 through 2.62).

(a) Space diagram

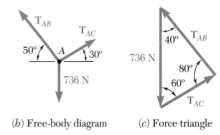

(b) Free-body diagram   (c) Force triangle

Fig. 2.29

Photo 2.1   As illustrated in the above example,
it is possible to determine the tensions in the
cables supporting the shaft shown by treating
the hook as a particle and then applying the
equations of equilibrium to the forces acting on
the hook.

SAMPLE PROBLEM 2.4

In a ship-unloading operation, a 3500-lb automobile is supported by a cable. A rope is tied to the cable at A and pulled in order to center the automobile over its intended position. The angle between the cable and the vertical is 2°, while the angle between the rope and the horizontal is 30°. What is the tension in the rope?

SOLUTION

Free-Body Diagram. Point A is chosen as a free body, and the complete free-body diagram is drawn. T_{AB} is the tension in the cable AB, and T_{AC} is the tension in the rope.

Equilibrium Condition. Since only three forces act on the free body, we draw a force triangle to express that it is in equilibrium. Using the law of sines, we write

$$\frac{T_{AB}}{\sin 120°} = \frac{T_{AC}}{\sin 2°} = \frac{3500 \text{ lb}}{\sin 58°}$$

With a calculator, we first compute and store the value of the last quotient. Multiplying this value successively by sin 120° and sin 2°, we obtain

$$T_{AB} = 3570 \text{ lb} \qquad T_{AC} = 144 \text{ lb} \quad \blacktriangleleft$$

SAMPLE PROBLEM 2.5

Determine the magnitude and direction of the smallest force $\mathbf{F}$ which will maintain the package shown in equilibrium. Note that the force exerted by the rollers on the package is perpendicular to the incline.

SOLUTION

Free-Body Diagram. We choose the package as a free body, assuming that it can be treated as a particle. We draw the corresponding free-body diagram.

Equilibrium Condition. Since only three forces act on the free body, we draw a force triangle to express that it is in equilibrium. Line 1-$1'$ represents the known direction of $\mathbf{P}$. In order to obtain the minimum value of the force $\mathbf{F}$, we choose the direction of $\mathbf{F}$ perpendicular to that of $\mathbf{P}$. From the geometry of the triangle obtained, we find

$$F = (294 \text{ N}) \sin 15° = 76.1 \text{ N} \qquad \alpha = 15°$$
$$\mathbf{F} = 76.1 \text{ N} \angle 15° \quad \blacktriangleleft$$

SAMPLE PROBLEM 2.6

As part of the design of a new sailboat, it is desired to determine the drag force which may be expected at a given speed. To do so, a model of the proposed hull is placed in a test channel and three cables are used to keep its bow on the centerline of the channel. Dynamometer readings indicate that for a given speed, the tension is 40 lb in cable AB and 60 lb in cable AE. Determine the drag force exerted on the hull and the tension in cable AC.

SOLUTION

Determination of the Angles. First, the angles α and β defining the direction of cables AB and AC are determined. We write

$$\tan \alpha = \frac{7 \text{ ft}}{4 \text{ ft}} = 1.75 \qquad \tan \beta = \frac{1.5 \text{ ft}}{4 \text{ ft}} = 0.375$$

$$\alpha = 60.26° \qquad\qquad \beta = 20.56°$$

Free-Body Diagram. Choosing the hull as a free body, we draw the free-body diagram shown. It includes the forces exerted by the three cables on the hull, as well as the drag force $\mathbf{F}_D$ exerted by the flow.

Equilibrium Condition. We express that the hull is in equilibrium by writing that the resultant of all forces is zero:

$$\mathbf{R} = \mathbf{T}_{AB} + \mathbf{T}_{AC} + \mathbf{T}_{AE} + \mathbf{F}_D = 0 \tag{1}$$

Since more than three forces are involved, we resolve the forces into x and y components:

$$\mathbf{T}_{AB} = -(40 \text{ lb}) \sin 60.26°\mathbf{i} + (40 \text{ lb}) \cos 60.26°\mathbf{j}$$
$$= -(34.73 \text{ lb})\mathbf{i} + (19.84 \text{ lb})\mathbf{j}$$
$$\mathbf{T}_{AC} = T_{AC} \sin 20.56°\mathbf{i} + T_{AC} \cos 20.56°\mathbf{j}$$
$$= 0.3512 T_{AC}\mathbf{i} + 0.9363 T_{AC}\mathbf{j}$$
$$\mathbf{T}_{AE} = -(60 \text{ lb})\mathbf{j}$$
$$\mathbf{F}_D = F_D\mathbf{i}$$

Substituting the expressions obtained into Eq. (1) and factoring the unit vectors $\mathbf{i}$ and $\mathbf{j}$, we have

$$(-34.73 \text{ lb} + 0.3512 T_{AC} + F_D)\mathbf{i} + (19.84 \text{ lb} + 0.9363 T_{AC} - 60 \text{ lb})\mathbf{j} = 0$$

This equation will be satisfied if, and only if, the coefficients of $\mathbf{i}$ and $\mathbf{j}$ are equal to zero. We thus obtain the following two equilibrium equations, which express, respectively, that the sum of the x components and the sum of the y components of the given forces must be zero.

$$(\Sigma F_x = 0:) \qquad -34.73 \text{ lb} + 0.3512 T_{AC} + F_D = 0 \tag{2}$$
$$(\Sigma F_y = 0:) \qquad 19.84 \text{ lb} + 0.9363 T_{AC} - 60 \text{ lb} = 0 \tag{3}$$

From Eq. (3) we find $\qquad\qquad\qquad\qquad\qquad T_{AC} = +42.9 \text{ lb}$ ◄

and, substituting this value into Eq. (2), $\qquad\qquad F_D = +19.66 \text{ lb}$ ◄

In drawing the free-body diagram, we assumed a sense for each unknown force. A positive sign in the answer indicates that the assumed sense is correct. The complete force polygon may be drawn to check the results.

SOLVING PROBLEMS ON YOUR OWN

When a particle is in *equilibrium,* the resultant of the forces acting on the particle must be zero. Expressing this fact in the case of a particle under *coplanar forces* will provide you with two relations among these forces. As you saw in the preceding sample problems, these relations may be used to determine two unknowns—such as the magnitude and direction of one force or the magnitudes of two forces.

Drawing a free-body diagram is the first step in the solution of a problem involving the equilibrium of a particle. This diagram shows the particle and all the forces acting on it. Indicate in your free-body diagram the magnitudes of known forces, as well as any angle or dimensions that define the direction of a force. Any unknown magnitude or angle should be denoted by an appropriate symbol. Nothing else should be included in the free-body diagram.

Drawing a clear and accurate free-body diagram is a must in the solution of any equilibrium problem. Skipping this step might save you pencil and paper, but is very likely to lead you to a wrong solution.

Case 1. If only three forces are involved in the free-body diagram, the rest of the solution is best carried out by drawing these forces in tip-to-tail fashion to form a *force triangle.* This triangle can be solved graphically or by trigonometry for no more than two unknowns [Sample Probs. 2.4 and 2.5].

Case 2. If more than three forces are involved, it is to your advantage to use an *analytic solution.* You select x and y axes and resolve each of the forces shown in the free-body diagram into x and y components. Expressing that the sum of the x components and the sum of the y components of all the forces are both zero, you will obtain two equations which you can solve for no more than two unknowns [Sample Prob. 2.6].

It is strongly recommended that when using an analytic solution the equations of equilibrium be written in the same form as Eqs. (2) and (3) of Sample Prob. 2.6. The practice adopted by some students of initially placing the unknowns on the left side of the equation and the known quantities on the right side may lead to confusion in assigning the appropriate sign to each term.

We have noted that regardless of the method used to solve a two-dimensional equilibrium problem we can determine at most two unknowns. If a two-dimensional problem involves more than two unknowns, one or more additional relations must be obtained from the information contained in the statement of the problem.

PROBLEMS

FREE BODY PRACTICE PROBLEMS

2.F1 Two cables are tied together at C and loaded as shown. Draw the free-body diagram needed to determine the tension in AC and BC.

2.F2 A chairlift has been stopped in the position shown. Knowing that each chair weighs 250 N and that the skier in chair E weighs 765 N, draw the free-body diagrams needed to determine the weight of the skier in chair F.

Fig. P2.F1

Fig. P2.F2

2.F3 Two cables are tied together at A and loaded as shown. Draw the free-body diagram needed to determine the tension in each cable.

2.F4 The 60-lb collar A can slide on a frictionless vertical rod and is connected as shown to a 65-lb counterweight C. Draw the free-body diagram needed to determine the value of h for which the system is in equilibrium.

Fig. P2.F3

Fig. P2.F4

END-OF-SECTION PROBLEMS

2.43 and 2.44   Two cables are tied together at C and are loaded as shown. Determine the tension (*a*) in cable AC, (*b*) in cable BC.

Fig. P2.43   **Fig. *P2.44***

2.45   Knowing that $\alpha = 20°$, determine the tension (*a*) in cable AC, (*b*) in rope BC.

Fig. P2.46

Fig. P2.45

2.46   Knowing that $\alpha = 55°$ and that boom AC exerts on pin C a force directed along line AC, determine (*a*) the magnitude of that force, (*b*) the tension in cable BC.

2.47   Two cables are tied together at C and loaded as shown. Determine the tension (*a*) in cable AC, (*b*) in cable BC.

Fig. *P2.47*

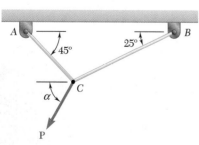

Fig. P2.48

2.48   Two cables are tied together at C and are loaded as shown. Knowing that **P** = 500 N and $\alpha = 60°$, determine the tension (*a*) in cable AC, (*b*) in cable BC.

2.49 Two forces of magnitude $T_A = 8$ kips and $T_B = 15$ kips are applied as shown to a welded connection. Knowing that the connection is in equilibrium, determine the magnitudes of the forces T_C and T_D.

2.50 Two forces of magnitude $T_A = 6$ kips and $T_C = 9$ kips are applied as shown to a welded connection. Knowing that the connection is in equilibrium, determine the magnitudes of the forces T_B and T_D.

2.51 Two cables are tied together at C and loaded as shown. Knowing that $P = 360$ N, determine the tension (*a*) in cable AC, (*b*) in cable BC.

2.52 Two cables are tied together at C and loaded as shown. Determine the range of values of P for which both cables remain taut.

2.53 A sailor is being rescued using a boatswain's chair that is suspended from a pulley that can roll freely on the support cable ACB and is pulled at a constant speed by cable CD. Knowing that $\alpha = 30°$ and $\beta = 10°$ and that the combined weight of the boatswain's chair and the sailor is 900 N, determine the tension (*a*) in the support cable ACB, (*b*) in the traction cable CD.

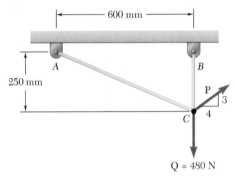

Fig. P2.49 and P2.50

Fig. P2.51 and P2.52

Fig. P2.53 and P2.54

2.54 A sailor is being rescued using a boatswain's chair that is suspended from a pulley that can roll freely on the support cable ACB and is pulled at a constant speed by cable CD. Knowing that $\alpha = 25°$ and $\beta = 15°$ and that the tension in cable CD is 80 N, determine (*a*) the combined weight of the boatswain's chair and the sailor, (*b*) the tension in the support cable ACB.

2.55 Two forces **P** and **Q** are applied as shown to an aircraft connection. Knowing that the connection is in equilibrium and that $P = 500$ lb and $Q = 650$ lb, determine the magnitudes of the forces exerted on the rods A and B.

2.56 Two forces **P** and **Q** are applied as shown to an aircraft connection. Knowing that the connection is in equilibrium and that the magnitudes of the forces exerted on rods A and B are $F_A = 750$ lb and $F_B = 400$ lb, determine the magnitudes of **P** and **Q**.

Fig. P2.55 and P2.56

Fig. P2.57 and P2.58

Fig. P2.62

2.57 Two cables tied together at C are loaded as shown. Knowing that the maximum allowable tension in each cable is 800 N, determine (*a*) the magnitude of the largest force **P** that can be applied at C, (*b*) the corresponding value of α.

2.58 Two cables tied together at C are loaded as shown. Knowing that the maximum allowable tension is 1200 N in cable AC and 600 N in cable BC, determine (*a*) the magnitude of the largest force **P** that can be applied at C, (*b*) the corresponding value of α.

2.59 For the situation described in Fig. P2.45, determine (*a*) the value of α for which the tension in rope BC is as small as possible, (*b*) the corresponding value of the tension.

2.60 For the structure and loading of Prob. 2.46, determine (*a*) the value of α for which the tension in cable BC is as small as possible, (*b*) the corresponding value of the tension.

2.61 For the cables of Prob. 2.48, it is known that the maximum allowable tension is 600 N in cable AC and 750 N in cable BC. Determine (*a*) the maximum force **P** that can be applied at C, (*b*) the corresponding value of α.

2.62 A movable bin and its contents have a combined weight of 2.8 kN. Determine the shortest chain sling ACB that can be used to lift the loaded bin if the tension in the chain is not to exceed 5 kN.

2.63 Collar A is connected as shown to a 50-lb load and can slide on a frictionless horizontal rod. Determine the magnitude of the force **P** required to maintain the equilibrium of the collar when (*a*) $x = 4.5$ in., (*b*) $x = 15$ in.

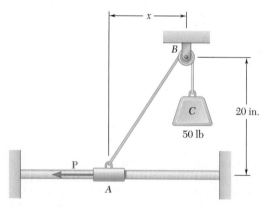

Fig. P2.63 and P2.64

2.64 Collar A is connected as shown to a 50-lb load and can slide on a frictionless horizontal rod. Determine the distance x for which the collar is in equilibrium when $P = 48$ lb.

2.65 Three forces are applied to a bracket as shown. The directions of the two 150-N forces may vary, but the angle between these forces is always 50°. Determine the range of values of α for which the magnitude of the resultant of the forces acting at A is less than 600 N.

500 N

30°

A

α

150 N

50°

150 N

Fig. P2.65

2.66 A 200-kg crate is to be supported by the rope-and-pulley arrangement shown. Determine the magnitude and direction of the force **P** that must be exerted on the free end of the rope to maintain equilibrium. (*Hint:* The tension in the rope is the same on each side of a simple pulley. This can be proved by the methods of Chap. 4.)

Fig. P2.66

2.67 A 600-lb crate is supported by several rope-and-pulley arrangements as shown. Determine for each arrangement the tension in the rope. (See the hint for Prob. 2.66.)

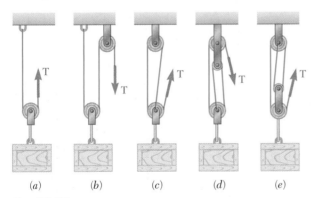

(*a*)   (*b*)   (*c*)   (*d*)   (*e*)

Fig. P2.67

2.68 Solve parts *b* and *d* of Prob. 2.67, assuming that the free end of the rope is attached to the crate.

2.69 A load **Q** is applied to the pulley *C*, which can roll on the cable *ACB*. The pulley is held in the position shown by a second cable *CAD*, which passes over the pulley *A* and supports a load **P**. Knowing that *P* = 750 N, determine (*a*) the tension in cable *ACB*, (*b*) the magnitude of load **Q**.

2.70 An 1800-N load **Q** is applied to the pulley *C*, which can roll on the cable *ACB*. The pulley is held in the position shown by a second cable *CAD*, which passes over the pulley *A* and supports a load **P**. Determine (*a*) the tension in cable *ACB*, (*b*) the magnitude of load **P**.

Fig. P2.69 and *P2.70*

(a)

(b)

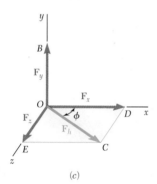

(c)

Fig. 2.30

2.12 RECTANGULAR COMPONENTS OF A FORCE IN SPACE

The problems considered in the first part of this chapter involved only two dimensions; they could be formulated and solved in a single plane. In this section and in the remaining sections of the chapter, we will discuss problems involving the three dimensions of space.

Consider a force $\mathbf{F}$ acting at the origin O of the system of rectangular coordinates x, y, z. To define the direction of $\mathbf{F}$, we draw the vertical plane $OBAC$ containing $\mathbf{F}$ (Fig. 2.30a). This plane passes through the vertical y axis; its orientation is defined by the angle ϕ it forms with the xy plane. The direction of $\mathbf{F}$ within the plane is defined by the angle θ_y that $\mathbf{F}$ forms with the y axis. The force $\mathbf{F}$ may be resolved into a vertical component $\mathbf{F}_y$ and a horizontal component $\mathbf{F}_h$; this operation, shown in Fig. 2.30b, is carried out in plane $OBAC$ according to the rules developed in the first part of the chapter. The corresponding scalar components are

$$F_y = F \cos \theta_y \qquad F_h = F \sin \theta_y \qquad (2.16)$$

But $\mathbf{F}_h$ may be resolved into two rectangular components $\mathbf{F}_x$ and $\mathbf{F}_z$ along the x and z axes, respectively. This operation, shown in Fig. 2.30c, is carried out in the xz plane. We obtain the following expressions for the corresponding scalar components:

$$F_x = F_h \cos \phi = F \sin \theta_y \cos \phi$$
$$F_z = F_h \sin \phi = F \sin \theta_y \sin \phi \qquad (2.17)$$

The given force $\mathbf{F}$ has thus been resolved into three rectangular vector components $\mathbf{F}_x$, $\mathbf{F}_y$, $\mathbf{F}_z$, which are directed along the three coordinate axes.

Applying the Pythagorean theorem to the triangles OAB and OCD of Fig. 2.30, we write

$$F^2 = (OA)^2 = (OB)^2 + (BA)^2 = F_y^2 + F_h^2$$
$$F_h^2 = (OC)^2 = (OD)^2 + (DC)^2 = F_x^2 + F_z^2$$

Eliminating F_h^2 from these two equations and solving for F, we obtain the following relation between the magnitude of $\mathbf{F}$ and its rectangular scalar components:

$$F = \sqrt{F_x^2 + F_y^2 + F_z^2} \qquad (2.18)$$

The relationship existing between the force $\mathbf{F}$ and its three components $\mathbf{F}_x$, $\mathbf{F}_y$, $\mathbf{F}_z$ is more easily visualized if a "box" having $\mathbf{F}_x$, $\mathbf{F}_y$, $\mathbf{F}_z$ for edges is drawn as shown in Fig. 2.31. The force $\mathbf{F}$ is then represented by the diagonal OA of this box. Figure 2.31b shows the right triangle OAB used to derive the first of the formulas (2.16): $F_y = F \cos \theta_y$. In Fig. 2.31a and c, two other right triangles have also been drawn: OAD and OAE. These triangles are seen to occupy in the box positions comparable with that of triangle OAB. Denoting by θ_x and θ_z, respectively, the angles that $\mathbf{F}$ forms

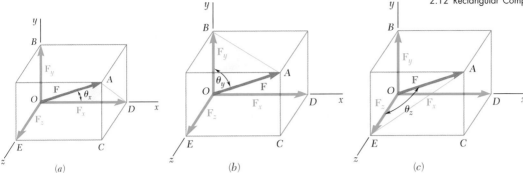

Fig. 2.31

with the x and z axes, we can derive two formulas similar to $F_y = F \cos \theta_y$. We thus write

$$F_x = F \cos \theta_x \qquad F_y = F \cos \theta_y \qquad F_z = F \cos \theta_z \quad (2.19)$$

The three angles θ_x, θ_y, θ_z define the direction of the force **F**; they are more commonly used for this purpose than the angles θ_y and ϕ introduced at the beginning of this section. The cosines of θ_x, θ_y, θ_z are known as the *direction cosines* of the force **F**.

Introducing the unit vectors **i**, **j**, and **k**, directed respectively along the x, y, and z axes (Fig. 2.32), we can express **F** in the form

$$\mathbf{F} = F_x\mathbf{i} + F_y\mathbf{j} + F_z\mathbf{k} \qquad (2.20)$$

where the scalar components F_x, F_y, F_z are defined by the relations (2.19).

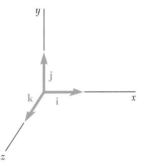

Fig. 2.32

EXAMPLE 1. A force of 500 N forms angles of 60°, 45°, and 120°, respectively, with the x, y, and z axes. Find the components F_x, F_y, and F_z of the force.

Substituting $F = 500$ N, $\theta_x = 60°$, $\theta_y = 45°$, $\theta_z = 120°$ into formulas (2.19), we write

$$F_x = (500 \text{ N}) \cos 60° = +250 \text{ N}$$
$$F_y = (500 \text{ N}) \cos 45° = +354 \text{ N}$$
$$F_z = (500 \text{ N}) \cos 120° = -250 \text{ N}$$

Carrying into Eq. (2.20) the values obtained for the scalar components of **F**, we have

$$\mathbf{F} = (250 \text{ N})\mathbf{i} + (354 \text{ N})\mathbf{j} - (250 \text{ N})\mathbf{k}$$

As in the case of two-dimensional problems, a plus sign indicates that the component has the same sense as the corresponding axis, and a minus sign indicates that it has the opposite sense. ■

The angle a force **F** forms with an axis should be measured from the positive side of the axis and will always be between 0 and 180°. An angle θ_x smaller than 90° (acute) indicates that **F** (assumed attached to O) is on the same side of the yz plane as the positive x axis; $\cos \theta_x$ and F_x will then be positive. An angle θ_x larger than 90° (obtuse) indicates

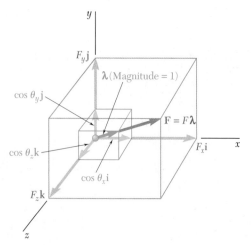

Fig. 2.33

that $\mathbf{F}$ is on the other side of the yz plane; $\cos \theta_x$ and F_x will then be negative. In Example 1 the angles θ_x and θ_y are acute, while θ_z is obtuse; consequently, F_x and F_y are positive, while F_z is negative.

Substituting into (2.20) the expressions obtained for F_x, F_y, F_z in (2.19), we write

$$\mathbf{F} = F(\cos \theta_x \mathbf{i} + \cos \theta_y \mathbf{j} + \cos \theta_z \mathbf{k}) \qquad (2.21)$$

which shows that the force $\mathbf{F}$ can be expressed as the product of the scalar F and the vector

$$\boldsymbol{\lambda} = \cos \theta_x \mathbf{i} + \cos \theta_y \mathbf{j} + \cos \theta_z \mathbf{k} \qquad (2.22)$$

Clearly, the vector $\boldsymbol{\lambda}$ is a vector whose magnitude is equal to 1 and whose direction is the same as that of $\mathbf{F}$ (Fig. 2.33). The vector $\boldsymbol{\lambda}$ is referred to as the *unit vector* along the line of action of $\mathbf{F}$. It follows from (2.22) that the components of the unit vector $\boldsymbol{\lambda}$ are respectively equal to the direction cosines of the line of action of $\mathbf{F}$:

$$\lambda_x = \cos \theta_x \qquad \lambda_y = \cos \theta_y \qquad \lambda_z = \cos \theta_z \qquad (2.23)$$

We should observe that the values of the three angles θ_x, θ_y, θ_z are not independent. Recalling that the sum of the squares of the components of a vector is equal to the square of its magnitude, we write

$$\lambda_x^2 + \lambda_y^2 + \lambda_z^2 = 1$$

or, substituting for λ_x, λ_y, λ_z from (2.23),

$$\cos^2 \theta_x + \cos^2 \theta_y + \cos^2 \theta_z = 1 \qquad (2.24)$$

In Example 1, for instance, once the values $\theta_x = 60°$ and $\theta_y = 45°$ have been selected, the value of θ_z *must* be equal to 60° or 120° in order to satisfy identity (2.24).

When the components F_x, F_y, F_z of a force $\mathbf{F}$ are given, the magnitude F of the force is obtained from (2.18).† The relations (2.19) can then be solved for the direction cosines,

$$\cos \theta_x = \frac{F_x}{F} \quad \cos \theta_y = \frac{F_y}{F} \quad \cos \theta_z = \frac{F_z}{F} \qquad (2.25)$$

and the angles θ_x, θ_y, θ_z characterizing the direction of $\mathbf{F}$ can be found.

EXAMPLE 2.   A force $\mathbf{F}$ has the components $F_x = 20$ lb, $F_y = -30$ lb, $F_z = 60$ lb. Determine its magnitude F and the angles θ_x, θ_y, θ_z it forms with the coordinate axes.

From formula (2.18) we obtain†

$$\begin{aligned}
F &= \sqrt{F_x^2 + F_y^2 + F_z^2} \\
&= \sqrt{(20 \text{ lb})^2 + (-30 \text{ lb})^2 + (60 \text{ lb})^2} \\
&= \sqrt{4900} \text{ lb} = 70 \text{ lb}
\end{aligned}$$

---

†With a calculator programmed to convert rectangular coordinates into polar coordinates, the following procedure will be found more expeditious for computing F: First determine F_h from its two rectangular components F_x and F_z (Fig. 2.30c), then determine F from its two rectangular components F_h and F_y (Fig. 2.30b). The actual order in which the three components F_x, F_y, F_z are entered is immaterial.

Substituting the values of the components and magnitude of **F** into Eqs. (2.25), we write

$$\cos \theta_x = \frac{F_x}{F} = \frac{20 \text{ lb}}{70 \text{ lb}} \qquad \cos \theta_y = \frac{F_y}{F} = \frac{-30 \text{ lb}}{70 \text{ lb}} \qquad \cos \theta_z = \frac{F_z}{F} = \frac{60 \text{ lb}}{70 \text{ lb}}$$

Calculating successively each quotient and its arc cosine, we obtain

$$\theta_x = 73.4° \qquad \theta_y = 115.4° \qquad \theta_z = 31.0°$$

These computations can be carried out easily with a calculator. ∎

2.13 FORCE DEFINED BY ITS MAGNITUDE AND TWO POINTS ON ITS LINE OF ACTION

In many applications, the direction of a force **F** is defined by the coordinates of two points, $M(x_1, y_1, z_1)$ and $N(x_2, y_2, z_2)$, located on its line of action (Fig. 2.34). Consider the vector $\overrightarrow{MN}$ joining M and N

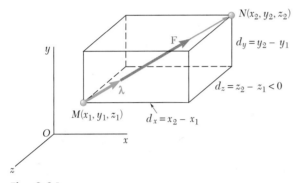

Fig. 2.34

and of the same sense as **F**. Denoting its scalar components by d_x, d_y, d_z, respectively, we write

$$\overrightarrow{MN} = d_x\mathbf{i} + d_y\mathbf{j} + d_z\mathbf{k} \qquad (2.26)$$

The unit vector $\boldsymbol{\lambda}$ along the line of action of **F** (i.e., along the line MN) may be obtained by dividing the vector $\overrightarrow{MN}$ by its magnitude MN. Substituting for $\overrightarrow{MN}$ from (2.26) and observing that MN is equal to the distance d from M to N, we write

$$\boldsymbol{\lambda} = \frac{\overrightarrow{MN}}{MN} = \frac{1}{d}(d_x\mathbf{i} + d_y\mathbf{j} + d_z\mathbf{k}) \qquad (2.27)$$

Recalling that **F** is equal to the product of F and λ, we have

$$\mathbf{F} = F\boldsymbol{\lambda} = \frac{F}{d}(d_x\mathbf{i} + d_y\mathbf{j} + d_z\mathbf{k}) \qquad (2.28)$$

from which it follows that the scalar components of **F** are, respectively,

$$F_x = \frac{Fd_x}{d} \qquad F_y = \frac{Fd_y}{d} \qquad F_z = \frac{Fd_z}{d} \qquad (2.29)$$

The relations (2.29) considerably simplify the determination of the components of a force $\mathbf{F}$ of given magnitude F when the line of action of $\mathbf{F}$ is defined by two points M and N. Subtracting the coordinates of M from those of N, we first determine the components of the vector $\overrightarrow{MN}$ and the distance d from M to N:

$$d_x = x_2 - x_1 \qquad d_y = y_2 - y_1 \qquad d_z = z_2 - z_1$$
$$d = \sqrt{d_x^2 + d_y^2 + d_z^2}$$

Substituting for F and for d_x, d_y, d_z, and d into the relations (2.29), we obtain the components F_x, F_y, F_z of the force.

The angles θ_x, θ_y, θ_z that $\mathbf{F}$ forms with the coordinate axes can then be obtained from Eqs. (2.25). Comparing Eqs. (2.22) and (2.27), we can also write

$$\cos \theta_x = \frac{d_x}{d} \qquad \cos \theta_y = \frac{d_y}{d} \qquad \cos \theta_z = \frac{d_z}{d} \qquad (2.30)$$

and determine the angles θ_x, θ_y, θ_z directly from the components and magnitude of the vector $\overrightarrow{MN}$.

2.14   ADDITION OF CONCURRENT FORCES IN SPACE

The resultant $\mathbf{R}$ of two or more forces in space will be determined by summing their rectangular components. Graphical or trigonometric methods are generally not practical in the case of forces in space.

The method followed here is similar to that used in Sec. 2.8 with coplanar forces. Setting

$$\mathbf{R} = \Sigma \mathbf{F}$$

we resolve each force into its rectangular components and write

$$R_x \mathbf{i} + R_y \mathbf{j} + R_z \mathbf{k} = \Sigma(F_x \mathbf{i} + F_y \mathbf{j} + F_z \mathbf{k})$$
$$= (\Sigma F_x)\mathbf{i} + (\Sigma F_y)\mathbf{j} + (\Sigma F_z)\mathbf{k}$$

from which it follows that

$$R_x = \Sigma F_x \qquad R_y = \Sigma F_y \qquad R_z = \Sigma F_z \qquad (2.31)$$

The magnitude of the resultant and the angles θ_x, θ_y, θ_z that the resultant forms with the coordinate axes are obtained using the method discussed in Sec. 2.12. We write

$$R = \sqrt{R_x^2 + R_y^2 + R_z^2} \qquad (2.32)$$

$$\cos \theta_x = \frac{R_x}{R} \qquad \cos \theta_y = \frac{R_y}{R} \qquad \cos \theta_z = \frac{R_z}{R} \qquad (2.33)$$

SAMPLE PROBLEM 2.7

A tower guy wire is anchored by means of a bolt at A. The tension in the wire is 2500 N. Determine (*a*) the components F_x, F_y, F_z of the force acting on the bolt, (*b*) the angles θ_x, θ_y, θ_z defining the direction of the force.

SOLUTION

a. Components of the Force. The line of action of the force acting on the bolt passes through A and B, and the force is directed from A to B. The components of the vector $\overrightarrow{AB}$, which has the same direction as the force, are

$$d_x = -40 \text{ m} \qquad d_y = +80 \text{ m} \qquad d_z = +30 \text{ m}$$

The total distance from A to B is

$$AB = d = \sqrt{d_x^2 + d_y^2 + d_z^2} = 94.3 \text{ m}$$

Denoting by $\mathbf{i}$, $\mathbf{j}$, $\mathbf{k}$ the unit vectors along the coordinate axes, we have

$$\overrightarrow{AB} = -(40 \text{ m})\mathbf{i} + (80 \text{ m})\mathbf{j} + (30 \text{ m})\mathbf{k}$$

Introducing the unit vector $\boldsymbol{\lambda} = \overrightarrow{AB}/AB$, we write

$$\mathbf{F} = F\boldsymbol{\lambda} = F\frac{\overrightarrow{AB}}{AB} = \frac{2500 \text{ N}}{94.3 \text{ m}}\overrightarrow{AB}$$

Substituting the expression found for $\overrightarrow{AB}$, we obtain

$$\mathbf{F} = \frac{2500 \text{ N}}{94.3 \text{ m}}[-(40 \text{ m})\mathbf{i} + (80 \text{ m})\mathbf{j} + (30 \text{ m})\mathbf{k}]$$

$$\mathbf{F} = -(1060 \text{ N})\mathbf{i} + (2120 \text{ N})\mathbf{j} + (795 \text{ N})\mathbf{k}$$

The components of $\mathbf{F}$, therefore, are

$$F_x = -1060 \text{ N} \qquad F_y = +2120 \text{ N} \qquad F_z = +795 \text{ N} \blacktriangleleft$$

b. Direction of the Force. Using Eqs. (2.25), we write

$$\cos\theta_x = \frac{F_x}{F} = \frac{-1060 \text{ N}}{2500 \text{ N}} \qquad \cos\theta_y = \frac{F_y}{F} = \frac{+2120 \text{ N}}{2500 \text{ N}}$$

$$\cos\theta_z = \frac{F_z}{F} = \frac{+795 \text{ N}}{2500 \text{ N}}$$

Calculating successively each quotient and its arc cosine, we obtain

$$\theta_x = 115.1° \qquad \theta_y = 32.0° \qquad \theta_z = 71.5° \blacktriangleleft$$

(*Note.* This result could have been obtained by using the components and magnitude of the vector $\overrightarrow{AB}$ rather than those of the force $\mathbf{F}$.)

SAMPLE PROBLEM 2.8

A wall section of precast concrete is temporarily held by the cables shown. Knowing that the tension is 840 lb in cable AB and 1200 lb in cable AC, determine the magnitude and direction of the resultant of the forces exerted by cables AB and AC on stake A.

SOLUTION

Components of the Forces. The force exerted by each cable on stake A will be resolved into x, y, and z components. We first determine the components and magnitude of the vectors $\overrightarrow{AB}$ and $\overrightarrow{AC}$, measuring them from A toward the wall section. Denoting by $\mathbf{i}$, $\mathbf{j}$, $\mathbf{k}$ the unit vectors along the coordinate axes, we write

$$\overrightarrow{AB} = -(16\text{ ft})\mathbf{i} + (8\text{ ft})\mathbf{j} + (11\text{ ft})\mathbf{k} \qquad AB = 21\text{ ft}$$
$$\overrightarrow{AC} = -(16\text{ ft})\mathbf{i} + (8\text{ ft})\mathbf{j} - (16\text{ ft})\mathbf{k} \qquad AC = 24\text{ ft}$$

Denoting by $\boldsymbol{\lambda}_{AB}$ the unit vector along AB, we have

$$\mathbf{T}_{AB} = T_{AB}\boldsymbol{\lambda}_{AB} = T_{AB}\frac{\overrightarrow{AB}}{AB} = \frac{840\text{ lb}}{21\text{ ft}}\overrightarrow{AB}$$

Substituting the expression found for $\overrightarrow{AB}$, we obtain

$$\mathbf{T}_{AB} = \frac{840\text{ lb}}{21\text{ ft}}[-(16\text{ ft})\mathbf{i} + (8\text{ ft})\mathbf{j} + (11\text{ ft})\mathbf{k}]$$
$$\mathbf{T}_{AB} = -(640\text{ lb})\mathbf{i} + (320\text{ lb})\mathbf{j} + (440\text{ lb})\mathbf{k}$$

Denoting by $\boldsymbol{\lambda}_{AC}$ the unit vector along AC, we obtain in a similar way

$$\mathbf{T}_{AC} = T_{AC}\boldsymbol{\lambda}_{AC} = T_{AC}\frac{\overrightarrow{AC}}{AC} = \frac{1200\text{ lb}}{24\text{ ft}}\overrightarrow{AC}$$
$$\mathbf{T}_{AC} = -(800\text{ lb})\mathbf{i} + (400\text{ lb})\mathbf{j} - (800\text{ lb})\mathbf{k}$$

Resultant of the Forces. The resultant $\mathbf{R}$ of the forces exerted by the two cables is

$$\mathbf{R} = \mathbf{T}_{AB} + \mathbf{T}_{AC} = -(1440\text{ lb})\mathbf{i} + (720\text{ lb})\mathbf{j} - (360\text{ lb})\mathbf{k}$$

The magnitude and direction of the resultant are now determined:

$$R = \sqrt{R_x^2 + R_y^2 + R_z^2} = \sqrt{(-1440)^2 + (720)^2 + (-360)^2}$$

$$R = 1650\text{ lb} \quad \blacktriangleleft$$

From Eqs. (2.33) we obtain

$$\cos\theta_x = \frac{R_x}{R} = \frac{-1440\text{ lb}}{1650\text{ lb}} \qquad \cos\theta_y = \frac{R_y}{R} = \frac{+720\text{ lb}}{1650\text{ lb}}$$

$$\cos\theta_z = \frac{R_z}{R} = \frac{-360\text{ lb}}{1650\text{ lb}}$$

Calculating successively each quotient and its arc cosine, we have

$$\theta_x = 150.8° \qquad \theta_y = 64.1° \qquad \theta_z = 102.6° \quad \blacktriangleleft$$

SOLVING PROBLEMS
ON YOUR OWN

In this lesson we saw that *forces in space* may be defined by their magnitude and direction or by the three rectangular components F_x, F_y, and F_z.

A. When a force is defined by its magnitude and direction, its rectangular components F_x, F_y, and F_z may be found as follows:

Case 1. If the direction of the force **F** is defined by the angles θ_y and ϕ shown in Fig. 2.30, projections of **F** through these angles or their complements will yield the components of **F** [Eqs. (2.17)]. Note that the x and z components of **F** are found by first projecting **F** onto the horizontal plane; the projection $\mathbf{F}_h$ obtained in this way is then resolved into the components $\mathbf{F}_x$ and $\mathbf{F}_z$ (Fig. 2.30c).

Case 2. If the direction of the force **F** is defined by the angles θ_x, θ_y, θ_z that **F** forms with the coordinate axes, each component can be obtained by multiplying the magnitude F of the force by the cosine of the corresponding angle [Example 1]:

$$F_x = F \cos \theta_x \qquad F_y = F \cos \theta_y \qquad F_z = F \cos \theta_z$$

Case 3. If the drection of the force **F** is defined by two points M and N located on its line of action (Fig. 2.34), you will first express the vector $\overrightarrow{MN}$ drawn from M to N in terms of its components d_x, d_y, d_z and the unit vectors **i**, **j**, **k**:

$$\overrightarrow{MN} = d_x\mathbf{i} + d_y\mathbf{j} + d_z\mathbf{k}$$

Next, you will determine the unit vector $\boldsymbol{\lambda}$ along the line of action of **F** by dividing the vector $\overrightarrow{MN}$ by its magnitude MN. Multiplying $\boldsymbol{\lambda}$ by the magnitude of **F**, you will obtain the desired expression for **F** in terms of its rectangular components [Sample Prob. 2.7]:

$$\mathbf{F} = F\boldsymbol{\lambda} = \frac{F}{d}(d_x\mathbf{i} + d_y\mathbf{j} + d_z\mathbf{k})$$

It is advantageous to use a consistent and meaningful system of notation when determining the rectangular components of a force. The method used in this text is illustrated in Sample Prob. 2.8 where, for example, the force $\mathbf{T}_{AB}$ acts from stake A toward point B. Note that the subscripts have been ordered to agree with the direction of the force. It is recommended that you adopt the same notation, as it will help you identify point 1 (the first subscript) and point 2 (the second subscript).

When forming the vector defining the line of action of a force, you may think of its scalar components as the number of steps you must take in each coordinate direction to go from point 1 to point 2. It is essential that you always remember to assign the correct sign to each of the components.

(continued)

B. When a force is defined by its rectangular components F_x, F_y, F_z, you can obtain its magnitude F by writing

$$F = \sqrt{F_x^2 + F_y^2 + F_z^2}$$

You can determine the direction cosines of the line of action of **F** by dividing the components of the force by F:

$$\cos \theta_x = \frac{F_x}{F} \quad \cos \theta_y = \frac{F_y}{F} \quad \cos \theta_z = \frac{F_z}{F}$$

From the direction cosines you can obtain the angles θ_x, θ_y, θ_z that **F** forms with the coordinate axes [Example 2].

C. To determine the resultant R of two or more forces in three-dimensional space, first determine the rectangular components of each force by one of the procedures described above. Adding these components will yield the components R_x, R_y, R_z of the resultant. The magnitude and direction of the resultant may then be obtained as indicated above for a force **F** [Sample Prob. 2.8].

PROBLEMS

2.71 Determine (*a*) the *x*, *y*, and *z* components of the 900-N force, (*b*) the angles θ_x, θ_y, and θ_z that the force forms with the coordinate axes.

2.72 Determine (*a*) the *x*, *y*, and *z* components of the 750-N force, (*b*) the angles θ_x, θ_y, and θ_z that the force forms with the coordinate axes.

2.73 A gun is aimed at a point *A* located 35° east of north. Knowing that the barrel of the gun forms an angle of 40° with the horizontal and that the maximum recoil force is 400 N, determine (*a*) the *x*, *y*, and *z* components of that force, (*b*) the values of the angles θ_x, θ_y, and θ_z defining the direction of the recoil force. (Assume that the *x*, *y*, and *z* axes are directed, respectively, east, up, and south.)

2.74 Solve Prob. 2.73, assuming that point *A* is located 15° north of west and that the barrel of the gun forms an angle of 25° with the horizontal.

2.75 Cable *AB* is 65 ft long, and the tension in that cable is 3900 lb. Determine (*a*) the *x*, *y*, and *z* components of the force exerted by the cable on the anchor *B*, (*b*) the angles θ_x, θ_y, and θ_z defining the direction of that force.

2.76 Cable *AC* is 70 ft long, and the tension in that cable is 5250 lb. Determine (*a*) the *x*, *y*, and *z* components of the force exerted by the cable on the anchor *C*, (*b*) the angles θ_x, θ_y, and θ_z defining the direction of that force.

2.77 The end of the coaxial cable *AE* is attached to the pole *AB*, which is strengthened by the guy wires *AC* and *AD*. Knowing that the tension in wire *AC* is 120 lb, determine (*a*) the components of the force exerted by this wire on the pole, (*b*) the angles θ_x, θ_y, and θ_z that the force forms with the coordinate axes.

2.78 The end of the coaxial cable *AE* is attached to the pole *AB*, which is strengthened by the guy wires *AC* and *AD*. Knowing that the tension in wire *AD* is 85 lb, determine (*a*) the components of the force exerted by this wire on the pole, (*b*) the angles θ_x, θ_y, and θ_z that the force forms with the coordinate axes.

2.79 Determine the magnitude and direction of the force $\mathbf{F} = (690\text{ lb})\mathbf{i} + (300\text{ lb})\mathbf{j} - (580\text{ lb})\mathbf{k}$.

2.80 Determine the magnitude and direction of the force $\mathbf{F} = (650\text{ N})\mathbf{i} - (320\text{ N})\mathbf{j} + (760\text{ N})\mathbf{k}$.

2.81 A force acts at the origin of a coordinate system in a direction defined by the angles $\theta_x = 75°$ and $\theta_z = 130°$. Knowing that the *y* component of the force is +300 lb, determine (*a*) the angle θ_y, (*b*) the other components and the magnitude of the force.

Fig. P2.71 and P2.72

Fig. P2.75 and P2.76

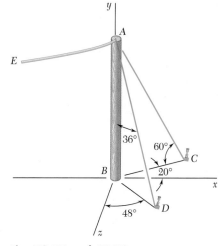

Fig. P2.77 and P2.78

2.82 A force acts at the origin of a coordinate system in a direction defined by the angles $\theta_y = 55°$ and $\theta_z = 45°$. Knowing that the x component of the force is -500 N, determine (a) the angle θ_x, (b) the other components and the magnitude of the force.

2.83 A force **F** of magnitude 230 N acts at the origin of a coordinate system. Knowing that $\theta_x = 32.5°$, $F_y = -60$ N, and $F_z > 0$, determine (a) the components F_x and F_z, (b) the angles θ_y and θ_z.

2.84 A force **F** of magnitude 210 N acts at the origin of a coordinate system. Knowing that $F_x = 80$ N, $\theta_z = 151.2°$, and $F_y < 0$, determine (a) the components F_y and F_z, (b) the angles θ_x and θ_y.

2.85 In order to move a wrecked truck, two cables are attached at A and pulled by winches B and C as shown. Knowing that the tension in cable AB is 2 kips, determine the components of the force exerted at A by the cable.

2.86 In order to move a wrecked truck, two cables are attached at A and pulled by winches B and C as shown. Knowing that the tension in cable AC is 1.5 kips, determine the components of the force exerted at A by the cable.

Fig. P2.85 and P2.86

2.87 Knowing that the tension in cable AB is 1425 N, determine the components of the force exerted on the plate at B.

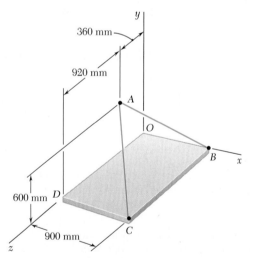
Fig. P2.87 and P2.88

2.88 Knowing that the tension in cable AC is 2130 N, determine the components of the force exerted on the plate at C.

2.89 A frame ABC is supported in part by cable DBE that passes through a frictionless ring at B. Knowing that the tension in the cable is 385 N, determine the components of the force exerted by the cable on the support at D.

2.90 For the frame and cable of Prob. 2.89, determine the components of the force exerted by the cable on the support at E.

Fig. P2.89

2.91 Find the magnitude and direction of the resultant of the two forces shown knowing that $P = 600$ N and $Q = 450$ N.

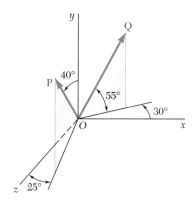

Fig. P2.91 and P2.92

2.92 Find the magnitude and direction of the resultant of the two forces shown knowing that $P = 450$ N and $Q = 600$ N.

2.93 Knowing that the tension is 425 lb in cable AB and 510 lb in cable AC, determine the magnitude and direction of the resultant of the forces exerted at A by the two cables.

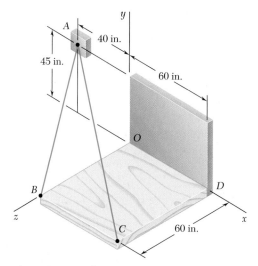

Fig. *P2.93* and P2.94

2.94 Knowing that the tension is 510 lb in cable AB and 425 lb in cable AC, determine the magnitude and direction of the resultant of the forces exerted at A by the two cables.

2.95 For the frame of Prob. 2.89, determine the magnitude and direction of the resultant of the forces exerted by the cable at B knowing that the tension in the cable is 385 N.

2.96   For the cables of Prob. 2.87, knowing that the tension is 1425 N in cable AB and 2130 N in cable AC, determine the magnitude and direction of the resultant of the forces exerted at A by the two cables.

2.97   The boom OA carries a load $\mathbf{P}$ and is supported by two cables as shown. Knowing that the tension in cable AB is 183 lb and that the resultant of the load $\mathbf{P}$ and of the forces exerted at A by the two cables must be directed along OA, determine the tension in cable AC.

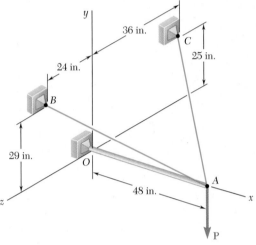

Fig. P2.97

2.98   For the boom and loading of Prob. 2.97, determine the magnitude of the load $\mathbf{P}$.

2.15   EQUILIBRIUM OF A PARTICLE IN SPACE

According to the definition given in Sec. 2.9, a particle A is in equilibrium if the resultant of all the forces acting on A is zero. The components R_x, R_y, R_z of the resultant are given by the relations (2.31); expressing that the components of the resultant are zero, we write

$$\Sigma F_x = 0 \qquad \Sigma F_y = 0 \qquad \Sigma F_z = 0 \qquad (2.34)$$

Equations (2.34) represent the necessary and sufficient conditions for the equilibrium of a particle in space. They can be used to solve problems dealing with the equilibrium of a particle involving no more than three unknowns.

To solve such problems, you first should draw a free-body diagram showing the particle in equilibrium and *all* the forces acting on it. You can then write the equations of equilibrium (2.34) and solve them for three unknowns. In the more common types of problems, these unknowns will represent (1) the three components of a single force or (2) the magnitude of three forces, each of known direction.

Photo 2.2   While the tension in the *four* cables supporting the car cannot be found using the *three* equations of (2.34), a relation between the tensions can be obtained by considering the equilibrium of the hook.

SAMPLE PROBLEM 2.9

A 200-kg cylinder is hung by means of two cables AB and AC, which are attached to the top of a vertical wall. A horizontal force $\mathbf{P}$ perpendicular to the wall holds the cylinder in the position shown. Determine the magnitude of $\mathbf{P}$ and the tension in each cable.

SOLUTION

Free-Body Diagram.   Point A is chosen as a free body; this point is subjected to four forces, three of which are of unknown magnitude.

Introducing the unit vectors $\mathbf{i}$, $\mathbf{j}$, $\mathbf{k}$, we resolve each force into rectangular components.

$$\mathbf{P} = P\mathbf{i}$$
$$\mathbf{W} = -mg\mathbf{j} = -(200 \text{ kg})(9.81 \text{ m/s}^2)\mathbf{j} = -(1962 \text{ N})\mathbf{j} \tag{1}$$

In the case of $\mathbf{T}_{AB}$ and $\mathbf{T}_{AC}$, it is necessary first to determine the components and magnitudes of the vectors $\overrightarrow{AB}$ and $\overrightarrow{AC}$. Denoting by $\boldsymbol{\lambda}_{AB}$ the unit vector along AB, we write

$$\overrightarrow{AB} = -(1.2 \text{ m})\mathbf{i} + (10 \text{ m})\mathbf{j} + (8 \text{ m})\mathbf{k} \qquad AB = 12.862 \text{ m}$$

$$\boldsymbol{\lambda}_{AB} = \frac{\overrightarrow{AB}}{12.862 \text{ m}} = -0.09330\mathbf{i} + 0.7775\mathbf{j} + 0.6220\mathbf{k}$$

$$\mathbf{T}_{AB} = T_{AB}\boldsymbol{\lambda}_{AB} = -0.09330T_{AB}\mathbf{i} + 0.7775T_{AB}\mathbf{j} + 0.6220T_{AB}\mathbf{k} \tag{2}$$

Denoting by $\boldsymbol{\lambda}_{AC}$ the unit vector along AC, we write in a similar way

$$\overrightarrow{AC} = -(1.2 \text{ m})\mathbf{i} + (10 \text{ m})\mathbf{j} - (10 \text{ m})\mathbf{k} \qquad AC = 14.193 \text{ m}$$

$$\boldsymbol{\lambda}_{AC} = \frac{\overrightarrow{AC}}{14.193 \text{ m}} = -0.08455\mathbf{i} + 0.7046\mathbf{j} - 0.7046\mathbf{k}$$

$$\mathbf{T}_{AC} = T_{AC}\boldsymbol{\lambda}_{AC} = -0.08455T_{AC}\mathbf{i} + 0.7046T_{AC}\mathbf{j} - 0.7046T_{AC}\mathbf{k} \tag{3}$$

Equilibrium Condition.   Since A is in equilibrium, we must have

$$\Sigma\mathbf{F} = 0: \qquad \mathbf{T}_{AB} + \mathbf{T}_{AC} + \mathbf{P} + \mathbf{W} = 0$$

or, substituting from (1), (2), (3) for the forces and factoring $\mathbf{i}$, $\mathbf{j}$, $\mathbf{k}$,

$$(-0.09330T_{AB} - 0.08455T_{AC} + P)\mathbf{i}$$
$$+ (0.7775T_{AB} + 0.7046T_{AC} - 1962 \text{ N})\mathbf{j}$$
$$+ (0.6220T_{AB} - 0.7046T_{AC})\mathbf{k} = 0$$

Setting the coefficients of $\mathbf{i}$, $\mathbf{j}$, $\mathbf{k}$ equal to zero, we write three scalar equations, which express that the sums of the x, y, and z components of the forces are respectively equal to zero.

$$(\Sigma F_x = 0:) \qquad -0.09330T_{AB} - 0.08455T_{AC} + P = 0$$
$$(\Sigma F_y = 0:) \qquad +0.7775T_{AB} + 0.7046T_{AC} - 1962 \text{ N} = 0$$
$$(\Sigma F_z = 0:) \qquad +0.6220T_{AB} - 0.7046T_{AC} = 0$$

Solving these equations, we obtain

$$P = 235 \text{ N} \qquad T_{AB} = 1402 \text{ N} \qquad T_{AC} = 1238 \text{ N} \quad \blacktriangleleft$$

SOLVING PROBLEMS
ON YOUR OWN

We saw earlier that when a particle is in *equilibrium*, the resultant of the forces acting on the particle must be zero. Expressing this fact in the case of the equilibrium of a *particle in three-dimensional space* will provide you with three relations among the forces acting on the particle. These relations may be used to determine three unknowns—usually the magnitudes of three forces.

Your solution will consist of the following steps:

1. Draw a free-body diagram of the particle. This diagram shows the particle and all the forces acting on it. Indicate on the diagram the magnitudes of known forces, as well as any angles or dimensions that define the direction of a force. Any unknown magnitude or angle should be denoted by an appropriate symbol. Nothing else should be included in your free-body diagram.

2. Resolve each of the forces into rectangular components. Following the method used in the preceding lesson, you will determine for each force **F** the unit vector **λ** defining the direction of that force and express **F** as the product of its magnitude F and the unit vector **λ**. You will obtain an expression of the form

$$\mathbf{F} = F\boldsymbol{\lambda} = \frac{F}{d}\,(d_x\mathbf{i} + d_y\mathbf{j} + d_z\mathbf{k})$$

where d, d_x, d_y, and d_z are dimensions obtained from the free-body diagram of the particle. If a force is known in magnitude as well as in direction, then F is known and the expression obtained for **F** is well defined; otherwise F is one of the three unknowns that should be determined.

3. Set the resultant, or sum, of the forces exerted on the particle equal to zero. You will obtain a vectorial equation consisting of terms containing the unit vectors **i**, **j**, or **k**. You will group the terms containing the same unit vector and factor that vector. For the vectorial equation to be satisfied, the coefficient of each of the unit vectors must be set equal to zero. This will yield three scalar equations that you can solve for no more than three unknowns [Sample Prob. 2.9].

PROBLEMS

FREE BODY PRACTICE PROBLEMS

2.F5 A 36-lb triangular plate is supported by three cables as shown. Draw the free-body diagram needed to determine the tension in each wire.

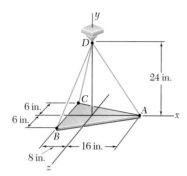

Fig. P2.F5

2.F6 A 70-kg cylinder is supported by two cables AC and BC, which are attached to the top of vertical posts. A horizontal force **P**, perpendicular to the plane containing the posts, holds the cylinder in the position shown. Draw the free-body diagram needed to determine the magnitude of **P** and the force in each cable.

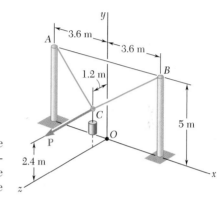

Fig. P2.F6

2.F7 Three cables are connected at point D, which is located 18 in. below the T-shaped pipe support ABC. The cables support a 180-lb cylinder as shown. Draw the free-body diagram needed to determine the tension in each cable.

2.F8 A 100-kg container is suspended from ring A, to which cables AC and AE are attached. A force P is applied to end F of a third cable that passes over a pulley at B and through ring A and then is attached to a support at D. Draw the free-body diagram needed to determine the magnitude of P. (*Hint:* The tension is the same in all portions of cable $FBAD$.)

Fig. P2.F7

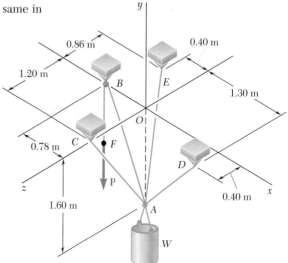

Fig. P2.F8

END-OF-SECTION PROBLEMS

2.99 A container is supported by three cables that are attached to a ceiling as shown. Determine the weight W of the container, knowing that the tension in cable AB is 6 kN.

Fig. P2.99 and *P2.100*

2.100 A container is supported by three cables that are attached to a ceiling as shown. Determine the weight W of the container, knowing that the tension in cable AD is 4.3 kN.

2.101 Three cables are used to tether a balloon as shown. Determine the vertical force **P** exerted by the balloon at A knowing that the tension in cable AD is 481 N.

Fig. P2.101 and *P2.102*

2.102 Three cables are used to tether a balloon as shown. Knowing that the balloon exerts an 800-N vertical force at A, determine the tension in each cable.

2.103 A crate is supported by three cables as shown. Determine the weight of the crate knowing that the tension in cable AB is 750 lb.

2.104 A crate is supported by three cables as shown. Determine the weight of the crate knowing that the tension in cable AD is 616 lb.

2.105 A crate is supported by three cables as shown. Determine the weight of the crate knowing that the tension in cable AC is 544 lb.

2.106 A 1600-lb crate is supported by three cables as shown. Determine the tension in each cable.

2.107 Three cables are connected at A, where the forces **P** and **Q** are applied as shown. Knowing that Q = 0, find the value of P for which the tension in cable AD is 305 N.

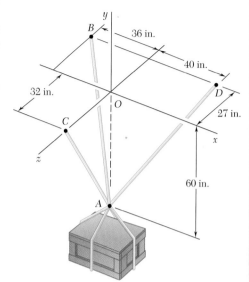

Fig. P2.103, P2.104, *P2.105*, and P2.106

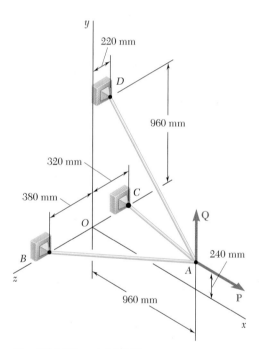

Fig. P2.107 and P2.108

2.108 Three cables are connected at A, where the forces **P** and **Q** are applied as shown. Knowing that P = 1200 N, determine the values of Q for which cable AD is taut.

2.109 A rectangular plate is supported by three cables as shown. Knowing that the tension in cable AC is 60 N, determine the weight of the plate.

2.110 A rectangular plate is supported by three cables as shown. Knowing that the tension in cable AD is 520 N, determine the weight of the plate.

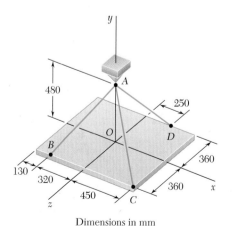

Dimensions in mm

Fig. P2.109 and P2.110

2.111 A transmission tower is held by three guy wires attached to a pin at A and anchored by bolts at B, C, and D. If the tension in wire AB is 630 lb, determine the vertical force **P** exerted by the tower on the pin at A.

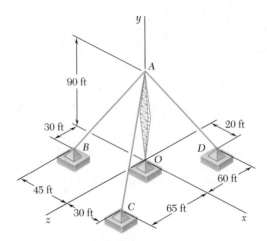

Fig. P2.111 and P2.112

2.112 A transmission tower is held by three guy wires attached to a pin at A and anchored by bolts at B, C, and D. If the tension in wire AC is 920 lb, determine the vertical force **P** exerted by the tower on the pin at A.

2.113 In trying to move across a slippery icy surface, a 180-lb man uses two ropes AB and AC. Knowing that the force exerted on the man by the icy surface is perpendicular to that surface, determine the tension in each rope.

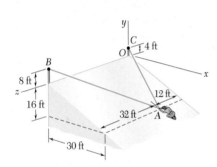

Fig. P2.113

2.114 Solve Prob. 2.113, assuming that a friend is helping the man at A by pulling on him with a force $\mathbf{P} = -(60 \text{ lb})\mathbf{k}$.

2.115 For the rectangular plate of Probs. 2.109 and 2.110, determine the tension in each of the three cables knowing that the weight of the plate is 792 N.

2.116 For the cable system of Probs. 2.107 and 2.108, determine the tension in each cable knowing that $P = 2880$ N and $Q = 0$.

2.117 For the cable system of Probs. 2.107 and 2.108, determine the tension in each cable knowing that $P = 2880$ N and $Q = 576$ N.

2.118 For the cable system of Probs. 2.107 and 2.108, determine the tension in each cable knowing that $P = 2880$ N and $Q = -576$ N (**Q** is directed downward).

2.119 For the transmission tower of Probs. 2.111 and 2.112, determine the tension in each guy wire knowing that the tower exerts on the pin at A an upward vertical force of 2100 lb.

2.120 A horizontal circular plate weighing 60 lb is suspended as shown from three wires that are attached to a support at D and form 30° angles with the vertical. Determine the tension in each wire.

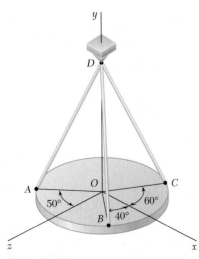

Fig. P2.120

2.121 Cable *BAC* passes through a frictionless ring *A* and is attached to fixed supports at *B* and *C*, while cables *AD* and *AE* are both tied to the ring and are attached, respectively, to supports at *D* and *E*. Knowing that a 200-lb vertical load **P** is applied to ring *A*, determine the tension in each of the three cables.

2.122 Knowing that the tension in cable *AE* of Prob. 2.121 is 75 lb, determine (*a*) the magnitude of the load **P**, (*b*) the tension in cables *BAC* and *AD*.

2.123 A container of weight *W* is suspended from ring *A*. Cable *BAC* passes through the ring and is attached to fixed supports at *B* and *C*. Two forces **P** = *P***i** and **Q** = *Q***k** are applied to the ring to maintain the container in the position shown. Knowing that *W* = 376 N, determine *P* and *Q*. (*Hint*: The tension is the same in both portions of cable *BAC*.)

Fig. P2.121

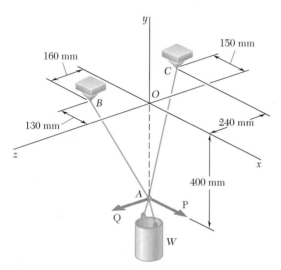

Fig. P2.123

2.124 For the system of Prob. 2.123, determine *W* and *Q* knowing that *P* = 164 N.

2.125 Collars *A* and *B* are connected by a 525-mm-long wire and can slide freely on frictionless rods. If a force **P** = (341 N)**j** is applied to collar *A*, determine (*a*) the tension in the wire when *y* = 155 mm, (*b*) the magnitude of the force **Q** required to maintain the equilibrium of the system.

2.126 Solve Prob. 2.125 assuming that *y* = 275 mm.

Fig. P2.125

REVIEW AND SUMMARY

In this chapter we have studied the effect of forces on particles, i.e., on bodies of such shape and size that all forces acting on them may be assumed applied at the same point.

Resultant of two forces

Fig. 2.35

Forces are *vector quantities;* they are characterized by a *point of application,* a *magnitude,* and a *direction,* and they add according to the *parallelogram law* (Fig. 2.35). The magnitude and direction of the resultant **R** of two forces **P** and **Q** can be determined either graphically or by trigonometry, using successively the law of cosines and the law of sines [Sample Prob. 2.1].

Components of a force

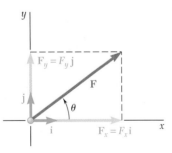

Fig. 2.36

Any given force acting on a particle can be resolved into two or more *components,* i.e., it can be replaced by two or more forces which have the same effect on the particle. A force **F** can be resolved into two components **P** and **Q** by drawing a parallelogram which has **F** for its diagonal; the components **P** and **Q** are then represented by the two adjacent sides of the parallelogram (Fig. 2.36) and can be determined either graphically or by trigonometry [Sec. 2.6].

A force **F** is said to have been resolved into two *rectangular components* if its components F_x and F_y are perpendicular to each other and are directed along the coordinate axes (Fig. 2.37). Introducing the *unit vectors* **i** and **j** along the x and y axes, respectively, we write [Sec. 2.7]

Rectangular components
Unit vectors

$$\mathbf{F}_x = F_x\mathbf{i} \qquad \mathbf{F}_y = F_y\mathbf{j} \tag{2.6}$$

and

$$\mathbf{F} = F_x\mathbf{i} + F_y\mathbf{j} \tag{2.7}$$

where F_x and F_y are the *scalar components* of **F**. These components, which can be positive or negative, are defined by the relations

$$F_x = F\cos\theta \qquad F_y = F\sin\theta \tag{2.8}$$

When the rectangular components F_x and F_y of a force **F** are given, the angle θ defining the direction of the force can be obtained by writing

$$\tan\theta = \frac{F_y}{F_x} \tag{2.9}$$

The magnitude F of the force can then be obtained by solving one of the equations (2.8) for F or by applying the Pythagorean theorem and writing

$$F = \sqrt{F_x^2 + F_y^2} \tag{2.10}$$

Fig. 2.37

When *three or more coplanar forces* act on a particle, the rectangular components of their resultant **R** can be obtained by adding algebraically the corresponding components of the given forces [Sec. 2.8]. We have

$$R_x = \Sigma F_x \qquad R_y = \Sigma F_y \qquad (2.13)$$

The magnitude and direction of **R** can then be determined from relations similar to Eqs. (2.9) and (2.10) [Sample Prob. 2.3].

A force **F** in *three-dimensional space* can be resolved into rectangular components $\mathbf{F}_x$, $\mathbf{F}_y$, and $\mathbf{F}_z$ [Sec. 2.12]. Denoting by θ_x, θ_y, and θ_z, respectively, the angles that **F** forms with the x, y, and z axes (Fig. 2.38), we have

$$F_x = F \cos \theta_x \qquad F_y = F \cos \theta_y \qquad F_z = F \cos \theta_z \qquad (2.19)$$

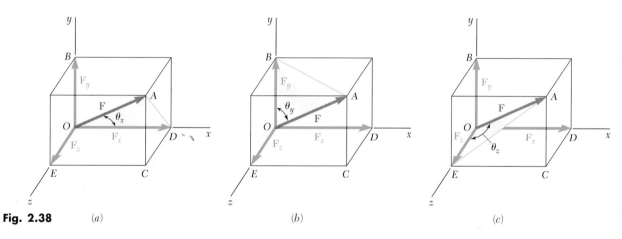

Fig. 2.38   (*a*)   (*b*)   (*c*)

The cosines of θ_x, θ_y, θ_z are known as the *direction cosines* of the force **F**. Introducing the unit vectors **i**, **j**, **k** along the coordinate axes, we write

$$\mathbf{F} = F_x\mathbf{i} + F_y\mathbf{j} + F_z\mathbf{k} \qquad (2.20)$$

or

$$\mathbf{F} = F(\cos \theta_x\mathbf{i} + \cos \theta_y\mathbf{j} + \cos \theta_z\mathbf{k}) \qquad (2.21)$$

which shows (Fig. 2.39) that **F** is the product of its magnitude F and the unit vector

$$\boldsymbol{\lambda} = \cos \theta_x\mathbf{i} + \cos \theta_y\mathbf{j} + \cos \theta_z\mathbf{k}$$

Since the magnitude of $\boldsymbol{\lambda}$ is equal to unity, we must have

$$\cos^2 \theta_x + \cos^2 \theta_y + \cos^2 \theta_z = 1 \qquad (2.24)$$

When the rectangular components F_x, F_y, F_z of a force **F** are given, the magnitude F of the force is found by writing

$$F = \sqrt{F_x^2 + F_y^2 + F_z^2} \qquad (2.18)$$

and the direction cosines of **F** are obtained from Eqs. (2.19). We have

$$\cos \theta_x = \frac{F_x}{F} \qquad \cos \theta_y = \frac{F_y}{F} \qquad \cos \theta_z = \frac{F_z}{F} \qquad (2.25)$$

Fig. 2.39

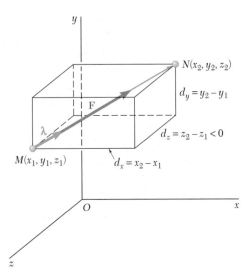

Fig. 2.40

When a force **F** is defined in three-dimensional space by its magnitude F and two points M and N on its line of action [Sec. 2.13], its rectangular components can be obtained as follows. We first express the vector $\overrightarrow{MN}$ joining points M and N in terms of its components d_x, d_y, and d_z (Fig. 2.40); we write

$$\overrightarrow{MN} = d_x\mathbf{i} + d_y\mathbf{j} + d_z\mathbf{k} \qquad (2.26)$$

We next determine the unit vector $\boldsymbol{\lambda}$ along the line of action of **F** by dividing $\overrightarrow{MN}$ by its magnitude $MN = d$:

$$\boldsymbol{\lambda} = \frac{\overrightarrow{MN}}{MN} = \frac{1}{d}(d_x\mathbf{i} + d_y\mathbf{j} + d_z\mathbf{k}) \qquad (2.27)$$

Recalling that **F** is equal to the product of F and $\boldsymbol{\lambda}$, we have

$$\mathbf{F} = F\boldsymbol{\lambda} = \frac{F}{d}(d_x\mathbf{i} + d_y\mathbf{j} + d_z\mathbf{k}) \qquad (2.28)$$

from which it follows [Sample Probs. 2.7 and 2.8] that the scalar components of **F** are, respectively,

$$F_x = \frac{Fd_x}{d} \qquad F_y = \frac{Fd_y}{d} \qquad F_z = \frac{Fd_z}{d} \qquad (2.29)$$

Resultant of forces in space

When *two or more forces* act on a particle in *three-dimensional space*, the rectangular components of their resultant **R** can be obtained by adding algebraically the corresponding components of the given forces [Sec. 2.14]. We have

$$R_x = \Sigma F_x \qquad R_y = \Sigma F_y \qquad R_z = \Sigma F_z \qquad (2.31)$$

The magnitude and direction of **R** can then be determined from relations similar to Eqs. (2.18) and (2.25) [Sample Prob. 2.8].

Equilibrium of a particle

A particle is said to be in *equilibrium* when the resultant of all the forces acting on it is zero [Sec. 2.9]. The particle will then remain at rest (if originally at rest) or move with constant speed in a straight line (if originally in motion) [Sec. 2.10].

Free-body diagram

To solve a problem involving a particle in equilibrium, one first should draw a *free-body diagram* of the particle showing all the forces acting on it [Sec. 2.11]. If *only three coplanar forces* act on the particle, a *force triangle* may be drawn to express that the particle is in equilibrium. Using graphical methods of trigonometry, this triangle can be solved for no more than two unknowns [Sample Prob. 2.4]. If *more than three coplanar forces* are involved, the equations of equilibrium

$$\Sigma F_x = 0 \qquad \Sigma F_y = 0 \qquad (2.15)$$

should be used. These equations can be solved for no more than two unknowns [Sample Prob. 2.6].

Equilibrium in space

When a particle is in *equilibrium in three-dimensional space* [Sec. 2.15], the three equations of equilibrium

$$\Sigma F_x = 0 \qquad \Sigma F_y = 0 \qquad \Sigma F_z = 0 \qquad (2.34)$$

should be used. These equations can be solved for no more than three unknowns [Sample Prob. 2.9].

REVIEW PROBLEMS

2.127 Two structural members A and B are bolted to a bracket as shown. Knowing that both members are in compression and that the force is 15 kN in member A and 10 kN in member B, determine by trigonometry the magnitude and direction of the resultant of the forces applied to the bracket by members A and B.

2.128 Member BD exerts on member ABC a force $\mathbf{P}$ directed along line BD. Knowing that $\mathbf{P}$ must have a 300-lb horizontal component, determine (a) the magnitude of the force $\mathbf{P}$, (b) its vertical component.

Fig. P2.127

Fig. P2.128

2.129 Determine (a) the required tension in cable AC, knowing that the resultant of the three forces exerted at point C of boom BC must be directed along BC, (b) the corresponding magnitude of the resultant.

Fig. P2.129

2.130 Two cables are tied together at C and are loaded as shown. Determine the tension (a) in cable AC, (b) in cable BC.

Fig. P2.130

2.131 A welded connection is in equilibrium under the action of the four forces shown. Knowing that $F_A = 8$ kN and $F_B = 16$ kN, determine the magnitudes of the other two forces.

Fig. P2.131

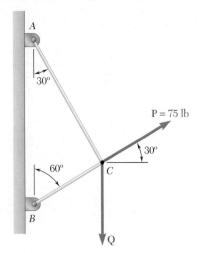

Fig. P2.132

2.132 Two cables tied together at C are loaded as shown. Determine the range of values of Q for which the tension will not exceed 60 lb in either cable.

2.133 A horizontal circular plate is suspended as shown from three wires that are attached to a support at D and form 30° angles with the vertical. Knowing that the x component of the force exerted by wire AD on the plate is 110.3 N, determine (a) the tension in wire AD, (b) the angles θ_x, θ_y, and θ_z that the force exerted at A forms with the coordinate axes.

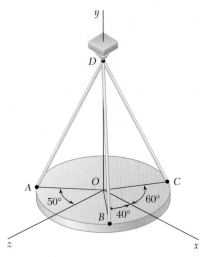

Fig. P2.133

2.134 A force acts at the origin of a coordinate system in a direction defined by the angles $\theta_y = 55°$ and $\theta_z = 45°$. Knowing that the x component of the force is -500 lb, determine (a) the angle θ_x, (b) the other components and the magnitude of the force.

2.135 Find the magnitude and direction of the resultant of the two forces shown knowing that $P = 300$ N and $Q = 400$ N.

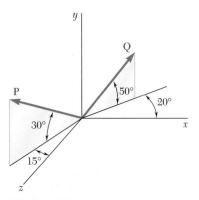

Fig. P2.135

2.136 Three cables are used to tether a balloon as shown. Determine the vertical force **P** exerted by the balloon at A knowing that the tension in cable AC is 444 N.

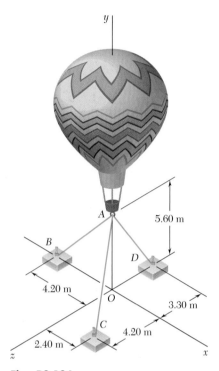

Fig. P2.136

2.137 Collars A and B are connected by a 25-in.-long wire and can slide freely on frictionless rods. If a 60-lb force **Q** is applied to collar B as shown, determine (*a*) the tension in the wire when $x = 9$ in., (*b*) the corresponding magnitude of the force **P** required to maintain the equilibrium of the system.

2.138 Collars A and B are connected by a 25-in.-long wire and can slide freely on frictionless rods. Determine the distances x and z for which the equilibrium of the system is maintained when $P = 120$ lb and $Q = 60$ lb.

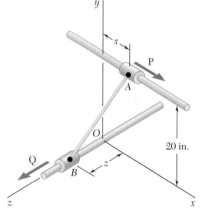

Fig. P2.137 and P2.138

COMPUTER PROBLEMS

2.C1 Write a computer program that can be used to determine the magnitude and direction of the resultant of n coplanar forces applied at a point A. Use this program to solve Probs. 2.32, 2.33, 2.35, and 2.38.

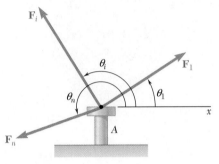

Fig. P2.C1

2.C2 A load **P** is supported by two cables as shown. Write a computer program that can be used to determine the tension in each cable for any given value of P and for values of θ ranging from $\theta_1 = \beta - 90°$ to $\theta_2 = 90° - \alpha$, using given increments $\Delta\theta$. Use this program to determine for the following three sets of numerical values (a) the tension in each cable for values of θ ranging from θ_1 to θ_2, (b) the value of θ for which the tension in the two cables is as small as possible, (c) the corresponding value of the tension:

(1) $\alpha = 35°$, $\beta = 75°$, $P = 400$ lb, $\Delta\theta = 5°$
(2) $\alpha = 50°$, $\beta = 30°$, $P = 600$ lb, $\Delta\theta = 10°$
(3) $\alpha = 40°$, $\beta = 60°$, $P = 250$ lb, $\Delta\theta = 5°$

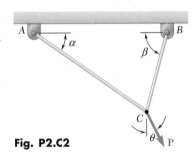

Fig. P2.C2

2.C3 An acrobat is walking on a tightrope of length $L = 20.1$ m attached to supports A and B at a distance of 20.0 m from each other. The combined weight of the acrobat and his balancing pole is 800 N, and the friction between his shoes and the rope is large enough to prevent him from slipping. Neglecting the weight of the rope and any elastic deformation, write a computer program to calculate the deflection y and the tension in portions AC and BC of the rope for values of x from 0.5 m to 10.0 m using 0.5-m increments. From the data obtained, determine (a) the maximum deflection of the rope, (b) the maximum tension in the rope, (c) the smallest values of the tension in portions AC and BC of the rope.

Fig. P2.C3

2.C4 Write a computer program that can be used to determine the magnitude and direction of the resultant of n forces $\mathbf{F}_i$, where $i = 1, 2, \ldots, n$, that are applied at point A_0 of coordinates x_0, y_0, and z_0, knowing that the line of action of $\mathbf{F}_i$ passes through point A_i of coordinates x_i, y_i, and z_i. Use this program to solve Probs. 2.93, 2.94, and 2.95.

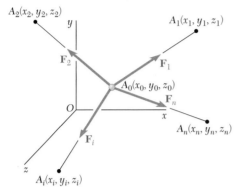

Fig. P2.C4

2.C5 Three cables are attached at points A_1, A_2, and A_3, respectively, and are connected at point A_0, to which a given load $\mathbf{P}$ is applied as shown. Write a computer program that can be used to determine the tension in each of the cables. Use this program to solve Probs. 2.102, 2.106, 2.107, and 2.115.

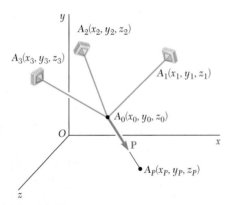

Fig. P2.C5

The battleship USS *New Jersey* is maneuvered by four tugboats at Bremerton Naval Shipyard. It will be shown in this chapter that the forces exerted on the ship by the tugboats could be replaced by an equivalent force exerted by a single, more powerful, tugboat.

CHAPTER 3

Rigid Bodies:
Equivalent Systems of Forces

3.1 INTRODUCTION

In the preceding chapter it was assumed that each of the bodies considered could be treated as a single particle. Such a view, however, is not always possible, and a body, in general, should be treated as a combination of a large number of particles. The size of the body will have to be taken into consideration, as well as the fact that forces will act on different particles and thus will have different points of application.

Most of the bodies considered in elementary mechanics are assumed to be *rigid*, a *rigid body* being defined as one which does not deform. Actual structures and machines, however, are never absolutely rigid and deform under the loads to which they are subjected. But these deformations are usually small and do not appreciably affect the conditions of equilibrium or motion of the structure under consideration. They are important, though, as far as the resistance of the structure to failure is concerned and are considered in the study of mechanics of materials.

In this chapter you will study the effect of forces exerted on a rigid body, and you will learn how to replace a given system of forces by a simpler equivalent system. This analysis will rest on the fundamental assumption that the effect of a given force on a rigid body remains unchanged if that force is moved along its line of action (*principle of transmissibility*). It follows that forces acting on a rigid body can be represented by *sliding vectors*, as indicated earlier in Sec. 2.3.

Two important concepts associated with the effect of a force on a rigid body are the *moment of a force about a point* (Sec. 3.6) and the *moment of a force about an axis* (Sec. 3.11). Since the determination of these quantities involves the computation of vector products and scalar products of two vectors, the fundamentals of vector algebra will be introduced in this chapter and applied to the solution of problems involving forces acting on rigid bodies.

Another concept introduced in this chapter is that of a *couple*, i.e., the combination of two forces which have the same magnitude, parallel lines of action, and opposite sense (Sec. 3.12). As you will see, any system of forces acting on a rigid body can be replaced by an equivalent system consisting of one force acting at a given point and one couple. This basic system is called a *force-couple system*. In the case of concurrent, coplanar, or parallel forces, the equivalent force-couple system can be further reduced to a single force, called the *resultant* of the system, or to a single couple, called the *resultant couple* of the system.

3.2 EXTERNAL AND INTERNAL FORCES

Forces acting on rigid bodies can be separated into two groups: (1) *external forces* and (2) *internal forces*.

1. The *external forces* represent the action of other bodies on the rigid body under consideration. They are entirely responsible for the external behavior of the rigid body. They will either cause it to move or ensure that it remains at rest. We shall be concerned only with external forces in this chapter and in Chaps. 4 and 5.

2. The *internal forces* are the forces which hold together the particles forming the rigid body. If the rigid body is structurally composed of several parts, the forces holding the component parts together are also defined as internal forces. Internal forces will be considered in Chaps. 6 and 7.

As an example of external forces, let us consider the forces acting on a disabled truck that three people are pulling forward by means of a rope attached to the front bumper (Fig. 3.1). The external forces acting on the truck are shown in a *free-body diagram* (Fig. 3.2). Let us first consider the *weight* of the truck. Although it embodies the effect of the earth's pull on each of the particles forming the truck, the weight can be represented by the single force **W**. The *point of application* of this force, i.e., the point at which the force acts, is defined as the *center of gravity* of the truck. It will be seen in Chap. 5 how centers of gravity can be determined. The weight **W** tends to make the truck move vertically downward. In fact, it would actually cause the truck to move downward, i.e., to fall, if it were not for the presence of the ground. The ground opposes the downward motion of the truck by means of the reactions **R**₁ and **R**₂. These forces are exerted *by* the ground *on* the truck and must therefore be included among the external forces acting on the truck.

Fig. 3.1

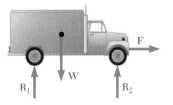

Fig. 3.2

The people pulling on the rope exert the force **F**. The point of application of **F** is on the front bumper. The force **F** tends to make the truck move forward in a straight line and does actually make it move, since no external force opposes this motion. (Rolling resistance has been neglected here for simplicity.) This forward motion of the truck, during which each straight line keeps its original orientation (the floor of the truck remains horizontal, and the walls remain vertical), is known as a *translation*. Other forces might cause the truck to move differently. For example, the force exerted by a jack placed under the front axle would cause the truck to pivot about its rear axle. Such a motion is a *rotation*. It can be concluded, therefore, that each of the *external forces* acting on a *rigid body* can, if unopposed, impart to the rigid body a motion of translation or rotation, or both.

3.3 PRINCIPLE OF TRANSMISSIBILITY. EQUIVALENT FORCES

The *principle of transmissibility* states that the conditions of equilibrium or motion of a rigid body will remain unchanged if a force **F** acting at a given point of the rigid body is replaced by a force **F**′ of the same magnitude and same direction, but acting at a different point, *provided that the two forces have the same line of action* (Fig. 3.3). The two forces **F** and **F**′ have the same effect on the rigid body and are said to be *equivalent*. This principle, which states that the action of a force may be *transmitted* along its line of action, is based on experimental evidence. It *cannot* be derived from the properties established so far in this text and must therefore be accepted as an experimental law. However, as you will see in Sec. 16.5, the principle of transmissibility can be derived from the study of the dynamics of rigid bodies, but this study requires the introduction of Newton's

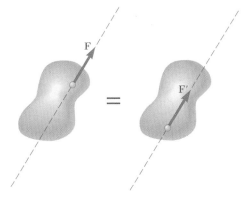

Fig. 3.3

second and third laws and of a number of other concepts as well. Therefore, our study of the statics of rigid bodies will be based on the three principles introduced so far, i.e., the parallelogram law of addition, Newton's first law, and the principle of transmissibility.

It was indicated in Chap. 2 that the forces acting on a particle could be represented by vectors. These vectors had a well-defined point of application, namely, the particle itself, and were therefore fixed, or bound, vectors. In the case of forces acting on a rigid body, however, the point of application of the force does not matter, as long as the line of action remains unchanged. Thus, forces acting on a rigid body must be represented by a different kind of vector, known as a *sliding vector*, since forces may be allowed to slide along their lines of action. We should note that all the properties which will be derived in the following sections for the forces acting on a rigid body will be valid more generally for any system of sliding vectors. In order to keep our presentation more intuitive, however, we will carry it out in terms of physical forces rather than in terms of mathematical sliding vectors.

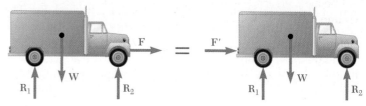

Fig. 3.4

Returning to the example of the truck, we first observe that the line of action of the force **F** is a horizontal line passing through both the front and the rear bumpers of the truck (Fig. 3.4). Using the principle of transmissibility, we can therefore replace **F** by an *equivalent force* **F′** acting on the rear bumper. In other words, the conditions of motion are unaffected, and all the other external forces acting on the truck (**W**, $\mathbf{R}_1$, $\mathbf{R}_2$) remain unchanged if the people push on the rear bumper instead of pulling on the front bumper.

The principle of transmissibility and the concept of equivalent forces have limitations, however. Consider, for example, a short bar *AB* acted upon by equal and opposite axial forces $\mathbf{P}_1$ and $\mathbf{P}_2$, as shown in Fig. 3.5*a*. According to the principle of transmissibility, the force $\mathbf{P}_2$ can be replaced by a force $\mathbf{P}_2'$ having the same magnitude, the same direction, and the same line of action but acting at *A* instead of *B* (Fig. 3.5*b*). The forces $\mathbf{P}_1$ and $\mathbf{P}_2'$ acting on the same particle

Fig. 3.5

can be added according to the rules of Chap. 2, and, as these forces are equal and opposite, their sum is equal to zero. Thus, in terms of the external behavior of the bar, the original system of forces shown in Fig. 3.5*a* is equivalent to no force at all (Fig. 3.5*c*).

Consider now the two equal and opposite forces $\mathbf{P}_1$ and $\mathbf{P}_2$ acting on the bar *AB* as shown in Fig. 3.5*d*. The force $\mathbf{P}_2$ can be replaced by a force $\mathbf{P}_2'$ having the same magnitude, the same direction, and the same line of action but acting at *B* instead of at *A* (Fig. 3.5*e*). The forces $\mathbf{P}_1$ and $\mathbf{P}_2'$ can then be added, and their sum is again zero (Fig. 3.5*f*). From the point of view of the mechanics of rigid bodies, the systems shown in Fig. 3.5*a* and *d* are thus equivalent. But the *internal forces* and *deformations* produced by the two systems are clearly different. The bar of Fig. 3.5*a* is in *tension* and, if not absolutely rigid, will increase in length slightly; the bar of Fig. 3.5*d* is in *compression* and, if not absolutely rigid, will decrease in length slightly. Thus, while the principle of transmissibility may be used freely to determine the conditions of motion or equilibrium of rigid bodies and to compute the external forces acting on these bodies, it should be avoided, or at least used with care, in determining internal forces and deformations.

3.4 VECTOR PRODUCT OF TWO VECTORS

In order to gain a better understanding of the effect of a force on a rigid body, a new concept, the concept of *a moment of a force about a point,* will be introduced at this time. This concept will be more clearly understood, and applied more effectively, if we first add to the mathematical tools at our disposal the *vector product* of two vectors.

The vector product of two vectors $\mathbf{P}$ and $\mathbf{Q}$ is defined as the vector $\mathbf{V}$ which satisfies the following conditions.

1. The line of action of $\mathbf{V}$ is perpendicular to the plane containing $\mathbf{P}$ and $\mathbf{Q}$ (Fig. 3.6*a*).
2. The magnitude of $\mathbf{V}$ is the product of the magnitudes of $\mathbf{P}$ and $\mathbf{Q}$ and of the sine of the angle θ formed by $\mathbf{P}$ and $\mathbf{Q}$ (the measure of which will always be 180° or less); we thus have

$$V = PQ \sin \theta \tag{3.1}$$

3. The direction of $\mathbf{V}$ is obtained from the *right-hand rule.* Close your right hand and hold it so that your fingers are curled in the same sense as the rotation through θ which brings the vector $\mathbf{P}$ in line with the vector $\mathbf{Q}$; your thumb will then indicate the direction of the vector $\mathbf{V}$ (Fig. 3.6*b*). Note that if $\mathbf{P}$ and $\mathbf{Q}$ do not have a common point of application, they should first be redrawn from the same point. The three vectors $\mathbf{P}$, $\mathbf{Q}$, and $\mathbf{V}$—taken in that order—are said to form a *right-handed triad.*†

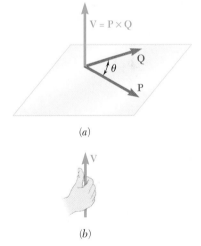

(*a*)

(*b*)

Fig. 3.6

---

†We should note that the x, y, and z axes used in Chap. 2 form a right-handed system of orthogonal axes and that the unit vectors $\mathbf{i}$, $\mathbf{j}$, $\mathbf{k}$ defined in Sec. 2.12 form a right-handed orthogonal triad.

As stated above, the vector **V** satisfying these three conditions (which define it uniquely) is referred to as the vector product of **P** and **Q**; it is represented by the mathematical expression

$$\mathbf{V} = \mathbf{P} \times \mathbf{Q} \qquad (3.2)$$

Because of the notation used, the vector product of two vectors **P** and **Q** is also referred to as the *cross product* of **P** and **Q**.

It follows from Eq. (3.1) that, when two vectors **P** and **Q** have either the same direction or opposite directions, their vector product is zero. In the general case when the angle θ formed by the two vectors is neither 0° nor 180°, Eq. (3.1) can be given a simple geometric interpretation: The magnitude V of the vector product of **P** and **Q** is equal to the area of the parallelogram which has **P** and **Q** for sides (Fig. 3.7). The vector product **P** × **Q** will therefore remain unchanged if we replace **Q** by a vector **Q**′ which is coplanar with **P** and **Q** and such that the line joining the tips of **Q** and **Q**′ is parallel to **P**. We write

$$\mathbf{V} = \mathbf{P} \times \mathbf{Q} = \mathbf{P} \times \mathbf{Q}' \qquad (3.3)$$

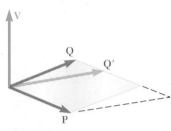

Fig. 3.7

From the third condition used to define the vector product **V** of **P** and **Q**, namely, the condition stating that **P**, **Q**, and **V** must form a right-handed triad, it follows that vector products *are not commutative*, i.e., **Q** × **P** is not equal to **P** × **Q**. Indeed, we can easily check that **Q** × **P** is represented by the vector −**V**, which is equal and opposite to **V**. We thus write

$$\mathbf{Q} \times \mathbf{P} = -(\mathbf{P} \times \mathbf{Q}) \qquad (3.4)$$

EXAMPLE   Let us compute the vector product **V** = **P** × **Q** where the vector **P** is of magnitude 6 and lies in the zx plane at an angle of 30° with the x axis, and where the vector **Q** is of magnitude 4 and lies along the x axis (Fig. 3.8).

It follows immediately from the definition of the vector product that the vector **V** must lie along the y axis, have the magnitude

$$V = PQ \sin \theta = (6)(4) \sin 30° = 12$$

and be directed upward. ▪

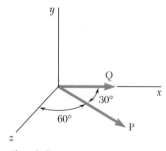

Fig. 3.8

We saw that the commutative property does not apply to vector products. We may wonder whether the *distributive* property holds, i.e., whether the relation

$$\mathbf{P} \times (\mathbf{Q}_1 + \mathbf{Q}_2) = \mathbf{P} \times \mathbf{Q}_1 + \mathbf{P} \times \mathbf{Q}_2 \qquad (3.5)$$

is valid. The answer is *yes*. Many readers are probably willing to accept without formal proof an answer which they intuitively feel is correct. However, since the entire structure of both vector algebra and statics depends upon the relation (3.5), we should take time out to derive it.

We can, without any loss of generality, assume that **P** is directed along the y axis (Fig. 3.9a). Denoting by **Q** the sum of **Q**$_1$ and **Q**$_2$, we drop perpendiculars from the tips of **Q**, **Q**$_1$, and **Q**$_2$ onto the zx plane, defining in this way the vectors **Q**′, **Q**′$_1$, and **Q**′$_2$. These vectors will be referred to, respectively, as the *projections* of **Q**, **Q**$_1$, and **Q**$_2$ on the zx plane. Recalling the property expressed by Eq. (3.3), we

note that the left-hand member of Eq. (3.5) can be replaced by
$\mathbf{P} \times \mathbf{Q}'$ and that, similarly, the vector products $\mathbf{P} \times \mathbf{Q}_1$ and $\mathbf{P} \times \mathbf{Q}_2$
can respectively be replaced by $\mathbf{P} \times \mathbf{Q}_1'$ and $\mathbf{P} \times \mathbf{Q}_2'$. Thus, the
relation to be proved can be written in the form

$$\mathbf{P} \times \mathbf{Q}' = \mathbf{P} \times \mathbf{Q}_1' + \mathbf{P} \times \mathbf{Q}_2' \tag{3.5'}$$

We now observe that $\mathbf{P} \times \mathbf{Q}'$ can be obtained from $\mathbf{Q}'$ by
multiplying this vector by the scalar P and rotating it counterclockwise through 90° in the zx plane (Fig. 3.9b); the other two vector

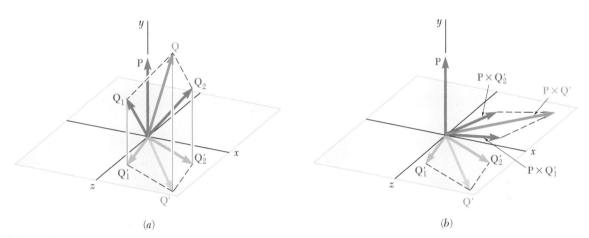

Fig. 3.9

products in (3.5′) can be obtained in the same manner from $\mathbf{Q}_1'$ and
$\mathbf{Q}_2'$, respectively. Now, since the projection of a parallelogram onto
an arbitrary plane is a parallelogram, the projection $\mathbf{Q}'$ of the sum
$\mathbf{Q}$ of $\mathbf{Q}_1$ and $\mathbf{Q}_2$ must be the sum of the projections $\mathbf{Q}_1'$ and $\mathbf{Q}_2'$ of
$\mathbf{Q}_1$ and $\mathbf{Q}_2$ on the same plane (Fig. 3.9a). This relation between the
vectors $\mathbf{Q}'$, $\mathbf{Q}_1'$, and $\mathbf{Q}_2'$ will still hold after the three vectors have
been multiplied by the scalar P and rotated through 90° (Fig. 3.9b).
Thus, the relation (3.5′) has been proved, and we can now be sure
that the distributive property holds for vector products.

A third property, the associative property, does not apply to
vector products; we have in general

$$(\mathbf{P} \times \mathbf{Q}) \times \mathbf{S} \neq \mathbf{P} \times (\mathbf{Q} \times \mathbf{S}) \tag{3.6}$$

3.5 VECTOR PRODUCTS EXPRESSED IN TERMS OF RECTANGULAR COMPONENTS

Let us now determine the vector product of any two of the unit
vectors $\mathbf{i}$, $\mathbf{j}$, and $\mathbf{k}$, which were defined in Chap. 2. Consider first the
product $\mathbf{i} \times \mathbf{j}$ (Fig. 3.10a). Since both vectors have a magnitude
equal to 1 and since they are at a right angle to each other, their
vector product will also be a unit vector. This unit vector must be $\mathbf{k}$,
since the vectors $\mathbf{i}$, $\mathbf{j}$, and $\mathbf{k}$ are mutually perpendicular and form a
right-handed triad. On the other hand, it follows from the right-hand
rule given on page 79 that the product $\mathbf{j} \times \mathbf{i}$ will be equal to $-\mathbf{k}$
(Fig. 3.10b). Finally, it should be observed that the vector product

Fig. 3.10

of a unit vector with itself, such as $\mathbf{i} \times \mathbf{i}$, is equal to zero, since both vectors have the same direction. The vector products of the various possible pairs of unit vectors are

$$
\begin{array}{lll}
\mathbf{i} \times \mathbf{i} = 0 & \mathbf{j} \times \mathbf{i} = -\mathbf{k} & \mathbf{k} \times \mathbf{i} = \mathbf{j} \\
\mathbf{i} \times \mathbf{j} = \mathbf{k} & \mathbf{j} \times \mathbf{j} = 0 & \mathbf{k} \times \mathbf{j} = -\mathbf{i} \\
\mathbf{i} \times \mathbf{k} = -\mathbf{j} & \mathbf{j} \times \mathbf{k} = \mathbf{i} & \mathbf{k} \times \mathbf{k} = 0
\end{array}
\tag{3.7}
$$

Fig. 3.11

By arranging in a circle and in counterclockwise order the three letters representing the unit vectors (Fig. 3.11), we can simplify the determination of the sign of the vector product of two unit vectors: The product of two unit vectors will be positive if they follow each other in counterclockwise order and will be negative if they follow each other in clockwise order.

We can now easily express the vector product $\mathbf{V}$ of two given vectors $\mathbf{P}$ and $\mathbf{Q}$ in terms of the rectangular components of these vectors. Resolving $\mathbf{P}$ and $\mathbf{Q}$ into components, we first write

$$
\mathbf{V} = \mathbf{P} \times \mathbf{Q} = (P_x\mathbf{i} + P_y\mathbf{j} + P_z\mathbf{k}) \times (Q_x\mathbf{i} + Q_y\mathbf{j} + Q_z\mathbf{k})
$$

Making use of the distributive property, we express $\mathbf{V}$ as the sum of vector products, such as $P_x\mathbf{i} \times Q_y\mathbf{j}$. Observing that each of the expressions obtained is equal to the vector product of two unit vectors, such as $\mathbf{i} \times \mathbf{j}$, multiplied by the product of two scalars, such as P_xQ_y, and recalling the identities (3.7), we obtain, after factoring out $\mathbf{i}$, $\mathbf{j}$, and $\mathbf{k}$,

$$
\mathbf{V} = (P_yQ_z - P_zQ_y)\mathbf{i} + (P_zQ_x - P_xQ_z)\mathbf{j} + (P_xQ_y - P_yQ_x)\mathbf{k} \tag{3.8}
$$

The rectangular components of the vector product $\mathbf{V}$ are thus found to be

$$
\begin{aligned}
V_x &= P_yQ_z - P_zQ_y \\
V_y &= P_zQ_x - P_xQ_z \\
V_z &= P_xQ_y - P_yQ_x
\end{aligned}
\tag{3.9}
$$

Returning to Eq. (3.8), we observe that its right-hand member represents the expansion of a determinant. The vector product $\mathbf{V}$ can thus be expressed in the following form, which is more easily memorized:†

$$
\mathbf{V} =
\begin{vmatrix}
\mathbf{i} & \mathbf{j} & \mathbf{k} \\
P_x & P_y & P_z \\
Q_x & Q_y & Q_z
\end{vmatrix}
\tag{3.10}
$$

†Any determinant consisting of three rows and three columns can be evaluated by repeating the first and second columns and forming products along each diagonal line. The sum of the products obtained along the red lines is then subtracted from the sum of the products obtained along the black lines.

3.6 MOMENT OF A FORCE ABOUT A POINT

Let us now consider a force **F** acting on a rigid body (Fig. 3.12a). As we know, the force **F** is represented by a vector which defines its magnitude and direction. However, the effect of the force on the rigid body depends also upon its point of application A. The position of A can be conveniently defined by the vector **r** which joins the fixed reference point O with A; this vector is known as the *position vector* of A.† The position vector **r** and the force **F** define the plane shown in Fig. 3.12a.

We will define the *moment of* **F** *about* O as the vector product of **r** and **F**:

$$M_O = r \times F \tag{3.11}$$

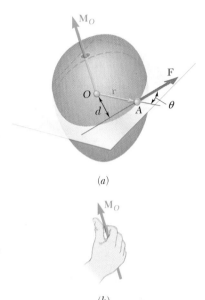

(a)

(b)

Fig. 3.12

According to the definition of the vector product given in Sec. 3.4, the moment **M**$_O$ must be perpendicular to the plane containing O and the force **F**. The sense of **M**$_O$ is defined by the sense of the rotation which will bring the vector **r** in line with the vector **F**; this rotation will be observed as *counterclockwise* by an observer located at the tip of **M**$_O$. Another way of defining the sense of **M**$_O$ is furnished by a variation of the right-hand rule: Close your right hand and hold it so that your fingers are curled in the sense of the rotation that **F** would impart to the rigid body about a fixed axis directed along the line of action of **M**$_O$; your thumb will indicate the sense of the moment **M**$_O$ (Fig. 3.12b).

Finally, denoting by θ the angle between the lines of action of the position vector **r** and the force **F**, we find that the magnitude of the moment of **F** about O is

$$M_O = rF \sin \theta = Fd \tag{3.12}$$

where d represents the perpendicular distance from O to the line of action of **F**. Since the tendency of a force **F** to make a rigid body rotate about a fixed axis perpendicular to the force depends upon the distance of **F** from that axis as well as upon the magnitude of **F**, we note that *the magnitude of* **M**$_O$ *measures the tendency of the force* **F** *to make the rigid body rotate about a fixed axis directed along* **M**$_O$.

In the SI system of units, where a force is expressed in newtons (N) and a distance in meters (m), the moment of a force is expressed in newton-meters (N · m). In the U.S. customary system of units, where a force is expressed in pounds and a distance in feet or inches, the moment of a force is expressed in lb · ft or lb · in.

We can observe that although the moment **M**$_O$ of a force about a point depends upon the magnitude, the line of action, and the sense of the force, it does *not* depend upon the actual position of the point of application of the force along its line of action. Conversely, the moment **M**$_O$ of a force **F** does not characterize the position of the point of application of **F**.

---

†We can easily verify that position vectors obey the law of vector addition and, thus, are truly vectors. Consider, for example, the position vectors **r** and **r'** of A with respect to two reference points O and O' and the position vector **s** of O with respect to O' (Fig. 3.40a, Sec. 3.16). We verify that the position vector **r'** = $\overrightarrow{O'A}$ can be obtained from the position vectors **s** = $\overrightarrow{O'O}$ and **r** = $\overrightarrow{OA}$ by applying the triangle rule for the addition of vectors.

However, as it will be seen presently, the moment $\mathbf{M}_O$ of a force $\mathbf{F}$ of given magnitude and direction *completely defines the line of action of* $\mathbf{F}$. Indeed, the line of action of $\mathbf{F}$ must lie in a plane through O perpendicular to the moment $\mathbf{M}_O$; its distance d from O must be equal to the quotient M_O/F of the magnitudes of $\mathbf{M}_O$ and $\mathbf{F}$; and the sense of $\mathbf{M}_O$ determines whether the line of action of $\mathbf{F}$ is to be drawn on one side or the other of the point O.

We recall from Sec. 3.3 that the principle of transmissibility states that two forces $\mathbf{F}$ and $\mathbf{F}'$ are equivalent (i.e., have the same effect on a rigid body) if they have the same magnitude, same direction, and same line of action. This principle can now be restated as follows: *Two forces* $\mathbf{F}$ *and* $\mathbf{F}'$ *are equivalent if, and only if, they are equal* (i.e., have the same magnitude and same direction) *and have equal moments about a given point* O. The necessary and sufficient conditions for two forces $\mathbf{F}$ and $\mathbf{F}'$ to be equivalent are thus

$$\mathbf{F} = \mathbf{F}' \qquad \text{and} \qquad \mathbf{M}_O = \mathbf{M}'_O \tag{3.13}$$

We should observe that it follows from this statement that if the relations (3.13) hold for a given point O, they will hold for any other point.

Problems Involving Only Two Dimensions.

Many applications deal with two-dimensional structures, i.e., structures which have length and breadth but only negligible depth and which are subjected to forces contained in the plane of the structure. Two-dimensional structures and the forces acting on them can be readily represented on a sheet of paper or on a blackboard. Their analysis is therefore considerably simpler than that of three-dimensional structures and forces.

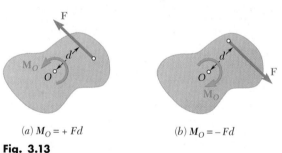

(a) $M_O = + Fd$ (b) $M_O = - Fd$

Fig. 3.13

Consider, for example, a rigid slab acted upon by a force $\mathbf{F}$ (Fig. 3.13). The moment of $\mathbf{F}$ about a point O chosen in the plane of the figure is represented by a vector $\mathbf{M}_O$ perpendicular to that plane and of magnitude Fd. In the case of Fig. 3.13a the vector $\mathbf{M}_O$ points *out of* the paper, while in the case of Fig. 3.13b it points *into* the paper. As we look at the figure, we observe in the first case that $\mathbf{F}$ tends to rotate the slab counterclockwise and in the second case that it tends to rotate the slab clockwise. Therefore, it is natural to refer to the sense of the moment of $\mathbf{F}$ about O in Fig. 3.13a as counterclockwise $\uparrow$, and in Fig. 3.13b as clockwise $\downarrow$.

Since the moment of a force $\mathbf{F}$ acting in the plane of the figure must be perpendicular to that plane, we need only specify the *magnitude* and the *sense* of the moment of $\mathbf{F}$ about O. This can be done by assigning to the magnitude M_O of the moment a positive or negative sign according to whether the vector $\mathbf{M}_O$ points out of or into the paper.

3.7 VARIGNON'S THEOREM

The distributive property of vector products can be used to determine the moment of the resultant of several *concurrent forces*. If several forces $\mathbf{F}_1$, $\mathbf{F}_2$, . . . are applied at the same point A (Fig. 3.14), and if we denote by $\mathbf{r}$ the position vector of A, it follows immediately from Eq. (3.5) of Sec. 3.4 that

$$\mathbf{r} \times (\mathbf{F}_1 + \mathbf{F}_2 + \cdots) = \mathbf{r} \times \mathbf{F}_1 + \mathbf{r} \times \mathbf{F}_2 + \cdots \quad (3.14)$$

In words, *the moment about a given point O of the resultant of several concurrent forces is equal to the sum of the moments of the various forces about the same point O*. This property, which was originally established by the French mathematician Varignon (1654–1722) long before the introduction of vector algebra, is known as *Varignon's theorem*.

The relation (3.14) makes it possible to replace the direct determination of the moment of a force $\mathbf{F}$ by the determination of the moments of two or more component forces. As you will see in the next section, $\mathbf{F}$ will generally be resolved into components parallel to the coordinate axes. However, it may be more expeditious in some instances to resolve $\mathbf{F}$ into components which are not parallel to the coordinate axes (see Sample Prob. 3.3).

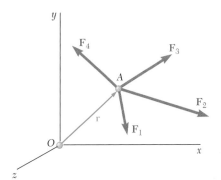

Fig. 3.14

3.8 RECTANGULAR COMPONENTS OF THE MOMENT OF A FORCE

In general, the determination of the moment of a force in space will be considerably simplified if the force and the position vector of its point of application are resolved into rectangular x, y, and z components. Consider, for example, the moment $\mathbf{M}_O$ about O of a force $\mathbf{F}$ whose components are F_x, F_y, and F_z and which is applied at a point A of coordinates x, y, and z (Fig. 3.15). Observing that the components of the position vector $\mathbf{r}$ are respectively equal to the coordinates x, y, and z of the point A, we write

$$\mathbf{r} = x\mathbf{i} + y\mathbf{j} + z\mathbf{k} \quad (3.15)$$
$$\mathbf{F} = F_x\mathbf{i} + F_y\mathbf{j} + F_z\mathbf{k} \quad (3.16)$$

Substituting for $\mathbf{r}$ and $\mathbf{F}$ from (3.15) and (3.16) into

$$\mathbf{M}_O = \mathbf{r} \times \mathbf{F} \quad (3.11)$$

and recalling the results obtained in Sec. 3.5, we write the moment $\mathbf{M}_O$ of $\mathbf{F}$ about O in the form

$$\mathbf{M}_O = M_x\mathbf{i} + M_y\mathbf{j} + M_z\mathbf{k} \quad (3.17)$$

where the components M_x, M_y, and M_z are defined by the relations

$$\begin{aligned} M_x &= yF_z - zF_y \\ M_y &= zF_x - xF_z \\ M_z &= xF_y - yF_x \end{aligned} \quad (3.18)$$

Fig. 3.15

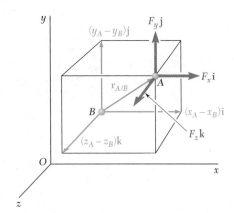

Fig. 3.16

As you will see in Sec. 3.11, the scalar components M_x, M_y, and M_z of the moment $\mathbf{M}_O$ measure the tendency of the force $\mathbf{F}$ to impart to a rigid body a motion of rotation about the x, y, and z axes, respectively. Substituting from (3.18) into (3.17), we can also write $\mathbf{M}_O$ in the form of the determinant

$$\mathbf{M}_O = \begin{vmatrix} \mathbf{i} & \mathbf{j} & \mathbf{k} \\ x & y & z \\ F_x & F_y & F_z \end{vmatrix} \qquad (3.19)$$

To compute the moment $\mathbf{M}_B$ about an arbitrary point B of a force $\mathbf{F}$ applied at A (Fig. 3.16), we must replace the position vector $\mathbf{r}$ in Eq. (3.11) by a vector drawn from B to A. This vector is the *position vector of A relative to B* and will be denoted by $\mathbf{r}_{A/B}$. Observing that $\mathbf{r}_{A/B}$ can be obtained by subtracting $\mathbf{r}_B$ from $\mathbf{r}_A$, we write

$$\mathbf{M}_B = \mathbf{r}_{A/B} \times \mathbf{F} = (\mathbf{r}_A - \mathbf{r}_B) \times \mathbf{F} \qquad (3.20)$$

or, using the determinant form,

$$\mathbf{M}_B = \begin{vmatrix} \mathbf{i} & \mathbf{j} & \mathbf{k} \\ x_{A/B} & y_{A/B} & z_{A/B} \\ F_x & F_y & F_z \end{vmatrix} \qquad (3.21)$$

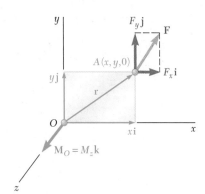

Fig. 3.17

where $x_{A/B}$, $y_{A/B}$, and $z_{A/B}$ denote the components of the vector $\mathbf{r}_{A/B}$:

$$x_{A/B} = x_A - x_B \qquad y_{A/B} = y_A - y_B \qquad z_{A/B} = z_A - z_B$$

In the case of *problems involving only two dimensions*, the force $\mathbf{F}$ can be assumed to lie in the xy plane (Fig. 3.17). Setting $z = 0$ and $F_z = 0$ in Eq. (3.19), we obtain

$$\mathbf{M}_O = (xF_y - yF_x)\mathbf{k}$$

We verify that the moment of $\mathbf{F}$ about O is perpendicular to the plane of the figure and that it is completely defined by the scalar

$$M_O = M_z = xF_y - yF_x \qquad (3.22)$$

As noted earlier, a positive value for M_O indicates that the vector $\mathbf{M}_O$ points out of the paper (the force $\mathbf{F}$ tends to rotate the body counterclockwise about O), and a negative value indicates that the vector $\mathbf{M}_O$ points into the paper (the force $\mathbf{F}$ tends to rotate the body clockwise about O).

To compute the moment about $B(x_B, y_B)$ of a force lying in the xy plane and applied at $A(x_A, y_A)$ (Fig. 3.18), we set $z_{A/B} = 0$ and $F_z = 0$ in the relations (3.21) and note that the vector $\mathbf{M}_B$ is perpendicular to the xy plane and is defined in magnitude and sense by the scalar

$$M_B = (x_A - x_B)F_y - (y_A - y_B)F_x \qquad (3.23)$$

Fig. 3.18

SAMPLE PROBLEM 3.1

A 100-lb vertical force is applied to the end of a lever which is attached to a shaft at O. Determine (a) the moment of the 100-lb force about O; (b) the horizontal force applied at A which creates the same moment about O; (c) the smallest force applied at A which creates the same moment about O; (d) how far from the shaft a 240-lb vertical force must act to create the same moment about O; (e) whether any one of the forces obtained in parts b, c, and d is equivalent to the original force.

SOLUTION

a. Moment about O. The perpendicular distance from O to the line of action of the 100-lb force is

$$d = (24 \text{ in.}) \cos 60° = 12 \text{ in.}$$

The magnitude of the moment about O of the 100-lb force is

$$M_O = Fd = (100 \text{ lb})(12 \text{ in.}) = 1200 \text{ lb} \cdot \text{in.}$$

Since the force tends to rotate the lever clockwise about O, the moment will be represented by a vector $\mathbf{M}_O$ perpendicular to the plane of the figure and pointing *into* the paper. We express this fact by writing

$$\mathbf{M}_O = 1200 \text{ lb} \cdot \text{in.} \downdownarrows \quad \blacktriangleleft$$

b. Horizontal Force. In this case, we have

$$d = (24 \text{ in.}) \sin 60° = 20.8 \text{ in.}$$

Since the moment about O must be 1200 lb · in., we write

$$M_O = Fd$$
$$1200 \text{ lb} \cdot \text{in.} = F(20.8 \text{ in.})$$
$$F = 57.7 \text{ lb} \qquad \mathbf{F} = 57.7 \text{ lb} \rightarrow \quad \blacktriangleleft$$

c. Smallest Force. Since $M_O = Fd$, the smallest value of F occurs when d is maximum. We choose the force perpendicular to OA and note that $d = 24$ in.; thus

$$M_O = Fd$$
$$1200 \text{ lb} \cdot \text{in.} = F(24 \text{ in.})$$
$$F = 50 \text{ lb} \qquad \mathbf{F} = 50 \text{ lb} \searrow 30° \quad \blacktriangleleft$$

d. 240-lb Vertical Force. In this case $M_O = Fd$ yields

$$1200 \text{ lb} \cdot \text{in.} = (240 \text{ lb})d \qquad d = 5 \text{ in.}$$

but $\qquad\qquad OB \cos 60° = d \qquad\qquad OB = 10 \text{ in.} \quad \blacktriangleleft$

e. None of the forces considered in parts b, c, and d is equivalent to the original 100-lb force. Although they have the same moment about O, they have different x and y components. In other words, although each force tends to rotate the shaft in the same manner, each causes the lever to pull on the shaft in a different way.

SAMPLE PROBLEM 3.2

A force of 800 N acts on a bracket as shown. Determine the moment of the force about B.

SOLUTION

The moment $\mathbf{M}_B$ of the force $\mathbf{F}$ about B is obtained by forming the vector product

$$\mathbf{M}_B = \mathbf{r}_{A/B} \times \mathbf{F}$$

where $\mathbf{r}_{A/B}$ is the vector drawn from B to A. Resolving $\mathbf{r}_{A/B}$ and $\mathbf{F}$ into rectangular components, we have

$$\mathbf{r}_{A/B} = -(0.2\ \text{m})\mathbf{i} + (0.16\ \text{m})\mathbf{j}$$
$$\mathbf{F} = (800\ \text{N})\cos 60°\mathbf{i} + (800\ \text{N})\sin 60°\mathbf{j}$$
$$= (400\ \text{N})\mathbf{i} + (693\ \text{N})\mathbf{j}$$

Recalling the relations (3.7) for the cross products of unit vectors (Sec. 3.5), we obtain

$$\mathbf{M}_B = \mathbf{r}_{A/B} \times \mathbf{F} = [-(0.2\ \text{m})\mathbf{i} + (0.16\ \text{m})\mathbf{j}] \times [(400\ \text{N})\mathbf{i} + (693\ \text{N})\mathbf{j}]$$
$$= -(138.6\ \text{N} \cdot \text{m})\mathbf{k} - (64.0\ \text{N} \cdot \text{m})\mathbf{k}$$
$$= -(202.6\ \text{N} \cdot \text{m})\mathbf{k} \qquad\qquad \mathbf{M}_B = 203\ \text{N} \cdot \text{m} \downarrow \blacktriangleleft$$

The moment $\mathbf{M}_B$ is a vector perpendicular to the plane of the figure and pointing *into* the paper.

SAMPLE PROBLEM 3.3

A 30-lb force acts on the end of the 3-ft lever as shown. Determine the moment of the force about O.

SOLUTION

The force is replaced by two components, one component $\mathbf{P}$ in the direction of OA and one component $\mathbf{Q}$ perpendicular to OA. Since O is on the line of action of $\mathbf{P}$, the moment of $\mathbf{P}$ about O is zero and the moment of the 30-lb force reduces to the moment of $\mathbf{Q}$, which is clockwise and, thus, is represented by a negative scalar.

$$Q = (30\ \text{lb}) \sin 20° = 10.26\ \text{lb}$$
$$M_O = -Q(3\ \text{ft}) = -(10.26\ \text{lb})(3\ \text{ft}) = -30.8\ \text{lb} \cdot \text{ft}$$

Since the value obtained for the scalar M_O is negative, the moment $\mathbf{M}_O$ points *into* the paper. We write

$$\mathbf{M}_O = 30.8\ \text{lb} \cdot \text{ft} \downarrow \blacktriangleleft$$

SAMPLE PROBLEM 3.4

A rectangular plate is supported by brackets at A and B and by a wire CD. Knowing that the tension in the wire is 200 N, determine the moment about A of the force exerted by the wire on point C.

SOLUTION

The moment $\mathbf{M}_A$ about A of the force $\mathbf{F}$ exerted by the wire on point C is obtained by forming the vector product

$$\mathbf{M}_A = \mathbf{r}_{C/A} \times \mathbf{F} \tag{1}$$

where $\mathbf{r}_{C/A}$ is the vector drawn from A to C,

$$\mathbf{r}_{C/A} = \overrightarrow{AC} = (0.3 \text{ m})\mathbf{i} + (0.08 \text{ m})\mathbf{k} \tag{2}$$

and $\mathbf{F}$ is the 200-N force directed along CD. Introducing the unit vector $\boldsymbol{\lambda} = \overrightarrow{CD}/CD$, we write

$$\mathbf{F} = F\boldsymbol{\lambda} = (200 \text{N}) \frac{\overrightarrow{CD}}{CD} \tag{3}$$

Resolving the vector $\overrightarrow{CD}$ into rectangular components, we have

$$\overrightarrow{CD} = -(0.3 \text{ m})\mathbf{i} + (0.24 \text{ m})\mathbf{j} - (0.32 \text{ m})\mathbf{k} \qquad CD = 0.50 \text{ m}$$

Substituting into (3), we obtain

$$\mathbf{F} = \frac{200 \text{N}}{0.50 \text{m}} [-(0.3 \text{ m})\mathbf{i} + (0.24 \text{ m})\mathbf{j} - (0.32 \text{ m})\mathbf{k}]$$
$$= -(120 \text{ N})\mathbf{i} + (96 \text{ N})\mathbf{j} - (128 \text{ N})\mathbf{k} \tag{4}$$

Substituting for $\mathbf{r}_{C/A}$ and $\mathbf{F}$ from (2) and (4) into (1) and recalling the relations (3.7) of Sec. 3.5, we obtain

$$\mathbf{M}_A = \mathbf{r}_{C/A} \times \mathbf{F} = (0.3\mathbf{i} + 0.08\mathbf{k}) \times (-120\mathbf{i} + 96\mathbf{j} - 128\mathbf{k})$$
$$= (0.3)(96)\mathbf{k} + (0.3)(-128)(-\mathbf{j}) + (0.08)(-120)\mathbf{j} + (0.08)(96)(-\mathbf{i})$$
$$\mathbf{M}_A = -(7.68 \text{ N} \cdot \text{m})\mathbf{i} + (28.8 \text{ N} \cdot \text{m})\mathbf{j} + (28.8 \text{ N} \cdot \text{m})\mathbf{k} \ \blacktriangleleft$$

Alternative Solution. As indicated in Sec. 3.8, the moment $\mathbf{M}_A$ can be expressed in the form of a determinant:

$$\mathbf{M}_A = \begin{vmatrix} \mathbf{i} & \mathbf{j} & \mathbf{k} \\ x_C - x_A & y_C - y_A & z_C - z_A \\ F_x & F_y & F_z \end{vmatrix} = \begin{vmatrix} \mathbf{i} & \mathbf{j} & \mathbf{k} \\ 0.3 & 0 & 0.08 \\ -120 & 96 & -128 \end{vmatrix}$$

$$\mathbf{M}_A = -(7.68 \text{N} \cdot \text{m})\mathbf{i} + (28.8 \text{ N} \cdot \text{m})\mathbf{j} + (28.8 \text{ N} \cdot \text{m})\mathbf{k} \ \blacktriangleleft$$

SOLVING PROBLEMS
ON YOUR OWN

In this lesson we introduced the *vector product* or *cross product* of two vectors. In the following problems, you may want to use the vector product to compute the *moment of a force about a point* and also to determine the *perpendicular distance* from a point to a line.

We defined the moment of the force $\mathbf{F}$ about the point O of a rigid body as

$$\mathbf{M}_O = \mathbf{r} \times \mathbf{F} \tag{3.11}$$

where $\mathbf{r}$ is the position vector *from O to any point* on the line of action of $\mathbf{F}$. Since the vector product is not commutative, it is absolutely necessary when computing such a product that you place the vectors in the proper order and that each vector have the correct sense. The moment $\mathbf{M}_O$ is important because its magnitude is a measure of the tendency of the force $\mathbf{F}$ to cause the rigid body to rotate about an axis directed along $\mathbf{M}_O$.

1. Computing the moment M_O of a force in two dimensions. You can use one of the following procedures:

a. Use Eq. (3.12), $M_O = Fd$, which expresses the magnitude of the moment as the product of the magnitude of $\mathbf{F}$ and the *perpendicular distance d* from O to the line of action of $\mathbf{F}$ [Sample Prob. 3.1].

b. Express $\mathbf{r}$ and $\mathbf{F}$ in component form and formally evaluate the vector product $\mathbf{M}_O = \mathbf{r} \times \mathbf{F}$ [Sample Prob. 3.2].

c. Resolve $\mathbf{F}$ into components respectively parallel and perpendicular to the position vector $\mathbf{r}$. Only the perpendicular component contributes to the moment of $\mathbf{F}$ [Sample Prob. 3.3].

d. Use Eq. (3.22), $M_O = M_z = xF_y - yF_x$. When applying this method, the simplest approach is to treat the scalar components of $\mathbf{r}$ and $\mathbf{F}$ as positive and then to assign, by observation, the proper sign to the moment produced by each force component. For example, applying this method to solve Sample Prob. 3.2, we observe that both force components tend to produce a clockwise rotation about B. Therefore, the moment of each force about B should be represented by a negative scalar. We then have for the total moment

$$M_B = -(0.16 \text{ m})(400 \text{ N}) - (0.20 \text{ m})(693 \text{ N}) = -202.6 \text{ N} \cdot \text{m}$$

2. Computing the moment M_O of a force F in three dimensions. Following the method of Sample Prob. 3.4, the first step in the process is to select the most convenient (simplest) position vector $\mathbf{r}$. You should next express $\mathbf{F}$ in terms of its rectangular components. The final step is to evaluate the vector product $\mathbf{r} \times \mathbf{F}$ to determine the moment. In most three-dimensional problems you will find it easiest to calculate the vector product using a determinant.

3. Determining the perpendicular distance d from a point A to a given line. First assume that a force $\mathbf{F}$ of known magnitude F lies along the given line. Next determine its moment about A by forming the vector product $\mathbf{M}_A = \mathbf{r} \times \mathbf{F}$, and calculate this product as indicated above. Then compute its magnitude M_A. Finally, substitute the values of F and M_A into the equation $M_A = Fd$ and solve for d.

PROBLEMS

3.1 A 20-lb force is applied to the control rod *AB* as shown. Knowing that the length of the rod is 9 in. and that $\alpha = 25°$, determine the moment of the force about point *B* by resolving the force into horizontal and vertical components.

3.2 A 20-lb force is applied to the control rod *AB* as shown. Knowing that the length of the rod is 9 in. and that $\alpha = 25°$, determine the moment of the force about point *B* by resolving the force into components along *AB* and in a direction perpendicular to *AB*.

3.3 A 20-lb force is applied to the control rod *AB* as shown. Knowing that the length of the rod is 9 in. and that the moment of the force about *B* is 120 lb · in. clockwise, determine the value of α.

Fig. P3.1, P3.2, and P3.3

3.4 A crate of mass 80 kg is held in the position shown. Determine (*a*) the moment produced by the weight **W** of the crate about *E*, (*b*) the smallest force applied at *B* that creates a moment of equal magnitude and opposite sense about *E*.

3.5 A crate of mass 80 kg is held in the position shown. Determine (*a*) the moment produced by the weight **W** of the crate about *E*, (*b*) the smallest force applied at *A* that creates a moment of equal magnitude and opposite sense about *E*, (*c*) the magnitude, sense, and point of application on the bottom of the crate of the smallest vertical force that creates a moment of equal magnitude and opposite sense about *E*.

Fig. P3.4 and P3.5

3.6 A 300-N force **P** is applied at point *A* of the bell crank shown. (*a*) Compute the moment of the force **P** about *O* by resolving it into horizontal and vertical components. (*b*) Using the result of part *a*, determine the perpendicular distance from *O* to the line of action of **P**.

Fig. P3.6 and P3.7

3.7 A 400-N force **P** is applied at point *A* of the bell crank shown. (*a*) Compute the moment of the force **P** about *O* by resolving it into components along line *OA* and in a direction perpendicular to that line. (*b*) Determine the magnitude and direction of the smallest force **Q** applied at *B* that has the same moment as **P** about *O*.

3.8 It is known that a vertical force of 200 lb is required to remove the nail at C from the board. As the nail first starts moving, determine (a) the moment about B of the force exerted on the nail, (b) the magnitude of the force **P** that creates the same moment about B if $\alpha = 10°$, (c) the smallest force **P** that creates the same moment about B.

3.9 and 3.10 It is known that the connecting rod AB exerts on the crank BC a 500-lb force directed down and to the left along the centerline of AB. Determine the moment of the force about C.

Fig. P3.8

Fig. P3.9   **Fig. P3.10**

3.11 A winch puller AB is used to straighten a fence post. Knowing that the tension in cable BC is 1040 N and length d is 1.90 m, determine the moment about D of the force exerted by the cable at C by resolving that force into horizontal and vertical components applied (a) at point C, (b) at point E.

3.12 It is known that a force with a moment of 960 N · m about D is required to straighten the fence post CD. If $d = 2.80$ m, determine the tension that must be developed in the cable of winch puller AB to create the required moment about point D.

3.13 It is known that a force with a moment of 960 N · m about D is required to straighten the fence post CD. If the capacity of winch puller AB is 2400 N, determine the minimum value of distance d to create the specified moment about point D.

Fig. P3.11, P3.12, and P3.13

3.14 A mechanic uses a piece of pipe *AB* as a lever when tightening an alternator belt. When he pushes down at *A*, a force of 485 N is exerted on the alternator at *B*. Determine the moment of that force about bolt *C* if its line of action passes through *O*.

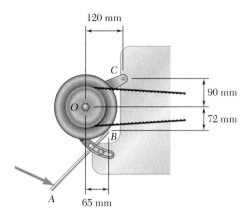

120 mm

C

90 mm

O

72 mm

B

A 65 mm

Fig. P3.14

3.15 Form the vector products **B** × **C** and **B′** × **C,** where *B* = *B′*, and use the results obtained to prove the identity

$$\sin \alpha \cos \beta = \tfrac{1}{2} \sin (\alpha + \beta) + \tfrac{1}{2} \sin (\alpha - \beta).$$

3.16 The vectors **P** and **Q** are two adjacent sides of a parallelogram. Determine the area of the parallelogram when (*a*) **P** = −7**i** + 3**j** − 3**k** and **Q** = 2**i** + 2**j** + 5**k**, (*b*) **P** = 6**i** − 5**j** − 2**k** and **Q** = −2**i** + 5**j** − **k**.

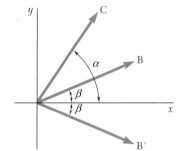

y   C

α   B

β
β   *x*

B′

Fig. P3.15

3.17 A plane contains the vectors **A** and **B**. Determine the unit vector normal to the plane when **A** and **B** are equal to, respectively, (*a*) **i** + 2**j** − 5**k** and 4**i** − 7**j** − 5**k**, (*b*) 3**i** − 3**j** + 2**k** and −2**i** + 6**j** − 4**k**.

3.18 A line passes through the points (20 m, 16 m) and (−1 m, −4 m). Determine the perpendicular distance *d* from the line to the origin *O* of the system of coordinates.

3.19 Determine the moment about the origin *O* of the force **F** = 4**i** − 3**j** + 5**k** that acts at a point *A*. Assume that the position vector of *A* is (*a*) **r** = 2**i** + 3**j** − 4**k**, (*b*) **r** = −8**i** + 6**j** − 10**k**, (*c*) **r** = 8**i** − 6**j** + 5**k**.

3.20 Determine the moment about the origin *O* of the force **F** = 2**i** + 3**j** − 4**k** that acts at a point *A*. Assume that the position vector of *A* is (*a*) **r** = 3**i** − 6**j** + 5**k**, (*b*) **r** = **i** − 4**j** − 2**k**, (*c*) **r** = 4**i** + 6**j** − 8**k**.

3.21 The wire *AE* is stretched between the corners *A* and *E* of a bent plate. Knowing that the tension in the wire is 435 N, determine the moment about *O* of the force exerted by the wire (*a*) on corner *A*, (*b*) on corner *E*.

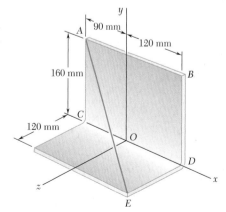

y
90 mm
A   120 mm
160 mm   *B*
C
120 mm   *O*
D
z   *x*
E

Fig. P3.21

Fig. P3.22

3.22 A small boat hangs from two davits, one of which is shown in the figure. The tension in line *ABAD* is 82 lb. Determine the moment about *C* of the resultant force $\mathbf{R}_A$ exerted on the davit at *A*.

3.23 A 6-ft-long fishing rod *AB* is securely anchored in the sand of a beach. After a fish takes the bait, the resulting force in the line is 6 lb. Determine the moment about *A* of the force exerted by the line at *B*.

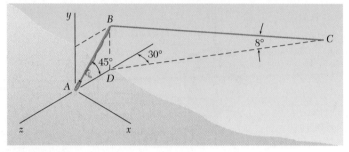

Fig. P3.23

3.24 A precast concrete wall section is temporarily held by two cables as shown. Knowing that the tension in cable *BD* is 900 N, determine the moment about point *O* of the force exerted by the cable at *B*.

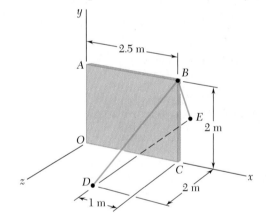

Fig. P3.24

3.25 A 200-N force is applied as shown to the bracket *ABC*. Determine the moment of the force about *A*.

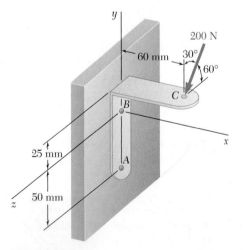

Fig. P3.25

3.26 The 6-m boom AB has a fixed end A. A steel cable is stretched from the free end B of the boom to a point C located on the vertical wall. If the tension in the cable is 2.5 kN, determine the moment about A of the force exerted by the cable at B.

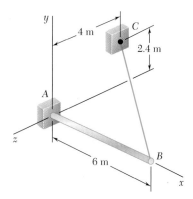

Fig. P3.26

3.27 In Prob. 3.21, determine the perpendicular distance from point O to wire AE.

3.28 In Prob. 3.21, determine the perpendicular distance from point B to wire AE.

3.29 In Prob. 3.22, determine the perpendicular distance from point C to portion AD of the line $ABAD$.

3.30 In Prob. 3.23, determine the perpendicular distance from point A to a line drawn through points B and C.

3.31 In Prob. 3.23, determine the perpendicular distance from point D to a line drawn through points B and C.

3.32 In Prob. 3.24, determine the perpendicular distance from point O to cable BD.

3.33 In Prob. 3.24, determine the perpendicular distance from point C to cable BD.

3.34 Determine the value of a that minimizes the perpendicular distance from point C to a section of pipeline that passes through points A and B.

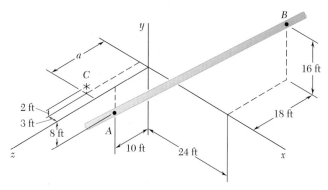

Fig. P3.34

3.9 SCALAR PRODUCT OF TWO VECTORS

Fig. 3.19

The *scalar product* of two vectors **P** and **Q** is defined as the product of the magnitudes of **P** and **Q** and of the cosine of the angle θ formed by **P** and **Q** (Fig. 3.19). The scalar product of **P** and **Q** is denoted by **P** · **Q**. We write therefore

$$\mathbf{P} \cdot \mathbf{Q} = PQ \cos \theta \qquad (3.24)$$

Note that the expression just defined is not a vector but a *scalar*, which explains the name *scalar product*; because of the notation used, **P** · **Q** is also referred to as the *dot product* of the vectors **P** and **Q**.

It follows from its very definition that the scalar product of two vectors is *commutative*, i.e., that

$$\mathbf{P} \cdot \mathbf{Q} = \mathbf{Q} \cdot \mathbf{P} \qquad (3.25)$$

To prove that the scalar product is also *distributive*, we must prove the relation

$$\mathbf{P} \cdot (\mathbf{Q}_1 + \mathbf{Q}_2) = \mathbf{P} \cdot \mathbf{Q}_1 + \mathbf{P} \cdot \mathbf{Q}_2 \qquad (3.26)$$

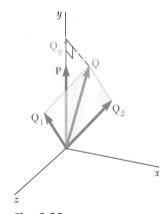

Fig. 3.20

We can, without any loss of generality, assume that **P** is directed along the y axis (Fig. 3.20). Denoting by **Q** the sum of **Q**$_1$ and **Q**$_2$ and by θ_y the angle **Q** forms with the y axis, we express the left-hand member of (3.26) as follows:

$$\mathbf{P} \cdot (\mathbf{Q}_1 + \mathbf{Q}_2) = \mathbf{P} \cdot \mathbf{Q} = PQ \cos \theta_y = PQ_y \qquad (3.27)$$

where Q_y is the y component of **Q**. We can, in a similar way, express the right-hand member of (3.26) as

$$\mathbf{P} \cdot \mathbf{Q}_1 + \mathbf{P} \cdot \mathbf{Q}_2 = P(Q_1)_y + P(Q_2)_y \qquad (3.28)$$

Since **Q** is the sum of **Q**$_1$ and **Q**$_2$, its y component must be equal to the sum of the y components of **Q**$_1$ and **Q**$_2$. Thus, the expressions obtained in (3.27) and (3.28) are equal, and the relation (3.26) has been proved.

As far as the third property—the associative property—is concerned, we note that this property cannot apply to scalar products. Indeed, (**P** · **Q**) · **S** has no meaning, since **P** · **Q** is not a vector but a scalar.

The scalar product of two vectors **P** and **Q** can be expressed in terms of their rectangular components. Resolving **P** and **Q** into components, we first write

$$\mathbf{P} \cdot \mathbf{Q} = (P_x\mathbf{i} + P_y\mathbf{j} + P_z\mathbf{k}) \cdot (Q_x\mathbf{i} + Q_y\mathbf{j} + Q_z\mathbf{k})$$

Making use of the distributive property, we express **P** · **Q** as the sum of scalar products, such as $P_x\mathbf{i} \cdot Q_x\mathbf{i}$ and $P_x\mathbf{i} \cdot Q_y\mathbf{j}$. However, from the definition of the scalar product it follows that the scalar products of the unit vectors are either zero or one.

$$
\begin{array}{lll}
\mathbf{i} \cdot \mathbf{i} = 1 & \mathbf{j} \cdot \mathbf{j} = 1 & \mathbf{k} \cdot \mathbf{k} = 1 \\
\mathbf{i} \cdot \mathbf{j} = 0 & \mathbf{j} \cdot \mathbf{k} = 0 & \mathbf{k} \cdot \mathbf{i} = 0
\end{array}
\tag{3.29}
$$

Thus, the expression obtained for $\mathbf{P} \cdot \mathbf{Q}$ reduces to

$$
\mathbf{P} \cdot \mathbf{Q} = P_x Q_x + P_y Q_y + P_z Q_z
\tag{3.30}
$$

In the particular case when $\mathbf{P}$ and $\mathbf{Q}$ are equal, we note that

$$
\mathbf{P} \cdot \mathbf{P} = P_x^2 + P_y^2 + P_z^2 = P^2
\tag{3.31}
$$

Applications

1. *Angle formed by two given vectors.* Let two vectors be given in terms of their components:

$$
\mathbf{P} = P_x \mathbf{i} + P_y \mathbf{j} + P_z \mathbf{k}
$$
$$
\mathbf{Q} = Q_x \mathbf{i} + Q_y \mathbf{j} + Q_z \mathbf{k}
$$

To determine the angle formed by the two vectors, we equate the expressions obtained in (3.24) and (3.30) for their scalar product and write

$$
PQ \cos \theta = P_x Q_x + P_y Q_y + P_z Q_z
$$

Solving for $\cos \theta$, we have

$$
\cos \theta = \frac{P_x Q_x + P_y Q_y + P_z Q_z}{PQ}
\tag{3.32}
$$

2. *Projection of a vector on a given axis.* Consider a vector $\mathbf{P}$ forming an angle θ with an axis, or directed line, OL (Fig. 3.21). The *projection of $\mathbf{P}$ on the axis OL* is defined as the scalar

$$
P_{OL} = P \cos \theta
\tag{3.33}
$$

We note that the projection P_{OL} is equal in absolute value to the length of the segment OA; it will be positive if OA has the same sense as the axis OL, that is, if θ is acute, and negative otherwise. If $\mathbf{P}$ and OL are at a right angle, the projection of $\mathbf{P}$ on OL is zero.

Consider now a vector $\mathbf{Q}$ directed along OL and of the same sense as OL (Fig. 3.22). The scalar product of $\mathbf{P}$ and $\mathbf{Q}$ can be expressed as

$$
\mathbf{P} \cdot \mathbf{Q} = PQ \cos \theta = P_{OL} Q
\tag{3.34}
$$

Fig. 3.21

Fig. 3.22

Fig. 3.23

from which it follows that

$$P_{OL} = \frac{\mathbf{P} \cdot \mathbf{Q}}{Q} = \frac{P_x Q_x + P_y Q_y + P_z Q_z}{Q} \tag{3.35}$$

In the particular case when the vector selected along OL is the unit vector $\boldsymbol{\lambda}$ (Fig. 3.23), we write

$$P_{OL} = \mathbf{P} \cdot \boldsymbol{\lambda} \tag{3.36}$$

Resolving $\mathbf{P}$ and $\boldsymbol{\lambda}$ into rectangular components and recalling from Sec. 2.12 that the components of $\boldsymbol{\lambda}$ along the coordinate axes are respectively equal to the direction cosines of OL, we express the projection of $\mathbf{P}$ on OL as

$$P_{OL} = P_x \cos \theta_x + P_y \cos \theta_y + P_z \cos \theta_z \tag{3.37}$$

where θ_x, θ_y, and θ_z denote the angles that the axis OL forms with the coordinate axes.

3.10 MIXED TRIPLE PRODUCT OF THREE VECTORS

We define the *mixed triple product* of the three vectors $\mathbf{S}$, $\mathbf{P}$, and $\mathbf{Q}$ as the scalar expression

$$\mathbf{S} \cdot (\mathbf{P} \times \mathbf{Q}) \tag{3.38}$$

Fig. 3.24

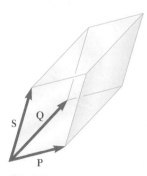

Fig. 3.25

obtained by forming the scalar product of $\mathbf{S}$ with the vector product of $\mathbf{P}$ and $\mathbf{Q}$.†

A simple geometrical interpretation can be given for the mixed triple product of $\mathbf{S}$, $\mathbf{P}$, and $\mathbf{Q}$ (Fig. 3.24). We first recall from Sec. 3.4 that the vector $\mathbf{P} \times \mathbf{Q}$ is perpendicular to the plane containing $\mathbf{P}$ and $\mathbf{Q}$ and that its magnitude is equal to the area of the parallelogram which has $\mathbf{P}$ and $\mathbf{Q}$ for sides. On the other hand, Eq. (3.34) indicates that the scalar product of $\mathbf{S}$ and $\mathbf{P} \times \mathbf{Q}$ can be obtained by multiplying the magnitude of $\mathbf{P} \times \mathbf{Q}$ (i.e., the area of the parallelogram defined by $\mathbf{P}$ and $\mathbf{Q}$) by the projection of $\mathbf{S}$ on the vector $\mathbf{P} \times \mathbf{Q}$ (i.e., by the projection of $\mathbf{S}$ on the normal to the plane containing the parallelogram). The mixed triple product is thus equal, in absolute value, to the volume of the parallelepiped having the vectors $\mathbf{S}$, $\mathbf{P}$, and $\mathbf{Q}$ for sides (Fig. 3.25). We note that the sign of the mixed triple product will be positive if $\mathbf{S}$, $\mathbf{P}$, and $\mathbf{Q}$ form a right-handed triad and negative if they form a left-handed triad [that is, $\mathbf{S} \cdot (\mathbf{P} \times \mathbf{Q})$ will be negative if the rotation which brings $\mathbf{P}$ into line with $\mathbf{Q}$ is observed as clockwise from the

---

†Another kind of triple product will be introduced later (Chap. 15): the *vector triple product* $\mathbf{S} \times (\mathbf{P} \times \mathbf{Q})$.

tip of **S**]. The mixed triple product will be zero if **S**, **P**, and **Q** are coplanar.

Since the parallelepiped defined in the preceding paragraph is independent of the order in which the three vectors are taken, the six mixed triple products which can be formed with **S**, **P**, and **Q** will all have the same absolute value, although not the same sign. It is easily shown that

$$
\begin{aligned}
\mathbf{S} \cdot (\mathbf{P} \times \mathbf{Q}) &= \mathbf{P} \cdot (\mathbf{Q} \times \mathbf{S}) = \mathbf{Q} \cdot (\mathbf{S} \times \mathbf{P}) \\
&= -\mathbf{S} \cdot (\mathbf{Q} \times \mathbf{P}) = -\mathbf{P} \cdot (\mathbf{S} \times \mathbf{Q}) = -\mathbf{Q} \cdot (\mathbf{P} \times \mathbf{S})
\end{aligned}
\tag{3.39}
$$

Arranging in a circle and in counterclockwise order the letters representing the three vectors (Fig. 3.26), we observe that the sign of the mixed triple product remains unchanged if the vectors are permuted in such a way that they are still read in counterclockwise order. Such a permutation is said to be a *circular permutation*. It also follows from Eq. (3.39) and from the commutative property of scalar products that the mixed triple product of **S**, **P**, and **Q** can be defined equally well as **S** · (**P** × **Q**) or (**S** × **P**) · **Q**.

Fig. 3.26

The mixed triple product of the vectors **S**, **P**, and **Q** can be expressed in terms of the rectangular components of these vectors. Denoting **P** × **Q** by **V** and using formula (3.30) to express the scalar product of **S** and **V**, we write

$$
\mathbf{S} \cdot (\mathbf{P} \times \mathbf{Q}) = \mathbf{S} \cdot \mathbf{V} = S_x V_x + S_y V_y + S_z V_z
$$

Substituting from the relations (3.9) for the components of **V**, we obtain

$$
\mathbf{S} \cdot (\mathbf{P} \times \mathbf{Q}) = S_x(P_y Q_z - P_z Q_y) + S_y(P_z Q_x - P_x Q_z) \\
+ S_z(P_x Q_y - P_y Q_x) \quad (3.40)
$$

This expression can be written in a more compact form if we observe that it represents the expansion of a determinant:

$$
\mathbf{S} \cdot (\mathbf{P} \times \mathbf{Q}) = \begin{vmatrix} S_x & S_y & S_z \\ P_x & P_y & P_z \\ Q_x & Q_y & Q_z \end{vmatrix}
\tag{3.41}
$$

By applying the rules governing the permutation of rows in a determinant, we could easily verify the relations (3.39) which were derived earlier from geometrical considerations.

3.11 MOMENT OF A FORCE ABOUT A GIVEN AXIS

Now that we have further increased our knowledge of vector algebra, we can introduce a new concept, the concept of *moment of a force about an axis*. Consider again a force **F** acting on a rigid body and the moment **M**$_O$ of that force about O (Fig. 3.27). Let *OL* be

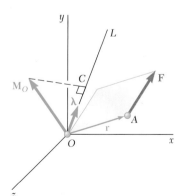

Fig. 3.27

an axis through O; *we define the moment M_{OL} of $\mathbf{F}$ about OL as the projection OC of the moment $\mathbf{M}_O$ onto the axis OL.* Denoting by $\boldsymbol{\lambda}$ the unit vector along OL and recalling from Secs. 3.9 and 3.6, respectively, the expressions (3.36) and (3.11) obtained for the projection of a vector on a given axis and for the moment $\mathbf{M}_O$ of a force $\mathbf{F}$, we write

$$M_{OL} = \boldsymbol{\lambda} \cdot \mathbf{M}_O = \boldsymbol{\lambda} \cdot (\mathbf{r} \times \mathbf{F}) \tag{3.42}$$

which shows that the moment M_{OL} of $\mathbf{F}$ about the axis OL is the scalar obtained by forming the mixed triple product of $\boldsymbol{\lambda}$, $\mathbf{r}$, and $\mathbf{F}$. Expressing M_{OL} in the form of a determinant, we write

$$M_{OL} = \begin{vmatrix} \lambda_x & \lambda_y & \lambda_z \\ x & y & z \\ F_x & F_y & F_z \end{vmatrix} \tag{3.43}$$

where $\lambda_x,\ \lambda_y,\ \lambda_z$ = direction cosines of axis OL

$\qquad\quad x,\ y,\ z$ = coordinates of point of application of $\mathbf{F}$

$\qquad F_x,\ F_y,\ F_z$ = components of force $\mathbf{F}$

The physical significance of the moment M_{OL} of a force $\mathbf{F}$ about a fixed axis OL becomes more apparent if we resolve $\mathbf{F}$ into two rectangular components $\mathbf{F}_1$ and $\mathbf{F}_2$, with $\mathbf{F}_1$ parallel to OL and $\mathbf{F}_2$ lying in a plane P perpendicular to OL (Fig. 3.28). Resolving $\mathbf{r}$ similarly into two components $\mathbf{r}_1$ and $\mathbf{r}_2$ and substituting for $\mathbf{F}$ and $\mathbf{r}$ into (3.42), we write

$$\begin{aligned} M_{OL} &= \boldsymbol{\lambda} \cdot [(\mathbf{r}_1 + \mathbf{r}_2) \times (\mathbf{F}_1 + \mathbf{F}_2)] \\ &= \boldsymbol{\lambda} \cdot (\mathbf{r}_1 \times \mathbf{F}_1) + \boldsymbol{\lambda} \cdot (\mathbf{r}_1 \times \mathbf{F}_2) + \boldsymbol{\lambda} \cdot (\mathbf{r}_2 \times \mathbf{F}_1) + \boldsymbol{\lambda} \cdot (\mathbf{r}_2 \times \mathbf{F}_2) \end{aligned}$$

Noting that all of the mixed triple products except the last one are equal to zero, since they involve vectors which are coplanar when drawn from a common origin (Sec. 3.10), we have

$$M_{OL} = \boldsymbol{\lambda} \cdot (\mathbf{r}_2 \times \mathbf{F}_2) \tag{3.44}$$

The vector product $\mathbf{r}_2 \times \mathbf{F}_2$ is perpendicular to the plane P and represents the moment of the component $\mathbf{F}_2$ of $\mathbf{F}$ about the point Q where OL intersects P. Therefore, the scalar M_{OL}, which will be positive if $\mathbf{r}_2 \times \mathbf{F}_2$ and OL have the same sense and negative otherwise, measures the tendency of $\mathbf{F}_2$ to make the rigid body rotate about the fixed axis OL. Since the other component $\mathbf{F}_1$ of $\mathbf{F}$ does not tend to make the body rotate about OL, we conclude that *the moment M_{OL} of $\mathbf{F}$ about OL measures the tendency of the force $\mathbf{F}$ to impart to the rigid body a motion of rotation about the fixed axis OL.*

It follows from the definition of the moment of a force about an axis that the moment of $\mathbf{F}$ about a coordinate axis is equal to the component of $\mathbf{M}_O$ along that axis. Substituting successively each

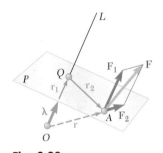

Fig. 3.28

of the unit vectors $\mathbf{i}$, $\mathbf{j}$, and $\mathbf{k}$ for $\boldsymbol{\lambda}$ in (3.42), we observe that the expressions thus obtained for the *moments of* $\mathbf{F}$ *about the coordinate axes* are respectively equal to the expressions obtained in Sec. 3.8 for the components of the moment $\mathbf{M}_O$ of $\mathbf{F}$ about O:

$$
\begin{aligned}
M_x &= yF_z - zF_y \\
M_y &= zF_x - xF_z \\
M_z &= xF_y - yF_x
\end{aligned}
\tag{3.18}
$$

We observe that just as the components F_x, F_y, and F_z of a force $\mathbf{F}$ acting on a rigid body measure, respectively, the tendency of $\mathbf{F}$ to move the rigid body in the x, y, and z directions, the moments M_x, M_y, and M_z of $\mathbf{F}$ about the coordinate axes measure the tendency of $\mathbf{F}$ to impart to the rigid body a motion of rotation about the x, y, and z axes, respectively.

More generally, the moment of a force $\mathbf{F}$ applied at A about an axis which does not pass through the origin is obtained by choosing an arbitrary point B on the axis (Fig. 3.29) and determining the projection on the axis BL of the moment $\mathbf{M}_B$ of $\mathbf{F}$ about B. We write

$$
M_{BL} = \boldsymbol{\lambda} \cdot \mathbf{M}_B = \boldsymbol{\lambda} \cdot (\mathbf{r}_{A/B} \times \mathbf{F})
\tag{3.45}
$$

where $\mathbf{r}_{A/B} = \mathbf{r}_A - \mathbf{r}_B$ represents the vector drawn from B to A. Expressing M_{BL} in the form of a determinant, we have

$$
M_{BL} = \begin{vmatrix} \lambda_x & \lambda_y & \lambda_z \\ x_{A/B} & y_{A/B} & z_{A/B} \\ F_x & F_y & F_z \end{vmatrix}
\tag{3.46}
$$

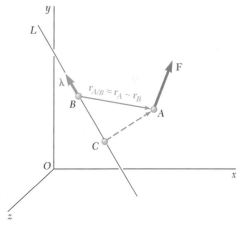

Fig. 3.29

where λ_x, λ_y, λ_z = direction cosines of axis BL
$$x_{A/B} = x_A - x_B \qquad y_{A/B} = y_A - y_B \qquad z_{A/B} = z_A - z_B$$
F_x, F_y, F_z = components of force $\mathbf{F}$

It should be noted that the result obtained is independent of the choice of the point B on the given axis. Indeed, denoting by M_{CL} the result obtained with a different point C, we have

$$
\begin{aligned}
M_{CL} &= \boldsymbol{\lambda} \cdot [(\mathbf{r}_A - \mathbf{r}_C) \times \mathbf{F}] \\
&= \boldsymbol{\lambda} \cdot [(\mathbf{r}_A - \mathbf{r}_B) \times \mathbf{F}] + \boldsymbol{\lambda} \cdot [(\mathbf{r}_B - \mathbf{r}_C) \times \mathbf{F}]
\end{aligned}
$$

But, since the vectors $\boldsymbol{\lambda}$ and $\mathbf{r}_B - \mathbf{r}_C$ lie in the same line, the volume of the parallelepiped having the vectors $\boldsymbol{\lambda}$, $\mathbf{r}_B - \mathbf{r}_C$, and $\mathbf{F}$ for sides is zero, as is the mixed triple product of these three vectors (Sec. 3.10). The expression obtained for M_{CL} thus reduces to its first term, which is the expression used earlier to define M_{BL}. In addition, it follows from Sec. 3.6 that, when computing the moment of $\mathbf{F}$ about the given axis, A can be any point on the line of action of $\mathbf{F}$.

SAMPLE PROBLEM 3.5

A cube of side a is acted upon by a force $\mathbf{P}$ as shown. Determine the moment of $\mathbf{P}$ (a) about A, (b) about the edge AB, (c) about the diagonal AG of the cube, (d). Using the result of part c, determine the perpendicular distance between AG and FC.

SOLUTION

a. Moment about A. Choosing x, y, and z axes as shown, we resolve into rectangular components the force $\mathbf{P}$ and the vector $\mathbf{r}_{F/A} = \overrightarrow{AF}$ drawn from A to the point of application F of $\mathbf{P}$.

$$\mathbf{r}_{F/A} = a\mathbf{i} - a\mathbf{j} = a(\mathbf{i} - \mathbf{j})$$
$$\mathbf{P} = (P/\sqrt{2})\mathbf{j} - (P/\sqrt{2})\mathbf{k} = (P/\sqrt{2})(\mathbf{j} - \mathbf{k})$$

The moment of $\mathbf{P}$ about A is

$$\mathbf{M}_A = \mathbf{r}_{F/A} \times \mathbf{P} = a(\mathbf{i} - \mathbf{j}) \times (P/\sqrt{2})(\mathbf{j} - \mathbf{k})$$
$$\mathbf{M}_A = (aP/\sqrt{2})(\mathbf{i} + \mathbf{j} + \mathbf{k}) \quad \blacktriangleleft$$

b. Moment about AB. Projecting $\mathbf{M}_A$ on AB, we write

$$M_{AB} = \mathbf{i} \cdot \mathbf{M}_A = \mathbf{i} \cdot (aP/\sqrt{2})(\mathbf{i} + \mathbf{j} + \mathbf{k})$$
$$M_{AB} = aP/\sqrt{2} \quad \blacktriangleleft$$

We verify that, since AB is parallel to the x axis, M_{AB} is also the x component of the moment $\mathbf{M}_A$.

c. Moment about Diagonal AG. The moment of $\mathbf{P}$ about AG is obtained by projecting $\mathbf{M}_A$ on AG. Denoting by $\boldsymbol{\lambda}$ the unit vector along AG, we have

$$\boldsymbol{\lambda} = \frac{\overrightarrow{AG}}{AG} = \frac{a\mathbf{i} - a\mathbf{j} - a\mathbf{k}}{a\sqrt{3}} = (1/\sqrt{3})(\mathbf{i} - \mathbf{j} - \mathbf{k})$$
$$M_{AG} = \boldsymbol{\lambda} \cdot \mathbf{M}_A = (1/\sqrt{3})(\mathbf{i} - \mathbf{j} - \mathbf{k}) \cdot (aP/\sqrt{2})(\mathbf{i} + \mathbf{j} + \mathbf{k})$$
$$M_{AG} = (aP/\sqrt{6})(1 - 1 - 1) \quad M_{AG} = -aP/\sqrt{6} \quad \blacktriangleleft$$

Alternative Method. The moment of $\mathbf{P}$ about AG can also be expressed in the form of a determinant:

$$M_{AG} = \begin{vmatrix} \lambda_x & \lambda_y & \lambda_z \\ x_{F/A} & y_{F/A} & z_{F/A} \\ F_x & F_y & F_z \end{vmatrix} = \begin{vmatrix} 1/\sqrt{3} & -1/\sqrt{3} & -1/\sqrt{3} \\ a & -a & 0 \\ 0 & P/\sqrt{2} & -P/\sqrt{2} \end{vmatrix} = -aP/\sqrt{6}$$

d. Perpendicular Distance between AG and FC. We first observe that $\mathbf{P}$ is perpendicular to the diagonal AG. This can be checked by forming the scalar product $\mathbf{P} \cdot \boldsymbol{\lambda}$ and verifying that it is zero:

$$\mathbf{P} \cdot \boldsymbol{\lambda} = (P/\sqrt{2})(\mathbf{j} - \mathbf{k}) \cdot (1/\sqrt{3})(\mathbf{i} - \mathbf{j} - \mathbf{k}) = (P\sqrt{6})(0 - 1 + 1) = 0$$

The moment M_{AG} can then be expressed as $-Pd$, where d is the perpendicular distance from AG to FC. (The negative sign is used since the rotation imparted to the cube by $\mathbf{P}$ appears as clockwise to an observer at G.) Recalling the value found for M_{AG} in part c,

$$M_{AG} = -Pd = -aP/\sqrt{6} \qquad d = a/\sqrt{6} \quad \blacktriangleleft$$

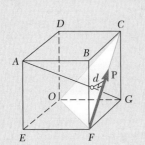

SOLVING PROBLEMS
ON YOUR OWN

In the problems for this lesson you will apply the *scalar product* or *dot product* of two vectors to determine the *angle formed by two given vectors* and the *projection of a force on a given axis*. You will also use the *mixed triple product* of three vectors to find the *moment of a force about a given axis* and the *perpendicular distance between two lines*.

1. Calculating the angle formed by two given vectors. First express the vectors in terms of their components and determine the magnitudes of the two vectors. The cosine of the desired angle is then obtained by dividing the scalar product of the two vectors by the product of their magnitudes [Eq. (3.32)].

2. Computing the projection of a vector P on a given axis *OL*. In general, begin by expressing $\mathbf{P}$ and the unit vector $\boldsymbol{\lambda}$, that defines the direction of the axis, in component form. Take care that $\boldsymbol{\lambda}$ has the correct sense (that is, $\boldsymbol{\lambda}$ is directed from O to L). The required projection is then equal to the scalar product $\mathbf{P} \cdot \boldsymbol{\lambda}$. However, if you know the angle θ formed by $\mathbf{P}$ and $\boldsymbol{\lambda}$, the projection is also given by $P \cos \theta$.

3. Determining the moment M_{OL} of a force about a given axis *OL*. We defined M_{OL} as

$$M_{OL} = \boldsymbol{\lambda} \cdot \mathbf{M}_O = \boldsymbol{\lambda} \cdot (\mathbf{r} \times \mathbf{F}) \qquad (3.42)$$

where $\boldsymbol{\lambda}$ is the unit vector along OL and $\mathbf{r}$ is a position vector *from any point* on the line OL *to any point* on the line of action of $\mathbf{F}$. As was the case for the moment of a force about a point, choosing the most convenient position vector will simplify your calculations. Also, recall the warning of the previous lesson: The vectors $\mathbf{r}$ and $\mathbf{F}$ must have the correct sense, and they must be placed in the proper order. The procedure you should follow when computing the moment of a force about an axis is illustrated in part c of Sample Prob. 3.5. The two essential steps in this procedure are to first express $\boldsymbol{\lambda}$, $\mathbf{r}$, and $\mathbf{F}$ in terms of their rectangular components and to then evaluate the mixed triple product $\boldsymbol{\lambda} \cdot (\mathbf{r} \times \mathbf{F})$ to determine the moment about the axis. In most three-dimensional problems the most convenient way to compute the mixed triple product is by using a determinant.

As noted in the text, when $\boldsymbol{\lambda}$ is directed along one of the coordinate axes, M_{OL} is equal to the scalar component of $\mathbf{M}_O$ along that axis.

(continued)

4. Determining the perpendicular distance between two lines. You should remember that it is the perpendicular component $\mathbf{F}_2$ of the force $\mathbf{F}$ that tends to make a body rotate about a given axis OL (Fig. 3.28). It then follows that

$$M_{OL} = F_2 d$$

where M_{OL} is the moment of $\mathbf{F}$ about axis OL and d is the perpendicular distance between OL and the line of action of $\mathbf{F}$. This last equation gives us a simple technique for determining d. First assume that a force $\mathbf{F}$ of known magnitude F lies along one of the given lines and that the unit vector $\boldsymbol{\lambda}$ lies along the other line. Next compute the moment M_{OL} of the force $\mathbf{F}$ about the second line using the method discussed above. The magnitude of the parallel component, F_1, of $\mathbf{F}$ is obtained using the scalar product:

$$F_1 = \mathbf{F} \cdot \boldsymbol{\lambda}$$

The value of F_2 is then determined from

$$F_2 = \sqrt{F^2 - F_1^2}$$

Finally, substitute the values of M_{OL} and F_2 into the equation $M_{OL} = F_2 d$ and solve for d.

You should now realize that the calculation of the perpendicular distance in part d of Sample Prob. 3.5 was simplified by $\mathbf{P}$ being perpendicular to the diagonal AG. In general, the two given lines will not be perpendicular, so that the technique just outlined will have to be used when determining the perpendicular distance between them.

PROBLEMS

3.35 Given the vectors $\mathbf{P} = 3\mathbf{i} - \mathbf{j} + 2\mathbf{k}$, $\mathbf{Q} = 4\mathbf{i} + 5\mathbf{j} - 3\mathbf{k}$, and $\mathbf{S} = -2\mathbf{i} + 3\mathbf{j} - \mathbf{k}$, compute the scalar products $\mathbf{P} \cdot \mathbf{Q}$, $\mathbf{P} \cdot \mathbf{S}$, and $\mathbf{Q} \cdot \mathbf{S}$.

3.36 Form the scalar product $\mathbf{B} \cdot \mathbf{C}$ and use the result obtained to prove the identity

$$\cos(\alpha - \beta) = \cos\alpha \cos\beta + \sin\alpha \sin\beta$$

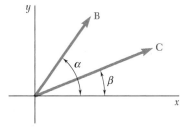

Fig. P3.36

3.37 Consider the volleyball net shown. Determine the angle formed by guy wires AB and AC.

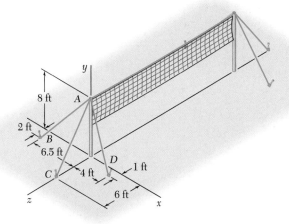

Fig. P3.37 and P3.38

3.38 Consider the volleyball net shown. Determine the angle formed by guy wires AC and AD.

3.39 Three cables are used to support a container as shown. Determine the angle formed by cables AB and AD.

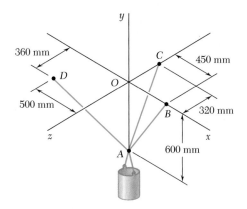

Fig. P3.39 and *P3.40*

3.40 Three cables are used to support a container as shown. Determine the angle formed by cables AC and AD.

Fig. P3.41

3.41 The 20-in. tube AB can slide along a horizontal rod. The ends A and B of the tube are connected by elastic cords to the fixed point C. For the position corresponding to $x = 11$ in., determine the angle formed by the two cords, (a) using Eq. (3.32), (b) applying the law of cosines to triangle ABC.

3.42 Solve Prob. 3.41 for the position corresponding to $x = 4$ in.

3.43 Ropes AB and BC are two of the ropes used to support a tent. The two ropes are attached to a stake at B. If the tension in rope AB is 540 N, determine (a) the angle between rope AB and the stake, (b) the projection on the stake of the force exerted by rope AB at point B.

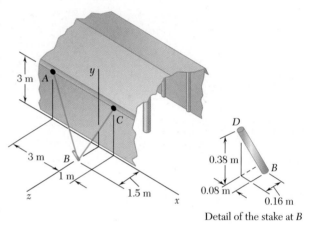

Fig. P3.43 and P3.44

3.44 Ropes AB and BC are two of the ropes used to support a tent. The two ropes are attached to a stake at B. If the tension in rope BC is 490 N, determine (a) the angle between rope BC and the stake, (b) the projection on the stake of the force exerted by rope BC at point B.

3.45 Given the vectors $\mathbf{P} = 4\mathbf{i} - 2\mathbf{j} + 3\mathbf{k}$, $\mathbf{Q} = 2\mathbf{i} + 4\mathbf{j} - 5\mathbf{k}$, and $\mathbf{S} = S_x\mathbf{i} - \mathbf{j} + 2\mathbf{k}$, determine the value of S_x for which the three vectors are coplanar.

3.46 Determine the volume of the parallelepiped of Fig. 3.25 when (a) $\mathbf{P} = 4\mathbf{i} - 3\mathbf{j} + 2\mathbf{k}$, $\mathbf{Q} = -2\mathbf{i} - 5\mathbf{j} + \mathbf{k}$, and $\mathbf{S} = 7\mathbf{i} + \mathbf{j} - \mathbf{k}$, (b) $\mathbf{P} = 5\mathbf{i} - \mathbf{j} + 6\mathbf{k}$, $\mathbf{Q} = 2\mathbf{i} + 3\mathbf{j} + \mathbf{k}$, and $\mathbf{S} = -3\mathbf{i} - 2\mathbf{j} + 4\mathbf{k}$.

3.47 Knowing that the tension in cable AB is 570 N, determine the moment about each of the coordinate axes of the force exerted on the plate at B.

3.48 Knowing that the tension in cable AC is 1065 N, determine the moment about each of the coordinate axes of the force exerted on the plate at C.

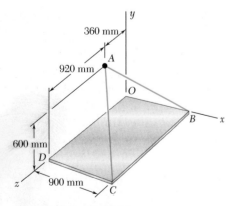

Fig. P3.47 and P3.48

3.49 A small boat hangs from two davits, one of which is shown in the figure. It is known that the moment about the z axis of the resultant force $\mathbf{R}_A$ exerted on the davit at A must not exceed 279 lb · ft in absolute value. Determine the largest allowable tension in line $ABAD$ when $x = 6$ ft.

Fig. P3.49

3.50 For the davit of Prob. 3.49, determine the largest allowable distance x when the tension in line $ABAD$ is 60 lb.

3.51 A farmer uses cables and winch pullers B and E to plumb one side of a small barn. If it is known that the sum of the moments about the x axis of the forces exerted by the cables on the barn at points A and D is equal to 4728 lb · ft, determine the magnitude of $\mathbf{T}_{DE}$ when $T_{AB} = 255$ lb.

3.52 Solve Prob. 3.51 when the tension in cable AB is 306 lb.

3.53 A single force $\mathbf{P}$ acts at C in a direction perpendicular to the handle BC of the crank shown. Knowing that $M_x = +20$ N · m and $M_y = -8.75$ N · m, and $M_z = -30$ N · m, determine the magnitude of $\mathbf{P}$ and the values of ϕ and θ.

Fig. P3.51

Fig. P3.53 and P3.54

3.54 A single force $\mathbf{P}$ acts at C in a direction perpendicular to the handle BC of the crank shown. Determine the moment M_x of $\mathbf{P}$ about the x axis when $\theta = 65°$, knowing that $M_y = -15$ N · m and $M_z = -36$ N · m.

3.55 The triangular plate ABC is supported by ball-and-socket joints at B and D and is held in the position shown by cables AE and CF. If the force exerted by cable AE at A is 55 N, determine the moment of that force about the line joining points D and B.

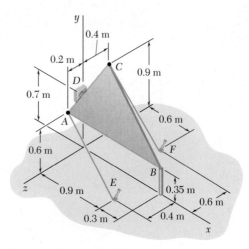

Fig. P3.55 and P3.56

3.56 The triangular plate ABC is supported by ball-and-socket joints at B and D and is held in the position shown by cables AE and CF. If the force exerted by cable CF at C is 33 N, determine the moment of that force about the line joining points D and B.

3.57 The 23-in. vertical rod CD is welded to the midpoint C of the 50-in. rod AB. Determine the moment about AB of the 235-lb force $\mathbf{P}$.

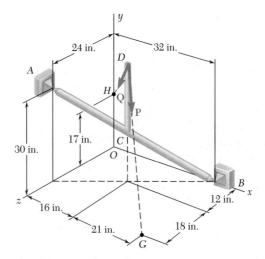

Fig. P3.57 and P3.58

3.58 The 23-in. vertical rod CD is welded to the midpoint C of the 50-in. rod AB. Determine the moment about AB of the 174-lb force $\mathbf{Q}$.

3.59 The frame *ACD* is hinged at *A* and *D* and is supported by a cable that passes through a ring at *B* and is attached to hooks at *G* and *H*. Knowing that the tension in the cable is 450 N, determine the moment about the diagonal *AD* of the force exerted on the frame by portion *BH* of the cable.

3.60 In Prob. 3.59, determine the moment about the diagonal *AD* of the force exerted on the frame by portion *BG* of the cable.

3.61 A regular tetrahedron has six edges of length *a*. A force **P** is directed as shown along edge *BC*. Determine the moment of **P** about edge *OA*.

Fig. P3.59

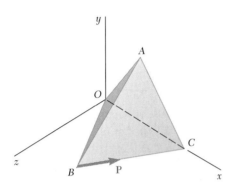

Fig. P3.61 and P3.62

3.62 A regular tetrahedron has six edges of length *a*. (*a*) Show that two opposite edges, such as *OA* and *BC*, are perpendicular to each other. (*b*) Use this property and the result obtained in Prob. 3.61 to determine the perpendicular distance between edges *OA* and *BC*.

3.63 Two forces $\mathbf{F}_1$ and $\mathbf{F}_2$ in space have the same magnitude *F*. Prove that the moment of $\mathbf{F}_1$ about the line of action of $\mathbf{F}_2$ is equal to the moment of $\mathbf{F}_2$ about the line of action of $\mathbf{F}_1$.

*3.64 In Prob. 3.55, determine the perpendicular distance between cable *AE* and the line joining points *D* and *B*.

*3.65 In Prob. 3.56, determine the perpendicular distance between cable *CF* and the line joining points *D* and *B*.

*3.66 In Prob. 3.57, determine the perpendicular distance between rod *AB* and the line of action of **P**.

*3.67 In Prob. 3.58, determine the perpendicular distance between rod *AB* and the line of action of **Q**.

*3.68 In Prob. 3.59, determine the perpendicular distance between portion *BH* of the cable and the diagonal *AD*.

*3.69 In Prob. 3.60, determine the perpendicular distance between portion *BG* of the cable and the diagonal *AD*.

Fig. 3.30

Fig. 3.31

Fig. 3.32

Photo 3.1 The parallel upward and downward forces of equal magnitude exerted on the arms of the lug nut wrench are an example of a couple.

3.12 MOMENT OF A COUPLE

Two forces **F** *and* −**F** *having the same magnitude, parallel lines of action, and opposite sense are said to form a* couple (Fig. 3.30). Clearly, the sum of the components of the two forces in any direction is zero. The sum of the moments of the two forces about a given point, however, is not zero. While the two forces will not translate the body on which they act, they will tend to make it rotate.

Denoting by $\mathbf{r}_A$ and $\mathbf{r}_B$, respectively, the position vectors of the points of application of **F** and −**F** (Fig. 3.31), we find that the sum of the moments of the two forces about O is

$$\mathbf{r}_A \times \mathbf{F} + \mathbf{r}_B \times (-\mathbf{F}) = (\mathbf{r}_A - \mathbf{r}_B) \times \mathbf{F}$$

Setting $\mathbf{r}_A - \mathbf{r}_B = \mathbf{r}$, where **r** is the vector joining the points of application of the two forces, we conclude that the sum of the moments of **F** and −**F** about O is represented by the vector

$$\mathbf{M} = \mathbf{r} \times \mathbf{F} \tag{3.47}$$

The vector **M** is called the *moment of the couple;* it is a vector perpendicular to the plane containing the two forces, and its magnitude is

$$M = rF \sin \theta = Fd \tag{3.48}$$

where d is the perpendicular distance between the lines of action of **F** and −**F**. The sense of **M** is defined by the right-hand rule.

Since the vector **r** in (3.47) is independent of the choice of the origin O of the coordinate axes, we note that the same result would have been obtained if the moments of **F** and −**F** had been computed about a different point O'. Thus, the moment **M** of a couple is a *free vector* (Sec. 2.3) which can be applied at any point (Fig. 3.32).

From the definition of the moment of a couple, it also follows that two couples, one consisting of the forces $\mathbf{F}_1$ and $-\mathbf{F}_1$, the other of the forces $\mathbf{F}_2$ and $-\mathbf{F}_2$ (Fig. 3.33), will have equal moments if

$$F_1 d_1 = F_2 d_2 \tag{3.49}$$

and if the two couples lie in parallel planes (or in the same plane) and have the same sense.

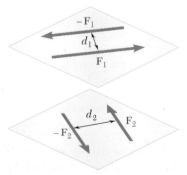

Fig. 3.33

3.13 EQUIVALENT COUPLES

Figure 3.34 shows three couples which act successively on the same rectangular box. As seen in the preceding section, the only motion a couple can impart to a rigid body is a rotation. Since each of the three couples shown has the same moment **M** (same direction and same magnitude $M = 120$ lb · in.), we can expect the three couples to have the same effect on the box.

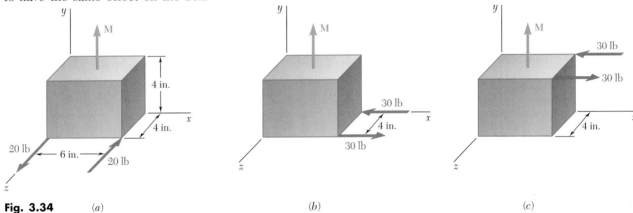

Fig. 3.34 (a) (b) (c)

As reasonable as this conclusion appears, we should not accept it hastily. While intuitive feeling is of great help in the study of mechanics, it should not be accepted as a substitute for logical reasoning. Before stating that two systems (or groups) of forces have the same effect on a rigid body, we should prove that fact on the basis of the experimental evidence introduced so far. This evidence consists of the parallelogram law for the addition of two forces (Sec. 2.2) and the principle of transmissibility (Sec. 3.3). Therefore, we will state that *two systems of forces are equivalent* (i.e., they have the same effect on a rigid body) *if we can transform one of them into the other by means of one or several of the following operations*: (1) replacing two forces acting on the same particle by their resultant; (2) resolving a force into two components; (3) canceling two equal and opposite forces acting on the same particle; (4) attaching to the same particle two equal and opposite forces; (5) moving a force along its line of action. Each of these operations is easily justified on the basis of the parallelogram law or the principle of transmissibility.

Let us now prove that *two couples having the same moment* **M** *are equivalent*. First consider two couples contained in the same plane, and assume that this plane coincides with the plane of the figure (Fig. 3.35). The first couple consists of the forces $\mathbf{F}_1$ and $-\mathbf{F}_1$ of magnitude F_1, which are located at a distance d_1 from each other (Fig. 3.35a), and the second couple consists of the forces $\mathbf{F}_2$ and $-\mathbf{F}_2$ of magnitude F_2, which are located at a distance d_2 from each other (Fig. 3.35d). Since the two couples have the same moment **M**, which is perpendicular to the plane of the figure, they must have the same sense (assumed here to be counterclockwise), and the relation

$$F_1 d_1 = F_2 d_2 \qquad (3.49)$$

must be satisfied. To prove that they are equivalent, we shall show that the first couple can be transformed into the second by means of the operations listed above.

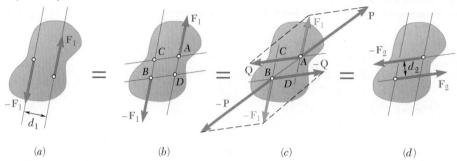

(a) (b) (c) (d)

Fig. 3.35

Denoting by A, B, C, and D the points of intersection of the lines of action of the two couples, we first slide the forces $\mathbf{F}_1$ and $-\mathbf{F}_1$ until they are attached, respectively, at A and B, as shown in Fig. 3.35b. The force $\mathbf{F}_1$ is then resolved into a component $\mathbf{P}$ along line AB and a component $\mathbf{Q}$ along AC (Fig. 3.35c); similarly, the force $-\mathbf{F}_1$ is resolved into $-\mathbf{P}$ along AB and $-\mathbf{Q}$ along BD. The forces $\mathbf{P}$ and $-\mathbf{P}$ have the same magnitude, the same line of action, and opposite sense; they can be moved along their common line of action until they are applied at the same point and may then be canceled. Thus the couple formed by $\mathbf{F}_1$ and $-\mathbf{F}_1$ reduces to a couple consisting of $\mathbf{Q}$ and $-\mathbf{Q}$.

We will now show that the forces $\mathbf{Q}$ and $-\mathbf{Q}$ are respectively equal to the forces $-\mathbf{F}_2$ and $\mathbf{F}_2$. The moment of the couple formed by $\mathbf{Q}$ and $-\mathbf{Q}$ can be obtained by computing the moment of $\mathbf{Q}$ about B; similarly, the moment of the couple formed by $\mathbf{F}_1$ and $-\mathbf{F}_1$ is the moment of $\mathbf{F}_1$ about B. But, by Varignon's theorem, the moment of $\mathbf{F}_1$ is equal to the sum of the moments of its components $\mathbf{P}$ and $\mathbf{Q}$. Since the moment of $\mathbf{P}$ about B is zero, the moment of the couple formed by $\mathbf{Q}$ and $-\mathbf{Q}$ must be equal to the moment of the couple formed by $\mathbf{F}_1$ and $-\mathbf{F}_1$. Recalling (3.49), we write

$$Qd_2 = F_1d_1 = F_2d_2 \qquad \text{and} \qquad Q = F_2$$

Thus the forces $\mathbf{Q}$ and $-\mathbf{Q}$ are respectively equal to the forces $-\mathbf{F}_2$ and $\mathbf{F}_2$, and the couple of Fig. 3.35a is equivalent to the couple of Fig. 3.35d.

Next consider two couples contained in parallel planes P_1 and P_2; we will prove that they are equivalent if they have the same moment. In view of the foregoing, we can assume that the couples consist of forces of the same magnitude F acting along parallel lines (Fig. 3.36a and d). We propose to show that the couple contained in plane P_1 can be transformed into the couple contained in plane P_2 by means of the standard operations listed above.

Let us consider the two planes defined respectively by the lines of action of $\mathbf{F}_1$ and $-\mathbf{F}_2$ and by those of $-\mathbf{F}_1$ and $\mathbf{F}_2$ (Fig. 3.36b). At a point on their line of intersection we attach two forces $\mathbf{F}_3$ and $-\mathbf{F}_3$, respectively equal to $\mathbf{F}_1$ and $-\mathbf{F}_1$. The couple formed by $\mathbf{F}_1$ and $-\mathbf{F}_3$ can be replaced by a couple consisting of $\mathbf{F}_3$ and $-\mathbf{F}_2$ (Fig. 3.36c), since both couples clearly have the same moment and are contained in the same plane. Similarly, the couple formed by $-\mathbf{F}_1$ and $\mathbf{F}_3$ can be replaced by a couple consisting of $-\mathbf{F}_3$ and $\mathbf{F}_2$. Canceling the two equal and opposite forces $\mathbf{F}_3$ and $-\mathbf{F}_3$, we obtain the desired couple in plane P_2 (Fig. 3.36d). Thus, we conclude that two couples having

Fig. 3.36

the same moment **M** are equivalent, whether they are contained in the same plane or in parallel planes.

The property we have just established is very important for the correct understanding of the mechanics of rigid bodies. It indicates that when a couple acts on a rigid body, it does not matter where the two forces forming the couple act or what magnitude and direction they have. The only thing which counts is the *moment* of the couple (magnitude and direction). Couples with the same moment will have the same effect on the rigid body.

3.14 ADDITION OF COUPLES

Consider two intersecting planes P_1 and P_2 and two couples acting respectively in P_1 and P_2. We can, without any loss of generality, assume that the couple in P_1 consists of two forces $\mathbf{F}_1$ and $-\mathbf{F}_1$ perpendicular to the line of intersection of the two planes and acting respectively at A and B (Fig. 3.37a). Similarly, we assume that the couple in P_2 consists of two forces $\mathbf{F}_2$ and $-\mathbf{F}_2$ perpendicular to AB and acting respectively at A and B. It is clear that the resultant $\mathbf{R}$ of $\mathbf{F}_1$ and $\mathbf{F}_2$ and the resultant $-\mathbf{R}$ of $-\mathbf{F}_1$ and $-\mathbf{F}_2$ form a couple. Denoting by $\mathbf{r}$ the vector joining B to A and recalling the definition of the moment of a couple (Sec. 3.12), we express the moment $\mathbf{M}$ of the resulting couple as follows:

$$\mathbf{M} = \mathbf{r} \times \mathbf{R} = \mathbf{r} \times (\mathbf{F}_1 + \mathbf{F}_2)$$

and, by Varignon's theorem,

$$\mathbf{M} = \mathbf{r} \times \mathbf{F}_1 + \mathbf{r} \times \mathbf{F}_2$$

But the first term in the expression obtained represents the moment $\mathbf{M}_1$ of the couple in P_1, and the second term represents the moment $\mathbf{M}_2$ of the couple in P_2. We have

$$\mathbf{M} = \mathbf{M}_1 + \mathbf{M}_2 \tag{3.50}$$

and we conclude that the sum of two couples of moments $\mathbf{M}_1$ and $\mathbf{M}_2$ is a couple of moment $\mathbf{M}$ equal to the vector sum of $\mathbf{M}_1$ and $\mathbf{M}_2$ (Fig. 3.37b).

3.15 COUPLES CAN BE REPRESENTED BY VECTORS

As we saw in Sec. 3.13, couples which have the same moment, whether they act in the same plane or in parallel planes, are equivalent. There is therefore no need to draw the actual forces forming a given couple in order to define its effect on a rigid body (Fig. 3.38a). It is sufficient to draw an arrow equal in magnitude and direction to the moment $\mathbf{M}$ of the couple (Fig. 3.38b). On the other hand, we saw in Sec. 3.14 that the sum of two couples is itself a couple and that the moment $\mathbf{M}$ of the resultant couple can be obtained by forming the vector sum of the moments $\mathbf{M}_1$ and $\mathbf{M}_2$ of the given couples. Thus, couples obey the law of addition of vectors, and the arrow used in Fig. 3.38b to represent the couple defined in Fig. 3.38a can truly be considered a vector.

(a)

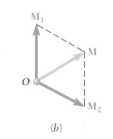

(b)

Fig. 3.37

The vector representing a couple is called a *couple vector*. Note that, in Fig. 3.38, a red arrow is used to distinguish the couple vector, *which represents the couple itself*, from the *moment* of the couple, which was represented by a green arrow in earlier figures. Also note that the symbol ↰ is added to this red arrow to avoid any confusion with vectors representing forces. A couple vector, like the moment of a couple, is a free vector. Its point of application, therefore, can be chosen at the origin of the system of coordinates, if so desired (Fig. 3.38c). Furthermore, the couple vector **M** can be resolved into component vectors $\mathbf{M}_x$, $\mathbf{M}_y$, and $\mathbf{M}_z$, which are directed along the coordinate axes (Fig. 3.38d). These component vectors represent couples acting, respectively, in the *yz*, *zx*, and *xy* planes.

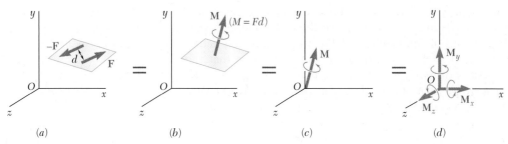

Fig. 3.38

3.16 RESOLUTION OF A GIVEN FORCE INTO A FORCE AT O AND A COUPLE

Consider a force **F** acting on a rigid body at a point *A* defined by the position vector **r** (Fig. 3.39a). Suppose that for some reason we would rather have the force act at point *O*. While we can move **F** along its line of action (principle of transmissibility), we cannot move it to a point *O* which does not lie on the original line of action without modifying the action of **F** on the rigid body.

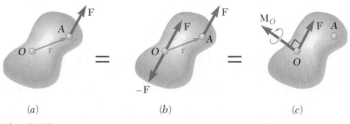

Fig. 3.39

We can, however, attach two forces at point *O*, one equal to **F** and the other equal to **−F**, without modifying the action of the original force on the rigid body (Fig. 3.39b). As a result of this transformation, a force **F** is now applied at *O*; the other two forces form a couple of moment $\mathbf{M}_O = \mathbf{r} \times \mathbf{F}$. Thus, *any force* **F** *acting on a rigid body can be moved to an arbitrary point O provided that a couple is added whose moment is equal to the moment of* **F** *about O.* The couple tends to impart to the rigid body the same rotational motion about O that the force $\mathbf{F}$ tended to produce before it was transferred to O. The couple is represented by a couple vector $\mathbf{M}_O$ perpendicular to the plane containing $\mathbf{r}$ and $\mathbf{F}$. Since $\mathbf{M}_O$ is a free vector, it may be applied anywhere; for convenience, however, the couple vector is usually attached at O, together with $\mathbf{F}$, and the combination obtained is referred to as a *force-couple system* (Fig. 3.39c).

If the force $\mathbf{F}$ had been moved from A to a different point O' (Fig. 3.40a and c), the moment $\mathbf{M}_{O'} = \mathbf{r}' \times \mathbf{F}$ of $\mathbf{F}$ about O' should have been computed, and a new force-couple system, consisting of $\mathbf{F}$ and of the couple vector $\mathbf{M}_{O'}$, would have been attached at O'. The relation existing between the moments of $\mathbf{F}$ about O and O' is obtained by writing

$$\mathbf{M}_{O'} = \mathbf{r}' \times \mathbf{F} = (\mathbf{r} + \mathbf{s}) \times \mathbf{F} = \mathbf{r} \times \mathbf{F} + \mathbf{s} \times \mathbf{F}$$

$$\mathbf{M}_{O'} = \mathbf{M}_O + \mathbf{s} \times \mathbf{F} \qquad (3.51)$$

where $\mathbf{s}$ is the vector joining O' to O. Thus, the moment $\mathbf{M}_{O'}$ of $\mathbf{F}$ about O' is obtained by adding to the moment $\mathbf{M}_O$ of $\mathbf{F}$ about O the vector product $\mathbf{s} \times \mathbf{F}$ representing the moment about O' of the force $\mathbf{F}$ applied at O.

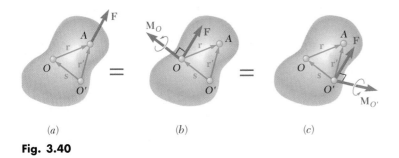

(a) = (b) = (c)

Fig. 3.40

This result could also have been established by observing that, in order to transfer to O' the force-couple system attached at O (Fig. 3.40b and c), the couple vector $\mathbf{M}_O$ can be freely moved to O'; to move the force $\mathbf{F}$ from O to O', however, it is necessary to add to $\mathbf{F}$ a couple vector whose moment is equal to the moment about O' of the force $\mathbf{F}$ applied at O. Thus, the couple vector $\mathbf{M}_{O'}$ must be the sum of $\mathbf{M}_O$ and the vector $\mathbf{s} \times \mathbf{F}$.

As noted above, the force-couple system obtained by transferring a force $\mathbf{F}$ from a point A to a point O consists of $\mathbf{F}$ and a couple vector $\mathbf{M}_O$ perpendicular to $\mathbf{F}$. Conversely, any force-couple system consisting of a force $\mathbf{F}$ and a couple vector $\mathbf{M}_O$ which are *mutually perpendicular* can be replaced by a single equivalent force. This is done by moving the force $\mathbf{F}$ in the plane perpendicular to $\mathbf{M}_O$ until its moment about O is equal to the moment of the couple to be eliminated.

Photo 3.2 The force exerted by each hand on the wrench could be replaced with an equivalent force-couple system acting on the nut.

SAMPLE PROBLEM 3.6

Determine the components of the single couple equivalent to the two couples shown.

SOLUTION

Our computations will be simplified if we attach two equal and opposite 20-lb forces at A. This enables us to replace the original 20-lb-force couple by two new 20-lb-force couples, one of which lies in the zx plane and the other in a plane parallel to the xy plane. The three couples shown in the adjoining sketch can be represented by three couple vectors $\mathbf{M}_x$, $\mathbf{M}_y$, and $\mathbf{M}_z$ directed along the coordinate axes. The corresponding moments are

$$M_x = -(30 \text{ lb})(18 \text{ in.}) = -540 \text{ lb} \cdot \text{in.}$$
$$M_y = +(20 \text{ lb})(12 \text{ in.}) = +240 \text{ lb} \cdot \text{in.}$$
$$M_z = +(20 \text{ lb})(9 \text{ in.}) = +180 \text{ lb} \cdot \text{in.}$$

These three moments represent the components of the single couple $\mathbf{M}$ equivalent to the two given couples. We write

$$\mathbf{M} = -(540 \text{ lb} \cdot \text{in.})\mathbf{i} + (240 \text{ lb} \cdot \text{in.})\mathbf{j} + (180 \text{ lb} \cdot \text{in.})\mathbf{k} \blacktriangleleft$$

Alternative Solution. The components of the equivalent single couple $\mathbf{M}$ can also be obtained by computing the sum of the moments of the four given forces about an arbitrary point. Selecting point D, we write

$$\mathbf{M} = \mathbf{M}_D = (18 \text{ in.})\mathbf{j} \times (-30 \text{ lb})\mathbf{k} + [(9 \text{ in.})\mathbf{j} - (12 \text{ in.})\mathbf{k}] \times (-20 \text{ lb})\mathbf{i}$$

and, after computing the various cross products,

$$\mathbf{M} = -(540 \text{ lb} \cdot \text{in.})\mathbf{i} + (240 \text{ lb} \cdot \text{in.})\mathbf{j} + (180 \text{ lb} \cdot \text{in.})\mathbf{k} \blacktriangleleft$$

SAMPLE PROBLEM 3.7

Replace the couple and force shown by an equivalent single force applied to the lever. Determine the distance from the shaft to the point of application of this equivalent force.

SOLUTION

First the given force and couple are replaced by an equivalent force-couple system at O. We move the force $\mathbf{F} = -(400\,\text{N})\mathbf{j}$ to O and at the same time add a couple of moment $\mathbf{M}_O$ equal to the moment about O of the force in its original position.

$$\mathbf{M}_O = \overrightarrow{OB} \times \mathbf{F} = [(0.150\,\text{m})\mathbf{i} + (0.260\,\text{m})\mathbf{j}] \times (-400\,\text{N})\mathbf{j}$$
$$= -(60\,\text{N} \cdot \text{m})\mathbf{k}$$

This couple is added to the couple of moment $-(24\,\text{N} \cdot \text{m})\mathbf{k}$ formed by the two 200-N forces, and a couple of moment $-(84\,\text{N} \cdot \text{m})\mathbf{k}$ is obtained. This last couple can be eliminated by applying $\mathbf{F}$ at a point C chosen in such a way that

$$-(84\,\text{N} \cdot \text{m})\mathbf{k} = \overrightarrow{OC} \times \mathbf{F}$$
$$= [(OC)\cos 60°\mathbf{i} + (OC)\sin 60°\mathbf{j}] \times (-400\,\text{N})\mathbf{j}$$
$$= -(OC)\cos 60°(400\,\text{N})\mathbf{k}$$

We conclude that

$$(OC)\cos 60° = 0.210\,\text{m} = 210\,\text{mm} \qquad OC = 420\,\text{mm} \quad \blacktriangleleft$$

Alternative Solution. Since the effect of a couple does not depend on its location, the couple of moment $-(24\,\text{N} \cdot \text{m})\mathbf{k}$ can be moved to B; we thus obtain a force-couple system at B. The couple can now be eliminated by applying $\mathbf{F}$ at a point C chosen in such a way that

$$-(24\,\text{N} \cdot \text{m})\mathbf{k} = \overrightarrow{BC} \times \mathbf{F}$$
$$= -(BC)\cos 60°(400\,\text{N})\mathbf{k}$$

We conclude that

$$(BC)\cos 60° = 0.060\,\text{m} = 60\,\text{mm} \qquad BC = 120\,\text{mm}$$
$$OC = OB + BC = 300\,\text{mm} + 120\,\text{mm} \qquad OC = 420\,\text{mm} \quad \blacktriangleleft$$

SOLVING PROBLEMS
ON YOUR OWN

In this lesson we discussed the properties of *couples*. To solve the problems which follow, you will need to remember that the net effect of a couple is to produce a moment **M**. Since this moment is independent of the point about which it is computed, **M** is a *free vector* and thus remains unchanged as it is moved from point to point. Also, two couples are *equivalent* (that is, they have the same effect on a given rigid body) if they produce the same moment.

When determining the moment of a couple, all previous techniques for computing moments apply. Also, since the moment of a couple is a free vector, it should be computed relative to the most convenient point.

Because the only effect of a couple is to produce a moment, it is possible to represent a couple with a vector, the *couple vector*, which is equal to the moment of the couple. The couple vector is a free vector and will be represented by a special symbol, $\not\!\!\nearrow$, to distinguish it from force vectors.

In solving the problems in this lesson, you will be called upon to perform the following operations:

1. Adding two or more couples. This results in a new couple, the moment of which is obtained by adding vectorially the moments of the given couples [Sample Prob. 3.6].

2. Replacing a force with an equivalent force-couple system at a specified point. As explained in Sec. 3.16, the force of the force-couple system is equal to the original force, while the required couple vector is equal to the moment of the original force about the given point. In addition, it is important to observe that the force and the couple vector are perpendicular to each other. Conversely, it follows that a force-couple system can be reduced to a single force only if the force and couple vector are mutually perpendicular (see the next paragraph).

3. Replacing a force-couple system (with F perpendicular to M) with a single equivalent force. Note that the requirement that **F** and **M** be mutually perpendicular will be satisfied in all two-dimensional problems. The single equivalent force is equal to **F** and is applied in such a way that its moment about the original point of application is equal to **M** [Sample Prob. 3.7].

PROBLEMS

3.70 A plate in the shape of a parallelogram is acted upon by two couples. Determine (*a*) the moment of the couple formed by the two 21-lb forces, (*b*) the perpendicular distance between the 12-lb forces if the resultant of the two couples is zero, (*c*) the value of α if the resultant couple is 72 lb · in. clockwise and *d* is 42 in.

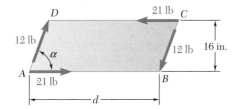

Fig. P3.70

3.71 Four 1-in.-diameter pegs are attached to a board as shown. Two strings are passed around the pegs and pulled with the forces indicated. (*a*) Determine the resultant couple acting on the board. (*b*) If only one string is used, around which pegs should it pass and in what directions should it be pulled to create the same couple with the minimum tension in the string? (*c*) What is the value of that minimum tension?

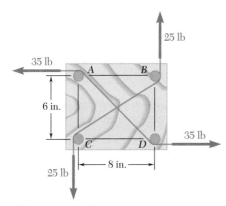

Fig. *P3.71* and P3.72

3.72 Four pegs of the same diameter are attached to a board as shown. Two strings are passed around the pegs and pulled with the forces indicated. Determine the diameter of the pegs knowing that the resultant couple applied to the board is 485 lb · in. counterclockwise.

3.73 A piece of plywood in which several holes are being drilled successively has been secured to a workbench by means of two nails. Knowing that the drill exerts a 12-N · m couple on the piece of plywood, determine the magnitude of the resulting forces applied to the nails if they are located (*a*) at *A* and *B*, (*b*) at *B* and *C*, (*c*) at *A* and *C*.

Fig. P3.73

Fig. P3.74

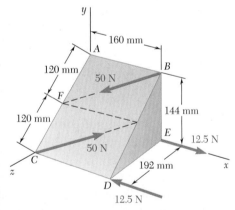

Fig. P3.76

3.74 Two parallel 40-N forces are applied to a lever as shown. Determine the moment of the couple formed by the two forces (*a*) by resolving each force into horizontal and vertical components and adding the moments of the two resulting couples, (*b*) by using the perpendicular distance between the two forces, (*c*) by summing the moments of the two forces about point *A*.

3.75 The two shafts of a speed-reducer unit are subjected to couples of magnitude $M_1 = 15$ lb · ft and $M_2 = 3$ lb · ft, respectively. Replace the two couples with a single equivalent couple, specifying its magnitude and the direction of its axis.

Fig. P3.75

3.76 Replace the two couples shown with a single equivalent couple, specifying its magnitude and the direction of its axis.

3.77 Solve Prob. 3.76, assuming that two 10-N vertical forces have been added, one acting upward at *C* and the other downward at *B*.

3.78 If $P = 0$, replace the two remaining couples with a single equivalent couple, specifying its magnitude and the direction of its axis.

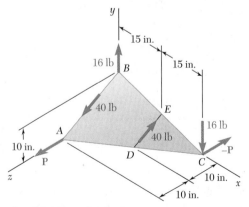

Fig. P3.78 and P3.79

3.79 If $P = 20$ lb, replace the three couples with a single equivalent couple, specifying its magnitude and the direction of its axis.

3.80 In a manufacturing operation, three holes are drilled simultaneously in a workpiece. If the holes are perpendicular to the surfaces of the workpiece, replace the couples applied to the drills with a single equivalent couple, specifying its magnitude and the direction of its axis.

3.81 A 260-lb force is applied at *A* to the rolled-steel section shown. Replace that force with an equivalent force-couple system at the center *C* of the section.

Fig. P3.81

Fig. P3.80

Fig. P3.82

3.82 A 30-lb vertical force **P** is applied at *A* to the bracket shown, which is held by screws at *B* and *C*. (*a*) Replace **P** with an equivalent force-couple system at *B*. (*b*) Find the two horizontal forces at *B* and *C* that are equivalent to the couple obtained in part *a*.

3.83 The force **P** has a magnitude of 250 N and is applied at the end *C* of a 500-mm rod *AC* attached to a bracket at *A* and *B*. Assuming $\alpha = 30°$ and $\beta = 60°$, replace **P** with (*a*) an equivalent force-couple system at *B*, (*b*) an equivalent system formed by two parallel forces applied at *A* and *B*.

3.84 Solve Prob. 3.83, assuming $\alpha = \beta = 25°$.

3.85 The 80-N horizontal force **P** acts on a bell crank as shown. (*a*) Replace **P** with an equivalent force-couple system at *B*. (*b*) Find the two vertical forces at *C* and *D* that are equivalent to the couple found in part *a*.

Fig. P3.83

Fig. *P3.85*

3.86 A dirigible is tethered by a cable attached to its cabin at *B*. If the tension in the cable is 1040 N, replace the force exerted by the cable at *B* with an equivalent system formed by two parallel forces applied at *A* and *C*.

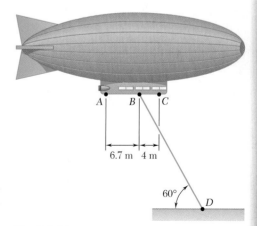

Fig. P3.86

3.87 Three control rods attached to a lever *ABC* exert on it the forces shown. (*a*) Replace the three forces with an equivalent force-couple system at *B*. (*b*) Determine the single force that is equivalent to the force-couple system obtained in part *a*, and specify its point of application on the lever.

Fig. P3.87

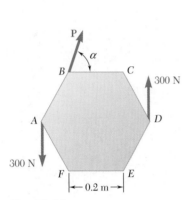

Fig. P3.88

3.88 A hexagonal plate is acted upon by the force **P** and the couple shown. Determine the magnitude and the direction of the smallest force **P** for which this system can be replaced with a single force at *E*.

3.89 A force and couple act as shown on a square plate of side *a* = 25 in. Knowing that *P* = 60 lb, *Q* = 40 lb, and *α* = 50°, replace the given force and couple with a single force applied at a point located (*a*) on line *AB*, (*b*) on line *AC*. In each case determine the distance from *A* to the point of application of the force.

Fig. P3.89 and P3.90

3.90 The force and couple shown are to be replaced by an equivalent single force. Knowing that *P* = 2*Q*, determine the required value of *α* if the line of action of the single equivalent force is to pass through (*a*) point *A*, (*b*) point *C*.

3.91 The shearing forces exerted on the cross section of a steel channel can be represented by a 900-N vertical force and two 250-N horizontal forces as shown. Replace this force and couple with a single force **F** applied at point *C*, and determine the distance *x* from *C* to line *BD*. (Point *C* is defined as the *shear center* of the section.)

Fig. P3.91

3.92 A force and a couple are applied as shown to the end of a cantilever beam. (*a*) Replace this system with a single force **F** applied at point *C*, and determine the distance *d* from *C* to a line drawn through points *D* and *E*. (*b*) Solve part *a* if the directions of the two 360-N forces are reversed.

3.93 An antenna is guyed by three cables as shown. Knowing that the tension in cable *AB* is 288 lb, replace the force exerted at *A* by cable *AB* with an equivalent force-couple system at the center *O* of the base of the antenna.

Fig. P3.92

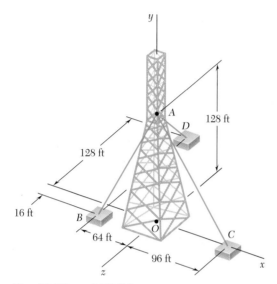

Fig. P3.93 and P3.94

3.94 An antenna is guyed by three cables as shown. Knowing that the tension in cable *AD* is 270 lb, replace the force exerted at *A* by cable *AD* with an equivalent force-couple system at the center *O* of the base of the antenna.

3.95 A 110-N force acting in a vertical plane parallel to the *yz* plane is applied to the 220-mm-long horizontal handle *AB* of a socket wrench. Replace the force with an equivalent force-couple system at the origin *O* of the coordinate system.

Fig. P3.95

3.96 An eccentric, compressive 1220-N force **P** is applied to the end of a cantilever beam. Replace **P** with an equivalent force-couple system at *G*.

Fig. P3.96

Fig. P3.97

3.97 To keep a door closed, a wooden stick is wedged between the floor and the doorknob. The stick exerts at B a 175-N force directed along line AB. Replace that force with an equivalent force-couple system at C.

3.98 A 46-lb force $\mathbf{F}$ and a 2120-lb · in. couple $\mathbf{M}$ are applied to corner A of the block shown. Replace the given force-couple system with an equivalent force-couple system at corner H.

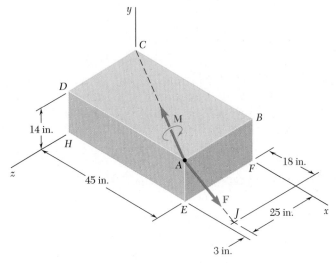

Fig. P3.98

3.99 A 77-N force $\mathbf{F}_1$ and a 31-N · m couple $\mathbf{M}_1$ are applied to corner E of the bent plate shown. If $\mathbf{F}_1$ and $\mathbf{M}_1$ are to be replaced with an equivalent force-couple system $(\mathbf{F}_2, \mathbf{M}_2)$ at corner B and if $(M_2)_z = 0$, determine (a) the distance d, (b) $\mathbf{F}_2$ and $\mathbf{M}_2$.

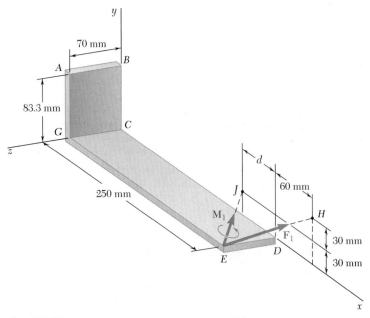

Fig. P3.99

3.100 A 2.6-kip force is applied at point D of the cast-iron post shown. Replace that force with an equivalent force-couple system at the center A of the base section.

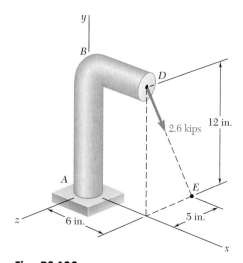

Fig. **P3.100**

3.17 REDUCTION OF A SYSTEM OF FORCES TO ONE FORCE AND ONE COUPLE

Consider a system of forces $\mathbf{F}_1$, $\mathbf{F}_2$, $\mathbf{F}_3$, . . . , acting on a rigid body at the points A_1, A_2, A_3, . . . , *defined by the position vectors* $\boldsymbol{r}_1$, $\boldsymbol{r}_2$, $\boldsymbol{r}_3$, *etc.* (Fig. 3.41a). As seen in the preceding section, $\mathbf{F}_1$ can be moved from A_1 to a given point O if a couple of moment $\mathbf{M}_1$ equal to the moment $\boldsymbol{r}_1 \times \mathbf{F}_1$ of $\mathbf{F}_1$ about O is added to the original system of forces. Repeating this procedure with $\mathbf{F}_2$, $\mathbf{F}_3$, . . . , we obtain the system shown in Fig. 3.41b, which consists of the original forces, now acting at O, and the added couple vectors. Since the forces are now concurrent, they can be added vectorially and replaced by their resultant $\mathbf{R}$. Similarly, the couple vectors $\mathbf{M}_1$, $\mathbf{M}_2$, $\mathbf{M}_3$, . . . , can be added vectorially and replaced by a single couple vector $\mathbf{M}_O^R$. Any system of forces, however complex, can thus be reduced to an *equivalent force-couple system acting at a given point O* (Fig. 3.41c). We

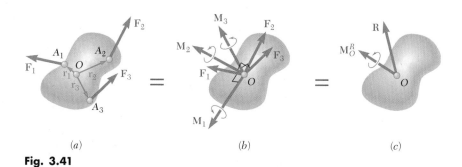

Fig. **3.41**

should note that while each of the couple vectors $\mathbf{M}_1$, $\mathbf{M}_2$, $\mathbf{M}_3$, . . . , in Fig. 3.41b is perpendicular to its corresponding force, the resultant force $\mathbf{R}$ and the resultant couple vector $\mathbf{M}_O^R$ in Fig. 3.41c will not, in general, be perpendicular to each other.

The equivalent force-couple system is defined by the equations

$$\mathbf{R} = \Sigma\mathbf{F} \qquad \mathbf{M}_O^R = \Sigma\mathbf{M}_O = \Sigma(\mathbf{r} \times \mathbf{F}) \qquad (3.52)$$

which express that the force $\mathbf{R}$ is obtained by adding all the forces of the system, while the moment of the resultant couple vector $\mathbf{M}_O^R$, called the *moment resultant* of the system, is obtained by adding the moments about O of all the forces of the system.

Once a given system of forces has been reduced to a force and a couple at a point O, it can easily be reduced to a force and a couple at another point O'. While the resultant force $\mathbf{R}$ will remain unchanged, the new moment resultant $\mathbf{M}_{O'}^R$ will be equal to the sum of $\mathbf{M}_O^R$ and the moment about O' of the force $\mathbf{R}$ attached at O (Fig. 3.42). We have

$$\mathbf{M}_{O'}^R = \mathbf{M}_O^R + \mathbf{s} \times \mathbf{R} \qquad (3.53)$$

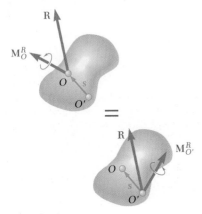

Fig. 3.42

In practice, the reduction of a given system of forces to a single force $\mathbf{R}$ at O and a couple vector $\mathbf{M}_O^R$ will be carried out in terms of components. Resolving each position vector $\mathbf{r}$ and each force $\mathbf{F}$ of the system into rectangular components, we write

$$\mathbf{r} = x\mathbf{i} + y\mathbf{j} + z\mathbf{k} \qquad (3.54)$$
$$\mathbf{F} = F_x\mathbf{i} + F_y\mathbf{j} + F_z\mathbf{k} \qquad (3.55)$$

Substituting for $\mathbf{r}$ and $\mathbf{F}$ in (3.52) and factoring out the unit vectors $\mathbf{i}$, $\mathbf{j}$, $\mathbf{k}$, we obtain $\mathbf{R}$ and $\mathbf{M}_O^R$ in the form

$$\mathbf{R} = R_x\mathbf{i} + R_y\mathbf{j} + R_z\mathbf{k} \qquad \mathbf{M}_O^R = M_x^R\mathbf{i} + M_y^R\mathbf{j} + M_z^R\mathbf{k} \quad (3.56)$$

The components R_x, R_y, R_z represent, respectively, the sums of the x, y, and z components of the given forces and measure the tendency of the system to impart to the rigid body a motion of translation in the x, y, or z direction. Similarly, the components M_x^R, M_y^R, M_z^R represent, respectively, the sum of the moments of the given forces about the x, y, and z axes and measure the tendency of the system to impart to the rigid body a motion of rotation about the x, y, or z axis.

If the magnitude and direction of the force $\mathbf{R}$ are desired, they can be obtained from the components R_x, R_y, R_z by means of the relations (2.18) and (2.19) of Sec. 2.12; similar computations will yield the magnitude and direction of the couple vector $\mathbf{M}_O^R$.

3.18 EQUIVALENT SYSTEMS OF FORCES

We saw in the preceding section that any system of forces acting on a rigid body can be reduced to a force-couple system at a given point O. This equivalent force-couple system characterizes completely the effect of the given force system on the rigid body. *Two systems of forces are equivalent, therefore, if they can be reduced to the same force-couple system at a given point O.* Recalling that the force-couple system at O is defined by the relations (3.52), we state that *two systems of forces, $\mathbf{F}_1$, $\mathbf{F}_2$, $\mathbf{F}_3$, . . . , and $\mathbf{F}'_1$, $\mathbf{F}'_2$, $\mathbf{F}'_3$, . . . , which act on the same rigid body are equivalent if, and only if, the sums of the forces and the sums of the moments about a given point O of the forces of the two systems are, respectively, equal.* Expressed mathematically, the necessary and sufficient conditions for the two systems of forces to be equivalent are

Photo 3.3 The forces exerted by the children upon the wagon can be replaced with an equivalent force-couple system when analyzing the motion of the wagon.

$$\Sigma\mathbf{F} = \Sigma\mathbf{F}' \qquad \text{and} \qquad \Sigma\mathbf{M}_O = \Sigma\mathbf{M}'_O \qquad (3.57)$$

Note that to prove that two systems of forces are equivalent, the second of the relations (3.57) must be established with respect to *only one point O.* It will hold, however, with respect to *any point* if the two systems are equivalent.

Resolving the forces and moments in (3.57) into their rectangular components, we can express the necessary and sufficient conditions for the equivalence of two systems of forces acting on a rigid body as follows:

$$
\begin{array}{ccc}
\Sigma F_x = \Sigma F'_x & \Sigma F_y = \Sigma F'_y & \Sigma F_z = \Sigma F'_z \\
\Sigma M_x = \Sigma M'_x & \Sigma M_y = \Sigma M'_y & \Sigma M_z = \Sigma M'_z
\end{array}
\qquad (3.58)
$$

These equations have a simple physical significance. They express that two systems of forces are equivalent if they tend to impart to the rigid body (1) the same translation in the x, y, and z directions, respectively, and (2) the same rotation about the x, y, and z axes, respectively.

3.19 EQUIPOLLENT SYSTEMS OF VECTORS

In general, when two systems of vectors satisfy Eqs. (3.57) or (3.58), i.e., when their resultants and their moment resultants about an arbitrary point O are respectively equal, the two systems are said to be *equipollent.* The result established in the preceding section can thus be restated as follows: *If two systems of forces acting on a rigid body are equipollent, they are also equivalent.*

It is important to note that this statement does not apply to *any* system of vectors. Consider, for example, a system of forces acting on a set of independent particles which do *not* form a rigid body. A different system of forces acting on the same particles may happen to be equipollent to the first one; i.e., it may have the same resultant and the same moment resultant. Yet, since different forces will now act on the various particles, their effects on these particles will be different; the two systems of forces, while equipollent, are *not equivalent.*

3.20 FURTHER REDUCTION OF A SYSTEM OF FORCES

We saw in Sec. 3.17 that any given system of forces acting on a rigid body can be reduced to an equivalent force-couple system at O consisting of a force $\mathbf{R}$ equal to the sum of the forces of the system and a couple vector $\mathbf{M}_O^R$ of moment equal to the moment resultant of the system.

When $\mathbf{R} = 0$, the force-couple system reduces to the couple vector $\mathbf{M}_O^R$. The given system of forces can then be reduced to a single couple, called the *resultant couple* of the system.

Let us now investigate the conditions under which a given system of forces can be reduced to a single force. It follows from Sec. 3.16 that the force-couple system at O can be replaced by a single force $\mathbf{R}$ acting along a new line of action if $\mathbf{R}$ and $\mathbf{M}_O^R$ are mutually perpendicular. The systems of forces which can be reduced to a single force, or *resultant*, are therefore the systems for which the force $\mathbf{R}$ and the couple vector $\mathbf{M}_O^R$ are mutually perpendicular. While this condition *is generally not satisfied* by systems of forces in space, it *will be satisfied* by systems consisting of (1) concurrent forces, (2) coplanar forces, or (3) parallel forces. These three cases will be discussed separately.

1. *Concurrent forces* are applied at the same point and can therefore be added directly to obtain their resultant $\mathbf{R}$. Thus, they always reduce to a single force. Concurrent forces were discussed in detail in Chap. 2.

2. *Coplanar forces* act in the same plane, which may be assumed to be the plane of the figure (Fig. 3.43a). The sum $\mathbf{R}$ of the forces of the system will also lie in the plane of the figure, while the moment of each force about O, and thus the moment resultant $\mathbf{M}_O^R$, will be perpendicular to that plane. The force-couple system at O consists, therefore, of a force $\mathbf{R}$ and a couple vector $\mathbf{M}_O^R$ which are mutually perpendicular (Fig. 3.43b).† They can be reduced to a single force $\mathbf{R}$ by moving $\mathbf{R}$ in the plane of the figure until its moment about O becomes equal to $\mathbf{M}_O^R$. The distance from O to the line of action of $\mathbf{R}$ is $d = M_O^R/R$ (Fig. 3.43c).

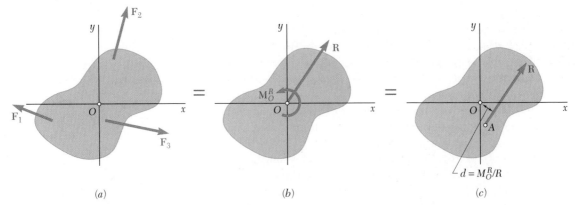

(a)   (b)   (c)

Fig. 3.43

†Since the couple vector $\mathbf{M}_O^R$ is perpendicular to the plane of the figure, it has been represented by the symbol ↰. A counterclockwise couple ↰ represents a vector pointing out of the paper, and a clockwise couple ↲ represents a vector pointing into the paper.

As noted in Sec. 3.17, the reduction of a system of forces is considerably simplified if the forces are resolved into rectangular components. The force-couple system at O is then characterized by the components (Fig. 3.44a)

$$R_x = \Sigma F_x \qquad R_y = \Sigma F_y \qquad M_z^R = M_O^R = \Sigma M_O \qquad (3.59)$$

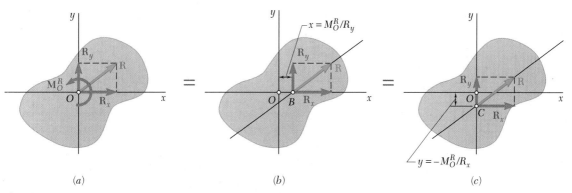

(a) (b) (c)

Fig. 3.44

To reduce the system to a single force $\mathbf{R}$, we express that the moment of $\mathbf{R}$ about O must be equal to $\mathbf{M}_O^R$. Denoting by x and y the coordinates of the point of application of the resultant and recalling formula (3.22) of Sec. 3.8, we write

$$xR_y - yR_x = M_O^R$$

which represents the equation of the line of action of $\mathbf{R}$. We can also determine directly the x and y intercepts of the line of action of the resultant by noting that $\mathbf{M}_O^R$ must be equal to the moment about O of the y component of $\mathbf{R}$ when $\mathbf{R}$ is attached at B (Fig. 3.44b) and to the moment of its x component when $\mathbf{R}$ is attached at C (Fig. 3.44c).

3. *Parallel forces* have parallel lines of action and may or may not have the same sense. Assuming here that the forces are parallel to the y axis (Fig. 3.45a), we note that their sum $\mathbf{R}$ will also be parallel to the y axis. On the other hand, since the moment of a given force must be perpendicular to that force, the moment about O of each force of the system, and thus the moment resultant $\mathbf{M}_O^R$, will lie in the zx plane. The force-couple system at O consists,

(a) (b) (c)

Fig. 3.45

Photo 3.4 The parallel wind forces acting on the highway signs can be reduced to a single equivalent force. Determining this force can simplify the calculation of the forces acting on the supports of the frame to which the signs are attached.

therefore, of a force $\mathbf{R}$ and a couple vector $\mathbf{M}_O^R$ which are mutually perpendicular (Fig. 3.45b). They can be reduced to a single force $\mathbf{R}$ (Fig. 3.45c) or, if $\mathbf{R} = 0$, to a single couple of moment $\mathbf{M}_O^R$.

In practice, the force-couple system at O will be characterized by the components

$$R_y = \Sigma F_y \qquad M_x^R = \Sigma M_x \qquad M_z^R = \Sigma M_z \qquad (3.60)$$

The reduction of the system to a single force can be carried out by moving $\mathbf{R}$ to a new point of application $A(x, 0, z)$ chosen so that the moment of $\mathbf{R}$ about O is equal to $\mathbf{M}_O^R$. We write

$$\mathbf{r} \times \mathbf{R} = \mathbf{M}_O^R$$
$$(x\mathbf{i} + z\mathbf{k}) \times R_y\mathbf{j} = M_x^R\mathbf{i} + M_z^R\mathbf{k}$$

By computing the vector products and equating the coefficients of the corresponding unit vectors in both members of the equation, we obtain two scalar equations which define the coordinates of A:

$$-zR_y = M_x^R \qquad xR_y = M_z^R$$

These equations express that the moments of $\mathbf{R}$ about the x and z axes must, respectively, be equal to M_x^R and M_z^R.

*3.21 REDUCTION OF A SYSTEM OF FORCES TO A WRENCH

In the general case of a system of forces in space, the equivalent force-couple system at O consists of a force $\mathbf{R}$ and a couple vector $\mathbf{M}_O^R$ which are not perpendicular, and neither of which is zero (Fig. 3.46a). Thus, the system of forces *cannot* be reduced to a single force or to a single couple. The couple vector, however, can be replaced by two other couple vectors obtained by resolving $\mathbf{M}_O^R$ into a component $\mathbf{M}_1$ along $\mathbf{R}$ and a component $\mathbf{M}_2$ in a plane perpendicular to $\mathbf{R}$ (Fig. 3.46b). The couple vector $\mathbf{M}_2$ and the force $\mathbf{R}$ can then be replaced by a single force $\mathbf{R}$ acting along a new line of action. The original system of forces thus reduces to $\mathbf{R}$ and to the couple vector $\mathbf{M}_1$ (Fig. 3.46c), i.e., to $\mathbf{R}$ and a couple acting in the plane perpendicular to $\mathbf{R}$. This particular force-couple system is called a *wrench* because the resulting combination of push and twist is the same as that which would be caused by an actual wrench. The line of action of $\mathbf{R}$ is known as the *axis of the wrench*, and the ratio $p = M_1/R$ is called the *pitch*

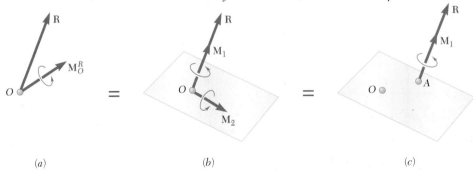

(a) (b) (c)

Fig. 3.46

of the wrench. A wrench, therefore, consists of two collinear vectors, namely, a force $\mathbf{R}$ and a couple vector

$$\mathbf{M}_1 = p\mathbf{R} \qquad (3.61)$$

Recalling the expression (3.35) obtained in Sec. 3.9 for the projection of a vector on the line of action of another vector, we note that the projection of $\mathbf{M}_O^R$ on the line of action of $\mathbf{R}$ is

$$M_1 = \frac{\mathbf{R} \cdot \mathbf{M}_O^R}{R}$$

Thus, the pitch of the wrench can be expressed as†

$$p = \frac{M_1}{R} = \frac{\mathbf{R} \cdot \mathbf{M}_O^R}{R^2} \qquad (3.62)$$

To define the axis of the wrench, we can write a relation involving the position vector $\mathbf{r}$ of an arbitrary point P located on that axis. Attaching the resultant force $\mathbf{R}$ and couple vector $\mathbf{M}_1$ at P (Fig. 3.47) and expressing that the moment about O of this force-couple system is equal to the moment resultant $\mathbf{M}_O^R$ of the original force system, we write

$$\mathbf{M}_1 + \mathbf{r} \times \mathbf{R} = \mathbf{M}_O^R \qquad (3.63)$$

or, recalling Eq. (3.61),

$$p\mathbf{R} + \mathbf{r} \times \mathbf{R} = \mathbf{M}_O^R \qquad (3.64)$$

Fig. 3.47

Photo 3.5 The pushing-turning action associated with the tightening of a screw illustrates the collinear lines of action of the force and couple vector that constitute a wrench.

†The expressions obtained for the projection of the couple vector on the line of action of $\mathbf{R}$ and for the pitch of the wrench are independent of the choice of point O. Using the relation (3.53) of Sec. 3.17, we note that if a different point O' had been used, the numerator in (3.62) would have been

$$\mathbf{R} \cdot \mathbf{M}_{O'}^R = \mathbf{R} \cdot (\mathbf{M}_O^R + \mathbf{s} \times \mathbf{R}) = \mathbf{R} \cdot \mathbf{M}_O^R + \mathbf{R} \cdot (\mathbf{s} \times \mathbf{R})$$

Since the mixed triple product $\mathbf{R} \cdot (\mathbf{s} \times \mathbf{R})$ is identically equal to zero, we have

$$\mathbf{R} \cdot \mathbf{M}_{O'}^R = \mathbf{R} \cdot \mathbf{M}_O^R$$

Thus, the scalar product $\mathbf{R} \cdot \mathbf{M}_O^R$ is independent of the choice of point O.

SAMPLE PROBLEM 3.8

A 4.80-m-long beam is subjected to the forces shown. Reduce the given system of forces to (a) an equivalent force-couple system at A, (b) an equivalent force-couple system at B, (c) a single force or resultant.

Note. Since the reactions at the supports are not included in the given system of forces, the given system will not maintain the beam in equilibrium.

SOLUTION

a. Force-Couple System at A. The force-couple system at A equivalent to the given system of forces consists of a force **R** and a couple $\mathbf{M}_A^R$ defined as follows:

$$\mathbf{R} = \Sigma\mathbf{F}$$
$$= (150\ \text{N})\mathbf{j} - (600\ \text{N})\mathbf{j} + (100\ \text{N})\mathbf{j} - (250\ \text{N})\mathbf{j} = -(600\ \text{N})\mathbf{j}$$
$$\mathbf{M}_A^R = \Sigma(\mathbf{r} \times \mathbf{F})$$
$$= (1.6\mathbf{i}) \times (-600\mathbf{j}) + (2.8\mathbf{i}) \times (100\mathbf{j}) + (4.8\mathbf{i}) \times (-250\mathbf{j})$$
$$= -(1880\ \text{N} \cdot \text{m})\mathbf{k}$$

The equivalent force-couple system at A is thus

$$\mathbf{R} = 600\ \text{N}\downarrow \qquad \mathbf{M}_A^R = 1880\ \text{N} \cdot \text{m}\ \downarrow \quad \blacktriangleleft$$

b. Force-Couple System at B. We propose to find a force-couple system at B equivalent to the force-couple system at A determined in part a. The force **R** is unchanged, but a new couple $\mathbf{M}_B^R$ must be determined, the moment of which is equal to the moment about B of the force-couple system determined in part a. Thus, we have

$$\mathbf{M}_B^R = \mathbf{M}_A^R + \overrightarrow{BA} \times \mathbf{R}$$
$$= -(1880\ \text{N} \cdot \text{m})\mathbf{k} + (-4.8\ \text{m})\mathbf{i} \times (-600\ \text{N})\mathbf{j}$$
$$= -(1880\ \text{N} \cdot \text{m})\mathbf{k} + (2880\ \text{N} \cdot \text{m})\mathbf{k} = +(1000\ \text{N} \cdot \text{m})\mathbf{k}$$

The equivalent force-couple system at B is thus

$$\mathbf{R} = 600\ \text{N}\downarrow \qquad \mathbf{M}_B^R = 1000\ \text{N} \cdot \text{m}\ \uparrow \quad \blacktriangleleft$$

c. Single Force or Resultant. The resultant of the given system of forces is equal to **R**, and its point of application must be such that the moment of **R** about A is equal to $\mathbf{M}_A^R$. We write

$$\mathbf{r} \times \mathbf{R} = \mathbf{M}_A^R$$
$$x\mathbf{i} \times (-600\ \text{N})\mathbf{j} = -(1880\ \text{N} \cdot \text{m})\mathbf{k}$$
$$-x(600\ \text{N})\mathbf{k} = -(1880\ \text{N} \cdot \text{m})\mathbf{k}$$

and conclude that $x = 3.13$ m. Thus, the single force equivalent to the given system is defined as

$$\mathbf{R} = 600\ \text{N}\downarrow \qquad x = 3.13\ \text{m} \quad \blacktriangleleft$$

SAMPLE PROBLEM 3.9

Four tugboats are used to bring an ocean liner to its pier. Each tugboat exerts a 5000-lb force in the direction shown. Determine (a) the equivalent force-couple system at the foremast O, (b) the point on the hull where a single, more powerful tugboat should push to produce the same effect as the original four tugboats.

SOLUTION

a. **Force-Couple System at O.** Each of the given forces is resolved into components in the diagram shown (kip units are used). The force-couple system at O equivalent to the given system of forces consists of a force **R** and a couple $\mathbf{M}_O^R$ defined as follows:

$$\mathbf{R} = \Sigma\mathbf{F}$$
$$= (2.50\mathbf{i} - 4.33\mathbf{j}) + (3.00\mathbf{i} - 4.00\mathbf{j}) + (-5.00\mathbf{j}) + (3.54\mathbf{i} + 3.54\mathbf{j})$$
$$= 9.04\mathbf{i} - 9.79\mathbf{j}$$

$$\mathbf{M}_O^R = \Sigma(\mathbf{r} \times \mathbf{F})$$
$$= (-90\mathbf{i} + 50\mathbf{j}) \times (2.50\mathbf{i} - 4.33\mathbf{j})$$
$$+ (100\mathbf{i} + 70\mathbf{j}) \times (3.00\mathbf{i} - 4.00\mathbf{j})$$
$$+ (400\mathbf{i} + 70\mathbf{j}) \times (-5.00\mathbf{j})$$
$$+ (300\mathbf{i} - 70\mathbf{j}) \times (3.54\mathbf{i} + 3.54\mathbf{j})$$
$$= (390 - 125 - 400 - 210 - 2000 + 1062 + 248)\mathbf{k}$$
$$= -1035\mathbf{k}$$

The equivalent force-couple system at O is thus

$$\mathbf{R} = (9.04 \text{ kips})\mathbf{i} - (9.79 \text{ kips})\mathbf{j} \qquad \mathbf{M}_O^R = -(1035 \text{ kip} \cdot \text{ft})\mathbf{k}$$

or $\qquad \mathbf{R} = 13.33 \text{ kips} \searrow 47.3° \qquad \mathbf{M}_O^R = 1035 \text{ kip} \cdot \text{ft} \downarrow$ ◀

Remark. Since all the forces are contained in the plane of the figure, we could have expected the sum of their moments to be perpendicular to that plane. Note that the moment of each force component could have been obtained directly from the diagram by first forming the product of its magnitude and perpendicular distance to O and then assigning to this product a positive or a negative sign depending upon the sense of the moment.

b. **Single Tugboat.** The force exerted by a single tugboat must be equal to **R**, and its point of application A must be such that the moment of **R** about O is equal to $\mathbf{M}_O^R$. Observing that the position vector of A is

$$\mathbf{r} = x\mathbf{i} + 70\mathbf{j}$$

we write

$$\mathbf{r} \times \mathbf{R} = \mathbf{M}_O^R$$
$$(x\mathbf{i} + 70\mathbf{j}) \times (9.04\mathbf{i} - 9.79\mathbf{j}) = -1035\mathbf{k}$$
$$-x(9.79)\mathbf{k} - 633\mathbf{k} = -1035\mathbf{k} \qquad x = 41.1 \text{ ft} ◀$$

E(150 mm, -50 mm, 100 mm)

SAMPLE PROBLEM 3.10

Three cables are attached to a bracket as shown. Replace the forces exerted by the cables with an equivalent force-couple system at A.

SOLUTION

We first determine the relative position vectors drawn from point A to the points of application of the various forces and resolve the forces into rectangular components. Observing that $\mathbf{F}_B = (700 \text{ N})\boldsymbol{\lambda}_{BE}$ where

$$\boldsymbol{\lambda}_{BE} = \frac{\overrightarrow{BE}}{BE} = \frac{75\mathbf{i} - 150\mathbf{j} + 50\mathbf{k}}{175}$$

we have, using meters and newtons,

$$\mathbf{r}_{B/A} = \overrightarrow{AB} = 0.075\mathbf{i} + 0.050\mathbf{k} \qquad \mathbf{F}_B = 300\mathbf{i} - 600\mathbf{j} + 200\mathbf{k}$$
$$\mathbf{r}_{C/A} = \overrightarrow{AC} = 0.075\mathbf{i} - 0.050\mathbf{k} \qquad \mathbf{F}_C = 707\mathbf{i} \qquad\quad - 707\mathbf{k}$$
$$\mathbf{r}_{D/A} = \overrightarrow{AD} = 0.100\mathbf{i} - 0.100\mathbf{j} \qquad \mathbf{F}_D = 600\mathbf{i} + 1039\mathbf{j}$$

The force-couple system at A equivalent to the given forces consists of a force $\mathbf{R} = \Sigma\mathbf{F}$ and a couple $\mathbf{M}_A^R = \Sigma(\mathbf{r} \times \mathbf{F})$. The force $\mathbf{R}$ is readily obtained by adding respectively the x, y, and z components of the forces:

$$\mathbf{R} = \Sigma\mathbf{F} = (1607 \text{ N})\mathbf{i} + (439 \text{ N})\mathbf{j} - (507 \text{ N})\mathbf{k} \quad \blacktriangleleft$$

The computation of $\mathbf{M}_A^R$ will be facilitated if we express the moments of the forces in the form of determinants (Sec. 3.8):

$$\mathbf{r}_{B/A} \times \mathbf{F}_B = \begin{vmatrix} \mathbf{i} & \mathbf{j} & \mathbf{k} \\ 0.075 & 0 & 0.050 \\ 300 & -600 & 200 \end{vmatrix} = 30\mathbf{i} \qquad\qquad -45\mathbf{k}$$

$$\mathbf{r}_{C/A} \times \mathbf{F}_C = \begin{vmatrix} \mathbf{i} & \mathbf{j} & \mathbf{k} \\ 0.075 & 0 & -0.050 \\ 707 & 0 & -707 \end{vmatrix} = \qquad 17.68\mathbf{j}$$

$$\mathbf{r}_{D/A} \times \mathbf{F}_D = \begin{vmatrix} \mathbf{i} & \mathbf{j} & \mathbf{k} \\ 0.100 & -0.100 & 0 \\ 600 & 1039 & 0 \end{vmatrix} = \qquad\qquad 163.9\mathbf{k}$$

Adding the expressions obtained, we have

$$\mathbf{M}_A^R = \Sigma(\mathbf{r} \times \mathbf{F}) = (30 \text{ N} \cdot \text{m})\mathbf{i} + (17.68 \text{ N} \cdot \text{m})\mathbf{j} + (118.9 \text{ N} \cdot \text{m})\mathbf{k} \quad \blacktriangleleft$$

The rectangular components of the force $\mathbf{R}$ and the couple $\mathbf{M}_A^R$ are shown in the adjoining sketch.

(17.68 N·m) j

(439 N) j

− (507 N) k

(118.9 N·m) k

(1607 N) i

(30 N·m) i

SAMPLE PROBLEM 3.11

A square foundation mat supports the four columns shown. Determine the magnitude and point of application of the resultant of the four loads.

SOLUTION

We first reduce the given system of forces to a force-couple system at the origin O of the coordinate system. This force-couple system consists of a force $\mathbf{R}$ and a couple vector $\mathbf{M}_O^R$ defined as follows:

$$\mathbf{R} = \Sigma\mathbf{F} \qquad \mathbf{M}_O^R = \Sigma(\mathbf{r} \times \mathbf{F})$$

The position vectors of the points of application of the various forces are determined, and the computations are arranged in tabular form.

r, ft	F, kips	r × F, kip · ft
0	$-40\mathbf{j}$	0
$10\mathbf{i}$	$-12\mathbf{j}$	$-120\mathbf{k}$
$10\mathbf{i} + 5\mathbf{k}$	$-8\mathbf{j}$	$40\mathbf{i} - 80\mathbf{k}$
$4\mathbf{i} + 10\mathbf{k}$	$-20\mathbf{j}$	$200\mathbf{i} - 80\mathbf{k}$
	$\mathbf{R} = -80\mathbf{j}$	$\mathbf{M}_O^R = 240\mathbf{i} - 280\mathbf{k}$

Since the force $\mathbf{R}$ and the couple vector $\mathbf{M}_O^R$ are mutually perpendicular, the force-couple system obtained can be reduced further to a single force $\mathbf{R}$. The new point of application of $\mathbf{R}$ will be selected in the plane of the mat and in such a way that the moment of $\mathbf{R}$ about O will be equal to $\mathbf{M}_O^R$. Denoting by $\mathbf{r}$ the position vector of the desired point of application, and by x and z its coordinates, we write

$$\mathbf{r} \times \mathbf{R} = \mathbf{M}_O^R$$
$$(x\mathbf{i} + z\mathbf{k}) \times (-80\mathbf{j}) = 240\mathbf{i} - 280\mathbf{k}$$
$$-80x\mathbf{k} + 80z\mathbf{i} = 240\mathbf{i} - 280\mathbf{k}$$

from which it follows that

$$-80x = -280 \qquad 80z = 240$$
$$x = 3.50 \text{ ft} \qquad z = 3.00 \text{ ft}$$

We conclude that the resultant of the given system of forces is

$$\mathbf{R} = 80 \text{ kips} \downarrow \qquad \text{at } x = 3.50 \text{ ft, } z = 3.00 \text{ ft} \quad \blacktriangleleft$$

SAMPLE PROBLEM 3.12

Two forces of the same magnitude P act on a cube of side a as shown. Replace the two forces by an equivalent wrench, and determine (a) the magnitude and direction of the resultant force $\mathbf{R}$, (b) the pitch of the wrench, (c) the point where the axis of the wrench intersects the yz plane.

SOLUTION

Equivalent Force-Couple System at O. We first determine the equivalent force-couple system at the origin O. We observe that the position vectors of the points of application E and D of the two given forces are $\mathbf{r}_E = a\mathbf{i} + a\mathbf{j}$ and $\mathbf{r}_D = a\mathbf{j} + a\mathbf{k}$. The resultant $\mathbf{R}$ of the two forces and their moment resultant $\mathbf{M}_O^R$ about O are

$$\mathbf{R} = \mathbf{F}_1 + \mathbf{F}_2 = P\mathbf{i} + P\mathbf{j} = P(\mathbf{i} + \mathbf{j}) \tag{1}$$
$$\mathbf{M}_O^R = \mathbf{r}_E \times \mathbf{F}_1 + \mathbf{r}_D \times \mathbf{F}_2 = (a\mathbf{i} + a\mathbf{j}) \times P\mathbf{i} + (a\mathbf{j} + a\mathbf{k}) \times P\mathbf{j}$$
$$= -Pa\mathbf{k} - Pa\mathbf{i} = -Pa(\mathbf{i} + \mathbf{k}) \tag{2}$$

a. Resultant Force R. It follows from Eq. (1) and the adjoining sketch that the resultant force $\mathbf{R}$ has the magnitude $R = P\sqrt{2}$, lies in the xy plane, and forms angles of 45° with the x and y axes. Thus

$$R = P\sqrt{2} \qquad \theta_x = \theta_y = 45° \qquad \theta_z = 90° \quad \blacktriangleleft$$

b. Pitch of Wrench. Recalling formula (3.62) of Sec. 3.21 and Eqs. (1) and (2) above, we write

$$p = \frac{\mathbf{R} \cdot \mathbf{M}_O^R}{R^2} = \frac{P(\mathbf{i} + \mathbf{j}) \cdot (-Pa)(\mathbf{i} + \mathbf{k})}{(P\sqrt{2})^2} = \frac{-P^2a(1 + 0 + 0)}{2P^2} \qquad p = -\frac{a}{2} \quad \blacktriangleleft$$

c. Axis of Wrench. It follows from the above and from Eq. (3.61) that the wrench consists of the force $\mathbf{R}$ found in (1) and the couple vector

$$\mathbf{M}_1 = p\mathbf{R} = -\frac{a}{2}P(\mathbf{i} + \mathbf{j}) = -\frac{Pa}{2}(\mathbf{i} + \mathbf{j}) \tag{3}$$

To find the point where the axis of the wrench intersects the yz plane, we express that the moment of the wrench about O is equal to the moment resultant $\mathbf{M}_O^R$ of the original system:

$$\mathbf{M}_1 + \mathbf{r} \times \mathbf{R} = \mathbf{M}_O^R$$

or, noting that $\mathbf{r} = y\mathbf{j} + z\mathbf{k}$ and substituting for $\mathbf{R}$, $\mathbf{M}_O^R$, and $\mathbf{M}_1$ from Eqs. (1), (2), and (3),

$$-\frac{Pa}{2}(\mathbf{i} + \mathbf{j}) + (y\mathbf{j} + z\mathbf{k}) \times P(\mathbf{i} + \mathbf{j}) = -Pa(\mathbf{i} + \mathbf{k})$$

$$-\frac{Pa}{2}\mathbf{i} - \frac{Pa}{2}\mathbf{j} - Py\mathbf{k} + Pz\mathbf{j} - Pz\mathbf{i} = -Pa\mathbf{i} - Pa\mathbf{k}$$

Equating the coefficients of $\mathbf{k}$, and then the coefficients of $\mathbf{j}$, we find

$$y = a \qquad z = a/2 \quad \blacktriangleleft$$

SOLVING PROBLEMS
ON YOUR OWN

This lesson was devoted to the reduction and simplification of force systems. In solving the problems which follow, you will be asked to perform the operations discussed below.

1. Reducing a force system to a force and a couple at a given point A. The force is the *resultant* **R** of the system and is obtained by adding the various forces; the moment of the couple is the *moment resultant* of the system and is obtained by adding the moments about A of the various forces. We have

$$\mathbf{R} = \Sigma\mathbf{F} \qquad \mathbf{M}_A^R = \Sigma(\mathbf{r} \times \mathbf{F})$$

where the position vector **r** is drawn from A to *any point* on the line of action of **F**.

2. Moving a force-couple system from point A to point B. If you wish to reduce a given force system to a force-couple system at point B after you have reduced it to a force-couple system at point A, you need not recompute the moments of the forces about B. The resultant **R** remains unchanged, and the new moment resultant $\mathbf{M}_B^R$ can be obtained by adding to $\mathbf{M}_A^R$ the moment about B of the force **R** applied at A [Sample Prob. 3.8]. Denoting by **s** the vector drawn from B to A, you can write

$$\mathbf{M}_B^R = \mathbf{M}_A^R + \mathbf{s} \times \mathbf{R}$$

3. Checking whether two force systems are equivalent. First reduce each force system to a force-couple system *at the same, but arbitrary, point A* (as explained in paragraph 1). The two systems are equivalent (that is, they have the same effect on the given rigid body) if the two force-couple systems you have obtained are identical, that is, if

$$\Sigma\mathbf{F} = \Sigma\mathbf{F}' \qquad \text{and} \qquad \Sigma\mathbf{M}_A = \Sigma\mathbf{M}_A'$$

You should recognize that if the first of these equations is not satisfied, that is, if the two systems do not have the same resultant **R**, the two systems cannot be equivalent and there is then no need to check whether or not the second equation is satisfied.

4. Reducing a given force system to a single force. First reduce the given system to a force-couple system consisting of the resultant **R** and the couple vector $\mathbf{M}_A^R$ at some convenient point A (as explained in paragraph 1). You will recall from the previous lesson that further reduction to a single force is possible *only if the*

(continued)

force **R** *and the couple vector* $\mathbf{M}_A^R$ *are mutually perpendicular.* This will certainly be the case for systems of forces which are either *concurrent, coplanar,* or *parallel.* The required single force can then be obtained by moving **R** until its moment about A is equal to $\mathbf{M}_A^R$, as you did in several problems of the preceding lesson. More formally, you can write that the position vector **r** drawn from A to any point on the line of action of the single force **R** must satisfy the equation

$$\mathbf{r} \times \mathbf{R} = \mathbf{M}_A^R$$

This procedure was used in Sample Probs. 3.8, 3.9, and 3.11.

5. Reducing a given force system to a wrench. If the given system is comprised of forces which are not concurrent, coplanar, or parallel, the equivalent force-couple system at a point A will consist of a force **R** and a couple vector $\mathbf{M}_A^R$ which, in general, *are not mutually perpendicular.* (To check whether **R** and $\mathbf{M}_A^R$ are mutually perpendicular, form their scalar product. If this product is zero, they are mutually perpendicular; otherwise, they are not.) If **R** and $\mathbf{M}_A^R$ are not mutually perpendicular, the force-couple system (and thus the given system of forces) *cannot be reduced to a single force.* However, the system can be reduced to a *wrench*— the combination of a force **R** and a couple vector $\mathbf{M}_1$ directed along a common line of action called the *axis of the wrench* (Fig. 3.47). The ratio $p = M_1/R$ is called the *pitch* of the wrench.

To reduce a given force system to a wrench, you should follow these steps:

   a. Reduce the given system to an equivalent force-couple system $(\mathbf{R}, \mathbf{M}_O^R)$, typically located at the origin O.
   b. Determine the pitch p from Eq. (3.62)

$$p = \frac{M_1}{R} = \frac{\mathbf{R} \cdot \mathbf{M}_O^R}{R^2} \tag{3.62}$$

and the couple vector from $\mathbf{M}_1 = p\mathbf{R}$.
   c. Express that the moment about O of the wrench is equal to the moment resultant $\mathbf{M}_O^R$ of the force-couple system at O:

$$\mathbf{M}_1 + \mathbf{r} \times \mathbf{R} = \mathbf{M}_O^R \tag{3.63}$$

This equation allows you to determine the point where the line of action of the wrench intersects a specified plane, since the position vector **r** is directed from O to that point.

These steps are illustrated in Sample Prob. 3.12. Although the determination of a wrench and the point where its axis intersects a plane may appear difficult, the process is simply the application of several of the ideas and techniques developed in this chapter. Thus, once you have mastered the wrench, you can feel confident that you understand much of Chap. 3.

PROBLEMS

Fig. P3.101

3.101 A 3-m-long beam is subjected to a variety of loadings. (*a*) Replace each loading with an equivalent force-couple system at end *A* of the beam. (*b*) Which of the loadings are equivalent?

3.102 A 3-m-long beam is loaded as shown. Determine the loading of Prob. 3.101 that is equivalent to this loading.

3.103 Determine the single equivalent force and the distance from point *A* to its line of action for the beam and loading of (*a*) Prob. 3.101*a*, (*b*) Prob. 3.101*b*, (*c*) Prob. 3.102.

3.104 Five separate force-couple systems act at the corners of a piece of sheet metal, which has been bent into the shape shown. Determine which of these systems is equivalent to a force $\mathbf{F} = (10 \text{ lb})\mathbf{i}$ and a couple of moment $\mathbf{M} = (15 \text{ lb} \cdot \text{ft})\mathbf{j} + (15 \text{ lb} \cdot \text{ft})\mathbf{k}$ located at the origin.

Fig. P3.102

Fig. P3.104

Fig. P3.105

3.105 Three horizontal forces are applied as shown to a vertical cast-iron arm. Determine the resultant of the forces and the distance from the ground to its line of action when (*a*) *P* = 200 N, (*b*) *P* = 2400 N, (*c*) *P* = 1000 N.

3.106 Three stage lights are mounted on a pipe as shown. The lights at *A* and *B* each weigh 4.1 lb, while the one at *C* weighs 3.5 lb. (*a*) If *d* = 25 in., determine the distance from *D* to the line of action of the resultant of the weights of the three lights. (*b*) Determine the value of *d* so that the resultant of the weights passes through the midpoint of the pipe.

Fig. P3.106

3.107 The weights of two children sitting at ends *A* and *B* of a seesaw are 84 lb and 64 lb, respectively. Where should a third child sit so that the resultant of the weights of the three children will pass through *C* if she weighs (*a*) 60 lb, (*b*) 52 lb?

3.108 A couple of magnitude *M* = 54 lb · in. and the three forces shown are applied to an angle bracket. (*a*) Find the resultant of this system of forces. (*b*) Locate the points where the line of action of the resultant intersects line *AB* and line *BC*.

Fig. P3.107

Fig. P3.108 and P3.109

3.109 A couple **M** and the three forces shown are applied to an angle bracket. Find the moment of the couple if the line of action of the resultant of the force system is to pass through (*a*) point *A*, (*b*) point *B*, (*c*) point *C*.

3.110 A 32-lb motor is mounted on the floor. Find the resultant of the weight and the forces exerted on the belt, and determine where the line of action of the resultant intersects the floor.

3.111 A machine component is subjected to the forces and couples shown. The component is to be held in place by a single rivet that can resist a force but not a couple. For $P = 0$, determine the location of the rivet hole if it is to be located (*a*) on line *FG*, (*b*) on line *GH*.

Fig. P3.110

Fig. P3.111

3.112 Solve Prob. 3.111, assuming that $P = 60$ N.

3.113 A truss supports the loading shown. Determine the equivalent force acting on the truss and the point of intersection of its line of action with a line drawn through points *A* and *G*.

3.114 Four ropes are attached to a crate and exert the forces shown. If the forces are to be replaced with a single equivalent force applied at a point on line *AB*, determine (*a*) the equivalent force and the distance from *A* to the point of application of the force when $\alpha = 30°$, (*b*) the value of α so that the single equivalent force is applied at point *B*.

Fig. P3.113

Fig. P3.114

3.115 Solve Prob. 3.114, assuming that the 90-lb force is removed.

3.116 Four forces act on a 700 × 375-mm plate as shown. (*a*) Find the resultant of these forces. (*b*) Locate the two points where the line of action of the resultant intersects the edge of the plate.

3.117 Solve Prob. 3.116, assuming that the 760-N force is directed to the right.

Fig. P3.116

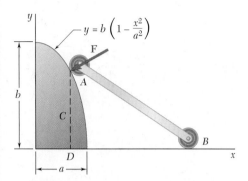

Fig. P3.118

3.118 As follower AB rolls along the surface of member C, it exerts a constant force **F** perpendicular to the surface. (*a*) Replace **F** with an equivalent force-couple system at the point D obtained by drawing the perpendicular from the point of contact to the x axis. (*b*) For $a = 1$ m and $b = 2$ m, determine the value of x for which the moment of the equivalent force-couple system at D is maximum.

3.119 As plastic bushings are inserted into a 60-mm-diameter cylindrical sheet metal enclosure, the insertion tools exert the forces shown on the enclosure. Each of the forces is parallel to one of the coordinate axes. Replace these forces with an equivalent force-couple system at C.

3.120 Two 150-mm-diameter pulleys are mounted on line shaft AD. The belts at B and C lie in vertical planes parallel to the yz plane. Replace the belt forces shown with an equivalent force-couple system at A.

Fig. P3.119

Fig. P3.120

3.121 Four forces are applied to the machine component $ABDE$ as shown. Replace these forces with an equivalent force-couple system at A.

Fig. P3.121

3.122 While using a pencil sharpener, a student applies the forces and couple shown. (*a*) Determine the forces exerted at *B* and *C* knowing that these forces and the couple are equivalent to a force-couple system at *A* consisting of the force $\mathbf{R} = (2.6\text{ lb})\mathbf{i} + R_y\mathbf{j} - (0.7\text{ lb})\mathbf{k}$ and the couple $\mathbf{M}_A^R = M_x\mathbf{i} + (1.0\text{ lb}\cdot\text{ft})\mathbf{j} - (0.72\text{ lb}\cdot\text{ft})\mathbf{k}$. (*b*) Find the corresponding values of R_y and M_x.

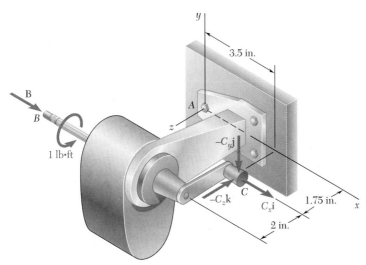

Fig. P3.122

3.123 A blade held in a brace is used to tighten a screw at *A*. (*a*) Determine the forces exerted at *B* and *C*, knowing that these forces are equivalent to a force-couple system at *A* consisting of $\mathbf{R} = -(30\text{ N})\mathbf{i} + R_y\mathbf{j} + R_z\mathbf{k}$ and $\mathbf{M}_A^R = -(12\text{ N}\cdot\text{m})\mathbf{i}$. (*b*) Find the corresponding values of R_y and R_z. (*c*) What is the orientation of the slot in the head of the screw for which the blade is least likely to slip when the brace is in the position shown?

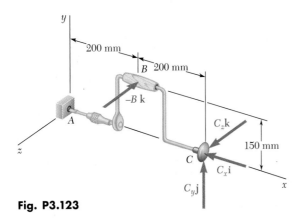

Fig. P3.123

3.124 In order to unscrew the tapped faucet A, a plumber uses two pipe wrenches as shown. By exerting a 40-lb force on each wrench, at a distance of 10 in. from the axis of the pipe and in a direction perpendicular to the pipe and to the wrench, he prevents the pipe from rotating, and thus avoids loosening or further tightening the joint between the pipe and the tapped elbow C. Determine (a) the angle θ that the wrench at A should form with the vertical if elbow C is not to rotate about the vertical, (b) the force-couple system at C equivalent to the two 40-lb forces when this condition is satisfied.

Fig. P3.124

3.125 Assuming $\theta = 60°$ in Prob. 3.124, replace the two 40-lb forces with an equivalent force-couple system at D and determine whether the plumber's action tends to tighten or loosen the joint between (a) pipe CD and elbow D, (b) elbow D and pipe DE. Assume all threads to be right-handed.

3.126 As an adjustable brace BC is used to bring a wall into plumb, the force-couple system shown is exerted on the wall. Replace this force-couple system with an equivalent force-couple system at A if $R = 21.2$ lb and $M = 13.25$ lb · ft.

3.127 Three children are standing on a 5 × 5-m raft. If the weights of the children at points A, B, and C are 375 N, 260 N, and 400 N, respectively, determine the magnitude and the point of application of the resultant of the three weights.

3.128 Three children are standing on a 5 × 5-m raft. The weights of the children at points A, B, and C are 375 N, 260 N, and 400 N, respectively. If a fourth child of weight 425 N climbs onto the raft, determine where she should stand if the other children remain in the positions shown and the line of action of the resultant of the four weights is to pass through the center of the raft.

Fig. P3.126

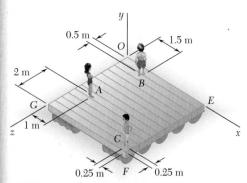

Fig. P3.127 and P3.128

3.129 Four signs are mounted on a frame spanning a highway, and the magnitudes of the horizontal wind forces acting on the signs are as shown. Determine the magnitude and the point of application of the resultant of the four wind forces when $a = 1$ ft and $b = 12$ ft.

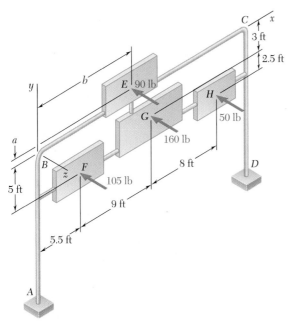

Fig. P3.129 and P3.130

3.130 Four signs are mounted on a frame spanning a highway, and the magnitudes of the horizontal wind forces acting on the signs are as shown. Determine a and b so that the point of application of the resultant of the four forces is at G.

*3.131 A group of students loads a 2×3.3-m flatbed trailer with two $0.66 \times 0.66 \times 0.66$-m boxes and one $0.66 \times 0.66 \times 1.2$-m box. Each of the boxes at the rear of the trailer is positioned so that it is aligned with both the back and a side of the trailer. Determine the smallest load the students should place in a second $0.66 \times 0.66 \times 1.2$-m box and where on the trailer they should secure it, without any part of the box overhanging the sides of the trailer, if each box is uniformly loaded and the line of action of the resultant of the weights of the four boxes is to pass through the point of intersection of the centerlines of the trailer and the axle. (*Hint:* Keep in mind that the box may be placed either on its side or on its end.)

*3.132 Solve Prob. 3.131 if the students want to place as much weight as possible in the fourth box and at least one side of the box must coincide with a side of the trailer.

*3.133 A piece of sheet metal is bent into the shape shown and is acted upon by three forces. If the forces have the same magnitude P, replace them with an equivalent wrench and determine (*a*) the magnitude and the direction of the resultant force **R**, (*b*) the pitch of the wrench, (*c*) the axis of the wrench.

Fig. P3.131

Fig. P3.133

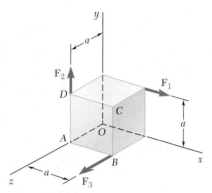

Fig. P3.134

*3.134 Three forces of the same magnitude P act on a cube of side a as shown. Replace the three forces with an equivalent wrench and determine (a) the magnitude and direction of the resultant force $\mathbf{R}$, (b) the pitch of the wrench, (c) the axis of the wrench.

*3.135 and *3.136 The forces and couples shown are applied to two screws as a piece of sheet metal is fastened to a block of wood. Reduce the forces and the couples to an equivalent wrench and determine (a) the resultant force $\mathbf{R}$, (b) the pitch of the wrench, (c) the point where the axis of the wrench intersects the xz plane.

Fig. P3.135

Fig. P3.136

*3.137 and *3.138 Two bolts at A and B are tightened by applying the forces and couples shown. Replace the two wrenches with a single equivalent wrench and determine (a) the resultant $\mathbf{R}$, (b) the pitch of the single equivalent wrench, (c) the point where the axis of the wrench intersects the xz plane.

Fig. P3.137

Fig. P3.138

*3.139 A flagpole is guyed by three cables. If the tensions in the cables have the same magnitude P, replace the forces exerted on the pole with an equivalent wrench and determine (a) the resultant force **R**, (b) the pitch of the wrench, (c) the point where the axis of the wrench intersects the xz plane.

*3.140 Two ropes attached at A and B are used to move the trunk of a fallen tree. Replace the forces exerted by the ropes with an equivalent wrench and determine (a) the resultant force **R**, (b) the pitch of the wrench, (c) the point where the axis of the wrench intersects the yz plane.

Fig. P3.139

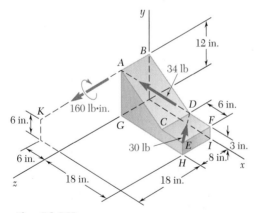

Fig. P3.140

*3.141 and *3.142 Determine whether the force-and-couple system shown can be reduced to a single equivalent force **R**. If it can, determine **R** and the point where the line of action of **R** intersects the yz plane. If it cannot be so reduced, replace the given system with an equivalent wrench and determine its resultant, its pitch, and the point where its axis intersects the yz plane.

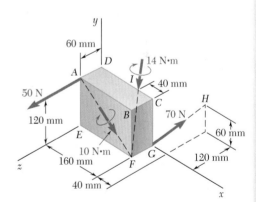

Fig. P3.141

Fig. P3.142

*3.143 Replace the wrench shown with an equivalent system consisting of two forces perpendicular to the y axis and applied respectively at A and B.

*3.144 Show that, in general, a wrench can be replaced with two forces chosen in such a way that one force passes through a given point while the other force lies in a given plane.

*3.145 Show that a wrench can be replaced with two perpendicular forces, one of which is applied at a given point.

*3.146 Show that a wrench can be replaced with two forces, one of which has a prescribed line of action.

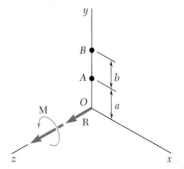

Fig. P3.143

REVIEW AND SUMMARY

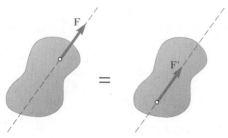

Fig. 3.48

In this chapter we studied the effect of forces exerted on a rigid body. We first learned to distinguish between *external* and *internal* forces [Sec. 3.2] and saw that, according to the *principle of transmissibility,* the effect of an external force on a rigid body remains unchanged if that force is moved along its line of action [Sec. 3.3]. In other words, two forces **F** and **F′** acting on a rigid body at two different points have the same effect on that body if they have the same magnitude, same direction, and same one of action (Fig. 3.48). Two such forces are said to be *equivalent.*

Before proceeding with the discussion of *equivalent systems of forces,* we introduced the concept of the *vector product of two vectors* [Sec. 3.4]. The vector product

$$\mathbf{V} = \mathbf{P} \times \mathbf{Q}$$

of the vectors **P** and **Q** was defined as a vector perpendicular to the plane containing **P** and **Q** (Fig. 3.49), of magnitude

$$V = PQ \sin \theta \tag{3.1}$$

Vector product of two vectors

(a)

(b)

Fig. 3.49

Fig. 3.50

and directed in such a way that a person located at the tip of **V** will observe as counterclockwise the rotation through θ which brings the vector **P** in line with the vector **Q**. The three vectors **P**, **Q**, and **V**—taken in that order—are said to form a *right-handed triad.* It follows that the vector products **Q** × **P** and **P** × **Q** are represented by equal and opposite vectors. We have

$$\mathbf{Q} \times \mathbf{P} = -(\mathbf{P} \times \mathbf{Q}) \tag{3.4}$$

It also follows from the definition of the vector product of two vectors that the vector products of the unit vectors **i**, **j**, and **k** are

$$\mathbf{i} \times \mathbf{i} = 0 \qquad \mathbf{i} \times \mathbf{j} = \mathbf{k} \qquad \mathbf{j} \times \mathbf{i} = -\mathbf{k}$$

and so on. The sign of the vector product of two unit vectors can be obtained by arranging in a circle and in counterclockwise order the three letters representing the unit vectors (Fig. 3.50): The vector product of two unit vectors will be positive if they follow each other in counterclockwise order and negative if they follow each other in clockwise order.

The *rectangular components of the vector product* **V** of two vectors **P** and **Q** were expressed [Sec. 3.5] as

Rectangular components of vector product

$$\begin{aligned} V_x &= P_y Q_z - P_z Q_y \\ V_y &= P_z Q_x - P_x Q_z \\ V_z &= P_x Q_y - P_y Q_x \end{aligned} \tag{3.9}$$

Using a determinant, we also wrote

$$\mathbf{V} = \begin{vmatrix} \mathbf{i} & \mathbf{j} & \mathbf{k} \\ P_x & P_y & P_z \\ Q_x & Q_y & Q_z \end{vmatrix} \quad (3.10)$$

The *moment of a force* **F** *about a point* O was defined [Sec. 3.6] as the vector product

$$\mathbf{M}_O = \mathbf{r} \times \mathbf{F} \quad (3.11)$$

where **r** is the *position vector* drawn from O to the point of application A of the force **F** (Fig. 3.51). Denoting by θ the angle between the lines of action of **r** and **F**, we found that the magnitude of the moment of **F** about O can be expressed as

$$M_O = rF \sin \theta = Fd \quad (3.12)$$

where d represents the perpendicular distance from O to the line of action of **F**.

The *rectangular components of the moment* $\mathbf{M}_O$ of a force **F** were expressed [Sec. 3.8] as

$$\begin{aligned} M_x &= yF_z - zF_y \\ M_y &= zF_x - xF_z \\ M_z &= xF_y - yF_x \end{aligned} \quad (3.18)$$

where x, y, z are the components of the position vector **r** (Fig. 3.52). Using a determinant form, we also wrote

$$\mathbf{M}_O = \begin{vmatrix} \mathbf{i} & \mathbf{j} & \mathbf{k} \\ x & y & z \\ F_x & F_y & F_z \end{vmatrix} \quad (3.19)$$

In the more general case of the moment about an arbitrary point B of a force **F** applied at A, we had

$$\mathbf{M}_B = \begin{vmatrix} \mathbf{i} & \mathbf{j} & \mathbf{k} \\ x_{A/B} & y_{A/B} & z_{A/B} \\ F_x & F_y & F_z \end{vmatrix} \quad (3.21)$$

where $x_{A/B}$, $y_{A/B}$, and $z_{A/B}$ denote the components of the vector $\mathbf{r}_{A/B}$:

$$x_{A/B} = x_A - x_B \qquad y_{A/B} = y_A - y_B \qquad z_{A/B} = z_A - z_B$$

In the case of *problems involving only two dimensions*, the force **F** can be assumed to lie in the xy plane. Its moment $\mathbf{M}_B$ about a point B in the same plane is perpendicular to that plane (Fig. 3.53) and is completely defined by the scalar

$$M_B = (x_A - x_B)F_y - (y_A - y_B)F_x \quad (3.23)$$

Various methods for the computation of the moment of a force about a point were illustrated in Sample Probs. 3.1 through 3.4.

The *scalar product* of two vectors **P** and **Q** [Sec. 3.9] was denoted by $\mathbf{P} \cdot \mathbf{Q}$ and was defined as the scalar quantity

$$\mathbf{P} \cdot \mathbf{Q} = PQ \cos \theta \quad (3.24)$$

Moment of a force about a point

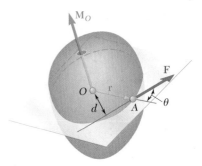

Fig. 3.51

Rectangular components of moment

Fig. 3.52

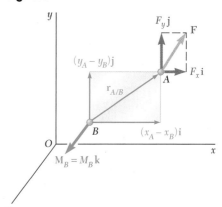

Fig. 3.53

Scalar product of two vectors

Fig. 3.54

Projection of a vector on an axis

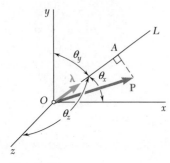

Fig. 3.55

Mixed triple product of three vectors

Moment of a force about an axis

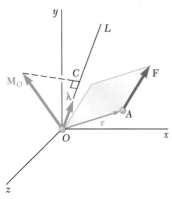

Fig. 3.56

where θ is the angle between $\mathbf{P}$ and $\mathbf{Q}$ (Fig. 3.54). By expressing the scalar product of $\mathbf{P}$ and $\mathbf{Q}$ in terms of the rectangular components of the two vectors, we determined that

$$\mathbf{P} \cdot \mathbf{Q} = P_x Q_x + P_y Q_y + P_z Q_z \tag{3.30}$$

The *projection of a vector* $\mathbf{P}$ *on an axis* OL (Fig. 3.55) can be obtained by forming the scalar product of $\mathbf{P}$ and the unit vector $\boldsymbol{\lambda}$ along OL. We have

$$P_{OL} = \mathbf{P} \cdot \boldsymbol{\lambda} \tag{3.36}$$

or, using rectangular components,

$$P_{OL} = P_x \cos \theta_x + P_y \cos \theta_y + P_z \cos \theta_z \tag{3.37}$$

where θ_x, θ_y, and θ_z denote the angles that the axis OL forms with the coordinate axes.

The *mixed triple product* of the three vectors $\mathbf{S}$, $\mathbf{P}$, and $\mathbf{Q}$ was defined as the scalar expression

$$\mathbf{S} \cdot (\mathbf{P} \times \mathbf{Q}) \tag{3.38}$$

obtained by forming the scalar product of $\mathbf{S}$ with the vector product of $\mathbf{P}$ and $\mathbf{Q}$ [Sec. 3.10]. It was shown that

$$\mathbf{S} \cdot (\mathbf{P} \times \mathbf{Q}) = \begin{vmatrix} S_x & S_y & S_z \\ P_x & P_y & P_z \\ Q_x & Q_y & Q_z \end{vmatrix} \tag{3.41}$$

where the elements of the determinant are the rectangular components of the three vectors.

The *moment of a force* $\mathbf{F}$ *about an axis* OL [Sec. 3.11] was defined as the projection OC on OL of the moment $\mathbf{M}_O$ of the force $\mathbf{F}$ (Fig. 3.56), i.e., as the mixed triple product of the unit vector $\boldsymbol{\lambda}$, the position vector $\mathbf{r}$, and the force $\mathbf{F}$:

$$M_{OL} = \boldsymbol{\lambda} \cdot \mathbf{M}_O = \boldsymbol{\lambda} \cdot (\mathbf{r} \times \mathbf{F}) \tag{3.42}$$

Using the determinant form for the mixed triple product, we have

$$M_{OL} = \begin{vmatrix} \lambda_x & \lambda_y & \lambda_z \\ x & y & z \\ F_x & F_y & F_z \end{vmatrix} \tag{3.43}$$

where $\lambda_x, \lambda_y, \lambda_z$ = direction cosines of axis OL
x, y, z = components of $\mathbf{r}$
F_x, F_y, F_z = components of $\mathbf{F}$

An example of the determination of the moment of a force about a skew axis was given in Sample Prob. 3.5.

*Two forces **F** and −**F** having the same magnitude, parallel lines of action, and opposite sense are said to form a couple* [Sec. 3.12]. It was shown that the moment of a couple is independent of the point about which it is computed; it is a vector **M** perpendicular to the plane of the couple and equal in magnitude to the product of the common magnitude F of the forces and the perpendicular distance d between their lines of action (Fig. 3.57).

Fig. 3.57

Two couples having the same moment **M** are *equivalent*, i.e., they have the same effect on a given rigid body [Sec. 3.13]. The sum of two couples is itself a couple [Sec. 3.14], and the moment **M** of the resultant couple can be obtained by adding vectorially the moments **M**$_1$ and **M**$_2$ of the original couples [Sample Prob. 3.6]. It follows that a couple can be represented by a vector, called a *couple vector*, equal in magnitude and direction to the moment **M** of the couple [Sec. 3.15]. A couple vector is a *free vector* which can be attached to the origin O if so desired and resolved into components (Fig. 3.58).

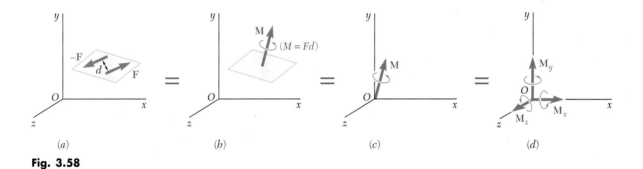

Fig. 3.58

Any force **F** acting at a point A of a rigid body can be replaced by a *force-couple system* at an arbitrary point O, consisting of the force **F** applied at O and a couple of moment **M**$_O$ equal to the moment about O of the force **F** in its original position [Sec. 3.16]; it should be noted that the force **F** and the couple vector **M**$_O$ are always perpendicular to each other (Fig. 3.59).

Force-couple system

Fig. 3.59

It follows [Sec. 3.17] that *any system of forces can be reduced to a force-couple system at a given point O* by first replacing each of the forces of the system by an equivalent force-couple system at O

Reduction of a system of forces to a force-couple system

(Fig. 3.60) and then adding all the forces and all the couples determined in this manner to obtain a resultant force **R** and a resultant couple vector $\mathbf{M}_O^R$ [Sample Probs. 3.8 through 3.11]. Note that, in general, the resultant **R** and the couple vector $\mathbf{M}_O^R$ will not be perpendicular to each other.

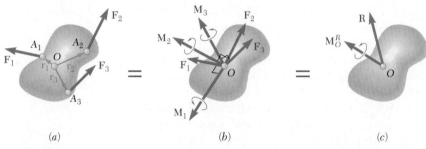

(a)   (b)   (c)

Fig. 3.60

Equivalent systems of forces

We concluded from the above [Sec. 3.18] that, as far as rigid bodies are concerned, *two systems of forces, $\mathbf{F}_1$, $\mathbf{F}_2$, $\mathbf{F}_3$, . . . and $\mathbf{F}_1'$, $\mathbf{F}_2'$, $\mathbf{F}_3'$, . . . , are equivalent if, and only if,*

$$\Sigma\mathbf{F} = \Sigma\mathbf{F}' \qquad \text{and} \qquad \Sigma\mathbf{M}_O = \Sigma\mathbf{M}_O' \qquad (3.57)$$

Further reduction of a system of forces

If the resultant force **R** and the resultant couple vector $\mathbf{M}_O^R$ are perpendicular to each other, the force-couple system at O can be further reduced to a single resultant force [Sec. 3.20]. This will be the case for systems consisting either of (a) concurrent forces (cf. Chap. 2), (b) coplanar forces [Sample Probs. 3.8 and 3.9], or (c) parallel forces [Sample Prob. 3.11]. If the resultant **R** and the couple vector $\mathbf{M}_O^R$ are *not* perpendicular to each other, the system *cannot* be reduced to a single force. It can, however, be reduced to a special type of force-couple system called a *wrench*, consisting of the resultant **R** and a couple vector $\mathbf{M}_1$ directed along **R** [Sec. 3.21 and Sample Prob. 3.12].

REVIEW PROBLEMS

3.147 A 300-N force is applied at A as shown. Determine (a) the moment of the 300-N force about D, (b) the smallest force applied at B that creates the same moment about D.

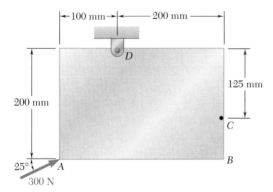

Fig. P3.147

3.148 The tailgate of a car is supported by the hydraulic lift BC. If the lift exerts a 125-lb force directed along its centerline on the ball and socket at B, determine the moment of the force about A.

3.149 The ramp $ABCD$ is supported by cables at corners C and D. The tension in each of the cables is 810 N. Determine the moment about A of the force exerted by (a) the cable at D, (b) the cable at C.

Fig. P3.148

Fig. P3.149

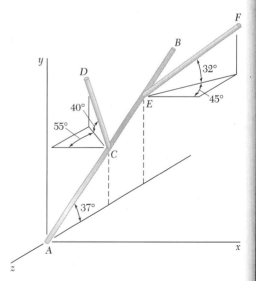

Fig. P3.150

3.150 Section AB of a pipeline lies in the yz plane and forms an angle of $37°$ with the z axis. Branch lines CD and EF join AB as shown. Determine the angle formed by pipes AB and CD.

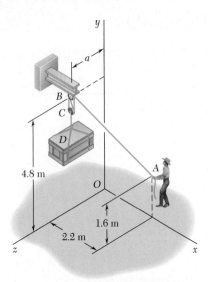

Fig. P3.151

3.151 To lift a heavy crate, a man uses a block and tackle attached to the bottom of an I-beam at hook B. Knowing that the moments about the y and the z axes of the force exerted at B by portion AB of the rope are, respectively, 120 N · m and –460 N · m, determine the distance a.

3.152 To loosen a frozen valve, a force $\mathbf{F}$ of magnitude 70 lb is applied to the handle of the valve. Knowing that $\theta = 25°$, $M_x = -61$ lb · ft, and $M_z = -43$ lb · ft, determine ϕ and d.

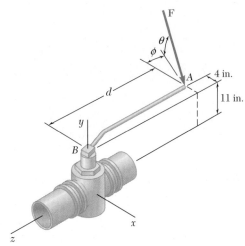

Fig. P3.152

3.153 The tension in the cable attached to the end C of an adjustable boom ABC is 560 lb. Replace the force exerted by the cable at C with an equivalent force-couple system (a) at A, (b) at B.

Fig. P3.153

Fig. P3.154

3.154 While tapping a hole, a machinist applies the horizontal forces shown to the handle of the tap wrench. Show that these forces are equivalent to a single force, and specify, if possible, the point of application of the single force on the handle.

3.155 Replace the 150-N force with an equivalent force-couple system at A.

3.156 A beam supports three loads of given magnitude and a fourth load whose magnitude is a function of position. If $b = 1.5$ m and the loads are to be replaced with a single equivalent force, determine (a) the value of a so that the distance from support A to the line of action of the equivalent force is maximum, (b) the magnitude of the equivalent force and its point of application on the beam.

Fig. P3.155

Fig. P3.156

3.157 A mechanic uses a crowfoot wrench to loosen a bolt at C. The mechanic holds the socket wrench handle at points A and B and applies forces at these points. Knowing that these forces are equivalent to a force-couple system at C consisting of the force $\mathbf{C} = -(8 \text{ lb})\mathbf{i} + (4 \text{ lb})\mathbf{k}$ and the couple $\mathbf{M}_C = (360 \text{ lb} \cdot \text{in.})\mathbf{i}$, determine the forces applied at A and at B when $A_z = 2$ lb.

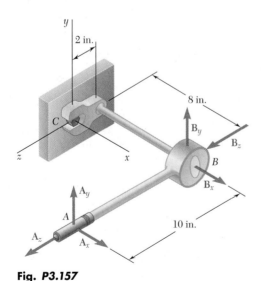

Fig. P3.157

3.158 A concrete foundation mat in the shape of a regular hexagon of side 12 ft supports four column loads as shown. Determine the magnitudes of the additional loads that must be applied at B and F if the resultant of all six loads is to pass through the center of the mat.

Fig. P3.158

COMPUTER PROBLEMS

Fig. P3.C1

3.C1 A beam AB is subjected to several vertical forces as shown. Write a computer program that can be used to determine the magnitude of the resultant of the forces and the distance x_C to point C, the point where the line of action of the resultant intersects AB. Use this program to solve (*a*) Sample Prob. 3.8*c*, (*b*) Prob. 3.106*a*.

3.C2 Write a computer program that can be used to determine the magnitude and the point of application of the resultant of the vertical forces $\mathbf{P}_1$, $\mathbf{P}_2$, . . . , $\mathbf{P}_n$ that act at points $\mathbf{A}_1$, $\mathbf{A}_2$, . . . , $\mathbf{A}_n$ that are located in the xz plane. Use this program to solve (*a*) Sample Prob. 3.11, (*b*) Prob. 3.127, (*c*) Prob. 3.129.

Fig. P3.C2

3.C3 A friend asks for your help in designing flower planter boxes. The boxes are to have 4, 5, 6, or 8 sides, which are to tilt outward at 10°, 20°, or 30°. Write a computer program that can be used to determine the bevel angle α for each of the 12 planter designs. (*Hint:* The bevel angle is equal to one-half of the angle formed by the inward normals of two adjacent sides.)

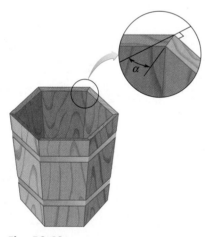

Fig. P3.C3

3.C4 The manufacturer of a spool for hoses wants to determine the moment of the force $\mathbf{F}$ about the axis AA'. The magnitude of the force, in newtons, is defined by the relation $F = 300(1 - x/L)$, where x is the length of hose wound on the 0.6-m-diameter drum and L is the total length of the hose. Write a computer program that can be used to calculate the required moment for a hose 30 m long and 50 mm in diameter. Beginning with $x = 0$, compute the moment after every revolution of the drum until the hose is wound on the drum.

Fig. P3.C4

3.C5 A body is acted upon by a system of n forces. Write a computer program that can be used to calculate the equivalent force-couple system at the origin of the coordinate axes and to determine, if the equivalent force and the equivalent couple are orthogonal, the magnitude and the point of application in the xz plane of the resultant of the original force system. Use this program to solve (*a*) Prob. 3.113, (*b*) Prob. 3.120, (*c*) Prob. 3.127.

3.C6 Two cylindrical ducts, *AB* and *CD*, enter a room through two parallel walls. The centerlines of the ducts are parallel to each other but are not perpendicular to the walls. The ducts are to be connected by two flexible elbows and a straight center portion. Write a computer program that can be used to determine the lengths of *AB* and *CD* that minimize the distance between the axis of the straight portion and a thermometer mounted on the wall at *E*. Assume that the elbows are of negligible length and that *AB* and *CD* have centerlines defined by $\lambda_{AB} = (7\mathbf{i} - 4\mathbf{j} + 4\mathbf{k})/9$ and $\lambda_{CD} = (-7\mathbf{i} + 4\mathbf{j} - 4\mathbf{k})/9$ and can vary in length from 9 in. to 36 in.

Fig. P3.C5

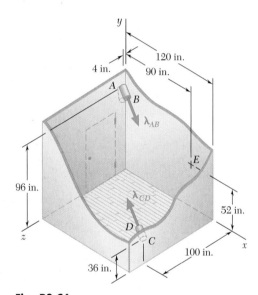

Fig. P3.C6

This telecommunications tower, constructed in the heart of the Barcelona Olympic complex to broadcast the 1992 games, was designed to remain in equilibrium under the vertical force of gravity and the lateral forces exerted by wind.

Equilibrium of Rigid Bodies

4.1 INTRODUCTION

We saw in the preceding chapter that the external forces acting on a rigid body can be reduced to a force-couple system at some arbitrary point O. When the force and the couple are both equal to zero, the external forces form a system equivalent to zero, and the rigid body is said to be in *equilibrium.*

The necessary and sufficient conditions for the equilibrium of a rigid body, therefore, can be obtained by setting $\mathbf{R}$ and $\mathbf{M}_O^R$ equal to zero in the relations (3.52) of Sec. 3.17:

$$\Sigma \mathbf{F} = 0 \qquad \Sigma \mathbf{M}_O = \Sigma(\mathbf{r} \times \mathbf{F}) = 0 \tag{4.1}$$

Resolving each force and each moment into its rectangular components, we can express the necessary and sufficient conditions for the equilibrium of a rigid body with the following six scalar equations:

$$\Sigma F_x = 0 \qquad \Sigma F_y = 0 \qquad \Sigma F_z = 0 \tag{4.2}$$
$$\Sigma M_x = 0 \qquad \Sigma M_y = 0 \qquad \Sigma M_z = 0 \tag{4.3}$$

The equations obtained can be used to determine unknown forces applied to the rigid body or unknown reactions exerted on it by its supports. We note that Eqs. (4.2) express the fact that the components of the external forces in the x, y, and z directions are balanced; Eqs. (4.3) express the fact that the moments of the external forces about the x, y, and z axes are balanced. Therefore, for a rigid body in equilibrium, the system of the external forces will impart no translational or rotational motion to the body considered.

In order to write the equations of equilibrium for a rigid body, it is essential to first identify all of the forces acting on that body and then to draw the corresponding *free-body diagram.* In this chapter we first consider the equilibrium of *two-dimensional structures* subjected to forces contained in their planes and learn how to draw their free-body diagrams. In addition to the forces *applied* to a structure, the *reactions* exerted on the structure by its supports will be considered. A specific reaction will be associated with each type of support. You will learn how to determine whether the structure is properly supported, so that you can know in advance whether the equations of equilibrium can be solved for the unknown forces and reactions.

Later in the chapter, the equilibrium of three-dimensional structures will be considered, and the same kind of analysis will be given to these structures and their supports.

4.2 FREE-BODY DIAGRAM

In solving a problem concerning the equilibrium of a rigid body, it is essential to consider *all* of the forces acting on the body; it is equally important to exclude any force which is not directly applied to the body. Omitting a force or adding an extraneous one would destroy the conditions of equilibrium. Therefore, the first step in the solution of the problem should be to draw a *free-body diagram* of the rigid body under consideration. Free-body diagrams have already been used on many occasions in Chap. 2. However, in view of their importance to the solution of equilibrium problems, we summarize here the various steps which must be followed in drawing a free-body diagram.

1. A clear decision should be made regarding the choice of the free body to be used. This body is then detached from the ground and is separated from all other bodies. The contour of the body thus isolated is sketched.
2. All external forces should be indicated on the free-body diagram. These forces represent the actions exerted *on* the free body *by* the ground and *by* the bodies which have been detached; they should be applied at the various points where the free body was supported by the ground or was connected to the other bodies. The *weight* of the free body should also be included among the external forces, since it represents the attraction exerted by the earth on the various particles forming the free body. As will be seen in Chap. 5, the weight should be applied at the center of gravity of the body. When the free body is made of several parts, the forces the various parts exert on each other should *not* be included among the external forces. These forces are internal forces as far as the free body is concerned.
3. The magnitudes and directions of the *known external forces* should be clearly marked on the free-body diagram. When indicating the directions of these forces, it must be remembered that the forces shown on the free-body diagram must be those which are exerted *on*, and not *by*, the free body. Known external forces generally include the *weight* of the free body and *forces applied* for a given purpose.
4. *Unknown external forces* usually consist of the *reactions*, through which the ground and other bodies oppose a possible motion of the free body. The reactions constrain the free body to remain in the same position, and, for that reason, are sometimes called *constraining forces*. Reactions are exerted at the points where the free body is *supported by* or *connected to* other bodies and should be clearly indicated. Reactions are discussed in detail in Secs. 4.3 and 4.8.
5. The free-body diagram should also include dimensions, since these may be needed in the computation of moments of forces. Any other detail, however, should be omitted.

Photo 4.1 A free-body diagram of the tractor shown would include all of the external forces acting on the tractor: the weight of the tractor, the weight of the load in the bucket, and the forces exerted by the ground on the tires.

Photo 4.2 In Chap. 6, we will discuss how to determine the internal forces in structures made of several connected pieces, such as the forces in the members that support the bucket of the tractor of Photo 4.1.

Photo 4.3 As the link of the awning window opening mechanism is extended, the force it exerts on the slider results in a normal force being applied to the rod, which causes the window to open.

Photo 4.4 The abutment-mounted rocker bearing shown is used to support the roadway of a bridge.

Photo 4.5 Shown is the rocker expansion bearing of a plate girder bridge. The convex surface of the rocker allows the support of the girder to move horizontally.

EQUILIBRIUM IN TWO DIMENSIONS

4.3 REACTIONS AT SUPPORTS AND CONNECTIONS FOR A TWO-DIMENSIONAL STRUCTURE

In the first part of this chapter, the equilibrium of a two-dimensional structure is considered; i.e., it is assumed that the structure being analyzed and the forces applied to it are contained in the same plane. Clearly, the reactions needed to maintain the structure in the same position will also be contained in this plane.

The reactions exerted on a two-dimensional structure can be divided into three groups corresponding to three types of *supports,* or *connections:*

1. *Reactions Equivalent to a Force with Known Line of Action.* Supports and connections causing reactions of this type include *rollers, rockers, frictionless surfaces, short links and cables, collars on frictionless rods,* and *frictionless pins in slots.* Each of these supports and connections can prevent motion in one direction only. They are shown in Fig. 4.1, together with the reactions they produce. Each of these reactions involves *one unknown,* namely, the magnitude of the reaction; this magnitude should be denoted by an appropriate letter. The line of action of the reaction is known and should be indicated clearly in the free-body diagram. The sense of the reaction must be as shown in Fig. 4.1 for the cases of a frictionless surface (toward the free body) or a cable (away from the free body). The reaction can be directed either way in the case of double-track rollers, links, collars on rods, and pins in slots. Single-track rollers and rockers are generally assumed to be reversible, and thus the corresponding reactions can also be directed either way.

2. *Reactions Equivalent to a Force of Unknown Direction and Magnitude.* Supports and connections causing reactions of this type include *frictionless pins in fitted holes, hinges,* and *rough surfaces.* They can prevent translation of the free body in all directions, but they cannot prevent the body from rotating about the connection. Reactions of this group involve *two unknowns* and are usually represented by their x and y components. In the case of a rough surface, the component normal to the surface must be directed away from the surface.

3. *Reactions Equivalent to a Force and a Couple.* These reactions are caused by *fixed supports,* which oppose any motion of the free body and thus constrain it completely. Fixed supports actually produce forces over the entire surface of contact; these forces, however, form a system which can be reduced to a force and a couple. Reactions of this group involve *three unknowns,* consisting usually of the two components of the force and the moment of the couple.

Support or Connection	Reaction	Number of Unknowns
Rollers Rocker Frictionless surface	Force with known line of action	1
Short cable Short link	Force with known line of action	1
Collar on frictionless rod Frictionless pin in slot	$90°$ Force with known line of action	1
Frictionless pin or hinge Rough surface	or α Force of unknown direction	2
Fixed support	or α Force and couple	3

Fig. 4.1 Reactions at supports and connections.

When the sense of an unknown force or couple is not readily apparent, no attempt should be made to determine it. Instead, the sense of the force or couple should be arbitrarily assumed; the sign of the answer obtained will indicate whether the assumption is correct or not.

4.4   EQUILIBRIUM OF A RIGID BODY IN TWO DIMENSIONS

The conditions stated in Sec. 4.1 for the equilibrium of a rigid body become considerably simpler for the case of a two-dimensional structure. Choosing the x and y axes to be in the plane of the structure, we have

$$F_z = 0 \qquad M_x = M_y = 0 \qquad M_z = M_O$$

for each of the forces applied to the structure. Thus, the six equations of equilibrium derived in Sec. 4.1 reduce to

$$\Sigma F_x = 0 \qquad \Sigma F_y = 0 \qquad \Sigma M_O = 0 \qquad (4.4)$$

and to three trivial identities, $0 = 0$. Since $\Sigma M_O = 0$ must be satisfied regardless of the choice of the origin O, we can write the equations of equilibrium for a two-dimensional structure in the more general form

$$\Sigma F_x = 0 \qquad \Sigma F_y = 0 \qquad \Sigma M_A = 0 \qquad (4.5)$$

where A is any point in the plane of the structure. The three equations obtained can be solved for no more than *three unknowns*.

We saw in the preceding section that unknown forces include reactions and that the number of unknowns corresponding to a given reaction depends upon the type of support or connection causing that reaction. Referring to Sec. 4.3, we observe that the equilibrium equations (4.5) can be used to determine the reactions associated with two rollers and one cable, one fixed support, or one roller and one pin in a fitted hole, etc.

Consider Fig. 4.2a, in which the truss shown is subjected to the given forces **P**, **Q**, and **S**. The truss is held in place by a pin at A and a roller at B. The pin prevents point A from moving by exerting on the truss a force which can be resolved into the components A_x and A_y; the roller keeps the truss from rotating about A by exerting the vertical force **B**. The free-body diagram of the truss is shown in Fig. 4.2b; it includes the reactions A_x, A_y, and **B** as well as the applied forces **P**, **Q**, **S** and the weight **W** of the truss. Expressing that the sum of the moments about A of all of the forces shown in Fig. 4.2b is zero, we write the equation $\Sigma M_A = 0$, which can be used to determine the magnitude B since it does not contain A_x or A_y. Next, expressing that the sum of the x components and the sum of the y components of the forces are zero, we write the equations $\Sigma F_x = 0$ and $\Sigma F_y = 0$, from which we can obtain the components A_x and A_y, respectively.

An additional equation could be obtained by expressing that the sum of the moments of the external forces about a point other than A is zero. We could write, for instance, $\Sigma M_B = 0$. Such a statement, however, does not contain any new information, since it has already been established that the system of the forces shown in Fig. 4.2b is equivalent to zero. The additional equation *is not independent* and cannot be used to determine a fourth unknown. It will be useful,

(a)

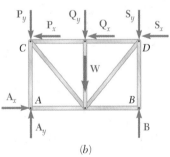

(b)

Fig. 4.2

however, for checking the solution obtained from the original three equations of equilibrium.

While the three equations of equilibrium cannot be *augmented* by additional equations, any of them can be *replaced* by another equation. Therefore, an alternative system of equations of equilibrium is

$$\Sigma F_x = 0 \qquad \Sigma M_A = 0 \qquad \Sigma M_B = 0 \qquad (4.6)$$

where the second point about which the moments are summed (in this case, point B) cannot lie on the line parallel to the y axis that passes through point A (Fig. 4.2b). These equations are sufficient conditions for the equilibrium of the truss. The first two equations indicate that the external forces must reduce to a single vertical force at A. Since the third equation requires that the moment of this force be zero about a point B which is not on its line of action, the force must be zero, and the rigid body is in equilibrium.

A third possible set of equations of equilibrium is

$$\Sigma M_A = 0 \qquad \Sigma M_B = 0 \qquad \Sigma M_C = 0 \qquad (4.7)$$

where the points A, B, and C do not lie in a straight line (Fig. 4.2b). The first equation requires that the external forces reduce to a single force at A; the second equation requires that this force pass through B; and the third equation requires that it pass through C. Since the points A, B, C do not lie in a straight line, the force must be zero, and the rigid body is in equilibrium.

The equation $\Sigma M_A = 0$, which expresses that the sum of the moments of the forces about pin A is zero, possesses a more definite physical meaning than either of the other two equations (4.7). These two equations express a similar idea of balance, but with respect to points about which the rigid body is not actually hinged. They are, however, as useful as the first equation, and our choice of equilibrium equations should not be unduly influenced by the physical meaning of these equations. Indeed, it will be desirable in practice to choose equations of equilibrium containing only one unknown, since this eliminates the necessity of solving simultaneous equations. Equations containing only one unknown can be obtained by summing moments about the point of intersection of the lines of action of two unknown forces or, if these forces are parallel, by summing components in a direction perpendicular to their common direction. For example, in Fig. 4.3, in which the truss shown is held by rollers at A and B and a short link at D, the reactions at A and B can be eliminated by summing x components. The reactions at A and D will be eliminated by summing moments about C, and the reactions at B and D by summing moments about D. The equations obtained are

$$\Sigma F_x = 0 \qquad \Sigma M_C = 0 \qquad \Sigma M_D = 0$$

Each of these equations contains only one unknown.

(a)

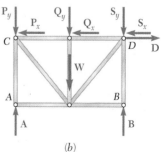

(b)

Fig. 4.3

4.5 STATICALLY INDETERMINATE REACTIONS. PARTIAL CONSTRAINTS

In the two examples considered in the preceding section (Figs. 4.2 and 4.3), the types of supports used were such that the rigid body could not possibly move under the given loads or under any other loading conditions. In such cases, the rigid body is said to be *completely constrained.* We also recall that the reactions corresponding to these supports involved *three unknowns* and could be determined by solving the three equations of equilibrium. When such a situation exists, the reactions are said to be *statically determinate.*

Consider Fig. 4.4*a*, in which the truss shown is held by pins at *A* and *B*. These supports provide more constraints than are necessary to keep the truss from moving under the given loads or under any other loading conditions. We also note from the free-body diagram of Fig. 4.4*b* that the corresponding reactions involve *four unknowns.* Since, as was pointed out in Sec. 4.4, only three independent equilibrium equations are available, there are *more unknowns than equations;* thus, all of the unknowns cannot be determined. While the equations $\Sigma M_A = 0$ and $\Sigma M_B = 0$ yield the vertical components B_y and A_y, respectively, the equation $\Sigma F_x = 0$ gives only the sum $A_x + B_x$ of the horizontal components of the reactions at *A* and *B*. The components A_x and B_x are said to be *statically indeterminate.* They could be determined by considering the deformations produced in the truss by the given loading, but this method is beyond the scope of statics and belongs to the study of mechanics of materials.

The supports used to hold the truss shown in Fig. 4.5*a* consist of rollers at *A* and *B*. Clearly, the constraints provided by these supports are not sufficient to keep the truss from moving. While any vertical motion is prevented, the truss is free to move horizontally. The truss is said to be *partially constrained.*[†] Turning our attention to Fig. 4.5*b*, we note that the reactions at *A* and *B* involve only *two unknowns.* Since three equations of equilibrium must still be satisfied, there are *fewer unknowns than equations,* and, in general, one of the equilibrium equations will not be satisfied. While the equations $\Sigma M_A = 0$ and $\Sigma M_B = 0$ can be satisfied by a proper choice of reactions at *A* and *B*, the equation $\Sigma F_x = 0$ will not be satisfied unless the sum of the horizontal components of the applied forces happens to be zero. We thus observe that the equilibrium of the truss of Fig. 4.5 cannot be maintained under general loading conditions.

It appears from the above that if a rigid body is to be completely constrained and if the reactions at its supports are to be statically determinate, *there must be as many unknowns as there are equations of equilibrium.* When this condition is *not* satisfied, we can be certain that either the rigid body is not completely constrained or that the reactions at its supports are not statically determinate; it is also possible that the rigid body is not completely constrained *and* that the reactions are statically indeterminate.

We should note, however, that, while *necessary,* the above condition is *not sufficient.* In other words, the fact that the number of

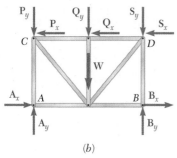

Fig. 4.4 Statically indeterminate reactions.

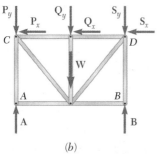

Fig. 4.5 Partial constraints.

[†]Partially constrained bodies are often referred to as *unstable.* However, to avoid confusion between this type of instability, due to insufficient constraints, and the type of instability considered in Chap. 10, which relates to the behavior of a rigid body when its equilibrium is disturbed, we shall restrict the use of the words *stable* and *unstable* to the latter case.

unknowns is equal to the number of equations is no guarantee that the body is completely constrained or that the reactions at its supports are statically determinate. Consider Fig. 4.6a, in which the truss shown is held by rollers at A, B, and E. While there are three unknown reactions, **A**, **B**, and **E** (Fig. 4.6b), the equation $\Sigma F_x = 0$ will not be satisfied unless the sum of the horizontal components of the applied forces happens to be zero. Although there are a sufficient number of constraints, these constraints are not properly arranged, and the truss is free to move horizontally. We say that the truss is *improperly constrained.* Since only two equilibrium equations are left for determining three unknowns, the reactions will be statically indeterminate. Thus, improper constraints also produce static indeterminacy.

Another example of improper constraints—and of static indeterminacy—is provided by the truss shown in Fig. 4.7. This truss is held by a pin at A and by rollers at B and C, which altogether involve four unknowns. Since only three independent equilibrium equations are available, the reactions at the supports are statically indeterminate. On the other hand, we note that the equation $\Sigma M_A = 0$ cannot be satisfied under general loading conditions, since the lines of action of the reactions **B** and **C** pass through A. We conclude that the truss can rotate about A and that it is improperly constrained.†

The examples of Figs. 4.6 and 4.7 lead us to conclude that *a rigid body is improperly constrained whenever the supports,* even though they may provide a sufficient number of reactions, *are arranged in such a way that the reactions must be either concurrent or parallel.*‡

In summary, to be sure that a two-dimensional rigid body is completely constrained and that the reactions at its supports are statically determinate, we should verify that the reactions involve three—and only three—unknowns and that the supports are arranged in such a way that they do not require the reactions to be either concurrent or parallel.

Supports involving statically indeterminate reactions should be used with care in the *design* of structures and only with a full knowledge of the problems they may cause. On the other hand, the *analysis* of structures possessing statically indeterminate reactions often can be partially carried out by the methods of statics. In the case of the truss of Fig. 4.4, for example, the vertical components of the reactions at A and B were obtained from the equilibrium equations.

For obvious reasons, supports producing partial or improper constraints should be avoided in the design of stationary structures. However, a partially or improperly constrained structure will not necessarily collapse; under particular loading conditions, equilibrium can be maintained. For example, the trusses of Figs. 4.5 and 4.6 will be in equilibrium if the applied forces **P**, **Q**, and **S** are vertical. Besides, structures which are designed to move *should* be only partially constrained. A railroad car, for instance, would be of little use if it were completely constrained by having its brakes applied permanently.

Fig. 4.6 Improper constraints.

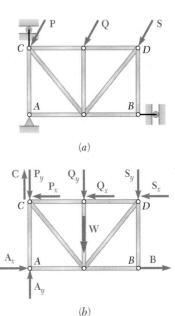

(a)

(b)

Fig. 4.7 Improper constraints.

---

†Rotation of the truss about A requires some "play" in the supports at B and C. In practice such play will always exist. In addition, we note that if the play is kept small, the displacements of the rollers B and C and, thus, the distances from A to the lines of action of the reactions **B** and **C** will also be small. The equation $\Sigma M_A = 0$ then requires that **B** and **C** be very large, a situation which can result in the failure of the supports at B and C.

‡Because this situation arises from an inadequate arrangement or *geometry* of the supports, it is often referred to as *geometric instability.*

SAMPLE PROBLEM 4.1

A fixed crane has a mass of 1000 kg and is used to lift a 2400-kg crate. It is held in place by a pin at A and a rocker at B. The center of gravity of the crane is located at G. Determine the components of the reactions at A and B.

SOLUTION

Free-Body Diagram. A free-body diagram of the crane is drawn. By multiplying the masses of the crane and of the crate by $g = 9.81 \text{ m/s}^2$, we obtain the corresponding weights, that is, 9810 N or 9.81 kN, and 23 500 N or 23.5 kN. The reaction at pin A is a force of unknown direction; it is represented by its components $\mathbf{A}_x$ and $\mathbf{A}_y$. The reaction at the rocker B is perpendicular to the rocker surface; thus, it is horizontal. We assume that $\mathbf{A}_x$, $\mathbf{A}_y$, and $\mathbf{B}$ act in the directions shown.

Determination of B. We express that the sum of the moments of all external forces about point A is zero. The equation obtained will contain neither A_x nor A_y, since the moments of $\mathbf{A}_x$ and $\mathbf{A}_y$ about A are zero. Multiplying the magnitude of each force by its perpendicular distance from A, we write

$$+\!\!\upuparrows \Sigma M_A = 0: \quad +B(1.5 \text{ m}) - (9.81 \text{ kN})(2 \text{ m}) - (23.5 \text{ kN})(6 \text{ m}) = 0$$
$$B = +107.1 \text{ kN} \qquad \mathbf{B} = 107.1 \text{ kN} \rightarrow \quad \blacktriangleleft$$

Since the result is positive, the reaction is directed as assumed.

Determination of A_x. The magnitude of $\mathbf{A}_x$ is determined by expressing that the sum of the horizontal components of all external forces is zero.

$$\xrightarrow{+} \Sigma F_x = 0: \quad A_x + B = 0$$
$$A_x + 107.1 \text{ kN} = 0$$
$$A_x = -107.1 \text{ kN} \qquad \mathbf{A}_x = 107.1 \text{ kN} \leftarrow \quad \blacktriangleleft$$

Since the result is negative, the sense of $\mathbf{A}_x$ is opposite to that assumed originally.

Determination of A_y. The sum of the vertical components must also equal zero.

$$+\!\!\uparrow \Sigma F_y = 0: \quad A_y - 9.81 \text{ kN} - 23.5 \text{ kN} = 0$$
$$A_y = +33.3 \text{ kN} \qquad \mathbf{A}_y = 33.3 \text{ kN} \uparrow \quad \blacktriangleleft$$

Adding vectorially the components $\mathbf{A}_x$ and $\mathbf{A}_y$, we find that the reaction at A is 112.2 kN ⟋17.3°.

Check. The values obtained for the reactions can be checked by recalling that the sum of the moments of all of the external forces about any point must be zero. For example, considering point B, we write

$$+\!\!\upuparrows \Sigma M_B = -(9.81 \text{ kN})(2 \text{ m}) - (23.5 \text{ kN})(6 \text{ m}) + (107.1 \text{ kN})(1.5 \text{ m}) = 0$$

SAMPLE PROBLEM 4.2

Three loads are applied to a beam as shown. The beam is supported by a roller at A and by a pin at B. Neglecting the weight of the beam, determine the reactions at A and B when $P = 15$ kips.

SOLUTION

Free-Body Diagram. A free-body diagram of the beam is drawn. The reaction at A is vertical and is denoted by $\mathbf{A}$. The reaction at B is represented by components $\mathbf{B}_x$ and $\mathbf{B}_y$. Each component is assumed to act in the direction shown.

Equilibrium Equations. We write the following three equilibrium equations and solve for the reactions indicated:

$$\xrightarrow{+}\Sigma F_x = 0: \qquad\qquad\qquad B_x = 0 \qquad\qquad \mathbf{B}_x = 0 \blacktriangleleft$$

$$+\curvearrowleft\Sigma M_A = 0:$$
$$-(15\text{ kips})(3\text{ ft}) + B_y(9\text{ ft}) - (6\text{ kips})(11\text{ ft}) - (6\text{ kips})(13\text{ ft}) = 0$$
$$B_y = +21.0\text{ kips} \qquad \mathbf{B}_y = 21.0\text{ kips}\uparrow \blacktriangleleft$$

$$+\curvearrowleft\Sigma M_B = 0:$$
$$-A(9\text{ ft}) + (15\text{ kips})(6\text{ ft}) - (6\text{ kips})(2\text{ ft}) - (6\text{ kips})(4\text{ ft}) = 0$$
$$A = +6.00\text{ kips} \qquad \mathbf{A} = 6.00\text{ kips}\uparrow \blacktriangleleft$$

Check. The results are checked by adding the vertical components of all of the external forces:

$$+\uparrow\Sigma F_y = +6.00\text{ kips} - 15\text{ kips} + 21.0\text{ kips} - 6\text{ kips} - 6\text{ kips} = 0$$

Remark. In this problem the reactions at both A and B are vertical; however, these reactions are vertical for different reasons. At A, the beam is supported by a roller; hence the reaction cannot have any horizontal component. At B, the horizontal component of the reaction is zero because it must satisfy the equilibrium equation $\Sigma F_x = 0$ and because none of the other forces acting on the beam has a horizontal component.

We could have noticed at first glance that the reaction at B was vertical and dispensed with the horizontal component $\mathbf{B}_x$. This, however, is a bad practice. In following it, we would run the risk of forgetting the component $\mathbf{B}_x$ when the loading conditions require such a component (i.e., when a horizontal load is included). Also, the component $\mathbf{B}_x$ was found to be zero by using and solving an equilibrium equation, $\Sigma F_x = 0$. By setting $\mathbf{B}_x$ equal to zero immediately, we might not realize that we actually make use of this equation and thus might lose track of the number of equations available for solving the problem.

SAMPLE PROBLEM 4.3

A loading car is at rest on a track forming an angle of 25° with the vertical. The gross weight of the car and its load is 5500 lb, and it is applied at a point 30 in. from the track, halfway between the two axles. The car is held by a cable attached 24 in. from the track. Determine the tension in the cable and the reaction at each pair of wheels.

SOLUTION

Free-Body Diagram. A free-body diagram of the car is drawn. The reaction at each wheel is perpendicular to the track, and the tension force **T** is parallel to the track. For convenience, we choose the x axis parallel to the track and the y axis perpendicular to the track. The 5500-lb weight is then resolved into x and y components.

$$W_x = +(5500 \text{ lb}) \cos 25° = +4980 \text{ lb}$$
$$W_y = -(5500 \text{ lb}) \sin 25° = -2320 \text{ lb}$$

Equilibrium Equations. We take moments about A to eliminate **T** and $\mathbf{R}_1$ from the computation.

$+\uparrow\Sigma M_A = 0$: $-(2320 \text{ lb})(25 \text{ in.}) - (4980 \text{ lb})(6 \text{ in.}) + R_2(50 \text{ in.}) = 0$
$R_2 = +1758 \text{ lb}$ $\mathbf{R}_2 = 1758 \text{ lb} \nearrow$ ◄

Now, taking moments about B to eliminate **T** and $\mathbf{R}_2$ from the computation, we write

$+\uparrow\Sigma M_B = 0$: $(2320 \text{ lb})(25 \text{ in.}) - (4980 \text{ lb})(6 \text{ in.}) - R_1(50 \text{ in.}) = 0$
$R_1 = +562 \text{ lb}$ $\mathbf{R}_1 = +562 \text{ lb} \nearrow$ ◄

The value of T is found by writing

$\searrow+\Sigma F_x = 0$: $+4980 \text{ lb} - T = 0$
$T = +4980 \text{ lb}$ $\mathbf{T} = 4980 \text{ lb} \nwarrow$ ◄

The computed values of the reactions are shown in the adjacent sketch.

Check. The computations are verified by writing

$$\nearrow+\Sigma F_y = +562 \text{ lb} + 1758 \text{ lb} - 2320 \text{ lb} = 0$$

The solution could also have been checked by computing moments about any point other than A or B.

SAMPLE PROBLEM 4.4

The frame shown supports part of the roof of a small building. Knowing that the tension in the cable is 150 kN, determine the reaction at the fixed end E.

SOLUTION

Free-Body Diagram. A free-body diagram of the frame and of the cable BDF is drawn. The reaction at the fixed end E is represented by the force components $\mathbf{E}_x$ and $\mathbf{E}_y$ and the couple $\mathbf{M}_E$. The other forces acting on the free body are the four 20-kN loads and the 150-kN force exerted at end F of the cable.

Equilibrium Equations. Noting that $DF = \sqrt{(4.5 \text{ m})^2 + (6 \text{ m})^2} = 7.5$ m, we write

$$\xrightarrow{+}\Sigma F_x = 0: \qquad E_x + \frac{4.5}{7.5}(150 \text{ kN}) = 0$$
$$E_x = -90.0 \text{ kN} \qquad \mathbf{E}_x = 90.0 \text{ kN} \leftarrow \quad \blacktriangleleft$$

$$+\uparrow\Sigma F_y = 0: \qquad E_y - 4(20 \text{ kN}) - \frac{6}{7.5}(150 \text{ kN}) = 0$$
$$E_y = +200 \text{ kN} \qquad \mathbf{E}_y = 200 \text{ kN} \uparrow \quad \blacktriangleleft$$

$$+\curvearrowleft\Sigma M_E = 0: \qquad (20 \text{ kN})(7.2 \text{ m}) + (20 \text{ kN})(5.4 \text{ m}) + (20 \text{ kN})(3.6 \text{ m})$$
$$+ (20 \text{ kN})(1.8 \text{ m}) - \frac{6}{7.5}(150 \text{ kN})(4.5 \text{ m}) + M_E = 0$$
$$M_E = +180.0 \text{ kN} \cdot \text{m} \qquad \mathbf{M}_E = 180.0 \text{ kN} \cdot \text{m} \curvearrowleft \quad \blacktriangleleft$$

SAMPLE PROBLEM 4.5

A 400-lb weight is attached at A to the lever shown. The constant of the spring BC is $k = 250$ lb/in., and the spring is unstretched when $\theta = 0$. Determine the position of equilibrium.

SOLUTION

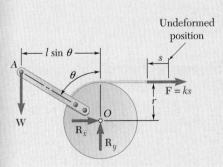

Free-Body Diagram. We draw a free-body diagram of the lever and cylinder. Denoting by s the deflection of the spring from its undeformed position, and noting that $s = r\theta$, we have $F = ks = kr\theta$.

Equilibrium Equation. Summing the moments of $\mathbf{W}$ and $\mathbf{F}$ about O, we write

$$+\curvearrowleft\Sigma M_O = 0: \qquad Wl \sin \theta - r(kr\theta) = 0 \qquad \sin \theta = \frac{kr^2}{Wl}\theta$$

Substituting the given data, we obtain

$$\sin \theta = \frac{(250 \text{ lb/in.})(3 \text{ in.})^2}{(400 \text{ lb})(8 \text{ in.})}\theta \qquad \sin \theta = 0.703\,\theta$$

Solving by trial and error, we find $\qquad \theta = 0 \qquad \theta = 80.3° \quad \blacktriangleleft$

171

SOLVING PROBLEMS ON YOUR OWN

You saw that the external forces acting on a rigid body in equilibrium form a system equivalent to zero. To solve an equilibrium problem your first task is to draw a neat, reasonably large *free-body diagram* on which you will show all external forces. Both known and unknown forces must be included.

For a two-dimensional rigid body, the reactions at the supports can involve one, two, or three unknowns depending on the type of support (Fig. 4.1). For the successful solution of a problem, a correct free-body diagram is essential. Never proceed with the solution of a problem until you are sure that your free-body diagram includes all loads, all reactions, and the weight of the body (if appropriate).

1. You can write three equilibrium equations and solve them for *three unknowns*. The three equations might be

$$\Sigma F_x = 0 \qquad \Sigma F_y = 0 \qquad \Sigma M_O = 0$$

However, there are usually several sets of equations that you can write, such as

$$\Sigma F_x = 0 \qquad \Sigma M_A = 0 \qquad \Sigma M_B = 0$$

where point B is chosen in such a way that the line AB is not parallel to the y axis, or

$$\Sigma M_A = 0 \qquad \Sigma M_B = 0 \qquad \Sigma M_C = 0$$

where the points A, B, and C do not lie in a straight line.

2. To simplify your solution, it may be helpful to use one of the following solution techniques if applicable.

   a. By summing moments about the point of intersection of the lines of action of two unknown forces, you will obtain an equation in a single unknown.

   b. By summing components in a direction perpendicular to two unknown parallel forces, you will obtain an equation in a single unknown.

3. After drawing your free-body diagram, you may find that one of the following special situations exists.

   a. The reactions involve fewer than three unknowns; the body is said to be *partially constrained* and motion of the body is possible.

   b. The reactions involve more than three unknowns; the reactions are said to be *statically indeterminate*. While you may be able to calculate one or two reactions, you cannot determine all of the reactions.

   c. The reactions pass through a single point or are parallel; the body is said to be *improperly constrained* and motion can occur under a general loading condition.

PROBLEMS

FREE BODY PRACTICE PROBLEMS

4.F1 For the frame and loading shown, draw the free-body diagram needed to determine the reactions at A and E when $\alpha = 30°$.

4.F2 Neglecting friction, draw the free-body diagram needed to determine the tension in cable ABD and the reaction at C when $\theta = 60°$.

Fig. P4.F1

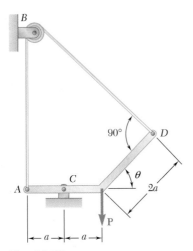

Fig. P4.F2

4.F3 Bar AC supports two 400-N loads as shown. Rollers at A and C rest against frictionless surfaces and a cable BD is attached at B. Draw the free-body diagram needed to determine the tension in cable BD and the reactions at A and C.

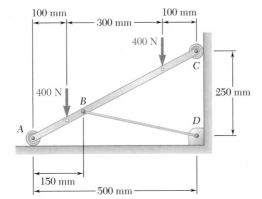

Fig. P4.F3

4.F4 Draw the free-body diagram needed to determine the tension in each cable and the reaction at D.

Fig. P4.F4

173

Fig. P4.1

Fig. P4.5

15 lb   20 lb   35 lb   20 lb   15 lb

Fig. P4.8

END-OF-SECTION PROBLEMS

4.1 Two crates, each of mass 350 kg, are placed as shown in the bed of a 1400-kg pickup truck. Determine the reactions at each of the two (*a*) rear wheels *A*, (*b*) front wheels *B*.

4.2 Solve Prob. 4.1, assuming that crate *D* is removed and that the position of crate *C* is unchanged.

4.3 A T-shaped bracket supports the four loads shown. Determine the reactions at *A* and *B* (*a*) if *a* = 10 in., (*b*) if *a* = 7 in.

40 lb   50 lb   30 lb   10 lb

6 in.   6 in.   *a*   8 in.

Fig. P4.3

4.4 For the bracket and loading of Prob. 4.3, determine the smallest distance *a* if the bracket is not to move.

4.5 A hand truck is used to move two kegs, each of mass 40 kg. Neglecting the mass of the hand truck, determine (*a*) the vertical force **P** that should be applied to the handle to maintain equilibrium when *α* = 35°, (*b*) the corresponding reaction at each of the two wheels.

4.6 Solve Prob. 4.5 when *α* = 40°.

4.7 A 3200-lb forklift truck is used to lift a 1700-lb crate. Determine the reaction at each of the two (*a*) front wheels *A*, (*b*) rear wheels *B*.

16 in.   24 in.   12 in.

Fig. P4.7

4.8 For the beam and loading shown, determine (*a*) the reaction at *A*, (*b*) the tension in cable *BC*.

4.9 For the beam and loading shown, determine the range of the distance a for which the reaction at B does not exceed 100 lb downward or 200 lb upward.

4.10 The maximum allowable value of each of the reactions is 180 N. Neglecting the weight of the beam, determine the range of the distance d for which the beam is safe.

Fig. P4.9

Fig. P4.10

4.11 Three loads are applied as shown to a light beam supported by cables attached at B and D. Neglecting the weight of the beam, determine the range of values of Q for which neither cable becomes slack when $P = 0$.

4.12 Three loads are applied as shown to a light beam supported by cables attached at B and D. Knowing that the maximum allowable tension in each cable is 4 kN and neglecting the weight of the beam, determine the range of values of Q for which the loading is safe when $P = 0$.

4.13 For the beam of Prob. 4.12, determine the range of values of Q for which the loading is safe when $P = 1$ kN.

4.14 For the beam of Sample Prob. 4.2, determine the range of values of P for which the beam will be safe, knowing that the maximum allowable value of each of the reactions is 30 kips and that the reaction at A must be directed upward.

Fig. P4.11 and P4.12

4.15 The bracket BCD is hinged at C and attached to a control cable at B. For the loading shown, determine (a) the tension in the cable, (b) the reaction at C.

Fig. P4.15

4.16 Solve Prob. 4.15, assuming that $a = 0.32$ m.

4.17 The lever BCD is hinged at C and attached to a control rod at B. If $P = 100$ lb, determine (a) the tension in rod AB, (b) the reaction at C.

4.18 The lever BCD is hinged at C and attached to a control rod at B. Determine the maximum force **P** that can be safely applied at D if the maximum allowable value of the reaction at C is 250 lb.

Fig. P4.17 and P4.18

4.19 Two links *AB* and *DE* are connected by a bell crank as shown. Knowing that the tension in link *AB* is 720 N, determine (*a*) the tension in link *DE*, (*b*) the reaction at *C*.

Fig. P4.19 and P4.20

4.20 Two links *AB* and *DE* are connected by a bell crank as shown. Determine the maximum force that can be safely exerted by link *AB* on the bell crank if the maximum allowable value for the reaction at *C* is 1600 N.

4.21 Determine the reactions at *A* and *C* when (*a*) $\alpha = 0$, (*b*) $\alpha = 30°$.

4.22 Determine the reactions at *A* and *B* when (*a*) $\alpha = 0$, (*b*) $\alpha = 90°$, (*c*) $\alpha = 30°$.

4.23 Determine the reactions at *A* and *B* when (*a*) $h = 0$, (*b*) $h = 200$ mm.

Fig. P4.21

Fig. P4.23

Fig. P4.22

4.24 A lever *AB* is hinged at *C* and attached to a control cable at *A*. If the lever is subjected to a 75-lb vertical force at *B*, determine (*a*) the tension in the cable, (*b*) the reaction at *C*.

Fig. P4.24

4.25 and 4.26 For each of the plates and loadings shown, determine the reactions at A and B.

Fig. P4.25

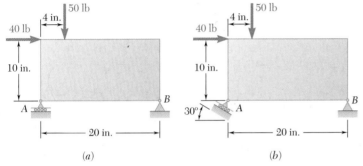

Fig. P4.26

4.27 A rod AB hinged at A and attached at B to cable BD supports the loads shown. Knowing that $d = 200$ mm, determine (a) the tension in cable BD, (b) the reaction at A.

Fig. P4.27 and P4.28

4.28 A rod AB, hinged at A and attached at B to cable BD, supports the loads shown. Knowing that $d = 150$ mm, determine (a) the tension in cable BD, (b) the reaction at A.

4.29 A force **P** of magnitude 90 lb is applied to member ACE, which is supported by a frictionless pin at D and by the cable ABE. Since the cable passes over a small pulley at B, the tension may be assumed to be the same in portions AB and BE of the cable. For the case when $a = 3$ in., determine (a) the tension in the cable, (b) the reaction at D.

4.30 Solve Prob. 4.29 for $a = 6$ in.

Fig. P4.29

Fig. P4.31

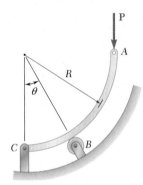

Fig. P4.33 and P4.34

4.31 Neglecting friction, determine the tension in cable *ABD* and the reaction at support *C*.

4.32 Neglecting friction and the radius of the pulley, determine (*a*) the tension in cable *ADB*, (*b*) the reaction at *C*.

Fig. P4.32

4.33 Rod *ABC* is bent in the shape of an arc of circle of radius *R*. Knowing the $\theta = 30°$, determine the reaction (*a*) at *B*, (*b*) at *C*.

4.34 Rod *ABC* is bent in the shape of an arc of circle of radius *R*. Knowing the $\theta = 60°$, determine the reaction (*a*) at *B*, (*b*) at *C*.

4.35 A movable bracket is held at rest by a cable attached at *C* and by frictionless rollers at *A* and *B*. For the loading shown, determine (*a*) the tension in the cable, (*b*) the reactions at *A* and *B*.

Fig. P4.35

4.36 A light bar *AB* supports a 15-kg block at its midpoint *C*. Rollers at *A* and *B* rest against frictionless surfaces, and a horizontal cable *AD* is attached at *A*. Determine (*a*) the tension in cable *AD*, (*b*) the reactions at *A* and *B*.

Fig. P4.36

4.37 A light bar *AD* is suspended from a cable *BE* and supports a 50-lb block at *C*. The ends *A* and *D* of the bar are in contact with frictionless vertical walls. Determine the tension in cable *BE* and the reactions at *A* and *D*.

4.38 A light rod *AD* is supported by frictionless pegs at *B* and *C* and rests against a frictionless wall at *A*. A vertical 120-lb force is applied at *D*. Determine the reactions at *A*, *B*, and *C*.

Fig. P4.37

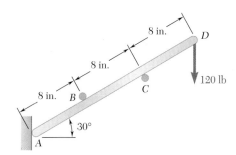

Fig. P4.38

4.39 Bar *AD* is attached at *A* and *C* to collars that can move freely on the rods shown. If the cord *BE* is vertical (α = 0), determine the tension in the cord and the reactions at *A* and *C*.

Fig. P4.39

4.40 Solve Prob. 4.39 if the cord *BE* is parallel to the rods (α = 30°).

4.41 The T-shaped bracket shown is supported by a small wheel at *E* and pegs at *C* and *D*. Neglecting the effect of friction, determine the reactions at *C*, *D*, and *E* when θ = 30°.

4.42 The T-shaped bracket shown is supported by a small wheel at *E* and pegs at *C* and *D*. Neglecting the effect of friction, determine (*a*) the smallest value of θ for which the equilibrium of the bracket is maintained, (*b*) the corresponding reactions at *C*, *D*, and *E*.

4.43 Beam *AD* carries the two 40-lb loads shown. The beam is held by a fixed support at *D* and by the cable *BE* that is attached to the counterweight *W*. Determine the reaction at *D* when (*a*) *W* = 100 lb, (*b*) *W* = 90 lb.

4.44 For the beam and loading shown, determine the range of values of *W* for which the magnitude of the couple at *D* does not exceed 40 lb · ft.

Fig. P4.41 and P4.42

Fig. P4.43 and P4.44

4.45 An 8-kg mass can be supported in the three different ways shown. Knowing that the pulleys have a 100-mm radius, determine the reaction at A in each case.

(a)   (b)   (c)

Fig. P4.45

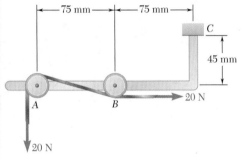

Fig. P4.46

4.46 A tension of 20 N is maintained in a tape as it passes through the support system shown. Knowing that the radius of each pulley is 10 mm, determine the reaction at C.

4.47 Solve Prob. 4.46, assuming that 15-mm-radius pulleys are used.

4.48 The rig shown consists of a 1200-lb horizontal member ABC and a vertical member DBE welded together at B. The rig is being used to raise a 3600-lb crate at a distance $x = 12$ ft from the vertical member DBE. If the tension in the cable is 4 kips, determine the reaction at E, assuming that the cable is (a) anchored at F as shown in the figure, (b) attached to the vertical member at a point located 1 ft above E.

Fig. P4.48

4.49 For the rig and crate of Prob. 4.48, and assuming that the cable is anchored at F as shown, determine (a) the required tension in cable $ADCF$ if the maximum value of the couple at E as x varies from 1.5 to 17.5 ft is to be as small as possible, (b) the corresponding maximum value of the couple.

4.50 A 6-m telephone pole weighing 1600 N is used to support the ends of two wires. The wires form the angles shown with the horizontal and the tensions in the wires are, respectively, $T_1 = 600$ N and $T_2 = 375$ N. Determine the reaction at the fixed end A.

4.51 and 4.52 A vertical load **P** is applied at end B of rod BC. (*a*) Neglecting the weight of the rod, express the angle θ corresponding to the equilibrium position in terms of P, l, and the counterweight W. (*b*) Determine the value of θ corresponding to equilibrium if $P = 2W$.

Fig. P4.50

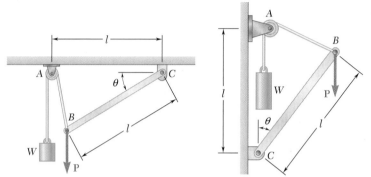

Fig. P4.51   **Fig. P4.52**

4.53 A slender rod AB, of weight W, is attached to blocks A and B, which move freely in the guides shown. The blocks are connected by an elastic cord that passes over a pulley at C. (*a*) Express the tension in the cord in terms of W and θ. (*b*) Determine the value of θ for which the tension in the cord is equal to $3W$.

Fig. P4.53

4.54 Rod AB is acted upon by a couple **M** and two forces, each of magnitude P. (*a*) Derive an equation in θ, P, M, and l that must be satisfied when the rod is in equilibrium. (*b*) Determine the value of θ corresponding to equilibrium when $M = 150$ N · m, $P = 200$ N, and $l = 600$ mm.

Fig. P4.54

4.55 Solve Sample Prob. 4.5, assuming that the spring is unstretched when $\theta = 90°$.

Fig. P4.56

4.56 A slender rod AB, of weight W, is attached to blocks A and B that move freely in the guides shown. The constant of the spring is k, and the spring is unstretched when $\theta = 0$. (a) Neglecting the weight of the blocks, derive an equation in W, k, l, and θ that must be satisfied when the rod is in equilibrium. (b) Determine the value of θ when $W = 75$ lb, $l = 30$ in., and $k = 3$ lb/in.

4.57 A vertical load **P** is applied at end B of rod BC. The constant of the spring is k, and the spring is unstretched when $\theta = 60°$. (a) Neglecting the weight of the rod, express the angle θ corresponding to the equilibrium position in terms of P, k, and l. (b) Determine the value of θ corresponding to equilibrium if $P = \frac{1}{4}kl$.

Fig. P4.57

Fig. P4.58

4.58 A collar B of weight W can move freely along the vertical rod shown. The constant of the spring is k, and the spring is unstretched when $\theta = 0$. (a) Derive an equation in θ, W, k, and l that must be satisfied when the collar is in equilibrium. (b) Knowing that $W = 300$ N, $l = 500$ mm, and $k = 800$ N/m, determine the value of θ corresponding to equilibrium.

4.59 Eight identical 500×750-mm rectangular plates, each of mass $m = 40$ kg, are held in a vertical plane as shown. All connections consist of frictionless pins, rollers, or short links. In each case, determine whether (a) the plate is completely, partially, or improperly constrained, (b) the reactions are statically determinate or indeterminate, (c) the equilibrium of the plate is maintained in the position shown. Also, wherever possible, compute the reactions.

Fig. P4.59

4.60 The bracket *ABC* can be supported in the eight different ways shown. All connections consist of smooth pins, rollers, or short links. For each case, answer the questions listed in Prob. 4.59, and, wherever possible, compute the reactions, assuming that the magnitude of the force **P** is 100 lb.

Fig. P4.60

4.6 EQUILIBRIUM OF A TWO-FORCE BODY

A particular case of equilibrium which is of considerable interest is that of a rigid body subjected to two forces. Such a body is commonly called a *two-force body*. It will be shown that *if a two-force body is in equilibrium, the two forces must have the same magnitude, the same line of action, and opposite sense.*

Consider a corner plate subjected to two forces $\mathbf{F}_1$ and $\mathbf{F}_2$ acting at *A* and *B*, respectively (Fig. 4.8*a*). If the plate is to be in equilibrium, the sum of the moments of $\mathbf{F}_1$ and $\mathbf{F}_2$ about any axis must be zero. First, we sum moments about *A*. Since the moment of $\mathbf{F}_1$ is obviously zero, the moment of $\mathbf{F}_2$ must also be zero and the line of action of $\mathbf{F}_2$ must pass through *A* (Fig. 4.8*b*). Summing moments about *B*, we prove similarly that the line of action of $\mathbf{F}_1$ must pass through *B* (Fig. 4.8*c*). Therefore, both forces have the same line of action (line *AB*). From either of the equations $\Sigma F_x = 0$ and $\Sigma F_y = 0$ it is seen that they must also have the same magnitude but opposite sense.

(a) *(b)* *(c)*

Fig. 4.8

Fig. 4.8 (repeated)

If several forces act at two points A and B, the forces acting at A can be replaced by their resultant $\mathbf{F}_1$ and those acting at B can be replaced by their resultant $\mathbf{F}_2$. Thus a two-force body can be more generally defined as *a rigid body subjected to forces acting at only two points.* The resultants $\mathbf{F}_1$ and $\mathbf{F}_2$ then must have the same line of action, the same magnitude, and opposite sense (Fig. 4.8).

In the study of structures, frames, and machines, you will see how the recognition of two-force bodies simplifies the solution of certain problems.

4.7 EQUILIBRIUM OF A THREE-FORCE BODY

Another case of equilibrium that is of great interest is that of a *three-force body,* i.e., a rigid body subjected to three forces or, more generally, *a rigid body subjected to forces acting at only three points.* Consider a rigid body subjected to a system of forces which can be reduced to three forces $\mathbf{F}_1$, $\mathbf{F}_2$, and $\mathbf{F}_3$ acting at A, B, and C, respectively (Fig. 4.9a). It will be shown that if the body is in equilibrium, *the lines of action of the three forces must be either concurrent or parallel.*

Since the rigid body is in equilibrium, the sum of the moments of $\mathbf{F}_1$, $\mathbf{F}_2$, and $\mathbf{F}_3$ about any axis must be zero. Assuming that the lines of action of $\mathbf{F}_1$ and $\mathbf{F}_2$ intersect and denoting their point of intersection by D, we sum moments about D (Fig. 4.9b). Since the moments of $\mathbf{F}_1$ and $\mathbf{F}_2$ about D are zero, the moment of $\mathbf{F}_3$ about D must also be zero, and the line of action of $\mathbf{F}_3$ must pass through D (Fig. 4.9c). Therefore, the three lines of action are concurrent. The only exception occurs when none of the lines intersect; the lines of action are then parallel.

Although problems concerning three-force bodies can be solved by the general methods of Secs. 4.3 to 4.5, the property just established can be used to solve them either graphically or mathematically from simple trigonometric or geometric relations.

Fig. 4.9

SAMPLE PROBLEM 4.6

A man raises a 10-kg joist, of length 4 m, by pulling on a rope. Find the tension T in the rope and the reaction at A.

SOLUTION

Free-Body Diagram. The joist is a three-force body, since it is acted upon by three forces: its weight $\mathbf{W}$, the force $\mathbf{T}$ exerted by the rope, and the reaction $\mathbf{R}$ of the ground at A. We note that

$$W = mg = (10 \text{ kg})(9.81 \text{ m/s}^2) = 98.1 \text{ N}$$

Three-Force Body. Since the joist is a three-force body, the forces acting on it must be concurrent. The reaction $\mathbf{R}$, therefore, will pass through the point of intersection C of the lines of action of the weight $\mathbf{W}$ and the tension force $\mathbf{T}$. This fact will be used to determine the angle α that $\mathbf{R}$ forms with the horizontal.

Drawing the vertical BF through B and the horizontal CD through C, we note that

$$AF = BF = (AB) \cos 45° = (4 \text{ m}) \cos 45° = 2.828 \text{ m}$$
$$CD = EF = AE = \tfrac{1}{2}(AF) = 1.414 \text{ m}$$
$$BD = (CD) \cot (45° + 25°) = (1.414 \text{ m}) \tan 20° = 0.515 \text{ m}$$
$$CE = DF = BF - BD = 2.828 \text{ m} - 0.515 \text{ m} = 2.313 \text{ m}$$

We write

$$\tan \alpha = \frac{CE}{AE} = \frac{2.313 \text{ m}}{1.414 \text{ m}} = 1.636$$

$$\alpha = 58.6° \quad \blacktriangleleft$$

We now know the direction of all the forces acting on the joist.

Force Triangle. A force triangle is drawn as shown, and its interior angles are computed from the known directions of the forces. Using the law of sines, we write

$$\frac{T}{\sin 31.4°} = \frac{R}{\sin 110°} = \frac{98.1 \text{ N}}{\sin 38.6°}$$

$$T = 81.9 \text{ N} \quad \blacktriangleleft$$
$$R = 147.8 \text{ N} \measuredangle 58.6° \quad \blacktriangleleft$$

SOLVING PROBLEMS
ON YOUR OWN

The preceding sections covered two particular cases of equilibrium of a rigid body.

1. A two-force body is a body subjected to forces at only two points. The resultants of the forces acting at each of these points must have the *same magnitude, the same line of action, and opposite sense.* This property will allow you to simplify the solutions of some problems by replacing the two unknown components of a reaction by a single force of unknown magnitude but of *known direction.*

2. A three-force body is subjected to forces at only three points. The resultants of the forces acting at each of these points must be *concurrent or parallel.* To solve a problem involving a three-force body with concurrent forces, draw your free-body diagram showing that these three forces pass through the same point. The use of simple geometry may then allow you to complete the solution by using a force triangle [Sample Prob. 4.6].

Although the principle noted above for the solution of problems involving three-force bodies is easily understood, it can be difficult to sketch the needed geometric constructions. If you encounter difficulty, first draw a reasonably large free-body diagram and then seek a relation between known or easily calculated lengths and a dimension that involves an unknown. This was done in Sample Prob. 4.6, where the easily calculated dimensions AE and CE were used to determine the angle α.

PROBLEMS

4.61 Determine the reactions at A and B when $a = 150$ mm.

4.62 Determine the value of a for which the magnitude of the reaction at B is equal to 800 N.

4.63 Using the method of Sec. 4.7, solve Prob. 4.22b.

4.64 A 500-lb cylindrical tank, 8 ft in diameter, is to be raised over a 2-ft obstruction. A cable is wrapped around the tank and pulled horizontally as shown. Knowing that the corner of the obstruction at A is rough, find the required tension in the cable and the reaction at A.

Fig. P4.61 and P4.62

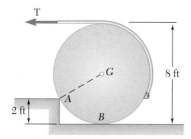

Fig. P4.64

4.65 For the frame and loading shown, determine the reactions at A and C.

4.66 For the frame and loading shown, determine the reactions at C and D.

Fig. P4.65

Fig. P4.66

4.67 Determine the reactions at B and D when $b = 60$ mm.

4.68 Determine the reactions at B and D when $b = 120$ mm.

Fig. P4.67 and P4.68

187

4.69 A T-shaped bracket supports a 300-N load as shown. Determine the reactions at A and C when $\alpha = 45°$.

Fig. P4.69 and P4.70

4.70 A T-shaped bracket supports a 300-N load as shown. Determine the reactions at A and C when $\alpha = 60°$.

4.71 A 40-lb roller, of diameter 8 in., which is to be used on a tile floor, is resting directly on the subflooring as shown. Knowing that the thickness of each tile is 0.3 in., determine the force **P** required to move the roller onto the tiles if the roller is (a) pushed to the left, (b) pulled to the right.

Fig. P4.71

4.72 One end of rod AB rests in the corner A and the other end is attached to cord BD. If the rod supports a 40-lb load at its mid-point C, find the reaction at A and the tension in the cord.

4.73 A 50-kg crate is attached to the trolley-beam system shown. Knowing that $a = 1.5$ m, determine (a) the tension in cable CD, (b) the reaction at B.

Fig. P4.72

Fig. P4.73

4.74 Solve Prob. 4.73, assuming that $a = 3$ m.

4.75 Determine the reactions at A and B when $\beta = 50°$.

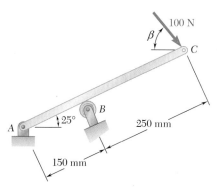

Fig. P4.75 and P4.76

4.76 Determine the reactions at A and B when $\beta = 80°$.

4.77 Knowing that $\theta = 30°$, determine the reaction (a) at B, (b) at C.

4.78 Knowing that $\theta = 60°$, determine the reaction (a) at B, (b) at C.

4.79 Using the method of Sec. 4.7, solve Prob. 4.23.

4.80 Using the method of Sec. 4.7, solve Prob. 4.24.

4.81 and 4.82 Member ABC is supported by a pin and bracket at B and by an inextensible cord attached at A and C and passing over a frictionless pulley at D. The tension may be assumed to be the same in portions AD and CD of the cord. For the loading shown and neglecting the size of the pulley, determine the tension in the cord and the reaction at B.

Fig. P4.77 and P4.78

Fig. P4.81

Fig. P4.82

4.83 A thin ring of mass 2 kg and radius $r = 140$ mm is held against a frictionless wall by a 125-mm string AB. Determine (a) the distance d, (b) the tension in the string, (c) the reaction at C.

Fig. P4.83

Fig. P4.84

Fig. P4.86

Fig. P4.89 and P4.90

4.84 A uniform rod AB of length $2R$ rests inside a hemispherical bowl of radius R as shown. Neglecting friction, determine the angle θ corresponding to equilibrium.

4.85 A slender rod BC of length L and weight W is held by two cables as shown. Knowing that cable AB is horizontal and that the rod forms an angle of 40° with the horizontal, determine (*a*) the angle θ that cable CD forms with the horizontal, (*b*) the tension in each cable.

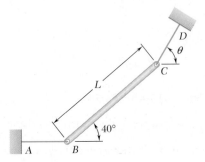

Fig. P4.85

4.86 A slender rod of length L and weight W is attached to a collar at A and is fitted with a small wheel at B. Knowing that the wheel rolls freely along a cylindrical surface of radius R, and neglecting friction, derive an equation in θ, L, and R that must be satisfied when the rod is in equilibrium.

4.87 Knowing that for the rod of Prob. 4.86, $L = 15$ in., $R = 20$ in., and $W = 10$ lb, determine (*a*) the angle θ corresponding to equilibrium, (*b*) the reactions at A and B.

4.88 Rod AB is bent into the shape of an arc of circle and is lodged between two pegs D and E. It supports a load $\mathbf{P}$ at end B. Neglecting friction and the weight of the rod, determine the distance c corresponding to equilibrium when $a = 20$ mm and $R = 100$ mm.

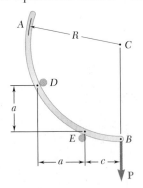

Fig. P4.88

4.89 A slender rod of length L is attached to collars that can slide freely along the guides shown. Knowing that the rod is in equilibrium, derive an expression for the angle θ in terms of the angle β.

4.90 An 8-kg slender rod of length L is attached to collars that can slide freely along the guides shown. Knowing that the rod is in equilibrium and that $\beta = 30°$, determine (*a*) the angle θ that the rod forms with the vertical, (*b*) the reactions at A and B.

4.8 EQUILIBRIUM OF A RIGID BODY IN THREE DIMENSIONS

We saw in Sec. 4.1 that six scalar equations are required to express the conditions for the equilibrium of a rigid body in the general three-dimensional case:

$$\Sigma F_x = 0 \qquad \Sigma F_y = 0 \qquad \Sigma F_z = 0 \qquad (4.2)$$
$$\Sigma M_x = 0 \qquad \Sigma M_y = 0 \qquad \Sigma M_z = 0 \qquad (4.3)$$

These equations can be solved for no more than *six unknowns*, which generally will represent reactions at supports or connections.

In most problems the scalar equations (4.2) and (4.3) will be more conveniently obtained if we first express in vector form the conditions for the equilibrium of the rigid body considered. We write

$$\Sigma \mathbf{F} = 0 \qquad \Sigma \mathbf{M}_O = \Sigma(\mathbf{r} \times \mathbf{F}) = 0 \qquad (4.1)$$

and express the forces **F** and position vectors **r** in terms of scalar components and unit vectors. Next, we compute all vector products, either by direct calculation or by means of determinants (see Sec. 3.8). We observe that as many as three unknown reaction components may be eliminated from these computations through a judicious choice of the point O. By equating to zero the coefficients of the unit vectors in each of the two relations (4.1), we obtain the desired scalar equations.†

4.9 REACTIONS AT SUPPORTS AND CONNECTIONS FOR A THREE-DIMENSIONAL STRUCTURE

The reactions on a three-dimensional structure range from the single force of known direction exerted by a frictionless surface to the force-couple system exerted by a fixed support. Consequently, in problems involving the equilibrium of a three-dimensional structure, there can be between one and six unknowns associated with the reaction at each support or connection. Various types of supports and

---

†In some problems, it will be found convenient to eliminate the reactions at two points A and B from the solution by writing the equilibrium equation $\Sigma M_{AB} = 0$, which involves the determination of the moments of the forces about the axis AB joining points A and B (see Sample Prob. 4.10).

Photo 4.6 Universal joints, easily seen on the drive shafts of rear-wheel-drive cars and trucks, allow rotational motion to be transferred between two noncollinear shafts.

Photo 4.7 The pillow block bearing shown supports the shaft of a fan used in an industrial facility.

connections are shown in Fig. 4.10 with their corresponding reactions. A simple way of determining the type of reaction corresponding to a given support or connection and the number of unknowns involved is to find which of the six fundamental motions (translation in the x, y, and z directions, rotation about the x, y, and z axes) are allowed and which motions are prevented.

Ball supports, frictionless surfaces, and cables, for example, prevent translation in one direction only and thus exert a single force whose line of action is known; each of these supports involves one unknown, namely, the magnitude of the reaction. Rollers on rough surfaces and wheels on rails prevent translation in two directions; the corresponding reactions consist of two unknown force components. Rough surfaces in direct contact and ball-and-socket supports prevent translation in three directions; these supports involve three unknown force components.

Some supports and connections can prevent rotation as well as translation; the corresponding reactions include couples as well as forces. For example, the reaction at a fixed support, which prevents any motion (rotation as well as translation), consists of three unknown forces and three unknown couples. A universal joint, which is designed to allow rotation about two axes, will exert a reaction consisting of three unknown force components and one unknown couple.

Other supports and connections are primarily intended to prevent translation; their design, however, is such that they also prevent some rotations. The corresponding reactions consist essentially of force components but *may* also include couples. One group of supports of this type includes hinges and bearings designed to support radial loads only (for example, journal bearings, roller bearings). The corresponding reactions consist of two force components but may also include two couples. Another group includes pin-and-bracket supports, hinges, and bearings designed to support an axial thrust as well as a radial load (for example, ball bearings). The corresponding reactions consist of three force components but may include two couples. However, these supports will not exert any appreciable couples under normal conditions of use. Therefore, *only* force components should be included in their analysis *unless* it is found that couples are necessary to maintain the equilibrium of the rigid body, or unless the support is known to have been specifically designed to exert a couple (see Probs. 4.119 through 4.122).

If the reactions involve more than six unknowns, there are more unknowns than equations, and some of the reactions are *statically indeterminate*. If the reactions involve fewer than six unknowns, there are more equations than unknowns, and some of the equations of equilibrium cannot be satisfied under general loading conditions; the rigid body is only *partially constrained*. Under the particular loading conditions corresponding to a given problem, however, the extra equations often reduce to trivial identities, such as $0 = 0$, and can be disregarded; although only partially constrained, the rigid body remains in equilibrium (see Sample Probs. 4.7 and 4.8). Even with six or more unknowns, it is possible that some equations of equilibrium will not be satisfied. This can occur when the reactions associated with the given supports either are parallel or intersect the same line; the rigid body is then *improperly constrained*.

Fig. 4.10 Reactions at supports and connections.

SAMPLE PROBLEM 4.7

A 20-kg ladder used to reach high shelves in a storeroom is supported by two flanged wheels A and B mounted on a rail and by an unflanged wheel C resting against a rail fixed to the wall. An 80-kg man stands on the ladder and leans to the right. The line of action of the combined weight $\mathbf{W}$ of the man and ladder intersects the floor at point D. Determine the reactions at A, B, and C.

SOLUTION

Free-Body Diagram. A free-body diagram of the ladder is drawn. The forces involved are the combined weight of the man and ladder,

$$\mathbf{W} = -mg\mathbf{j} = -(80 \text{ kg} + 20 \text{ kg})(9.81 \text{ m/s}^2)\mathbf{j} = -(981 \text{ N})\mathbf{j}$$

and five unknown reaction components, two at each flanged wheel and one at the unflanged wheel. The ladder is thus only partially constrained; it is free to roll along the rails. It is, however, in equilibrium under the given load since the equation $\Sigma F_x = 0$ is satisfied.

Equilibrium Equations. We express that the forces acting on the ladder form a system equivalent to zero:

$$\Sigma \mathbf{F} = 0: \quad A_y\mathbf{j} + A_z\mathbf{k} + B_y\mathbf{j} + B_z\mathbf{k} - (981 \text{ N})\mathbf{j} + C\mathbf{k} = 0$$
$$(A_y + B_y - 981 \text{ N})\mathbf{j} + (A_z + B_z + C)\mathbf{k} = 0 \quad (1)$$
$$\Sigma \mathbf{M}_A = \Sigma(\mathbf{r} \times \mathbf{F}) = 0: \quad 1.2\mathbf{i} \times (B_y\mathbf{j} + B_z\mathbf{k}) + (0.9\mathbf{i} - 0.6\mathbf{k}) \times (-981\mathbf{j})$$
$$+ (0.6\mathbf{i} + 3\mathbf{j} - 1.2\mathbf{k}) \times C\mathbf{k} = 0$$

Computing the vector products, we have[†]

$$1.2B_y\mathbf{k} - 1.2B_z\mathbf{j} - 882.9\mathbf{k} - 588.6\mathbf{i} - 0.6C\mathbf{j} + 3C\mathbf{i} = 0$$
$$(3C - 588.6)\mathbf{i} - (1.2B_z + 0.6C)\mathbf{j} + (1.2B_y - 882.9)\mathbf{k} = 0 \quad (2)$$

Setting the coefficients of $\mathbf{i}$, $\mathbf{j}$, $\mathbf{k}$ equal to zero in Eq. (2), we obtain the following three scalar equations, which express that the sum of the moments about each coordinate axis must be zero:

$$3C - 588.6 = 0 \qquad C = +196.2 \text{ N}$$
$$1.2B_z + 0.6C = 0 \qquad B_z = -98.1 \text{ N}$$
$$1.2B_y - 882.9 = 0 \qquad B_y = +736 \text{ N}$$

The reactions at B and C are therefore

$$\mathbf{B} = +(736 \text{ N})\mathbf{j} - (98.1 \text{ N})\mathbf{k} \qquad \mathbf{C} = +(196.2 \text{ N})\mathbf{k} \quad \blacktriangleleft$$

Setting the coefficients of $\mathbf{j}$ and $\mathbf{k}$ equal to zero in Eq. (1), we obtain two scalar equations expressing that the sums of the components in the y and z directions are zero. Substituting for B_y, B_z, and C the values obtained above, we write

$$A_y + B_y - 981 = 0 \qquad A_y + 736 - 981 = 0 \qquad A_y = +245 \text{ N}$$
$$A_z + B_z + C = 0 \qquad A_z - 98.1 + 196.2 = 0 \qquad A_z = -98.1 \text{ N}$$

We conclude that the reaction at A is $\quad \mathbf{A} = +(245 \text{ N})\mathbf{j} - (98.1 \text{ N})\mathbf{k} \quad \blacktriangleleft$

[†]The moments in this sample problem and in Sample Probs. 4.8 and 4.9 can also be expressed in the form of determinants (see Sample Prob. 3.10).

SAMPLE PROBLEM 4.8

A 5 × 8-ft sign of uniform density weighs 270 lb and is supported by a ball-and-socket joint at A and by two cables. Determine the tension in each cable and the reaction at A.

SOLUTION

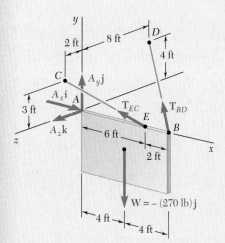

Free-Body Diagram. A free-body diagram of the sign is drawn. The forces acting on the free body are the weight $\mathbf{W} = -(270 \text{ lb})\mathbf{j}$ and the reactions at A, B, and E. The reaction at A is a force of unknown direction and is represented by three unknown components. Since the directions of the forces exerted by the cables are known, these forces involve only one unknown each, namely, the magnitudes T_{BD} and T_{EC}. Since there are only five unknowns, the sign is partially constrained. It can rotate freely about the x axis; it is, however, in equilibrium under the given loading, since the equation $\Sigma M_x = 0$ is satisfied.

The components of the forces $\mathbf{T}_{BD}$ and $\mathbf{T}_{EC}$ can be expressed in terms of the unknown magnitudes T_{BD} and T_{EC} by writing

$$\vec{BD} = -(8 \text{ ft})\mathbf{i} + (4 \text{ ft})\mathbf{j} - (8 \text{ ft})\mathbf{k} \qquad BD = 12 \text{ ft}$$
$$\vec{EC} = -(6 \text{ ft})\mathbf{i} + (3 \text{ ft})\mathbf{j} + (2 \text{ ft})\mathbf{k} \qquad EC = 7 \text{ ft}$$
$$\mathbf{T}_{BD} = T_{BD}\left(\frac{\vec{BD}}{BD}\right) = T_{BD}(-\tfrac{2}{3}\mathbf{i} + \tfrac{1}{3}\mathbf{j} - \tfrac{2}{3}\mathbf{k})$$
$$\mathbf{T}_{EC} = T_{EC}\left(\frac{\vec{EC}}{EC}\right) = T_{EC}(-\tfrac{6}{7}\mathbf{i} + \tfrac{3}{7}\mathbf{j} - \tfrac{2}{7}\mathbf{k})$$

Equilibrium Equations. We express that the forces acting on the sign form a system equivalent to zero:

$$\Sigma\mathbf{F} = 0: \qquad A_x\mathbf{i} + A_y\mathbf{j} + A_z\mathbf{k} + \mathbf{T}_{BD} + \mathbf{T}_{EC} - (270 \text{ lb})\mathbf{j} = 0$$
$$(A_x - \tfrac{2}{3}T_{BD} - \tfrac{6}{7}T_{EC})\mathbf{i} + (A_y + \tfrac{1}{3}T_{BD} + \tfrac{3}{7}T_{EC} - 270 \text{ lb})\mathbf{j}$$
$$+ (A_z - \tfrac{2}{3}T_{BD} + \tfrac{2}{7}T_{EC})\mathbf{k} = 0 \quad (1)$$

$$\Sigma\mathbf{M}_A = \Sigma(\mathbf{r} \times \mathbf{F}) = 0:$$
$$(8 \text{ ft})\mathbf{i} \times T_{BD}(-\tfrac{2}{3}\mathbf{i} + \tfrac{1}{3}\mathbf{j} - \tfrac{2}{3}\mathbf{k}) + (6 \text{ ft})\mathbf{i} \times T_{EC}(-\tfrac{6}{7}\mathbf{i} + \tfrac{3}{7}\mathbf{j} + \tfrac{2}{7}\mathbf{k})$$
$$+ (4 \text{ ft})\mathbf{i} \times (-270 \text{ lb})\mathbf{j} = 0$$
$$(2.667T_{BD} + 2.571T_{EC} - 1080 \text{ lb})\mathbf{k} + (5.333T_{BD} - 1.714T_{EC})\mathbf{j} = 0 \quad (2)$$

Setting the coefficients of $\mathbf{j}$ and $\mathbf{k}$ equal to zero in Eq. (2), we obtain two scalar equations which can be solved for T_{BD} and T_{EC}:

$$T_{BD} = 101.3 \text{ lb} \qquad T_{EC} = 315 \text{ lb} \quad \blacktriangleleft$$

Setting the coefficients of $\mathbf{i}$, $\mathbf{j}$, and $\mathbf{k}$ equal to zero in Eq. (1), we obtain three more equations, which yield the components of $\mathbf{A}$. We have

$$\mathbf{A} = +(338 \text{ lb})\mathbf{i} + (101.2 \text{ lb})\mathbf{j} - (22.5 \text{ lb})\mathbf{k} \quad \blacktriangleleft$$

A uniform pipe cover of radius $r = 240$ mm and mass 30 kg is held in a horizontal position by the cable CD. Assuming that the bearing at B does not exert any axial thrust, determine the tension in the cable and the reactions at A and B.

SOLUTION

Free-Body Diagram. A free-body diagram is drawn with the coordinate axes shown. The forces acting on the free body are the weight of the cover,

$$\mathbf{W} = -mg\mathbf{j} = -(30 \text{ kg})(9.81 \text{ m/s}^2)\mathbf{j} = -(294 \text{ N})\mathbf{j}$$

and reactions involving six unknowns, namely, the magnitude of the force $\mathbf{T}$ exerted by the cable, three force components at hinge A, and two at hinge B. The components of $\mathbf{T}$ are expressed in terms of the unknown magnitude T by resolving the vector $\overrightarrow{DC}$ into rectangular components and writing

$$\overrightarrow{DC} = -(480 \text{ mm})\mathbf{i} + (240 \text{ mm})\mathbf{j} - (160 \text{ mm})\mathbf{k} \qquad DC = 560 \text{ mm}$$

$$\mathbf{T} = T\frac{\overrightarrow{DC}}{DC} = -\tfrac{6}{7}T\mathbf{i} + \tfrac{3}{7}T\mathbf{j} - \tfrac{2}{7}T\mathbf{k}$$

Equilibrium Equations. We express that the forces acting on the pipe cover form a system equivalent to zero:

$$\Sigma\mathbf{F} = 0: \qquad A_x\mathbf{i} + A_y\mathbf{j} + A_z\mathbf{k} + B_x\mathbf{i} + B_y\mathbf{j} + \mathbf{T} - (294 \text{ N})\mathbf{j} = 0$$
$$(A_x + B_x - \tfrac{6}{7}T)\mathbf{i} + (A_y + B_y + \tfrac{3}{7}T - 294 \text{ N})\mathbf{j} + (A_z - \tfrac{2}{7}T)\mathbf{k} = 0 \quad (1)$$

$$\Sigma\mathbf{M}_B = \Sigma(\mathbf{r} \times \mathbf{F}) = 0:$$
$$2r\mathbf{k} \times (A_x\mathbf{i} + A_y\mathbf{j} + A_z\mathbf{k})$$
$$+ (2r\mathbf{i} + r\mathbf{k}) \times (-\tfrac{6}{7}T\mathbf{i} + \tfrac{3}{7}T\mathbf{j} - \tfrac{2}{7}T\mathbf{k})$$
$$+ (r\mathbf{i} + r\mathbf{k}) \times (-294 \text{ N})\mathbf{j} = 0$$
$$(-2A_y - \tfrac{3}{7}T + 294 \text{ N})r\mathbf{i} + (2A_x - \tfrac{2}{7}T)r\mathbf{j} + (\tfrac{6}{7}T - 294 \text{ N})r\mathbf{k} = 0 \quad (2)$$

Setting the coefficients of the unit vectors equal to zero in Eq. (2), we write three scalar equations, which yield

$$A_x = +49.0 \text{ N} \qquad A_y = +73.5 \text{ N} \qquad T = 343 \text{ N} \quad \blacktriangleleft$$

Setting the coefficients of the unit vectors equal to zero in Eq. (1), we obtain three more scalar equations. After substituting the values of T, A_x, and A_y into these equations, we obtain

$$A_z = +98.0 \text{ N} \qquad B_x = +245 \text{ N} \qquad B_y = +73.5 \text{ N}$$

The reactions at A and B are therefore

$$\mathbf{A} = +(49.0 \text{ N})\mathbf{i} + (73.5 \text{ N})\mathbf{j} + (98.0 \text{ N})\mathbf{k} \quad \blacktriangleleft$$
$$\mathbf{B} = +(245 \text{ N})\mathbf{i} + (73.5 \text{ N})\mathbf{j} \quad \blacktriangleleft$$

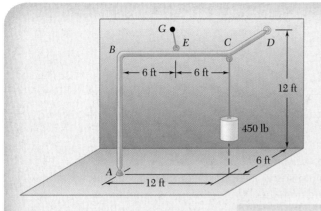

SAMPLE PROBLEM 4.10

A 450-lb load hangs from the corner C of a rigid piece of pipe $ABCD$ which has been bent as shown. The pipe is supported by the ball-and-socket joints A and D, which are fastened, respectively, to the floor and to a vertical wall, and by a cable attached at the midpoint E of the portion BC of the pipe and at a point G on the wall. Determine (a) where G should be located if the tension in the cable is to be minimum, (b) the corresponding minimum value of the tension.

SOLUTION

Free-Body Diagram. The free-body diagram of the pipe includes the load $\mathbf{W} = (-450 \text{ lb})\mathbf{j}$, the reactions at A and D, and the force $\mathbf{T}$ exerted by the cable. To eliminate the reactions at A and D from the computations, we express that the sum of the moments of the forces about AD is zero. Denoting by $\boldsymbol{\lambda}$ the unit vector along AD, we write

$$\Sigma M_{AD} = 0: \qquad \boldsymbol{\lambda} \cdot (\overrightarrow{AE} \times \mathbf{T}) + \boldsymbol{\lambda} \cdot (\overrightarrow{AC} \times \mathbf{W}) = 0 \qquad (1)$$

The second term in Eq. (1) can be computed as follows:

$$\overrightarrow{AC} \times \mathbf{W} = (12\mathbf{i} + 12\mathbf{j}) \times (-450\mathbf{j}) = -5400\mathbf{k}$$

$$\boldsymbol{\lambda} = \frac{\overrightarrow{AD}}{AD} = \frac{12\mathbf{i} + 12\mathbf{j} - 6\mathbf{k}}{18} = \tfrac{2}{3}\mathbf{i} + \tfrac{2}{3}\mathbf{j} - \tfrac{1}{3}\mathbf{k}$$

$$\boldsymbol{\lambda} \cdot (\overrightarrow{AC} \times \mathbf{W}) = (\tfrac{2}{3}\mathbf{i} + \tfrac{2}{3}\mathbf{j} - \tfrac{1}{3}\mathbf{k}) \cdot (-5400\mathbf{k}) = +1800$$

Substituting the value obtained into Eq. (1), we write

$$\boldsymbol{\lambda} \cdot (\overrightarrow{AE} \times \mathbf{T}) = -1800 \text{ lb} \cdot \text{ft} \qquad (2)$$

Minimum Value of Tension. Recalling the commutative property for mixed triple products, we rewrite Eq. (2) in the form

$$\mathbf{T} \cdot (\boldsymbol{\lambda} \times \overrightarrow{AE}) = -1800 \text{ lb} \cdot \text{ft} \qquad (3)$$

which shows that the projection of $\mathbf{T}$ on the vector $\boldsymbol{\lambda} \times \overrightarrow{AE}$ is a constant. It follows that $\mathbf{T}$ is minimum when parallel to the vector

$$\boldsymbol{\lambda} \times \overrightarrow{AE} = (\tfrac{2}{3}\mathbf{i} + \tfrac{2}{3}\mathbf{j} - \tfrac{1}{3}\mathbf{k}) \times (6\mathbf{i} + 12\mathbf{j}) = 4\mathbf{i} - 2\mathbf{j} + 4\mathbf{k}$$

Since the corresponding unit vector is $\tfrac{2}{3}\mathbf{i} - \tfrac{1}{3}\mathbf{j} + \tfrac{2}{3}\mathbf{k}$, we write

$$\mathbf{T}_{\min} = T(\tfrac{2}{3}\mathbf{i} - \tfrac{1}{3}\mathbf{j} + \tfrac{2}{3}\mathbf{k}) \qquad (4)$$

Substituting for $\mathbf{T}$ and $\boldsymbol{\lambda} \times \overrightarrow{AE}$ in Eq. (3) and computing the dot products, we obtain $6T = -1800$ and, thus, $T = -300$. Carrying this value into (4), we obtain

$$\mathbf{T}_{\min} = -200\mathbf{i} + 100\mathbf{j} - 200\mathbf{k} \qquad T_{\min} = 300 \text{ lb} \quad \blacktriangleleft$$

Location of G. Since the vector $\overrightarrow{EG}$ and the force $\mathbf{T}_{\min}$ have the same direction, their components must be proportional. Denoting the coordinates of G by $x, y, 0$, we write

$$\frac{x - 6}{-200} = \frac{y - 12}{+100} = \frac{0 - 6}{-200} \qquad x = 0 \qquad y = 15 \text{ ft} \quad \blacktriangleleft$$

SOLVING PROBLEMS
ON YOUR OWN

The equilibrium of a *three-dimensional body* was considered in the sections you just completed. It is again most important that you draw a complete *free-body diagram* as the first step of your solution.

1. As you draw the free-body diagram, pay particular attention to the reactions at the supports. The number of unknowns at a support can range from one to six (Fig. 4.10). To decide whether an unknown reaction or reaction component exists at a support, ask yourself whether the support prevents motion of the body in a certain direction or about a certain axis.

  a. If motion is prevented in a certain direction, include in your free-body diagram an unknown *reaction* or *reaction component* that acts in the *same direction.*

  b. If a support prevents rotation about a certain axis, include in your free-body diagram a *couple* of unknown magnitude that acts about the *same axis.*

2. The external forces acting on a three-dimensional body form a system equivalent to zero. Writing $\Sigma \mathbf{F} = 0$ and $\Sigma \mathbf{M}_A = 0$ about an appropriate point A, and setting the coefficients of $\mathbf{i}$, $\mathbf{j}$, $\mathbf{k}$ in both equations equal to zero will provide you with six scalar equations. In general, these equations will contain six unknowns and may be solved for these unknowns.

3. After completing your free-body diagram, you may want to seek equations involving as few unknowns as possible. The following strategies may help you.

  a. By summing moments about a ball-and-socket support or a hinge, you will obtain equations from which three unknown reaction components have been eliminated [Sample Probs. 4.8 and 4.9].

  b. If you can draw an axis through the points of application of all but one of the unknown reactions, summing moments about that axis will yield an equation in a single unknown [Sample Prob. 4.10].

4. After drawing your free-body diagram, you may find that one of the following situations exists.

  a. The reactions involve fewer than six unknowns; the body is said to be *partially constrained* and motion of the body is possible. However, you may be able to determine the reactions for a given loading condition [Sample Prob. 4.7].

  b. The reactions involve more than six unknowns; the reactions are said to be *statically indeterminate.* Although you may be able to calculate one or two reactions, you cannot determine all of the reactions [Sample Prob. 4.10].

  c. The reactions are parallel or intersect the same line; the body is said to be *improperly constrained,* and motion can occur under a general loading condition.

PROBLEMS

FREE BODY PRACTICE PROBLEMS

4.F5 A 4 × 8-ft sheet of plywood weighing 34 lb has been temporarily placed among three pipe supports. The lower edge of the sheet rests on small collars at A and B and its upper edge leans against pipe C. Neglecting friction on all surfaces, draw the free-body diagram needed to determine the reactions at A, B, and C.

Fig. P4.F5

4.F6 Two transmission belts pass over sheaves welded to an axle supported by bearings at B and D. The sheave at A has a radius of 2.5 in. and the sheave at C has a radius of 2 in. Knowing that the system rotates at a constant rate, draw the free-body diagram needed to determine the tension T and the reactions at B and D. Assume that the bearing at D does not exert any axial thrust and neglect the weights of the sheaves and axle.

Fig. P4.F6

4.F7 The 6-m pole ABC is acted upon by a 455-N force as shown. The pole is held by a ball-and-socket joint at A and by two cables BD and BE. Draw the free-body diagram needed to determine the tension in each cable and the reaction at A.

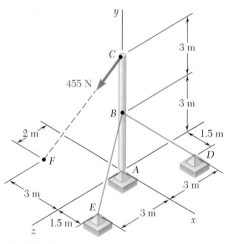

Fig. P4.F7

199

END-OF-SECTION PROBLEMS

4.91 A 200-mm lever and a 240-mm-diameter pulley are welded to the axle *BE* that is supported by bearings at *C* and *D*. If a 720-N vertical load is applied at *A* when the lever is horizontal, determine (*a*) the tension in the cord, (*b*) the reactions at *C* and *D*. Assume that the bearing at *D* does not exert any axial thrust.

Fig. P4.91

4.92 Solve Prob. 4.91, assuming that the axle has been rotated clockwise in its bearings by 30° and that the 720-N load remains vertical.

4.93 A 4 × 8-ft sheet of plywood weighing 40 lb has been temporarily propped against column *CD*. It rests at *A* and *B* on small wooden blocks and against protruding nails. Neglecting friction at all surfaces of contact, determine the reactions at *A*, *B*, and *C*.

Fig. P4.93

Fig. P4.94

4.94 Two tape spools are attached to an axle supported by bearings at *A* and *D*. The radius of spool *B* is 1.5 in. and the radius of spool *C* is 2 in. Knowing that $T_B = 20$ lb and that the system rotates at a constant rate, determine the reactions at *A* and *D*. Assume that the bearing at *A* does not exert any axial thrust and neglect the weights of the spools and axle.

4.95 Two transmission belts pass over a double-sheaved pulley that is attached to an axle supported by bearings at A and D. The radius of the inner sheave is 125 mm and the radius of the outer sheave is 250 mm. Knowing that when the system is at rest, the tension is 90 N in both portions of belt B and 150 N in both portions of belt C, determine the reactions at A and D. Assume that the bearing at D does not exert any axial thrust.

Fig. P4.95

4.96 Solve Prob. 4.95, assuming that the pulley rotates at a constant rate and that $T_B = 104$ N, $T'_B = 84$ N, and $T_C = 175$ N.

4.97 Two steel pipes AB and BC, each having a mass per unit length of 8 kg/m, are welded together at B and supported by three vertical wires. Knowing that $a = 0.4$ m, determine the tension in each wire.

4.98 For the pipe assembly of Prob. 4.97, determine (*a*) the largest permissible value of a if the assembly is not to tip, (*b*) the corresponding tension in each wire.

4.99 The 45-lb square plate shown is supported by three vertical wires. Determine the tension in each wire.

Fig. P4.97

Fig. P4.99

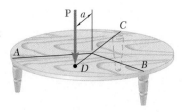

Fig. P4.100

4.100 The table shown weighs 30 lb and has a diameter of 4 ft. It is supported by three legs equally spaced around the edge. A vertical load **P** of magnitude 100 lb is applied to the top of the table at D. Determine the maximum value of a if the table is not to tip over. Show, on a sketch, the area of the table over which **P** can act without tipping the table.

4.101 An opening in a floor is covered by a 1×1.2-m sheet of plywood of mass 18 kg. The sheet is hinged at A and B and is maintained in a position slightly above the floor by a small block C. Determine the vertical component of the reaction (a) at A, (b) at B, (c) at C.

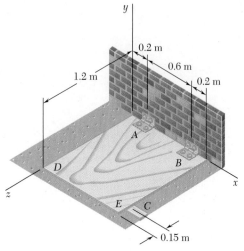

Fig. P4.101

4.102 Solve Prob. 4.101, assuming that the small block C is moved and placed under edge DE at a point 0.15 m from corner E.

4.103 The rectangular plate shown weighs 80 lb and is supported by three vertical wires. Determine the tension in each wire.

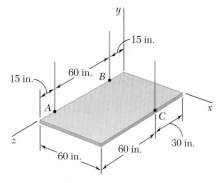

Fig. P4.103 and P4.104

4.104 The rectangular plate shown weighs 80 lb and is supported by three vertical wires. Determine the weight and location of the lightest block that should be placed on the plate if the tensions in the three wires are to be equal.

4.105 A 2.4-m boom is held by a ball-and-socket joint at C and by two cables AD and AE. Determine the tension in each cable and the reaction at C.

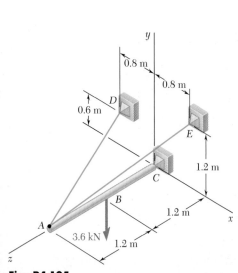

Fig. P4.105

4.106 Solve Prob. 4.105, assuming that the 3.6-kN load is applied at point A.

4.107 A 10-ft boom is acted upon by the 840-lb force shown. Determine the tension in each cable and the reaction at the ball-and-socket joint at A.

4.108 A 12-m pole supports a horizontal cable CD and is held by a ball and socket at A and two cables BE and BF. Knowing that the tension in cable CD is 14 kN and assuming that CD is parallel to the x axis ($\phi = 0$), determine the tension in cables BE and BF and the reaction at A.

Fig. P4.107

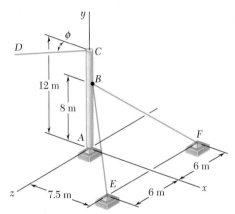

Fig. P4.108

4.109 Solve Prob. 4.108, assuming that cable CD forms an angle $\phi = 25°$ with the vertical xy plane.

4.110 A 48-in. boom is held by a ball-and-socket joint at C and by two cables BF and DAE; cable DAE passes around a frictionless pulley at A. For the loading shown, determine the tension in each cable and the reaction at C.

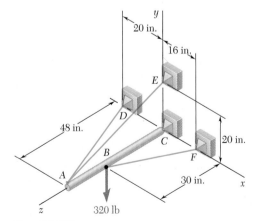

Fig. P4.110

4.111 Solve Prob. 4.110, assuming that the 320-lb load is applied at A.

4.112 A 600-lb crate hangs from a cable that passes over a pulley B and is attached to a support at H. The 200-lb boom AB is supported by a ball-and-socket joint at A and by two cables DE and DF. The center of gravity of the boom is located at G. Determine (a) the tension in cables DE and DF, (b) the reaction at A.

Fig. P4.112

4.113 A 100-kg uniform rectangular plate is supported in the position shown by hinges A and B and by cable DCE that passes over a frictionless hook at C. Assuming that the tension is the same in both parts of the cable, determine (a) the tension in the cable, (b) the reactions at A and B. Assume that the hinge at B does not exert any axial thrust.

4.114 Solve Prob. 4.113, assuming that cable DCE is replaced by a cable attached to point E and hook C.

4.115 The rectangular plate shown weighs 75 lb and is held in the position shown by hinges at A and B and by cable EF. Assuming that the hinge at B does not exert any axial thrust, determine (a) the tension in the cable, (b) the reactions at A and B.

Fig. P4.113

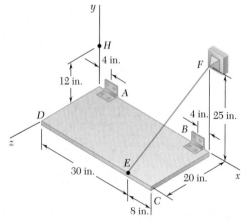

Fig. P4.115

4.116 Solve Prob. 4.115, assuming that cable EF is replaced by a cable attached at points E and H.

4.117 A 20-kg cover for a roof opening is hinged at corners A and B. The roof forms an angle of 30° with the horizontal, and the cover is maintained in a horizontal position by the brace CE. Determine (*a*) the magnitude of the force exerted by the brace, (*b*) the reactions at the hinges. Assume that the hinge at A does not exert any axial thrust.

4.118 The bent rod $ABEF$ is supported by bearings at C and D and by wire AH. Knowing that portion AB of the rod is 250 mm long, determine (*a*) the tension in wire AH, (*b*) the reactions at C and D. Assume that the bearing at D does not exert any axial thrust.

Fig. P4.117

Fig. *P4.118*

4.119 Solve Prob. 4.115, assuming that the hinge at B is removed and that the hinge at A can exert couples about axes parallel to the y and z axes.

4.120 Solve Prob. 4.118, assuming that the bearing at D is removed and that the bearing at C can exert couples about axes parallel to the y and z axes.

4.121 The assembly shown is welded to collar A that fits on the vertical pin shown. The pin can exert couples about the x and z axes but does not prevent motion about or along the y axis. For the loading shown, determine the tension in each cable and the reaction at A.

Fig. P4.121

Fig. P4.122

4.122 The assembly shown is used to control the tension T in a tape that passes around a frictionless spool at E. Collar C is welded to rods ABC and CDE. It can rotate about shaft FG but its motion along the shaft is prevented by a washer S. For the loading shown, determine (a) the tension T in the tape, (b) the reaction at C.

4.123 The rigid L-shaped member ABF is supported by a ball-and-socket joint at A and by three cables. For the loading shown, determine the tension in each cable and the reaction at A.

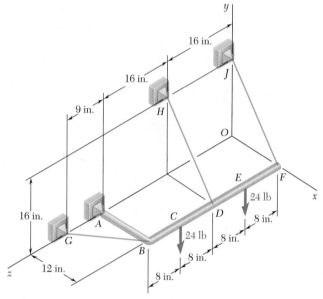

Fig. P4.123

4.124 Solve Prob. 4.123, assuming that the load at C has been removed.

4.125 The rigid L-shaped member ABC is supported by a ball-and-socket joint at A and by three cables. If a 1.8-kN load is applied at F, determine the tension in each cable.

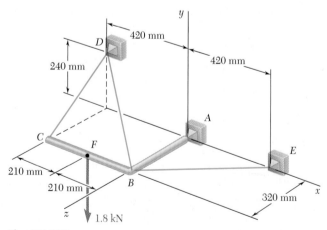

Fig. P4.125

4.126 Solve Prob. 4.125, assuming that the 1.8-kN load is applied at C.

4.127 The assembly shown consists of an 80-mm rod AF that is welded to a cross consisting of four 200-mm arms. The assembly is supported by a ball-and-socket joint at F and by three short links, each of which forms an angle of 45° with the vertical. For the loading shown, determine (a) the tension in each link, (b) the reaction at F.

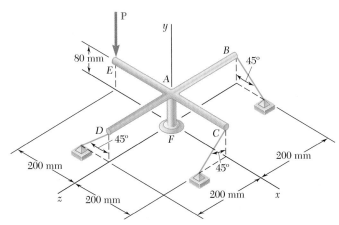

Fig. P4.127

4.128 The uniform 10-kg rod AB is supported by a ball-and-socket joint at A and by the cord CG that is attached to the midpoint G of the rod. Knowing that the rod leans against a frictionless vertical wall at B, determine (a) the tension in the cord, (b) the reactions at A and B.

Fig. P4.128

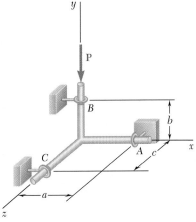

Fig. P4.129

4.129 Three rods are welded together to form a "corner" that is supported by three eyebolts. Neglecting friction, determine the reactions at A, B, and C when $P = 240$ lb, $a = 12$ in., $b = 8$ in., and $c = 10$ in.

4.130 Solve Prob. 4.129, assuming that the force $\mathbf{P}$ is removed and is replaced by a couple $\mathbf{M} = +(600 \text{ lb} \cdot \text{in.})\mathbf{j}$ acting at B.

4.131 In order to clean the clogged drainpipe AE, a plumber has disconnected both ends of the pipe and inserted a power snake through the opening at A. The cutting head of the snake is connected by a heavy cable to an electric motor that rotates at a constant speed as the plumber forces the cable into the pipe. The forces exerted by the plumber and the motor on the end of the cable can be represented by the wrench $\mathbf{F} = -(48 \text{ N})\mathbf{k}$, $\mathbf{M} = -(90 \text{ N} \cdot \text{m})\mathbf{k}$. Determine the additional reactions at B, C, and D caused by the cleaning operation. Assume that the reaction at each support consists of two force components perpendicular to the pipe.

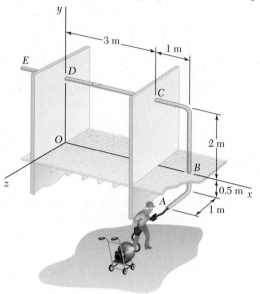

Fig. P4.131

4.132 Solve Prob. 4.131, assuming that the plumber exerts a force $\mathbf{F} = -(48 \text{ N})\mathbf{k}$ and that the motor is turned off ($\mathbf{M} = 0$).

4.133 The 50-kg plate $ABCD$ is supported by hinges along edge AB and by wire CE. Knowing that the plate is uniform, determine the tension in the wire.

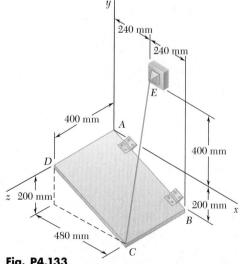

Fig. P4.133

4.134 Solve Prob. 4.133, assuming that wire CE is replaced by a wire connecting E and D.

4.135 Two rectangular plates are welded together to form the assembly shown. The assembly is supported by ball-and-socket joints at B and D and by a ball on a horizontal surface at C. For the loading shown, determine the reaction at C.

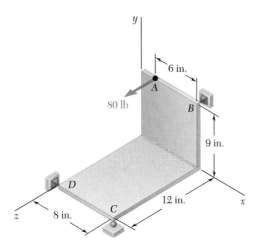

Fig. P4.135

4.136 Two 2 × 4-ft plywood panels, each of weight 12 lb, are nailed together as shown. The panels are supported by ball-and-socket joints at A and F and by the wire BH. Determine (a) the location of H in the xy plane if the tension in the wire is to be minimum, (b) the corresponding minimum tension.

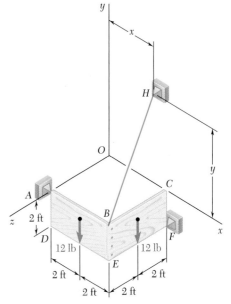

Fig. P4.136

4.137 Solve Prob. 4.136, subject to the restriction that H must lie on the y axis.

4.138 The frame ACD is supported by ball-and-socket joints at A and D and by a cable that passes through a ring at B and is attached to hooks at G and H. Knowing that the frame supports at point C a load of magnitude $P = 268$ N, determine the tension in the cable.

4.139 Solve Prob. 4.138, assuming that cable GBH is replaced by a cable GB attached at G and B.

4.140 The bent rod $ABDE$ is supported by ball-and-socket joints at A and E and by the cable DF. If a 60-lb load is applied at C as shown, determine the tension in the cable.

Fig. P4.138

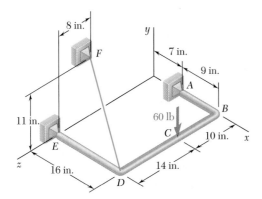

Fig. P4.140

4.141 Solve Prob. 4.140, assuming that cable DF is replaced by a cable connecting B and F.

REVIEW AND SUMMARY

Equilibrium equations

This chapter was devoted to the study of the *equilibrium of rigid bodies*, i.e., to the situation when the external forces acting on a rigid body *form a system equivalent to zero* [Sec. 4.1]. We then have

$$\Sigma\mathbf{F} = 0 \qquad \Sigma\mathbf{M}_O = \Sigma(\mathbf{r} \times \mathbf{F}) = 0 \qquad (4.1)$$

Resolving each force and each moment into its rectangular components, we can express the necessary and sufficient conditions for the equilibrium of a rigid body with the following six scalar equations:

$$\Sigma F_x = 0 \qquad \Sigma F_y = 0 \qquad \Sigma F_z = 0 \qquad (4.2)$$
$$\Sigma M_x = 0 \qquad \Sigma M_y = 0 \qquad \Sigma M_z = 0 \qquad (4.3)$$

These equations can be used to determine unknown forces applied to the rigid body or unknown reactions exerted by its supports.

Free-body diagram

When solving a problem involving the equilibrium of a rigid body, it is essential to consider *all* of the forces acting on the body. Therefore, the first step in the solution of the problem should be to draw a *free-body diagram* showing the body under consideration and all of the unknown as well as known forces acting on it [Sec. 4.2].

Equilibrium of a two-dimensional structure

In the first part of the chapter, we considered the *equilibrium of a two-dimensional structure;* i.e., we assumed that the structure considered and the forces applied to it were contained in the same plane. We saw that each of the reactions exerted on the structure by its supports could involve one, two, or three unknowns, depending upon the type of support [Sec. 4.3].

In the case of a two-dimensional structure, Eqs. (4.1), or Eqs. (4.2) and (4.3), reduce to *three equilibrium equations*, namely

$$\Sigma F_x = 0 \qquad \Sigma F_y = 0 \qquad \Sigma M_A = 0 \qquad (4.5)$$

where A is an arbitrary point in the plane of the structure [Sec. 4.4]. These equations can be used to solve for three unknowns. While the three equilibrium equations (4.5) cannot be *augmented* with additional equations, any of them can be *replaced* by another equation. Therefore, we can write alternative sets of equilibrium equations, such as

$$\Sigma F_x = 0 \qquad \Sigma M_A = 0 \qquad \Sigma M_B = 0 \qquad (4.6)$$

where point B is chosen in such a way that the line AB is not parallel to the y axis, or

$$\Sigma M_A = 0 \qquad \Sigma M_B = 0 \qquad \Sigma M_C = 0 \qquad (4.7)$$

where the points A, B, and C do not lie in a straight line.

Since any set of equilibrium equations can be solved for only three unknowns, the reactions at the supports of a rigid two-dimensional structure cannot be completely determined if they involve *more than three unknowns*; they are said to be *statically indeterminate* [Sec. 4.5]. On the other hand, if the reactions involve *fewer than three unknowns*, equilibrium will not be maintained under general loading conditions; the structure is said to be *partially constrained*. The fact that the reactions involve exactly three unknowns is no guarantee that the equilibrium equations can be solved for all three unknowns. If the supports are arranged in such a way that the reactions are *either concurrent or parallel*, the reactions are statically indeterminate, and the structure is said to be *improperly constrained*.

Two particular cases of equilibrium of a rigid body were given special attention. In Sec. 4.6, a *two-force body* was defined as a rigid body subjected to forces at only two points, and it was shown that the resultants $\mathbf{F}_1$ and $\mathbf{F}_2$ of these forces must have the *same magnitude, the same line of action, and opposite sense* (Fig. 4.11), a property which will simplify the solution of certain problems in later chapters. In Sec. 4.7, a *three-force body* was defined as a rigid body subjected to forces at only three points, and it was shown that the resultants $\mathbf{F}_1$, $\mathbf{F}_2$, and $\mathbf{F}_3$ of these forces must be *either concurrent* (Fig. 4.12) or *parallel*. This property provides us with an alternative approach to the solution of problems involving a three-force body [Sample Prob. 4.6].

Two-force body

Three-force body

Fig. 4.11

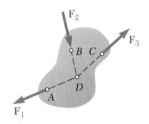

Fig. 4.12

In the second part of the chapter, we considered the *equilibrium of a three-dimensional body* and saw that each of the reactions exerted on the body by its supports could involve between one and six unknowns, depending upon the type of support [Sec. 4.8].

Equilibrium of a three-dimensional body

In the general case of the equilibrium of a three-dimensional body, all of the six scalar equilibrium equations (4.2) and (4.3) listed at the beginning of this review should be used and solved for *six unknowns* [Sec. 4.9]. In most problems, however, these equations will be more conveniently obtained if we first write

$$\Sigma \mathbf{F} = 0 \qquad \Sigma \mathbf{M}_O = \Sigma(\mathbf{r} \times \mathbf{F}) = 0 \qquad (4.1)$$

and express the forces $\mathbf{F}$ and position vectors $\mathbf{r}$ in terms of scalar components and unit vectors. The vector products can then be computed either directly or by means of determinants, and the desired scalar equations obtained by equating to zero the coefficients of the unit vectors [Sample Probs. 4.7 through 4.9].

We noted that as many as three unknown reaction components may be eliminated from the computation of $\Sigma \mathbf{M}_O$ in the second of the relations (4.1) through a judicious choice of point O. Also, the reactions at two points A and B can be eliminated from the solution of some problems by writing the equation $\Sigma M_{AB} = 0$, which involves the computation of the moments of the forces about an axis AB joining points A and B [Sample Prob. 4.10].

If the reactions involve more than six unknowns, some of the reactions are *statically indeterminate;* if they involve fewer than six unknowns, the rigid body is only *partially constrained.* Even with six or more unknowns, the rigid body will be *improperly constrained* if the reactions associated with the given supports either are parallel or intersect the same line.

REVIEW PROBLEMS

4.142 A gardener uses a 60-N wheelbarrow to transport a 250-N bag of fertilizer. What force must she exert on each handle?

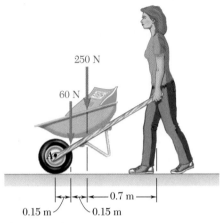

Fig. P4.142

4.143 The required tension in cable AB is 200 lb. Determine (*a*) the vertical force **P** that must be applied to the pedal, (*b*) the corresponding reaction at C.

Fig. P4.143

4.144 A lever AB is hinged at C and attached to a control cable at A. If the lever is subjected to a 500-N horizontal force at B, determine (*a*) the tension in the cable, (*b*) the reaction at C.

Fig. P4.144

4.145 A force **P** of magnitude 280 lb is applied to member $ABCD$, which is supported by a pin at A and by the cable CED. Neglecting friction and considering the case when $a = 3$ in., determine (*a*) the tension in the cable, (*b*) the reaction at A.

Fig. P4.145

4.146 Two slots have been cut in plate *DEF*, and the plate has been placed so that the slots fit two fixed, frictionless pins *A* and *B*. Knowing that *P* = 15 lb, determine (*a*) the force each pin exerts on the plate, (*b*) the reaction at *F*.

Fig. P4.146

4.147 Knowing that the tension in wire *BD* is 1300 N, determine the reaction at the fixed support *C* of the frame shown.

4.148 The spanner shown is used to rotate a shaft. A pin fits in a hole at *A*, while a flat, frictionless surface rests against the shaft at *B*. If a 60-lb force **P** is exerted on the spanner at *D*, find the reactions at *A* and *B*.

Fig. P4.147

Fig. P4.148

4.149 Rod *AB* is supported by a pin and bracket at *A* and rests against a frictionless peg at *C*. Determine the reactions at *A* and *C* when a 170-N vertical force is applied at *B*.

4.150 The 24-lb square plate shown is supported by three vertical wires. Determine (*a*) the tension in each wire when *a* = 10 in., (*b*) the value of *a* for which the tension in each wire is 8 lb.

Fig. P4.149

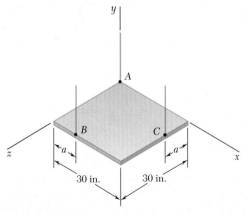

Fig. P4.150

4.151 Frame *ABCD* is supported by a ball-and-socket joint at *A* and by three cables. For $a = 150$ mm, determine the tension in each cable and the reaction at *A*.

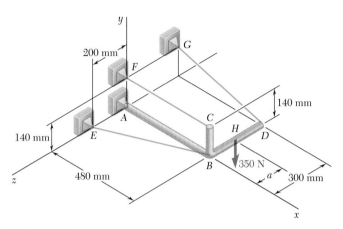

Fig. P4.151

4.152 The pipe *ACDE* is supported by ball-and-socket joints at *A* and *E* and by the wire *DF*. Determine the tension in the wire when a 640-N load is applied at *B* as shown.

Fig. P4.152

4.153 A force **P** is applied to a bent rod *ABC*, which may be supported in four different ways as shown. In each case, if possible, determine the reactions at the supports.

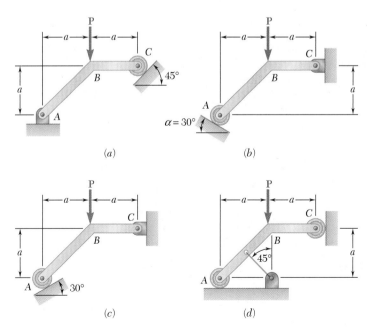

Fig. P4.153

COMPUTER PROBLEMS

4.C1 The position of the L-shaped rod shown is controlled by a cable attached at B. Knowing that the rod supports a load of magnitude $P = 50$ lb, write a computer program that can be used to calculate the tension T in the cable for values of θ from 0 to 120° using 10° increments. Using appropriate smaller increments, calculate the maximum tension T and the corresponding value of θ.

Fig. P4.C1

Fig. P4.C2

4.C2 The position of the 10-kg rod AB is controlled by the block shown, which is slowly moved to the left by the force **P**. Neglecting the effect of friction, write a computer program that can be used to calculate the magnitude P of the force for values of x decreasing from 750 mm to 0 using 50-mm increments. Using appropriate smaller increments, determine the maximum value of P and the corresponding value of x.

4.C3 and 4.C4 The constant of spring AB is k, and the spring is unstretched when $\theta = 0$. Knowing that $R = 10$ in., $a = 20$ in., and $k = 5$ lb/in., write a computer program that can be used to calculate the weight W corresponding to equilibrium for values of θ from 0 to 90° using 10° increments. Using appropriate smaller increments, determine the value of θ corresponding to equilibrium when $W = 5$ lb.

Fig. P4.C3

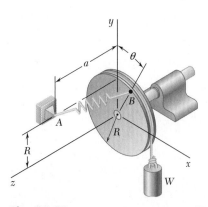

Fig. P4.C4

4.C5 A 200 × 250-mm panel of mass 20 kg is supported by hinges along edge *AB*. Cable *CDE* is attached to the panel at *C*, passes over a small pulley at *D*, and supports a cylinder of mass *m*. Neglecting the effect of friction, write a computer program that can be used to calculate the mass of the cylinder corresponding to equilibrium for values of θ from 0 to 90° using 10° increments. Using appropriate smaller increments, determine the value of θ corresponding to equilibrium when *m* = 10 kg.

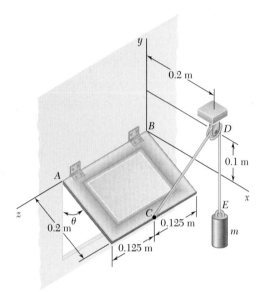

Fig. P4.C5

4.C6 The derrick shown supports a 2000-kg crate. It is held by a ball-and-socket joint at *A* and by two cables attached at *D* and *E*. Knowing that the derrick stands in a vertical plane forming an angle ϕ with the *xy* plane, write a computer program that can be used to calculate the tension in each cable for values of ϕ from 0 to 60° using 5° increments. Using appropriate smaller increments, determine the value of ϕ for which the tension in cable *BE* is maximum.

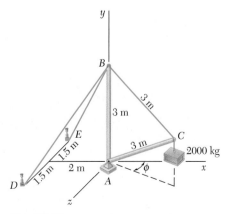

Fig. P4.C6

The Revelstoke Dam, located on the Columbia River in British Columbia, is subjected to three different kinds of distributed forces: the weights of its constituent elements, the pressure forces exerted by the water of its submerged face, and the pressure forces exerted by the ground on its base.

CHAPTER 5

Distributed Forces: Centroids and Centers of Gravity

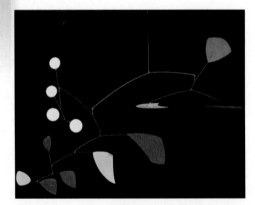

Photo 5.1   The precise balancing of the components of a mobile requires an understanding of centers of gravity and centroids, the main topics of this chapter.

5.1   INTRODUCTION

We have assumed so far that the attraction exerted by the earth on a rigid body could be represented by a single force $\mathbf{W}$. This force, called the force of gravity or the weight of the body, was to be applied at the *center of gravity* of the body (Sec. 3.2). Actually, the earth exerts a force on each of the particles forming the body. The action of the earth on a rigid body should thus be represented by a large number of small forces distributed over the entire body. You will learn in this chapter, however, that all of these small forces can be replaced by a single equivalent force $\mathbf{W}$. You will also learn how to determine the center of gravity, i.e., the point of application of the resultant $\mathbf{W}$, for bodies of various shapes.

In the first part of the chapter, two-dimensional bodies, such as flat plates and wires contained in a given plane, are considered. Two concepts closely associated with the determination of the center of gravity of a plate or a wire are introduced: the concept of the *centroid* of an area or a line and the concept of the *first moment* of an area or a line with respect to a given axis.

You will also learn that the computation of the area of a surface of revolution or of the volume of a body of revolution is directly related to the determination of the centroid of the line or area used to generate that surface or body of revolution (theorems of Pappus-Guldinus). And, as is shown in Secs. 5.8 and 5.9, the determination of the centroid of an area simplifies the analysis of beams subjected to distributed loads and the computation of the forces exerted on submerged rectangular surfaces, such as hydraulic gates and portions of dams.

In the last part of the chapter, you will learn how to determine the center of gravity of a three-dimensional body as well as the centroid of a volume and the first moments of that volume with respect to the coordinate planes.

AREAS AND LINES

5.2   CENTER OF GRAVITY OF A TWO-DIMENSIONAL BODY

Let us first consider a flat horizontal plate (Fig. 5.1). We can divide the plate into n small elements. The coordinates of the first element

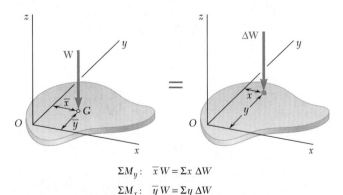

$$\Sigma M_y: \quad \overline{x}\,W = \Sigma x\,\Delta W$$

$$\Sigma M_x: \quad \overline{y}\,W = \Sigma y\,\Delta W$$

Fig. 5.1   Center of gravity of a plate.

are denoted by x_1 and y_1, those of the second element by x_2 and y_2, etc. The forces exerted by the earth on the elements of the plate will be denoted, respectively, by $\Delta\mathbf{W}_1$, $\Delta\mathbf{W}_2$, ... , $\Delta\mathbf{W}_n$. These forces or weights are directed toward the center of the earth; however, for all practical purposes they can be assumed to be parallel. Their resultant is therefore a single force in the same direction. The magnitude W of this force is obtained by adding the magnitudes of the elemental weights.

$$\Sigma F_z: \qquad W = \Delta W_1 + \Delta W_2 + \cdots + \Delta W_n$$

To obtain the coordinates $\bar{x}$ and $\bar{y}$ of the point G where the resultant $\mathbf{W}$ should be applied, we write that the moments of $\mathbf{W}$ about the y and x axes are equal to the sum of the corresponding moments of the elemental weights,

$$\Sigma M_y: \qquad \bar{x}W = x_1\,\Delta W_1 + x_2\,\Delta W_2 + \cdots + x_n\,\Delta W_n$$
$$\Sigma M_x: \qquad \bar{y}W = y_1\,\Delta W_1 + y_2\,\Delta W_2 + \cdots + y_n\,\Delta W_n \qquad (5.1)$$

If we now increase the number of elements into which the plate is divided and simultaneously decrease the size of each element, we obtain in the limit the following expressions:

$$W = \int dW \qquad \bar{x}W = \int x\,dW \qquad \bar{y}W = \int y\,dW \qquad (5.2)$$

These equations define the weight $\mathbf{W}$ and the coordinates $\bar{x}$ and $\bar{y}$ of the center of gravity G of a flat plate. The same equations can be derived for a wire lying in the xy plane (Fig. 5.2). We note that the center of gravity G of a wire is usually not located on the wire.

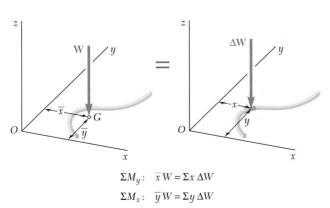

$$\Sigma M_y: \quad \bar{x}\,W = \Sigma x\,\Delta W$$
$$\Sigma M_x: \quad \bar{y}\,W = \Sigma y\,\Delta W$$

Fig. 5.2 Center of gravity of a wire.

5.3 CENTROIDS OF AREAS AND LINES

In the case of a flat homogeneous plate of uniform thickness, the
magnitude ΔW of the weight of an element of the plate can be
expressed as

$$\Delta W = \gamma t \, \Delta A$$

where γ = specific weight (weight per unit volume) of the material
t = thickness of the plate
ΔA = area of the element

Similarly, we can express the magnitude W of the weight of the entire
plate as

$$W = \gamma t A$$

where A is the total area of the plate.

If U.S. customary units are used, the specific weight γ should
be expressed in lb/ft^3, the thickness t in feet, and the areas ΔA and A
in square feet. We observe that ΔW and W will then be expressed
in pounds. If SI units are used, γ should be expressed in N/m^3, t in
meters, and the areas ΔA and A in square meters; the weights ΔW
and W will then be expressed in newtons.†

Substituting for ΔW and W in the moment equations (5.1) and
dividing throughout by γt, we obtain

$$\Sigma M_y: \qquad \bar{x}A = x_1 \Delta A_1 + x_2 \Delta A_2 + \cdots + x_n \Delta A_n$$
$$\Sigma M_x: \qquad \bar{y}A = y_1 \Delta A_1 + y_2 \Delta A_2 + \cdots + y_n \Delta A_n$$

If we increase the number of elements into which the area A is
divided and simultaneously decrease the size of each element, we
obtain in the limit

$$\bar{x}A = \int x \, dA \qquad \bar{y}A = \int y \, dA \qquad (5.3)$$

These equations define the coordinates $\bar{x}$ and $\bar{y}$ of the center of
gravity of a homogeneous plate. The point whose coordinates are $\bar{x}$
and $\bar{y}$ is also known as the *centroid C of the area A* of the plate
(Fig. 5.3). If the plate is not homogeneous, these equations cannot
be used to determine the center of gravity of the plate; they still
define, however, the centroid of the area.

In the case of a homogeneous wire of uniform cross section, the
magnitude ΔW of the weight of an element of wire can be expressed as

$$\Delta W = \gamma a \, \Delta L$$

where γ = specific weight of the material
a = cross-sectional area of the wire
ΔL = length of the element

---

†It should be noted that in the SI system of units a given material is generally characterized by its density ρ (mass per unit volume) rather than by its specific weight γ. The
specific weight of the material can then be obtained from the relation

$$\gamma = \rho g$$

where g = 9.81 m/s^2. Since ρ is expressed in kg/m^3, we observe that γ will be expressed in
(kg/m^3)(m/s^2), that is, in N/m^3.

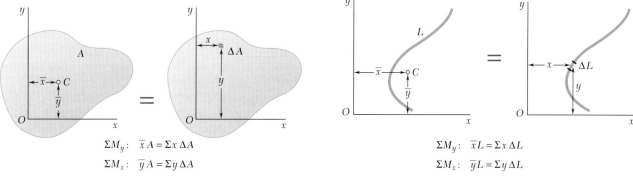

$$\Sigma M_y: \quad \bar{x}A = \Sigma x \, \Delta A$$
$$\Sigma M_x: \quad \bar{y}A = \Sigma y \, \Delta A$$

Fig. 5.3 Centroid of an area.

$$\Sigma M_y: \quad \bar{x}L = \Sigma x \, \Delta L$$
$$\Sigma M_x: \quad \bar{y}L = \Sigma y \, \Delta L$$

Fig. 5.4 Centroid of a line.

The center of gravity of the wire then coincides with the *centroid C of the line L* defining the shape of the wire (Fig. 5.4). The coordinates $\bar{x}$ and $\bar{y}$ of the centroid of the line L are obtained from the equations

$$\bar{x}L = \int x \, dL \qquad \bar{y}L = \int y \, dL \qquad (5.4)$$

5.4 FIRST MOMENTS OF AREAS AND LINES

The integral $\int x \, dA$ in Eqs. (5.3) of the preceding section is known as the *first moment of the area A with respect to the y axis* and is denoted by Q_y. Similarly, the integral $\int y \, dA$ defines the *first moment of A with respect to the x axis* and is denoted by Q_x. We write

$$Q_y = \int x \, dA \qquad Q_x = \int y \, dA \qquad (5.5)$$

Comparing Eqs. (5.3) with Eqs. (5.5), we note that the first moments of the area A can be expressed as the products of the area and the coordinates of its centroid:

$$Q_y = \bar{x}A \qquad Q_x = \bar{y}A \qquad (5.6)$$

It follows from Eqs. (5.6) that the coordinates of the centroid of an area can be obtained by dividing the first moments of that area by the area itself. The first moments of the area are also useful in mechanics of materials for determining the shearing stresses in beams under transverse loadings. Finally, we observe from Eqs. (5.6) that if the centroid of an area is located on a coordinate axis, the first moment of the area with respect to that axis is zero. Conversely, if the first moment of an area with respect to a coordinate axis is zero, then the centroid of the area is located on that axis.

Relations similar to Eqs. (5.5) and (5.6) can be used to define the first moments of a line with respect to the coordinate axes and

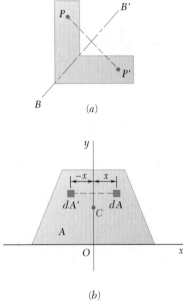

Fig. 5.5

to express these moments as the products of the length L of the line and the coordinates $\bar{x}$ and $\bar{y}$ of its centroid.

An area A is said to be *symmetric with respect to an axis BB′* if for every point P of the area there exists a point P' of the same area such that the line PP' is perpendicular to BB' and is divided into two equal parts by that axis (Fig. 5.5a). A line L is said to be symmetric with respect to an axis BB' if it satisfies similar conditions. When an area A or a line L possesses an axis of symmetry BB', its first moment with respect to BB' is zero, and its centroid is located on that axis. For example, in the case of the area A of Fig. 5.5b, which is symmetric with respect to the y axis, we observe that for every element of area dA of abscissa x there exists an element dA' of equal area and with abscissa $-x$. It follows that the integral in the first of Eqs. (5.5) is zero and, thus, that $Q_y = 0$. It also follows from the first of the relations (5.3) that $\bar{x} = 0$. Thus, if an area A or a line L possesses an axis of symmetry, its centroid C is located on that axis.

We further note that if an area or line possesses two axes of symmetry, its centroid C must be located at the intersection of the two axes (Fig. 5.6). This property enables us to determine immediately the centroid of areas such as circles, ellipses, squares, rectangles, equilateral triangles, or other symmetric figures as well as the centroid of lines in the shape of the circumference of a circle, the perimeter of a square, etc.

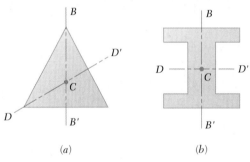

Fig. 5.6

An area A is said to be *symmetric with respect to a center O* if for every element of area dA of coordinates x and y there exists an element dA' of equal area with coordinates $-x$ and $-y$ (Fig. 5.7). It then follows that the integrals in Eqs. (5.5) are both zero and that $Q_x = Q_y = 0$. It also follows from Eqs. (5.3) that $\bar{x} = \bar{y} = 0$, that is, that the centroid of the area coincides with its center of symmetry O. Similarly, if a line possesses a center of symmetry O, the centroid of the line will coincide with the center O.

It should be noted that a figure possessing a center of symmetry does not necessarily possess an axis of symmetry (Fig. 5.7), while a figure possessing two axes of symmetry does not necessarily possess a center of symmetry (Fig. 5.6a). However, if a figure possesses two axes of symmetry at a right angle to each other, the point of intersection of these axes is a center of symmetry (Fig. 5.6b).

Determining the centroids of unsymmetrical areas and lines and of areas and lines possessing only one axis of symmetry will be discussed in Secs. 5.6 and 5.7. Centroids of common shapes of areas and lines are shown in Fig. 5.8A and B.

Fig. 5.7

Shape		$\bar{x}$	$\bar{y}$	Area
Triangular area			$\dfrac{h}{3}$	$\dfrac{bh}{2}$
Quarter-circular area		$\dfrac{4r}{3\pi}$	$\dfrac{4r}{3\pi}$	$\dfrac{\pi r^2}{4}$
Semicircular area		0	$\dfrac{4r}{3\pi}$	$\dfrac{\pi r^2}{2}$
Quarter-elliptical area		$\dfrac{4a}{3\pi}$	$\dfrac{4b}{3\pi}$	$\dfrac{\pi ab}{4}$
Semielliptical area		0	$\dfrac{4b}{3\pi}$	$\dfrac{\pi ab}{2}$
Semiparabolic area		$\dfrac{3a}{8}$	$\dfrac{3h}{5}$	$\dfrac{2ah}{3}$
Parabolic area		0	$\dfrac{3h}{5}$	$\dfrac{4ah}{3}$
Parabolic spandrel		$\dfrac{3a}{4}$	$\dfrac{3h}{10}$	$\dfrac{ah}{3}$
General spandrel		$\dfrac{n+1}{n+2}a$	$\dfrac{n+1}{4n+2}h$	$\dfrac{ah}{n+1}$
Circular sector		$\dfrac{2r\sin\alpha}{3\alpha}$	0	αr^2

Fig. 5.8A Centroids of common shapes of areas.

Shape		$\bar{x}$	$\bar{y}$	Length
Quarter-circular arc		$\dfrac{2r}{\pi}$	$\dfrac{2r}{\pi}$	$\dfrac{\pi r}{2}$
Semicircular arc		0	$\dfrac{2r}{\pi}$	πr
Arc of circle		$\dfrac{r \sin \alpha}{\alpha}$	0	$2\alpha r$

Fig. 5.8B Centroids of common shapes of lines.

5.5 COMPOSITE PLATES AND WIRES

In many instances, a flat plate can be divided into rectangles, triangles, or the other common shapes shown in Fig. 5.8A. The abscissa $\bar{X}$ of its center of gravity G can be determined from the abscissas $\bar{x}_1, \bar{x}_2, \ldots, \bar{x}_n$ of the centers of gravity of the various parts by expressing that the moment of the weight of the whole plate about the y axis is equal to the sum of the moments of the weights of the various parts about the same axis (Fig. 5.9). The ordinate $\bar{Y}$ of the center of gravity of the plate is found in a similar way by equating moments about the x axis. We write

$$\Sigma M_y: \quad \bar{X}(W_1 + W_2 + \cdots + W_n) = \bar{x}_1 W_1 + \bar{x}_2 W_2 + \cdots + \bar{x}_n W_n$$
$$\Sigma M_x: \quad \bar{Y}(W_1 + W_2 + \cdots + W_n) = \bar{y}_1 W_1 + \bar{y}_2 W_2 + \cdots + \bar{y}_n W_n$$

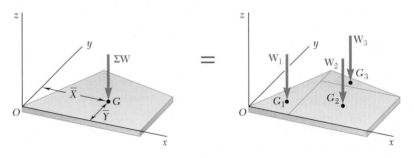

$$\Sigma M_y: \quad \bar{X}\,\Sigma W = \Sigma\,\bar{x}\,W$$
$$\Sigma M_x: \quad \bar{Y}\,\Sigma W = \Sigma\,\bar{y}\,W$$

Fig. 5.9 Center of gravity of a composite plate.

or, for short,

$$\overline{X}\Sigma W = \Sigma \overline{x}W \qquad \overline{Y}\Sigma W = \Sigma \overline{y}W \qquad (5.7)$$

These equations can be solved for the coordinates $\overline{X}$ and $\overline{Y}$ of the center of gravity of the plate.

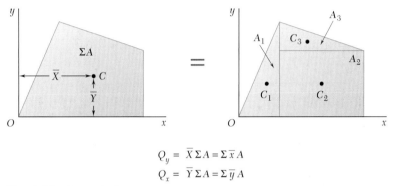

$$Q_y = \overline{X}\Sigma A = \Sigma \overline{x}A$$
$$Q_x = \overline{Y}\Sigma A = \Sigma \overline{y}A$$

Fig. 5.10 Centroid of a composite area.

If the plate is homogeneous and of uniform thickness, the center of gravity coincides with the centroid C of its area. The abscissa $\overline{X}$ of the centroid of the area can be determined by noting that the first moment Q_y of the composite area with respect to the y axis can be expressed both as the product of $\overline{X}$ and the total area and as the sum of the first moments of the elementary areas with respect to the y axis (Fig. 5.10). The ordinate $\overline{Y}$ of the centroid is found in a similar way by considering the first moment Q_x of the composite area. We have

$$Q_y = \overline{X}(A_1 + A_2 + \cdots + A_n) = \overline{x}_1A_1 + \overline{x}_2A_2 + \cdots + \overline{x}_nA_n$$
$$Q_x = \overline{Y}(A_1 + A_2 + \cdots + A_n) = \overline{y}_1A_1 + \overline{y}_2A_2 + \cdots + \overline{y}_nA_n$$

or, for short,

$$Q_y = \overline{X}\Sigma A = \Sigma \overline{x}A \qquad Q_x = \overline{Y}\Sigma A = \Sigma \overline{y}A \qquad (5.8)$$

These equations yield the first moments of the composite area, or they can be used to obtain the coordinates $\overline{X}$ and $\overline{Y}$ of its centroid.

Care should be taken to assign the appropriate sign to the moment of each area. First moments of areas, like moments of forces, can be positive or negative. For example, an area whose centroid is located to the left of the y axis will have a negative first moment with respect to that axis. Also, the area of a hole should be assigned a negative sign (Fig. 5.11).

Similarly, it is possible in many cases to determine the center of gravity of a composite wire or the centroid of a composite line by dividing the wire or line into simpler elements (see Sample Prob. 5.2).

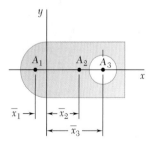

	$\overline{x}$	A	$\overline{x}A$
A_1 Semicircle	−	+	−
A_2 Full rectangle	+	+	+
A_3 Circular hole	+	−	−

Fig. 5.11

SAMPLE PROBLEM 5.1

For the plane area shown, determine (a) the first moments with respect to the x and y axes, (b) the location of the centroid.

SOLUTION

Components of Area. The area is obtained by adding a rectangle, a triangle, and a semicircle and by then subtracting a circle. Using the coordinate axes shown, the area and the coordinates of the centroid of each of the component areas are determined and entered in the table below. The area of the circle is indicated as negative, since it is to be subtracted from the other areas. We note that the coordinate $\bar{y}$ of the centroid of the triangle is negative for the axes shown. The first moments of the component areas with respect to the coordinate axes are computed and entered in the table.

Component	A, mm²	$\bar{x}$, mm	$\bar{y}$, mm	$\bar{x}A$, mm³	$\bar{y}A$, mm³
Rectangle	$(120)(80) = 9.6 \times 10^3$	60	40	$+576 \times 10^3$	$+384 \times 10^3$
Triangle	$\frac{1}{2}(120)(60) = 3.6 \times 10^3$	40	-20	$+144 \times 10^3$	-72×10^3
Semicircle	$\frac{1}{2}\pi(60)^2 = 5.655 \times 10^3$	60	105.46	$+339.3 \times 10^3$	$+596.4 \times 10^3$
Circle	$-\pi(40)^2 = -5.027 \times 10^3$	60	80	-301.6×10^3	-402.2×10^3
	$\Sigma A = 13.828 \times 10^3$			$\Sigma\bar{x}A = +757.7 \times 10^3$	$\Sigma\bar{y}A = +506.2 \times 10^3$

a. First Moments of the Area. Using Eqs. (5.8), we write

$$Q_x = \Sigma\bar{y}A = 506.2 \times 10^3 \text{ mm}^3 \qquad Q_x = 506 \times 10^3 \text{ mm}^3 \blacktriangleleft$$
$$Q_y = \Sigma\bar{x}A = 757.7 \times 10^3 \text{ mm}^3 \qquad Q_y = 758 \times 10^3 \text{ mm}^3 \blacktriangleleft$$

b. Location of Centroid. Substituting the values given in the table into the equations defining the centroid of a composite area, we obtain

$$\bar{X}\Sigma A = \Sigma\bar{x}A: \quad \bar{X}(13.828 \times 10^3 \text{ mm}^2) = 757.7 \times 10^3 \text{ mm}^3$$
$$\bar{X} = 54.8 \text{ mm} \blacktriangleleft$$

$$\bar{Y}\Sigma A = \Sigma\bar{y}A: \quad \bar{Y}(13.828 \times 10^3 \text{ mm}^2) = 506.2 \times 10^3 \text{ mm}^3$$
$$\bar{Y} = 36.6 \text{ mm} \blacktriangleleft$$

SAMPLE PROBLEM 5.2

The figure shown is made from a piece of thin, homogeneous wire. Determine the location of its center of gravity.

SOLUTION

Since the figure is formed of homogeneous wire, its center of gravity coincides with the centroid of the corresponding line. Therefore, that centroid will be determined. Choosing the coordinate axes shown, with origin at A, we determine the coordinates of the centroid of each line segment and compute the first moments with respect to the coordinate axes.

Segment	L, in.	$\bar{x}$, in.	$\bar{y}$, in.	$\bar{x}L$, in^2	$\bar{y}L$, in^2
AB	24	12	0	288	0
BC	26	12	5	312	130
CA	10	0	5	0	50
	$\Sigma L = 60$			$\Sigma\bar{x}L = 600$	$\Sigma\bar{y}L = 180$

Substituting the values obtained from the table into the equations defining the centroid of a composite line, we obtain

$\bar{X}\Sigma L = \Sigma\bar{x}L:$ $\bar{X}(60 \text{ in.}) = 600 \text{ in}^2$ $\bar{X} = 10 \text{ in.}$ ◀

$\bar{Y}\Sigma L = \Sigma\bar{y}L:$ $\bar{Y}(60 \text{ in.}) = 180 \text{ in}^2$ $\bar{Y} = 3 \text{ in.}$ ◀

SAMPLE PROBLEM 5.3

A uniform semicircular rod of weight W and radius r is attached to a pin at A and rests against a frictionless surface at B. Determine the reactions at A and B.

SOLUTION

Free-Body Diagram. A free-body diagram of the rod is drawn. The forces acting on the rod are its weight **W**, which is applied at the center of gravity G (whose position is obtained from Fig. 5.8B); a reaction at A, represented by its components $\mathbf{A}_x$ and $\mathbf{A}_y$; and a horizontal reaction at B.

Equilibrium Equations

$$+\!\!\uparrow \Sigma M_A = 0: \qquad B(2r) - W\!\left(\frac{2r}{\pi}\right) = 0$$

$$B = +\frac{W}{\pi} \qquad\qquad \mathbf{B} = \frac{W}{\pi} \rightarrow \;\blacktriangleleft$$

$$\xrightarrow{+} \Sigma F_x = 0: \qquad A_x + B = 0$$

$$A_x = -B = -\frac{W}{\pi} \qquad \mathbf{A}_x = \frac{W}{\pi} \leftarrow$$

$$+\!\!\uparrow\Sigma F_y = 0: \qquad A_y - W = 0 \qquad \mathbf{A}_y = W \uparrow$$

Adding the two components of the reaction at A:

$$A = \left[W^2 + \left(\frac{W}{\pi}\right)^2 \right]^{1/2} \qquad\qquad A = W\!\left(1 + \frac{1}{\pi^2}\right)^{1/2} \;\blacktriangleleft$$

$$\tan \alpha = \frac{W}{W/\pi} = \pi \qquad\qquad \alpha = \tan^{-1}\pi \;\blacktriangleleft$$

The answers can also be expressed as follows:

$$\mathbf{A} = 1.049W \;\text{⦥}\,72.3° \qquad \mathbf{B} = 0.318W \rightarrow \;\blacktriangleleft$$

SOLVING PROBLEMS
ON YOUR OWN

In this lesson we developed the general equations for locating the centers of gravity of two-dimensional bodies and wires [Eqs. (5.2)] and the centroids of plane areas [Eqs. (5.3)] and lines [Eqs. (5.4)]. In the following problems, you will have to locate the centroids of composite areas and lines or determine the first moments of the area for composite plates [Eqs. (5.8)].

1. Locating the centroids of composite areas and lines. Sample Problems 5.1 and 5.2 illustrate the procedure you should follow when solving problems of this type. There are, however, several points that should be emphasized.

a. The first step in your solution should be to decide how to construct the given area or line from the common shapes of Fig. 5.8. You should recognize that for plane areas it is often possible to construct a particular shape in more than one way. Also, showing the different components (as is done in Sample Prob. 5.1) will help you to correctly establish their centroids and areas or lengths. Do not forget that you can subtract areas as well as add them to obtain a desired shape.

b. We strongly recommend that for each problem you construct a table containing the areas or lengths and the respective coordinates of the centroids. It is essential for you to remember that areas which are "removed" (for example, holes) are treated as negative. Also, the sign of negative coordinates must be included. Therefore, you should always carefully note the location of the origin of the coordinate axes.

c. When possible, use symmetry [Sec. 5.4] to help you determine the location of a centroid.

d. In the formulas for the circular sector and for the arc of a circle in Fig. 5.8, the angle α must always be expressed in radians.

2. Calculating the first moments of an area. The procedures for locating the centroid of an area and for determining the first moments of an area are similar; however, for the latter it is not necessary to compute the total area. Also, as noted in Sec. 5.4, you should recognize that the first moment of an area relative to a centroidal axis is zero.

3. Solving problems involving the center of gravity. The bodies considered in the following problems are homogeneous; thus, their centers of gravity and centroids coincide. In addition, when a body that is suspended from a single pin is in equilibrium, the pin and the body's center of gravity must lie on the same vertical line.

It may appear that many of the problems in this lesson have little to do with the study of mechanics. However, being able to locate the centroid of composite shapes will be essential in several topics that you will soon encounter.

PROBLEMS

5.1 through 5.9 Locate the centroid of the plane area shown.

Fig. P5.1

Fig. P5.2

Fig. P5.3

Fig. P5.4

Fig. P5.5

Fig. P5.6

Fig. P5.7

Fig. P5.8

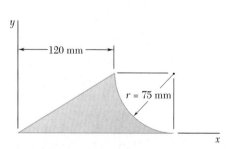

Fig. P5.9

5.10 through 5.15 Locate the centroid of the plane area shown.

Fig. P5.10

Fig. P5.11

Fig. *P5.12*

Fig. P5.13

Fig. P5.14

Fig. *P5.15*

5.16 Determine the x coordinate of the centroid of the trapezoid shown in terms of h_1, h_2, and a.

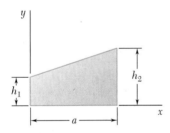

Fig. *P5.16*

5.17 For the plane area of Prob. 5.5, determine the ratio a/r so that the centroid of the area is located at point B.

5.18 Determine the y coordinate of the centroid of the shaded area in terms of r_1, r_2, and α.

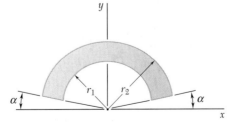

Fig. P5.18 and P5.19

5.19 Show that as r_1 approaches r_2, the location of the centroid approaches that for an arc of circle of radius $(r_1 + r_2)/2$.

5.20 and 5.21 The horizontal x axis is drawn through the centroid C of the area shown, and it divides the area into two component areas A_1 and A_2. Determine the first moment of each component area with respect to the x axis, and explain the results obtained.

Fig. P5.20   **Fig. P5.21**

5.22 A composite beam is constructed by bolting four plates to four $60 \times 60 \times 12$-mm angles as shown. The bolts are equally spaced along the beam, and the beam supports a vertical load. As proved in mechanics of materials, the shearing forces exerted on the bolts at A and B are proportional to the first moments with respect to the centroidal x axis of the red shaded areas shown, respectively, in parts a and b of the figure. Knowing that the force exerted on the bolt at A is 280 N, determine the force exerted on the bolt at B.

Fig. P5.22

5.23 The first moment of the shaded area with respect to the x axis is denoted by Q_x. (*a*) Express Q_x in terms of b, c, and the distance y from the base of the shaded area to the x axis. (*b*) For what value of y is Q_x maximum, and what is that maximum value?

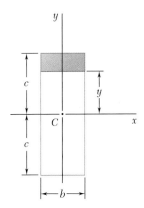

Fig. P5.23

5.24 through 5.27   A thin, homogeneous wire is bent to form the perimeter of the figure indicated. Locate the center of gravity of the wire figure thus formed.

      5.24   Fig. P5.2.
      5.25   Fig. P5.3.
      5.26   Fig. P5.4.
      5.27   Fig. P5.5.

5.28 The homogeneous wire *ABCD* is bent as shown and is attached to a hinge at *C*. Determine the length *L* for which portion *BCD* of the wire is horizontal.

5.29 The homogeneous wire *ABCD* is bent as shown and is attached to a hinge at *C*. Determine the length *L* for which portion *AB* of the wire is horizontal.

Fig. P5.28 and P5.29

5.30 The homogeneous wire *ABC* is bent into a semicircular arc and a straight section as shown and is attached to a hinge at *A*. Determine the value of θ for which the wire is in equilibrium for the indicated position.

Fig. P5.30

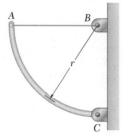

Fig. P5.31

5.31 A uniform circular rod of weight 8 lb and radius 10 in. is attached to a pin at C and to the cable AB. Determine (a) the tension in the cable, (b) the reaction at C.

5.32 Determine the distance h for which the centroid of the shaded area is as far above line BB' as possible when (a) $k = 0.10$, (b) $k = 0.80$.

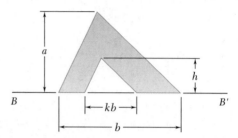

Fig. P5.32 and P5.33

5.33 Knowing that the distance h has been selected to maximize the distance $\bar{y}$ from line BB' to the centroid of the shaded area, show that $\bar{y} = 2h/3$.

5.6 DETERMINATION OF CENTROIDS BY INTEGRATION

The centroid of an area bounded by analytical curves (i.e., curves defined by algebraic equations) is usually determined by evaluating the integrals in Eqs. (5.3) of Sec. 5.3:

$$\bar{x}A = \int x \, dA \qquad \bar{y}A = \int y \, dA \qquad (5.3)$$

If the element of area dA is a small rectangle of sides dx and dy, the evaluation of each of these integrals requires a *double integration* with respect to x and y. A double integration is also necessary if polar coordinates are used for which dA is a small element of sides dr and $r \, d\theta$.

In most cases, however, it is possible to determine the coordinates of the centroid of an area by performing a single integration. This is achieved by choosing dA to be a thin rectangle or strip or a thin sector or pie-shaped element (Fig. 5.12); the centroid of the thin rectangle is located at its center, and the centroid of the thin sector is located at a distance $\frac{2}{3}r$ from its vertex (as it is for a triangle). The coordinates of the centroid of the area under consideration are then obtained by expressing that the first moment of the entire area with respect to each of the coordinate axes is equal to the sum (or integral) of the corresponding moments of the elements of area.

Denoting by $\bar{x}_{el}$ and $\bar{y}_{el}$ the coordinates of the centroid of the element dA, we write

$$Q_y = \bar{x}A = \int \bar{x}_{el}\, dA$$

$$Q_x = \bar{y}A = \int \bar{y}_{el}\, dA$$

(5.9)

If the area A is not already known, it can also be computed from these elements.

The coordinates $\bar{x}_{el}$ and $\bar{y}_{el}$ of the centroid of the element of area dA should be expressed in terms of the coordinates of a point located on the curve bounding the area under consideration. Also, the area of the element dA should be expressed in terms of the coordinates of that point and the appropriate differentials. This has been done in Fig. 5.12 for three common types of elements; the pie-shaped element of part c should be used when the equation of the curve bounding the area is given in polar coordinates. The appropriate expressions should be substituted into formulas (5.9), and the equation of the bounding curve should be used to express one of the coordinates in terms of the other. The integration is thus reduced to a single integration. Once the area has been determined and the integrals in Eqs. (5.9) have been evaluated, these equations can be solved for the coordinates $\bar{x}$ and $\bar{y}$ of the centroid of the area.

When a line is defined by an algebraic equation, its centroid can be determined by evaluating the integrals in Eqs. (5.4) of Sec. 5.3:

$$\bar{x}L = \int x\, dL \qquad \bar{y}L = \int y\, dL$$

(5.4)

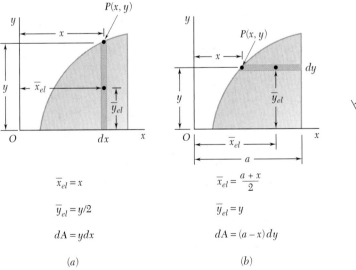

$$\bar{x}_{el} = x$$

$$\bar{y}_{el} = y/2$$

$$dA = y\, dx$$

(a)

$$\bar{x}_{el} = \frac{a + x}{2}$$

$$\bar{y}_{el} = y$$

$$dA = (a - x)\, dy$$

(b)

$$\bar{x}_{el} = \frac{2r}{3}\cos\theta$$

$$\bar{y}_{el} = \frac{2r}{3}\sin\theta$$

$$dA = \frac{1}{2}\, r^2\, d\theta$$

(c)

Fig. 5.12 Centroids and areas of differential elements.

The differential length dL should be replaced by one of the following expressions, depending upon which coordinate, x, y, or θ, is chosen as the independent variable in the equation used to define the line (these expressions can be derived using the Pythagorean theorem):

$$dL = \sqrt{1 + \left(\frac{dy}{dx}\right)^2}\,dx \qquad dL = \sqrt{1 + \left(\frac{dx}{dy}\right)^2}\,dy$$

$$dL = \sqrt{r^2 + \left(\frac{dr}{d\theta}\right)^2}\,d\theta$$

After the equation of the line has been used to express one of the coordinates in terms of the other, the integration can be performed, and Eqs. (5.4) can be solved for the coordinates $\bar{x}$ and $\bar{y}$ of the centroid of the line.

Photo 5.2 The storage tanks shown are all bodies of revolution. Thus, their surface areas and volumes can be determined using the theorems of Pappus-Guldinus.

5.7 THEOREMS OF PAPPUS-GULDINUS

These theorems, which were first formulated by the Greek geometer Pappus during the third century A.D. and later restated by the Swiss mathematician Guldinus, or Guldin, (1577–1643) deal with surfaces and bodies of revolution.

A *surface of revolution* is a surface which can be generated by rotating a plane curve about a fixed axis. For example (Fig. 5.13), the

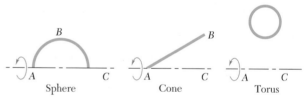

Fig. 5.13

surface of a sphere can be obtained by rotating a semicircular arc ABC about the diameter AC, the surface of a cone can be produced by rotating a straight line AB about an axis AC, and the surface of a torus or ring can be generated by rotating the circumference of a circle about a nonintersecting axis. A *body of revolution* is a body which can be generated by rotating a plane area about a fixed axis. As shown in Fig. 5.14, a sphere, a cone, and a torus can each be generated by rotating the appropriate shape about the indicated axis.

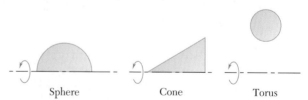

Fig. 5.14

THEOREM I. *The area of a surface of revolution is equal to the length of the generating curve times the distance traveled by the centroid of the curve while the surface is being generated.*

Proof. Consider an element dL of the line L (Fig. 5.15), which is revolved about the x axis. The area dA generated by the element

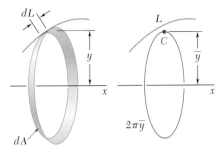

Fig. 5.15

dL is equal to $2\pi y\, dL$. Thus, the entire area generated by L is $A = \int 2\pi y\, dL$. Recalling that we found in Sec. 5.3 that the integral $\int y\, dL$ is equal to $\bar{y}L$, we therefore have

$$A = 2\pi\bar{y}L \tag{5.10}$$

where $2\pi\bar{y}$ is the distance traveled by the centroid of L (Fig. 5.15). It should be noted that the generating curve must not cross the axis about which it is rotated; if it did, the two sections on either side of the axis would generate areas having opposite signs, and the theorem would not apply.

THEOREM II. *The volume of a body of revolution is equal to the generating area times the distance traveled by the centroid of the area while the body is being generated.*

Proof. Consider an element dA of the area A which is revolved about the x axis (Fig. 5.16). The volume dV generated by the element dA is equal to $2\pi y\, dA$. Thus, the entire volume generated by A is $V = \int 2\pi y\, dA$, and since the integral $\int y\, dA$ is equal to $\bar{y}A$ (Sec. 5.3), we have

$$V = 2\pi\bar{y}A \tag{5.11}$$

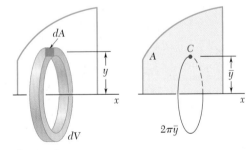

Fig. 5.16

where $2\pi\bar{y}$ is the distance traveled by the centroid of A. Again, it should be noted that the theorem does not apply if the axis of rotation intersects the generating area.

The theorems of Pappus-Guldinus offer a simple way to compute the areas of surfaces of revolution and the volumes of bodies of revolution. Conversely, they can also be used to determine the centroid of a plane curve when the area of the surface generated by the curve is known or to determine the centroid of a plane area when the volume of the body generated by the area is known (see Sample Prob. 5.8).

SAMPLE PROBLEM 5.4

Determine by direct integration the location of the centroid of a parabolic spandrel.

SOLUTION

Determination of the Constant k. The value of k is determined by substituting $x = a$ and $y = b$ into the given equation. We have $b = ka^2$ or $k = b/a^2$. The equation of the curve is thus

$$y = \frac{b}{a^2}x^2 \qquad \text{or} \qquad x = \frac{a}{b^{1/2}}y^{1/2}$$

Vertical Differential Element. We choose the differential element shown and find the total area of the figure.

$$A = \int dA = \int y \, dx = \int_0^a \frac{b}{a^2}x^2 \, dx = \left[\frac{b}{a^2}\frac{x^3}{3}\right]_0^a = \frac{ab}{3}$$

The first moment of the differential element with respect to the y axis is $\bar{x}_{el} \, dA$; hence, the first moment of the entire area with respect to this axis is

$$Q_y = \int \bar{x}_{el} \, dA = \int xy \, dx = \int_0^a x\left(\frac{b}{a^2}x^2\right)dx = \left[\frac{b}{a^2}\frac{x^4}{4}\right]_0^a = \frac{a^2 b}{4}$$

Since $Q_y = \bar{x}A$, we have

$$\bar{x}A = \int \bar{x}_{el} \, dA \qquad \bar{x}\frac{ab}{3} = \frac{a^2 b}{4} \qquad \bar{x} = \tfrac{3}{4}a \quad \blacktriangleleft$$

Likewise, the first moment of the differential element with respect to the x axis is $\bar{y}_{el} \, dA$, and the first moment of the entire area is

$$Q_x = \int \bar{y}_{el} \, dA = \int \frac{y}{2}y \, dx = \int_0^a \frac{1}{2}\left(\frac{b}{a^2}x^2\right)^2 dx = \left[\frac{b^2}{2a^4}\frac{x^5}{5}\right]_0^a = \frac{ab^2}{10}$$

Since $Q_x = \bar{y}A$, we have

$$\bar{y}A = \int \bar{y}_{el} \, dA \qquad \bar{y}\frac{ab}{3} = \frac{ab^2}{10} \qquad \bar{y} = \tfrac{3}{10}b \quad \blacktriangleleft$$

Horizontal Differential Element. The same results can be obtained by considering a horizontal element. The first moments of the area are

$$Q_y = \int \bar{x}_{el} \, dA = \int \frac{a + x}{2}(a - x)dy = \int_0^b \frac{a^2 - x^2}{2}dy$$

$$= \frac{1}{2}\int_0^b \left(a^2 - \frac{a^2}{b}y\right)dy = \frac{a^2 b}{4}$$

$$Q_x = \int \bar{y}_{el} \, dA = \int y(a - x)dy = \int y\left(a - \frac{a}{b^{1/2}}y^{1/2}\right)dy$$

$$= \int_0^b \left(ay - \frac{a}{b^{1/2}}y^{3/2}\right)dy = \frac{ab^2}{10}$$

To determine $\bar{x}$ and $\bar{y}$, the expressions obtained are again substituted into the equations defining the centroid of the area.

SAMPLE PROBLEM 5.5

Determine the location of the centroid of the arc of circle shown.

SOLUTION

Since the arc is symmetrical with respect to the x axis, $\bar{y} = 0$. A differential element is chosen as shown, and the length of the arc is determined by integration.

$$L = \int dL = \int_{-\alpha}^{\alpha} r\, d\theta = r \int_{-\alpha}^{\alpha} d\theta = 2r\alpha$$

The first moment of the arc with respect to the y axis is

$$Q_y = \int x\, dL = \int_{-\alpha}^{\alpha} (r\cos\theta)(r\, d\theta) = r^2 \int_{-\alpha}^{\alpha} \cos\theta\, d\theta$$
$$= r^2 [\sin\theta]_{-\alpha}^{\alpha} = 2r^2 \sin\alpha$$

Since $Q_y = \bar{x}L$, we write

$$\bar{x}(2r\alpha) = 2r^2 \sin\alpha \qquad \bar{x} = \frac{r\sin\alpha}{\alpha} \quad \blacktriangleleft$$

SAMPLE PROBLEM 5.6

Determine the area of the surface of revolution shown, which is obtained by rotating a quarter-circular arc about a vertical axis.

SOLUTION

According to Theorem I of Pappus-Guldinus, the area generated is equal to the product of the length of the arc and the distance traveled by its centroid. Referring to Fig. 5.8B, we have

$$\bar{x} = 2r - \frac{2r}{\pi} = 2r\left(1 - \frac{1}{\pi}\right)$$

$$A = 2\pi \bar{x}L = 2\pi \left[2r\left(1 - \frac{1}{\pi}\right)\right]\left(\frac{\pi r}{2}\right)$$

$$A = 2\pi r^2(\pi - 1) \quad \blacktriangleleft$$

SAMPLE PROBLEM 5.7

The outside diameter of a pulley is 0.8 m, and the cross section of its rim is as shown. Knowing that the pulley is made of steel and that the density of steel is $\rho = 7.85 \times 10^3$ kg/m^3, determine the mass and the weight of the rim.

SOLUTION

The volume of the rim can be found by applying Theorem II of Pappus-Guldinus, which states that the volume equals the product of the given cross-sectional area and the distance traveled by its centroid in one complete revolution. However, the volume can be more easily determined if we observe that the cross section can be formed from rectangle I, whose area is positive, and rectangle II, whose area is negative.

	Area, mm^2	$\bar{y}$, mm	Distance Traveled by C, mm	Volume, mm^3
I	+5000	375	$2\pi(375) = 2356$	$(5000)(2356) = 11.78 \times 10^6$
II	−1800	365	$2\pi(365) = 2293$	$(-1800)(2293) = -4.13 \times 10^6$
				Volume of rim $= 7.65 \times 10^6$

Since 1 mm $= 10^{-3}$ m, we have 1 mm^3 $= (10^{-3}$ m$)^3 = 10^{-9}$ m^3, and we obtain $V = 7.65 \times 10^6$ mm^3 $= (7.65 \times 10^6)(10^{-9}$ m$^3) = 7.65 \times 10^{-3}$ m^3.

$$m = \rho V = (7.85 \times 10^3 \text{ kg/m}^3)(7.65 \times 10^{-3} \text{ m}^3) \qquad m = 60.0 \text{ kg} \blacktriangleleft$$
$$W = mg = (60.0 \text{ kg})(9.81 \text{ m/s}^2) = 589 \text{ kg} \cdot \text{m/s}^2 \qquad W = 589 \text{ N} \blacktriangleleft$$

SAMPLE PROBLEM 5.8

Using the theorems of Pappus-Guldinus, determine (a) the centroid of a semicircular area, (b) the centroid of a semicircular arc. We recall that the volume and the surface area of a sphere are $\frac{4}{3}\pi r^3$ and $4\pi r^2$, respectively.

SOLUTION

The volume of a sphere is equal to the product of the area of a semicircle and the distance traveled by the centroid of the semicircle in one revolution about the x axis.

$$V = 2\pi\bar{y}A \qquad \tfrac{4}{3}\pi r^3 = 2\pi\bar{y}(\tfrac{1}{2}\pi r^2) \qquad \bar{y} = \frac{4r}{3\pi} \blacktriangleleft$$

Likewise, the area of a sphere is equal to the product of the length of the generating semicircle and the distance traveled by its centroid in one revolution.

$$A = 2\pi\bar{y}L \qquad 4\pi r^2 = 2\pi\bar{y}(\pi r) \qquad \bar{y} = \frac{2r}{\pi} \blacktriangleleft$$

SOLVING PROBLEMS
ON YOUR OWN

In the problems for this lesson, you will use the equations

$$\bar{x}A = \int x \, dA \qquad \bar{y}A = \int y \, dA \qquad (5.3)$$

$$\bar{x}L = \int x \, dL \qquad \bar{y}L = \int y \, dL \qquad (5.4)$$

to locate the centroids of plane areas and lines, respectively. You will also apply the theorems of Pappus-Guldinus (Sec. 5.7) to determine the areas of surfaces of revolution and the volumes of bodies of revolution.

1. Determining by direct integration the centroids of areas and lines. When solving problems of this type, you should follow the method of solution shown in Sample Probs. 5.4 and 5.5: compute A or L, determine the first moments of the area or the line, and solve Eqs. (5.3) or (5.4) for the coordinates of the centroid. In addition, you should pay particular attention to the following points.

   a. Begin your solution by carefully defining or determining each term in the applicable integral formulas. We strongly encourage you to show on your sketch of the given area or line your choice for dA or dL and the distances to its centroid.

   b. As explained in Sec. 5.6, the x and the y in the above equations represent the *coordinates of the centroid* of the differential elements dA and dL. It is important to recognize that the coordinates of the centroid of dA are not equal to the coordinates of a point located on the curve bounding the area under consideration. You should carefully study Fig. 5.12 until you fully understand this important point.

   c. To possibly simplify or minimize your computations, always examine the shape of the given area or line before defining the differential element that you will use. For example, sometimes it may be preferable to use horizontal rectangular elements instead of vertical ones. Also, it will usually be advantageous to use polar coordinates when a line or an area has circular symmetry.

   d. Although most of the integrations in this lesson are straightforward, at times it may be necessary to use more advanced techniques, such as trigonometric substitution or integration by parts. Of course, using a table of integrals is the fastest method to evaluate difficult integrals.

2. Applying the theorems of Pappus-Guldinus. As shown in Sample Probs. 5.6 through 5.8, these simple, yet very useful theorems allow you to apply your knowledge of centroids to the computation of areas and volumes. Although the theorems refer to the distance traveled by the centroid and to the length of the generating curve or to the generating area, the resulting equations [Eqs. (5.10) and (5.11)] contain the products of these quantities, which are simply the first moments of a line $(\bar{y}L)$ and an area $(\bar{y}A)$, respectively. Thus, for those problems for which the generating line or area consists of more than one common shape, you need only determine $\bar{y}L$ or $\bar{y}A$; you do not have to calculate the length of the generating curve or the generating area.

PROBLEMS

5.34 through 5.36 Determine by direct integration the centroid of the area shown. Express your answer in terms of a and h.

Fig. P5.34

Fig. P5.35

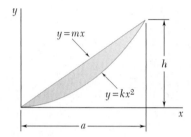

Fig. P5.36

5.37 through 5.39 Determine by direct integration the centroid of the area shown.

Fig. P5.37

Fig. P5.38

Fig. P5.39

5.40 and 5.41 Determine by direct integration the centroid of the area shown. Express your answer in terms of a and b.

Fig. P5.40

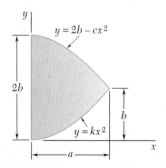

Fig. P5.41

244

5.42 Determine by direct integration the centroid of the area shown.

5.43 and 5.44 Determine by direct integration the centroid of the area shown. Express your answer in terms of a and b.

Fig. P5.42

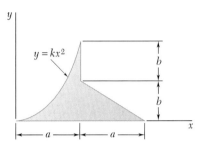

Fig. P5.43

Fig. P5.44

5.45 and 5.46 A homogeneous wire is bent into the shape shown. Determine by direct integration the x coordinate of its centroid.

Fig. P5.45

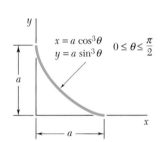

Fig. P5.46

*5.47 A homogeneous wire is bent into the shape shown. Determine by direct integration the x coordinate of its centroid. Express your answer in terms of a.

*5.48 and *5.49 Determine by direct integration the centroid of the area shown.

Fig. P5.47

Fig. P5.48

Fig. P5.49

5.50 Determine the centroid of the area shown when $a = 2$ in.

5.51 Determine the value of a for which the ratio $\bar{x}/\bar{y}$ is 9.

5.52 Determine the volume and the surface area of the solid obtained by rotating the area of Prob. 5.1 about (a) the x axis, (b) the y axis.

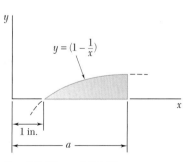

Fig. P5.50 and P5.51

5.53 Determine the volume and the surface area of the solid obtained by rotating the area of Prob. 5.2 about (*a*) the line $y = 72$ mm, (*b*) the *x* axis.

5.54 Determine the volume and the surface area of the solid obtained by rotating the area of Prob. 5.8 about (*a*) the line $x = -60$ mm, (*b*) the line $y = 120$ mm.

5.55 Determine the volume of the solid generated by rotating the parabolic area shown about (*a*) the *x* axis, (*b*) the axis *AA'*.

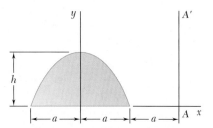

Fig. P5.55

5.56 Determine the volume and the surface area of the chain link shown, which is made from a 6-mm-diameter bar, if $R = 10$ mm and $L = 30$ mm.

Fig. P5.56

5.57 Verify that the expressions for the volumes of the first four shapes in Fig. 5.21 on page 260 are correct.

5.58 Determine the volume and weight of the solid brass knob shown, knowing that the specific weight of brass is 0.306 lb/in^3.

5.59 Determine the total surface area of the solid brass knob shown.

Fig. P5.58 and P5.59

5.60 The aluminum shade for the small high-intensity lamp shown has a uniform thickness of 1 mm. Knowing that the density of aluminum is 2800 kg/m^3, determine the mass of the shade.

Fig. P5.60

5.61 The escutcheon (a decorative plate placed on a pipe where the pipe exits from a wall) shown is cast from brass. Knowing that the density of brass is 8470 kg/m³, determine the mass of the escutcheon.

Fig. P5.61

5.62 A $\frac{3}{4}$-in.-diameter hole is drilled in a piece of 1-in.-thick steel; the hole is then countersunk as shown. Determine the volume of steel removed during the countersinking process.

Fig. P5.62

5.63 Knowing that two equal caps have been removed from a 10-in.-diameter wooden sphere, determine the total surface area of the remaining portion.

Fig. P5.63

5.64 Determine the capacity, in liters, of the punch bowl shown if $R = 250$ mm.

Fig. P5.64

*5.65 The shade for a wall-mounted light is formed from a thin sheet of translucent plastic. Determine the surface area of the outside of the shade, knowing that it has the parabolic cross section shown.

Fig. P5.65

(a)

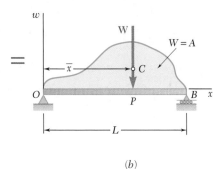

(b)

Fig. 5.17

*5.8 DISTRIBUTED LOADS ON BEAMS

The concept of the centroid of an area can be used to solve other problems besides those dealing with the weights of flat plates. Consider, for example, a beam supporting a *distributed load;* this load may consist of the weight of materials supported directly or indirectly by the beam, or it may be caused by wind or hydrostatic pressure. The distributed load can be represented by plotting the load w supported per unit length (Fig. 5.17); this load is expressed in N/m or in lb/ft. The magnitude of the force exerted on an element of beam of length dx is $dW = w\,dx$, and the total load supported by the beam is

$$W = \int_0^L w\,dx$$

We observe that the product $w\,dx$ is equal in magnitude to the element of area dA shown in Fig. 5.17a. The load W is thus equal in magnitude to the total area A under the load curve:

$$W = \int dA = A$$

We now determine where a *single concentrated load* **W**, of the same magnitude W as the total distributed load, should be applied on the beam if it is to produce the same reactions at the supports (Fig. 5.17b). However, this concentrated load **W**, which represents the resultant of the given distributed loading, is equivalent to the loading only when considering the free-body diagram of the entire beam. The point of application P of the equivalent concentrated load **W** is obtained by expressing that the moment of **W** about point O is equal to the sum of the moments of the elemental loads d**W** about O:

$$(OP)W = \int x\,dW$$

or, since $dW = w\,dx = dA$ and $W = A$,

$$(OP)A = \int_0^L x\,dA \qquad (5.12)$$

Since the integral represents the first moment with respect to the w axis of the area under the load curve, it can be replaced by the product $\bar{x}A$. We therefore have $OP = \bar{x}$, where $\bar{x}$ is the distance from the w axis to the centroid C of the area A (this is *not* the centroid of the beam).

A *distributed load on a beam can thus be replaced by a concentrated load; the magnitude of this single load is equal to the area under the load curve, and its line of action passes through the centroid of that area.* It should be noted, however, that the concentrated load is equivalent to the given loading only as far as external forces are concerned. It can be used to determine reactions but should not be used to compute internal forces and deflections.

Photo 5.3 The roofs of the buildings shown must be able to support not only the total weight of the snow but also the nonsymmetric distributed loads resulting from drifting of the snow.

*5.9 FORCES ON SUBMERGED SURFACES

The approach used in the preceding section can be used to determine the resultant of the hydrostatic pressure forces exerted on a *rectangular surface* submerged in a liquid. Consider the rectangular plate shown in Fig. 5.18, which is of length L and width b, where b is measured perpendicular to the plane of the figure. As noted in Sec. 5.8, the load exerted on an element of the plate of length dx is $w\,dx$, where w is the load per unit length. However, this load can also be expressed as $p\,dA = pb\,dx$, where p is the gage pressure in the liquid† and b is the width of the plate; thus, $w = bp$. Since the gage pressure in a liquid is $p = \gamma h$, where γ is the specific weight of the liquid and h is the vertical distance from the free surface, it follows that

$$w = bp = b\gamma h \tag{5.13}$$

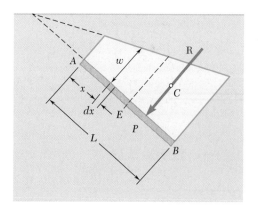

Fig. 5.18

which shows that the load per unit length w is proportional to h and, thus, varies linearly with x.

Recalling the results of Sec. 5.8, we observe that the resultant **R** of the hydrostatic forces exerted on one side of the plate is equal in magnitude to the trapezoidal area under the load curve and that its line of action passes through the centroid C of that area. The point P of the plate where **R** is applied is known as the *center of pressure*.‡

Next, we consider the forces exerted by a liquid on a curved surface of constant width (Fig. 5.19*a*). Since the determination of the resultant **R** of these forces by direct integration would not be easy, we consider the free body obtained by detaching the volume of liquid ABD bounded by the curved surface AB and by the two plane surfaces AD and DB shown in Fig. 5.19*b*. The forces acting on the free body ABD are the weight **W** of the detached volume of liquid, the resultant $\mathbf{R}_1$ of the forces exerted on AD, the resultant $\mathbf{R}_2$ of the forces exerted on BD, and the resultant $-\mathbf{R}$ of the forces exerted *by the curved surface on the liquid*. The resultant $-\mathbf{R}$ is equal and opposite to, and has the same line of action as, the resultant **R** of the forces exerted *by the liquid on the curved surface*. The forces **W**, $\mathbf{R}_1$, and $\mathbf{R}_2$ can be determined by standard methods; after their values have been found, the force $-\mathbf{R}$ is obtained by solving the equations of equilibrium for the free body of Fig. 5.19*b*. The resultant **R** of the hydrostatic forces exerted on the curved surface is then obtained by reversing the sense of $-\mathbf{R}$.

The methods outlined in this section can be used to determine the resultant of the hydrostatic forces exerted on the surfaces of dams and rectangular gates and vanes. The resultants of forces on submerged surfaces of variable width will be determined in Chap. 9.

(a)

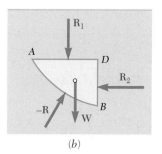

(b)

Fig. 5.19

---

†The pressure p, which represents a load per unit area, is expressed in N/m^2 or in lb/ft^2. The derived SI unit N/m^2 is called a *pascal* (Pa).

‡Noting that the area under the load curve is equal to $w_E L$, where w_E is the load per unit length at the center E of the plate, and recalling Eq. (5.13), we can write

$$R = w_E L = (bp_E)L = p_E(bL) = p_E A$$

where A denotes the area of the *plate*. Thus, the magnitude of **R** can be obtained by multiplying the area of the plate by the pressure at its center E. The resultant **R**, however, *should be applied at P, not at E.*

$w_B = 4500$ N/m

$w_A = 1500$ N/m

A B

$L = 6$ m

SAMPLE PROBLEM 5.9

A beam supports a distributed load as shown. (*a*) Determine the equivalent concentrated load. (*b*) Determine the reactions at the supports.

$\bar{x} = 4$ m

II   4.5 kN/m

1.5 kN/m   I

x

$\bar{x} = 2$ m

6 m

SOLUTION

a. Equivalent Concentrated Load.

The magnitude of the resultant of the load is equal to the area under the load curve, and the line of action of the resultant passes through the centroid of the same area. We divide the area under the load curve into two triangles and construct the table below. To simplify the computations and tabulation, the given loads per unit length have been converted into kN/m.

Component	A, kN	$\bar{x}$, m	$\bar{x}A$, kN · m
Triangle I	4.5	2	9
Triangle II	13.5	4	54
	$\Sigma A = 18.0$		$\Sigma\bar{x}A = 63$

Thus, $\bar{X}\Sigma A = \Sigma\bar{x}A$: $\bar{X}(18 \text{ kN}) = 63 \text{ kN} \cdot \text{m}$ $\bar{X} = 3.5$ m

The equivalent concentrated load is

$$\mathbf{W} = 18 \text{ kN} \downarrow \quad \blacktriangleleft$$

18 kN

$\bar{X} = 3.5$ m

A B

and its line of action is located at a distance

$$\bar{X} = 3.5 \text{ m to the right of } A \quad \blacktriangleleft$$

b. Reactions.

The reaction at A is vertical and is denoted by $\mathbf{A}$; the reaction at B is represented by its components $\mathbf{B}_x$ and $\mathbf{B}_y$. The given load can be considered to be the sum of two triangular loads as shown. The resultant of each triangular load is equal to the area of the triangle and acts at its centroid. We write the following equilibrium equations for the free body shown:

4.5 kN   13.5 kN

B_x

A B_y

2 m

4 m

6 m

$\xrightarrow{+}\Sigma F_x = 0:$ $\mathbf{B}_x = 0$ $\blacktriangleleft$

$+\uparrow \Sigma M_A = 0:$ $-(4.5 \text{ kN})(2 \text{ m}) - (13.5 \text{ kN})(4 \text{ m}) + B_y(6 \text{ m}) = 0$

$$\mathbf{B}_y = 10.5 \text{ kN} \uparrow \quad \blacktriangleleft$$

$+\uparrow \Sigma M_B = 0:$ $+(4.5 \text{ kN})(4 \text{ m}) + (13.5 \text{ kN})(2 \text{ m}) - A(6 \text{ m}) = 0$

$$\mathbf{A} = 7.5 \text{ kN} \uparrow \quad \blacktriangleleft$$

Alternative Solution. The given distributed load can be replaced by its resultant, which was found in part *a*. The reactions can be determined by writing the equilibrium equations $\Sigma F_x = 0$, $\Sigma M_A = 0$, and $\Sigma M_B = 0$. We again obtain

$$\mathbf{B}_x = 0 \quad \mathbf{B}_y = 10.5 \text{ kN} \uparrow \quad \mathbf{A} = 7.5 \text{ kN} \uparrow \quad \blacktriangleleft$$

SAMPLE PROBLEM 5.10

The cross section of a concrete dam is as shown. Consider a 1-ft-thick section of the dam, and determine (a) the resultant of the reaction forces exerted by the ground on the base AB of the dam, (b) the resultant of the pressure forces exerted by the water on the face BC of the dam. The specific weights of concrete and water are 150 lb/ft³ and 62.4 lb/ft³, respectively.

SOLUTION

a. Ground Reaction. We choose as a free body the 1-ft-thick section AEFCDB of the dam and water. The reaction forces exerted by the ground on the base AB are represented by an equivalent force-couple system at A. Other forces acting on the free body are the weight of the dam, represented by the weights of its components $\mathbf{W}_1$, $\mathbf{W}_2$, and $\mathbf{W}_3$; the weight of the water $\mathbf{W}_4$; and the resultant $\mathbf{P}$ of the pressure forces exerted on section BD by the water to the right of section BD. We have

$$W_1 = \tfrac{1}{2}(9 \text{ ft})(22 \text{ ft})(1 \text{ ft})(150 \text{ lb/ft}^3) = 14{,}850 \text{ lb}$$
$$W_2 = (5 \text{ ft})(22 \text{ ft})(1 \text{ ft})(150 \text{ lb/ft}^3) = 16{,}500 \text{ lb}$$
$$W_3 = \tfrac{1}{3}(10 \text{ ft})(18 \text{ ft})(1 \text{ ft})(150 \text{ lb/ft}^3) = 9000 \text{ lb}$$
$$W_4 = \tfrac{2}{3}(10 \text{ ft})(18 \text{ ft})(1 \text{ ft})(62.4 \text{ lb/ft}^3) = 7488 \text{ lb}$$
$$P = \tfrac{1}{2}(18 \text{ ft})(1 \text{ ft})(18 \text{ ft})(62.4 \text{ lb/ft}^3) = 10{,}109 \text{ lb}$$

Equilibrium Equations

$\xrightarrow{+} \Sigma F_x = 0: \qquad H - 10{,}109 \text{ lb} = 0 \qquad\qquad \mathbf{H} = 10{,}110 \text{ lb} \rightarrow$ ◄

$+\uparrow \Sigma F_y = 0: \qquad V - 14{,}850 \text{ lb} - 16{,}500 \text{ lb} - 9000 \text{ lb} - 7488 \text{ lb} = 0$

$\qquad\qquad\qquad\qquad\qquad\qquad\qquad\qquad \mathbf{V} = 47{,}840 \text{ lb} \uparrow$ ◄

$+\! \uparrow \Sigma M_A = 0: \qquad -(14{,}850 \text{ lb})(6 \text{ ft}) - (16{,}500 \text{ lb})(11.5 \text{ ft})$
$\qquad\qquad - (9000 \text{ lb})(17 \text{ ft}) - (7488 \text{ lb})(20 \text{ ft}) + (10{,}109 \text{ lb})(6 \text{ ft}) + M = 0$

$\qquad\qquad\qquad\qquad\qquad\qquad\qquad \mathbf{M} = 520{,}960 \text{ lb} \cdot \text{ft} \!\uparrow$ ◄

We can replace the force-couple system obtained by a single force acting at a distance d to the right of A, where

$$d = \frac{520{,}960 \text{ lb} \cdot \text{ft}}{47{,}840 \text{ lb}} = 10.89 \text{ ft}$$

b. Resultant R of Water Forces. The parabolic section of water BCD is chosen as a free body. The forces involved are the resultant $-\mathbf{R}$ of the forces exerted by the dam on the water, the weight $\mathbf{W}_4$, and the force $\mathbf{P}$. Since these forces must be concurrent, $-\mathbf{R}$ passes through the point of intersection G of $\mathbf{W}_4$ and $\mathbf{P}$. A force triangle is drawn from which the magnitude and direction of $-\mathbf{R}$ are determined. The resultant $\mathbf{R}$ of the forces exerted by the water on the face BC is equal and opposite:

$$\mathbf{R} = 12{,}580 \text{ lb} \;\nearrow 36.5°$$ ◄

SOLVING PROBLEMS ON YOUR OWN

The problems in this lesson involve two common and very important types of loading: distributed loads on beams and forces on submerged surfaces of constant width. As we discussed in Secs. 5.8 and 5.9 and illustrated in Sample Probs. 5.9 and 5.10, determining the single equivalent force for each of these loadings requires a knowledge of centroids.

1. Analyzing beams subjected to distributed loads. In Sec. 5.8, we showed that a distributed load on a beam can be replaced by a single equivalent force. The magnitude of this force is equal to the area under the distributed load curve and its line of action passes through the centroid of that area. Thus, you should begin your solution by replacing the various distributed loads on a given beam by their respective single equivalent forces. The reactions at the supports of the beam can then be determined by using the methods of Chap. 4.

When possible, complex distributed loads should be divided into the common-shape areas shown in Fig. 5.8A [Sample Prob. 5.9]. Each of these areas can then be replaced by a single equivalent force. If required, the system of equivalent forces can be reduced further to a single equivalent force. As you study Sample Prob. 5.9, note how we have used the analogy between force and area and the techniques for locating the centroid of a composite area to analyze a beam subjected to a distributed load.

2. Solving problems involving forces on submerged bodies. The following points and techniques should be remembered when solving problems of this type.

   a. The pressure p at a depth h below the free surface of a liquid is equal to γh or $\rho g h$, where γ and ρ are the specific weight and the density of the liquid, respectively. The load per unit length w acting on a submerged surface of constant width b is then

$$w = bp = b\gamma h = b\rho g h$$

   b. The line of action of the resultant force $\mathbf{R}$ acting on a submerged plane surface is perpendicular to the surface.

   c. For a vertical or inclined plane rectangular surface of width b, the loading on the surface can be represented by a linearly distributed load which is trapezoidal in shape (Fig. 5.18). Further, the magnitude of $\mathbf{R}$ is given by

$$R = \gamma h_E A$$

where h_E is the vertical distance to the center of the surface and A is the area of the surface.

d. The load curve will be triangular (rather than trapezoidal) when the top edge of a plane rectangular surface coincides with the free surface of the liquid, since the pressure of the liquid at the free surface is zero. For this case, the line of action of **R** is easily determined, for it passes through the centroid of a *triangular* distributed load.

e. For the general case, rather than analyzing a trapezoid, we suggest that you use the method indicated in part *b* of Sample Prob. 5.9. First divide the trapezoidal distributed load into two triangles, and then compute the magnitude of the resultant of each triangular load. (The magnitude is equal to the area of the triangle times the width of the plate.) Note that the line of action of each resultant force passes through the centroid of the corresponding triangle and that the sum of these forces is equivalent to **R**. Thus, rather than using **R**, you can use the two equivalent resultant forces, whose points of application are easily calculated. Of course, the equation given for R in paragraph c should be used when only the magnitude of **R** is needed.

f. When the submerged surface of constant width is curved, the resultant force acting on the surface is obtained by considering the equilibrium of the volume of liquid bounded by the curved surface and by horizontal and vertical planes (Fig. 5.19). Observe that the force $\mathbf{R}_1$ of Fig. 5.19 is equal to the weight of the liquid lying above the plane AD. The method of solution for problems involving curved surfaces is shown in part b of Sample Prob. 5.10.

In subsequent mechanics courses (in particular, mechanics of materials and fluid mechanics), you will have ample opportunity to use the ideas introduced in this lesson.

PROBLEMS

5.66 and 5.67 For the beam and loading shown, determine (*a*) the magnitude and location of the resultant of the distributed load, (*b*) the reactions at the beam supports.

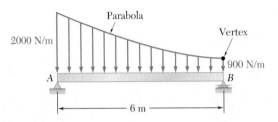

Fig. P5.66

Fig. P5.67

5.68 through 5.73 Determine the reactions at the beam supports for the given loading.

Fig. P5.68

Fig. P5.69

Fig. P5.70

Fig. P5.71

Fig. P5.72

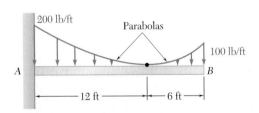

Fig. P5.73

5.74 Determine the reactions at the beam supports for the given loading when $w_0 = 400$ lb/ft.

Fig. P5.74 and P5.75

5.75 Determine (a) the distributed load w_0 at the end A of the beam ABC for which the reaction at C is zero, (b) the corresponding reaction at B.

5.76 Determine (a) the distance a so that the vertical reactions at supports A and B are equal, (b) the corresponding reactions at the supports.

5.77 Determine (a) the distance a so that the reaction at support B is minimum, (b) the corresponding reactions at the supports.

5.78 A beam is subjected to a linearly distributed downward load and rests on two wide supports BC and DE, which exert uniformly distributed upward loads as shown. Determine the values of w_{BC} and w_{DE} corresponding to equilibrium when $w_A = 600$ N/m.

5.79 A beam is subjected to a linearly distributed downward load and rests on two wide supports BC and DE, which exert uniformly distributed upward loads as shown. Determine (a) the value of w_A so that $w_{BC} = w_{DE}$, (b) the corresponding values of w_{BC} and w_{DE}.

In the following problems, use $\gamma = 62.4$ lb/ft^3 for the specific weight of fresh water and $\gamma_c = 150$ lb/ft^3 for the specific weight of concrete if U.S. customary units are used. With SI units, use $\rho = 10^3$ kg/m^3 for the density of fresh water and $\rho_c = 2.40 \times 10^3$ kg/m^3 for the density of concrete. (See the footnote on page 222 for how to determine the specific weight of a material given its density.)

5.80 and 5.81 The cross section of a concrete dam is as shown. For a 1-m-wide dam section determine (a) the resultant of the reaction forces exerted by the ground on the base AB of the dam, (b) the point of application of the resultant of part a, (c) the resultant of the pressure forces exerted by the water on the face BC of the dam.

Fig. P5.76 and P5.77

Fig. P5.78 and P5.79

Fig. P5.80

Fig. P5.81

Fig. P5.82 and P5.83

5.82 An automatic valve consists of a 9 × 9-in. square plate that is pivoted about a horizontal axis through A located at a distance $h = 3.6$ in. above the lower edge. Determine the depth of water d for which the valve will open.

5.83 An automatic valve consists of a 9 × 9-in. square plate that is pivoted about a horizontal axis through A. If the valve is to open when the depth of water is $d = 18$ in., determine the distance h from the bottom of the valve to the pivot A.

5.84 The 3 × 4-m side AB of a tank is hinged at its bottom A and is held in place by a thin rod BC. The maximum tensile force the rod can withstand without breaking is 200 kN, and the design specifications require the force in the rod not to exceed 20 percent of this value. If the tank is slowly filled with water, determine the maximum allowable depth of water d in the tank.

Fig. *P5.84* and P5.85

5.85 The 3 × 4-m side of an open tank is hinged at its bottom A and is held in place by a thin rod BC. The tank is to be filled with glycerine, whose density is 1263 kg/m³. Determine the force **T** in the rod and the reactions at the hinge after the tank is filled to a depth of 2.9 m.

Fig. P5.86

5.86 The friction force between a 6 × 6-ft square sluice gate AB and its guides is equal to 10 percent of the resultant of the pressure forces exerted by the water on the face of the gate. Determine the initial force needed to lift the gate if it weighs 1000 lb.

5.87 A tank is divided into two sections by a 1 × 1-m square gate that is hinged at A. A couple of magnitude 490 N · m is required for the gate to rotate. If one side of the tank is filled with water at the rate of 0.1 m³/min and the other side is filled simultaneously with methyl alcohol (density $\rho_{ma} = 789$ kg/m³) at the rate of 0.2 m³/min, determine at what time and in which direction the gate will rotate.

Fig. *P5.87*

5.88 A prismatically shaped gate placed at the end of a freshwater channel is supported by a pin and bracket at A and rests on a frictionless support at B. The pin is located at a distance $h = 0.10$ m below the center of gravity C of the gate. Determine the depth of water d for which the gate will open.

5.89 A prismatically shaped gate placed at the end of a freshwater channel is supported by a pin and bracket at A and rests on a frictionless support at B. The pin is located at a distance h below the center of gravity C of the gate. Determine the distance h if the gate is to open when $d = 0.75$ m.

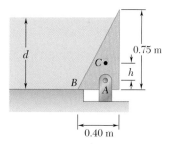

Fig. P5.88 and P5.89

5.90 The square gate AB is held in the position shown by hinges along its top edge A and by a shear pin at B. For a depth of water $d = 3.5$ ft, determine the force exerted on the gate by the shear pin.

5.91 A long trough is supported by a continuous hinge along its lower edge and by a series of horizontal cables attached to its upper edge. Determine the tension in each of the cables, at a time when the trough is completely full of water.

Fig. P5.90

Fig. P5.91

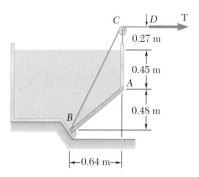

Fig. P5.92 and P5.93

5.92 A 0.5 × 0.8-m gate AB is located at the bottom of a tank filled with water. The gate is hinged along its top edge A and rests on a frictionless stop at B. Determine the reactions at A and B when cable BCD is slack.

5.93 A 0.5 × 0.8-m gate AB is located at the bottom of a tank filled with water. The gate is hinged along its top edge A and rests on a frictionless stop at B. Determine the minimum tension required in cable BCD to open the gate.

5.94 A 4 × 2-ft gate is hinged at A and is held in position by rod CD. End D rests against a spring whose constant is 828 lb/ft. The spring is undeformed when the gate is vertical. Assuming that the force exerted by rod CD on the gate remains horizontal, determine the minimum depth of water d for which the bottom B of the gate will move to the end of the cylindrical portion of the floor.

5.95 Solve Prob. 5.94 if the gate weighs 1000 lb.

Fig. P5.94

Photo 5.4 To predict the flight characteristics of the modified Boeing 747 when used to transport a space shuttle, the center of gravity of each craft had to be determined.

VOLUMES

5.10 CENTER OF GRAVITY OF A THREE-DIMENSIONAL BODY. CENTROID OF A VOLUME

The *center of gravity* G of a three-dimensional body is obtained by dividing the body into small elements and by then expressing that the weight **W** of the body acting at G is equivalent to the system of distributed forces Δ**W** representing the weights of the small elements. Choosing the y axis to be vertical with positive sense upward (Fig. 5.20) and denoting by $\bar{\mathbf{r}}$ the position vector of G, we write that

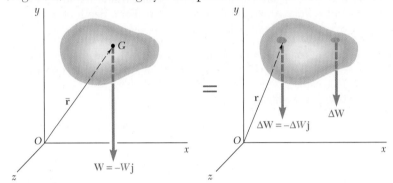

Fig. 5.20

W is equal to the sum of the elemental weights Δ**W** and that its moment about O is equal to the sum of the moments about O of the elemental weights:

$$\Sigma\mathbf{F}: \qquad -W\mathbf{j} = \Sigma(-\Delta W\mathbf{j})$$
$$\Sigma\mathbf{M}_O: \qquad \bar{\mathbf{r}} \times (-W\mathbf{j}) = \Sigma[\mathbf{r} \times (-\Delta W\mathbf{j})] \qquad (5.14)$$

Rewriting the last equation in the form

$$\bar{\mathbf{r}}W \times (-\mathbf{j}) = (\Sigma\mathbf{r}\,\Delta W) \times (-\mathbf{j}) \qquad (5.15)$$

we observe that the weight **W** of the body is equivalent to the system of the elemental weights Δ**W** if the following conditions are satisfied:

$$W = \Sigma\,\Delta W \qquad \bar{\mathbf{r}}W = \Sigma\mathbf{r}\,\Delta W$$

Increasing the number of elements and simultaneously decreasing the size of each element, we obtain in the limit

$$W = \int dW \qquad \bar{\mathbf{r}}W = \int \mathbf{r}\,dW \qquad (5.16)$$

We note that the relations obtained are independent of the orientation of the body. For example, if the body and the coordinate axes were rotated so that the z axis pointed upward, the unit vector $-\mathbf{j}$ would be replaced by $-\mathbf{k}$ in Eqs. (5.14) and (5.15), but the relations (5.16) would remain unchanged. Resolving the vectors $\bar{\mathbf{r}}$ and $\mathbf{r}$ into rectangular components, we note that the second of the relations (5.16) is equivalent to the three scalar equations

$$\bar{x}W = \int x\,dW \qquad \bar{y}W = \int y\,dW \qquad \bar{z}W = \int z\,dW \qquad (5.17)$$

If the body is made of a homogeneous material of specific weight γ, the magnitude dW of the weight of an infinitesimal element can be expressed in terms of the volume dV of the element, and the magnitude W of the total weight can be expressed in terms of the total volume V. We write

$$dW = \gamma \, dV \qquad W = \gamma V$$

Substituting for dW and W in the second of the relations (5.16), we write

$$\bar{\mathbf{r}}V = \int \mathbf{r} \, dV \tag{5.18}$$

or, in scalar form,

$$\bar{x}V = \int x \, dV \qquad \bar{y}V = \int y \, dV \qquad \bar{z}V = \int z \, dV \tag{5.19}$$

The point whose coordinates are $\bar{x}, \bar{y}, \bar{z}$ is also known as the *centroid C of the volume* V of the body. If the body is not homogeneous, Eqs. (5.19) cannot be used to determine the center of gravity of the body; however, Eqs. (5.19) still define the centroid of the volume.

The integral $\int x \, dV$ is known as the *first moment of the volume with respect to the yz plane*. Similarly, the integrals $\int y \, dV$ and $\int z \, dV$ define the first moments of the volume with respect to the zx plane and the xy plane, respectively. It is seen from Eqs. (5.19) that if the centroid of a volume is located in a coordinate plane, the first moment of the volume with respect to that plane is zero.

A volume is said to be symmetrical with respect to a given plane if for every point P of the volume there exists a point P' of the same volume, such that the line PP' is perpendicular to the given plane and is bisected by that plane. The plane is said to be a *plane of symmetry* for the given volume. When a volume V possesses a plane of symmetry, the first moment of V with respect to that plane is zero, and the centroid of the volume is located in the plane of symmetry. When a volume possesses two planes of symmetry, the centroid of the volume is located on the line of intersection of the two planes. Finally, when a volume possesses three planes of symmetry which intersect at a well-defined point (i.e., not along a common line), the point of intersection of the three planes coincides with the centroid of the volume. This property enables us to determine immediately the locations of the centroids of spheres, ellipsoids, cubes, rectangular parallelepipeds, etc.

The centroids of unsymmetrical volumes or of volumes possessing only one or two planes of symmetry should be determined by integration (Sec. 5.12). The centroids of several common volumes are shown in Fig. 5.21. It should be observed that in general the centroid of a volume of revolution *does not coincide* with the centroid of its cross section. Thus, the centroid of a hemisphere is different from that of a semicircular area, and the centroid of a cone is different from that of a triangle.

Shape		$\bar{x}$	Volume
Hemisphere		$\dfrac{3a}{8}$	$\dfrac{2}{3}\pi a^3$
Semiellipsoid of revolution		$\dfrac{3h}{8}$	$\dfrac{2}{3}\pi a^2 h$
Paraboloid of revolution		$\dfrac{h}{3}$	$\dfrac{1}{2}\pi a^2 h$
Cone		$\dfrac{h}{4}$	$\dfrac{1}{3}\pi a^2 h$
Pyramid		$\dfrac{h}{4}$	$\dfrac{1}{3}abh$

Fig. 5.21 Centroids of common shapes and volumes.

5.11 COMPOSITE BODIES

If a body can be divided into several of the common shapes shown
in Fig. 5.21, its center of gravity G can be determined by expressing
that the moment about O of its total weight is equal to the sum of
the moments about O of the weights of the various component parts.
Proceeding as in Sec. 5.10, we obtain the following equations defining the coordinates $\bar{X}, \bar{Y}, \bar{Z}$ of the center of gravity G.

$$\bar{X}\Sigma W = \Sigma\bar{x}W \quad \bar{Y}\Sigma W = \Sigma\bar{y}W \quad \bar{Z}\Sigma W = \Sigma\bar{z}W \quad (5.20)$$

If the body is made of a homogeneous material, its center of
gravity coincides with the centroid of its volume, and we obtain:

$$\bar{X}\Sigma V = \Sigma\bar{x}V \quad \bar{Y}\Sigma V = \Sigma\bar{y}V \quad \bar{Z}\Sigma V = \Sigma\bar{z}V \quad (5.21)$$

5.12 DETERMINATION OF CENTROIDS OF VOLUMES BY INTEGRATION

The centroid of a volume bounded by analytical surfaces can be
determined by evaluating the integrals given in Sec. 5.10:

$$\bar{x}V = \int x\,dV \quad \bar{y}V = \int y\,dV \quad \bar{z}V = \int z\,dV \quad (5.22)$$

If the element of volume dV is chosen to be equal to a small cube
of sides dx, dy, and dz, the evaluation of each of these integrals
requires a *triple integration*. However, it is possible to determine the
coordinates of the centroid of most volumes by *double integration* if
dV is chosen to be equal to the volume of a thin filament (Fig. 5.22).
The coordinates of the centroid of the volume are then obtained by
rewriting Eqs. (5.22) as

$$\bar{x}V = \int \bar{x}_{el}\,dV \quad \bar{y}V = \int \bar{y}_{el}\,dV \quad \bar{z}V = \int \bar{z}_{el}\,dV \quad (5.23)$$

and by then substituting the expressions given in Fig. 5.22 for the
volume dV and the coordinates $\bar{x}_{el}, \bar{y}_{el}, \bar{z}_{el}$. By using the equation of
the surface to express z in terms of x and y, the integration is reduced
to a double integration in x and y.

If the volume under consideration possesses *two planes of symmetry*, its centroid must be located on the line of intersection of the
two planes. Choosing the x axis to lie along this line, we have

$$\bar{y} = \bar{z} = 0$$

and the only coordinate to determine is $\bar{x}$. This can be done with a *single
integration* by dividing the given volume into thin slabs parallel to the
yz plane and expressing dV in terms of x and dx in the equation

$$\bar{x}V = \int \bar{x}_{el}\,dV \quad (5.24)$$

For a body of revolution, the slabs are circular and their volume is
given in Fig. 5.23.

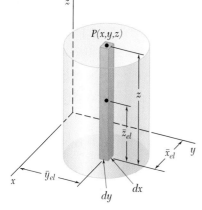

$$\bar{x}_{el} = x, \quad \bar{y}_{el} = y, \quad \bar{z}_{el} = \frac{z}{2}$$
$$dV = z\,dx\,dy$$

Fig. 5.22 Determination of the centroid
of a volume by double integration.

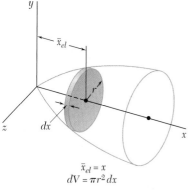

$$\bar{x}_{el} = x$$
$$dV = \pi r^2\,dx$$

Fig. 5.23 Determination of the
centroid of a body of revolution.

SAMPLE PROBLEM 5.11

Determine the location of the center of gravity of the homogeneous body of revolution shown, which was obtained by joining a hemisphere and a cylinder and carving out a cone.

SOLUTION

Because of symmetry, the center of gravity lies on the x axis. As shown in the figure below, the body can be obtained by adding a hemisphere to a cylinder and then subtracting a cone. The volume and the abscissa of the centroid of each of these components are obtained from Fig. 5.21 and are entered in the table below. The total volume of the body and the first moment of its volume with respect to the yz plane are then determined.

Component	Volume, mm^3		$\bar{x}$, mm	$\bar{x}V$, mm^4
Hemisphere	$\dfrac{1}{2}\dfrac{4\pi}{3}(60)^3 =$	0.4524×10^6	-22.5	-10.18×10^6
Cylinder	$\pi(60)^2(100) =$	1.1310×10^6	$+50$	$+56.55 \times 10^6$
Cone	$-\dfrac{\pi}{3}(60)^2(100) =$	-0.3770×10^6	$+75$	-28.28×10^6
	$\Sigma V =$	1.206×10^6		$\Sigma \bar{x}V = +18.09 \times 10^6$

Thus,

$$\bar{X}\Sigma V = \Sigma \bar{x}V: \qquad \bar{X}(1.206 \times 10^6 \text{ mm}^3) = 18.09 \times 10^6 \text{ mm}^4$$

$$\bar{X} = 15 \text{ mm} \blacktriangleleft$$

SAMPLE PROBLEM 5.12

Locate the center of gravity of the steel machine element shown. The diameter of each hole is 1 in.

SOLUTION

The machine element can be obtained by adding a rectangular parallelepiped (I) to a quarter cylinder (II) and then subtracting two 1-in.-diameter cylinders (III and IV). The volume and the coordinates of the centroid of each component are determined and are entered in the table below. Using the data in the table, we then determine the total volume and the moments of the volume with respect to each of the coordinate planes.

	V, in^3	$\bar{x}$, in.	$\bar{y}$, in.	$\bar{z}$, in.	$\bar{x}V$, in^4	$\bar{y}V$, in^4	$\bar{z}V$, in^4
I	$(4.5)(2)(0.5) = 4.5$	0.25	-1	2.25	1.125	-4.5	10.125
II	$\frac{1}{4}\pi(2)^2(0.5) = 1.571$	1.3488	-0.8488	0.25	2.119	-1.333	0.393
III	$-\pi(0.5)^2(0.5) = -0.3927$	0.25	-1	3.5	-0.098	0.393	-1.374
IV	$-\pi(0.5)^2(0.5) = -0.3927$	0.25	-1	1.5	-0.098	0.393	-0.589
	$\Sigma V = 5.286$				$\Sigma\bar{x}V = 3.048$	$\Sigma\bar{y}V = -5.047$	$\Sigma\bar{z}V = 8.555$

Thus,

$$\bar{X}\Sigma V = \Sigma\bar{x}V: \qquad \bar{X}(5.286 \text{ in}^3) = 3.048 \text{ in}^4 \qquad \bar{X} = \quad 0.577 \text{ in.} \blacktriangleleft$$

$$\bar{Y}\Sigma V = \Sigma\bar{y}V: \qquad \bar{Y}(5.286 \text{ in}^3) = -5.047 \text{ in}^4 \qquad \bar{Y} = -0.955 \text{ in.} \blacktriangleleft$$

$$\bar{Z}\Sigma V = \Sigma\bar{z}V: \qquad \bar{Z}(5.286 \text{ in}^3) = 8.555 \text{ in}^4 \qquad \bar{Z} = \quad 1.618 \text{ in.} \blacktriangleleft$$

SAMPLE PROBLEM 5.13

Determine the location of the centroid of the half right circular cone shown.

SOLUTION

Since the xy plane is a plane of symmetry, the centroid lies in this plane and $\bar{z} = 0$. A slab of thickness dx is chosen as a differential element. The volume of this element is

$$dV = \tfrac{1}{2}\pi r^2 \, dx$$

The coordinates $\bar{x}_{el}$ and $\bar{y}_{el}$ of the centroid of the element are obtained from Fig. 5.8 (semicircular area).

$$\bar{x}_{el} = x \qquad \bar{y}_{el} = \frac{4r}{3\pi}$$

We observe that r is proportional to x and write

$$\frac{r}{x} = \frac{a}{h} \qquad r = \frac{a}{h}x$$

The volume of the body is

$$V = \int dV = \int_0^h \tfrac{1}{2}\pi r^2 \, dx = \int_0^h \tfrac{1}{2}\pi \left(\frac{a}{h}x\right)^2 dx = \frac{\pi a^2 h}{6}$$

The moment of the differential element with respect to the yz plane is $\bar{x}_{el} \, dV$; the total moment of the body with respect to this plane is

$$\int \bar{x}_{el} \, dV = \int_0^h x(\tfrac{1}{2}\pi r^2) \, dx = \int_0^h x(\tfrac{1}{2}\pi)\left(\frac{a}{h}x\right)^2 dx = \frac{\pi a^2 h^2}{8}$$

Thus,

$$\bar{x}V = \int \bar{x}_{el} \, dV \qquad \bar{x}\frac{\pi a^2 h}{6} = \frac{\pi a^2 h^2}{8} \qquad \bar{x} = \tfrac{3}{4}h \quad \blacktriangleleft$$

Likewise, the moment of the differential element with respect to the zx plane is $\bar{y}_{el} \, dV$; the total moment is

$$\int \bar{y}_{el} \, dV = \int_0^h \frac{4r}{3\pi}(\tfrac{1}{2}\pi r^2) \, dx = \frac{2}{3}\int_0^h \left(\frac{a}{h}x\right)^3 dx = \frac{a^3 h}{6}$$

Thus,

$$\bar{y}V = \int \bar{y}_{el} \, dV \qquad \bar{y}\frac{\pi a^2 h}{6} = \frac{a^3 h}{6} \qquad \bar{y} = \frac{a}{\pi} \quad \blacktriangleleft$$

SOLVING PROBLEMS
ON YOUR OWN

In the problems for this lesson, you will be asked to locate the centers of gravity of three-dimensional bodies or the centroids of their volumes. All of the techniques we previously discussed for two-dimensional bodies—using symmetry, dividing the body into common shapes, choosing the most efficient differential element, etc.—may also be applied to the general three-dimensional case.

1. Locating the centers of gravity of composite bodies. In general, Eqs. (5.20) must be used:

$$\overline{X}\Sigma W = \Sigma \overline{x}W \qquad \overline{Y}\Sigma W = \Sigma \overline{y}W \qquad \overline{Z}\Sigma W = \Sigma \overline{z}W \qquad (5.20)$$

However, for the case of a *homogeneous body*, the center of gravity of the body coincides with the *centroid of its volume*. Therefore, for this special case, the center of gravity of the body can also be located using Eqs. (5.21):

$$\overline{X}\Sigma V = \Sigma \overline{x}V \qquad \overline{Y}\Sigma V = \Sigma \overline{y}V \qquad \overline{Z}\Sigma V = \Sigma \overline{z}V \qquad (5.21)$$

You should realize that these equations are simply an extension of the equations used for the two-dimensional problems considered earlier in the chapter. As the solutions of Sample Probs. 5.11 and 5.12 illustrate, the methods of solution for two- and three-dimensional problems are identical. Thus, we once again strongly encourage you to construct appropriate diagrams and tables when analyzing composite bodies. Also, as you study Sample Prob. 5.12, observe how the x and y coordinates of the centroid of the quarter cylinder were obtained using the equations for the centroid of a quarter circle.

We note that *two special cases* of interest occur when the given body consists of either uniform wires or uniform plates made of the same material.

    a. For a body made of *several wire elements* of the *same uniform cross section*, the cross-sectional area A of the wire elements will factor out of Eqs. (5.21) when V is replaced with the product AL, where L is the length of a given element. Equations (5.21) thus reduce in this case to

$$\overline{X}\Sigma L = \Sigma \overline{x}L \qquad \overline{Y}\Sigma L = \Sigma \overline{y}L \qquad \overline{Z}\Sigma L = \Sigma \overline{z}L$$

    b. For a body made of *several plates* of the *same uniform thickness*, the thickness t of the plates will factor out of Eqs. (5.21) when V is replaced with the product tA, where A is the area of a given plate. Equations (5.21) thus reduce in this case to

$$\overline{X}\Sigma A = \Sigma \overline{x}A \qquad \overline{Y}\Sigma A = \Sigma \overline{y}A \qquad \overline{Z}\Sigma A = \Sigma \overline{z}A$$

2. Locating the centroids of volumes by direct integration. As explained in Sec. 5.12, evaluating the integrals of Eqs. (5.22) can be simplified by choosing either a thin filament (Fig. 5.22) or a thin slab (Fig. 5.23) for the element of volume dV. Thus, you should begin your solution by identifying, if possible, the dV which produces the single or double integrals that are the easiest to compute. For bodies of revolution, this may be a thin slab (as in Sample Prob. 5.13) or a thin cylindrical shell. However, it is important to remember that the relationship that you establish among the variables (like the relationship between r and x in Sample Prob. 5.13) will directly affect the complexity of the integrals you will have to compute. Finally, we again remind you that $\overline{x}_{el}$, $\overline{y}_{el}$, and $\overline{z}_{el}$ in Eqs. (5.23) are the coordinates of the centroid of dV.

5.96 A hemisphere and a cone are attached as shown. Determine the location of the centroid of the composite body when (a) h = 1.5a, (b) h = 2a.

Fig. P5.96

5.97 Consider the composite body shown. Determine (a) the value of $\bar{x}$ when h = L/2, (b) the ratio h/L for which $\bar{x}$ = L.

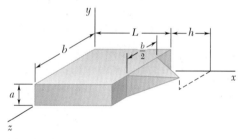

Fig. P5.97

5.98 Determine the y coordinate of the centroid of the body shown.

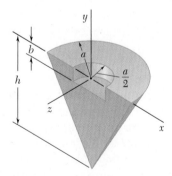

Fig. P5.98 and P5.99

5.99 Determine the z coordinate of the centroid of the body shown. (*Hint:* Use the result of Sample Prob. 5.13.)

5.100 and 5.101 For the machine element shown, locate the y coordinate of the center of gravity.

5.102 For the machine element shown, locate the x coordinate of the center of gravity.

5.103 For the machine element shown, locate the z coordinate of the center of gravity.

Fig. P5.100 and P5.103

Fig. P5.101 and *P5.102*

5.104 For the machine element shown, locate the x coordinate of the center of gravity.

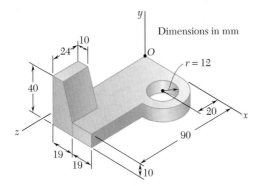

Fig. P5.104 and *P5.105*

5.105 For the machine element shown, locate the z coordinate of the center of gravity.

Fig. P5.106

5.106 and 5.107 Locate the center of gravity of the sheet-metal form shown.

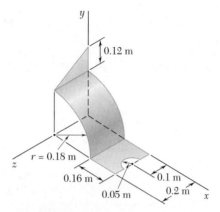

Fig. P5.107

5.108 A window awning is fabricated from sheet metal of uniform thickness. Locate the center of gravity of the awning.

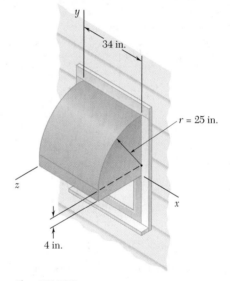

Fig. P5.108

5.109 A thin sheet of plastic of uniform thickness is bent to form a desk organizer. Locate the center of gravity of the organizer.

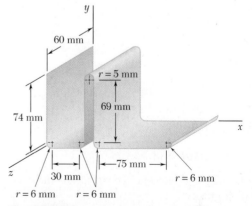

Fig. P5.109

5.110 A wastebasket, designed to fit in the corner of a room, is 16 in. high and has a base in the shape of a quarter circle of radius 10 in. Locate the center of gravity of the wastebasket, knowing that it is made of sheet metal of uniform thickness.

5.111 A mounting bracket for electronic components is formed from sheet metal of uniform thickness. Locate the center of gravity of the bracket.

Fig. P5.110

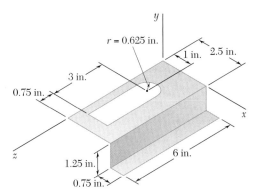

Fig. *P5.111*

5.112 An 8-in.-diameter cylindrical duct and a 4 × 8-in. rectangular duct are to be joined as indicated. Knowing that the ducts were fabricated from the same sheet metal, which is of uniform thickness, locate the center of gravity of the assembly.

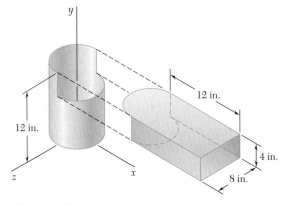

Fig. *P5.112*

5.113 An elbow for the duct of a ventilating system is made of sheet metal of uniform thickness. Locate the center of gravity of the elbow.

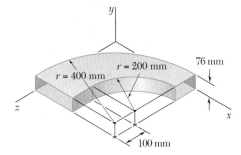

Fig. P5.113

5.114 and 5.115 Locate the center of gravity of the figure shown, knowing that it is made of thin brass rods of uniform diameter.

Fig. P5.114

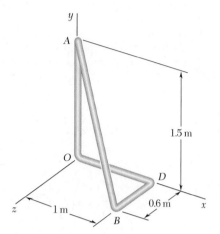

Fig. P5.115

5.116 A thin steel wire of uniform cross section is bent into the shape shown. Locate its center of gravity.

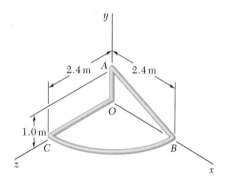

Fig. P5.116

5.117 The frame of a greenhouse is constructed from uniform aluminum channels. Locate the center of gravity of the portion of the frame shown.

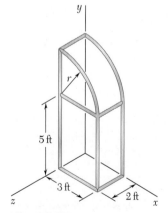

Fig. P5.117

5.118 Three brass plates are brazed to a steel pipe to form the flagpole base shown. Knowing that the pipe has a wall thickness of 8 mm and that each plate is 6 mm thick, determine the location of the center of gravity of the base. (Densities: brass = 8470 kg/m³; steel = 7860 kg/m³.)

5.119 A brass collar, of length 2.5 in., is mounted on an aluminum rod of length 4 in. Locate the center of gravity of the composite body. (Specific weights: brass = 0.306 lb/in³, aluminum = 0.101 lb/in³.)

Fig. P5.118

Fig. *P5.119*

5.120 A bronze bushing is mounted inside a steel sleeve. Knowing that the specific weight of bronze is 0.318 lb/in³ and of steel is 0.284 lb/in³, determine the location of the center of gravity of the assembly.

5.121 A scratch awl has a plastic handle and a steel blade and shank. Knowing that the density of plastic is 1030 kg/m³ and of steel is 7860 kg/m³, locate the center of gravity of the awl.

Fig. P5.120

Fig. P5.121

5.122 through 5.124 Determine by direct integration the values of $\bar{x}$ for the two volumes obtained by passing a vertical cutting plane through the given shape of Fig. 5.21. The cutting plane is parallel to the base of the given shape and divides the shape into two volumes of equal height.

    5.122 A hemisphere.
    5.123 A semiellipsoid of revolution.
    5.124 A paraboloid of revolution.

5.125 and 5.126 Locate the centroid of the volume obtained by rotating the shaded area about the x axis.

Fig. P5.125

Fig. P5.126

5.127 Locate the centroid of the volume obtained by rotating the shaded area about the line $x = h$.

Fig. P5.127

*5.128 Locate the centroid of the volume generated by revolving the portion of the sine curve shown about the x axis.

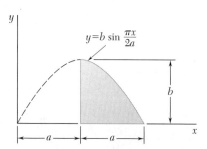

Fig. P5.128 and P5.129

*5.129 Locate the centroid of the volume generated by revolving the portion of the sine curve shown about the y axis. (*Hint:* Use a thin cylindrical shell of radius r and thickness dr as the element of volume.)

*5.130 Show that for a regular pyramid of height h and n sides ($n = 3$, 4, …) the centroid of the volume of the pyramid is located at a distance $h/4$ above the base.

5.131 Determine by direct integration the location of the centroid of one-half of a thin, uniform hemispherical shell of radius R.

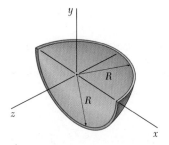

Fig. P5.131

5.132 The sides and the base of a punch bowl are of uniform thickness t. If $t \ll R$ and $R = 250$ mm, determine the location of the center of gravity of (*a*) the bowl, (*b*) the punch.

Fig. P5.132

5.133 Locate the centroid of the section shown, which was cut from a thin circular pipe by two oblique planes.

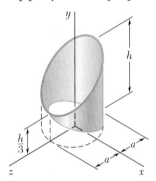

Fig. *P5.133*

*5.134 Locate the centroid of the section shown, which was cut from an elliptical cylinder by an oblique plane.

5.135 After grading a lot, a builder places four stakes to designate the corners of the slab for a house. To provide a firm, level base for the slab, the builder places a minimum of 3 in. of gravel beneath the slab. Determine the volume of gravel needed and the x coordinate of the centroid of the volume of the gravel. (*Hint:* The bottom surface of the gravel is an oblique plane, which can be represented by the equation $y = a + bx + cz$.)

Fig. P5.134

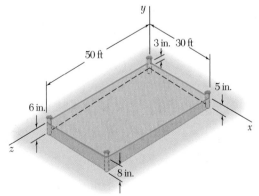

Fig. P5.135

5.136 Determine by direct integration the location of the centroid of the volume between the xz plane and the portion shown of the surface $y = 16h(ax - x^2)(bz - z^2)/a^2b^2$.

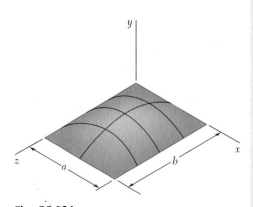

Fig. P5.136

This chapter was devoted chiefly to the determination of the *center of gravity* of a rigid body, i.e., to the determination of the point G where a single force $\mathbf{W}$, called the *weight* of the body, can be applied to represent the effect of the earth's attraction on the body.

Center of gravity of a two-dimensional body

In the first part of the chapter, we considered *two-dimensional bodies*, such as flat plates and wires contained in the xy plane. By adding force components in the vertical z direction and moments about the horizontal y and x axes [Sec. 5.2], we derived the relations

$$W = \int dW \qquad \bar{x}W = \int x\, dW \qquad \bar{y}W = \int y\, dW \qquad (5.2)$$

which define the weight of the body and the coordinates $\bar{x}$ and $\bar{y}$ of its center of gravity.

Centroid of an area or line

In the case of a *homogeneous flat plate of uniform thickness* [Sec. 5.3], the center of gravity G of the plate coincides with the *centroid C of the area A* of the plate, the coordinates of which are defined by the relations

$$\bar{x}A = \int x\, dA \qquad \bar{y}A = \int y\, dA \qquad (5.3)$$

Similarly, the determination of the center of gravity of a *homogeneous wire of uniform cross section* contained in a plane reduces to the determination of the *centroid C of the line L* representing the wire; we have

$$\bar{x}L = \int x\, dL \qquad \bar{y}L = \int y\, dL \qquad (5.4)$$

First moments

The integrals in Eqs. (5.3) are referred to as the *first moments* of the area A with respect to the y and x axes and are denoted by Q_y and Q_x, respectively [Sec. 5.4]. We have

$$Q_y = \bar{x}A \qquad Q_x = \bar{y}A \qquad (5.6)$$

The first moments of a line can be defined in a similar way.

Properties of symmetry

The determination of the centroid C of an area or line is simplified when the area or line possesses certain *properties of symmetry*. If the area or line is symmetric with respect to an axis, its centroid C

lies on that axis; if it is symmetric with respect to two axes, C is located at the intersection of the two axes; if it is symmetric with respect to a center O, C coincides with O.

The *areas and the centroids of various common shapes* are tabulated in Fig. 5.8. When a flat plate can be divided into several of these shapes, the coordinates $\overline{X}$ and $\overline{Y}$ of its center of gravity G can be determined from the coordinates $\overline{x}_1, \overline{x}_2, \ldots$ and $\overline{y}_1, \overline{y}_2, \ldots$ of the centers of gravity $G_1, G_2, \ldots$ of the various parts [Sec. 5.5]. Equating moments about the y and x axes, respectively (Fig. 5.24), we have

$$\overline{X}\Sigma W = \Sigma\overline{x}W \qquad \overline{Y}\Sigma W = \Sigma\overline{y}W \qquad (5.7)$$

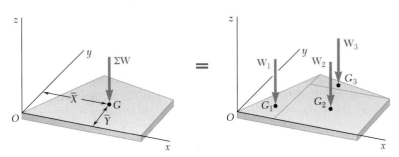

Fig. 5.24

If the plate is homogeneous and of uniform thickness, its center of gravity coincides with the centroid C of the area of the plate, and Eqs. (5.7) reduce to

$$Q_y = \overline{X}\Sigma A = \Sigma\overline{x}A \qquad Q_x = \overline{Y}\Sigma A = \Sigma\overline{y}A \qquad (5.8)$$

These equations yield the first moments of the composite area, or they can be solved for the coordinates $\overline{X}$ and $\overline{Y}$ of its centroid [Sample Prob. 5.1]. The determination of the center of gravity of a composite wire is carried out in a similar fashion [Sample Prob. 5.2].

When an area is bounded by analytical curves, the coordinates of its centroid can be determined by *integration* [Sec. 5.6]. This can be done by evaluating either the double integrals in Eqs. (5.3) or a *single integral* which uses one of the thin rectangular or pie-shaped elements of area shown in Fig. 5.12. Denoting by $\overline{x}_{el}$ and $\overline{y}_{el}$ the coordinates of the centroid of the element dA, we have

$$Q_y = \overline{x}A = \int \overline{x}_{el}\,dA \qquad Q_x = \overline{y}A = \int \overline{y}_{el}\,dA \qquad (5.9)$$

It is advantageous to use the same element of area to compute both of the first moments Q_y and Q_x; the same element can also be used to determine the area A [Sample Prob. 5.4].

Theorems of Pappus-Guldinus

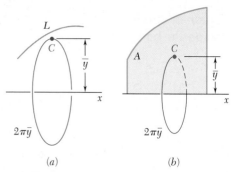

(a)

(b)

Fig. 5.25

Distributed loads

The *theorems of Pappus-Guldinus* relate the determination of the area of a surface of revolution or the volume of a body of revolution to the determination of the centroid of the generating curve or area [Sec. 5.7]. The area A of the surface generated by rotating a curve of length L about a fixed axis (Fig. 5.25a) is

$$A = 2\pi\bar{y}L \tag{5.10}$$

where $\bar{y}$ represents the distance from the centroid C of the curve to the fixed axis. Similarly, the volume V of the body generated by rotating an area A about a fixed axis (Fig. 5.25b) is

$$V = 2\pi\bar{y}A \tag{5.11}$$

where $\bar{y}$ represents the distance from the centroid C of the area to the fixed axis.

The concept of centroid of an area can also be used to solve problems other than those dealing with the weight of flat plates. For example, to determine the reactions at the supports of a beam [Sec. 5.8], we can replace a *distributed load* w by a concentrated load $\mathbf{W}$ equal in magnitude to the area A under the load curve and passing through the centroid C of that area (Fig. 5.26). The same approach can be used to determine the resultant of the hydrostatic forces exerted on a *rectangular plate submerged in a liquid* [Sec. 5.9].

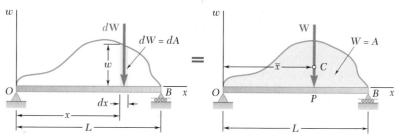

Fig. 5.26

Center of gravity of a three-dimensional body

The last part of the chapter was devoted to the determination of the *center of gravity G of a three-dimensional body*. The coordinates $\bar{x}$, $\bar{y}$, $\bar{z}$ of G were defined by the relations

$$\bar{x}W = \int x\, dW \qquad \bar{y}W = \int y\, dW \qquad \bar{z}W = \int z\, dW \tag{5.17}$$

Centroid of a volume

In the case of a *homogeneous body*, the center of gravity G coincides with the *centroid C of the volume V* of the body; the coordinates of C are defined by the relations

$$\bar{x}V = \int x\, dV \qquad \bar{y}V = \int y\, dV \qquad \bar{z}V = \int z\, dV \tag{5.19}$$

If the volume possesses a *plane of symmetry*, its centroid C will lie in that plane; if it possesses two planes of symmetry, C will be located on the line of intersection of the two planes; if it possesses three planes of symmetry which intersect at only one point, C will coincide with that point [Sec. 5.10].

The *volumes and centroids of various common three-dimensional shapes* are tabulated in Fig. 5.21. When a body can be divided into several of these shapes, the coordinates $\overline{X}$, $\overline{Y}$, $\overline{Z}$ of its center of gravity G can be determined from the corresponding coordinates of the centers of gravity of its various parts [Sec. 5.11]. We have

$$\overline{X}\Sigma W = \Sigma\overline{x}\,W \qquad \overline{Y}\Sigma W = \Sigma\overline{y}\,W \qquad \overline{Z}\Sigma W = \Sigma\overline{z}\,W \qquad (5.20)$$

If the body is made of a homogeneous material, its center of gravity coincides with the centroid C of its volume, and we write [Sample Probs. 5.11 and 5.12]

$$\overline{X}\Sigma V = \Sigma\overline{x}\,V \qquad \overline{Y}\Sigma V = \Sigma\overline{y}\,V \qquad \overline{Z}\Sigma V = \Sigma\overline{z}\,V \qquad (5.21)$$

When a volume is bounded by analytical surfaces, the coordinates of its centroid can be determined by *integration* [Sec. 5.12]. To avoid the computation of the triple integrals in Eqs. (5.19), we can use elements of volume in the shape of thin filaments, as shown in Fig. 5.27.

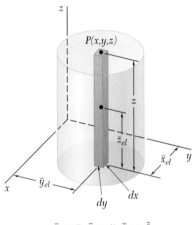

$$\overline{x}_{el} = x, \ \overline{y}_{el} = y, \ \overline{z}_{el} = \tfrac{z}{2}$$
$$dV = z\,dx\,dy$$

Fig. 5.27

Denoting by $\overline{x}_{el}$, $\overline{y}_{el}$, and $\overline{z}_{el}$ the coordinates of the centroid of the element dV, we rewrite Eqs. (5.19) as

$$\overline{x}V = \int \overline{x}_{el}\,dV \qquad \overline{y}V = \int \overline{y}_{el}\,dV \qquad \overline{z}V = \int \overline{z}_{el}\,dV \qquad (5.23)$$

which involve only double integrals. If the volume possesses *two planes of symmetry*, its centroid C is located on their line of intersection. Choosing the x axis to lie along that line and dividing the volume into thin slabs parallel to the yz plane, we can determine C from the relation

$$\overline{x}V = \int \overline{x}_{el}\,dV \qquad (5.24)$$

with a *single integration* [Sample Prob. 5.13]. For a body of revolution, these slabs are circular and their volume is given in Fig. 5.28.

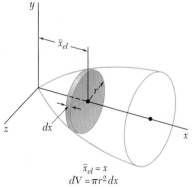

$$\overline{x}_{el} = x$$
$$dV = \pi r^2\,dx$$

Fig. 5.28

REVIEW PROBLEMS

5.137 and 5.138 Locate the centroid of the plane area shown.

Fig. P5.137

Fig. P5.138

5.139 The frame for a sign is fabricated from thin, flat steel bar stock of mass per unit length 4.73 kg/m. The frame is supported by a pin at C and by a cable AB. Determine (a) the tension in the cable, (b) the reaction at C.

Fig. P5.139

5.140 Determine by direct integration the centroid of the area shown. Express your answer in terms of a and h.

5.141 Determine by direct integration the centroid of the area shown.

Fig. P5.140

Fig. P5.141

278

5.142 Three different drive belt profiles are to be studied. If at any given time each belt makes contact with one-half of the circumference of its pulley, determine the *contact area* between the belt and the pulley for each design.

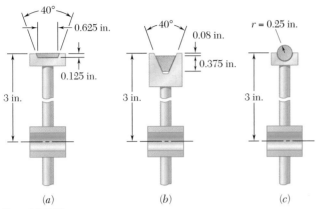

(a)   (b)   (c)

Fig. P5.142

5.143 Determine the reactions at the beam supports for the given loading.

5.144 The beam *AB* supports two concentrated loads and rests on soil that exerts a linearly distributed upward load as shown. Determine the values of w_A and w_B corresponding to equilibrium.

5.145 The base of a dam for a lake is designed to resist up to 120 percent of the horizontal force of the water. After construction, it is found that silt (that is equivalent to a liquid of density $\rho_s = 1.76 \times 10^3$ kg/m^3) is settling on the lake bottom at the rate of 12 mm/year. Considering a 1-m-wide section of dam, determine the number of years until the dam becomes unsafe.

Fig. P5.145

5.146 Determine the location of the centroid of the composite body shown when (a) $h = 2b$, (b) $h = 2.5b$.

5.147 Locate the center of gravity of the sheet-metal form shown.

5.148 Locate the centroid of the volume obtained by rotating the shaded area about the *x* axis.

Fig. P5.148

Fig. P5.143

Fig. P5.144

Fig. P5.146

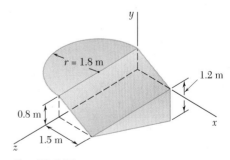

Fig. P5.147

COMPUTER PROBLEMS

5.C1 A beam is to carry a series of uniform and uniformly varying distributed loads as shown in part a of the figure. Divide the area under each portion of the load curve into two triangles (see Sample Prob. 5.9), and then write a computer program that can be used to calculate the reactions at A and B. Use this program to calculate the reactions at the supports for the beams shown in parts b and c of the figure.

(a)

Fig. P5.C1

(b)

(c)

5.C2 The three-dimensional structure shown is fabricated from five thin steel rods of equal diameter. Write a computer program that can be used to calculate the coordinates of the center of gravity of the structure. Use this program to locate the center of gravity when (a) $h = 12$ m, $R = 5$ m, $\alpha = 90°$; (b) $h = 570$ mm, $R = 760$ mm, $\alpha = 30°$; (c) $h = 21$ m, $R = 20$ m, $\alpha = 135°$.

Fig. P5.C2

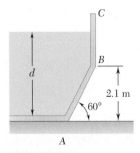

Fig. P5.C3

5.C3 An open tank is to be slowly filled with water. (The density of water is 10^3 kg/m³.) Write a computer program that can be used to determine the resultant of the pressure forces exerted by the water on a 1-m-wide section of side ABC of the tank. Determine the resultant of the pressure forces for values of d from 0 to 3 m using 0.25-m increments.

5.C4 Approximate the curve shown using 10 straight-line segments, and then write a computer program that can be used to determine the location of the centroid of the curve. Use this program to determine the location of the centroid when (a) $a = 1$ in., $L = 11$ in., $h = 2$ in.; (b) $a = 2$ in., $L = 17$ in., $h = 4$ in.; (c) $a = 5$ in., $L = 12$ in., $h = 1$ in.

Fig. P5.C4

5.C5 Approximate the general spandrel shown using a series of n rectangles, each of width Δa and of the form $bcc'b'$, and then write a computer program that can be used to calculate the coordinates of the centroid of the area. Use this program to locate the centroid when (a) $m = 2$, $a = 80$ mm, $h = 80$ mm; (b) $m = 2$, $a = 80$ mm, $h = 500$ mm; (c) $m = 5$, $a = 80$ mm, $h = 80$ mm; (d) $m = 5$, $a = 80$ mm, $h = 500$ mm. In each case, compare the answers obtained to the exact values of $\bar{x}$ and $\bar{y}$ computed from the formulas given in Fig. 5.8A and determine the percentage error.

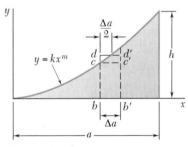

Fig. P5.C5

5.C6 Solve Prob. 5.C5, using rectangles of the form $bdd'b'$.

*5.C7 A farmer asks a group of engineering students to determine the volume of water in a small pond. Using cord, the students first establish a 2×2-ft grid across the pond and then record the depth of the water, in feet, at each intersection point of the grid (see the accompanying table). Write a computer program that can be used to determine (a) the volume of water in the pond, (b) the location of the center of gravity of the water. Approximate the depth of each 2×2-ft element of water using the average of the water depths at the four corners of the element.

				Cord						
	1	2	3	4	5	6	7	8	9	10
1	...	...	...	...	0	0	0	...	...	...
2	...	...	0	0	0	1	0	0	0	...
3	...	0	0	1	3	3	3	1	0	0
4	0	0	1	3	6	6	6	3	1	0
5	0	1	3	6	8	8	6	3	1	0
6	0	1	3	6	8	7	7	3	0	0
7	0	3	4	6	6	6	4	1	0	...
8	0	3	3	3	3	3	1	0	0	...
9	0	0	0	1	1	0	0	0	...	...
10	...	...	0	0	0	0	...	...	...	...

Cord

Trusses, such as this Pratt-style cantilever arch bridge in New York State, provide both a practical and an economical solution to many engineering problems.

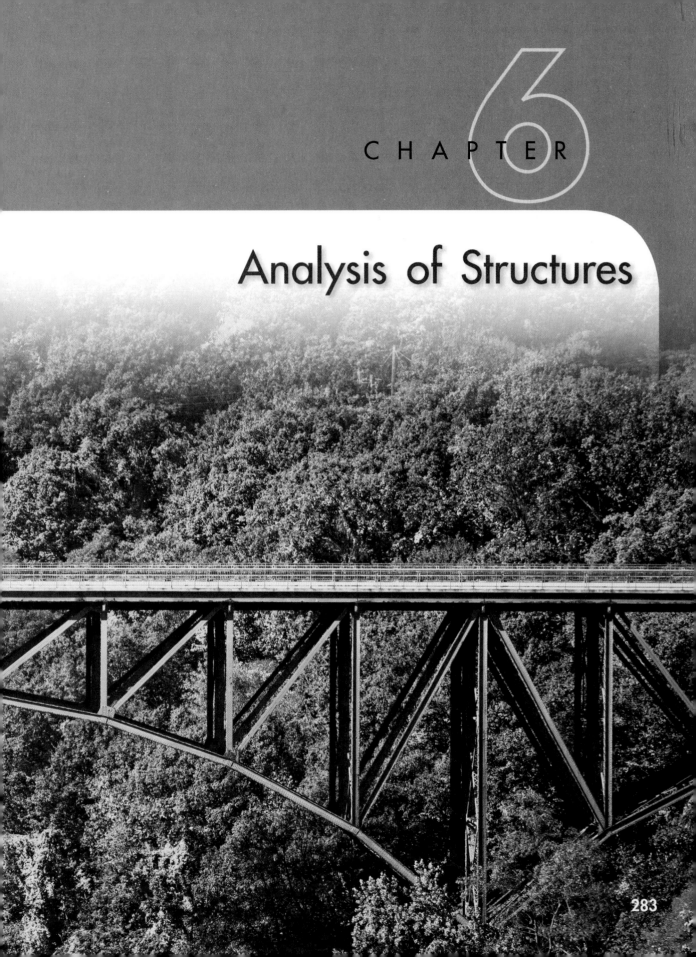

CHAPTER

6

Analysis of Structures

6.1 INTRODUCTION

The problems considered in the preceding chapters concerned the equilibrium of a single rigid body, and all forces involved were external to the rigid body. We now consider problems dealing with the equilibrium of structures made of several connected parts. These problems call for the determination not only of the external forces acting on the structure but also of the forces which hold together the various parts of the structure. From the point of view of the structure as a whole, these forces are *internal forces*.

Consider, for example, the crane shown in Fig. 6.1*a*, which carries a load W. The crane consists of three beams AD, CF, and BE connected by frictionless pins; it is supported by a pin at A and by a cable DG. The free-body diagram of the crane has been drawn in Fig. 6.1*b*. The external forces, which are shown in the diagram, include the weight $\mathbf{W}$, the two components $\mathbf{A}_x$ and $\mathbf{A}_y$ of the reaction at A, and the force $\mathbf{T}$ exerted by the cable at D. The internal forces holding the various parts of the crane together do not appear in the diagram. If, however, the crane is dismembered and if a free-body diagram is drawn for each of its component parts, the forces holding the three beams together will also be represented, since these forces are external forces from the point of view of each component part (Fig. 6.1*c*).

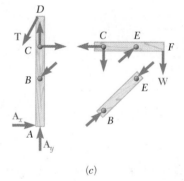

(a)   (b)   (c)

Fig. 6.1

It will be noted that the force exerted at B by member BE on member AD has been represented as equal and opposite to the force exerted at the same point by member AD on member BE; the force exerted at E by BE on CF is shown equal and opposite to the force exerted by CF on BE; and the components of the force exerted at C by CF on AD are shown equal and opposite to the components of the force exerted by AD on CF. This is in conformity with Newton's third law, which states that *the forces of action and reaction between bodies in contact have the same magnitude, same line of action, and opposite sense*. As pointed out in Chap. 1, this law, which is based on experimental evidence, is one of the six fundamental principles of elementary mechanics, and its application is essential to the solution of problems involving connected bodies.

In this chapter, three broad categories of engineering structures will be considered:

1. *Trusses,* which are designed to support loads and are usually stationary, fully constrained structures. Trusses consist exclusively of straight members connected at joints located at the ends of each member. Members of a truss, therefore, are *two-force members,* i.e., members acted upon by two equal and opposite forces directed along the member.

2. *Frames,* which are also designed to support loads and are also usually stationary, fully constrained structures. However, like the crane of Fig. 6.1, frames always contain at least one *multiforce member,* i.e., a member acted upon by three or more forces which, in general, are not directed along the member.

3. *Machines,* which are designed to transmit and modify forces and are structures containing moving parts. Machines, like frames, always contain at least one multiforce member.

Photo 6.1 Shown is a pin-jointed connection on the approach span to the San Francisco–Oakland Bay Bridge.

TRUSSES

6.2 DEFINITION OF A TRUSS

The truss is one of the major types of engineering structures. It provides both a practical and an economical solution to many engineering situations, especially in the design of bridges and buildings. A typical truss is shown in Fig. 6.2a. A truss consists of straight members connected at joints. Truss members are connected at their extremities only; thus no member is continuous through a joint. In Fig. 6.2a, for example, there is no member *AB*; there are instead two distinct members *AD* and *DB*. Most actual structures are made of several trusses joined together to form a space framework. Each truss is designed to carry those loads which act in its plane and thus may be treated as a two-dimensional structure.

In general, the members of a truss are slender and can support little lateral load; all loads, therefore, must be applied to the various joints, and not to the members themselves. When a concentrated load is to be applied between two joints, or when a distributed load is to be supported by the truss, as in the case of a bridge truss, a floor system must be provided which, through the use of stringers and floor beams, transmits the load to the joints (Fig. 6.3).

The weights of the members of the truss are also assumed to be applied to the joints, half of the weight of each member being applied to each of the two joints the member connects. Although the members are actually joined together by means of welded, bolted, or riveted connections, it is customary to assume that the members are pinned together; therefore, the forces acting at each end of a member reduce to a single force and no couple. Thus, the only forces assumed to be applied to a truss member are a single

(a)

(b)

Fig. 6.2

Fig. 6.3

(a)   (b)

Fig. 6.4

force at each end of the member. Each member can then be treated as a two-force member, and the entire truss can be considered as a group of pins and two-force members (Fig. 6.2b). An individual member can be acted upon as shown in either of the two sketches of Fig. 6.4. In Fig. 6.4a, the forces tend to pull the member apart, and the member is in tension; in Fig. 6.4b, the forces tend to compress the member, and the member is in compression. A number of typical trusses are shown in Fig. 6.5.

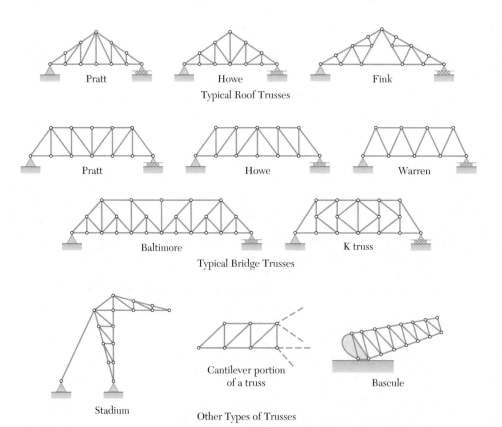

Pratt   Howe   Fink

Typical Roof Trusses

Pratt   Howe   Warren

Baltimore   K truss

Typical Bridge Trusses

Stadium   Cantilever portion
of a truss   Bascule

Other Types of Trusses

Fig. 6.5

6.3 SIMPLE TRUSSES

Consider the truss of Fig. 6.6*a*, which is made of four members connected by pins at *A*, *B*, *C*, and *D*. If a load is applied at *B*, the truss will greatly deform, completely losing its original shape. In contrast, the truss of Fig. 6.6*b*, which is made of three members connected by pins at *A*, *B*, and *C*, will deform only slightly under a load applied at *B*. The only possible deformation for this truss is one involving small changes in the length of its members. The truss of Fig. 6.6*b* is said to be a *rigid truss*, the term rigid being used here to indicate that the truss *will not collapse*.

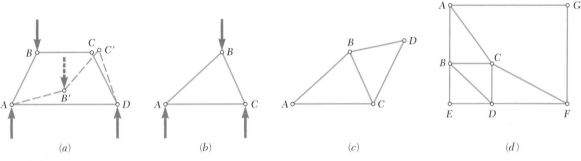

Fig. 6.6

As shown in Fig. 6.6*c*, a larger rigid truss can be obtained by adding two members *BD* and *CD* to the basic triangular truss of Fig. 6.6*b*. This procedure can be repeated as many times as desired, and the resulting truss will be rigid if each time two new members are added, they are attached to two existing joints and connected at a new joint.† A truss which can be constructed in this manner is called a *simple truss*.

It should be noted that a simple truss is not necessarily made only of triangles. The truss of Fig. 6.6*d*, for example, is a simple truss which was constructed from triangle *ABC* by adding successively the joints *D*, *E*, *F*, and *G*. On the other hand, rigid trusses are not always simple trusses, even when they appear to be made of triangles. The Fink and Baltimore trusses shown in Fig. 6.5, for instance, are not simple trusses, since they cannot be constructed from a single triangle in the manner described above. All the other trusses shown in Fig. 6.5 are simple trusses, as may be easily checked. (For the K truss, start with one of the central triangles.)

Returning to Fig. 6.6, we note that the basic triangular truss of Fig. 6.6*b* has three members and three joints. The truss of Fig. 6.6*c* has two more members and one more joint, i.e., five members and four joints altogether. Observing that every time two new members are added, the number of joints is increased by one, we find that in a simple truss the total number of members is $m = 2n - 3$, where n is the total number of joints.

Photo 6.2 Two K trusses were used as the main components of the movable bridge shown which moved above a large stockpile of ore. The bucket below the trusses picked up ore and redeposited it until the ore was thoroughly mixed. The ore was then sent to the mill for processing into steel.

---

†The three joints must not be in a straight line.

(a)

(b)

Fig. 6.7

Photo 6.3   Because roof trusses, such as those shown, require support only at their ends, it is possible to construct buildings with large, unobstructed floor areas.

6.4 ANALYSIS OF TRUSSES BY THE METHOD OF JOINTS

We saw in Sec. 6.2 that a truss can be considered as a group of pins and two-force members. The truss of Fig. 6.2, whose free-body diagram is shown in Fig. 6.7a, can thus be dismembered, and a free-body diagram can be drawn for each pin and each member (Fig. 6.7b). Each member is acted upon by two forces, one at each end; these forces have the same magnitude, same line of action, and opposite sense (Sec. 4.6). Furthermore, Newton's third law indicates that the forces of action and reaction between a member and a pin are equal and opposite. Therefore, the forces exerted by a member on the two pins it connects must be directed along that member and be equal and opposite. The common magnitude of the forces exerted by a member on the two pins it connects is commonly referred to as the *force in the member* considered, even though this quantity is actually a scalar. Since the lines of action of all the internal forces in a truss are known, the analysis of a truss reduces to computing the forces in its various members and to determining whether each of its members is in tension or in compression.

Since the entire truss is in equilibrium, each pin must be in equilibrium. The fact that a pin is in equilibrium can be expressed by drawing its free-body diagram and writing two equilibrium equations (Sec. 2.9). If the truss contains n pins, there will, therefore, be $2n$ equations available, which can be solved for $2n$ unknowns. In the case of a simple truss, we have $m = 2n - 3$, that is, $2n = m + 3$, and the number of unknowns which can be determined from the free-body diagrams of the pins is thus $m + 3$. This means that the forces in all the members, the two components of the reaction $\mathbf{R}_A$, and the reaction $\mathbf{R}_B$ can be found by considering the free-body diagrams of the pins.

The fact that the entire truss is a rigid body in equilibrium can be used to write three more equations involving the forces shown in the free-body diagram of Fig. 6.7a. Since they do not contain any new information, these equations are not independent of the equations associated with the free-body diagrams of the pins. Nevertheless, they can be used to determine the components of the reactions at the supports. The arrangement of pins and members in a simple truss is such that it will then always be possible to find a joint involving only two unknown forces. These forces can be determined by the methods of Sec. 2.11 and their values transferred to the adjacent joints and treated as known quantities at these joints. This procedure can be repeated until all unknown forces have been determined.

As an example, the truss of Fig. 6.7 will be analyzed by considering the equilibrium of each pin successively, starting with a joint at which only two forces are unknown. In the truss considered, all pins are subjected to at least three unknown forces. Therefore, the reactions at the supports must first be determined by considering the entire truss as a free body and using the equations of equilibrium of a rigid body. We find in this way that $\mathbf{R}_A$ is vertical and determine the magnitudes of $\mathbf{R}_A$ and $\mathbf{R}_B$.

The number of unknown forces at joint A is thus reduced to two, and these forces can be determined by considering the equilibrium of pin A. The reaction $\mathbf{R}_A$ and the forces $\mathbf{F}_{AC}$ and $\mathbf{F}_{AD}$ exerted on pin A by members AC and AD, respectively, must form a force triangle. First we draw $\mathbf{R}_A$ (Fig. 6.8); noting that $\mathbf{F}_{AC}$ and $\mathbf{F}_{AD}$ are directed along AC and AD, respectively, we complete the triangle and determine the magnitude and sense of $\mathbf{F}_{AC}$ and $\mathbf{F}_{AD}$. The magnitudes F_{AC} and F_{AD} represent the forces in members AC and AD. Since $\mathbf{F}_{AC}$ is directed down and to the left, that is, *toward* joint A, member AC pushes on pin A and is in compression. Since $\mathbf{F}_{AD}$ is directed *away* from joint A, member AD pulls on pin A and is in tension.

Fig. 6.8

We can now proceed to joint D, where only two forces, $\mathbf{F}_{DC}$ and $\mathbf{F}_{DB}$, are still unknown. The other forces are the load $\mathbf{P}$, which is given, and the force $\mathbf{F}_{DA}$ exerted on the pin by member AD. As indicated above, this force is equal and opposite to the force $\mathbf{F}_{AD}$ exerted by the same member on pin A. We can draw the force polygon corresponding to joint D, as shown in Fig. 6.8, and determine the forces

$\mathbf{F}_{DC}$ and $\mathbf{F}_{DB}$ from that polygon. However, when more than three forces are involved, it is usually more convenient to solve the equations of equilibrium $\Sigma F_x = 0$ and $\Sigma F_y = 0$ for the two unknown forces. Since both of these forces are found to be directed away from joint D, members DC and DB pull on the pin and are in tension.

Next, joint C is considered; its free-body diagram is shown in Fig. 6.8. It is noted that both $\mathbf{F}_{CD}$ and $\mathbf{F}_{CA}$ are known from the analysis of the preceding joints and that only $\mathbf{F}_{CB}$ is unknown. Since the equilibrium of each pin provides sufficient information to determine two unknowns, a check of our analysis is obtained at this joint. The force triangle is drawn, and the magnitude and sense of $\mathbf{F}_{CB}$ are determined. Since $\mathbf{F}_{CB}$ is directed toward joint C, member CB pushes on pin C and is in compression. The check is obtained by verifying that the force $\mathbf{F}_{CB}$ and member CB are parallel.

At joint B, all of the forces are known. Since the corresponding pin is in equilibrium, the force triangle must close and an additional check of the analysis is obtained.

It should be noted that the force polygons shown in Fig. 6.8 are not unique. Each of them could be replaced by an alternative configuration. For example, the force triangle corresponding to joint A could be drawn as shown in Fig. 6.9. The triangle actually shown in Fig. 6.8 was obtained by drawing the three forces $\mathbf{R}_A$, $\mathbf{F}_{AC}$, and $\mathbf{F}_{AD}$ in tip-to-tail fashion in the order in which their lines of action are encountered when moving clockwise around joint A. The other force polygons in Fig. 6.8, having been drawn in the same way, can be made to fit into a single diagram, as shown in Fig. 6.10. Such a diagram, known as *Maxwell's diagram*, greatly facilitates the *graphical analysis* of truss problems.

Fig. 6.9

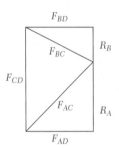

Fig. 6.10

*6.5 JOINTS UNDER SPECIAL LOADING CONDITIONS

Consider Fig. 6.11a, in which the joint shown connects four members lying in two intersecting straight lines. The free-body diagram of Fig. 6.11b shows that pin A is subjected to two pairs of directly opposite forces. The corresponding force polygon, therefore, must be a parallelogram (Fig. 6.11c), and *the forces in opposite members must be equal*.

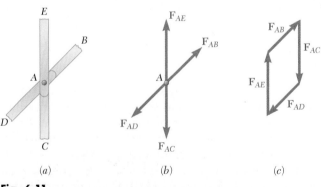

(a) (b) (c)

Fig. 6.11

Consider next Fig. 6.12*a*, in which the joint shown connects three members and supports a load **P**. Two of the members lie in the same line, and the load **P** acts along the third member. The free-body diagram of pin *A* and the corresponding force polygon will be as shown in Fig. 6.11*b* and *c*, with **F**$_{AE}$ replaced by the load **P**. Thus, *the forces in the two opposite members must be equal, and the force in the other member must equal P.* A particular case of special interest is shown in Fig. 6.12*b*. Since, in this case, no external load is applied to the joint, we have *P* = 0, and the force in member *AC* is zero. Member *AC* is said to be a *zero-force member*.

Consider now a joint connecting two members only. From Sec. 2.9, we know that a particle which is acted upon by two forces will be in equilibrium if the two forces have the same magnitude, same line of action, and opposite sense. In the case of the joint of Fig. 6.13*a*, which connects two members *AB* and *AD* lying in the same line, *the forces in the two members must be equal* for pin *A* to be in equilibrium. In the case of the joint of Fig. 6.13*b*, pin *A* cannot be in equilibrium unless the forces in both members are zero. Members connected as shown in Fig. 6.13*b*, therefore, must be *zero-force members*.

Spotting the joints which are under the special loading conditions listed above will expedite the analysis of a truss. Consider, for example, a Howe truss loaded as shown in Fig. 6.14. All of the members represented by green lines will be recognized as zero-force members. Joint *C* connects three members, two of which lie in the same line, and is not subjected to any external load; member *BC* is thus a zero-force member. Applying the same reasoning to joint *K*, we find that member *JK* is also a zero-force member. But joint *J* is now in the same situation as joints *C* and *K*, and member *IJ* must be a zero-force member. The examination of joints *C*, *J*, and *K* also shows that the forces in members *AC* and *CE* are equal, that the forces in members *HJ* and *JL* are equal, and that the forces in members *IK* and *KL* are equal. Turning our attention to joint *I*, where the 20-kN load and member *HI* are collinear, we note that the force in member *HI* is 20 kN (tension) and that the forces in members *GI* and *IK* are equal. Hence, the forces in members *GI*, *IK*, and *KL* are equal.

Fig. 6.12

Fig. 6.13

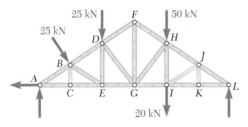

Fig. 6.14

Note that the conditions described above do not apply to joints *B* and *D* in Fig. 6.14, and it would be wrong to assume that the force in member *DE* is 25 kN or that the forces in members *AB* and *BD* are equal. The forces in these members and in all remaining members should be found by carrying out the analysis of joints *A*, *B*, *D*, *E*, *F*, *G*, *H*, and *L* in the usual manner. Thus, until you have become thoroughly familiar with the conditions under which the rules established in this

Photo 6.4 Three-dimensional or space trusses are used for broadcast and power transmission line towers, roof framing, and spacecraft applications, such as components of the *International Space Station.*

section can be applied, you should draw the free-body diagrams of all pins and write the corresponding equilibrium equations (or draw the corresponding force polygons) whether or not the joints being considered are under one of the special loading conditions described above.

A final remark concerning zero-force members: These members are not useless. For example, although the zero-force members of Fig. 6.14 do not carry any loads under the loading conditions shown, the same members would probably carry loads if the loading conditions were changed. Besides, even in the case considered, these members are needed to support the weight of the truss and to maintain the truss in the desired shape.

*6.6 SPACE TRUSSES

When several straight members are joined together at their extremities to form a three-dimensional configuration, the structure obtained is called a *space truss*.

We recall from Sec. 6.3 that the most elementary two-dimensional rigid truss consisted of three members joined at their extremities to form the sides of a triangle; by adding two members at a time to this basic configuration, and connecting them at a new joint, it was possible to obtain a larger rigid structure which was defined as a simple truss. Similarly, the most elementary rigid space truss consists of six members joined at their extremities to form the edges of a tetrahedron *ABCD* (Fig. 6.15a). By adding three members at a time to this basic configuration, such as *AE, BE,* and *CE,* attaching them to three existing joints, and connecting them at a new joint,† we can obtain a larger rigid structure which is defined as a *simple space truss* (Fig. 6.15b). Observing that the basic tetrahedron has six members and four joints and that every time three members are added, the number of joints is increased by one, we conclude that in a simple space truss the total number of members is $m = 3n - 6$, where n is the total number of joints.

If a space truss is to be completely constrained and if the reactions at its supports are to be statically determinate, the supports should consist of a combination of balls, rollers, and balls and sockets which provides six unknown reactions (see Sec. 4.9). These unknown reactions may be readily determined by solving the six equations expressing that the three-dimensional truss is in equilibrium.

Although the members of a space truss are actually joined together by means of bolted or welded connections, it is assumed that each joint consists of a ball-and-socket connection. Thus, no couple will be applied to the members of the truss, and each member can be treated as a two-force member. The conditions of equilibrium for each joint will be expressed by the three equations $\Sigma F_x = 0$, $\Sigma F_y = 0$, and $\Sigma F_z = 0$. In the case of a simple space truss containing n joints, writing the conditions of equilibrium for each joint will thus yield $3n$ equations. Since $m = 3n - 6$, these equations suffice to determine all unknown forces (forces in m members and six reactions at the supports). However, to avoid the necessity of solving simultaneous equations, care should be taken to select joints in such an order that no selected joint will involve more than three unknown forces.

(a)

(b)

Fig. 6.15

---

†The four joints must not lie in a plane.

SAMPLE PROBLEM 6.1

Using the method of joints, determine the force in each member of the truss shown.

SOLUTION

Free-Body: Entire Truss. A free-body diagram of the entire truss is drawn; external forces acting on this free body consist of the applied loads and the reactions at C and E. We write the following equilibrium equations.

$+\uparrow\Sigma M_C = 0$: $\quad$ $(2000\text{ lb})(24\text{ ft}) + (1000\text{ lb})(12\text{ ft}) - E(6\text{ ft}) = 0$
$$E = +10{,}000\text{ lb} \qquad\qquad \mathbf{E} = 10{,}000\text{ lb}\uparrow$$

$\xrightarrow{+}\Sigma F_x = 0$: $\qquad\qquad\qquad\qquad\qquad\qquad\qquad\qquad \mathbf{C}_x = 0$

$+\uparrow\Sigma F_y = 0$: $\quad$ $-2000\text{ lb} - 1000\text{ lb} + 10{,}000\text{ lb} + C_y = 0$
$$C_y = -7000\text{ lb} \qquad\qquad \mathbf{C}_y = 7000\text{ lb}\downarrow$$

Free-Body: Joint A. This joint is subjected to only two unknown forces, namely, the forces exerted by members AB and AD. A force triangle is used to determine $\mathbf{F}_{AB}$ and $\mathbf{F}_{AD}$. We note that member AB pulls on the joint and thus is in tension and that member AD pushes on the joint and thus is in compression. The magnitudes of the two forces are obtained from the proportion

$$\frac{2000\text{ lb}}{4} = \frac{F_{AB}}{3} = \frac{F_{AD}}{5}$$

$$F_{AB} = 1500\text{ lb } T \quad \blacktriangleleft$$
$$F_{AD} = 2500\text{ lb } C \quad \blacktriangleleft$$

Free-Body: Joint D. Since the force exerted by member AD has been determined, only two unknown forces are now involved at this joint. Again, a force triangle is used to determine the unknown forces in members DB and DE.

$$F_{DB} = F_{DA} \qquad\qquad F_{DB} = 2500\text{ lb } T \quad \blacktriangleleft$$
$$F_{DE} = 2(\tfrac{3}{5})F_{DA} \qquad F_{DE} = 3000\text{ lb } C \quad \blacktriangleleft$$

Free-Body: Joint B. Since more than three forces act at this joint, we determine the two unknown forces $\mathbf{F}_{BC}$ and $\mathbf{F}_{BE}$ by solving the equilibrium equations $\Sigma F_x = 0$ and $\Sigma F_y = 0$. We arbitrarily assume that both unknown forces act away from the joint, i.e., that the members are in tension. The positive value obtained for F_{BC} indicates that our assumption was correct; member BC is in tension. The negative value of F_{BE} indicates that our assumption was wrong; member BE is in compression.

$$+\uparrow\Sigma F_y = 0: \qquad -1000 - \tfrac{4}{5}(2500) - \tfrac{4}{5}F_{BE} = 0$$
$$F_{BE} = -3750 \text{ lb} \qquad\qquad F_{BE} = 3750 \text{ lb } C \quad \blacktriangleleft$$

$$\xrightarrow{+}\Sigma F_x = 0: \qquad F_{BC} - 1500 - \tfrac{3}{5}(2500) - \tfrac{3}{5}(3750) = 0$$
$$F_{BC} = +5250 \text{ lb} \qquad\qquad F_{BC} = 5250 \text{ lb } T \quad \blacktriangleleft$$

Free-Body: Joint E. The unknown force $\mathbf{F}_{EC}$ is assumed to act away from the joint. Summing x components, we write

$$\xrightarrow{+}\Sigma F_x = 0: \qquad \tfrac{3}{5}F_{EC} + 3000 + \tfrac{3}{5}(3750) = 0$$
$$F_{EC} = -8750 \text{ lb} \qquad\qquad F_{EC} = 8750 \text{ lb } C \quad \blacktriangleleft$$

Summing y components, we obtain a check of our computations:

$$+\uparrow\Sigma F_y = 10,000 - \tfrac{4}{5}(3750) - \tfrac{4}{5}(8750)$$
$$= 10,000 - 3000 - 7000 = 0 \qquad\qquad \text{(checks)}$$

Free-Body: Joint C. Using the computed values of $\mathbf{F}_{CB}$ and $\mathbf{F}_{CE}$, we can determine the reactions $\mathbf{C}_x$ and $\mathbf{C}_y$ by considering the equilibrium of this joint. Since these reactions have already been determined from the equilibrium of the entire truss, we will obtain two checks of our computations. We can also simply use the computed values of all forces acting on the joint (forces in members and reactions) and check that the joint is in equilibrium:

$$\xrightarrow{+}\Sigma F_x = -5250 + \tfrac{3}{5}(8750) = -5250 + 5250 = 0 \qquad \text{(checks)}$$
$$+\uparrow\Sigma F_y = -7000 + \tfrac{4}{5}(8750) = -7000 + 7000 = 0 \qquad \text{(checks)}$$

SOLVING PROBLEMS
ON YOUR OWN

In this lesson you learned to use the *method of joints* to determine the forces in the members of a *simple truss*, that is, a truss that can be constructed from a basic triangular truss by adding to it two new members at a time and connecting them at a new joint.

Your solution will consist of the following steps:

1. Draw a free-body diagram of the entire truss, and use this diagram to determine the reactions at the supports.

2. Locate a joint connecting only two members, and draw the free-body diagram of its pin. Use this free-body diagram to determine the unknown force in each of the two members. If only three forces are involved (the two unknown forces and a known one), you will probably find it more convenient to draw and solve the corresponding force triangle. If more than three forces are involved, you should write and solve the equilibrium equations for the pin, $\Sigma F_x = 0$ and $\Sigma F_y = 0$, assuming that the members are in tension. A positive answer means that the member is in tension, a negative answer that the member is in compression. Once the forces have been found, enter their values on a sketch of the truss, with T for tension and C for compression.

3. Next, locate a joint where the forces in only two of the connected members are still unknown. Draw the free-body diagram of the pin and use it as indicated above to determine the two unknown forces.

4. Repeat this procedure until the forces in all the members of the truss have been found. Since you previously used the three equilibrium equations associated with the free-body diagram of the entire truss to determine the reactions at the supports, you will end up with three extra equations. These equations can be used to check your computations.

5. Note that the choice of the first joint is not unique. Once you have determined the reactions at the supports of the truss, you can choose either of two joints as a starting point for your analysis. In Sample Prob. 6.1, we started at joint A and proceeded through joints D, B, E, and C, but we could also have started at joint C and proceeded through joints E, B, D, and A. On the other hand, having selected a first joint, you may in some cases reach a point in your analysis beyond which you cannot proceed. You must then start again from another joint to complete your solution.

Keep in mind that the analysis of a *simple truss* can always be carried out by the method of joints. Also remember that it is helpful to outline your solution *before* starting any computations.

PROBLEMS

6.1 through 6.8 Using the method of joints, determine the force in each member of the truss shown. State whether each member is in tension or compression.

Fig. P6.1

Fig. P6.2

Fig. P6.3

Fig. P6.4

Fig. P6.5

Fig. P6.6

Fig. P6.7

Fig. P6.8

6.9 Determine the force in each member of the Gambrel roof truss shown. State whether each member is in tension or compression.

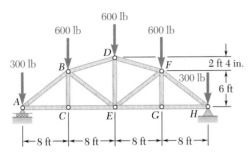

Fig. P6.9

6.10 Determine the force in each member of the Howe roof truss shown. State whether each member is in tension or compression.

6.11 Determine the force in each member of the Pratt roof truss shown. State whether each member is in tension or compression.

Fig. P6.10

Fig. P6.11

6.12 Determine the force in each member of the Fink roof truss shown. State whether each member is in tension or compression.

6.13 Determine the force in each member of the double-pitch roof truss shown. State whether each member is in tension or compression.

Fig. P6.12

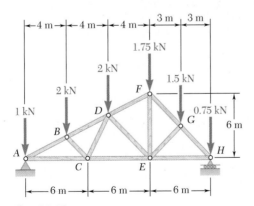

Fig. P6.13

6.14 The truss shown is one of several supporting an advertising panel. Determine the force in each member of the truss for a wind load equivalent to the two forces shown. State whether each member is in tension or compression.

Fig. P6.14

6.15 Determine the force in each of the members located to the left of line *FGH* for the studio roof truss shown. State whether each member is in tension or compression.

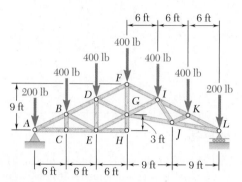

Fig. P6.15 and P6.16

6.16 Determine the force in member *FG* and in each of the members located to the right of *FG* for the studio roof truss shown. State whether each member is in tension or compression.

6.17 Determine the force in each of the members located to the left of *FG* for the scissors roof truss shown. State whether each member is in tension or compression.

6.18 Determine the force in member *FG* and in each of the members located to the right of *FG* for the scissors roof truss shown. State whether each member is in tension or compression.

6.19 Determine the force in each member of the Warren bridge truss shown. State whether each member is in tension or compression.

6.20 Solve Prob. 6.19 assuming that the load applied at *E* has been removed.

6.21 Determine the force in each member of the Pratt bridge truss shown. State whether each member is in tension or compression.

Fig. P6.17 and P6.18

Fig. P6.19

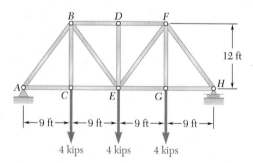

Fig. P6.21

6.22 Solve Prob. 6.21 assuming that the load applied at *G* has been removed.

6.23 The portion of truss shown represents the upper part of a power transmission line tower. For the given loading, determine the force in each of the members located above *HJ*. State whether each member is in tension or compression.

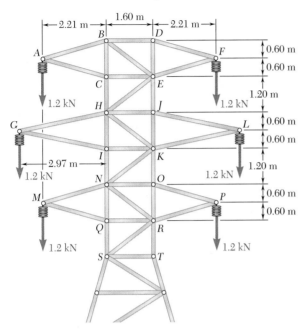

Fig. P6.23

6.24 For the tower and loading of Prob. 6.23 and knowing that $F_{CH} = F_{EJ} = 1.2$ kN C and $F_{EH} = 0$, determine the force in member *HJ* and in each of the members located between *HJ* and *NO*. State whether each member is in tension or compression.

6.25 Solve Prob. 6.23 assuming that the cables hanging from the right side of the tower have fallen to the ground.

6.26 Determine the force in each of the members connecting joints *A* through *F* of the vaulted roof truss shown. State whether each member is in tension or compression.

6.27 Determine the force in each member of the truss shown. State whether each member is in tension or compression.

Fig. P6.26

Fig. P6.27

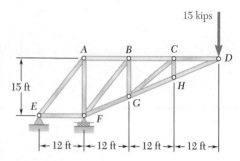

15 kips

15 ft

← 12 ft → ← 12 ft → ← 12 ft → ← 12 ft →

Fig. P6.28

6.28 Determine the force in each member of the truss shown. State whether each member is in tension or compression.

6.29 Determine whether the trusses of Probs. 6.31a, 6.32a, and 6.33a are simple trusses.

6.30 Determine whether the trusses of Probs. 6.31b, 6.32b, and 6.33b are simple trusses.

6.31 For the given loading, determine the zero-force members in each of the two trusses shown.

(a)

(b)

Fig. P6.31

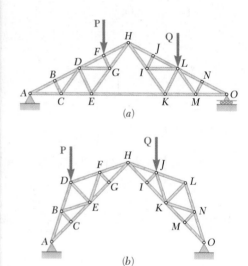

(a)

(b)

Fig. P6.32

6.32 For the given loading, determine the zero-force members in each of the two trusses shown.

6.33 For the given loading, determine the zero-force members in each of the two trusses shown.

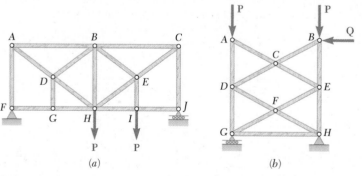

(a)

(b)

Fig. P6.33

6.34 Determine the zero-force members in the truss of (a) Prob. 6.26, (b) Prob. 6.28.

*6.35 The truss shown consists of six members and is supported by a short link at A, two short links at B, and a ball and socket at D. Determine the force in each of the members for the given loading.

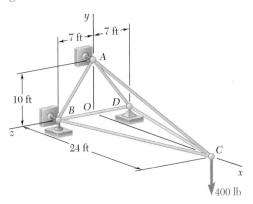

Fig. P6.35

*6.36 The truss shown consists of six members and is supported by a ball and socket at B, a short link at C, and two short links at D. Determine the force in each of the members for $\mathbf{P} = (-2184 \text{ N})\mathbf{j}$ and $\mathbf{Q} = 0$.

*6.37 The truss shown consists of six members and is supported by a ball and socket at B, a short link at C, and two short links at D. Determine the force in each of the members for $\mathbf{P} = 0$ and $\mathbf{Q} = (2968 \text{ N})\mathbf{i}$.

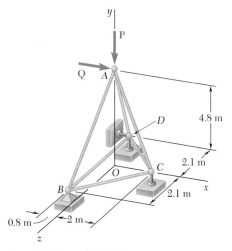

Fig. P6.36 and P6.37

*6.38 The truss shown consists of nine members and is supported by a ball and socket at A, two short links at B, and a short link at C. Determine the force in each of the members for the given loading.

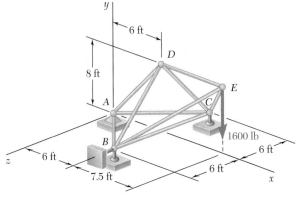

Fig. P6.38

*6.39 The truss shown consists of nine members and is supported by a ball and socket at B, a short link at C, and two short links at D. (a) Check that this truss is a simple truss, that it is completely constrained, and that the reactions at its supports are statically determinate. (b) Determine the force in each member for $\mathbf{P} = (-1200 \text{ N})\mathbf{j}$ and $\mathbf{Q} = 0$.

*6.40 Solve Prob. 6.39 for $\mathbf{P} = 0$ and $\mathbf{Q} = (-900 \text{ N})\mathbf{k}$.

Fig. P6.39

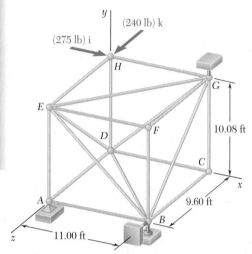

(275 lb) i

(240 lb) k

y

H

G

10.08 ft

E

D

F

C

x

A

B

9.60 ft

z

11.00 ft

Fig. P6.41 and P6.42

*6.41 The truss shown consists of 18 members and is supported by a ball and socket at A, two short links at B, and one short link at G. (a) Check that this truss is a simple truss, that it is completely constrained, and that the reactions at its supports are statically determinate. (b) For the given loading, determine the force in each of the six members joined at E.

*6.42 The truss shown consists of 18 members and is supported by a ball and socket at A, two short links at B, and one short link at G. (a) Check that this truss is a simple truss, that it is completely constrained, and that the reactions at its supports are statically determinate. (b) For the given loading, determine the force in each of the six members joined at G.

6.7 ANALYSIS OF TRUSSES BY THE METHOD OF SECTIONS

The method of joints is most effective when the forces in all the members of a truss are to be determined. If, however, the force in only one member or the forces in a very few members are desired, another method, the method of sections, is more efficient.

Assume, for example, that we want to determine the force in member BD of the truss shown in Fig. 6.16a. To do this, we must determine the force with which member BD acts on either joint B or joint D. If we were to use the method of joints, we would choose either joint B or joint D as a free body. However, we can also choose as a free body a larger portion of the truss, composed of several joints and members, provided that the desired force is one of the external forces acting on that portion. If, in addition, the portion of the truss is chosen so that there is a total of only three unknown forces acting upon it, the desired force can be obtained by solving the equations of equilibrium for this portion of the truss. In practice, the portion of the truss to be utilized is obtained by *passing a section* through three members of the truss, one of which is the desired member, i.e., by drawing a line which divides the truss into two completely separate parts but does not intersect more than three members. Either of the two portions of the truss obtained after the intersected members have been removed can then be used as a free body.†

In Fig. 6.16a, the section nn has been passed through members BD, BE, and CE, and the portion ABC of the truss is chosen as the free body (Fig. 6.16b). The forces acting on the free body are the loads $\mathbf{P}_1$ and $\mathbf{P}_2$ at points A and B and the three unknown forces $\mathbf{F}_{BD}$, $\mathbf{F}_{BE}$, and $\mathbf{F}_{CE}$. Since it is not known whether the members removed were in tension or compression, the three forces have been arbitrarily drawn away from the free body as if the members were in tension.

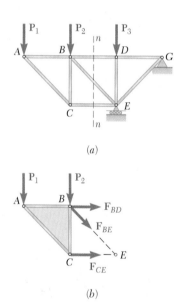

P_1 P_2 P_3

n

A B D G

C E

n

(a)

P_1 P_2

A B

F_{BD}

F_{BE}

C E

F_{CE}

(b)

Fig. 6.16

†In the analysis of certain trusses, sections are passed which intersect more than three members; the forces in one, or possibly two, of the intersected members may be obtained if equilibrium equations can be found, each of which involves only one unknown (see Probs. 6.61 through 6.64).

The fact that the rigid body ABC is in equilibrium can be expressed by writing three equations which can be solved for the three unknown forces. If only the force $\mathbf{F}_{BD}$ is desired, we need write only one equation, provided that the equation does not contain the other unknowns. Thus the equation $\Sigma M_E = 0$ yields the value of the magnitude F_{BD} of the force $\mathbf{F}_{BD}$ (Fig. 6.16b). A positive sign in the answer will indicate that our original assumption regarding the sense of $\mathbf{F}_{BD}$ was correct and that member BD is in tension; a negative sign will indicate that our assumption was incorrect and that BD is in compression.

On the other hand, if only the force $\mathbf{F}_{CE}$ is desired, an equation which does not involve $\mathbf{F}_{BD}$ or $\mathbf{F}_{BE}$ should be written; the appropriate equation is $\Sigma M_B = 0$. Again a positive sign for the magnitude F_{CE} of the desired force indicates a correct assumption, that is, tension; and a negative sign indicates an incorrect assumption, that is, compression.

If only the force $\mathbf{F}_{BE}$ is desired, the appropriate equation is $\Sigma F_y = 0$. Whether the member is in tension or compression is again determined from the sign of the answer.

When the force in only one member is determined, no independent check of the computation is available. However, when all the unknown forces acting on the free body are determined, the computations can be checked by writing an additional equation. For instance, if $\mathbf{F}_{BD}$, $\mathbf{F}_{BE}$, and $\mathbf{F}_{CE}$ are determined as indicated above, the computation can be checked by verifying that $\Sigma F_x = 0$.

*6.8 TRUSSES MADE OF SEVERAL SIMPLE TRUSSES

Consider two simple trusses ABC and DEF. If they are connected by three bars BD, BE, and CE as shown in Fig. 6.17a, they will form together a rigid truss $ABDF$. The trusses ABC and DEF can also be combined into a single rigid truss by joining joints B and D into a single joint B and by connecting joints C and E by a bar CE (Fig. 6.17b). The truss thus obtained is known as a *Fink truss*. It should be noted that the trusses of Fig. 6.17a and b are *not* simple trusses; they cannot be constructed from a triangular truss by adding successive pairs of members as prescribed in Sec. 6.3. They are rigid trusses, however, as we can check by comparing the systems of connections used to hold the simple trusses ABC and DEF together (three bars in Fig. 6.17a, one pin and one bar in Fig. 6.17b) with the systems of supports discussed in Secs. 4.4 and 4.5. Trusses made of several simple trusses rigidly connected are known as *compound trusses*.

(a)

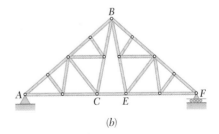

(b)

Fig. 6.17

In a compound truss the number of members m and the number of joints n are still related by the formula $m = 2n - 3$. This can be verified by observing that, if a compound truss is supported by a frictionless pin and a roller (involving three unknown reactions), the total number of unknowns is $m + 3$, and this number must be equal to the number $2n$ of equations obtained by expressing that the n pins are in equilibrium; it follows that $m = 2n - 3$. Compound trusses supported by a pin and a roller, or by an equivalent system of supports, are *statically determinate, rigid,* and *completely constrained.* This means that all of the unknown reactions and the forces in all the members can be determined by the methods of statics, and that the truss will neither collapse nor move. The forces in the members, however, cannot all be determined by the method of joints, except by solving a large number of simultaneous equations. In the case of the compound truss of Fig. 6.17*a*, for example, it is more efficient to pass a section through members *BD*, *BE*, and *CE* to determine the forces in these members.

Suppose, now, that the simple trusses *ABC* and *DEF* are connected by *four* bars *BD*, *BE*, *CD*, or *CE* (Fig. 6.18). The number of members m is now larger than $2n - 3$; the truss obtained is *overrigid,* and one of the four members *BD*, *BE*, *CD*, or *CE* is said to be *redundant.* If the truss is supported by a pin at *A* and a roller at *F*, the total number of unknowns is $m + 3$. Since $m > 2n - 3$, the number $m + 3$ of unknowns is now larger than the number $2n$ of available independent equations; the truss is *statically indeterminate.*

Finally, let us assume that the two simple trusses *ABC* and *DEF* are joined by a pin as shown in Fig. 6.19*a*. The number of members m is smaller than $2n - 3$. If the truss is supported by a pin at *A* and a roller at *F*, the total number of unknowns is $m + 3$. Since $m < 2n - 3$, the number $m + 3$ of unknowns is now smaller than the number $2n$ of equilibrium equations which should be satisfied; the truss is *nonrigid* and will collapse under its own weight. However, if two pins are used to support it, the truss becomes *rigid* and will not collapse (Fig. 6.19*b*). We note that the total number of unknowns is now $m + 4$ and is equal to the number $2n$ of equations. More generally, if the reactions at the supports involve r unknowns, the condition for a compound truss to be statically determinate, rigid, and completely constrained is $m + r = 2n$. However, while necessary this condition is not sufficient for the equilibrium of a structure which ceases to be rigid when detached from its supports (see Sec. 6.11).

Fig. 6.17 (*repeated*)

Fig. 6.18

(*a*)   (*b*)

Fig. 6.19

Determine the force in members EF and GI of the truss shown.

SOLUTION

Free-Body: Entire Truss. A free-body diagram of the entire truss is drawn; external forces acting on this free body consist of the applied loads and the reactions at B and J. We write the following equilibrium equations.

$+\uparrow\Sigma M_B = 0$:

$$-(28 \text{ kips})(8 \text{ ft}) - (28 \text{ kips})(24 \text{ ft}) - (16 \text{ kips})(10 \text{ ft}) + J(32 \text{ ft}) = 0$$
$$J = +33 \text{ kips} \qquad \mathbf{J} = 33 \text{ kips}\uparrow$$

$\xrightarrow{+}\Sigma F_x = 0$: $\qquad B_x + 16 \text{ kips} = 0$

$$B_x = -16 \text{ kips} \qquad \mathbf{B}_x = 16 \text{ kips}\leftarrow$$

$+\uparrow\Sigma M_J = 0$:

$$(28 \text{ kips})(24 \text{ ft}) + (28 \text{ kips})(8 \text{ ft}) - (16 \text{ kips})(10 \text{ ft}) - B_y(32 \text{ ft}) = 0$$
$$B_y = +23 \text{ kips} \qquad \mathbf{B}_y = 23 \text{ kips}\uparrow$$

Force in Member EF. Section nn is passed through the truss so that it intersects member EF and only two additional members. After the intersected members have been removed, the left-hand portion of the truss is chosen as a free body. Three unknowns are involved; to eliminate the two horizontal forces, we write

$+\uparrow\Sigma F_y = 0$: $\qquad +23 \text{ kips} - 28 \text{ kips} - F_{EF} = 0$
$$F_{EF} = -5 \text{ kips}$$

The sense of $\mathbf{F}_{EF}$ was chosen assuming member EF to be in tension; the negative sign obtained indicates that the member is in compression.

$$F_{EF} = 5 \text{ kips } C \quad \blacktriangleleft$$

Force in Member GI. Section mm is passed through the truss so that it intersects member GI and only two additional members. After the intersected members have been removed, we choose the right-hand portion of the truss as a free body. Three unknown forces are again involved; to eliminate the two forces passing through point H, we write

$+\uparrow\Sigma M_H = 0$: $\qquad (33 \text{ kips})(8 \text{ ft}) - (16 \text{ kips})(10 \text{ ft}) + F_{GI}(10 \text{ ft}) = 0$
$$F_{GI} = -10.4 \text{ kips} \qquad F_{GI} = 10.4 \text{ kips } C \quad \blacktriangleleft$$

Determine the force in members FH, GH, and GI of the roof truss shown.

SOLUTION

Free Body: Entire Truss. From the free-body diagram of the entire truss, we find the reactions at A and L:

$$\mathbf{A} = 12.50 \text{ kN}\uparrow \qquad \mathbf{L} = 7.50 \text{ kN}\uparrow$$

We note that

$$\tan \alpha = \frac{FG}{GL} = \frac{8 \text{ m}}{15 \text{ m}} = 0.5333 \quad \alpha = 28.07°$$

Force in Member GI. Section nn is passed through the truss as shown. Using the portion HLI of the truss as a free body, the value of F_{GI} is obtained by writing

$$+\circlearrowleft\Sigma M_H = 0: \qquad (7.50 \text{ kN})(10 \text{ m}) - (1 \text{ kN})(5 \text{ m}) - F_{GI}(5.33 \text{ m}) = 0$$
$$F_{GI} = +13.13 \text{ kN} \qquad F_{GI} = 13.13 \text{ kN } T \quad \blacktriangleleft$$

Force in Member FH. The value of F_{FH} is obtained from the equation $\Sigma M_G = 0$. We move $\mathbf{F}_{FH}$ along its line of action until it acts at point F, where it is resolved into its x and y components. The moment of $\mathbf{F}_{FH}$ with respect to point G is now equal to $(F_{FH} \cos \alpha)(8 \text{ m})$.

$$+\circlearrowleft\Sigma M_G = 0:$$
$$(7.50 \text{ kN})(15 \text{ m}) - (1 \text{ kN})(10 \text{ m}) - (1 \text{ kN})(5 \text{ m}) + (F_{FH} \cos \alpha)(8 \text{ m}) = 0$$
$$F_{FH} = -13.81 \text{ kN} \qquad F_{FH} = 13.81 \text{ kN } C \quad \blacktriangleleft$$

Force in Member GH. We first note that

$$\tan \beta = \frac{GI}{HI} = \frac{5\text{m}}{\frac{2}{3}(8\text{m})} = 0.9375 \quad \beta = 43.15°$$

The value of F_{GH} is then determined by resolving the force $\mathbf{F}_{GH}$ into x and y components at point G and solving the equation $\Sigma M_L = 0$.

$$+\circlearrowleft\Sigma M_L = 0: \qquad (1 \text{ kN})(10 \text{ m}) + (1 \text{ kN})(5 \text{ m}) + (F_{GH} \cos \beta)(15 \text{ m}) = 0$$
$$F_{GH} = -1.371 \text{ kN} \qquad F_{GH} = 1.371 \text{ kN } C \quad \blacktriangleleft$$

SOLVING PROBLEMS ON YOUR OWN

The *method of joints* that you studied earlier is usually the best method to use when the forces *in all the members* of a simple truss are to be found. However, the method of sections, which was covered in this lesson, is more effective when the force *in only one member* or the forces *in a very few members* of a simple truss are desired. The method of sections must also be used when the truss *is not a simple truss.*

A. To determine the force in a given truss member by the method of sections, you should follow these steps:

1. Draw a free-body diagram of the entire truss, and use this diagram to determine the reactions at the supports.

2. Pass a section through three members of the truss, one of which is the desired member. After you have removed these members, you will obtain two separate portions of truss.

3. Select one of the two portions of truss you have obtained, and draw its free-body diagram. This diagram should include the external forces applied to the selected portion as well as the forces exerted on it by the intersected members before these members were removed.

4. You can now write three equilibrium equations which can be solved for the forces in the three intersected members.

5. An alternative approach is to write a single equation, which can be solved for the force in the desired member. To do so, first observe whether the forces exerted by the other two members on the free body are parallel or whether their lines of action intersect.

   a. If these forces are parallel, they can be eliminated by writing an equilibrium equation involving *components in a direction perpendicular* to these two forces.

   b. If their lines of action intersect at a point *H*, they can be eliminated by writing an equilibrium equation involving *moments about H*.

6. Keep in mind that the section you use must intersect three members only. This is because the equilibrium equations in step 4 can be solved for three unknowns only. However, you can pass a section through more than three members to find the force in one of those members if you can write an equilibrium equation containing only that force as an unknown. Such special situations are found in Probs. 6.61 through 6.64.

(continued)

B. About completely constrained and determinate trusses:

1. First note that any simple truss which is simply supported is a completely constrained and determinate truss.

2. To determine whether any other truss is or is not completely constrained and determinate, you first count the number m of its members, the number n of its joints, and the number r of the reaction components at its supports. You then compare the sum $m + r$ representing the number of unknowns and the product $2n$ representing the number of available independent equilibrium equations.

    a. If $m + r < 2n$, there are fewer unknowns than equations. Thus, some of the equations cannot be satisfied; the truss is only *partially constrained*.

    b. If $m + r > 2n$, there are more unknowns than equations. Thus, some of the unknowns cannot be determined; the truss is *indeterminate*.

    c. If $m + r = 2n$, there are as many unknowns as there are equations. This, however, does not mean that all the unknowns can be determined and that all the equations can be satisfied. To find out whether the truss is *completely* or *improperly constrained*, you should try to determine the reactions at its supports and the forces in its members. If all can be found, the truss is *completely constrained and determinate*.

PROBLEMS

6.43 Determine the force in members *CD* and *DF* of the truss shown.

12 kN   12 kN

1.8 m

4 panels @ 2.4 m = 9.6 m

Fig. P6.43 and P6.44

6.44 Determine the force in members *FG* and *FH* of the truss shown.

6.45 A Warren bridge truss is loaded as shown. Determine the force in members *CE*, *DE*, and *DF*.

6.46 A Warren bridge truss is loaded as shown. Determine the force in members *EG*, *FG*, and *FH*.

6.47 Determine the force in members *DF*, *EF*, and *EG* of the truss shown.

6.25 ft   12.5 ft   12.5 ft   12.5 ft   12.5 ft

15 ft

12.5 ft   12.5 ft   12.5 ft   12.5 ft   12.5 ft

6000 lb   6000 lb

Fig. P6.45 and P6.46

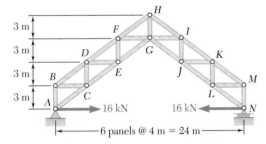

3 m

3 m

3 m

3 m

16 kN   16 kN

6 panels @ 4 m = 24 m

Fig. P6.47 and P6.48

6.48 Determine the force in members *GI*, *GJ*, and *HI* of the truss shown.

6.49 Determine the force in members *AD*, *CD*, and *CE* of the truss shown.

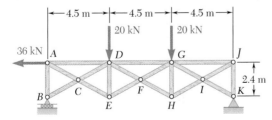

4.5 m   4.5 m   4.5 m

20 kN   20 kN

36 kN

2.4 m

Fig. P6.49 and P6.50

6.50 Determine the force in members *DG*, *FG*, and *FH* of the truss shown.

309

6.51 A stadium roof truss is loaded as shown. Determine the force in members *AB*, *AG*, and *FG*.

Fig. P6.51 and P6.52

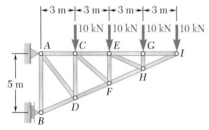

Fig. P6.53 and P6.54

6.52 A stadium roof truss is loaded as shown. Determine the force in members *AE*, *EF*, and *FJ*.

6.53 Determine the force in members *CD* and *DF* of the truss shown.

6.54 Determine the force in members *CE* and *EF* of the truss shown.

6.55 The truss shown was designed to support the roof of a food market. For the given loading, determine the force in members *FG*, *EG*, and *EH*.

Fig. P6.55 and P6.56

6.56 The truss shown was designed to support the roof of a food market. For the given loading, determine the force in members *KM*, *LM*, and *LN*.

6.57 A Polynesian, or duopitch, roof truss is loaded as shown. Determine the force in members *DF*, *EF*, and *EG*.

Fig. P6.57 and P6.58

6.58 A Polynesian, or duopitch, roof truss is loaded as shown. Determine the force in members *HI*, *GI*, and *GJ*.

6.59 Determine the force in members *DE* and *DF* of the truss shown when *P* = 20 kips.

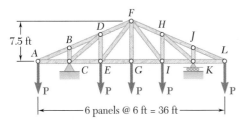

Fig. P6.59 and P6.60

6.60 Determine the force in members *EG* and *EF* of the truss shown when *P* = 20 kips.

6.61 Determine the force in members *EH* and *GI* of the truss shown. (*Hint:* Use section *aa*.)

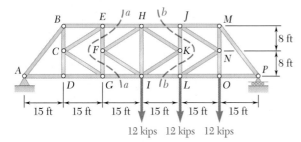

Fig. P6.61 and P6.62

6.62 Determine the force in members *HJ* and *IL* of the truss shown. (*Hint:* Use section *bb*.)

6.63 Determine the force in members *DG* and *FI* of the truss shown. (*Hint:* Use section *aa*.)

6.64 Determine the force in members *GJ* and *IK* of the truss shown. (*Hint:* Use section *bb*.)

Fig. P6.63 and P6.64

Fig. P6.65

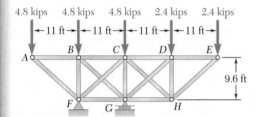

Fig. P6.66

6.65 and 6.66 The diagonal members in the center panels of the truss shown are very slender and can act only in tension; such members are known as *counters*. Determine the forces in the counters that are acting under the given loading.

6.67 and 6.68 The diagonal members in the center panels of the power transmission line tower shown are very slender and can act only in tension; such members are known as *counters*. For the given loading, determine (*a*) which of the two counters listed below is acting, (*b*) the force in that counter.
6.67 Counters *CJ* and *HE*.
6.68 Counters *IO* and *KN*.

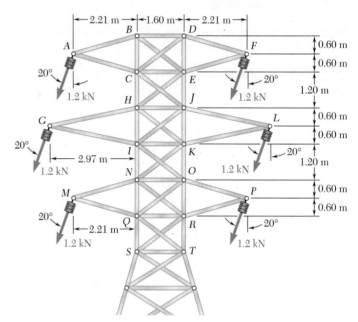

Fig. P6.67 and P6.68

6.69 Classify each of the structures shown as completely, partially, or improperly constrained; if completely constrained, further classify as determinate or indeterminate. (All members can act both in tension and in compression.)

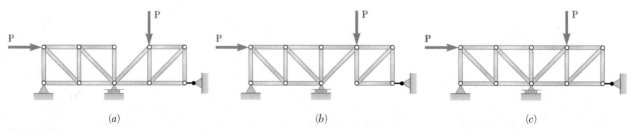

(*a*)   (*b*)   (*c*)

Fig. P6.69

6.70 through 6.74 Classify each of the structures shown as completely, partially, or improperly constrained; if completely constrained, further classify as determinate or indeterminate. (All members can act both in tension and in compression.)

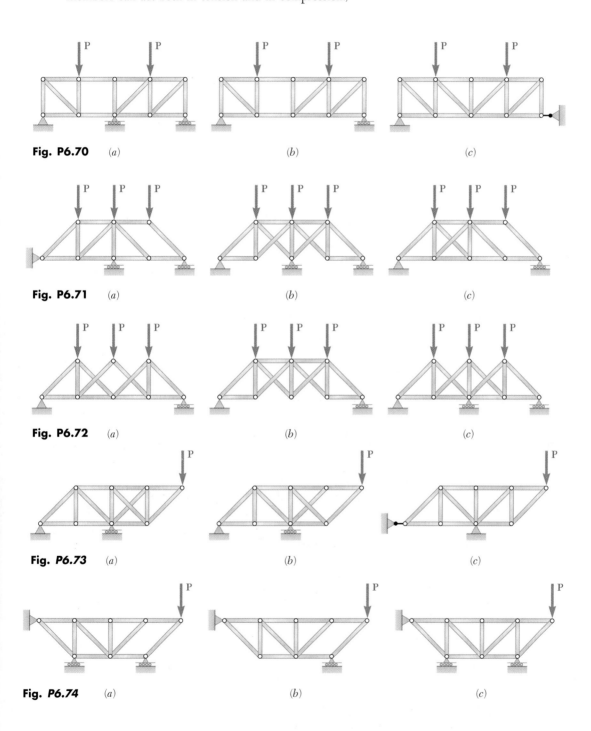

Fig. P6.70   (a)   (b)   (c)

Fig. P6.71   (a)   (b)   (c)

Fig. P6.72   (a)   (b)   (c)

Fig. P6.73   (a)   (b)   (c)

Fig. P6.74   (a)   (b)   (c)

6.9 STRUCTURES CONTAINING MULTIFORCE MEMBERS

Under trusses, we have considered structures consisting entirely of pins and straight two-force members. The forces acting on the two-force members were known to be directed along the members themselves. We now consider structures in which at least one of the members is a *multiforce* member, i.e., a member acted upon by three or more forces. These forces will generally not be directed along the members on which they act; their direction is unknown, and they should be represented therefore by two unknown components.

Frames and machines are structures containing multiforce members. *Frames* are designed to support loads and are usually stationary, fully constrained structures. *Machines* are designed to transmit and modify forces; they may or may not be stationary and will always contain moving parts.

6.10 ANALYSIS OF A FRAME

As a first example of analysis of a frame, the crane described in Sec. 6.1, which carries a given load W (Fig. 6.20a), will again be considered. The free-body diagram of the entire frame is shown in Fig. 6.20b. This diagram can be used to determine the external forces acting on the frame. Summing moments about A, we first determine the force $\mathbf{T}$ exerted by the cable; summing x and y components, we then determine the components $\mathbf{A}_x$ and $\mathbf{A}_y$ of the reaction at the pin A.

In order to determine the internal forces holding the various parts of a frame together, we must dismember the frame and draw a free-body diagram for each of its component parts (Fig. 6.20c). First, the two-force members should be considered. In this frame, member BE is the only two-force member. The forces acting at each end of this member must have the same magnitude, same line of action, and opposite sense (Sec. 4.6). They are therefore directed along BE and will be denoted, respectively, by $\mathbf{F}_{BE}$ and $-\mathbf{F}_{BE}$. Their sense will be arbitrarily assumed as shown in Fig. 6.20c; later the sign obtained for the common magnitude F_{BE} of the two forces will confirm or deny this assumption.

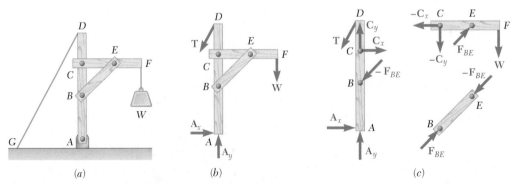

(a) (b) (c)

Fig. 6.20

Next, we consider the multiforce members, i.e., the members which are acted upon by three or more forces. According to Newton's third law, the force exerted at B by member BE on member AD must be equal and opposite to the force $\mathbf{F}_{BE}$ exerted by AD on BE. Similarly, the force exerted at E by member BE on member CF must be equal and opposite to the force $-\mathbf{F}_{BE}$ exerted by CF on BE. Thus the forces that the two-force member BE exerts on AD and CF are, respectively, equal to $-\mathbf{F}_{BE}$ and $\mathbf{F}_{BE}$; they have the same magnitude F_{BE} and opposite sense, and should be directed as shown in Fig. 6.20c.

At C two multiforce members are connected. Since neither the direction nor the magnitude of the forces acting at C is known, these forces will be represented by their x and y components. The components $\mathbf{C}_x$ and $\mathbf{C}_y$ of the force acting on member AD will be arbitrarily directed to the right and upward. Since, according to Newton's third law, the forces exerted by member CF on AD and by member AD on CF are equal and opposite, the components of the force acting on member CF *must* be directed to the left and downward; they will be denoted, respectively, by $-\mathbf{C}_x$ and $-\mathbf{C}_y$. Whether the force $\mathbf{C}_x$ is actually directed to the right and the force $-\mathbf{C}_x$ is actually directed to the left will be determined later from the sign of their common magnitude C_x, a plus sign indicating that the assumption made was correct, and a minus sign that it was wrong. The free-body diagrams of the multiforce members are completed by showing the external forces acting at A, D, and F.†

The internal forces can now be determined by considering the free-body diagram of either of the two multiforce members. Choosing the free-body diagram of CF, for example, we write the equations $\Sigma M_C = 0$, $\Sigma M_E = 0$, and $\Sigma F_x = 0$, which yield the values of the magnitudes F_{BE}, C_y, and C_x, respectively. These values can be checked by verifying that member AD is also in equilibrium.

It should be noted that the pins in Fig. 6.20 were assumed to form an integral part of one of the two members they connected and so it was not necessary to show their free-body diagram. This assumption can always be used to simplify the analysis of frames and machines. When a pin connects three or more members, however, or when a pin connects a support and two or more members, or when a load is applied to a pin, a clear decision must be made in choosing the member to which the pin will be assumed to belong. (If multiforce members are involved, the pin should be attached to one of these members.) The various forces exerted on the pin should then be clearly identified. This is illustrated in Sample Prob. 6.6.

---

†It is not strictly necessary to use a minus sign to distinguish the force exerted by one member on another from the equal and opposite force exerted by the second member on the first, since the two forces belong to different free-body diagrams and thus cannot easily be confused. In the Sample Problems, the same symbol is used to represent equal and opposite forces which are applied to different free bodies. It should be noted that, under these conditions, the sign obtained for a given force component will not directly relate the sense of that component to the sense of the corresponding coordinate axis. Rather, a positive sign will indicate that *the sense assumed for that component in the free-body diagram* is correct, and a negative sign will indicate that it is wrong.

6.11 FRAMES WHICH CEASE TO BE RIGID WHEN DETACHED FROM THEIR SUPPORTS

The crane analyzed in Sec. 6.10 was so constructed that it could keep the same shape without the help of its supports; it was therefore considered as a rigid body. Many frames, however, will collapse if detached from their supports; such frames cannot be considered as rigid bodies. Consider, for example, the frame shown in Fig. 6.21a, which consists of two members AC and CB carrying loads $\mathbf{P}$ and $\mathbf{Q}$ at their midpoints; the members are supported by pins at A and B and are connected by a pin at C. If detached from its supports, this frame will not maintain its shape; it should therefore be considered as made of *two distinct rigid parts AC and CB*.

The equations $\Sigma F_x = 0$, $\Sigma F_y = 0$, $\Sigma M = 0$ (about any given point) express the conditions for the *equilibrium of a rigid body* (Chap. 4); we should use them, therefore, in connection with the free-body diagrams of rigid bodies, namely, the free-body diagrams of members AC and CB (Fig. 6.21b). Since these members are multiforce members, and since pins are used at the supports and at the connection, the reactions at A and B and the forces at C will each be represented by two components. In accordance with Newton's third law, the components of the force exerted by CB on AC and the components of the force exerted by AC on CB will be represented by vectors of the same magnitude and opposite sense; thus, if the first pair of components consists of $\mathbf{C}_x$ and $\mathbf{C}_y$, the second pair will be represented by $-\mathbf{C}_x$ and $-\mathbf{C}_y$. We note that four unknown force components act on free body AC, while only three independent equations can be used to express that the body is in equilibrium; similarly, four unknowns, but only three equations, are associated with CB. However, only six different unknowns are involved in the analysis of the two members, and altogether six equations are available to express that the members are in equilibrium. Writing $\Sigma M_A = 0$ for free body AC and $\Sigma M_B = 0$ for CB, we obtain two simultaneous equations which may be solved for the common magnitude C_x of the components $\mathbf{C}_x$ and $-\mathbf{C}_x$, and for the common magnitude C_y of the components $\mathbf{C}_y$ and $-\mathbf{C}_y$. We then write $\Sigma F_x = 0$ and $\Sigma F_y = 0$ for each of the two free bodies, obtaining, successively, the magnitudes A_x, A_y, B_x, and B_y.

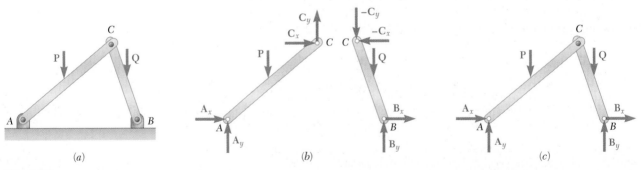

(a)   (b)   (c)

Fig. 6.21

It can now be observed that since the equations of equilibrium $\Sigma F_x = 0$, $\Sigma F_y = 0$, and $\Sigma M = 0$ (about any given point) are satisfied by the forces acting on free body AC, and since they are also satisfied by the forces acting on free body CB, they must be satisfied when the forces acting on the two free bodies are considered simultaneously. Since the internal forces at C cancel each other, we find that the equations of equilibrium must be satisfied by the external forces shown on the free-body diagram of the frame ACB itself (Fig. 6.21c), although the frame is not a rigid body. These equations can be used to determine some of the components of the reactions at A and B. We will also find, however, that *the reactions cannot be completely determined from the free-body diagram of the whole frame*. It is thus necessary to dismember the frame and to consider the free-body diagrams of its component parts (Fig. 6.21b), even when we are interested in determining external reactions only. This is because the equilibrium equations obtained for free body ACB *are necessary conditions* for the equilibrium of a nonrigid structure, *but are not sufficient conditions*.

The method of solution outlined in the second paragraph of this section involved simultaneous equations. A more efficient method is now presented, which utilizes the free body ACB as well as the free bodies AC and CB. Writing $\Sigma M_A = 0$ and $\Sigma M_B = 0$ for free body ACB, we obtain B_y and A_y. Writing $\Sigma M_C = 0$, $\Sigma F_x = 0$, and $\Sigma F_y = 0$ for free body AC, we obtain, successively, A_x, C_x, and C_y. Finally, writing $\Sigma F_x = 0$ for ACB, we obtain B_x.

We noted above that the analysis of the frame of Fig. 6.21 involves six unknown force components and six independent equilibrium equations. (The equilibrium equations for the whole frame were obtained from the original six equations and, therefore, are not independent.) Moreover, we checked that all unknowns could be actually determined and that all equations could be satisfied. The frame considered is *statically determinate and rigid*.† In general, to determine whether a structure is statically determinate and rigid, we should draw a free-body diagram for each of its component parts and count the reactions and internal forces involved. We should also determine the number of independent equilibrium equations (excluding equations expressing the equilibrium of the whole structure or of groups of component parts already analyzed). If there are more unknowns than equations, the structure is *statically indeterminate*. If there are fewer unknowns than equations, the structure is *nonrigid*. If there are as many unknowns as equations, *and if all unknowns can be determined and all equations satisfied* under general loading conditions, the structure is *statically determinate and rigid*. If, however, due to an *improper arrangement* of members and supports, all unknowns cannot be determined and all equations cannot be satisfied, the structure is *statically indeterminate and nonrigid*.

---

†The word "rigid" is used here to indicate that the frame will maintain its shape as long as it remains attached to its supports.

SAMPLE PROBLEM 6.4

In the frame shown, members *ACE* and *BCD* are connected by a pin at *C* and by the link *DE*. For the loading shown, determine the force in link *DE* and the components of the force exerted at *C* on member *BCD*.

SOLUTION

Free Body: Entire Frame. Since the external reactions involve only three unknowns, we compute the reactions by considering the free-body diagram of the entire frame.

$+\uparrow \Sigma F_y = 0$: $A_y - 480 \text{ N} = 0$ $A_y = +480 \text{ N}$ $\mathbf{A}_y = 480 \text{ N}\uparrow$

$+\gamma \Sigma M_A = 0$: $-(480 \text{ N})(100 \text{ mm}) + B(160 \text{ mm}) = 0$
$\qquad\qquad\qquad\qquad\qquad B = +300 \text{ N}$ $\mathbf{B} = 300 \text{ N}\rightarrow$

$\xrightarrow{+} \Sigma F_x = 0$: $B + A_x = 0$
$\qquad\qquad\qquad 300 \text{ N} + A_x = 0$ $A_x = -300 \text{ N}$ $\mathbf{A}_x = 300 \text{ N}\leftarrow$

Members. We now dismember the frame. Since only two members are connected at *C*, the components of the unknown forces acting on *ACE* and *BCD* are, respectively, equal and opposite and are assumed directed as shown. We assume that link *DE* is in tension and exerts equal and opposite forces at *D* and *E*, directed as shown.

Free Body: Member BCD. Using the free body *BCD*, we write

$+\gamma \Sigma M_C = 0$:
$\qquad (F_{DE} \sin \alpha)(250 \text{ mm}) + (300 \text{ N})(80 \text{ mm}) + (480 \text{ N})(100 \text{ mm}) = 0$
$\qquad F_{DE} = -561 \text{ N}$ $\qquad\qquad\qquad\qquad F_{DE} = 561 \text{ N } C$ ◀

$\xrightarrow{+} \Sigma F_x = 0$: $C_x - F_{DE} \cos \alpha + 300 \text{ N} = 0$
$\qquad\qquad C_x - (-561 \text{ N}) \cos 28.07° + 300 \text{ N} = 0$ $C_x = -795 \text{ N}$

$+\uparrow \Sigma F_y = 0$: $C_y - F_{DE} \sin \alpha - 480 \text{ N} = 0$
$\qquad\qquad C_y - (-561 \text{ N}) \sin 28.07° - 480 \text{ N} = 0$ $C_y = +216 \text{ N}$

From the signs obtained for C_x and C_y we conclude that the force components $\mathbf{C}_x$ and $\mathbf{C}_y$ exerted on member *BCD* are directed, respectively, to the left and up. We have

$$C_x = 795 \text{ N}\leftarrow, C_y = 216 \text{ N}\uparrow$$ ◀

Free Body: Member ACE (Check). The computations are checked by considering the free body *ACE*. For example,

$+\gamma \Sigma M_A = (F_{DE} \cos \alpha)(300 \text{ mm}) + (F_{DE} \sin \alpha)(100 \text{ mm}) - C_x(220 \text{ mm})$
$\qquad\qquad = (-561 \cos \alpha)(300) + (-561 \sin \alpha)(100) - (-795)(220) = 0$

SAMPLE PROBLEM 6.5

Determine the components of the forces acting on each member of the frame shown.

SOLUTION

Free Body: Entire Frame. Since the external reactions involve only three unknowns, we compute the reactions by considering the free-body diagram of the entire frame.

$+\curvearrowright\Sigma M_E = 0$: $\quad -(2400 \text{ N})(3.6 \text{ m}) + F(4.8 \text{ m}) = 0$
$$F = +1800 \text{ N} \qquad\qquad \mathbf{F} = 1800 \text{ N}\uparrow \quad \blacktriangleleft$$
$+\uparrow\Sigma F_y = 0$: $\quad -2400 \text{ N} + 1800 \text{ N} + E_y = 0$
$$E_y = +600 \text{ N} \qquad\qquad \mathbf{E}_y = 600 \text{ N}\uparrow \quad \blacktriangleleft$$
$\xrightarrow{+}\Sigma F_x = 0$: $\qquad\qquad\qquad\qquad\qquad\qquad\qquad \mathbf{E}_x = 0 \quad \blacktriangleleft$

Members. The frame is now dismembered; since only two members are connected at each joint, equal and opposite components are shown on each member at each joint.

Free Body: Member BCD

$+\curvearrowright\Sigma M_B = 0$: $\quad -(2400 \text{ N})(3.6 \text{ m}) + C_y(2.4 \text{ m}) = 0 \quad C_y = +3600 \text{ N} \quad \blacktriangleleft$
$+\curvearrowright\Sigma M_C = 0$: $\quad -(2400 \text{ N})(1.2 \text{ m}) + B_y(2.4 \text{ m}) = 0 \quad B_y = +1200 \text{ N} \quad \blacktriangleleft$
$\xrightarrow{+}\Sigma F_x = 0$: $\quad -B_x + C_x = 0$

We note that neither B_x nor C_x can be obtained by considering only member *BCD*. The positive values obtained for B_y and C_y indicate that the force components $\mathbf{B}_y$ and $\mathbf{C}_y$ are directed as assumed.

Free Body: Member ABE

$+\curvearrowright\Sigma M_A = 0$: $\quad B_x(2.7 \text{ m}) = 0 \qquad\qquad\qquad\qquad B_x = 0 \quad \blacktriangleleft$
$\xrightarrow{+}\Sigma F_x = 0$: $\quad +B_x - A_x = 0 \qquad\qquad\qquad\qquad A_x = 0 \quad \blacktriangleleft$
$+\uparrow\Sigma F_y = 0$: $\quad -A_y + B_y + 600 \text{ N} = 0$
$$-A_y + 1200 \text{ N} + 600 \text{ N} = 0 \qquad A_y = +1800 \text{ N} \quad \blacktriangleleft$$

Free Body: Member BCD. Returning now to member *BCD*, we write

$\xrightarrow{+}\Sigma F_x = 0$: $\quad -B_x + C_x = 0 \qquad 0 + C_x = 0 \qquad\qquad C_x = 0 \quad \blacktriangleleft$

Free Body: Member ACF (Check). All unknown components have now been found; to check the results, we verify that member *ACF* is in equilibrium.

$+\curvearrowright\Sigma M_C = (1800 \text{ N})(2.4 \text{ m}) - A_y(2.4 \text{ m}) - A_x(2.7 \text{ m})$
$$= (1800 \text{ N})(2.4 \text{ m}) - (1800 \text{ N})(2.4 \text{ m}) - 0 = 0 \qquad \text{(checks)}$$

SAMPLE PROBLEM 6.6

A 600-lb horizontal force is applied to pin A of the frame shown. Determine the forces acting on the two vertical members of the frame.

SOLUTION

Free Body: Entire Frame. The entire frame is chosen as a free body; although the reactions involve four unknowns, $\mathbf{E}_y$ and $\mathbf{F}_y$ may be determined by writing

$+\circlearrowleft\Sigma M_E = 0:$ $-(600\text{ lb})(10\text{ ft}) + F_y(6\text{ ft}) = 0$
            $F_y = +1000\text{ lb}$ $\mathbf{F}_y = 1000\text{ lb}\uparrow$ ◀

$+\uparrow\Sigma F_y = 0:$ $E_y + F_y = 0$
            $E_y = -1000\text{ lb}$ $\mathbf{E}_y = 1000\text{ lb}\downarrow$ ◀

Members. The equations of equilibrium of the entire frame are not sufficient to determine $\mathbf{E}_x$ and $\mathbf{F}_x$. The free-body diagrams of the various members must now be considered in order to proceed with the solution. In dismembering the frame we will assume that pin A is attached to the multiforce member ACE and, thus, that the 600-lb force is applied to that member. We also note that AB and CD are two-force members.

Free Body: Member ACE

$+\uparrow\Sigma F_y = 0:$ $-\frac{5}{13}F_{AB} + \frac{5}{13}F_{CD} - 1000\text{ lb} = 0$
$+\circlearrowleft\Sigma M_E = 0:$ $-(600\text{ lb})(10\text{ ft}) - (\frac{12}{13}F_{AB})(10\text{ ft}) - (\frac{12}{13}F_{CD})(2.5\text{ ft}) = 0$

Solving these equations simultaneously, we find

$$F_{AB} = -1040\text{ lb} \qquad F_{CD} = +1560\text{ lb}$$ ◀

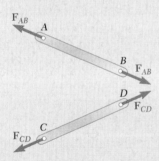

The signs obtained indicate that the sense assumed for F_{CD} was correct and the sense for F_{AB} incorrect. Summing now x components,

$\xrightarrow{+}\Sigma F_x = 0:$ $600\text{ lb} + \frac{12}{13}(-1040\text{ lb}) + \frac{12}{13}(+1560\text{ lb}) + E_x = 0$
        $E_x = -1080\text{ lb}$ $\mathbf{E}_x = 1080\text{ lb}\leftarrow$ ◀

Free Body: Entire Frame. Since $\mathbf{E}_x$ has been determined, we can return to the free-body diagram of the entire frame and write

$\xrightarrow{+}\Sigma F_x = 0:$ $600\text{ lb} - 1080\text{ lb} + F_x = 0$
        $F_x = +480\text{ lb}$ $\mathbf{F}_x = 480\text{ lb}\rightarrow$ ◀

Free Body: Member BDF (Check). We can check our computations by verifying that the equation $\Sigma M_B = 0$ is satisfied by the forces acting on member BDF.

$+\circlearrowleft\Sigma M_B = -(\frac{12}{13}F_{CD})(2.5\text{ ft}) + (F_x)(7.5\text{ ft})$
           $= -\frac{12}{13}(1560\text{ lb})(2.5\text{ ft}) + (480\text{ lb})(7.5\text{ ft})$
           $= -3600\text{ lb}\cdot\text{ft} + 3600\text{ lb}\cdot\text{ft} = 0$ (checks)

SOLVING PROBLEMS ON YOUR OWN

In this lesson you learned to analyze *frames containing one or more multiforce members*. In the problems that follow you will be asked to determine the external reactions exerted on the frame and the internal forces that hold together the members of the frame.

In solving problems involving frames containing one or more multiforce members, follow these steps:

1. Draw a free-body diagram of the entire frame. Use this free-body diagram to calculate, to the extent possible, the reactions at the supports. (In Sample Prob. 6.6 only two of the four reaction components could be found from the free body of the entire frame.)

2. Dismember the frame, and draw a free-body diagram of each member.

3. Considering first the two-force members, apply equal and opposite forces to each two-force member at the points where it is connected to another member. If the two-force member is a straight member, these forces will be directed along the axis of the member. If you cannot tell at this point whether the member is in tension or compression, just *assume* that the member is in tension and *direct both of the forces away from the member*. Since these forces have the same unknown magnitude, give them both the *same name* and, to avoid any confusion later, *do not use a plus sign or a minus sign*.

4. Next, consider the multiforce members. For each of these members, show all the forces acting on the member, including *applied loads, reactions, and internal forces at connections*. The magnitude and direction of any reaction or reaction component found earlier from the free-body diagram of the entire frame should be clearly indicated.

   a. Where a multiforce member is connected to a two-force member, apply to the multiforce member a force *equal and opposite* to the force drawn on the free-body diagram of the two-force member, *giving it the same name*.

   b. Where a multiforce member is connected to another multiforce member, use *horizontal and vertical components* to represent the internal forces at that point, since neither the direction nor the magnitude of these forces is known. The direction you choose for each of the two force components exerted on the first multiforce member is arbitary, but *you must apply equal and opposite force components of the same name* to the other multiforce member. Again, *do not use a plus sign or a minus sign*.

(continued)

5. The internal forces may now be determined, as well as any *reactions* that you have not already found.

   a. The free-body diagram of each of the multiforce members can provide you with *three equilibrium equations.*

   b. To simplify your solution, you should seek a way to write an equation involving a single unknown. If you can locate *a point where all but one of the unknown force components intersect,* you will obtain an equation in a single unknown by summing moments about that point. *If all unknown forces except one are parallel,* you will obtain an equation in a single unknown by summing force components in a direction perpendicular to the parallel forces.

   c. Since you arbitrarily chose the direction of each of the unknown forces, you cannot determine until the solution is completed whether your guess was correct. To do that, consider the *sign* of the value found for each of the unknowns: a *positive* sign means that the direction you selected was *correct;* a *negative* sign means that the direction is *opposite* to the direction you assumed.

6. To be more effective and efficient as you proceed through your solution, observe the following rules:

   a. If an equation involving only one unknown can be found, write that equation and *solve it for that unknown.* Immediately *replace* that unknown wherever it appears on other free-body diagrams *by the value you have found.* Repeat this process by seeking equilibrium equations involving only one unknown until you have found all of the internal forces and unknown reactions.

   b. If an equation involving only one unknown cannot be found, you may have to *solve a pair of simultaneous equations.* Before doing so, check that you have shown the values of all of the reactions that were obtained from the free-body diagram of the entire frame.

   c. The total number of equations of equilibrium for the entire frame and for the individual members *will be larger than the number of unknown forces and reactions.* After you have found all the reactions and all the internal forces, you can use the remaining equations to check the accuracy of your computations.

PROBLEMS

FREE BODY PRACTICE PROBLEMS

6.F1 For the frame and loading shown, draw the free-body diagram(s) needed to determine the forces acting on member *ABC* at *B* and *C*.

Fig. P6.F1

6.F2 For the frame and loading shown, draw the free-body diagram(s) needed to determine all forces acting on member *GBEH*.

6.F3 For the frame and loading shown, draw the free-body diagram(s) needed to determine the reactions at *B* and *F*.

Dimensions in mm

Fig. P6.F2

Fig. P6.F3

6.F4 Knowing that the surfaces at *A* and *D* are frictionless, draw the free-body diagram(s) needed to determine the forces exerted at *B* and *C* on member *BCE*.

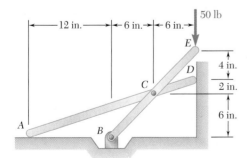

Fig. P6.F4

6.75 and 6.76 Determine the force in member BD and the component of the reaction at C.

Fig. P6.75

Fig. P6.76

6.77 Determine the components of all forces acting on member $ABCD$ of the assembly shown.

Fig. P6.77

Fig. P6.78

6.78 Determine the components of all forces acting on member ABD of the frame shown.

6.79 For the frame and loading shown, determine the components of all forces acting on member ABC.

6.80 Solve Prob. 6.79 assuming that the 20-kip load is replaced by a clockwise couple of magnitude 100 kip · ft applied to member EDC at point D.

6.81 Determine the components of all forces acting on member $ABCD$ when $\theta = 0$.

Fig. P6.79

Fig. P6.81 and P6.82

6.82 Determine the components of all forces acting on member $ABCD$ when $\theta = 90°$.

6.83 and 6.84 Determine the components of the reactions at A and E if a 750-N force directed vertically downward is applied (*a*) at B, (*b*) at D.

6.85 and 6.86 Determine the components of the reactions at A and E if the frame is loaded by a clockwise couple of magnitude 36 N · m applied (*a*) at B, (*b*) at D.

6.87 Determine all the forces exerted on member AI if the frame is loaded by a clockwise couple of magnitude 1200 lb · in. applied (*a*) at point D, (*b*) at point E.

Fig. P6.83 and P6.85

Fig. P6.87 and P6.88

Fig. P6.84 and P6.86

6.88 Determine all the forces exerted on member AI if the frame is loaded by a 40-lb force directed horizontally to the right and applied (*a*) at point D, (*b*) at point E.

6.89 Determine the components of the reactions at A and B, (*a*) if the 100-lb load is applied as shown, (*b*) if the 100-lb load is moved along its line of action and is applied at point F.

6.90 (*a*) Show that when a frame supports a pulley at A, an equivalent loading of the frame and of each of its component parts can be obtained by removing the pulley and applying at A two forces equal and parallel to the forces that the cable exerted on the pulley. (*b*) Show that if one end of the cable is attached to the frame at a point B, a force of magnitude equal to the tension in the cable should also be applied at B.

Fig. P6.89

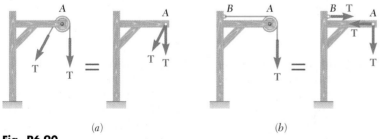

Fig. P6.90

6.91 A 3-ft-diameter pipe is supported every 16 ft by a small frame like that shown. Knowing that the combined weight of the pipe and its contents is 500 lb/ft and assuming frictionless surfaces, determine the components (*a*) of the reaction at E, (*b*) of the force exerted at C on member CDE.

6.92 Solve Prob. 6.91 for a frame where $h = 6$ ft.

Fig. P6.91

6.93 Knowing that the pulley has a radius of 0.5 m, determine the components of the reactions at A and E.

Fig. P6.93

Fig. P6.94

6.94 Knowing that the pulley has a radius of 50 mm, determine the components of the reactions at B and E.

6.95 A trailer weighing 2400 lb is attached to a 2900-lb pickup truck by a ball-and-socket truck hitch at D. Determine (*a*) the reactions at each of the six wheels when the truck and trailer are at rest, (*b*) the additional load on each of the truck wheels due to the trailer.

Fig. P6.95

Fig. P6.96

6.96 In order to obtain a better weight distribution over the four wheels of the pickup truck of Prob. 6.95, a compensating hitch of the type shown is used to attach the trailer to the truck. The hitch consists of two bar springs (only one is shown in the figure) that fit into bearings inside a support rigidly attached to the truck. The springs are also connected by chains to the trailer frame, and specially designed hooks make it possible to place both chains in tension. (*a*) Determine the tension T required in each of the two chains if the additional load due to the trailer is to be evenly distributed over the four wheels of the truck. (*b*) What are the resulting reactions at each of the six wheels of the trailer-truck combination?

6.97 The cab and motor units of the front-end loader shown are connected by a vertical pin located 2 m behind the cab wheels. The distance from C to D is 1 m. The center of gravity of the 300-kN motor unit is located at G_m, while the centers of gravity of the 100-kN cab and 75-kN load are located, respectively, at G_c and G_l. Knowing that the machine is at rest with its brakes released, determine (a) the reactions at each of the four wheels, (b) the forces exerted on the motor unit at C and D.

Fig. P6.97

6.98 Solve Prob. 6.97 assuming that the 75-kN load has been removed.

6.99 and 6.100 For the frame and loading shown, determine the components of all forces acting on member ABE.

6.101 For the frame and loading shown, determine the components of all forces acting on member ABD.

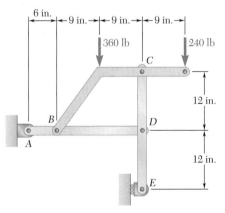

Fig. P6.101

6.102 Solve Prob. 6.101 assuming that the 360-lb load has been removed.

6.103 For the frame and loading shown, determine the components of the forces acting on member CDE at C and D.

Fig. P6.99

Fig. P6.100

Fig. P6.103

Fig. P6.104

Fig. P6.107

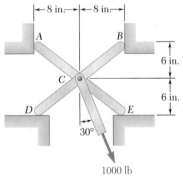

Fig. P6.108

6.104 For the frame and loading shown, determine the components of the forces acting on member *CFE* at *C* and *F*.

6.105 For the frame and loading shown, determine the components of all forces acting on member *ABD*.

Dimensions in mm

Fig. P6.105

6.106 Solve Prob. 6.105 assuming that the 3-kN load has been removed.

6.107 Determine the reaction at *F* and the force in members *AE* and *BD*.

6.108 For the frame and loading shown, determine the reactions at *A*, *B*, *D*, and *E*. Assume that the surface at each support is frictionless.

6.109 The axis of the three-hinge arch *ABC* is a parabola with vertex at *B*. Knowing that $P = 112$ kN and $Q = 140$ kN, determine (*a*) the components of the reaction at *A*, (*b*) the components of the force exerted at *B* on segment *AB*.

Fig. P6.109 and P6.110

6.110 The axis of the three-hinge arch *ABC* is a parabola with vertex at *B*. Knowing that $P = 140$ kN and $Q = 112$ kN, determine (*a*) the components of the reaction at *A*, (*b*) the components of the force exerted at *B* on segment *AB*.

6.111, 6.112, and 6.113 Members *ABC* and *CDE* are pin-connected at *C* and supported by four links. For the loading shown, determine the force in each link.

Fig. P6.111

Fig. P6.112

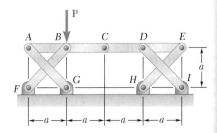

Fig. P6.113

6.114 Members *ABC* and *CDE* are pin-connected at *C* and supported by the four links *AF*, *BG*, *DG*, and *EH*. For the loading shown, determine the force in each link.

6.115 Solve Prob. 6.112 assuming that the force **P** is replaced by a clockwise couple of moment $\mathbf{M}_0$ applied to member *CDE* at *D*.

6.116 Solve Prob. 6.114 assuming that the force **P** is replaced by a clockwise couple of moment $\mathbf{M}_0$ applied at the same point.

6.117 Four beams, each of length 3*a*, are held together by single nails at *A*, *B*, *C*, and *D*. Each beam is attached to a support located at a distance *a* from an end of the beam as shown. Assuming that only vertical forces are exerted at the connections, determine the vertical reactions at *E*, *F*, *G*, and *H*.

Fig. P6.114

Fig. P6.117

6.118 Four beams, each of length 2*a*, are nailed together at their midpoints to form the support system shown. Assuming that only vertical forces are exerted at the connections, determine the vertical reactions at *A*, *D*, *E*, and *H*.

Fig. P6.118

6.119 through 6.121 Each of the frames shown consists of two L-shaped members connected by two rigid links. For each frame, determine the reactions at the supports and indicate whether the frame is rigid.

(a)

(b)

(c)

Fig. P6.119

(a)

(b)

(c)

Fig. P6.120

(a)

(b)

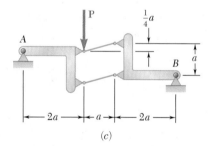

(c)

Fig. P6.121

6.12 MACHINES

Machines are structures designed to transmit and modify forces. Whether they are simple tools or include complicated mechanisms, their main purpose is to transform *input forces* into *output forces*. Consider, for example, a pair of cutting pliers used to cut a wire (Fig. 6.22a). If we apply two equal and opposite forces **P** and −**P** on their handles, they will exert two equal and opposite forces **Q** and −**Q** on the wire (Fig. 6.22b).

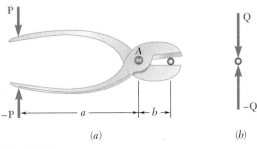

Fig. 6.22

To determine the magnitude Q of the output forces when the magnitude P of the input forces is known (or, conversely, to determine P when Q is known), we draw a free-body diagram of the pliers *alone*, showing the input forces **P** and $-$**P** and the *reactions* $-$**Q** and **Q** that the wire exerts on the pliers (Fig. 6.23). However, since a pair of pliers forms a nonrigid structure, we must use one of the component parts as a free body in order to determine the unknown forces. Considering Fig. 6.24a, for example, and taking moments about A, we obtain the relation $Pa = Qb$, which defines the magnitude Q in terms of P or P in terms of Q. The same free-body diagram can be used to determine the components of the internal force at A; we find $A_x = 0$ and $A_y = P + Q$.

In the case of more complicated machines, it generally will be necessary to use several free-body diagrams and, possibly, to solve simultaneous equations involving various internal forces. The free bodies should be chosen to include the input forces and the reactions to the output forces, and the total number of unknown force components involved should not exceed the number of available independent equations. It is advisable, before attempting to solve a problem, to determine whether the structure considered is determinate. There is no point, however, in discussing the rigidity of a machine, since a machine includes moving parts and thus *must* be nonrigid.

Photo 6.5 The lamp shown can be placed in many positions. By considering various free bodies, the force in the springs and the internal forces at the joints can be determined.

Fig. 6.23

Fig. 6.24

SAMPLE PROBLEM 6.7

A hydraulic-lift table is used to raise a 1000-kg crate. It consists of a platform and two identical linkages on which hydraulic cylinders exert equal forces. (Only one linkage and one cylinder are shown.) Members EDB and CG are each of length $2a$, and member AD is pinned to the midpoint of EDB. If the crate is placed on the table, so that half of its weight is supported by the system shown, determine the force exerted by each cylinder in raising the crate for $\theta = 60°$, $a = 0.70$ m, and $L = 3.20$ m. Show that the result obtained is independent of the distance d.

SOLUTION

The machine considered consists of the platform and of the linkage. Its free-body diagram includes an input force $\mathbf{F}_{DH}$ exerted by the cylinder, the weight $\frac{1}{2}\mathbf{W}$, equal and opposite to the output force, and reactions at E and G that we assume to be directed as shown. Since more than three unknowns are involved, this diagram will not be used. The mechanism is dismembered and a free-body diagram is drawn for each of its component parts. We note that AD, BC, and CG are two-force members. We already assumed member CG to be in compression; we now assume that AD and BC are in tension and direct as shown the forces exerted on them. Equal and opposite vectors will be used to represent the forces exerted by the two-force members on the platform, on member BDE, and on roller C.

Free Body: Platform ABC.

$$\xrightarrow{+}\Sigma F_x = 0: \qquad F_{AD} \cos\theta = 0 \qquad F_{AD} = 0$$
$$+\uparrow\Sigma F_y = 0: \qquad B + C - \tfrac{1}{2}W = 0 \qquad B + C = \tfrac{1}{2}W \qquad (1)$$

Free Body: Roller C. We draw a force triangle and obtain $F_{BC} = C \cot\theta$.

Free Body: Member BDE. Recalling that $F_{AD} = 0$,

$$+\curvearrowleft\Sigma M_E = 0: \quad F_{DH} \cos(\phi - 90°)a - B(2a\cos\theta) - F_{BC}(2a\sin\theta) = 0$$
$$F_{DH}a\sin\phi - B(2a\cos\theta) - (C\cot\theta)(2a\sin\theta) = 0$$
$$F_{DH}\sin\phi - 2(B + C)\cos\theta = 0$$

Recalling Eq. (1), we have

$$F_{DH} = W\frac{\cos\theta}{\sin\phi} \qquad (2)$$

and we observe that *the result obtained is independent of d.* ◀

Applying first the law of sines to triangle EDH, we write

$$\frac{\sin\phi}{EH} = \frac{\sin\theta}{DH} \qquad \sin\phi = \frac{EH}{DH}\sin\theta \qquad (3)$$

Using now the law of cosines, we have

$$(DH)^2 = a^2 + L^2 - 2aL\cos\theta$$
$$= (0.70)^2 + (3.20)^2 - 2(0.70)(3.20)\cos 60°$$
$$(DH)^2 = 8.49 \qquad DH = 2.91 \text{ m}$$

We also note that

$$W = mg = (1000 \text{ kg})(9.81 \text{ m/s}^2) = 9810 \text{ N} = 9.81 \text{ kN}$$

Substituting for $\sin\phi$ from (3) into (2) and using the numerical data, we write

$$F_{DH} = W\frac{DH}{EH}\cot\theta = (9.81 \text{ kN})\frac{2.91 \text{ m}}{3.20 \text{ m}}\cot 60°$$

$$F_{DH} = 5.15 \text{ kN} \quad ◀$$

SOLVING PROBLEMS
ON YOUR OWN

This lesson was devoted to the analysis of *machines*. Since machines are designed to transmit or modify forces, they always contain moving parts. However, the machines considered here will always be at rest, and you will be working with the set of *forces required to maintain the equilibrium of the machine*.

Known forces that act on a machine are called *input forces*. *A machine transforms the input forces into output forces*, such as the cutting forces applied by the pliers of Fig. 6.22. You will determine the output forces by finding the forces equal and opposite to the output forces that should be applied to the machine to maintain its equilibrium.

In the preceding lesson you analyzed frames; you will now use almost the same procedure to analyze machines:

1. Draw a free-body diagram of the whole machine, and use it to determine as many as possible of the unknown forces exerted on the machine.

2. Dismember the machine, and draw a free-body diagram of each member.

3. Considering first the two-force members, apply equal and opposite forces to each two-force member at the points where it is connected to another member. If you cannot tell at this point whether the member is in tension or in compression just *assume* that the member is in tension and *direct both of the forces away from the member*. Since these forces have the same unknown magnitude, *give them both the same name*.

4. Next consider the multiforce members. For each of these members, show all the forces acting on the member, including applied loads and forces, reactions, and internal forces at connections.

   a. Where a multiforce member is connected to a two-force member, apply to the multiforce member a force *equal and opposite* to the force drawn on the free-body diagram of the two-force member, *giving it the same name*.

   b. Where a multiforce member is connected to another multiforce member, use *horizontal and vertical components* to represent the internal forces at that point. The directions you choose for each of the two force components exerted on the first multiforce member are arbitrary, but *you must apply equal and opposite force components of the same name* to the other multiforce member.

5. Equilibrium equations can be written after you have completed the various free-body diagrams.

   a. To simplify your solution, you should, whenever possible, write and solve equilibrium equations involving single unknowns.

   b. Since you arbitrarily chose the direction of each of the unknown forces, you must determine at the end of the solution whether your guess was correct. To that effect, *consider the sign* of the value found for each of the unknowns. A *positive* sign indicates that your guess was correct, and a *negative* sign indicates that it was not.

6. Finally, you should check your solution by substituting the results obtained into an equilibrium equation that you have not previously used.

PROBLEMS

FREE BODY PRACTICE PROBLEMS

6.F5 The position of member *ABC* is controlled by the hydraulic cylinder *CD*. Knowing that $\theta = 30°$, draw the free-body diagram(s) needed to determine the force exerted by the hydraulic cylinder on pin *C*, and the reaction at *B*.

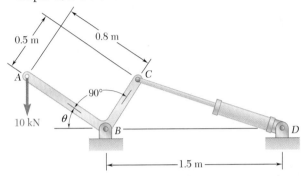

Fig. P6.F5

6.F6 Arm *ABC* is connected by pins to a collar at *B* and to crank *CD* at *C*. Neglecting the effect of friction, draw the free-body diagram(s) needed to determine the couple **M** to hold the system in equilibrium when $\theta = 30°$.

Fig. P6.F6

6.F7 Since the brace shown must remain in position even when the magnitude of **P** is very small, a single safety spring is attached at *D* and *E*. The spring *DE* has a constant of 50 lb/in. and an unstretched length of 7 in. Knowing that $l = 10$ in. and that the magnitude of **P** is 800 lb, draw the free-body diagram(s) needed to determine the force **Q** required to release the brace.

Fig. P6.F7

6.F8 A log weighing 800 lb is lifted by a pair of tongs as shown. Draw the free-body diagram(s) needed to determine the forces exerted at *E* and *F* on tong *DEF*.

Fig. P6.F8

Fig. P6.122

6.122 The shear shown is used to cut and trim electronic-circuit-board laminates. For the position shown, determine (*a*) the vertical component of the force exerted on the shearing blade at *D*, (*b*) the reaction at *C*.

6.123 The press shown is used to emboss a small seal at *E*. Knowing that *P* = 250 N, determine (*a*) the vertical component of the force exerted on the seal, (*b*) the reaction at *A*.

Fig. P6.123 and P6.124

6.124 The press shown is used to emboss a small seal at *E*. Knowing that the vertical component of the force exerted on the seal must be 900 N, determine (*a*) the required vertical force **P**, (*b*) the corresponding reaction at *A*.

6.125 Water pressure in the supply system exerts a downward force of 135 N on the vertical plug at *A*. Determine the tension in the fusible link *DE* and the force exerted on member *BCE* at *B*.

Fig. P6.125

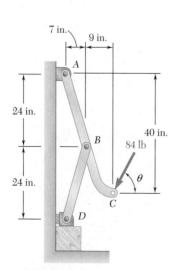

Fig. P6.126

6.126 An 84-lb force is applied to the toggle vise at *C*. Knowing that *θ* = 90°, determine (*a*) the vertical force exerted on the block at *D*, (*b*) the force exerted on member *ABC* at *B*.

6.127 Solve Prob. 6.126 when *θ* = 0.

6.128 For the system and loading shown, determine (a) the force **P** required for equilibrium, (b) the corresponding force in member BD, (c) the corresponding reaction at C.

6.129 The Whitworth mechanism shown is used to produce a quick-return motion of point D. The block at B is pinned to the crank AB and is free to slide in a slot cut in member CD. Determine the couple **M** that must be applied to the crank AB to hold the mechanism in equilibrium when (a) $\alpha = 0$, (b) $\alpha = 30°$.

Fig. P6.128

Fig. P6.129

6.130 Solve Prob. 6.129 when (a) $\alpha = 60°$, (b) $\alpha = 90°$.

6.131 A couple **M** of magnitude 1.5 kN · m is applied to the crank of the engine system shown. For each of the two positions shown, determine the force **P** required to hold the system in equilibrium.

(a)   (b)

Fig. P6.131 and P6.132

6.132 A force **P** of magnitude 16 kN is applied to the piston of the engine system shown. For each of the two positions shown, determine the couple **M** required to hold the system in equilibrium.

6.133 The pin at B is attached to member ABC and can slide freely along the slot cut in the fixed plate. Neglecting the effect of friction, determine the couple **M** required to hold the system in equilibrium when $\theta = 30°$.

6.134 The pin at B is attached to member ABC and can slide freely along the slot cut in the fixed plate. Neglecting the effect of friction, determine the couple **M** required to hold the system in equilibrium when $\theta = 60°$.

Fig. P6.133 and P6.134

Fig. P6.135

6.135 and 6.136 Rod CD is attached to the collar D and passes through a collar welded to end B of lever AB. Neglecting the effect of friction, determine the couple $\mathbf{M}$ required to hold the system in equilibrium when $\theta = 30°$.

Fig. P6.136

6.137 and 6.138 Two rods are connected by a frictionless collar B. Knowing that the magnitude of the couple $\mathbf{M}_A$ is 500 lb · in., determine (a) the couple $\mathbf{M}_C$ required for equilibrium, (b) the corresponding components of the reaction at C.

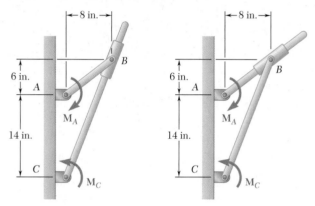

Fig. P6.137   **Fig. P6.138**

6.139 Two hydraulic cylinders control the position of the robotic arm ABC. Knowing that in the position shown the cylinders are parallel, determine the force exerted by each cylinder when $P = 160$ N and $Q = 80$ N.

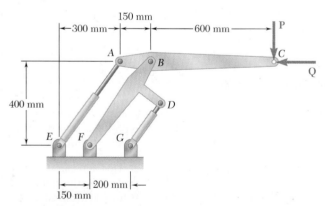

Fig. P6.139 and P6.140

6.140 Two hydraulic cylinders control the position of the robotic arm ABC. In the position shown, the cylinders are parallel and both are in tension. Knowing that $F_{AE} = 600$ N and $F_{DG} = 50$ N, determine the forces $\mathbf{P}$ and $\mathbf{Q}$ applied at C to arm ABC.

6.141 The tongs shown are used to apply a total upward force of 45 kN on a pipe cap. Determine the forces exerted at D and F on tong ADF.

Fig. P6.141

6.142 If the toggle shown is added to the tongs of Prob. 6.141 and a single vertical force is applied at G, determine the forces exerted at D and F on tong ADF.

Fig. P6.142

6.143 A small barrel weighing 60 lb is lifted by a pair of tongs as shown. Knowing that $a = 5$ in., determine the forces exerted at B and D on tong ABD.

6.144 A 39-ft length of railroad rail of weight 44 lb/ft is lifted by the tongs shown. Determine the forces exerted at D and F on tong BDF.

Fig. P6.143

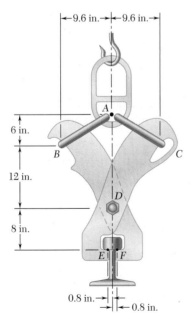

Fig. P6.144

6.145 Determine the magnitude of the gripping forces produced when two 300-N forces are applied as shown.

Fig. P6.145

6.146 The compound-lever pruning shears shown can be adjusted by placing pin A at various ratchet positions on blade ACE. Knowing that 300-lb vertical forces are required to complete the pruning of a small branch, determine the magnitude P of the forces that must be applied to the handles when the shears are adjusted as shown.

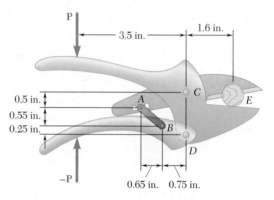

Fig. P6.146

6.147 The pliers shown are used to grip a 0.3-in.-diameter rod. Knowing that two 60-lb forces are applied to the handles, determine (a) the magnitude of the forces exerted on the rod, (b) the force exerted by the pin at A on portion AB of the pliers.

Fig. P6.147

6.148 In using the bolt cutter shown, a worker applies two 300-N forces to the handles. Determine the magnitude of the forces exerted by the cutter on the bolt.

Fig. P6.148

6.149 The specialized plumbing wrench shown is used in confined areas (e.g., under a basin or sink). It consists essentially of a jaw BC pinned at B to a long rod. Knowing that the forces exerted on the nut are equivalent to a clockwise (when viewed from above) couple of magnitude 135 lb · in., determine (a) the magnitude of the force exerted by pin B on jaw BC, (b) the couple $\mathbf{M}_0$ that is applied to the wrench.

6.150 and 6.151 Determine the force $\mathbf{P}$ that must be applied to the toggle CDE to maintain bracket ABC in the position shown.

Fig. P6.149

Fig. P6.150

Fig. P6.151

6.152 A 45-lb shelf is held horizontally by a self-locking brace that consists of two parts *EDC* and *CDB* hinged at *C* and bearing against each other at *D*. Determine the force **P** required to release the brace.

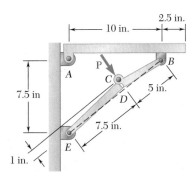

Fig. P6.152

6.153 The telescoping arm *ABC* is used to provide an elevated platform for construction workers. The workers and the platform together have a mass of 200 kg and have a combined center of gravity located directly above *C*. For the position when $\theta = 20°$, determine (*a*) the force exerted at *B* by the single hydraulic cylinder *BD*, (*b*) the force exerted on the supporting carriage at *A*.

6.154 The telescoping arm *ABC* of Prob. 6.153 can be lowered until end *C* is close to the ground, so that workers can easily board the platform. For the position when $\theta = -20°$, determine (*a*) the force exerted at *B* by the single hydraulic cylinder *BD*, (*b*) the force exerted on the supporting carriage at *A*.

6.155 The bucket of the front-end loader shown carries a 3200-lb load. The motion of the bucket is controlled by two identical mechanisms, only one of which is shown. Knowing that the mechanism shown supports one-half of the 3200-lb load, determine the force exerted (*a*) by cylinder *CD*, (*b*) by cylinder *FH*.

Fig. P6.153

Dimensions in inches

Fig. P6.155

6.156 The motion of the bucket of the front-end loader shown is controlled by two arms and a linkage that are pin-connected at D. The arms are located symmetrically with respect to the central, vertical, and longitudinal plane of the loader; one arm AFJ and its control cylinder EF are shown. The single linkage $GHDB$ and its control cylinder BC are located in the plane of symmetry. For the position and loading shown, determine the force exerted (a) by cylinder BC, (b) by cylinder EF.

Fig. P6.156

6.157 The motion of the backhoe bucket shown is controlled by the hydraulic cylinders AD, CG, and EF. As a result of an attempt to dislodge a portion of a slab, a 2-kip force **P** is exerted on the bucket teeth at J. Knowing that $\theta = 45°$, determine the force exerted by each cylinder.

Fig. P6.157

6.158 Solve Prob. 6.157 assuming that the 2-kip force **P** acts horizontally to the right ($\theta = 0$).

6.159 In the planetary gear system shown, the radius of the central gear A is $a = 18$ mm, the radius of each planetary gear is b, and the radius of the outer gear E is $(a + 2b)$. A clockwise couple of magnitude $M_A = 10$ N $\cdot$ m is applied to the central gear A and a counterclockwise couple of magnitude $M_S = 50$ N $\cdot$ m is applied to the spider BCD. If the system is to be in equilibrium, determine (a) the required radius b of the planetary gears, (b) the magnitude M_E of the couple that must be applied to the outer gear E.

Fig. P6.159

6.160 The gears D and G are rigidly attached to shafts that are held by frictionless bearings. If $r_D = 90$ mm and $r_G = 30$ mm, determine (a) the couple $\mathbf{M}_0$ that must be applied for equilibrium, (b) the reactions at A and B.

Fig. P6.160

*6.161 Two shafts AC and CF, which lie in the vertical xy plane, are connected by a universal joint at C. The bearings at B and D do not exert any axial force. A couple of magnitude 500 lb · in. (clockwise when viewed from the positive x axis) is applied to shaft CF at F. At a time when the arm of the crosspiece attached to shaft CF is horizontal, determine (a) the magnitude of the couple that must be applied to shaft AC at A to maintain equilibrium, (b) the reactions at B, D, and E. (*Hint:* The sum of the couples exerted on the crosspiece must be zero.)

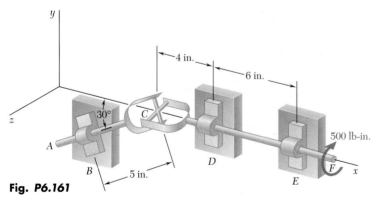

Fig. P6.161

*6.162 Solve Prob. 6.161 assuming that the arm of the crosspiece attached to shaft CF is vertical.

*6.163 The large mechanical tongs shown are used to grab and lift a thick 7500-kg steel slab HJ. Knowing that slipping does not occur between the tong grips and the slab at H and J, determine the components of all forces acting on member EFH. (*Hint:* Consider the symmetry of the tongs to establish relationships between the components of the force acting at E on EFH and the components of the force acting at D on DGJ.)

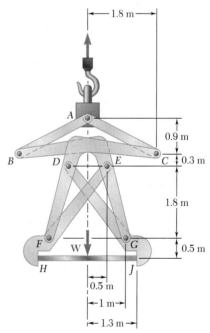

Fig. P6.163

REVIEW AND SUMMARY

In this chapter you learned to determine the *internal forces* holding together the various parts of a structure.

Analysis of trusses

The first half of the chapter was devoted to the analysis of *trusses*, i.e., to the analysis of structures consisting of *straight members connected at their extremities only*. The members being slender and unable to support lateral loads, all the loads must be applied at the joints; a truss may thus be assumed to consist of *pins and two-force members* [Sec. 6.2].

Simple trusses

A truss is said to be *rigid* if it is designed in such a way that it will not greatly deform or collapse under a small load. A triangular truss consisting of three members connected at three joints is clearly a rigid truss (Fig. 6.25a) and so will be the truss obtained by adding two new members to the first one and connecting them at a new joint (Fig. 6.25b). Trusses obtained by repeating this procedure are called *simple trusses*. We may check that in a simple truss the total number of members is $m = 2n - 3$, where n is the total number of joints [Sec. 6.3].

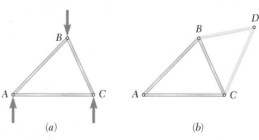

(a)   (b)

Fig. 6.25

Method of joints

The forces in the various members of a simple truss can be determined by the *method of joints* [Sec. 6.4]. First, the reactions at the supports can be obtained by considering the entire truss as a free body. The free-body diagram of each pin is then drawn, showing the forces exerted on the pin by the members or supports it connects. Since the members are straight two-force members, the force exerted by a member on the pin is directed along that member, and only the magnitude of the force is unknown. It is always possible in the case of a simple truss to draw the free-body diagrams of the pins in such an order that only two unknown forces are included in each diagram. These forces can be obtained from the corresponding two equilibrium equations or—if only three forces are involved—from the corresponding force triangle. If the force exerted by a member on a pin is directed toward that pin, the member is in *compression*;

if it is directed away from the pin, the member is in *tension* [Sample Prob. 6.1]. The analysis of a truss is sometimes expedited by first recognizing *joints under special loading conditions* [Sec. 6.5]. The method of joints can also be extended to the analysis of three-dimensional or *space trusses* [Sec. 6.6].

The *method of sections* is usually preferred to the method of joints when the force in only one member—or very few members—of a truss is desired [Sec. 6.7]. To determine the force in member *BD* of the truss of Fig. 6.26a, for example, we *pass a section* through members *BD*, *BE*, and *CE*, remove these members, and use the portion *ABC* of the truss as a free body (Fig. 6.26b). Writing $\Sigma M_E = 0$, we determine the magnitude of the force $\mathbf{F}_{BD}$, which represents the force in member *BD*. A positive sign indicates that the member is in *tension;* a negative sign indicates that it is in *compression* [Sample Probs. 6.2 and 6.3].

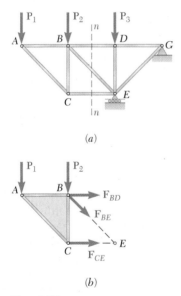

Fig. 6.26

The method of sections is particularly useful in the analysis of *compound trusses*, i.e., trusses which cannot be constructed from the basic triangular truss of Fig. 6.25a but which can be obtained by rigidly connecting several simple trusses [Sec. 6.8]. If the component trusses have been properly connected (e.g., one pin and one link, or three nonconcurrent and nonparallel links) and if the resulting structure is properly supported (e.g., one pin and one roller), the compound truss is *statically determinate, rigid, and completely constrained*. The following necessary—but not sufficient—condition is then satisfied: $m + r = 2n$, where m is the number of members, r is the number of unknowns representing the reactions at the supports, and n is the number of joints.

Frames and machines

The second part of the chapter was devoted to the analysis of *frames and machines.* Frames and machines are structures which contain *multiforce members,* i.e., members acted upon by three or more forces. Frames are designed to support loads and are usually stationary, fully constrained structures. Machines are designed to transmit or modify forces and always contain moving parts [Sec. 6.9].

Analysis of a frame

To *analyze a frame,* we first consider the *entire frame as a free body* and write three equilibrium equations [Sec. 6.10]. If the frame remains rigid when detached from its supports, the reactions involve only three unknowns and may be determined from these equations [Sample Probs. 6.4 and 6.5]. On the other hand, if the frame ceases to be rigid when detached from its supports, the reactions involve more than three unknowns and cannot be completely determined from the equilibrium equations of the frame [Sec. 6.11; Sample Prob. 6.6].

Multiforce members

We then *dismember the frame* and identify the various members as either two-force members or multiforce members; pins are assumed to form an integral part of one of the members they connect. We draw the free-body diagram of each of the multiforce members, noting that when two multiforce members are connected to the same two-force member, they are acted upon by that member with *equal and opposite forces of unknown magnitude but known direction.* When two multiforce members are connected by a pin, they exert on each other *equal and opposite forces of unknown direction,* which should be represented by *two unknown components.* The equilibrium equations obtained from the free-body diagrams of the multiforce members can then be solved for the various internal forces [Sample Probs. 6.4 and 6.5]. The equilibrium equations can also be used to complete the determination of the reactions at the supports [Sample Prob. 6.6]. Actually, if the frame is *statically determinate and rigid,* the free-body diagrams of the multiforce members could provide as many equations as there are unknown forces (including the reactions) [Sec. 6.11]. However, as suggested above, it is advisable to first consider the free-body diagram of the entire frame to minimize the number of equations that must be solved simultaneously.

Analysis of a machine

To *analyze a machine,* we dismember it and, following the same procedure as for a frame, draw the free-body diagram of each of the multiforce members. The corresponding equilibrium equations yield the *output forces* exerted by the machine in terms of the *input forces* applied to it, as well as the *internal forces* at the various connections [Sec. 6.12; Sample Prob. 6.7].

REVIEW PROBLEMS

6.164 Using the method of joints, determine the force in each member of the truss shown. State whether each member is in tension or compression.

6.165 Using the method of joints, determine the force in each member of the roof truss shown. State whether each member is in tension or compression.

Fig. P6.164

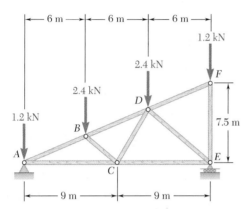

Fig. P6.165

6.166 A Howe scissors roof truss is loaded as shown. Determine the force in members *DF*, *DG*, and *EG*.

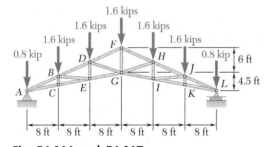

Fig. P6.166 and P6.167

6.167 A Howe scissors roof truss is loaded as shown. Determine the force in members *GI*, *HI*, and *HJ*.

6.168 Rod *CD* is fitted with a collar at *D* that can be moved along rod *AB*, which is bent in the shape of an arc of circle. For the position when $\theta = 30°$, determine (*a*) the force in rod *CD*, (*b*) the reaction at *B*.

Fig. P6.168

347

Fig. P6.169

6.169 For the frame and loading shown, determine the components of all forces acting on member ABC.

6.170 Knowing that each pulley has a radius of 250 mm, determine the components of the reactions at D and E.

Fig. P6.170

6.171 For the frame and loading shown, determine the components of the forces acting on member $DABC$ at B and D.

Fig. P6.172

Fig. P6.171

6.172 For the frame and loading shown, determine (a) the reaction at C, (b) the force in member AD.

6.173 The control rod CE passes through a horizontal hole in the body of the toggle system shown. Knowing that link BD is 250 mm long, determine the force **Q** required to hold the system in equilibrium when $\beta = 20°$.

Fig. P6.173

6.174 Determine the magnitude of the gripping forces exerted along line *aa* on the nut when two 50-lb forces are applied to the handles as shown. Assume that pins *A* and *D* slide freely in slots cut in the jaws.

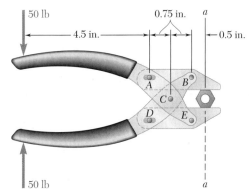

Fig. P6.174

6.175 Knowing that the frame shown has a sag at *B* of *a* = 1 in., determine the force **P** required to maintain equilibrium in the position shown.

Fig. P6.175

COMPUTER PROBLEMS

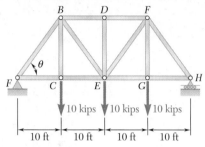

Fig. P6.C1

6.C1 A Pratt steel truss is to be designed to support three 10-kip loads as shown. The length of the truss is to be 40 ft. The height of the truss and thus the angle θ, as well as the cross-sectional areas of the various members, are to be selected to obtain the most economical design. Specifically, the cross-sectional area of each member is to be chosen so that the stress (force divided by area) in that member is equal to 20 kips/in^2, the allowable stress for the steel used; the total weight of the steel, and thus its cost, must be as small as possible. (*a*) Knowing that the specific weight of the steel used is 0.284 lb/in^3, write a computer program that can be used to calculate the weight of the truss and the cross-sectional area of each load-bearing member located to the left of *DE* for values of θ from 20° to 80° using 5° increments. (*b*) Using appropriate smaller increments, determine the optimum value of θ and the corresponding values of the weight of the truss and of the cross-sectional areas of the various members. Ignore the weight of any zero-force member in your computations.

6.C2 The floor of a bridge will rest on stringers that will be simply supported by transverse floor beams, as in Fig. 6.3. The ends of the beams will be connected to the upper joints of two trusses, one of which is shown in Fig. P6.C2. As part of the design of the bridge, it is desired to simulate the effect on this truss of driving a 12-kN truck over the bridge. Knowing that the distance between the truck's axles is $b = 2.25$ m and assuming that the weight of the truck is equally distributed over its four wheels, write a computer program that can be used to calculate the forces created by the truck in members *BH* and *GH* for values of *x* from 0 to 17.25 m using 0.75-m increments. From the results obtained, determine (*a*) the maximum tensile force in *BH*, (*b*) the maximum compressive force in *BH*, (*c*) the maximum tensile force in *GH*. Indicate in each case the corresponding value of *x*. (*Note:* The increments have been selected so that the desired values are among those that will be tabulated.)

Fig. P6.C2

6.C3 In the mechanism shown the position of boom AC is controlled by arm BD. For the loading shown, write a computer program and use it to determine the couple $\mathbf{M}$ required to hold the system in equilibrium for values of θ from $-30°$ to $90°$ using $10°$ increments. Also, for the same values of θ, determine the reaction at A. As a part of the design process of the mechanism, use appropriate smaller increments and determine (a) the value of θ for which M is maximum and the corresponding value of M, (b) the value of θ for which the reaction at A is maximum and the corresponding magnitude of this reaction.

6.C4 The design of a robotic system calls for the two-rod mechanism shown. Rods AC and BD are connected by a slider block D as shown. Neglecting the effect of friction, write a computer program and use it to determine the couple $\mathbf{M}_A$ required to hold the rods in equilibrium for values of θ from 0 to $120°$ using $10°$ increments. For the same values of θ, determine the magnitude of the force $\mathbf{F}$ exerted by rod AC on the slider block.

6.C5 The compound-lever pruning shears shown can be adjusted by placing pin A at various ratchet positions on blade ACE. Knowing that the length AB is 0.85 in., write a computer program and use it to determine the magnitude of the vertical forces applied to the small branch for values of d from 0.4 in. to 0.6 in. using 0.025-in. increments. As a part of the design of the shears, use appropriate smaller increments and determine the smallest allowable value of d if the force in link AB is not to exceed 500 lb.

Fig. P6.C3

Fig. P6.C4

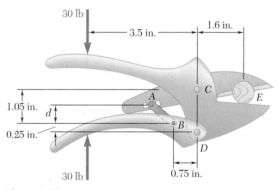

Fig. P6.C5

6.C6 Rod CD is attached to collar D and passes through a collar welded to end B of lever AB. As an initial step in the design of lever AB, write a computer program and use it to calculate the magnitude M of the couple required to hold the system in equilibrium for values of θ from $15°$ to $90°$ using $5°$ increments. Using appropriate smaller increments, determine the value of θ for which M is minimum and the corresponding value of M.

Fig. P6.C6

The George Washington Bridge connects Manhattan, New York, and Fort Lee, New Jersey. This suspension bridge carries traffic on two levels over roadways that are supported by a system of beams. Trusses are used both to connect these roadways to complete the overall bridge span as well as to form the towers. The bridge span itself is supported by the cable system.

Forces in Beams and Cables

*7.1 INTRODUCTION

In preceding chapters, two basic problems involving structures were considered: (1) determining the external forces acting on a structure (Chap. 4) and (2) determining the forces which hold together the various members forming a structure (Chap. 6). The problem of determining the internal forces which hold together the various parts of a given member will now be considered.

We will first analyze the internal forces in the members of a frame, such as the crane considered in Secs. 6.1 and 6.10, noting that whereas the internal forces in a straight two-force member can produce only *tension* or *compression* in that member, the internal forces in any other type of member usually produce *shear* and *bending* as well.

Most of this chapter will be devoted to the analysis of the internal forces in two important types of engineering structures, namely,

1. *Beams,* which are usually long, straight prismatic members designed to support loads applied at various points along the member.
2. *Cables,* which are flexible members capable of withstanding only tension, designed to support either concentrated or distributed loads. Cables are used in many engineering applications, such as suspension bridges and transmission lines.

*7.2 INTERNAL FORCES IN MEMBERS

Let us first consider a *straight two-force member AB* (Fig. 7.1a). From Sec. 4.6, we know that the forces **F** and −**F** acting at *A* and *B*, respectively, must be directed along *AB* in opposite sense and have the same magnitude *F*. Now, let us cut the member at *C*. To maintain the equilibrium of the free bodies *AC* and *CB* thus obtained, we must apply to *AC* a force −**F** equal and opposite to **F**, and to *CB* a force **F** equal and opposite to −**F** (Fig. 7.1b). These new forces are directed along *AB* in opposite sense and have the same magnitude *F*. Since the two parts *AC* and *CB* were in equilibrium before the member was cut, *internal forces* equivalent to these new forces must have existed in the member itself. We conclude that in the case of a straight two-force member, the internal forces that the two portions of the member exert on each other are equivalent to *axial forces*. The common magnitude *F* of these forces does not depend upon the location of the section *C* and is referred to as the *force in member AB*. In the case considered, the member is in tension and will elongate under the action of the internal forces. In the case represented in Fig. 7.2, the member is in compression and will decrease in length under the action of the internal forces.

Next, let us consider a *multiforce member*. Take, for instance, member *AD* of the crane analyzed in Sec. 6.10. This crane is shown again in Fig. 7.3a, and the free-body diagram of member *AD* is drawn in Fig. 7.3b. We now cut member *AD* at *J* and draw a free-body diagram for each of the portions *JD* and *AJ* of the member

Fig. 7.1

Fig. 7.2

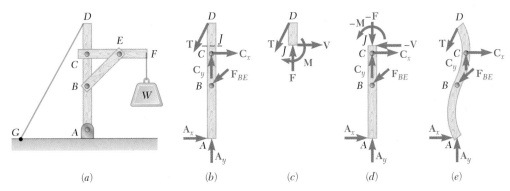

Fig. 7.3

(Fig. 7.3*c* and *d*). Considering the free body *JD*, we find that its equilibrium will be maintained if we apply at *J* a force **F** to balance the vertical component of **T**, a force **V** to balance the horizontal component of **T**, and a couple **M** to balance the moment of **T** about *J*. Again we conclude that internal forces must have existed at *J* before member *AD* was cut. The internal forces acting on the portion *JD* of member *AD* are equivalent to the force-couple system shown in Fig. 7.3*c*. According to Newton's third law, the internal forces acting on *AJ* must be equivalent to an equal and opposite force-couple system, as shown in Fig. 7.3*d*. It is clear that the action of the internal forces in member *AD* *is not limited to producing tension or compression* as in the case of straight two-force members; the internal forces *also produce shear and bending*. The force **F** is an *axial force;* the force **V** is called a *shearing force;* and the moment **M** of the couple is known as the *bending moment at J.* We note that when determining internal forces in a member, we should clearly indicate on which portion of the member the forces are supposed to act. The deformation which will occur in member *AD* is sketched in Fig. 7.3*e*. The actual analysis of such a deformation is part of the study of mechanics of materials.

It should be noted that in a *two-force member which is not straight*, the internal forces are also equivalent to a force-couple system. This is shown in Fig. 7.4, where the two-force member *ABC* has been cut at *D*.

Photo 7.1 The design of the shaft of a circular saw must account for the internal forces resulting from the forces applied to the teeth of the blade. At a given point in the shaft, these internal forces are equivalent to a force-couple system consisting of axial and shearing forces and a couple representing the bending and torsional moments.

Fig. 7.4

SAMPLE PROBLEM 7.1

In the frame shown, determine the internal forces (a) in member *ACF* at point *J*, (b) in member *BCD* at point *K*. This frame has been previously considered in Sample Prob. 6.5.

SOLUTION

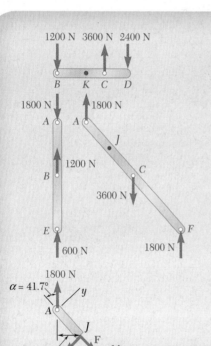

Reactions and Forces at Connections. The reactions and the forces acting on each member of the frame are determined; this has been previously done in Sample Prob. 6.5, and the results are repeated here.

a. Internal Forces at J. Member *ACF* is cut at point *J*, and the two parts shown are obtained. The internal forces at *J* are represented by an equivalent force-couple system and can be determined by considering the equilibrium of either part. Considering the *free body AJ*, we write

$$+\gamma \; \Sigma M_J = 0: \qquad -(1800 \text{ N})(1.2 \text{ m}) + M = 0$$
$$M = +2160 \text{ N} \cdot \text{m} \qquad \mathbf{M} = 2160 \text{ N} \cdot \text{m} \; \gamma \; \blacktriangleleft$$
$$+\searrow \; \Sigma F_x = 0: \qquad F - (1800 \text{ N}) \cos 41.7° = 0$$
$$F = +1344 \text{ N} \qquad \mathbf{F} = 1344 \text{ N} \searrow \; \blacktriangleleft$$
$$+\nearrow \; \Sigma F_y = 0: \qquad -V + (1800 \text{ N}) \sin 41.7° = 0$$
$$V = +1197 \text{ N} \qquad \mathbf{V} = 1197 \text{ N} \swarrow \; \blacktriangleleft$$

The internal forces at *J* are therefore equivalent to a couple **M**, an axial force **F**, and a shearing force **V**. The internal force-couple system acting on part *JCF* is equal and opposite.

b. Internal Forces at K. We cut member *BCD* at *K* and obtain the two parts shown. Considering the *free body BK*, we write

$$+\gamma \; \Sigma M_K = 0: \qquad (1200 \text{ N})(1.5 \text{ m}) + M = 0$$
$$M = -1800 \text{ N} \cdot \text{m} \qquad \mathbf{M} = 1800 \text{ N} \cdot \text{m} \; \downdownarrows \; \blacktriangleleft$$
$$\xrightarrow{+} \Sigma F_x = 0: \qquad F = 0 \qquad \mathbf{F} = 0 \; \blacktriangleleft$$
$$+\uparrow \Sigma F_y = 0: \qquad -1200 \text{ N} - V = 0$$
$$V = -1200 \text{ N} \qquad \mathbf{V} = 1200 \text{ N} \uparrow \; \blacktriangleleft$$

SOLVING PROBLEMS
ON YOUR OWN

In this lesson you learned to determine the internal forces in the member of a frame. The internal forces at a given point in a *straight two-force member* reduce to an axial force, but in all other cases, they are equivalent to a *force-couple system* consisting of an *axial force* **F**, a *shearing force* **V**, and a couple **M** representing the *bending moment* at that point.

To determine the internal forces at a given point J of the member of a frame, you should take the following steps.

1. Draw a free-body diagram of the entire frame, and use it to determine as many of the reactions at the supports as you can.

2. Dismember the frame, and draw a free-body diagram of each of its members. Write as many equilibrium equations as are necessary to find all the forces acting on the member on which point J is located.

3. Cut the member at point J, and draw a free-body diagram of each of the two portions of the member that you have obtained, applying to each portion at point J the force components and couple representing the internal forces exerted by the other portion. Note that these force components and couples are equal in magnitude and opposite in sense.

4. Select one of the two free-body diagrams you have drawn and use it to write three equilibrium equations for the corresponding portion of member.
   a. Summing moments about J and equating them to zero will yield the bending moment at point J.
   b. Summing components in directions parallel and perpendicular to the member at J and equating them to zero will yield, respectively, the axial and shearing force.

5. When recording your answers, be sure to specify the portion of the member you have used, since the forces and couples acting on the two portions have opposite senses.

Since the solutions of the problems in this lesson require the determination of the forces exerted on each other by the various members of a frame, be sure to review the methods used in Chap. 6 to solve this type of problem. When frames involve pulleys and cables, for instance, remember that the forces exerted by a pulley on the member of the frame to which it is attached have the same magnitude and direction as the forces exerted by the cable on the pulley [Prob. 6.90].

PROBLEMS

7.1 and 7.2 Determine the internal forces (axial force, shearing force, and bending moment) at point J of the structure indicated.

    7.1 Frame and loading of Prob. 6.75

    7.2 Frame and loading of Prob. 6.78

7.3 Determine the internal forces at point J when $\alpha = 90°$.

Fig. P7.3 and P7.4

7.4 Determine the internal forces at point J when $\alpha = 0$.

7.5 and 7.6 Knowing that the turnbuckle has been tightened until the tension in wire AD is 850 N, determine the internal forces at the point indicated:

    7.5 Point J

    7.6 Point K

7.7 Two members, each consisting of a straight and a quarter-circular portion of rod, are connected as shown and support a 75-lb load at A. Determine the internal forces at point J.

Fig. P7.5 and P7.6

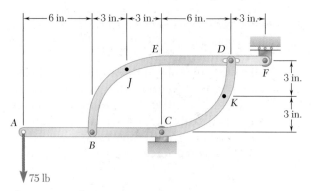

Fig. P7.7 and P7.8

7.8 Two members, each consisting of a straight and a quarter-circular portion of rod, are connected as shown and support a 75-lb load at A. Determine the internal forces at point K.

7.9 A semicircular rod is loaded as shown. Determine the internal forces at point *J*.

7.10 A semicircular rod is loaded as shown. Determine the internal forces at point *K*.

7.11 A semicircular rod is loaded as shown. Determine the internal forces at point *J* knowing that $\theta = 30°$.

Fig. P7.9 and P7.10

Fig. P7.11 and P7.12

7.12 A semicircular rod is loaded as shown. Determine the magnitude and location of the maximum bending moment in the rod.

7.13 The axis of the curved member *AB* is a parabola with vertex at *A*. If a vertical load **P** of magnitude 450 lb is applied at *A*, determine the internal forces at *J* when $h = 12$ in., $L = 40$ in., and $a = 24$ in.

7.14 Knowing that the axis of the curved member *AB* is a parabola with vertex at *A*, determine the magnitude and location of the maximum bending moment.

7.15 Knowing that the radius of each pulley is 200 mm and neglecting friction, determine the internal forces at point *J* of the frame shown.

Fig. P7.13 and P7.14

Fig. P7.15 and P7.16

7.16 Knowing that the radius of each pulley is 200 mm and neglecting friction, determine the internal forces at point *K* of the frame shown.

7.17 A 5-in.-diameter pipe is supported every 9 ft by a small frame consisting of two members as shown. Knowing that the combined weight of the pipe and its contents is 10 lb/ft and neglecting the effect of friction, determine the magnitude and location of the maximum bending moment in member AC.

Fig. P7.17

7.18 For the frame of Prob. 7.17, determine the magnitude and location of the maximum bending moment in member BC.

7.19 Knowing that the radius of each pulley is 150 mm, that $\alpha = 20°$, and neglecting friction, determine the internal forces at (a) point J, (b) point K.

7.20 Knowing that the radius of each pulley is 150 mm, that $\alpha = 30°$, and neglecting friction, determine the internal forces at (a) point J, (b) point K.

7.21 and 7.22 A force **P** is applied to a bent rod that is supported by a roller and a pin and bracket. For each of the three cases shown, determine the internal forces at point J.

Fig. P7.19 and P7.20

Fig. P7.21

Fig. P7.22

7.23 and 7.24   A quarter-circular rod of weight W and uniform cross section is supported as shown. Determine the bending moment at point J when $\theta = 30°$.

Fig. P7.23   **Fig. P7.24**

7.25   For the rod of Prob. 7.23, determine the magnitude and location of the maximum bending moment.

7.26   For the rod of Prob. 7.24, determine the magnitude and location of the maximum bending moment.

7.27 and 7.28   A half section of pipe rests on a frictionless horizontal surface as shown. If the half section of pipe has a mass of 9 kg and a diameter of 300 mm, determine the bending moment at point J when $\theta = 90°$.

Fig. P7.27   **Fig. P7.28**

BEAMS

*7.3   VARIOUS TYPES OF LOADING AND SUPPORT

A structural member designed to support loads applied at various points along the member is known as a *beam*. In most cases, the loads are perpendicular to the axis of the beam and will cause only shear and bending in the beam. When the loads are not at a right angle to the beam, they will also produce axial forces in the beam.

Beams are usually long, straight prismatic bars. Designing a beam for the most effective support of the applied loads is a two-part

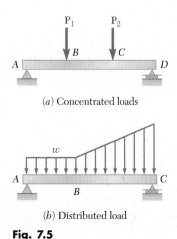

(a) Concentrated loads

(b) Distributed load

Fig. 7.5

process: (1) determining the shearing forces and bending moments produced by the loads and (2) selecting the cross section best suited to resist the shearing forces and bending moments determined in the first part. Here we are concerned with the first part of the problem of beam design. The second part belongs to the study of mechanics of materials.

A beam can be subjected to *concentrated loads* $\mathbf{P}_1$, $\mathbf{P}_2$, . . ., expressed in newtons, pounds, or their multiples kilonewtons and kips (Fig. 7.5a), to a *distributed load w*, expressed in N/m, kN/m, lb/ft, or kips/ft (Fig. 7.5b), or to a combination of both. When the load w per unit length has a constant value over part of the beam (as between A and B in Fig. 7.5b), the load is said to be *uniformly distributed* over that part of the beam. The determination of the reactions at the supports is considerably simplified if distributed loads are replaced by equivalent concentrated loads, as explained in Sec. 5.8. This substitution, however, should not be performed, or at least should be performed with care, when internal forces are being computed (see Sample Prob. 7.3).

Beams are classified according to the way in which they are supported. Several types of beams frequently used are shown in Fig. 7.6. The distance L between supports is called the *span*. It should be noted that the reactions will be determinate if the supports involve only three unknowns. If more unknowns are involved, the reactions will be statically indeterminate and the methods of statics will not be sufficient to determine the reactions; the properties of the beam with regard to its resistance to bending must then be taken into consideration. Beams supported by two rollers are not shown here; they are only partially constrained and will move under certain loadings.

Sometimes two or more beams are connected by hinges to form a single continuous structure. Two examples of beams hinged at a point H are shown in Fig. 7.7. It will be noted that the reactions at the supports involve four unknowns and cannot be

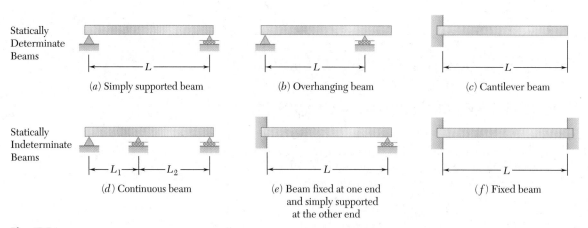

Statically Determinate Beams

(a) Simply supported beam   (b) Overhanging beam   (c) Cantilever beam

Statically Indeterminate Beams

(d) Continuous beam   (e) Beam fixed at one end and simply supported at the other end   (f) Fixed beam

Fig. 7.6

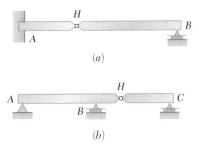

(a)

(b)

Fig. 7.7

determined from the free-body diagram of the two-beam system. They can be determined, however, by considering the free-body diagram of each beam separately; six unknowns are involved (including two force components at the hinge), and six equations are available.

*7.4 SHEAR AND BENDING MOMENT IN A BEAM

Consider a beam AB subjected to various concentrated and distributed loads (Fig. 7.8a). We propose to determine the shearing force and bending moment at any point of the beam. In the example considered here, the beam is simply supported, but the method used could be applied to any type of statically determinate beam.

First we determine the reactions at A and B by choosing the entire beam as a free body (Fig. 7.8b); writing $\Sigma M_A = 0$ and $\Sigma M_B = 0$, we obtain, respectively, $\mathbf{R}_B$ and $\mathbf{R}_A$.

Photo 7.2 The internal forces in the beams of the overpass shown vary as the truck crosses the overpass.

Fig. 7.8

To determine the internal forces at C, we cut the beam at C and draw the free-body diagrams of the portions AC and CB of the beam (Fig. 7.8c). Using the free-body diagram of AC, we can determine the shearing force **V** at C by equating to zero the sum of the vertical components of all forces acting on AC. Similarly, the bending moment **M** at C can be found by equating to zero the sum of the moments about C of all forces and couples acting on AC. Alternatively, we could use the free-body diagram of CB† and determine the shearing force **V′** and the bending moment **M′** by equating to zero the sum of the vertical components and the sum of the moments about C of all forces and couples acting on CB. While this choice of free bodies may facilitate the computation of the numerical values of the shearing force and bending moment, it makes it necessary to indicate on which portion of the beam the internal forces considered are acting. If the shearing force and bending moment are to be computed at every point of the beam and efficiently recorded, we must find a way to avoid having to specify every time which portion of the beam is used as a free body. We shall adopt, therefore, the following conventions:

In determining the shearing force in a beam, *it will always be assumed* that the internal forces **V** and **V′** are directed as shown in Fig. 7.8c. A positive value obtained for their common magnitude V will indicate that this assumption was correct and that the shearing forces are actually directed as shown. A negative value obtained for V will indicate that the assumption was wrong and that the shearing forces are directed in the opposite way. Thus, only the magnitude V, together with a plus or minus sign, needs to be recorded to define completely the shearing forces at a given point of the beam. The scalar V is commonly referred to as the *shear* at the given point of the beam.

Similarly, *it will always be assumed* that the internal couples **M** and **M′** are directed as shown in Fig. 7.8c. A positive value obtained for their magnitude M, commonly referred to as the bending moment, will indicate that this assumption was correct, and a negative value will indicate that it was wrong. Summarizing the sign conventions we have presented, we state:

The shear V and the bending moment M at a given point of a beam are said to be positive when the internal forces and couples acting on each portion of the beam are directed as shown in Fig. 7.9a.

These conventions can be more easily remembered if we note that:

1. *The shear at C is positive when the* **external** *forces (loads and reactions) acting on the beam tend to shear off the beam at C as indicated in Fig. 7.9b.*
2. *The bending moment at C is positive when the* **external** *forces acting on the beam tend to bend the beam at C as indicated in Fig. 7.9c.*

(a) Internal forces at section
(positive shear and positive bending moment)

(b) Effect of external forces
(positive shear)

(c) Effect of external forces
(positive bending moment)

Fig. 7.9

†The force and couple representing the internal forces acting on CB will now be denoted by **V′** and **M′**, rather than by −**V** and −**M** as done earlier, in order to avoid confusion when applying the sign convention which we are about to introduce.

It may also help to note that the situation described in Fig. 7.9, in which the values of the shear and of the bending moment are positive, is precisely the situation which occurs in the left half of a simply supported beam carrying a single concentrated load at its midpoint. This particular example is fully discussed in the following section.

*7.5 SHEAR AND BENDING-MOMENT DIAGRAMS

Now that shear and bending moment have been clearly defined in sense as well as in magnitude, we can easily record their values at any point of a beam by plotting these values against the distance x measured from one end of the beam. The graphs obtained in this way are called, respectively, the *shear diagram* and the *bending-moment diagram*. As an example, consider a simply supported beam AB of span L subjected to a single concentrated load $\mathbf{P}$ applied at its midpoint D (Fig. 7.10a). We first determine the reactions at the supports from the free-body diagram of the entire beam (Fig. 7.10b); we find that the magnitude of each reaction is equal to $P/2$.

Next we cut the beam at a point C between A and D and draw the free-body diagrams of AC and CB (Fig. 7.10c). *Assuming that shear and bending moment are positive*, we direct the internal forces $\mathbf{V}$ and $\mathbf{V}'$ and the internal couples $\mathbf{M}$ and $\mathbf{M}'$ as indicated in Fig. 7.9a. Considering the free body AC and writing that the sum of the vertical components and the sum of the moments about C of the forces acting on the free body are zero, we find $V = +P/2$ and $M = +Px/2$. Both shear and bending moment are therefore positive; this can be checked by observing that the reaction at A tends to shear off and to bend the beam at C as indicated in Fig. 7.9b and c. We can plot V and M between A and D (Fig. 7.10e and f); the shear has a constant value $V = P/2$, while the bending moment increases linearly from $M = 0$ at $x = 0$ to $M = PL/4$ at $x = L/2$.

Cutting, now, the beam at a point E between D and B and considering the free body EB (Fig. 7.10d), we write that the sum of the vertical components and the sum of the moments about E of the forces acting on the free body are zero. We obtain $V = -P/2$ and $M = P(L - x)/2$. The shear is therefore negative and the bending moment positive; this can be checked by observing that the reaction at B bends the beam at E as indicated in Fig. 7.9c but tends to shear it off in a manner opposite to that shown in Fig. 7.9b. We can complete, now, the shear and bending-moment diagrams of Fig. 7.10e and f; the shear has a constant value $V = -P/2$ between D and B, while the bending moment decreases linearly from $M = PL/4$ at $x = L/2$ to $M = 0$ at $x = L$.

It should be noted that when a beam is subjected to concentrated loads only, the shear is of constant value between loads and the bending moment varies linearly between loads, but when a beam is subjected to distributed loads, the shear and bending moment vary quite differently (see Sample Prob. 7.3).

Fig. 7.10

SAMPLE PROBLEM 7.2

Draw the shear and bending-moment diagrams for the beam and loading shown.

SOLUTION

Free-Body: Entire Beam. From the free-body diagram of the entire beam, we find the reactions at B and D:

$$\mathbf{R}_B = 46 \text{ kN} \uparrow \qquad \mathbf{R}_D = 14 \text{ kN} \uparrow$$

Shear and Bending Moment. We first determine the internal forces just to the right of the 20-kN load at A. Considering the stub of beam to the left of section 1 as a free body and assuming V and M to be positive (according to the standard convention), we write

$$+\uparrow\Sigma F_y = 0: \qquad -20 \text{ kN} - V_1 = 0 \qquad\qquad V_1 = -20 \text{ kN}$$
$$+\curvearrowleft \Sigma M_1 = 0: \qquad (20 \text{ kN})(0 \text{ m}) + M_1 = 0 \qquad M_1 = 0$$

We next consider as a free body the portion of the beam to the left of section 2 and write

$$+\uparrow\Sigma F_y = 0: \qquad -20 \text{ kN} - V_2 = 0 \qquad\qquad V_2 = -20 \text{ kN}$$
$$+\curvearrowleft \Sigma M_2 = 0: \qquad (20 \text{ kN})(2.5 \text{ m}) + M_2 = 0 \qquad M_2 = -50 \text{ kN} \cdot \text{m}$$

The shear and bending moment at sections *3, 4, 5,* and *6* are determined in a similar way from the free-body diagrams shown. We obtain

$$
\begin{aligned}
V_3 &= +26 \text{ kN} & M_3 &= -50 \text{ kN} \cdot \text{m} \\
V_4 &= +26 \text{ kN} & M_4 &= +28 \text{ kN} \cdot \text{m} \\
V_5 &= -14 \text{ kN} & M_5 &= +28 \text{ kN} \cdot \text{m} \\
V_6 &= -14 \text{ kN} & M_6 &= 0
\end{aligned}
$$

For several of the latter sections, the results are more easily obtained by considering as a free body the portion of the beam to the right of the section. For example, considering the portion of the beam to the right of section 4, we write

$$+\uparrow\Sigma F_y = 0: \qquad V_4 - 40 \text{ kN} + 14 \text{ kN} = 0 \qquad V_4 = +26 \text{ kN}$$
$$+\curvearrowleft \Sigma M_4 = 0: \qquad -M_4 + (14 \text{ kN})(2 \text{ m}) = 0 \qquad M_4 = +28 \text{ kN} \cdot \text{m}$$

Shear and Bending-Moment Diagrams. We can now plot the six points shown on the shear and bending-moment diagrams. As indicated in Sec. 7.5, the shear is of constant value between concentrated loads, and the bending moment varies linearly; we therefore obtain the shear and bending-moment diagrams shown.

Draw the shear and bending-moment diagrams for the beam AB. The distributed load of 40 lb/in. extends over 12 in. of the beam, from A to C, and the 400-lb load is applied at E.

SOLUTION

Free-Body: Entire Beam. The reactions are determined by considering the entire beam as a free body.

$$+\uparrow \Sigma M_A = 0: \quad B_y(32 \text{ in.}) - (480 \text{ lb})(6 \text{ in.}) - (400 \text{ lb})(22 \text{ in.}) = 0$$
$$B_y = +365 \text{ lb} \qquad \mathbf{B}_y = 365 \text{ lb} \uparrow$$
$$+\uparrow \Sigma M_B = 0: \quad (480 \text{ lb})(26 \text{ in.}) + (400 \text{ lb})(10 \text{ in.}) - A(32 \text{ in.}) = 0$$
$$A = +515 \text{ lb} \qquad \mathbf{A} = 515 \text{ lb} \uparrow$$
$$\overset{+}{\rightarrow}\Sigma F_x = 0: \quad B_x = 0 \qquad \mathbf{B}_x = 0$$

The 400-lb load is now replaced by an equivalent force-couple system acting on the beam at point D.

Shear and Bending Moment. *From A to C.* We determine the internal forces at a distance x from point A by considering the portion of the beam to the left of section 1. That part of the distributed load acting on the free body is replaced by its resultant, and we write

$$+\uparrow \Sigma F_y = 0: \quad 515 - 40x - V = 0 \qquad V = 515 - 40x$$
$$+\uparrow \Sigma M_1 = 0: \quad -515x + 40x(\tfrac{1}{2}x) + M = 0 \qquad M = 515x - 20x^2$$

Since the free-body diagram shown can be used for all values of x smaller than 12 in., the expressions obtained for V and M are valid throughout the region $0 < x < 12$ in.

From C to D. Considering the portion of the beam to the left of section 2 and again replacing the distributed load by its resultant, we obtain

$$+\uparrow \Sigma F_y = 0: \quad 515 - 480 - V = 0 \quad V = 35 \text{ lb}$$
$$+\uparrow \Sigma M_2 = 0: \quad -515x + 480(x - 6) + M = 0 \qquad M = (2880 + 35x) \text{ lb} \cdot \text{in.}$$

These expressions are valid in the region $12 \text{ in.} < x < 18 \text{ in.}$

From D to B. Using the portion of the beam to the left of section 3, we obtain for the region $18 \text{ in.} < x < 32 \text{ in.}$

$$+\uparrow \Sigma F_y = 0: \quad 515 - 480 - 400 - V = 0 \quad V = -365 \text{ lb}$$
$$+\uparrow \Sigma M_3 = 0: \quad -515x + 480(x - 6) - 1600 + 400(x - 18) + M = 0$$
$$M = (11{,}680 - 365x) \text{ lb} \cdot \text{in.}$$

Shear and Bending-Moment Diagrams. The shear and bending-moment diagrams for the entire beam can now be plotted. We note that the couple of moment 1600 lb · in. applied at point D introduces a discontinuity into the bending-moment diagram.

SOLVING PROBLEMS ON YOUR OWN

In this lesson you learned to determine the shear V and the *bending moment* M at any point in a beam. You also learned to draw the *shear diagram* and the *bending-moment diagram* for the beam by plotting, respectively, V and M against the distance x measured along the beam.

A. Determining the shear and bending moment in a beam. To determine the shear V and the bending moment M at a given point C of a beam, you should take the following steps.

1. Draw a free-body diagram of the entire beam, and use it to determine the reactions at the beam supports.

2. Cut the beam at point C, and, using the original loading, select one of the two portions of the beam you have obtained.

3. Draw the free-body diagram of the portion of the beam you have selected, showing:

    a. The loads and the reaction exerted on that portion of the beam, replacing each distributed load by an equivalent concentrated load as explained earlier in Sec. 5.8.

    b. The shearing force and the bending couple representing the internal forces at C. To facilitate recording the shear V and the bending moment M after they have been determined, follow the convention indicated in Figs. 7.8 and 7.9. Thus, if you are using the portion of the beam located to the *left of C*, apply at C a *shearing force* **V** *directed downward* and a *bending couple* **M** *directed counterclockwise*. If you are using the portion of the beam located to the *right of C*, apply at C a *shearing force* **V′** *directed upward* and a *bending couple* **M′** *directed clockwise* [Sample Prob. 7.2].

4. Write the equilibrium equations for the portion of the beam you have selected. Solve the equation $\Sigma F_y = 0$ for V and the equation $\Sigma M_C = 0$ for M.

5. Record the values of V and M with the sign obtained for each of them. A positive sign for V means that the shearing forces exerted at C on each of the two portions of the beam are directed as shown in Figs. 7.8 and 7.9; a negative sign means that they have the opposite sense. Similarly, a positive sign for M means that the bending couples at C are directed as shown in these figures, and a negative sign means that they have the opposite sense. In addition, a positive sign for M means that the concavity of the beam at C is directed upward, and a negative sign means that it is directed downward.

B. Drawing the shear and bending-moment diagrams for a beam. These diagrams are obtained by plotting, respectively, V and M against the distance x measured along the beam. However, in most cases the values of V and M need to be computed only at a few points.

1. For a beam supporting only concentrated loads, we note [Sample Prob. 7.2] that

    a. The shear diagram consists of segments of horizontal lines. Thus, to draw the shear diagram of the beam you will need to compute V only just to the left or just to the right of the points where the loads or the reactions are applied.

    b. The bending-moment diagram consists of segments of oblique straight lines. Thus, to draw the bending-moment diagram of the beam you will need to compute M only at the points where the loads or the reactions are applied.

2. For a beam supporting uniformly distributed loads, we note [Sample Prob. 7.3] that under each of the distributed loads:

    a. The shear diagram consists of a segment of an oblique straight line. Thus, you will need to compute V only where the distributed load begins and where it ends.

    b. The bending-moment diagram consists of an arc of parabola. In most cases you will need to compute M only where the distributed load begins and where it ends.

3. For a beam with a more complicated loading, it is necessary to consider the free-body diagram of a portion of the beam of arbitrary length x and determine V and M as functions of x. This procedure may have to be repeated several times, since V and M are often represented by different functions in various parts of the beam [Sample Prob. 7.3].

4. When a couple is applied to a beam, the shear has the same value on both sides of the point of application of the couple, but the bending-moment diagram will show a discontinuity at that point, rising or falling by an amount equal to the magnitude of the couple. Note that a couple can either be applied directly to the beam, or result from the application of a load on a curved member rigidly attached to the beam [Sample Prob. 7.3].

PROBLEMS

Fig. P7.29

Fig. P7.30

7.29 through 7.32 For the beam and loading shown, (*a*) draw the shear and bending-moment diagrams, (*b*) determine the maximum absolute values of the shear and bending moment.

Fig. P7.31

Fig. P7.32

7.33 and 7.34 For the beam and loading shown, (*a*) draw the shear and bending-moment diagrams, (*b*) determine the maximum absolute values of the shear and bending moment.

Fig. *P7.33*

Fig. *P7.34*

7.35 and 7.36 For the beam and loading shown, (*a*) draw the shear and bending-moment diagrams, (*b*) determine the maximum absolute values of the shear and bending moment.

Fig. P7.35

Fig. P7.36

7.37 and 7.38 For the beam and loading shown, (*a*) draw the shear and bending-moment diagrams, (*b*) determine the maximum absolute values of the shear and bending moment.

Fig. *P7.37*

Fig. *P7.38*

7.39 through 7.42 For the beam and loading shown, (*a*) draw the shear and bending-moment diagrams, (*b*) determine the maximum absolute values of the shear and bending moment.

Fig. P7.39

Fig. P7.40

Fig. P7.41

7.43 Assuming the upward reaction of the ground on beam *AB* to be uniformly distributed and knowing that $P = wa$, (*a*) draw the shear and bending-moment diagrams, (*b*) determine the maximum absolute values of the shear and bending moment.

Fig. P7.42

Fig. *P7.43*

7.44 Solve Prob. 7.43 knowing that $P = 3wa$.

7.45 Assuming the upward reaction of the ground on beam *AB* to be uniformly distributed and knowing that $a = 0.3$ m, (*a*) draw the shear and bending-moment diagrams, (*b*) determine the maximum absolute values of the shear and bending moment.

7.46 Solve Prob. 7.45 knowing that $a = 0.5$ m.

7.47 and 7.48 Assuming the upward reaction of the ground on beam *AB* to be uniformly distributed, (*a*) draw the shear and bending-moment diagrams, (*b*) determine the maximum absolute values of the shear and bending moment.

7.49 and 7.50 Draw the shear and bending-moment diagrams for the beam *AB*, and determine the maximum absolute values of the shear and bending moment.

Fig. P7.45

Fig. P7.47

Fig. P7.48

Fig. P7.50

Fig. P7.49

Fig. P7.51

Fig. P7.53

Fig. P7.58

Fig. P7.59

Fig. P7.60

7.51 and 7.52 Draw the shear and bending-moment diagrams for the beam AB, and determine the maximum absolute values of the shear and bending moment.

Fig. P7.52

7.53 Two small channel sections DF and EH have been welded to the uniform beam AB of weight $W = 3$ kN to form the rigid structural member shown. This member is being lifted by two cables attached at D and E. Knowing that $\theta = 30°$ and neglecting the weight of the channel sections, (a) draw the shear and bending-moment diagrams for beam AB, (b) determine the maximum absolute values of the shear and bending moment in the beam.

7.54 Solve Prob. 7.53 when $\theta = 60°$.

7.55 For the structural member of Prob. 7.53, determine (a) the angle θ for which the maximum absolute value of the bending moment in beam AB is as small as possible, (b) the corresponding value of $|M|_{max}$. (*Hint:* Draw the bending-moment diagram and then equate the absolute values of the largest positive and negative bending moments obtained.)

7.56 For the beam of Prob. 7.43, determine (a) the ratio $k = P/wa$ for which the maximum absolute value of the bending moment in the beam is as small as possible, (b) the corresponding value of $|M|_{max}$. (See hint for Prob. 7.55.)

7.57 For the beam of Prob. 7.45, determine (a) the distance a for which the maximum absolute value of the bending moment in the beam is as small as possible, (b) the corresponding value of $|M|_{max}$. (See hint for Prob. 7.55.)

7.58 For the beam and loading shown, determine (a) the distance a for which the maximum absolute value of the bending moment in the beam is as small as possible, (b) the corresponding value of $|M|_{max}$. (See hint for Prob. 7.55.)

7.59 A uniform beam is to be picked up by crane cables attached at A and B. Determine the distance a from the ends of the beam to the points where the cables should be attached if the maximum absolute value of the bending moment in the beam is to be as small as possible. (*Hint:* Draw the bending-moment diagram in terms of a, L, and the weight per unit length w, and then equate the absolute values of the largest positive and negative bending moments obtained.)

7.60 Knowing that $P = Q = 150$ lb, determine (a) the distance a for which the maximum absolute value of the bending moment in beam AB is as small as possible, (b) the corresponding value of $|M|_{max}$. (See hint for Prob. 7.55.)

7.61 Solve Prob. 7.60 assuming that $P = 300$ lb and $Q = 150$ lb.

*7.62 In order to reduce the bending moment in the cantilever beam AB, a cable and counterweight are permanently attached at end B. Determine the magnitude of the counterweight for which the maximum absolute value of the bending moment in the beam is as small as possible and the corresponding value of $|M|_{max}$. Consider (*a*) the case when the distributed load is permanently applied to the beam, (*b*) the more general case when the distributed load may either be applied or removed.

Fig. P7.62

*7.6 RELATIONS AMONG LOAD, SHEAR, AND BENDING MOMENT

When a beam carries more than two or three concentrated loads, or when it carries distributed loads, the method outlined in Sec. 7.5 for plotting shear and bending moment is likely to be quite cumbersome. The construction of the shear diagram and, especially, of the bending-moment diagram will be greatly facilitated if certain relations existing among load, shear, and bending moment are taken into consideration.

Let us consider a simply supported beam AB carrying a distributed load w per unit length (Fig. 7.11*a*), and let C and C' be two points of the beam at a distance Δx from each other. The shear and bending moment at C will be denoted by V and M, respectively, and will be assumed positive; the shear and bending moment at C' will be denoted by $V + \Delta V$ and $M + \Delta M$.

Let us now detach the portion of beam CC' and draw its free-body diagram (Fig. 7.11*b*). The forces exerted on the free body include a load of magnitude $w \Delta x$ and internal forces and couples at C and C'. Since shear and bending moment have been assumed positive, the forces and couples will be directed as shown in the figure.

Relations Between Load and Shear. We write that the sum of the vertical components of the forces acting on the free body CC' is zero:

$$V - (V + \Delta V) - w \Delta x = 0$$
$$\Delta V = -w \Delta x$$

Dividing both members of the equation by Δx and then letting Δx approach zero, we obtain

$$\frac{dV}{dx} = -w \tag{7.1}$$

Formula (7.1) indicates that for a beam loaded as shown in Fig. 7.11*a*, the slope dV/dx of the shear curve is negative; the numerical value of the slope at any point is equal to the load per unit length at that point.

(*a*)

(*b*)

Fig. 7.11

Fig. 7.11 (repeated)

Integrating (7.1) between points C and D, we obtain

$$V_D - V_C = -\int_{x_C}^{x_D} w \, dx \qquad (7.2)$$

$$V_D - V_C = -(\text{area under load curve between } C \text{ and } D) \qquad (7.2')$$

Note that this result could also have been obtained by considering the equilibrium of the portion of beam CD, since the area under the load curve represents the total load applied between C and D.

It should be observed that formula (7.1) *is not valid* at a point where a concentrated load is applied; the shear curve is discontinuous at such a point, as seen in Sec. 7.5. Similarly, formulas (7.2) and (7.2′) cease to be valid when concentrated loads are applied between C and D, since they do not take into account the sudden change in shear caused by a concentrated load. Formulas (7.2) and (7.2′), therefore, should be applied only between successive concentrated loads.

Relations Between Shear and Bending Moment. Returning to the free-body diagram of Fig. 7.11b, and writing now that the sum of the moments about C' is zero, we obtain

$$(M + \Delta M) - M - V \, \Delta x + w \Delta x \frac{\Delta x}{2} = 0$$

$$\Delta M = V \, \Delta x - \tfrac{1}{2} w (\Delta x)^2$$

Dividing both members of the equation by Δx and then letting Δx approach zero, we obtain

$$\frac{dM}{dx} = V \qquad (7.3)$$

Formula (7.3) indicates that the slope dM/dx of the bending-moment curve is equal to the value of the shear. This is true at any point where the shear has a well-defined value, i.e., at any point where no concentrated load is applied. Formula (7.3) also shows that the shear is zero at points where the bending moment is maximum. This property facilitates the determination of the points where the beam is likely to fail under bending.

Integrating (7.3) between points C and D, we obtain

$$M_D - M_C = \int_{x_C}^{x_D} V \, dx \qquad (7.4)$$

$$M_D - M_C = \text{area under shear curve between } C \text{ and } D \qquad (7.4')$$

Note that the area under the shear curve should be considered positive where the shear is positive and negative where the shear is negative. Formulas (7.4) and (7.4′) are valid even when concentrated loads are applied between C and D, as long as the shear curve has been correctly drawn. The formulas cease to be valid, however, if a *couple* is applied at a point between C and D, since they do not take into account the sudden change in bending moment caused by a couple (see Sample Prob. 7.7).

EXAMPLE Let us consider a simply supported beam AB of span L carrying a uniformly distributed load w (Fig. 7.12a). From the free-body diagram of the entire beam we determine the magnitude of the reactions at the supports: $R_A = R_B = wL/2$ (Fig. 7.12b). Next, we draw the shear diagram. Close to the end A of the beam, the shear is equal to R_A, that is, to $wL/2$, as we can check by considering a very small portion of the beam as a free body. Using formula (7.2), we can then determine the shear V at any distance x from A. We write

$$V - V_A = -\int_0^x w\, dx = -wx$$

$$V = V_A - wx = \frac{wL}{2} - wx = w\left(\frac{L}{2} - x\right)$$

The shear curve is thus an oblique straight line which crosses the x axis at $x = L/2$ (Fig. 7.12c). Considering, now, the bending moment, we first observe that $M_A = 0$. The value M of the bending moment at any distance x from A can then be obtained from formula (7.4); we have

$$M - M_A = \int_0^x V\, dx$$

$$M = \int_0^x w\left(\frac{L}{2} - x\right) dx = \frac{w}{2}(Lx - x^2)$$

The bending-moment curve is a parabola. The maximum value of the bending moment occurs when $x = L/2$, since V (and thus dM/dx) is zero for that value of x. Substituting $x = L/2$ in the last equation, we obtain $M_{max} = wL^2/8$. ∎

In most engineering applications, the value of the bending moment needs to be known only at a few specific points. Once the shear diagram has been drawn, and after M has been determined at one of the ends of the beam, the value of the bending moment can then be obtained at any given point by computing the area under the shear curve and using formula (7.4′). For instance, since $M_A = 0$ for the beam of Fig. 7.12, the maximum value of the bending moment for that beam can be obtained simply by measuring the area of the shaded triangle in the shear diagram:

$$M_{max} = \frac{1}{2}\frac{L}{2}\frac{wL}{2} = \frac{wL^2}{8}$$

In this example, the load curve is a horizontal straight line, the shear curve is an oblique straight line, and the bending-moment curve is a parabola. If the load curve had been an oblique straight line (first degree), the shear curve would have been a parabola (second degree), and the bending-moment curve would have been a cubic (third degree). The shear and bending-moment curves will always be, respectively, one and two degrees higher than the load curve. Thus, once a few values of the shear and bending moment have been computed, we should be able to sketch the shear and bending-moment diagrams without actually determining the functions $V(x)$ and $M(x)$. The sketches obtained will be more accurate if we make use of the fact that at any point where the curves are continuous, the slope of the shear curve is equal to $-w$ and the slope of the bending-moment curve is equal to V.

Fig. 7.12

SAMPLE PROBLEM 7.4

Draw the shear and bending-moment diagrams for the beam and loading shown.

SOLUTION

Free-Body: Entire Beam. Considering the entire beam as a free body, we determine the reactions:

$+\Sigma \uparrow M_A = 0$:

$$D(24 \text{ ft}) - (20 \text{ kips})(6 \text{ ft}) - (12 \text{ kips})(14 \text{ ft}) - (12 \text{ kips})(28 \text{ ft}) = 0$$

$$D = +26 \text{ kips} \qquad \mathbf{D} = 26 \text{ kips} \uparrow$$

$+\uparrow \Sigma F_y = 0$: $\quad A_y - 20 \text{ kips} - 12 \text{ kips} + 26 \text{ kips} - 12 \text{ kips} = 0$

$$A_y = +18 \text{ kips} \qquad \mathbf{A}_y = 18 \text{ kips} \uparrow$$

$\xrightarrow{+} \Sigma F_x = 0$: $\quad A_x = 0 \qquad\qquad\qquad \mathbf{A}_x = 0$

We also note that at both A and E the bending moment is zero; thus two points (indicated by small circles) are obtained on the bending-moment diagram.

Shear Diagram. Since $dV/dx = -w$, we find that between concentrated loads and reactions the slope of the shear diagram is zero (i.e., the shear is constant). The shear at any point is determined by dividing the beam into two parts and considering either part as a free body. For example, using the portion of beam to the left of section *1*, we obtain the shear between B and C:

$+\uparrow \Sigma F_y = 0$: $\quad +18 \text{ kips} - 20 \text{ kips} - V = 0 \qquad V = -2 \text{ kips}$

We also find that the shear is $+12$ kips just to the right of D and zero at end E. Since the slope $dV/dx = -w$ is constant between D and E, the shear diagram between these two points is a straight line.

Bending-Moment Diagram. We recall that the area under the shear curve between two points is equal to the change in bending moment between the same two points. For convenience, the area of each portion of the shear diagram is computed and is indicated on the diagram. Since the bending moment M_A at the left end is known to be zero, we write

$$\begin{aligned} M_B - M_A &= +108 & M_B &= +108 \text{ kip} \cdot \text{ft} \\ M_C - M_B &= -16 & M_C &= +92 \text{ kip} \cdot \text{ft} \\ M_D - M_C &= -140 & M_D &= -48 \text{ kip} \cdot \text{ft} \\ M_E - M_D &= +48 & M_E &= 0 \end{aligned}$$

Since M_E is known to be zero, a check of the computations is obtained.

Between the concentrated loads and reactions the shear is constant; thus the slope dM/dx is constant, and the bending-moment diagram is drawn by connecting the known points with straight lines. Between D and E, where the shear diagram is an oblique straight line, the bending-moment diagram is a parabola.

From the V and M diagrams we note that $V_{\max} = 18$ kips and $M_{\max} = 108$ kip $\cdot$ ft.

Draw the shear and bending-moment diagrams for the beam and loading shown and determine the location and magnitude of the maximum bending moment.

SOLUTION

Free-Body: Entire Beam. Considering the entire beam as a free body, we obtain the reactions

$$\mathbf{R}_A = 80 \text{ kN} \uparrow \qquad \mathbf{R}_C = 40 \text{ kN} \uparrow$$

Shear Diagram. The shear just to the right of A is $V_A = +80$ kN. Since the change in shear between two points is equal to *minus* the area under the load curve between the same two points, we obtain V_B by writing

$$V_B - V_A = -(20 \text{ kN/m})(6 \text{ m}) = -120 \text{ kN}$$
$$V_B = -120 + V_A = -120 + 80 = -40 \text{ kN}$$

Since the slope $dV/dx = -w$ is constant between A and B, the shear diagram between these two points is represented by a straight line. Between B and C, the area under the load curve is zero; therefore,

$$V_C - V_B = 0 \qquad V_C = V_B = -40 \text{ kN}$$

and the shear is constant between B and C.

Bending-Moment Diagram. We note that the bending moment at each end of the beam is zero. In order to determine the maximum bending moment, we locate the section D of the beam where $V = 0$. We write

$$V_D - V_A = -wx$$
$$0 - 80 \text{ kN} = -(20 \text{ kN/m})x$$

and, solving for x: $\qquad\qquad x = 4 \text{ m}$ ◀

The maximum bending moment occurs at point D, where we have $dM/dx = V = 0$. The areas of the various portions of the shear diagram are computed and are given (in parentheses) on the diagram. Since the area of the shear diagram between two points is equal to the change in bending moment between the same two points, we write

$$
\begin{array}{ll}
M_D - M_A = +160 \text{ kN} \cdot \text{m} & M_D = +160 \text{ kN} \cdot \text{m} \\
M_B - M_D = -\ 40 \text{ kN} \cdot \text{m} & M_B = +120 \text{ kN} \cdot \text{m} \\
M_C - M_B = -120 \text{ kN} \cdot \text{m} & M_C = 0
\end{array}
$$

The bending-moment diagram consists of an arc of parabola followed by a segment of straight line; the slope of the parabola at A is equal to the value of V at that point.

The maximum bending moment is

$$M_{\text{max}} = M_D = +160 \text{ kN} \cdot \text{m}$$ ◀

SAMPLE PROBLEM 7.6

Sketch the shear and bending-moment diagrams for the cantilever beam shown.

SOLUTION

Shear Diagram. At the free end of the beam, we find $V_A = 0$. Between A and B, the area under the load curve is $\frac{1}{2}w_0a$; we find V_B by writing

$$V_B - V_A = -\tfrac{1}{2}w_0a \qquad V_B = -\tfrac{1}{2}w_0a$$

Between B and C, the beam is not loaded; thus $V_C = V_B$. At A, we have $w = w_0$, and, according to Eq. (7.1), the slope of the shear curve is $dV/dx = -w_0$, while at B the slope is $dV/dx = 0$. Between A and B, the loading decreases linearly, and the shear diagram is parabolic. Between B and C, $w = 0$, and the shear diagram is a horizontal line.

Bending-Moment Diagram. We note that $M_A = 0$ at the free end of the beam. We compute the area under the shear curve and write

$$M_B - M_A = -\tfrac{1}{3}w_0a^2 \qquad M_B = -\tfrac{1}{3}w_0a^2$$
$$M_C - M_B = -\tfrac{1}{2}w_0a(L - a)$$
$$M_C = -\tfrac{1}{6}w_0a(3L - a)$$

The sketch of the bending-moment diagram is completed by recalling that $dM/dx = V$. We find that between A and B the diagram is represented by a cubic curve with zero slope at A, and between B and C the diagram is represented by a straight line.

SAMPLE PROBLEM 7.7

The simple beam AC is loaded by a couple of magnitude T applied at point B. Draw the shear and bending-moment diagrams for the beam.

SOLUTION

Free-Body: Entire Beam. The entire beam is taken as a free body, and we obtain

$$\mathbf{R}_A = \frac{T}{L}\uparrow \qquad \mathbf{R}_C = \frac{T}{L}\downarrow$$

Shear and Bending-Moment Diagrams. The shear at any section is constant and equal to T/L. Since a couple is applied at B, the bending-moment diagram is discontinuous at B; the bending moment decreases suddenly by an amount equal to T.

SOLVING PROBLEMS ON YOUR OWN

In this lesson you learned how to use the relations existing among load, shear, and bending moment to simplify the drawing of the shear and bending-moment diagrams. These relations are

$$\frac{dV}{dx} = -w \qquad (7.1)$$

$$\frac{dM}{dx} = V \qquad (7.3)$$

$$V_D - V_C = -(\text{area under load curve between } C \text{ and } D) \qquad (7.2')$$
$$M_D - M_C = (\text{area under shear curve between } C \text{ and } D) \qquad (7.4')$$

Taking into account these relations, you can use the following procedure to draw the shear and bending-moment diagrams for a beam.

1. Draw a free-body diagram of the entire beam, and use it to determine the reactions at the beam supports.

2. Draw the shear diagram. This can be done as in the preceding lesson by cutting the beam at various points and considering the free-body diagram of one of the two portions of the beam that you have obtained [Sample Prob. 7.3]. You can, however, consider one of the following alternative procedures.

   a. The shear V at any point of the beam is the sum of the reactions and loads to the left of that point; an upward force is counted as positive, and a downward force is counted as negative.

   b. For a beam carrying a distributed load, you can start from a point where you know V and use Eq. (7.2') repeatedly to find V at all the other points of interest.

3. Draw the bending-moment diagram, using the following procedure.

   a. Compute the area under each portion of the shear curve, assigning a positive sign to areas located above the x axis and a negative sign to areas located below the x axis.

   b. Apply Eq. (7.4') repeatedly [Sample Probs. 7.4 and 7.5], starting from the left end of the beam, where $M = 0$ (except if a couple is applied at that end, or if the beam is a cantilever beam with a fixed left end).

   c. Where a couple is applied to the beam, be careful to show a discontinuity in the bending-moment diagram by *increasing* the value of M at that point by an amount equal to the magnitude of the couple if the couple is *clockwise*, or *decreasing* the value of M by that amount if the couple is *counterclockwise* [Sample Prob. 7.7].

(continued)

4. Determine the location and magnitude of $|M|_{max}$. The maximum absolute value of the bending moment occurs at one of the points where $dM/dx = 0$, that is, according to Eq. (7.3), at a point where V is equal to zero or changes sign. You should, therefore:

a. Determine from the shear diagram the value of $|M|$ where V changes sign; this will occur under the concentrated loads [Sample Prob. 7.4].

b. Determine the points where $V = 0$ and the corresponding values of $|M|$; this will occur under a distributed load. To find the distance x between point C, where the distributed load starts, and point D, where the shear is zero, use Eq. (7.2′); for V_C use the known value of the shear at point C, for V_D use zero, and express the area under the load curve as a function of x [Sample Prob. 7.5].

5. You can improve the quality of your drawings by keeping in mind that at any given point, according to Eqs. (7.1) and (7.3), the slope of the V curve is equal to $-w$ and the slope of the M curve is equal to V.

6. Finally, for beams supporting a distributed load expressed as a function $w(x)$, remember that the shear V can be obtained by integrating the function $-w(x)$, and the bending moment M can be obtained by integrating $V(x)$ [Eqs. (7.3) and (7.4)].

PROBLEMS

7.63 Using the method of Sec. 7.6, solve Prob. 7.29.

7.64 Using the method of Sec. 7.6, solve Prob. 7.30.

7.65 Using the method of Sec. 7.6, solve Prob. 7.31.

7.66 Using the method of Sec. 7.6, solve Prob. 7.32.

7.67 Using the method of Sec. 7.6, solve Prob. 7.33.

7.68 Using the method of Sec. 7.6, solve Prob. 7.34.

7.69 and 7.70 For the beam and loading shown, (a) draw the shear and bending-moment diagrams, (b) determine the maximum absolute values of the shear and bending moment.

7.71 Using the method of Sec. 7.6, solve Prob. 7.39.

7.72 Using the method of Sec. 7.6, solve Prob. 7.40.

7.73 Using the method of Sec. 7.6, solve Prob. 7.41.

7.74 Using the method of Sec. 7.6, solve Prob. 7.42.

7.75 and 7.76 For the beam and loading shown, (a) draw the shear and bending-moment diagrams, (b) determine the maximum absolute values of the shear and bending moment.

Fig. P7.69

Fig. P7.70

Fig. P7.75

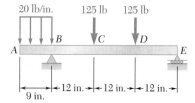

Fig. P7.76

7.77 through 7.79 For the beam and loading shown, (a) draw the shear and bending-moment diagrams, (b) determine the magnitude and location of the maximum absolute value of the bending moment.

Fig. P7.77   **Fig. P7.78**

Fig. P7.79

7.80 Solve Prob. 7.79 assuming that the 20-kN · m couple applied at B is counterclockwise.

381

7.81 For the beam shown, draw the shear and bending-moment diagrams, and determine the magnitude and location of the maximum absolute value of the bending moment, knowing that (a) M = 0, (b) M = 24 kip · ft.

Fig. P7.81

7.82 For the beam shown, draw the shear and bending-moment diagrams, and determine the magnitude and location of the maximum absolute value of the bending moment, knowing that (a) P = 6 kips, (b) P = 3 kips.

Fig. P7.82

7.83 (a) Draw the shear and bending-moment diagrams for beam AB, (b) determine the magnitude and location of the maximum absolute value of the bending moment.

7.84 Solve Prob. 7.83 assuming that the 300-lb force applied at D is directed upward.

7.85 through 7.87 For the beam and loading shown, (a) write the equations of the shear and bending-moment curves, (b) determine the magnitude and location of the maximum bending moment.

Fig. P7.83

Fig. P7.85   **Fig. P7.86**

7.88 For the beam and loading shown, (a) write the equations of the shear and bending-moment curves, (b) determine the maximum bending moment.

Fig. P7.87

Fig. P7.88

7.89 The beam AB is subjected to the uniformly distributed load shown and to two unknown forces **P** and **Q**. Knowing that it has been experimentally determined that the bending moment is +800 N · m at D and +1300 N · m at E, (a) determine **P** and **Q**, (b) draw the shear and bending-moment diagrams for the beam.

7.90 Solve Prob. 7.89 assuming that the bending moment was found to be +650 N · m at D and +1450 N · m at E.

Fig. P7.89

*7.91 The beam AB is subjected to the uniformly distributed load shown and to two unknown forces **P** and **Q**. Knowing that it has been experimentally determined that the bending moment is $+6.10$ kip · ft at D and $+5.50$ kip · ft at E, (a) determine **P** and **Q**, (b) draw the shear and bending-moment diagrams for the beam.

*7.92 Solve Prob. 7.91 assuming that the bending moment was found to be $+5.96$ kip · ft at D and $+6.84$ kip · ft at E.

Fig. P7.91

CABLES

*7.7 CABLES WITH CONCENTRATED LOADS

Cables are used in many engineering applications, such as suspension bridges, transmission lines, aerial tramways, guy wires for high towers, etc. Cables may be divided into two categories, according to their loading: (1) cables supporting concentrated loads, (2) cables supporting distributed loads. In this section, cables of the first category are examined.

Consider a cable attached to two fixed points A and B and supporting n vertical concentrated loads $\mathbf{P}_1, \mathbf{P}_2, \ldots, \mathbf{P}_n$ (Fig. 7.13a). We assume that the cable is *flexible*, i.e., that its resistance to bending is small and can be neglected. We further assume that the *weight of the cable is negligible* compared with the loads supported by the cable. Any portion of cable between successive loads can therefore be considered as a two-force member, and the internal forces at any point in the cable reduce to a *force of tension directed along the cable*.

We assume that each of the loads lies in a given vertical line, i.e., that the horizontal distance from support A to each of the loads is known; we also assume that the horizontal and vertical distances between the supports are known. We propose to determine the shape of the cable, i.e., the vertical distance from support A to each of the points $C_1, C_2, \ldots, C_n$, and also the tension T in each portion of the cable.

Photo 7.3 Since the weight of the cable of the chairlift shown is negligible compared to the weights of the chairs and skiers, the methods of this section can be used to determine the force at any point in the cable.

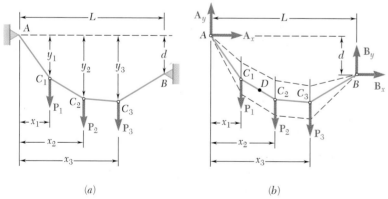

(a)　　　　　　　　(b)

Fig. 7.13

We first draw the free-body diagram of the entire cable (Fig. 7.13b). Since the slope of the portions of cable attached at A and B is not known, the reactions at A and B must be represented by two components each. Thus, four unknowns are involved, and the three equations of equilibrium are not sufficient to determine the reactions at A and B.† We must

†Clearly, the cable is not a rigid body; the equilibrium equations represent, therefore, *necessary but not sufficient conditions* (see Sec. 6.11).

Fig. 7.13 (repeated)

Fig. 7.14

therefore obtain an additional equation by considering the equilibrium of a portion of the cable. This is possible if we know the coordinates x and y of a point D of the cable. Drawing the free-body diagram of the portion of cable AD (Fig. 7.14a) and writing $\Sigma M_D = 0$, we obtain an additional relation between the scalar components A_x and A_y and can determine the reactions at A and B. The problem would remain indeterminate, however, if we did not know the coordinates of D, unless some other relation between A_x and A_y (or between B_x and B_y) were given. The cable might hang in any of various possible ways, as indicated by the dashed lines in Fig. 7.13b.

Once A_x and A_y have been determined, the vertical distance from A to any point of the cable can easily be found. Considering point C_2, for example, we draw the free-body diagram of the portion of cable AC_2 (Fig. 7.14b). Writing $\Sigma M_{C_2} = 0$, we obtain an equation which can be solved for y_2. Writing $\Sigma F_x = 0$ and $\Sigma F_y = 0$, we obtain the components of the force $\mathbf{T}$ representing the tension in the portion of cable to the right of C_2. We observe that $T \cos \theta = -A_x$; *the horizontal component of the tension force is the same at any point of the cable.* It follows that the tension T is maximum when $\cos \theta$ is minimum, i.e., in the portion of cable which has the largest angle of inclination θ. Clearly, this portion of cable must be adjacent to one of the two supports of the cable.

*7.8 CABLES WITH DISTRIBUTED LOADS

Consider a cable attached to two fixed points A and B and carrying a *distributed load* (Fig. 7.15a). We saw in the preceding section that for a cable supporting concentrated loads, the internal force at any point is a force of tension directed along the cable. In the case of a cable carrying a distributed load, the cable hangs in the shape of a curve, and the internal force at a point D is a force of tension $\mathbf{T}$ *directed along the tangent to the curve.* In this section, you will learn to determine the tension at any point of a cable supporting a given distributed load. In the following sections, the shape of the cable will be determined for two particular types of distributed loads.

Considering the most general case of distributed load, we draw the free-body diagram of the portion of cable extending from the lowest point C to a given point D of the cable (Fig. 7.15b). The

Fig. 7.15

forces acting on the free body are the tension force T_0 at C, which is horizontal, the tension force $\mathbf{T}$ at D, directed along the tangent to the cable at D, and the resultant $\mathbf{W}$ of the distributed load supported by the portion of cable CD. Drawing the corresponding force triangle (Fig. 7.15c), we obtain the following relations:

$$T \cos \theta = T_0 \qquad T \sin \theta = W \tag{7.5}$$

$$T = \sqrt{T_0^2 + W^2} \qquad \tan \theta = \frac{W}{T_0} \tag{7.6}$$

From the relations (7.5), it appears that the horizontal component of the tension force $\mathbf{T}$ is the same at any point and that the vertical component of $\mathbf{T}$ is equal to the magnitude W of the load measured from the lowest point. Relations (7.6) show that the tension T is minimum at the lowest point and maximum at one of the two points of support.

*7.9 PARABOLIC CABLE

Let us assume, now, that the cable AB carries a load *uniformly distributed along the horizontal* (Fig. 7.16a). Cables of suspension bridges may be assumed loaded in this way, since the weight of the cables is small compared with the weight of the roadway. We denote by w the load per unit length (*measured horizontally*) and express it in N/m or in lb/ft. Choosing coordinate axes with origin at the lowest point C of the cable, we find that the magnitude W of the total load carried by the portion of cable extending from C to the point D of coordinates x and y is $W = wx$. The relations (7.6) defining the magnitude and direction of the tension force at D become

$$T = \sqrt{T_0^2 + w^2 x^2} \qquad \tan \theta = \frac{wx}{T_0} \tag{7.7}$$

Moreover, the distance from D to the line of action of the resultant $\mathbf{W}$ is equal to half the horizontal distance from C to D (Fig. 7.16b). Summing moments about D, we write

$$+\!\!\uparrow \Sigma M_D = 0: \qquad wx\frac{x}{2} - T_0 y = 0$$

(a)

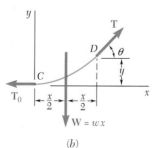

(b)

Fig. 7.16

and, solving for y,

$$y = \frac{wx^2}{2T_0} \tag{7.8}$$

This is the equation of a *parabola* with a vertical axis and its vertex at the origin of coordinates. The curve formed by cables loaded uniformly along the horizontal is thus a parabola.[†]

When the supports A and B of the cable have the same elevation, the distance L between the supports is called the *span* of the cable and the vertical distance h from the supports to the lowest point is called the *sag* of the cable (Fig. 7.17a). If the span and sag of a cable are known, and if the load w per unit horizontal length is given, the minimum tension T_0 may be found by substituting $x = L/2$ and $y = h$ in Eq. (7.8). Equations (7.7) will then yield the tension and the slope at any point of the cable and Eq. (7.8) will define the shape of the cable.

When the supports have different elevations, the position of the lowest point of the cable is not known and the coordinates x_A, y_A and x_B, y_B of the supports must be determined. To this effect, we express that the coordinates of A and B satisfy Eq. (7.8) and that $x_B - x_A = L$ and $y_B - y_A = d$, where L and d denote, respectively, the horizontal and vertical distances between the two supports (Fig. 7.17b and c).

The length of the cable from its lowest point C to its support B can be obtained from the formula

$$s_B = \int_0^{x_B} \sqrt{1 + \left(\frac{dy}{dx}\right)^2}\, dx \tag{7.9}$$

Differentiating (7.8), we obtain the derivative $dy/dx = wx/T_0$; substituting into (7.9) and using the binomial theorem to expand the radical in an infinite series, we have

$$s_B = \int_0^{x_B} \sqrt{1 + \frac{w^2 x^2}{T_0^2}}\, dx = \int_0^{x_B} \left(1 + \frac{w^2 x^2}{2T_0^2} - \frac{w^4 x^4}{8T_0^4} + \cdots\right) dx$$

$$s_B = x_B\left(1 + \frac{w^2 x_B^2}{6T_0^2} - \frac{w^4 x_B^4}{40T_0^4} + \cdots\right)$$

and, since $wx_B^2/2T_0 = y_B$,

$$s_B = x_B\left[1 + \frac{2}{3}\left(\frac{y_B}{x_B}\right)^2 - \frac{2}{5}\left(\frac{y_B}{x_B}\right)^4 + \cdots\right] \tag{7.10}$$

The series converges for values of the ratio y_B/x_B less than 0.5; in most cases, this ratio is much smaller, and only the first two terms of the series need be computed.

(a)

(b)

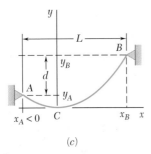

(c)

Fig. 7.17

[†]Cables hanging under their own weight are not loaded uniformly along the horizontal, and they do not form a parabola. The error introduced by assuming a parabolic shape for cables hanging under their weight, however, is small when the cable is sufficiently taut. A complete discussion of cables hanging under their own weight is given in the next section.

SAMPLE PROBLEM 7.8

The cable AE supports three vertical loads from the points indicated. If point C is 5 ft below the left support, determine (a) the elevation of points B and D, (b) the maximum slope and the maximum tension in the cable.

SOLUTION

Reactions at Supports. The reaction components $\mathbf{A}_x$ and $\mathbf{A}_y$ are determined as follows:

Free Body: Entire Cable

$+\circlearrowleft \Sigma M_E = 0$:
$$A_x(20 \text{ ft}) - A_y(60 \text{ ft}) + (6 \text{ kips})(40 \text{ ft}) + (12 \text{ kips})(30 \text{ ft}) + (4 \text{ kips})(15 \text{ ft}) = 0$$
$$20A_x - 60A_y + 660 = 0$$

Free Body: ABC

$+\circlearrowleft \Sigma M_C = 0$: $-A_x(5 \text{ ft}) - A_y(30 \text{ ft}) + (6 \text{ kips})(10 \text{ ft}) = 0$
$$-5A_x - 30A_y + 60 = 0$$

Solving the two equations simultaneously, we obtain

$$A_x = -18 \text{ kips} \qquad \mathbf{A}_x = 18 \text{ kips} \leftarrow$$
$$A_y = +5 \text{ kips} \qquad \mathbf{A}_y = 5 \text{ kips} \uparrow$$

a. Elevation of Points B and D.

Free Body: AB Considering the portion of cable AB as a free body, we write

$+\circlearrowleft \Sigma M_B = 0$: $(18 \text{ kips})y_B - (5 \text{ kips})(20 \text{ ft}) = 0$
$$y_B = 5.56 \text{ ft below } A \quad \blacktriangleleft$$

Free Body: ABCD Using the portion of cable $ABCD$ as a free body, we write

$+\circlearrowleft \Sigma M_D = 0$:
$$-(18 \text{ kips})y_D - (5 \text{ kips})(45 \text{ ft}) + (6 \text{ kips})(25 \text{ ft}) + (12 \text{ kips})(15 \text{ ft}) = 0$$
$$y_D = 5.83 \text{ ft above } A \quad \blacktriangleleft$$

b. Maximum Slope and Maximum Tension.
We observe that the maximum slope occurs in portion DE. Since the horizontal component of the tension is constant and equal to 18 kips, we write

$$\tan \theta = \frac{14.17}{15 \text{ ft}} \qquad \theta = 43.4° \quad \blacktriangleleft$$

$$T_{max} = \frac{18 \text{ kips}}{\cos \theta} \qquad T_{max} = 24.8 \text{ kips} \quad \blacktriangleleft$$

SAMPLE PROBLEM 7.9

A light cable is attached to a support at A, passes over a small pulley at B, and supports a load **P**. Knowing that the sag of the cable is 0.5 m and that the mass per unit length of the cable is 0.75 kg/m, determine (a) the magnitude of the load **P**, (b) the slope of the cable at B, (c) the total length of the cable from A to B. Since the ratio of the sag to the span is small, assume the cable to be parabolic. Also, neglect the weight of the portion of cable from B to D.

SOLUTION

a. Load P. We denote by C the lowest point of the cable and draw the free-body diagram of the portion CB of cable. Assuming the load to be uniformly distributed along the horizontal, we write

$$w = (0.75 \text{ kg/m})(9.81 \text{ m/s}^2) = 7.36 \text{ N/m}$$

The total load for the portion CB of cable is

$$W = wx_B = (7.36 \text{ N/m})(20 \text{ m}) = 147.2 \text{ N}$$

and is applied halfway between C and B. Summing moments about B, we write

$$+\text{↰} \ \Sigma M_B = 0: \quad (147.2 \text{ N})(10 \text{ m}) - T_0(0.5 \text{ m}) = 0 \qquad T_0 = 2944 \text{ N}$$

From the force triangle we obtain

$$T_B = \sqrt{T_0^2 + W^2}$$
$$= \sqrt{(2944 \text{ N})^2 + (147.2 \text{ N})^2} = 2948 \text{ N}$$

Since the tension on each side of the pulley is the same, we find

$$P = T_B = 2948 \text{ N} \quad \blacktriangleleft$$

b. Slope of Cable at B. We also obtain from the force triangle

$$\tan \theta = \frac{W}{T_0} = \frac{147.2 \text{ N}}{2944 \text{ N}} = 0.05$$

$$\theta = 2.9° \quad \blacktriangleleft$$

c. Length of Cable. Applying Eq. (7.10) between C and B, we write

$$s_B = x_B \left[1 + \frac{2}{3}\left(\frac{y_B}{x_B}\right)^2 + \cdots \right]$$
$$= (20 \text{ m}) \left[1 + \frac{2}{3}\left(\frac{0.5 \text{ m}}{20 \text{ m}}\right)^2 + \cdots \right] = 20.00833 \text{ m}$$

The total length of the cable between A and B is twice this value,

$$\text{Length} = 2s_B = 40.0167 \text{ m} \quad \blacktriangleleft$$

SOLVING PROBLEMS ON YOUR OWN

In the problems of this section you will apply the equations of equilibrium to *cables that lie in a vertical plane*. We assume that a cable cannot resist bending, so that the force of tension in the cable is always directed along the cable.

A. In the first part of this lesson we considered cables subjected to concentrated loads. Since the weight of the cable is neglected, the cable is straight between loads.

Your solution will consist of the following steps:

1. Draw a free-body diagram of the entire cable showing the loads and the horizontal and vertical components of the reaction at each support. Use this free-body diagram to write the corresponding equilibrium equations.

2. You will be confronted with four unknown components and only three equations of equilibrium (see Fig. 7.13). You must therefore find an additional piece of information, such as the *position* of a point on the cable or the *slope* of the cable at a given point.

3. After you have identified the point of the cable where the additional information exists, cut the cable at that point, and draw a free-body diagram of one of the two portions of the cable you have obtained.
   a. If you know the position of the point where you have cut the cable, writing $\Sigma M = 0$ about that point for the new free body will yield the additional equation required to solve for the four unknown components of the reactions [Sample Prob. 7.8].
   b. If you know the slope of the portion of the cable you have cut, writing $\Sigma F_x = 0$ and $\Sigma F_y = 0$ for the new free body will yield two equilibrium equations which, together with the original three, can be solved for the four reaction components and for the tension in the cable where it has been cut.

4. To find the elevation of a given point of the cable and the slope and tension at that point once the reactions at the supports have been found, you should cut the cable at that point and draw a free-body diagram of one of the two portions of the cable you have obtained. Writing $\Sigma M = 0$ about the given point yields its elevation. Writing $\Sigma F_x = 0$ and $\Sigma F_y = 0$ yields the components of the tension force, from which its magnitude and direction can easily be found.

(continued)

5. For a cable supporting vertical loads only, you will observe that *the horizontal component of the tension force is the same at any point*. It follows that, for such a cable, the *maximum tension occurs in the steepest portion of the cable*.

B. In the second portion of this lesson we considered cables carrying a load uniformly distributed along the horizontal. The shape of the cable is then parabolic.

Your solution will use one or more of the following concepts:

1. Placing the origin of coordinates at the lowest point of the cable and directing the x and y axes to the right and upward, respectively, we find that *the equation of the parabola* is

$$y = \frac{wx^2}{2T_0} \tag{7.8}$$

The minimum cable tension occurs at the origin, where the cable is horizontal, and the maximum tension is at the support where the slope is maximum.

2. If the supports of the cable have the same elevation, the sag h of the cable is the vertical distance from the lowest point of the cable to the horizontal line joining the supports. To solve a problem involving such a parabolic cable, you should write Eq. (7.8) for one of the supports; this equation can be solved for one unknown.

3. If the supports of the cable have different elevations, you will have to write Eq. (7.8) for each of the supports (see Fig. 7.17).

4. To find the length of the cable from the lowest point to one of the supports, you can use Eq. (7.10). In most cases, you will need to compute only the first two terms of the series.

PROBLEMS

7.93 Three loads are suspended as shown from the cable $ABCDE$. Knowing that $d_C = 3$ m, determine (a) the components of the reaction at E, (b) the maximum tension in the cable.

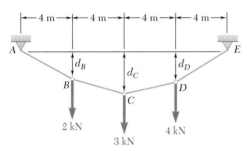

Fig. P7.93 and P7.94

7.94 Knowing that the maximum tension in cable $ABCDE$ is 13 kN, determine the distance d_C.

7.95 If $d_C = 8$ ft, determine (a) the reaction at A, (b) the reaction at E.

7.96 If $d_C = 4.5$ ft, determine (a) the reaction at A, (b) the reaction at E.

7.97 Knowing that $d_C = 3$ m, determine (a) the distances d_B and d_D, (b) the reaction at E.

7.98 Determine (a) distance d_C for which portion DE of the cable is horizontal, (b) the corresponding reactions at A and E.

7.99 An oil pipeline is supported at 6-ft intervals by vertical hangers attached to the cable shown. Due to the combined weight of the pipe and its contents, the tension in each hanger is 400 lb. Knowing that $d_C = 12$ ft, determine (a) the maximum tension in the cable, (b) the distance d_D.

Fig. P7.95 and P7.96

Fig. P7.97 and P7.98

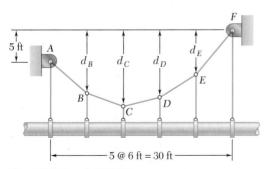

Fig. P7.99 and P7.100

7.100 Solve Prob. 7.99 assuming that $d_C = 9$ ft.

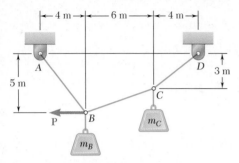

Fig. P7.101 and P7.102

7.101 Knowing that $m_B = 70$ kg and $m_C = 25$ kg, determine the magnitude of the force **P** required to maintain equilibrium.

7.102 Knowing that $m_B = 18$ kg and $m_C = 10$ kg, determine the magnitude of the force **P** required to maintain equilibrium.

7.103 Cable ABC supports two loads as shown. Knowing that $b = 21$ ft, determine (a) the required magnitude of the horizontal force **P**, (b) the corresponding distance a.

Fig. P7.103 and P7.104

7.104 Cable ABC supports two loads as shown. Determine the distances a and b when a horizontal force **P** of magnitude 200 lb is applied at C.

7.105 If $a = 3$ m, determine the magnitudes of **P** and **Q** required to maintain the cable in the shape shown.

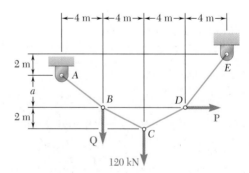

Fig. P7.105 and P7.106

7.106 If $a = 4$ m, determine the magnitudes of **P** and **Q** required to maintain the cable in the shape shown.

7.107 A transmission cable having a mass per unit length of 0.8 kg/m is strung between two insulators at the same elevation that are 75 m apart. Knowing that the sag of the cable is 2 m, determine (a) the maximum tension in the cable, (b) the length of the cable.

7.108 The total mass of cable ACB is 20 kg. Assuming that the mass of the cable is distributed uniformly along the horizontal, determine (a) the sag h, (b) the slope of the cable at A.

7.109 The center span of the Verrazano-Narrows Bridge consists of two uniform roadways suspended from four cables. The uniform load supported by each cable is $w = 10.8$ kips/ft along the horizontal. Knowing that the span L is 4260 ft and that the sag h is 390 ft, determine (a) the maximum tension in each cable, (b) the length of each cable.

7.110 The center span of the Verrazano-Narrows Bridge consists of two uniform roadways suspended from four cables. The design of the bridge allows for the effect of extreme temperature changes that cause the sag of the center span to vary from $h_w = 386$ ft in winter to $h_s = 394$ ft in summer. Knowing that the span is $L = 4260$ ft, determine the change in length of the cables due to extreme temperature changes.

7.111 Each cable of the Golden Gate Bridge supports a load $w = 11.1$ kips/ft along the horizontal. Knowing that the span L is 4150 ft and that the sag h is 464 ft, determine (a) the maximum tension in each cable, (b) the length of each cable.

7.112 Two cables of the same gauge are attached to a transmission tower at B. Since the tower is slender, the horizontal component of the resultant of the forces exerted by the cables at B is to be zero. Knowing that the mass per unit length of the cables is 0.4 kg/m, determine (a) the required sag h, (b) the maximum tension in each cable.

Fig. P7.108

Fig. P7.112

7.113 A 50.5-m length of wire having a mass per unit length of 0.75 kg/m is used to span a horizontal distance of 50 m. Determine (a) the approximate sag of the wire, (b) the maximum tension in the wire. [*Hint:* Use only the first two terms of Eq. (7.10).]

7.114 A cable of length $L + \Delta$ is suspended between two points that are at the same elevation and a distance L apart. (a) Assuming that Δ is small compared to L and that the cable is parabolic, determine the approximate sag in terms of L and Δ. (b) If $L = 100$ ft and $\Delta = 4$ ft, determine the approximate sag. [*Hint:* Use only the first two terms of Eq. (7.10).]

7.115 The total mass of cable AC is 25 kg. Assuming that the mass of the cable is distributed uniformly along the horizontal, determine the sag h and the slope of the cable at A and C.

Fig. P7.115

7.116 Cable *ACB* supports a load uniformly distributed along the horizontal as shown. The lowest point *C* is located 9 m to the right of *A*. Determine (*a*) the vertical distance *a*, (*b*) the length of the cable, (*c*) the components of the reaction at *A*.

Fig. P7.116

Fig. P7.117

7.117 Each cable of the side spans of the Golden Gate Bridge supports a load $w = 10.2$ kips/ft along the horizontal. Knowing that for the side spans the maximum vertical distance *h* from each cable to the chord *AB* is 30 ft and occurs at midspan, determine (*a*) the maximum tension in each cable, (*b*) the slope at *B*.

7.118 A steam pipe weighing 45 lb/ft that passes between two buildings 40 ft apart is supported by a system of cables as shown. Assuming that the weight of the cable system is equivalent to a uniformly distributed loading of 5 lb/ft, determine (*a*) the location of the lowest point *C* of the cable, (*b*) the maximum tension in the cable.

Fig. P7.118

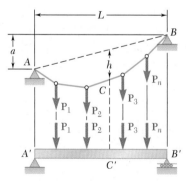

Fig. P7.119

*7.119 A cable *AB* of span *L* and a simple beam *A'B'* of the same span are subjected to identical vertical loadings as shown. Show that the magnitude of the bending moment at a point *C'* in the beam is equal to the product $T_0 h$, where T_0 is the magnitude of the horizontal component of the tension force in the cable and *h* is the vertical distance between point *C* and the chord joining the points of support *A* and *B*.

7.120 through 7.123 Making use of the property established in Prob. 7.119, solve the problem indicated by first solving the corresponding beam problem.

7.120	Prob. 7.94.
7.121	Prob. 7.97*a*.
7.122	Prob. 7.99*b*.
7.123	Prob. 7.100*b*.

*7.124 Show that the curve assumed by a cable that carries a distributed load $w(x)$ is defined by the differential equation $d^2y/dx^2 = w(x)/T_0$, where T_0 is the tension at the lowest point.

*7.125 Using the property indicated in Prob. 7.124, determine the curve assumed by a cable of span L and sag h carrying a distributed load $w = w_0 \cos(\pi x/L)$, where x is measured from mid-span. Also determine the maximum and minimum values of the tension in the cable.

*7.126 If the weight per unit length of the cable AB is $w_0/\cos^2 \theta$, prove that the curve formed by the cable is a circular arc. (*Hint:* Use the property indicated in Prob. 7.124.)

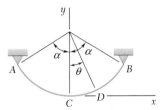

Fig. P7.126

*7.10 CATENARY

Let us now consider a cable AB carrying a load *uniformly distributed along the cable itself* (Fig. 7.18a). Cables hanging under their own weight are loaded in this way. We denote by w the load per unit length (*measured along the cable*) and express it in N/m or in lb/ft. The magnitude W of the total load carried by a portion of cable of length s extending from the lowest point C to a point D is $W = ws$. Substituting this value for W in formula (7.6), we obtain the tension at D:

$$T = \sqrt{T_0^2 + w^2s^2}$$

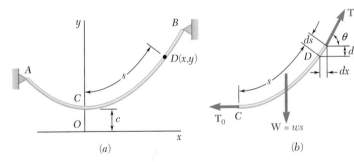

Fig. 7.18

In order to simplify the subsequent computations, we introduce the constant $c = T_0/w$. We thus write

$$T_0 = wc \qquad W = ws \qquad T = w\sqrt{c^2 + s^2} \qquad (7.11)$$

The free-body diagram of the portion of cable CD is shown in Fig. 7.18b. This diagram, however, cannot be used to obtain directly the equation of the curve assumed by the cable, since we do not know the horizontal distance from D to the line of action of the resultant $\mathbf{W}$ of the load. To obtain this equation, we first write that the horizontal projection of a small element of cable of length ds is

Photo 7.4 The forces on the supports and the internal forces in the cables of the power line shown are discussed in this section.

(a)

(b)

(c)

Fig. 7.18 *(continued)*

$dx = ds \cos \theta$. Observing from Fig. 7.18c that $\cos \theta = T_0/T$ and using (7.11), we write

$$dx = ds \cos \theta = \frac{T_0}{T}\,ds = \frac{wc\,ds}{w\sqrt{c^2 + s^2}} = \frac{ds}{\sqrt{1 + s^2/c^2}}$$

Selecting the origin O of the coordinates at a distance c directly below C (Fig. 7.18a) and integrating from $C(0, c)$ to $D(x, y)$, we obtain†

$$x = \int_0^s \frac{ds}{\sqrt{1 + s^2/c^2}} = c\left[\sinh^{-1}\frac{s}{c}\right]_0^s = c\sinh^{-1}\frac{s}{c}$$

This equation, which relates the length s of the portion of cable CD and the horizontal distance x, can be written in the form

$$s = c\sinh\frac{x}{c} \tag{7.15}$$

The relation between the coordinates x and y can now be obtained by writing $dy = dx\tan\theta$. Observing from Fig. 7.18c that $\tan\theta = W/T_0$ and using (7.11) and (7.15), we write

$$dy = dx\tan\theta = \frac{W}{T_0}dx = \frac{s}{c}dx = \sinh\frac{x}{c}dx$$

Integrating from $C(0, c)$ to $D(x, y)$ and using (7.12) and (7.13), we obtain

$$y - c = \int_0^x \sinh\frac{x}{c}dx = c\left[\cosh\frac{x}{c}\right]_0^x = c\left(\cosh\frac{x}{c} - 1\right)$$

$$y - c = c\cosh\frac{x}{c} - c$$

†This integral can be found in all standard integral tables. The function

$$z = \sinh^{-1} u$$

(read "arc hyperbolic sine u") is the *inverse* of the function $u = \sinh z$ (read "hyperbolic sine z"). This function and the function $v = \cosh z$ (read "hyperbolic cosine z") are defined as follows:

$$u = \sinh z = \tfrac{1}{2}(e^z - e^{-z}) \qquad v = \cosh z = \tfrac{1}{2}(e^z + e^{-z})$$

Numerical values of the functions $\sinh z$ and $\cosh z$ are found in *tables of hyperbolic functions*. They may also be computed on most calculators either directly or from the above definitions. The student is referred to any calculus text for a complete description of the properties of these functions. In this section, we use only the following properties, which are easily derived from the above definitions:

$$\frac{d\sinh z}{dz} = \cosh z \qquad \frac{d\cosh z}{dz} = \sinh z \tag{7.12}$$

$$\sinh 0 = 0 \qquad \cosh 0 = 1 \tag{7.13}$$

$$\cosh^2 z - \sinh^2 z = 1 \tag{7.14}$$

which reduces to

$$y = c \cosh \frac{x}{c} \qquad (7.16)$$

This is the equation of a *catenary* with vertical axis. The ordinate c of the lowest point C is called the *parameter* of the catenary. Squaring both sides of Eqs. (7.15) and (7.16), subtracting, and taking (7.14) into account, we obtain the following relation between y and s:

$$y^2 - s^2 = c^2 \qquad (7.17)$$

Solving (7.17) for s^2 and carrying into the last of the relations (7.11), we write these relations as follows:

$$T_0 = wc \qquad W = ws \qquad T = wy \qquad (7.18)$$

The last relation indicates that the tension at any point D of the cable is proportional to the vertical distance from D to the horizontal line representing the x axis.

When the supports A and B of the cable have the same elevation, the distance L between the supports is called the *span* of the cable and the vertical distance h from the supports to the lowest point C is called the *sag* of the cable. These definitions are the same as those given in the case of parabolic cables, but it should be noted that because of our choice of coordinate axes, the sag h is now

$$h = y_A - c \qquad (7.19)$$

It should also be observed that certain catenary problems involve transcendental equations which must be solved by successive approximations (see Sample Prob. 7.10). When the cable is fairly taut, however, the load can be assumed uniformly distributed *along the horizontal* and the catenary can be replaced by a parabola. This greatly simplifies the solution of the problem, and the error introduced is small.

When the supports A and B have different elevations, the position of the lowest point of the cable is not known. The problem can then be solved in a manner similar to that indicated for parabolic cables, by expressing that the cable must pass through the supports and that $x_B - x_A = L$ and $y_B - y_A = d$, where L and d denote, respectively, the horizontal and vertical distances between the two supports.

SAMPLE PROBLEM 7.10

A uniform cable weighing 3 lb/ft is suspended between two points A and B as shown. Determine (a) the maximum and minimum values of the tension in the cable, (b) the length of the cable.

SOLUTION

Equation of Cable. The origin of coordinates is placed at a distance c below the lowest point of the cable. The equation of the cable is given by Eq. (7.16),

$$y = c \cosh \frac{x}{c}$$

The coordinates of point B are

$$x_B = 250 \text{ ft} \qquad y_B = 100 + c$$

Substituting these coordinates into the equation of the cable, we obtain

$$100 + c = c \cosh \frac{250}{c}$$

$$\frac{100}{c} + 1 = \cosh \frac{250}{c}$$

The value of c is determined by assuming successive trial values, as shown in the following table:

c	$\dfrac{250}{c}$	$\dfrac{100}{c}$	$\dfrac{100}{c} + 1$	$\cosh \dfrac{250}{c}$
300	0.833	0.333	1.333	1.367
350	0.714	0.286	1.286	1.266
330	0.758	0.303	1.303	1.301
328	0.762	0.305	1.305	1.305

Taking $c = 328$, we have

$$y_B = 100 + c = 428 \text{ ft}$$

a. Maximum and Minimum Values of the Tension. Using Eqs. (7.18), we obtain

$$T_{\min} = T_0 = wc = (3 \text{ lb/ft})(328 \text{ ft}) \qquad T_{\min} = 984 \text{ lb} \blacktriangleleft$$
$$T_{\max} = T_B = wy_B = (3 \text{ lb/ft})(428 \text{ ft}) \qquad T_{\max} = 1284 \text{ lb} \blacktriangleleft$$

b. Length of Cable. One-half the length of the cable is found by solving Eq. (7.17):

$$y_B^2 - s_{CB}^2 = c^2 \qquad s_{CB}^2 = y_B^2 - c^2 = (428)^2 - (328)^2 \qquad s_{CB} = 275 \text{ ft}$$

The total length of the cable is therefore

$$s_{AB} = 2s_{CB} = 2(275 \text{ ft}) \qquad s_{AB} = 550 \text{ ft} \blacktriangleleft$$

SOLVING PROBLEMS ON YOUR OWN

In the last section of this chapter you learned to solve problems involving a *cable carrying a load uniformly distributed along the cable*. The shape assumed by the cable is a catenary and is defined by the equation:

$$y = c \cosh \frac{x}{c} \tag{7.16}$$

1. You should keep in mind that the origin of coordinates for a catenary is located at a distance c directly below the lowest point of the catenary. The length of the cable from the origin to any point is expressed as

$$s = c \sinh \frac{x}{c} \tag{7.15}$$

2. You should first identify all of the known and unknown quantities. Then consider each of the equations listed in the text (Eqs. 7.15 through 7.19), and solve an equation that contains only one unknown. Substitute the value found into another equation, and solve that equation for another unknown.

3. If the sag h is given, use Eq. (7.19) to replace y by $h + c$ in Eq. (7.16) if x is known [Sample Prob. 7.10], or in Eq. (7.17) if s is known, and solve the equation obtained for the constant c.

4. Many of the problems that you will encounter will involve the solution by trial and error of an equation involving a hyperbolic sine or cosine. You can make your work easier by keeping track of your calculations in a table, as in Sample Prob. 7.10, or by applying a numerical methods approach using a computer or calculator.

PROBLEMS

7.127 A 20-m chain of mass 12 kg is suspended between two points at the same elevation. Knowing that the sag is 8 m, determine (a) the distance between the supports, (b) the maximum tension in the chain.

7.128 A 600-ft-long aerial tramway cable having a weight per unit length of 3.0 lb/ft is suspended between two points at the same elevation. Knowing that the sag is 150 ft, find (a) the horizontal distance between the supports, (b) the maximum tension in the cable.

7.129 A 40-m cable is strung as shown between two buildings. The maximum tension is found to be 350 N, and the lowest point of the cable is observed to be 6 m above the ground. Determine (a) the horizontal distance between the buildings, (b) the total mass of the cable.

Fig. P7.129

7.130 A 200-ft steel surveying tape weighs 4 lb. If the tape is stretched between two points at the same elevation and pulled until the tension at each end is 16 lb, determine the horizontal distance between the ends of the tape. Neglect the elongation of the tape due to the tension.

7.131 A 20-m length of wire having a mass per unit length of 0.2 kg/m is attached to a fixed support at A and to a collar at B. Neglecting the effect of friction, determine (a) the force $\mathbf{P}$ for which $h = 8$ m, (b) the corresponding span L.

7.132 A 20-m length of wire having a mass per unit length of 0.2 kg/m is attached to a fixed support at A and to a collar at B. Knowing that the magnitude of the horizontal force applied to the collar is $P = 20$ N, determine (a) the sag h, (b) the span L.

7.133 A 20-m length of wire having a mass per unit length of 0.2 kg/m is attached to a fixed support at A and to a collar at B. Neglecting the effect of friction, determine (a) the sag h for which $L = 15$ m, (b) the corresponding force $\mathbf{P}$.

Fig. P7.131, P7.132, and P7.133

7.134 Determine the sag of a 30-ft chain that is attached to two points at the same elevation that are 20 ft apart.

7.135 A 10-ft rope is attached to two supports A and B as shown. Determine (a) the span of the rope for which the span is equal to the sag, (b) the corresponding angle θ_B.

7.136 A 90-m wire is suspended between two points at the same elevation that are 60 m apart. Knowing that the maximum tension is 300 N, determine (a) the sag of the wire, (b) the total mass of the wire.

7.137 A cable weighing 2 lb/ft is suspended between two points at the same elevation that are 160 ft apart. Determine the smallest allowable sag of the cable if the maximum tension is not to exceed 400 lb.

7.138 A uniform cord 50 in. long passes over a pulley at B and is attached to a pin support at A. Knowing that $L = 20$ in. and neglecting the effect of friction, determine the smaller of the two values of h for which the cord is in equilibrium.

Fig. P7.135

Fig. P7.138

7.139 A motor M is used to slowly reel in the cable shown. Knowing that the mass per unit length of the cable is 0.4 kg/m, determine the maximum tension in the cable when $h = 5$ m.

7.140 A motor M is used to slowly reel in the cable shown. Knowing that the mass per unit length of the cable is 0.4 kg/m, determine the maximum tension in the cable when $h = 3$ m.

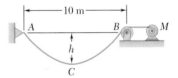

Fig. P7.139 and P7.140

7.141 The cable ACB has a mass per unit length of 0.45 kg/m. Knowing that the lowest point of the cable is located at a distance $a = 0.6$ m below the support A, determine (a) the location of the lowest point C, (b) the maximum tension in the cable.

7.142 The cable ACB has a mass per unit length of 0.45 kg/m. Knowing that the lowest point of the cable is located at a distance $a = 2$ m below the support A, determine (a) the location of the lowest point C, (b) the maximum tension in the cable.

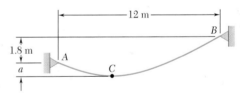

Fig. P7.141 and P7.142

7.143 A uniform cable weighing 3 lb/ft is held in the position shown by a horizontal force $\mathbf{P}$ applied at B. Knowing that $P = 180$ lb and $\theta_A = 60°$, determine (a) the location of point B, (b) the length of the cable.

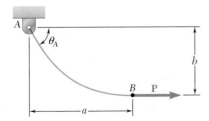

Fig. P7.143 and P7.144

7.144 A uniform cable weighing 3 lb/ft is held in the position shown by a horizontal force $\mathbf{P}$ applied at B. Knowing that $P = 150$ lb and $\theta_A = 60°$, determine (a) the location of point B, (b) the length of the cable.

7.145 To the left of point B the long cable $ABDE$ rests on the rough horizontal surface shown. Knowing that the mass per unit length of the cable is 2 kg/m, determine the force $\mathbf{F}$ when $a = 3.6$ m.

Fig. P7.145 and P7.146

7.146 To the left of point B the long cable $ABDE$ rests on the rough horizontal surface shown. Knowing that the mass per unit length of the cable is 2 kg/m, determine the force $\mathbf{F}$ when $a = 6$ m.

*7.147 The 10-ft cable AB is attached to two collars as shown. The collar at A can slide freely along the rod; a stop attached to the rod prevents the collar at B from moving on the rod. Neglecting the effect of friction and the weight of the collars, determine the distance a.

Fig. P7.147

*7.148 Solve Prob. 7.147 assuming that the angle θ formed by the rod and the horizontal is 45°.

7.149 Denoting by θ the angle formed by a uniform cable and the horizontal, show that at any point (a) $s = c \tan \theta$, (b) $y = c \sec \theta$.

*7.150 (a) Determine the maximum allowable horizontal span for a uniform cable of weight per unit length w if the tension in the cable is not to exceed a given value T_m. (b) Using the result of part a, determine the maximum span of a steel wire for which $w = 0.25$ lb/ft and $T_m = 8000$ lb.

*7.151 A cable has a mass per unit length of 3 kg/m and is supported as shown. Knowing that the span L is 6 m, determine the *two* values of the sag h for which the maximum tension is 350 N.

*7.152 Determine the sag-to-span ratio for which the maximum tension in the cable is equal to the total weight of the entire cable AB.

Fig. P7.151, P7.152, and P7.153

*7.153 A cable of weight per unit length w is suspended between two points at the same elevation that are a distance L apart. Determine (a) the sag-to-span ratio for which the maximum tension is as small as possible, (b) the corresponding values of θ_B and T_m.

In this chapter you learned to determine the internal forces which hold together the various parts of a given member in a structure.

Considering first a *straight two-force member AB* [Sec. 7.2], we recall that such a member is subjected at A and B to equal and opposite forces $\mathbf{F}$ and $-\mathbf{F}$ directed along AB (Fig. 7.19a). Cutting member AB at C and drawing the free-body diagram of portion AC, we conclude that the internal forces which existed at C in member AB are equivalent to an *axial force* $-\mathbf{F}$ equal and opposite to $\mathbf{F}$ (Fig. 7.19b). We note that in the case of a two-force member which is not straight, the internal forces reduce to a force-couple system and not to a single force.

Fig. 7.19

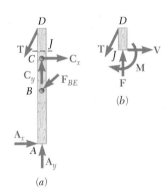

Fig. 7.20

Considering next a *multiforce member AD* (Fig. 7.20a), cutting it at J, and drawing the free-body diagram of portion JD, we conclude that the internal forces at J are equivalent to a force-couple system consisting of the *axial force* $\mathbf{F}$, the *shearing force* $\mathbf{V}$, and a couple $\mathbf{M}$ (Fig. 7.20b). The magnitude of the shearing force measures the *shear* at point J, and the moment of the couple is referred to as the *bending moment* at J. Since an equal and opposite force-couple system would have been obtained by considering the free-body diagram of portion AJ, it is necessary to specify which portion of member AD was used when recording the answers [Sample Prob. 7.1].

Most of the chapter was devoted to the analysis of the internal forces in two important types of engineering structures: *beams* and *cables*. *Beams* are usually long, straight prismatic members designed to support loads applied at various points along the member. In general the loads are perpendicular to the axis of the beam and produce only *shear and bending* in the beam. The loads may be either *concentrated* at specific points, or *distributed* along the entire length or a portion of the beam. The beam itself may be supported in various ways; since only statically determinate beams are considered in this text, we limited our analysis to that of *simply supported beams, overhanging beams,* and *cantilever beams* [Sec. 7.3].

Shear and bending moment in a beam

Internal forces at section
(positive shear and positive bending moment)
Fig. 7.21

To obtain the *shear V* and *bending moment M* at a given point *C* of a beam, we first determine the reactions at the supports by considering the entire beam as a free body. We then cut the beam at *C* and use the free-body diagram of one of the two portions obtained in this fashion to determine *V* and *M*. In order to avoid any confusion regarding the sense of the shearing force **V** and couple **M** (which act in opposite directions on the two portions of the beam), the sign convention illustrated in Fig. 7.21 was adopted [Sec. 7.4]. Once the values of the shear and bending moment have been determined at a few selected points of the beam, it is usually possible to draw a *shear diagram* and a *bending-moment diagram* representing, respectively, the shear and bending moment at any point of the beam [Sec. 7.5]. When a beam is subjected to concentrated loads only, the shear is of constant value between loads and the bending moment varies linearly between loads [Sample Prob. 7.2]. On the other hand, when a beam is subjected to distributed loads, the shear and bending moment vary quite differently [Sample Prob. 7.3].

The construction of the shear and bending-moment diagrams is facilitated if the following relations are taken into account. Denoting by w the distributed load per unit length (assumed positive if directed downward), we have [Sec. 7.5]:

Relations among load, shear, and bending moment

$$\frac{dV}{dx} = -w \tag{7.1}$$

$$\frac{dM}{dx} = V \tag{7.3}$$

or, in integrated form,

$$V_D - V_C = -(\text{area under load curve between } C \text{ and } D) \tag{7.2'}$$
$$M_D - M_C = \text{area under shear curve between } C \text{ and } D \tag{7.4'}$$

Equation (7.2′) makes it possible to draw the shear diagram of a beam from the curve representing the distributed load on that beam and the value of *V* at one end of the beam. Similarly, Eq. (7.4′) makes it possible to draw the bending-moment diagram from the shear diagram and the value of *M* at one end of the beam. However, concentrated loads introduce discontinuities in the shear diagram and concentrated couples in the bending-moment diagram, none of which are accounted for in these equations [Sample Probs. 7.4 and 7.7]. Finally, we note from Eq. (7.3) that the points of the beam where the bending moment is maximum or minimum are also the points where the shear is zero [Sample Prob. 7.5].

Cables with concentrated loads

The second half of the chapter was devoted to the analysis of *flexible cables*. We first considered a cable of negligible weight supporting *concentrated loads* [Sec. 7.7]. Using the entire cable *AB* as a free body (Fig. 7.22), we noted that the three available equilibrium equations were not sufficient to determine the four unknowns representing the reactions at the supports A and B. However, if the coordinates of a point D of the cable are known, an additional equation can be obtained by considering the free-body diagram of the portion AD or DB of the cable. Once the reactions at the supports have been determined, the elevation of any point of the cable and the tension in any portion of the cable can be found from the appropriate free-body diagram [Sample Prob. 7.8]. It was noted that the horizontal component of the force $\mathbf{T}$ representing the tension is the same at any point of the cable.

We next considered cables carrying *distributed loads* [Sec. 7.8]. Using as a free body a portion of cable CD extending from the lowest point C to an arbitrary point D of the cable (Fig. 7.23), we observed that the horizontal component of the tension force $\mathbf{T}$ at D is constant and equal to the tension T_0 at C, while its vertical component is equal to the weight W of the portion of cable CD. The magnitude and direction of $\mathbf{T}$ were obtained from the force triangle:

$$T = \sqrt{T_0^2 + W^2} \qquad \tan \theta = \frac{W}{T_0} \qquad (7.6)$$

In the case of a load *uniformly distributed along the horizontal*—as in a suspension bridge (Fig. 7.24)—the load supported by portion CD is $W = wx$, where w is the constant load per unit horizontal length [Sec. 7.9]. We also found that the curve formed by the cable is a *parabola* of equation

$$y = \frac{wx^2}{2T_0} \qquad (7.8)$$

and that the length of the cable can be found by using the expansion in series given in Eq. (7.10) [Sample Prob. 7.9].

In the case of a load *uniformly distributed along the cable itself*—e.g., a cable hanging under its own weight (Fig. 7.25)—the load supported by portion CD is $W = ws$, where s is the length measured along the cable and w is the constant load per unit length [Sec. 7.10]. Choosing the origin O of the coordinate axes at a distance $c = T_0/w$ below C, we derived the relations

$$s = c \sinh \frac{x}{c} \qquad (7.15)$$

$$y = c \cosh \frac{x}{c} \qquad (7.16)$$

$$y^2 - s^2 = c^2 \qquad (7.17)$$

$$T_0 = wc \qquad W = ws \qquad T = wy \qquad (7.18)$$

which can be used to solve problems involving cables hanging under their own weight [Sample Prob. 7.10]. Equation (7.16), which defines the shape of the cable, is the equation of a *catenary*.

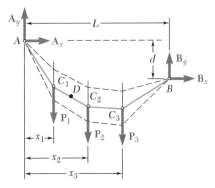

Fig. 7.22

Cables with distributed loads

Fig. 7.23

Parabolic cable

Fig. 7.24

Catenary

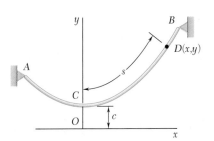

Fig. 7.25

REVIEW PROBLEMS

Fig. P7.154 and P7.155

240 mm

225 mm

225 mm

60 mm
60 mm

135 mm

400 N

C   D

J

B

K

A

7.154 Determine the internal forces at point *J* of the structure shown.

7.155 Determine the internal forces at point *K* of the structure shown.

7.156 An archer aiming at a target is pulling with a 45-lb force on the bowstring. Assuming that the shape of the bow can be approximated by a parabola, determine the internal forces at point *J*.

8 in.

24 in.

B

16 in.

J

16 in.

A

C

32 in.

D

Fig. P7.156

7.157 Knowing that the radius of each pulley is 200 mm and neglecting friction, determine the internal forces at point *J* of the frame shown.

1 m

B

C

1.8 m

A   D

J

K   E

0.2 m

0.8 m 0.8 m 0.8 m

360 N

Fig. P7.157

60 kips   60 kips

P   P

C   D   E   F

A   B

2 ft 2 ft 2 ft 2 ft 2 ft

Fig. P7.158

7.158 For the beam shown, determine (*a*) the magnitude *P* of the two upward forces for which the maximum absolute value of the bending moment in the beam is as small as possible, (*b*) the corresponding value of $|M|_{max}$.

7.159 and 7.160   For the beam and loading shown, (a) draw the shear and bending-moment diagrams, (b) determine the maximum absolute values of the shear and bending moment.

Fig. P7.159

Fig. P7.160

7.161   For the beam and loading shown, (a) draw the shear and bending-moment diagrams, (b) determine the magnitude and location of the maximum absolute value of the bending moment.

7.162   The beam AB, which lies on the ground, supports the parabolic load shown. Assuming the upward reaction of the ground to be uniformly distributed, (a) write the equations of the shear and bending-moment curves, (b) determine the maximum bending moment.

Fig. P7.161

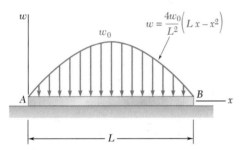

Fig. P7.162

7.163   Two loads are suspended as shown from the cable $ABCD$. Knowing that $d_B = 1.8$ m, determine (a) the distance d_C, (b) the components of the reaction at D, (c) the maximum tension in the cable.

7.164   A wire having a mass per unit length of 0.65 kg/m is suspended from two supports at the same elevation that are 120 m apart. If the sag is 30 m, determine (a) the total length of the wire, (b) the maximum tension in the wire.

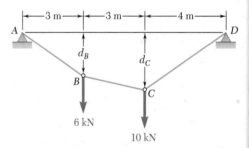

Fig. P7.163

7.165   A counterweight D is attached to a cable that passes over a small pulley at A and is attached to a support at B. Knowing that $L = 45$ ft and $h = 15$ ft, determine (a) the length of the cable from A to B, (b) the weight per unit length of the cable. Neglect the weight of the cable from A to D.

Fig. P7.165

COMPUTER PROBLEMS

Fig. P7.C1

7.C1 An overhanging beam is to be designed to support several concentrated loads. One of the first steps in the design of the beam is to determine the values of the bending moment that can be expected at the supports A and B and under each of the concentrated loads. Write a computer program that can be used to calculate those values for the arbitrary beam and loading shown. Use this program for the beam and loading of (*a*) Prob. 7.36, (*b*) Prob. 7.37, (*c*) Prob. 7.38.

7.C2 Several concentrated loads and a uniformly distributed load are to be applied to a simply supported beam AB. As a first step in the design of the beam, write a computer program that can be used to calculate the shear and bending moment in the beam for the arbitrary loading shown using given increments Δx. Use this program for the beam of (*a*) Prob. 7.39, with $\Delta x = 0.25$ m; (*b*) Prob. 7.41, with $\Delta x = 0.5$ ft; (*c*) Prob. 7.42, with $\Delta x = 0.5$ ft.

Fig. P7.C2

7.C3 A beam AB hinged at B and supported by a roller at D is to be designed to carry a load uniformly distributed from its end A to its midpoint C with maximum efficiency. As part of the design process, write a computer program that can be used to determine the distance a from end A to the point D where the roller should be placed to minimize the absolute value of the bending moment M in the beam. (*Note:* A short preliminary analysis will show that the roller should be placed under the load and that the largest negative value of M will occur at D, while its largest positive value will occur somewhere between D and C. Also see the hint for Prob. 7.55.)

Fig. P7.C3

7.C4 The floor of a bridge will consist of narrow planks resting on two simply supported beams, one of which is shown in the figure. As part of the design of the bridge, it is desired to simulate the effect that driving a 3000-lb truck over the bridge will have on this beam. The distance between the truck's axles is 6 ft, and it is assumed that the weight of the truck is equally distributed over its four wheels. (*a*) Write a computer program that can be used to calculate the magnitude and location of the maximum bending moment in the beam for values of *x* from −3 ft to 10 ft using 0.5-ft increments. (*b*) Using smaller increments if necessary, determine the largest value of the bending moment that occurs in the beam as the truck is driven over the bridge and determine the corresponding value of *x*.

Fig. P7.C4

*7.C5 Write a computer program that can be used to plot the shear and bending-moment diagrams for the beam of Prob. 7.C1. Using this program and a plotting increment $\Delta x \leq L/100$, plot the *V* and *M* diagrams for the beam and loading of (*a*) Prob. 7.36, (*b*) Prob. 7.37, (*c*) Prob. 7.38.

*7.C6 Write a computer program that can be used to plot the shear and bending-moment diagrams for the beam of Prob. 7.C2. Using this program and a plotting increment $\Delta x \leq L/100$, plot the *V* and *M* diagrams for the beam and loading of (*a*) Prob. 7.39, (*b*) Prob. 7.41, (*c*) Prob. 7.42.

7.C7 Write a computer program that can be used in the design of cable supports to calculate the horizontal and vertical components of the reaction at the support A_n from values of the loads $P_1, P_2, \ldots, P_{n-1}$, the horizontal distances $d_1, d_2, \ldots, d_n$, and the two vertical distances h_0 and h_k. Use this program to solve Probs. 7.95*b*, 7.96*b*, and 7.97*b*.

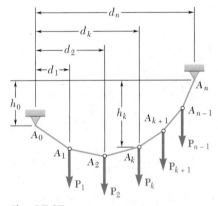

Fig. P7.C7

7.C8 A typical transmission-line installation consists of a cable of length s_{AB} and weight *w* per unit length suspended as shown between two points at the same elevation. Write a computer program and use it to develop a table that can be used in the design of future installations. The table should present the dimensionless quantities h/L, s_{AB}/L, T_0/wL, and T_{max}/wL for values of c/L from 0.2 to 0.5 using 0.025 increments and from 0.5 to 4 using 0.5 increments.

Fig. P7.C8

7.C9 Write a computer program and use it to solve Prob. 7.132 for values of *P* from 0 to 50 N using 5-N increments.

The tractive force that a railroad locomotive can develop depends upon the frictional resistance between the drive wheels and the rails. When the potential exists for wheel slip to occur, such as when a train travels upgrade over wet rails, sand is deposited on top of the railhead to increase this friction.

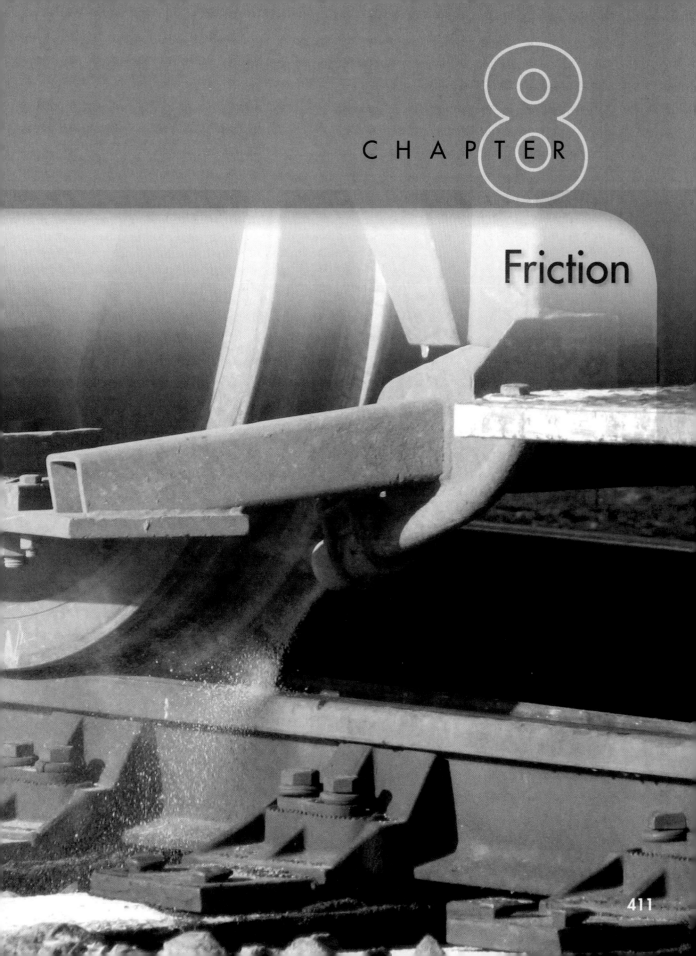

411

Chapter 8 Friction

8.1   INTRODUCTION

In the preceding chapters, it was assumed that surfaces in contact were either *frictionless* or *rough*. If they were frictionless, the force each surface exerted on the other was normal to the surfaces and the two surfaces could move freely with respect to each other. If they were rough, it was assumed that tangential forces could develop to prevent the motion of one surface with respect to the other.

This view was a simplified one. Actually, no perfectly frictionless surface exists. When two surfaces are in contact, tangential forces, called *friction forces,* will always develop if one attempts to move one surface with respect to the other. On the other hand, these friction forces are limited in magnitude and will not prevent motion if sufficiently large forces are applied. The distinction between frictionless and rough surfaces is thus a matter of degree. This will be seen more clearly in the present chapter, which is devoted to the study of friction and of its applications to common engineering situations.

There are two types of friction: *dry friction,* sometimes called *Coulomb friction,* and *fluid friction.* Fluid friction develops between layers of fluid moving at different velocities. Fluid friction is of great importance in problems involving the flow of fluids through pipes and orifices or dealing with bodies immersed in moving fluids. It is also basic in the analysis of the motion of *lubricated mechanisms.* Such problems are considered in texts on fluid mechanics. The present study is limited to dry friction, i.e., to problems involving rigid bodies which are in contact along *nonlubricated* surfaces.

In the first part of this chapter, the equilibrium of various rigid bodies and structures, assuming dry friction at the surfaces of contact, is analyzed. Later a number of specific engineering applications where dry friction plays an important role are considered: wedges, square-threaded screws, journal bearings, thrust bearings, rolling resistance, and belt friction.

8.2   THE LAWS OF DRY FRICTION. COEFFICIENTS OF FRICTION

The laws of dry friction are exemplified by the following experiment. A block of weight **W** is placed on a horizontal plane surface (Fig. 8.1a). The forces acting on the block are its weight **W** and the reaction of the surface. Since the weight has no horizontal component, the reaction of the surface also has no horizontal component; the reaction is therefore *normal* to the surface and is represented by **N** in Fig. 8.1a. Suppose, now, that a horizontal force **P** is applied to the block (Fig. 8.1b). If **P** is small, the block will not move; some other horizontal force must therefore exist, which balances **P**. This other force is the *static-friction* force **F**, which is actually the resultant of a great number of forces acting over the entire surface of contact between the block and the plane. The nature of these forces is not known exactly, but it is generally assumed that these forces are due to the irregularities of the surfaces in contact and, to a certain extent, to molecular attraction.

If the force **P** is increased, the friction force **F** also increases, continuing to oppose **P**, until its magnitude reaches a certain *maximum value* F_m (Fig. 8.1c). If **P** is further increased, the friction force

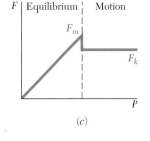

(a)   (b)   (c)

Fig. 8.1

cannot balance it any more and the block starts sliding.† As soon as the block has been set in motion, the magnitude of **F** drops from F_m to a lower value F_k. This is because there is less interpenetration between the irregularities of the surfaces in contact when these surfaces move with respect to each other. From then on, the block keeps sliding with increasing velocity while the friction force, denoted by $\mathbf{F}_k$ and called the *kinetic-friction force*, remains approximately constant.

Experimental evidence shows that the maximum value F_m of the static-friction force is proportional to the normal component N of the reaction of the surface. We have

$$F_m = \mu_s N \qquad (8.1)$$

where μ_s is a constant called the *coefficient of static friction*. Similarly, the magnitude F_k of the kinetic-friction force may be put in the form

$$F_k = \mu_k N \qquad (8.2)$$

where μ_k is a constant called the *coefficient of kinetic friction*. The coefficients of friction μ_s and μ_k do not depend upon the area of

†It should be noted that, as the magnitude F of the friction force increases from 0 to F_m, the point of application A of the resultant **N** of the normal forces of contact moves to the right, so that the couples formed, respectively, by **P** and **F** and by **W** and **N** remain balanced. If **N** reaches B before F reaches its maximum value F_m, the block will tip about B before it can start sliding (see Probs. 8.15 through 8.18).

the surfaces in contact. Both coefficients, however, depend strongly on the *nature* of the surfaces in contact. Since they also depend upon the exact condition of the surfaces, their value is seldom known with an accuracy greater than 5 percent. Approximate values of coefficients of static friction for various dry surfaces are given in Table 8.1. The corresponding values of the coefficient of kinetic friction would be about 25 percent smaller. Since coefficients of friction are dimensionless quantities, the values given in Table 8.1 can be used with both SI units and U.S. customary units.

(*a*) No friction ($P_x = 0$)

(*b*) No motion ($P_x < F_m$)

(*c*) Motion impending ⟶ ($P_x = F_m$)

(*d*) Motion ⟶ ($P_x > F_m$)

Fig. 8.2

TABLE 8.1 Approximate Values of Coefficient of Static Friction for Dry Surfaces

Metal on metal	0.15–0.60
Metal on wood	0.20–0.60
Metal on stone	0.30–0.70
Metal on leather	0.30–0.60
Wood on wood	0.25–0.50
Wood on leather	0.25–0.50
Stone on stone	0.40–0.70
Earth on earth	0.20–1.00
Rubber on concrete	0.60–0.90

From the description given above, it appears that four different situations can occur when a rigid body is in contact with a horizontal surface:

1. The forces applied to the body do not tend to move it along the surface of contact; there is no friction force (Fig. 8.2*a*).
2. The applied forces tend to move the body along the surface of contact but are not large enough to set it in motion. The friction force **F** which has developed can be found by solving the equations of equilibrium for the body. Since there is no evidence that **F** has reached its maximum value, the equation $F_m = \mu_s N$ *cannot be used* to determine the friction force (Fig. 8.2*b*).
3. The applied forces are such that the body is just about to slide. We say that *motion is impending*. The friction force **F** has reached its maximum value F_m and, together with the normal force **N**, balances the applied forces. Both the equations of equilibrium and the equation $F_m = \mu_s N$ *can be used*. We also note that the friction force has a sense opposite to the sense of impending motion (Fig. 8.2*c*).
4. The body is sliding under the action of the applied forces, and the equations of equilibrium do not apply any more. However, **F** is now equal to $\mathbf{F}_k$ and the equation $F_k = \mu_k N$ may be used. The sense of $\mathbf{F}_k$ is opposite to the sense of motion (Fig. 8.2*d*).

8.3 ANGLES OF FRICTION

It is sometimes convenient to replace the normal force **N** and the friction force **F** by their resultant **R**. Let us consider again a block of weight **W** resting on a horizontal plane surface. If no horizontal force is applied to the block, the resultant **R** reduces to the normal force **N** (Fig. 8.3*a*). However, if the applied force **P** has a horizontal component $\mathbf{P}_x$ which tends to move the block, the force **R** will have a horizontal component **F** and, thus, will form an angle ϕ with the normal to the surface (Fig. 8.3*b*). If $\mathbf{P}_x$ is increased until motion becomes impending, the angle between **R** and the vertical grows and reaches a maximum value (Fig. 8.3*c*). This value is called the *angle of static friction* and is denoted by ϕ_s. From the geometry of Fig. 8.3*c*, we note that

$$\tan \phi_s = \frac{F_m}{N} = \frac{\mu_s N}{N}$$

$$\tan \phi_s = \mu_s \tag{8.3}$$

If motion actually takes place, the magnitude of the friction force drops to F_k; similarly, the angle ϕ between **R** and **N** drops to a lower value ϕ_k, called the *angle of kinetic friction* (Fig. 8.3*d*). From the geometry of Fig. 8.3*d*, we write

$$\tan \phi_k = \frac{F_k}{N} = \frac{\mu_k N}{N}$$

$$\tan \phi_k = \mu_k \tag{8.4}$$

Another example will show how the angle of friction can be used to advantage in the analysis of certain types of problems. Consider a block resting on a board and subjected to no other force than its weight **W** and the reaction **R** of the board. The board can be given any desired inclination. If the board is horizontal, the force **R** exerted by the board on the block is perpendicular to the board and balances the weight **W** (Fig. 8.4*a*). If the board is given a small angle of inclination θ, the force **R** will deviate from the perpendicular to the board by the angle θ and will keep balancing **W** (Fig. 8.4*b*); it will then have a normal component **N** of magnitude $N = W \cos \theta$ and a tangential component **F** of magnitude $F = W \sin \theta$.

If we keep increasing the angle of inclination, motion will soon become impending. At that time, the angle between **R** and the normal will have reached its maximum value ϕ_s (Fig. 8.4*c*). The value of the angle of inclination corresponding to impending motion is called the *angle of repose*. Clearly, the angle of repose is equal to the angle of static friction ϕ_s. If the angle of inclination θ is further increased, motion starts and the angle between **R** and the normal drops to the lower value ϕ_k (Fig. 8.4*d*). The reaction **R** is not vertical any more, and the forces acting on the block are unbalanced.

(*a*) No friction

(*b*) No motion

(*c*) Motion impending ⟶

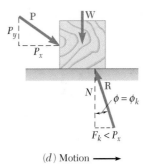

(*d*) Motion ⟶

Fig. 8.3

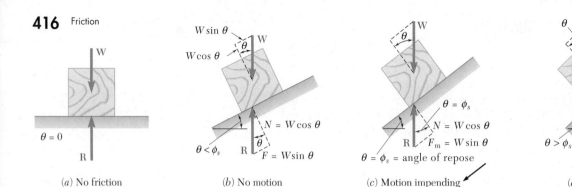

(a) No friction   (b) No motion   (c) Motion impending   (d) Motion

Fig. 8.4

Photo 8.1 The coefficient of static friction between a package and the inclined conveyer belt must be sufficiently large to enable the package to be transported without slipping.

8.4 PROBLEMS INVOLVING DRY FRICTION

Problems involving dry friction are found in many engineering applications. Some deal with simple situations such as the block sliding on a plane described in the preceding sections. Others involve more complicated situations as in Sample Prob. 8.3; many deal with the stability of rigid bodies in accelerated motion and will be studied in dynamics. Also, a number of common machines and mechanisms can be analyzed by applying the laws of dry friction. These include wedges, screws, journal and thrust bearings, and belt transmissions. They will be studied in the following sections.

The *methods* which should be used to solve problems involving dry friction are the same that were used in the preceding chapters. If a problem involves only a motion of translation, with no possible rotation, the body under consideration can usually be treated as a particle, and the methods of Chap. 2 used. If the problem involves a possible rotation, the body must be considered as a rigid body, and the methods of Chap. 4 should be used. If the structure considered is made of several parts, the principle of action and reaction must be used as was done in Chap. 6.

If the body considered is acted upon by more than three forces (including the reactions at the surfaces of contact), the reaction at each surface will be represented by its components **N** and **F** and the problem will be solved from the equations of equilibrium. If only three forces act on the body under consideration, it may be more convenient to represent each reaction by the single force **R** and to solve the problem by drawing a force triangle.

Most problems involving friction fall into one of the following *three groups:* In the *first group* of problems, all applied forces are given and the coefficients of friction are known; we are to determine whether the body considered will remain at rest or slide. The friction force **F** *required to maintain equilibrium* is unknown (its magnitude is *not* equal to $\mu_s N$) and should be determined, together with the normal force **N**, by drawing a free-body diagram and *solving the equations of equilibrium* (Fig. 8.5a). The value found for the magnitude F of the friction force is then compared with the maximum value $F_m = \mu_s N$. If F is smaller than or equal to F_m, the body remains at rest. If the value found for F is larger than F_m, equilibrium cannot be maintained and motion takes place; the actual magnitude of the friction force is then $F_k = \mu_k N$.

In problems of the *second group*, all applied forces are given and the motion is known to be impending; we are to determine the value of the coefficient of static friction. Here again, we determine the friction force and the normal force by drawing a free-body diagram and solving the equations of equilibrium (Fig. 8.5*b*). Since we know that the value found for F is the maximum value F_m, the coefficient of friction may be found by writing and solving the equation $F_m = \mu_s N$.

In problems of the *third group*, the coefficient of static friction is given, and it is known that the motion is impending in a given direction; we are to determine the magnitude or the direction of one of the applied forces. The friction force should be shown in the free-body diagram with a *sense opposite to that of the impending motion* and with a magnitude $F_m = \mu_s N$ (Fig. 8.5*c*). The equations of equilibrium can then be written, and the desired force determined.

As noted above, when only three forces are involved it may be more convenient to represent the reaction of the surface by a single force **R** and to solve the problem by drawing a force triangle. Such a solution is used in Sample Prob. 8.2.

When two bodies A and B are in contact (Fig. 8.6*a*), the forces of friction exerted, respectively, by A on B and by B on A are equal and opposite (Newton's third law). In drawing the free-body diagram of one of the bodies, it is important to include the appropriate friction force with its correct sense. The following rule should then be observed: *The sense of the friction force acting on A is opposite to that of the motion (or impending motion) of A as observed from B* (Fig. 8.6*b*).† The sense of the friction force acting on B is determined in a similar way (Fig. 8.6*c*). Note that the motion of A as observed from B is a *relative motion*. For example, if body A is fixed and body B moves, body A will have a relative motion with respect to B. Also, if both B and A are moving down but B is moving faster than A, body A will be observed, from B, to be moving up.

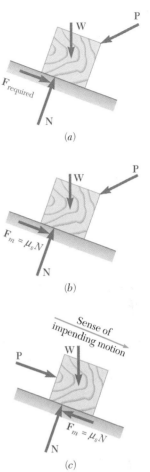

Fig. 8.5

†It is therefore *the same as that of the motion of B as observed from A.*

Fig. 8.6

SAMPLE PROBLEM 8.1

A 100-lb force acts as shown on a 300-lb block placed on an inclined plane. The coefficients of friction between the block and the plane are $\mu_s = 0.25$ and $\mu_k = 0.20$. Determine whether the block is in equilibrium, and find the value of the friction force.

SOLUTION

Force Required for Equilibrium. We first determine the value of the friction force *required to maintain equilibrium*. Assuming that **F** is directed down and to the left, we draw the free-body diagram of the block and write

$$+\nearrow \Sigma F_x = 0: \qquad 100 \text{ lb} - \tfrac{3}{5}(300 \text{ lb}) - F = 0$$
$$F = -80 \text{ lb} \qquad \mathbf{F} = 80 \text{ lb} \nearrow$$

$$+\nwarrow \Sigma F_y = 0: \qquad N - \tfrac{4}{5}(300 \text{ lb}) = 0$$
$$N = +240 \text{ lb} \qquad \mathbf{N} = 240 \text{ lb} \nwarrow$$

The force **F** required to maintain equilibrium is an 80-lb force directed up and to the right; the tendency of the block is thus to move down the plane.

Maximum Friction Force. The magnitude of the maximum friction force which may be developed is

$$F_m = \mu_s N \qquad F_m = 0.25(240 \text{ lb}) = 60 \text{ lb}$$

Since the value of the force required to maintain equilibrium (80 lb) is larger than the maximum value which may be obtained (60 lb), equilibrium will not be maintained and *the block will slide down the plane.*

Actual Value of Friction Force. The magnitude of the actual friction force is obtained as follows:

$$F_{\text{actual}} = F_k = \mu_k N$$
$$= 0.20(240 \text{ lb}) = 48 \text{ lb}$$

The sense of this force is opposite to the sense of motion; the force is thus directed up and to the right:

$$\mathbf{F}_{\text{actual}} = 48 \text{ lb} \nearrow \quad \blacktriangleleft$$

It should be noted that the forces acting on the block are not balanced; the resultant is

$$\tfrac{3}{5}(300 \text{ lb}) - 100 \text{ lb} - 48 \text{ lb} = 32 \text{ lb} \swarrow$$

800 N

SAMPLE PROBLEM 8.2

A support block is acted upon by two forces as shown. Knowing that the coefficients of friction between the block and the incline are $\mu_s = 0.35$ and $\mu_k = 0.25$, determine the force **P** required (a) to start the block moving up the incline, (b) to keep it moving up, (c) to prevent it from sliding down.

SOLUTION

Free-Body Diagram. For each part of the problem we draw a free-body diagram of the block and a force triangle including the 800-N vertical force, the horizontal force **P**, and the force **R** exerted on the block by the incline. The direction of **R** must be determined in each separate case. We note that since **P** is perpendicular to the 800-N force, the force triangle is a right triangle, which can easily be solved for **P**. In most other problems, however, the force triangle will be an oblique triangle and should be solved by applying the law of sines.

a. Force P to Start Block Moving Up

$\tan \phi_s = \mu_s$
$= 0.35$
$\phi_s = 19.29°$
$25° + 19.29° = 44.29°$

$P = (800 \text{ N}) \tan 44.29°$ $\mathbf{P} = 780 \text{ N} \leftarrow$ ◀

b. Force P to Keep Block Moving Up

$\tan \phi_k = \mu_k$
$= 0.25$
$\phi_k = 14.04°$
$25° + 14.04° = 39.04°$

$P = (800 \text{ N}) \tan 39.04°$ $\mathbf{P} = 649 \text{ N} \leftarrow$ ◀

c. Force P to Prevent Block from Sliding Down

$\phi_s = 19.29°$
$25° - 19.29° = 5.71°$

$P = (800 \text{ N}) \tan 5.71°$ $\mathbf{P} = 80.0 \text{ N} \leftarrow$ ◀

SAMPLE PROBLEM 8.3

The movable bracket shown may be placed at any height on the 3-in.-diameter pipe. If the coefficient of static friction between the pipe and bracket is 0.25, determine the minimum distance x at which the load **W** can be supported. Neglect the weight of the bracket.

SOLUTION

Free-Body Diagram. We draw the free-body diagram of the bracket. When **W** is placed at the minimum distance x from the axis of the pipe, the bracket is just about to slip, and the forces of friction at A and B have reached their maximum values:

$$F_A = \mu_s N_A = 0.25\ N_A$$
$$F_B = \mu_s N_B = 0.25\ N_B$$

Equilibrium Equations

$$\xrightarrow{+}\Sigma F_x = 0: \qquad N_B - N_A = 0$$
$$N_B = N_A$$

$$+\uparrow\Sigma F_y = 0: \qquad F_A + F_B - W = 0$$
$$0.25N_A + 0.25N_B = W$$

And, since N_B has been found equal to N_A,

$$0.50N_A = W$$
$$N_A = 2W$$

$$+\gamma\ \Sigma M_B = 0: \qquad N_A(6\text{ in.}) - F_A(3\text{ in.}) - W(x - 1.5\text{ in.}) = 0$$
$$6N_A - 3(0.25N_A) - Wx + 1.5W = 0$$
$$6(2W) - 0.75(2W) - Wx + 1.5W = 0$$

Dividing through by W and solving for x,

$$x = 12\text{ in.} \quad \blacktriangleleft$$

SOLVING PROBLEMS
ON YOUR OWN

In this lesson you studied and applied the *laws of dry friction*. Previously you had encountered only (*a*) frictionless surfaces that could move freely with respect to each other, (*b*) rough surfaces that allowed no motion relative to each other.

A. In solving problems involving dry friction, you should keep the following in mind.

1. The reaction R exerted by a surface on a free body can be resolved into a component **N** and a tangential component **F**. The tangential component is known as the *friction force*. When a body is in contact with a fixed surface the direction of the friction force **F** is opposite to that of the actual or impending motion of the body.

    a. No motion will occur as long as F does not exceed the maximum value $F_m = \mu_s N$, where μ_s is the *coefficient of static friction*.

    b. Motion will occur if a value of F larger than F_m is required to maintain equilibrium. As motion takes place, the actual value of F drops to $F_k = \mu_k N$, where μ_k is the *coefficient of kinetic friction* [Sample Prob. 8.1].

2. When only three forces are involved an alternative approach to the analysis of friction may be preferred [Sample Prob. 8.2]. The reaction **R** is defined by its magnitude R and the angle ϕ it forms with the normal to the surface. No motion will occur as long as ϕ does not exceed the maximum value ϕ_s, where $\tan \phi_s = \mu_s$. Motion will occur if a value of ϕ larger than ϕ_s is required to maintain equilibrium, and the actual value of ϕ will drop to ϕ_k, where $\tan \phi_k = \mu_k$.

3. When two bodies are in contact the sense of the actual or impending relative motion at the point of contact must be determined. On each of the two bodies a friction force **F** should be shown in a direction opposite to that of the actual or impending motion of the body as seen from the other body.

(continued)

B. Methods of solution. The first step in your solution is to *draw a free-body diagram* of the body under consideration, resolving the force exerted on each surface where friction exists into a normal component **N** and a friction force **F**. If several bodies are involved, draw a free-body diagram of each of them, labeling and directing the forces at each surface of contact as you learned to do when analyzing frames in Chap. 6.

The problem you have to solve may fall in one of the following three categories:

1. All the applied forces and the coefficients of friction are known, and you must determine whether equilibrium is maintained. Note that in this situation the friction force is unknown and *cannot be assumed to be equal* to $\mu_s N$.

   a. Write the equations of equilibrium to determine N and F.

   b. Calculate the maximum allowable friction force, $F_m = \mu_s N$. If $F \leq F_m$, equilibrium is maintained. If $F > F_m$, motion occurs, and the magnitude of the friction force is $F_k = \mu_k N$ [Sample Prob. 8.1].

2. All the applied forces are known, and you must find the smallest allowable value of μ_s for which equilibrium is maintained. You will assume that motion is impending and determine the corresponding value of μ_s.

   a. Write the equations of equilibrium to determine N and F.

   b. Since motion is impending, $F = F_m$. Substitute the values found for N and F into the equation $F_m = \mu_s N$ and solve for μ_s.

3. The motion of the body is impending and μ_s is known; you must find some unknown quantity, such as a distance, an angle, the magnitude of a force, or the direction of a force.

   a. Assume a possible motion of the body and, on the free-body diagram, draw the friction force in a direction opposite to that of the assumed motion.

   b. Since motion is impending, $F = F_m = \mu_s N$. Substituting for μ_s its known value, you can express F in terms of N on the free-body diagram, thus eliminating one unknown.

   c. Write and solve the equilibrium equations for the unknown you seek [Sample Prob. 8.3].

PROBLEMS

FREE BODY PRACTICE PROBLEMS

8.F1 Draw the free-body diagram needed to determine the smallest force **P** for which equilibrium of the 7.5-kg block is maintained.

8.F2 Two blocks A and B are connected by a cable as shown. Knowing that the coefficient of static friction at all surfaces of contact is 0.30 and neglecting the friction of the pulleys, draw the free-body diagrams needed to determine the smallest force **P** required to move the blocks.

Fig. P8.F1

Fig. P8.F2

8.F3 The cylinder shown is of weight W and radius r, and the coefficient of static friction μ_s is the same at A and B. Draw the free-body diagram needed to determine the largest couple **M** that can be applied to the cylinder if it is not to rotate.

Fig. P8.F3

8.F4 A uniform crate of mass 30 kg must be moved up along the 15° incline without tipping. Knowing that the force **P** is horizontal, draw the free-body diagram needed to determine the largest allowable coefficient of static friction between the crate and the incline, and the corresponding force **P**.

Fig. P8.F4

423

END-OF-SECTION PROBLEMS

8.1 Determine whether the block shown is in equilibrium and find the magnitude and direction of the friction force when $\theta = 25°$ and $P = 750$ N.

Fig. P8.1 and P8.2

8.2 Determine whether the block shown is in equilibrium and find the magnitude and direction of the friction force when $\theta = 30°$ and $P = 150$ N.

Fig. P8.3, P8.4, and P8.5

8.3 Determine whether the block shown is in equilibrium and find the magnitude and direction of the friction force when $P = 100$ lb.

8.4 Determine whether the block shown is in equilibrium and find the magnitude and direction of the friction force when $P = 60$ lb.

8.5 Determine the smallest value of P required to (*a*) start the block up the incline, (*b*) keep it moving up, (*c*) prevent it from moving down.

8.6 Knowing that the coefficient of friction between the 25-kg block and the incline is $\mu_s = 0.25$, determine (*a*) the smallest value of P required to start the block moving up the incline, (*b*) the corresponding value of β.

Fig. P8.6

8.7 The 80-lb block is attached to link AB and rests on a moving belt. Knowing that $\mu_s = 0.25$ and $\mu_k = 0.20$, determine the magnitude of the horizontal force **P** that should be applied to the belt to maintain its motion (*a*) to the right, (*b*) to the left.

Fig. P8.7

Fig. P8.8

8.8 The coefficients of friction between the block and the rail are $\mu_s = 0.30$ and $\mu_k = 0.25$. Knowing that $\theta = 65°$, determine the smallest value of P required (*a*) to start the block moving up the rail, (*b*) to keep it from moving down.

8.9 Considering only values of θ less than 90°, determine the smallest value of θ required to start the block moving to the right when (*a*) W = 75 lb, (*b*) W = 100 lb.

8.10 Determine the range of values of P for which equilibrium of the block shown is maintained.

$\mu_s = 0.25$
$\mu_k = 0.20$

30°

P

500 N

Fig. P8.10

8.11 The 20-lb block A and the 30-lb block B are supported by an incline that is held in the position shown. Knowing that the coefficient of static friction is 0.15 between the two blocks and zero between block B and the incline, determine the value of θ for which motion is impending.

8.12 The 20-lb block A and the 30-lb block B are supported by an incline that is held in the position shown. Knowing that the coefficient of static friction is 0.15 between all surfaces of contact, determine the value of θ for which motion is impending.

8.13 and 8.14 The coefficients of friction are $\mu_s = 0.40$ and $\mu_k = 0.30$ between all surfaces of contact. Determine the smallest force **P** required to start the 30-kg block moving if cable AB (*a*) is attached as shown, (*b*) is removed.

8.15 A 40-kg packing crate must be moved to the left along the floor without tipping. Knowing that the coefficient of static friction between the crate and the floor is 0.35, determine (*a*) the largest allowable value of α, (*b*) the corresponding magnitude of the force **P**.

P
α
A
B
0.5 m
C
D
0.8 m

Fig. P8.15 and P8.16

8.16 A 40-kg packing crate is pulled by a rope as shown. The coefficient of static friction between the crate and the floor is 0.35. If $\alpha = 40°$, determine (*a*) the magnitude of the force **P** required to move the crate, (*b*) whether the crate will slide or tip.

$\mu_s = 0.25$
$\mu_k = 0.20$

W

θ   30 lb

Fig. P8.9

A
B
θ

Fig. P8.11 and P8.12

20 kg   A
P
30 kg   B

Fig. P8.13

20 kg   A   B
P
30 kg

Fig. P8.14

Fig. P8.17 and P8.18

8.17 A 120-lb cabinet is mounted on casters that can be locked to prevent their rotation. The coefficient of static friction between the floor and each caster is 0.30. If $h = 32$ in., determine the magnitude of the force **P** required to move the cabinet to the right (*a*) if all casters are locked, (*b*) if the casters at *B* are locked and the casters at *A* are free to rotate, (*c*) if the casters at *A* are locked and the casters at *B* are free to rotate.

8.18 A 120-lb cabinet is mounted on casters that can be locked to prevent their rotation. The coefficient of static friction between the floor and each caster is 0.30. Assuming that the casters at both *A* and *B* are locked, determine (*a*) the force **P** required to move the cabinet to the right, (*b*) the largest allowable value of h if the cabinet is not to tip over.

8.19 Wire is being drawn at a constant rate from a spool by applying a vertical force **P** to the wire as shown. The spool and the wire wrapped on the spool have a combined weight of 20 lb. Knowing that the coefficients of friction at both *A* and *B* are $\mu_s = 0.40$ and $\mu_k = 0.30$, determine the required magnitude of the force **P**.

8.20 Solve Prob. 8.19 assuming that the coefficients of friction at *B* are zero.

Fig. P8.19

8.21 The hydraulic cylinder shown exerts a force of 3 kN directed to the right on point *B* and to the left on point *E*. Determine the magnitude of the couple **M** required to rotate the drum clockwise at a constant speed.

Fig. P8.21 and P8.22

8.22 A couple **M** of magnitude 100 N · m is applied to the drum as shown. Determine the smallest force that must be exerted by the hydraulic cylinder on joints *B* and *E* if the drum is not to rotate.

8.23 A slender rod of length *L* is lodged between peg *C* and the vertical wall, and supports a load **P** at end *A*. Knowing that the coefficient of static friction between the peg and the rod is 0.15 and neglecting friction at the roller, determine the range of values of the ratio *L/a* for which equilibrium is maintained.

8.24 Solve Prob. 8.23 assuming that the coefficient of static friction between the peg and the rod is 0.60.

Fig. P8.23

8.25 A 6.5-m ladder AB leans against a wall as shown. Assuming that the coefficient of static friction μ_s is zero at B, determine the smallest value of μ_s at A for which equilibrium is maintained.

8.26 A 6.5-m ladder AB leans against a wall as shown. Assuming that the coefficient of static friction μ_s is the same at A and B, determine the smallest value of μ_s for which equilibrium is maintained.

8.27 The press shown is used to emboss a small seal at E. Knowing that the coefficient of static friction between the vertical guide and the embossing die D is 0.30, determine the force exerted by the die on the seal.

Fig. P8.25 and P8.26

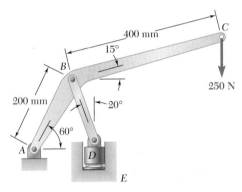

Fig. P8.27

8.28 The machine base shown has a mass of 75 kg and is fitted with skids at A and B. The coefficient of static friction between the skids and the floor is 0.30. If a force **P** of magnitude 500 N is applied at corner C, determine the range of values of θ for which the base will not move.

8.29 The 50-lb plate $ABCD$ is attached at A and D to collars that can slide on the vertical rod. Knowing that the coefficient of static friction is 0.40 between both collars and the rod, determine whether the plate is in equilibrium in the position shown when the magnitude of the vertical force applied at E is (a) $P = 0$, (b) $P = 20$ lb.

Fig. P8.28

Fig. P8.29

8.30 In Prob. 8.29, determine the range of values of the magnitude P of the vertical force applied at E for which the plate will move downward.

8.31 A rod *DE* and a small cylinder are placed between two guides as shown. The rod is not to slip downward, however large the force **P** may be; i.e., the arrangement is said to be self-locking. Neglecting the weight of the cylinder, determine the minimum allowable coefficients of static friction at *A*, *B*, and *C*.

Fig. P8.31

8.32 A 500-N concrete block is to be lifted by the pair of tongs shown. Determine the smallest allowable value of the coefficient of static friction between the block and the tongs at *F* and *G*.

8.33 The 100-mm-radius cam shown is used to control the motion of the plate *CD*. Knowing that the coefficient of static friction between the cam and the plate is 0.45 and neglecting friction at the roller supports, determine (*a*) the force **P** required to maintain the motion of the plate, knowing that the plate is 20 mm thick, (*b*) the largest thickness of the plate for which the mechanism is self-locking (i.e., for which the plate cannot be moved however large the force **P** may be).

Fig. P8.32

Fig. P8.33

8.34 A safety device used by workers climbing ladders fixed to high structures consists of a rail attached to the ladder and a sleeve that can slide on the flange of the rail. A chain connects the worker's belt to the end of an eccentric cam that can be rotated about an axle attached to the sleeve at *C*. Determine the smallest allowable common value of the coefficient of static friction between the flange of the rail, the pins at *A* and *B*, and the eccentric cam if the sleeve is not to slide down when the chain is pulled vertically downward.

Fig. P8.34

8.35 To be of practical use, the safety sleeve described in Prob. 8.34 must be free to slide along the rail when pulled upward. Determine the largest allowable value of the coefficient of static friction between the flange of the rail and the pins at A and B if the sleeve is to be free to slide when pulled as shown in the figure, assuming (a) $\theta = 60°$, (b) $\theta = 50°$, (c) $\theta = 40°$.

8.36 Two 10-lb blocks A and B are connected by a slender rod of negligible weight. The coefficient of static friction is 0.30 between all surfaces of contact, and the rod forms an angle $\theta = 30°$ with the vertical. (a) Show that the system is in equilibrium when $P = 0$. (b) Determine the largest value of P for which equilibrium is maintained.

Fig. P8.35

Fig. P8.36

8.37 Bar AB is attached to collars that can slide on the inclined rods shown. A force $\mathbf{P}$ is applied at point D located at a distance a from end A. Knowing that the coefficient of static friction μ_s between each collar and the rod upon which it slides is 0.30 and neglecting the weights of the bar and of the collars, determine the smallest value of the ratio a/L for which equilibrium is maintained.

Fig. P8.37

8.38 Two identical uniform boards, each of weight 40 lb, are temporarily leaned against each other as shown. Knowing that the coefficient of static friction between all surfaces is 0.40, determine (a) the largest magnitude of the force $\mathbf{P}$ for which equilibrium will be maintained, (b) the surface at which motion will impend.

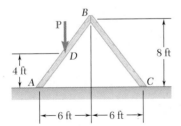

Fig. P8.38

8.39 Knowing that the coefficient of static friction between the collar and the rod is 0.35, determine the range of values of P for which equilibrium is maintained when $\theta = 50°$ and $M = 20$ N · m.

8.40 Knowing that the coefficient of static friction between the collar and the rod is 0.40, determine the range of values of M for which equilibrium is maintained when $\theta = 60°$ and $P = 200$ N.

Fig. P8.39 and P8.40

Fig. P8.41

Fig. P8.43

Fig. P8.46 and P8.47

8.41 A 10-ft beam, weighing 1200 lb, is to be moved to the left onto the platform. A horizontal force **P** is applied to the dolly, which is mounted on frictionless wheels. The coefficients of friction between all surfaces are $\mu_s = 0.30$ and $\mu_k = 0.25$, and initially $x = 2$ ft. Knowing that the top surface of the dolly is slightly higher than the platform, determine the force **P** required to start moving the beam. (*Hint:* The beam is supported at A and D.)

8.42 (*a*) Show that the beam of Prob. 8.41 *cannot* be moved if the top surface of the dolly is slightly *lower* than the platform. (*b*) Show that the beam *can* be moved if two 175-lb workers stand on the beam at B and determine how far to the left the beam can be moved.

8.43 Two 8-kg blocks A and B resting on shelves are connected by a rod of negligible mass. Knowing that the magnitude of a horizontal force **P** applied at C is slowly increased from zero, determine the value of P for which motion occurs, and what that motion is, when the coefficient of static friction between all surfaces is (*a*) $\mu_s = 0.40$, (*b*) $\mu_s = 0.50$.

8.44 A slender steel rod of length 225 mm is placed inside a pipe as shown. Knowing that the coefficient of static friction between the rod and the pipe is 0.20, determine the largest value of θ for which the rod will not fall into the pipe.

Fig. P8.44

8.45 In Prob. 8.44, determine the smallest value of θ for which the rod will not fall out of the pipe.

8.46 Two slender rods of negligible weight are pin-connected at C and attached to blocks A and B, each of weight W. Knowing that $\theta = 80°$ and that the coefficient of static friction between the blocks and the horizontal surface is 0.30, determine the largest value of P for which equilibrium is maintained.

8.47 Two slender rods of negligible weight are pin-connected at C and attached to blocks A and B, each of weight W. Knowing that $P = 1.260W$ and that the coefficient of static friction between the blocks and the horizontal surface is 0.30, determine the range of values of θ, between 0 and 180°, for which equilibrium is maintained.

8.5 WEDGES

Wedges are simple machines used to raise large stone blocks and other heavy loads. These loads can be raised by applying to the wedge a force usually considerably smaller than the weight of the load. In addition, because of the friction between the surfaces in contact, a properly shaped wedge will remain in place after being forced under the load. Wedges can thus be used advantageously to make small adjustments in the position of heavy pieces of machinery.

Consider the block A shown in Fig. 8.7a. This block rests against a vertical wall B and is to be raised slightly by forcing a wedge C between block A and a second wedge D. We want to find the minimum value of the force **P** which must be applied to the wedge C to move the block. It will be assumed that the weight **W** of the block is known, either given in pounds or determined in newtons from the mass of the block expressed in kilograms.

The free-body diagrams of block A and of wedge C have been drawn in Fig. 8.7b and c. The forces acting on the block include its weight and the normal and friction forces at the surfaces of contact with wall B and wedge C. The magnitudes of the friction forces F_1 and F_2 are equal, respectively, to $\mu_s N_1$ and $\mu_s N_2$ since the motion of the block must be started. It is important to show the friction forces with their correct sense. Since the block will move upward, the force F_1 exerted by the wall on the block must be directed downward. On the other hand, since the wedge C moves to the right, the relative motion of A with respect to C is to the left and the force F_2 exerted by C on A must be directed to the right.

Considering now the free body C in Fig. 8.7c, we note that the forces acting on C include the applied force **P** and the normal and friction forces at the surfaces of contact with A and D. The weight of the wedge is small compared with the other forces involved and can be neglected. The forces exerted by A on C are equal and opposite to the forces N_2 and F_2 exerted by C on A and are denoted, respectively, by $-N_2$ and $-F_2$; the friction force $-F_2$ must therefore be directed to the left. We check that the force F_3 exerted by D is also directed to the left.

The total number of unknowns involved in the two free-body diagrams can be reduced to four if the friction forces are expressed in terms of the normal forces. Expressing that block A and wedge C are in equilibrium will provide four equations which can be solved to obtain the magnitude of **P**. It should be noted that in the example considered here, it will be more convenient to replace each pair of normal and friction forces by their resultant. Each free body is then subjected to only three forces, and the problem can be solved by drawing the corresponding force triangles (see Sample Prob. 8.4).

8.6 SQUARE-THREADED SCREWS

Square-threaded screws are frequently used in jacks, presses, and other mechanisms. Their analysis is similar to the analysis of a block sliding along an inclined plane.

Consider the jack shown in Fig. 8.8. The screw carries a load **W** and is supported by the base of the jack. Contact between screw and base takes place along a portion of their threads. By applying a force **P** on the handle, the screw can be made to turn and to raise the load **W**.

(a)

(b)

(c)

Fig. 8.7

Fig. 8.8

Photo 8.2 Wedges are used as shown to split tree trunks because the normal forces exerted by the wedges on the wood are much larger than the forces required to insert the wedges.

The thread of the base has been unwrapped and shown as a straight line in Fig. 8.9a. The correct slope was obtained by plotting horizontally the product $2\pi r$, where r is the mean radius of the thread, and vertically the *lead L* of the screw, i.e., the distance through which the screw advances in one turn. The angle θ this line forms with the horizontal is the *lead angle*. Since the force of friction between two surfaces in contact does not depend upon the area of contact, a much smaller than actual area of contact between the two threads can be assumed, and the screw can be represented by the block shown in Fig. 8.9a. It should be noted, however, that in this analysis of the jack, the friction between cap and screw is neglected.

The free-body diagram of the block should include the load **W**, the reaction **R** of the base thread, and a horizontal force **Q** having the same effect as the force **P** exerted on the handle. The force **Q** should have the same moment as **P** about the axis of the screw and its magnitude should thus be $Q = Pa/r$. The force **Q**, and thus the force **P** required to raise the load **W**, can be obtained from the free-body diagram shown in Fig. 8.9a. The friction angle is taken equal to ϕ_s since the load will presumably be raised through a succession of short strokes. In mechanisms providing for the continuous rotation of a screw, it may be desirable to distinguish between the force required to start motion (using ϕ_s) and that required to maintain motion (using ϕ_k).

(a) Impending motion upward

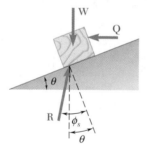

(b) Impending motion downward with $\phi_s > \theta$

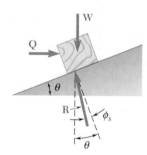

(c) Impending motion downward with $\phi_s < \theta$

Fig. 8.9 Block-and-incline analysis of a screw.

If the friction angle ϕ_s is larger than the lead angle θ, the screw is said to be *self-locking*; it will remain in place under the load. To lower the load, we must then apply the force shown in Fig. 8.9b. If ϕ_s is smaller than θ, the screw will unwind under the load; it is then necessary to apply the force shown in Fig. 8.9c to maintain equilibrium.

The lead of a screw should not be confused with its *pitch*. The lead was defined as the distance through which the screw advances in one turn; the pitch is the distance measured between two consecutive threads. While lead and pitch are equal in the case of *single-threaded* screws, they are different in the case of *multiple-threaded* screws, i.e., screws having several independent threads. It is easily verified that for double-threaded screws, the lead is twice as large as the pitch; for triple-threaded screws, it is three times as large as the pitch; etc.

400 lb

SAMPLE PROBLEM 8.4

The position of the machine block B is adjusted by moving the wedge A. Knowing that the coefficient of static friction is 0.35 between all surfaces of contact, determine the force $\mathbf{P}$ required (a) to raise block B, (b) to lower block B.

SOLUTION

For each part, the free-body diagrams of block B and wedge A are drawn, together with the corresponding force triangles, and the law of sines is used to find the desired forces. We note that since $\mu_s = 0.35$, the angle of friction is

$$\phi_s = \tan^{-1} 0.35 = 19.3°$$

a. Force P to Raise Block

Free Body: Block B

$$\frac{R_1}{\sin 109.3°} = \frac{400 \text{ lb}}{\sin 43.4°}$$
$$R_1 = 549 \text{ lb}$$

Free Body: Wedge A

$$\frac{P}{\sin 46.6°} = \frac{549 \text{ lb}}{\sin 70.7°}$$
$$P = 423 \text{ lb} \qquad \mathbf{P} = 423 \text{ lb} \leftarrow \blacktriangleleft$$

b. Force P to Lower Block

Free Body: Block B

$$\frac{R_1}{\sin 70.7°} = \frac{400 \text{ lb}}{\sin 98.0°}$$
$$R_1 = 381 \text{ lb}$$

Free Body: Wedge A

$$\frac{P}{\sin 30.6°} = \frac{381 \text{ lb}}{\sin 70.7°}$$
$$P = 206 \text{ lb} \qquad \mathbf{P} = 206 \text{ lb} \rightarrow \blacktriangleleft$$

433

SAMPLE PROBLEM 8.5

A clamp is used to hold two pieces of wood together as shown. The clamp has a double square thread of mean diameter equal to 10 mm with a pitch of 2 mm. The coefficient of friction between threads is $\mu_s = 0.30$. If a maximum couple of 40 N · m is applied in tightening the clamp, determine (a) the force exerted on the pieces of wood, (b) the couple required to loosen the clamp.

SOLUTION

a. Force Exerted by Clamp. The mean radius of the screw is $r = 5$ mm. Since the screw is double-threaded, the lead L is equal to twice the pitch: $L = 2(2 \text{ mm}) = 4$ mm. The lead angle θ and the friction angle ϕ_s are obtained by writing

$$\tan \theta = \frac{L}{2\pi r} = \frac{4 \text{ mm}}{10\pi \text{ mm}} = 0.1273 \qquad \theta = 7.3°$$
$$\tan \phi_s = \mu_s = 0.30 \qquad\qquad \phi_s = 16.7°$$

The force $\mathbf{Q}$ which should be applied to the block representing the screw is obtained by expressing that its moment Qr about the axis of the screw is equal to the applied couple.

$$Q(5 \text{ mm}) = 40 \text{ N} \cdot \text{m}$$
$$Q = \frac{40 \text{ N} \cdot \text{m}}{5 \text{ mm}} = \frac{40 \text{ N} \cdot \text{m}}{5 \times 10^{-3} \text{ m}} = 8000 \text{ N} = 8 \text{ kN}$$

The free-body diagram and the corresponding force triangle can now be drawn for the block; the magnitude of the force $\mathbf{W}$ exerted on the pieces of wood is obtained by solving the triangle.

$$W = \frac{Q}{\tan(\theta + \phi_s)} = \frac{8 \text{ kN}}{\tan 24.0°}$$

$$W = 17.97 \text{ kN} \blacktriangleleft$$

b. Couple Required to Loosen Clamp. The force $\mathbf{Q}$ required to loosen the clamp and the corresponding couple are obtained from the free-body diagram and force triangle shown.

$$Q = W \tan (\phi_s - \theta) = (17.97 \text{ kN}) \tan 9.4°$$
$$= 2.975 \text{ kN}$$
$$\text{Couple} = Qr = (2.975 \text{ kN})(5 \text{ mm})$$
$$= (2.975 \times 10^3 \text{ N})(5 \times 10^{-3} \text{ m}) = 14.87 \text{ N} \cdot \text{m}$$

$$\text{Couple} = 14.87 \text{ N} \cdot \text{m} \blacktriangleleft$$

SOLVING PROBLEMS
ON YOUR OWN

In this lesson you learned to apply the laws of friction to the solution of problems involving *wedges* and *square-threaded screws*.

1. Wedges. Keep the following in mind when solving a problem involving a wedge:

a. First draw a free-body diagram of the wedge and of all the other bodies involved. Carefully note the sense of the relative motion of all surfaces of contact and show each friction force acting in *a direction opposite* to the direction of that relative motion.

b. Show the maximum static friction force F_m at each surface if the wedge is to be inserted or removed, *since motion will be impending in each of these cases.*

c. The reaction R and the angle of friction, rather than the normal force and the friction force, can be used in many applications. You can then draw one or more force triangles and determine the unknown quantities either graphically or by trigonometry [Sample Prob. 8.4].

2. Square-Threaded Screws. The analysis of a square-threaded screw is equivalent to the analysis of a block sliding on an incline. To draw the appropriate incline, you should unwrap the thread of the screw and represent it by a straight line [Sample Prob. 8.5]. When solving a problem involving a square-threaded screw, keep the following in mind:

a. Do not confuse the pitch of a screw with the lead of a screw. The *pitch* of a screw is the distance between two consecutive threads, while the *lead* of a screw is the distance the screw advances in one full turn. The lead and the pitch are equal only in single-threaded screws. In a double-threaded screw, the lead is twice the pitch.

b. The couple required to tighten a screw is different from the couple required to loosen it. Also, screws used in jacks and clamps are usually *self-locking;* that is, the screw will remain stationary as long as no couple is applied to it, and a couple must be applied to the screw to loosen it [Sample Prob. 8.5].

PROBLEMS

Fig. P8.48

Fig. P8.50

Fig. P8.51

Fig. P8.54, P8.55, and P8.56

8.48 The machine part *ABC* is supported by a frictionless hinge at *B* and a 10° wedge at *C*. Knowing that the coefficient of static friction at both surfaces of the wedge is 0.20, determine (*a*) the force **P** required to move the wedge, (*b*) the components of the corresponding reaction at *B*.

8.49 Solve Prob. 8.48 assuming that the force **P** is directed to the right.

8.50 and 8.51 The elevation of the end of the steel beam supported by a concrete floor is adjusted by means of the steel wedges *E* and *F*. The base plate *CD* has been welded to the lower flange of the beam, and the end reaction of the beam is known to be 100 kN. The coefficient of static friction is 0.30 between two steel surfaces and 0.60 between steel and concrete. If the horizontal motion of the beam is prevented by the force **Q**, determine (*a*) the force **P** required to raise the beam, (*b*) the corresponding force **Q**.

8.52 and 8.53 Two 10° wedges of negligible weight are used to move and position the 400-lb block. Knowing that the coefficient of static friction is 0.25 at all surfaces of contact, determine the smallest force **P** that should be applied as shown to one of the wedges.

Fig. P8.52   **Fig. P8.53**

8.54 Block *A* supports a pipe column and rests as shown on wedge *B*. Knowing that the coefficient of static friction at all surfaces of contact is 0.25 and that *θ* = 45°, determine the smallest force **P** required to raise block *A*.

8.55 Block *A* supports a pipe column and rests as shown on wedge *B*. Knowing that the coefficient of static friction at all surfaces of contact is 0.25 and that *θ* = 45°, determine the smallest force **P** for which equilibrium is maintained.

8.56 Block *A* supports a pipe column and rests as shown on wedge *B*. The coefficient of static friction at all surfaces of contact is 0.25. If **P** = 0, determine (*a*) the angle *θ* for which sliding is impending, (*b*) the corresponding force exerted on the block by the vertical wall.

8.57 A wedge A of negligible weight is to be driven between two 100-lb plates B and C. The coefficient of static friction between all surfaces of contact is 0.35. Determine the magnitude of the force **P** required to start moving the wedge (*a*) if the plates are equally free to move, (*b*) if plate C is securely bolted to the surface.

Fig. P8.57

8.58 A 10° wedge is used to split a section of a log. The coefficient of static friction between the wedge and the log is 0.35. Knowing that a force **P** of magnitude 600 lb was required to insert the wedge, determine the magnitude of the forces exerted on the wood by the wedge after insertion.

Fig. P8.58

8.59 A 10° wedge is to be forced under end B of the 5-kg rod AB. Knowing that the coefficient of static friction is 0.40 between the wedge and the rod and 0.20 between the wedge and the floor, determine the smallest force **P** required to raise end B of the rod.

8.60 The spring of the door latch has a constant of 1.8 lb/in. and in the position shown exerts a 0.6-lb force on the bolt. The coefficient of static friction between the bolt and the strike plate is 0.40; all other surfaces are well lubricated and may be assumed frictionless. Determine the magnitude of the force **P** required to start closing the door.

Fig. P8.59

Fig. P8.60

8.61 In Prob. 8.60, determine the angle that the face of the bolt should form with the line BC if the force **P** required to close the door is to be the same for both the position shown and the position when B is almost at the strike plate.

8.62 A 5° wedge is to be forced under a 1400-lb machine base at A. Knowing that the coefficient of static friction at all surfaces is 0.20, (*a*) determine the force **P** required to move the wedge, (*b*) indicate whether the machine base will move.

8.63 Solve Prob. 8.62 assuming that the wedge is to be forced under the machine base at B instead of A.

Fig. P8.62

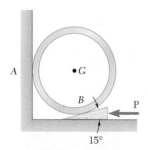

Fig. P8.64 and P8.65

8.64 A 15° wedge is forced under a 50-kg pipe as shown. The coefficient of static friction at all surfaces is 0.20. (*a*) Show that slipping will occur between the pipe and the vertical wall. (*b*) Determine the force **P** required to move the wedge.

8.65 A 15° wedge is forced under a 50-kg pipe as shown. Knowing that the coefficient of static friction at both surfaces of the wedge is 0.20, determine the largest coefficient of static friction between the pipe and the vertical wall for which slipping will occur at *A*.

*8.66 A 200-N block rests as shown on a wedge of negligible weight. The coefficient of static friction μ_s is the same at both surfaces of the wedge, and friction between the block and the vertical wall may be neglected. For $P = 100$ N, determine the value of μ_s for which motion is impending. (*Hint:* Solve the equation obtained by trial and error.)

Fig. P8.66

*8.67 Solve Prob. 8.66 assuming that the rollers are removed and that μ_s is the coefficient of friction at all surfaces of contact.

8.68 Derive the following formulas relating the load **W** and the force **P** exerted on the handle of the jack discussed in Sec. 8.6. (*a*) $P = (Wr/a) \tan (\theta + \phi_s)$, to raise the load; (*b*) $P = (Wr/a) \tan (\phi_s - \theta)$, to lower the load if the screw is self-locking; (*c*) $P = (Wr/a) \tan (\theta - \phi_s)$, to hold the load if the screw is not self-locking.

8.69 The square-threaded worm gear shown has a mean radius of 2 in. and a lead of 0.5 in. The large gear is subjected to a constant clockwise couple of 9.6 kip · in. Knowing that the coefficient of static friction between the two gears is 0.12, determine the couple that must be applied to shaft *AB* in order to rotate the large gear counterclockwise. Neglect friction in the bearings at *A*, *B*, and *C*.

Fig. P8.69

8.70 In Prob. 8.69, determine the couple that must be applied to shaft *AB* in order to rotate the large gear clockwise.

8.71 High-strength bolts are used in the construction of many steel structures. For a 24-mm-nominal-diameter bolt, the required minimum bolt tension is 210 kN. Assuming the coefficient of friction to be 0.40, determine the required couple that should be applied to the bolt and nut. The mean diameter of the thread is 22.6 mm, and the lead is 3 mm. Neglect friction between the nut and washer, and assume the bolt to be square-threaded.

Fig. P8.71

8.72 The position of the automobile jack shown is controlled by a screw *ABC* that is single-threaded at each end (right-handed thread at *A*, left-handed thread at *C*). Each thread has a pitch of 0.1 in. and a mean diameter of 0.375 in. If the coefficient of static friction is 0.15, determine the magnitude of the couple **M** that must be applied to raise the automobile.

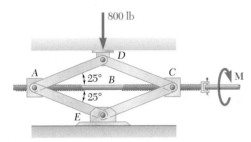

Fig. P8.72

8.73 For the jack of Prob. 8.72, determine the magnitude of the couple **M** that must be applied to lower the automobile.

8.74 In the gear-pulling assembly shown, the square-threaded screw *AB* has a mean radius of 15 mm and a lead of 4 mm. Knowing that the coefficient of static friction is 0.10, determine the couple that must be applied to the screw in order to produce a force of 3 kN on the gear. Neglect friction at end *A* of the screw.

Fig. P8.74

8.75 The ends of two fixed rods *A* and *B* are each made in the form of a single-threaded screw of mean radius 6 mm and pitch 2 mm. Rod *A* has a right-handed thread and rod *B* has a left-handed thread. The coefficient of static friction between the rods and the threaded sleeve is 0.12. Determine the magnitude of the couple that must be applied to the sleeve in order to draw the rods closer together.

Fig. P8.75

8.76 Assuming that in Prob. 8.75 a right-handed thread is used on *both* rods *A* and *B*, determine the magnitude of the couple that must be applied to the sleeve in order to rotate it.

*8.7 JOURNAL BEARINGS. AXLE FRICTION

Journal bearings are used to provide lateral support to rotating shafts and axles. Thrust bearings, which will be studied in the next section, are used to provide axial support to shafts and axles. If the journal bearing is fully lubricated, the frictional resistance depends upon the speed of rotation, the clearance between axle and bearing, and the viscosity of the lubricant. As indicated in Sec. 8.1, such problems are studied in fluid mechanics. The methods of this chapter, however, can be applied to the study of axle friction when the bearing is not lubricated or only partially lubricated. It can then be assumed that the axle and the bearing are in direct contact along a single straight line.

(a)   *(b)*

(c)   *(d)*   *(e)*

Fig. 8.10

Consider two wheels, each of weight **W**, rigidly mounted on an axle supported symmetrically by two journal bearings (Fig. 8.10*a*). If the wheels rotate, we find that to keep them rotating at constant speed, it is necessary to apply to each of them a couple **M**. The free-body diagram in Fig. 8.10*c* represents one of the wheels and the corresponding half axle in projection on a plane perpendicular to the axle. The forces acting on the free body include the weight **W** of the wheel, the couple **M** required to maintain its motion, and a force **R** representing the reaction of the bearing. This force is vertical, equal, and opposite to **W** but does not pass through the center *O* of the axle; **R** is located to the right of *O* at a distance such that its moment about *O* balances the moment **M** of the couple. Therefore, contact between the axle and bearing does not take place at the lowest point *A* when the axle rotates. It takes place at point *B* (Fig. 8.10*b*) or, rather, along a straight line intersecting the plane of the figure at *B*. Physically, this is explained by the fact that when the wheels are set in motion, the axle "climbs" in the bearings until slippage occurs. After sliding back slightly, the axle settles more or less in the position shown. This position is such that the angle between the reaction **R** and the normal to the surface of the bearing is equal to the angle of kinetic friction ϕ_k. The distance from *O* to the line of action of **R** is thus $r \sin \phi_k$, where r is the radius of the axle. Writing that $\Sigma M_O = 0$ for the forces acting on the free body considered, we obtain the magnitude of the couple **M** required to overcome the frictional resistance of one of the bearings:

$$M = Rr \sin \phi_k \qquad (8.5)$$

Observing that, for small values of the angle of friction, $\sin \phi_k$ can be replaced by $\tan \phi_k$, that is, by μ_k, we write the approximate formula

$$M \approx Rr\mu_k \tag{8.6}$$

In the solution of certain problems, it may be more convenient to let the line of action of **R** pass through O, as it does when the axle does not rotate. A couple $-\mathbf{M}$ of the same magnitude as the couple **M** but of opposite sense must then be added to the reaction **R** (Fig. 8.10d). This couple represents the frictional resistance of the bearing.

In case a graphical solution is preferred, the line of action of **R** can be readily drawn (Fig. 8.10e) if we note that it must be tangent to a circle centered at O and of radius

$$r_f = r \sin \phi_k \approx r\mu_k \tag{8.7}$$

This circle is called the *circle of friction* of the axle and bearing and is independent of the loading conditions of the axle.

*8.8 THRUST BEARINGS. DISK FRICTION

Two types of thrust bearings are used to provide axial support to rotating shafts and axles: (1) *end bearings* and (2) *collar bearings* (Fig. 8.11). In the case of collar bearings, friction forces develop between the two ring-shaped areas which are in contact. In the case of end bearings, friction takes place over full circular areas, or over ring-shaped areas when the end of the shaft is hollow. Friction between circular areas, called *disk friction,* also occurs in other mechanisms, such as *disk clutches.*

(a) End bearing　　　　　(b) Collar bearing

Fig. 8.11 Thrust bearings.

To obtain a formula which is valid in the most general case of disk friction, let us consider a rotating hollow shaft. A couple **M** keeps the shaft rotating at constant speed while a force **P** maintains it in contact with a fixed bearing (Fig. 8.12). Contact between the shaft and

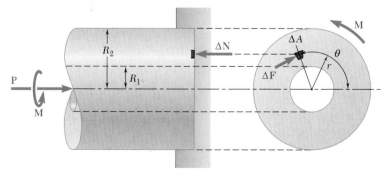

Fig. 8.12

the bearing takes place over a ring-shaped area of inner radius R_1 and outer radius R_2. Assuming that the pressure between the two surfaces in contact is uniform, we find that the magnitude of the normal force $\Delta \mathbf{N}$ exerted on an element of area ΔA is $\Delta N = P\,\Delta A/A$, where $A = \pi(R_2^2 - R_1^2)$, and that the magnitude of the friction force $\Delta \mathbf{F}$ acting on ΔA is $\Delta F = \mu_k\,\Delta N$. Denoting by r the distance from the axis of the shaft to the element of area ΔA, we express the magnitude ΔM of the moment of $\Delta \mathbf{F}$ about the axis of the shaft as follows:

$$\Delta M = r\,\Delta F = \frac{r\mu_k P\,\Delta A}{\pi(R_2^2 - R_1^2)}$$

The equilibrium of the shaft requires that the moment $\mathbf{M}$ of the couple applied to the shaft be equal in magnitude to the sum of the moments of the friction forces $\Delta \mathbf{F}$. Replacing ΔA by the infinitesimal element $dA = r\,d\theta\,dr$ used with polar coordinates, and integrating over the area of contact, we thus obtain the following expression for the magnitude of the couple $\mathbf{M}$ required to overcome the frictional resistance of the bearing:

$$M = \frac{\mu_k P}{\pi(R_2^2 - R_1^2)} \int_0^{2\pi} \int_{R_1}^{R_2} r^2\,dr\,d\theta$$

$$= \frac{\mu_k P}{\pi(R_2^2 - R_1^2)} \int_0^{2\pi} \tfrac{1}{3}(R_2^3 - R_1^3)\,d\theta$$

$$M = \tfrac{2}{3}\mu_k P \frac{R_2^3 - R_1^3}{R_2^2 - R_1^2} \tag{8.8}$$

When contact takes place over a full circle of radius R, formula (8.8) reduces to

$$M = \tfrac{2}{3}\mu_k PR \tag{8.9}$$

The value of M is then the same as would be obtained if contact between shaft and bearing took place at a single point located at a distance $2R/3$ from the axis of the shaft.

The largest couple which can be transmitted by a disk clutch without causing slippage is given by a formula similar to (8.9), where μ_k has been replaced by the coefficient of static friction μ_s.

*8.9 WHEEL FRICTION. ROLLING RESISTANCE

The wheel! is one of the most important inventions of our civilization. Its use makes it possible to move heavy loads with relatively little effort. Because the point of the wheel in contact with the ground at any given instant has no relative motion with respect to the ground, the wheel eliminates the large friction forces which would arise if the load were in direct contact with the ground. However, some resistance to the wheel's motion exists. This resistance has two distinct causes. It is due (1) to a combined effect of axle friction and friction at the rim and (2) to the fact that the wheel and the ground deform, with the result that contact between wheel and ground takes place over a certain area, rather than at a single point.

To understand better the first cause of resistance to the motion of a wheel, let us consider a railroad car supported by eight wheels mounted on axles and bearings. The car is assumed to be moving to the right at constant speed along a straight horizontal track. The free-body diagram of one of the wheels is shown in Fig. 8.13*a*. The forces acting on the free body include the load **W** supported by the wheel and the normal reaction **N** of the track. Since **W** is drawn through the center *O* of the axle, the frictional resistance of the bearing should be represented by a counterclockwise couple **M** (see Sec. 8.7). To keep the free body in equilibrium, we must add two equal and opposite forces **P** and **F**, forming a clockwise couple of moment −**M**. The force **F** is the friction force exerted by the track on the wheel, and **P** represents the force which should be applied to the wheel to keep it rolling at constant speed. Note that the forces **P** and **F** would not exist if there were no friction between wheel and track. The couple **M** representing the axle friction would then be zero; the wheel would slide on the track without turning in its bearing.

The couple **M** and the forces **P** and **F** also reduce to zero when there is no axle friction. For example, a wheel which is not held in bearings and rolls freely and at constant speed on horizontal ground (Fig. 8.13*b*) will be subjected to only two forces: its own weight **W** and the normal reaction **N** of the ground. Regardless of the value of the coefficient of friction between wheel and ground no friction force will act on the wheel. A wheel rolling freely on horizontal ground should thus keep rolling indefinitely.

Experience, however, indicates that the wheel will slow down and eventually come to rest. This is due to the second type of resistance mentioned at the beginning of this section, known as the *rolling resistance*. Under the load **W**, both the wheel and the ground deform slightly, causing the contact between wheel and ground to take place over a certain area. Experimental evidence shows that the resultant of the forces exerted by the ground on the wheel over this area is a force **R** applied at a point *B*, which is not located directly under the center *O* of the wheel, but slightly in front of it (Fig. 8.13*c*). To balance the moment of **W** about *B* and to keep the wheel rolling at constant speed, it is necessary to apply a horizontal force **P** at the center of the wheel. Writing $\Sigma M_B = 0$, we obtain

$$Pr = Wb \qquad (8.10)$$

where r = radius of wheel
b = horizontal distance between O and B

The distance b is commonly called the *coefficient of rolling resistance*. It should be noted that b is not a dimensionless coefficient since it represents a length; b is usually expressed in inches or in millimeters. The value of b depends upon several parameters in a manner which has not yet been clearly established. Values of the coefficient of rolling resistance vary from about 0.01 in. or 0.25 mm for a steel wheel on a steel rail to 5.0 in. or 125 mm for the same wheel on soft ground.

(*a*) Effect of axle friction

(*b*) Free wheel

(*c*) Rolling resistance

Fig. 8.13

SAMPLE PROBLEM 8.6

A pulley of diameter 4 in. can rotate about a fixed shaft of diameter 2 in. The coefficient of static friction between the pulley and shaft is 0.20. Determine (a) the smallest vertical force **P** required to start raising a 500-lb load, (b) the smallest vertical force **P** required to hold the load, (c) the smallest horizontal force **P** required to start raising the same load.

SOLUTION

W = 500 lb   R
← 2.20 in. → ← 1.80 in. →

W = 500 lb   R
← 1.80 in. → ← 2.20 in. →

a. Vertical Force P Required to Start Raising the Load. When the forces in both parts of the rope are equal, contact between the pulley and shaft takes place at A. When **P** is increased, the pulley rolls around the shaft slightly and contact takes place at B. The free-body diagram of the pulley when motion is impending is drawn. The perpendicular distance from the center O of the pulley to the line of action of **R** is

$$r_f = r \sin \phi_s \approx r\mu_s \qquad r_f \approx (1 \text{ in.})0.20 = 0.20 \text{ in.}$$

Summing moments about B, we write

$$+\!\uparrow \Sigma M_B = 0: \qquad (2.20 \text{ in.})(500 \text{ lb}) - (1.80 \text{ in.})P = 0$$
$$P = 611 \text{ lb} \qquad\qquad \mathbf{P} = 611 \text{ lb} \downarrow \quad \blacktriangleleft$$

b. Vertical Force P to Hold the Load. As the force **P** is decreased, the pulley rolls around the shaft and contact takes place at C. Considering the pulley as a free body and summing moments about C, we write

$$+\!\uparrow \Sigma M_C = 0: \qquad (1.80 \text{ in.})(500 \text{ lb}) - (2.20 \text{ in.})P = 0$$
$$P = 409 \text{ lb} \qquad\qquad \mathbf{P} = 409 \text{ lb} \downarrow \quad \blacktriangleleft$$

c. Horizontal Force P to Start Raising the Load. Since the three forces **W**, **P**, and **R** are not parallel, they must be concurrent. The direction of **R** is thus determined from the fact that its line of action must pass through the point of intersection D of **W** and **P**, and must be tangent to the circle of friction. Recalling that the radius of the circle of friction is $r_f = 0.20$ in., we write

$$\sin \theta = \frac{OE}{OD} = \frac{0.20 \text{ in.}}{(2 \text{ in.})\sqrt{2}} = 0.0707 \qquad \theta = 4.1°$$

From the force triangle, we obtain

$$P = W \cot (45° - \theta) = (500 \text{ lb}) \cot 40.9°$$
$$= 577 \text{ lb} \qquad\qquad \mathbf{P} = 577 \text{ lb} \rightarrow \quad \blacktriangleleft$$

SOLVING PROBLEMS
ON YOUR OWN

In this lesson you learned about several additional engineering applications of the laws of friction.

1. Journal bearings and axle friction. In journal bearings, the *reaction does not pass through the center of the shaft or axle* which is being supported. The distance from the center of the shaft or axle to the line of action of the reaction (Fig. 8.10) is defined by the equation.

$$r_f = r \sin \phi_k \approx r\mu_k$$

if motion is actually taking place, and by the equation

$$r_f = r \sin \phi_s \approx r\mu_s$$

if the motion is impending.

Once you have determined the line of action of the reaction, you can draw a *free-body diagram* and use the corresponding equations of equilibrium to complete your solution [Sample Prob. 8.6]. In some problems, it is useful to observe that the line of action of the reaction must be tangent to a circle of radius $r_f \approx r\mu_k$, or $r_f \approx r\mu_s$, known as the *circle of friction* [Sample Prob. 8.6, part *c*].

2. Thrust bearings and disk friction. In a *thrust bearing* the magnitude of the couple required to overcome frictional resistance is equal to the sum of the moments of the *kinetic friction forces* exerted on the elements of the end of the shaft [Eqs. (8.8) and (8.9)].

An example of disk friction is the *disk clutch*. It is analyzed in the same way as a thrust bearing, except that to determine the largest couple that can be transmitted, you must compute the sum of the moments of the *maximum static friction forces* exerted on the disk.

3. Wheel friction and rolling resistance. You saw that the rolling resistance of a wheel is caused by deformations of both the wheel and the ground. The line of action of the reaction **R** of the ground on the wheel intersects the ground at a horizontal distance *b* from the center of the wheel. The distance *b* is known as the *coefficient of rolling resistance* and is expressed in inches or millimeters.

4. In problems involving both rolling resistance and axle friction, your free-body diagram should show that the line of action of the reaction **R** of the ground on the wheel is tangent to the friction circle of the axle and intersects the ground at a horizontal distance from the center of the wheel equal to the coefficient of rolling resistance.

8.77 A lever of negligible weight is loosely fitted onto a 30-mm-radius fixed shaft as shown. Knowing that a force **P** of magnitude 275 N will just start the lever rotating clockwise, determine (*a*) the coefficient of static friction between the shaft and the lever, (*b*) the smallest force **P** for which the lever does not start rotating counterclockwise.

Fig. P8.77

Fig. P8.78

8.78 A hot-metal ladle and its contents weigh 130 kips. Knowing that the coefficient of static friction between the hooks and the pinion is 0.30, determine the tension in cable *AB* required to start tipping the ladle.

8.79 and 8.80 The double pulley shown is attached to a 10-mm-radius shaft that fits loosely in a fixed bearing. Knowing that the coefficient of static friction between the shaft and the poorly lubricated bearing is 0.40, determine the magnitude of the force **P** required to start raising the load.

Fig. P8.79 and P8.81

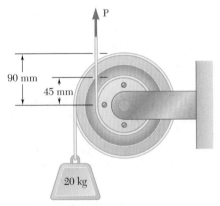

Fig. P8.80 and P8.82

8.81 and 8.82 The double pulley shown is attached to a 10-mm-radius shaft that fits loosely in a fixed bearing. Knowing that the coefficient of static friction between the shaft and the poorly lubricated bearing is 0.40, determine the magnitude of the smallest force **P** required to maintain equilibrium.

8.83 The block and tackle shown are used to raise a 150-lb load. Each of the 3-in.-diameter pulleys rotates on a 0.5-in.-diameter axle. Knowing that the coefficient of static friction is 0.20, determine the tension in each portion of the rope as the load is slowly raised.

8.84 The block and tackle shown are used to lower a 150-lb load. Each of the 3-in.-diameter pulleys rotates on a 0.5-in.-diameter axle. Knowing that the coefficient of static friction is 0.20, determine the tension in each portion of the rope as the load is slowly lowered.

8.85 A scooter is to be designed to roll down a 2 percent slope at a constant speed. Assuming that the coefficient of kinetic friction between the 25-mm-diameter axles and the bearings is 0.10, determine the required diameter of the wheels. Neglect the rolling resistance between the wheels and the ground.

8.86 The link arrangement shown is frequently used in highway bridge construction to allow for expansion due to changes in temperature. At each of the 60-mm-diameter pins A and B the coefficient of static friction is 0.20. Knowing that the vertical component of the force exerted by BC on the link is 200 kN, determine (a) the horizontal force that should be exerted on beam BC to just move the link, (b) the angle that the resulting force exerted by beam BC on the link will form with the vertical.

8.87 and 8.88 A lever AB of negligible weight is loosely fitted onto a 2.5-in.-diameter fixed shaft. Knowing that the coefficient of static friction between the fixed shaft and the lever is 0.15, determine the force $\mathbf{P}$ required to start the lever rotating counterclockwise.

Fig. **P8.83** and **P8.84**

Fig. **P8.86**

Fig. **P8.87** and **P8.89**

Fig. **P8.88** and **P8.90**

8.89 and 8.90 A lever AB of negligible weight is loosely fitted onto a 2.5-in.-diameter fixed shaft. Knowing that the coefficient of static friction between the fixed shaft and the lever is 0.15, determine the force $\mathbf{P}$ required to start the lever rotating clockwise.

8.91 A loaded railroad car has a mass of 30 Mg and is supported by eight 800-mm-diameter wheels with 125-mm-diameter axles. Knowing that the coefficients of friction are $\mu_s = 0.020$ and $\mu_k = 0.015$, determine the horizontal force required (a) to start the car moving, (b) to keep the car moving at a constant speed. Neglect rolling resistance between the wheels and the rails.

4 kN

M

50 mm

120 mm

Fig. P8.92

P

M

θ θ

R_1

R_2

Fig. P8.96

8.92 Knowing that a couple of magnitude 30 N · m is required to start the vertical shaft rotating, determine the coefficient of static friction between the annular surfaces of contact.

8.93 A 50-lb electric floor polisher is operated on a surface for which the coefficient of kinetic friction is 0.25. Assuming that the normal force per unit area between the disk and the floor is uniformly distributed, determine the magnitude Q of the horizontal forces required to prevent motion of the machine.

−Q

20 in.

Q

18 in.

Fig. P8.93

*8.94 The frictional resistance of a thrust bearing decreases as the shaft and bearing surfaces wear out. It is generally assumed that the wear is directly proportional to the distance traveled by any given point of the shaft and thus to the distance r from the point to the axis of the shaft. Assuming, then, that the normal force per unit area is inversely proportional to r, show that the magnitude M of the couple required to overcome the frictional resistance of a worn-out end bearing (with contact over the full circular area) is equal to 75 percent of the value given by Eq. (8.9) for a new bearing.

*8.95 Assuming that bearings wear out as indicated in Prob. 8.94, show that the magnitude M of the couple required to overcome the frictional resistance of a worn-out collar bearing is

$$M = \tfrac{1}{2}\mu_k P(R_1 + R_2)$$

where P = magnitude of the total axial force
R_1, R_2 = inner and outer radii of the collar

*8.96 Assuming that the pressure between the surfaces of contact is uniform, show that the magnitude M of the couple required to overcome frictional resistance for the conical bearing shown is

$$M = \frac{2}{3}\frac{\mu_k P}{\sin\theta}\frac{R_2^3 - R_1^3}{R_2^2 - R_1^2}$$

8.97 Solve Prob. 8.93 assuming that the normal force per unit area between the disk and the floor varies linearly from a maximum at the center to zero at the circumference of the disk.

8.98 Determine the horizontal force required to move a 2500-lb automobile with 23-in.-diameter tires along a horizontal road at a constant speed. Neglect all forms of friction except rolling resistance, and assume the coefficient of rolling resistance to be 0.05 in.

8.99 Knowing that a 6-in.-diameter disk rolls at a constant velocity down a 2 percent incline, determine the coefficient of rolling resistance between the disk and the incline.

8.100 A 900-kg machine base is rolled along a concrete floor using a series of steel pipes with outside diameters of 100 mm. Knowing that the coefficient of rolling resistance is 0.5 mm between the pipes and the base and 1.25 mm between the pipes and the concrete floor, determine the magnitude of the force **P** required to slowly move the base along the floor.

8.101 Solve Prob. 8.85 including the effect of a coefficient of rolling resistance of 1.75 mm.

8.102 Solve Prob. 8.91 including the effect of a coefficient of rolling resistance of 0.5 mm.

Fig. P8.100

8.10 BELT FRICTION

Consider a flat belt passing over a fixed cylindrical drum (Fig. 8.14a). We propose to determine the relation existing between the values T_1 and T_2 of the tension in the two parts of the belt when the belt is just about to slide toward the right.

Let us detach from the belt a small element PP' subtending an angle $\Delta\theta$. Denoting by T the tension at P and by $T + \Delta T$ the tension at P', we draw the free-body diagram of the element of the belt (Fig. 8.14b). Besides the two forces of tension, the forces acting on the free body are the normal component $\Delta\mathbf{N}$ of the reaction of the drum and the friction force $\Delta\mathbf{F}$. Since motion is assumed to be impending, we have $\Delta F = \mu_s \Delta N$. It should be noted that if $\Delta\theta$ is made to approach zero, the magnitudes ΔN and ΔF, and the *difference* ΔT between the tension at P and the tension at P', will also approach zero; the value T of the tension at P, however, will remain unchanged. This observation helps in understanding our choice of notations.

Choosing the coordinate axes shown in Fig. 8.14b, we write the equations of equilibrium for the element PP':

$$\Sigma F_x = 0: \quad (T + \Delta T) \cos \frac{\Delta\theta}{2} - T \cos \frac{\Delta\theta}{2} - \mu_s \Delta N = 0 \quad (8.11)$$

$$\Sigma F_y = 0: \quad \Delta N - (T + \Delta T) \sin \frac{\Delta\theta}{2} - T \sin \frac{\Delta\theta}{2} = 0 \quad (8.12)$$

Solving Eq. (8.12) for ΔN and substituting into (8.11), we obtain after reductions

$$\Delta T \cos \frac{\Delta\theta}{2} - \mu_s (2T + \Delta T) \sin \frac{\Delta\theta}{2} = 0$$

Both terms are now divided by $\Delta\theta$. For the first term, this is done simply by dividing ΔT by $\Delta\theta$. The division of the second term is

(a)

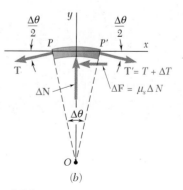

(b)

Fig. 8.14

carried out by dividing the terms in the parentheses by 2 and the sine by $\Delta\theta/2$. We write

$$\frac{\Delta T}{\Delta\theta}\cos\frac{\Delta\theta}{2} - \mu_s\left(T + \frac{\Delta T}{2}\right)\frac{\sin(\Delta\theta/2)}{\Delta\theta/2} = 0$$

If we now let $\Delta\theta$ approach 0, the cosine approaches 1 and $\Delta T/2$ approaches zero, as noted above. The quotient of $\sin(\Delta\theta/2)$ over $\Delta\theta/2$ approaches 1, according to a lemma derived in all calculus textbooks. Since the limit of $\Delta T/\Delta\theta$ is by definition equal to the derivative $dT/d\theta$, we write

$$\frac{dT}{d\theta} - \mu_s T = 0 \qquad \frac{dT}{T} = \mu_s\,d\theta$$

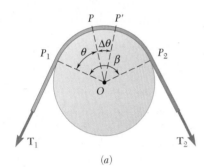

Both members of the last equation (Fig. 8.14a) will now be integrated from P_1 to P_2. At P_1, we have $\theta = 0$ and $T = T_1$; at P_2, we have $\theta = \beta$ and $T = T_2$. Integrating between these limits, we write

$$\int_{T_1}^{T_2}\frac{dT}{T} = \int_0^{\beta}\mu_s\,d\theta$$

$$\ln T_2 - \ln T_1 = \mu_s\beta$$

Fig. 8.14a (repeated)

or, noting that the left-hand member is equal to the natural logarithm of the quotient of T_2 and T_1,

$$\ln\frac{T_2}{T_1} = \mu_s\beta \tag{8.13}$$

This relation can also be written in the form

$$\frac{T_2}{T_1} = e^{\mu_s\beta} \tag{8.14}$$

Photo 8.3 By wrapping the rope around the bollard, the force exerted by the worker to control the rope is much smaller than the tension in the taut portion of the rope.

The formulas we have derived apply equally well to problems involving flat belts passing over fixed cylindrical drums and to problems involving ropes wrapped around a post or capstan. They can also be used to solve problems involving band brakes. In such problems, it is the drum which is about to rotate, while the band remains fixed. The formulas can also be applied to problems involving belt drives. In these problems, both the pulley and the belt rotate; our concern is then to find whether the belt will slip, i.e., whether it will move *with respect* to the pulley.

Formulas (8.13) and (8.14) should be used only if the belt, rope, or brake is *about to slip*. Formula (8.14) will be used if T_1 or T_2 is desired; formula (8.13) will be preferred if either μ_s or the angle of contact β is desired. We should note that T_2 is always larger than T_1; T_2 therefore represents the tension in that part of the belt or rope which *pulls*, while T_1 is the tension in the part which *resists*. We should also observe that the angle of contact β must be expressed in *radians*. The angle β may be larger than 2π; for example, if a rope is wrapped n times around a post, β is equal to $2\pi n$.

If the belt, rope, or brake is actually slipping, formulas similar to (8.13) and (8.14), but involving the coefficient of kinetic friction μ_k, should be used. If the belt, rope, or brake is not slipping and is not about to slip, none of these formulas can be used.

The belts used in belt drives are often V-shaped. In the V belt shown in Fig. 8.15a contact between belt and pulley takes place

(a)   (b)   (c)

Fig. 8.15

along the sides of the groove. The relation existing between the values T_1 and T_2 of the tension in the two parts of the belt when the belt is just about to slip can again be obtained by drawing the free-body diagram of an element of belt (Fig. 8.15b and c). Equations similar to (8.11) and (8.12) are derived, but the magnitude of the total friction force acting on the element is now $2\,\Delta F$, and the sum of the y components of the normal forces is $2\,\Delta N \sin{(\alpha/2)}$. Proceeding as above, we obtain

$$\ln\frac{T_2}{T_1} = \frac{\mu_s\beta}{\sin{(\alpha/2)}} \tag{8.15}$$

or,

$$\frac{T_2}{T_1} = e^{\mu_s\beta/\sin{(\alpha/2)}} \tag{8.16}$$

150 N   7500 N

SAMPLE PROBLEM 8.7

A hawser thrown from a ship to a pier is wrapped two full turns around a bollard. The tension in the hawser is 7500 N; by exerting a force of 150 N on its free end, a dockworker can just keep the hawser from slipping. (*a*) Determine the coefficient of friction between the hawser and the bollard. (*b*) Determine the tension in the hawser that could be resisted by the 150-N force if the hawser were wrapped three full turns around the bollard.

SOLUTION

a. Coefficient of Friction. Since slipping of the hawser is impending, we use Eq. (8.13):

$$\ln \frac{T_2}{T_1} = \mu_s \beta$$

Since the hawser is wrapped two full turns around the bollard, we have

$$\beta = 2(2\pi \text{ rad}) = 12.57 \text{ rad}$$
$$T_1 = 150 \text{ N} \qquad T_2 = 7500 \text{ N}$$

Therefore,

$$\mu_s \beta = \ln \frac{T_2}{T_1}$$
$$\mu_s (12.57 \text{ rad}) = \ln \frac{7500 \text{ N}}{150 \text{ N}} = \ln 50 = 3.91$$
$$\mu_s = 0.311 \qquad\qquad \mu_s = 0.311 \blacktriangleleft$$

b. Hawser Wrapped Three Turns Around Bollard. Using the value of μ_s obtained in part *a*, we now have

$$\beta = 3(2\pi \text{ rad}) = 18.85 \text{ rad}$$
$$T_1 = 150 \text{ N} \qquad \mu_s = 0.311$$

Substituting these values into Eq. (8.14), we obtain

$$\frac{T_2}{T_1} = e^{\mu_s \beta}$$
$$\frac{T_2}{150 \text{ N}} = e^{(0.311)(18.85)} = e^{5.862} = 351.5$$
$$T_2 = 52\ 725 \text{ N}$$
$$T_2 = 52.7 \text{ kN} \blacktriangleleft$$

$T_1 = 150$ N T_2

SAMPLE PROBLEM 8.8

A flat belt connects pulley A, which drives a machine tool, to pulley B, which is attached to the shaft of an electric motor. The coefficients of friction are $\mu_s = 0.25$ and $\mu_k = 0.20$ between both pulleys and the belt. Knowing that the maximum allowable tension in the belt is 600 lb, determine the largest torque which can be exerted by the belt on pulley A.

SOLUTION

Since the resistance to slippage depends upon the angle of contact β between pulley and belt, as well as upon the coefficient of static friction μ_s, and since μ_s is the same for both pulleys, slippage will occur first on pulley B, for which β is smaller.

Pulley B. Using Eq. (8.14) with $T_2 = 600$ lb, $\mu_s = 0.25$, and $\beta = 120° = 2\pi/3$ rad, we write

$$\frac{T_2}{T_1} = e^{\mu_s \beta} \qquad \frac{600 \text{ lb}}{T_1} = e^{0.25(2\pi/3)} = 1.688$$

$$T_1 = \frac{600 \text{ lb}}{1.688} = 355.4 \text{ lb}$$

Pulley A. We draw the free-body diagram of pulley A. The couple $\mathbf{M}_A$ is applied to the pulley by the machine tool to which it is attached and is equal and opposite to the torque exerted by the belt. We write

$$+\curvearrowleft \Sigma M_A = 0: \quad M_A - (600 \text{ lb})(8 \text{ in.}) + (355.4 \text{ lb})(8 \text{ in.}) = 0$$
$$M_A = 1957 \text{ lb} \cdot \text{in.} \qquad M_A = 163.1 \text{ lb} \cdot \text{ft} \blacktriangleleft$$

Note. We may check that the belt does not slip on pulley A by computing the value of μ_s required to prevent slipping at A and verifying that it is smaller than the actual value of μ_s. From Eq. (8.13) we have

$$\mu_s \beta = \ln \frac{T_2}{T_1} = \ln \frac{600 \text{ lb}}{355.4 \text{ lb}} = 0.524$$

and, since $\beta = 240° = 4\pi/3$ rad,

$$\frac{4\pi}{3} \mu_s = 0.524 \qquad \mu_s = 0.125 < 0.25$$

SOLVING PROBLEMS ON YOUR OWN

In the preceding section you learned about *belt friction*. The problems you will solve include belts passing over fixed drums, band brakes in which the drum rotates while the band remains fixed, and belt drives.

1. Problems involving belt friction fall into one of the following two categories:

a. Problems in which slipping is impending. One of the following formulas, involving the *coefficient of static friction* μ_s, may then be used,

$$\ln\frac{T_2}{T_1} = \mu_s\beta \tag{8.13}$$

or

$$\frac{T_2}{T_1} = e^{\mu_s\beta} \tag{8.14}$$

b. Problems in which slipping is occurring. The formulas to be used can be obtained from Eqs. (8.13) and (8.14) by replacing μ_s with the *coefficient of kinetic friction* μ_k.

2. As you start solving a belt-friction problem, be sure to remember the following:

a. The angle β must be expressed in radians. In a belt-and-drum problem, this is the angle subtending the arc of the drum on which the belt is wrapped.

b. The larger tension is always denoted by T_2 and the *smaller tension is denoted by T_1.*

c. The larger tension occurs at the end of the belt which is in the direction of the motion, or impending motion, of the belt relative to the drum.

3. In each of the problems you will be asked to solve, three of the four quantities T_1, T_2, β, and μ_s (or μ_k) will either be given or readily found, and you will then solve the appropriate equation for the fourth quantity. Here are two kinds of problems that you will encounter:

a. Find μ_s between belt and drum, knowing that slipping is impending. From the given data, determine T_1, T_2, and β; substitute these values into Eq. (8.13) and solve for μ_s [Sample Prob. 8.7, part *a*]. Follow the same procedure to find the *smallest value* of μ_s for which slipping will not occur.

b. Find the magnitude of a force or couple applied to the belt or drum, knowing that slipping is impending. The given data should include μ_s and β. If it also includes T_1 or T_2, use Eq. (8.14) to find the other tension. If neither T_1 nor T_2 is known but some other data is given, use the free-body diagram of the belt-drum system to write an equilibrium equation that you will solve simultaneously with Eq. (8.14) for T_1 and T_2. You will then be able to find the magnitude of the specified force or couple from the free-body diagram of the system. Follow the same procedure to determine the *largest value* of a force or couple which can be applied to the belt or drum if no slipping is to occur [Sample Prob. 8.8].

PROBLEMS

8.103 A 300-lb block is supported by a rope that is wrapped $1\frac{1}{2}$ times around a horizontal rod. Knowing that the coefficient of static friction between the rope and the rod is 0.15, determine the range of values of P for which equilibrium is maintained.

8.104 A hawser is wrapped two full turns around a bollard. By exerting an 80-lb force on the free end of the hawser, a dockworker can resist a force of 5000 lb on the other end of the hawser. Determine (*a*) the coefficient of static friction between the hawser and the bollard, (*b*) the number of times the hawser should be wrapped around the bollard if a 20,000-lb force is to be resisted by the same 80-lb force.

Fig. P8.103

8.105 A rope *ABCD* is looped over two pipes as shown. Knowing that the coefficient of static friction is 0.25, determine (*a*) the smallest value of the mass *m* for which equilibrium is possible, (*b*) the corresponding tension in portion *BC* of the rope.

8.106 A rope *ABCD* is looped over two pipes as shown. Knowing that the coefficient of static friction is 0.25, determine (*a*) the largest value of the mass *m* for which equilibrium is possible, (*b*) the corresponding tension in portion *BC* of the rope.

Fig. P8.105 and P8.106

8.107 Knowing that the coefficient of static friction is 0.25 between the rope and the horizontal pipe and 0.20 between the rope and the vertical pipe, determine the range of values of P for which equilibrium is maintained.

Fig. *P8.107* and *P8.108*

8.108 Knowing that the coefficient of static friction is 0.30 between the rope and the horizontal pipe and that the smallest value of P for which equilibrium is maintained is 80 N, determine (*a*) the largest value of P for which equilibrium is maintained, (*b*) the coefficient of static friction between the rope and the vertical pipe.

8.109 A band brake is used to control the speed of a flywheel as shown. The coefficients of friction are $\mu_s = 0.30$ and $\mu_k = 0.25$. Determine the magnitude of the couple being applied to the flywheel, knowing that $P = 45$ N and that the flywheel is rotating counterclockwise at a constant speed.

Fig. P8.109

Fig. P8.110 and P8.111

8.110 The setup shown is used to measure the output of a small turbine. When the flywheel is at rest, the reading of each spring scale is 14 lb. If a 105-lb · in. couple must be applied to the flywheel to keep it rotating clockwise at a constant speed, determine (a) the reading of each scale at that time, (b) the coefficient of kinetic friction. Assume that the length of the belt does not change.

8.111 The setup shown is used to measure the output of a small turbine. The coefficient of kinetic friction is 0.20 and the reading of each spring scale is 16 lb when the flywheel is at rest. Determine (a) the reading of each scale when the flywheel is rotating clockwise at a constant speed, (b) the couple that must be applied to the flywheel. Assume that the length of the belt does not change.

8.112 A flat belt is used to transmit a couple from drum B to drum A. Knowing that the coefficient of static friction is 0.40 and that the allowable belt tension is 450 N, determine the largest couple that can be exerted on drum A.

Fig. P8.112

8.113 A flat belt is used to transmit a couple from pulley A to pulley B. The radius of each pulley is 60 mm, and a force of magnitude $P = 900$ N is applied as shown to the axle of pulley A. Knowing that the coefficient of static friction is 0.35, determine (a) the largest couple that can be transmitted, (b) the corresponding maximum value of the tension in the belt.

Fig. P8.113

8.114 Solve Prob. 8.113 assuming that the belt is looped around the pulleys in a figure eight.

8.115 The speed of the brake drum shown is controlled by a belt attached to the control bar AD. A force **P** of magnitude 25 lb is applied to the control bar at A. Determine the magnitude of the couple being applied to the drum, knowing that the coefficient of kinetic friction between the belt and the drum is 0.25, that $a = 4$ in., and that the drum is rotating at a constant speed (*a*) counterclockwise, (*b*) clockwise.

8.116 The speed of the brake drum shown is controlled by a belt attached to the control bar AD. Knowing that $a = 4$ in., determine the maximum value of the coefficient of static friction for which the brake is not self-locking when the drum rotates counterclockwise.

8.117 The speed of the brake drum shown is controlled by a belt attached to the control bar AD. Knowing that the coefficient of static friction is 0.30 and that the brake drum is rotating counterclockwise, determine the minimum value of a for which the brake is not self-locking.

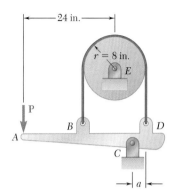

Fig. P8.115, P8.116, and P8.117

8.118 Bucket A and block C are connected by a cable that passes over drum B. Knowing that drum B rotates slowly counterclockwise and that the coefficients of friction at all surfaces are $\mu_s = 0.35$ and $\mu_k = 0.25$, determine the smallest combined mass m of the bucket and its contents for which block C will (*a*) remain at rest, (*b*) start moving up the incline, (*c*) continue moving up the incline at a constant speed.

Fig. P8.118

8.119 Solve Prob. 8.118 assuming that drum B is frozen and cannot rotate.

8.120 and 8.122 A cable is placed around three parallel pipes. Knowing that the coefficients of friction are $\mu_s = 0.25$ and $\mu_k = 0.20$, determine (*a*) the smallest weight W for which equilibrium is maintained, (*b*) the largest weight W that can be raised if pipe B is slowly rotated counterclockwise while pipes A and C remain fixed.

Fig. P8.120 and P8.121

Fig. P8.122 and P8.123

8.121 and 8.123 A cable is placed around three parallel pipes. Two of the pipes are fixed and do not rotate; the third pipe is slowly rotated. Knowing that the coefficients of friction are $\mu_s = 0.25$ and $\mu_k = 0.20$, determine the largest weight W that can be raised (*a*) if only pipe A is rotated counterclockwise, (*b*) if only pipe C is rotated clockwise.

8.124 A recording tape passes over the 20-mm-radius drive drum B and under the idler drum C. Knowing that the coefficients of friction between the tape and the drums are $\mu_s = 0.40$ and $\mu_k = 0.30$ and that drum C is free to rotate, determine the smallest allowable value of P if slipping of the tape on drum B is not to occur.

Fig. P8.124

8.125 Solve Prob. 8.124 assuming that the idler drum C is frozen and cannot rotate.

8.126 The strap wrench shown is used to grip the pipe firmly without marring the external surface of the pipe. Knowing that the coefficient of static friction is the same for all surfaces of contact, determine the smallest value of μ_s for which the wrench will be self-locking when $a = 200$ mm, $r = 30$ mm, and $\theta = 65°$.

Fig. P8.126

8.127 Solve Prob. 8.126 assuming that $\theta = 75°$.

8.128 The 10-lb bar AE is suspended by a cable that passes over a 5-in.-radius drum. Vertical motion of end E of the bar is prevented by the two stops shown. Knowing that $\mu_s = 0.30$ between the cable and the drum, determine (a) the largest counterclockwise couple $\mathbf{M}_0$ that can be applied to the drum if slipping is not to occur, (b) the corresponding force exerted on end E of the bar.

Fig. P8.128

8.129 Solve Prob. 8.128 assuming that a clockwise couple $\mathbf{M}_0$ is applied to the drum.

8.130 Prove that Eqs. (8.13) and (8.14) are valid for any shape of surface provided that the coefficient of friction is the same at all points of contact.

Fig. P8.130

8.131 Complete the derivation of Eq. (8.15), which relates the tension in both parts of a V belt.

8.132 Solve Prob. 8.112 assuming that the flat belt and drums are replaced by a V belt and V pulleys with $\alpha = 36°$. (The angle α is as shown in Fig. 8.15*a*.)

8.133 Solve Prob. 8.113 assuming that the flat belt and pulleys are replaced by a V belt and V pulleys with $\alpha = 36°$. (The angle α is as shown in Fig. 8.15*a*.)

REVIEW AND SUMMARY

This chapter was devoted to the study of *dry friction,* i.e., to problems involving rigid bodies which are in contact along *nonlubricated surfaces.*

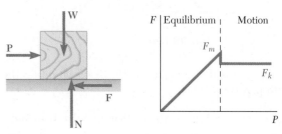

Fig. 8.16

Static and kinetic friction

Applying a horizontal force **P** to a block resting on a horizontal surface [Sec. 8.2], we note that the block at first does not move. This shows that a *friction force* **F** must have developed to balance **P** (Fig. 8.16). As the magnitude of **P** is increased, the magnitude of **F** also increases until it reaches a maximum value F_m. If **P** is further increased, the block starts sliding and the magnitude of **F** drops from F_m to a lower value F_k. Experimental evidence shows that F_m and F_k are proportional to the normal component N of the reaction of the surface. We have

$$F_m = \mu_s N \qquad F_k = \mu_k N \qquad (8.1, 8.2)$$

where μ_s and μ_k are called, respectively, the *coefficient of static friction* and the *coefficient of kinetic friction.* These coefficients depend on the nature and the condition of the surfaces in contact. Approximate values of the coefficients of static friction were given in Table 8.1.

Angles of friction

Fig. 8.17

It is sometimes convenient to replace the normal force **N** and the friction force **F** by their resultant **R** (Fig. 8.17). As the friction force increases and reaches its maximum value $F_m = \mu_s N$, the angle ϕ that **R** forms with the normal to the surface increases and reaches a maximum value ϕ_s, called the *angle of static friction.* If motion actually takes place, the magnitude of **F** drops to F_k; similarly the angle ϕ drops to a lower value ϕ_k, called the *angle of kinetic friction.* As shown in Sec. 8.3, we have

$$\tan \phi_s = \mu_s \qquad \tan \phi_k = \mu_k \qquad (8.3, 8.4)$$

When solving equilibrium problems involving friction, we should keep in mind that the magnitude F of the friction force is equal to $F_m = \mu_s N$ only if the body is about to slide [Sec. 8.4]. *If motion is not impending,* F and N should be considered as *independent unknowns* to be determined from the equilibrium equations (Fig. 8.18a). We

Problems involving friction

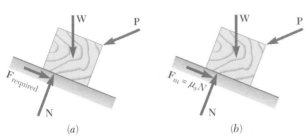

(a)   (b)

Fig. 8.18

should also check that the value of F required to maintain equilibrium is not larger than F_m; if it were, the body would move and the magnitude of the friction force would be $F_k = \mu_k N$ [Sample Prob. 8.1]. On the other hand, *if motion is known to be impending,* F has reached its maximum value $F_m = \mu_s N$ (Fig. 8.18b), and this expression may be substituted for F in the equilibrium equations [Sample Prob. 8.3]. When only three forces are involved in a free-body diagram, including the reaction **R** of the surface in contact with the body, it is usually more convenient to solve the problem by drawing a force triangle [Sample Prob. 8.2].

When a problem involves the analysis of the forces exerted on each other by *two bodies A and B*, it is important to show the friction forces with their correct sense. The correct sense for the friction force exerted by B on A, for instance, is opposite to that of the *relative motion* (or impending motion) of A with respect to B [Fig. 8.6].

In the second part of the chapter we considered a number of specific engineering applications where dry friction plays an important role. In the case of *wedges,* which are simple machines used to raise heavy loads [Sec. 8.5], two or more free-body diagrams were drawn and care was taken to show each friction force with its correct sense [Sample Prob. 8.4]. The analysis of *square-threaded screws,* which are frequently used in jacks, presses, and other mechanisms, was reduced to the analysis of a block sliding on an incline by unwrapping the thread of the screw and showing it as a straight line [Sec. 8.6]. This is done again in Fig. 8.19, where r denotes the *mean radius* of the thread, L is the *lead* of the screw, i.e., the distance through which the screw advances in one turn, **W** is the load, and Qr is equal to the couple exerted on the screw. It was noted that in the case of multiple-threaded screws the lead L of the screw is *not* equal to its pitch, which is the distance measured between two consecutive threads.

Wedges and screws

Fig. 8.19

Other engineering applications considered in this chapter were *journal bearings* and *axle friction* [Sec. 8.7], *thrust bearings* and *disk friction* [Sec. 8.8], *wheel friction* and *rolling resistance* [Sec. 8.9], and *belt friction* [Sec. 8.10].

Belt friction
In solving a problem involving a *flat belt* passing over a fixed cylinder, it is important to first determine the direction in which the belt slips or is about to slip. If the drum is rotating, the motion or impending motion of the belt should be determined *relative* to the rotating drum. For instance, if the belt shown in Fig. 8.20 is about to slip to

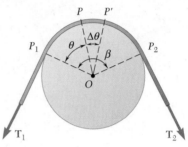

Fig. 8.20

the right relative to the drum, the friction forces exerted by the drum on the belt will be directed to the left and the tension will be larger in the right-hand portion of the belt than in the left-hand portion. Denoting the larger tension by T_2, the smaller tension by T_1, the coefficient of static friction by μ_s, and the angle (in radians) subtended by the belt by β, we derived in Sec. 8.10 the formulas

$$\ln\frac{T_2}{T_1} = \mu_s\beta \tag{8.13}$$

$$\frac{T_2}{T_1} = e^{\mu_s\beta} \tag{8.14}$$

which were used in solving Sample Probs. 8.7 and 8.8. If the belt actually slips on the drum, the coefficient of static friction μ_s should be replaced by the coefficient of kinetic friction μ_k in both of these formulas.

8.134 Determine whether the block shown is in equilibrium and find the magnitude and direction of the friction force when $\theta = 35°$ and $P = 200$ N.

8.135 Three 4-kg packages A, B, and C are placed on a conveyor belt that is at rest. Between the belt and both packages A and C the coefficients of friction are $\mu_s = 0.30$ and $\mu_k = 0.20$; between package B and the belt the coefficients are $\mu_s = 0.10$ and $\mu_k = 0.08$. The packages are placed on the belt so that they are in contact with each other and at rest. Determine which, if any, of the packages will move and the friction force acting on each package.

Fig. P8.134

Fig. P8.135

8.136 The cylinder shown is of weight W and radius r. Express in terms W and r the magnitude of the largest couple **M** that can be applied to the cylinder if it is not to rotate, assuming the coefficient of static friction to be (a) zero at A and 0.30 at B, (b) 0.25 at A and 0.30 at B.

8.137 End A of a slender, uniform rod of length L and weight W bears on a surface as shown, while end B is supported by a cord BC. Knowing that the coefficients of friction are $\mu_s = 0.40$ and $\mu_k = 0.30$, determine (a) the largest value of θ for which motion is impending, (b) the corresponding value of the tension in the cord.

Fig. P8.136

Fig. P8.137

Fig. P8.138

8.138 A worker slowly moves a 50-kg crate to the left along a loading dock by applying a force **P** at corner B as shown. Knowing that the crate starts to tip about the edge E of the loading dock when $a = 200$ mm, determine (a) the coefficient of kinetic friction between the crate and the loading dock, (b) the corresponding magnitude P of the force.

8.139 A window sash weighing 10 lb is normally supported by two 5-lb sash weights. Knowing that the window remains open after one sash cord has broken, determine the smallest possible value of the coefficient of static friction. (Assume that the sash is slightly smaller than the frame and will bind only at points A and D.)

Fig. P8.139

8.140 The slender rod AB of length $l = 600$ mm is attached to a collar at B and rests on a small wheel located at a horizontal distance $a = 80$ mm from the vertical rod on which the collar slides. Knowing that the coefficient of static friction between the collar and the vertical rod is 0.25 and neglecting the radius of the wheel, determine the range of values of P for which equilibrium is maintained when $Q = 100$ N and $\theta = 30°$.

Fig. P8.140

8.141 The machine part ABC is supported by a frictionless hinge at B and a 10° wedge at C. Knowing that the coefficient of static friction is 0.20 at both surfaces of the wedge, determine (a) the force **P** required to move the wedge to the left, (b) the components of the corresponding reaction at B.

Fig. P8.141

8.142 A conical wedge is placed between two horizontal plates that are then slowly moved toward each other. Indicate what will happen to the wedge (*a*) if $\mu_s = 0.20$, (*b*) if $\mu_s = 0.30$.

Fig. P8.142

8.143 In the machinist's vise shown, the movable jaw *D* is rigidly attached to the tongue *AB* that fits loosely into the fixed body of the vise. The screw is single-threaded into the fixed base and has a mean diameter of 0.75 in. and a pitch of 0.25 in. The coefficient of static friction is 0.25 between the threads and also between the tongue and the body. Neglecting bearing friction between the screw and the movable head, determine the couple that must be applied to the handle in order to produce a clamping force of 1 kip.

Fig. P8.143

8.144 A lever of negligible weight is loosely fitted onto a 75-mm-diameter fixed shaft. It is observed that the lever will just start rotating if a 3-kg mass is added at *C*. Determine the coefficient of static friction between the shaft and the lever.

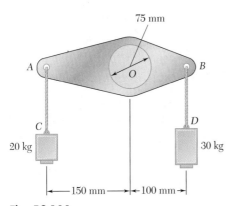

Fig. P8.144

8.145 In the pivoted motor mount shown, the weight **W** of the 175-lb motor is used to maintain tension in the drive belt. Knowing that the coefficient of static friction between the flat belt and drums *A* and *B* is 0.40, and neglecting the weight of platform *CD*, determine the largest couple that can be transmitted to drum *B* when the drive drum *A* is rotating clockwise.

Fig. P8.145

COMPUTER PROBLEMS

Fig. P8.C1

8.C1 The position of the 10-kg rod *AB* is controlled by the 2-kg block shown, which is slowly moved to the left by the force **P**. Knowing that the coefficient of kinetic friction between all surfaces of contact is 0.25, write a computer program and use it to calculate the magnitude *P* of the force for values of *x* from 900 to 100 mm, using 50-mm decrements. Using appropriate smaller decrements, determine the maximum value of *P* and the corresponding value of *x*.

8.C2 Blocks *A* and *B* are supported by an incline that is held in the position shown. Knowing that block *A* weighs 20 lb and that the coefficient of static friction between all surfaces of contact is 0.15, write a computer program and use it to calculate the value of θ for which motion is impending for weights of block *B* from 0 to 100 lb, using 10-lb increments.

Fig. P8.C2

Fig. P8.C3

8.C3 A 300-g cylinder *C* rests on cylinder *D* as shown. Knowing that the coefficient of static friction μ_s is the same at *A* and *B*, write a computer program and use it to determine, for values of μ_s from 0 to 0.40 and using 0.05 increments, the largest counterclockwise couple **M** that can be applied to cylinder *D* if it is not to rotate.

8.C4 Two rods are connected by a slider block *D* and are held in equilibrium by the couple **M**$_A$ as shown. Knowing that the coefficient of static friction between rod *AC* and the slider block is 0.40, write a computer program and use it to determine, for values of θ from 0 to 120° and using 10° increments, the range of values of M_A for which equilibrium is maintained.

Fig. P8.C4

8.C5 The 10-lb block A is slowly moved up the circular cylindrical surface by a cable that passes over a small fixed cylindrical drum at B. The coefficient of kinetic friction is known to be 0.30 between the block and the surface and between the cable and the drum. Write a computer program and use it to calculate the force $\mathbf{P}$ required to maintain the motion for values of θ from 0 to 90°, using 10° increments. For the same values of θ calculate the magnitude of the reaction between the block and the surface. [Note that the angle of contact between the cable and the fixed drum is $\beta = \pi - (\theta/2)$.]

Fig. P8.C5

8.C6 A flat belt is used to transmit a couple from drum A to drum B. The radius of each drum is 80 mm, and the system is fitted with an idler wheel C that is used to increase the contact between the belt and the drums. The allowable belt tension is 200 N, and the coefficient of static friction between the belt and the drums is 0.30. Write a computer program and use it to calculate the largest couple that can be transmitted for values of θ from 0 to 30°, using 5° increments.

Fig. P8.C6

8.C7 Two collars A and B that slide on vertical rods with negligible friction are connected by a 30-in. cord that passes over a fixed shaft at C. The coefficient of static friction between the cord and the fixed shaft is 0.30. Knowing that the weight of collar B is 8 lb, write a computer program and use it to determine, for values of θ from 0 to 60° and using 10° increments, the largest and smallest weight of collar A for which equilibrium is maintained.

Fig. P8.C7

8.C8 The end B of a uniform beam of length L is being pulled by a stationary crane. Initially the beam lies on the ground with end A directly below pulley C. As the cable is slowly pulled in, the beam first slides to the left with $\theta = 0$ until it has moved through a distance x_0. In a second phase, end B is raised, while end A keeps sliding to the left until x reaches its maximum value x_m and θ the corresponding value θ_1. The beam then rotates about A' while θ keeps increasing. As θ reaches the value θ_2, end A starts sliding to the right and keeps sliding in an irregular manner until B reaches C. Knowing that the coefficients of friction between the beam and the ground are $\mu_s = 0.50$ and $\mu_k = 0.40$, (a) write a program to compute x for any value of θ while the beam is sliding to the left and use this program to determine x_0, x_m, and θ_1, (b) modify the program to compute for any θ the value of x for which sliding would be impending to the right and use this new program to determine the value θ_2 of θ corresponding to $x = x_m$.

Fig. P8.C8

The strength of structural members used in the construction of buildings depends to a large extent on the properties of their cross sections. This includes the second moments of area, or moments of inertia, of these cross sections.

CHAPTER 9

Distributed Forces: Moments of Inertia

9.1 INTRODUCTION

In Chap. 5, we analyzed various systems of forces distributed over an area or volume. The three main types of forces considered were (1) weights of homogeneous plates of uniform thickness (Secs. 5.3 through 5.6), (2) distributed loads on beams (Sec. 5.8) and hydrostatic forces (Sec. 5.9), and (3) weights of homogeneous three-dimensional bodies (Secs. 5.10 and 5.11). In the case of homogeneous plates, the magnitude ΔW of the weight of an element of a plate was proportional to the area ΔA of the element. For distributed loads on beams, the magnitude ΔW of each elemental weight was represented by an element of area $\Delta A = \Delta W$ under the load curve; in the case of hydrostatic forces on submerged rectangular surfaces, a similar procedure was followed. In the case of homogeneous three-dimensional bodies, the magnitude ΔW of the weight of an element of the body was proportional to the volume ΔV of the element. Thus, in all cases considered in Chap. 5, the distributed forces were proportional to the elemental areas or volumes associated with them. The resultant of these forces, therefore, could be obtained by summing the corresponding areas or volumes, and the moment of the resultant about any given axis could be determined by computing the first moments of the areas or volumes about that axis.

In the first part of this chapter, we consider distributed forces $\Delta \mathbf{F}$ whose magnitudes depend not only upon the elements of area ΔA on which these forces act but also upon the distance from ΔA to some given axis. More precisely, the magnitude of the force per unit area $\Delta F/\Delta A$ is assumed to vary linearly with the distance to the axis. As indicated in the next section, forces of this type are found in the study of the bending of beams and in problems involving submerged nonrectangular surfaces. Assuming that the elemental forces involved are distributed over an area A and vary linearly with the distance y to the x axis, it will be shown that while the magnitude of their resultant $\mathbf{R}$ depends upon the first moment $Q_x = \int y \, dA$ of the area A, the location of the point where $\mathbf{R}$ is applied depends upon the *second moment, or moment of inertia,* $I_x = \int y^2 \, dA$ of the same area with respect to the x axis. You will learn to compute the moments of inertia of various areas with respect to given x and y axes. Also introduced in the first part of this chapter is the *polar moment of inertia* $J_O = \int r^2 \, dA$ of an area, where r is the distance from the element of area dA to the point O. To facilitate your computations, a relation will be established between the moment of inertia I_x of an area A with respect to a given x axis and the moment of inertia $I_{x'}$ of the same area with respect to the parallel centroidal x' axis (parallel-axis theorem). You will also study the transformation of the moments of inertia of a given area when the coordinate axes are rotated (Secs. 9.9 and 9.10).

In the second part of the chapter, you will learn how to determine the moments of inertia of various *masses* with respect to a given axis. As you will see in Sec. 9.11, the moment of inertia of a given mass about an axis AA' is defined as $I = \int r^2 \, dm$, where r is the distance from the axis AA' to the element of mass dm. Moments of inertia of masses are encountered in dynamics in problems involving the rotation of a rigid body about an axis. To facilitate the computation of mass moments of inertia, the parallel-axis theorem will be introduced (Sec. 9.12). Finally, you will learn to analyze the transformation of moments of inertia of masses when the coordinate axes are rotated (Secs. 9.16 through 9.18).

MOMENTS OF INERTIA OF AREAS

9.2 SECOND MOMENT, OR MOMENT OF INERTIA, OF AN AREA

In the first part of this chapter, we consider distributed forces $\Delta \mathbf{F}$ whose magnitudes ΔF are proportional to the elements of area ΔA on which the forces act and at the same time vary linearly with the distance from ΔA to a given axis.

Consider, for example, a beam of uniform cross section which is subjected to two equal and opposite couples applied at each end of the beam. Such a beam is said to be in *pure bending*, and it is shown in mechanics of materials that the internal forces in any section of the beam are distributed forces whose magnitudes $\Delta F = ky \, \Delta A$ vary linearly with the distance y between the element of area ΔA and an axis passing through the centroid of the section. This axis, represented by the x axis in Fig. 9.1, is known as the *neutral axis* of the section. The forces on one side of the neutral axis are forces of compression, while those on the other side are forces of tension; on the neutral axis itself the forces are zero.

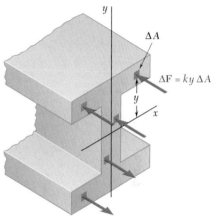

Fig. 9.1

The magnitude of the resultant $\mathbf{R}$ of the elemental forces $\Delta \mathbf{F}$ which act over the entire section is

$$R = \int ky \, dA = k \int y \, dA$$

The last integral obtained is recognized as the *first moment* Q_x of the section about the x axis; it is equal to $\bar{y}A$ and is thus equal to zero, since the centroid of the section is located on the x axis. The system of the forces $\Delta \mathbf{F}$ thus reduces to a couple. The magnitude M of this couple (bending moment) must be equal to the sum of the moments $\Delta M_x = y \, \Delta F = ky^2 \, \Delta A$ of the elemental forces. Integrating over the entire section, we obtain

$$M = \int ky^2 \, dA = k \int y^2 \, dA$$

The last integral is known as the *second moment*, or *moment of inertia*,† of the beam section with respect to the x axis and is denoted by I_x. It is obtained by multiplying each element of area dA by the *square of its distance* from the x axis and integrating over the beam section. Since each product $y^2 \, dA$ is positive, regardless of the sign of y, or zero (if y is zero), the integral I_x will always be positive.

Another example of a second moment, or moment of inertia, of an area is provided by the following problem from hydrostatics: A

---

†The term *second moment* is more proper than the term *moment of inertia*, since, logically, the latter should be used only to denote integrals of mass (see Sec. 9.11). In engineering practice, however, moment of inertia is used in connection with areas as well as masses.

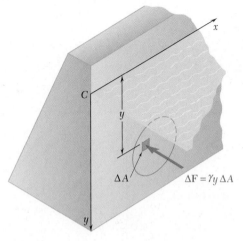

Fig. 9.2

vertical circular gate used to close the outlet of a large reservoir is submerged under water as shown in Fig. 9.2. What is the resultant of the forces exerted by the water on the gate, and what is the moment of the resultant about the line of intersection of the plane of the gate and the water surface (x axis)?

If the gate were rectangular, the resultant of the forces of pressure could be determined from the pressure curve, as was done in Sec. 5.9. Since the gate is circular, however, a more general method must be used. Denoting by y the depth of an element of area ΔA and by γ the specific weight of water, the pressure at the element is $p = \gamma y$, and the magnitude of the elemental force exerted on ΔA is $\Delta F = p\,\Delta A = \gamma y\,\Delta A$. The magnitude of the resultant of the elemental forces is thus

$$R = \int \gamma y\,dA = \gamma \int y\,dA$$

and can be obtained by computing the first moment $Q_x = \int y\,dA$ of the area of the gate with respect to the x axis. The moment M_x of the resultant must be equal to the sum of the moments $\Delta M_x = y\,\Delta F = \gamma y^2\,\Delta A$ of the elemental forces. Integrating over the area of the gate, we have

$$M_x = \int \gamma y^2\,dA = \gamma \int y^2\,dA$$

Here again, the integral obtained represents the second moment, or moment of inertia, I_x of the area with respect to the x axis.

9.3 DETERMINATION OF THE MOMENT OF INERTIA OF AN AREA BY INTEGRATION

We defined in the preceding section the second moment, or moment of inertia, of an area A with respect to the x axis. Defining in a similar way the moment of inertia I_y of the area A with respect to the y axis, we write (Fig. 9.3a)

$$I_x = \int y^2\,dA \qquad I_y = \int x^2\,dA \tag{9.1}$$

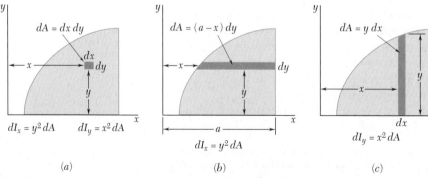

(a)   (b)   (c)

Fig. 9.3

These integrals, known as the *rectangular moments of inertia* of the area A, can be more easily evaluated if we choose dA to be a thin strip parallel to one of the coordinate axes. To compute I_x, the strip is chosen parallel to the x axis, so that all of the points of the strip are at the same distance y from the x axis (Fig. 9.3b); the moment of inertia dI_x of the strip is then obtained by multiplying the area dA of the strip by y^2. To compute I_y, the strip is chosen parallel to the y axis so that all of the points of the strip are at the same distance x from the y axis (Fig. 9.3c); the moment of inertia dI_y of the strip is $x^2\, dA$.

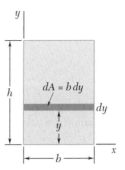

Fig. 9.4

Moment of Inertia of a Rectangular Area. As an example, let us determine the moment of inertia of a rectangle with respect to its base (Fig. 9.4). Dividing the rectangle into strips parallel to the x axis, we obtain

$$dA = b\, dy \qquad dI_x = y^2 b\, dy$$

$$I_x = \int_0^h by^2\, dy = \tfrac{1}{3}bh^3 \tag{9.2}$$

Computing I_x and I_y Using the Same Elemental Strips. The formula just derived can be used to determine the moment of inertia dI_x with respect to the x axis of a rectangular strip which is parallel to the y axis, such as the strip shown in Fig. 9.3c. Setting $b = dx$ and $h = y$ in formula (9.2), we write

$$dI_x = \tfrac{1}{3}y^3\, dx$$

On the other hand, we have

$$dI_y = x^2\, dA = x^2 y\, dx$$

The same element can thus be used to compute the moments of inertia I_x and I_y of a given area (Fig. 9.5).

$$dI_x = \tfrac{1}{3}y^3\, dx$$
$$dI_y = x^2 y\, dx$$

Fig. 9.5

9.4 POLAR MOMENT OF INERTIA

An integral of great importance in problems concerning the torsion of cylindrical shafts and in problems dealing with the rotation of slabs is

$$J_O = \int r^2\, dA \tag{9.3}$$

where r is the distance from O to the element of area dA (Fig. 9.6). This integral is the *polar moment of inertia* of the area A with respect to *the "pole" O*.

The polar moment of inertia of a given area can be computed from the rectangular moments of inertia I_x and I_y of the area if these quantities are already known. Indeed, noting that $r^2 = x^2 + y^2$, we write

$$J_O = \int r^2\, dA = \int (x^2 + y^2)\, dA = \int y^2\, dA + \int x^2\, dA$$

Fig. 9.6

(a)

(b)

(c)

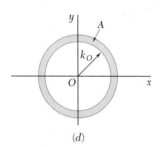

(d)

Fig. 9.7

that is,

$$J_O = I_x + I_y \tag{9.4}$$

9.5 RADIUS OF GYRATION OF AN AREA

Consider an area A which has a moment of inertia I_x with respect to the x axis (Fig. 9.7a). Let us imagine that we concentrate this area into a thin strip parallel to the x axis (Fig. 9.7b). If the area A, thus concentrated, is to have the same moment of inertia with respect to the x axis, the strip should be placed at a distance k_x from the x axis, where k_x is defined by the relation

$$I_x = k_x^2 A$$

Solving for k_x, we write

$$k_x = \sqrt{\frac{I_x}{A}} \tag{9.5}$$

The distance k_x is referred to as the *radius of gyration* of the area with respect to the x axis. In a similar way, we can define the radii of gyration k_y and k_O (Fig. 9.7c and d); we write

$$I_y = k_y^2 A \qquad k_y = \sqrt{\frac{I_y}{A}} \tag{9.6}$$

$$J_O = k_O^2 A \qquad k_O = \sqrt{\frac{J_O}{A}} \tag{9.7}$$

If we rewrite Eq. (9.4) in terms of the radii of gyration, we find that

$$k_O^2 = k_x^2 + k_y^2 \tag{9.8}$$

EXAMPLE For the rectangle shown in Fig. 9.8, let us compute the radius of gyration k_x with respect to its base. Using formulas (9.5) and (9.2), we write

$$k_x^2 = \frac{I_x}{A} = \frac{\frac{1}{3}bh^3}{bh} = \frac{h^2}{3} \qquad k_x = \frac{h}{\sqrt{3}}$$

The radius of gyration k_x of the rectangle is shown in Fig. 9.8. It should not be confused with the ordinate $\bar{y} = h/2$ of the centroid of the area. While k_x depends upon the *second moment*, or moment of inertia, of the area, the ordinate $\bar{y}$ is related to the *first moment* of the area. ■

Fig. 9.8

SAMPLE PROBLEM 9.1

Determine the moment of inertia of a triangle with respect to its base.

SOLUTION

A triangle of base b and height h is drawn; the x axis is chosen to coincide with the base. A differential strip parallel to the x axis is chosen to be dA. Since all portions of the strip are at the same distance from the x axis, we write

$$dI_x = y^2 \, dA \qquad dA = l \, dy$$

Using similar triangles, we have

$$\frac{l}{b} = \frac{h-y}{h} \qquad l = b\frac{h-y}{h} \qquad dA = b\frac{h-y}{h}dy$$

Integrating dI_x from $y = 0$ to $y = h$, we obtain

$$I_x = \int y^2 \, dA = \int_0^h y^2 b\frac{h-y}{h}dy = \frac{b}{h}\int_0^h (hy^2 - y^3)\,dy$$

$$= \frac{b}{h}\left[h\frac{y^3}{3} - \frac{y^4}{4} \right]_0^h \qquad\qquad I_x = \frac{bh^3}{12} \quad\blacktriangleleft$$

SAMPLE PROBLEM 9.2

(a) Determine the centroidal polar moment of inertia of a circular area by direct integration. (b) Using the result of part a, determine the moment of inertia of a circular area with respect to a diameter.

SOLUTION

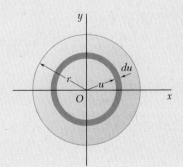

a. Polar Moment of Inertia. An annular differential element of area is chosen to be dA. Since all portions of the differential area are at the same distance from the origin, we write

$$dJ_O = u^2 \, dA \qquad dA = 2\pi u \, du$$

$$J_O = \int dJ_O = \int_0^r u^2(2\pi u \, du) = 2\pi \int_0^r u^3 \, du$$

$$J_O = \frac{\pi}{2}r^4 \quad\blacktriangleleft$$

b. Moment of Inertia with Respect to a Diameter. Because of the symmetry of the circular area, we have $I_x = I_y$. We then write

$$J_O = I_x + I_y = 2I_x \qquad \frac{\pi}{2}r^4 = 2I_x \qquad I_{\text{diameter}} = I_x = \frac{\pi}{4}r^4 \quad\blacktriangleleft$$

SAMPLE PROBLEM 9.3

(a) Determine the moment of inertia of the shaded area shown with respect to each of the coordinate axes. (Properties of this area were considered in Sample Prob. 5.4.) (b) Using the results of part a, determine the radius of gyration of the shaded area with respect to each of the coordinate axes.

SOLUTION

Referring to Sample Prob. 5.4, we obtain the following expressions for the equation of the curve and the total area:

$$y = \frac{b}{a^2}x^2 \qquad A = \tfrac{1}{3}ab$$

Moment of Inertia I_x. A vertical differential element of area is chosen to be dA. Since all portions of this element are *not* at the same distance from the x axis, we must treat the element as a thin rectangle. The moment of inertia of the element with respect to the x axis is then

$$dI_x = \tfrac{1}{3}y^3\,dx = \frac{1}{3}\left(\frac{b}{a^2}x^2\right)^3 dx = \frac{1}{3}\frac{b^3}{a^6}x^6\,dx$$

$$I_x = \int dI_x = \int_0^a \frac{1}{3}\frac{b^3}{a^6}x^6\,dx = \left[\frac{1}{3}\frac{b^3}{a^6}\frac{x^7}{7}\right]_0^a$$

$$I_x = \frac{ab^3}{21} \quad \blacktriangleleft$$

Moment of Inertia I_y. The same vertical differential element of area is used. Since all portions of the element are at the same distance from the y axis, we write

$$dI_y = x^2\,dA = x^2(y\,dx) = x^2\left(\frac{b}{a^2}x^2\right)dx = \frac{b}{a^2}x^4\,dx$$

$$I_y = \int dI_y = \int_0^a \frac{b}{a^2}x^4\,dx = \left[\frac{b}{a^2}\frac{x^5}{5}\right]_0^a$$

$$I_y = \frac{a^3b}{5} \quad \blacktriangleleft$$

Radii of Gyration k_x and k_y. We have, by definition,

$$k_x^2 = \frac{I_x}{A} = \frac{ab^3/21}{ab/3} = \frac{b^2}{7} \qquad k_x = \sqrt{\tfrac{1}{7}}\,b \quad \blacktriangleleft$$

and

$$k_y^2 = \frac{I_y}{A} = \frac{a^3b/5}{ab/3} = \tfrac{3}{5}a^2 \qquad k_y = \sqrt{\tfrac{3}{5}}\,a \quad \blacktriangleleft$$

SOLVING PROBLEMS ON YOUR OWN

The purpose of this lesson was to introduce the *rectangular and polar moments of inertia of areas* and the corresponding *radii of gyration*. Although the problems you are about to solve may appear to be more appropriate for a calculus class than for one in mechanics, we hope that our introductory comments have convinced you of the relevance of the moments of inertia to your study of a variety of engineering topics.

1. **Calculating the rectangular moments of inertia I_x and I_y.** We defined these quantities as

$$I_x = \int y^2 \, dA \qquad I_y = \int x^2 \, dA \qquad (9.1)$$

where dA is a differential element of area $dx \, dy$. The moments of inertia are *the second moments of the area;* it is for that reason that I_x, for example, depends on the perpendicular distance y to the area dA. As you study Sec. 9.3, you should recognize the importance of carefully defining the shape and the orientation of dA. Further, you should note the following points.

   a. The moments of inertia of most areas can be obtained by means of a single integration. The expressions given in Figs. 9.3*b* and *c* and Fig. 9.5 can be used to calculate I_x and I_y. Regardless of whether you use a single or a double integration, be sure to show on your sketch the element dA that you have chosen.

   b. The moment of inertia of an area is always positive, regardless of the location of the area with respect to the coordinate axes. This is because it is obtained by integrating the product of dA and the *square* of distance. (Note how this differs from the results for the first moment of the area.) Only when an area is *removed* (as in the case for a hole) will its moment of inertia be entered in your computations with a minus sign.

   c. As a partial check of your work, observe that the moments of inertia are equal to an area times the square of a length. Thus, every term in an expression for a moment of inertia must be a *length to the fourth power*.

2. **Computing the polar moment of inertia J_O.** We defined J_O as

$$J_O = \int r^2 \, dA \qquad (9.3)$$

where $r^2 = x^2 + y^2$. If the given area has circular symmetry (as in Sample Prob. 9.2), it is possible to express dA as a function of r and to compute J_O with a single integration. When the area lacks circular symmetry, it is usually easier first to calculate I_x and I_y and then to determine J_O from

$$J_O = I_x + I_y \qquad (9.4)$$

Lastly, if the equation of the curve that bounds the given area is expressed in polar coordinates, then $dA = r \, dr \, d\theta$ and a double integration is required to compute the integral for J_O [see Prob. 9.27].

3. **Determining the radii of gyration k_x and k_y and the polar radius of gyration k_O.** These quantities were defined in Sec. 9.5, and you should realize that they can be determined only after the area and the appropriate moments of inertia have been computed. It is important to remember that k_x is measured in the y direction, while k_y is measured in the x direction; you should carefully study Sec. 9.5 until you understand this point.

PROBLEMS

Fig. P9.1 and *P9.5*

9.1 through 9.4 Determine by direct integration the moment of inertia of the shaded area with respect to the y axis.

9.5 through 9.8 Determine by direct integration the moment of inertia of the shaded area with respect to the x axis.

Fig. P9.2 and P9.6

Fig. P9.3 and *P9.7*

Fig. P9.4 and P9.8

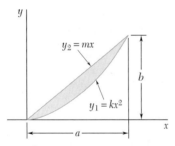

Fig. P9.9 and P9.12

9.9 through 9.11 Determine by direct integration the moment of inertia of the shaded area with respect to the x axis.

9.12 through *9.14* Determine by direct integration the moment of inertia of the shaded area with respect to the y axis.

Fig. P9.10 and *P9.13*

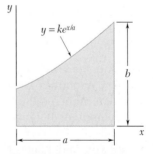

Fig. P9.11 and *P9.14*

9.15 and 9.16 Determine the moment of inertia and the radius of gyration of the shaded area shown with respect to the x axis.

Fig. P9.15 and P9.17

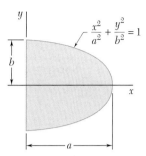

Fig. P9.16 and P9.18

9.17 and 9.18 Determine the moment of inertia and the radius of gyration of the shaded area shown with respect to the y axis.

9.19 Determine the moment of inertia and the radius of gyration of the shaded area shown with respect to the x axis.

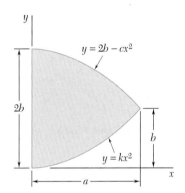

Fig. P9.19 and P9.20

9.20 Determine the moment of inertia and the radius of gyration of the shaded area shown with respect to the y axis.

9.21 and 9.22 Determine the polar moment of inertia and the polar radius of gyration of the shaded area shown with respect to point P.

Fig. P9.21

Fig. P9.22

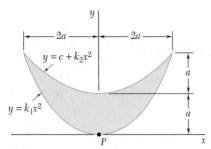

Fig. P9.23

9.23 and 9.24 Determine the polar moment of inertia and the polar radius of gyration of the shaded area shown with respect to point P.

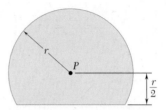

Fig. P9.24

9.25 (a) Determine by direct integration the polar moment of inertia of the semiannular area shown with respect to point O. (b) Using the result of part a, determine the moments of inertia of the given area with respect to the x and y axes.

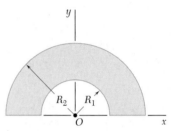

Fig. P9.25 and P9.26

9.26 (a) Show that the polar radius of gyration k_O of the semiannular area shown is approximately equal to the mean radius $R_m = (R_1 + R_2)/2$ for small values of the thickness $t = R_2 - R_1$. (b) Determine the percentage error introduced by using R_m in place of k_O for the following values of t/R_m: 1, $\frac{1}{2}$, and $\frac{1}{10}$.

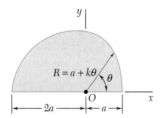

Fig. P9.27

9.27 Determine the polar moment of inertia and the polar radius of gyration of the shaded area shown with respect to the point O.

9.28 Determine the polar moment of inertia and the polar radius of gyration of the isosceles triangle shown with respect to the point O.

*9.29 Using the polar moment of inertia of the isosceles triangle of Prob. 9.28, show that the centroidal polar moment of inertia of a circular area of radius r is $\pi r^4/2$. (*Hint:* As a circular area is divided into an increasing number of equal circular sectors, what is the approximate shape of each circular sector?)

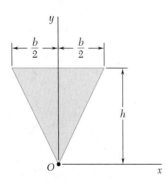

Fig. P9.28

*9.30 Prove that the centroidal polar moment of inertia of a given area A cannot be smaller than $A^2/2\pi$. (*Hint:* Compare the moment of inertia of the given area with the moment of inertia of a circle that has the same area and the same centroid.)

9.6 PARALLEL-AXIS THEOREM

Consider the moment of inertia I of an area A with respect to an axis AA' (Fig. 9.9). Denoting by y the distance from an element of area dA to AA', we write

$$I = \int y^2\, dA$$

Let us now draw through the centroid C of the area an axis BB' parallel to AA'; this axis is called a *centroidal axis*. Denoting by y'

Fig. 9.9

the distance from the element dA to BB', we write $y = y' + d$, where d is the distance between the axes AA' and BB'. Substituting for y in the above integral, we write

$$I = \int y^2\, dA = \int (y' + d)^2\, dA$$

$$= \int y'^2\, dA + 2d \int y'\, dA + d^2 \int dA$$

The first integral represents the moment of inertia $\bar{I}$ of the area with respect to the centroidal axis BB'. The second integral represents the first moment of the area with respect to BB'; since the centroid C of the area is located on that axis, the second integral must be zero. Finally, we observe that the last integral is equal to the total area A. Therefore, we have

$$I = \bar{I} + Ad^2 \qquad (9.9)$$

This formula expresses that the moment of inertia I of an area with respect to any given axis AA' is equal to the moment of inertia $\bar{I}$ of the area with respect to a centroidal axis BB' parallel to AA' *plus* the product of the area A and the square of the distance d between the two axes. This theorem is known as the *parallel-axis theorem*. Substituting k^2A for I and $\bar{k}^2A$ for $\bar{I}$, the theorem can also be expressed as

$$k^2 = \bar{k}^2 + d^2 \qquad (9.10)$$

A similar theorem can be used to relate the polar moment of inertia J_O of an area about a point O to the polar moment of inertia $\bar{J}_C$ of the same area about its centroid C. Denoting by d the distance between O and C, we write

$$J_O = \bar{J}_C + Ad^2 \qquad \text{or} \qquad k_O^2 = \bar{k}_C^2 + d^2 \qquad (9.11)$$

Fig. 9.10

Fig. 9.11

EXAMPLE 1 As an application of the parallel-axis theorem, let us determine the moment of inertia I_T of a circular area with respect to a line tangent to the circle (Fig. 9.10). We found in Sample Prob. 9.2 that the moment of inertia of a circular area about a centroidal axis is $\bar{I} = \frac{1}{4}\pi r^4$. We can write, therefore,

$$I_T = \bar{I} + Ad^2 = \tfrac{1}{4}\pi r^4 + (\pi r^2)r^2 = \tfrac{5}{4}\pi r^4 \ \blacksquare$$

EXAMPLE 2 The parallel-axis theorem can also be used to determine the centroidal moment of inertia of an area when the moment of inertia of the area with respect to a parallel axis is known. Consider, for instance, a triangular area (Fig. 9.11). We found in Sample Prob. 9.1 that the moment of inertia of a triangle with respect to its base AA' is equal to $\frac{1}{12}bh^3$. Using the parallel-axis theorem, we write

$$I_{AA'} = \bar{I}_{BB'} + Ad^2$$
$$\bar{I}_{BB'} = I_{AA'} - Ad^2 = \tfrac{1}{12}bh^3 - \tfrac{1}{2}bh(\tfrac{1}{3}h)^2 = \tfrac{1}{36}bh^3$$

It should be observed that the product Ad^2 was *subtracted* from the given moment of inertia in order to obtain the centroidal moment of inertia of the triangle. Note that this product is *added* when transferring *from* a centroidal axis to a parallel axis, but it should be *subtracted* when transferring *to* a centroidal axis. In other words, the moment of inertia of an area is always smaller with respect to a centroidal axis than with respect to any parallel axis.

Returning to Fig. 9.11, we observe that the moment of inertia of the triangle with respect to the line DD' (which is drawn through a vertex) can be obtained by writing

$$I_{DD'} = \bar{I}_{BB'} + Ad'^2 = \tfrac{1}{36}bh^3 + \tfrac{1}{2}bh(\tfrac{2}{3}h)^2 = \tfrac{1}{4}bh^3$$

Note that $I_{DD'}$ could not have been obtained directly from $I_{AA'}$. The parallel-axis theorem can be applied only if one of the two parallel axes passes through the centroid of the area. $\blacksquare$

9.7 MOMENTS OF INERTIA OF COMPOSITE AREAS

Consider a composite area A made of several component areas A_1, A_2, A_3, . . . Since the integral representing the moment of inertia of A can be subdivided into integrals evaluated over A_1, A_2, A_3, . . . , the moment of inertia of A with respect to a given axis is obtained by adding the moments of inertia of the areas A_1, A_2, A_3, . . . , with respect to the same axis. The moment of inertia of an area consisting of several of the common shapes shown in Fig. 9.12 can thus be obtained by using the formulas given in that figure. Before adding the moments of inertia of the component areas, however, the parallel-axis theorem may have to be used to transfer each moment of inertia to the desired axis. This is shown in Sample Probs. 9.4 and 9.5.

The properties of the cross sections of various structural shapes are given in Fig. 9.13. As noted in Sec. 9.2, the moment of inertia of a beam section about its neutral axis is closely related to the computation of the bending moment in that section of the beam. The

Photo 9.1 Figure 9.13 tabulates data for a small sample of the rolled-steel shapes that are readily available. Shown above are two examples of wide-flange shapes that are commonly used in the construction of buildings.

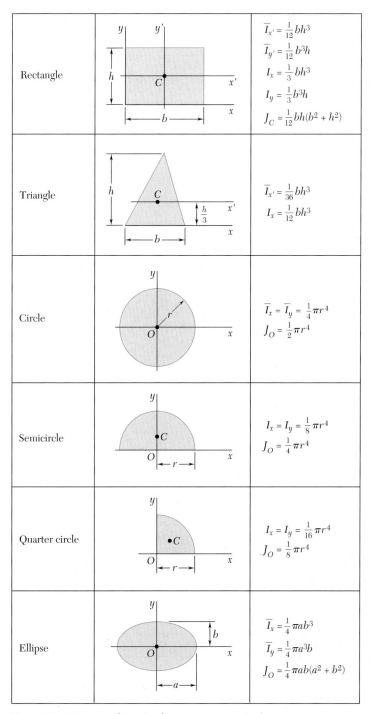

Rectangle		$\overline{I}_{x'} = \frac{1}{12}bh^3$ $\overline{I}_{y'} = \frac{1}{12}b^3h$ $I_x = \frac{1}{3}bh^3$ $I_y = \frac{1}{3}b^3h$ $J_C = \frac{1}{12}bh(b^2 + h^2)$
Triangle		$\overline{I}_{x'} = \frac{1}{36}bh^3$ $I_x = \frac{1}{12}bh^3$
Circle		$\overline{I}_x = \overline{I}_y = \frac{1}{4}\pi r^4$ $J_O = \frac{1}{2}\pi r^4$
Semicircle		$I_x = I_y = \frac{1}{8}\pi r^4$ $J_O = \frac{1}{4}\pi r^4$
Quarter circle		$I_x = I_y = \frac{1}{16}\pi r^4$ $J_O = \frac{1}{8}\pi r^4$
Ellipse		$\overline{I}_x = \frac{1}{4}\pi ab^3$ $\overline{I}_y = \frac{1}{4}\pi a^3b$ $J_O = \frac{1}{4}\pi ab(a^2 + b^2)$

Fig. 9.12 Moments of inertia of common geometric shapes.

determination of moments of inertia is thus a prerequisite to the analysis and design of structural members.

It should be noted that the radius of gyration of a composite area is *not* equal to the sum of the radii of gyration of the component areas. In order to determine the radius of gyration of a composite area, it is first necessary to compute the moment of inertia of the area.

	Designation	Area in²	Depth in.	Width in.	Axis X-X			Axis Y-Y		
					$\overline{I}_x$, in⁴	$\overline{k}_x$, in.	$\overline{y}$, in.	$\overline{I}_y$, in⁴	$\overline{k}_y$, in.	$\overline{x}$, in.
W Shapes (Wide-Flange Shapes)	W18 × 76†	22.3	18.2	11.0	1330	7.73		152	2.61	
	W16 × 57	16.8	16.4	7.12	758	6.72		43.1	1.60	
	W14 × 38	11.2	14.1	6.77	385	5.87		26.7	1.55	
	W8 × 31	9.12	8.00	8.00	110	3.47		37.1	2.02	
S Shapes (American Standard Shapes)	S18 × 54.7†	16.0	18.0	6.00	801	7.07		20.7	1.14	
	S12 × 31.8	9.31	12.0	5.00	217	4.83		9.33	1.00	
	S10 × 25.4	7.45	10.0	4.66	123	4.07		6.73	0.950	
	S6 × 12.5	3.66	6.00	3.33	22.0	2.45		1.80	0.702	
C Shapes (American Standard Channels)	C12 × 20.7†	6.08	12.0	2.94	129	4.61		3.86	0.797	0.698
	C10 × 15.3	4.48	10.0	2.60	67.3	3.87		2.27	0.711	0.634
	C8 × 11.5	3.37	8.00	2.26	32.5	3.11		1.31	0.623	0.572
	C6 × 8.2	2.39	6.00	1.92	13.1	2.34		0.687	0.536	0.512
Angles	L6 × 6 × 1‡	11.0			35.4	1.79	1.86	35.4	1.79	1.86
	L4 × 4 × ½	3.75			5.52	1.21	1.18	5.52	1.21	1.18
	L3 × 3 × ¼	1.44			1.23	0.926	0.836	1.23	0.926	0.836
	L6 × 4 × ½	4.75			17.3	1.91	1.98	6.22	1.14	0.981
	L5 × 3 × ½	3.75			9.43	1.58	1.74	2.55	0.824	0.746
	L3 × 2 × ¼	1.19			1.09	0.953	0.980	0.390	0.569	0.487

Fig. 9.13A Properties of rolled-steel shapes (U.S. customary units).*

*Courtesy of the American Institute of Steel Construction, Chicago, Illinois
†Nominal depth in inches and weight in pounds per foot
‡Depth, width, and thickness in inches

			Area	Depth	Width	Axis X-X			Axis Y-Y		
	Designation		mm²	mm	mm	$\overline{I}_x$ 10^6 mm⁴	$\overline{k}_x$ mm	$\overline{y}$ mm	$\overline{I}_y$ 10^6 mm⁴	$\overline{k}_y$ mm	$\overline{x}$ mm
W Shapes (Wide-Flange Shapes)	W460 × 113† W410 × 85 W360 × 57.8 W200 × 46.1		14400 10800 7230 5880	462 417 358 203	279 181 172 203	554 316 160 45.8	196 171 149 88.1		63.3 17.9 11.1 15.4	66.3 40.6 39.4 51.3	
S Shapes (American Standard Shapes)	S460 × 81.4† S310 × 47.3 S250 × 37.8 S150 × 18.6		10300 6010 4810 2360	457 305 254 152	152 127 118 84.6	333 90.3 51.2 9.16	180 123 103 62.2		8.62 3.88 2.80 0.749	29.0 25.4 24.1 17.8	
C Shapes (American Standard Channels)	C310 × 30.8† C250 × 22.8 C200 × 17.1 C150 × 12.2		3920 2890 2170 1540	305 254 203 152	74.7 66.0 57.4 48.8	53.7 28.0 13.5 5.45	117 98.3 79.0 59.4		1.61 0.945 0.545 0.286	20.2 18.1 15.8 13.6	17.7 16.1 14.5 13.0
Angles	L152 × 152 × 25.4‡ L102 × 102 × 12.7 L76 × 76 × 6.4 L152 × 102 × 12.7 L127 × 76 × 12.7 L76 × 51 × 6.4		7100 2420 929 3060 2420 768			14.7 2.30 0.512 7.20 3.93 0.454	45.5 30.7 23.5 48.5 40.1 24.2	47.2 30.0 21.2 50.3 44.2 24.9	14.7 2.30 0.512 2.59 1.06 0.162	45.5 30.7 23.5 29.0 20.9 14.5	47.2 30.0 21.2 24.9 18.9 12.4

Fig. 9.13B Properties of rolled-steel shapes (SI units).

†Nominal depth in millimeters and mass in kilograms per meter
‡Depth, width, and thickness in millimeters

SAMPLE PROBLEM 9.4

The strength of a W14 × 38 rolled-steel beam is increased by attaching a $9 \times \frac{3}{4}$-in. plate to its upper flange as shown. Determine the moment of inertia and the radius of gyration of the composite section with respect to an axis which is parallel to the plate and passes through the centroid C of the section.

SOLUTION

The origin O of the coordinates is placed at the centroid of the wide-flange shape, and the distance $\overline{Y}$ to the centroid of the composite section is computed using the methods of Chap. 5. The area of the wide-flange shape is found by referring to Fig. 9.13A. The area and the y coordinate of the centroid of the plate are

$$A = (9 \text{ in.})(0.75 \text{ in.}) = 6.75 \text{ in}^2$$
$$\overline{y} = \tfrac{1}{2}(14.1 \text{ in.}) + \tfrac{1}{2}(0.75 \text{ in.}) = 7.425 \text{ in.}$$

Section	Area, in^2	$\overline{y}$, in.	$\overline{y}A$, in^3
Plate	6.75	7.425	50.12
Wide-flange shape	11.2	0	0
	$\Sigma A = 17.95$		$\Sigma \overline{y}A = 50.12$

$$\overline{Y}\Sigma A = \Sigma \overline{y}A \qquad \overline{Y}(17.95) = 50.12 \qquad \overline{Y} = 2.792 \text{ in.}$$

Moment of Inertia. The parallel-axis theorem is used to determine the moments of inertia of the wide-flange shape and the plate with respect to the x' axis. This axis is a centroidal axis for the composite section but *not* for either of the elements considered separately. The value of $\overline{I}_x$ for the wide-flange shape is obtained from Fig. 9.13A.

For the wide-flange shape,

$$I_{x'} = \overline{I}_x + A\overline{Y}^2 = 385 + (11.2)(2.792)^2 = 472.3 \text{ in}^4$$

For the plate,

$$I_{x'} = \overline{I}_x + Ad^2 = (\tfrac{1}{12})(9)(\tfrac{3}{4})^3 + (6.75)(7.425 - 2.792)^2 = 145.2 \text{ in}^4$$

For the composite area,

$$I_{x'} = 472.3 + 145.2 = 617.5 \text{ in}^4 \qquad I_{x'} = 618 \text{ in}^4 \;\blacktriangleleft$$

Radius of Gyration. We have

$$k_{x'}^2 = \frac{I_{x'}}{A} = \frac{617.5 \text{ in}^4}{17.95 \text{ in}^2} \qquad k_{x'} = 5.87 \text{ in.} \;\blacktriangleleft$$

SAMPLE PROBLEM 9.5

Determine the moment of inertia of the shaded area with respect to the x axis.

SOLUTION

The given area can be obtained by subtracting a half circle from a rectangle. The moments of inertia of the rectangle and the half circle will be computed separately.

Moment of Inertia of Rectangle. Referring to Fig. 9.12, we obtain

$$I_x = \tfrac{1}{3}bh^3 = \tfrac{1}{3}(240 \text{ mm})(120 \text{ mm})^3 = 138.2 \times 10^6 \text{ mm}^4$$

Moment of Inertia of Half Circle. Referring to Fig. 5.8, we determine the location of the centroid C of the half circle with respect to diameter AA'.

$$a = \frac{4r}{3\pi} = \frac{(4)(90 \text{ mm})}{3\pi} = 38.2 \text{ mm}$$

The distance b from the centroid C to the x axis is

$$b = 120 \text{ mm} - a = 120 \text{ mm} - 38.2 \text{ mm} = 81.8 \text{ mm}$$

Referring now to Fig. 9.12, we compute the moment of inertia of the half circle with respect to diameter AA'; we also compute the area of the half circle.

$$I_{AA'} = \tfrac{1}{8}\pi r^4 = \tfrac{1}{8}\pi(90 \text{ mm})^4 = 25.76 \times 10^6 \text{ mm}^4$$
$$A = \tfrac{1}{2}\pi r^2 = \tfrac{1}{2}\pi(90 \text{ mm})^2 = 12.72 \times 10^3 \text{ mm}^2$$

Using the parallel-axis theorem, we obtain the value of $\bar{I}_{x'}$:

$$I_{AA'} = \bar{I}_{x'} + Aa^2$$
$$25.76 \times 10^6 \text{ mm}^4 = \bar{I}_{x'} + (12.72 \times 10^3 \text{ mm}^2)(38.2 \text{ mm})^2$$
$$\bar{I}_{x'} = 7.20 \times 10^6 \text{ mm}^4$$

Again using the parallel-axis theorem, we obtain the value of I_x:

$$I_x = \bar{I}_{x'} + Ab^2 = 7.20 \times 10^6 \text{ mm}^4 + (12.72 \times 10^3 \text{ mm}^2)(81.8 \text{ mm})^2$$
$$= 92.3 \times 10^6 \text{ mm}^4$$

Moment of Inertia of Given Area. Subtracting the moment of inertia of the half circle from that of the rectangle, we obtain

$$I_x = 138.2 \times 10^6 \text{ mm}^4 - 92.3 \times 10^6 \text{ mm}^4$$
$$I_x = 45.9 \times 10^6 \text{ mm}^4 \quad \blacktriangleleft$$

SOLVING PROBLEMS
ON YOUR OWN

In this lesson we introduced the *parallel-axis theorem* and illustrated how it can be used to simplify the computation of moments and polar moments of inertia of composite areas. The areas that you will consider in the following problems will consist of common shapes and rolled-steel shapes. You will also use the parallel-axis theorem to locate the point of application (the center of pressure) of the resultant of the hydrostatic forces acting on a submerged plane area.

1. Applying the parallel-axis theorem. In Sec. 9.6 we derived the parallel-axis theorem

$$I = \bar{I} + Ad^2 \tag{9.9}$$

which states that the moment of inertia I of an area A with respect to a given axis is equal to the sum of the moment of inertia $\bar{I}$ of that area with respect to a *parallel centroidal axis* and the product Ad^2, where d is the distance between the two axes. It is important that you remember the following points as you use the parallel-axis theorem.

a. The centroidal moment of inertia $\bar{I}$ of an area A can be obtained by subtracting the product Ad^2 from the moment of inertia I of the area with respect to a parallel axis. It follows that the moment of inertia $\bar{I}$ is *smaller* than the moment of inertia I of the same area with respect to any parallel axis.

b. The parallel-axis theorem can be applied only if one of the two axes involved is a centroidal axis. Therefore, as we noted in Example 2, to compute the moment of inertia of an area with respect to a *noncentroidal axis* when the moment of inertia of the area is known with respect to *another noncentroidal axis*, it is necessary to *first compute* the moment of inertia of the area with respect to a *centroidal axis parallel to the two given axes*.

2. Computing the moments and polar moments of inertia of composite areas. Sample Probs. 9.4 and 9.5 illustrate the steps you should follow to solve problems of this type. As with all composite-area problems, you should show on your sketch the common shapes or rolled-steel shapes that constitute the various elements of the given area, as well as the distances between the centroidal axes of the elements and the axes about which the moments of inertia are to be computed. In addition, it is important that the following points be noted.

a. The moment of inertia of an area is always positive, regardless of the location of the axis with respect to which it is computed. As pointed out in the comments for the preceding lesson, it is only when an area is *removed* (as in the case of a hole) that its moment of inertia should be entered in your computations with a minus sign.

b. The moments of inertia of a semiellipse and a quarter ellipse can be determined by dividing the moment of inertia of an ellipse by 2 and 4, respectively. It should be noted, however, that the moments of inertia obtained in this manner are *with respect to the axes of symmetry of the ellipse*. To obtain the *centroidal* moments of inertia of these shapes, the parallel-axis theorem should be used. Note that this remark also applies to a semicircle and to a quarter circle and that the expressions given for these shapes in Fig. 9.12 are *not* centroidal moments of inertia.

c. To calculate the polar moment of inertia of a composite area, you can use either the expressions given in Fig. 9.12 for J_O or the relationship

$$J_O = I_x + I_y \tag{9.4}$$

depending on the shape of the given area.

d. Before computing the centroidal moments of inertia of a given area, you may find it necessary to first locate the centroid of the area using the methods of Chap. 5.

3. Locating the point of application of the resultant of a system of hydrostatic forces. In Sec. 9.2 we found that

$$R = \gamma \int y \, dA = \gamma \bar{y} A$$

$$M_x = \gamma \int y^2 \, dA = \gamma I_x$$

where $\bar{y}$ is the distance from the x axis to the centroid of the submerged plane area. Since **R** is equivalent to the system of elemental hydrostatic forces, it follows that

$$\Sigma M_x: \qquad y_P R = M_x$$

where y_P is the depth of the point of application of **R**. Then

$$y_P(\gamma \bar{y} A) = \gamma I_x \qquad \text{or} \qquad y_P = \frac{I_x}{\bar{y} A}$$

In closing, we encourage you to carefully study the notation used in Fig. 9.13 for the rolled-steel shapes, as you will likely encounter it again in subsequent engineering courses.

PROBLEMS

9.31 and 9.32 Determine the moment of inertia and the radius of gyration of the shaded area with respect to the x axis.

Fig. P9.31 and P9.33

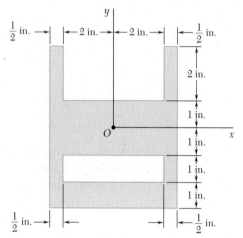

Fig. P9.32 and P9.34

9.33 and 9.34 Determine the moment of inertia and the radius of gyration of the shaded area with respect to the y axis.

9.35 and 9.36 Determine the moments of inertia of the shaded area shown with respect to the x and y axes when $a = 20$ mm.

Fig. P9.35

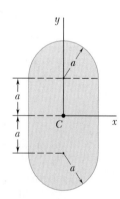

Fig. P9.36

9.37 The shaded area is equal to 50 in². Determine its centroidal moments of inertia $\bar{I}_x$ and $\bar{I}_y$, knowing that $\bar{I}_y = 2\bar{I}_x$ and that the polar moment of inertia of the area about point A is $J_A = 2250$ in⁴.

9.38 The polar moments of inertia of the shaded area with respect to points A, B, and D are, respectively, $J_A = 2880$ in⁴, $J_B = 6720$ in⁴, and $J_D = 4560$ in⁴. Determine the shaded area, its centroidal moment of inertia $\bar{J}_C$, and the distance d from C to D.

Fig. P9.37 and P9.38

9.39 Determine the shaded area and its moment of inertia with respect to the centroidal axis parallel to AA', knowing that $d_1 = 30$ mm and $d_2 = 10$ mm, and that the moments of inertia with respect to AA' and BB' are 4.1×10^6 mm^4 and 6.9×10^6 mm^4, respectively.

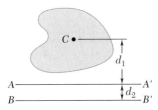

Fig. P9.39 and P9.40

9.40 Knowing that the shaded area is equal to 7500 mm^2 and that its moment of inertia with respect to AA' is 31×10^6 mm^4, determine its moment of inertia with respect to BB', for $d_1 = 60$ mm and $d_2 = 15$ mm.

9.41 through 9.44 Determine the moments of inertia $\bar{I}_x$ and $\bar{I}_y$ of the area shown with respect to centroidal axes respectively parallel and perpendicular to side AB.

Fig. P9.41

Fig. P9.42

Fig. P9.43

Fig. P9.44

9.45 and 9.46 Determine the polar moment of inertia of the area shown with respect to (a) point O, (b) the centroid of the area.

Fig. P9.45

Fig. P9.46

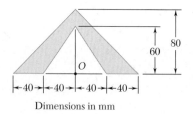

Fig. P9.47

Dimensions in mm

Fig. P9.49

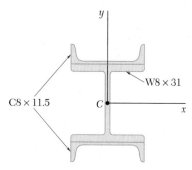

Fig. P9.51

9.47 and 9.48 Determine the polar moment of inertia of the area shown with respect to (*a*) point *O*, (*b*) the centroid of the area.

Fig. P9.48

9.49 Two channels and two plates are used to form the column section shown. For $b = 200$ mm, determine the moments of inertia and the radii of gyration of the combined section with respect to the centroidal *x* and *y* axes.

9.50 Two L6 × 4 × $\frac{1}{2}$-in. angles are welded together to form the section shown. Determine the moments of inertia and the radii of gyration of the combined section with respect to the centroidal *x* and *y* axes.

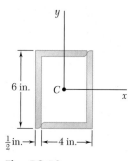

Fig. P9.50

9.51 Two channels are welded to a rolled W section as shown. Determine the moments of inertia and the radii of gyration of the combined section with respect to the centroidal *x* and *y* axes.

9.52 Two 20-mm steel plates are welded to a rolled S section as shown. Determine the moments of inertia and the radii of gyration of the combined section with respect to the centroidal *x* and *y* axes.

Fig. P9.52

9.53 A channel and a plate are welded together as shown to form a section that is symmetrical with respect to the y axis. Determine the moments of inertia of the combined section with respect to its centroidal x and y axes.

C8 × 11.5

0.5 in.

12 in.

Fig. P9.53

C250 × 22.8

W460 × 113

Fig. P9.54

9.54 The strength of the rolled W section shown is increased by welding a channel to its upper flange. Determine the moments of inertia of the combined section with respect to its centroidal x and y axes.

9.55 Two L76 × 76 × 6.4-mm angles are welded to a C250 × 22.8 channel. Determine the moments of inertia of the combined section with respect to centroidal axes respectively parallel and perpendicular to the web of the channel.

L76 × 76 × 6.4

C250 × 22.8

Fig. P9.55

$\frac{1}{2}$ in.

10 in.

L4 × 4 × $\frac{1}{2}$

Fig. P9.56

9.56 Two L4 × 4 × $\frac{1}{2}$-in. angles are welded to a steel plate as shown. Determine the moments of inertia of the combined section with respect to centroidal axes respectively parallel and perpendicular to the plate.

9.57 and 9.58 The panel shown forms the end of a trough that is filled with water to the line AA'. Referring to Sec. 9.2, determine the depth of the point of application of the resultant of the hydrostatic forces acting on the panel (the center of pressure).

A — a — A'

h

Fig. P9.57

a

A — A'

h

b

Fig. P9.58

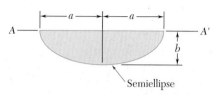

Fig. P9.59

9.59 and *9.60 The panel shown forms the end of a trough that is filled with water to the line AA'. Referring to Sec. 9.2, determine the depth of the point of application of the resultant of the hydrostatic forces acting on the panel (the center of pressure).

Fig. P9.60

9.61 A vertical trapezoidal gate that is used as an automatic valve is held shut by two springs attached to hinges located along edge AB. Knowing that each spring exerts a couple of magnitude $1470 \text{ N} \cdot \text{m}$, determine the depth d of water for which the gate will open.

9.62 The cover for a 0.5-m-diameter access hole in a water storage tank is attached to the tank with four equally spaced bolts as shown. Determine the additional force on each bolt due to the water pressure when the center of the cover is located 1.4 m below the water surface.

Fig. P9.61

Fig. P9.62

*9.63 Determine the x coordinate of the centroid of the volume shown. (*Hint:* The height y of the volume is proportional to the x coordinate; consider an analogy between this height and the water pressure on a submerged surface.)

*9.64 Determine the x coordinate of the centroid of the volume shown; this volume was obtained by intersecting an elliptic cylinder with an oblique plane. (See hint of Prob. 9.63.)

Fig. P9.63

Fig. P9.64

*9.65 Show that the system of hydrostatic forces acting on a submerged plane area A can be reduced to a force $\mathbf{P}$ at the centroid C of the area and two couples. The force $\mathbf{P}$ is perpendicular to the area and is of magnitude $P = \gamma A \bar{y} \sin \theta$, where γ is the specific weight of the liquid, and the couples are $\mathbf{M}_{x'} = (\gamma \bar{I}_{x'} \sin \theta)\mathbf{i}$ and $\mathbf{M}_{y'} = (\gamma \bar{I}_{x'y'} \sin \theta)\mathbf{j}$, where $\bar{I}_{x'y'} = \int x'y' \, dA$ (see Sec. 9.8). Note that the couples are independent of the depth at which the area is submerged.

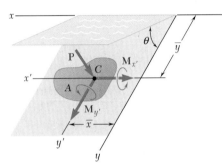

Fig. P9.65

*9.66 Show that the resultant of the hydrostatic forces acting on a submerged plane area A is a force $\mathbf{P}$ perpendicular to the area and of magnitude $P = \gamma A \bar{y} \sin \theta = \bar{p} A$, where γ is the specific weight of the liquid and $\bar{p}$ is the pressure at the centroid C of the area. Show that $\mathbf{P}$ is applied at a point C_P, called the center of pressure, whose coordinates are $x_p = I_{xy}/A\bar{y}$ and $y_p = I_x/A\bar{y}$, where $I_{xy} = \int xy \, dA$ (see Sec. 9.8). Show also that the difference of ordinates $y_p - \bar{y}$ is equal to $\bar{k}_x^2/\bar{y}$ and thus depends upon the depth at which the area is submerged.

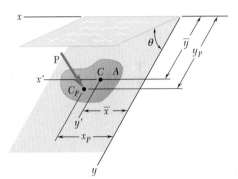

Fig. P9.66

*9.8 PRODUCT OF INERTIA

The integral

$$I_{xy} = \int xy \, dA \tag{9.12}$$

which is obtained by multiplying each element dA of an area A by its coordinates x and y and integrating over the area (Fig. 9.14), is known as the *product of inertia* of the area A with respect to the x and y axes. Unlike the moments of inertia I_x and I_y, the product of inertia I_{xy} can be positive, negative, or zero.

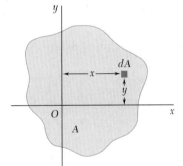

Fig. 9.14

When one or both of the x and y axes are axes of symmetry for the area A, the product of inertia I_{xy} is zero. Consider, for example, the channel section shown in Fig. 9.15. Since this section is symmetrical with respect to the x axis, we can associate with each element dA of coordinates x and y an element dA' of coordinates x and $-y$. Clearly, the contributions to I_{xy} of any pair of elements chosen in this way cancel out, and the integral (9.12) reduces to zero.

A parallel-axis theorem similar to the one established in Sec. 9.6 for moments of inertia can be derived for products of inertia. Consider an area A and a system of rectangular coordinates x and y

Fig. 9.15

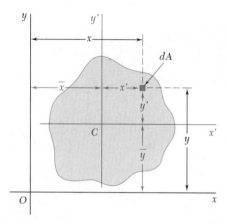

Fig. 9.16

(Fig. 9.16). Through the centroid C of the area, of coordinates $\bar{x}$ and $\bar{y}$, we draw two *centroidal axes* x' and y' which are parallel, respectively, to the x and y axes. Denoting by x and y the coordinates of an element of area dA with respect to the original axes, and by x' and y' the coordinates of the same element with respect to the centroidal axes, we write $x = x' + \bar{x}$ and $y = y' + \bar{y}$. Substituting into (9.12), we obtain the following expression for the product of inertia I_{xy}:

$$I_{xy} = \int xy\, dA = \int (x' + \bar{x})(y' + \bar{y})\, dA$$

$$= \int x'y'\, dA + \bar{y} \int x'\, dA + \bar{x} \int y'\, dA + \bar{x}\bar{y} \int dA$$

The first integral represents the product of inertia $\bar{I}_{x'y'}$ of the area A with respect to the centroidal axes x' and y'. The next two integrals represent first moments of the area with respect to the centroidal axes; they reduce to zero, since the centroid C is located on these axes. Finally, we observe that the last integral is equal to the total area A. Therefore, we have

$$I_{xy} = \bar{I}_{x'y'} + \bar{x}\bar{y}A \tag{9.13}$$

*9.9 PRINCIPAL AXES AND PRINCIPAL MOMENTS OF INERTIA

Consider the area A and the coordinate axes x and y (Fig. 9.17). Assuming that the moments and product of inertia

$$I_x = \int y^2\, dA \qquad I_y = \int x^2\, dA \qquad I_{xy} = \int xy\, dA \tag{9.14}$$

of the area A are known, we propose to determine the moments and product of inertia $I_{x'}$, $I_{y'}$, and $I_{x'y'}$ of A with respect to new axes x' and y' which are obtained by rotating the original axes about the origin through an angle θ.

We first note the following relations between the coordinates x', y' and x, y of an element of area dA:

$$x' = x \cos \theta + y \sin \theta \qquad y' = y \cos \theta - x \sin \theta$$

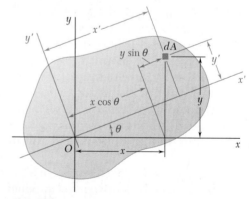

Fig. 9.17

Substituting for y' in the expression for $I_{x'}$, we write

$$I_{x'} = \int (y')^2 \, dA = \int (y \cos \theta - x \sin \theta)^2 \, dA$$

$$= \cos^2 \theta \int y^2 \, dA - 2 \sin \theta \cos \theta \int xy \, dA + \sin^2 \theta \int x^2 \, dA$$

Using the relations (9.14), we write

$$I_{x'} = I_x \cos^2 \theta - 2I_{xy} \sin \theta \cos \theta + I_y \sin^2 \theta \qquad (9.15)$$

Similarly, we obtain for $I_{y'}$ and $I_{x'y'}$ the expressions

$$I_{y'} = I_x \sin^2 \theta + 2I_{xy} \sin \theta \cos \theta + I_y \cos^2 \theta \qquad (9.16)$$

$$I_{x'y'} = (I_x - I_y) \sin \theta \cos \theta + I_{xy}(\cos^2 \theta - \sin^2 \theta) \qquad (9.17)$$

Recalling the trigonometric relations

$$\sin 2\theta = 2 \sin \theta \cos \theta \qquad \cos 2\theta = \cos^2 \theta - \sin^2 \theta$$

and

$$\cos^2 \theta = \frac{1 + \cos 2\theta}{2} \qquad \sin^2 \theta = \frac{1 - \cos 2\theta}{2}$$

we can write (9.15), (9.16), and (9.17) as follows:

$$I_{x'} = \frac{I_x + I_y}{2} + \frac{I_x - I_y}{2} \cos 2\theta - I_{xy} \sin 2\theta \qquad (9.18)$$

$$I_{y'} = \frac{I_x + I_y}{2} - \frac{I_x - I_y}{2} \cos 2\theta + I_{xy} \sin 2\theta \qquad (9.19)$$

$$I_{x'y'} = \frac{I_x - I_y}{2} \sin 2\theta + I_{xy} \cos 2\theta \qquad (9.20)$$

Adding (9.18) and (9.19) we observe that

$$I_{x'} + I_{y'} = I_x + I_y \qquad (9.21)$$

This result could have been anticipated, since both members of (9.21) are equal to the polar moment of inertia J_O.

Equations (9.18) and (9.20) are the parametric equations of a circle. This means that if we choose a set of rectangular axes and plot a point M of abscissa $I_{x'}$ and ordinate $I_{x'y'}$ for any given value of the parameter θ, all of the points thus obtained will lie on a circle. To establish this property, we eliminate θ from Eqs. (9.18) and (9.20); this is done by transposing $(I_x + I_y)/2$ in Eq. (9.18), squaring both members of Eqs. (9.18) and (9.20), and adding. We write

$$\left(I_{x'} - \frac{I_x + I_y}{2}\right)^2 + I_{x'y'}^2 = \left(\frac{I_x - I_y}{2}\right)^2 + I_{xy}^2 \qquad (9.22)$$

Setting

$$I_{\text{ave}} = \frac{I_x + I_y}{2} \qquad \text{and} \qquad R = \sqrt{\left(\frac{I_x - I_y}{2}\right)^2 + I_{xy}^2} \qquad (9.23)$$

we write the identity (9.22) in the form

$$(I_{x'} - I_{\text{ave}})^2 + I_{x'y'}^2 = R^2 \qquad (9.24)$$

(a)

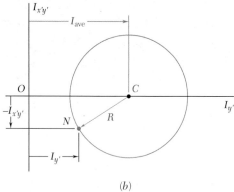

(b)

Fig. 9.18

which is the equation of a circle of radius R centered at the point C whose x and y coordinates are I_{ave} and 0, respectively (Fig. 9.18a). We observe that Eqs. (9.19) and (9.20) are the parametric equations of the same circle. Furthermore, because of the symmetry of the circle about the horizontal axis, the same result would have been obtained if instead of plotting M, we had plotted a point N of coordinates $I_{y'}$ and $-I_{x'y'}$ (Fig. 9.18b). This property will be used in Sec. 9.10.

The two points A and B where the above circle intersects the horizontal axis (Fig. 9.18a) are of special interest: Point A corresponds to the maximum value of the moment of inertia $I_{x'}$, while point B corresponds to its minimum value. In addition, both points correspond to a zero value of the product of inertia $I_{x'y'}$. Thus, the values θ_m of the parameter θ which correspond to the points A and B can be obtained by setting $I_{x'y'} = 0$ in Eq. (9.20). We obtain†

$$\tan 2\theta_m = -\frac{2I_{xy}}{I_x - I_y} \qquad (9.25)$$

This equation defines two values $2\theta_m$ which are 180° apart and thus two values θ_m which are 90° apart. One of these values corresponds to point A in Fig. 9.18a and to an axis through O in Fig. 9.17 with respect to which the moment of inertia of the given area is maximum; the other value corresponds to point B and to an axis through O with respect to which the moment of inertia of the area is minimum. The two axes thus defined, which are perpendicular to each other, are called the *principal axes of the area about O*, and the corresponding values I_{max} and I_{min} of the moment of inertia are called the *principal moments of inertia of the area about O*. Since the two values θ_m defined by Eq. (9.25) were obtained by setting $I_{x'y'} = 0$ in Eq. (9.20), it is clear that the product of inertia of the given area with respect to its principal axes is zero.

We observe from Fig. 9.18a that

$$I_{max} = I_{ave} + R \qquad I_{min} = I_{ave} - R \qquad (9.26)$$

Using the values for I_{ave} and R from formulas (9.23), we write

$$I_{max,min} = \frac{I_x + I_y}{2} \pm \sqrt{\left(\frac{I_x - I_y}{2}\right)^2 + I_{xy}^2} \qquad (9.27)$$

Unless it is possible to tell by inspection which of the two principal axes corresponds to I_{max} and which corresponds to I_{min}, it is necessary to substitute one of the values of θ_m into Eq. (9.18) in order to determine which of the two corresponds to the maximum value of the moment of inertia of the area about O.

Referring to Sec. 9.8, we note that if an area possesses an axis of symmetry through a point O, this axis must be a principal axis of the area about O. On the other hand, a principal axis does not need to be an axis of symmetry; whether or not an area possesses any axes of symmetry, it will have two principal axes of inertia about any point O.

The properties we have established hold for any point O located inside or outside the given area. If the point O is chosen to coincide with the centroid of the area, any axis through O is a centroidal axis; the two principal axes of the area about its centroid are referred to as the *principal centroidal axes of the area*.

†This relation can also be obtained by differentiating $I_{x'}$ in Eq. (9.18) and setting $dI_{x'}/d\theta = 0$.

SAMPLE PROBLEM 9.6

Determine the product of inertia of the right triangle shown (a) with respect to the x and y axes and (b) with respect to centroidal axes parallel to the x and y axes.

SOLUTION

a. Product of Inertia I_{xy}. A vertical rectangular strip is chosen as the differential element of area. Using the parallel-axis theorem, we write

$$dI_{xy} = dI_{x'y'} + \bar{x}_{el}\bar{y}_{el}\, dA$$

Since the element is symmetrical with respect to the x' and y' axes, we note that $dI_{x'y'} = 0$. From the geometry of the triangle, we obtain

$$y = h\left(1 - \frac{x}{b}\right) \qquad dA = y\, dx = h\left(1 - \frac{x}{b}\right)dx$$

$$\bar{x}_{el} = x \qquad \bar{y}_{el} = \tfrac{1}{2}y = \tfrac{1}{2}h\left(1 - \frac{x}{b}\right)$$

Integrating dI_{xy} from $x = 0$ to $x = b$, we obtain

$$I_{xy} = \int dI_{xy} = \int \bar{x}_{el}\bar{y}_{el}\, dA = \int_0^b x(\tfrac{1}{2})h^2\left(1 - \frac{x}{b}\right)^2 dx$$

$$= h^2\int_0^b \left(\frac{x}{2} - \frac{x^2}{b} + \frac{x^3}{2b^2}\right)dx = h^2\left[\frac{x^2}{4} - \frac{x^3}{3b} + \frac{x^4}{8b^2}\right]_0^b$$

$$I_{xy} = \tfrac{1}{24}b^2h^2 \quad \blacktriangleleft$$

b. Product of Inertia $\bar{I}_{x''y''}$. The coordinates of the centroid of the triangle relative to the x and y axes are

$$\bar{x} = \tfrac{1}{3}b \qquad \bar{y} = \tfrac{1}{3}h$$

Using the expression for I_{xy} obtained in part a, we apply the parallel-axis theorem and write

$$I_{xy} = \bar{I}_{x''y''} + \bar{x}\,\bar{y}A$$

$$\tfrac{1}{24}b^2h^2 = \bar{I}_{x''y''} + (\tfrac{1}{3}b)(\tfrac{1}{3}h)(\tfrac{1}{2}bh)$$

$$\bar{I}_{x''y''} = \tfrac{1}{24}b^2h^2 - \tfrac{1}{18}b^2h^2$$

$$\bar{I}_{x''y''} = -\tfrac{1}{72}b^2h^2 \quad \blacktriangleleft$$

499

SAMPLE PROBLEM 9.7

For the section shown, the moments of inertia with respect to the x and y axes have been computed and are known to be

$$I_x = 10.38 \text{ in}^4 \qquad I_y = 6.97 \text{ in}^4$$

Determine (a) the orientation of the principal axes of the section about O, (b) the values of the principal moments of inertia of the section about O.

SOLUTION

We first compute the product of inertia with respect to the x and y axes. The area is divided into three rectangles as shown. We note that the product of inertia $\bar{I}_{x'y'}$ with respect to centroidal axes parallel to the x and y axes is zero for each rectangle. Using the parallel-axis theorem $I_{xy} = \bar{I}_{x'y'} + \bar{x}\,\bar{y}A$, we find that I_{xy} reduces to $\bar{x}\,\bar{y}A$ for each rectangle.

Rectangle	Area, in^2	$\bar{x}$, in.	$\bar{y}$, in.	$\bar{x}\bar{y}A$, in^4
I	1.5	−1.25	+1.75	−3.28
II	1.5	0	0	0
III	1.5	+1.25	−1.75	−3.28
				$\Sigma\bar{x}\bar{y}A = -6.56$

$$I_{xy} = \Sigma\bar{x}\,\bar{y}A = -6.56 \text{ in}^4$$

a. Principal Axes. Since the magnitudes of I_x, I_y, and I_{xy} are known, Eq. (9.25) is used to determine the values of θ_m:

$$\tan 2\theta_m = -\frac{2I_{xy}}{I_x - I_y} = -\frac{2(-6.56)}{10.38 - 6.97} = +3.85$$

$$2\theta_m = 75.4° \text{ and } 255.4°$$

$$\theta_m = 37.7° \qquad \text{and} \qquad \theta_m = 127.7° \quad \blacktriangleleft$$

b. Principal Moments of Inertia. Using Eq. (9.27), we write

$$I_{max,min} = \frac{I_x + I_y}{2} \pm \sqrt{\left(\frac{I_x - I_y}{2}\right)^2 + I_{xy}^2}$$

$$= \frac{10.38 + 6.97}{2} \pm \sqrt{\left(\frac{10.38 - 6.97}{2}\right)^2 + (-6.56)^2}$$

$$I_{max} = 15.45 \text{ in}^4 \qquad I_{min} = 1.897 \text{ in}^4 \quad \blacktriangleleft$$

Noting that the elements of the area of the section are more closely distributed about the b axis than about the a axis, we conclude that $I_a = I_{max} = 15.45 \text{ in}^4$ and $I_b = I_{min} = 1.897 \text{ in}^4$. This conclusion can be verified by substituting $\theta = 37.7°$ into Eqs. (9.18) and (9.19).

SOLVING PROBLEMS
ON YOUR OWN

In the problems for this lesson, you will continue your work with *moments of inertia* and will utilize various techniques for computing *products of inertia*. Although the problems are generally straightforward, several items are worth noting.

1. Calculating the product of inertia I_{xy} by integration. We defined this quantity as

$$I_{xy} = \int xy \, dA \qquad (9.12)$$

and stated that its value can be positive, negative, or zero. The product of inertia can be computed directly from the above equation using double integration, or it can be determined using single integration as shown in Sample Prob. 9.6. When applying the latter technique and using the parallel-axis theorem, it is important to remember that $\bar{x}_{el}$ and $\bar{y}_{el}$ in the equation

$$dI_{xy} = dI_{x'y'} + \bar{x}_{el}\bar{y}_{el} \, dA$$

are the coordinates of the centroid of the element of area dA. Thus, if dA is not in the first quadrant, one or both of these coordinates will be negative.

2. Calculating the products of inertia of composite areas. They can easily be computed from the products of inertia of their component parts by using the parallel-axis theorem

$$I_{xy} = \bar{I}_{x'y'} + \bar{x}\bar{y}A \qquad (9.13)$$

The proper technique to use for problems of this type is illustrated in Sample Probs. 9.6 and 9.7. In addition to the usual rules for composite-area problems, it is essential that you remember the following points.

    a. If either of the centroidal axes of a component area is an axis of symmetry for that area, the product of inertia $\bar{I}_{x'y'}$ for that area is zero. Thus, $\bar{I}_{x'y'}$ is zero for component areas such as circles, semicircles, rectangles, and isosceles triangles which possess an axis of symmetry parallel to one of the coordinate axes.

    b. Pay careful attention to the signs of the coordinates $\bar{x}$ and $\bar{y}$ of each component area when you use the parallel-axis theorem [Sample Prob. 9.7].

3. Determining the moments of inertia and the product of inertia for rotated coordinate axes. In Sec. 9.9 we derived Eqs. (9.18), (9.19), and (9.20), from which the moments of inertia and the product of inertia can be computed for coordinate axes which have been rotated about the origin O. To apply these equations, you must know a set of values I_x, I_y, and I_{xy} for a given orientation of the axes, and you must remember that θ is positive for counterclockwise rotations of the axes and negative for clockwise rotations of the axes.

4. Computing the principal moments of inertia. We showed in Sec. 9.9 that there is a particular orientation of the coordinate axes for which the moments of inertia attain their maximum and minimum values, I_{max} and I_{min}, and for which the product of inertia is zero. Equation (9.27) can be used to compute these values, known as the *principal moments of inertia* of the area about O. The corresponding axes are referred to as the *principal axes* of the area about O, and their orientation is defined by Eq. (9.25). *To determine which of the principal axes corresponds to I_{max} and which corresponds to I_{min}*, you can either follow the procedure outlined in the text after Eq. (9.27) or observe about which of the two principal axes the area is more closely distributed; that axis corresponds to I_{min} [Sample Prob. 9.7].

PROBLEMS

9.67 through 9.70 Determine by direct integration the product of inertia of the given area with respect to the x and y axes.

Fig. P9.67

Fig. P9.68

Fig. P9.69

Fig. P9.70

9.71 through 9.74 Using the parallel-axis theorem, determine the product of inertia of the area shown with respect to the centroidal x and y axes.

Fig. P9.71

Fig. P9.72

Fig. P9.73

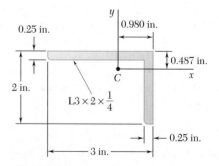

Fig. P9.74

9.75 through 9.78 Using the parallel-axis theorem, determine the product of inertia of the area shown with respect to the centroidal x and y axes.

Fig. P9.75

Fig. P9.76

Fig. P9.77

Fig. P9.78

9.79 Determine for the quarter ellipse of Prob. 9.67 the moments of inertia and the product of inertia with respect to new axes obtained by rotating the x and y axes about O (*a*) through 45° counterclockwise, (*b*) through 30° clockwise.

9.80 Determine the moments of inertia and the product of inertia of the area of Prob. 9.72 with respect to new centroidal axes obtained by rotating the x and y axes 30° counterclockwise.

9.81 Determine the moments of inertia and the product of inertia of the area of Prob. 9.73 with respect to new centroidal axes obtained by rotating the x and y axes 60° counterclockwise.

9.82 Determine the moments of inertia and the product of inertia of the area of Prob. 9.75 with respect to new centroidal axes obtained by rotating the x and y axes 45° clockwise.

9.83 Determine the moments of inertia and the product of inertia of the L3 × 2 × $\frac{1}{4}$-in. angle cross section of Prob. 9.74 with respect to new centroidal axes obtained by rotating the x and y axes 30° clockwise.

9.84 Determine the moments of inertia and the product of inertia of the L127 × 76 × 12.7-mm angle cross section of Prob. 9.78 with respect to new centroidal axes obtained by rotating the x and y axes 45° counterclockwise.

9.85 For the quarter ellipse of Prob. 9.67, determine the orientation of the principal axes at the origin and the corresponding values of the moments of inertia.

9.86 through 9.88 For the area indicated, determine the orientation of the principal axes at the origin and the corresponding values of the moments of inertia.
    9.86 Area of Prob. 9.72
    9.87 Area of Prob. 9.73
    9.88 Area of Prob. 9.75

9.89 and 9.90 For the angle cross section indicated, determine the orientation of the principal axes at the origin and the corresponding values of the moments of inertia.
    9.89 The L3 × 2 × $\frac{1}{4}$-in. angle cross section of Prob. 9.74
    9.90 The L127 × 76 × 12.7-mm angle cross section of Prob. 9.78

*9.10 MOHR'S CIRCLE FOR MOMENTS AND PRODUCTS OF INERTIA

The circle used in the preceding section to illustrate the relations existing between the moments and products of inertia of a given area with respect to axes passing through a fixed point O was first introduced by the German engineer Otto Mohr (1835–1918) and is known as *Mohr's circle*. It will be shown that if the moments and product of inertia of an area A are known with respect to two rectangular x and y axes which pass through a point O, Mohr's circle can be used to graphically determine (*a*) the principal axes and principal moments of inertia of the area about O and (*b*) the moments and product of inertia of the area with respect to any other pair of rectangular axes x' and y' through O.

Consider a given area A and two rectangular coordinate axes x and y (Fig. 9.19*a*). Assuming that the moments of inertia I_x and I_y and the product of inertia I_{xy} are known, we will represent them on a diagram by plotting a point X of coordinates I_x and I_{xy} and a point Y of coordinates I_y and $-I_{xy}$ (Fig. 9.19*b*). If I_{xy} is positive, as assumed in Fig. 9.19*a*, point X is located above the horizontal axis and point Y is located below, as shown in Fig. 9.19*b*. If I_{xy} is negative, X is located below the horizontal axis and Y is located above. Joining X and Y with a straight line, we denote by C the point of intersection of line XY with the horizontal axis and draw the circle of center C and diameter XY. Noting that the abscissa of C and the radius of the circle are respectively equal to the quantities I_{ave} and R defined by the formula (9.23), we conclude that the circle obtained is Mohr's circle for the given area about point O. Thus, the abscissas of the points A and B where the circle intersects the horizontal axis represent, respectively, the principal moments of inertia $I_{\max}$ and $I_{\min}$ of the area.

We also note that, since $\tan (XCA) = 2I_{xy}/(I_x - I_y)$, the angle XCA is equal in magnitude to one of the angles $2\theta_m$ which satisfy Eq. (9.25);

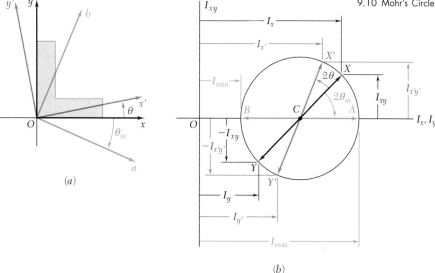

(a)

(b)

Fig. 9.19

thus, the angle θ_m, which defines in Fig. 9.19a the principal axis Oa corresponding to point A in Fig. 9.19b, is equal to half of the angle XCA of Mohr's circle. We further observe that if $I_x > I_y$ and $I_{xy} > 0$, as in the case considered here, the rotation which brings CX into CA is clockwise. Also, under these conditions, the angle θ_m obtained from Eq. (9.25), which defines the principal axis Oa in Fig. 9.19a, is negative; thus, the rotation which brings Ox into Oa is also clockwise. We conclude that the senses of rotation in both parts of Fig. 9.19 are the same. If a clockwise rotation through $2\theta_m$ is required to bring CX into CA on Mohr's circle, a clockwise rotation through θ_m will bring Ox into the corresponding principal axis Oa in Fig. 9.19a.

Since Mohr's circle is uniquely defined, the same circle can be obtained by considering the moments and product of inertia of the area A with respect to the rectangular axes x' and y' (Fig. 9.19a). The point X' of coordinates $I_{x'}$ and $I_{x'y'}$ and the point Y' of coordinates $I_{y'}$ and $-I_{x'y'}$ are thus located on Mohr's circle, and the angle $X'CA$ in Fig. 9.19b must be equal to twice the angle $x'Oa$ in Fig. 9.19a. Since, as noted above, the angle XCA is twice the angle xOa, it follows that the angle XCX' in Fig. 9.19b is twice the angle xOx' in Fig. 9.19a. The diameter $X'Y'$, which defines the moments and product of inertia $I_{x'}$, $I_{y'}$, and $I_{x'y'}$ of the given area with respect to rectangular axes x' and y' forming an angle θ with the x and y axes can be obtained by rotating through an angle 2θ the diameter XY which corresponds to the moments and product of inertia I_x, I_y, and I_{xy}. We note that the rotation which brings the diameter XY into the diameter $X'Y'$ in Fig. 9.19b has the same sense as the rotation which brings the x and y axes into the x' and y' axes in Fig. 9.19a.

It should be noted that the use of Mohr's circle is not limited to graphical solutions, i.e., to solutions based on the careful drawing and measuring of the various parameters involved. By merely sketching Mohr's circle and using trigonometry, one can easily derive the various relations required for a numerical solution of a given problem (see Sample Prob. 9.8).

L152 × 102 × 12.7

$\theta = 60°$

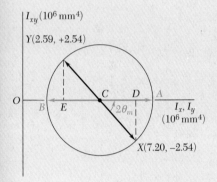

$I_{xy}(10^6 \text{ mm}^4)$

$Y(2.59, +2.54)$

$X(7.20, -2.54)$

$2\theta_m$

I_x, I_y (10^6 mm^4)

$\theta_m = 23.9°$

I_{xy}

$4.895 \times 10^6 \text{ mm}^4$

$3.430 \times 10^6 \text{ mm}^4$

$2\theta = 120°$

$2\theta_m = 47.8°$

I_x, I_y

SAMPLE PROBLEM 9.8

For the section shown, the moments and product of inertia with respect to the x and y axes are known to be

$$I_x = 7.20 \times 10^6 \text{ mm}^4 \qquad I_y = 2.59 \times 10^6 \text{ mm}^4 \qquad I_{xy} = -2.54 \times 10^6 \text{ mm}^4$$

Using Mohr's circle, determine (a) the principal axes of the section about O, (b) the values of the principal moments of inertia of the section about O, (c) the moments and product of inertia of the section with respect to the x' and y' axes which form an angle of 60° with the x and y axes.

SOLUTION

Drawing Mohr's Circle. We first plot point X of coordinates $I_x = 7.20$, $I_{xy} = -2.54$, and point Y of coordinates $I_y = 2.59$, $-I_{xy} = +2.54$. Joining X and Y with a straight line, we define the center C of Mohr's circle. The abscissa of C, which represents I_{ave}, and the radius R of the circle can be measured directly or calculated as follows:

$$I_{ave} = OC = \tfrac{1}{2}(I_x + I_y) = \tfrac{1}{2}(7.20 \times 10^6 + 2.59 \times 10^6) = 4.895 \times 10^6 \text{ mm}^4$$
$$CD = \tfrac{1}{2}(I_x - I_y) = \tfrac{1}{2}(7.20 \times 10^6 - 2.59 \times 10^6) = 2.305 \times 10^6 \text{ mm}^4$$
$$R = \sqrt{(CD)^2 + (DX)^2} = \sqrt{(2.305 \times 10^6)^2 + (2.54 \times 10^6)^2}$$
$$= 3.430 \times 10^6 \text{ mm}^4$$

a. Principal Axes. The principal axes of the section correspond to points A and B on Mohr's circle, and the angle through which we should rotate CX to bring it into CA defines $2\theta_m$. We have

$$\tan 2\theta_m = \frac{DX}{CD} = \frac{2.54}{2.305} = 1.102 \qquad 2\theta_m = 47.8° \text{↰} \qquad \theta_m = 23.9° \text{↰} \quad \blacktriangleleft$$

Thus, the principal axis Oa corresponding to the maximum value of the moment of inertia is obtained by rotating the x axis through 23.9° counterclockwise; the principal axis Ob corresponding to the minimum value of the moment of inertia can be obtained by rotating the y axis through the same angle.

b. Principal Moments of Inertia. The principal moments of inertia are represented by the abscissas of A and B. We have

$$I_{max} = OA = OC + CA = I_{ave} + R = (4.895 + 3.430)10^6 \text{ mm}^4$$
$$I_{max} = 8.33 \times 10^6 \text{ mm}^4 \quad \blacktriangleleft$$
$$I_{min} = OB = OC - BC = I_{ave} - R = (4.895 - 3.430)10^6 \text{ mm}^4$$
$$I_{min} = 1.47 \times 10^6 \text{ mm}^4 \quad \blacktriangleleft$$

c. Moments and Product of Inertia with Respect to the x' and y' Axes. On Mohr's circle, the points X' and Y', which correspond to the x' and y' axes, are obtained by rotating CX and CY through an angle $2\theta = 2(60°) = 120°$ counterclockwise. The coordinates of X' and Y' yield the desired moments and product of inertia. Noting that the angle that CX' forms with the horizontal axis is $\phi = 120° - 47.8° = 72.2°$, we write

$$I_{x'} = OF = OC + CF = 4.895 \times 10^6 \text{ mm}^4 + (3.430 \times 10^6 \text{ mm}^4) \cos 72.2°$$
$$I_{x'} = 5.94 \times 10^6 \text{ mm}^4 \quad \blacktriangleleft$$
$$I_{y'} = OG = OC - GC = 4.895 \times 10^6 \text{ mm}^4 - (3.430 \times 10^6 \text{ mm}^4) \cos 72.2°$$
$$I_{y'} = 3.85 \times 10^6 \text{ mm}^4 \quad \blacktriangleleft$$
$$I_{x'y'} = FX' = (3.430 \times 10^6 \text{ mm}^4) \sin 72.2°$$
$$I_{x'y'} = 3.27 \times 10^6 \text{ mm}^4 \quad \blacktriangleleft$$

SOLVING PROBLEMS
ON YOUR OWN

In the problems for this lesson, you will use *Mohr's circle* to determine the moments and products of inertia of a given area for different orientations of the coordinate axes. Although in some cases using Mohr's circle may not be as direct as substituting into the appropriate equations [Eqs. (9.18) through (9.20)], this method of solution has the advantage of providing a visual representation of the relationships among the various variables. Further, Mohr's circle shows all of the values of the moments and products of inertia which are possible for a given problem.

Using Mohr's circle. The underlying theory was presented in Sec. 9.9, and we discussed the application of this method in Sec. 9.10 and in Sample Prob. 9.8. In the same problem, we presented the steps you should follow to determine the *principal axes*, the *principal moments of inertia,* and the *moments and product of inertia with respect to a specified orientation of the coordinates axes.* When you use Mohr's circle to solve problems, it is important that you remember the following points.

a. Mohr's circle is completely defined by the quantities R and I_{ave}, which represent, respectively, the radius of the circle and the distance from the origin O to the center C of the circle. These quantities can be obtained from Eqs. (9.23) if the moments and product of inertia are known for a given orientation of the axes. However, Mohr's circle can be defined by other combinations of known values [Probs. 9.103, 9.106, and 9.107]. For these cases, it may be necessary to first make one or more assumptions, such as choosing an arbitrary location for the center when I_{ave} is unknown, assigning relative magnitudes to the moments of inertia (for example, $I_x > I_y$), or selecting the sign of the product of inertia.

b. Point X of coordinates (I_x, I_{xy}) and point Y of coordinates $(I_y, -I_{xy})$ are both located on Mohr's circle and are diametrically opposite.

c. Since moments of inertia must be positive, the entire Mohr's circle must lie to the right of the I_{xy} axis; it follows that $I_{ave} > R$ for all cases.

d. As the coordinate axes are rotated through an angle θ, the associated rotation of the diameter of Mohr's circle is equal to 2θ and is in the same sense (clockwise or counterclockwise). We strongly suggest that the known points on the circumference of the circle be labeled with the appropriate capital letter, as was done in Fig. 9.19*b* and for the Mohr circles of Sample Prob. 9.8. This will enable you to determine, for each value of θ, the sign of the corresponding product of inertia and to determine which moment of inertia is associated with each of the coordinate axes [Sample Prob. 9.8, parts *a* and *c*].

Although we have introduced Mohr's circle within the specific context of the study of moments and products of inertia, the Mohr circle technique is also applicable to the solution of analogous but physically different problems in mechanics of materials. This multiple use of a specific technique is not unique, and as you pursue your engineering studies, you will encounter several methods of solution which can be applied to a variety of problems.

PROBLEMS

9.91 Using Mohr's circle, determine for the quarter ellipse of Prob. 9.67 the moments of inertia and the product of inertia with respect to new axes obtained by rotating the x and y axes about O (a) through 45° counterclockwise, (b) through 30° clockwise.

9.92 Using Mohr's circle, determine the moments of inertia and the product of inertia of the area of Prob. 9.72 with respect to new centroidal axes obtained by rotating the x and y axes 30° counterclockwise.

9.93 Using Mohr's circle, determine the moments of inertia and the product of inertia of the area of Prob. 9.73 with respect to new centroidal axes obtained by rotating the x and y axes 60° counterclockwise.

9.94 Using Mohr's circle, determine the moments of inertia and the product of inertia of the area of Prob. 9.75 with respect to new centroidal axes obtained by rotating the x and y axes 45° clockwise.

9.95 Using Mohr's circle, determine the moments of inertia and the product of inertia of the L3 × 2 × $\frac{1}{4}$-in. angle cross section of Prob. 9.74 with respect to new centroidal axes obtained by rotating the x and y axes 30° clockwise.

9.96 Using Mohr's circle, determine the moments of inertia and the product of inertia of the L127 × 76 × 12.7-mm angle cross section of Prob. 9.78 with respect to new centroidal axes obtained by rotating the x and y axes 45° counterclockwise.

9.97 For the quarter ellipse of Prob. 9.67, use Mohr's circle to determine the orientation of the principal axes at the origin and the corresponding values of the moments of inertia.

9.98 through 9.102 Using Mohr's circle, determine for the area indicated the orientation of the principal centroidal axes and the corresponding values of the moments of inertia.

    9.98 Area of Prob. 9.72
    9.99 Area of Prob. 9.76
    9.100 Area of Prob. 9.73
    9.101 Area of Prob. 9.74
    9.102 Area of Prob. 9.77

(The moments of inertia $\bar{I}_x$ and $\bar{I}_y$ of the area of Prob. 9.102 were determined in Prob. 9.44.)

9.103 The moments and product of inertia of an L4 × 3 × $\frac{1}{4}$-in. angle cross section with respect to two rectangular axes x and y through C are, respectively, $\bar{I}_x = 1.33$ in⁴, $\bar{I}_y = 2.75$ in⁴, and $\bar{I}_{xy} < 0$, with the minimum value of the moment of inertia of the area with respect to any axis through C being $\bar{I}_{min} = 0.692$ in⁴. Using Mohr's circle, determine (a) the product of inertia $\bar{I}_{xy}$ of the area, (b) the orientation of the principal axes, (c) the value of $\bar{I}_{max}$.

9.104 and 9.105   Using Mohr's circle, determine for the cross section of the rolled-steel angle shown the orientation of the principal centroidal axes and the corresponding values of the moments of inertia. (Properties of the cross sections are given in Fig. 9.13.)

Fig. P9.104

Fig. P9.105

*9.106   For a given area the moments of inertia with respect to two rectangular centroidal x and y axes are $\bar{I}_x = 1200$ in^4 and $\bar{I}_y = 300$ in^4, respectively. Knowing that after rotating the x and y axes about the centroid 30° counterclockwise, the moment of inertia relative to the rotated x axis is 1450 in^4, use Mohr's circle to determine (*a*) the orientation of the principal axes, (*b*) the principal centroidal moments of inertia.

9.107   It is known that for a given area $\bar{I}_y = 48 \times 10^6$ mm^4 and $\bar{I}_{xy} = -20 \times 10^6$ mm^4, where the x and y axes are rectangular centroidal axes. If the axis corresponding to the maximum product of inertia is obtained by rotating the x axis 67.5° counterclockwise about C, use Mohr's circle to determine (*a*) the moment of inertia $\bar{I}_x$ of the area, (*b*) the principal centroidal moments of inertia.

9.108   Using Mohr's circle, show that for any regular polygon (such as a pentagon) (*a*) the moment of inertia with respect to every axis through the centroid is the same, (*b*) the product of inertia with respect to every pair of rectangular axes through the centroid is zero.

9.109   Using Mohr's circle, prove that the expression $I_{x'}I_{y'} - I_{x'y'}^2$ is independent of the orientation of the x' and y' axes, where $I_{x'}$, $I_{y'}$, and $I_{x'y'}$ represent the moments and product of inertia, respectively, of a given area with respect to a pair of rectangular axes x' and y' through a given point O. Also show that the given expression is equal to the square of the length of the tangent drawn from the origin of the coordinate system to Mohr's circle.

9.110   Using the invariance property established in the preceding problem, express the product of inertia I_{xy} of an area A with respect to a pair of rectangular axes through O in terms of the moments of inertia I_x and I_y of A and the principal moments of inertia I_{min} and I_{max} of A about O. Use the formula obtained to calculate the product of inertia I_{xy} of the L3 × 2 × $\frac{1}{4}$-in. angle cross section shown in Fig. 9.13A, knowing that its maximum moment of inertia is 1.257 in^4.

9.11 MOMENT OF INERTIA OF A MASS

Consider a small mass Δm mounted on a rod of negligible mass which can rotate freely about an axis AA' (Fig. 9.20a). If a couple is applied to the system, the rod and mass, assumed to be initially at rest, will start rotating about AA'. The details of this motion will be studied later in dynamics. At present, we wish only to indicate that the time required for the system to reach a given speed of rotation is proportional to the mass Δm and to the square of the distance r. The product $r^2 \Delta m$ provides, therefore, a measure of the *inertia* of the system, i.e., a measure of the resistance the system offers when we try to set it in motion. For this reason, the product $r^2 \Delta m$ is called the *moment of inertia* of the mass Δm with respect to the axis AA'.

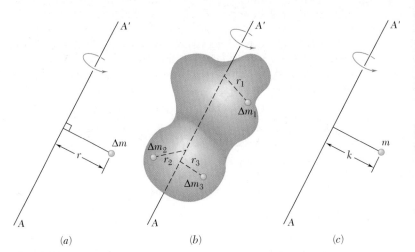

(a)   (b)   (c)

Fig. 9.20

Consider now a body of mass m which is to be rotated about an axis AA' (Fig. 9.20b). Dividing the body into elements of mass Δm_1, Δm_2, etc., we find that the body's resistance to being rotated is measured by the sum $r_1^2 \Delta m_1 + r_2^2 \Delta m_2 + \cdots$. This sum defines, therefore, the moment of inertia of the body with respect to the axis AA'. Increasing the number of elements, we find that the moment of inertia is equal, in the limit, to the integral

$$I = \int r^2 \, dm \tag{9.28}$$

The *radius of gyration* k of the body with respect to the axis AA' is defined by the relation

$$I = k^2 m \qquad \text{or} \qquad k = \sqrt{\dfrac{I}{m}} \qquad (9.29)$$

The radius of gyration k represents, therefore, the distance at which the entire mass of the body should be concentrated if its moment of inertia with respect to AA' is to remain unchanged (Fig. 9.20c). Whether it is kept in its original shape (Fig. 9.20b) or whether it is concentrated as shown in Fig. 9.20c, the mass m will react in the same way to a rotation, or *gyration*, about AA'.

If SI units are used, the radius of gyration k is expressed in meters and the mass m in kilograms, and thus the unit used for the moment of inertia of a mass is $kg \cdot m^2$. If U.S. customary units are used, the radius of gyration is expressed in feet and the mass in slugs (i.e., in $lb \cdot s^2/ft$), and thus the derived unit used for the moment of inertia of a mass is $lb \cdot ft \cdot s^2$.[†]

The moment of inertia of a body with respect to a coordinate axis can easily be expressed in terms of the coordinates x, y, z of the element of mass dm (Fig. 9.21). Noting, for example, that the square of the distance r from the element dm to the y axis is $z^2 + x^2$, we express the moment of inertia of the body with respect to the y axis as

$$I_y = \int r^2 \, dm = \int (z^2 + x^2) \, dm$$

Similar expressions can be obtained for the moments of inertia with respect to the x and z axes. We write

$$I_x = \int (y^2 + z^2) \, dm$$

$$I_y = \int (z^2 + x^2) \, dm \qquad (9.30)$$

$$I_z = \int (x^2 + y^2) \, dm$$

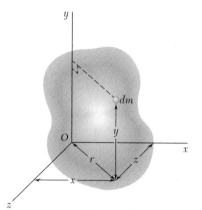

Fig. 9.21

[†]It should be kept in mind when converting the moment of inertia of a mass from U.S. customary units to SI units that the base unit *pound* used in the derived unit $lb \cdot ft \cdot s^2$ is a unit of force (*not* of mass) and should therefore be converted into newtons. We have

$$1 \ lb \cdot ft \cdot s^2 = (4.45 \ N)(0.3048 \ m)(1 \ s)^2 = 1.356 \ N \cdot m \cdot s^2$$

or, since $1 \ N = 1 \ kg \cdot m/s^2$,

$$1 \ lb \cdot ft \cdot s^2 = 1.356 \ kg \cdot m^2$$

Photo 9.2 As you will discuss in your dynamics course, the rotational behavior of the camshaft shown is dependent upon the mass moment of inertia of the camshaft with respect to its axis of rotation.

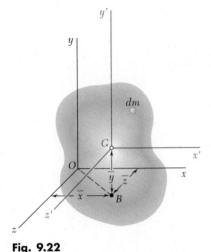

Fig. 9.22

9.12 PARALLEL-AXIS THEOREM

Consider a body of mass m. Let $Oxyz$ be a system of rectangular coordinates whose origin is at the arbitrary point O, and $Gx'y'z'$ a system of parallel *centroidal axes*, i.e., a system whose origin is at the center of gravity G of the body† and whose axes x', y', z' are parallel to the x, y, and z axes, respectively (Fig. 9.22). Denoting by $\bar{x}, \bar{y}, \bar{z}$ the coordinates of G with respect to $Oxyz$, we write the following relations between the coordinates x, y, z of the element dm with respect to $Oxyz$ and its coordinates x', y', z' with respect to the centroidal axes $Gx'y'z'$:

$$x = x' + \bar{x} \qquad y = y' + \bar{y} \qquad z = z' + \bar{z} \qquad (9.31)$$

Referring to Eqs. (9.30), we can express the moment of inertia of the body with respect to the x axis as follows:

$$I_x = \int (y^2 + z^2)\, dm = \int [(y' + \bar{y})^2 + (z' + \bar{z})^2]\, dm$$

$$= \int (y'^2 + z'^2)\, dm + 2\bar{y}\int y'\, dm + 2\bar{z}\int z'\, dm + (\bar{y}^2 + \bar{z}^2)\int dm$$

The first integral in this expression represents the moment of inertia $\bar{I}_{x'}$ of the body with respect to the centroidal axis x'; the second and third integrals represent the first moment of the body with respect to the $z'x'$ and $x'y'$ planes, respectively, and, since both planes contain G, the two integrals are zero; the last integral is equal to the total mass m of the body. We write, therefore,

$$I_x = \bar{I}_{x'} + m(\bar{y}^2 + \bar{z}^2) \qquad (9.32)$$

and, similarly,

$$I_y = \bar{I}_{y'} + m(\bar{z}^2 + \bar{x}^2) \qquad I_z = \bar{I}_{z'} + m(\bar{x}^2 + \bar{y}^2) \qquad (9.32')$$

We easily verify from Fig. 9.22 that the sum $\bar{z}^2 + \bar{x}^2$ represents the square of the distance OB between the y and y' axes. Similarly, $\bar{y}^2 + \bar{z}^2$ and $\bar{x}^2 + \bar{y}^2$ represent the squares of the distance between the x and x' axes and the z and z' axes, respectively. Denoting by d the distance between an arbitrary axis AA' and a parallel centroidal axis BB' (Fig. 9.23), we can, therefore, write the following general relation between the moment of inertia I of the body with respect to AA' and its moment of inertia $\bar{I}$ with respect to BB':

$$I = \bar{I} + md^2 \qquad (9.33)$$

Expressing the moments of inertia in terms of the corresponding radii of gyration, we can also write

$$k^2 = \bar{k}^2 + d^2 \qquad (9.34)$$

where k and $\bar{k}$ represent the radii of gyration of the body about AA' and BB', respectively.

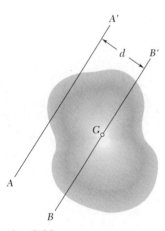

Fig. 9.23

†Note that the term *centroidal* is used here to define an axis passing through the center of gravity G of the body, whether or not G coincides with the centroid of the volume of the body.

9.13 MOMENTS OF INERTIA OF THIN PLATES

Consider a thin plate of uniform thickness t, which is made of a homogeneous material of density ρ (density = mass per unit volume). The mass moment of inertia of the plate with respect to an axis AA' *contained in the plane* of the plate (Fig. 9.24a) is

$$I_{AA', \text{ mass}} = \int r^2 \, dm$$

Since $dm = \rho t \, dA$, we write

$$I_{AA', \text{ mass}} = \rho t \int r^2 \, dA$$

But r represents the distance of the element of area dA to the axis

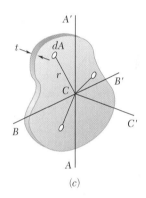

(a)   (b)   (c)

Fig. 9.24

AA'; the integral is therefore equal to the moment of inertia of the area of the plate with respect to AA'. We have

$$I_{AA', \text{ mass}} = \rho t I_{AA', \text{ area}} \qquad (9.35)$$

Similarly, for an axis BB' which is contained in the plane of the plate and is perpendicular to AA' (Fig. 9.24b), we have

$$I_{BB', \text{ mass}} = \rho t I_{BB', \text{ area}} \qquad (9.36)$$

Considering now the axis CC' which is *perpendicular* to the plate and passes through the point of intersection C of AA' and BB' (Fig. 9.24c), we write

$$I_{CC', \text{ mass}} = \rho t J_{C, \text{ area}} \qquad (9.37)$$

where J_C is the *polar* moment of inertia of the area of the plate with respect to point C.

Recalling the relation $J_C = I_{AA'} + I_{BB'}$ which exists between polar and rectangular moments of inertia of an area, we write the following relation between the mass moments of inertia of a thin plate:

$$I_{CC'} = I_{AA'} + I_{BB'} \qquad (9.38)$$

Fig. 9.25

Fig. 9.26

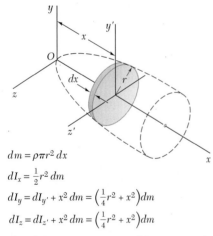

$$dm = \rho \pi r^2\, dx$$

$$dI_x = \tfrac{1}{2}r^2\, dm$$

$$dI_y = dI_{y'} + x^2\, dm = \left(\tfrac{1}{4}r^2 + x^2\right)dm$$

$$dI_z = dI_{z'} + x^2\, dm = \left(\tfrac{1}{4}r^2 + x^2\right)dm$$

Fig. 9.27 Determination of the moment of inertia of a body of revolution.

Rectangular Plate. In the case of a rectangular plate of sides a and b (Fig. 9.25), we obtain the following mass moments of inertia with respect to axes through the center of gravity of the plate:

$$I_{AA',\,\text{mass}} = \rho t I_{AA',\,\text{area}} = \rho t\left(\tfrac{1}{12}a^3 b\right)$$
$$I_{BB',\,\text{mass}} = \rho t I_{BB',\,\text{area}} = \rho t\left(\tfrac{1}{12}ab^3\right)$$

Observing that the product ρabt is equal to the mass m of the plate, we write the mass moments of inertia of a thin rectangular plate as follows:

$$I_{AA'} = \tfrac{1}{12}ma^2 \qquad I_{BB'} = \tfrac{1}{12}mb^2 \tag{9.39}$$
$$I_{CC'} = I_{AA'} + I_{BB'} = \tfrac{1}{12}m(a^2 + b^2) \tag{9.40}$$

Circular Plate. In the case of a circular plate, or disk, of radius r (Fig. 9.26), we write

$$I_{AA',\,\text{mass}} = \rho t I_{AA',\,\text{area}} = \rho t\left(\tfrac{1}{4}\pi r^4\right)$$

Observing that the product $\rho \pi r^2 t$ is equal to the mass m of the plate and that $I_{AA'} = I_{BB'}$, we write the mass moments of inertia of a circular plate as follows:

$$I_{AA'} = I_{BB'} = \tfrac{1}{4}mr^2 \tag{9.41}$$
$$I_{CC'} = I_{AA'} + I_{BB'} = \tfrac{1}{2}mr^2 \tag{9.42}$$

9.14 DETERMINATION OF THE MOMENT OF INERTIA OF A THREE-DIMENSIONAL BODY BY INTEGRATION

The moment of inertia of a three-dimensional body is obtained by evaluating the integral $I = \int r^2\, dm$. If the body is made of a homogeneous material of density ρ, the element of mass dm is equal to $\rho\, dV$ and we can write $I = \rho\int r^2\, dV$. This integral depends only upon the shape of the body. Thus, in order to compute the moment of inertia of a three-dimensional body, it will generally be necessary to perform a triple, or at least a double, integration.

However, if the body possesses two planes of symmetry, it is usually possible to determine the body's moment of inertia with a single integration by choosing as the element of mass dm a thin slab which is perpendicular to the planes of symmetry. In the case of bodies of revolution, for example, the element of mass would be a thin disk (Fig. 9.27). Using formula (9.42), the moment of inertia of the disk with respect to the axis of revolution can be expressed as indicated in Fig. 9.27. Its moment of inertia with respect to each of the other two coordinate axes is obtained by using formula (9.41) and the parallel-axis theorem. Integration of the expression obtained yields the desired moment of inertia of the body.

9.15 MOMENTS OF INERTIA OF COMPOSITE BODIES

The moments of inertia of a few common shapes are shown in Fig. 9.28. For a body consisting of several of these simple shapes, the moment of inertia of the body with respect to a given axis can be obtained by first computing the moments of inertia of its component parts about the desired axis and then adding them together. As was the case for areas, the radius of gyration of a composite body *cannot* be obtained by adding the radii of gyration of its component parts.

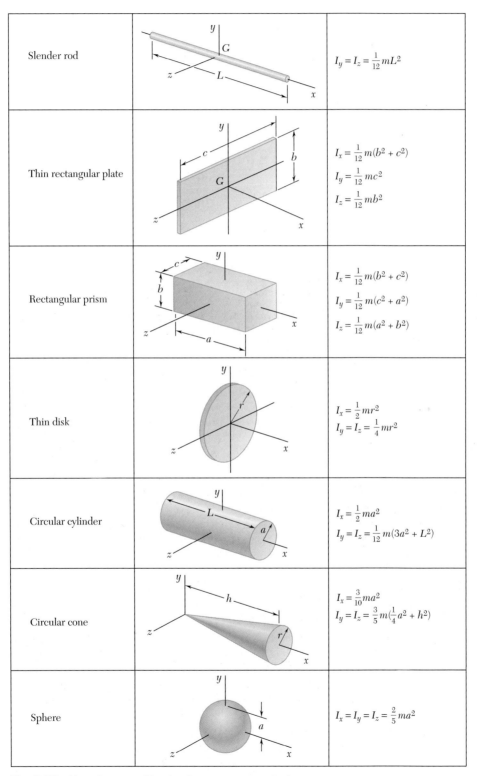

Slender rod		$I_y = I_z = \frac{1}{12}mL^2$
Thin rectangular plate		$I_x = \frac{1}{12}m(b^2 + c^2)$ $I_y = \frac{1}{12}mc^2$ $I_z = \frac{1}{12}mb^2$
Rectangular prism		$I_x = \frac{1}{12}m(b^2 + c^2)$ $I_y = \frac{1}{12}m(c^2 + a^2)$ $I_z = \frac{1}{12}m(a^2 + b^2)$
Thin disk		$I_x = \frac{1}{2}mr^2$ $I_y = I_z = \frac{1}{4}mr^2$
Circular cylinder		$I_x = \frac{1}{2}ma^2$ $I_y = I_z = \frac{1}{12}m(3a^2 + L^2)$
Circular cone		$I_x = \frac{3}{10}ma^2$ $I_y = I_z = \frac{3}{5}m(\frac{1}{4}a^2 + h^2)$
Sphere		$I_x = I_y = I_z = \frac{2}{5}ma^2$

Fig. 9.28 Mass moments of inertia of common geometric shapes.

SAMPLE PROBLEM 9.9

Determine the moment of inertia of a slender rod of length L and mass m with respect to an axis which is perpendicular to the rod and passes through one end of the rod.

SOLUTION

Choosing the differential element of mass shown, we write

$$dm = \frac{m}{L}dx$$

$$I_y = \int x^2\, dm = \int_0^L x^2\frac{m}{L}dx = \left[\frac{m}{L}\frac{x^3}{3}\right]_0^L \qquad I_y = \tfrac{1}{3}mL^2 \quad \blacktriangleleft$$

SAMPLE PROBLEM 9.10

For the homogeneous rectangular prism shown, determine the moment of inertia with respect to the z axis.

SOLUTION

We choose as the differential element of mass the thin slab shown; thus

$$dm = \rho bc\, dx$$

Referring to Sec. 9.13, we find that the moment of inertia of the element with respect to the z' axis is

$$dI_{z'} = \tfrac{1}{12}b^2\, dm$$

Applying the parallel-axis theorem, we obtain the mass moment of inertia of the slab with respect to the z axis.

$$dI_z = dI_{z'} + x^2\, dm = \tfrac{1}{12}b^2\, dm + x^2\, dm = (\tfrac{1}{12}b^2 + x^2)\rho bc\, dx$$

Integrating from $x = 0$ to $x = a$, we obtain

$$I_z = \int dI_z = \int_0^a (\tfrac{1}{12}b^2 + x^2)\rho bc\, dx = \rho abc(\tfrac{1}{12}b^2 + \tfrac{1}{3}a^2)$$

Since the total mass of the prism is $m = \rho abc$, we can write

$$I_z = m(\tfrac{1}{12}b^2 + \tfrac{1}{3}a^2) \qquad I_z = \tfrac{1}{12}m(4a^2 + b^2) \quad \blacktriangleleft$$

We note that if the prism is thin, b is small compared to a, and the expression for I_z reduces to $\tfrac{1}{3}ma^2$, which is the result obtained in Sample Prob. 9.9 when $L = a$.

SAMPLE PROBLEM 9.11

Determine the moment of inertia of a right circular cone with respect to (*a*) its longitudinal axis, (*b*) an axis through the apex of the cone and perpendicular to its longitudinal axis, (*c*) an axis through the centroid of the cone and perpendicular to its longitudinal axis.

SOLUTION

We choose the differential element of mass shown.

$$r = a\frac{x}{h} \qquad dm = \rho\pi r^2\, dx = \rho\pi \frac{a^2}{h^2}x^2\, dx$$

a. Moment of Inertia I_x. Using the expression derived in Sec. 9.13 for a thin disk, we compute the mass moment of inertia of the differential element with respect to the x axis.

$$dI_x = \tfrac{1}{2}r^2\, dm = \tfrac{1}{2}\left(a\frac{x}{h}\right)^2\left(\rho\pi\frac{a^2}{h^2}x^2\, dx\right) = \tfrac{1}{2}\rho\pi\frac{a^4}{h^4}x^4\, dx$$

Integrating from $x = 0$ to $x = h$, we obtain

$$I_x = \int dI_x = \int_0^h \tfrac{1}{2}\rho\pi\frac{a^4}{h^4}x^4\, dx = \tfrac{1}{2}\rho\pi\frac{a^4}{h^4}\frac{h^5}{5} = \tfrac{1}{10}\rho\pi a^4 h$$

Since the total mass of the cone is $m = \tfrac{1}{3}\rho\pi a^2 h$, we can write

$$I_x = \tfrac{1}{10}\rho\pi a^4 h = \tfrac{3}{10}a^2(\tfrac{1}{3}\rho\pi a^2 h) = \tfrac{3}{10}ma^2 \qquad I_x = \tfrac{3}{10}ma^2 \quad \blacktriangleleft$$

b. Moment of Inertia I_y. The same differential element is used. Applying the parallel-axis theorem and using the expression derived in Sec. 9.13 for a thin disk, we write

$$dI_y = dI_{y'} + x^2\, dm = \tfrac{1}{4}r^2\, dm + x^2\, dm = (\tfrac{1}{4}r^2 + x^2)\, dm$$

Substituting the expressions for r and dm into the equation, we obtain

$$dI_y = \left(\frac{1}{4}\frac{a^2}{h^2}x^2 + x^2\right)\left(\rho\pi\frac{a^2}{h^2}x^2\, dx\right) = \rho\pi\frac{a^2}{h^2}\left(\frac{a^2}{4h^2} + 1\right)x^4\, dx$$

$$I_y = \int dI_y = \int_0^h \rho\pi\frac{a^2}{h^2}\left(\frac{a^2}{4h^2} + 1\right)x^4\, dx = \rho\pi\frac{a^2}{h^2}\left(\frac{a^2}{4h^2} + 1\right)\frac{h^5}{5}$$

Introducing the total mass of the cone m, we rewrite I_y as follows:

$$I_y = \tfrac{3}{5}(\tfrac{1}{4}a^2 + h^2)\tfrac{1}{3}\rho\pi a^2 h \qquad I_y = \tfrac{3}{5}m(\tfrac{1}{4}a^2 + h^2) \quad \blacktriangleleft$$

c. Moment of Inertia $I_{y''}$. We apply the parallel-axis theorem and write

$$I_y = \bar{I}_{y''} + m\bar{x}^2$$

Solving for $\bar{I}_{y''}$ and recalling that $\bar{x} = \tfrac{3}{4}h$, we have

$$\bar{I}_{y''} = I_y - m\bar{x}^2 = \tfrac{3}{5}m(\tfrac{1}{4}a^2 + h^2) - m(\tfrac{3}{4}h)^2$$

$$\bar{I}_{y''} = \tfrac{3}{20}m(a^2 + \tfrac{1}{4}h^2) \quad \blacktriangleleft$$

SAMPLE PROBLEM 9.12

A steel forging consists of a $6 \times 2 \times 2$-in. rectangular prism and two cylinders of diameter 2 in. and length 3 in. as shown. Determine the moments of inertia of the forging with respect to the coordinate axes, knowing that the specific weight of steel is 490 lb/ft³.

SOLUTION

Computation of Masses

Prism

$$V = (2 \text{ in.})(2 \text{ in.})(6 \text{ in.}) = 24 \text{ in}^3$$

$$W = \frac{(24 \text{ in}^3)(490 \text{ lb/ft}^3)}{1728 \text{ in}^3/\text{ft}^3} = 6.81 \text{ lb}$$

$$m = \frac{6.81 \text{ lb}}{32.2 \text{ ft/s}^2} = 0.211 \text{ lb} \cdot \text{s}^2/\text{ft}$$

Each Cylinder

$$V = \pi(1 \text{ in.})^2(3 \text{ in.}) = 9.42 \text{ in}^3$$

$$W = \frac{(9.42 \text{ in}^3)(490 \text{ lb/ft}^3)}{1728 \text{ in}^3/\text{ft}^3} = 2.67 \text{ lb}$$

$$m = \frac{2.67 \text{ lb}}{32.2 \text{ ft/s}^2} = 0.0829 \text{ lb} \cdot \text{s}^2/\text{ft}$$

Moments of Inertia. The moments of inertia of each component are computed from Fig. 9.28, using the parallel-axis theorem when necessary. Note that all lengths should be expressed in feet.

Prism

$I_x = I_z = \frac{1}{12}(0.211 \text{ lb} \cdot \text{s}^2/\text{ft})[(\frac{6}{12} \text{ ft})^2 + (\frac{2}{12} \text{ ft})^2] = 4.88 \times 10^{-3} \text{ lb} \cdot \text{ft} \cdot \text{s}^2$

$I_y = \frac{1}{12}(0.211 \text{ lb} \cdot \text{s}^2/\text{ft})[(\frac{2}{12} \text{ ft})^2 + (\frac{2}{12} \text{ ft})^2] = 0.977 \times 10^{-3} \text{ lb} \cdot \text{ft} \cdot \text{s}^2$

Each Cylinder

$I_x = \frac{1}{2}ma^2 + m\bar{y}^2 = \frac{1}{2}(0.0829 \text{ lb} \cdot \text{s}^2/\text{ft})(\frac{1}{12} \text{ ft})^2$
$\qquad\qquad + (0.0829 \text{ lb} \cdot \text{s}^2/\text{ft})(\frac{2}{12} \text{ ft})^2 = 2.59 \times 10^{-3} \text{ lb} \cdot \text{ft} \cdot \text{s}^2$

$I_y = \frac{1}{12}m(3a^2 + L^2) = m\bar{x}^2 = \frac{1}{12}(0.0829 \text{ lb} \cdot \text{s}^2/\text{ft})[3(\frac{1}{12} \text{ ft})^2 + (\frac{3}{12} \text{ ft})^2]$
$\qquad\qquad + (0.0829 \text{ lb} \cdot \text{s}^2/\text{ft})(\frac{2.5}{12} \text{ ft})^2 = 4.17 \times 10^{-3} \text{ lb} \cdot \text{ft} \cdot \text{s}^2$

$I_z = \frac{1}{12}m(3a^2 + L^2) + m(\bar{x}^2 + \bar{y}^2) = \frac{1}{12}(0.0829 \text{ lb} \cdot \text{s}^2/\text{ft})[3(\frac{1}{12} \text{ ft})^2 + (\frac{3}{12} \text{ ft})^2]$
$\qquad\qquad + (0.0829 \text{ lb} \cdot \text{s}^2/\text{ft})[(\frac{2.5}{12} \text{ ft})^2 + (\frac{2}{12} \text{ ft})^2] = 6.48 \times 10^{-3} \text{ lb} \cdot \text{ft} \cdot \text{s}^2$

Entire Body. Adding the values obtained,

$I_x = 4.88 \times 10^{-3} + 2(2.59 \times 10^{-3}) \qquad I_x = 10.06 \times 10^{-3} \text{ lb} \cdot \text{ft} \cdot \text{s}^2 \quad \blacktriangleleft$

$I_y = 0.977 \times 10^{-3} + 2(4.17 \times 10^{-3}) \qquad I_y = 9.32 \times 10^{-3} \text{ lb} \cdot \text{ft} \cdot \text{s}^2 \quad \blacktriangleleft$

$I_z = 4.88 \times 10^{-3} + 2(6.48 \times 10^{-3}) \qquad I_z = 17.84 \times 10^{-3} \text{ lb} \cdot \text{ft} \cdot \text{s}^2 \quad \blacktriangleleft$

Dimensions in mm

SAMPLE PROBLEM 9.13

A thin steel plate which is 4 mm thick is cut and bent to form the machine part shown. Knowing that the density of steel is 7850 kg/m^3, determine the moments of inertia of the machine part with respect to the coordinate axes.

SOLUTION

We observe that the machine part consists of a semicircular plate and a rectangular plate from which a circular plate has been removed.

Computation of Masses. *Semicircular Plate*

$$V_1 = \tfrac{1}{2}\pi r^2 t = \tfrac{1}{2}\pi(0.08\text{ m})^2(0.004\text{ m}) = 40.21 \times 10^{-6}\text{ m}^3$$
$$m_1 = \rho V_1 = (7.85 \times 10^3\text{ kg/m}^3)(40.21 \times 10^{-6}\text{ m}^3) = 0.3156\text{ kg}$$

Rectangular Plate

$$V_2 = (0.200\text{ m})(0.160\text{ m})(0.004\text{ m}) = 128 \times 10^{-6}\text{ m}^3$$
$$m_2 = \rho V_2 = (7.85 \times 10^3\text{ kg/m}^3)(128 \times 10^{-6}\text{ m}^3) = 1.005\text{ kg}$$

Circular Plate

$$V_3 = \pi a^2 t = \pi(0.050\text{ m})^2(0.004\text{ m}) = 31.42 \times 10^{-6}\text{ m}^3$$
$$m_3 = \rho V_3 = (7.85 \times 10^3\text{ kg/m}^3)(31.42 \times 10^{-6}\text{ m}^3) = 0.2466\text{ kg}$$

Moments of Inertia. Using the method presented in Sec. 9.13, we compute the moments of inertia of each component.

Semicircular Plate. From Fig. 9.28, we observe that for a circular plate of mass m and radius r

$$I_x = \tfrac{1}{2}mr^2 \qquad I_y = I_z = \tfrac{1}{4}mr^2$$

Because of symmetry, we note that for a semicircular plate

$$I_x = \tfrac{1}{2}(\tfrac{1}{2}mr^2) \qquad I_y = I_z = \tfrac{1}{2}(\tfrac{1}{4}mr^2)$$

Since the mass of the semicircular plate is $m_1 = \tfrac{1}{2}m$, we have

$$I_x = \tfrac{1}{2}m_1 r^2 = \tfrac{1}{2}(0.3156\text{ kg})(0.08\text{ m})^2 = 1.010 \times 10^{-3}\text{ kg} \cdot \text{m}^2$$
$$I_y = I_z = \tfrac{1}{4}(\tfrac{1}{2}mr^2) = \tfrac{1}{4}m_1 r^2 = \tfrac{1}{4}(0.3156\text{ kg})(0.08\text{ m})^2 = 0.505 \times 10^{-3}\text{ kg} \cdot \text{m}^2$$

Rectangular Plate

$$I_x = \tfrac{1}{12}m_2 c^2 = \tfrac{1}{12}(1.005\text{ kg})(0.16\text{ m})^2 = 2.144 \times 10^{-3}\text{ kg} \cdot \text{m}^2$$
$$I_z = \tfrac{1}{3}m_2 b^2 = \tfrac{1}{3}(1.005\text{ kg})(0.2\text{ m})^2 = 13.400 \times 10^{-3}\text{ kg} \cdot \text{m}^2$$
$$I_y = I_x + I_z = (2.144 + 13.400)(10^{-3}) = 15.544 \times 10^{-3}\text{ kg} \cdot \text{m}^2$$

Circular Plate

$$I_x = \tfrac{1}{4}m_3 a^2 = \tfrac{1}{4}(0.2466\text{ kg})(0.05\text{ m})^2 = 0.154 \times 10^{-3}\text{ kg} \cdot \text{m}^2$$
$$I_y = \tfrac{1}{2}m_3 a^2 + m_3 d^2$$
$$= \tfrac{1}{2}(0.2466\text{ kg})(0.05\text{ m})^2 + (0.2466\text{ kg})(0.1\text{ m})^2 = 2.774 \times 10^{-3}\text{ kg} \cdot \text{m}^2$$
$$I_z = \tfrac{1}{4}m_3 a^2 + m_3 d^2 = \tfrac{1}{4}(0.2466\text{ kg})(0.05\text{ m})^2 + (0.2466\text{ kg})(0.1\text{ m})^2$$
$$= 2.620 \times 10^{-3}\text{ kg} \cdot \text{m}^2$$

Entire Machine Part

$$I_x = (1.010 + 2.144 - 0.154)(10^{-3})\text{ kg} \cdot \text{m}^2 \qquad I_x = 3.00 \times 10^{-3}\text{ kg} \cdot \text{m}^2 \;\blacktriangleleft$$
$$I_y = (0.505 + 15.544 - 2.774)(10^{-3})\text{ kg} \cdot \text{m}^2 \qquad I_y = 13.28 \times 10^{-3}\text{ kg} \cdot \text{m}^2 \;\blacktriangleleft$$
$$I_z = (0.505 + 13.400 - 2.620)(10^{-3})\text{ kg} \cdot \text{m}^2 \qquad I_z = 11.29 \times 10^{-3}\text{ kg} \cdot \text{m}^2 \;\blacktriangleleft$$

SOLVING PROBLEMS
ON YOUR OWN

In this lesson we introduced the *mass moment of inertia* and the *radius of gyration* of a three-dimensional body with respect to a given axis [Eqs. (9.28) and (9.29)]. We also derived a *parallel-axis theorem* for use with mass moments of inertia and discussed the computation of the mass moments of inertia of thin plates and three-dimensional bodies.

1. Computing mass moments of inertia. The mass moment of inertia I of a body with respect to a given axis can be calculated directly from the definition given in Eq. (9.28) for simple shapes [Sample Prob. 9.9]. In most cases, however, it is necessary to divide the body into thin slabs, compute the moment of inertia of a typical slab with respect to the given axis—using the parallel-axis theorem if necessary—and integrate the expression obtained.

2. Applying the parallel-axis theorem. In Sec. 9.12 we derived the parallel-axis theorem for mass moments of inertia

$$I = \bar{I} + md^2 \tag{9.33}$$

which states that the moment of inertia I of a body of mass m with respect to a given axis is equal to the sum of the moment of inertia $\bar{I}$ of that body with respect to a *parallel centroidal axis* and the product md^2, where d is the distance between the two axes. When the moment of inertia of a three-dimensional body is calculated with respect to one of the coordinate axes, d^2 can be replaced by the sum of the squares of distances measured along the other two coordinate axes [Eqs. (9.32) and (9.32′)].

3. Avoiding unit-related errors. To avoid errors, it is essential that you be consistent in your use of units. Thus, all lengths should be expressed in meters or feet, as appropriate, and for problems using U.S. customary units, masses should be given in lb $\cdot$ s^2/ft. In addition, we strongly recommend that you include units as you perform your calculations [Sample Probs. 9.12 and 9.13].

4. Calculating the mass moment of inertia of thin plates. We showed in Sec. 9.13 that the mass moment of inertia of a thin plate with respect to a given axis can be obtained by multiplying the corresponding moment of inertia of the area of the plate by the density ρ and the thickness t of the plate [Eqs. (9.35) through (9.37)]. Note that since the axis CC' in Fig. 9.24c is *perpendicular to the plate*, $I_{CC', \text{mass}}$ is associated with the *polar* moment of inertia $J_{C, \text{area}}$.

Instead of calculating directly the moment of inertia of a thin plate with respect to a specified axis, you may sometimes find it convenient to first compute its moment of inertia with respect to an axis parallel to the specified axis and then apply the parallel-axis theorem. Further, to determine the moment of inertia of a thin plate with respect to an axis perpendicular to the plate, you may wish to first determine its moments of inertia with respect to two perpendicular in-plane axes and then use Eq. (9.38). Finally, remember that the mass of a plate of area A, thickness t, and density ρ is $m = \rho t A$.

5. Determining the moment of inertia of a body by direct single integration. We discussed in Sec. 9.14 and illustrated in Sample Probs. 9.10 and 9.11 how single integration can be used to compute the moment of inertia of a body that can be divided into a series of thin, parallel slabs. For such cases, you will often need to express the mass of the body in terms of the body's density and dimensions. Assuming that the body has been divided, as in the sample problems, into thin slabs perpendicular to the x axis, you will need to express the dimensions of each slab as functions of the variable x.

　　a. In the special case of a body of revolution, the elemental slab is a thin disk, and the equations given in Fig. 9.27 should be used to determine the moments of inertia of the body [Sample Prob. 9.11].

　　b. In the general case, when the body is not of revolution, the differential element is not a disk, but a thin slab of a different shape, and the equations of Fig. 9.27 cannot be used. See, for example, Sample Prob. 9.10, where the element was a thin, rectangular slab. For more complex configurations, you may want to use one or more of the following equations, which are based on Eqs. (9.32) and (9.32′) of Sec. 9.12.

$$dI_x = dI_{x'} + (\overline{y}_{el}^2 + \overline{z}_{el}^2)\, dm$$
$$dI_y = dI_{y'} + (\overline{z}_{el}^2 + \overline{x}_{el}^2)\, dm$$
$$dI_z = dI_{z'} + (\overline{x}_{el}^2 + \overline{y}_{el}^2)\, dm$$

where the primes denote the centroidal axes of each elemental slab, and where $\overline{x}_{el}$, $\overline{y}_{el}$, and $\overline{z}_{el}$ represent the coordinates of its centroid. The centroidal moments of inertia of the slab are determined in the manner described earlier for a thin plate: Referring to Fig. 9.12 on page 483, calculate the corresponding moments of inertia of the area of the slab and multiply the result by the density ρ and the thickness t of the slab. Also, assuming that the body has been divided into thin slabs perpendicular to the x axis, remember that you can obtain $dI_{x'}$ by adding $dI_{y'}$ and $dI_{z'}$ instead of computing it directly. Finally, using the geometry of the body, express the result obtained in terms of the single variable x and integrate in x.

6. Computing the moment of inertia of a composite body. As stated in Sec. 9.15, the moment of inertia of a composite body with respect to a specified axis is equal to the sum of the moments of its components with respect to that axis. Sample Probs. 9.12 and 9.13 illustrate the appropriate method of solution. You must also remember that the moment of inertia of a component will be negative only if the component is *removed* (as in the case of a hole).

Although the composite-body problems in this lesson are relatively straightforward, you will have to work carefully to avoid computational errors. In addition, if some of the moments of inertia that you need are not given in Fig. 9.28, you will have to derive your own formulas, using the techniques of this lesson.

PROBLEMS

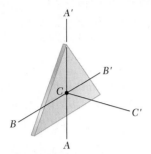

Fig. P9.111

9.111 A thin plate of mass m is cut in the shape of an equilateral triangle of side a. Determine the mass moment of inertia of the plate with respect to (a) the centroidal axes AA' and BB', (b) the centroidal axis CC' that is perpendicular to the plate.

9.112 The elliptical ring shown was cut from a thin, uniform plate. Denoting the mass of the ring by m, determine its mass moment of inertia with respect to (a) the centroidal axis BB', (b) the centroidal axis CC' that is perpendicular to the plane of the ring.

Fig. P9.112

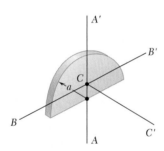

Fig. P9.113

9.113 A thin semicircular plate has a radius a and a mass m. Determine the mass moment of inertia of the plate with respect to (a) the centroidal axis BB', (b) the centroidal axis CC' that is perpendicular to the plate.

9.114 The quarter ring shown has a mass m and was cut from a thin, uniform plate. Knowing that $r_1 = \frac{3}{4}r_2$, determine the mass moment of inertia of the quarter ring with respect to (a) the axis AA', (b) the centroidal axis CC' that is perpendicular to the plane of the quarter ring.

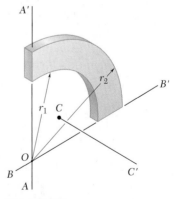

Fig. P9.114

9.115 A piece of thin, uniform sheet metal is cut to form the machine component shown. Denoting the mass of the component by m, determine its mass moment of inertia with respect to (a) the x axis, (b) the y axis.

9.116 A piece of thin, uniform sheet metal is cut to form the machine component shown. Denoting the mass of the component by m, determine its mass moment of inertia with respect to (a) the axis AA', (b) the axis BB', where the AA' and BB' axes are parallel to the x axis and lie in a plane parallel to and at a distance a above the xz plane.

9.117 A thin plate of mass m was cut in the shape of a parallelogram as shown. Determine the mass moment of inertia of the plate with respect to (a) the x axis, (b) the axis BB', which is perpendicular to the plate.

9.118 A thin plate of mass m was cut in the shape of a parallelogram as shown. Determine the mass moment of inertia of the plate with respect to (a) the y axis, (b) the axis AA', which is perpendicular to the plate.

9.119 Determine by direct integration the mass moment of inertia with respect to the z axis of the right circular cylinder shown, assuming that it has a uniform density and a mass m.

Fig. P9.115 and P9.116

Fig. P9.117 and P9.118

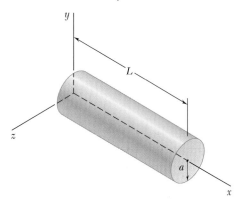

Fig. P9.119

9.120 The area shown is revolved about the x axis to form a homogeneous solid of revolution of mass m. Using direct integration, express the mass moment of inertia of the solid with respect to the x axis in terms of m and h.

9.121 The area shown is revolved about the x axis to form a homogeneous solid of revolution of mass m. Determine by direct integration the mass moment of inertia of the solid with respect to (a) the x axis, (b) the y axis. Express your answers in terms of m and the dimensions of the solid.

Fig. P9.120

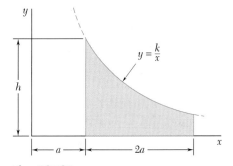

Fig. P9.121

9.122 Determine by direct integration the mass moment of inertia with respect to the x axis of the pyramid shown, assuming that it has a uniform density and a mass m.

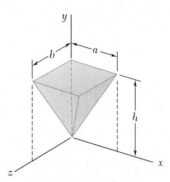

Fig. P9.122 and P9.123

9.123 Determine by direct integration the mass moment of inertia with respect to the y axis of the pyramid shown, assuming that it has a uniform density and a mass m.

9.124 Determine by direct integration the mass moment of inertia with respect to the y axis of the paraboloid shown, assuming that it has a uniform density and a mass m.

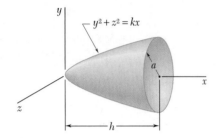

Fig. P9.124

9.125 A thin rectangular plate of mass m is welded to a vertical shaft AB as shown. Knowing that the plate forms an angle θ with the y axis, determine by direct integration the mass moment of inertia of the plate with respect to (a) the y axis, (b) the z axis.

Fig. P9.125

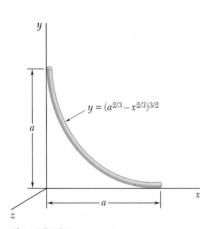

Fig. P9.126

*9.126 A thin steel wire is bent into the shape shown. Denoting the mass per unit length of the wire by m', determine by direct integration the mass moment of inertia of the wire with respect to each of the coordinate axes.

9.127 Shown is the cross section of an idler roller. Determine its mass moment of inertia and its radius of gyration with respect to the axis AA'. (The specific weight of bronze is 0.310 lb/in^3; of aluminum, 0.100 lb/in^3; and of neoprene, 0.0452 lb/in^3.)

9.128 Shown is the cross section of a molded flat-belt pulley. Determine its mass moment of inertia and its radius of gyration with respect to the axis AA'. (The density of brass is 8650 kg/m^3 and the density of the fiber-reinforced polycarbonate used is 1250 kg/m^3.)

Fig. P9.127

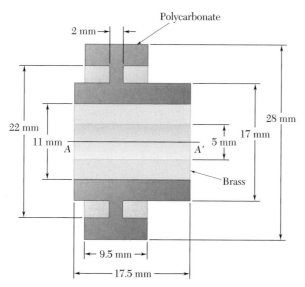

Fig. P9.128

9.129 The machine part shown is formed by machining a conical surface into a circular cylinder. For $b = \frac{1}{2}h$, determine the mass moment of inertia and the radius of gyration of the machine part with respect to the y axis.

Fig. P9.129

9.130 Given the dimensions and the mass m of the thin conical shell shown, determine the mass moment of inertia and the radius of gyration of the shell with respect to the x axis. (*Hint:* Assume that the shell was formed by removing a cone with a circular base of radius a from a cone with a circular base of radius $a + t$, where t is the thickness of the wall. In the resulting expressions, neglect terms containing t^2, t^3, etc. Do not forget to account for the difference in the heights of the two cones.)

Fig. P9.130

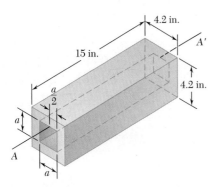

Fig. P9.131

9.131 A square hole is centered in and extends through the aluminum machine component shown. Determine (a) the value of a for which the mass moment of inertia of the component with respect to the axis AA', which bisects the top surface of the hole, is maximum, (b) the corresponding values of the mass moment of inertia and the radius of gyration with respect to the axis AA'. (The specific weight of aluminum is 0.100 lb/in³.)

9.132 The cups and the arms of an anemometer are fabricated from a material of density ρ. Knowing that the mass moment of inertia of a thin, hemispherical shell of mass m and thickness t with respect to its centroidal axis GG' is $5ma^2/12$, determine (a) the mass moment of inertia of the anemometer with respect to the axis AA', (b) the ratio of a to l for which the centroidal moment of inertia of the cups is equal to 1 percent of the moment of inertia of the cups with respect to the axis AA'.

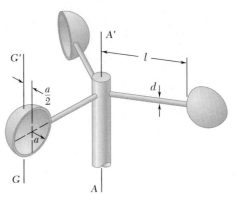

Fig. P9.132

9.133 After a period of use, one of the blades of a shredder has been worn to the shape shown and is of mass 0.18 kg. Knowing that the mass moments of inertia of the blade with respect to the AA' and BB' axes are 0.320 g · m² and 0.680 g · m², respectively, determine (a) the location of the centroidal axis GG', (b) the radius of gyration with respect to axis GG'.

Fig. P9.133

Fig. P9.134

9.134 Determine the mass moment of inertia of the 0.9-lb machine component shown with respect to the axis AA'.

9.135 and 9.136 A 2-mm-thick piece of sheet steel is cut and bent into the machine component shown. Knowing that the density of steel is 7850 kg/m³, determine the mass moment of inertia of the component with respect to each of the coordinate axes.

Fig. P9.135

Fig. P9.136

9.137 A subassembly for a model airplane is fabricated from three pieces of 1.5-mm plywood. Neglecting the mass of the adhesive used to assemble the three pieces, determine the mass moment of inertia of the subassembly with respect to each of the coordinate axes. (The density of the plywood is 780 kg/m³.)

9.138 The cover for an electronic device is formed from sheet aluminum that is 0.05 in. thick. Determine the mass moment of inertia of the cover with respect to each of the coordinate axes. (The specific weight of aluminum is 0.100 lb/in³.)

Fig. P9.137

Fig. P9.138

9.139 A framing anchor is formed of 0.05-in.-thick galvanized steel. Determine the mass moment of inertia of the anchor with respect to each of the coordinate axes. (The specific weight of galvanized steel is 470 lb/ft³.)

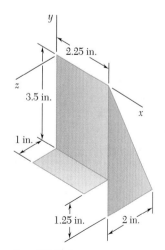

Fig. P9.139

*9.140 A farmer constructs a trough by welding a rectangular piece of 2-mm-thick sheet steel to half of a steel drum. Knowing that the density of steel is 7850 kg/m³ and that the thickness of the walls of the drum is 1.8 mm, determine the mass moment of inertia of the trough with respect to each of the coordinate axes. Neglect the mass of the welds.

Fig. P9.140

9.141 The machine element shown is fabricated from steel. Determine the mass moment of inertia of the assembly with respect to (a) the x axis, (b) the y axis, (c) the z axis. (The density of steel is 7850 kg/m³.)

Fig. P9.141

9.142 Determine the mass moments of inertia and the radii of gyration of the steel machine element shown with respect to the x and y axes. (The density of steel is 7850 kg/m³.)

Dimensions in mm

Fig. P9.142

9.143 Determine the mass moment of inertia of the steel machine element shown with respect to the y axis. (The specific weight of steel is 490 lb/ft³.)

Fig. P9.143 and *P9.144*

9.144 Determine the mass moment of inertia of the steel machine element shown with respect to the z axis. (The specific weight of steel is 490 lb/ft³.)

9.145 Determine the mass moment of inertia of the steel fixture shown with respect to (*a*) the x axis, (*b*) the y axis, (*c*) the z axis. (The density of steel is 7850 kg/m³.)

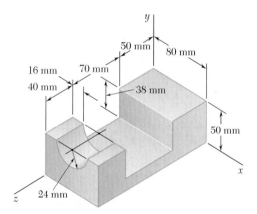

Fig. P9.145

9.146 Aluminum wire with a weight per unit length of 0.033 lb/ft is used to form the circle and the straight members of the figure shown. Determine the mass moment of inertia of the assembly with respect to each of the coordinate axes.

Fig. *P9.146*

9.147 The figure shown is formed of $\frac{1}{8}$-in.-diameter steel wire. Knowing that the specific weight of the steel is 490 lb/ft^3, determine the mass moment of inertia of the wire with respect to each of the coordinate axes.

Fig. P9.147

9.148 A homogeneous wire with a mass per unit length of 0.056 kg/m is used to form the figure shown. Determine the mass moment of inertia of the wire with respect to each of the coordinate axes.

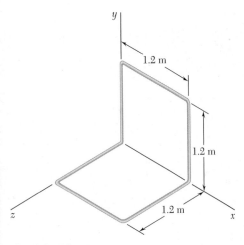

Fig. P9.148

*9.16   MOMENT OF INERTIA OF A BODY WITH RESPECT TO AN ARBITRARY AXIS THROUGH O. MASS PRODUCTS OF INERTIA

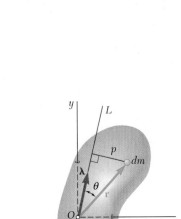

Fig. 9.29

In this section you will see how the moment of inertia of a body can be determined with respect to an arbitrary axis OL through the origin (Fig. 9.29) if its moments of inertia with respect to the three coordinate axes, as well as certain other quantities to be defined below, have already been determined.

The moment of inertia I_{OL} of the body with respect to OL is equal to $\int p^2 dm$, where p denotes the perpendicular distance from the element of mass dm to the axis OL. If we denote by λ the unit vector along OL and by $\mathbf{r}$ the position vector of the element dm, we observe that the perpendicular distance p is equal to $r \sin \theta$, which is the magnitude of the vector product $\lambda \times \mathbf{r}$. We therefore write

$$I_{OL} = \int p^2 \, \mathrm{dm} = \int |\lambda \times \mathbf{r}|^2 \, dm \qquad (9.43)$$

Expressing $|\lambda \times \mathbf{r}|^2$ in terms of the rectangular components of the vector product, we have

$$I_{OL} = \int [(\lambda_x y - \lambda_y x)^2 + (\lambda_y z - \lambda_z y)^2 + (\lambda_z x - \lambda_x z)^2] \, dm$$

where the components λ_x, λ_y, λ_z of the unit vector λ represent the direction cosines of the axis OL and the components x, y, z of $\mathbf{r}$ represent the coordinates of the element of mass dm. Expanding the squares and rearranging the terms, we write

$$I_{OL} = \lambda_x^2 \int (y^2 + z^2) \, dm + \lambda_y^2 \int (z^2 + x^2) \, dm + \lambda_z^2 \int (x^2 + y^2) \, dm$$

$$- 2\lambda_x \lambda_y \int xy \, dm - 2\lambda_y \lambda_z \int yz \, dm - 2\lambda_z \lambda_x \int zx \, dm \qquad (9.44)$$

Referring to Eqs. (9.30), we note that the first three integrals in (9.44) represent, respectively, the moments of inertia I_x, I_y, and I_z of the body with respect to the coordinate axes. The last three integrals in (9.44), which involve products of coordinates, are called the *products of inertia* of the body with respect to the x and y axes, the y and z axes, and the z and x axes, respectively. We write

$$I_{xy} = \int xy \, dm \qquad I_{yz} = \int yz \, dm \qquad I_{zx} = \int zx \, dm \qquad (9.45)$$

Rewriting Eq. (9.44) in terms of the integrals defined in Eqs. (9.30) and (9.45), we have

$$I_{OL} = I_x\lambda_x^2 + I_y\lambda_y^2 + I_z\lambda_z^2 - 2I_{xy}\lambda_x\lambda_y - 2I_{yz}\lambda_y\lambda_z - 2I_{zx}\lambda_z\lambda_x \qquad (9.46)$$

We note that the definition of the products of inertia of a mass given in Eqs. (9.45) is an extension of the definition of the product of inertia of an area (Sec. 9.8). Mass products of inertia reduce to zero under the same conditions of symmetry as do products of inertia of areas, and the parallel-axis theorem for mass products of inertia is expressed by relations similar to the formula derived for the product of inertia of an area. Substituting the expressions for x, y, and z given in Eqs. (9.31) into Eqs. (9.45), we find that

$$\begin{aligned} I_{xy} &= \bar{I}_{x'y'} + m\bar{x}\bar{y} \\ I_{yz} &= \bar{I}_{y'z'} + m\bar{y}\bar{z} \\ I_{zx} &= \bar{I}_{z'x'} + m\bar{z}\bar{x} \end{aligned} \qquad (9.47)$$

where $\bar{x}, \bar{y}, \bar{z}$ are the coordinates of the center of gravity G of the body and $\bar{I}_{x'y'}, \bar{I}_{y'z'}, \bar{I}_{z'x'}$ denote the products of inertia of the body with respect to the centroidal axes x', y', z' (See Fig. 9.22).

*9.17 ELLIPSOID OF INERTIA. PRINCIPAL AXES OF INERTIA

Let us assume that the moment of inertia of the body considered in the preceding section has been determined with respect to a large number of axes OL through the fixed point O and that a point Q has been plotted on each axis OL at a distance $OQ = 1/\sqrt{I_{OL}}$ from O. The locus of the points Q thus obtained forms a surface (Fig. 9.30). The equation of that surface can be obtained by substituting $1/(OQ)^2$ for I_{OL} in (9.46) and then multiplying both sides of the equation by $(OQ)^2$. Observing that

$$(OQ)\lambda_x = x \qquad (OQ)\lambda_y = y \qquad (OQ)\lambda_z = z$$

where x, y, z denote the rectangular coordinates of Q, we write

$$I_x x^2 + I_y y^2 + I_z z^2 - 2I_{xy}xy - 2I_{yz}yz - 2I_{zx}zx = 1 \qquad (9.48)$$

The equation obtained is the equation of a *quadric surface*. Since the moment of inertia I_{OL} is different from zero for every axis OL, no point Q can be at an infinite distance from O. Thus, the quadric surface obtained is an *ellipsoid*. This ellipsoid, which defines the

Fig. 9.30

Fig. 9.31

Fig. 9.32

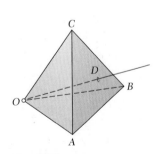

Fig. 9.33

moment of inertia of the body with respect to any axis through O, is known as the *ellipsoid of inertia* of the body at O.

We observe that if the axes in Fig. 9.30 are rotated, the coefficients of the equation defining the ellipsoid change, since they are equal to the moments and products of inertia of the body with respect to the rotated coordinate axes. However, the *ellipsoid itself remains unaffected,* since its shape depends only upon the distribution of mass in the given body. Suppose that we choose as coordinate axes the principal axes x', y', z' of the ellipsoid of inertia (Fig. 9.31). The equation of the ellipsoid with respect to these coordinate axes is known to be of the form

$$I_{x'}x'^2 + I_{y'}y'^2 + I_{z'}z'^2 = 1 \tag{9.49}$$

which does not contain any products of the coordinates. Comparing Eqs. (9.48) and (9.49), we observe that the products of inertia of the body with respect to the x', y', z' axes must be zero. The x', y', z' axes are known as the *principal axes of inertia* of the body at O, and the coefficients $I_{x'}$, $I_{y'}$, $I_{z'}$ are referred to as the *principal moments of inertia* of the body at O. Note that, given a body of arbitrary shape and a point O, it is always possible to find axes which are the principal axes of inertia of the body at O, that is, axes with respect to which the products of inertia of the body are zero. Indeed, whatever the shape of the body, the moments and products of inertia of the body with respect to x, y, and z axes through O will define an ellipsoid, and this ellipsoid will have principal axes which, by definition, are the principal axes of inertia of the body at O.

If the principal axes of inertia x', y', z' are used as coordinate axes, the expression obtained in Eq. (9.46) for the moment of inertia of a body with respect to an arbitrary axis through O reduces to

$$I_{OL} = I_{x'}\lambda_{x'}^2 + I_{y'}\lambda_{y'}^2 + I_{z'}\lambda_{z'}^2 \tag{9.50}$$

The determination of the principal axes of inertia of a body of arbitrary shape is somewhat involved and will be discussed in the next section. There are many cases, however, where these axes can be spotted immediately. Consider, for instance, the homogeneous cone of elliptical base shown in Fig. 9.32; this cone possesses two mutually perpendicular planes of symmetry OAA' and OBB'. From the definition (9.45), we observe that if the $x'y'$ and $y'z'$ planes are chosen to coincide with the two planes of symmetry, all of the products of inertia are zero. The x', y', and z' axes thus selected are therefore the principal axes of inertia of the cone at O. In the case of the homogeneous regular tetrahedron $OABC$ shown in Fig. 9.33, the line joining the corner O to the center D of the opposite face is a principal axis of inertia at O, and any line through O perpendicular to OD is also a principal axis of inertia at O. This property is apparent if we observe that rotating the tetrahedron through 120° about OD leaves its shape and mass distribution unchanged. It follows that the ellipsoid of inertia at O also remains unchanged under this rotation. The ellipsoid, therefore, is a body of revolution whose axis of revolution is OD, and the line OD, as well as any perpendicular line through O, must be a principal axis of the ellipsoid.

9.18 Determination of the Principal Axes and **533**
Principal Moments of Inertia of a
Body of Arbitrary Shape

*9.18 DETERMINATION OF THE PRINCIPAL AXES AND PRINCIPAL MOMENTS OF INERTIA OF A BODY OF ARBITRARY SHAPE

The method of analysis described in this section should be used when the body under consideration has no obvious property of symmetry.

Consider the ellipsoid of inertia of the body at a given point O (Fig. 9.34); let $\mathbf{r}$ be the radius vector of a point P on the surface of the ellipsoid and let $\mathbf{n}$ be the unit vector along the normal to that surface at P. We observe that the only points where $\mathbf{r}$ and $\mathbf{n}$ are collinear are the points P_1, P_2, and P_3, where the principal axes intersect the visible portion of the surface of the ellipsoid, and the corresponding points on the other side of the ellipsoid.

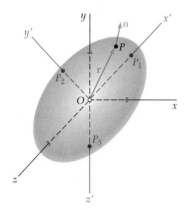

Fig. 9.34

We now recall from calculus that the direction of the normal to a surface of equation $f(x, y, z) = 0$ at a point $P(x, y, z)$ is defined by the gradient ∇f of the function f at that point. To obtain the points where the principal axes intersect the surface of the ellipsoid of inertia, we must therefore write that $\mathbf{r}$ and ∇f are collinear,

$$\nabla f = (2K)\mathbf{r} \qquad (9.51)$$

where K is a constant, $\mathbf{r} = x\mathbf{i} + y\mathbf{j} + z\mathbf{k}$, and

$$\nabla f = \frac{\partial f}{\partial x}\mathbf{i} + \frac{\partial f}{\partial y}\mathbf{j} + \frac{\partial f}{\partial x}\mathbf{k}$$

Recalling Eq. (9.48), we note that the function $f(x, y, z)$ corresponding to the ellipsoid of inertia is

$$f(x, y, z) = I_x x^2 + I_y y^2 + I_z z^2 - 2I_{xy}xy - 2I_{yz}yz - 2I_{zx}zx - 1$$

Substituting for $\mathbf{r}$ and ∇f into Eq. (9.51) and equating the coefficients of the unit vectors, we write

$$\begin{aligned}
I_x x &- I_{xy}y - I_{zx}z = Kx \\
-I_{xy}x &+ I_y y - I_{yz}z = Ky \\
-I_{zx}x &- I_{yz}y + I_z z = Kz
\end{aligned} \qquad (9.52)$$

Dividing each term by the distance r from O to P, we obtain similar equations involving the direction cosines λ_x, λ_y, and λ_z:

$$
\begin{aligned}
I_x\lambda_x &- I_{xy}\lambda_y - I_{zx}\lambda_z = K\lambda_x \\
-I_{xy}\lambda_x &+ I_y\lambda_y - I_{yz}\lambda_z = K\lambda_y \\
-I_{zx}\lambda_x &- I_{yz}\lambda_y + I_z\lambda_z = K\lambda_z
\end{aligned}
\tag{9.53}
$$

Transposing the right-hand members leads to the following homogeneous linear equations:

$$
\begin{aligned}
(I_x - K)\lambda_x &- I_{xy}\lambda_y - I_{zx}\lambda_z = 0 \\
-I_{xy}\lambda_x &+ (I_y - K)\lambda_y - I_{yz}\lambda_z = 0 \\
-I_{zx}\lambda_x &- I_{yz}\lambda_y + (I_z - K)\lambda_z = 0
\end{aligned}
\tag{9.54}
$$

For this system of equations to have a solution different from $\lambda_x = \lambda_y = \lambda_z = 0$, its discriminant must be zero:

$$
\begin{vmatrix}
I_x - K & -I_{xy} & -I_{zx} \\
-I_{xy} & I_y - K & -I_{yz} \\
-I_{zx} & -I_{yz} & I_z - K
\end{vmatrix}
= 0
\tag{9.55}
$$

Expanding this determinant and changing signs, we write

$$
\begin{aligned}
K^3 - (I_x + I_y + I_z)K^2 &+ (I_xI_y + I_yI_z + I_zI_x - I_{xy}^2 - I_{yz}^2 - I_{zx}^2)K \\
&- (I_xI_yI_z - I_xI_{yz}^2 - I_yI_{zx}^2 - I_zI_{xy}^2 - 2I_{xy}I_{yz}I_{zx}) = 0
\end{aligned}
\tag{9.56}
$$

This is a cubic equation in K, which yields three real, positive roots K_1, K_2, and K_3.

To obtain the direction cosines of the principal axis corresponding to the root K_1 we substitute K_1 for K in Eqs. (9.54). Since these equations are now linearly dependent, only two of them may be used to determine λ_x, λ_y, and λ_z. An additional equation may be obtained, however, by recalling from Sec. 2.12 that the direction cosines must satisfy the relation

$$
\lambda_x^2 + \lambda_y^2 + \lambda_z^2 = 1
\tag{9.57}
$$

Repeating this procedure with K_2 and K_3, we obtain the direction cosines of the other two principal axes.

We will now show that *the roots K_1, K_2, and K_3 of Eq. (9.56) are the principal moments of inertia of the given body*. Let us substitute for K in Eqs. (9.53) the root K_1, and for λ_x, λ_y, and λ_z the corresponding values $(\lambda_x)_1$, $(\lambda_y)_1$, and $(\lambda_z)_1$ of the direction cosines; the three equations will be satisfied. We now multiply by $(\lambda_x)_1$, $(\lambda_y)_1$, and $(\lambda_z)_1$, respectively, each term in the first, second, and third equation and add the equations obtained in this way. We write

$$
\begin{aligned}
I_x^2(\lambda_x)_1^2 + I_y^2(\lambda_y)_1^2 &+ I_z^2(\lambda_z)_1^2 - 2I_{xy}(\lambda_x)_1(\lambda_y)_1 \\
&- 2I_{yz}(\lambda_y)_1(\lambda_z)_1 - 2I_{zx}(\lambda_z)_1(\lambda_x)_1 = K_1[(\lambda_x)_1^2 + (\lambda_y)_1^2 + (\lambda_z)_1^2]
\end{aligned}
$$

Recalling Eq. (9.46), we observe that the left-hand member of this equation represents the moment of inertia of the body with respect to the principal axis corresponding to K_1; it is thus the principal moment of inertia corresponding to that root. On the other hand, recalling Eq. (9.57), we note that the right-hand member reduces to K_1. Thus K_1 itself is the principal moment of inertia. We can show in the same fashion that K_2 and K_3 are the other two principal moments of inertia of the body.

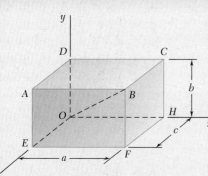

SAMPLE PROBLEM 9.14

Consider a rectangular prism of mass m and sides a, b, c. Determine (a) the moments and products of inertia of the prism with respect to the coordinate axes shown, (b) its moment of inertia with respect to the diagonal OB.

SOLUTION

a. Moments and Products of Inertia with Respect to the Coordinate Axes. *Moments of Inertia.* Introducing the centroidal axes x', y', z', with respect to which the moments of inertia are given in Fig. 9.28, we apply the parallel-axis theorem:

$$I_x = \bar{I}_{x'} + m(\bar{y}^2 + \bar{z}^2) = \tfrac{1}{12}m(b^2 + c^2) + m(\tfrac{1}{4}b^2 + \tfrac{1}{4}c^2)$$
$$I_x = \tfrac{1}{3}m(b^2 + c^2) \quad \blacktriangleleft$$

Similarly, $\qquad I_y = \tfrac{1}{3}m(c^2 + a^2) \qquad I_z = \tfrac{1}{3}m(a^2 + b^2) \quad \blacktriangleleft$

Products of Inertia. Because of symmetry, the products of inertia with respect to the centroidal axes x', y', z' are zero, and these axes are principal axes of inertia. Using the parallel-axis theorem, we have

$$I_{xy} = \bar{I}_{x'y'} + m\bar{x}\bar{y} = 0 + m(\tfrac{1}{2}a)(\tfrac{1}{2}b) \qquad I_{xy} = \tfrac{1}{4}mab \quad \blacktriangleleft$$

Similarly, $\qquad\qquad I_{yz} = \tfrac{1}{4}mbc \qquad I_{zx} = \tfrac{1}{4}mca \quad \blacktriangleleft$

b. Moment of Inertia with Respect to OB. We recall Eq. (9.46):

$$I_{OB} = I_x\lambda_x^2 + I_y\lambda_y^2 + I_z\lambda_z^2 - 2I_{xy}\lambda_x\lambda_y - 2I_{yz}\lambda_y\lambda_z - 2I_{zx}\lambda_z\lambda_x$$

where the direction cosines of OB are

$$\lambda_x = \cos\theta_x = \frac{OH}{OB} = \frac{a}{(a^2 + b^2 + c^2)^{1/2}}$$

$$\lambda_y = \frac{b}{(a^2 + b^2 + c^2)^{1/2}} \qquad \lambda_z = \frac{c}{(a^2 + b^2 + c^2)^{1/2}}$$

Substituting the values obtained for the moments and products of inertia and for the direction cosines into the equation for I_{OB}, we have

$$I_{OB} = \frac{1}{a^2 + b^2 + c^2}\left[\tfrac{1}{3}m(b^2 + c^2)a^2 + \tfrac{1}{3}m(c^2 + a^2)b^2 + \tfrac{1}{3}m(a^2 + b^2)c^2\right.$$

$$\left. -\tfrac{1}{2}ma^2b^2 - \tfrac{1}{2}mb^2c^2 - \tfrac{1}{2}mc^2a^2\right]$$

$$I_{OB} = \frac{m}{6}\frac{a^2b^2 + b^2c^2 + c^2a^2}{a^2 + b^2 + c^2} \quad \blacktriangleleft$$

Alternative Solution. The moment of inertia I_{OB} can be obtained directly from the principal moments of inertia $\bar{I}_{x'}$, $\bar{I}_{y'}$, $\bar{I}_{z'}$, since the line OB passes through the centroid O'. Since the x', y', z' axes are principal axes of inertia, we use Eq. (9.50) to write

$$I_{OB} = \bar{I}_{x'}\lambda_x^2 + \bar{I}_{y'}\lambda_y^2 + \bar{I}_{z'}\lambda_z^2$$

$$= \frac{1}{a^2 + b^2 + c^2}\left[\frac{m}{12}(b^2 + c^2)a^2 + \frac{m}{12}(c^2 + a^2)b^2 + \frac{m}{12}(a^2 + b^2)c^2\right]$$

$$I_{OB} = \frac{m}{6}\frac{a^2b^2 + b^2c^2 + c^2a^2}{a^2 + b^2 + c^2} \quad \blacktriangleleft$$

If $a = 3c$ and $b = 2c$ for the rectangular prism of Sample Prob. 9.14, determine (a) the principal moments of inertia at the origin O, (b) the principal axes of inertia at O.

SOLUTION

a. Principal Moments of Inertia at the Origin O. Substituting $a = 3c$ and $b = 2c$ into the solution to Sample Prob. 9.14, we have

$$I_x = \tfrac{5}{3}mc^2 \qquad I_y = \tfrac{10}{3}mc^2 \qquad I_z = \tfrac{13}{3}mc^2$$
$$I_{xy} = \tfrac{3}{2}mc^2 \qquad I_{yz} = \tfrac{1}{2}mc^2 \qquad I_{zx} = \tfrac{3}{4}mc^2$$

Substituting the values of the moments and products of inertia into Eq. (9.56) and collecting terms yields

$$K^3 - (\tfrac{28}{3}mc^2)K^2 + (\tfrac{3479}{144}m^2c^4)K - \tfrac{589}{54}m^3c^6 = 0$$

We then solve for the roots of this equation; from the discussion in Sec. 9.18, it follows that these roots are the principal moments of inertia of the body at the origin.

$$K_1 = 0.568867mc^2 \qquad K_2 = 4.20885mc^2 \qquad K_3 = 4.55562mc^2$$
$$K_1 = 0.569mc^2 \qquad K_2 = 4.21mc^2 \qquad K_3 = 4.56mc^2 \qquad \blacktriangleleft$$

b. Principal Axes of Inertia at O. To determine the direction of a principal axis of inertia, we first substitute the corresponding value of K into two of the equations (9.54); the resulting equations together with Eq. (9.57) constitute a system of three equations from which the direction cosines of the corresponding principal axis can be determined. Thus, we have for the first principal moment of inertia K_1:

$$(\tfrac{5}{3} - 0.568867)mc^2(\lambda_x)_1 - \tfrac{3}{2}mc^2(\lambda_y)_1 - \tfrac{3}{4}mc^2(\lambda_z)_1 = 0$$
$$-\tfrac{3}{2}mc^2(\lambda_x)_1 + (\tfrac{10}{3} - 0.568867)mc^2(\lambda_y)_1 - \tfrac{1}{2}mc^2(\lambda_z)_1 = 0$$
$$(\lambda_x)_1^2 + (\lambda_y)_1^2 + (\lambda_z)_1^2 = 1$$

Solving yields

$$(\lambda_x)_1 = 0.836600 \qquad (\lambda_y)_1 = 0.496001 \qquad (\lambda_z)_1 = 0.232557$$

The angles that the first principal axis of inertia forms with the coordinate axes are then

$$(\theta_x)_1 = 33.2° \qquad (\theta_y)_1 = 60.3° \qquad (\theta_z)_1 = 76.6° \qquad \blacktriangleleft$$

Using the same set of equations successively with K_2 and K_3, we find that the angles associated with the second and third principal moments of inertia at the origin are, respectively,

$$(\theta_x)_2 = 57.8° \qquad (\theta_y)_2 = 146.6° \qquad (\theta_z)_2 = 98.0° \qquad \blacktriangleleft$$

and

$$(\theta_x)_3 = 82.8° \qquad (\theta_y)_3 = 76.1° \qquad (\theta_z)_3 = 164.3° \qquad \blacktriangleleft$$

SOLVING PROBLEMS
ON YOUR OWN

In this lesson we defined the *mass products of inertia* I_{xy}, I_{yz}, and I_{zx} of a body and showed you how to determine the moments of inertia of that body with respect to an arbitrary axis passing through the origin O. You also learned how to determine at the origin O the *principal axes of inertia* of a body and the corresponding *principal moments of inertia*.

1. Determining the mass products of inertia of a composite body. The mass products of inertia of a composite body with respect to the coordinate axes can be expressed as the sums of the products of inertia of its component parts with respect to those axes. For each component part, we can use the parallel-axis theorem and write Eqs. (9.47)

$$I_{xy} = \bar{I}_{x'y'} + m\bar{x}\,\bar{y} \qquad I_{yz} = \bar{I}_{y'z'} + m\bar{y}\,\bar{z} \qquad I_{zx} = \bar{I}_{z'x'} + m\bar{z}\,\bar{x}$$

where the primes denote the centroidal axes of each component part and where $\bar{x}$, $\bar{y}$, and $\bar{z}$ represent the coordinates of its center of gravity. Keep in mind that the mass products of inertia can be positive, negative, or zero, and be sure to take into account the signs of $\bar{x}$, $\bar{y}$, and $\bar{z}$.

a. From the properties of symmetry of a component part, you can deduce that two or all three of its centroidal mass products of inertia are zero. For instance, you can verify that for a thin plate parallel to the xy plane; a wire lying in a plane parallel to the xy plane; a body with a plane of symmetry parallel to the xy plane; and a body with an axis of symmetry parallel to the z axis, *the products of inertia $\bar{I}_{y'z'}$ and $\bar{I}_{z'x'}$ are zero.*

For rectangular, circular, or semicircular plates with axes of symmetry parallel to the coordinate axes; straight wires parallel to a coordinate axis; circular and semicircular wires with axes of symmetry parallel to the coordinate axes; and rectangular prisms with axes of symmetry parallel to the coordinate axes, *the products of inertia $\bar{I}_{x'y'}$, $\bar{I}_{y'z'}$, and $\bar{I}_{z'x'}$ are all zero.*

b. Mass products of inertia which are different from zero can be computed from Eqs. (9.45). Although, in general, a triple integration is required to determine a mass product of inertia, a single integration can be used if the given body can be divided into a series of thin, parallel slabs. The computations are then similar to those discussed in the previous lesson for moments of inertia.

(continued)

2. Computing the moment of inertia of a body with respect to an arbitrary axis OL. An expression for the moment of inertia I_{OL} was derived in Sec. 9.16 and is given in Eq. (9.46). Before computing I_{OL}, you must first determine the mass moments and products of inertia of the body with respect to the given coordinate axes as well as the direction cosines of the unit vector $\boldsymbol{\lambda}$ along OL.

3. Calculating the principal moments of inertia of a body and determining its principal axes of inertia. You saw in Sec. 9.17 that it is always possible to find an orientation of the coordinate axes for which the mass products of inertia are zero. These axes are referred to as the *principal axes of inertia* and the corresponding moments of inertia are known as the *principal moments of inertia* of the body. In many cases, the principal axes of inertia of a body can be determined from its properties of symmetry. The procedure required to determine the principal moments and principal axes of inertia of a body with no obvious property of symmetry was discussed in Sec. 9.18 and was illustrated in Sample Prob. 9.15. It consists of the following steps.

a. Expand the determinant in Eq. (9.55) and solve the resulting cubic equation. The solution can be obtained by trial and error or, preferably, with an advanced scientific calculator or with the appropriate computer software. The roots K_1, K_2, and K_3 of this equation are the principal moments of inertia of the body.

b. To determine the direction of the principal axis corresponding to K_1, substitute this value for K in two of the equations (9.54) and solve these equations together with Eq. (9.57) for the direction cosines of the principal axis corresponding to K_1.

c. Repeat this procedure with K_2 and K_3 to determine the directions of the other two principal axes. As a check of your computations, you may wish to verify that the scalar product of any two of the unit vectors along the three axes you have obtained is zero and, thus, that these axes are perpendicular to each other.

PROBLEMS

9.149 Determine the mass products of inertia I_{xy}, I_{yz}, and I_{zx} of the steel fixture shown. (The density of steel is 7850 kg/m³.)

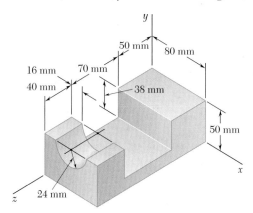

Fig. P9.149

9.150 Determine the mass products of inertia I_{xy}, I_{yz}, and I_{zx} of the steel machine element shown. (The density of steel is 7850 kg/m³.)

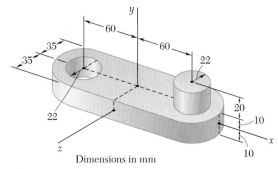

Dimensions in mm

Fig. P9.150

9.151 and 9.152 Determine the mass products of inertia I_{xy}, I_{yz}, and I_{zx} of the cast aluminum machine component shown. (The specific weight of aluminum is 0.100 lb/in³.)

Fig. P9.151

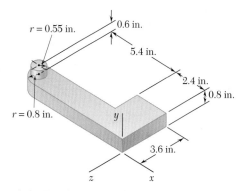

Fig. P9.152

9.153 through 9.156   A section of sheet steel 2 mm thick is cut and bent into the machine component shown. Knowing that the density of steel is 7850 kg/m^3, determine the mass products of inertia I_{xy}, I_{yz}, and I_{zx} of the component.

Fig. P9.153

Fig. P9.154

Fig. P9.155

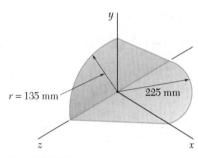

Fig. P9.156

9.157   The figure shown is formed of 1.5-mm-diameter aluminum wire. Knowing that the density of aluminum is 2800 kg/m^3, determine the mass products of inertia I_{xy}, I_{yz}, and I_{zx} of the wire figure.

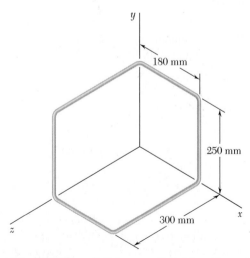

Fig. P9.157

9.158 Thin aluminum wire of uniform diameter is used to form the figure shown. Denoting by m' the mass per unit length of the wire, determine the mass products of inertia I_{xy}, I_{yz}, and I_{zx} of the wire figure.

9.159 and 9.160 Brass wire with a weight per unit length w is used to form the figure shown. Determine the mass products of inertia I_{xy}, I_{yz}, and I_{zx} of the wire figure.

Fig. P9.158

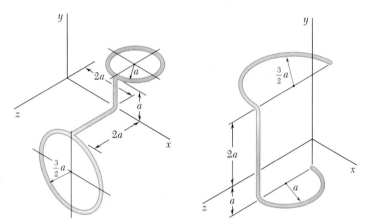

Fig. P9.159   **Fig. P9.160**

9.161 Complete the derivation of Eqs. (9.47), which express the parallel-axis theorem for mass products of inertia.

9.162 For the homogeneous tetrahedron of mass m shown, (a) determine by direct integration the mass product of inertia I_{zx}, (b) deduce I_{yz} and I_{xy} from the result obtained in part a.

9.163 The homogeneous circular cone shown has a mass m. Determine the mass moment of inertia of the cone with respect to the line joining the origin O and point A.

Fig. P9.162

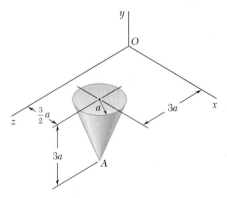

Fig. P9.163

9.164 The homogeneous circular cylinder shown has a mass m. Determine the mass moment of inertia of the cylinder with respect to the line joining the origin O and point A that is located on the perimeter of the top surface of the cylinder.

Fig. P9.164

9.165 Shown is the machine element of Prob. 9.141. Determine its mass moment of inertia with respect to the line joining the origin O and point A.

Fig. P9.165

9.166 Determine the mass moment of inertia of the steel fixture of Probs. 9.145 and 9.149 with respect to the axis through the origin that forms equal angles with the x, y, and z axes.

9.167 The thin bent plate shown is of uniform density and weight W. Determine its mass moment of inertia with respect to the line joining the origin O and point A.

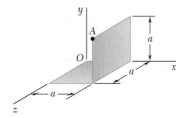

Fig. P9.167

9.168 A piece of sheet steel of thickness t and specific weight γ is cut and bent into the machine component shown. Determine the mass moment of inertia of the component with respect to the line joining the origin O and point A.

9.169 Determine the mass moment of inertia of the machine component of Probs. 9.136 and 9.155 with respect to the axis through the origin characterized by the unit vector $\lambda = (-4\mathbf{i} + 8\mathbf{j} + \mathbf{k})/9$.

9.170 through 9.172 For the wire figure of the problem indicated, determine the mass moment of inertia of the figure with respect to the axis through the origin characterized by the unit vector $\lambda = (-3\mathbf{i} - 6\mathbf{j} + 2\mathbf{k})/7$.
    9.170 Prob. 9.148
    9.171 Prob. 9.147
    9.172 Prob. 9.146

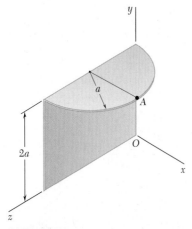

Fig. P9.168

9.173 For the homogeneous circular cylinder shown, of radius a and length L, determine the value of the ratio a/L for which the ellipsoid of inertia of the cylinder is a sphere when computed (a) at the centroid of the cylinder, (b) at point A.

9.174 For the rectangular prism shown, determine the values of the ratios b/a and c/a so that the ellipsoid of inertia of the prism is a sphere when computed (a) at point A, (b) at point B.

Fig. P9.173

Fig. P9.174

9.175 For the right circular cone of Sample Prob. 9.11, determine the value of the ratio a/h for which the ellipsoid of inertia of the cone is a sphere when computed (a) at the apex of the cone, (b) at the center of the base of the cone.

9.176 Given an arbitrary body and three rectangular axes x, y, and z, prove that the mass moment of inertia of the body with respect to any one of the three axes cannot be larger than the sum of the mass moments of inertia of the body with respect to the other two axes. That is, prove that the inequality $I_x \leq I_y + I_z$ and the two similar inequalities are satisfied. Further, prove that $I_y \geq \frac{1}{2} I_x$ if the body is a homogeneous solid of revolution, where x is the axis of revolution and y is a transverse axis.

9.177 Consider a cube of mass m and side a. (a) Show that the ellipsoid of inertia at the center of the cube is a sphere, and use this property to determine the moment of inertia of the cube with respect to one of its diagonals. (b) Show that the ellipsoid of inertia at one of the corners of the cube is an ellipsoid of revolution, and determine the principal moments of inertia of the cube at that point.

9.178 Given a homogeneous body of mass m and of arbitrary shape and three rectangular axes x, y, and z with origin at O, prove that the sum $I_x + I_y + I_z$ of the mass moments of inertia of the body cannot be smaller than the similar sum computed for a sphere of the same mass and the same material centered at O. Further, using the result of Prob. 9.176, prove that if the body is a solid of revolution, where x is the axis of revolution, its mass moment of inertia I_y about a transverse axis y cannot be smaller than $3ma^2/10$, where a is the radius of the sphere of the same mass and the same material.

*9.179 The homogeneous circular cylinder shown has a mass m, and the diameter OB of its top surface forms 45° angles with the x and z axes. (*a*) Determine the principal mass moments of inertia of the cylinder at the origin O. (*b*) Compute the angles that the principal axes of inertia at O form with the coordinate axes. (*c*) Sketch the cylinder, and show the orientation of the principal axes of inertia relative to the x, y, and z axes.

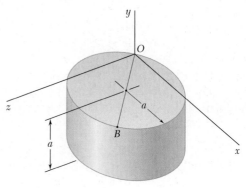

Fig. P9.179

9.180 through 9.184 For the component described in the problem indicated, determine (*a*) the principal mass moments of inertia at the origin, (*b*) the principal axes of inertia at the origin. Sketch the body and show the orientation of the principal axes of inertia relative to the x, y, and z axes.

  *9.180 Prob. 9.165
  *9.181 Probs. 9.145 and 9.149
  *9.182 Prob. 9.167
  *9.183 Prob. 9.168
  *9.184 Probs. 9.148 and 9.170

REVIEW AND SUMMARY

In the first half of this chapter, we discussed the determination of the resultant $\mathbf{R}$ of forces $\Delta\mathbf{F}$ distributed over a plane area A when the magnitudes of these forces are proportional to both the areas ΔA of the elements on which they act and the distances y from these elements to a given x axis; we thus had $\Delta F = ky\,\Delta A$. We found that the magnitude of the resultant $\mathbf{R}$ is proportional to the first moment $Q_x = \int y\,dA$ of the area A, while the moment of $\mathbf{R}$ about the x axis is proportional to the *second moment*, or *moment of inertia*, $I_x = \int y^2\,dA$ of A with respect to the same axis [Sec. 9.2].

The *rectangular moments of inertia* I_x *and* I_y *of an area* [Sec. 9.3] were obtained by evaluating the integrals

Rectangular moments of inertia

$$I_x = \int y^2\,dA \qquad I_y = \int x^2\,dA \qquad (9.1)$$

These computations can be reduced to single integrations by choosing dA to be a thin strip parallel to one of the coordinate axes. We also recall that it is possible to compute I_x and I_y from the same elemental strip (Fig. 9.35) using the formula for the moment of inertia of a rectangular area [Sample Prob. 9.3].

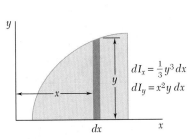

$$dI_x = \tfrac{1}{3}y^3\,dx$$
$$dI_y = x^2y\,dx$$

Fig. 9.35

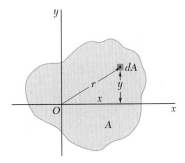

Fig. 9.36

The *polar moment of inertia of an area* A *with respect to the pole* O [Sec. 9.4] was defined as

Polar moment of inertia

$$J_O = \int r^2\,dA \qquad (9.3)$$

where r is the distance from O to the element of area dA (Fig. 9.36). Observing that $r^2 = x^2 + y^2$, we established the relation

$$J_O = I_x + I_y \qquad (9.4)$$

Radius of gyration

The *radius of gyration of an area A* with respect to the x axis [Sec. 9.5] was defined as the distance k_x, where $I_x = K_x^2 A$. With similar definitions for the radii of gyration of A with respect to the y axis and with respect to O, we had

$$k_x = \sqrt{\frac{I_x}{A}} \qquad k_y = \sqrt{\frac{I_y}{A}} \qquad k_O = \sqrt{\frac{J_O}{A}} \qquad (9.5\text{--}9.7)$$

Parallel-axis theorem

The parallel-axis theorem was presented in Sec. 9.6. It states that the moment of inertia I of an area with respect to any given axis AA' (Fig. 9.37) is equal to the moment of inertia $\bar{I}$ of the area with respect to the centroidal axis BB' that is parallel to AA' *plus* the product of the area A and the square of the distance d between the two axes:

$$I = \bar{I} + Ad^2 \qquad (9.9)$$

Fig. 9.37

This formula can also be used to determine the moment of inertia $\bar{I}$ of an area with respect to a centroidal axis BB' when its moment of inertia I with respect to a parallel axis AA' is known. In this case, however, the product Ad^2 should be *subtracted* from the known moment of inertia I.

A similar relation holds between the polar moment of inertia J_O of an area about a point O and the polar moment of inertia $\bar{J}_C$ of the same area about its centroid C. Letting d be the distance between O and C, we have

$$J_O = \bar{J}_C + Ad^2 \qquad (9.11)$$

Composite areas

The parallel-axis theorem can be used very effectively to compute the *moment of inertia of a composite area* with respect to a given axis [Sec. 9.7]. Considering each component area separately, we first compute the moment of inertia of each area with respect to its centroidal axis, using the data provided in Figs. 9.12 and 9.13 whenever possible. The parallel-axis theorem is then applied to determine the moment of inertia of each component area with respect to the desired axis, and the various values obtained are added [Sample Probs. 9.4 and 9.5].

Product of inertia

Sections 9.8 through 9.10 were devoted to the transformation of the moments of inertia of an area *under a rotation of the coordinate axes*. First, we defined the *product of inertia of an area A* as

$$I_{xy} = \int xy \, dA \qquad (9.12)$$

and showed that $I_{xy} = 0$ if the area A is symmetrical with respect to either or both of the coordinate axes. We also derived the *parallel-axis theorem for products of inertia*. We had

$$I_{xy} = \bar{I}_{x'y'} + \bar{x}\,\bar{y}A \qquad (9.13)$$

where $\bar{I}_{x'y'}$ is the product of inertia of the area with respect to the centroidal axes x' and y' which are parallel to the x and y axis and $\bar{x}$ and $\bar{y}$ are the coordinates of the centroid of the area [Sec. 9.8].

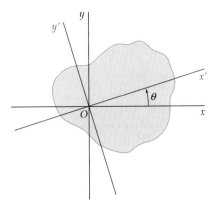

Fig. 9.38

In Sec. 9.9 we determined the moments and product of inertia $I_{x'}$, $I_{y'}$, and $I_{x'y'}$ of an area with respect to x' and y' axes obtained by rotating the original x and y coordinate axes through an angle θ counterclockwise (Fig. 9.38). We expressed $I_{x'}$, $I_{y'}$, and $I_{x'y'}$ in terms of the moments and product of inertia I_x, I_y, and I_{xy} computed with respect to the original x and y axes. We had

Rotation of axes

$$I_{x'} = \frac{I_x + I_y}{2} + \frac{I_x - I_y}{2}\cos 2\theta - I_{xy}\sin 2\theta \qquad (9.18)$$

$$I_{y'} = \frac{I_x + I_y}{2} - \frac{I_x - I_y}{2}\cos 2\theta + I_{xy}\sin 2\theta \qquad (9.19)$$

$$I_{x'y'} = \frac{I_x - I_y}{2}\sin 2\theta + I_{xy}\cos 2\theta \qquad (9.20)$$

The *principal axes of the area about O* were defined as the two axes perpendicular to each other, with respect to which the moments of inertia of the area are maximum and minimum. The corresponding values of θ, denoted by θ_m, were obtained from the formula

Principal axes

$$\tan 2\theta_m = -\frac{2I_{xy}}{I_x - I_y} \qquad (9.25)$$

The corresponding maximum and minimum values of I are called the *principal moments of inertia* of the area about O; we had

Principal moments of inertia

$$I_{\max,\min} = \frac{I_x + I_y}{2} \pm \sqrt{\left(\frac{I_x - I_y}{2}\right)^2 + I_{xy}^2} \qquad (9.27)$$

We also noted that the corresponding value of the product of inertia is zero.

The transformation of the moments and product of inertia of an area under a rotation of axes can be represented graphically by drawing *Mohr's circle* [Sec. 9.10]. Given the moments and product of inertia I_x, I_y, and I_{xy} of the area with respect to the x and y coordinate axes, we

Mohr's circle

Fig. 9.39

plot points X (I_x, I_{xy}) and Y (I_y, $-I_{xy}$) and draw the line joining these two points (Fig. 9.39). This line is a diameter of Mohr's circle and thus defines this circle. As the coordinate axes are rotated through θ, the diameter rotates through *twice that angle*, and the coordinates of X' and Y' yield the new values $I_{x'}$, $I_{y'}$, and $I_{x'y'}$ of the moments and product of inertia of the area. Also, the angle θ_m and the coordinates of points A and B define the principal axes a and b and the principal moments of inertia of the area [Sample Prob. 9.8].

Moments of inertia of masses The second half of the chapter was devoted to the determination of *moments of inertia of masses*, which are encountered in dynamics in problems involving the rotation of a rigid body about an axis. The mass moment of inertia of a body with respect to an axis AA' (Fig. 9.40) was defined as

$$I = \int r^2 \, dm \qquad (9.28)$$

where r is the distance from AA' to the element of mass [Sec. 9.11]. The *radius of gyration* of the body was defined as

$$k = \sqrt{\frac{I}{m}} \qquad (9.29)$$

The moments of inertia of a body with respect to the coordinates axes were expressed as

$$I_x = \int (y^2 + z^2) \, dm$$

$$I_y = \int (z^2 + x^2) \, dm \qquad (9.30)$$

$$I_z = \int (x^2 + y^2) \, dm$$

Fig. 9.40

We saw that the *parallel-axis theorem* also applies to mass moments of inertia [Sec. 9.12]. Thus, the moment of inertia I of a body with respect to an arbitrary axis AA' (Fig. 9.41) can be expressed as

Parallel-axis theorem

$$I = \bar{I} + md^2 \qquad (9.33)$$

where $\bar{I}$ is the moment of inertia of the body with respect to the centroidal axis BB' which is parallel to the axis AA', m is the mass of the body, and d is the distance between the two axes.

Fig. 9.41

Fig. 9.42

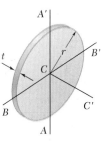

Fig. 9.43

The moments of inertia of *thin plates* can be readily obtained from the moments of inertia of their areas [Sec. 9.13]. We found that for a *rectangular plate* the moments of inertia with respect to the axes shown (Fig. 9.42) are

Moments of inertia of thin plates

$$I_{AA'} = \tfrac{1}{12}ma^2 \qquad I_{BB'} = \tfrac{1}{12}mb^2 \qquad (9.39)$$

$$I_{CC'} = I_{AA'} + I_{BB'} = \tfrac{1}{12}m(a^2 + b^2) \qquad (9.40)$$

while for a *circular plate* (Fig. 9.43) they are

$$I_{AA'} = I_{BB'} = \tfrac{1}{4}mr^2 \qquad (9.41)$$

$$I_{CC'} = I_{AA'} + I_{BB'} = \tfrac{1}{2}mr^2 \qquad (9.42)$$

When a body possesses *two planes of symmetry*, it is usually possible to use a single integration to determine its moment of inertia with respect to a given axis by selecting the element of mass dm to be a thin plate [Sample Probs. 9.10 and 9.11]. On the other hand, when a body consists of *several common geometric shapes*, its moment of inertia with respect to a given axis can be obtained by using the formulas given in Fig. 9.28 together with the parallel-axis theorem [Sample Probs. 9.12 and 9.13].

Composite bodies

In the last portion of the chapter, we learned to determine the moment of inertia of a body *with respect to an arbitrary axis OL* which is drawn through the origin O [Sec. 9.16]. Denoting by λ_x, λ_y,

Moment of inertia with respect to an arbitrary axis

λ_z the components of the unit vector $\boldsymbol{\lambda}$ along OL (Fig. 9.44) and introducing the *products of inertia*

$$I_{xy} = \int xy \, dm \qquad I_{yz} = \int yz \, dm \qquad I_{zx} = \int zx \, dm \qquad (9.45)$$

we found that the moment of inertia of the body with respect to OL could be expressed as

$$I_{OL} = I_x\lambda_x^2 + I_y\lambda_y^2 + I_z\lambda_z^2 - 2I_{xy}\lambda_x\lambda_y - 2I_{yz}\lambda_y\lambda_z - 2I_{zx}\lambda_z\lambda_x \qquad (9.46)$$

Fig. 9.44

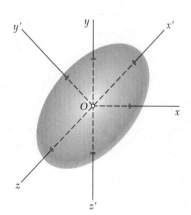

Fig. 9.45

Ellipsoid of inertia

By plotting a point Q along each axis OL at a distance $OQ = 1\sqrt{I_{OL}}$ from O [Sec. 9.17], we obtained the surface of an ellipsoid, known as the *ellipsoid of inertia* of the body at point O. The principal axes x', y', z' of this ellipsoid (Fig. 9.45) are the *principal axes of inertia*

Principal axes of inertia
Principal moments of inertia

of the body; that is, the products of inertia $I_{x'y'}$, $I_{y'z'}$, $I_{z'x'}$ of the body with respect to these axes are all zero. There are many situations when the principal axes of inertia of a body can be deduced from properties of symmetry of the body. Choosing these axes to be the coordinate axes, we can then express I_{OL} as

$$I_{OL} = I_{x'}\lambda_{x'}^2 + I_{y'}\lambda_{y'}^2 + I_{z'}\lambda_{z'}^2 \qquad (9.50)$$

where $I_{x'}$, $I_{y'}$, $I_{z'}$ are the *principal moments of inertia* of the body at O.

When the principal axes of inertia cannot be obtained by observation [Sec. 9.17], it is necessary to solve the cubic equation

$$K^3 - (I_x + I_y + I_z)K^2 + (I_xI_y + I_yI_z + I_zI_x - I_{xy}^2 - I_{yz}^2 - I_{zx}^2)K$$
$$- (I_xI_yI_z - I_xI_{yz}^2 - I_yI_{zx}^2 - I_zI_{xy}^2 - 2I_{xy}I_{yz}I_{zx}) = 0 \qquad (9.56)$$

We found [Sec. 9.18] that the roots K_1, K_2, and K_3 of this equation are the principal moments of inertia of the given body. The direction cosines $(\lambda_x)_1$, $(\lambda_y)_1$, and $(\lambda_z)_1$ of the principal axis corresponding to the principal moment of inertia K_1 are then determined by substituting K_1 into Eqs. (9.54) and solving two of these equations and Eq. (9.57) simultaneously. The same procedure is then repeated using K_2 and K_3 to determine the direction cosines of the other two principal axes [Sample Prob. 9.15].

REVIEW PROBLEMS

9.185 Determine by direct integration the moments of inertia of the shaded area with respect to the x and y axes.

Fig. P9.185

9.186 Determine the moment of inertia and the radius of gyration of the shaded area shown with respect to the y axis.

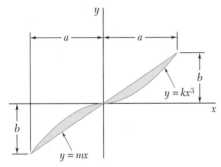

Fig. P9.186

9.187 Determine the moment of inertia and the radius of gyration of the shaded area shown with respect to the x axis.

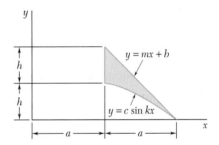

Fig. P9.187

9.188 Determine the moments of inertia $\bar{I}_x$ and $\bar{I}_y$ of the area shown with respect to centroidal axes respectively parallel and perpendicular to side AB.

Fig. P9.188

Fig. P9.189

9.189 Determine the polar moment of inertia of the area shown with respect to (a) point O, (b) the centroid of the area.

9.190 Two L5 × 3 × $\frac{1}{2}$-in. angles are welded to a $\frac{1}{2}$-in. steel plate. Determine the distance b and the centroidal moments of inertia $\bar{I}_x$ and $\bar{I}_y$ of the combined section, knowing that $\bar{I}_y = 4\bar{I}_x$.

Fig. P9.190

9.191 Using the parallel-axis theorem, determine the product of inertia of the L5 × 3 × $\frac{1}{2}$-in. angle cross section shown with respect to the centroidal x and y axes.

9.192 For the L5 × 3 × $\frac{1}{2}$-in. angle cross section shown, use Mohr's circle to determine (a) the moments of inertia and the product of inertia with respect to new centroidal axes obtained by rotating the x and y axes 30° clockwise, (b) the orientation of the principal axes through the centroid and the corresponding values of the moments of inertia.

Fig. P9.191 and P9.192

9.193 A thin plate of mass m has the trapezoidal shape shown. Determine the mass moment of inertia of the plate with respect to (a) the x axis, (b) the y axis.

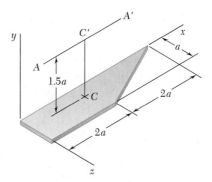

Fig. P9.193 and P9.194

9.194 A thin plate of mass m has the trapezoidal shape shown. Determine the mass moment of inertia of the plate with respect to (a) the centroidal axis CC' that is perpendicular to the plate, (b) the axis AA' that is parallel to the x axis and is located at a distance $1.5a$ from the plate.

9.195 A 2-mm-thick piece of sheet steel is cut and bent into the machine component shown. Knowing that the density of steel is 7850 kg/m^3, determine the mass moment of inertia of the component with respect to each of the coordinate axes.

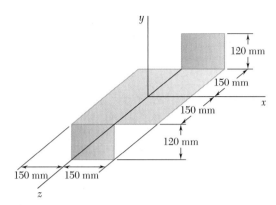

Fig. P9.195

9.196 Determine the mass moment of inertia and the radius of gyration of the steel machine element shown with respect to the x axis. (The density of steel is 7850 kg/m^3.)

Fig. P9.196

COMPUTER PROBLEMS

9.C1 Write a computer program that, for an area with known moments and product of inertia I_x, I_y, and I_{xy}, can be used to calculate the moments and product of inertia $I_{x'}$, $I_{y'}$, and $I_{x'y'}$ of the area with respect to axes x' and y' obtained by rotating the original axes counterclockwise through an angle θ. Use this program to compute $I_{x'}$, $I_{y'}$, and $I_{x'y'}$ for the section of Sample Prob. 9.7 for values of θ from 0 to 90° using 5° increments.

9.C2 Write a computer program that, for an area with known moments and product of inertia I_x, I_y, and I_{xy}, can be used to calculate the orientation of the principal axes of the area and the corresponding values of the principal moments of inertia. Use this program to solve (a) Prob. 9.89, (b) Sample Prob. 9.7.

9.C3 Many cross sections can be approximated by a series of rectangles as shown. Write a computer program that can be used to calculate the moments of inertia and the radii of gyration of cross sections of this type with respect to horizontal and vertical centroidal axes. Apply this program to the cross sections shown in (a) Figs. P9.31 and P9.33, (b) Figs. P9.32 and P9.34, (c) Fig. P9.43, (d) Fig. P9.44.

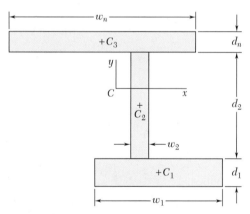

Fig. P9.C3 and P9.C4

9.C4 Many cross sections can be approximated by a series of rectangles as shown. Write a computer program that can be used to calculate the products of inertia of cross sections of this type with respect to horizontal and vertical centroidal axes. Use this program to solve (a) Prob. 9.71, (b) Prob. 9.75, (c) Prob. 9.77.

9.C5 The area shown is revolved about the x axis to form a homogeneous solid of mass m. Approximate the area using a series of 400 rectangles of the form $bcc'b'$, each of width Δl, and then write a computer program that can be used to determine the mass moment of inertia of the solid with respect to the x axis. Use this program to solve part a of (a) Sample Prob. 9.11, (b) Prob. 9.121, assuming that in these problems $m = 2$ kg, $a = 100$ mm, and $h = 400$ mm.

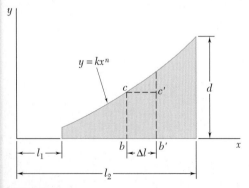

Fig. P9.C5

554

9.C6 A homogeneous wire with a weight per unit length of 0.04 lb/ft is used to form the figure shown. Approximate the figure using 10 straight line segments, and then write a computer program that can be used to determine the mass moment of inertia I_x of the wire with respect to the x axis. Use this program to determine I_x when (a) $a = 1$ in., $L = 11$ in., $h = 4$ in., (b) $a = 2$ in., $L = 17$ in., $h = 10$ in., (c) $a = 5$ in., $L = 25$ in., $h = 6$ in.

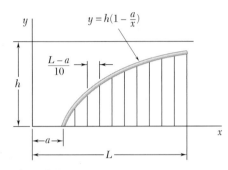

Fig. P9.C6

*9.C7 Write a computer program that, for a body with known mass moments and products of inertia I_x, I_y, I_z, I_{xy}, I_{yz}, and I_{zx}, can be used to calculate the principal mass moments of inertia K_1, K_2, and K_3 of the body at the origin. Use this program to solve part a of (a) Prob. 9.180, (b) Prob. 9.181, (c) Prob. 9.184.

*9.C8 Extend the computer program of Prob. 9.C7 to include the computation of the angles that the principal axes of inertia at the origin form with the coordinate axes. Use this program to solve (a) Prob. 9.180, (b) Prob. 9.181, (c) Prob. 9.184.

The method of virtual work is particularly effective when a simple relation can be found among the displacements of the points of application of the various forces involved. This is the case for the scissor lift platform being used by workers to gain access to a highway bridge under construction.

CHAPTER

Method of Virtual Work

*10.1 INTRODUCTION

In the preceding chapters, problems involving the equilibrium of rigid bodies were solved by expressing that the external forces acting on the bodies were balanced. The equations of equilibrium $\Sigma F_x = 0$, $\Sigma F_y = 0$, $\Sigma M_A = 0$ were written and solved for the desired unknowns. A different method, which will prove more effective for solving certain types of equilibrium problems, will now be considered. This method is based on the *principle of virtual work* and was first formally used by the Swiss mathematician Jean Bernoulli in the eighteenth century.

As you will see in Sec. 10.3, the principle of virtual work states that if a particle or rigid body, or, more generally, a system of connected rigid bodies, which is in equilibrium under various external forces, is given an arbitrary displacement from that position of equilibrium, the total work done by the external forces during the displacement is zero. This principle is particularly effective when applied to the solution of problems involving the equilibrium of machines or mechanisms consisting of several connected members.

In the second part of the chapter, the method of virtual work will be applied in an alternative form based on the concept of *potential energy*. It will be shown in Sec. 10.8 that if a particle, rigid body, or system of rigid bodies is in equilibrium, then the derivative of its potential energy with respect to a variable defining its position must be zero.

In this chapter, you will also learn to evaluate the mechanical efficiency of a machine (Sec. 10.5) and to determine whether a given position of equilibrium is stable, unstable, or neutral (Sec. 10.9).

*10.2 WORK OF A FORCE

Let us first define the terms *displacement* and *work* as they are used in mechanics. Consider a particle which moves from a point A to a neighboring point A' (Fig. 10.1). If $\mathbf{r}$ denotes the position vector corresponding to point A, the small vector joining A and A' may be denoted by the differential $d\mathbf{r}$; the vector $d\mathbf{r}$ is called the *displacement* of the particle. Now let us assume that a force $\mathbf{F}$ is acting on the particle. The *work of the force $\mathbf{F}$ corresponding to the displacement $d\mathbf{r}$* is defined as the quantity

Fig. 10.1

$$dU = \mathbf{F} \cdot d\mathbf{r} \tag{10.1}$$

obtained by forming the scalar product of the force $\mathbf{F}$ and the displacement $d\mathbf{r}$. Denoting respectively by F and ds the magnitudes of the force and of the displacement, and by α the angle formed by $\mathbf{F}$ and $d\mathbf{r}$, and recalling the definition of the scalar product of two vectors (Sec. 3.9), we write

$$dU = F \, ds \cos \alpha \tag{10.1'}$$

Being a *scalar quantity*, work has a magnitude and a sign, but no direction. We also note that work should be expressed in units obtained by multiplying units of length by units of force. Thus, if U.S. customary units are used, work should be expressed in ft · lb or in · lb. If SI units are used, work should be expressed in N · m. The unit of work N · m is called a *joule* (J).†

It follows from (10.1′) that the work dU is positive if the angle α is acute and negative if α is obtuse. Three particular cases are of special interest. If the force **F** has the same direction as $d\mathbf{r}$, the work dU reduces to $F\,ds$. If **F** has a direction opposite to that of $d\mathbf{r}$, the work is $dU = -F\,ds$. Finally, if **F** is perpendicular to $d\mathbf{r}$, the work dU is zero.

The work dU of a force **F** during a displacement $d\mathbf{r}$ can also be considered as the product of F and the component $ds\cos\alpha$ of the displacement $d\mathbf{r}$ along **F** (Fig. 10.2a). This view is particularly

Fig. 10.2

Photo 10.1 The forces exerted by the hydraulic cylinders to position the bucket lift shown can be effectively determined using the method of virtual work since a simple relation exists among the displacements of the points of application of the forces acting on the members of the lift.

useful in the computation of the work done by the weight **W** of a body (Fig. 10.2b). The work of **W** is equal to the product of W and the vertical displacement dy of the center of gravity G of the body. If the displacement is downward, the work is positive; if it is upward, the work is negative.

A number of forces frequently encountered in statics *do no work:* forces applied to fixed points ($ds = 0$) or acting in a direction perpendicular to the displacement ($\cos\alpha = 0$). Among these forces are the reaction at a frictionless pin when the body supported rotates about the pin; the reaction at a frictionless surface when the body in contact moves along the surface; the reaction at a roller moving along its track; the weight of a body when its center of gravity moves horizontally; and the friction force acting on a wheel rolling without slipping (since at any instant the point of contact does not move). Examples of forces which *do work* are the weight of a body (except in the case considered above), the friction force acting on a body sliding on a rough surface, and most forces applied on a moving body.

---

†The joule is the SI unit of *energy,* whether in mechanical form (work, potential energy, kinetic energy) or in chemical, electrical, or thermal form. We should note that even though N · m = J, the moment of a force must be expressed in N · m, and not in joules, since the moment of a force is not a form of energy.

In certain cases, the sum of the work done by several forces is zero. Consider, for example, two rigid bodies AC and BC connected at C by a *frictionless pin* (Fig. 10.3a). Among the forces acting on AC is the force **F** exerted at C by BC. In general, the work of this

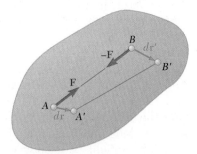

Fig. 10.3

force will not be zero, but it will be equal in magnitude and opposite in sign to the work of the force $-$**F** exerted by AC on BC, since these forces are equal and opposite and are applied to the same particle. Thus, when the total work done by all the forces acting on AB and BC is considered, the work of the two internal forces at C cancels out. A similar result is obtained if we consider a system consisting of two blocks connected by an *inextensible cord AB* (Fig. 10.3b). The work of the tension force **T** at A is equal in magnitude to the work of the tension force **T'** at B, since these forces have the same magnitude and the points A and B move through the same distance; but in one case the work is positive, and in the other it is negative. Thus, the work of the internal forces again cancels out.

It can be shown that the total work of the internal forces holding together the particles of a rigid body is zero. Consider two particles A and B of a rigid body and the two equal and opposite forces **F** and $-$**F** they exert on each other (Fig. 10.4). While, in general,

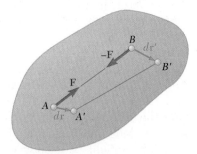

Fig. 10.4

small displacements d**r** and d**r'** of the two particles are different, the components of these displacements along AB must be equal; otherwise, the particles would not remain at the same distance from each other, and the body would not be rigid. Therefore, the work of **F** is equal in magnitude and opposite in sign to the work of $-$**F**, and their sum is zero.

In computing the work of the external forces acting on a rigid body, it is often convenient to determine the work of a couple without considering separately the work of each of the two forces forming the couple. Consider the two forces **F** and $-$**F** forming a couple of moment **M** and acting on a rigid body (Fig. 10.5). Any small displacement of the rigid body bringing A and B, respectively, into A' and B'' can be divided into two parts, one in which points A and B undergo equal displacements $d\mathbf{r}_1$, the other in which A' remains fixed while B' moves into B'' through a displacement $d\mathbf{r}_2$ of magnitude $ds_2 = r\, d\theta$. In the first part of the motion, the work of **F** is equal in magnitude and opposite in sign to the work of $-\mathbf{F}$, and their sum is zero. In the second part of the motion, only force **F** works, and its work is $dU = F\, ds_2 = Fr\, d\theta$. But the product Fr is equal to the magnitude M of the moment of the couple. Thus, the work of a couple of moment **M** acting on a rigid body is

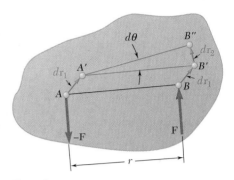

Fig. 10.5

$$dU = M\, d\theta \tag{10.2}$$

where $d\theta$ is the small angle expressed in radians through which the body rotates. We again note that work should be expressed in units obtained by multiplying units of force by units of length.

*10.3 PRINCIPLE OF VIRTUAL WORK

Consider a particle acted upon by several forces $\mathbf{F}_1, \mathbf{F}_2, \ldots, \mathbf{F}_n$ (Fig. 10.6). We can imagine that the particle undergoes a small displacement from A to A'. This displacement is possible, but it will not necessarily take place. The forces may be balanced and the particle at rest, or the particle may move under the action of the given forces in a direction different from that of AA'. Since the displacement considered does not actually occur, it is called a *virtual displacement* and is denoted by $\delta\mathbf{r}$. The symbol $\delta\mathbf{r}$ represents a differential of the first order; it is used to distinguish the virtual displacement from the displacement $d\mathbf{r}$ which would take place under actual motion. As you will see, virtual displacements can be used to determine whether the conditions of equilibrium of a particle are satisfied.

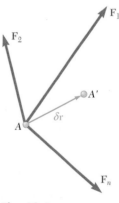

Fig. 10.6

The work of each of the forces $\mathbf{F}_1, \mathbf{F}_2, \ldots, \mathbf{F}_n$ during the virtual displacement $\delta\mathbf{r}$ is called *virtual work*. The virtual work of all the forces acting on the particle of Fig. 10.6 is

$$\delta U = \mathbf{F}_1 \cdot \delta\mathbf{r} + \mathbf{F}_2 \cdot \delta\mathbf{r} + \cdots + \mathbf{F}_n \cdot \delta\mathbf{r}$$
$$= (\mathbf{F}_1 + \mathbf{F}_2 + \cdots + \mathbf{F}_n) \cdot \delta\mathbf{r}$$

or

$$\delta U = \mathbf{R} \cdot \delta\mathbf{r} \tag{10.3}$$

where **R** is the resultant of the given forces. Thus, the total virtual work of the forces $\mathbf{F}_1, \mathbf{F}_2, \ldots, \mathbf{F}_n$ is equal to the virtual work of their resultant **R**.

The principle of virtual work for a particle states that *if a particle is in equilibrium, the total virtual work of the forces acting on the particle is zero for any virtual displacement of the particle*. This condition is necessary: if the particle is in equilibrium, the resultant **R** of the forces is zero, and it follows from (10.3) that the total virtual work δU is zero. The condition is also sufficient: if the total virtual work δU is zero for any virtual displacement, the scalar product $\mathbf{R} \cdot \delta\mathbf{r}$ is zero for any $\delta\mathbf{r}$, and the resultant **R** must be zero.

In the case of a rigid body, the principle of virtual work states that *if a rigid body is in equilibrium, the total virtual work of the external forces acting on the rigid body is zero for any virtual displacement of the body.* The condition is necessary: if the body is in equilibrium, all the particles forming the body are in equilibrium and the total virtual work of the forces acting on all the particles must be zero; but we have seen in the preceding section that the total work of the internal forces is zero; the total work of the external forces must therefore also be zero. The condition can also be proved to be sufficient.

The principle of virtual work can be extended to the case of a *system of connected rigid bodies.* If the system remains connected during the virtual displacement, *only the work of the forces external to the system need be considered,* since the total work of the internal forces at the various connections is zero.

*10.4 APPLICATIONS OF THE PRINCIPLE OF VIRTUAL WORK

The principle of virtual work is particularly effective when applied to the solution of problems involving machines or mechanisms consisting of several connected rigid bodies. Consider, for instance, the toggle vise *ACB* of Fig. 10.7*a*, used to compress a wooden block. We

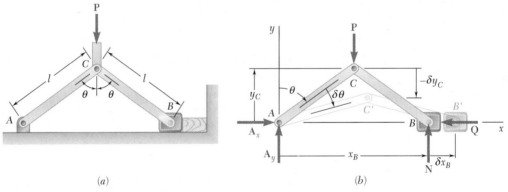

(a)   *(b)*

Fig. 10.7

wish to determine the force exerted by the vise on the block when a given force **P** is applied at *C*, assuming that there is no friction. Denoting by **Q** the reaction of the block on the vise, we draw the free-body diagram of the vise and consider the virtual displacement obtained by giving a positive increment $\delta\theta$ to the angle θ (Fig. 10.7*b*). Choosing a system of coordinate axes with origin at *A*, we note that x_B increases while y_C decreases. This is indicated in the figure, where a positive increment δx_B and a negative increment $-\delta y_C$ are shown. The reactions $\mathbf{A}_x$, $\mathbf{A}_y$, and **N** will do no work during the virtual displacement considered, and we need only compute the work of **P** and **Q**. Since **Q** and δx_B have opposite senses, the virtual work of **Q** is $\delta U_Q = -Q\,\delta x_B$. Since **P** and the increment shown $(-\delta y_C)$ have the same sense, the virtual work of **P** is $\delta U_P = +P(-\delta y_C) = -P\,\delta y_C$. The minus signs obtained could have been predicted by simply noting that the forces **Q** and **P** are directed opposite to the positive x and y axes, respectively. Expressing the coordinates x_B and y_C in terms of the angle θ and differentiating, we obtain

$$x_B = 2l \sin \theta \qquad y_C = l \cos \theta$$
$$\delta x_B = 2l \cos \theta \, \delta\theta \qquad \delta y_C = -l \sin \theta \, \delta\theta \qquad (10.4)$$

The total virtual work of the forces **Q** and **P** is thus

$$\delta U = \delta U_Q + \delta U_P = -Q \, \delta x_B - P \, \delta y_C$$
$$= -2Ql \cos \theta \, \delta\theta + Pl \sin \theta \, \delta\theta$$

Making $\delta U = 0$, we obtain

$$2Ql \cos \theta \, \delta\theta = Pl \sin \theta \, \delta\theta \qquad (10.5)$$
$$Q = \tfrac{1}{2}P \tan \theta \qquad (10.6)$$

The superiority of the method of virtual work over the conventional equilibrium equations in the problem considered here is clear: by using the method of virtual work, we were able to eliminate all unknown reactions, while the equation $\Sigma M_A = 0$ would have eliminated only two of the unknown reactions. This property of the method of virtual work can be used in solving many problems involving machines and mechanisms. *If the virtual displacement considered is consistent with the constraints imposed by the supports and connections, all reactions and internal forces are eliminated and only the work of the loads, applied forces, and friction forces need be considered.*

The method of virtual work can also be used to solve problems involving completely constrained structures, although the virtual displacements considered will never actually take place. Consider, for example, the frame *ACB* shown in Fig. 10.8*a*. If point *A* is kept fixed, while *B* is given a horizontal virtual displacement (Fig. 10.8*b*), we need consider only the work of **P** and **B**$_x$. We can thus determine

Photo 10.2 The clamping force of the toggle clamp shown can be expressed as a function of the force applied to the handle by first establishing the geometric relations among the members of the clamp and then applying the method of virtual work.

(a)

Fig. 10.8

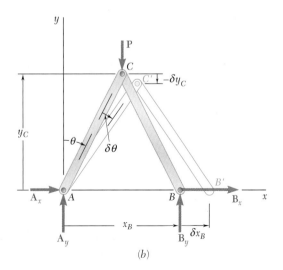

(b)

the reaction component $\mathbf{B}_x$ in the same way as the force $\mathbf{Q}$ of the preceding example (Fig. 10.7b); we have

$$B_x = -\tfrac{1}{2}P \tan \theta$$

Keeping B fixed and giving to A a horizontal virtual displacement, we can similarly determine the reaction component $\mathbf{A}_x$. The components $\mathbf{A}_y$ and $\mathbf{B}_y$ can be determined by rotating the frame ACB as a rigid body about B and A, respectively.

The method of virtual work can also be used to determine the configuration of a system in equilibrium under given forces. For example, the value of the angle θ for which the linkage of Fig. 10.7 is in equilibrium under two given forces $\mathbf{P}$ and $\mathbf{Q}$ can be obtained by solving Eq. (10.6) for $\tan \theta$.

It should be noted, however, that the attractiveness of the method of virtual work depends to a large extent upon the existence of simple geometric relations between the various virtual displacements involved in the solution of a given problem. When no such simple relations exist, it is usually advisable to revert to the conventional method of Chap. 6.

*10.5 REAL MACHINES. MECHANICAL EFFICIENCY

In analyzing the toggle vise in the preceding section, we assumed that no friction forces were involved. Thus, the virtual work consisted only of the work of the applied force $\mathbf{P}$ and of the reaction $\mathbf{Q}$. But the work of the reaction $\mathbf{Q}$ is equal in magnitude and opposite in sign to the work of the force exerted by the vise on the block. Equation (10.5), therefore, expresses that the *output work* $2Ql \cos \theta \, \delta\theta$ is equal to the *input work* $Pl \sin \theta \, \delta\theta$. A machine in which input and output work are equal is said to be an "ideal" machine. In a "real" machine, friction forces will always do some work, and the output work will be smaller than the input work.

Consider, for example, the toggle vise of Fig. 10.7a, and assume now that a friction force $\mathbf{F}$ develops between the sliding block B and the horizontal plane (Fig. 10.9). Using the conventional methods of statics and summing moments about A, we find $N = P/2$. Denoting by μ the coefficient of friction between block B and the horizontal

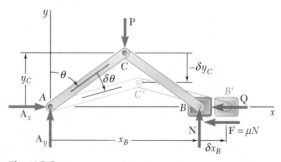

Fig. 10.9

plane, we have $F = \mu N = \mu P/2$. Recalling formulas (10.4), we find that the total virtual work of the forces $\mathbf{Q}$, $\mathbf{P}$, and $\mathbf{F}$ during the virtual displacement shown in Fig. 10.9 is

$$\begin{aligned} \delta U &= -Q\,\delta x_B - P\,\delta y_C - F\,\delta x_B \\ &= -2Ql\cos\theta\,\delta\theta + Pl\sin\theta\,\delta\theta - \mu Pl\cos\theta\,\delta\theta \end{aligned}$$

Making $\delta U = 0$, we obtain

$$2Ql\cos\theta\,\delta\theta = Pl\sin\theta\,\delta\theta - \mu Pl\cos\theta\,\delta\theta \qquad (10.7)$$

which expresses that the output work is equal to the input work minus the work of the friction force. Solving for Q, we have

$$Q = \tfrac{1}{2}P(\tan\theta - \mu) \qquad (10.8)$$

We note that $Q = 0$ when $\tan\theta = \mu$, that is, when θ is equal to the angle of friction ϕ, and that $Q < 0$ when $\theta < \phi$. The toggle vise may thus be used only for values of θ larger than the angle of friction.

The *mechanical efficiency* of a machine is defined as the ratio

$$\eta = \frac{\text{output work}}{\text{input work}} \qquad (10.9)$$

Clearly, the mechanical efficiency of an ideal machine is $\eta = 1$, since input and output work are then equal, while the mechanical efficiency of a real machine will always be less than 1.

In the case of the toggle vise we have just analyzed, we write

$$\eta = \frac{\text{output work}}{\text{input work}} = \frac{2Ql\cos\theta\,\delta\theta}{Pl\sin\theta\,\delta\theta}$$

Substituting from (10.8) for Q, we obtain

$$\eta = \frac{P(\tan\theta - \mu)l\cos\theta\,\delta\theta}{Pl\sin\theta\,\delta\theta} = 1 - \mu\cot\theta \qquad (10.10)$$

We check that in the absence of friction forces, we would have $\mu = 0$ and $\eta = 1$. In the general case, when μ is different from zero, the efficiency η becomes zero for $\mu\cot\theta = 1$, that is, for $\tan\theta = \mu$, or $\theta = \tan^{-1}\mu = \phi$. We note again that the toggle vise can be used only for values of θ larger than the angle of friction ϕ.

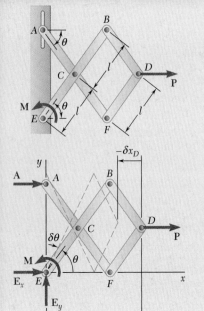

SAMPLE PROBLEM 10.1

Using the method of virtual work, determine the magnitude of the couple **M** required to maintain the equilibrium of the mechanism shown.

SOLUTION

Choosing a coordinate system with origin at E, we write

$$x_D = 3l \cos \theta \qquad \delta x_D = -3l \sin \theta \; \delta\theta$$

Principle of Virtual Work. Since the reactions **A**, $\mathbf{E}_x$, and $\mathbf{E}_y$ will do no work during the virtual displacement, the total virtual work done by **M** and **P** must be zero. Noting that **P** acts in the positive x direction and **M** acts in the positive θ direction, we write

$$\delta U = 0: \qquad +M \; \delta\theta + P \; \delta x_D = 0$$
$$+M \; \delta\theta + P(-3l \sin \theta \; \delta\theta) = 0$$

$$M = 3Pl \sin \theta \quad \blacktriangleleft$$

SAMPLE PROBLEM 10.2

Determine the expressions for θ and for the tension in the spring which correspond to the equilibrium position of the mechanism. The unstretched length of the spring is h, and the constant of the spring is k. Neglect the weight of the mechanism.

SOLUTION

With the coordinate system shown

$$y_B = l \sin \theta \qquad y_C = 2l \sin \theta$$
$$\delta y_B = l \cos \theta \; \delta\theta \qquad \delta y_C = 2l \cos \theta \; \delta\theta$$

The elongation of the spring is $\quad s = y_C - h = 2l \sin \theta - h$

The magnitude of the force exerted at C by the spring is

$$F = ks = k(2l \sin \theta - h) \tag{1}$$

Principle of Virtual Work. Since the reactions $\mathbf{A}_x$, $\mathbf{A}_y$, and **C** do no work, the total virtual work done by **P** and **F** must be zero.

$$\delta U = 0: \qquad P \; \delta y_B - F \; \delta y_C = 0$$
$$P(l \cos \theta \; \delta\theta) - k(2l \sin \theta - h)(2l \cos \theta \; \delta\theta) = 0$$

$$\sin \theta = \frac{P + 2kh}{4kl} \quad \blacktriangleleft$$

Substituting this expression into (1), we obtain

$$F = \tfrac{1}{2}P \quad \blacktriangleleft$$

566

SAMPLE PROBLEM 10.3

A hydraulic-lift table is used to raise a 1000-kg crate. It consists of a platform and of two identical linkages on which hydraulic cylinders exert equal forces. (Only one linkage and one cylinder are shown.) Members EDB and CG are each of length $2a$, and member AD is pinned to the midpoint of EDB. If the crate is placed on the table, so that half of its weight is supported by the system shown, determine the force exerted by each cylinder in raising the crate for $\theta = 60°$, $a = 0.70$ m, and $L = 3.20$ m. This mechanism has been previously considered in Sample Prob. 6.7.

SOLUTION

The machine considered consists of the platform and of the linkage, with an input force $\mathbf{F}_{DH}$ exerted by the cylinder and an output force equal and opposite to $\frac{1}{2}\mathbf{W}$.

Principle of Virtual Work. We first observe that the reactions at E and G do no work. Denoting by y the elevation of the platform above the base, and by s the length DH of the cylinder-and-piston assembly, we write

$$\delta U = 0: \qquad -\tfrac{1}{2}W\,\delta y + F_{DH}\,\delta s = 0 \tag{1}$$

The vertical displacement δy of the platform is expressed in terms of the angular displacement $\delta\theta$ of EDB as follows:

$$y = (EB)\sin\theta = 2a\sin\theta$$
$$\delta y = 2a\cos\theta\,\delta\theta$$

To express δs similarly in terms of $\delta\theta$, we first note that by the law of cosines,

$$s^2 = a^2 + L^2 - 2aL\cos\theta$$

Differentiating,

$$2s\,\delta s = -2aL(-\sin\theta)\,\delta\theta$$
$$\delta s = \frac{aL\sin\theta}{s}\,\delta\theta$$

Substituting for δy and δs into (1), we write

$$(-\tfrac{1}{2}W)2a\cos\theta\,\delta\theta + F_{DH}\frac{aL\sin\theta}{s}\,\delta\theta = 0$$

$$F_{DH} = W\frac{s}{L}\cot\theta$$

With the given numerical data, we have

$$W = mg = (1000 \text{ kg})(9.81 \text{ m/s}^2) = 9810 \text{ N} = 9.81 \text{ kN}$$
$$s^2 = a^2 + L^2 - 2aL\cos\theta$$
$$= (0.70)^2 + (3.20)^2 - 2(0.70)(3.20)\cos 60° = 8.49$$
$$s = 2.91 \text{ m}$$

$$F_{DH} = W\frac{s}{L}\cot\theta = (9.81 \text{ kN})\frac{2.91 \text{ m}}{3.20 \text{ m}}\cot 60°$$

$$F_{DH} = 5.15 \text{ kN} \quad \blacktriangleleft$$

SOLVING PROBLEMS
ON YOUR OWN

In this lesson you learned to use the *method of virtual work*, which is a different way of solving problems involving the equilibrium of rigid bodies.

The work done by a force during a displacement of its point of application or by a couple during a rotation is found by using Eqs. (10.1) and (10.2), respectively:

$$dU = F \, ds \cos \alpha \qquad (10.1)$$
$$dU = M \, d\theta \qquad (10.2)$$

Principle of virtual work. In its more general and more useful form, this principle can be stated as follows: *If a system of connected rigid bodies is in equilibrium, the total virtual work of the external forces applied to the system is zero for any virtual displacement of the system.*

As you apply the principle of virtual work, keep in mind the following:

1. Virtual displacement. A machine or mechanism in equilibrium has no tendency to move. However, *we can cause, or imagine, a small displacement.* Since it does not actually occur, such a displacement is called a *virtual displacement*.

2. Virtual work. The work done by a force or couple during a virtual displacement is called *virtual work*.

3. You need consider only the forces which do work during the virtual displacement.

4. Forces which do no work during a virtual displacement that is consistent with the constraints imposed on the system are:
   a. Reactions at supports
   b. Internal forces at connections
   c. Forces exerted by inextensible cords and cables
None of these forces need be considered when you use the method of virtual work.

5. Be sure to express the various virtual displacements involved in your computations in terms of a *single virtual displacement*. This is done in each of the three preceding sample problems, where the virtual displacements are all expressed in terms of $\delta\theta$.

6. Remember that the method of virtual work is effective only in those cases where the geometry of the system makes it relatively easy to relate the displacements involved.

PROBLEMS

10.1 and 10.2   Determine the vertical force **P** that must be applied at *G* to maintain the equilibrium of the linkage.

10.3 and 10.4   Determine the couple **M** that must be applied to member *DEFG* to maintain the equilibrium of the linkage.

Fig. P10.1 and P10.3

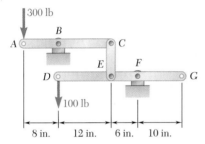

Fig. P10.2 and P10.4

10.5 Determine the force **P** required to maintain the equilibrium of the linkage shown. All members are of the same length and the wheels at *A* and *B* roll freely on the horizontal rod.

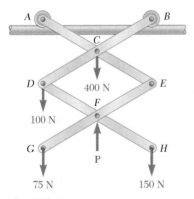

Fig. P10.5

10.6 Solve Prob. 10.5 assuming that the vertical force **P** is applied at point *E*.

10.7 The two-bar linkage shown is supported by a pin and bracket at *B* and a collar at *D* that slides freely on a vertical rod. Determine the force **P** required to maintain the equilibrium of the linkage.

Fig. P10.7

Fig. P10.8

10.8 Knowing that the maximum friction force exerted by the bottle on the cork is 60 lb, determine (a) the force **P** that must be applied to the corkscrew to open the bottle, (b) the maximum force exerted by the base of the corkscrew on the top of the bottle.

10.9 Rod *AD* is acted upon by a vertical force **P** at end *A* and by two equal and opposite horizontal forces of magnitude *Q* at points *B* and *C*. Derive an expression for the magnitude *Q* of the horizontal forces required for equilibrium.

Fig. P10.9

10.10 and 10.11 The slender rod *AB* is attached to a collar *A* and rests on a small wheel at *C*. Neglecting the radius of the wheel and the effect of friction, derive an expression for the magnitude of the force **Q** required to maintain the equilibrium of the rod.

Fig. P10.10   **Fig. P10.11**

10.12 Knowing that the line of action of the force **Q** passes through point *C*, derive an expression for the magnitude of **Q** required to maintain equilibrium.

10.13 Solve Prob. 10.12 assuming that the force **P** applied at point *A* acts horizontally to the left.

Fig. P10.12

10.14 The mechanism shown is acted upon by the force **P**; derive an expression for the magnitude of the force **Q** required to maintain equilibrium.

10.15 and 10.16 Derive an expression for the magnitude of the couple **M** required to maintain the equilibrium of the linkage shown.

Fig. P10.14

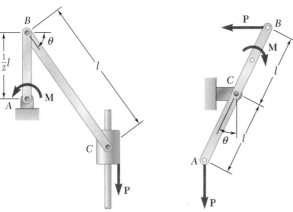

Fig. P10.15

Fig. P10.16

10.17 A uniform rod AB of length l and weight W is suspended from two cords AC and BC of equal length. Derive an expression for the magnitude of the couple **M** required to maintain equilibrium of the rod in the position shown.

10.18 Collar B can slide along rod AC and is attached by a pin to a block that can slide in the vertical slot shown. Derive an expression for the magnitude of the couple **M** required to maintain equilibrium.

Fig. P10.17

Fig. P10.18

10.19 For the linkage shown, determine the couple **M** required for equilibrium when $l = 1.8$ ft, $Q = 40$ lb, and $\theta = 65°$.

10.20 For the linkage shown, determine the force **Q** required for equilibrium when $l = 18$ in., $M = 600$ lb · in., and $\theta = 70°$.

Fig. P10.19 and P10.20

Fig. P10.21 and P10.22

10.21 A 4-kN force **P** is applied as shown to the piston of the engine system. Knowing that $AB = 50$ mm and $BC = 200$ mm, determine the couple **M** required to maintain the equilibrium of the system when (a) $\theta = 30°$, (b) $\theta = 150°$.

10.22 A couple **M** of magnitude 100 N · m is applied as shown to the crank of the engine system. Knowing that $AB = 50$ mm and $BC = 200$ mm, determine the force **P** required to maintain the equilibrium of the system when (a) $\theta = 60°$, (b) $\theta = 120°$.

10.23 A slender rod of length l is attached to a collar at B and rests on a portion of a circular cylinder of radius r. Neglecting the effect of friction, determine the value of θ corresponding to the equilibrium position of the mechanism when $l = 200$ mm, $r = 60$ mm, $P = 40$ N, and $Q = 80$ N.

Fig. P10.23 and P10.24

10.24 A slender rod of length l is attached to a collar at B and rests on a portion of a circular cylinder of radius r. Neglecting the effect of friction, determine the value of θ corresponding to the equilibrium position of the mechanism when $l = 14$ in., $r = 5$ in., $P = 75$ lb, and $Q = 150$ lb.

10.25 Determine the value of θ corresponding to the equilibrium position of the rod of Prob. 10.10 when $l = 30$ in., $a = 5$ in., $P = 25$ lb, and $Q = 40$ lb.

10.26 Determine the values of θ corresponding to the equilibrium position of the rod of Prob. 10.11 when $l = 600$ mm, $a = 100$ mm, $P = 50$ N, and $Q = 90$ N.

10.27 Determine the value of θ corresponding to the equilibrium position of the mechanism of Prob. 10.12 when $P = 80$ N and $Q = 100$ N.

10.28 Determine the value of θ corresponding to the equilibrium position of the mechanism of Prob. 10.14 when $P = 270$ N and $Q = 960$ N.

10.29 A load **W** of magnitude 600 N is applied to the linkage at B. The constant of the spring is $k = 2.5$ kN/m, and the spring is unstretched when AB and BC are horizontal. Neglecting the weight of the linkage and knowing that $l = 300$ mm, determine the value of θ corresponding to equilibrium.

10.30 A vertical load **W** is applied to the linkage at B. The constant of the spring is k, and the spring is unstretched when AB and BC are horizontal. Neglecting the weight of the linkage, derive an equation in θ, W, l, and k that must be satisfied when the linkage is in equilibrium.

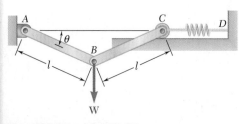

Fig. P10.29 and P10.30

10.31 Two bars AD and DG are connected by a pin at D and by a spring AG. Knowing that the spring is 12 in. long when unstretched and that the constant of the spring is 125 lb/in., determine the value of x corresponding to equilibrium when a 900-lb load is applied at E as shown.

10.32 Solve Prob. 10.31 assuming that the 900-lb vertical force is applied at C instead of E.

10.33 Two 5-kg bars AB and BC are connected by a pin at B and by a spring DE. Knowing that the spring is 150 mm long when unstretched and that the constant of the spring is 1 kN/m, determine the value of x corresponding to equilibrium.

Fig. P10.31

Fig. P10.33

10.34 Rod ABC is attached to blocks A and B that can move freely in the guides shown. The constant of the spring attached at A is $k = 3$ kN/m, and the spring is unstretched when the rod is vertical. For the loading shown, determine the value of θ corresponding to equilibrium.

Fig. P10.34

10.35 A vertical force **P** of magnitude 150 N is applied to end E of cable CDE, which passes over a small pulley D and is attached to the mechanism at C. The constant of the spring is $k = 4$ kN/m, and the spring is unstretched when $\theta = 0$. Neglecting the weight of the mechanism and the radius of the pulley, determine the value of θ corresponding to equilibrium.

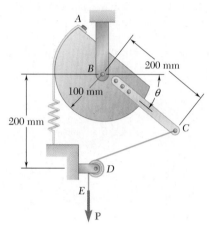

Fig. P10.35

10.36 A horizontal force **P** of magnitude 40 lb is applied to the mechanism at C. The constant of the spring is $k = 9$ lb/in., and the spring is unstretched when $\theta = 0$. Neglecting the weight of the mechanism, determine the value of θ corresponding to equilibrium.

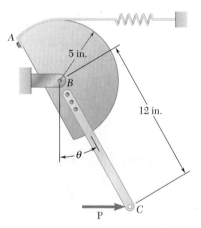

Fig. P10.36

10.37 and 10.38 Knowing that the constant of spring CD is k and that the spring is unstretched when rod ABC is horizontal, determine the value of θ corresponding to equilibrium for the data indicated.

10.37 $P = 300$ N, $l = 400$ mm, $k = 5$ kN/m
10.38 $P = 75$ lb, $l = 15$ in., $k = 20$ lb/in.

Fig. P10.37 and P10.38

10.39 The lever AB is attached to the horizontal shaft BC that passes through a bearing and is welded to a fixed support at C. The torsional spring constant of the shaft BC is K; that is, a couple of magnitude K is required to rotate end B through 1 rad. Knowing that the shaft is untwisted when AB is horizontal, determine the value of θ corresponding to the position of equilibrium when $P = 100$ N, $l = 250$ mm, and $K = 12.5$ N $\cdot$ m/rad.

10.40 Solve Prob. 10.39 assuming that $P = 350$ N, $l = 250$ mm, and $K = 12.5$ N $\cdot$ m/rad. Obtain answers in each of the following quadrants: $0 < \theta < 90°$, $270° < \theta < 360°$, $360° < \theta < 450°$.

Fig. P10.39

10.41 The position of boom ABC is controlled by the hydraulic cylinder BD. For the loading shown, determine the force exerted by the hydraulic cylinder on pin B when $\theta = 65°$.

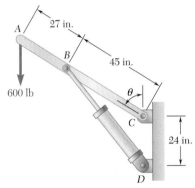

Fig. P10.41 and P10.42

10.42 The position of boom ABC is controlled by the hydraulic cylinder BD. For the loading shown, (a) express the force exerted by the hydraulic cylinder on pin B as a function of the length BD, (b) determine the smallest possible value of the angle θ if the maximum force that the cylinder can exert on pin B is 2.5 kips.

10.43 The position of member ABC is controlled by the hydraulic cylinder CD. For the loading shown, determine the force exerted by the hydraulic cylinder on pin C when $\theta = 55°$.

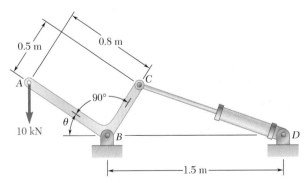

Fig. P10.43 and P10.44

10.44 The position of member ABC is controlled by the hydraulic cylinder CD. Determine the angle θ knowing that the hydraulic cylinder exerts a 15-kN force on pin C.

10.45 The telescoping arm ABC is used to provide an elevated platform for construction workers. The workers and the platform together weigh 500 lb and their combined center of gravity is located directly above C. For the position when $\theta = 20°$, determine the force exerted on pin B by the single hydraulic cylinder BD.

Fig. P10.45

10.46 Solve Prob. 10.45 assuming that the workers are lowered to a point near the ground so that $\theta = -20°$.

10.47 Denoting by μ_s the coefficient of static friction between collar C and the vertical rod, derive an expression for the magnitude of the largest couple $\mathbf{M}$ for which equilibrium is maintained in the position shown. Explain what happens if $\mu_s \geq \tan \theta$.

Fig. P10.47 and P10.48

10.48 Knowing that the coefficient of static friction between collar C and the vertical rod is 0.40, determine the magnitude of the largest and smallest couple $\mathbf{M}$ for which equilibrium is maintained in the position shown, when $\theta = 35°$, $l = 600$ mm, and $P = 300$ N.

10.49 A block of weight W is pulled up a plane forming an angle α with the horizontal by a force $\mathbf{P}$ directed along the plane. If μ is the coefficient of friction between the block and the plane, derive an expression for the mechanical efficiency of the system. Show that the mechanical efficiency cannot exceed $\frac{1}{2}$ if the block is to remain in place when the force $\mathbf{P}$ is removed.

10.50 Derive an expression for the mechanical efficiency of the jack discussed in Sec. 8.6. Show that if the jack is to be self-locking, the mechanical efficiency cannot exceed $\frac{1}{2}$.

10.51 Denoting by μ_s the coefficient of static friction between the block attached to rod *ACE* and the horizontal surface, derive expressions in terms of *P*, μ_s, and θ for the largest and smallest magnitude of the force **Q** for which equilibrium is maintained.

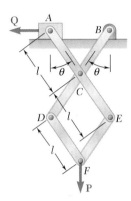

Fig. P10.51 and P10.52

10.52 Knowing that the coefficient of static friction between the block attached to rod *ACE* and the horizontal surface is 0.15, determine the magnitude of the largest and smallest force **Q** for which equilibrium is maintained when $\theta = 30°$, $l = 0.2$ m, and $P = 40$ N.

10.53 Using the method of virtual work, determine the reaction at *E*.

10.54 Using the method of virtual work, determine separately the force and couple representing the reaction at *H*.

Fig. P10.53 and P10.54

10.55 Referring to Prob. 10.43 and using the value found for the force exerted by the hydraulic cylinder *CD*, determine the change in the length of *CD* required to raise the 10-kN load by 15 mm.

10.56 Referring to Prob. 10.45 and using the value found for the force exerted by the hydraulic cylinder *BD*, determine the change in the length of *BD* required to raise the platform attached at *C* by 2.5 in.

10.57 Determine the vertical movement of joint *D* if the length of member *BF* is increased by 1.5 in. (*Hint:* Apply a vertical load at joint *D*, and, using the methods of Chap. 6, compute the force exerted by member *BF* on joints *B* and *F*. Then apply the method of virtual work for a virtual displacement resulting in the specified increase in length of member *BF*. This method should be used only for small changes in the lengths of members.)

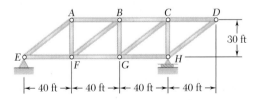

Fig. P10.57 and P10.58

10.58 Determine the horizontal movement of joint *D* if the length of member *BF* is increased by 1.5 in. (See the hint for Prob. 10.57.)

*10.6 WORK OF A FORCE DURING A FINITE DISPLACEMENT

Consider a force $\mathbf{F}$ acting on a particle. The work of $\mathbf{F}$ corresponding to an infinitesimal displacement $d\mathbf{r}$ of the particle was defined in Sec. 10.2 as

$$dU = \mathbf{F} \cdot d\mathbf{r} \qquad (10.1)$$

The work of $\mathbf{F}$ corresponding to a finite displacement of the particle from A_1 to A_2 (Fig. 10.10a) is denoted by $U_{1 \to 2}$ and is obtained by integrating (10.1) along the curve described by the particle:

$$U_{1 \to 2} = \int_{A_1}^{A_2} \mathbf{F} \cdot d\mathbf{r} \qquad (10.11)$$

Using the alternative expression

$$dU = F \, ds \, \cos \alpha \qquad (10.1')$$

given in Sec. 10.2 for the elementary work dU, we can also express the work $U_{1 \to 2}$ as

$$U_{1 \to 2} = \int_{s_1}^{s_2} (F \cos \alpha) \, ds \qquad (10.11')$$

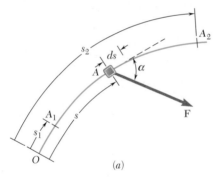

(a)

(b)

Fig. 10.10

where the variable of integration s measures the distance along the path traveled by the particle. The work $U_{1 \to 2}$ is represented by the area under the curve obtained by plotting $F \cos \alpha$ against s (Fig. 10.10b). In the case of a force $\mathbf{F}$ of constant magnitude acting in the direction of motion, formula (10.11') yields $U_{1 \to 2} = F(s_2 - s_1)$.

Recalling from Sec. 10.2 that the work of a couple of moment $\mathbf{M}$ during an infinitesimal rotation $d\theta$ of a rigid body is

$$dU = M \, d\theta \qquad (10.2)$$

we express as follows the work of the couple during a finite rotation of the body:

$$U_{1 \to 2} = \int_{\theta_1}^{\theta_2} M \, d\theta \qquad (10.12)$$

In the case of a constant couple, formula (10.12) yields

$$U_{1 \to 2} = M(\theta_2 - \theta_1)$$

Work of a Weight. It was stated in Sec. 10.2 that the work of the weight **W** of a body during an infinitesimal displacement of the body is equal to the product of W and the vertical displacement of the center of gravity of the body. With the y axis pointing upward, the work of **W** during a finite displacement of the body (Fig. 10.11) is obtained by writing

$$dU = -W \, dy$$

Integrating from A_1 to A_2, we have

$$U_{1 \to 2} = -\int_{y_1}^{y_2} W \, dy = Wy_1 - Wy_2 \qquad (10.13)$$

or

$$U_{1 \to 2} = -W(y_2 - y_1) = -W \, \Delta y \qquad (10.13')$$

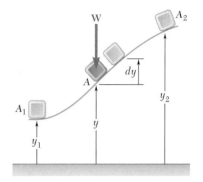

Fig. 10.11

where Δy is the vertical displacement from A_1 to A_2. The work of the weight **W** is thus equal to *the product of W and the vertical displacement of the center of gravity of the body.* The work is *positive* when $\Delta y < 0$, that is, *when the body moves down.*

Work of the Force Exerted by a Spring. Consider a body A attached to a fixed point B by a spring; it is assumed that the spring is undeformed when the body is at A_0 (Fig. 10.12a). Experimental evidence shows that the magnitude of the force **F** exerted by the spring on a body A is proportional to the deflection x of the spring measured from the position A_0. We have

$$F = kx \qquad (10.14)$$

where k is the *spring constant,* expressed in N/m if SI units are used and expressed in lb/ft or lb/in. if U.S. customary units are used. The work of the force **F** exerted by the spring during a finite displacement of the body from $A_1(x = x_1)$ to $A_2(x = x_2)$ is obtained by writing

$$dU = -F \, dx = -kx \, dx$$

$$U_{1 \to 2} = -\int_{x_1}^{x_2} kx \, dx = \tfrac{1}{2}kx_1^2 - \tfrac{1}{2}kx_2^2 \qquad (10.15)$$

Care should be taken to express k and x in consistent units. For example, if U.S. customary units are used, k should be expressed in lb/ft and x expressed in feet, or k in lb/in. and x in inches; in the first case, the work is obtained in ft · lb; in the second case, in in · lb. We note that the work of the force **F** exerted by the spring on the body is *positive* when $x_2 < x_1$, that is, *when the spring is returning to its undeformed position.*

Since Eq. (10.14) is the equation of a straight line of slope k passing through the origin, the work $U_{1 \to 2}$ of **F** during the displacement from A_1 to A_2 can be obtained by evaluating the area of the trapezoid shown in Fig. 10.12b. This is done by computing the values F_1 and F_2 and multiplying the base Δx of the trapezoid by its mean height $\tfrac{1}{2}(F_1 + F_2)$. Since the work of the force **F** exerted by the spring is positive for a negative value of Δx, we write

$$U_{1 \to 2} = -\tfrac{1}{2}(F_1 + F_2) \, \Delta x \qquad (10.16)$$

Formula (10.16) is usually more convenient to use than (10.15) and affords fewer chances of confusing the units involved.

Fig. 10.12

Fig. 10.11 (repeated)

Fig. 10.12a (repeated)

*10.7 POTENTIAL ENERGY

Considering again the body of Fig. 10.11, we note from Eq. (10.13) that the work of the weight **W** during a finite displacement is obtained by subtracting the value of the function Wy corresponding to the second position of the body from its value corresponding to the first position. The work of **W** is thus independent of the actual path followed; it depends only upon the initial and final values of the function Wy. This function is called the *potential energy* of the body with respect to the *force of gravity* **W** and is denoted by V_g. We write

$$U_{1 \to 2} = (V_g)_1 - (V_g)_2 \quad \text{with } V_g = Wy \qquad (10.17)$$

We note that if $(V_g)_2 > (V_g)_1$, that is, *if the potential energy increases* during the displacement (as in the case considered here), *the work $U_{1 \to 2}$ is negative.* If, on the other hand, the work of **W** is positive, the potential energy decreases. Therefore, the potential energy V_g of the body provides a measure of *the work which can be done* by its weight **W**. Since only the *change* in potential energy, and not the actual value of V_g, is involved in formula (10.17), an arbitrary constant can be added to the expression obtained for V_g. In other words, the level from which the elevation y is measured can be chosen arbitrarily. Note that potential energy is expressed in the same units as work, i.e., in joules (J) if SI units are used† and in ft · lb or in · lb if U.S. customary units are used.

Considering now the body of Fig. 10.12a, we note from Eq. (10.15) that the work of the elastic force **F** is obtained by subtracting the value of the function $\frac{1}{2}kx^2$ corresponding to the second position of the body from its value corresponding to the first position. This function is denoted by V_e and is called the *potential energy* of the body with respect to the *elastic force* **F**. We write

$$U_{1 \to 2} = (V_e)_1 - (V_e)_2 \quad \text{with } V_e = \tfrac{1}{2}kx^2 \qquad (10.18)$$

and observe that during the displacement considered, the work of the force **F** exerted by the spring on the body is negative and the potential energy V_e increases. We should note that the expression obtained for V_e is valid only if the deflection of the spring is measured from its undeformed position.

The concept of potential energy can be used when forces other than gravity forces and elastic forces are involved. It remains valid as long as the elementary work dU of the force considered is an *exact differential.* It is then possible to find a function V, called potential energy, such that

$$dU = -dV \qquad (10.19)$$

Integrating (10.19) over a finite displacement, we obtain the general formula

$$U_{1 \to 2} = V_1 - V_2 \qquad (10.20)$$

which expresses that *the work of the force is independent of the path followed and is equal to minus the change in potential energy.* A force which satisfies Eq. (10.20) is said to be a *conservative force.*‡

†See footnote, page 559.

‡A detailed discussion of conservative forces is given in Sec. 13.7 of *Dynamics.*

*10.8 POTENTIAL ENERGY AND EQUILIBRIUM

The application of the principle of virtual work is considerably simplified when the potential energy of a system is known. In the case of a virtual displacement, formula (10.19) becomes $\delta U = -\delta V$. Moreover, if the position of the system is defined by a single independent variable θ, we can write $\delta V = (dV/d\theta)\,\delta\theta$. Since $\delta\theta$ must be different from zero, the condition $\delta U = 0$ for the equilibrium of the system becomes

$$\frac{dV}{d\theta} = 0 \qquad (10.21)$$

In terms of potential energy, therefore, the principle of virtual work states that *if a system is in equilibrium, the derivative of its total potential energy is zero.* If the position of the system depends upon several independent variables (the system is then said to possess *several degrees of freedom*), the partial derivatives of V with respect to each of the independent variables must be zero.

Consider, for example, a structure made of two members AC and CB and carrying a load W at C. The structure is supported by a pin at A and a roller at B, and a spring BD connects B to a fixed point D (Fig. 10.13a). The constant of the spring is k, and it is assumed that the natural length of the spring is equal to AD and thus that the spring is undeformed when B coincides with A. Neglecting the friction forces and the weight of the members, we find that the only forces which work during a displacement of the structure are the weight $\mathbf{W}$ and the force $\mathbf{F}$ exerted by the spring at point B (Fig. 10.13b). The total potential energy of the system will thus be obtained by adding the potential energy V_g corresponding to the gravity force $\mathbf{W}$ and the potential energy V_e corresponding to the elastic force $\mathbf{F}$.

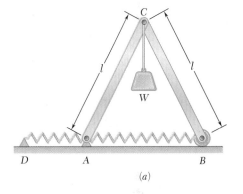

(a)

Choosing a coordinate system with origin at A and noting that the deflection of the spring, measured from its undeformed position, is $AB = x_B$, we write

$$V_e = \tfrac{1}{2}kx_B^2 \qquad V_g = Wy_C$$

Expressing the coordinates x_B and y_C in terms of the angle θ, we have

$$
\begin{aligned}
x_B &= 2l\sin\theta & y_C &= l\cos\theta \\
V_e &= \tfrac{1}{2}k(2l\sin\theta)^2 & V_g &= W(l\cos\theta) \\
V &= V_e + V_g = 2kl^2\sin^2\theta + Wl\cos\theta & & (10.22)
\end{aligned}
$$

The positions of equilibrium of the system are obtained by equating to zero the derivative of the potential energy V. We write

$$\frac{dV}{d\theta} = 4kl^2\sin\theta\cos\theta - Wl\sin\theta = 0$$

or, factoring $l\sin\theta$,

$$\frac{dV}{d\theta} = l\sin\theta(4kl\cos\theta - W) = 0$$

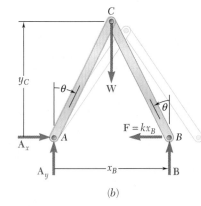

(b)

Fig. 10.13

There are therefore two positions of equilibrium, corresponding to the values $\theta = 0$ and $\theta = \cos^{-1}(W/4kl)$, respectively.[†]

[†]The second position does not exist if $W > 4kl$.

*10.9 STABILITY OF EQUILIBRIUM

Consider the three uniform rods of length $2a$ and weight **W** shown in Fig. 10.14. While each rod is in equilibrium, there is an important difference between the three cases considered. Suppose that each rod is slightly disturbed from its position of equilibrium and then released: rod a will move back toward its original position, rod b will keep moving away from its original position, and rod c will remain in its new position. In case a, the equilibrium of the rod is said to be *stable;* in case b, it is said to be *unstable;* and, in case c, it is said to be *neutral*.

(*a*) Stable equilibrium

(*b*) Unstable equilibrium

(*c*) Neutral equilibrium

Fig. 10.14

Recalling from Sec. 10.7 that the potential energy V_g with respect to gravity is equal to Wy, where y is the elevation of the point of application of **W** measured from an arbitrary level, we observe that the potential energy of rod a is minimum in the position of equilibrium considered, that the potential energy of rod b is maximum, and that the potential energy of rod c is constant. Equilibrium is thus *stable, unstable,* or *neutral* according to whether the potential energy is *minimum, maximum,* or *constant* (Fig. 10.15).

That the result obtained is quite general can be seen as follows: We first observe that a force always tends to do positive work and thus to decrease the potential energy of the system on which it is applied. Therefore, when a system is disturbed from its position of equilibrium, the forces acting on the system will tend to bring it back to its original position if V is minimum (Fig. 10.15a) and to move it farther away if V is maximum (Fig. 10.15b). If V is constant (Fig. 10.15c), the forces will not tend to move the system either way.

Recalling from calculus that a function is minimum or maximum according to whether its second derivative is positive or negative, we can summarize the conditions for the equilibrium of a system with one degree of freedom (i.e., a system the position of which is defined by a single independent variable θ) as follows:

$$\frac{dV}{d\theta} = 0 \qquad \frac{d^2V}{d\theta^2} > 0: \text{stable equilibrium}$$

$$\frac{dV}{d\theta} = 0 \qquad \frac{d^2V}{d\theta^2} < 0: \text{unstable equilibrium}$$
(10.23)

(a) Stable equilibrium $\qquad$ (b) Unstable equilibrium $\qquad$ (c) Neutral equilibrium

Fig. 10.15

If both the first and the second derivatives of V are zero, it is necessary to examine derivatives of a higher order to determine whether the equilibrium is stable, unstable, or neutral. The equilibrium will be neutral if all derivatives are zero, since the potential energy V is then a constant. The equilibrium will be stable if the first derivative found to be different from zero is of even order and positive. In all other cases the equilibrium will be unstable.

If the system considered possesses *several degrees of freedom*, the potential energy V depends upon several variables, and it is thus necessary to apply the theory of functions of several variables to determine whether V is minimum. It can be verified that a system with 2 degrees of freedom will be stable, and the corresponding potential energy $V(\theta_1, \theta_2)$ will be minimum, if the following relations are satisfied simultaneously:

$$\frac{\partial V}{\partial \theta_1} = \frac{\partial V}{\partial \theta_2} = 0$$

$$\left(\frac{\partial^2 V}{\partial \theta_1 \partial \theta_2}\right)^2 - \frac{\partial^2 V}{\partial \theta_1^2}\frac{\partial^2 V}{\partial \theta_2^2} < 0$$
(10.24)

$$\frac{\partial^2 V}{\partial \theta_1^2} > 0 \qquad \text{or} \qquad \frac{\partial^2 V}{\partial \theta_2^2} > 0$$

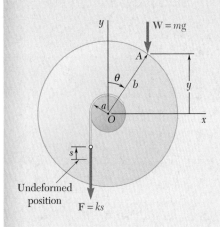

SAMPLE PROBLEM 10.4

A 10-kg block is attached to the rim of a 300-mm-radius disk as shown. Knowing that spring BC is unstretched when $\theta = 0$, determine the position or positions of equilibrium, and state in each case whether the equilibrium is stable, unstable, or neutral.

SOLUTION

Potential Energy. Denoting by s the deflection of the spring from its undeformed position and placing the origin of coordinates at O, we obtain

$$V_e = \tfrac{1}{2}ks^2 \qquad V_g = Wy = mgy$$

Measuring θ in radians, we have

$$s = a\theta \qquad y = b \cos \theta$$

Substituting for s and y in the expressions for V_e and V_g, we write

$$V_e = \tfrac{1}{2}ka^2\theta^2 \qquad V_g = mgb \cos \theta$$
$$V = V_e + V_g = \tfrac{1}{2}ka^2\theta^2 + mgb \cos \theta$$

Positions of Equilibrium. Setting $dV/d\theta = 0$, we write

$$\frac{dV}{d\theta} = ka^2\theta - mgb \sin \theta = 0$$

$$\sin \theta = \frac{ka^2}{mgb}\theta$$

Substituting $a = 0.08$ m, $b = 0.3$ m, $k = 4$ kN/m, and $m = 10$ kg, we obtain

$$\sin \theta = \frac{(4 \text{ kN/m})(0.08 \text{ m})^2}{(10 \text{ kg})(9.81 \text{ m/s}^2)(0.3 \text{ m})}\theta$$

$$\sin \theta = 0.8699\,\theta$$

where θ is expressed in radians. Solving by trial and error for θ, we find

$$\theta = 0 \qquad \text{and} \qquad \theta = 0.902 \text{ rad}$$
$$\theta = 0 \qquad \text{and} \qquad \theta = 51.7° \qquad \blacktriangleleft$$

Stability of Equilibrium. The second derivative of the potential energy V with respect to θ is

$$\frac{d^2V}{d\theta^2} = ka^2 - mgb \cos \theta$$
$$= (4 \text{ kN/m})(0.08 \text{ m})^2 - (10 \text{ kg})(9.81 \text{ m/s}^2)(0.3 \text{ m}) \cos \theta$$
$$= 25.6 - 29.43 \cos \theta$$

For $\theta = 0$: $\qquad \dfrac{d^2V}{d\theta^2} = 25.6 - 29.43 \cos 0° = -3.83 < 0$

The equilibrium is unstable for $\theta = 0$ $\qquad \blacktriangleleft$

For $\theta = 51.7°$: $\qquad \dfrac{d^2V}{d\theta^2} = 25.6 - 29.43 \cos 51.7° = +7.36 > 0$

The equilibrium is stable for $\theta = 51.7°$ $\qquad \blacktriangleleft$

SOLVING PROBLEMS
ON YOUR OWN

In this lesson we defined the *work of a force during a finite displacement* and the *potential energy* of a rigid body or a system of rigid bodies. You learned to use the concept of potential energy to determine the *equilibrium position* of a rigid body or a system of rigid bodies.

1. The potential energy V of a system is the sum of the potential energies associated with the various forces acting on the system that *do work* as the system moves. In the problems of this lesson you will determine the following:

a. Potential energy of a weight. This is the potential energy due to *gravity*, $V_g = Wy$, where y is the elevation of the weight W measured from some arbitrary reference level. Note that the potential energy V_g may be used with any vertical force **P** of constant magnitude directed downward; we write $V_g = Py$.

b. Potential energy of a spring. This is the potential energy due to the *elastic* force exerted by a spring, $V_e = \frac{1}{2}kx^2$, where k is the constant of the spring and x is the deformation of the spring *measured from its unstretched position*.

Reactions at fixed supports, internal forces at connections, forces exerted by inextensible cords and cables, and other forces which do no work do not contribute to the potential energy of the system.

2. Express all distances and angles in terms of a single variable, such as an angle θ, when computing the potential energy V of a system. This is necessary, since the determination of the equilibrium position of the system requires the computation of the derivative $dV/d\theta$.

3. When a system is in equilibrium, the first derivative of its potential energy is zero. Therefore:

a. To determine a position of equilibrium of a system, once its potential energy V has been expressed in terms of the single variable θ, compute its derivative and solve the equation $dV/d\theta = 0$ for θ.

b. To determine the force or couple required to maintain a system in a given position of equilibrium, substitute the known value of θ in the equation $dV/d\theta = 0$ and solve this equation for the desired force or couple.

4. Stability of equilibrium. The following rules generally apply:

a. Stable equilibrium occurs when the potential energy of the system is *minimum*, that is, when $dV/d\theta = 0$ and $d^2V/d\theta^2 > 0$ (Figs. 10.14*a* and 10.15*a*).

b. Unstable equilibrium occurs when the potential energy of the system is *maximum*, that is, when $dV/d\theta = 0$ and $d^2V/d\theta^2 < 0$ (Figs. 10.14*b* and 10.15*b*).

c. Neutral equilibrium occurs when the potential energy of the system is *constant; $dV/d\theta$, $d^2V/d\theta^2$*, and all the successive derivatives of V are then equal to zero (Figs. 10.14*c* and 10.15*c*).

See page 583 for a discussion of the case when $dV/d\theta$, $d^2V/d\theta^2$ but *not all* of the successive derivatives of V are equal to zero.

PROBLEMS

10.59 Using the method of Sec. 10.8, solve Prob. 10.29.

10.60 Using the method of Sec. 10.8, solve Prob. 10.30.

10.61 Using the method of Sec. 10.8, solve Prob. 10.31.

10.62 Using the method of Sec. 10.8, solve Prob. 10.32.

10.63 Using the method of Sec. 10.8, solve Prob. 10.33.

10.64 Using the method of Sec. 10.8, solve Prob. 10.35.

10.65 Using the method of Sec. 10.8, solve Prob. 10.37.

10.66 Using the method of Sec. 10.8, solve Prob. 10.38.

10.67 Show that equilibrium is neutral in Prob. 10.1.

10.68 Show that equilibrium is neutral in Prob. 10.7.

10.69 Two uniform rods, each of mass m, are attached to gears of equal radii as shown. Determine the positions of equilibrium of the system and state in each case whether the equilibrium is stable, unstable, or neutral.

10.70 Two uniform rods, AB and CD, are attached to gears of equal radii as shown. Knowing that $W_{AB} = 8$ lb and $W_{CD} = 4$ lb, determine the positions of equilibrium of the system and state in each case whether the equilibrium is stable, unstable, or neutral.

10.71 Two uniform rods, each of mass m and length l, are attached to gears as shown. For the range $0 \leq \theta \leq 180°$, determine the positions of equilibrium of the system and state in each case whether the equilibrium is stable, unstable, or neutral.

Fig. P10.69 and P10.70

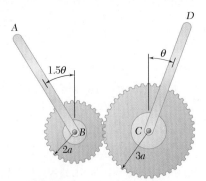

Fig. P10.71

10.72 Two uniform rods, each of mass m and length l, are attached to drums that are connected by a belt as shown. Assuming that no slipping occurs between the belt and the drums, determine the positions of equilibrium of the system and state in each case whether the equilibrium is stable, unstable, or neutral.

10.73 Using the method of Sec. 10.8, solve Prob. 10.39. Determine whether the equilibrium is stable, unstable, or neutral. (*Hint:* The potential energy corresponding to the couple exerted by a torsion spring is $\frac{1}{2}K\theta^2$, where K is the torsional spring constant and θ is the angle of twist.)

10.74 In Prob. 10.40, determine whether each of the positions of equilibrium is stable, unstable, or neutral. (See hint for Prob. 10.73.)

10.75 A load **W** of magnitude 100 lb is applied to the mechanism at C. Knowing that the spring is unstretched when $\theta = 15°$, determine that value of θ corresponding to equilibrium and check that the equilibrium is stable.

Fig. P10.72

Fig. *P10.75* and *P10.76*

10.76 A load **W** of magnitude 100 lb is applied to the mechanism at C. Knowing that the spring is unstretched when $\theta = 30°$, determine that value of θ corresponding to equilibrium and check that the equilibrium is stable.

10.77 A slender rod AB, of weight W, is attached to two blocks A and B that can move freely in the guides shown. Knowing that the spring is unstretched when $y = 0$, determine the value of y corresponding to equilibrium when $W = 80$ N, $l = 500$ mm, and $k = 600$ N/m.

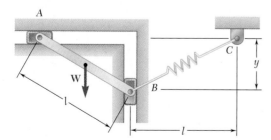

Fig. P10.77

10.78 A slender rod AB, of weight W, is attached to two blocks A and B that can move freely in the guides shown. Knowing that both springs are unstretched when $y = 0$, determine the value of y corresponding to equilibrium when $W = 80$ N, $l = 500$ mm, and $k = 600$ N/m.

Fig. P10.78

10.79 A slender rod AB, of weight W, is attached to two blocks A and B that can move freely in the guides shown. The constant of the spring is k, and the spring is unstretched when AB is horizontal. Neglecting the weight of the blocks, derive an equation in θ, W, l, and k that must be satisfied when the rod is in equilibrium.

Fig. *P10.79* and P10.80

10.80 A slender rod AB, of weight W, is attached to two blocks A and B that can move freely in the guides shown. Knowing that the spring is unstretched when AB is horizontal, determine three values of θ corresponding to equilibrium when $W = 300$ lb, $l = 16$ in., and $k = 75$ lb/in. State in each case whether the equilibrium is stable, unstable, or neutral.

10.81 A spring AB of constant k is attached to two identical gears as shown. Knowing that the spring is undeformed when $\theta = 0$, determine two values of the angle θ corresponding to equilibrium when $P = 30$ lb, $a = 4$ in., $b = 3$ in., $r = 6$ in., and $k = 5$ lb/in. State in each case whether the equilibrium is stable, unstable, or neutral.

10.82 A spring AB of constant k is attached to two identical gears as shown. Knowing that the spring is undeformed when $\theta = 0$, and given that $a = 60$ mm, $b = 45$ mm, $r = 90$ mm, and $k = 6$ kN/m, determine (a) the range of values of P for which a position of equilibrium exists, (b) two values of θ corresponding to equilibrium if the value of P is equal to half the upper limit of the range found in part a.

Fig. P10.81 and *P10.82*

10.83 A slender rod AB is attached to two collars A and B that can move freely along the guide rods shown. Knowing that $\beta = 30°$ and $P = Q = 400$ N, determine the value of the angle θ corresponding to equilibrium.

10.84 A slender rod AB is attached to two collars A and B that can move freely along the guide rods shown. Knowing that $\beta = 30°$, $P = 100$ N, and $Q = 25$ N, determine the value of the angle θ corresponding to equilibrium.

10.85 and 10.86 Cart B, which weighs 75 kN, rolls along a sloping track that forms an angle β with the horizontal. The spring constant is 5 kN/m, and the spring is unstretched when $x = 0$. Determine the distance x corresponding to equilibrium for the angle β indicated.

> **10.85** Angle $\beta = 30°$
> **10.86** Angle $\beta = 60°$

Fig. P10.83 and P10.84

Fig. P10.85 and P10.86

10.87 and 10.88 Collar A can slide freely on the semicircular rod shown. Knowing that the constant of the spring is k and that the unstretched length of the spring is equal to the radius r, determine the value of θ corresponding to equilibrium when $W = 50$ lb, $r = 9$ in., and $k = 15$ lb/in.

Fig. P10.87

Fig. P10.88

10.89 Two bars AB and BC of negligible weight are attached to a single spring of constant k that is unstretched when the bars are horizontal. Determine the range of values of the magnitude P of two equal and opposite forces **P** and $-$**P** for which the equilibrium of the system is stable in the position shown.

Fig. P10.89

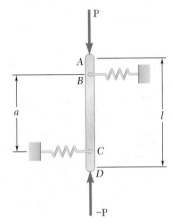

10.90 A vertical bar AD is attached to two springs of constant k and is in equilibrium in the position shown. Determine the range of values of the magnitude P of two equal and opposite vertical forces **P** and $-$**P** for which the equilibrium position is stable if (a) $AB = CD$, (b) $AB = 2CD$.

Fig. P10.90

10.91 Rod AB is attached to a hinge at A and to two springs, each of constant k. If $h = 25$ in., $d = 12$ in., and $W = 80$ lb, determine the range of values of k for which the equilibrium of the rod is stable in the position shown. Each spring can act in either tension or compression.

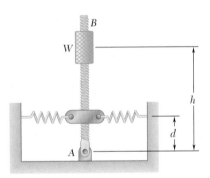

Fig. P10.91 and P10.92

10.92 Rod AB is attached to a hinge at A and to two springs, each of constant k. If $h = 45$ in., $k = 6$ lb/in., and $W = 60$ lb, determine the smallest distance d for which the equilibrium of the rod is stable in the position shown. Each spring can act in either tension or compression.

10.93 and 10.94 Two bars are attached to a single spring of constant k that is unstretched when the bars are vertical. Determine the range of values of P for which the equilibrium of the system is stable in the position shown.

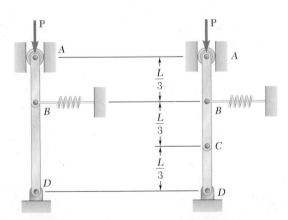

Fig. P10.93   **Fig. P10.94**

10.95 The horizontal bar *BEH* is connected to three vertical bars. The collar at *E* can slide freely on bar *DF*. Determine the range of values of *Q* for which the equilibrium of the system is stable in the position shown when $a = 24$ in., $b = 20$ in., and $P = 150$ lb.

10.96 The horizontal bar *BEH* is connected to three vertical bars. The collar at *E* can slide freely on bar *DF*. Determine the range of values of *P* for which the equilibrium of the system is stable in the position shown when $a = 150$ mm, $b = 200$ mm, and $Q = 45$ N.

*10.97 Bars *AB* and *BC*, each of length *l* and of negligible weight, are attached to two springs, each of constant *k*. The springs are undeformed and the system is in equilibrium when $\theta_1 = \theta_2 = 0$. Determine the range of values of *P* for which the equilibrium position is stable.

Fig. P10.95 and P10.96

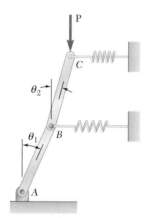

Fig. P10.97

*10.98 Solve Prob. 10.97 knowing that $l = 800$ mm and $k = 2.5$ kN/m.

*10.99 Two rods of negligible weight are attached to drums of radius *r* that are connected by a belt and spring of constant *k*. Knowing that the spring is undeformed when the rods are vertical, determine the range of values of *P* for which the equilibrium position $\theta_1 = \theta_2 = 0$ is stable.

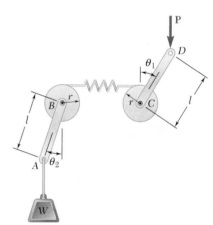

Fig. P10.99

*10.100 Solve Prob. 10.99 knowing that $k = 20$ lb/in., $r = 3$ in., $l = 6$ in., and (*a*) $W = 15$ lb, (*b*) $W = 60$ lb.

REVIEW AND SUMMARY

Work of a force

The first part of this chapter was devoted to the *principle of virtual work* and to its direct application to the solution of equilibrium problems. We first defined the *work of a force* **F** *corresponding to the small displacement* d**r** [Sec. 10.2] as the quantity

$$dU = \mathbf{F} \cdot d\mathbf{r} \tag{10.1}$$

Fig. 10.16

obtained by forming the scalar product of the force **F** and the displacement d**r** (Fig. 10.16). Denoting respectively by F and ds the magnitudes of the force and of the displacement, and by α the angle formed by **F** and d**r**, we wrote

$$dU = F \, ds \cos \alpha \tag{10.1'}$$

The work dU is positive if $\alpha < 90°$, zero if $\alpha = 90°$, and negative if $\alpha > 90°$. We also found that the *work of a couple of moment* **M** acting on a rigid body is

$$dU = M \, d\theta \tag{10.2}$$

where $d\theta$ is the small angle expressed in radians through which the body rotates.

Virtual displacement

Considering a particle located at A and acted upon by several forces $\mathbf{F}_1, \mathbf{F}_2, \ldots, \mathbf{F}_n$ [Sec. 10.3], we imagined that the particle moved to a new position A' (Fig. 10.17). Since this displacement did not actually take place, it was referred to as a *virtual displacement* and denoted by $\delta\mathbf{r}$, while the corresponding work of the forces was called *virtual work* and denoted by δU. We had

$$\delta U = \mathbf{F}_1 \cdot \delta\mathbf{r} + \mathbf{F}_2 \cdot \delta\mathbf{r} + \cdots + \mathbf{F}_n \cdot \delta\mathbf{r}$$

Principle of virtual work

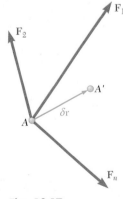

The *principle of virtual work* states that *if a particle is in equilibrium, the total virtual work* δU *of the forces acting on the particle is zero for any virtual displacement of the particle.*

The principle of virtual work can be extended to the case of rigid bodies and systems of rigid bodies. Since it involves *only forces which do work,* its application provides a useful alternative to the use of the equilibrium equations in the solution of many engineering problems. It is particularly effective in the case of machines and mechanisms consisting of connected rigid bodies, since the work of the reactions at the supports is zero and the work of the internal forces at the pin connections cancels out [Sec. 10.4; Sample Probs. 10.1, 10.2, and 10.3].

Fig. 10.17

In the case of *real machines*, however [Sec. 10.5], the work of the friction forces should be taken into account, with the result that the *output work will be less than the input work.* Defining the *mechanical efficiency* of a machine as the ratio

$$\eta = \frac{\text{output work}}{\text{input work}} \tag{10.9}$$

we also noted that for an ideal machine (no friction) $\eta = 1$, while for a real machine $\eta < 1$.

In the second part of the chapter we considered the *work of forces corresponding to finite displacements* of their points of application. The work $U_{1 \rightarrow 2}$ of the force **F** corresponding to a displacement of the particle A from A_1 to A_2 (Fig. 10.18) was obtained by integrating the right-hand member of Eq. (10.1) or (10.1′) along the curve described by the particle [Sec. 10.6]:

$$U_{1 \rightarrow 2} = \int_{A_1}^{A_2} \mathbf{F} \cdot d\mathbf{r} \tag{10.11}$$

or

$$U_{1 \rightarrow 2} = \int_{s_1}^{s_2} (F \cos \alpha)\, ds \tag{10.11′}$$

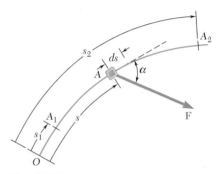

Fig. 10.18

Similarly, the work of a couple of moment **M** corresponding to a finite rotation from θ_1 to θ_2 of a rigid body was expressed as

$$U_{1 \rightarrow 2} = \int_{\theta_1}^{\theta_2} M\, d\theta \tag{10.12}$$

The *work of the weight* **W** *of a body* as its center of gravity moves from the elevation y_1 to y_2 (Fig. 10.19) can be obtained by making $F = W$ and $\alpha = 180°$ in Eq. (10.11′):

$$U_{1 \rightarrow 2} = - \int_{y_1}^{y_2} W\, dy = Wy_1 - Wy_2 \tag{10.13}$$

The work of **W** is therefore positive *when the elevation y decreases.*

Fig. 10.19

Work of the force exerted by a spring

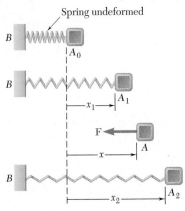

Spring undeformed

Fig. 10.20

The *work of the force* **F** *exerted by a spring* on a body A as the spring is stretched from x_1 to x_2 (Fig. 10.20) can be obtained by making $F = kx$, where k is the constant of the spring, and $\alpha = 180°$ in Eq. (10.11′):

$$U_{1 \to 2} = -\int_{x_1}^{x_2} kx \, dx = \tfrac{1}{2}kx_1^2 - \tfrac{1}{2}kx_2^2 \qquad (10.15)$$

The work of **F** is therefore positive *when the spring is returning to its undeformed position.*

When the work of a force **F** is independent of the path actually followed between A_1 and A_2, the force is said to be a *conservative force,* and its work can be expressed as

$$U_{1 \to 2} = V_1 - V_2 \qquad (10.20)$$

Potential energy

where V is the *potential energy* associated with **F**, and V_1 and V_2 represent the values of V at A_1 and A_2, respectively [Sec. 10.7]. The potential energies associated, respectively, with the *force of gravity* **W** and the *elastic force* **F** exerted by a spring were found to be

$$V_g = Wy \quad \text{and} \quad V_e = \tfrac{1}{2}kx^2 \qquad (10.17, 10.18)$$

Alternative expression for the principle of virtual work

When the position of a mechanical system depends upon a single independent variable θ, the potential energy of the system is a function $V(\theta)$ of that variable, and it follows from Eq. (10.20) that $\delta U = -\delta V = -(dV/d\theta)\,\delta\theta$. The condition $\delta U = 0$ required by the principle of virtual work for the equilibrium of the system can thus be replaced by the condition

$$\frac{dV}{d\theta} = 0 \qquad (10.21)$$

When all the forces involved are conservative, it may be preferable to use Eq. (10.21) rather than apply the principle of virtual work directly [Sec. 10.8; Sample Prob. 10.4].

Stability of equilibrium

This approach presents another advantage, since it is possible to determine from the sign of the second derivative of V whether the equilibrium of the system is *stable, unstable,* or *neutral* [Sec. 10.9]. If $d^2V/d\theta^2 > 0$, V is *minimum* and the equilibrium is *stable;* if $d^2V/d\theta^2 < 0$, V is *maximum* and the equilibrium is *unstable;* if $d^2V/d\theta^2 = 0$, it is necessary to examine derivatives of a higher order.

REVIEW PROBLEMS

10.101 Determine the horizontal force **P** that must be applied at A to maintain the equilibrium of the linkage.

Fig. P10.101 and P10.102

10.102 Determine the couple **M** that must be applied to member ABC to maintain the equilibrium of the linkage.

10.103 A spring of constant 15 kN/m connects points C and F of the linkage shown. Neglecting the weight of the spring and linkage, determine the force in the spring and the vertical motion of point G when a vertical downward 120-N force is applied (*a*) at point C, (*b*) at points C and H.

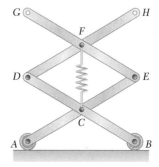

Fig. P10.103

10.104 Derive an expression for the magnitude of the force **Q** required to maintain the equilibrium of the mechanism shown.

10.105 Derive an expression for the magnitude of the couple **M** required to maintain the equilibrium of the linkage shown.

Fig. P10.104

Fig. P10.105

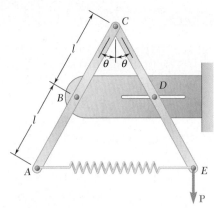

Fig. P10.106

10.106 Two rods AC and CE are connected by a pin at C and by a spring AE. The constant of the spring is k, and the spring is unstretched when $\theta = 30°$. For the loading shown, derive an equation in P, θ, l, and k that must be satisfied when the system is in equilibrium.

10.107 A force $\mathbf{P}$ of magnitude 240 N is applied to end E of cable CDE, which passes under pulley D and is attached to the mechanism at C. Neglecting the weight of the mechanism and the radius of the pulley, determine the value of θ corresponding to equilibrium. The constant of the spring is $k = 4$ kN/m, and the spring is unstretched when $\theta = 90°$.

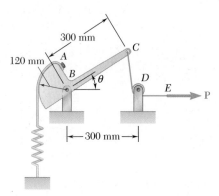

Fig. P10.107

10.108 Two identical rods ABC and DBE are connected by a pin at B and by a spring CE. Knowing that the spring is 4 in. long when unstretched and that the constant of the spring is 8 lb/in., determine the distance x corresponding to equilibrium when a 24-lb load is applied at E as shown.

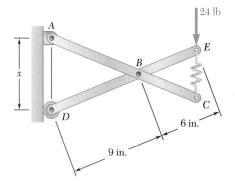

Fig. P10.108

10.109 Solve Prob. 10.108 assuming that the 24-lb load is applied at C instead of E.

10.110 Two bars AB and BC are attached to a single spring of constant k that is unstretched when the bars are vertical. Determine the range of values of P for which the equilibrium of the system is stable in the position shown.

Fig. P10.110

10.111 A homogeneous hemisphere of radius r is placed on an incline as shown. Assuming that friction is sufficient to prevent slipping between the hemisphere and the incline, determine the angle θ corresponding to equilibrium when $\beta = 10°$.

Fig. *P10.111* and P10.112

10.112 A homogeneous hemisphere of radius r is placed on an incline as shown. Assuming that friction is sufficient to prevent slipping between the hemisphere and the incline, determine (a) the largest angle β for which a position of equilibrium exists, (b) the angle θ corresponding to equilibrium when the angle β is equal to half the value found in part a.

COMPUTER PROBLEMS

10.C1 A couple **M** is applied to crank AB in order to maintain the equilibrium of the engine system shown when a force **P** is applied to the piston. Knowing that $b = 2.4$ in. and $l = 7.5$ in., write a computer program that can be used to calculate the ratio M/P for values of θ from 0 to 180° using 10° increments. Using appropriate smaller increments, determine the value of θ for which the ratio M/P is maximum, and the corresponding value of M/P.

Fig. P10.C1

10.C2 Knowing that $a = 500$ mm, $b = 150$ mm, $L = 500$ mm, and $P = 100$ N, write a computer program that can be used to calculate the force in member BD for values of θ from 30° to 150° using 10° increments. Using appropriate smaller increments, determine the range of values of θ for which the absolute value of the force in member BD is less than 400 N.

Fig. P10.C2

10.C3 Solve Prob. 10.C2 assuming that the force **P** applied at A is directed horizontally to the right.

10.C4 The constant of spring AB is k, and the spring is unstretched when $\theta = 0$. (*a*) Neglecting the weight of the member BCD, write a computer program that can be used to calculate the potential energy of the system and its derivative $dV/d\theta$. (*b*) For $W = 150$ lb, $a = 10$ in., and $k = 75$ lb/in., calculate and plot the potential energy versus θ for values of θ from 0 to 165° using 15° increments. (*c*) Using appropriate smaller increments, determine the values of θ for which the system is in equilibrium and state in each case whether the equilibrium is stable, unstable, or neutral.

Fig. P10.C4

10.C5 Two rods, AC and DE, each of length L, are connected by a collar that is attached to rod AC at its midpoint B. (a) Write a computer program that can be used to calculate the potential energy V of the system and its derivative $dV/d\theta$. (b) For $W = 75$ N, $P = 200$ N, and $L = 500$ mm, calculate V and $dV/d\theta$ for values of θ from 0 to 70° using 5° increments. (c) Using appropriate smaller increments, determine the values of θ for which the system is in equilibrium and state in each case whether the equilibrium is stable, unstable, or neutral.

10.C6 A slender rod ABC is attached to blocks A and B that can move freely in the guides shown. The constant of the spring is k, and the spring is unstretched when the rod is vertical. (a) Neglecting the weights of the rod and of the blocks, write a computer program that can be used to calculate the potential energy V of the system and its derivative $dV/d\theta$. (b) For $P = 150$ N, $l = 200$ mm, and $k = 3$ kN/m, calculate and plot the potential energy versus θ for values of θ from 0 to 75° using 5° increments. (c) Using appropriate smaller increments, determine any positions of equilibrium in the range $0 \le \theta \le 75°$ and state in each case whether the equilibrium is stable, unstable, or neutral.

Fig. P10.C5

Fig. P10.C6

10.C7 Solve Prob. 10.C6 assuming that the force $\mathbf{P}$ applied at C is directed horizontally to the right.

The motion of the space shuttle can be described in terms of its *position, velocity,* and *acceleration.* When landing, the pilot of the shuttle needs to consider the wind velocity and the *relative motion* of the shuttle with respect to the wind. The study of motion is known as *kinematics* and is the subject of this chapter.

Kinematics of Particles

11.1   INTRODUCTION TO DYNAMICS

Chapters 1 to 10 were devoted to *statics*, i.e., to the analysis of bodies at rest. We now begin the study of *dynamics*, the part of mechanics that deals with the analysis of bodies in motion.

While the study of statics goes back to the time of the Greek philosophers, the first significant contribution to dynamics was made by Galileo (1564–1642). Galileo's experiments on uniformly accelerated bodies led Newton (1642–1727) to formulate his fundamental laws of motion.

Dynamics includes:

1. *Kinematics*, which is the study of the geometry of motion. Kinematics is used to relate displacement, velocity, acceleration, and time, without reference to the cause of the motion.
2. *Kinetics*, which is the study of the relation existing between the forces acting on a body, the mass of the body, and the motion of the body. Kinetics is used to predict the motion caused by given forces or to determine the forces required to produce a given motion.

Chapters 11 to 14 are devoted to the *dynamics of particles;* in Chap. 11 the *kinematics of particles* will be considered. The use of the word *particles* does not mean that our study will be restricted to small corpuscles; rather, it indicates that in these first chapters the motion of bodies—possibly as large as cars, rockets, or airplanes—will be considered without regard to their size. By saying that the bodies are analyzed as particles, we mean that only their motion as an entire unit will be considered; any rotation about their own mass center will be neglected. There are cases, however, when such a rotation is not negligible; the bodies cannot then be considered as particles. Such motions will be analyzed in later chapters, dealing with the *dynamics of rigid bodies*.

In the first part of Chap. 11, the rectilinear motion of a particle will be analyzed; that is, the position, velocity, and acceleration of a particle will be determined at every instant as it moves along a straight line. First, general methods of analysis will be used to study the motion of a particle; then two important particular cases will be considered, namely, the uniform motion and the uniformly accelerated motion of a particle (Secs. 11.4 and 11.5). In Sec. 11.6 the simultaneous motion of several particles will be considered, and the concept of the relative motion of one particle with respect to another will be introduced. The first part of this chapter concludes with a study of graphical methods of analysis and their application to the solution of various problems involving the rectilinear motion of particles (Secs. 11.7 and 11.8).

In the second part of this chapter, the motion of a particle as it moves along a curved path will be analyzed. Since the position, velocity, and acceleration of a particle will be defined as vector quantities, the concept of the derivative of a vector function will be introduced in Sec. 11.10 and added to our mathematical tools. Applications in which the motion of a particle is defined by the rectangular components of its velocity and acceleration will then be considered; at this point, the motion of a projectile will be analyzed (Sec. 11.11). In Sec. 11.12, the motion of a particle relative to a reference frame in translation will be considered. Finally, the curvilinear motion of a particle will be analyzed in terms of components other than rectangular. The tangential and normal components of a particular velocity and an acceleration will be introduced in Sec. 11.13 and the radial and transverse components of its velocity and acceleration in Sec. 11.14.

RECTILINEAR MOTION OF PARTICLES

11.2 POSITION, VELOCITY, AND ACCELERATION

A particle moving along a straight line is said to be in *rectilinear motion*. At any given instant t, the particle will occupy a certain position on the straight line. To define the position P of the particle, we choose a fixed origin O on the straight line and a positive direction along the line. We measure the distance x from O to P and record it with a plus or minus sign, according to whether P is reached from O by moving along the line in the positive or the negative direction. The distance x, with the appropriate sign, completely defines the position of the particle; it is called the *position coordinate* of the particle considered. For example, the position coordinate corresponding to P in Fig. 11.1a is $x = +5$ m; the coordinate corresponding to P' in Fig. 11.1b is $x' = -2$ m.

When the position coordinate x of a particle is known for every value of time t, we say that the motion of the particle is known. The "timetable" of the motion can be given in the form of an equation in x and t, such as $x = 6t^2 - t^3$, or in the form of a graph of x versus t as shown in Fig. 11.6. The units most often used to measure the position coordinate x are the meter (m) in the SI system of units† and the foot (ft) in the U.S. customary system of units. Time t is usually measured in seconds (s).

Consider the position P occupied by the particle at time t and the corresponding coordinate x (Fig. 11.2). Consider also the position P' occupied by the particle at a later time $t + \Delta t$; the position coordinate of P' can be obtained by adding to the coordinate x of P the small displacement Δx, which will be positive or negative according to whether P' is to the right or to the left of P. The *average velocity* of the particle over the time interval Δt is defined as the quotient of the displacement Δx and the time interval Δt:

$$\text{Average velocity} = \frac{\Delta x}{\Delta t}$$

† Cf. Sec. 1.3.

Fig. 11.1

Fig. 11.2

Photo 11.1 The motion of this solar car can be described by its position, velocity, and acceleration.

If SI units are used, Δx is expressed in meters and Δt in seconds; the average velocity will thus be expressed in meters per second (m/s). If U.S. customary units are used, Δx is expressed in feet and Δt in seconds; the average velocity will then be expressed in feet per second (ft/s).

The *instantaneous velocity* v of the particle at the instant t is obtained from the average velocity by choosing shorter and shorter time intervals Δt and displacements Δx:

$$\text{Instantaneous velocity} = v = \lim_{\Delta t \to 0} \frac{\Delta x}{\Delta t}$$

The instantaneous velocity will also be expressed in m/s or ft/s. Observing that the limit of the quotient is equal, by definition, to the derivative of x with respect to t, we write

$$v = \frac{dx}{dt} \tag{11.1}$$

The velocity v is represented by an algebraic number which can be positive or negative.[†] A positive value of v indicates that x increases, i.e., that the particle moves in the positive direction (Fig. 11.3a); a negative value of v indicates that x decreases, i.e., that the particle moves in the negative direction (Fig. 11.3b). The magnitude of v is known as the *speed* of the particle.

Consider the velocity v of the particle at time t and also its velocity $v + \Delta v$ at a later time $t + \Delta t$ (Fig. 11.4). The *average acceleration* of the particle over the time interval Δt is defined as the quotient of Δv and Δt:

$$\text{Average acceleration} = \frac{\Delta v}{\Delta t}$$

If SI units are used, Δv is expressed in m/s and Δt in seconds; the average acceleration will thus be expressed in m/s². If U.S. customary units are used, Δv is expressed in ft/s and Δt in seconds; the average acceleration will then be expressed in ft/s².

The *instantaneous acceleration* a of the particle at the instant t is obtained from the average acceleration by choosing smaller and smaller values for Δt and Δv:

$$\text{Instantaneous acceleration} = a = \lim_{\Delta t \to 0} \frac{\Delta v}{\Delta t}$$

The instantaneous acceleration will also be expressed in m/s² or ft/s². The limit of the quotient, which is by definition the derivative of v

Fig. 11.3

Fig. 11.4

[†]As you will see in Sec. 11.9, the velocity is actually a vector quantity. However, since we are considering here the rectilinear motion of a particle, where the velocity of the particle has a known and fixed direction, we need only specify the sense and magnitude of the velocity; this can be conveniently done by using a scalar quantity with a plus or minus sign. The same is true of the acceleration of a particle in rectilinear motion.

with respect to t, measures the rate of change of the velocity. We write

$$a = \frac{dv}{dt} \qquad (11.2)$$

or, substituting for v from (11.1),

$$a = \frac{d^2x}{dt^2} \qquad (11.3)$$

The acceleration a is represented by an algebraic number which can be positive or negative.† A positive value of a indicates that the velocity (i.e., the algebraic number v) increases. This may mean that the particle is moving faster in the positive direction (Fig. 11.5a) or that it is moving more slowly in the negative direction (Fig. 11.5b); in both cases, Δv is positive. A negative value of a indicates that the velocity decreases; either the particle is moving more slowly in the positive direction (Fig. 11.5c) or it is moving faster in the negative direction (Fig. 11.5d).

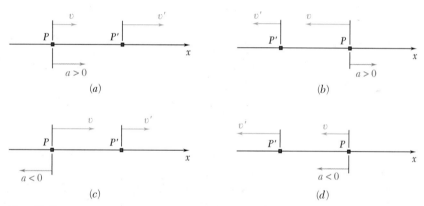

Fig. 11.5

The term *deceleration* is sometimes used to refer to a when the speed of the particle (i.e., the magnitude of v) decreases; the particle is then moving more slowly. For example, the particle of Fig. 11.5 is decelerated in parts b and c; it is truly accelerated (i.e., moves faster) in parts a and d.

Another expression for the acceleration can be obtained by eliminating the differential dt in Eqs. (11.1) and (11.2). Solving (11.1) for dt, we obtain $dt = dx/v$; substituting into (11.2), we write

$$a = v\frac{dv}{dx} \qquad (11.4)$$

†See footnote, page 604.

EXAMPLE Consider a particle moving in a straight line, and assume that its position is defined by the equation

$$x = 6t^2 - t^3$$

where t is expressed in seconds and x in meters. The velocity v at any time t is obtained by differentiating x with respect to t:

$$v = \frac{dx}{dt} = 12t - 3t^2$$

The acceleration a is obtained by differentiating again with respect to t:

$$a = \frac{dv}{dt} = 12 - 6t$$

Fig. 11.6

The position coordinate, the velocity, and the acceleration have been plotted against t in Fig. 11.6. The curves obtained are known as *motion curves*. Keep in mind, however, that the particle does not move along any of these curves; the particle moves in a straight line. Since the derivative of a function measures the slope of the corresponding curve, the slope of the x–t curve at any given time is equal to the value of v at that time and the slope of the v–t curve is equal to the value of a. Since $a = 0$ at $t = 2$ s, the slope of the v–t curve must be zero at $t = 2$ s; the velocity reaches a maximum at this instant. Also, since $v = 0$ at $t = 0$ and at $t = 4$ s, the tangent to the x–t curve must be horizontal for both of these values of t.

A study of the three motion curves of Fig. 11.6 shows that the motion of the particle from $t = 0$ to $t = \infty$ can be divided into four phases:

1. The particle starts from the origin, $x = 0$, with no velocity but with a positive acceleration. Under this acceleration, the particle gains a positive velocity and moves in the positive direction. From $t = 0$ to $t = 2$ s, x, v, and a are all positive.
2. At $t = 2$ s, the acceleration is zero; the velocity has reached its maximum value. From $t = 2$ s to $t = 4$ s, v is positive, but a is negative; the particle still moves in the positive direction but more and more slowly; the particle is decelerating.
3. At $t = 4$ s, the velocity is zero; the position coordinate x has reached its maximum value. From then on, both v and a are negative; the particle is accelerating and moves in the negative direction with increasing speed.
4. At $t = 6$ s, the particle passes through the origin; its coordinate x is then zero, while the total distance traveled since the beginning of the motion is 64 m. For values of t larger than 6 s, x, v, and a will all be negative. The particle keeps moving in the negative direction, away from O, faster and faster. ■

11.3 DETERMINATION OF THE MOTION OF A PARTICLE

We saw in the preceding section that the motion of a particle is said to be known if the position of the particle is known for every value of the time t. In practice, however, a motion is seldom defined by a relation between x and t. More often, the conditions of the motion will be specified by the type of acceleration that the particle possesses. For example, a freely falling body will have a constant acceleration, directed downward and equal to 9.81 m/s², or 32.2 ft/s²; a mass attached to a spring which has been stretched will have an acceleration proportional to the instantaneous elongation of the spring measured from the equilibrium position, etc. In general, the acceleration of the particle can be expressed as a function of one or more of the variables x, v, and t. In order to determine the position coordinate x in terms of t, it will thus be necessary to perform two successive integrations.

Let us consider three common classes of motion:

1. $a = f(t)$. *The Acceleration Is a Given Function of t.* Solving (11.2) for dv and substituting $f(t)$ for a, we write

$$dv = a\,dt$$
$$dv = f(t)\,dt$$

Integrating both members, we obtain the equation

$$\int dv = \int f(t)\,dt$$

which defines v in terms of t. It should be noted, however, that an arbitrary constant will be introduced as a result of the integration. This is due to the fact that there are many motions which correspond to the given acceleration $a = f(t)$. In order to uniquely define the motion of the particle, it is necessary to specify the *initial conditions* of the motion, i.e., the value v_0 of the velocity and the value x_0 of the position coordinate at $t = 0$. Replacing the indefinite integrals by *definite integrals* with lower limits corresponding to the initial conditions $t = 0$ and $v = v_0$ and upper limits corresponding to $t = t$ and $v = v$, we write

$$\int_{v_0}^{v} dv = \int_{0}^{t} f(t)\,dt$$

$$v - v_0 = \int_{0}^{t} f(t)\,dt$$

which yields v in terms of t.

Equation (11.1) can now be solved for dx,

$$dx = v\,dt$$

and the expression just obtained substituted for v. Both members are then integrated, the left-hand member with respect to x from $x = x_0$ to $x = x$, and the right-hand member with respect to t from $t = 0$ to $t = t$. The position coordinate x is thus obtained in terms of t; the motion is completely determined.

Two important particular cases will be studied in greater detail in Secs. 11.4 and 11.5: the case when $a = 0$, corresponding to a *uniform motion*, and the case when $a =$ constant, corresponding to a *uniformly accelerated motion*.

2. $a = f(x)$. *The Acceleration Is a Given Function of x*. Rearranging Eq. (11.4) and substituting $f(x)$ for a, we write

$$v \, dv = a \, dx$$
$$v \, dv = f(x) \, dx$$

Since each member contains only one variable, we can integrate the equation. Denoting again by v_0 and x_0, respectively, the initial values of the velocity and of the position coordinate, we obtain

$$\int_{v_0}^{v} v \, dv = \int_{x_0}^{x} f(x) \, dx$$

$$\tfrac{1}{2}v^2 - \tfrac{1}{2}v_0^2 = \int_{x_0}^{x} f(x) \, dx$$

which yields v in terms of x. We now solve (11.1) for dt,

$$dt = \frac{dx}{v}$$

and substitute for v the expression just obtained. Both members can then be integrated to obtain the desired relation between x and t. However, in most cases this last integration cannot be performed analytically and one must resort to a numerical method of integration.

3. $a = f(v)$. *The Acceleration Is a Given Function of v*. We can now substitute $f(v)$ for a in either (11.2) or (11.4) to obtain either of the following relations:

$$f(v) = \frac{dv}{dt} \qquad f(v) = v\frac{dv}{dx}$$
$$dt = \frac{dv}{f(v)} \qquad dx = \frac{v \, dv}{f(v)}$$

Integration of the first equation will yield a relation between v and t; integration of the second equation will yield a relation between v and x. Either of these relations can be used in conjunction with Eq. (11.1) to obtain the relation between x and t which characterizes the motion of the particle.

SAMPLE PROBLEM 11.1

The position of a particle which moves along a straight line is defined by the relation $x = t^3 - 6t^2 - 15t + 40$, where x is expressed in feet and t in seconds. Determine (a) the time at which the velocity will be zero, (b) the position and distance traveled by the particle at that time, (c) the acceleration of the particle at that time, (d) the distance traveled by the particle from $t = 4$ s to $t = 6$ s.

SOLUTION

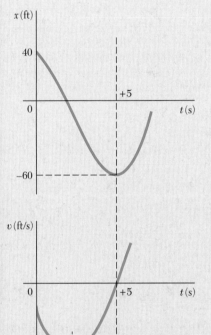

The equations of motion are

$$x = t^3 - 6t^2 - 15t + 40 \tag{1}$$

$$v = \frac{dx}{dt} = 3t^2 - 12t - 15 \tag{2}$$

$$a = \frac{dv}{dt} = 6t - 12 \tag{3}$$

a. Time at Which $v = 0$. We set $v = 0$ in (2):

$$3t^2 - 12t - 15 = 0 \qquad t = -1 \text{ s} \quad \text{and} \quad t = +5 \text{ s} \blacktriangleleft$$

Only the root $t = +5$ s corresponds to a time after the motion has begun: for $t < 5$ s, $v < 0$, the particle moves in the negative direction; for $t > 5$ s, $v > 0$, the particle moves in the positive direction.

b. Position and Distance Traveled When $v = 0$. Carrying $t = +5$ s into (1), we have

$$x_5 = (5)^3 - 6(5)^2 - 15(5) + 40 \qquad x_5 = -60 \text{ ft} \blacktriangleleft$$

The initial position at $t = 0$ was $x_0 = +40$ ft. Since $v \neq 0$ during the interval $t = 0$ to $t = 5$ s, we have

$$\text{Distance traveled} = x_5 - x_0 = -60 \text{ ft} - 40 \text{ ft} = -100 \text{ ft}$$

$$\text{Distance traveled} = 100 \text{ ft in the negative direction} \blacktriangleleft$$

c. Acceleration When $v = 0$. We substitute $t = +5$ s into (3):

$$a_5 = 6(5) - 12 \qquad a_5 = +18 \text{ ft/s}^2 \blacktriangleleft$$

d. Distance Traveled from $t = 4$ s to $t = 6$ s. The particle moves in the negative direction from $t = 4$ s to $t = 5$ s and in the positive direction from $t = 5$ s to $t = 6$ s; therefore, the distance traveled during each of these time intervals will be computed separately.

From $t = 4$ s to $t = 5$ s: $\qquad x_5 = -60$ ft

$$x_4 = (4)^3 - 6(4)^2 - 15(4) + 40 = -52 \text{ ft}$$

$$\text{Distance traveled} = x_5 - x_4 = -60 \text{ ft} - (-52 \text{ ft}) = -8 \text{ ft}$$
$$= 8 \text{ ft in the negative direction}$$

From $t = 5$ s to $t = 6$ s: $\qquad x_5 = -60$ ft

$$x_6 = (6)^3 - 6(6)^2 - 15(6) + 40 = -50 \text{ ft}$$
$$\text{Distance traveled} = x_6 - x_5 = -50 \text{ ft} - (-60 \text{ ft}) = +10 \text{ ft}$$
$$= 10 \text{ ft in the positive direction}$$

Total distance traveled from $t = 4$ s to $t = 6$ s is 8 ft + 10 ft $= 18$ ft $\blacktriangleleft$

SAMPLE PROBLEM 11.2

A ball is tossed with a velocity of 10 m/s directed vertically upward from a window located 20 m above the ground. Knowing that the acceleration of the ball is constant and equal to 9.81 m/s² downward, determine (a) the velocity v and elevation y of the ball above the ground at any time t, (b) the highest elevation reached by the ball and the corresponding value of t, (c) the time when the ball will hit the ground and the corresponding velocity. Draw the v–t and y–t curves.

SOLUTION

a. Velocity and Elevation. The y axis measuring the position coordinate (or elevation) is chosen with its origin O on the ground and its positive sense upward. The value of the acceleration and the initial values of v and y are as indicated. Substituting for a in $a = dv/dt$ and noting that at $t = 0$, $v_0 = +10$ m/s, we have

$$\frac{dv}{dt} = a = -9.81 \,\text{m/s}^2$$

$$\int_{v_0=10}^{v} dv = -\int_{0}^{t} 9.81 \, dt$$

$$[v]_{10}^{v} = -[9.81t]_{0}^{t}$$

$$v - 10 = -9.81t$$

$$v = 10 - 9.81t \quad (1) \quad \blacktriangleleft$$

Substituting for v in $v = dy/dt$ and noting that at $t = 0$, $y_0 = 20$ m, we have

$$\frac{dy}{dt} = v = 10 - 9.81t$$

$$\int_{y_0=20}^{y} dy = \int_{0}^{t} (10 - 9.81t) \, dt$$

$$[y]_{20}^{y} = [10t - 4.905t^2]_{0}^{t}$$

$$y - 20 = 10t - 4.905t^2$$

$$y = 20 + 10t - 4.905t^2 \quad (2) \quad \blacktriangleleft$$

b. Highest Elevation. When the ball reaches its highest elevation, we have $v = 0$. Substituting into (1), we obtain

$$10 - 9.81t = 0 \qquad t = 1.019 \,\text{s} \quad \blacktriangleleft$$

Carrying $t = 1.019$ s into (2), we have

$$y = 20 + 10(1.019) - 4.905(1.019)^2 \quad y = 25.1 \,\text{m} \quad \blacktriangleleft$$

c. Ball Hits the Ground. When the ball hits the ground, we have $y = 0$. Substituting into (2), we obtain

$$20 + 10t - 4.905t^2 = 0 \qquad t = -1.243 \,\text{s} \quad \text{and} \qquad t = +3.28 \,\text{s} \quad \blacktriangleleft$$

Only the root $t = +3.28$ s corresponds to a time after the motion has begun. Carrying this value of t into (1), we have

$$v = 10 - 9.81(3.28) = -22.2 \,\text{m/s} \qquad v = 22.2 \,\text{m/s} \downarrow \quad \blacktriangleleft$$

SAMPLE PROBLEM 11.3

The brake mechanism used to reduce recoil in certain types of guns consists essentially of a piston attached to the barrel and moving in a fixed cylinder filled with oil. As the barrel recoils with an initial velocity v_0, the piston moves and oil is forced through orifices in the piston, causing the piston and the barrel to decelerate at a rate proportional to their velocity; that is, $a = -kv$. Express (a) v in terms of t, (b) x in terms of t, (c) v in terms of x. Draw the corresponding motion curves.

SOLUTION

a. v in Terms of t. Substituting $-kv$ for a in the fundamental formula defining acceleration, $a = dv/dt$, we write

$$-kv = \frac{dv}{dt} \qquad \frac{dv}{v} = -k\,dt \qquad \int_{v_0}^{v} \frac{dv}{v} = -k\int_0^t dt$$

$$\ln\frac{v}{v_0} = -kt \qquad\qquad v = v_0 e^{-kt} \quad \blacktriangleleft$$

b. x in Terms of t. Substituting the expression just obtained for v into $v = dx/dt$, we write

$$v_0 e^{-kt} = \frac{dx}{dt}$$

$$\int_0^x dx = v_0 \int_0^t e^{-kt}\,dt$$

$$x = -\frac{v_0}{k}[e^{-kt}]_0^t = -\frac{v_0}{k}(e^{-kt} - 1)$$

$$x = \frac{v_0}{k}(1 - e^{-kt}) \quad \blacktriangleleft$$

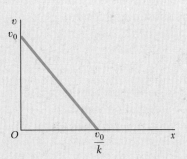

c. v in Terms of x. Substituting $-kv$ for a in $a = v\,dv/dx$, we write

$$-kv = v\frac{dv}{dx}$$

$$dv = -k\,dx$$

$$\int_{v_0}^{v} dv = -k\int_0^x dx$$

$$v - v_0 = -kx \qquad\qquad v = v_0 - kx \quad \blacktriangleleft$$

Check. Part c could have been solved by eliminating t from the answers obtained for parts a and b. This alternative method can be used as a check. From part a we obtain $e^{-kt} = v/v_0$; substituting into the answer of part b, we obtain

$$x = \frac{v_0}{k}(1 - e^{-kt}) = \frac{v_0}{k}\left(1 - \frac{v}{v_0}\right) \qquad v = v_0 - kx \qquad \text{(checks)}$$

SOLVING PROBLEMS
ON YOUR OWN

In the problems for this lesson, you will be asked to determine the *position,* the *velocity,* or the *acceleration* of a particle in *rectilinear motion.* As you read each problem, it is important that you identify both the independent variable (typically t or x) and what is required (for example, the need to express v as a function of x). You may find it helpful to start each problem by writing down both the given information and a simple statement of what is to be determined.

1. Determining v(t) and a(t) for a given x(t). As explained in Sec. 11.2, the first and the second derivatives of x with respect to t are respectively equal to the velocity and the acceleration of the particle [Eqs. (11.1) and (11.2)]. If the velocity and the acceleration have opposite signs, the particle can come to rest and then move in the opposite direction [Sample Prob. 11.1]. Thus, when computing the total distance traveled by a particle, you should first determine if the particle will come to rest during the specified interval of time. Constructing a diagram similar to that of Sample Prob. 11.1 that shows the position and the velocity of the particle at each critical instant ($v = v_{max}$, $v = 0$, etc.) will help you to visualize the motion.

2. Determining v(t) and x(t) for a given a(t). The solution of problems of this type was discussed in the first part of Sec. 11.3. We used the initial conditions, $t = 0$ and $v = v_0$, for the lower limits of the integrals in t and v, but any other known state (for example, $t = t_1$, $v = v_1$) could have been used instead. Also, if the given function $a(t)$ contains an unknown constant (for example, the constant k if $a = kt$), you will first have to determine that constant by substituting a set of known values of t and a in the equation defining $a(t)$.

3. Determining v(x) and x(t) for a given a(x). This is the second case considered in Sec. 11.3. We again note that the lower limits of integration can be any known state (for example, $x = x_1$, $v = v_1$). In addition, since $v = v_{max}$ when $a = 0$, the positions where the maximum values of the velocity occur are easily determined by writing $a(x) = 0$ and solving for x.

4. Determining v(x), v(t), and x(t) for a given a(v). This is the last case treated in Sec. 11.3; the appropriate solution techniques for problems of this type are illustrated in Sample Prob. 11.3. All of the general comments for the preceding cases once again apply. Note that Sample Prob. 11.3 provides a summary of how and when to use the equations $v = dx/dt$, $a = dv/dt$, and $a = v\, dv/dx$.

PROBLEMS†

CONCEPT QUESTIONS

11.CQ1 A bus travels the 100 miles between A and B at 50 mi/h and then another 100 miles between B and C at 70 mi/h. The average speed of the bus for the entire 200-mile trip is:

    a. More than 60 mi/h.

    b. Equal to 60 mi/h.

    c. Less than 60 mi/h.

Fig. P11.CQ1

11.CQ2 Two cars A and B race each other down a straight road. The position of each car as a function of time is shown. Which of the following statements are true (more than one answer can be correct)?

    a. At time t_2 both cars have traveled the same distance.

    b. At time t_1 both cars have the same speed.

    c. Both cars have the same speed at some time $t < t_1$.

    d. Both cars have the same acceleration at some time $t < t_1$.

    e. Both cars have the same acceleration at some time $t_1 < t < t_2$.

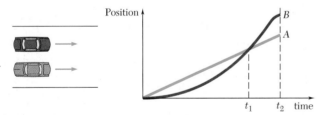

Fig. P11.CQ2

END-OF-SECTION PROBLEMS

11.1 The motion of a particle is defined by the relation $x = t^4 - 10t^2 + 8t + 12$, where x and t are expressed in inches and seconds, respectively. Determine the position, the velocity, and the acceleration of the particle when $t = 1$ s.

11.2 The motion of a particle is defined by the relation $x = 2t^3 - 9t^2 + 12t + 10$, where x and t are expressed in feet and seconds, respectively. Determine the time, the position, and the acceleration of the particle when $v = 0$.

†Answers to all problems set in straight type (such as **11.1**) are given at the end of the book. Answers to problems with a number set in italic type (such as **11.7**) are not given.

Fig. P11.3

11.3 The vertical motion of mass A is defined by the relation $x = 10 \sin 2t + 15 \cos 2t + 100$, where x and t are expressed in millimeters and seconds, respectively. Determine (a) the position, velocity, and acceleration of A when $t = 1$ s, (b) the maximum velocity and acceleration of A.

11.4 A loaded railroad car is rolling at a constant velocity when it couples with a spring and dashpot bumper system. After the coupling, the motion of the car is defined by the relation $x = 60e^{-4.8t} \sin 16t$, where x and t are expressed in millimeters and seconds, respectively. Determine the position, the velocity, and the acceleration of the railroad car when (a) $t = 0$, (b) $t = 0.3$ s.

Fig. P11.4

11.5 The motion of a particle is defined by the relation $x = 6t^4 - 2t^3 - 12t^2 + 3t + 3$, where x and t are expressed in meters and seconds, respectively. Determine the time, the position, and the velocity when $a = 0$.

11.6 The motion of a particle is defined by the relation $x = t^3 - 9t^2 + 24t - 8$, where x and t are expressed in inches and seconds, respectively. Determine (a) when the velocity is zero, (b) the position and the total distance traveled when the acceleration is zero.

11.7 The motion of a particle is defined by the relation $x = 2t^3 - 15t^2 + 24t + 4$, where x is expressed in meters and t in seconds. Determine (a) when the velocity is zero, (b) the position and the total distance traveled when the acceleration is zero.

11.8 The motion of a particle is defined by the relation $x = t^3 - 6t^2 - 36t - 40$, where x and t are expressed in feet and seconds, respectively. Determine (a) when the velocity is zero, (b) the velocity, the acceleration, and the total distance traveled when $x = 0$.

Fig. P11.9

11.9 The brakes of a car are applied, causing it to slow down at a rate of 10 ft/s². Knowing that the car stops in 300 ft, determine (a) how fast the car was traveling immediately before the brakes were applied, (b) the time required for the car to stop.

11.10 The acceleration of a particle is directly proportional to the time t. At $t = 0$, the velocity of the particle is $v = 16$ in./s. Knowing that $v = 15$ in./s and that $x = 20$ in. when $t = 1$ s, determine the velocity, the position, and the total distance traveled when $t = 7$ s.

11.11 The acceleration of a particle is directly proportional to the square of the time t. When $t = 0$, the particle is at $x = 24$ m. Knowing that at $t = 6$ s, $x = 96$ m and $v = 18$ m/s, express x and v in terms of t.

11.12 The acceleration of a particle is defined by the relation $a = kt^2$. (a) Knowing that $v = -8$ m/s when $t = 0$ and that $v = +8$ m/s when $t = 2$ s, determine the constant k. (b) Write the equations of motion, knowing also that $x = 0$ when $t = 2$ s.

11.13 The acceleration of point A is defined by the relation $a = -1.8 \sin kt$, where a and t are expressed in m/s^2 and seconds, respectively, and $k = 3$ rad/s. Knowing that $x = 0$ and $v = 0.6$ m/s when $t = 0$, determine the velocity and position of point A when $t = 0.5$ s.

11.14 The acceleration of point A is defined by the relation $a = -1.08 \sin kt - 1.44 \cos kt$, where a and t are expressed in m/s^2 and seconds, respectively, and $k = 3$ rad/s. Knowing that $x = 0.16$ m and $v = 0.36$ m/s when $t = 0$, determine the velocity and position of point A when $t = 0.5$ s.

Fig. P11.13 and P11.14

11.15 A piece of electronic equipment that is surrounded by packing material is dropped so that it hits the ground with a speed of 4 m/s. After contact the equipment experiences an acceleration of $a = -kx$, where k is a constant and x is the compression of the packing material. If the packing material experiences a maximum compression of 20 mm, determine the maximum acceleration of the equipment.

Fig. P11.15

11.16 A projectile enters a resisting medium at $x = 0$ with an initial velocity $v_0 = 900$ ft/s and travels 4 in. before coming to rest. Assuming that the velocity of the projectile is defined by the relation $v = v_0 - kx$, where v is expressed in ft/s and x is in feet, determine (a) the initial acceleration of the projectile, (b) the time required for the projectile to penetrate 3.9 in. into the resisting medium.

Fig. P11.16

11.17 The acceleration of a particle is defined by the relation $a = -k/x$. It has been experimentally determined that $v = 15$ ft/s when $x = 0.6$ ft and that $v = 9$ ft/s when $x = 1.2$ ft. Determine (a) the velocity of the particle when $x = 1.5$ ft, (b) the position of the particle at which its velocity is zero.

11.18 A brass (nonmagnetic) block A and a steel magnet B are in equilibrium in a brass tube under the magnetic repelling force of another steel magnet C located at a distance $x = 0.004$ m from B. The force is inversely proportional to the square of the distance between B and C. If block A is suddenly removed, the acceleration of block B is $a = -9.81 + k/x^2$, where a and x are expressed in m/s^2 and meters, respectively, and $k = 4 \times 10^{-4}$ m^3/s^2. Determine the maximum velocity and acceleration of B.

Fig. P11.18

Fig. P11.20

Fig. P11.23

Fig. P11.27

11.19 Based on experimental observations, the acceleration of a particle is defined by the relation $a = -(0.1 + \sin x/b)$, where a and x are expressed in m/s^2 and meters, respectively. Knowing that $b = 0.8$ m and that $v = 1$ m/s when $x = 0$, determine (a) the velocity of the particle when $x = -1$ m, (b) the position where the velocity is maximum, (c) the maximum velocity.

11.20 A spring AB is attached to a support at A and to a collar. The unstretched length of the spring is l. Knowing that the collar is released from rest at $x = x_0$ and has an acceleration defined by the relation $a = -100(x - lx/\sqrt{l^2 + x^2})$, determine the velocity of the collar as it passes through point C.

11.21 The acceleration of a particle is defined by the relation $a = -0.8v$, where a is expressed in m/s^2 and v in m/s. Knowing that at $t = 0$ the velocity is 1 m/s, determine (a) the distance the particle will travel before coming to rest, (b) the time required for the particle's velocity to be reduced by 50 percent of its initial value.

11.22 Starting from $x = 0$ with no initial velocity, a particle is given an acceleration $a = 0.1\sqrt{v^2 + 16}$, where a and v are expressed in ft/s^2 and ft/s, respectively. Determine (a) the position of the particle when $v = 3$ ft/s, (b) the speed and acceleration of the particle when $x = 4$ ft.

11.23 A ball is dropped from a boat so that it strikes the surface of a lake with a speed of 16.5 ft/s. While in the water the ball experiences an acceleration of $a = 10 - 0.8v$, where a and v are expressed in ft/s^2 and ft/s, respectively. Knowing the ball takes 3 s to reach the bottom of the lake, determine (a) the depth of the lake, (b) the speed of the ball when it hits the bottom of the lake.

11.24 The acceleration of a particle is defined by the relation $a = -k\sqrt{v}$, where k is a constant. Knowing that $x = 0$ and $v = 81$ m/s at $t = 0$ and that $v = 36$ m/s when $x = 18$ m, determine (a) the velocity of the particle when $x = 20$ m, (b) the time required for the particle to come to rest.

11.25 A particle is projected to the right from the position $x = 0$ with an initial velocity of 9 m/s. If the acceleration of the particle is defined by the relation $a = -0.6v^{3/2}$, where a and v are expressed in m/s^2 and m/s, respectively, determine (a) the distance the particle will have traveled when its velocity is 4 m/s, (b) the time when $v = 1$ m/s, (c) the time required for the particle to travel 6 m.

11.26 The acceleration of a particle is defined by the relation $a = 0.4(1 - kv)$, where k is a constant. Knowing that at $t = 0$ the particle starts from rest at $x = 4$ m and that when $t = 15$ s, $v = 4$ m/s, determine (a) the constant k, (b) the position of the particle when $v = 6$ m/s, (c) the maximum velocity of the particle.

11.27 Experimental data indicate that in a region downstream of a given louvered supply vent the velocity of the emitted air is defined by $v = 0.18v_0/x$, where v and x are expressed in m/s and meters, respectively, and v_0 is the initial discharge velocity of the air. For $v_0 = 3.6$ m/s, determine (a) the acceleration of the air at $x = 2$ m, (b) the time required for the air to flow from $x = 1$ to $x = 3$ m.

11.28 Based on observations, the speed of a jogger can be approximated by the relation $v = 7.5(1 - 0.04x)^{0.3}$, where v and x are expressed in mi/h and miles, respectively. Knowing that $x = 0$ at $t = 0$, determine (a) the distance the jogger has run when $t = 1$ h, (b) the jogger's acceleration in ft/s² at $t = 0$, (c) the time required for the jogger to run 6 mi.

Fig. P11.28

11.29 The acceleration due to gravity at an altitude y above the surface of the earth can be expressed as

$$a = \frac{-32.2}{[1 + (y/20.9 \times 10^6)]^2}$$

where a and y are expressed in ft/s² and feet, respectively. Using this expression, compute the height reached by a projectile fired vertically upward from the surface of the earth if its initial velocity is (a) 1800 ft/s, (b) 3000 ft/s, (c) 36,700 ft/s.

Fig. P11.29

11.30 The acceleration due to gravity of a particle falling toward the earth is $a = -gR^2/r^2$, where r is the distance from the *center* of the earth to the particle, R is the radius of the earth, and g is the acceleration due to gravity at the surface of the earth. If $R = 3960$ mi, calculate the *escape velocity*, that is, the minimum velocity with which a particle must be projected vertically upward from the surface of the earth if it is not to return to the earth. (*Hint:* $v = 0$ for $r = \infty$.)

Fig. P11.30

11.31 The velocity of a particle is $v = v_0[1 - \sin(\pi t/T)]$. Knowing that the particle starts from the origin with an initial velocity v_0, determine (a) its position and its acceleration at $t = 3T$, (b) its average velocity during the interval $t = 0$ to $t = T$.

11.32 The velocity of a slider is defined by the relation $v = v'\sin(\omega_n t + \phi)$. Denoting the velocity and the position of the slider at $t = 0$ by v_0 and x_0, respectively, and knowing that the maximum displacement of the slider is $2x_0$, show that (a) $v' = (v_0^2 + x_0^2\omega_n^2)/2x_0\omega_n$, (b) the maximum value of the velocity occurs when $x = x_0[3 - (v_0/x_0\omega_n)^2]/2$.

11.4 UNIFORM RECTILINEAR MOTION

Uniform rectilinear motion is a type of straight-line motion which is frequently encountered in practical applications. In this motion, the acceleration a of the particle is zero for every value of t. The velocity v is therefore constant, and Eq. (11.1) becomes

$$\frac{dx}{dt} = v = \text{constant}$$

The position coordinate x is obtained by integrating this equation. Denoting by x_0 the initial value of x, we write

$$\int_{x_0}^{x} dv = v \int_{0}^{t} dt$$

$$x - x_0 = vt$$

$$x = x_0 + vt \tag{11.5}$$

This equation can be used *only if the velocity of the particle is known to be constant.*

11.5 UNIFORMLY ACCELERATED RECTILINEAR MOTION

Uniformly accelerated rectilinear motion is another common type of motion. In this motion, the acceleration a of the particle is constant, and Eq. (11.2) becomes

$$\frac{dv}{dt} = a = \text{constant}$$

The velocity v of the particle is obtained by integrating this equation:

$$\int_{v_0}^{v} dv = a \int_{0}^{t} dt$$

$$v - v_0 = at$$

$$v = v_0 + at \tag{11.6}$$

where v_0 is the initial velocity. Substituting for v in (11.1), we write

$$\frac{dx}{dt} = v_0 + at$$

Denoting by x_0 the initial value of x and integrating, we have

$$\int_{x_0}^{x} dx = \int_{0}^{t} (v_0 + at)\,dt$$

$$x - x_0 = v_0 t + \tfrac{1}{2}at^2$$

$$x = x_0 + v_0 t + \tfrac{1}{2}at^2 \tag{11.7}$$

We can also use Eq. (11.4) and write

$$v \frac{dv}{dx} = a = \text{constant}$$

$$v \, dv = a \, dx$$

Integrating both sides, we obtain

$$\int_{v_0}^{v} v \, dv = a \int_{x_0}^{x} dx$$

$$\tfrac{1}{2}(v^2 - v_0^2) = a(x - x_0)$$

$$v^2 = v_0^2 + 2a(x - x_0) \qquad (11.8)$$

The three equations we have derived provide useful relations among position coordinate, velocity, and time in the case of a uniformly accelerated motion, as soon as appropriate values have been substituted for a, v_0, and x_0. The origin O of the x axis should first be defined and a positive direction chosen along the axis; this direction will be used to determine the signs of a, v_0, and x_0. Equation (11.6) relates v and t and should be used when the value of v corresponding to a given value of t is desired, or inversely. Equation (11.7) relates x and t; Eq. (11.8) relates v and x. An important application of uniformly accelerated motion is the motion of a *freely falling body*. The acceleration of a freely falling body (usually denoted by g) is equal to 9.81 m/s^2 or 32.2 ft/s^2.

It is important to keep in mind that the three equations can be used *only when the acceleration of the particle is known to be constant*. If the acceleration of the particle is variable, its motion should be determined from the fundamental equations (11.1) to (11.4) according to the methods outlined in Sec. 11.3.

11.6 MOTION OF SEVERAL PARTICLES

When several particles move independently along the same line, independent equations of motion can be written for each particle. Whenever possible, time should be recorded from the same initial instant for all particles, and displacements should be measured from the same origin and in the same direction. In other words, a single clock and a single measuring tape should be used.

Relative Motion of Two Particles. Consider two particles A and B moving along the same straight line (Fig. 11.7). If the position coordinates x_A and x_B are measured from the same origin, the difference $x_B - x_A$ defines the *relative position coordinate of B with respect to A* and is denoted by $x_{B/A}$. We write

$$x_{B/A} = x_B - x_A \qquad \text{or} \qquad x_B = x_A + x_{B/A} \qquad (11.9)$$

Regardless of the positions of A and B with respect to the origin, a positive sign for $x_{B/A}$ means that B is to the right of A, and a negative sign means that B is to the left of A.

Fig. 11.7

Photo 11.2 Multiple cables and pulleys are used by this shipyard crane.

Fig. 11.8

Fig. 11.9

The rate of change of $x_{B/A}$ is known as the *relative velocity of B with respect to A* and is denoted by $v_{B/A}$. Differentiating (11.9), we write

$$v_{B/A} = v_B - v_A \qquad \text{or} \qquad v_B = v_A + v_{B/A} \qquad (11.10)$$

A positive sign for $v_{B/A}$ means that B is *observed from A* to move in the positive direction; a negative sign means that it is observed to move in the negative direction.

The rate of change of $v_{B/A}$ is known as the *relative acceleration of B with respect to A* and is denoted by $a_{B/A}$. Differentiating (11.10), we obtain†

$$a_{B/A} = a_B - a_A \qquad \text{or} \qquad a_B = a_A + a_{B/A} \qquad (11.11)$$

Dependent Motions. Sometimes, the position of a particle will depend upon the position of another particle or of several other particles. The motions are then said to be *dependent*. For example, the position of block B in Fig. 11.8 depends upon the position of block A. Since the rope $ACDEFG$ is of constant length, and since the lengths of the portions of rope CD and EF wrapped around the pulleys remain constant, it follows that the sum of the lengths of the segments AC, DE, and FG is constant. Observing that the length of the segment AC differs from x_A only by a constant and that, similarly, the lengths of the segments DE and FG differ from x_B only by a constant, we write

$$x_A + 2x_B = \text{constant}$$

Since only one of the two coordinates x_A and x_B can be chosen arbitrarily, we say that the system shown in Fig. 11.8 has *one degree of freedom*. From the relation between the position coordinates x_A and x_B, it follows that if x_A is given an increment Δx_A, that is, if block A is lowered by an amount Δx_A, the coordinate x_B will receive an increment $\Delta x_B = -\frac{1}{2}\Delta x_A$. In other words, block B will rise by half the same amount; this can easily be checked directly from Fig. 11.8.

In the case of the three blocks of Fig. 11.9, we can again observe that the length of the rope which passes over the pulleys is constant, and thus the following relation must be satisfied by the position coordinates of the three blocks:

$$2x_A + 2x_B + x_C = \text{constant}$$

Since two of the coordinates can be chosen arbitrarily, we say that the system shown in Fig. 11.9 has *two degrees of freedom*.

When the relation existing between the position coordinates of several particles is *linear*, a similar relation holds between the velocities and between the accelerations of the particles. In the case of the blocks of Fig. 11.9, for instance, we differentiate twice the equation obtained and write

$$2\frac{dx_A}{dt} + 2\frac{dx_B}{dt} + \frac{dx_C}{dt} = 0 \qquad \text{or} \qquad 2v_A + 2v_B + v_C = 0$$

$$2\frac{dv_A}{dt} + 2\frac{dv_B}{dt} + \frac{dv_C}{dt} = 0 \qquad \text{or} \qquad 2a_A + 2a_B + a_C = 0$$

†Note that the product of the subscripts A and B/A used in the right-hand member of Eqs. (11.9), (11.10), and (11.11) is equal to the subscript B used in their left-hand member.

SAMPLE PROBLEM 11.4

A ball is thrown vertically upward from the 12-m level in an elevator shaft with an initial velocity of 18 m/s. At the same instant an open-platform elevator passes the 5-m level, moving upward with a constant velocity of 2 m/s. Determine (*a*) when and where the ball will hit the elevator, (*b*) the relative velocity of the ball with respect to the elevator when the ball hits the elevator.

SOLUTION

Motion of Ball. Since the ball has a constant acceleration, its motion is *uniformly accelerated*. Placing the origin O of the y axis at ground level and choosing its positive direction upward, we find that the initial position is $y_0 = +12$ m, the initial velocity is $v_0 = +18$ m/s, and the acceleration is $a = -9.81$ m/s^2. Substituting these values in the equations for uniformly accelerated motion, we write

$$v_B = v_0 + at \qquad v_B = 18 - 9.81t \qquad (1)$$
$$y_B = y_0 + v_0 t + \tfrac{1}{2}at^2 \qquad y_B = 12 + 18t - 4.905t^2 \qquad (2)$$

Motion of Elevator. Since the elevator has a constant velocity, its motion is *uniform*. Again placing the origin O at the ground level and choosing the positive direction upward, we note that $y_0 = +5$ m and write

$$v_E = +2 \text{ m/s} \qquad (3)$$
$$y_E = y_0 + v_E t \qquad y_E = 5 + 2t \qquad (4)$$

Ball Hits Elevator. We first note that the same time t and the same origin O were used in writing the equations of motion of both the ball and the elevator. We see from the figure that when the ball hits the elevator,

$$y_E = y_B \qquad (5)$$

Substituting for y_E and y_B from (2) and (4) into (5), we have

$$5 + 2t = 12 + 18t - 4.905t^2$$
$$t = -0.39 \text{ s} \qquad \text{and} \qquad t = 3.65 \text{ s} \blacktriangleleft$$

Only the root $t = 3.65$ s corresponds to a time after the motion has begun. Substituting this value into (4), we have

$$y_E = 5 + 2(3.65) = 12.30 \text{ m}$$
$$\text{Elevation from ground} = 12.30 \text{ m} \blacktriangleleft$$

The relative velocity of the ball with respect to the elevator is

$$v_{B/E} = v_B - v_E = (18 - 9.81t) - 2 = 16 - 9.81t$$

When the ball hits the elevator at time $t = 3.65$ s, we have

$$v_{B/E} = 16 - 9.81(3.65) \qquad v_{B/E} = -19.81 \text{ m/s} \blacktriangleleft$$

The negative sign means that the ball is observed from the elevator to be moving in the negative sense (downward).

SAMPLE PROBLEM 11.5

Collar A and block B are connected by a cable passing over three pulleys C, D, and E as shown. Pulleys C and E are fixed, while D is attached to a collar which is pulled downward with a constant velocity of 3 in./s. At $t = 0$, collar A starts moving downward from position K with a constant acceleration and no initial velocity. Knowing that the velocity of collar A is 12 in./s as it passes through point L, determine the change in elevation, the velocity, and the acceleration of block B when collar A passes through L.

SOLUTION

Motion of Collar A. We place the origin O at the upper horizontal surface and choose the positive direction downward. We observe that when $t = 0$, collar A is at the position K and $(v_A)_0 = 0$. Since $v_A = 12$ in./s and $x_A - (x_A)_0 = 8$ in. when the collar passes through L, we write

$$v_A^2 = (v_A)_0^2 + 2a_A[x_A - (x_A)_0] \qquad (12)^2 = 0 + 2a_A(8)$$
$$a_A = 9 \text{ in./s}^2$$

The time at which collar A reaches point L is obtained by writing

$$v_A = (v_A)_0 + a_A t \qquad 12 = 0 + 9t \qquad t = 1.333 \text{ s}$$

Motion of Pulley D. Recalling that the positive direction is downward, we write

$$a_D = 0 \qquad v_D = 3 \text{ in./s} \qquad x_D = (x_D)_0 + v_D t = (x_D)_0 + 3t$$

When collar A reaches L, at $t = 1.333$ s, we have

$$x_D = (x_D)_0 + 3(1.333) = (x_D)_0 + 4$$

Thus, $$x_D - (x_D)_0 = 4 \text{ in.}$$

Motion of Block B. We note that the total length of cable $ACDEB$ differs from the quantity $(x_A + 2x_D + x_B)$ only by a constant. Since the cable length is constant during the motion, this quantity must also remain constant. Thus, considering the times $t = 0$ and $t = 1.333$ s, we write

$$x_A + 2x_D + x_B = (x_A)_0 + 2(x_D)_0 + (x_B)_0 \qquad (1)$$
$$[x_A - (x_A)_0] + 2[x_D - (x_D)_0] + [x_B - (x_B)_0] = 0 \qquad (2)$$

But we know that $x_A - (x_A)_0 = 8$ in. and $x_D - (x_D)_0 = 4$ in.; substituting these values in (2), we find

$$8 + 2(4) + [x_B - (x_B)_0] = 0 \qquad x_B - (x_B)_0 = -16 \text{ in.}$$

Thus: Change in elevation of $B = 16$ in. ↑ ◄

Differentiating (1) twice, we obtain equations relating the velocities and the accelerations of A, B, and D. Substituting for the velocities and accelerations of A and D at $t = 1.333$ s, we have

$$v_A + 2v_D + v_B = 0: \qquad 12 + 2(3) + v_B = 0$$
$$v_B = -18 \text{ in./s} \qquad v_B = 18 \text{ in./s} ↑ \quad ◄$$
$$a_A + 2a_D + a_B = 0: \qquad 9 + 2(0) + a_B = 0$$
$$a_B = -9 \text{ in./s}^2 \qquad a_B = 9 \text{ in./s}^2 ↑ \quad ◄$$

SOLVING PROBLEMS
ON YOUR OWN

In this lesson we derived the equations that describe *uniform rectilinear motion* (constant velocity) and *uniformly accelerated rectilinear motion* (constant acceleration). We also introduced the concept of *relative motion*. The equations for relative motion [Eqs. (11.9) to (11.11)] can be applied to the independent or dependent motions of any two particles moving along the same straight line.

A. Independent motion of one or more particles. The solution of problems of this type should be organized as follows:

1. Begin your solution by listing the given information, sketching the system, and selecting the origin and the positive direction of the coordinate axis [Sample Prob. 11.4]. It is always advantageous to have a visual representation of problems of this type.

2. Write the equations that describe the motions of the various particles as well as those that describe how these motions are related [Eq. (5) of Sample Prob. 11.4].

3. Define the initial conditions, i.e., specify the state of the system corresponding to $t = 0$. This is especially important if the motions of the particles begin at different times. In such cases, either of two approaches can be used.

    a. Let $t = 0$ be the time when the last particle begins to move. You must then determine the initial position x_0 and the initial velocity v_0 of each of the other particles.

    b. Let $t = 0$ be the time when the first particle begins to move. You must then, in each of the equations describing the motion of another particle, replace t with $t - t_0$, where t_0 is the time at which that specific particle begins to move. It is important to recognize that the equations obtained in this way are valid only for $t \geq t_0$.

B. Dependent motion of two or more particles. In problems of this type the particles of the system are connected to each other, typically by ropes or by cables. The method of solution of these problems is similar to that of the preceding group of problems, except that it will now be necessary to describe the *physical connections* between the particles. In the following problems, the connection is provided by one or more cables. For each cable, you will have to write equations similar to the last three equations of Sec. 11.6. We suggest that you use the following procedure:

1. Draw a sketch of the system and select a coordinate system, indicating clearly a positive sense for each of the coordinate axes. For example, in Sample Prob. 11.5 lengths are measured downward from the upper horizontal support. It thus follows that those displacements, velocities, and accelerations which have positive values are directed downward.

2. Write the equation describing the constraint imposed by each cable on the motion of the particles involved. Differentiating this equation twice, you will obtain the corresponding relations among velocities and accelerations.

3. If several directions of motion are involved, you must select a coordinate axis and a positive sense for each of these directions. You should also try to locate the origins of your coordinate axes so that the equations of constraints will be as simple as possible. For example, in Sample Prob. 11.5 it is easier to define the various coordinates by measuring them downward from the upper support than by measuring them upward from the bottom support.

Finally, keep in mind that the method of analysis described in this lesson and the corresponding equations can be used only for particles moving with *uniform* or *uniformly accelerated rectilinear motion.*

PROBLEMS

11.33 A stone is thrown vertically upward from a point on a bridge located 40 m above the water. Knowing that it strikes the water 4 s after release, determine (a) the speed with which the stone was thrown upward, (b) the speed with which the stone strikes the water.

11.34 A motorist is traveling at 54 km/h when she observes that a traffic light 240 m ahead of her turns red. The traffic light is timed to stay red for 24 s. If the motorist wishes to pass the light without stopping just as it turns green again, determine (a) the required uniform deceleration of the car, (b) the speed of the car as it passes the light.

Fig. P11.34

11.35 A motorist enters a freeway at 30 mi/h and accelerates uniformly to 60 mi/h. From the odometer in the car, the motorist knows that she traveled 550 ft while accelerating. Determine (a) the acceleration of the car, (b) the time required to reach 60 mi/h.

Fig. P11.35

11.36 A group of students launches a model rocket in the vertical direction. Based on tracking data, they determine that the altitude of the rocket was 89.6 ft at the end of the powered portion of the flight and that the rocket landed 16 s later. Knowing that the descent parachute failed to deploy so that the rocket fell freely to the ground after reaching its maximum altitude and assuming that $g = 32.2$ ft/s^2, determine (a) the speed v_1 of the rocket at the end of powered flight, (b) the maximum altitude reached by the rocket.

Fig. P11.36

11.37 A small package is released from rest at A and moves along the skate wheel conveyor $ABCD$. The package has a uniform acceleration of 4.8 m/s² as it moves down sections AB and CD, and its velocity is constant between B and C. If the velocity of the package at D is 7.2 m/s, determine (a) the distance d between C and D, (b) the time required for the package to reach D.

Fig. P11.37

Fig. P11.38

11.38 A sprinter in a 100-m race accelerates uniformly for the first 35 m and then runs with constant velocity. If the sprinter's time for the first 35 m is 5.4 s, determine (a) his acceleration, (b) his final velocity, (c) his time for the race.

Fig. P11.39

11.39 As relay runner A enters the 20-m-long exchange zone with a speed of 12.9 m/s, he begins to slow down. He hands the baton to runner B 1.82 s later as they leave the exchange zone with the same velocity. Determine (a) the uniform acceleration of each of the runners, (b) when runner B should begin to run.

11.40 In a boat race, boat A is leading boat B by 50 m and both boats are traveling at a constant speed of 180 km/h. At $t = 0$, the boats accelerate at constant rates. Knowing that when B passes A, $t = 8$ s and $v_A = 225$ km/h, determine (a) the acceleration of A, (b) the acceleration of B.

Fig. P11.40

11.41 A police officer in a patrol car parked in a 45 mi/h speed zone observes a passing automobile traveling at a slow, constant speed. Believing that the driver of the automobile might be intoxicated, the officer starts his car, accelerates uniformly to 60 mi/h in 8 s, and, maintaining a constant velocity of 60 mi/h, overtakes the motorist 42 s after the automobile passed him. Knowing that 18 s elapsed before the officer began pursuing the motorist, determine (*a*) the distance the officer traveled before overtaking the motorist, (*b*) the motorist's speed.

11.42 Automobiles *A* and *B* are traveling in adjacent highway lanes and at *t* = 0 have the positions and speeds shown. Knowing that automobile *A* has a constant acceleration of 1.8 ft/s² and that *B* has a constant deceleration of 1.2 ft/s², determine (*a*) when and where *A* will overtake *B*, (*b*) the speed of each automobile at that time.

Fig. P11.42

11.43 Two automobiles *A* and *B* are approaching each other in adjacent highway lanes. At *t* = 0, *A* and *B* are 3200 ft apart, their speeds are v_A = 65 mi/h and v_B = 40 mi/h, and they are at points *P* and *Q*, respectively. Knowing that *A* passes point *Q* 40 s after *B* was there and that *B* passes point *P* 42 s after *A* was there, determine (*a*) the uniform accelerations of *A* and *B*, (*b*) when the vehicles pass each other, (*c*) the speed of *B* at that time.

Fig. P11.43

11.44 An elevator is moving upward at a constant speed of 4 m/s. A man standing 10 m above the top of the elevator throws a ball upward with a speed of 3 m/s. Determine (*a*) when the ball will hit the elevator, (*b*) where the ball will hit the elevator with respect to the location of the man.

Fig. P11.44

Fig. P11.45

11.45 Two rockets are launched at a fireworks display. Rocket A is launched with an initial velocity $v_0 = 100$ m/s and rocket B is launched t_1 s later with the same initial velocity. The two rockets are timed to explode simultaneously at a height of 300 m as A is falling and B is rising. Assuming a constant acceleration $g = 9.81$ m/s², determine (a) the time t_1, (b) the velocity of B relative to A at the time of the explosion.

11.46 Car A is parked along the northbound lane of a highway, and car B is traveling in the southbound lane at a constant speed of 60 mi/h. At $t = 0$, A starts and accelerates at a constant rate a_A, while at $t = 5$ s, B begins to slow down with a constant deceleration of magnitude $a_A/6$. Knowing that when the cars pass each other $x = 294$ ft and $v_A = v_B$, determine (a) the acceleration a_A, (b) when the vehicles pass each other, (c) the distance d between the vehicles at $t = 0$.

Fig. P11.46

11.47 The elevator shown in the figure moves downward with a constant velocity of 4 m/s. Determine (a) the velocity of the cable C, (b) the velocity of the counterweight W, (c) the relative velocity of the cable C with respect to the elevator, (d) the relative velocity of the counterweight W with respect to the elevator.

Fig. P11.47 and P11.48

11.48 The elevator shown starts from rest and moves upward with a constant acceleration. If the counterweight W moves through 30 ft in 5 s, determine (a) the acceleration of the elevator and the cable C, (b) the velocity of the elevator after 5 s.

11.49 Slider block A moves to the left with a constant velocity of 6 m/s. Determine (a) the velocity of block B, (b) the velocity of portion D of the cable, (c) the relative velocity of portion C of the cable with respect to portion D.

Fig. P11.49 and P11.50

11.50 Block B starts from rest and moves downward with a constant acceleration. Knowing that after slider block A has moved 9 in. its velocity is 6 ft/s, determine (a) the accelerations of A and B, (b) the velocity and the change in position of B after 2 s.

11.51 Slider block B moves to the right with a constant velocity of 300 mm/s. Determine (a) the velocity of slider block A, (b) the velocity of portion C of the cable, (c) the velocity of portion D of the cable, (d) the relative velocity of portion C of the cable with respect to slider block A.

Fig. P11.51 and P11.52

11.52 At the instant shown, slider block B is moving with a constant acceleration, and its speed is 150 mm/s. Knowing that after slider block A has moved 240 mm to the right its velocity is 60 mm/s, determine (a) the accelerations of A and B, (b) the acceleration of portion D of the cable, (c) the velocity and the change in position of slider block B after 4 s.

11.53 Collar A starts from rest and moves upward with a constant acceleration. Knowing that after 8 s the relative velocity of collar B with respect to collar A is 24 in./s, determine (a) the accelerations of A and B, (b) the velocity and the change in position of B after 6 s.

Fig. P11.53

11.54 The motor M reels in the cable at a constant rate of 100 mm/s. Determine (a) the velocity of load L, (b) the velocity of pulley B with respect to load L.

Fig. P11.54

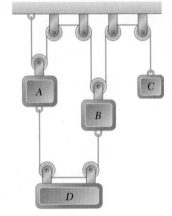

Fig. P11.55

11.55 Block C starts from rest at $t = 0$ and moves downward with a constant acceleration of 4 in./s². Knowing that block B has a constant velocity of 3 in./s upward, determine (a) the time when the velocity of block A is zero, (b) the time when the velocity of block A is equal to the velocity of block D, (c) the change in position of block A after 5 s.

11.56 Block A starts from rest at $t = 0$ and moves downward with a constant acceleration of 6 in./s². Knowing that block B moves up with a constant velocity of 3 in./s, determine (a) the time when the velocity of block C is zero, (b) the corresponding position of block C.

Fig. P11.56

11.57 Block B starts from rest, block A moves with a constant acceleration, and slider block C moves to the right with a constant acceleration of 75 mm/s². Knowing that at $t = 2$ s the velocities of B and C are 480 mm/s downward and 280 mm/s to the right, respectively, determine (a) the accelerations of A and B, (b) the initial velocities of A and C, (c) the change in position of slider block C after 3 s.

11.58 Block B moves downward with a constant velocity of 20 mm/s. At $t = 0$, block A is moving upward with a constant acceleration, and its velocity is 30 mm/s. Knowing that at $t = 3$ s slider block C has moved 57 mm to the right, determine (a) the velocity of slider block C at $t = 0$, (b) the accelerations of A and C, (c) the change in position of block A after 5 s.

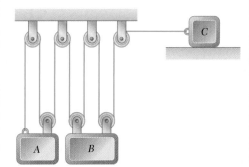

Fig. P11.57 and P11.58

11.59 The system shown starts from rest, and each component moves with a constant acceleration. If the relative acceleration of block C with respect to collar B is 60 mm/s² upward and the relative acceleration of block D with respect to block A is 110 mm/s² downward, determine (a) the velocity of block C after 3 s, (b) the change in position of block D after 5 s.

Fig. P11.59 and P11.60

***11.60** The system shown starts from rest, and the length of the upper cord is adjusted so that A, B, and C are initially at the same level. Each component moves with a constant acceleration, and after 2 s the relative change in position of block C with respect to block A is 280 mm upward. Knowing that when the relative velocity of collar B with respect to block A is 80 mm/s downward, the displacements of A and B are 160 mm downward and 320 mm downward, respectively, determine (a) the accelerations of A and B if $a_B > 10$ mm/s², (b) the change in position of block D when the velocity of block C is 600 mm/s upward.

*11.7 GRAPHICAL SOLUTION OF RECTILINEAR-MOTION PROBLEMS

It was observed in Sec. 11.2 that the fundamental formulas

$$v = \frac{dx}{dt} \quad \text{and} \quad a = \frac{dv}{dt}$$

have a geometrical significance. The first formula expresses that the velocity at any instant is equal to the slope of the x–t curve at the same instant (Fig. 11.10). The second formula expresses that the accel-

Fig. 11.10

eration is equal to the slope of the v–t curve. These two properties can be used to determine graphically the v–t and a–t curves of a motion when the x–t curve is known.

Integrating the two fundamental formulas from a time t_1 to a time t_2, we write

$$x_2 - x_1 = \int_{t_1}^{t_2} v \, dt \quad \text{and} \quad v_2 - v_1 = \int_{t_1}^{t_2} a \, dt \quad (11.12)$$

The first formula expresses that the area measured under the v–t curve from t_1 to t_2 is equal to the change in x during that time interval (Fig. 11.11). Similarly, the second formula expresses that the area measured under the a–t curve from t_1 to t_2 is equal to the change in v during that time interval. These two properties can be used to determine graphically the x–t curve of a motion when its v–t curve or its a–t curve is known (see Sample Prob. 11.6).

Graphical solutions are particularly useful when the motion considered is defined from experimental data and when x, v, and a are not analytical functions of t. They can also be used to advantage when the motion consists of distinct parts and when its analysis requires writing a different equation for each of its parts. When using a graphical solution, however, one should be careful to note that (1) the area under the v–t curve measures the *change in* x, not x itself, and similarly, that the area under the a–t curve measures the change in v; (2) an area above the t axis corresponds to an *increase* in x or v, while an area located below the t axis measures a *decrease* in x or v.

It will be useful to remember in drawing motion curves that if the velocity is constant, it will be represented by a horizontal straight line; the position coordinate x will then be a linear function of t and will be represented by an oblique straight line. If the acceleration is

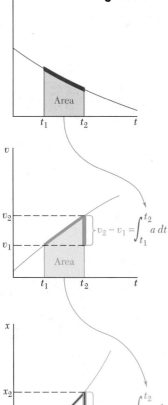

Fig. 11.11

constant and different from zero, it will be represented by a horizontal straight line; v will then be a linear function of t, represented by an oblique straight line, and x will be expressed as a second-degree polynomial in t, represented by a parabola. If the acceleration is a linear function of t, the velocity and the position coordinate will be equal, respectively, to second-degree and third-degree polynomials; a will then be represented by an oblique straight line, v by a parabola, and x by a cubic. In general, if the acceleration is a polynomial of degree n in t, the velocity will be a polynomial of degree $n + 1$ and the position coordinate a polynomial of degree $n + 2$; these polynomials are represented by motion curves of a corresponding degree.

*11.8 OTHER GRAPHICAL METHODS

An alternative graphical solution can be used to determine the position of a particle at a given instant directly from the a–t curve. Denoting the values of x and v at $t = 0$ as x_0 and v_0 and their values at $t = t_1$ as x_1 and v_1, and observing that the area under the v–t curve can be divided into a rectangle of area $v_0 t_1$ and horizontal differential elements of area $(t_1 - t)\, dv$ (Fig. 11.12a), we write

$$x_1 - x_0 = \text{area under } v\text{–}t \text{ curve} = v_0 t_1 + \int_{v_0}^{v_1} (t_1 - t)\, dv$$

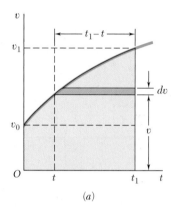

(a)

Substituting $dv = a\, dt$ in the integral, we obtain

$$x_1 - x_0 = v_0 t_1 + \int_{0}^{t_1} (t_1 - t)\, a\, dt$$

Referring to Fig. 11.12b, we note that the integral represents the first moment of the area under the a–t curve with respect to the line $t = t_1$ bounding the area on the right. This method of solution is known, therefore, as the *moment-area method*. If the abscissa $\bar{t}$ of the centroid C of the area is known, the position coordinate x_1 can be obtained by writing

$$x_1 = x_0 + v_0 t_1 + (\text{area under } a\text{–}t \text{ curve})(t_1 - \bar{t}) \qquad (11.13)$$

If the area under the a–t curve is a composite area, the last term in (11.13) can be obtained by multiplying each component area by the distance from its centroid to the line $t = t_1$. Areas above the t axis should be considered as positive and areas below the t axis as negative.

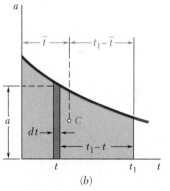

(b)

Fig. 11.12

Another type of motion curve, the v–x curve, is sometimes used. If such a curve has been plotted (Fig. 11.13), the acceleration a can be obtained at any time by drawing the normal AC to the curve and *measuring the subnormal* BC. Indeed, observing that the angle between AC and AB is equal to the angle θ between the horizontal and the tangent at A (the slope of which is $\tan \theta = dv/dx$), we write

$$BC = AB \tan \theta = v \frac{dv}{dx}$$

and thus, recalling formula (11.4),

$$BC = a$$

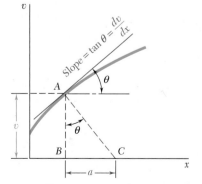

Fig. 11.13

SAMPLE PROBLEM 11.6

A subway car leaves station A; it gains speed at the rate of 4 ft/s^2 for 6 s and then at the rate of 6 ft/s^2 until it has reached the speed of 48 ft/s. The car maintains the same speed until it approaches station B; brakes are then applied, giving the car a constant deceleration and bringing it to a stop in 6 s. The total running time from A to B is 40 s. Draw the a–t, v–t, and x–t curves, and determine the distance between stations A and B.

SOLUTION

Acceleration-Time Curve. Since the acceleration is either constant or zero, the a–t curve is made of horizontal straight-line segments. The values of t_2 and a_4 are determined as follows:

$0 < t < 6$:   Change in v = area under a–t curve
$$v_6 - 0 = (6 \text{ s})(4 \text{ ft/s}^2) = 24 \text{ ft/s}$$

$6 < t < t_2$:   Since the velocity increases from 24 to 48 ft/s,

Change in v = area under a–t curve
$$48 \text{ ft/s} - 24 \text{ ft/s} = (t_2 - 6)(6 \text{ ft/s}^2) \qquad t_2 = 10 \text{ s}$$

$t_2 < t < 34$:   Since the velocity is constant, the acceleration is zero.

$34 < t < 40$:   Change in v = area under a–t curve
$$0 - 48 \text{ ft/s} = (6 \text{ s})a_4 \qquad a_4 = -8 \text{ ft/s}^2$$

The acceleration being negative, the corresponding area is below the t axis; this area represents a decrease in velocity.

Velocity-Time Curve. Since the acceleration is either constant or zero, the v–t curve is made of straight-line segments connecting the points determined above.

Change in x = area under v–t curve

$0 < t < 6$: $x_6 - 0 = \frac{1}{2}(6)(24) = 72 \text{ ft}$

$6 < t < 10$: $x_{10} - x_6 = \frac{1}{2}(4)(24 + 48) = 144 \text{ ft}$

$10 < t < 34$: $x_{34} - x_{10} = (24)(48) = 1152 \text{ ft}$

$34 < t < 40$: $x_{40} - x_{34} = \frac{1}{2}(6)(48) = 144 \text{ ft}$

Adding the changes in x, we obtain the distance from A to B:

$$d = x_{40} - 0 = 1512 \text{ ft}$$

$$d = 1512 \text{ ft} \blacktriangleleft$$

Position-Time Curve. The points determined above should be joined by three arcs of parabola and one straight-line segment. In constructing the x–t curve, keep in mind that for any value of t the slope of the tangent to the x–t curve is equal to the value of v at that instant.

SOLVING PROBLEMS ON YOUR OWN

In this lesson (Secs. 11.7 and 11.8), we reviewed and developed several *graphical techniques for the solution of problems involving rectilinear motion*. These techniques can be used to solve problems directly or to complement analytical methods of solution by providing a visual description, and thus a better understanding, of the motion of a given body. We suggest that you sketch one or more motion curves for several of the problems in this lesson, even if these problems are not part of your homework assignment.

1. Drawing *x–t*, *v–t*, and *a–t* curves and applying graphical methods. The following properties were indicated in Sec. 11.7 and should be kept in mind as you use a graphical method of solution.

   a. The slopes of the *x–t* and *v–t* curves at a time t_1 are respectively equal to the *velocity* and the *acceleration* at time t_1.

   b. The areas under the *a–t* and *v–t* curves between the times t_1 and t_2 are respectively equal to the change Δv in the velocity and to the change Δx in the position coordinate during that time interval.

   c. If one of the motion curves is known, the fundamental properties we have summarized in paragraphs *a* and *b* will enable you to construct the other two curves. However, when using the properties of paragraph *b*, the velocity and the position coordinate at time t_1 must be known in order to determine the velocity and the position coordinate at time t_2. Thus, in Sample Prob. 11.6, knowing that the initial value of the velocity was zero allowed us to find the velocity at $t = 6$ s: $v_6 = v_0 + \Delta v = 0 + 24$ ft/s $= 24$ ft/s.

If you have previously studied the shear and bending-moment diagrams for a beam, you should recognize the analogy that exists between the three motion curves and the three diagrams representing respectively the distributed load, the shear, and the bending moment in the beam. Thus, any techniques that you learned regarding the construction of these diagrams can be applied when drawing the motion curves.

2. Using approximate methods. When the *a–t* and *v–t* curves are not represented by analytical functions or when they are based on experimental data, it is often necessary to use approximate methods to calculate the areas under these curves. In those cases, the given area is approximated by a series of rectangles of width Δt. The smaller the value of Δt, the smaller the error introduced by the approximation. The velocity and the position coordinate are obtained by writing

$$v = v_0 + \Sigma a_{\text{ave}} \, \Delta t \qquad x = x_0 + \Sigma v_{\text{ave}} \, \Delta t$$

where a_{ave} and v_{ave} are the heights of an acceleration rectangle and a velocity rectangle, respectively.

(continued)

3. Applying the moment-area method.

This graphical technique is used when the a–t curve is given and the change in the position coordinate is to be determined. We found in Sec. 11.8 that the position coordinate x_1 can be expressed as

$$x_1 = x_0 + v_0 t_1 + (\text{area under } a\text{–}t \text{ curve})(t_1 - \bar{t}) \qquad (11.13)$$

Keep in mind that when the area under the a–t curve is a composite area, the same value of t_1 should be used for computing the contribution of each of the component areas.

4. Determining the acceleration from a v–x curve.

You saw in Sec. 11.8 that it is possible to determine the acceleration from a v–x curve by direct measurement. It is important to note, however, that this method is applicable only if the same linear scale is used for the v and x axes (for example, 1 in. = 10 ft and 1 in. = 10 ft/s). When this condition is not satisfied, the acceleration can still be determined from the equation

$$a = v \frac{dv}{dx}$$

where the slope dv/dx is obtained as follows: First, draw the tangent to the curve at the point of interest. Next, using appropriate scales, measure along that tangent corresponding increments Δx and Δv. The desired slope is equal to the ratio $\Delta v / \Delta x$.

PROBLEMS

11.61 A particle moves in a straight line with the acceleration shown in the figure. Knowing that it starts from the origin with $v_0 = -14$ ft/s, plot the v–t and x–t curves for $0 < t < 15$ s and determine (a) the maximum value of the velocity of the particle, (b) the maximum value of its position coordinate.

11.62 For the particle and motion of Prob. 11.61, plot the v–t and x–t curves for $0 < t < 15$ s and determine the velocity of the particle, its position, and the total distance traveled after 10 s.

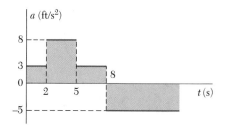

Fig. P11.61 and P11.62

11.63 A particle moves in a straight line with the velocity shown in the figure. Knowing that $x = -540$ m at $t = 0$, (a) construct the a–t and x–t curves for $0 < t < 50$ s, and determine (b) the total distance traveled by the particle when $t = 50$ s, (c) the two times at which $x = 0$.

Fig. P11.63 and P11.64

11.64 A particle moves in a straight line with the velocity shown in the figure. Knowing that $x = -540$ m at $t = 0$, (a) construct the a–t and x–t curves for $0 < t < 50$ s, and determine (b) the maximum value of the position coordinate of the particle, (c) the values of t for which the particle is at $x = 100$ m.

11.65 During a finishing operation the bed of an industrial planer moves alternately 30 in. to the right and 30 in. to the left. The velocity of the bed is limited to a maximum value of 6 in./s to the right and 12 in./s to the left; the acceleration is successively equal to 6 in./s^2 to the right, zero, 6 in./s^2 to the left, zero, etc. Determine the time required for the bed to complete a full cycle, and draw the v–t and x–t curves.

11.66 A parachutist is in free fall at a rate of 200 km/h when he opens his parachute at an altitude of 600 m. Following a rapid and constant deceleration, he then descends at a constant rate of 50 km/h from 586 m to 30 m, where he maneuvers the parachute into the wind to further slow his descent. Knowing that the parachutist lands with a negligible downward velocity, determine (a) the time required for the parachutist to land after opening his parachute, (b) the initial deceleration.

Fig. P11.66

637

11.67 A commuter train traveling at 40 mi/h is 3 mi from a station. The train then decelerates so that its speed is 20 mi/h when it is 0.5 mi from the station. Knowing that the train arrives at the station 7.5 min after beginning to decelerate and assuming constant decelerations, determine (a) the time required for the train to travel the first 2.5 mi, (b) the speed of the train as it arrives at the station, (c) the final constant deceleration of the train.

Fig. P11.67

Fig. P11.68

11.68 A temperature sensor is attached to slider AB which moves back and forth through 60 in. The maximum velocities of the slider are 12 in./s to the right and 30 in./s to the left. When the slider is moving to the right, it accelerates and decelerates at a constant rate of 6 in./s^2; when moving to the left, the slider accelerates and decelerates at a constant rate of 20 in./s^2. Determine the time required for the slider to complete a full cycle, and construct the v–t and x–t curves of its motion.

11.69 In a water-tank test involving the launching of a small model boat, the model's initial horizontal velocity is 6 m/s and its horizontal acceleration varies linearly from -12 m/s^2 at $t = 0$ to -2 m/s^2 at $t = t_1$ and then remains equal to -2 m/s^2 until $t = 1.4$ s. Knowing that $v = 1.8$ m/s when $t = t_1$, determine (a) the value of t_1, (b) the velocity and the position of the model at $t = 1.4$ s.

Fig. P11.69

11.70 The acceleration record shown was obtained for a small airplane traveling along a straight course. Knowing that $x = 0$ and $v = 60$ m/s when $t = 0$, determine (a) the velocity and position of the plane at $t = 20$ s, (b) its average velocity during the interval 6 s $< t < 14$ s.

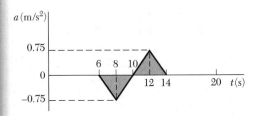

Fig. P11.70

11.71 In a 400-m race, runner A reaches her maximum velocity v_A in 4 s with constant acceleration and maintains that velocity until she reaches the halfway point with a split time of 25 s. Runner B reaches her maximum velocity v_B in 5 s with constant acceleration and maintains that velocity until she reaches the halfway point with a split time of 25.2 s. Both runners then run the second half of the race with the same constant deceleration of 0.1 m/s^2. Determine (a) the race times for both runners, (b) the position of the winner relative to the loser when the winner reaches the finish line.

Fig. P11.71

11.72 A car and a truck are both traveling at the constant speed of 35 mi/h; the car is 40 ft behind the truck. The driver of the car wants to pass the truck, i.e., he wishes to place his car at B, 40 ft in front of the truck, and then resume the speed of 35 mi/h. The maximum acceleration of the car is 5 ft/s² and the maximum deceleration obtained by applying the brakes is 20 ft/s². What is the shortest time in which the driver of the car can complete the passing operation if he does not at any time exceed a speed of 50 mi/h? Draw the v–t curve.

Fig. P11.72

11.73 Solve Prob. 11.72, assuming that the driver of the car does not pay any attention to the speed limit while passing and concentrates on reaching position B and resuming a speed of 35 mi/h in the shortest possible time. What is the maximum speed reached? Draw the v–t curve.

11.74 Car A is traveling on a highway at a constant speed $(v_A)_0 = 60$ mi/h and is 380 ft from the entrance of an access ramp when car B enters the acceleration lane at that point at a speed $(v_B)_0 = 15$ mi/h. Car B accelerates uniformly and enters the main traffic lane after traveling 200 ft in 5 s. It then continues to accelerate at the same rate until it reaches a speed of 60 mi/h, which it then maintains. Determine the final distance between the two cars.

Fig. P11.74

Fig. P11.75

11.75 An elevator starts from rest and moves upward, accelerating at a rate of 1.2 m/s² until it reaches a speed of 7.8 m/s, which it then maintains. Two seconds after the elevator begins to move, a man standing 12 m above the initial position of the top of the elevator throws a ball upward with an initial velocity of 20 m/s. Determine when the ball will hit the elevator.

11.76 Car A is traveling at 40 mi/h when it enters a 30 mi/h speed zone. The driver of car A decelerates at a rate of 16 ft/s² until reaching a speed of 30 mi/h, which she then maintains. When car B, which was initially 60 ft behind car A and traveling at a constant speed of 45 mi/h, enters the speed zone, its driver decelerates at a rate of 20 ft/s² until reaching a speed of 28 mi/h. Knowing that the driver of car B maintains a speed of 28 mi/h, determine (a) the closest that car B comes to car A, (b) the time at which car A is 70 ft in front of car B.

Fig. P11.76

11.77 An accelerometer record for the motion of a given part of a mechanism is approximated by an arc of a parabola for 0.2 s and a straight line for the next 0.2 s as shown in the figure. Knowing that $v = 0$ when $t = 0$ and $x = 0.8$ ft when $t = 0.4$ s, (a) construct the v–t curve for $0 \le t \le 0.4$ s, (b) determine the position of the part at $t = 0.3$ s and $t = 0.2$ s.

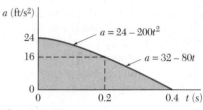

Fig. P11.77

11.78 A car is traveling at a constant speed of 54 km/h when its driver sees a child run into the road. The driver applies her brakes until the child returns to the sidewalk and then accelerates to resume her original speed of 54 km/h; the acceleration record of the car is shown in the figure. Assuming $x = 0$ when $t = 0$, determine (a) the time t_1 at which the velocity is again 54 km/h, (b) the position of the car at that time, (c) the average velocity of the car during the interval 1 s $\le t \le t_1$.

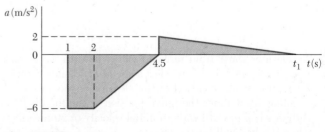

Fig. P11.78

11.79 An airport shuttle train travels between two terminals that are 1.6 mi apart. To maintain passenger comfort, the acceleration of the train is limited to ± 4 ft/s^2, and the jerk, or rate of change of acceleration, is limited to ± 0.8 ft/s^2 per second. If the shuttle has a maximum speed of 20 mi/h, determine (a) the shortest time for the shuttle to travel between the two terminals, (b) the corresponding average velocity of the shuttle.

11.80 During a manufacturing process, a conveyor belt starts from rest and travels a total of 1.2 ft before temporarily coming to rest. Knowing that the jerk, or rate of change of acceleration, is limited to ± 4.8 ft/s^2 per second, determine (a) the shortest time required for the belt to move 1.2 ft, (b) the maximum and average values of the velocity of the belt during that time.

11.81 Two seconds are required to bring the piston rod of an air cylinder to rest; the acceleration record of the piston rod during the 2 s is as shown. Determine by approximate means (a) the initial velocity of the piston rod, (b) the distance traveled by the piston rod as it is brought to rest.

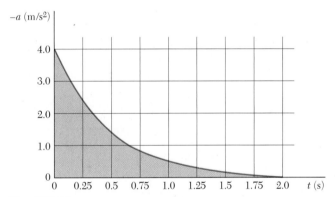

Fig. P11.81

11.82 The acceleration record shown was obtained during the speed trials of a sports car. Knowing that the car starts from rest, determine by approximate means (a) the velocity of the car at $t = 8$ s, (b) the distance the car has traveled at $t = 20$ s.

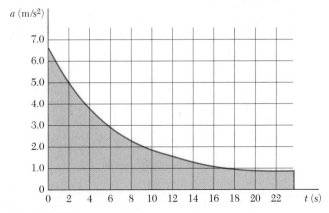

Fig. P11.82

11.83 A training airplane has a velocity of 126 ft/s when it lands on an aircraft carrier. As the arresting gear of the carrier brings the airplane to rest, the velocity and the acceleration of the airplane are recorded; the results are shown (solid curve) in the figure. Determine by approximate means (a) the time required for the airplane to come to rest, (b) the distance traveled in that time.

Fig. P11.83

11.84 Shown in the figure is a portion of the experimentally determined v–x curve for a shuttle cart. Determine by approximate means the acceleration of the cart when (a) $x = 10$ in., (b) $v = 80$ in./s.

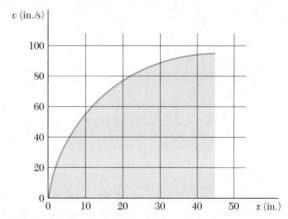

Fig. P11.84

11.85 Using the method of Sec. 11.8, derive the formula $x = x_0 + v_0 t + \frac{1}{2}at^2$ for the position coordinate of a particle in uniformly accelerated rectilinear motion.

11.86 Using the method of Sec. 11.8, determine the position of the particle of Prob. 11.61 when $t = 8$ s.

11.87 The acceleration of an object subjected to the pressure wave of a large explosion is defined approximately by the curve shown. The object is initially at rest and is again at rest at time t_1. Using the method of Sec. 11.8, determine (a) the time t_1, (b) the distance through which the object is moved by the pressure wave.

11.88 For the particle of Prob. 11.63, draw the a–t curve and determine, using the method of Sec. 11.8, (a) the position of the particle when $t = 52$ s, (b) the maximum value of its position coordinate.

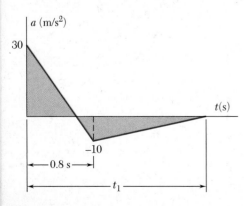

Fig. P11.87

CURVILINEAR MOTION OF PARTICLES

11.9 POSITION VECTOR, VELOCITY, AND ACCELERATION

When a particle moves along a curve other than a straight line, we say that the particle is in *curvilinear motion*. To define the position P occupied by the particle at a given time t, we select a fixed reference system, such as the x, y, z axes shown in Fig. 11.14a, and draw the vector $\mathbf{r}$ joining the origin O and point P. Since the vector $\mathbf{r}$ is characterized by its magnitude r and its direction with respect to the reference axes, it completely defines the position of the particle with respect to those axes; the vector $\mathbf{r}$ is referred to as the *position vector* of the particle at time t.

Consider now the vector $\mathbf{r}'$ defining the position P' occupied by the same particle at a later time $t + \Delta t$. The vector $\Delta \mathbf{r}$ joining P and P' represents the change in the position vector during the time interval Δt since, as we can easily check from Fig. 11.14a, the vector $\mathbf{r}'$ is obtained by adding the vectors $\mathbf{r}$ and $\Delta \mathbf{r}$ according to the triangle rule. We note that $\Delta \mathbf{r}$ represents a change in *direction* as well as a change in *magnitude* of the position vector $\mathbf{r}$. The *average velocity* of the particle over the time interval Δt is defined as the quotient of $\Delta \mathbf{r}$ and Δt. Since $\Delta \mathbf{r}$ is a vector and Δt is a scalar, the quotient $\Delta \mathbf{r}/\Delta t$ is a vector attached at P, of the same direction as $\Delta \mathbf{r}$ and of magnitude equal to the magnitude of $\Delta \mathbf{r}$ divided by Δt (Fig. 11.14b).

The *instantaneous velocity* of the particle at time t is obtained by choosing shorter and shorter time intervals Δt and, correspondingly, shorter and shorter vector increments $\Delta \mathbf{r}$. The instantaneous velocity is thus represented by the vector

$$\mathbf{v} = \lim_{\Delta t \to 0} \frac{\Delta \mathbf{r}}{\Delta t} \qquad (11.14)$$

As Δt and $\Delta \mathbf{r}$ become shorter, the points P and P' get closer; the vector $\mathbf{v}$ obtained in the limit must therefore be tangent to the path of the particle (Fig. 11.14c).

Since the position vector $\mathbf{r}$ depends upon the time t, we can refer to it as a *vector function* of the scalar variable t and denote it by $\mathbf{r}(t)$. Extending the concept of derivative of a scalar function introduced in elementary calculus, we will refer to the limit of the quotient $\Delta \mathbf{r}/\Delta t$ as the *derivative* of the vector function $\mathbf{r}(t)$. We write

$$\mathbf{v} = \frac{d\mathbf{r}}{dt} \qquad (11.15)$$

The magnitude v of the vector $\mathbf{v}$ is called the *speed* of the particle. It can be obtained by substituting for the vector $\Delta \mathbf{r}$ in formula (11.14) the magnitude of this vector represented by the straight-line segment PP'. But the length of the segment PP' approaches the length Δs of the arc PP' as Δt decreases (Fig. 11.14a), and we can write

$$v = \lim_{\Delta t \to 0} \frac{PP'}{\Delta t} = \lim_{\Delta t \to 0} \frac{\Delta s}{\Delta t} \qquad v = \frac{ds}{dt} \qquad (11.16)$$

Fig. 11.14

(a)

(b)

(c)

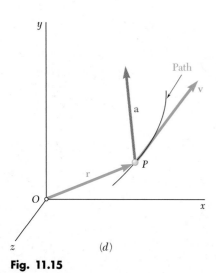

(d)

Fig. 11.15

The speed v can thus be obtained by differentiating with respect to t the length s of the arc described by the particle.

Consider the velocity $\mathbf{v}$ of the particle at time t and its velocity $\mathbf{v}'$ at a later time $t + \Delta t$ (Fig. 11.15a). Let us draw both vectors $\mathbf{v}$ and $\mathbf{v}'$ from the same origin O' (Fig. 11.15b). The vector $\Delta\mathbf{v}$ joining Q and Q' represents the change in the velocity of the particle during the time interval Δt, since the vector $\mathbf{v}'$ can be obtained by adding the vectors $\mathbf{v}$ and $\Delta\mathbf{v}$. We should note that $\Delta\mathbf{v}$ represents a change in the *direction* of the velocity as well as a change in *speed*. The *average acceleration* of the particle over the time interval Δt is defined as the quotient of $\Delta\mathbf{v}$ and Δt. Since $\Delta\mathbf{v}$ is a vector and Δt a scalar, the quotient $\Delta\mathbf{v}/\Delta t$ is a vector of the same direction as $\Delta\mathbf{v}$.

The *instantaneous acceleration* of the particle at time t is obtained by choosing smaller and smaller values for Δt and $\Delta\mathbf{v}$. The instantaneous acceleration is thus represented by the vector

$$\mathbf{a} = \lim_{\Delta t \to 0} \frac{\Delta\mathbf{v}}{\Delta t} \tag{11.17}$$

Noting that the velocity $\mathbf{v}$ is a vector function $\mathbf{v}(t)$ of the time t, we can refer to the limit of the quotient $\Delta\mathbf{v}/\Delta t$ as the derivative of $\mathbf{v}$ with respect to t. We write

$$\mathbf{a} = \frac{d\mathbf{v}}{dt} \tag{11.18}$$

We observe that the acceleration $\mathbf{a}$ is tangent to the curve described by the tip Q of the vector $\mathbf{v}$ when the latter is drawn from a fixed origin O' (Fig. 11.15c) and that, in general, the acceleration is *not* tangent to the path of the particle (Fig. 11.15d). The curve described by the tip of $\mathbf{v}$ and shown in Fig. 11.15c is called the *hodograph* of the motion.

11.10 DERIVATIVES OF VECTOR FUNCTIONS

We saw in the preceding section that the velocity **v** of a particle in curvilinear motion can be represented by the derivative of the vector function **r**(t) characterizing the position of the particle. Similarly, the acceleration **a** of the particle can be represented by the derivative of the vector function **v**(t). In this section, we will give a formal definition of the derivative of a vector function and establish a few rules governing the differentiation of sums and products of vector functions.

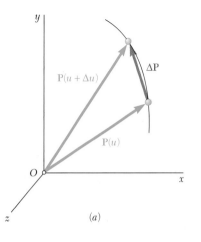

(a)

Let **P**(u) be a vector function of the scalar variable u. By that we mean that the scalar u completely defines the magnitude and direction of the vector **P**. If the vector **P** is drawn from a fixed origin O and the scalar u is allowed to vary, the tip of **P** will describe a given curve in space. Consider the vectors **P** corresponding, respectively, to the values u and u + Δu of the scalar variable (Fig. 11.16a). Let Δ**P** be the vector joining the tips of the two given vectors; we write

$$\Delta \mathbf{P} = \mathbf{P}(u + \Delta u) - \mathbf{P}(u)$$

Dividing through by Δu and letting Δu approach zero, *we define the derivative of the vector function* **P**(u):

$$\frac{d\mathbf{P}}{du} = \lim_{\Delta u \to 0} \frac{\Delta \mathbf{P}}{\Delta u} = \lim_{\Delta u \to 0} \frac{\mathbf{P}(u + \Delta u) - \mathbf{P}(u)}{\Delta u} \qquad (11.19)$$

As Δu approaches zero, the line of action of Δ**P** becomes tangent to the curve of Fig. 11.16a. Thus, the derivative d**P**/du of the vector function **P**(u) *is tangent to the curve described by the tip of* **P**(u) (Fig. 11.16b).

The standard rules for the differentiation of the sums and products of scalar functions can be extended to vector functions. Consider first the *sum of two vector functions* **P**(u) and **Q**(u) of the same scalar variable u. According to the definition given in (11.19), the derivative of the vector **P** + **Q** is

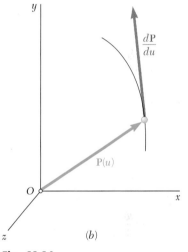

(b)

Fig. 11.16

$$\frac{d(\mathbf{P} + \mathbf{Q})}{du} = \lim_{\Delta u \to 0} \frac{\Delta(\mathbf{P} + \mathbf{Q})}{\Delta u} = \lim_{\Delta u \to 0} \left(\frac{\Delta \mathbf{P}}{\Delta u} + \frac{\Delta \mathbf{Q}}{\Delta u} \right)$$

or since the limit of a sum is equal to the sum of the limits of its terms,

$$\frac{d(\mathbf{P} + \mathbf{Q})}{du} = \lim_{\Delta u \to 0} \frac{\Delta \mathbf{P}}{\Delta u} + \lim_{\Delta u \to 0} \frac{\Delta \mathbf{Q}}{\Delta u}$$

$$\frac{d(\mathbf{P} + \mathbf{Q})}{du} = \frac{d\mathbf{P}}{du} + \frac{d\mathbf{Q}}{du} \qquad (11.20)$$

The *product of a scalar function f(u) and a vector function* **P**(u) of the same scalar variable u will now be considered. The derivative of the vector f **P** is

$$\frac{d(f\mathbf{P})}{du} = \lim_{\Delta u \to 0} \frac{(f + \Delta f)(\mathbf{P} + \Delta \mathbf{P}) - f\mathbf{P}}{\Delta u} = \lim_{\Delta u \to 0} \left(\frac{\Delta f}{\Delta u} \mathbf{P} + f \frac{\Delta \mathbf{P}}{\Delta u} \right)$$

or recalling the properties of the limits of sums and products,

$$\frac{d(f\mathbf{P})}{du} = \frac{df}{du}\mathbf{P} + f\frac{d\mathbf{P}}{du} \tag{11.21}$$

The derivatives of the *scalar product* and the *vector product* of two vector functions $\mathbf{P}(u)$ and $\mathbf{Q}(u)$ can be obtained in a similar way. We have

$$\frac{d(\mathbf{P} \cdot \mathbf{Q})}{du} = \frac{d\mathbf{P}}{du} \cdot \mathbf{Q} + \mathbf{P} \cdot \frac{d\mathbf{Q}}{du} \tag{11.22}$$

$$\frac{d(\mathbf{P} \times \mathbf{Q})}{du} = \frac{d\mathbf{P}}{du} \times \mathbf{Q} + \mathbf{P} \times \frac{d\mathbf{Q}}{du} \tag{11.23}\dagger$$

The properties established above can be used to determine the *rectangular components of the derivative of a vector function* $\mathbf{P}(u)$. Resolving $\mathbf{P}$ into components along fixed rectangular axes x, y, z, we write

$$\mathbf{P} = P_x\mathbf{i} + P_y\mathbf{j} + P_z\mathbf{k} \tag{11.24}$$

where P_x, P_y, P_z are the rectangular scalar components of the vector $\mathbf{P}$, and $\mathbf{i}$, $\mathbf{j}$, $\mathbf{k}$ the unit vectors corresponding, respectively, to the x, y, and z axes (Sec. 2.12). By (11.20), the derivative of $\mathbf{P}$ is equal to the sum of the derivatives of the terms in the right-hand member. Since each of these terms is the product of a scalar and a vector function, we should use (11.21). But the unit vectors $\mathbf{i}$, $\mathbf{j}$, $\mathbf{k}$ have a constant magnitude (equal to 1) and fixed directions. Their derivatives are therefore zero, and we write

$$\frac{d\mathbf{P}}{du} = \frac{dP_x}{du}\mathbf{i} + \frac{dP_y}{du}\mathbf{j} + \frac{dP_z}{du}\mathbf{k} \tag{11.25}$$

Noting that the coefficients of the unit vectors are, by definition, the scalar components of the vector $d\mathbf{P}/du$, we conclude that *the rectangular scalar components of the derivative $d\mathbf{P}/du$ of the vector function* $\mathbf{P}(u)$ *are obtained by differentiating the corresponding scalar components of* $\mathbf{P}$.

Rate of Change of a Vector. When the vector $\mathbf{P}$ is a function of the time t, its derivative $d\mathbf{P}/dt$ represents the *rate of change* of $\mathbf{P}$ with respect to the frame $Oxyz$. Resolving $\mathbf{P}$ into rectangular components, we have, by (11.25),

$$\frac{d\mathbf{P}}{dt} = \frac{dP_x}{dt}\mathbf{i} + \frac{dP_y}{dt}\mathbf{j} + \frac{dP_z}{dt}\mathbf{k}$$

or, using dots to indicate differentiation with respect to t,

$$\dot{\mathbf{P}} = \dot{P}_x\mathbf{i} + \dot{P}_y\mathbf{j} + \dot{P}_z\mathbf{k} \tag{11.25'}$$

---

†Since the vector product is not commutative (Sec. 3.4), the order of the factors in Eq. (11.23) must be maintained.

As you will see in Sec. 15.10, the rate of change of a vector as observed from a *moving frame of reference* is, in general, different from its rate of change as observed from a fixed frame of reference. However, if the moving frame $O'x'y'z'$ is in *translation*, i.e., if its axes remain parallel to the corresponding axes of the fixed frame $Oxyz$ (Fig. 11.17), the same unit vectors $\mathbf{i}$, $\mathbf{j}$, $\mathbf{k}$ are used in both frames, and at any given instant the vector $\mathbf{P}$ has the same components P_x, P_y, P_z in both frames. It follows from (11.25′) that the rate of change $\mathbf{P}$ is the same with respect to the frames $Oxyz$ and $O'x'y'z'$. We state, therefore: *The rate of change of a vector is the same with respect to a fixed frame and with respect to a frame in translation.* This property will greatly simplify our work, since we will be concerned mainly with frames in translation.

Fig. 11.17

11.11 RECTANGULAR COMPONENTS OF VELOCITY AND ACCELERATION

When the position of a particle P is defined at any instant by its rectangular coordinates x, y, and z, it is convenient to resolve the velocity $\mathbf{v}$ and the acceleration $\mathbf{a}$ of the particle into rectangular components (Fig. 11.18).

Resolving the position vector $\mathbf{r}$ of the particle into rectangular components, we write

$$\mathbf{r} = x\mathbf{i} + y\mathbf{j} + z\mathbf{k} \tag{11.26}$$

where the coordinates x, y, z are functions of t. Differentiating twice, we obtain

$$\mathbf{v} = \frac{d\mathbf{r}}{dt} = \dot{x}\mathbf{i} + \dot{y}\mathbf{j} + \dot{z}\mathbf{k} \tag{11.27}$$

$$\mathbf{a} = \frac{d\mathbf{v}}{dt} = \ddot{x}\mathbf{i} + \ddot{y}\mathbf{j} + \ddot{z}\mathbf{k} \tag{11.28}$$

where $\dot{x}, \dot{y}, \dot{z}$ and $\ddot{x}, \ddot{y}, \ddot{z}$ represent, respectively, the first and second derivatives of x, y, and z with respect to t. It follows from (11.27) and (11.28) that the scalar components of the velocity and acceleration are

$$v_x = \dot{x} \qquad v_y = \dot{y} \qquad v_z = \dot{z} \tag{11.29}$$
$$a_x = \ddot{x} \qquad a_y = \ddot{y} \qquad a_z = \ddot{z} \tag{11.30}$$

A positive value for v_x indicates that the vector component $\mathbf{v}_x$ is directed to the right, and a negative value indicates that it is directed to the left. The sense of each of the other vector components can be determined in a similar way from the sign of the corresponding scalar component. If desired, the magnitudes and directions of the velocity and acceleration can be obtained from their scalar components by the methods of Secs. 2.7 and 2.12.

The use of rectangular components to describe the position, the velocity, and the acceleration of a particle is particularly effective when the component a_x of the acceleration depends only upon t, x, and/or v_x, and when, similarly, a_y depends only upon t, y, and/or v_y,

(a)

(b)

Fig. 11.18

Photo 11.3 The motion of this snowboarder in the air will be a parabola assuming we can neglect air resistance.

(a) Motion of a projectile

(b) Equivalent rectilinear motions

Fig. 11.19

and a_z upon t, z, and/or v_z. Equations (11.30) can then be integrated independently, and so can Eqs. (11.29). In other words, the motion of the particle in the x direction, its motion in the y direction, and its motion in the z direction can be considered separately.

In the case of the *motion of a projectile*, for example, it can be shown (see Sec. 12.5) that the components of the acceleration are

$$a_x = \ddot{x} = 0 \qquad a_y = \ddot{y} = -g \qquad a_z = \ddot{z} = 0$$

if the resistance of the air is neglected. Denoting by x_0, y_0, and z_0 the coordinates of a gun, and by $(v_x)_0$, $(v_y)_0$, and $(v_z)_0$ the components of the initial velocity $\mathbf{v}_0$ of the projectile (a bullet), we integrate twice in t and obtain

$$v_x = \dot{x} = (v_x)_0 \qquad v_y = \dot{y} = (v_y)_0 - gt \qquad v_z = \dot{z} = (v_z)_0$$
$$x = x_0 + (v_x)_0 t \qquad y = y_0 + (v_y)_0 t - \tfrac{1}{2}gt^2 \qquad z = z_0 + (v_z)_0 t$$

If the projectile is fired in the xy plane from the origin O, we have $x_0 = y_0 = z_0 = 0$ and $(v_z)_0 = 0$, and the equations of motion reduce to

$$v_x = (v_x)_0 \qquad v_y = (v_y)_0 - gt \qquad v_z = 0$$
$$x = (v_x)_0 t \qquad y = (v_y)_0 t - \tfrac{1}{2}gt^2 \qquad z = 0$$

These equations show that the projectile remains in the xy plane, that its motion in the horizontal direction is uniform, and that its motion in the vertical direction is uniformly accelerated. The motion of a projectile can thus be replaced by two independent rectilinear motions, which are easily visualized if we assume that the projectile is fired vertically with an initial velocity $(\mathbf{v}_y)_0$ from a platform moving with a constant horizontal velocity $(\mathbf{v}_x)_0$ (Fig. 11.19). The coordinate x of the projectile is equal at any instant to the distance traveled by the platform, and its coordinate y can be computed as if the projectile were moving along a vertical line.

It can be observed that the equations defining the coordinates x and y of a projectile at any instant are the parametric equations of a parabola. Thus, the trajectory of a projectile is *parabolic*. This result, however, ceases to be valid when the resistance of the air or the variation with altitude of the acceleration of gravity is taken into account.

11.12 MOTION RELATIVE TO A FRAME IN TRANSLATION

In the preceding section, a single frame of reference was used to describe the motion of a particle. In most cases this frame was attached to the earth and was considered as fixed. Situations in which it is convenient to use several frames of reference simultaneously will now be analyzed. If one of the frames is attached to the earth, it will be called a *fixed frame of reference*, and the other frames will be referred to as *moving frames of reference*. It should be understood, however, that the selection of a fixed frame of reference is purely arbitrary. Any frame can be designated as "fixed"; all other frames not rigidly attached to this frame will then be described as "moving."

Consider two particles A and B moving in space (Fig. 11.20); the vectors $\mathbf{r}_A$ and $\mathbf{r}_B$ define their positions at any given instant with respect to the fixed frame of reference $Oxyz$. Consider now a system of axes x', y', z' centered at A and parallel to the x, y, z axes. While the origin of these axes moves, their orientation remains the same; the frame of reference $Ax'y'z'$ is in *translation* with respect to $Oxyz$. The vector $\mathbf{r}_{B/A}$ joining A and B defines *the position of B relative to the moving frame $Ax'y'z'$* (or, for short, *the position of B relative to A*).

We note from Fig. 11.20 that the position vector $\mathbf{r}_B$ of particle B is the sum of the position vector $\mathbf{r}_A$ of particle A and of the position vector $\mathbf{r}_{B/A}$ of B relative to A; we write

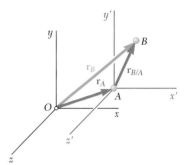

Fig. 11.20

$$\mathbf{r}_B = \mathbf{r}_A + \mathbf{r}_{B/A} \qquad (11.31)$$

Differentiating (11.31) with respect to t within the fixed frame of reference, and using dots to indicate time derivatives, we have

$$\dot{\mathbf{r}}_B = \dot{\mathbf{r}}_A + \dot{\mathbf{r}}_{B/A} \qquad (11.32)$$

The derivatives $\dot{\mathbf{r}}_A$ and $\dot{\mathbf{r}}_B$ represent, respectively, the velocities $\mathbf{v}_A$ and $\mathbf{v}_B$ of the particles A and B. Since $Ax'y'z'$ is in translation, the derivative $\dot{\mathbf{r}}_{B/A}$ represents the rate of change of $\mathbf{r}_{B/A}$ with respect to the frame $Ax'y'z'$ as well as with respect to the fixed frame (Sec. 11.10). This derivative, therefore, defines *the velocity $\mathbf{v}_{B/A}$ of B relative to the frame $Ax'y'z'$* (or, for short, *the velocity $\mathbf{v}_{B/A}$ of B relative to A*). We write

$$\mathbf{v}_B = \mathbf{v}_A + \mathbf{v}_{B/A} \qquad (11.33)$$

Differentiating Eq. (11.33) with respect to t, and using the derivative $\dot{\mathbf{v}}_{B/A}$ to define *the acceleration $\mathbf{a}_{B/A}$ of B relative to the frame $Ax'y'z'$* (or, for short, *the acceleration $\mathbf{a}_{B/A}$ of B relative to A*), we write

$$\mathbf{a}_B = \mathbf{a}_A + \mathbf{a}_{B/A} \qquad (11.34)$$

Photo 11.4 The pilot of a helicopter must take into account the relative motion of the ship when landing.

The motion of B with respect to the fixed frame $Oxyz$ is referred to as the *absolute motion of B*. The equations derived in this section show that *the absolute motion of B can be obtained by combining the motion of A and the relative motion of B with respect to the moving frame attached to A.* Equation (11.33), for example, expresses that the absolute velocity $\mathbf{v}_B$ of particle B can be obtained by adding vectorially the velocity of A and the velocity of B relative to the frame $Ax'y'z'$. Equation (11.34) expresses a similar property in terms of the accelerations.† We should keep in mind, however, that *the frame $Ax'y'z'$ is in translation;* that is, while it moves with A, it maintains the same orientation. As you will see later (Sec. 15.14), different relations must be used in the case of a rotating frame of reference.

†Note that the product of the subscripts A and B/A used in the right-hand member of Eqs. (11.31) through (11.34) is equal to the subscript B used in their left-hand member.

SAMPLE PROBLEM 11.7

A projectile is fired from the edge of a 150-m cliff with an initial velocity of 180 m/s at an angle of 30° with the horizontal. Neglecting air resistance, find (a) the horizontal distance from the gun to the point where the projectile strikes the ground, (b) the greatest elevation above the ground reached by the projectile.

SOLUTION

The vertical and the horizontal motion will be considered separately.

Vertical Motion. *Uniformly Accelerated Motion.* Choosing the positive sense of the y axis upward and placing the origin O at the gun, we have

$$(v_y)_0 = (180 \text{ m/s}) \sin 30° = +90 \text{ m/s}$$
$$a = -9.81 \text{ m/s}^2$$

Substituting into the equations of uniformly accelerated motion, we have

$$v_y = (v_y)_0 + at \qquad v_y = 90 - 9.81t \qquad (1)$$
$$y = (v_y)_0 t + \tfrac{1}{2}at^2 \qquad y = 90t - 4.90t^2 \qquad (2)$$
$$v_y^2 = (v_y)_0^2 + 2ay \qquad v_y^2 = 8100 - 19.62y \qquad (3)$$

Horizontal Motion. *Uniform Motion.* Choosing the positive sense of the x axis to the right, we have

$$(v_x)_0 = (180 \text{ m/s}) \cos 30° = +155.9 \text{ m/s}$$

Substituting into the equation of uniform motion, we obtain

$$x = (v_x)_0 t \qquad x = 155.9t \qquad (4)$$

a. Horizontal Distance. When the projectile strikes the ground, we have

$$y = -150 \text{ m}$$

Carrying this value into Eq. (2) for the vertical motion, we write

$$-150 = 90t - 4.90t^2 \qquad t^2 - 18.37t - 30.6 = 0 \qquad t = 19.91 \text{ s}$$

Carrying $t = 19.91$ s into Eq. (4) for the horizontal motion, we obtain

$$x = 155.9(19.91) \qquad x = 3100 \text{ m} \quad \blacktriangleleft$$

b. Greatest Elevation. When the projectile reaches its greatest elevation, we have $v_y = 0$; carrying this value into Eq. (3) for the vertical motion, we write

$$0 = 8100 - 19.62y \qquad y = 413 \text{ m}$$
$$\text{Greatest elevation above ground} = 150 \text{ m} + 413 \text{ m} = 563 \text{ m} \quad \blacktriangleleft$$

SAMPLE PROBLEM 11.8

A projectile is fired with an initial velocity of 800 ft/s at a target B located 2000 ft above the gun A and at a horizontal distance of 12,000 ft. Neglecting air resistance, determine the value of the firing angle α.

SOLUTION

The horizontal and the vertical motion will be considered separately.

Horizontal Motion. Placing the origin of the coordinate axes at the gun, we have

$$(v_x)_0 = 800 \cos \alpha$$

Substituting into the equation of uniform horizontal motion, we obtain

$$x = (v_x)_0 t \qquad x = (800 \cos \alpha)t$$

The time required for the projectile to move through a horizontal distance of 12,000 ft is obtained by setting x equal to 12,000 ft.

$$12{,}000 = (800 \cos \alpha)t$$
$$t = \frac{12{,}000}{800 \cos \alpha} = \frac{15}{\cos \alpha}$$

Vertical Motion

$$(v_y)_0 = 800 \sin \alpha \qquad a = -32.2 \text{ ft/s}^2$$

Substituting into the equation of uniformly accelerated vertical motion, we obtain

$$y = (v_y)_0 t + \tfrac{1}{2}at^2 \qquad y = (800 \sin \alpha)t - 16.1t^2$$

Projectile Hits Target. When $x = 12{,}000$ ft, we must have $y = 2000$ ft. Substituting for y and setting t equal to the value found above, we write

$$2000 = 800 \sin \alpha \frac{15}{\cos \alpha} - 16.1\left(\frac{15}{\cos \alpha}\right)^2$$

Since $1/\cos^2 \alpha = \sec^2 \alpha = 1 + \tan^2 \alpha$, we have

$$2000 = 800(15) \tan \alpha - 16.1(15^2)(1 + \tan^2 \alpha)$$
$$3622 \tan^2 \alpha - 12{,}000 \tan \alpha + 5622 = 0$$

Solving this quadratic equation for $\tan \alpha$, we have

$$\tan \alpha = 0.565 \qquad \text{and} \qquad \tan \alpha = 2.75$$
$$\alpha = 29.5° \qquad \text{and} \qquad \alpha = 70.0° \quad \blacktriangleleft$$

The target will be hit if either of these two firing angles is used (see figure).

SAMPLE PROBLEM 11.9

Automobile A is traveling east at the constant speed of 36 km/h. As automobile A crosses the intersection shown, automobile B starts from rest 35 m north of the intersection and moves south with a constant acceleration of 1.2 m/s². Determine the position, velocity, and acceleration of B relative to A 5 s after A crosses the intersection.

SOLUTION

We choose x and y axes with origin at the intersection of the two streets and with positive senses directed respectively east and north.

Motion of Automobile A. First the speed is expressed in m/s:

$$v_A = \left(36\frac{\text{km}}{\text{h}}\right)\left(\frac{1000 \text{ m}}{1 \text{ km}}\right)\left(\frac{1 \text{ h}}{3600 \text{ s}}\right) = 10 \text{ m/s}$$

Noting that the motion of A is uniform, we write, for any time t,

$$a_A = 0$$
$$v_A = +10 \text{ m/s}$$
$$x_A = (x_A)_0 + v_A t = 0 + 10t$$

For $t = 5$ s, we have

$$a_A = 0 \qquad\qquad \mathbf{a}_A = 0$$
$$v_A = +10 \text{ m/s} \qquad\qquad \mathbf{v}_A = 10 \text{ m/s} \rightarrow$$
$$x_A = +(10 \text{ m/s})(5 \text{ s}) = +50 \text{ m} \qquad \mathbf{r}_A = 50 \text{ m} \rightarrow$$

Motion of Automobile B. We note that the motion of B is uniformly accelerated and write

$$a_B = -1.2 \text{ m/s}^2$$
$$v_B = (v_B)_0 + at = 0 - 1.2\,t$$
$$y_B = (y_B)_0 + (v_B)_0 t + \tfrac{1}{2}a_B t^2 = 35 + 0 - \tfrac{1}{2}(1.2)t^2$$

For $t = 5$ s, we have

$$a_B = -1.2 \text{ m/s}^2 \qquad\qquad \mathbf{a}_B = 1.2 \text{ m/s}^2 \downarrow$$
$$v_B = -(1.2 \text{ m/s}^2)(5 \text{ s}) = -6 \text{ m/s} \qquad \mathbf{v}_B = 6 \text{ m/s} \downarrow$$
$$y_B = 35 - \tfrac{1}{2}(1.2 \text{ m/s}^2)(5 \text{ s})^2 = +20 \text{ m} \qquad \mathbf{r}_B = 20 \text{ m} \uparrow$$

Motion of B Relative to A. We draw the triangle corresponding to the vector equation $\mathbf{r}_B = \mathbf{r}_A + \mathbf{r}_{B/A}$ and obtain the magnitude and direction of the position vector of B relative to A.

$$r_{B/A} = 53.9 \text{ m} \qquad \alpha = 21.8° \qquad \mathbf{r}_{B/A} = 53.9 \text{ m} \;\angle\; 21.8° \quad \blacktriangleleft$$

Proceeding in a similar fashion, we find the velocity and acceleration of B relative to A.

$$\mathbf{v}_B = \mathbf{v}_A + \mathbf{v}_{B/A}$$
$$v_{B/A} = 11.66 \text{ m/s} \qquad \beta = 31.0° \qquad \mathbf{v}_{B/A} = 11.66 \text{ m/s} \;\angle\; 31.0° \quad \blacktriangleleft$$

$$\mathbf{a}_B = \mathbf{a}_A + \mathbf{a}_{B/A} \qquad\qquad\qquad \mathbf{a}_{B/A} = 1.2 \text{ m/s}^2 \downarrow \quad \blacktriangleleft$$

SOLVING PROBLEMS
ON YOUR OWN

$\mathbf{I}$n the problems for this lesson, you will analyze the *two- and three-dimensional motion* of a particle. While the physical interpretations of the velocity and acceleration are the same as in the first lessons of the chapter, you should remember that these quantities are vectors. In addition, you should understand from your experience with vectors in statics that it will often be advantageous to express position vectors, velocities, and accelerations in terms of their rectangular scalar components [Eqs. (11.27) and (11.28)]. Furthermore, given two vectors $\mathbf{A}$ and $\mathbf{B}$, recall that $\mathbf{A} \cdot \mathbf{B} = 0$ if $\mathbf{A}$ and $\mathbf{B}$ are perpendicular to each other, while $\mathbf{A} \times \mathbf{B} = 0$ if $\mathbf{A}$ and $\mathbf{B}$ are parallel.

A. Analyzing the motion of a projectile. Many of the following problems deal with the two-dimensional motion of a projectile, where the resistance of the air can be neglected. In Sec. 11.11, we developed the equations which describe this type of motion, and we observed that the horizontal component of the velocity remained constant (uniform motion) while the vertical component of the acceleration was constant (uniformly accelerated motion). We were able to consider separately the horizontal and the vertical motions of the particle. Assuming that the projectile is fired from the origin, we can write the two equations

$$x = (v_x)_0 t \qquad y = (v_y)_0 t - \tfrac{1}{2}gt^2$$

1. If the initial velocity and firing angle are known, the value of y corresponding to any given value of x (or the value of x for any value of y) can be obtained by solving one of the above equations for t and substituting for t into the other, equation [Sample Prob. 11.7].

2. If the initial velocity and the coordinates of a point of the trajectory are known, and you wish to *determine the firing angle* α, begin your solution by expressing the components $(v_x)_0$ and $(v_y)_0$ of the initial velocity as functions of the angle α. These expressions and the known values of x and y are then substituted into the above equations. Finally, solve the first equation for t and substitute that value of t into the second equation to obtain a trigonometric equation in α, which you can solve for that unknown [Sample Prob. 11.8].

(continued)

B. Solving translational two-dimensional relative-motion problems. You saw in Sec. 11.12 that the absolute motion of a particle B can be obtained by combining the motion of a particle A and the *relative motion* of B with respect to a frame attached to A which is in *translation*. The velocity and acceleration of B can then be expressed as shown in Eqs. (11.33) and (11.34), respectively.

1. To visualize the relative motion of B with respect to A, imagine that you are attached to particle A as you observe the motion of particle B. For example, to a passenger in automobile A of Sample Prob. 11.9, automobile B appears to be heading in a southwesterly direction (*south* should be obvious; and *west* is due to the fact that automobile A is moving to the east—automobile B then appears to travel to the west). Note that this conclusion is consistent with the direction of $\mathbf{v}_{B/A}$.

2. To solve a relative-motion problem, first write the vector equations (11.31), (11.33), and (11.34), which relate the motions of particles A and B. You may then use either of the following methods:

    a. Construct the corresponding vector triangles and solve them for the desired position vector, velocity, and acceleration [Sample Prob. 11.9].

    b. Express all vectors in terms of their rectangular components and solve the two independent sets of scalar equations obtained in that way. If you choose this approach, be sure to select the same positive direction for the displacement, velocity, and acceleration of each particle.

PROBLEMS

CONCEPT QUESTIONS

11.CQ3 Two model rockets are fired simultaneously from a ledge and follow the trajectories shown. Neglecting air resistance, which of the rockets will hit the ground first?

   a. A.

   b. B.

   c. They hit at the same time.

   d. The answer depends on h.

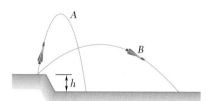

Fig. P11.CQ3

11.CQ4 Ball A is thrown straight up. Which of the following statements about the ball are true at the highest point in its path?

   a. The velocity and acceleration are both zero.

   b. The velocity is zero, but the acceleration is not zero.

   c. The velocity is not zero, but the acceleration is zero.

   d. Neither the velocity nor the acceleration is zero.

Fig. P11.CQ4

11.CQ5 Ball A is thrown straight up with an initial speed v_0 and reaches a maximum elevation h before falling back down. When A reaches its maximum elevation, a second ball is thrown straight upward with the same initial speed v_0. At what height, y, will the balls cross paths?

   a. $y = h$

   b. $y > h/2$

   c. $y = h/2$

   d. $y < h/2$

   e. $y = 0$

11.CQ6 Two cars are approaching an intersection at constant speeds as shown. What velocity will car B appear to have to an observer in car A?

   a. →   **b.** ↘   **c.** ↖   **d.** ↗   **e.** ↙

Fig. P11.CQ6

11.CQ7 Blocks A and B are released from rest in the positions shown. Neglecting friction between all surfaces, which figure best indicates the direction α of the acceleration of block B?

Fig. P11.CQ7

655

END-OF-SECTION PROBLEMS

11.89 A ball is thrown so that the motion is defined by the equations $x = 5t$ and $y = 2 + 6t - 4.9t^2$, where x and y are expressed in meters and t is expressed in seconds. Determine (a) the velocity at $t = 1$ s, (b) the horizontal distance the ball travels before hitting the ground.

Fig. P11.89

11.90 The motion of a vibrating particle is defined by the position vector $\mathbf{r} = 10(1 - e^{-3t})\mathbf{i} + (4e^{-2t} \sin 15t)\mathbf{j}$, where $\mathbf{r}$ and t are expressed in millimeters and seconds, respectively. Determine the velocity and acceleration when (a) $t = 0$, (b) $t = 0.5$ s.

11.91 The motion of a vibrating particle is defined by the position vector $\mathbf{r} = (4 \sin \pi t)\mathbf{i} - (\cos 2\pi t)\mathbf{j}$, where r is expressed in inches and t in seconds. (a) Determine the velocity and acceleration when $t = 1$ s. (b) Show that the path of the particle is parabolic.

11.92 The motion of a particle is defined by the equations $x = 10t - 5 \sin t$ and $y = 10 - 5 \cos t$, where x and y are expressed in feet and t is expressed in seconds. Sketch the path of the particle for the time interval $0 \le t \le 2\pi$, and determine (a) the magnitudes of the smallest and largest velocities reached by the particle, (b) the corresponding times, positions, and directions of the velocities.

11.93 The damped motion of a vibrating particle is defined by the position vector $\mathbf{r} = x_1[1 - 1/(t + 1)]\mathbf{i} + (y_1 e^{-\pi t/2} \cos 2\pi t)\mathbf{j}$, where t is expressed in seconds. For $x_1 = 30$ mm and $y_1 = 20$ mm, determine the position, the velocity, and the acceleration of the particle when (a) $t = 0$, (b) $t = 1.5$ s.

Fig. P11.90

Fig. P11.91

Fig. P11.94

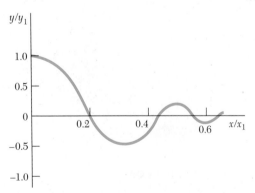

Fig. P11.93

11.94 The motion of a particle is defined by the position vector $\mathbf{r} = A(\cos t + t \sin t)\mathbf{i} + A(\sin t - t \cos t)\mathbf{j}$, where t is expressed in seconds. Determine the values of t for which the position vector and the acceleration are (a) perpendicular, (b) parallel.

11.95 The three-dimensional motion of a particle is defined by the position vector $\mathbf{r} = (Rt \cos \omega_n t)\mathbf{i} + ct\mathbf{j} + (Rt \sin \omega_n t)\mathbf{k}$. Determine the magnitudes of the velocity and acceleration of the particle. (The space curve described by the particle is a conic helix.)

*11.96 The three-dimensional motion of a particle is defined by the position vector $\mathbf{r} = (At \cos t)\mathbf{i} + (A\sqrt{t^2 + 1})\mathbf{j} + (Bt \sin t)\mathbf{k}$, where r and t are expressed in feet and seconds, respectively. Show that the curve described by the particle lies on the hyperboloid $(y/A)^2 - (x/A)^2 - (z/B)^2 = 1$. For $A = 3$ and $B = 1$, determine (a) the magnitudes of the velocity and acceleration when $t = 0$, (b) the smallest nonzero value of t for which the position vector and the velocity are perpendicular to each other.

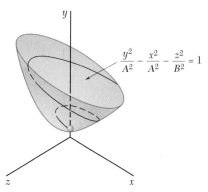

$$\frac{y^2}{A^2} - \frac{x^2}{A^2} - \frac{z^2}{B^2} = 1$$

Fig. P11.96

11.97 An airplane used to drop water on brushfires is flying horizontally in a straight line at 180 mi/h at an altitude of 300 ft. Determine the distance d at which the pilot should release the water so that it will hit the fire at B.

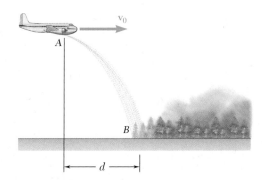

Fig. P11.97

11.98 A helicopter is flying with a constant horizontal velocity of 180 km/h and is directly above point A when a loose part begins to fall. The part lands 6.5 s later at point B on an inclined surface. Determine (a) the distance d between points A and B. (b) the initial height h.

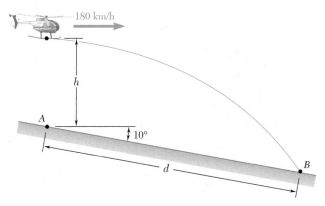

Fig. P11.98

11.99 A baseball pitching machine "throws" baseballs with a horizontal velocity $\mathbf{v}_0$. Knowing that height h varies between 788 mm and 1068 mm, determine (a) the range of values of v_0, (b) the values of α corresponding to $h = 788$ mm and $h = 1068$ mm.

Fig. P11.99

11.100 While delivering newspapers, a girl throws a newspaper with a horizontal velocity $\mathbf{v}_0$. Determine the range of values of v_0 if the newspaper is to land between points B and C.

Fig. P11.100

11.101 Water flows from a drain spout with an initial velocity of 2.5 ft/s at an angle of 15° with the horizontal. Determine the range of values of the distance d for which the water will enter the trough BC.

Fig. P11.101

11.102 Milk is poured into a glass of height 140 mm and inside diameter 66 mm. If the initial velocity of the milk is 1.2 m/s at an angle of 40° with the horizontal, determine the range of values of the height h for which the milk will enter the glass.

11.103 A volleyball player serves the ball with an initial velocity $\mathbf{v}_0$ of magnitude 13.40 m/s at an angle of 20° with the horizontal. Determine (a) if the ball will clear the top of the net, (b) how far from the net the ball will land.

Fig. P11.102

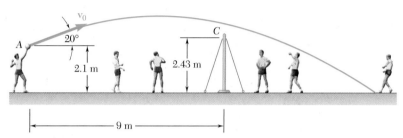

Fig. P11.103

11.104 A golfer hits a golf ball with an initial velocity of 160 ft/s at an angle of 25° with the horizontal. Knowing that the fairway slopes downward at an average angle of 5°, determine the distance d between the golfer and point B where the ball first lands.

Fig. P11.104

11.105 A homeowner uses a snowblower to clear his driveway. Knowing that the snow is discharged at an average angle of 40° with the horizontal, determine the initial velocity v_0 of the snow.

Fig. P11.105

11.106 At halftime of a football game souvenir balls are thrown to the spectators with a velocity $\mathbf{v}_0$. Determine the range of values of v_0 if the balls are to land between points B and C.

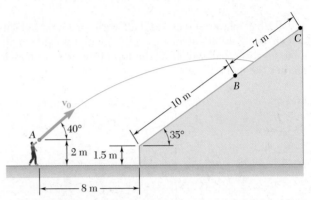

Fig. P11.106

11.107 A basketball player shoots when she is 16 ft from the backboard. Knowing that the ball has an initial velocity $\mathbf{v}_0$ at an angle of 30° with the horizontal, determine the value of v_0 when d is equal to (a) 9 in., (b) 17 in.

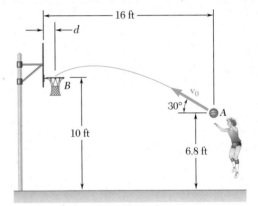

Fig. P11.107

11.108 A tennis player serves the ball at a height $h = 2.5$ m with an initial velocity of $\mathbf{v}_0$ at an angle of 5° with the horizontal. Determine the range of v_0 for which the ball will land in the service area that extends to 6.4 m beyond the net.

Fig. P11.108

11.109 The nozzle at A discharges cooling water with an initial velocity v_0 at an angle of 6° with the horizontal onto a grinding wheel 350 mm in diameter. Determine the range of values of the initial velocity for which the water will land on the grinding wheel between points B and C.

Fig. P11.109

11.110 While holding one of its ends, a worker lobs a coil of rope over the lowest limb of a tree. If he throws the rope with an initial velocity v_0 at an angle of 65° with the horizontal, determine the range of values of v_0 for which the rope will go over only the lowest limb.

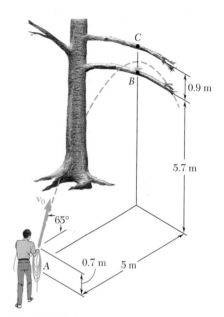

Fig. P11.110

11.111 The pitcher in a softball game throws a ball with an initial velocity v_0 of 72 km/h at an angle α with the horizontal. If the height of the ball at point B is 0.68 m, determine (*a*) the angle α, (*b*) the angle θ that the velocity of the ball at point B forms with the horizontal.

Fig. P11.111

11.112 A model rocket is launched from point A with an initial velocity v_0 of 75 m/s. If the rocket's descent parachute does not deploy and the rocket lands a distance $d = 100$ m from A, determine (*a*) the angle α that v_0 forms with the vertical, (*b*) the maximum height above point A reached by the rocket, (*c*) the duration of the flight.

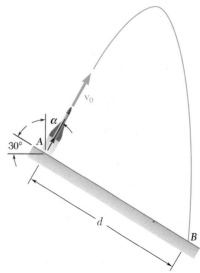

Fig. P11.112

11.113 The initial velocity $\mathbf{v}_0$ of a hockey puck is 105 mi/h. Determine (a) the largest value (less than 45°) of the angle α for which the puck will enter the net, (b) the corresponding time required for the puck to reach the net.

Fig. P11.113

11.114 A worker uses high-pressure water to clean the inside of a long drainpipe. If the water is discharged with an initial velocity $\mathbf{v}_0$ of 11.5 m/s, determine (a) the distance d to the farthest point B on the top of the pipe that the worker can wash from his position at A, (b) the corresponding angle α.

Fig. P11.114

11.115 An oscillating garden sprinkler which discharges water with an initial velocity $\mathbf{v}_0$ of 8 m/s is used to water a vegetable garden. Determine the distance d to the farthest point B that will be watered and the corresponding angle α when (a) the vegetables are just beginning to grow, (b) the height h of the corn is 1.8 m.

Fig. P11.116

Fig. P11.115

*11.116 A mountain climber plans to jump from A to B over a crevasse. Determine the smallest value of the climber's initial velocity $\mathbf{v}_0$ and the corresponding value of angle α so that he lands at B.

11.117 The velocities of skiers *A* and *B* are as shown. Determine the velocity of *A* with respect to *B*.

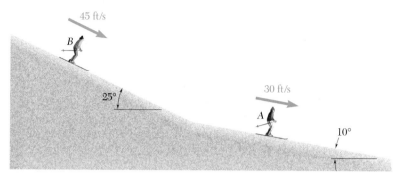

Fig. P11.117

11.118 The three blocks shown move with constant velocities. Find the velocity of each block, knowing that the relative velocity of *A* with respect to *C* is 300 mm/s upward and that the relative velocity of *B* with respect to *A* is 200 mm/s downward.

Fig. P11.118

11.119 Three seconds after automobile *B* passes through the intersection shown, automobile *A* passes through the same intersection. Knowing that the speed of each automobile is constant, determine (*a*) the relative velocity of *B* with respect to *A*, (*b*) the change in position of *B* with respect to *A* during a 4-s interval, (*c*) the distance between the two automobiles 2 s after *A* has passed through the intersection.

11.120 Shore-based radar indicates that a ferry leaves its slip with a velocity $\mathbf{v} = 18$ km/h ⬈ 70°, while instruments aboard the ferry indicate a speed of 18.4 km/h and a heading of 30° west of south relative to the river. Determine the velocity of the river.

Fig. P11.119

Fig. P11.120

Fig. *P11.121*

11.121 Airplanes A and B are flying at the same altitude and are tracking the eye of hurricane C. The relative velocity of C with respect to A is $\mathbf{v}_{C/A} = 350$ km/h ⤢75°, and the relative velocity of C with respect to B is $\mathbf{v}_{C/B} = 400$ km/h ⤡40°. Determine (*a*) the relative velocity of B with respect to A, (*b*) the velocity of A if ground-based radar indicates that the hurricane is moving at a speed of 30 km/h due north, (*c*) the change in position of C with respect to B during a 15-min interval.

11.122 Pin P moves at a constant speed of 150 mm/s in a counterclockwise sense along a circular slot which has been milled in the slider block A shown. Knowing that the block moves downward at a constant speed of 100 mm/s, determine the velocity of pin P when (*a*) $\theta = 30°$, (*b*) $\theta = 120°$.

Fig. *P11.122*

$\theta = 50°$

Fig. P11.123

11.123 Knowing that at the instant shown assembly A has a velocity of 9 in./s and an acceleration of 15 in./s² both directed downward, determine (*a*) the velocity of block B, (*b*) the acceleration of block B.

11.124 Knowing that at the instant shown block A has a velocity of 8 in./s and an acceleration of 6 in./s² both directed down the incline, determine (*a*) the velocity of block B, (*b*) the acceleration of block B.

Fig. P11.124

11.125 A boat is moving to the right with a constant deceleration of 0.3 m/s^2 when a boy standing on the deck D throws a ball with an initial velocity relative to the deck which is vertical. The ball rises to a maximum height of 8 m above the release point and the boy must step forward a distance d to catch it at the same height as the release point. Determine (a) the distance d, (b) the relative velocity of the ball with respect to the deck when the ball is caught.

Fig. P11.125

11.126 The assembly of rod A and wedge B starts from rest and moves to the right with a constant acceleration of 2 mm/s^2. Determine (a) the acceleration of wedge C, (b) the velocity of wedge C when $t = 10$ s.

Fig. P11.126

11.127 Determine the required velocity of the belt B if the relative velocity with which the sand hits belt B is to be (a) vertical, (b) as small as possible.

Fig. P11.127

11.128 Conveyor belt A, which forms a 20° angle with the horizontal, moves at a constant speed of 4 ft/s and is used to load an airplane. Knowing that a worker tosses duffel bag B with an initial velocity of 2.5 ft/s at an angle of 30° with the horizontal, determine the velocity of the bag relative to the belt as it lands on the belt.

Fig. P11.128

11.129 During a rainstorm the paths of the raindrops appear to form an angle of 30° with the vertical and to be directed to the left when observed from a side window of a train moving at a speed of 15 km/h. A short time later, after the speed of the train has increased to 24 km/h, the angle between the vertical and the paths of the drops appears to be 45°. If the train were stopped, at what angle and with what velocity would the drops be observed to fall?

11.130 As observed from a ship moving due east at 9 km/h, the wind appears to blow from the south. After the ship has changed course and speed, and as it is moving north at 6 km/h, the wind appears to blow from the southwest. Assuming that the wind velocity is constant during the period of observation, determine the magnitude and direction of the true wind velocity.

11.131 When a small boat travels north at 5 km/h, a flag mounted on its stern forms an angle $\theta = 50°$ with the centerline of the boat as shown. A short time later, when the boat travels east at 20 km/h, angle θ is again 50°. Determine the speed and the direction of the wind.

Fig. P11.131

11.132 As part of a department store display, a model train D runs on a slight incline between the store's up and down escalators. When the train and shoppers pass point A, the train appears to a shopper on the up escalator B to move downward at an angle of 22° with the horizontal, and to a shopper on the down escalator C to move upward at an angle of 23° with the horizontal and to travel to the left. Knowing that the speed of the escalators is 3 ft/s, determine the speed and the direction of the train.

Fig. P11.132

11.13 TANGENTIAL AND NORMAL COMPONENTS

We saw in Sec. 11.9 that the velocity of a particle is a vector tangent to the path of the particle but that, in general, the acceleration is not tangent to the path. It is sometimes convenient to resolve the acceleration into components directed, respectively, along the tangent and the normal to the path of the particle.

Plane Motion of a Particle. First, let us consider a particle which moves along a curve contained in the plane of the figure. Let P be the position of the particle at a given instant. We attach at P a unit vector $\mathbf{e}_t$ tangent to the path of the particle and pointing in the direction of motion (Fig. 11.21a). Let $\mathbf{e}'_t$ be the unit vector corresponding to the position P' of the particle at a later instant. Drawing both vectors from the same origin O', we define the vector $\Delta\mathbf{e}_t = \mathbf{e}'_t - \mathbf{e}_t$ (Fig. 11.21b). Since $\mathbf{e}_t$ and $\mathbf{e}'_t$ are of unit length, their tips lie on a circle of radius 1. Denoting by $\Delta\theta$ the angle formed by $\mathbf{e}_t$ and $\mathbf{e}'_t$, we find that the magnitude of $\Delta\mathbf{e}_t$ is $2\sin(\Delta\theta/2)$. Considering now the vector $\Delta\mathbf{e}_t/\Delta\theta$, we note that as $\Delta\theta$ approaches zero, this vector becomes tangent to the unit circle of Fig. 11.21b, i.e., perpendicular to $\mathbf{e}_t$, and that its magnitude approaches

$$\lim_{\Delta\theta \to 0} \frac{2\sin(\Delta\theta/2)}{\Delta\theta} = \lim_{\Delta\theta \to 0} \frac{\sin(\Delta\theta/2)}{\Delta\theta/2} = 1$$

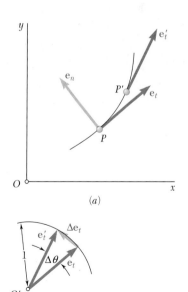

Fig. 11.21

Thus, the vector obtained in the limit is a unit vector along the normal to the path of the particle, in the direction toward which $\mathbf{e}_t$ turns. Denoting this vector by $\mathbf{e}_n$, we write

$$\mathbf{e}_n = \lim_{\Delta\theta\to 0} \frac{\Delta\mathbf{e}_t}{\Delta\theta}$$

$$\mathbf{e}_n = \frac{d\mathbf{e}_t}{d\theta} \tag{11.35}$$

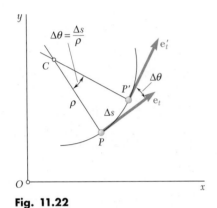

Fig. 11.22

Since the velocity $\mathbf{v}$ of the particle is tangent to the path, it can be expressed as the product of the scalar v and the unit vector $\mathbf{e}_t$. We have

$$\mathbf{v} = v\mathbf{e}_t \tag{11.36}$$

To obtain the acceleration of the particle, (11.36) will be differentiated with respect to t. Applying the rule for the differentiation of the product of a scalar and a vector function (Sec. 11.10), we write

$$\mathbf{a} = \frac{d\mathbf{v}}{dt} = \frac{dv}{dt}\mathbf{e}_t + v\frac{d\mathbf{e}_t}{dt} \tag{11.37}$$

But

$$\frac{d\mathbf{e}_t}{dt} = \frac{d\mathbf{e}_t}{d\theta}\frac{d\theta}{ds}\frac{ds}{dt}$$

Recalling from (11.16) that $ds/dt = v$, from (11.35) that $d\mathbf{e}_t/d\theta = \mathbf{e}_n$, and from elementary calculus that $d\theta/ds$ is equal to $1/\rho$, where ρ is the radius of curvature of the path at P (Fig. 11.22), we have

$$\frac{d\mathbf{e}_t}{dt} = \frac{v}{\rho}\mathbf{e}_n \tag{11.38}$$

Substituting into (11.37), we obtain

$$\mathbf{a} = \frac{dv}{dt}\mathbf{e}_t + \frac{v^2}{\rho}\mathbf{e}_n \tag{11.39}$$

Thus, the scalar components of the acceleration are

$$a_t = \frac{dv}{dt} \qquad a_n = \frac{v^2}{\rho} \tag{11.40}$$

Photo 11.5 The passengers in a train traveling around a curve will experience a normal acceleration toward the center of curvature of the path.

The relations obtained express that the *tangential component of the acceleration is equal to the rate of change of the speed of the particle*, while the *normal component is equal to the square of the speed divided by the radius of curvature of the path at P*. If the speed of the particle increases, a_t is positive and the vector component $\mathbf{a}_t$ points in the direction of motion. If the speed of the particle decreases, a_t is negative and $\mathbf{a}_t$ points against the direction of motion. The vector component $\mathbf{a}_n$, on the other hand, *is always directed toward the center of curvature C of the path* (Fig. 11.23).

We conclude from the above that the tangential component of the acceleration reflects a change in the speed of the particle, while

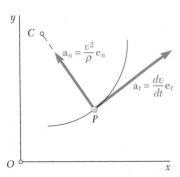

Fig. 11.23

its normal component reflects a change in the direction of motion of the particle. The acceleration of a particle will be zero only if both its components are zero. Thus, the acceleration of a particle moving with constant speed along a curve will not be zero unless the particle happens to pass through a point of inflection of the curve (where the radius of curvature is infinite) or unless the curve is a straight line.

The fact that the normal component of the acceleration depends upon the radius of curvature of the path followed by the particle is taken into account in the design of structures or mechanisms as widely different as airplane wings, railroad tracks, and cams. In order to avoid sudden changes in the acceleration of the air particles flowing past a wing, wing profiles are designed without any sudden change in curvature. Similar care is taken in designing railroad curves, to avoid sudden changes in the acceleration of the cars (which would be hard on the equipment and unpleasant for the passengers). A straight section of track, for instance, is never directly followed by a circular section. Special transition sections are used to help pass smoothly from the infinite radius of curvature of the straight section to the finite radius of the circular track. Likewise, in the design of high-speed cams, abrupt changes in acceleration are avoided by using transition curves which produce a continuous change in acceleration.

Motion of a Particle in Space. The relations (11.39) and (11.40) still hold in the case of a particle moving along a space curve. However, since there are an infinite number of straight lines which are perpendicular to the tangent at a given point P of a space curve, it is necessary to define more precisely the direction of the unit vector $\mathbf{e}_n$.

Let us consider again the unit vectors $\mathbf{e}_t$ and $\mathbf{e}'_t$ tangent to the path of the particle at two neighboring points P and P' (Fig. 11.24a) and the vector $\Delta\mathbf{e}_t$ representing the difference between $\mathbf{e}_t$ and $\mathbf{e}'_t$ (Fig. 11.24b). Let us now imagine a plane through P (Fig. 11.24a) parallel to the plane defined by the vectors $\mathbf{e}_t$, $\mathbf{e}'_t$, and $\Delta\mathbf{e}_t$ (Fig. 11.24b). This plane contains the tangent to the curve at P and is parallel to the tangent at P'. If we let P' approach P, we obtain in the limit the plane which fits the curve most closely in the neighborhood of P. This plane is called the *osculating plane* at P.† It follows from this

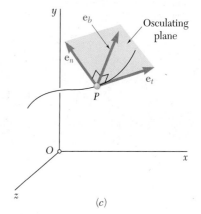

Fig. 11.24

†From the Latin *osculari*, to kiss.

definition that the osculating plane contains the unit vector $\mathbf{e}_n$, since this vector represents the limit of the vector $\Delta\mathbf{e}_t/\Delta\theta$. The normal defined by $\mathbf{e}_n$ is thus contained in the osculating plane; it is called the *principal normal* at P. The unit vector $\mathbf{e}_b = \mathbf{e}_t \times \mathbf{e}_n$ which completes the right-handed triad $\mathbf{e}_t$, $\mathbf{e}_n$, $\mathbf{e}_b$ (Fig. 11.24c) defines the *binormal* at P. The binormal is thus perpendicular to the osculating plane. We conclude that the acceleration of the particle at P can be resolved into two components, one along the tangent, the other along the principal normal at P, as indicated in Eq. (11.39). Note that the acceleration has no component along the binormal.

11.14 RADIAL AND TRANSVERSE COMPONENTS

In certain problems of plane motion, the position of the particle P is defined by its polar coordinates r and θ (Fig. 11.25a). It is then convenient to resolve the velocity and acceleration of the particle into components parallel and perpendicular, respectively, to the line OP. These components are called *radial* and *transverse components*.

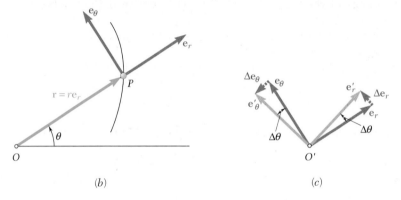

(a)   (b)   (c)

Fig. 11.25

Photo 11.6 The footpads on an elliptical trainer undergo curvilinear motion.

We attach at P two unit vectors, $\mathbf{e}_r$ and $\mathbf{e}_\theta$ (Fig. 11.25b). The vector $\mathbf{e}_r$ is directed along OP and the vector $\mathbf{e}_\theta$ is obtained by rotating $\mathbf{e}_r$ through 90° counterclockwise. The unit vector $\mathbf{e}_r$ defines the *radial* direction, i.e., the direction in which P would move if r were increased and θ were kept constant; the unit vector $\mathbf{e}_\theta$ defines the *transverse* direction, i.e., the direction in which P would move if θ were increased and r were kept constant. A derivation similar to the one we used in Sec. 11.13 to determine the derivative of the unit vector $\mathbf{e}_t$ leads to the relations

$$\frac{d\mathbf{e}_r}{d\theta} = \mathbf{e}_\theta \qquad \frac{d\mathbf{e}_\theta}{d\theta} = -\mathbf{e}_r \qquad (11.41)$$

where $-\mathbf{e}_r$ denotes a unit vector of sense opposite to that of $\mathbf{e}_r$ (Fig. 11.25c). Using the chain rule of differentiation, we express the time derivatives of the unit vectors $\mathbf{e}_r$ and $\mathbf{e}_\theta$ as follows:

$$\frac{d\mathbf{e}_r}{dt} = \frac{d\mathbf{e}_r}{d\theta}\frac{d\theta}{dt} = \mathbf{e}_\theta\frac{d\theta}{dt} \qquad \frac{d\mathbf{e}_\theta}{dt} = \frac{d\mathbf{e}_\theta}{d\theta}\frac{d\theta}{dt} = -\mathbf{e}_r\frac{d\theta}{dt}$$

or, using dots to indicate differentiation with respect to t,

$$\dot{\mathbf{e}}_r = \dot{\theta}\mathbf{e}_\theta \qquad \dot{\mathbf{e}}_\theta = -\dot{\theta}\mathbf{e}_r \qquad (11.42)$$

To obtain the velocity **v** of the particle P, we express the position vector **r** of P as the product of the scalar r and the unit vector $\mathbf{e}_r$ and differentiate with respect to t:

$$\mathbf{v} = \frac{d}{dt}(r\mathbf{e}_r) = \dot{r}\mathbf{e}_r + r\dot{\mathbf{e}}_r$$

or, recalling the first of the relations (11.42),

$$\mathbf{v} = \dot{r}\mathbf{e}_r + r\dot{\theta}\mathbf{e}_\theta \tag{11.43}$$

Differentiating again with respect to t to obtain the acceleration, we write

$$\mathbf{a} = \frac{d\mathbf{v}}{dt} = \ddot{r}\mathbf{e}_r + \dot{r}\dot{\mathbf{e}}_r + \dot{r}\dot{\theta}\mathbf{e}_\theta + r\ddot{\theta}\mathbf{e}_\theta + r\dot{\theta}\dot{\mathbf{e}}_\theta$$

or, substituting for $\dot{\mathbf{e}}_r$ and $\dot{\mathbf{e}}_\theta$ from (11.42) and factoring $\mathbf{e}_r$ and $\mathbf{e}_\theta$,

$$\boldsymbol{a} = (\ddot{r} - r\dot{\theta}^2)\mathbf{e}_r + (r\ddot{\theta} + 2\dot{r}\dot{\theta})\mathbf{e}_\theta \tag{11.44}$$

The scalar components of the velocity and the acceleration in the radial and transverse directions are, therefore,

$$v_r = \dot{r} \qquad\qquad v_\theta = r\dot{\theta} \tag{11.45}$$

$$a_r = \ddot{r} - r\dot{\theta}^2 \qquad a_\theta = r\ddot{\theta} + 2\dot{r}\dot{\theta} \tag{11.46}$$

It is important to note that a_r is *not* equal to the time derivative of v_r and that a_θ is *not* equal to the time derivative of v_θ.

In the case of a particle moving along a circle of center O, we have $r = $ constant and $\dot{r} = \ddot{r} = 0$, and the formulas (11.43) and (11.44) reduce, respectively, to

$$\mathbf{v} = r\dot{\theta}\mathbf{e}_\theta \qquad \mathbf{a} = -r\dot{\theta}^2\mathbf{e}_r + r\ddot{\theta}\mathbf{e}_\theta \tag{11.47}$$

Extension to the Motion of a Particle in Space: Cylindrical Coordinates.

The position of a particle P in space is sometimes defined by its cylindrical coordinates R, θ, and z (Fig. 11.26a). It is then convenient to use the unit vectors $\mathbf{e}_R$, $\mathbf{e}_\theta$, and **k** shown in Fig. 11.26b. Resolving the position vector **r** of the particle P into components along the unit vectors, we write

$$\mathbf{r} = R\mathbf{e}_R + z\mathbf{k} \tag{11.48}$$

Observing that $\mathbf{e}_R$ and $\mathbf{e}_\theta$ define, respectively, the radial and transverse directions in the horizontal xy plane, and that the vector **k**, which defines the *axial* direction, is constant in direction as well as in magnitude, we easily verify that

$$\mathbf{v} = \frac{d\mathbf{r}}{dt} = \dot{R}\mathbf{e}_R + R\dot{\theta}\mathbf{e}_\theta + \dot{z}\mathbf{k} \tag{11.49}$$

$$\mathbf{a} = \frac{d\mathbf{v}}{dt} = (\ddot{R} - R\dot{\theta}^2)\mathbf{e}_R + (R\ddot{\theta} + 2\dot{R}\dot{\theta})\mathbf{e}_\theta + \ddot{z}\mathbf{k} \tag{11.50}$$

(a)

(b)

Fig. 11.26

$v_A = 60$ mi/h

A

2500 ft

SAMPLE PROBLEM 11.10

A motorist is traveling on a curved section of highway of radius 2500 ft at the speed of 60 mi/h. The motorist suddenly applies the brakes, causing the automobile to slow down at a constant rate. Knowing that after 8 s the speed has been reduced to 45 mi/h, determine the acceleration of the automobile immediately after the brakes have been applied.

SOLUTION

Tangential Component of Acceleration. First the speeds are expressed in ft/s.

$$60 \text{ mi/h} = \left(60\frac{\text{mi}}{\text{h}}\right)\left(\frac{5280 \text{ ft}}{1 \text{ mi}}\right)\left(\frac{1 \text{ h}}{3600 \text{ s}}\right) = 88 \text{ ft/s}$$

$$45 \text{ mi/h} = 66 \text{ ft/s}$$

Since the automobile slows down at a constant rate, we have

$$a_t = \text{average } a_t = \frac{\Delta v}{\Delta t} = \frac{66 \text{ ft/s} - 88 \text{ ft/s}}{8 \text{ s}} = -2.75 \text{ ft/s}^2$$

Normal Component of Acceleration. Immediately after the brakes have been applied, the speed is still 88 ft/s, and we have

$$a_n = \frac{v^2}{\rho} = \frac{(88 \text{ ft/s})^2}{2500 \text{ ft}} = 3.10 \text{ ft/s}^2$$

Magnitude and Direction of Acceleration. The magnitude and direction of the resultant $\mathbf{a}$ of the components $\mathbf{a}_n$ and $\mathbf{a}_t$ are

$$\tan \alpha = \frac{a_n}{a_t} = \frac{3.10 \text{ ft/s}^2}{2.75 \text{ ft/s}^2} \qquad \alpha = 48.4° \blacktriangleleft$$

$$a = \frac{a_n}{\sin \alpha} = \frac{3.10 \text{ ft/s}^2}{\sin 48.4°} \qquad \mathbf{a} = 4.14 \text{ ft/s}^2 \blacktriangleleft$$

$a_t = 2.75$ ft/s²

A α

Motion

a

$a_n = 3.10$ ft/s²

SAMPLE PROBLEM 11.11

Determine the minimum radius of curvature of the trajectory described by the projectile considered in Sample Prob. 11.7.

SOLUTION

$v = v_x$

$a = a_n$

Since $a_n = v^2/\rho$, we have $\rho = v^2/a_n$. The radius will be small when v is small or when a_n is large. The speed v is minimum at the top of the trajectory since $v_y = 0$ at that point; a_n is maximum at that same point, since the direction of the vertical coincides with the direction of the normal. Therefore, the minimum radius of curvature occurs at the top of the trajectory. At this point, we have

$$v = v_x = 155.9 \text{ m/s} \qquad a_n = a = 9.81 \text{ m/s}^2$$

$$\rho = \frac{v^2}{a_n} = \frac{(155.9 \text{ m/s})^2}{9.81 \text{ m/s}^2} \qquad \rho = 2480 \text{ m} \blacktriangleleft$$

SAMPLE PROBLEM 11.12

The rotation of the 0.9-m arm OA about O is defined by the relation $\theta = 0.15t^2$, where θ is expressed in radians and t in seconds. Collar B slides along the arm in such a way that its distance from O is $r = 0.9 - 0.12t^2$, where r is expressed in meters and t in seconds. After the arm OA has rotated through 30°, determine (a) the total velocity of the collar, (b) the total acceleration of the collar, (c) the relative acceleration of the collar with respect to the arm.

SOLUTION

$$\mathbf{v} = v_r\mathbf{e}_r + v_\theta\mathbf{e}_\theta$$
$$\mathbf{a} = a_r\mathbf{e}_r + a_\theta\mathbf{e}_\theta$$

Time t at which $\theta = 30°$. Substituting $\theta = 30° = 0.524$ rad into the expression for θ, we obtain

$$\theta = 0.15t^2 \qquad 0.524 = 0.15t^2 \qquad t = 1.869 \text{ s}$$

Equations of Motion. Substituting $t = 1.869$ s in the expressions for r, θ, and their first and second derivatives, we have

$$r = 0.9 - 0.12t^2 = 0.481 \text{ m} \qquad \theta = 0.15t^2 = 0.524 \text{ rad}$$
$$\dot{r} = -0.24t = -0.449 \text{ m/s} \qquad \dot{\theta} = 0.30t = 0.561 \text{ rad/s}$$
$$\ddot{r} = -0.24 = -0.240 \text{ m/s}^2 \qquad \ddot{\theta} = 0.30 = 0.300 \text{ rad/s}^2$$

a. Velocity of B. Using Eqs. (11.45), we obtain the values of v_r and v_θ when $t = 1.869$ s.

$$v_r = \dot{r} = -0.449 \text{ m/s}$$
$$v_\theta = r\dot{\theta} = 0.481(0.561) = 0.270 \text{ m/s}$$

Solving the right triangle shown, we obtain the magnitude and direction of the velocity,

$$v = 0.524 \text{ m/s} \qquad \beta = 31.0° \quad \blacktriangleleft$$

b. Acceleration of B. Using Eqs. (11.46), we obtain

$$a_r = \ddot{r} - r\dot{\theta}^2$$
$$= -0.240 - 0.481(0.561)^2 = -0.391 \text{ m/s}^2$$
$$a_\theta = r\ddot{\theta} + 2\dot{r}\dot{\theta}$$
$$= 0.481(0.300) + 2(-0.449)(0.561) = -0.359 \text{ m/s}^2$$
$$a = 0.531 \text{ m/s}^2 \qquad \gamma = 42.6° \quad \blacktriangleleft$$

c. Acceleration of B with Respect to Arm OA. We note that the motion of the collar with respect to the arm is rectilinear and defined by the coordinate r. We write

$$a_{B/OA} = \ddot{r} = -0.240 \text{ m/s}^2$$
$$a_{B/OA} = 0.240 \text{ m/s}^2 \text{ toward } O. \quad \blacktriangleleft$$

SOLVING PROBLEMS ON YOUR OWN

You will be asked in the following problems to express the velocity and the acceleration of particles in terms of either their *tangential and normal components* or their *radial and transverse components*. Although those components may not be as familiar to you as the rectangular components, you will find that they can simplify the solution of many problems and that certain types of motion are more easily described when they are used.

1. Using tangential and normal components. These components are most often used when the particle of interest travels along a circular path or when the radius of curvature of the path is to be determined. Remember that the unit vector $\mathbf{e}_t$ is tangent to the path of the particle (and thus aligned with the velocity) while the unit vector $\mathbf{e}_n$ is directed along the normal to the path and always points toward its center of curvature. It follows that, as the particle moves, the directions of the two unit vectors are constantly changing.

2. Expressing the acceleration in terms of its tangential and normal components. We derived in Sec. 11.13 the following equation, applicable to both the two-dimensional and the three-dimensional motion of a particle:

$$\mathbf{a} = \frac{dv}{dt}\mathbf{e}_t + \frac{v^2}{\rho}\mathbf{e}_n \qquad (11.39)$$

The following observations may help you in solving the problems of this lesson.

a. The tangential component of the acceleration measures the rate of change of the speed: $a_t = dv/dt$. It follows that when a_t is constant, the equations for uniformly accelerated motion can be used with the acceleration equal to a_t. Furthermore, when a particle moves at a constant speed, we have $a_t = 0$ and the acceleration of the particle reduces to its normal component.

b. The normal component of the acceleration is always directed toward the center of curvature of the path of the particle, and its magnitude is $a_n = v^2/\rho$. Thus, the normal component can be easily determined if the speed of the particle and the radius of curvature ρ of the path are known. Conversely, when the speed and normal acceleration of the particle are known, the radius of curvature of the path can be obtained by solving this equation for ρ [Sample Prob. 11.11].

c. In three-dimensional motion, a third unit vector is used, $\mathbf{e}_b = \mathbf{e}_t \times \mathbf{e}_n$, which defines the direction of the *binormal*. Since this vector is perpendicular to both the velocity and the acceleration, it can be obtained by writing

$$\mathbf{e}_b = \frac{\mathbf{v} \times \mathbf{a}}{|\mathbf{v} \times \mathbf{a}|}$$

3. Using radial and transverse components. These components are used to analyze the plane motion of a particle P, when the position of P is defined by its polar coordinates r and θ. As shown in Fig. 11.25, the unit vector $\mathbf{e}_r$, which defines the *radial* direction, is attached to P and points away from the fixed point O, while the unit vector $\mathbf{e}_\theta$, which defines the *transverse* direction, is obtained by rotating $\mathbf{e}_r$ *counterclockwise* through 90°. The velocity and the acceleration of a particle were expressed in terms of their radial and transverse components in Eqs. (11.43) and (11.44), respectively. You will note that the expressions obtained contain the first and second derivatives with respect to t of both coordinates r and θ.

In the problems of this lesson, you will encounter the following types of problems involving radial and transverse components:

  a. Both r and θ are known functions of t. In this case, you will compute the first and second derivatives of r and θ and substitute the expressions obtained into Eqs. (11.43) and (11.44).

  b. A certain relationship exists between r and θ. First, you should determine this relationship from the geometry of the given system and use it to express r as a function of θ. Once the function $r = f(\theta)$ is known, you can apply the chain rule to determine $\dot{r}$ in terms of θ and $\dot{\theta}$, and $\ddot{r}$ in terms of θ, $\dot{\theta}$, $\ddot{\theta}$:

$$\dot{r} = f'(\theta)\dot{\theta}$$
$$\ddot{r} = f''(\theta)\dot{\theta}^2 + f'(\theta)\ddot{\theta}$$

The expressions obtained can then be substituted into Eqs. (11.43) and (11.44).

  c. The three-dimensional motion of a particle, as indicated at the end of Sec. 11.14, can often be effectively described in terms of the *cylindrical coordinates* R, θ, and z (Fig. 11.26). The unit vectors then should consist of $\mathbf{e}_R$, $\mathbf{e}_\theta$, and $\mathbf{k}$. The corresponding components of the velocity and the acceleration are given in Eqs. (11.49) and (11.50). Please note that the radial distance R is always measured in a plane parallel to the xy plane, and be careful not to confuse the position vector $\mathbf{r}$ with its radial component $R\mathbf{e}_R$.

PROBLEMS

Fig. P11.CQ8

CONCEPT QUESTIONS

11.CQ8 The Ferris wheel is rotating with a constant angular velocity ω. What is the direction of the acceleration of point A?

 a. $\rightarrow$ **b.** $\uparrow$ **c.** $\downarrow$ **d.** $\leftarrow$ **e.** The acceleration is zero.

11.CQ9 A race car travels around the track shown at a constant speed. At which point will the race car have the largest acceleration?

 a. A. **b.** B. **c.** C. **d.** D. **e.** The acceleration will be zero at all the points.

Fig. P11.CQ9

Fig. P11.CQ10

11.CQ10 A child walks across merry-go-round A with a constant speed u relative to A. The merry-go-round undergoes fixed-axis rotation about its center with a constant angular velocity ω counterclockwise. When the child is at the center of A, as shown, what is the direction of his acceleration when viewed from above?

 a. $\rightarrow$ **b.** $\leftarrow$ **c.** $\uparrow$ **d.** $\downarrow$ **e.** The acceleration is zero.

END-OF-SECTION PROBLEMS

11.133 Determine the smallest radius that should be used for a highway if the normal component of the acceleration of a car traveling at 72 km/h is not to exceed 0.8 m/s^2.

Fig. P11.133

11.134 Determine the maximum speed that the cars of the roller-coaster can reach along the circular portion AB of the track if $\rho = 25$ m and the normal component of their acceleration cannot exceed 3g.

Fig. P11.134

11.135 A bull-roarer is a piece of wood that produces a roaring sound when attached to the end of a string and whirled around in a circle. Determine the magnitude of the normal acceleration of a bull-roarer when it is spun in a circle of radius 0.9 m at a speed of 20 m/s.

Fig. P11.135

11.136 To test its performance, an automobile is driven around a circular test track of diameter d. Determine (a) the value of d if when the speed of the automobile is 45 mi/h, the normal component of the acceleration is 11 ft/s^2, (b) the speed of the automobile if $d = 600$ ft and the normal component of the acceleration is measured to be 0.6g.

11.137 An outdoor track is 420 ft in diameter. A runner increases her speed at a constant rate from 14 to 24 ft/s over a distance of 95 ft. Determine the magnitude of the total acceleration of the runner 2 s after she begins to increase her speed.

Fig. P11.137

11.138 A robot arm moves so that P travels in a circle about point B, which is not moving. Knowing that P starts from rest, and its speed increases at a constant rate of 10 mm/s^2, determine (a) the magnitude of the acceleration when $t = 4$ s, (b) the time for the magnitude of the acceleration to be 80 mm/s^2.

Fig. P11.138

11.139 A monorail train starts from rest on a curve of radius 400 m and accelerates at the constant rate a_t. If the maximum total acceleration of the train must not exceed 1.5 m/s^2, determine (a) the shortest distance in which the train can reach a speed of 72 km/h, (b) the corresponding constant rate of acceleration a_t.

11.140 A motorist starts from rest at point A on a circular entrance ramp when $t = 0$, increases the speed of her automobile at a constant rate and enters the highway at point B. Knowing that her speed continues to increase at the same rate until it reaches 100 km/h at point C, determine (a) the speed at point B, (b) the magnitude of the total acceleration when $t = 20$ s.

Fig. P11.140

Fig. P11.141

11.141 Race car A is traveling on a straight portion of the track while race car B is traveling on a circular portion of the track. At the instant shown, the speed of A is increasing at the rate of 10 m/s², and the speed of B is decreasing at the rate of 6 m/s². For the position shown, determine (*a*) the velocity of B relative to A, (*b*) the acceleration of B relative to A.

11.142 At a given instant in an airplane race, airplane A is flying horizontally in a straight line, and its speed is being increased at the rate of 8 m/s². Airplane B is flying at the same altitude as airplane A and, as it rounds a pylon, is following a circular path of 300-m radius. Knowing that at the given instant the speed of B is being decreased at the rate of 3 m/s², determine, for the positions shown, (*a*) the velocity of B relative to A, (*b*) the acceleration of B relative to A.

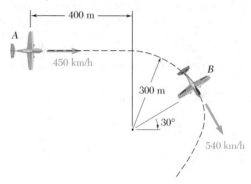

Fig. *P11.142*

11.143 From a photograph of a homeowner using a snowblower, it is determined that the radius of curvature of the trajectory of the snow was 30 ft as the snow left the discharge chute at A. Determine (*a*) the discharge velocity $\mathbf{v}_A$ of the snow, (*b*) the radius of curvature of the trajectory at its maximum height.

Fig. P11.143

Fig. P11.144

Fig. P11.145

11.144 A basketball is bounced on the ground at point A and rebounds with a velocity $\mathbf{v}_A$ of magnitude 7.5 ft/s as shown. Determine the radius of curvature of the trajectory described by the ball (*a*) at point A, (*b*) at the highest point of the trajectory.

11.145 A golfer hits a golf ball from point A with an initial velocity of 50 m/s at an angle of 25° with the horizontal. Determine the radius of curvature of the trajectory described by the ball (*a*) at point A, (*b*) at the highest point of the trajectory.

11.146 Three children are throwing snowballs at each other. Child A throws a snowball with a horizontal velocity $\mathbf{v}_0$. If the snowball just passes over the head of child B and hits child C, determine the radius of curvature of the trajectory described by the snowball (*a*) at point B, (*b*) at point C.

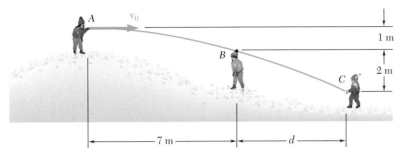

Fig. P11.146

11.147 Coal is discharged from the tailgate A of a dump truck with an initial velocity $\mathbf{v}_A = 2$ m/s $\nearrow 50°$. Determine the radius of curvature of the trajectory described by the coal (*a*) at point A, (*b*) at the point of the trajectory 1 m below point A.

11.148 From measurements of a photograph, it has been found that as the stream of water shown left the nozzle at A, it had a radius of curvature of 25 m. Determine (*a*) the initial velocity $\mathbf{v}_A$ of the stream, (*b*) the radius of curvature of the stream as it reaches its maximum height at B.

Fig. P11.147

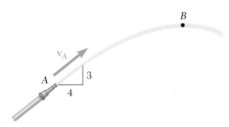

Fig. *P11.148*

11.149 A child throws a ball from point A with an initial velocity $\mathbf{v}_A$ of 20 m/s at an angle of 25° with the horizontal. Determine the velocity of the ball at the points of the trajectory described by the ball where the radius of curvature is equal to three-quarters of its value at A.

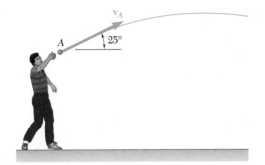

Fig. P11.149

11.150 A projectile is fired from point A with an initial velocity $\mathbf{v}_0$. (a) Show that the radius of curvature of the trajectory of the projectile reaches its minimum value at the highest point B of the trajectory. (b) Denoting by θ the angle formed by the trajectory and the horizontal at a given point C, show that the radius of curvature of the trajectory at C is $\rho = \rho_{\text{min}}/\cos^3\theta$.

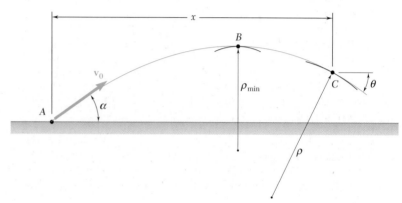

Fig. P11.150

*11.151 Determine the radius of curvature of the path described by the particle of Prob. 11.95 when $t = 0$.

*11.152 Determine the radius of curvature of the path described by the particle of Prob. 11.96 when $t = 0$, $A = 3$, and $B = 1$.

11.153 and 11.154 A satellite will travel indefinitely in a circular orbit around a planet if the normal component of the acceleration of the satellite is equal to $g(R/r)^2$, where g is the acceleration of gravity at the surface of the planet, R is the radius of the planet, and r is the distance from the center of the planet to the satellite. Knowing that the diameter of the sun is 1.39 Gm and that the acceleration of gravity at its surface is 274 m/s², determine the radius of the orbit of the indicated planet around the sun assuming that the orbit is circular.

  11.153 Earth: $(v_{\text{mean}})_{\text{orbit}} = 107$ Mm/h.
  11.154 Saturn: $(v_{\text{mean}})_{\text{orbit}} = 34.7$ Mm/h.

11.155 through 11.157 Determine the speed of a satellite relative to the indicated planet if the satellite is to travel indefinitely in a circular orbit 100 mi above the surface of the planet. (See information given in Probs. 11.153–11.154.)

  11.155 Venus: $g = 29.20$ ft/s², $R = 3761$ mi.
  11.156 Mars: $g = 12.17$ ft/s², $R = 2102$ mi.
  11.157 Jupiter: $g = 75.35$ ft/s², $R = 44{,}432$ mi.

11.158 A satellite is traveling in a circular orbit around Mars at an altitude of 300 km. After the altitude of the satellite is adjusted, it is found that the time of one orbit has increased by 10 percent. Knowing that the radius of Mars is 3382 km, determine the new altitude of the satellite. (See information given in Probs. 11.153–11.154).

11.159 Knowing that the radius of the earth is 6370 km, determine the time of one orbit of the Hubble Space Telescope knowing that the telescope travels in a circular orbit 590 km above the surface of the earth. (See information given in Probs. 11.153–11.154.)

11.160 Satellites A and B are traveling in the same plane in circular orbits around the earth at altitudes of 120 and 200 mi, respectively. If at $t = 0$ the satellites are aligned as shown and knowing that the radius of the earth is $R = 3960$ mi, determine when the satellites will next be radially aligned. (See information given in Probs. 11.153–11.154.)

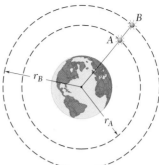

Fig. P11.160

11.161 The oscillation of rod OA about O is defined by the relation $\theta = (3/\pi)(\sin \pi t)$, where θ and t are expressed in radians and seconds, respectively. Collar B slides along the rod so that its distance from O is $r = 6(1 - e^{-2t})$ where r and t are expressed in inches and seconds, respectively. When $t = 1$ s, determine (a) the velocity of the collar, (b) the acceleration of the collar, (c) the acceleration of the collar relative to the rod.

Fig. P11.161 and P11.162

11.162 The rotation of rod OA about O is defined by the relation $\theta = t^3 - 4t$, where θ and t are expressed in radians and seconds, respectively. Collar B slides along the rod so that its distance from O is $r = 2.5t^3 - 5t^2$, where r and t are expressed in inches and seconds, respectively. When $t = 1$ s, determine (a) the velocity of the collar, (b) the acceleration of the collar, (c) the radius of curvature of the path of the collar.

11.163 The path of particle P is the ellipse defined by the relations $r = 2/(2 - \cos \pi t)$ and $\theta = \pi t$, where r is expressed in meters, t is in seconds, and θ is in radians. Determine the velocity and the acceleration of the particle when (a) $t = 0$, (b) $t = 0.5$ s.

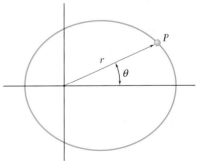

Fig. P11.163

11.164 The two-dimensional motion of a particle is defined by the relations $r = 2a \cos \theta$ and $\theta = bt^2/2$, where a and b are constants. Determine (a) the magnitudes of the velocity and acceleration at any instant, (b) the radius of curvature of the path. What conclusion can you draw regarding the path of the particle?

Fig. P11.165

11.165 As rod OA rotates, pin P moves along the parabola BCD. Knowing that the equation of this parabola is $r = 2b/(1 + \cos\theta)$ and that $\theta = kt$, determine the velocity and acceleration of P when (a) $\theta = 0$, (b) $\theta = 90°$.

11.166 The pin at B is free to slide along the circular slot DE and along the rotating rod OC. Assuming that the rod OC rotates at a constant rate $\dot{\theta}$, (a) show that the acceleration of pin B is of constant magnitude, (b) determine the direction of the acceleration of pin B.

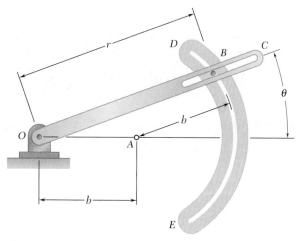

Fig. P11.166

11.167 To study the performance of a race car, a high-speed camera is positioned at point A. The camera is mounted on a mechanism which permits it to record the motion of the car as the car travels on straightaway BC. Determine (a) the speed of the car in terms of b, θ, and $\dot{\theta}$, (b) the magnitude of the acceleration in terms of b, θ, $\dot{\theta}$, and $\ddot{\theta}$.

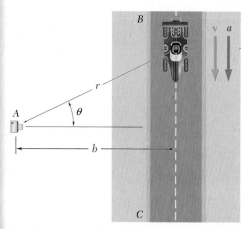

Fig. P11.167

11.168 After taking off, a helicopter climbs in a straight line at a constant angle β. Its flight is tracked by radar from point A. Determine the speed of the helicopter in terms of d, β, θ, and $\dot{\theta}$.

Fig. P11.168

11.169 At the bottom of a loop in the vertical plane an airplane has a horizontal velocity of 315 mi/h and is speeding up at a rate of 10 ft/s². The radius of curvature of the loop is 1 mi. The plane is being tracked by radar at O. What are the recorded values of $\dot{r}$, $\ddot{r}$, $\dot{\theta}$, and $\ddot{\theta}$ for this instant?

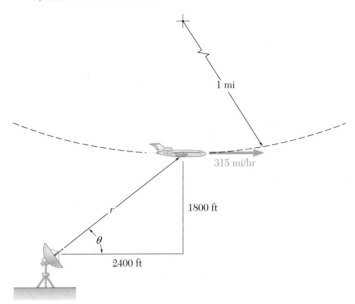

Fig. P11.169

11.170 Pin C is attached to rod BC and slides freely in the slot of rod OA which rotates at the constant rate ω. At the instant when $\beta = 60°$, determine (a) $\dot{r}$ and $\dot{\theta}$, (b) $\ddot{r}$ and $\ddot{\theta}$. Express your answers in terms of d and ω.

Fig. P11.170

11.171 For the race car of Prob. 11.167, it was found that it took 0.5 s for the car to travel from the position $\theta = 60°$ to the position $\theta = 35°$. Knowing that $b = 25$ m, determine the average speed of the car during the 0.5-s interval.

11.172 For the helicopter of Prob. 11.168, it was found that when the helicopter was at B, the distance and the angle of elevation of the helicopter were $r = 3000$ ft and $\theta = 20°$, respectively. Four seconds later, the radar station sighted the helicopter at $r = 3320$ ft and $\theta = 23.1°$. Determine the average speed and the angle of climb β of the helicopter during the 4-s interval.

11.173 and 11.174 A particle moves along the spiral shown; determine the magnitude of the velocity of the particle in terms of b, θ, and $\dot{\theta}$.

Hyperbolic spiral $r\theta = b$ Logarithmic spiral $r = e^{b\theta}$

Fig. P11.173 and P11.175   **Fig. P11.174 and P11.176**

11.175 and 11.176 A particle moves along the spiral shown. Knowing that $\dot{\theta}$ is constant and denoting this constant by ω, determine the magnitude of the acceleration of the particle in terms of b, θ, and ω.

11.177 The motion of a particle on the surface of a right circular cylinder is defined by the relations $R = A$, $\theta = 2\pi t$, and $z = B \sin 2\pi nt$, where A and B are constants and n is an integer. Determine the magnitudes of the velocity and acceleration of the particle at any time t.

Fig. P11.177

11.178 Show that $\dot{r} = h\dot{\phi} \sin \theta$ knowing that at the instant shown, step AB of the step exerciser is rotating counterclockwise at a constant rate $\dot{\phi}$.

11.179 The three-dimensional motion of a particle is defined by the relations $R = A(1 - e^{-t})$, $\theta = 2\pi t$, and $z = B(1 - e^{-t})$. Determine the magnitudes of the velocity and acceleration when (a) $t = 0$, (b) $t = \infty$.

*11.180 For the conic helix of Prob. 11.95, determine the angle that the osculating plane forms with the y axis.

*11.181 Determine the direction of the binormal of the path described by the particle of Prob. 11.96 when (a) $t = 0$, (b) $t = \pi/2$ s.

Fig. P11.178

REVIEW AND SUMMARY

In the first half of the chapter, we analyzed the *rectilinear motion of a particle,* i.e., the motion of a particle along a straight line. To define the position P of the particle on that line, we chose a fixed origin O and a positive direction (Fig. 11.27). The distance x from O to P, with the appropriate sign, completely defines the position of the particle on the line and is called the *position coordinate* of the particle [Sec. 11.2].

Position coordinate of a particle in rectilinear motion

Fig. 11.27

The *velocity v* of the particle was shown to be equal to the time derivative of the position coordinate x,

$$v = \frac{dx}{dt} \qquad (11.1)$$

and the *acceleration a* was obtained by differentiating v with respect to t,

$$a = \frac{dv}{dt} \qquad (11.2)$$

or

$$a = \frac{d^2x}{dt^2} \qquad (11.3)$$

We also noted that a could be expressed as

$$a = v\frac{dv}{dx} \qquad (11.4)$$

Velocity and acceleration in rectilinear motion

We observed that the velocity v and the acceleration a were represented by algebraic numbers which can be positive or negative. A positive value for v indicates that the particle moves in the positive direction, and a negative value that it moves in the negative direction. A positive value for a, however, may mean that the particle is truly accelerated (i.e., moves faster) in the positive direction, or that it is decelerated (i.e., moves more slowly) in the negative direction. A negative value for a is subject to a similar interpretation [Sample Prob. 11.1].

In most problems, the conditions of motion of a particle are defined by the type of acceleration that the particle possesses and by the initial conditions [Sec. 11.3]. The velocity and position of the particle can then be obtained by integrating two of the equations (11.1) to (11.4). Which of these equations should be selected depends upon the type of acceleration involved [Sample Probs. 11.2 and 11.3].

Determination of the velocity and acceleration by integration

685

Two types of motion are frequently encountered: the *uniform rectilinear motion* [Sec. 11.4], in which the velocity v of the particle is constant and

Uniform rectilinear motion

$$x = x_0 + vt \qquad (11.5)$$

Uniformly accelerated rectilinear motion

and the *uniformly accelerated rectilinear motion* [Sec. 11.5], in which the acceleration a of the particle is constant and we have

$$v = v_0 + at \qquad (11.6)$$
$$x = x_0 + v_0 t + \tfrac{1}{2}at^2 \qquad (11.7)$$
$$v^2 = v_0^2 + 2a(x - x_0) \qquad (11.8)$$

Relative motion of two particles

When two particles A and B move along the same straight line, we may wish to consider the *relative motion* of B with respect to A

Fig. 11.28

[Sec. 11.6]. Denoting by $x_{B/A}$ the *relative position coordinate* of B with respect to A (Fig. 11.28), we had

$$x_B = x_A + x_{B/A} \qquad (11.9)$$

Differentiating Eq. (11.9) twice with respect to t, we obtained successively

$$v_B = v_A + v_{B/A} \qquad (11.10)$$
$$a_B = a_A + a_{B/A} \qquad (11.11)$$

where $v_{B/A}$ and $a_{B/A}$ represent, respectively, the *relative velocity* and the *relative acceleration* of B with respect to A.

Blocks connected by inextensible cords

When several blocks are *connected by inextensible cords,* it is possible to write a *linear relation* between their position coordinates. Similar relations can then be written between their velocities and between their accelerations and can be used to analyze their motion [Sample Prob. 11.5].

Graphical solutions

It is sometimes convenient to use a *graphical solution* for problems involving the rectilinear motion of a particle [Secs. 11.7 and 11.8]. The graphical solution most commonly used involves the x–t, v–t, and a–t curves [Sec. 11.7; Sample Prob. 11.6]. It was shown that, at any given time t,

$$v = \text{slope of } x\text{–}t \text{ curve}$$
$$a = \text{slope of } v\text{–}t \text{ curve}$$

while, over any given time interval from t_1 to t_2,

$$v_2 - v_1 = \text{area under } a\text{–}t \text{ curve}$$
$$x_2 - x_1 = \text{area under } v\text{–}t \text{ curve}$$

Position vector and velocity in curvilinear motion

In the second half of the chapter, we analyzed the *curvilinear motion of a particle,* i.e., the motion of a particle along a curved path. The position P of the particle at a given time [Sec. 11.9] was defined by the *position vector* **r** joining the O of the coordinates and point P (Fig. 11.29). The *velocity* **v** of the particle was defined by the relation

$$\mathbf{v} = \frac{d\mathbf{r}}{dt} \tag{11.15}$$

and was found to be a *vector tangent to the path of the particle* and of magnitude v (called the *speed* of the particle) equal to the time derivative of the length s of the arc described by the particle:

$$v = \frac{ds}{dt} \tag{11.16}$$

The *acceleration* **a** of the particle was defined by the relation

$$\mathbf{a} = \frac{d\mathbf{v}}{dt} \tag{11.18}$$

and we noted that, in general, *the acceleration is not tangent to the path of the particle.*

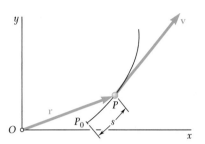

Fig. 11.29
Acceleration in curvilinear motion

Before proceeding to the consideration of the components of velocity and acceleration, we reviewed the formal definition of the derivative of a vector function and established a few rules governing the differentiation of sums and products of vector functions. We then showed that the rate of change of a vector is the same with respect to a fixed frame and with respect to a frame in translation [Sec. 11.10].

Derivative of a vector function

Denoting by x, y, and z the rectangular coordinates of a particle P, we found that the rectangular components of the velocity and acceleration of P equal, respectively, the first and second derivatives with respect to t of the corresponding coordinates:

Rectangular components of velocity and acceleration

$$v_x = \dot{x} \qquad v_y = \dot{y} \qquad v_z = \dot{z} \tag{11.29}$$
$$a_x = \ddot{x} \qquad a_y = \ddot{y} \qquad a_z = \ddot{z} \tag{11.30}$$

When the component a_x of the acceleration depends only upon t, x, and/or v_x, and when similarly a_y depends only upon t, y, and/or v_y, and a_z upon t, z, and/or v_z, Eq. (11.30) can be integrated independently. The analysis of the given curvilinear motion can thus be reduced to the analysis of three independent rectilinear component motions [Sec. 11.11]. This approach is particularly effective in the study of the motion of projectiles [Sample Probs. 11.7 and 11.8].

Component motions

For two particles A and B moving in space (Fig. 11.30), we considered the relative motion of B with respect to A, or more precisely, with respect to a moving frame attached to A and in translation with A [Sec. 11.12]. Denoting by $\mathbf{r}_{B/A}$ the *relative position vector* of B with respect to A (Fig. 11.30), we had

Relative motion of two particles

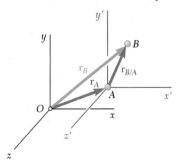

$$\mathbf{r}_B = \mathbf{r}_A + \mathbf{r}_{B/A} \tag{11.31}$$

Denoting by $\mathbf{v}_{B/A}$ and $\mathbf{a}_{B/A}$, respectively, the *relative velocity* and the *relative acceleration* of B with respect to A, we also showed that

$$\mathbf{v}_B = \mathbf{v}_A + \mathbf{v}_{B/A} \tag{11.33}$$

and

$$\mathbf{a}_B = \mathbf{a}_A + \mathbf{a}_{B/A} \tag{11.34}$$

Fig. 11.30

Tangential and normal components

Fig. 11.31

It is sometimes convenient to resolve the velocity and acceleration of a particle P into components other than the rectangular x, y, and z components. For a particle P moving along a path contained in a plane, we attached to P unit vectors $\mathbf{e}_t$ tangent to the path and $\mathbf{e}_n$ normal to the path and directed toward the center of curvature of the path [Sec. 11.13]. We then expressed the velocity and acceleration of the particle in terms of tangential and normal components. We wrote

$$\mathbf{v} = v\mathbf{e}_t \tag{11.36}$$

and

$$\mathbf{a} = \frac{dv}{dt}\mathbf{e}_t + \frac{v^2}{\rho}\mathbf{e}_n \tag{11.39}$$

where v is the speed of the particle and ρ the radius of curvature of its path [Sample Probs. 11.10 and 11.11]. We observed that while the velocity $\mathbf{v}$ is directed along the tangent to the path, the acceleration $\mathbf{a}$ consists of a component $\mathbf{a}_t$ directed along the tangent to the path and a component $\mathbf{a}_n$ directed toward the center of curvature of the path (Fig. 11.31).

Motion along a space curve

For a particle P moving along a space curve, we defined the plane which most closely fits the curve in the neighborhood of P as the *osculating plane*. This plane contains the unit vectors $\mathbf{e}_t$ and $\mathbf{e}_n$ which define, respectively, the tangent and principal normal to the curve. The unit vector $\mathbf{e}_b$ which is perpendicular to the osculating plane defines the *binormal*.

Radial and transverse components

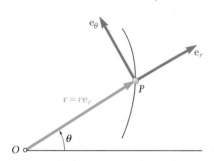

Fig. 11.32

When the position of a particle P moving in a plane is defined by its polar coordinates r and θ, it is convenient to use radial and transverse components directed, respectively, along the position vector $\mathbf{r}$ of the particle and in the direction obtained by rotating $\mathbf{r}$ through 90° counterclockwise [Sec. 11.14]. We attached to P unit vectors $\mathbf{e}_r$ and $\mathbf{e}_\theta$ directed, respectively, in the radial and transverse directions (Fig. 11.32). We then expressed the velocity and acceleration of the particle in terms of radial and transverse components

$$\mathbf{v} = \dot{r}\mathbf{e}_r + r\dot{\theta}\mathbf{e}_\theta \tag{11.43}$$
$$\mathbf{a} = (\ddot{r} - r\dot{\theta}^2)\mathbf{e}_r + (r\ddot{\theta} + 2\dot{r}\dot{\theta})\mathbf{e}_\theta \tag{11.44}$$

where dots are used to indicate differentiation with respect to time. The scalar components of the velocity and acceleration in the radial and transverse directions are therefore

$$v_r = \dot{r} \qquad v_\theta = r\dot{\theta} \tag{11.45}$$
$$a_r = \ddot{r} - r\dot{\theta}^2 \qquad a_\theta = r\ddot{\theta} + 2\dot{r}\dot{\theta} \tag{11.46}$$

It is important to note that a_r is *not* equal to the time derivative of v_r, and that a_θ is *not* equal to the time derivative of v_θ [Sample Prob. 11.12].

The chapter ended with a discussion of the use of cylindrical coordinates to define the position and motion of a particle in space.

REVIEW PROBLEMS

11.182 The motion of a particle is defined by the relation $x = 2t^3 - 15t^2 + 24t + 4$, where x and t are expressed in meters and seconds, respectively. Determine (a) when the velocity is zero, (b) the position and the total distance traveled when the acceleration is zero.

11.183 A particle starting from rest at $x = 1$ m is accelerated so that its velocity doubles in magnitude between $x = 2$ m and $x = 8$ m. Knowing that the acceleration of the particle is defined by the relation $a = k[x - (A/x)]$, determine the values of the constants A and k if the particle has a velocity of 29 m/s when $x = 16$ m.

11.184 A particle moves in a straight line with the acceleration shown in the figure. Knowing that the particle starts from the origin with $v_0 = -2$ m/s, (a) construct the v–t and x–t curves for $0 < t < 18$ s, (b) determine the position and the velocity of the particle and the total distance traveled when $t = 18$ s.

Fig. P11.184

11.185 The velocities of commuter trains A and B are as shown. Knowing that the speed of each train is constant and that B reaches the crossing 10 min after A passed through the same crossing, determine (a) the relative velocity of B with respect to A, (b) the distance between the fronts of the engines 3 min after A passed through the crossing.

Fig. P11.185

Fig. P11.186

Fig. P11.187

Fig. P11.189

11.186 Slider block B starts from rest and moves to the right with a constant acceleration of 1 ft/s^2. Determine (*a*) the relative acceleration of portion C of the cable with respect to slider block A, (*b*) the velocity of portion C of the cable after 2 s.

11.187 Collar A starts from rest at $t = 0$ and moves downward with a constant acceleration of 7 in./s^2. Collar B moves upward with a constant acceleration, and its initial velocity is 8 in./s. Knowing that collar B moves through 20 in. between $t = 0$ and $t = 2$ s, determine (*a*) the accelerations of collar B and block C, (*b*) the time at which the velocity of block C is zero, (*c*) the distance through which block C will have moved at that time.

11.188 A golfer hits a ball with an initial velocity of magnitude v_0 at an angle α with the horizontal. Knowing that the ball must clear the tops of two trees and land as close as possible to the flag, determine v_0 and the distance d when the golfer uses (*a*) a six-iron with $\alpha = 31°$, (*b*) a five-iron with $\alpha = 27°$.

Fig. P11.188

11.189 As the truck shown begins to back up with a constant acceleration of 4 ft/s^2, the outer section B of its boom starts to retract with a constant acceleration of 1.6 ft/s^2 relative to the truck. Determine (*a*) the acceleration of section B, (*b*) the velocity of section B when $t = 2$ s.

11.190 A motorist traveling along a straight portion of a highway is decreasing the speed of his automobile at a constant rate before exiting from the highway onto a circular exit ramp with a radius of 560 ft. He continues to decelerate at the same constant rate so that 10 s after entering the ramp, his speed has decreased to 20 mi/h, a speed which he then maintains. Knowing that at this constant speed the total acceleration of the automobile is equal to one-quarter of its value prior to entering the ramp, determine the maximum value of the total acceleration of the automobile.

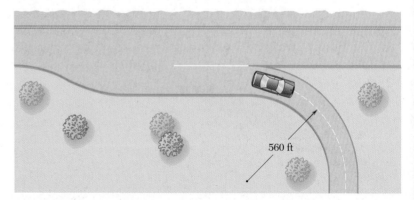

Fig. P11.190

11.191 Sand is discharged at A from a conveyor belt and falls onto the top of a stockpile at B. Knowing that the conveyor belt forms an angle $\alpha = 25°$ with the horizontal, determine (a) the speed v_0 of the belt, (b) the radius of curvature of the trajectory described by the sand at point B.

11.192 The end point B of a boom is originally 5 m from fixed point A when the driver starts to retract the boom with a constant radial acceleration of $\ddot{r} = -1.0$ m/s^2 and lower it with a constant angular acceleration $\ddot{\theta} = -0.5$ rad/s^2. At $t = 2$ s, determine (a) the velocity of point B, (b) the acceleration of point B, (c) the radius of curvature of the path.

Fig. P11.191

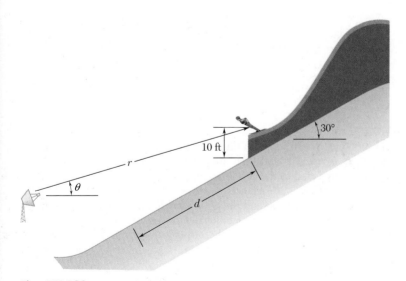

Fig. P11.192

11.193 A telemetry system is used to quantify kinematic values of a ski jumper immediately before she leaves the ramp. According to the system $r = 500$ ft, $\dot{r} = -105$ ft/s, $\ddot{r} = -10$ ft/s^2, $\theta = 25°$, $\dot{\theta} = 0.07$ rad/s, $\ddot{\theta} = 0.06$ rad/s^2. Determine (a) the velocity of the skier immediately before she leaves the jump, (b) the acceleration of the skier at this instant, (c) the distance of the jump d neglecting lift and air resistance.

Fig. P11.193

COMPUTER PROBLEMS

Fig. P11.C1

11.C1 The mechanism shown is known as a Whitworth quick-return mechanism. The input rod AP rotates at a constant rate $\dot{\phi}$, and the pin P is free to slide in the slot of the output rod BD. Plot θ versus ϕ and $\dot{\theta}$ versus ϕ for one revolution of rod AP. Assume $\dot{\phi} = 1$ rad/s, $l = 4$ in., and (a) $b = 2.5$ in., (b) $b = 3$ in., (c) $b = 3.5$ in.

11.C2 A ball is dropped with a velocity $\mathbf{v}_0$ at an angle α with the vertical onto the top step of a flight of stairs consisting of 8 steps. The ball rebounds and bounces down the steps as shown. Each time the ball bounces, at points $A, B, C, \ldots$, the horizontal component of its velocity remains constant and the magnitude of the vertical component of its velocity is reduced by k percent. Use computational software to determine (a) if the ball bounces down the steps without skipping any step, (b) if the ball bounces down the steps without bouncing twice on the same step, (c) the first step on which the ball bounces twice. Use values of v_0 from 1.8 m/s to 3.0 m/s in 0.6-m/s increments, values of α from 18° to 26° in 4° increments, and values of k equal to 40 and 50.

Fig. P11.C2

Fig. P11.C3

11.C3 In an amusement park ride, "airplane" A is attached to the 10-m-long rigid member OB. To operate the ride, the airplane and OB are rotated so that $70° \le \theta_0 \le 130°$ and then are allowed to swing freely about O. The airplane is subjected to the acceleration of gravity and to a deceleration due to air resistance, $-kv^2$, which acts in a direction opposite to that of its velocity $\mathbf{v}$. Neglecting the mass and the aerodynamic drag of OB and the friction in the bearing at O, use computational software or write a computer program to determine the speed of the airplane for given values of θ_0 and θ and the value of θ at which the airplane first comes to rest after being released. Use values of θ_0 from 70° to 130° in 30° increments, and determine the maximum speed of the airplane and the first two values of θ at which $v = 0$. For each value of θ_0, let (a) $k = 0$, (b) $k = 2 \times 10^{-4}$ m^{-1}, (c) $k = 4 \times 10^{-2}$ m^{-1}. (*Hint:* Express the tangential acceleration of the airplane in terms of g, k, and θ. Recall that $v_\theta = r\dot{\theta}$.)

11.C4 A motorist traveling on a highway at a speed of 60 mi/h exits onto an ice-covered exit ramp. Wishing to stop, he applies his brakes until his automobile comes to rest. Knowing that the magnitude of the total acceleration of the automobile cannot exceed 10 ft/s², use computational software to determine the minimum time required for the automobile to come to rest and the distance it travels on the exit ramp during that time if the exit ramp (*a*) is straight, (*b*) has a constant radius of curvature of 800 ft. Solve each part assuming that the driver applies his brakes so that *dv/dt*, during each time interval, (1) remains constant, (2) varies linearly.

11.C5 An oscillating garden sprinkler discharges water with an initial velocity **v**₀ of 10 m/s. (*a*) Knowing that the sides but not the top of arbor *BCDE* are open, use computational software to calculate the distance *d* to the point *F* that will be watered for values of α from 20° to 80°. (*b*) Determine the maximum value of *d* and the corresponding value of α.

Fig. P11.C5

The forces experienced by the passengers on a roller coaster will depend on whether the roller-coaster car is traveling up a hill or down a hill, in a straight line, or along a horizontal or vertical curved path. The relation existing among force, mass, and acceleration will be studied in this chapter.

694

CHAPTER

12

Kinetics of Particles:
Newton's Second Law

12.1 INTRODUCTION

Newton's first and third laws of motion were used extensively in statics to study bodies at rest and the forces acting upon them. These two laws are also used in dynamics; in fact, they are sufficient for the study of the motion of bodies which have no acceleration. However, when bodies are accelerated, i.e., when the magnitude or the direction of their velocity changes, it is necessary to use Newton's second law of motion to relate the motion of the body with the forces acting on it.

In this chapter we will discuss Newton's second law and apply it to the analysis of the motion of particles. As we state in Sec. 12.2, if the resultant of the forces acting on a particle is not zero, the particle will have an acceleration proportional to the magnitude of the resultant and in the direction of this resultant force. Moreover, the ratio of the magnitudes of the resultant force and of the acceleration can be used to define the *mass* of the particle.

In Sec. 12.3, the *linear momentum* of a particle is defined as the product $\mathbf{L} = m\mathbf{v}$ of the mass m and velocity $\mathbf{v}$ of the particle, and it is demonstrated that Newton's second law can be expressed in an alternative form relating the rate of change of the linear momentum with the resultant of the forces acting on that particle.

Section 12.4 stresses the need for consistent units in the solution of dynamics problems and provides a review of the International System of Units (SI units) and the system of U.S. customary units.

In Secs. 12.5 and 12.6 and in the Sample Problems which follow, Newton's second law is applied to the solution of engineering problems, using either rectangular components or tangential and normal components of the forces and accelerations involved. We recall that an actual body—including bodies as large as a car, rocket, or airplane—can be considered as a particle for the purpose of analyzing its motion as long as the effect of a rotation of the body about its mass center can be ignored.

The second part of the chapter is devoted to the solution of problems in terms of radial and transverse components, with particular emphasis on the motion of a particle under a central force. In Sec. 12.7, the *angular momentum* $\mathbf{H}_O$ of a particle about a point O is defined as the moment about O of the linear momentum of the particle: $\mathbf{H}_O = \mathbf{r} \times m\mathbf{v}$. It then follows from Newton's second law that the rate of change of the angular momentum $\mathbf{H}_O$ of a particle is equal to the sum of the moments about O of the forces acting on that particle.

Section 12.9 deals with the motion of a particle under a *central force*, i.e., under a force directed toward or away from a fixed point O. Since such a force has zero moment about O, it follows that the angular momentum of the particle about O is conserved. This property greatly simplifies the analysis of the motion of a particle under a central force; in Sec. 12.10 it is applied to the solution of problems involving the orbital motion of bodies under gravitational attraction.

Sections 12.11 through 12.13 are optional. They present a more extensive discussion of orbital motion and contain a number of problems related to space mechanics.

12.2 NEWTON'S SECOND LAW OF MOTION

Newton's second law can be stated as follows:

If the resultant force acting on a particle is not zero, the particle will have an acceleration proportional to the magnitude of the resultant and in the direction of this resultant force.

Newton's second law of motion is best understood by imagining the following experiment: A particle is subjected to a force $\mathbf{F}_1$ of constant direction and constant magnitude F_1. Under the action of that force, the particle is observed to move in a straight line and *in the direction of the force* (Fig. 12.1a). By determining the position of the particle at various instants, we find that its acceleration has a constant magnitude a_1. If the experiment is repeated with forces $\mathbf{F}_2$, $\mathbf{F}_3$, . . . , of different magnitude or direction (Fig. 12.1b and c), we find each time that the particle moves in the direction of the force acting on it and that the magnitudes a_1, a_2, a_3, . . . , of the accelerations are proportional to the magnitudes F_1, F_2, F_3, . . . , of the corresponding forces:

$$\frac{F_1}{a_1} = \frac{F_2}{a_2} = \frac{F_3}{a_3} = \cdots = \text{constant}$$

The constant value obtained for the ratio of the magnitudes of the forces and accelerations is a characteristic of the particle under consideration; it is called the *mass* of the particle and is denoted by m. When a particle of mass m is acted upon by a force $\mathbf{F}$, the force $\mathbf{F}$ and the acceleration $\mathbf{a}$ of the particle must therefore satisfy the relation

$$\mathbf{F} = m\mathbf{a} \qquad (12.1)$$

This relation provides a complete formulation of Newton's second law; it expresses not only that the magnitudes of $\mathbf{F}$ and $\mathbf{a}$ are proportional but also (since m is a positive scalar) that the vectors $\mathbf{F}$ and $\mathbf{a}$ have the same direction (Fig. 12.2). We should note that Eq. (12.1) still holds when $\mathbf{F}$ is not constant but varies with time in magnitude or direction. The magnitudes of $\mathbf{F}$ and $\mathbf{a}$ remain proportional, and the two vectors have the same direction at any given instant. However, they will not, in general, be tangent to the path of the particle.

When a particle is subjected simultaneously to several forces, Eq. (12.1) should be replaced by

$$\Sigma\mathbf{F} = m\mathbf{a} \qquad (12.2)$$

where $\Sigma\mathbf{F}$ represents the sum, or resultant, of all the forces acting on the particle.

It should be noted that the system of axes with respect to which the acceleration $\mathbf{a}$ is determined is not arbitrary. These axes must have a constant orientation with respect to the stars, and their origin must either be attached to the sun† or move with a constant velocity

†More accurately, to the mass center of the solar system.

(a)

(b)

(c)

Fig. 12.1

Fig. 12.2

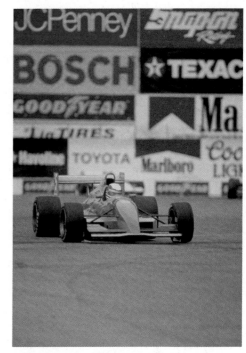

Photo 12.1 When the racecar accelerates forward the rear tires have a friction force acting on them in the direction the car is moving.

with respect to the sun. Such a system of axes is called a *newtonian frame of reference.*[†] A system of axes attached to the earth does *not* constitute a newtonian frame of reference, since the earth rotates with respect to the stars and is accelerated with respect to the sun. However, in most engineering applications, the acceleration **a** can be determined with respect to axes attached to the earth and Eqs. (12.1) and (12.2) used without any appreciable error. On the other hand, these equations do not hold if **a** represents a relative acceleration measured with respect to moving axes, such as axes attached to an accelerated car or to a rotating piece of machinery.

We observe that if the resultant $\Sigma\mathbf{F}$ of the forces acting on the particle is zero, it follows from Eq. (12.2) that the acceleration **a** of the particle is also zero. If the particle is initially at rest ($\mathbf{v}_0 = 0$) with respect to the newtonian frame of reference used, it will thus remain at rest ($\mathbf{v} = 0$). If originally moving with a velocity $\mathbf{v}_0$, the particle will maintain a constant velocity $\mathbf{v} = \mathbf{v}_0$; that is, it will move with the constant speed v_0 in a straight line. This, we recall, is the statement of Newton's first law (Sec. 2.10). Thus, Newton's first law is a particular case of Newton's second law and can be omitted from the fundamental principles of mechanics.

12.3 LINEAR MOMENTUM OF A PARTICLE. RATE OF CHANGE OF LINEAR MOMENTUM

Replacing the acceleration **a** by the derivative $d\mathbf{v}/dt$ in Eq. (12.2), we write

$$\Sigma\mathbf{F} = m\frac{d\mathbf{v}}{dt}$$

or, since the mass m of the particle is constant,

$$\Sigma\mathbf{F} = \frac{d}{dt}(m\mathbf{v}) \tag{12.3}$$

Fig. 12.3

The vector $m\mathbf{v}$ is called the *linear momentum,* or simply the *momentum,* of the particle. It has the same direction as the velocity of the particle, and its magnitude is equal to the product of the mass m and the speed v of the particle (Fig. 12.3). Equation (12.3) expresses that *the resultant of the forces acting on the particle is equal to the rate of change of the linear momentum of the particle.* It is in this form that the second law of motion was originally stated by Newton. Denoting by **L** the linear momentum of the particle,

$$\mathbf{L} = m\mathbf{v} \tag{12.4}$$

and by $\dot{\mathbf{L}}$ its derivative with respect to t, we can write Eq. (12.3) in the alternative form

$$\Sigma\mathbf{F} = \dot{\mathbf{L}} \tag{12.5}$$

[†]Since stars are not actually fixed, a more rigorous definition of a newtonian frame of reference (also called an *inertial system*) is one with respect to which Eq. (12.2) holds.

It should be noted that the mass m of the particle is assumed to be constant in Eqs. (12.3) to (12.5). Equation (12.3) or (12.5) should therefore not be used to solve problems involving the motion of bodies, such as rockets, which gain or lose mass. Problems of that type will be considered in Sec. 14.12.†

It follows from Eq. (12.3) that the rate of change of the linear momentum $m\mathbf{v}$ is zero when $\Sigma\mathbf{F} = 0$. Thus, *if the resultant force acting on a particle is zero, the linear momentum of the particle remains constant, in both magnitude and direction.* This is the principle of *conservation of linear momentum* for a particle, which can be recognized as an alternative statement of Newton's first law (Sec. 2.10).

12.4 SYSTEMS OF UNITS

In using the fundamental equation $\mathbf{F} = m\mathbf{a}$, the units of force, mass, length, and time cannot be chosen arbitrarily. If they are, the magnitude of the force $\mathbf{F}$ required to give an acceleration $\mathbf{a}$ to the mass m will *not* be numerically equal to the product ma; it will be only proportional to this product. Thus, we can choose three of the four units arbitrarily but must choose the fourth unit so that the equation $\mathbf{F} = m\mathbf{a}$ is satisfied. The units are then said to form a system of consistent kinetic units.

Two systems of consistent kinetic units are currently used by American engineers, the International System of Units (SI units‡) and the system of U.S. customary units. Both systems were discussed in detail in Sec. 1.3 and are described only briefly in this section.

International System of Units (SI Units). In this system, the base units are the units of length, mass, and time, and are called, respectively, the *meter* (m), the *kilogram* (kg), and the *second* (s). All three are arbitrarily defined (Sec. 1.3). The unit of force is a derived unit. It is called the *newton* (N) and is defined as the force which gives an acceleration of 1 m/s^2 to a mass of 1 kg (Fig. 12.4). From Eq. (12.1) we write

$$1 \text{ N} = (1 \text{ kg})(1 \text{ m/s}^2) = 1 \text{ kg} \cdot \text{m/s}^2$$

Fig. 12.4

The SI units are said to form an *absolute* system of units. This means that the three base units chosen are independent of the location where measurements are made. The meter, the kilogram, and the second may be used anywhere on the earth; they may even be used on another planet. They will always have the same significance.

The *weight* $\mathbf{W}$ of a body, or *force of gravity* exerted on that body, should, like any other force, be expressed in newtons. Since a body subjected to its own weight acquires an acceleration equal to the acceleration of gravity g, it follows from Newton's second law that the magnitude W of the weight of a body of mass m is

$$W = mg \qquad (12.6)$$

†On the other hand, Eqs. (12.3) and (12.5) do hold in *relativistic mechanics*, where the mass m of the particle is assumed to vary with the speed of the particle.

‡SI stands for *Système International d'Unités* (French).

Fig. 12.5

Recalling that $g = 9.81$ m/s^2, we find that the weight of a body of mass 1 kg (Fig. 12.5) is

$$W = (1 \text{ kg})(9.81 \text{ m/s}^2) = 9.81 \text{ N}$$

Multiples and submultiples of the units of length, mass, and force are frequently used in engineering practice. They are, respectively, the *kilometer* (km) and the *millimeter* (mm); the *megagram*† (Mg) and the *gram* (g); and the *kilonewton* (kN). By definition,

$$1 \text{ km} = 1000 \text{ m} \qquad 1 \text{ mm} = 0.001 \text{ m}$$
$$1 \text{ Mg} = 1000 \text{ kg} \qquad 1 \text{ g} = 0.001 \text{ kg}$$
$$1 \text{ kN} = 1000 \text{ N}$$

The conversion of these units to meters, kilograms, and newtons, respectively, can be effected simply by moving the decimal point three places to the right or to the left.

Units other than the units of mass, length, and time can all be expressed in terms of these three base units. For example, the unit of linear momentum can be obtained by recalling the definition of linear momentum and writing

$$mv = (\text{kg})(\text{m/s}) = \text{kg} \cdot \text{m/s}$$

U.S. Customary Units. Most practicing American engineers still commonly use a system in which the base units are the units of length, force, and time. These units are, respectively, the *foot* (ft), the *pound* (lb), and the *second* (s). The second is the same as the corresponding SI unit. The foot is defined as 0.3048 m. The pound is defined as the *weight* of a platinum standard, called the *standard pound,* which is kept at the National Institute of Standards and Technology outside Washington and the mass of which is 0.453 592 43 kg. Since the weight of a body depends upon the gravitational attraction of the earth, which varies with location, it is specified that the standard pound should be placed at sea level and at a latitude of 45° to properly define a force of 1 lb. Clearly, the U.S. customary units do not form an absolute system of units. Because of their dependence upon the gravitational attraction of the earth, they are said to form a *gravitational* system of units.

While the standard pound also serves as the unit of mass in commercial transactions in the United States, it cannot be so used in engineering computations since such a unit would not be consistent with the base units defined in the preceding paragraph. Indeed, when acted upon by a force of 1 lb, that is, when subjected to its own weight, the standard pound receives the acceleration of gravity, $g = 32.2$ ft/s^2 (Fig. 12.6), and not the unit acceleration required by Eq. (12.1). The unit of mass consistent with the foot, the pound, and the second is the mass which receives an acceleration of 1 ft/s^2 when a force of 1 lb is applied to it (Fig. 12.7). This unit, sometimes called a *slug,* can be derived from the equation $F = ma$ after substituting 1 lb and 1 ft/s^2 for F and a, respectively. We write

$$F = ma \qquad 1 \text{ lb} = (1 \text{ slug})(1 \text{ ft/s}^2)$$

a = 32.2 ft/s^2

m = 1 lb

F = 1 lb

Fig. 12.6

a = 1 ft/s^2

m = 1 slug
(= 1 lb·s^2/ft)

F = 1 lb

Fig. 12.7

†Also known as a *metric ton.*

and obtain

$$1 \text{ slug} = \frac{1 \text{ lb}}{1 \text{ ft/s}^2} = 1 \text{ lb} \cdot \text{s}^2/\text{ft}$$

Comparing Figs. 12.6 and 12.7, we conclude that the slug is a mass 32.2 times larger than the mass of the standard pound.

The fact that bodies are characterized in the U.S. customary system of units by their weight in pounds rather than by their mass in slugs was a convenience in the study of statics, where we were dealing for the most part with weights and other forces and only seldom with masses. However, in the study of kinetics, which involves forces, masses, and accelerations, it will be repeatedly necessary to express in slugs the mass m of a body, the weight W of which has been given in pounds. Recalling Eq. (12.6), we will write

$$m = \frac{W}{g} \tag{12.7}$$

where g is the acceleration of gravity ($g = 32.2 \text{ ft/s}^2$).

Units other than the units of force, length, and time can all be expressed in terms of these three base units. For example, the unit of linear momentum can be obtained by using the definition of linear momentum to write

$$mv = (\text{lb} \cdot \text{s}^2/\text{ft})(\text{ft/s}) = \text{lb} \cdot \text{s}$$

Conversion from One System of Units to Another. The conversion from U.S. customary units to SI units, and vice versa, was discussed in Sec. 1.4. You will recall that the conversion factors obtained for the units of length, force, and mass are, respectively,

Length: 1 ft = 0.3048 m
Force: 1 lb = 4.448 N
Mass: 1 slug = 1 lb · s²/ft = 14.59 kg

Although it cannot be used as a consistent unit of mass, the mass of the standard pound is, by definition,

1 pound-mass = 0.4536 kg

This constant can be used to determine the *mass* in SI units (kilograms) of a body which has been characterized by its *weight* in U.S. customary units (pounds).

12.5 EQUATIONS OF MOTION

Consider a particle of mass m acted upon by several forces. We recall from Sec. 12.2 that Newton's second law can be expressed by the equation

$$\Sigma \mathbf{F} = m\mathbf{a} \tag{12.2}$$

which relates the forces acting on the particle and the vector $m\mathbf{a}$ (Fig. 12.8). In order to solve problems involving the motion of a particle, however, it will be found more convenient to replace Eq. (12.2) by equivalent equations involving scalar quantities.

Fig. 12.8

Photo 12.2 The pilot of a fighter aircraft will experience very large normal forces when taking a sharp turn.

Rectangular Components. Resolving each force **F** and the acceleration **a** into rectangular components, we write

$$\Sigma(F_x\mathbf{i} + F_y\mathbf{j} + F_z\mathbf{k}) = m(a_x\mathbf{i} + a_y\mathbf{j} + a_z\mathbf{k})$$

from which it follows that

$$\Sigma F_x = ma_x \qquad \Sigma F_y = ma_y \qquad \Sigma F_z = ma_z \qquad (12.8)$$

Recalling from Sec. 11.11 that the components of the acceleration are equal to the second derivatives of the coordinates of the particle, we have

$$\Sigma F_x = m\ddot{x} \qquad \Sigma F_y = m\ddot{y} \qquad \Sigma F_z = m\ddot{z} \qquad (12.8')$$

Consider, as an example, the motion of a projectile. If the resistance of the air is neglected, the only force acting on the projectile after it has been fired is its weight $\mathbf{W} = -W\mathbf{j}$. The equations defining the motion of the projectile are therefore

$$m\ddot{x} = 0 \qquad m\ddot{y} = -W \qquad m\ddot{z} = 0$$

and the components of the acceleration of the projectile are

$$\ddot{x} = 0 \qquad \ddot{y} = -\frac{W}{m} = -g \qquad \ddot{z} = 0$$

where g is 9.81 m/s^2 or 32.2 ft/s^2. The equations obtained can be integrated independently, as shown in Sec. 11.11, to obtain the velocity and displacement of the projectile at any instant.

When a problem involves two or more bodies, equations of motion should be written for each of the bodies (see Sample Probs. 12.3 and 12.4). You will recall from Sec. 12.2 that all accelerations should be measured with respect to a newtonian frame of reference. In most engineering applications, accelerations can be determined with respect to axes attached to the earth, but relative accelerations measured with respect to moving axes, such as axes attached to an accelerated body, cannot be substituted for **a** in the equations of motion.

Tangential and Normal Components. Resolving the forces and the acceleration of the particle into components along the tangent to the path (in the direction of motion) and the normal (toward the inside of

Fig. 12.9

the path) (Fig. 12.9), and substituting into Eq. (12.2), we obtain the two scalar equations

$$\Sigma F_t = ma_t \qquad \Sigma F_n = ma_n \qquad (12.9)$$

Substituting for a_t and a_n from Eqs. (11.40), we have

$$\Sigma F_t = m\frac{dv}{dt} \qquad \Sigma F_n = m\frac{v^2}{\rho} \qquad (12.9')$$

The equations obtained may be solved for two unknowns.

12.6 DYNAMIC EQUILIBRIUM

Returning to Eq. (12.2) and transposing the right-hand member, we write Newton's second law in the alternative form

$$\Sigma \mathbf{F} - m\mathbf{a} = 0 \qquad (12.10)$$

which expresses that if we add the vector $-m\mathbf{a}$ to the forces acting on the particle, *we obtain a system of vectors equivalent to zero* (Fig. 12.10). The vector $-m\mathbf{a}$, of magnitude ma and of *direction opposite* to that of the acceleration, is called an *inertia vector*. The particle may thus be considered to be in equilibrium under the given forces and the inertia vector. The particle is said to be in *dynamic equilibrium*, and the problem under consideration can be solved by the methods developed earlier in statics.

In the case of coplanar forces, all the vectors shown in Fig. 12.10, *including the inertia vector,* can be drawn tip-to-tail to form a closed-vector polygon. Or the sums of the components of all the vectors in Fig. 12.10, again including the inertia vector, can be equated to zero. Using rectangular components, we therefore write

$$\Sigma F_x = 0 \qquad \Sigma F_y = 0 \qquad \textit{including inertia vector} \qquad (12.11)$$

Fig. 12.10

When tangential and normal components are used, it is more convenient to represent the inertia vector by its two components $-m\mathbf{a}_t$ and $-m\mathbf{a}_n$ in the sketch itself (Fig. 12.11). The tangential component of the inertia vector provides a measure of the resistance the particle offers to a change in speed, while its normal component (also called *centrifugal force*) represents the tendency of the particle to leave its curved path. We should note that either of these two components may be zero under special conditions: (1) If the particle starts from rest, its initial velocity is zero and the normal component of the inertia vector is zero at $t = 0$; (2) if the particle moves at constant speed along its path, the tangential component of the inertia vector is zero and only its normal component needs to be considered.

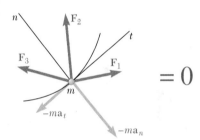

Fig. 12.11

Because they measure the resistance that particles offer when we try to set them in motion or when we try to change the conditions of their motion, inertia vectors are often called *inertia forces*. The inertia forces, however, are not forces like the forces found in statics, which are either contact forces or gravitational forces (weights). Many people, therefore, object to the use of the word "force" when referring to the vector $-m\mathbf{a}$ or even avoid altogether the concept of dynamic equilibrium. Others point out that inertia forces and actual forces, such as gravitational forces, affect our senses in the same way and cannot be distinguished by physical measurements. A man riding in an elevator which is accelerated upward will have the feeling that his weight has suddenly increased; and no measurement made within the elevator could establish whether the elevator is truly accelerated or whether the force of attraction exerted by the earth has suddenly increased.

Sample problems have been solved in this text by the direct application of Newton's second law, as illustrated in Figs. 12.8 and 12.9, rather than by the method of dynamic equilibrium.

Photo 12.3 The angle each rider is with respect to the horizontal will depend on the weight of the rider and the speed of rotation.

SAMPLE PROBLEM 12.1

A 200-lb block rests on a horizontal plane. Find the magnitude of the force **P** required to give the block an acceleration of 10 ft/s² to the right. The coefficient of kinetic friction between the block and the plane is $\mu_k = 0.25$.

SOLUTION

The mass of the block is

$$m = \frac{W}{g} = \frac{200 \text{ lb}}{32.2 \text{ ft/s}^2} = 6.21 \text{ lb} \cdot \text{s}^2/\text{ft}$$

We note that $F = \mu_k N = 0.25N$ and that $a = 10$ ft/s². Expressing that the forces acting on the block are equivalent to the vector **ma**, we write

$\xrightarrow{+}\Sigma F_x = ma:$ $P \cos 30° - 0.25N = (6.21 \text{ lb} \cdot \text{s}^2/\text{ft})(10 \text{ ft/s}^2)$
       $P \cos 30° - 0.25N = 62.1 \text{ lb}$ (1)

$+\uparrow\Sigma F_y = 0:$ $N - P \sin 30° - 200 \text{ lb} = 0$ (2)

Solving (2) for N and substituting the result into (1), we obtain

$$N = P \sin 30° + 200 \text{ lb}$$
$$P \cos 30° - 0.25(P \sin 30° + 200 \text{ lb}) = 62.1 \text{ lb} \qquad P = 151 \text{ lb} \quad \blacktriangleleft$$

SAMPLE PROBLEM 12.2

An 80-kg block rests on a horizontal plane. Find the magnitude of the force **P** required to give the block an acceleration of 2.5 m/s² to the right. The coefficient of kinetic friction between the block and the plane is $\mu_k = 0.25$.

SOLUTION

The weight of the block is

$$W = mg = (80 \text{ kg})(9.81 \text{ m/s}^2) = 785 \text{ N}$$

We note that $F = \mu_k N = 0.25N$ and that $a = 2.5$ m/s². Expressing that the forces acting on the block are equivalent to the vector **ma**, we write

$\xrightarrow{+}\Sigma F_x = ma:$ $P \cos 30° - 0.25N = (80 \text{ kg})(2.5 \text{ m/s}^2)$
       $P \cos 30° - 0.25N = 200 \text{ N}$ (1)

$+\uparrow\Sigma F_y = 0:$ $N - P \sin 30° - 785 \text{ N} = 0$ (2)

Solving (2) for N and substituting the result into (1), we obtain

$$N = P \sin 30° + 785 \text{ N}$$
$$P \cos 30° - 0.25(P \sin 30° + 785 \text{ N}) = 200 \text{ N} \qquad P = 535 \text{ N} \quad \blacktriangleleft$$

SAMPLE PROBLEM 12.3

The two blocks shown start from rest. The horizontal plane and the pulley are frictionless, and the pulley is assumed to be of negligible mass. Determine the acceleration of each block and the tension in each cord.

SOLUTION

Kinematics. We note that if block A moves through x_A to the right, block B moves down through

$$x_B = \tfrac{1}{2}x_A$$

Differentiating twice with respect to t, we have

$$a_B = \tfrac{1}{2}a_A \tag{1}$$

Kinetics. We apply Newton's second law successively to block A, block B, and pulley C.

Block A. Denoting by T_1 the tension in cord ACD, we write

$$\xrightarrow{+}\Sigma F_x = m_A a_A: \qquad T_1 = 100a_A \tag{2}$$

Block B. Observing that the weight of block B is

$$W_B = m_B g = (300 \text{ kg})(9.81 \text{ m/s}^2) = 2940 \text{ N}$$

and denoting by T_2 the tension in cord BC, we write

$$+\downarrow\Sigma F_y = m_B a_B: \qquad 2940 - T_2 = 300a_B$$

or, substituting for a_B from (1),

$$2940 - T_2 = 300(\tfrac{1}{2}a_A)$$
$$T_2 = 2940 - 150a_A \tag{3}$$

Pulley C. Since m_C is assumed to be zero, we have

$$+\downarrow\Sigma F_y = m_C a_C = 0: \qquad T_2 - 2T_1 = 0 \tag{4}$$

Substituting for T_1 and T_2 from (2) and (3), respectively, into (4) we write

$$2940 - 150a_A - 2(100a_A) = 0$$
$$2940 - 350a_A = 0 \qquad a_A = 8.40 \text{ m/s}^2 \blacktriangleleft$$

Substituting the value obtained for a_A into (1) and (2), we have

$$a_B = \tfrac{1}{2}a_A = \tfrac{1}{2}(8.40 \text{ m/s}^2) \qquad a_B = 4.20 \text{ m/s}^2 \blacktriangleleft$$
$$T_1 = 100a_A = (100 \text{ kg})(8.40 \text{ m/s}^2) \qquad T_1 = 840 \text{ N} \blacktriangleleft$$

Recalling (4), we write

$$T_2 = 2T_1 \qquad T_2 = 2(840 \text{ N}) \qquad T_2 = 1680 \text{ N} \blacktriangleleft$$

We note that the value obtained for T_2 is *not* equal to the weight of block B.

SAMPLE PROBLEM 12.4

The 12-lb block B starts from rest and slides on the 30-lb wedge A, which is supported by a horizontal surface. Neglecting friction, determine (a) the acceleration of the wedge, (b) the acceleration of the block relative to the wedge.

SOLUTION

Kinematics. We first examine the acceleration of the wedge and the acceleration of the block.

Wedge A. Since the wedge is constrained to move on the horizontal surface, its acceleration $\mathbf{a}_A$ is horizontal. We will assume that it is directed to the right.

Block B. The acceleration $\mathbf{a}_B$ of block B can be expressed as the sum of the acceleration of A and the acceleration of B relative to A. We have

$$\mathbf{a}_B = \mathbf{a}_A + \mathbf{a}_{B/A}$$

where $\mathbf{a}_{B/A}$ is directed along the inclined surface of the wedge.

Kinetics. We draw the free-body diagrams of the wedge and of the block and apply Newton's second law.

Wedge A. We denote the forces exerted by the block and the horizontal surface on wedge A by $\mathbf{N}_1$ and $\mathbf{N}_2$, respectively.

$$\xrightarrow{+} \Sigma F_x = m_A a_A: \qquad N_1 \sin 30° = m_A a_A$$
$$0.5 N_1 = (W_A/g)a_A \qquad (1)$$

Block B. Using the coordinate axes shown and resolving $\mathbf{a}_B$ into its components $\mathbf{a}_A$ and $\mathbf{a}_{B/A}$, we write

$$+\nearrow \Sigma F_x = m_B a_x: \qquad -W_B \sin 30° = m_B a_A \cos 30° - m_B a_{B/A}$$
$$-W_B \sin 30° = (W_B/g)(a_A \cos 30° - a_{B/A})$$
$$a_{B/A} = a_A \cos 30° + g \sin 30° \qquad (2)$$
$$+\nwarrow \Sigma F_y = m_B a_y: \qquad N_1 - W_B \cos 30° = -m_B a_A \sin 30°$$
$$N_1 - W_B \cos 30° = -(W_B/g)a_A \sin 30° \qquad (3)$$

a. Acceleration of Wedge A. Substituting for N_1 from Eq. (1) into Eq. (3), we have

$$2(W_A/g)a_A - W_B \cos 30° = -(W_B/g)a_A \sin 30°$$

Solving for a_A and substituting the numerical data, we write

$$a_A = \frac{W_B \cos 30°}{2W_A + W_B \sin 30°}g = \frac{(12 \text{ lb}) \cos 30°}{2(30 \text{ lb}) + (12 \text{ lb}) \sin 30°}(32.2 \text{ ft/s}^2)$$
$$a_A = +5.07 \text{ ft/s}^2 \qquad \mathbf{a}_A = 5.07 \text{ ft/s}^2 \rightarrow \blacktriangleleft$$

b. Acceleration of Block B Relative to A. Substituting the value obtained for a_A into Eq. (2), we have

$$a_{B/A} = (5.07 \text{ ft/s}^2) \cos 30° + (32.2 \text{ ft/s}^2) \sin 30°$$
$$a_{B/A} = +20.5 \text{ ft/s}^2 \qquad \mathbf{a}_{B/A} = 20.5 \text{ ft/s}^2 \; \measuredangle 30° \; \blacktriangleleft$$

SAMPLE PROBLEM 12.5

The bob of a 2-m pendulum describes an arc of circle in a vertical plane. If the tension in the cord is 2.5 times the weight of the bob for the position shown, find the velocity and the acceleration of the bob in that position.

SOLUTION

The weight of the bob is $W = mg$; the tension in the cord is thus $2.5\ mg$. Recalling that $\mathbf{a}_n$ is directed toward O and assuming $\mathbf{a}_t$ as shown, we apply Newton's second law and obtain

$+\swarrow \Sigma F_t = ma_t:$
$$mg \sin 30° = ma_t$$
$$a_t = g \sin 30° = +4.90 \text{ m/s}^2 \qquad \mathbf{a}_t = 4.90 \text{ m/s}^2 \swarrow \blacktriangleleft$$

$+\nwarrow \Sigma F_n = ma_n:$
$$2.5\ mg - mg \cos 30° = ma_n$$
$$a_n = 1.634\ g = +16.03 \text{ m/s}^2 \qquad \mathbf{a}_n = 16.03 \text{ m/s}^2 \nwarrow \blacktriangleleft$$

Since $a_n = v^2/\rho$, we have $v^2 = \rho a_n = (2 \text{ m})(16.03 \text{ m/s}^2)$
$$v = \pm 5.66 \text{ m/s} \qquad \mathbf{v} = 5.66 \text{ m/s} \nearrow \text{(up or down)} \blacktriangleleft$$

SAMPLE PROBLEM 12.6

Determine the rated speed of a highway curve of radius $\rho = 400$ ft banked through an angle $\theta = 18°$. The *rated speed* of a banked highway curve is the speed at which a car should travel if no lateral friction force is to be exerted on its wheels.

SOLUTION

The car travels in a *horizontal* circular path of radius ρ. The normal component $\mathbf{a}_n$ of the acceleration is directed toward the center of the path; its magnitude is $a_n = v^2/\rho$, where v is the speed of the car in ft/s. The mass m of the car is W/g, where W is the weight of the car. Since no lateral friction force is to be exerted on the car, the reaction $\mathbf{R}$ of the road is shown perpendicular to the roadway. Applying Newton's second law, we write

$+\uparrow \Sigma F_y = 0:$
$$R \cos \theta - W = 0 \qquad R = \frac{W}{\cos \theta} \qquad (1)$$

$\xrightarrow{+} \Sigma F_n = ma_n:$
$$R \sin \theta = \frac{W}{g} a_n \qquad (2)$$

Substituting for R from (1) into (2), and recalling that $a_n = v^2/\rho$,
$$\frac{W}{\cos \theta} \sin \theta = \frac{W}{g} \frac{v^2}{\rho} \qquad v^2 = g\rho \tan \theta$$

Substituting $\rho = 400$ ft and $\theta = 18°$ into this equation, we obtain
$$v^2 = (32.2 \text{ ft/s}^2)(400 \text{ ft}) \tan 18°$$
$$v = 64.7 \text{ ft/s} \qquad\qquad v = 44.1 \text{ mi/h} \blacktriangleleft$$

SOLVING PROBLEMS
ON YOUR OWN

In the problems for this lesson, you will apply *Newton's second law of motion*, $\Sigma\mathbf{F} = m\mathbf{a}$, to relate the forces acting on a particle to the motion of the particle.

1. Writing the equations of motion. When applying Newton's second law to the types of motion discussed in this lesson, you will find it most convenient to express the vectors $\mathbf{F}$ and $\mathbf{a}$ in terms of either their rectangular components or their tangential and normal components.

    a. When using rectangular components, and recalling from Sec. 11.11 the expressions found for a_x, a_y, and a_z, you will write

$$\Sigma F_x = m\ddot{x} \qquad \Sigma F_y = m\ddot{y} \qquad \Sigma F_z = m\ddot{z}$$

    b. When using tangential and normal components, and recalling from Sec. 11.13 the expressions found for a_t and a_n, you will write

$$\Sigma F_t = m\frac{dv}{dt} \qquad \Sigma F_n = m\frac{v^2}{\rho}$$

2. Drawing a free-body diagram showing the applied forces *and an equivalent diagram* showing the vector $m\mathbf{a}$ or its components will provide you with a pictorial representation of Newton's second law [Sample Probs. 12.1 through 12.6]. These diagrams will be of great help to you when writing the equations of motion. Note that when a problem involves two or more bodies, it is usually best to consider each body separately.

3. Applying Newton's second law. As we observed in Sec. 12.2, the acceleration used in the equation $\Sigma\mathbf{F} = m\mathbf{a}$ should always be the *absolute acceleration* of the particle (that is, it should be measured with respect to a newtonian frame of reference). Also, *if the sense of the acceleration* $\mathbf{a}$ *is unknown* or is not easily deduced, assume an arbitrary sense for $\mathbf{a}$ (usually the positive direction of a coordinate axis) and then let the solution provide the correct sense. Finally, note how the solutions of Sample Probs. 12.3 and 12.4 were divided into a *kinematics* portion and a *kinetics* portion, and how in Sample Prob. 12.4 we used two systems of coordinate axes to simplify the equations of motion.

4. When a problem involves dry friction, be sure to review the relevant sections of *Statics* [Secs. 8.1 to 8.3] before attempting to solve that problem. In particular, you should know when each of the equations $F = \mu_s N$ and $F = \mu_k N$ may be used.

You should also recognize that if the motion of a system is not specified, it is necessary first to assume a possible motion and then to check the validity of that assumption.

5. Solving problems involving relative motion. When a body B moves with respect to a body A, as in Sample Prob. 12.4, it is often convenient to express the acceleration of B as

$$\mathbf{a}_B = \mathbf{a}_A + \mathbf{a}_{B/A}$$

where $\mathbf{a}_{B/A}$ is the acceleration of B relative to A, that is, the acceleration of B as observed from a frame of reference attached to A and in translation. If B is observed to move in a straight line, $\mathbf{a}_{B/A}$ will be directed along that line. On the other hand, if B is observed to move along a circular path, the relative acceleration $\mathbf{a}_{B/A}$ should be resolved into components tangential and normal to that path.

6. Finally, always consider the implications of any assumption you make. Thus, in a problem involving two cords, if you assume that the tension in one of the cords is equal to its maximum allowable value, check whether any requirements set for the other cord will then be satisfied. For instance, will the tension T in that cord satisfy the relation $0 \leq T \leq T_{max}$? That is, will the cord remain taut and will its tension be less than its maximum allowable value?

PROBLEMS

Fig. P12.CQ1

CONCEPT QUESTIONS

12.CQ1 A 1000-lb boulder B is resting on a 200-lb platform A when truck C accelerates to the left with a constant acceleration. Which of the following statements are true (more than one may be true)?

  a. The tension in the cord connected to the truck is 200 lb.

  b. The tension in the cord connected to the truck is 1200 lb.

  c. The tension in the cord connected to the truck is greater than 1200 lb.

  d. The normal force between A and B is 1000 lb.

  e. The normal force between A and B is 1200 lb.

  f. None of the above are true.

12.CQ2 Marble A is placed in a hollow tube, and the tube is swung in a horizontal plane causing the marble to be thrown out. As viewed from the top, which of the following choices best describes the path of the marble after leaving the tube?

  a. 1 **b.** 2 **c.** 3 **d.** 4 **e.** 5

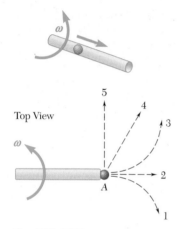

Fig. P12.CQ2

12.CQ3 The two systems shown start from rest. On the left, two 40-lb weights are connected by an inextensible cord, and on the right, a constant 40-lb force pulls on the cord. Neglecting all frictional forces, which of the following statements is true?

  a. Blocks A and C will have the same acceleration.

  b. Block C will have a larger acceleration than block A.

  c. Block A will have a larger acceleration than block C.

  d. Block A will not move.

  e. None of the above are true.

Fig. P12.CQ3

12.CQ4 Blocks A and B are released from rest in the position shown. Neglecting friction, the normal force between block A and the ground is:

  a. Less than the weight of A plus the weight of B.

  b. Equal to the weight of A plus the weight of B.

  c. Greater than the weight of A plus the weight of B.

Fig. P12.CQ4

12.CQ5 People sit on a Ferris wheel at points A, B, C, and D. The Ferris wheel travels at a constant angular velocity. At the instant shown, which person experiences the largest force from his or her chair (back and seat)? Assume you can neglect the size of the chairs—that is, the people are located the same distance from the axis of rotation.

    a. A

    b. B

    c. C

    d. D

    e. The force is the same for all the passengers.

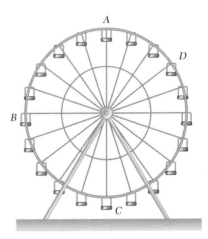

Fig. P12.CQ5

FREE BODY PRACTICE PROBLEMS

12.F1 Crate A is gently placed with zero initial velocity onto a moving conveyor belt. The coefficient of kinetic friction between the crate and the belt is μ_k. Draw the FBD and KD for A immediately after it contacts the belt.

Fig. P12.F1

12.F2 Two blocks weighing W_A and W_B are at rest on a conveyor that is initially at rest. The belt is suddenly started in an upward direction so that slipping occurs between the belt and the boxes. Assuming the coefficient of friction between the boxes and the belt is μ_k, draw the FBDs and KDs for blocks A and B. How would you determine if A and B remain in contact?

Fig. P12.F2

12.F3 Objects A, B, and C have masses m_A, m_B, and m_C, respectively. The coefficient of kinetic friction between A and B is μ_k, and the friction between A and the ground is negligible and the pulleys are massless and frictionless. Assuming B slides on A, draw the FBD and KD for each of the three masses A, B, and C.

Fig. P12.F3

Fig. P12.F4

12.F4 Blocks A and B have masses m_A and m_B, respectively. Neglecting friction between all surfaces, draw the FBD and KD for each mass.

12.F5 Blocks A and B have masses m_A and m_B, respectively. Neglecting friction between all surfaces, draw the FBD and KD for the two systems shown.

Fig. P12.F5

12.F6 A pilot of mass m flies a jet in a half-vertical loop of radius R so that the speed of the jet, v, remains constant. Draw a FBD and KD of the pilot at points A, B, and C.

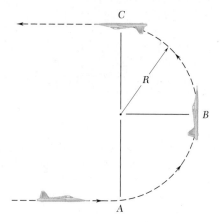

Fig. P12.F6

12.F7 Wires AC and BC are attached to a sphere which revolves at a constant speed v in the horizontal circle of radius r as shown. Draw a FBD and KD of C.

12.F8 A collar of mass m is attached to a spring and slides without friction along a circular rod in a vertical plane. The spring has an undeformed length of 5 in. and a constant k. Knowing that the collar has a speed v at point B, draw the FBD and KD of the collar at this point.

Fig. P12.F7

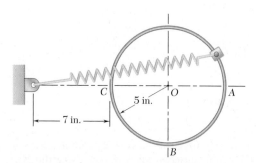

Fig. P12.F8

12.1 Astronauts who landed on the moon during the Apollo 15, 16, and 17 missions brought back a large collection of rocks to the earth. Knowing the rocks weighed 139 lb when they were on the moon, determine (a) the weight of the rocks on the earth, (b) the mass of the rocks in slugs. The acceleration due to gravity on the moon is 5.30 ft/s^2.

12.2 The value of g at any latitude ϕ may be obtained from the formula

$$g = 32.09(1 + 0.0053 \sin^2 \phi)\text{ft/s}^2$$

which takes into account the effect of the rotation of the earth, as well as the fact that the earth is not truly spherical. Determine to four significant figures (a) the weight in pounds, (b) the mass in pounds, (c) the mass in lb · s^2/ft, at the latitudes of 0°, 45°, 60°, of a silver bar, the mass of which has been officially designated as 5 lb.

12.3 A 400-kg satellite has been placed in a circular orbit 1500 km above the surface of the earth. The acceleration of gravity at this elevation is 6.43 m/s^2. Determine the linear momentum of the satellite, knowing that its orbital speed is 25.6 × 10^3 km/h.

12.4 A spring scale A and a lever scale B having equal lever arms are fastened to the roof of an elevator, and identical packages are attached to the scales as shown. Knowing that when the elevator moves downward with an acceleration of 1 m/s^2 the spring scale indicates a load of 60 N, determine (a) the weight of the packages, (b) the load indicated by the spring scale and the mass needed to balance the lever scale when the elevator moves upward with an acceleration of 1 m/s^2.

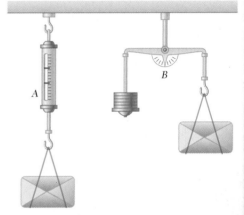

Fig. P12.4

12.5 In anticipation of a long 7° upgrade, a bus driver accelerates at a constant rate of 3 ft/s^2 while still on a level section of the highway. Knowing that the speed of the bus is 60 mi/h as it begins to climb the grade and that the driver does not change the setting of his throttle or shift gears, determine the distance traveled by the bus up the grade when its speed has decreased to 50 mi/h.

12.6 A hockey player hits a puck so that it comes to rest in 10 s after sliding 100 ft on the ice. Determine (a) the initial velocity of the puck, (b) the coefficient of friction between the puck and the ice.

12.7 The acceleration of a package sliding at point A is 3 m/s^2. Assuming that the coefficient of kinetic friction is the same for each section, determine the acceleration of the package at point B.

Fig. P12.7

12.8 Determine the maximum theoretical speed that may be achieved over a distance of 60 m by a car starting from rest, knowing that the coefficient of static friction is 0.80 between the tires and the pavement and that 60 percent of the weight of the car is distributed over its front wheels and 40 percent over its rear wheels. Assume (a) four-wheel drive, (b) front-wheel drive, (c) rear-wheel drive.

12.9 If an automobile's braking distance from 90 km/h is 45 m on level pavement, determine the automobile's braking distance from 90 km/h when it is (*a*) going up a 5° incline, (*b*) going down a 3-percent incline. Assume the braking force is independent of grade.

12.10 A mother and her child are skiing together, and the mother is holding the end of a rope tied to the child's waist. They are moving at a speed of 7.2 km/h on a gently sloping portion of the ski slope when the mother observes that they are approaching a steep descent. She pulls on the rope with an average force of 7 N. Knowing the coefficient of friction between the child and the ground is 0.1 and the angle of the rope does not change, determine (*a*) the time required for the child's speed to be cut in half, (*b*) the distance traveled in this time.

Fig. P12.10

12.11 The coefficients of friction between the load and the flatbed trailer shown are $\mu_s = 0.40$ and $\mu_k = 0.30$. Knowing that the speed of the rig is 72 km/h, determine the shortest distance in which the rig can be brought to a stop if the load is not to shift.

12.12 A light train made up of two cars is traveling at 90 km/h when the brakes are applied to both cars. Knowing that car A has a mass of 25 Mg and car B a mass of 20 Mg, and that the braking force is 30 kN on each car, determine (*a*) the distance traveled by the train before it comes to a stop, (*b*) the force in the coupling between the cars while the train is slowing down.

Fig. P12.12

12.13 The two blocks shown are originally at rest. Neglecting the masses of the pulleys and the effect of friction in the pulleys and between block A and the incline, determine (*a*) the acceleration of each block, (*b*) the tension in the cable.

Fig. P12.11

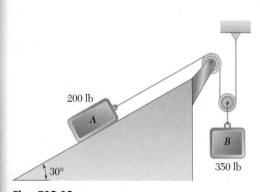

Fig. P12.13

12.14 Solve Prob. 12.13, assuming that the coefficients of friction between block A and the incline are $\mu_s = 0.25$ and $\mu_k = 0.20$.

12.15 Each of the systems shown is initially at rest. Neglecting axle friction and the masses of the pulleys, determine for each system (a) the acceleration of block A, (b) the velocity of block A after it has moved through 10 ft, (c) the time required for block A to reach a velocity of 20 ft/s.

100 lb

A

200 lb

(1)

100 lb

A

200 lb

(2)

A   2100 lb

2200 lb

(3)

Fig. P12.15

12.16 Boxes A and B are at rest on a conveyor belt that is initially at rest. The belt is suddenly started in an upward direction so that slipping occurs between the belt and the boxes. Knowing that the coefficients of kinetic friction between the belt and the boxes are $(\mu_k)_A = 0.30$ and $(\mu_k)_B = 0.32$, determine the initial acceleration of each box.

Fig. P12.16

12.17 A 5000-lb truck is being used to lift a 1000-lb boulder B that is on a 200-lb pallet A. Knowing the acceleration of the truck is 1 ft/s², determine (a) the horizontal force between the tires and the ground, (b) the force between the boulder and the pallet.

Fig. P12.17

12.18 Block A has a mass of 40 kg, and block B has a mass of 8 kg. The coefficients of friction between all surfaces of contact are $\mu_s = 0.20$ and $\mu_k = 0.15$. If $P = 0$, determine (a) the acceleration of block B, (b) the tension in the cord.

12.19 Block A has a mass of 40 kg, and block B has a mass of 8 kg. The coefficients of friction between all surfaces of contact are $\mu_s = 0.20$ and $\mu_k = 0.15$. If $P = 40$ N →, determine (a) the acceleration of block B, (b) the tension in the cord.

Fig. P12.18 and P12.19

Fig. P12.20

Fig. P12.21

Fig. P12.22

12.20 A package is at rest on a conveyor belt which is initially at rest. The belt is started and moves to the right for 1.3 s with a constant acceleration of 2 m/s². The belt then moves with a constant deceleration $\mathbf{a}_2$ and comes to a stop after a total displacement of 2.2 m. Knowing that the coefficients of friction between the package and the belt are $\mu_s = 0.35$ and $\mu_k = 0.25$, determine (a) the deceleration $\mathbf{a}_2$ of the belt, (b) the displacement of the package relative to the belt as the belt comes to a stop.

12.21 A baggage conveyor is used to unload luggage from an airplane. The 10-kg duffel bag A is sitting on top of the 20-kg suitcase B. The conveyor is moving the bags down at a constant speed of 0.5 m/s when the belt suddenly stops. Knowing that the coefficient of friction between the belt and B is 0.3 and that bag A does not slip on suitcase B, determine the smallest allowable coefficient of static friction between the bags.

12.22 To unload a bound stack of plywood from a truck, the driver first tilts the bed of the truck and then accelerates from rest. Knowing that the coefficients of friction between the bottom sheet of plywood and the bed are $\mu_s = 0.40$ and $\mu_k = 0.30$, determine (a) the smallest acceleration of the truck which will cause the stack of plywood to slide, (b) the acceleration of the truck which causes corner A of the stack to reach the end of the bed in 0.9 s.

12.23 To transport a series of bundles of shingles A to a roof, a contractor uses a motor-driven lift consisting of a horizontal platform BC which rides on rails attached to the sides of a ladder. The lift starts from rest and initially moves with a constant acceleration $\mathbf{a}_1$ as shown. The lift then decelerates at a constant rate $\mathbf{a}_2$ and comes to rest at D, near the top of the ladder. Knowing that the coefficient of static friction between a bundle of shingles and the horizontal platform is 0.30, determine the largest allowable acceleration $\mathbf{a}_1$ and the largest allowable deceleration $\mathbf{a}_2$ if the bundle is not to slide on the platform.

Fig. P12.23

12.24 An airplane has a mass of 25 Mg and its engines develop a total thrust of 40 kN during take-off. If the drag **D** exerted on the plane has a magnitude $D = 2.25 \, v^2$, where v is expressed in meters per second and D in newtons, and if the plane becomes airborne at a speed of 240 km/h, determine the length of runway required for the plane to take off.

12.25 The propellers of a ship of weight W can produce a propulsive force $\mathbf{F}_0$; they produce a force of the same magnitude but of opposite direction when the engines are reversed. Knowing that the ship was proceeding forward at its maximum speed v_0 when the engines were put into reverse, determine the distance the ship travels before coming to a stop. Assume that the frictional resistance of the water varies directly with the square of the velocity.

12.26 A constant force **P** is applied to a piston and rod of total mass m to make them move in a cylinder filled with oil. As the piston moves, the oil is forced through orifices in the piston and exerts on the piston a force of magnitude kv in a direction opposite to the motion of the piston. Knowing that the piston starts from rest at $t = 0$ and $x = 0$, show that the equation relating x, v, and t, where x is the distance traveled by the piston and v is the speed of the piston, is linear in each of these variables.

Fig. P12.26

12.27 A spring AB of constant k is attached to a support at A and to a collar of mass m. The unstretched length of the spring is l. Knowing that the collar is released from rest at $x = x_0$ and neglecting friction between the collar and the horizontal rod, determine the magnitude of the velocity of the collar as it passes through point C.

12.28 Block A has a mass of 10 kg, and blocks B and C have masses of 5 kg each. Knowing that the blocks are initially at rest and that B moves through 3 m in 2 s, determine (*a*) the magnitude of the force **P**, (*b*) the tension in the cord AD. Neglect the masses of the pulleys and axle friction.

Fig. P12.27

Fig. P12.28

12.29 A 40-lb sliding panel is supported by rollers at B and C. A 25-lb counterweight A is attached to a cable as shown and, in cases a and c, is initially in contact with a vertical edge of the panel. Neglecting friction, determine in each case shown the acceleration of the panel and the tension in the cord immediately after the system is released from rest.

(a)   (b)   (c)

Fig. P12.29

12.30 The coefficients of friction between blocks A and C and the horizontal surfaces are $\mu_s = 0.24$ and $\mu_k = 0.20$. Knowing that $m_A = 5$ kg, $m_B = 10$ kg, and $m_C = 10$ kg, determine (a) the tension in the cord, (b) the acceleration of each block.

Fig. P12.30

12.31 A 10-lb block B rests as shown on a 20-lb bracket A. The coefficients of friction are $\mu_s = 0.30$ and $\mu_k = 0.25$ between block B and bracket A, and there is no friction in the pulley or between the bracket and the horizontal surface. (a) Determine the maximum weight of block C if block B is not to slide on bracket A. (b) If the weight of block C is 10 percent larger than the answer found in a, determine the accelerations of A, B, and C.

Fig. P12.31

12.32 The masses of blocks A, B, C, and D are 9 kg, 9 kg, 6 kg, and 7 kg, respectively. Knowing that a downward force of magnitude 120 N is applied to block D, determine (*a*) the acceleration of each block, (*b*) the tension in cord ABC. Neglect the weights of the pulleys and the effect of friction.

12.33 The masses of blocks A, B, C, and D are 9 kg, 9 kg, 6 kg, and 7 kg, respectively. Knowing that a downward force of magnitude 50 N is applied to block B and that the system starts from rest, determine at $t = 3$ s the velocity (*a*) of D relative to A, (*b*) of C relative to D. Neglect the weights of the pulleys and the effect of friction.

Fig. P12.32 and P12.33

12.34 The 15-kg block B is supported by the 25-kg block A and is attached to a cord to which a 225-N horizontal force is applied as shown. Neglecting friction, determine (*a*) the acceleration of block A, (*b*) the acceleration of block B relative to A.

Fig. P12.34

12.35 Block B of mass 10 kg rests as shown on the upper surface of a 22-kg wedge A. Knowing that the system is released from rest and neglecting friction, determine (*a*) the acceleration of B, (*b*) the velocity of B relative to A at $t = 0.5$ s.

12.36 A 450-g tetherball A is moving along a horizontal circular path at a constant speed of 4 m/s. Determine (*a*) the angle θ that the cord forms with pole BC, (*b*) the tension in the cord.

Fig. P12.35

Fig. P12.36

12.37 During a hammer thrower's practice swings, the 7.1-kg head A of the hammer revolves at a constant speed v in a horizontal circle as shown. If $\rho = 0.93$ m and $\theta = 60°$, determine (a) the tension in wire BC, (b) the speed of the hammer's head.

Fig. P12.37

12.38 A single wire ACB passes through a ring at C attached to a sphere which revolves at a constant speed v in the horizontal circle shown. Knowing that the tension is the same in both portions of the wire, determine the speed v.

12.39 Two wires AC and BC are tied at C to a sphere which revolves at a constant speed v in the horizontal circle shown. Determine the range of values of v for which both wires remain taut.

*12.40 Two wires AC and BC are tied at C to a sphere which revolves at a constant speed v in the horizontal circle shown. Determine the range of the allowable values of v if both wires are to remain taut and if the tension in either of the wires is not to exceed 60 N.

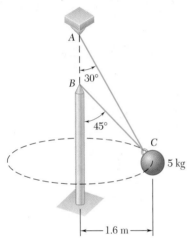

Fig. P12.38, P12.39, and P12.40

12.41 A 100-g sphere D is at rest relative to drum ABC which rotates at a constant rate. Neglecting friction, determine the range of the allowable values of the velocity v of the sphere if neither of the normal forces exerted by the sphere on the inclined surfaces of the drum is to exceed 1.1 N.

Fig. P12.41

*12.42 As part of an outdoor display, a 12-lb model C of the earth is attached to wires AC and BC and revolves at a constant speed v in the horizontal circle shown. Determine the range of the allowable values of v if both wires are to remain taut and if the tension in either of the wires is not to exceed 26 lb.

Fig. P12.42

*12.43 The 1.2-lb flyballs of a centrifugal governor revolve at a constant speed v in the horizontal circle of 6-in. radius shown. Neglecting the weights of links AB, BC, AD, and DE and requiring that the links support only tensile forces, determine the range of the allowable values of v so that the magnitudes of the forces in the links do not exceed 17 lb.

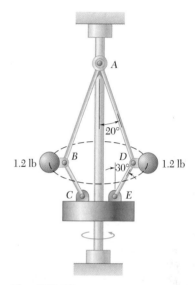

Fig. P12.43

12.44 A 130-lb wrecking ball B is attached to a 45-ft-long steel cable AB and swings in the vertical arc shown. Determine the tension in the cable (*a*) at the top C of the swing, (*b*) at the bottom D of the swing, where the speed of B is 13.2 ft/s.

Fig. P12.44

12.45 During a high-speed chase, a 2400-lb sports car traveling at a speed of 100 mi/h just loses contact with the road as it reaches the crest A of a hill. (*a*) Determine the radius of curvature ρ of the vertical profile of the road at A. (*b*) Using the value of ρ found in part *a*, determine the force exerted on a 160-lb driver by the seat of his 3100-lb car as the car, traveling at a constant speed of 50 mi/h, passes through A.

Fig. P12.45

Fig. P12.46

12.46 A child having a mass of 22 kg sits on a swing and is held in the position shown by a second child. Neglecting the mass of the swing, determine the tension in rope AB (a) while the second child holds the swing with his arms outstretched horizontally, (b) immediately after the swing is released.

12.47 The roller-coaster track shown is contained in a vertical plane. The portion of track between A and B is straight and horizontal, while the portions to the left of A and to the right of B have radii of curvature as indicated. A car is traveling at a speed of 72 km/h when the brakes are suddenly applied, causing the wheels of the car to slide on the track ($\mu_k = 0.20$). Determine the initial deceleration of the car if the brakes are applied as the car (a) has almost reached A, (b) is traveling between A and B, (c) has just passed B.

Fig. P12.47

12.48 A 250-g block fits inside a small cavity cut in arm OA, which rotates in the vertical plane at a constant rate such that $v = 3$ m/s. Knowing that the spring exerts on block B a force of magnitude $P = 1.5$ N and neglecting the effect of friction, determine the range of values of θ for which block B is in contact with the face of the cavity closest to the axis of rotation O.

12.49 A series of small packages, each with a mass of 0.5 kg, are discharged from a conveyor belt as shown. Knowing that the coefficient of static friction between each package and the conveyor belt is 0.4, determine (a) the force exerted by the belt on the package just after it has passed point A, (b) the angle θ defining the point B where the packages first *slip* relative to the belt.

Fig. P12.48

Fig. P12.49

12.50 A 54-kg pilot flies a jet trainer in a half-vertical loop of 1200-m radius so that the speed of the trainer decreases at a constant rate. Knowing that the pilot's apparent weights at points A and C are 1680 N and 350 N, respectively, determine the force exerted on her by the seat of the trainer when the trainer is at point B.

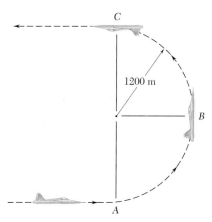

Fig. P12.50

12.51 A carnival ride is designed to allow the general public to experience high-acceleration motion. The ride rotates about point O in a horizontal circle such that the rider has a speed v_0. The rider reclines on a platform A which rides on rollers such that friction is negligible. A mechanical stop prevents the platform from rolling down the incline. Determine (*a*) the speed v_0 at which the platform A begins to roll upward, (*b*) the normal force experienced by an 80-kg rider at this speed.

Fig. P12.51

12.52 A curve in a speed track has a radius of 1000 ft and a rated speed of 120 mi/h. (See Sample Prob. 12.6 for the definition of rated speed.) Knowing that a racing car starts skidding on the curve when traveling at a speed of 180 mi/h, determine (*a*) the banking angle θ, (*b*) the coefficient of static friction between the tires and the track under the prevailing conditions, (*c*) the minimum speed at which the same car could negotiate the curve.

Fig. *P12.52*

Fig. P12.53 and P12.54

12.53 Tilting trains, such as the *American Flyer* which will run from Washington to New York and Boston, are designed to travel safely at high speeds on curved sections of track which were built for slower, conventional trains. As it enters a curve, each car is tilted by hydraulic actuators mounted on its trucks. The tilting feature of the cars also increases passenger comfort by eliminating or greatly reducing the side force $\mathbf{F}_s$ (parallel to the floor of the car) to which passengers feel subjected. For a train traveling at 100 mi/h on a curved section of track banked through an angle $\theta = 6°$ and with a rated speed of 60 mi/h, determine (*a*) the magnitude of the side force felt by a passenger of weight W in a standard car with no tilt ($\phi = 0$), (*b*) the required angle of tilt ϕ if the passenger is to feel no side force. (See Sample Prob. 12.6 for the definition of rated speed.)

12.54 Tests carried out with the tilting trains described in Prob. 12.53 revealed that passengers feel queasy when they see through the car windows that the train is rounding a curve at high speed, yet do not feel any side force. Designers, therefore, prefer to reduce, but not eliminate that force. For the train of Prob. 12.53, determine the required angle of tilt ϕ if passengers are to feel side forces equal to 10 percent of their weights.

12.55 A 3-kg block is at rest relative to a parabolic dish which rotates at a constant rate about a vertical axis. Knowing that the coefficient of static friction is 0.5 and that $r = 2$ m, determine the maximum allowable velocity v of the block.

Fig. P12.55

12.56 Three seconds after a polisher is started from rest, small tufts of fleece from along the circumference of the 225-mm-diameter polishing pad are observed to fly free of the pad. If the polisher is started so that the fleece along the circumference undergoes a constant tangential acceleration of 4 m/s², determine (*a*) the speed v of a tuft as it leaves the pad, (*b*) the magnitude of the force required to free a tuft if the average mass of a tuft is 1.6 mg.

Fig. P12.56

12.57 A turntable A is built into a stage for use in a theatrical production. It is observed during a rehearsal that a trunk B starts to slide on the turntable 10 s after the turntable begins to rotate. Knowing that the trunk undergoes a constant tangential acceleration of 0.24 m/s², determine the coefficient of static friction between the trunk and the turntable.

Fig. P12.57

12.58 A small, 300-g collar D can slide on portion AB of a rod which is bent as shown. Knowing that $\alpha = 40°$ and that the rod rotates about the vertical AC at a constant rate of 5 rad/s, determine the value of r for which the collar will not slide on the rod if the effect of friction between the rod and the collar is neglected.

Fig. P12.58 and P12.59

12.59 A small, 200-g collar D can slide on portion AB of a rod which is bent as shown. Knowing that the rod rotates about the vertical AC at a constant rate and that $\alpha = 30°$ and $r = 600$ mm, determine the range of values of the speed v for which the collar will not slide on the rod if the coefficient of static friction between the rod and the collar is 0.30.

12.60 A semicircular slot of 10-in. radius is cut in a flat plate which rotates about the vertical AD at a constant rate of 14 rad/s. A small, 0.8-lb block E is designed to slide in the slot as the plate rotates. Knowing that the coefficients of friction are $\mu_s = 0.35$ and $\mu_k = 0.25$, determine whether the block will slide in the slot if it is released in the position corresponding to (a) $\theta = 80°$, (b) $\theta = 40°$. Also determine the magnitude and the direction of the friction force exerted on the block immediately after it is released.

Fig. P12.60

Fig. P12.61

12.61 A small block B fits inside a slot cut in arm OA which rotates in a vertical plane at a constant rate. The block remains in contact with the end of the slot closest to A and its speed is 1.4 m/s for $0 \leq \theta \leq 150°$. Knowing that the block begins to slide when $\theta = 150°$, determine the coefficient of static friction between the block and the slot.

12.62 The parallel-link mechanism $ABCD$ is used to transport a component I between manufacturing processes at stations E, F, and G by picking it up at a station when $\theta = 0$ and depositing it at the next station when $\theta = 180°$. Knowing that member BC remains horizontal throughout its motion and that links AB and CD rotate at a constant rate in a vertical plane in such a way that $v_B = 2.2$ ft/s, determine (a) the minimum value of the coefficient of static friction between the component and BC if the component is not to slide on BC while being transferred, (b) the values of θ for which sliding is impending.

Fig. P12.62

12.63 Knowing that the coefficients of friction between the component I and member BC of the mechanism of Prob. 12.62 are $\mu_s = 0.35$ and $\mu_k = 0.25$, determine (a) the maximum allowable constant speed v_B if the component is not to slide on BC while being transferred, (b) the values of θ for which sliding is impending.

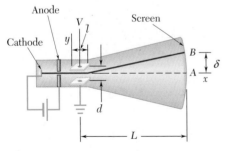

Fig. P12.64

12.64 In the cathode-ray tube shown, electrons emitted by the cathode and attracted by the anode pass through a small hole in the anode and then travel in a straight line with a speed v_0 until they strike the screen at A. However, if a difference of potential V is established between the two parallel plates, the electrons will be subjected to a force $\mathbf{F}$ perpendicular to the plates while they travel between the plates and will strike the screen at point B, which is at a distance δ from A. The magnitude of the force $\mathbf{F}$ is $F = eV/d$, where $-e$ is the charge of an electron and d is the distance between the plates. Derive an expression for the deflection d in terms of V, v_0, the charge $-e$ and the mass m of an electron, and the dimensions d, l, and L.

12.65 In Prob. 12.64, determine the smallest allowable value of the ratio d/l in terms of e, m, v_0, and V if at $x = l$ the minimum permissible distance between the path of the electrons and the positive plate is $0.05d$.

12.7 ANGULAR MOMENTUM OF A PARTICLE. RATE OF CHANGE OF ANGULAR MOMENTUM

Consider a particle P of mass m moving with respect to a newtonian frame of reference $Oxyz$. As we saw in Sec. 12.3, the linear momentum of the particle at a given instant is defined as the vector $m\mathbf{v}$ obtained by multiplying the velocity $\mathbf{v}$ of the particle by its mass m. The moment about O of the vector $m\mathbf{v}$ is called the *moment of momentum*, or the *angular momentum*, of the particle about O at that instant and is denoted by $\mathbf{H}_O$. Recalling the definition of the moment of a vector (Sec. 3.6) and denoting by $\mathbf{r}$ the position vector of P, we write

$$\mathbf{H}_O = \mathbf{r} \times m\mathbf{v} \qquad (12.12)$$

and note that $\mathbf{H}_O$ is a vector perpendicular to the plane containing $\mathbf{r}$ and $m\mathbf{v}$ and of magnitude

$$H_O = rmv \sin \phi \qquad (12.13)$$

where ϕ is the angle between $\mathbf{r}$ and $m\mathbf{v}$ (Fig. 12.12). The sense of $\mathbf{H}_O$ can be determined from the sense of $m\mathbf{v}$ by applying the right-hand rule. The unit of angular momentum is obtained by multiplying the units of length and of linear momentum (Sec. 12.4). With SI units, we have

$$(m)(kg \cdot m/s) = kg \cdot m^2/s$$

With U.S. customary units, we write

$$(ft)(lb \cdot s) = ft \cdot lb \cdot s$$

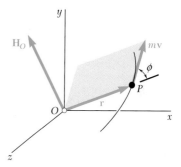

Fig. 12.12

Resolving the vectors $\mathbf{r}$ and $m\mathbf{v}$ into components and applying formula (3.10), we write

$$\mathbf{H}_O = \begin{vmatrix} \mathbf{i} & \mathbf{j} & \mathbf{k} \\ x & y & z \\ mv_x & mv_y & mv_z \end{vmatrix} \qquad (12.14)$$

The components of $\mathbf{H}_O$, which also represent the moments of the linear momentum $m\mathbf{v}$ about the coordinate axes, can be obtained by expanding the determinant in (12.14). We have

$$\begin{aligned} H_x &= m(yv_z - zv_y) \\ H_y &= m(zv_x - xv_z) \\ H_z &= m(xv_y - yv_x) \end{aligned} \qquad (12.15)$$

In the case of a particle moving in the xy plane, we have $z = v_z = 0$ and the components H_x and H_y reduce to zero. The angular momentum is thus perpendicular to the xy plane; it is then completely defined by the scalar

$$H_O = H_z = m(xv_y - yv_x) \qquad (12.16)$$

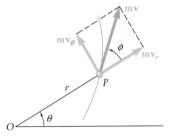

Fig. 12.13

which will be positive or negative according to the sense in which the particle is observed to move from O. If polar coordinates are used, we resolve the linear momentum of the particle into radial and transverse components (Fig. 12.13) and write

$$H_O = rmv \sin \phi = rmv_\theta \tag{12.17}$$

or, recalling from (11.45) that $v_\theta = r\dot{\theta}$,

$$H_O = mr^2\dot{\theta} \tag{12.18}$$

Let us now compute the derivative with respect to t of the angular momentum $\mathbf{H}_O$ of a particle P moving in space. Differentiating both members of Eq. (12.12), and recalling the rule for the differentiation of a vector product (Sec. 11.10), we write

$$\dot{\mathbf{H}}_O = \dot{\mathbf{r}} \times m\mathbf{v} + \mathbf{r} \times m\dot{\mathbf{v}} = \mathbf{v} \times m\mathbf{v} + \mathbf{r} \times m\mathbf{a}$$

Since the vectors $\mathbf{v}$ and $m\mathbf{v}$ are collinear, the first term of the expression obtained is zero; and, by Newton's second law, $m\mathbf{a}$ is equal to the sum $\Sigma\mathbf{F}$ of the forces acting on P. Noting that $\mathbf{r} \times \Sigma\mathbf{F}$ represents the sum $\Sigma\mathbf{M}_O$ of the moments about O of these forces, we write

$$\Sigma\mathbf{M}_O = \dot{\mathbf{H}}_O \tag{12.19}$$

Equation (12.19), which results directly from Newton's second law, states that *the sum of the moments about O of the forces acting on the particle is equal to the rate of change of the moment of momentum, or angular momentum, of the particle about O.*

12.8 EQUATIONS OF MOTION IN TERMS OF RADIAL AND TRANSVERSE COMPONENTS

Consider a particle P, of polar coordinates r and θ, which moves in a plane under the action of several forces. Resolving the forces and the acceleration of the particle into radial and transverse components (Fig. 12.14) and substituting into Eq. (12.2), we obtain the two scalar equations

$$\Sigma F_r = ma_r \qquad \Sigma F_\theta = ma_\theta \tag{12.20}$$

Substituting for a_r and a_θ from Eqs. (11.46), we have

$$\Sigma F_r = m(\ddot{r} - r\dot{\theta}^2) \tag{12.21}$$

$$\Sigma F_\theta = m(r\ddot{\theta} + 2\dot{r}\dot{\theta}) \tag{12.22}$$

The equations obtained can be solved for two unknowns.

Photo 12.4 The forces on the specimens used in a high speed centrifuge can be described in terms of radial and transverse components.

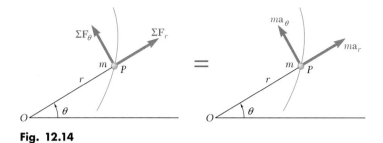

Fig. 12.14

Equation (12.22) could have been derived from Eq. (12.19). Recalling (12.18) and noting that $\Sigma M_O = r\Sigma F_\theta$, Eq. (12.19) yields

$$r\Sigma F_\theta = \frac{d}{dt}(mr^2\dot{\theta})$$
$$= m(r^2\ddot{\theta} + 2r\dot{r}\dot{\theta})$$

and, after dividing both members by r,

$$\Sigma F_\theta = m(r\ddot{\theta} + 2\dot{r}\dot{\theta}) \tag{12.22}$$

12.9 MOTION UNDER A CENTRAL FORCE. CONSERVATION OF ANGULAR MOMENTUM

When the only force acting on a particle P is a force $\mathbf{F}$ directed toward or away from a fixed point O, the particle is said to be *moving under a central force,* and the point O is referred to as the *center of force* (Fig. 12.15). Since the line of action of $\mathbf{F}$ passes through O, we must have $\Sigma \mathbf{M}_O = 0$ at any given instant. Substituting into Eq. (12.19), we therefore obtain

$$\dot{\mathbf{H}}_O = 0$$

for all values of t and, integrating in t,

$$\mathbf{H}_O = \text{constant} \tag{12.23}$$

We thus conclude that *the angular momentum of a particle moving under a central force is constant, in both magnitude and direction.*

Recalling the definition of the angular momentum of a particle (Sec. 12.7), we write

$$\mathbf{r} \times m\mathbf{v} = \mathbf{H}_O = \text{constant} \tag{12.24}$$

from which it follows that the position vector $\mathbf{r}$ of the particle P must be perpendicular to the constant vector $\mathbf{H}_O$. Thus, a particle under

Fig. 12.15

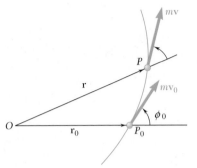

Fig. 12.16

a central force moves in a fixed plane perpendicular to $\mathbf{H}_O$. The vector $\mathbf{H}_O$ and the fixed plane are defined by the initial position vector $\mathbf{r}_0$ and the initial velocity $\mathbf{v}_0$ of the particle. For convenience, let us assume that the plane of the figure coincides with the fixed plane of motion (Fig. 12.16).

Since the magnitude H_O of the angular momentum of the particle P is constant, the right-hand member in Eq. (12.13) must be constant. We therefore write

$$rmv \sin \phi = r_0 m v_0 \sin \phi_0 \qquad (12.25)$$

This relation applies to the motion of any particle under a central force. Since the gravitational force exerted by the sun on a planet is a central force directed toward the center of the sun, Eq. (12.25) is fundamental to the study of planetary motion. For a similar reason, it is also fundamental to the study of the motion of space vehicles in orbit about the earth.

Alternatively, recalling Eq. (12.18), we can express the fact that the magnitude H_O of the angular momentum of the particle P is constant by writing

$$mr^2 \dot{\theta} = H_O = \text{constant} \qquad (12.26)$$

or, dividing by m and denoting by h the angular momentum per unit mass H_O/m,

$$r^2 \dot{\theta} = h \qquad (12.27)$$

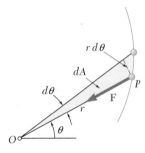

Fig. 12.17

Equation (12.27) can be given an interesting geometric interpretation. Observing from Fig. 12.17 that the radius vector OP sweeps an infinitesimal area $dA = \frac{1}{2} r^2 \, d\theta$ as it rotates through an angle $d\theta$, and defining the *areal velocity* of the particle as the quotient dA/dt, we note that the left-hand member of Eq. (12.27) represents twice the areal velocity of the particle. We thus conclude that *when a particle moves under a central force, its areal velocity is constant.*

12.10 NEWTON'S LAW OF GRAVITATION

As you saw in the preceding section, the gravitational force exerted by the sun on a planet or by the earth on an orbiting satellite is an important example of a central force. In this section, you will learn how to determine the magnitude of a gravitational force.

In his *law of universal gravitation*, Newton states that two particles of masses M and m at a distance r from each other attract each other with equal and opposite forces $\mathbf{F}$ and $-\mathbf{F}$ directed along the line joining the particles (Fig. 12.18). The common magnitude F of the two forces is

$$F = G \frac{Mm}{r^2} \qquad (12.28)$$

Fig. 12.18

where G is a universal constant, called the *constant of gravitation*. Experiments show that the value of G is $(66.73 \pm 0.03) \times 10^{-12}$ m^3/kg $\cdot$ s^2 in SI units or approximately 34.4×10^{-9} ft^4/lb $\cdot$ s^4 in U.S. customary units. Gravitational forces exist between any pair of bodies, but their effect is appreciable only when one of the bodies has a very large mass. The effect of gravitational forces is apparent in the cases of the motion of a planet about the sun, of satellites orbiting about the earth, or of bodies falling on the surface of the earth.

Since the force exerted by the earth on a body of mass m located on or near its surface is defined as the weight $\mathbf{W}$ of the body, we can substitute the magnitude $W = mg$ of the weight for F, and the radius R of the earth for r, in Eq. (12.28). We obtain

$$W = mg = \frac{GM}{R^2}m \qquad \text{or} \qquad g = \frac{GM}{R^2} \qquad (12.29)$$

where M is the mass of the earth. Since the earth is not truly spherical, the distance R from the center of the earth depends upon the point selected on its surface, and the values of W and g will thus vary with the altitude and latitude of the point considered. Another reason for the variation of W and g with latitude is that a system of axes attached to the earth does not constitute a newtonian frame of reference (see Sec. 12.2). A more accurate definition of the weight of a body should therefore include a component representing the centrifugal force due to the rotation of the earth. Values of g at sea level vary from 9.781 m/s^2, or 32.09 ft/s^2, at the equator to 9.833 m/s^2, or 32.26 ft/s^2, at the poles.†

The force exerted by the earth on a body of mass m located in space at a distance r from its center can be found from Eq. (12.28). The computations will be somewhat simplified if we note that according to Eq. (12.29), the product of the constant of gravitation G and the mass M of the earth can be expressed as

$$GM = gR^2 \qquad (12.30)$$

where g and the radius R of the earth will be given their average values $g = 9.81$ m/s^2 and $R = 6.37 \times 10^6$ m in SI units‡ and $g = 32.2$ ft/s^2 and $R = (3960$ mi$)(5280$ ft/mi$)$ in U.S. customary units.

The discovery of the law of universal gravitation has often been attributed to the belief that, after observing an apple falling from a tree, Newton had reflected that the earth must attract an apple and the moon in much the same way. While it is doubtful that this incident actually took place, it may be said that Newton would not have formulated his law if he had not first perceived that the acceleration of a falling body must have the same cause as the acceleration which keeps the moon in its orbit. This basic concept of the continuity of gravitational attraction is more easily understood today, when the gap between the apple and the moon is being filled with artificial earth satellites.

---

†A formula expressing g in terms of the latitude ϕ was given in Prob. 12.2.

‡The value of R is easily found if one recalls that the circumference of the earth is $2\pi R = 40 \times 10^6$ m.

SAMPLE PROBLEM 12.7

A block B of mass m can slide freely on a frictionless arm OA which rotates in a horizontal plane at a constant rate $\dot{\theta}_0$. Knowing that B is released at a distance r_0 from O, express as a function of r, (a) the component v_r of the velocity of B along OA, (b) the magnitude of the horizontal force $\mathbf{F}$ exerted on B by the arm OA.

SOLUTION

Since all other forces are perpendicular to the plane of the figure, the only force shown acting on B is the force $\mathbf{F}$ perpendicular to OA.

Equations of Motion. Using radial and transverse components,

$$+\nearrow \ \Sigma F_r = ma_r: \qquad\qquad 0 = m(\ddot{r} - r\dot{\theta}^2) \qquad\qquad (1)$$
$$+\nwarrow \ \Sigma F_\theta = ma_\theta: \qquad\qquad F = m(r\ddot{\theta} + 2\dot{r}\dot{\theta}) \qquad\qquad (2)$$

a. Component v_r of Velocity. Since $v_r = \dot{r}$, we have

$$\ddot{r} = \dot{v}_r = \frac{dv_r}{dt} = \frac{dv_r}{dr}\frac{dr}{dt} = v_r\frac{dv_r}{dr}$$

Substituting for $\ddot{r}$ in (1), recalling that $\dot{\theta} = \dot{\theta}_0$, and separating the variables,

$$v_r\,dv_r = \dot{\theta}_0^2 r\,dr$$

Multiplying by 2, and integrating from 0 to v_r and from r_0 to r,

$$v_r^2 = \dot{\theta}_0^2(r^2 - r_0^2) \qquad v_r = \dot{\theta}_0(r^2 - r_0^2)^{1/2} \ \blacktriangleleft$$

b. Horizontal Force F. Setting $\dot{\theta} = \dot{\theta}_0$, $\ddot{\theta} = 0$, $\dot{r} = v_r$ in Eq. (2), and substituting for v_r the expression obtained in part a,

$$F = 2m\dot{\theta}_0(r^2 - r_0^2)^{1/2}\dot{\theta}_0 \qquad F = 2m\dot{\theta}_0^2(r^2 - r_0^2)^{1/2} \ \blacktriangleleft$$

SAMPLE PROBLEM 12.8

A satellite is launched in a direction parallel to the surface of the earth with a velocity of 18,820 mi/h from an altitude of 240 mi. Determine the velocity of the satellite as it reaches its maximum altitude of 2340 mi. It is recalled that the radius of the earth is 3960 mi.

SOLUTION

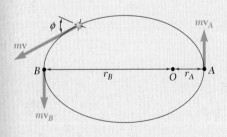

Since the satellite is moving under a central force directed toward the center O of the earth, its angular momentum $\mathbf{H}_O$ is constant. From Eq. (12.13) we have

$$rmv \sin \phi = H_O = \text{constant}$$

which shows that v is minimum at B, where both r and $\sin \phi$ are maximum. Expressing conservation of angular momentum between A and B,

$$r_A mv_A = r_B mv_B$$

$$v_B = v_A\frac{r_A}{r_B} = (18{,}820 \text{ mi/h})\frac{3960 \text{ mi} + 240 \text{ mi}}{3960 \text{ mi} + 2340 \text{ mi}}$$

$$v_B = 12{,}550 \text{ mi/h} \ \blacktriangleleft$$

SOLVING PROBLEMS
ON YOUR OWN

In this lesson we continued our study of Newton's second law by expressing the force and the acceleration in terms of their *radial and transverse components*, where the corresponding equations of motion are

$$\Sigma F_r = ma_r: \qquad\qquad \Sigma F_r = m(\ddot{r} - r\dot{\theta}^2)$$
$$\Sigma F_\theta = ma_\theta: \qquad\qquad \Sigma F_\theta = m(r\ddot{\theta} + 2\dot{r}\dot{\theta})$$

We introduced the *moment of the momentum*, or the *angular momentum*, $\mathbf{H}_O$ of a particle about O:

$$\mathbf{H}_O = \mathbf{r} \times m\mathbf{v} \qquad\qquad (12.12)$$

and found that $\mathbf{H}_O$ is constant when the particle moves under a *central force* with its center located at O.

1. Using radial and transverse components. Radial and transverse components were introduced in the last lesson of Chap. 11 [Sec. 11.14]; you should review that material before attempting to solve the following problems. Also, our comments in the preceding lesson regarding the application of Newton's second law (drawing a free-body diagram and a $m\mathbf{a}$ diagram, etc.) still apply [Sample Prob. 12.7]. Finally, note that the solution of that sample problem depends on the application of techniques developed in Chap. 11—you will need to use similar techniques to solve some of the problems of this lesson.

2. Solving problems involving the motion of a particle under a central force. In problems of this type, the angular momentum $\mathbf{H}_O$ of the particle about the center of force O is conserved. You will find it convenient to introduce the constant $h = H_O/m$ representing the angular momentum per unit mass. Conservation of the angular momentum of the particle P about O can then be expressed by either of the following equations

$$rv \sin \phi = h \qquad \text{or} \qquad r^2\dot{\theta} = h$$

where r and θ are the polar coordinates of P, and ϕ is the angle that the velocity $\mathbf{v}$ of the particle forms with the line OP (Fig. 12.16). The constant h can be determined from the initial conditions and either of the above equations can be solved for one unknown.

(continued)

3. In space-mechanics problems involving the orbital motion of a planet about the sun, or a satellite about the earth, the moon, or some other planet, the central force $\mathbf{F}$ is the force of gravitational attraction; it is directed *toward* the center of force O and has the magnitude

$$F = G\frac{Mm}{r^2} \tag{12.28}$$

Note that in the particular case of the gravitational force exerted by the earth, the product GM can be replaced by gR^2, where R is the radius of the earth [Eq. 12.30].

The following two cases of orbital motion are frequently encountered:

a. For a satellite in a circular orbit, the force $\mathbf{F}$ is normal to the orbit and you can write $F = ma_n$; substituting for F from Eq. (12.28) and observing that $a_n = v^2/\rho = v^2/r$, you will obtain

$$G\frac{Mm}{r^2} = m\frac{v^2}{r} \qquad \text{or} \qquad v^2 = \frac{GM}{r}$$

b. For a satellite in an elliptic orbit, the radius vector $\mathbf{r}$ and the velocity $\mathbf{v}$ of the satellite are perpendicular to each other at the points A and B which are, respectively, farthest and closest to the center of force O [Sample Prob. 12.8]. Thus, conservation of angular momentum of the satellite between these two points can be expressed as

$$r_A m v_A = r_B m v_B$$

PROBLEMS

FREE-BODY PRACTICE PROBLEMS

12.F9 Four pins slide in four separate slots cut in a horizontal circular plate as shown. When the plate is at rest, each pin has a velocity directed as shown and of the same constant magnitude u. Each pin has a mass m and maintains the same velocity relative to the plate when the plate rotates about O with a constant counterclockwise angular velocity ω. Draw the FBDs and KDs to determine the forces on pins P_1 and P_2.

Fig. P12.F9

12.F10 At the instant shown, the length of the boom AB is being *decreased* at the constant rate of 0.2 m/s, and the boom is being lowered at the constant rate of 0.08 rad/s. If the mass of the men and lift connected to the boom at point B is m, draw the FBD and KD that could be used to determine the horizontal and vertical forces at B.

Fig. P12.F10

12.F11 Disk A rotates in a horizontal plane about a vertical axis at the constant rate $\dot{\theta}_0$. Slider B has a mass m and moves in a frictionless slot cut in the disk. The slider is attached to a spring of constant k, which is undeformed when $r = 0$. Knowing that the slider is released with no radial velocity in the position $r = r_0$, draw a FBD and KD at an arbitrary distance r from O.

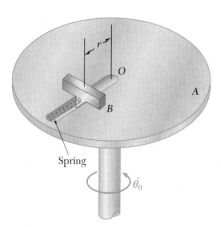

Fig. P12.F11

12.F12 Pin B has a mass m and slides along the slot in the rotating arm OC and along the slot DE which is cut in a fixed horizontal plate. Neglecting friction and knowing that rod OC rotates at the constant rate $\dot{\theta}_0$, draw a FBD and KD that can be used to determine the forces **P** and **Q** exerted on pin B by rod OC and the wall of slot DE, respectively.

Fig. P12.F12

735

Fig. P12.66 and P12.67

Fig. P12.68

Fig. P12.70

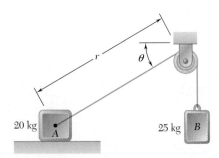

Fig. P12.71 and P12.72

END-OF-SECTION PROBLEMS

12.66 Rod *OA* rotates about *O* in a horizontal plane. The motion of the 0.5-lb collar *B* is defined by the relations $r = 10 + 6 \cos \pi t$ and $\theta = \pi(4t^2 - 8t)$, where *r* is expressed in inches, *t* in seconds, and θ in radians. Determine the radial and transverse components of the force exerted on the collar when (*a*) $t = 0$, (*b*) $t = 0.5$ s.

12.67 Rod *OA* oscillates about *O* in a horizontal plane. The motion of the 2-lb collar *B* is defined by the relations $r = 6(1 - e^{-2t})$ and $\theta = (3/\pi)(\sin \pi t)$, where *r* is expressed in inches, *t* in seconds, and θ in radians. Determine the radial and transverse components of the force exerted on the collar when (*a*) $t = 1$ s, (*b*) $t = 1.5$ s.

12.68 The 3-kg collar *B* slides on the frictionless arm *AA'*. The arm is attached to drum *D* and rotates about *O* in a horizontal plane at the rate $\dot\theta = 0.75t$, where $\dot\theta$ and *t* are expressed in rad/s and seconds, respectively. As the arm-drum assembly rotates, a mechanism within the drum releases cord so that the collar moves outward from *O* with a constant speed of 0.5 m/s. Knowing that at $t = 0$, $r = 0$, determine the time at which the tension in the cord is equal to the magnitude of the horizontal force exerted on *B* by arm *AA'*.

12.69 The horizontal rod *OA* rotates about a vertical shaft according to the relation $\dot\theta = 10t$, where $\dot\theta$ and *t* are expressed in rad/s and seconds, respectively. A 250-g collar *B* is held by a cord with a breaking strength of 18 N. Neglecting friction, determine, immediately after the cord breaks, (*a*) the relative acceleration of the collar with respect to the rod, (*b*) the magnitude of the horizontal force exerted on the collar by the rod.

Fig. P12.69

12.70 Pin *B* weighs 4 oz and is free to slide in a horizontal plane along the rotating arm *OC* and along the circular slot *DE* of radius $b = 20$ in. Neglecting friction and assuming that $\dot\theta = 15$ rad/s and $\ddot\theta = 250$ rad/s² for the position $\theta = 20°$, determine for that position (*a*) the radial and transverse components of the resultant force exerted on pin *B*, (*b*) the forces **P** and **Q** exerted on pin *B*, respectively, by rod *OC* and the wall of slot *DE*.

12.71 The two blocks are released from rest when $r = 0.8$ m and $\theta = 30°$. Neglecting the mass of the pulley and the effect of friction in the pulley and between block *A* and the horizontal surface, determine (*a*) the initial tension in the cable, (*b*) the initial acceleration of block *A*, (*c*) the initial acceleration of block *B*.

12.72 The velocity of block *A* is 2 m/s to the right at the instant when $r = 0.8$ m and $\theta = 30°$. Neglecting the mass of the pulley and the effect of friction in the pulley and between block *A* and the horizontal surface, determine, at this instant, (*a*) the tension in the cable, (*b*) the acceleration of block *A*, (*c*) the acceleration of block *B*.

*12.73 Slider C has a weight of 0.5 lb and may move in a slot cut in arm AB, which rotates at the constant rate $\dot{\theta}_0 = 10$ rad/s in a horizontal plane. The slider is attached to a spring of constant $k = 2.5$ lb/ft, which is unstretched when $r = 0$. Knowing that the slider is released from rest with no radial velocity in the position $r = 18$ in. and neglecting friction, determine for the position $r = 12$ in. (a) the radial and transverse components of the velocity of the slider, (b) the radial and transverse components of its acceleration, (c) the horizontal force exerted on the slider by arm AB.

Fig. P12.73

12.74 A particle of mass m is projected from point A with an initial velocity $\mathbf{v}_0$ perpendicular to line OA and moves under a central force $\mathbf{F}$ directed away from the center of force O. Knowing that the particle follows a path defined by the equation $r = r_0/\sqrt{\cos 2\theta}$ and using Eq. (12.27), express the radial and transverse components of the velocity $\mathbf{v}$ of the particle as functions of θ.

Fig. P12.74

12.75 For the particle of Prob. 12.74, show (a) that the velocity of the particle and the central force $\mathbf{F}$ are proportional to the distance r from the particle to the center of force O, (b) that the radius of curvature of the path is proportional to r^3.

12.76 A particle of mass m is projected from point A with an initial velocity $\mathbf{v}_0$ perpendicular to line OA and moves under a central force $\mathbf{F}$ along a semicircular path of diameter OA. Observing that $r = r_0 \cos \theta$ and using Eq. (12.27), show that the speed of the particle is $v = v_0/\cos^2 \theta$.

Fig. P12.76

12.77 For the particle of Prob. 12.76, determine the tangential component F_t of the central force $\mathbf{F}$ along the tangent to the path of the particle for (a) $\theta = 0$, (b) $\theta = 45°$.

12.78 Determine the mass of the earth knowing that the mean radius of the moon's orbit about the earth is 238,910 mi and that the moon requires 27.32 days to complete one full revolution about the earth.

12.79 Show that the radius r of the moon's orbit can be determined from the radius R of the earth, the acceleration of gravity g at the surface of the earth, and the time τ required for the moon to complete one full revolution about the earth. Compute r knowing that $\tau = 27.3$ days, giving the answer in both SI and U.S. customary units.

12.80 Communication satellites are placed in a geosynchronous orbit, i.e., in a circular orbit such that they complete one full revolution about the earth in one sidereal day (23.934 h), and thus appear stationary with respect to the ground. Determine (a) the altitude of these satellites above the surface of the earth, (b) the velocity with which they describe their orbit. Give the answers in both SI and U.S. customary units.

12.81 Show that the radius r of the orbit of a moon of a given planet can be determined from the radius R of the planet, the acceleration of gravity at the surface of the planet, and the time τ required by the moon to complete one full revolution about the planet. Determine the acceleration of gravity at the surface of the planet Jupiter knowing that $R = 71\ 492$ km and that $\tau = 3.551$ days and $r = 670.9 \times 10^3$ km for its moon Europa.

12.82 The orbit of the planet Venus is nearly circular with an orbital velocity of 126.5×10^3 km/h. Knowing that the mean distance from the center of the sun to the center of Venus is 108×10^6 km and that the radius of the sun is 695.5×10^3 km, determine (a) the mass of the sun, (b) the acceleration of gravity at the surface of the sun.

12.83 A satellite is placed into a circular orbit about the planet Saturn at an altitude of 2100 mi. The satellite describes its orbit with a velocity of 54.7×10^3 mi/h. Knowing that the radius of the orbit about Saturn and the periodic time of Atlas, one of Saturn's moons, are 85.54×10^3 mi and 0.6017 days, respectively, determine (a) the radius of Saturn, (b) the mass of Saturn. (The *periodic time* of a satellite is the time it requires to complete one full revolution about the planet.)

12.84 The periodic times (see Prob. 12.83) of the planet Uranus's moons Juliet and Titania have been observed to be 0.4931 days and 8.706 days, respectively. Knowing that the radius of Juliet's orbit is 40,000 mi, determine (a) the mass of Uranus, (b) the radius of Titania's orbit.

12.85 A 500-kg spacecraft first is placed into a circular orbit about the earth at an altitude of 4500 km and then is transferred to a circular orbit about the moon. Knowing that the mass of the moon is 0.01230 times the mass of the earth and that the radius of the moon is 1737 km, determine (a) the gravitational force exerted on the spacecraft as it was orbiting the earth, (b) the required radius of the orbit of the spacecraft about the moon if the periodic times (see Prob. 12.83) of the two orbits are to be equal, (c) the acceleration of gravity at the surface of the moon.

12.86 A space vehicle is in a circular orbit of 2200-km radius around the moon. To transfer it to a smaller circular orbit of 2080-km radius, the vehicle is first placed on an elliptic path AB by reducing its speed by 26.3 m/s as it passes through A. Knowing that the mass of the moon is 73.49×10^{21} kg, determine (a) the speed of the vehicle as it approaches B on the elliptic path, (b) the amount by which its speed should be reduced as it approaches B to insert it into the smaller circular orbit.

Fig. P12.86

12.87 Plans for an unmanned landing mission on the planet Mars called for the earth-return vehicle to first describe a circular orbit at an altitude d_A = 2200 km above the surface of the planet with a velocity of 2771 m/s. As it passed through point A, the vehicle was to be inserted into an elliptic transfer orbit by firing its engine and increasing its speed by Δv_A = 1046 m/s. As it passed through point B, at an altitude d_B = 100 000 km, the vehicle was to be inserted into a second transfer orbit located in a slightly different plane, by changing the direction of its velocity and reducing its speed by Δv_B = −22.0 m/s. Finally, as the vehicle passed through point C, at an altitude d_C = 1000 km, its speed was to be increased by Δv_C = 660 m/s to insert it into its return trajectory. Knowing that the radius of the planet Mars is R = 3400 km, determine the velocity of the vehicle after completion of the last maneuver.

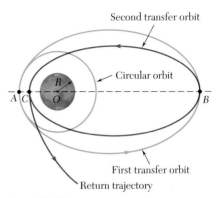

Fig. P12.87

12.88 To place a communications satellite into a geosynchronous orbit (see Prob. 12.80) at an altitude of 22,240 mi above the surface of the earth, the satellite first is released from a space shuttle, which is in a circular orbit at an altitude of 185 mi, and then is propelled by an upper-stage booster to its final altitude. As the satellite passes through A, the booster's motor is fired to insert the satellite into an elliptic transfer orbit. The booster is again fired at B to insert the satellite into a geosynchronous orbit. Knowing that the second firing increases the speed of the satellite by 4810 ft/s, determine (a) the speed of the satellite as it approaches B on the elliptic transfer orbit, (b) the increase in speed resulting from the first firing at A.

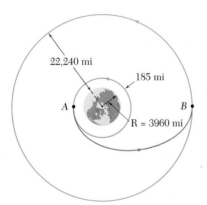

Fig. P12.88

12.89 A space shuttle S and a satellite A are in the circular orbits shown. In order for the shuttle to recover the satellite, the shuttle is first placed in an elliptic path BC by increasing its speed by Δv_B = 280 ft/s as it passes through B. As the shuttle approaches C, its speed is increased by Δv_C = 260 ft/s to insert it into a second elliptic transfer orbit CD. Knowing that the distance from O to C is 4289 mi, determine the amount by which the speed of the shuttle should be increased as it approaches D to insert it into the circular orbit of the satellite.

Fig. P12.89

Fig. P12.90

Fig. P12.91

Fig. P12.93

12.90 A 1-kg collar can slide on a horizontal rod which is free to rotate about a vertical shaft. The collar is initially held at A by a cord attached to the shaft. A spring of constant 30 N/m is attached to the collar and to the shaft and is undeformed when the collar is at A. As the rod rotates at the rate $\dot{\theta} = 16$ rad/s, the cord is cut and the collar moves out along the rod. Neglecting friction and the mass of the rod, determine (a) the radial and transverse components of the acceleration of the collar at A, (b) the acceleration of the collar relative to the rod at A, (c) the transverse component of the velocity of the collar at B.

12.91 A 1-lb ball A and a 2-lb ball B are mounted on a horizontal rod which rotates freely about a vertical shaft. The balls are held in the positions shown by pins. The pin holding B is suddenly removed and the ball moves to position C as the rod rotates. Neglecting friction and the mass of the rod and knowing that the initial speed of A is $v_A = 8$ ft/s, determine (a) the radial and transverse components of the acceleration of ball B immediately after the pin is removed, (b) the acceleration of ball B relative to the rod at that instant, (c) the speed of ball A after ball B has reached the stop at C.

12.92 Two 2.6-lb collars A and B can slide without friction on a frame, consisting of the horizontal rod OE and the vertical rod CD, which is free to rotate about CD. The two collars are connected by a cord running over a pulley that is attached to the frame at O and a stop prevents collar B from moving. The frame is rotating at the rate $\dot{\theta} = 12$ rad/s and $r = 0.6$ ft when the stop is removed allowing collar A to move out along rod OE. Neglecting friction and the mass of the frame, determine, for the position $r = 1.2$ ft, (a) the transverse component of the velocity of collar A, (b) the tension in the cord and the acceleration of collar A relative to the rod OE.

Fig. P12.92

12.93 A small ball swings in a horizontal circle at the end of a cord of length l_1, which forms an angle θ_1 with the vertical. The cord is then slowly drawn through the support at O until the length of the free end is l_2. (a) Derive a relation among l_1, l_2, θ_1, and θ_2. (b) If the ball is set in motion so that initially $l_1 = 0.8$ m and $\theta_1 = 35°$, determine the angle θ_2 when $l_2 = 0.6$ m.

*12.11 TRAJECTORY OF A PARTICLE UNDER A CENTRAL FORCE

Consider a particle P moving under a central force $\mathbf{F}$. We propose to obtain the differential equation which defines its trajectory.

Assuming that the force $\mathbf{F}$ is directed toward the center of force O, we note that ΣF_r and ΣF_θ reduce, respectively, to $-F$ and zero in Eqs. (12.21) and (12.22). We therefore write

$$m(\ddot{r} - r\dot{\theta}^2) = -F \qquad (12.31)$$

$$m(r\ddot{\theta} + 2\dot{r}\dot{\theta}) = 0 \qquad (12.32)$$

These equations define the motion of P. We will, however, replace Eq. (12.32) by Eq. (12.27), which is equivalent to Eq. (12.32), as can easily be checked by differentiating it with respect to t, but which is more convenient to use. We write

$$r^2\dot{\theta} = h \qquad \text{or} \qquad r^2\frac{d\theta}{dt} = h \qquad (12.33)$$

Equation (12.33) can be used to eliminate the independent variable t from Eq. (12.31). Solving Eq. (12.33) for $\dot{\theta}$ or $d\theta/dt$, we have

$$\dot{\theta} = \frac{d\theta}{dt} = \frac{h}{r^2} \qquad (12.34)$$

from which it follows that

$$\dot{r} = \frac{dr}{dt} = \frac{dr}{d\theta}\frac{d\theta}{dt} = \frac{h}{r^2}\frac{dr}{d\theta} = -h\frac{d}{d\theta}\left(\frac{1}{r}\right) \qquad (12.35)$$

$$\ddot{r} = \frac{d\dot{r}}{dt} = \frac{d\dot{r}}{d\theta}\frac{d\theta}{dt} = \frac{h}{r^2}\frac{d\dot{r}}{d\theta}$$

or, substituting for $\dot{r}$ from (12.35),

$$\ddot{r} = \frac{h}{r^2}\frac{d}{d\theta}\left[-h\frac{d}{d\theta}\left(\frac{1}{r}\right)\right]$$

$$\ddot{r} = -\frac{h^2}{r^2}\frac{d^2}{d\theta^2}\left(\frac{1}{r}\right) \qquad (12.36)$$

Substituting for $\dot{\theta}$ and $\ddot{r}$ from (12.34) and (12.36), respectively, in Eq. (12.31) and introducing the function $u = 1/r$, we obtain after reductions

$$\frac{d^2u}{d\theta^2} + u = \frac{F}{mh^2u^2} \qquad (12.37)$$

In deriving Eq. (12.37), the force $\mathbf{F}$ was assumed directed toward O. The magnitude F should therefore be positive if $\mathbf{F}$ is actually directed toward O (attractive force) and negative if $\mathbf{F}$ is directed away from O (repulsive force). If F is a known function of r and thus of u, Eq. (12.37) is a differential equation in u and θ. This differential equation defines the trajectory followed by the particle under the central force $\mathbf{F}$. The equation of the trajectory can be obtained by solving the differential equation (12.37) for u as a function of θ and determining the constants of integration from the initial conditions.

*12.12 APPLICATION TO SPACE MECHANICS

After the last stages of their launching rockets have burned out, earth satellites and other space vehicles are subjected to only the gravitational pull of the earth. Their motion can therefore be determined from Eqs. (12.33) and (12.37), which govern the motion of a particle under a central force, after F has been replaced by the expression obtained for the force of gravitational attraction.† Setting in Eq. (12.37)

$$F = \frac{GMm}{r^2} = GMmu^2$$

where M = mass of earth
m = mass of space vehicle
r = distance from center of earth to vehicle
u = 1/r

we obtain the differential equation

$$\frac{d^2u}{d\theta^2} + u = \frac{GM}{h^2} \qquad (12.38)$$

where the right-hand member is observed to be a constant.

The solution of the differential equation (12.38) is obtained by adding the particular solution $u = GM/h^2$ to the general solution $u = C \cos(\theta - \theta_0)$ of the corresponding homogeneous equation (i.e., the equation obtained by setting the right-hand member equal to zero). Choosing the polar axis so that $\theta_0 = 0$, we write

$$\frac{1}{r} = u = \frac{GM}{h^2} + C \cos \theta \qquad (12.39)$$

Equation (12.39) is the equation of a *conic section* (ellipse, parabola, or hyperbola) in the polar coordinates r and θ. The origin O of the coordinates, which is located at the center of the earth, is a *focus* of this conic section, and the polar axis is one of its axes of symmetry (Fig. 12.19).

The ratio of the constants C and GM/h^2 defines the *eccentricity* ε of the conic section; letting

$$\varepsilon = \frac{C}{GM/h^2} = \frac{Ch^2}{GM} \qquad (12.40)$$

we can write Eq. (12.39) in the form

$$\frac{1}{r} = \frac{GM}{h^2}(1 + \varepsilon \cos \theta) \qquad (12.39')$$

This equation represents three possible trajectories.

1. $\varepsilon > 1$, or $C > GM/h^2$: There are two values θ_1 and $-\theta_1$ of the polar angle, defined by $\cos \theta_1 = -GM/Ch^2$, for which the

Photo 12.5 The Hubble telescope was carried into orbit by the space shuttle in 1990 (first geosynchronous from NASA).

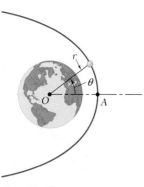

Fig. 12.19

---

†It is assumed that the space vehicles considered here are attracted by the earth only and that their mass is negligible compared with the mass of the earth. If a vehicle moves very far from the earth, its path may be affected by the attraction of the sun, the moon, or another planet.

right-hand member of Eq. (12.39) becomes zero. For both these values, the radius vector r becomes infinite; the conic section is a *hyperbola* (Fig. 12.20).

2. $\varepsilon = 1$, or $C = GM/h^2$: The radius vector becomes infinite for $\theta = 180°$; the conic section is a *parabola*.
3. $\varepsilon < 1$, or $C < GM/h^2$: The radius vector remains finite for every value of θ; the conic section is an *ellipse*. In the particular case when $\varepsilon = C = 0$, the length of the radius vector is constant; the conic section is a circle.

Let us now see how the constants C and GM/h^2, which characterize the trajectory of a space vehicle, can be determined from the vehicle's position and velocity at the beginning of its free flight. We will assume that, as is generally the case, the powered phase of its flight has been programmed in such a way that as the last stage of the launching rocket burns out, the vehicle has a velocity parallel to the surface of the earth (Fig. 12.21). In other words, we will assume that the space vehicle begins its free flight at the vertex A of its trajectory.[†]

Denoting the radius vector and speed of the vehicle at the beginning of its free flight by r_0 and v_0, respectively, we observe that the velocity reduces to its transverse component and, thus, that $v_0 = r_0\dot{\theta}_0$. Recalling Eq. (12.27), we express the angular momentum per unit mass h as

$$h = r_0^2\dot{\theta}_0 = r_0 v_0 \qquad (12.41)$$

The value obtained for h can be used to determine the constant GM/h^2. We also note that the computation of this constant will be simplified if we use the relation obtained in Sec. 12.10.

$$GM = gR^2 \qquad (12.30)$$

where R is the radius of the earth ($R = 6.37 \times 10^6$ m or 3960 mi) and g is the acceleration of gravity at the surface of the earth.

The constant C is obtained by setting $\theta = 0$, $r = r_0$ in (12.39):

$$C = \frac{1}{r_0} - \frac{GM}{h^2} \qquad (12.42)$$

Substituting for h from (12.41), we can then easily express C in terms of r_0 and v_0.

Let us now determine the initial conditions corresponding to each of the three fundamental trajectories indicated above. Considering first the parabolic trajectory, we set C equal to GM/h^2 in Eq. (12.42) and eliminate h between Eqs. (12.41) and (12.42). Solving for v_0, we obtain

$$v_0 = \sqrt{\frac{2GM}{r_0}}$$

We can easily check that a larger value of the initial velocity corresponds to a hyperbolic trajectory and a smaller value corresponds to an elliptic orbit. Since the value of v_0 obtained for the parabolic trajectory

Fig. 12.20

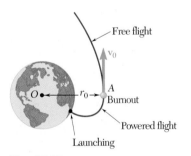

Fig. 12.21

[†]Problems involving oblique launchings will be considered in Sec. 13.9.

is the smallest value for which the space vehicle does not return to its starting point, it is called the *escape velocity*. We write therefore

$$v_{esc} = \sqrt{\frac{2GM}{r_0}} \qquad \text{or} \qquad v_{esc} = \sqrt{\frac{2gR^2}{r_0}} \qquad (12.43)$$

if we make use of Eq. (12.30). We note that the trajectory will be (1) hyperbolic if $v_0 > v_{esc}$, (2) parabolic if $v_0 = v_{esc}$, and (3) elliptic if $v_0 < v_{esc}$.

Among the various possible elliptic orbits, the one obtained when $C = 0$, the *circular orbit*, is of special interest. The value of the initial velocity corresponding to a circular orbit is easily found to be

$$v_{circ} = \sqrt{\frac{GM}{r_0}} \qquad \text{or} \qquad v_{circ} = \sqrt{\frac{gR^2}{r_0}} \qquad (12.44)$$

if Eq. (12.30) is taken into account. We note from Fig. 12.22 that for values of v_0 larger than v_{circ} but smaller than v_{esc}, point A where free flight begins is the point of the orbit closest to the earth; this point is called the *perigee*, while point A', which is farthest away from the earth, is known as the *apogee*. For values of v_0 smaller than v_{circ}, point A is the apogee, while point A'', on the other side of the orbit, is the perigee. For values of v_0 much smaller than v_{circ}, the trajectory of the space vehicle intersects the surface of the earth; in such a case, the vehicle does not go into orbit.

Ballistic missiles, which were designed to hit the surface of the earth, also travel along elliptic trajectories. In fact, we should now realize that any object projected in vacuum with an initial velocity v_0 smaller than v_{esc} will move along an elliptic path. It is only when the distances involved are small that the gravitational field of the earth can be assumed uniform and that the elliptic path can be approximated by a parabolic path, as was done earlier (Sec. 11.11) in the case of conventional projectiles.

Periodic Time. An important characteristic of the motion of an earth satellite is the time required by the satellite to describe its orbit. This time, known as the *periodic time* of the satellite, is denoted by τ. We first observe, in view of the definition of areal velocity (Sec. 12.9), that τ can be obtained by dividing the area inside the orbit by the areal velocity. Noting that the area of an ellipse is equal to πab, where a and b denote the semimajor and semiminor axes, respectively, and that the areal velocity is equal to $h/2$, we write

$$\tau = \frac{2\pi ab}{h} \qquad (12.45)$$

While h can be readily determined from r_0 and v_0 in the case of a satellite launched in a direction parallel to the surface of the earth, the semiaxes a and b are not directly related to the initial conditions. Since, on the other hand, the values r_0 and r_1 of r corresponding to the perigee and apogee of the orbit can easily be determined from Eq. (12.39), we will express the semiaxes a and b in terms of r_0 and r_1.

Consider the elliptic orbit shown in Fig. 12.23. The earth's center is located at O and coincides with one of the two foci of the

Fig. 12.22

Fig. 12.23

ellipse, while the points A and A' represent, respectively, the perigee and apogee of the orbit. We easily check that

$$r_0 + r_1 = 2a$$

and thus

$$a = \tfrac{1}{2}(r_0 + r_1) \tag{12.46}$$

Recalling that the sum of the distances from each of the foci to any point of the ellipse is constant, we write

$$O'B + BO = O'A + OA = 2a \qquad \text{or} \qquad BO = a$$

On the other hand, we have $CO = a - r_0$. We can therefore write

$$b^2 = (BC)^2 = (BO)^2 - (CO)^2 = a^2 - (a - r_0)^2$$
$$b^2 = r_0(2a - r_0) = r_0 r_1$$

and thus

$$b = \sqrt{r_0 r_1} \tag{12.47}$$

Fig. 12.23 (repeated)

Formulas (12.46) and (12.47) indicate that the semimajor and semiminor axes of the orbit are equal, respectively, to the arithmetic and geometric means of the maximum and minimum values of the radius vector. Once r_0 and r_1 have been determined, the lengths of the semiaxes can be easily computed and substituted for a and b in formula (12.45).

*12.13 KEPLER'S LAWS OF PLANETARY MOTION

The equations governing the motion of an earth satellite can be used to describe the motion of the moon around the earth. In that case, however, the mass of the moon is not negligible compared with the mass of the earth, and the results obtained are not entirely accurate.

The theory developed in the preceding sections can also be applied to the study of the motion of the planets around the sun. Although another error is introduced by neglecting the forces exerted by the planets on one another, the approximation obtained is excellent. Indeed, even before Newton had formulated his fundamental theory, the properties expressed by Eq. (12.39), where M now represents the mass of the sun, and by Eq. (12.33) had been discovered by the German astronomer Johann Kepler (1571–1630) from astronomical observations of the motion of the planets.

Kepler's three *laws of planetary motion* can be stated as follows:

1. Each planet describes an ellipse, with the sun located at one of its foci.
2. The radius vector drawn from the sun to a planet sweeps equal areas in equal times.
3. The squares of the periodic times of the planets are proportional to the cubes of the semimajor axes of their orbits.

The first law states a particular case of the result established in Sec. 12.12, and the second law expresses that the areal velocity of each planet is constant (see Sec. 12.9). Kepler's third law can also be derived from the results obtained in Sec. 12.12.[†]

†See Prob. 12.120.

SAMPLE PROBLEM 12.9

A satellite is launched in a direction parallel to the surface of the earth with a velocity of 36 900 km/h from an altitude of 500 km. Determine (a) the maximum altitude reached by the satellite, (b) the periodic time of the satellite.

SOLUTION

a. Maximum Altitude. After the satellite is launched, it is subjected only to the gravitational attraction of the earth; its motion is thus governed by Eq. (12.39),

$$\frac{1}{r} = \frac{GM}{h^2} + C \cos \theta \tag{1}$$

Since the radial component of the velocity is zero at the point of launching A, we have $h = r_0 v_0$. Recalling that for the earth $R = 6370$ km, we compute

$$r_0 = 6370 \text{ km} + 500 \text{ km} = 6870 \text{ km} = 6.87 \times 10^6 \text{ m}$$

$$v_0 = 36\,900 \text{ km/h} = \frac{36.9 \times 10^6 \text{ m}}{3.6 \times 10^3 \text{ s}} = 10.25 \times 10^3 \text{ m/s}$$

$$h = r_0 v_0 = (6.87 \times 10^6 \text{ m})(10.25 \times 10^3 \text{ m/s}) = 70.4 \times 10^9 \text{ m}^2/\text{s}$$

$$h^2 = 4.96 \times 10^{21} \text{ m}^4/\text{s}^2$$

Since $GM = gR^2$, where R is the radius of the earth, we have

$$GM = gR^2 = (9.81 \text{ m/s}^2)(6.37 \times 10^6 \text{ m})^2 = 398 \times 10^{12} \text{ m}^3/\text{s}^2$$

$$\frac{GM}{h^2} = \frac{398 \times 10^{12} \text{ m}^3/\text{s}^2}{4.96 \times 10^{21} \text{ m}^4/\text{s}^2} = 80.3 \times 10^{-9} \text{ m}^{-1}$$

Substituting this value into (1), we obtain

$$\frac{1}{r} = 80.3 \times 10^{-9} \text{ m}^{-1} + C \cos \theta \tag{2}$$

Noting that at point A we have $\theta = 0$ and $r = r_0 = 6.87 \times 10^6$ m, we compute the constant C:

$$\frac{1}{6.87 \times 10^6 \text{ m}} = 80.3 \times 10^{-9} \text{ m}^{-1} + C \cos 0° \qquad C = 65.3 \times 10^{-9} \text{ m}^{-1}$$

At A', the point on the orbit farthest from the earth, we have $\theta = 180°$. Using (2), we compute the corresponding distance r_1:

$$\frac{1}{r_1} = 80.3 \times 10^{-9} \text{ m}^{-1} + (65.3 \times 10^{-9} \text{ m}^{-1}) \cos 180°$$

$$r_1 = 66.7 \times 10^6 \text{ m} = 66\,700 \text{ km}$$

$$\textit{Maximum altitude} = 66\,700 \text{ km} - 6370 \text{ km} = 60\,300 \text{ km} \quad \blacktriangleleft$$

b. Periodic Time. Since A and A' are the perigee and apogee, respectively, of the elliptic orbit, we use Eqs. (12.46) and (12.47) and compute the semimajor and semiminor axes of the orbit:

$$a = \tfrac{1}{2}(r_0 + r_1) = \tfrac{1}{2}(6.87 + 66.7)(10^6) \text{ m} = 36.8 \times 10^6 \text{ m}$$

$$b = \sqrt{r_0 r_1} = \sqrt{(6.87)(66.7)} \times 10^6 \text{ m} = 21.4 \times 10^6 \text{ m}$$

$$\tau = \frac{2\pi ab}{h} = \frac{2\pi (36.8 \times 10^6 \text{ m})(21.4 \times 10^6 \text{ m})}{70.4 \times 10^9 \text{ m}^2/\text{s}}$$

$$\tau = 70.3 \times 10^3 \text{ s} = 1171 \text{ min} = 19 \text{ h } 31 \text{ min} \quad \blacktriangleleft$$

SOLVING PROBLEMS
ON YOUR OWN

In this lesson, we continued our study of the motion of a particle under a central force and applied the results to problems in space mechanics. We found that the trajectory of a particle under a central force is defined by the differential equation

$$\frac{d^2u}{d\theta^2} + u = \frac{F}{mh^2u^2} \tag{12.37}$$

where u is the reciprocal of the distance r of the particle to the center of force ($u = 1/r$), F is the magnitude of the central force $\mathbf{F}$, and h is a constant equal to the angular momentum per unit mass of the particle. In space-mechanics problems, $\mathbf{F}$ is the force of gravitational attraction exerted on the satellite or spacecraft by the sun, earth, or other planet about which it travels. Substituting $F = GMm/r^2 = GMmu^2$ into Eq. (12.37), we obtain for that case

$$\frac{d^2u}{d\theta^2} + u = \frac{GM}{h^2} \tag{12.38}$$

where the right-hand member is a constant.

1. Analyzing the motion of satellites and spacecraft. The solution of the differential equation (12.38) defines the trajectory of a satellite or spacecraft. It was obtained in Sec. 12.12 and was given in the alternative forms

$$\frac{1}{r} = \frac{GM}{h^2} + C\cos\theta \quad \text{or} \quad \frac{1}{r} = \frac{GM}{h^2}(1 + \varepsilon\cos\theta) \quad \text{(12.39, 12.39')}$$

Remember when applying these equations that $\theta = 0$ always corresponds to the perigee (the point of closest approach) of the trajectory (Fig. 12.19) and that h is a constant for a given trajectory. Depending on the value of the eccentricity ε, the trajectory will be a hyperbola, a parabola, or an ellipse.

a. $\varepsilon > 1$: The trajectory is a hyperbola, so that for this case the spacecraft never returns to its starting point.

b. $\varepsilon = 1$: The trajectory is a parabola. This is the limiting case between open (hyperbolic) and closed (elliptic) trajectories. We had observed for this case that the velocity v_0 at the perigee is equal to the escape velocity v_{esc},

$$v_0 = v_{esc} = \sqrt{\frac{2GM}{r_0}} \tag{12.43}$$

Note that the escape velocity is the smallest velocity for which the spacecraft does not return to its starting point.

c. $\varepsilon < 1$: The trajectory is an elliptic orbit. For problems involving elliptic orbits, you may find that the relation derived in Prob. 12.102,

$$\frac{1}{r_0} + \frac{1}{r_1} = \frac{2GM}{h^2}$$

(continued)

will be useful in the solution of subsequent problems. When you apply this equation, remember that r_0 and r_1 are the distances from the center of force to the perigee ($\theta = 0$) and apogee ($\theta = 180°$), respectively; that $h = r_0v_0 = r_1v_1$; and that, for a satellite orbiting the earth, $GM_{\text{earth}} = gR^2$, where R is the radius of the earth. Also recall that the trajectory is a circle when $\varepsilon = 0$.

2. Determining the point of impact of a descending spacecraft. For problems of this type, you may assume that the trajectory is elliptic and that the initial point of the descent trajectory is the apogee of the path (Fig. 12.22). Note that at the point of impact, the distance r in Eqs. (12.39) and (12.39′) is equal to the radius R of the body on which the spacecraft lands or crashes. In addition, we have $h = Rv_I \sin \phi_I$, where v_I is the speed of the spacecraft at impact and ϕ_I is the angle that its path forms with the vertical at the point of impact.

3. Calculating the time to travel between two points on a trajectory. For central force motion, the time t required for a particle to travel along a portion of its trajectory can be determined by recalling from Sec. 12.9 that the rate at which area is swept per unit time by the position vector $\mathbf{r}$ is equal to one-half of the angular momentum per unit mass h of the particle: $dA/dt = h/2$. It follows, since h is a constant for a given trajectory, that

$$t = \frac{2A}{h}$$

where A is the total area swept in the time t.

    a. In the case of an elliptic trajectory, the time required to complete one orbit is called the *periodic time* and is expressed as

$$\tau = \frac{2(\pi ab)}{h} \tag{12.45}$$

where a and b are the semimajor and semiminor axes, respectively, of the ellipse and are related to the distances r_0 and r_1 by

$$a = \tfrac{1}{2}(r_0 + r_1) \qquad \text{and} \qquad b = \sqrt{r_0r_1} \tag{12.46, 12.47}$$

    b. Kepler's third law provides a convenient relation between the periodic times of two satellites describing elliptic orbits about the same body [Sec. 12.13]. Denoting the semimajor axes of the two orbits by a_1 and a_2, respectively, and the corresponding periodic times by τ_1 and τ_2, we have

$$\frac{\tau_1^2}{\tau_2^2} = \frac{a_1^3}{a_2^3}$$

    c. In the case of a parabolic trajectory, you may be able to use the expression given on the inside of the front cover of the book for a parabolic or a semiparabolic area to calculate the time required to travel between two points of the trajectory.

PROBLEMS

CONCEPTS QUESTIONS

12.CQ6 A uniform crate C with mass m_C is being transported to the left by a forklift with a constant speed v_1. What is the magnitude of the angular momentum of the crate about point D, that is, the upper left corner of the crate?

  a. 0

  b. mv_1a

  c. mv_1b

  d. $mv_1\sqrt{a^2 + b^2}$

Fig. P12.CQ6 and P12.CQ7

12.CQ7 A uniform crate C with mass m_C is being transported to the left by a forklift with a constant speed v_1. What is the magnitude of the angular momentum of the crate about point A, that is, the point of contact between the front tire of the forklift and the ground?

  a. 0

  b. mv_1d

  c. $3mv_1$

  d. $mv_1\sqrt{3^2 + d^2}$

END-OF-SECTION PROBLEMS

12.94 A particle of mass m is projected from point A with an initial velocity $\mathbf{v}_0$ perpendicular to OA and moves under a central force $\mathbf{F}$ along an elliptic path defined by the equation $r = r_0/(2 - \cos\theta)$. Using Eq. (12.37), show that $\mathbf{F}$ is inversely proportional to the square of the distance r from the particle to the center of force O.

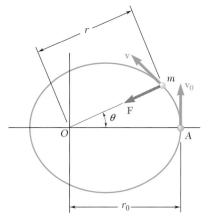

Fig. P12.94

12.95 A particle of mass m describes the logarithmic spiral $r = r_0 e^{b\theta}$ under a central force **F** directed toward the center of force O. Using Eq. (12.37), show that **F** is inversely proportional to the cube of the distance r from the particle to O.

12.96 For the particle of Prob. 12.74, and using Eq. (12.37), show that the central force **F** is proportional to the distance r from the particle to the center of force O.

12.97 A particle of mass m describes the path defined by the equation $r = r_0 \sin \theta$ under a central force **F** directed toward the center of force O. Using Eq. (12.37), show that **F** is inversely proportional to the fifth power of the distance r from the particle to O.

12.98 It was observed that during its second flyby of the earth, the Galileo spacecraft had a velocity of 14.1 km/s as it reached its minimum altitude of 303 km above the surface of the earth. Determine the eccentricity of the trajectory of the spacecraft during this portion of its flight.

12.99 It was observed that during the Galileo spacecraft's first flyby of the earth, its minimum altitude was 600 mi above the surface of the earth. Assuming that the trajectory of the spacecraft was parabolic, determine the maximum velocity of Galileo during its first flyby of the earth.

12.100 As a space probe approaching the planet Venus on a parabolic trajectory reaches point A closest to the planet, its velocity is decreased to insert it into a circular orbit. Knowing that the mass and the radius of Venus are 4.87×10^{24} kg and 6052 km, respectively, determine (a) the velocity of the probe as it approaches A, (b) the decrease in velocity required to insert it into the circular orbit.

280 km

Fig. P12.100

12.101 It was observed that as the Voyager I spacecraft reached the point of its trajectory closest to the planet Saturn, it was at a distance of 185×10^3 km from the center of the planet and had a velocity of 21.0 km/s. Knowing that Tethys, one of Saturn's moons, describes a circular orbit of radius 295×10^3 km at a speed of 11.35 km/s, determine the eccentricity of the trajectory of Voyager I on its approach to Saturn.

12.102 A satellite describes an elliptic orbit about a planet of mass M. Denoting by r_0 and r_1, respectively, the minimum and maximum values of the distance r from the satellite to the center of the planet, derive the relation

$$\frac{1}{r_0} + \frac{1}{r_1} = \frac{2GM}{h^2}$$

where h is the angular momentum per unit mass of the satellite.

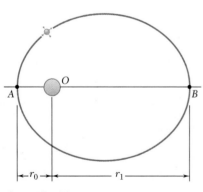

Fig. P12.102

12.103 A space probe is describing a circular orbit about a planet of radius R. The altitude of the probe above the surface of the planet is αR and its speed is v_0. To place the probe in an elliptic orbit which will bring it closer to the planet, its speed is reduced from v_0 to βv_0, where $\beta < 1$, by firing its engine for a short interval of time. Determine the smallest permissible value of β if the probe is not to crash on the surface of the planet.

12.104 At main engine cutoff of its thirteenth flight, the space shuttle Discovery was in an elliptic orbit of minimum altitude 60 km and maximum altitude 500 km above the surface of the earth. Knowing that at point A the shuttle had a velocity $\mathbf{v}_0$ parallel to the surface of the earth and that the shuttle was transferred to a circular orbit as it passed through point B, determine (a) the speed v_0 of the shuttle at A, (b) the increase in speed required at B to insert the shuttle into the circular orbit.

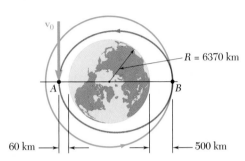

Fig. P12.104

12.105 A space probe is to be placed in a circular orbit of 5600-mi radius about the planet Venus in a specified plane. As the probe reaches A, the point of its original trajectory closest to Venus, it is inserted in a first elliptic transfer orbit by reducing its speed of Δv_A. This orbit brings it to point B with a much reduced velocity. There the probe is inserted in a second transfer orbit located in the specified plane by changing the direction of its velocity and further reducing its speed by Δv_B. Finally, as the probe reaches point C, it is inserted in the desired circular orbit by reducing its speed by Δv_C. Knowing that the mass of Venus is 0.82 times the mass of the earth, that $r_A = 9.3 \times 10^3$ mi and $r_B = 190 \times 10^3$ mi, and that the probe approaches A on a parabolic trajectory, determine by how much the velocity of the probe should be reduced (a) at A, (b) at B, (c) at C.

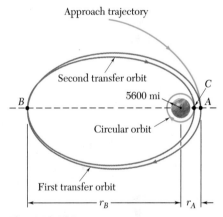

Fig. P12.105

12.106 For the space probe of Prob. 12.105, it is known that $r_A = 9.3 \times 10^3$ mi and that the velocity of the probe is reduced to 20,000 ft/s as it passes through A. Determine (a) the distance from the center of Venus to point B, (b) the amounts by which the velocity of the probe should be reduced at B and C, respectively.

12.107 As it describes an elliptic orbit about the sun, a spacecraft reaches a maximum distance of 202×10^6 mi from the center of the sun at point A (called the aphelion) and a minimum distance of 92×10^6 mi at point B (called the perihelion). To place the spacecraft in a smaller elliptic orbit with aphelion at A' and perihelion at B', where A' and B' are located 164.5×10^6 mi and 85.5×10^6 mi, respectively, from the center of the sun, the speed of the spacecraft is first reduced as it passes through A and then is further reduced as it passes through B'. Knowing that the mass of the sun is 332.8×10^3 times the mass of the earth, determine (a) the speed of the spacecraft at A, (b) the amounts by which the speed of the spacecraft should be reduced at A and B' to insert it into the desired elliptic orbit.

Fig. P12.107

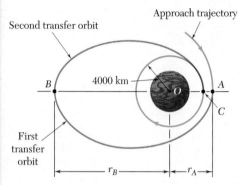

Fig. P12.110

12.108 Halley's comet travels in an elongated elliptic orbit for which the minimum distance from the sun is approximately $\frac{1}{2} r_E$, where $r_E = 150 \times 10^6$ km is the mean distance from the sun to the earth. Knowing that the periodic time of Halley's comet is about 76 years, determine the maximum distance from the sun reached by the comet.

12.109 Based on observations made during the 1996 sighting of comet Hyakutake, it was concluded that the trajectory of the comet is a highly elongated ellipse for which the eccentricity is approximately $\varepsilon = 0.999887$. Knowing that for the 1996 sighting the minimum distance between the comet and the sun was $0.230 R_E$, where R_E is the mean distance from the sun to the earth, determine the periodic time of the comet.

12.110 A space probe is to be placed in a circular orbit of radius 4000 km about the planet Mars. As the probe reaches A, the point of its original trajectory closest to Mars, it is inserted into a first elliptic transfer orbit by reducing its speed. This orbit brings it to point B with a much-reduced velocity. There the probe is inserted into a second transfer orbit by further reducing its speed. Knowing that the mass of Mars is 0.1074 times the mass of the earth, that $r_A = 9000$ km and $r_B = 180\ 000$ km, and that the probe approaches A on a parabolic trajectory, determine the time needed for the space probe to travel from A to B on its first transfer orbit.

12.111 A space shuttle is in an elliptic orbit of eccentricity 0.0356 and a minimum altitude of 300 km above the surface of the earth. Knowing that the radius of the earth is 6370 km, determine the periodic time for the orbit.

12.112 The Clementine spacecraft described an elliptic orbit of minimum altitude $h_A = 400$ km and maximum altitude $h_B = 2940$ km above the surface of the moon. Knowing that the radius of the moon is 1737 km and that the mass of the moon is 0.01230 times the mass of the earth, determine the periodic time of the spacecraft.

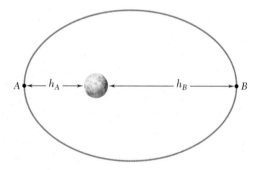

Fig. P12.112

12.113 Determine the time needed for the space probe of Prob. 12.100 to travel from B to C.

12.114 A space probe is describing a circular orbit of radius nR with a velocity v_0 about a planet of radius R and center O. As the probe passes through point A, its velocity is reduced from v_0 to βv_0, where $\beta < 1$, to place the probe on a crash trajectory. Express in terms of n and β the angle AOB, where B denotes the point of impact of the probe on the planet.

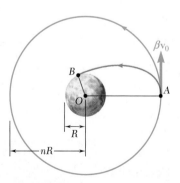

Fig. P12.114

12.115 A long-range ballistic trajectory between points A and B on the earth's surface consists of a portion of an ellipse with the apogee at point C. Knowing that point C is 1500 km above the surface of the earth and the range $R\phi$ of the trajectory is 6000 km, determine (a) the velocity of the projectile at C, (b) the eccentricity ε of the trajectory.

12.116 A space shuttle is describing a circular orbit at an altitude of 563 km above the surface of the earth. As it passes through point A, it fires its engine for a short interval of time to reduce its speed by 152 m/s and begin its descent toward the earth. Determine the angle AOB so that the altitude of the shuttle at point B is 121 km. (*Hint:* Point A is the apogee of the elliptic descent trajectory.)

Fig. P12.115

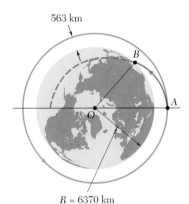

Fig. *P12.116*

12.117 As a spacecraft approaches the planet Jupiter, it releases a probe which is to enter the planet's atmosphere at point B at an altitude of 280 mi above the surface of the planet. The trajectory of the probe is a hyperbola of eccentricity $\varepsilon = 1.031$. Knowing that the radius and the mass of Jupiter are 44,423 mi and 1.30×10^{26} slug, respectively, and that the velocity $\mathbf{v}_B$ of the probe at B forms an angle of 82.9° with the direction of OA, determine (a) the angle AOB, (b) the speed v_B of the probe at B.

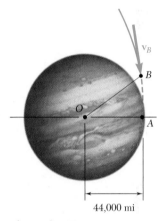

Fig. *P12.117*

12.118 A satellite describes an elliptic orbit about a planet. Denoting by r_0 and r_1 the distances corresponding, respectively, to the perigee and apogee of the orbit, show that the curvature of the orbit at each of these two points can be expressed as

$$\frac{1}{\rho} = \frac{1}{2}\left(\frac{1}{r_0} + \frac{1}{r_1}\right)$$

12.119 (a) Express the eccentricity ε of the elliptic orbit described by a satellite about a planet in terms of the distances r_0 and r_1 corresponding, respectively, to the perigee and apogee of the orbit. (b) Use the result obtained in part a and the data given in Prob. 12.109, where $R_E = 149.6 \times 10^6$ km, to determine the approximate maximum distance from the sun reached by comet Hyakutake.

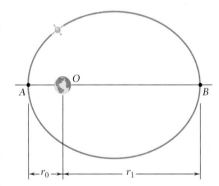

Fig. P12.118 and *P12.119*

12.120 Derive Kepler's third law of planetary motion from Eqs. (12.39) and (12.45).

12.121 Show that the angular momentum per unit mass h of a satellite describing an elliptic orbit of semimajor axis a and eccentricity ε about a planet of mass M can be expressed as

$$h = \sqrt{GMa(1 - \varepsilon^2)}$$

REVIEW AND SUMMARY

This chapter was devoted to Newton's second law and its application to the analysis of the motion of particles.

Newton's second law

Denoting by m the mass of a particle, by $\Sigma\mathbf{F}$ the sum, or resultant, of the forces acting on the particle, and by $\mathbf{a}$ the acceleration of the particle relative to a *newtonian frame of reference* [Sec. 12.2], we wrote

$$\Sigma\mathbf{F} = m\mathbf{a} \qquad (12.2)$$

Linear momentum

Introducing the *linear momentum* of a particle, $\mathbf{L} = m\mathbf{v}$ [Sec. 12.3], we saw that Newton's second law can also be written in the form

$$\Sigma\mathbf{F} = \dot{\mathbf{L}} \qquad (12.5)$$

which expresses that *the resultant of the forces acting on a particle is equal to the rate of change of the linear momentum of the particle.*

Consistent systems of units

Equation (12.2) holds only if a consistent system of units is used. With SI units, the forces should be expressed in newtons, the masses in kilograms, and the accelerations in m/s^2; with U.S. customary units, the forces should be expressed in pounds, the masses in $lb \cdot s^2/ft$ (also referred to as *slugs*), and the accelerations in ft/s^2 [Sec. 12.4].

Equations of motion for a particle

To solve a problem involving the motion of a particle, Eq. (12.2) should be replaced by equations containing scalar quantities [Sec. 12.5]. Using *rectangular components* of $\mathbf{F}$ and $\mathbf{a}$, we wrote

$$\Sigma F_x = ma_x \qquad \Sigma F_y = ma_y \qquad \Sigma F_z = ma_z \qquad (12.8)$$

Using *tangential and normal components*, we had

$$\Sigma F_t = m\frac{dv}{dt} \qquad \Sigma F_n = m\frac{v^2}{\rho} \qquad (12.9')$$

Dynamic equilibrium

We also noted [Sec. 12.6] that the equations of motion of a particle can be replaced by equations similar to the equilibrium equations used in statics if a vector $-m\mathbf{a}$ of magnitude ma but of sense opposite to that of the acceleration is added to the forces applied to the particle; the particle is then said to be in *dynamic equilibrium*. For the sake of uniformity, however, all the Sample Problems were solved by using the equations of motion, first with rectangular components [Sample Probs. 12.1 through 12.4], then with tangential and normal components [Sample Probs. 12.5 and 12.6].

In the second part of the chapter, we defined the *angular momentum* $\mathbf{H}_O$ of a particle about a point O as the moment about O of the linear momentum $m\mathbf{v}$ of that particle [Sec. 12.7]. We wrote

$$\mathbf{H}_O = \mathbf{r} \times m\mathbf{v} \tag{12.12}$$

and noted that $\mathbf{H}_O$ is a vector perpendicular to the plane containing $\mathbf{r}$ and $m\mathbf{v}$ (Fig. 12.24) and of magnitude

$$H_O = rmv \sin \phi \tag{12.13}$$

Resolving the vectors $\mathbf{r}$ and $m\mathbf{v}$ into rectangular components, we expressed the angular momentum $\mathbf{H}_O$ in the determinant form

$$\mathbf{H}_O = \begin{vmatrix} \mathbf{i} & \mathbf{j} & \mathbf{k} \\ x & y & z \\ mv_x & mv_y & mv_z \end{vmatrix} \tag{12.14}$$

In the case of a particle moving in the xy plane, we have $z = v_z = 0$. The angular momentum is perpendicular to the xy plane and is completely defined by its magnitude. We wrote

$$H_O = H_z = m(xv_y - yv_x) \tag{12.16}$$

Computing the rate of change $\dot{\mathbf{H}}_O$ of the angular momentum $\mathbf{H}_O$, and applying Newton's second law, we wrote the equation

$$\Sigma \mathbf{M}_O = \dot{\mathbf{H}}_O \tag{12.19}$$

which states that *the sum of the moments about O of the forces acting on a particle is equal to the rate of change of the angular momentum of the particle about O.*

In many problems involving the plane motion of a particle, it is found convenient to use *radial and transverse components* [Sec. 12.8, Sample Prob. 12.7] and to write the equations

$$\Sigma F_r = m(\ddot{r} - r\dot{\theta}^2) \tag{12.21}$$
$$\Sigma F_\theta = m(r\ddot{\theta} + 2\dot{r}\dot{\theta}) \tag{12.22}$$

When the only force acting on a particle P is a force $\mathbf{F}$ directed toward or away from a fixed point O, the particle is said to be moving *under a central force* [Sec. 12.9]. Since $\Sigma \mathbf{M}_O = 0$ at any given instant, it follows from Eq. (12.19) that $\dot{\mathbf{H}}_O = 0$ for all values of t and, thus, that

$$\mathbf{H}_O = \text{constant} \tag{12.23}$$

We concluded that *the angular momentum of a particle moving under a central force is constant, both in magnitude and direction,* and that the particle moves in a plane perpendicular to the vector $\mathbf{H}_O$.

Angular momentum

Fig. 12.24

Rate of change of angular momentum

Radial and transverse components

Motion under a central force

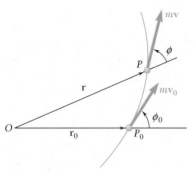

Fig. 12.25

Recalling Eq. (12.13), we wrote the relation

$$rmv \sin \phi = r_0 mv_0 \sin \phi_0 \qquad (12.25)$$

for the motion of any particle under a central force (Fig. 12.25). Using polar coordinates and recalling Eq. (12.18), we also had

$$r^2\dot{\theta} = h \qquad (12.27)$$

where h is a constant representing the angular momentum per unit mass, H_O/m, of the particle. We observed (Fig. 12.26) that the infinitesimal area dA swept by the radius vector OP as it rotates through $d\theta$ is equal to $\frac{1}{2}r^2 d\theta$ and, thus, that the left-hand member of Eq. (12.27) represents twice the *areal velocity* dA/dt of the particle. Therefore, *the areal velocity of a particle moving under a central force is constant.*

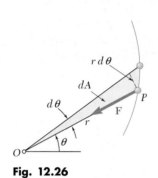

Fig. 12.26

Newton's law of universal gravitation

Fig. 12.27

An important application of the motion under a central force is provided by the orbital motion of bodies under gravitational attraction [Sec. 12.10]. According to *Newton's law of universal gravitation*, two particles at a distance r from each other and of masses M and m, respectively, attract each other with equal and opposite forces $\mathbf{F}$ and $-\mathbf{F}$ directed along the line joining the particles (Fig. 12.27). The common magnitude F of the two forces is

$$F = G\frac{Mm}{r^2} \qquad (12.28)$$

where G is the *constant of gravitation*. In the case of a body of mass m subjected to the gravitational attraction of the earth, the product GM, where M is the mass of the earth, can be expressed as

$$GM = gR^2 \qquad (12.30)$$

where $g = 9.81 \text{ m/s}^2 = 32.2 \text{ ft/s}^2$ and R is the radius of the earth.

Orbital motion

It was shown in Sec. 12.11 that a particle moving under a central force describes a trajectory defined by the differential equation

$$\frac{d^2u}{d\theta^2} + u = \frac{F}{mh^2u^2} \qquad (12.37)$$

where $F > 0$ corresponds to an attractive force and $u = 1/r$. In the case of a particle moving under a force of gravitational attraction [Sec. 12.12], we substituted for F the expression given in Eq. (12.28). Measuring θ from the axis OA joining the focus O to the point A of the trajectory closest to O (Fig. 12.28), we found that the solution to Eq. (12.37) was

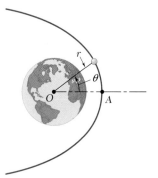

Fig. 12.28

$$\frac{1}{r} = u = \frac{GM}{h^2} + C \cos \theta \qquad (12.39)$$

This is the equation of a conic of eccentricity $\varepsilon = Ch^2/GM$. The conic is an *ellipse* if $\varepsilon < 1$, a *parabola* if $\varepsilon = 1$, and a hyperbola if $\varepsilon > 1$. The constants C and h can be determined from the initial conditions; if the particle is projected from point A ($\theta = 0$, $r = r_0$) with an initial velocity $\mathbf{v}_0$ perpendicular to OA, we have $h = r_0 v_0$ [Sample Prob. 12.9].

It was also shown that the values of the initial velocity corresponding, respectively, to a parabolic and a circular trajectory were

Escape velocity

$$v_{esc} = \sqrt{\frac{2GM}{r_0}} \qquad (12.43)$$

$$v_{circ} = \sqrt{\frac{GM}{r_0}} \qquad (12.44)$$

and that the first of these values, called the *escape velocity*, is the smallest value of v_0 for which the particle will not return to its starting point.

The *periodic time* τ of a planet or satellite was defined as the time required by that body to describe its orbit. It was shown that

Periodic time

$$\tau = \frac{2\pi ab}{h} \qquad (12.45)$$

where $h = r_0 v_0$ and where a and b represent the semimajor and semiminor axes of the orbit. It was further shown that these semiaxes are respectively equal to the arithmetic and geometric means of the maximum and minimum values of the radius vector r.

The last section of the chapter [Sec. 12.13] presented *Kepler's laws of planetary motion* and showed that these empirical laws, obtained from early astronomical observations, confirm Newton's laws of motion as well as his law of gravitation.

Kepler's laws

REVIEW PROBLEMS

Fig. P12.123

Fig. P12.124

Fig. P12.126

12.122 In the braking test of a sports car its velocity is reduced from 70 mi/h to zero in a distance of 170 ft with slipping impending. Knowing that the coefficient of kinetic friction is 80 percent of the coefficient of static friction, determine (a) the coefficient of static friction, (b) the stopping distance for the same initial velocity if the car skids. Ignore air resistance and rolling resistance.

12.123 A bucket is attached to a rope of length $L = 1.2$ m and is made to revolve in a horizontal circle. Drops of water leaking from the bucket fall and strike the floor along the perimeter of a circle of radius a. Determine the radius a when $\theta = 30°$.

12.124 A 12-lb block B rests as shown on the upper surface of a 30-lb wedge A. Neglecting friction, determine immediately after the system is released from rest (a) the acceleration of A, (b) the acceleration of B relative to A.

12.125 A 500-lb crate B is suspended from a cable attached to a 40-lb trolley A which rides on an inclined I-beam as shown. Knowing that at the instant shown the trolley has an acceleration of 1.2 ft/s² up and to the right, determine (a) the acceleration of B relative to A, (b) the tension in cable CD.

Fig. P12.125

12.126 The roller-coaster track shown is contained in a vertical plane. The portion of track between A and B is straight and horizontal, while the portions to the left of A and to the right of B have radii of curvature as indicated. A car is traveling at a speed of 72 km/h when the brakes are suddenly applied, causing the wheels of the car to slide on the track ($\mu_k = 0.25$). Determine the initial deceleration of the car if the brakes are applied as the car (a) has almost reached A, (b) is traveling between A and B, (c) has just passed B.

12.127 The 100-g pin *B* slides along the slot in the rotating arm *OC* and along the slot *DE* which is cut in a fixed horizontal plate. Neglecting friction and knowing that rod *OC* rotates at the constant rate $\dot{\theta}_0 = 12$ rad/s, determine for any given value of θ (*a*) the radial and transverse components of the resultant force **F** exerted on pin *B*, (*b*) the forces **P** and **Q** exerted on pin *B* by rod *OC* and the wall of slot *DE*, respectively.

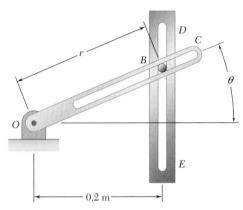

Fig. P12.127

12.128 A small 200-g collar *C* can slide on a semicircular rod which is made to rotate about the vertical *AB* at the constant rate of 6 rad/s. Determine the minimum required value of the coefficient of static friction between the collar and the rod if the collar is not to slide when (*a*) $\theta = 90°$, (*b*) $\theta = 75°$, (*c*) $\theta = 45°$. Indicate in each case the direction of the impending motion.

12.129 Telemetry technology is used to quantify kinematic values of a 200-kg roller-coaster cart as it passes overhead. According to the system, $r = 25$ m, $\dot{r} = -10$ m/s, $\ddot{r} = -2$ m/s², $\theta = 90°$, $\dot{\theta} = -0.4$ rad/s, $\ddot{\theta} = -0.32$ rad/s². At this instant, determine (*a*) the normal force between the cart and the track, (*b*) the radius of curvature of the track.

Fig. P12.128

Fig. P12.129

12.130 The radius of the orbit of a moon of a given planet is equal to twice the radius of that planet. Denoting by ρ the mean density of the planet, show that the time required by the moon to complete one full revolution about the planet is $(24\pi/G\rho)^{1/2}$, where *G* is the constant of gravitation.

12.131 At engine burnout on a mission, a shuttle had reached point A at an altitude of 40 mi above the surface of the earth and had a horizontal velocity $\mathbf{v}_0$. Knowing that its first orbit was elliptic and that the shuttle was transferred to a circular orbit as it passed through point B at an altitude of 170 mi, determine (a) the time needed for the shuttle to travel from A to B on its original elliptic orbit, (b) the periodic time of the shuttle on its final circular orbit.

Fig. P12.131

12.132 It was observed that as the Galileo spacecraft reached the point on its trajectory closest to Io, a moon of the planet Jupiter, it was at a distance of 1750 mi from the center of Io and had a velocity of 49.4×10^3 ft/s. Knowing that the mass of Io is 0.01496 times the mass of the earth, determine the eccentricity of the trajectory of the spacecraft as it approached Io.

*12.133 Disk A rotates in a horizontal plane about a of vertical axis at the constant rate $\dot{\theta}_0 = 10$ rad/s. Slider B has mass 1 kg and moves in a frictionless slot cut in the disk. The slider is attached to a spring of constant k, which is undeformed when $r = 0$. Knowing that the slider is released with no radial velocity in the position $r = 500$ mm, determine the position of the slider and the horizontal force exerted on it by the disk at $t = 0.1$ s for (a) $k = 100$ N/m, (b) $k = 200$ N/m.

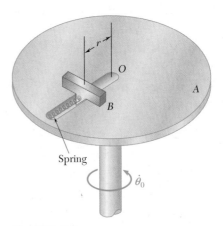

Fig. P12.133

COMPUTER PROBLEMS

12.C1 Block B of mass 10 kg is initially at rest as shown on the upper surface of a 20-kg wedge A which is supported by a horizontal surface. A 2-kg block C is connected to block B by a cord which passes over a pulley of negligible mass. Using computational software and denoting by μ the coefficient of friction at all surfaces, use this program to determine the accelerations for values of $\mu \geq 0$. Use 0.01 increments for μ until the wedge does not move and then use 0.1 increments until no motion occurs.

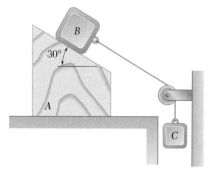

Fig. P12.C1

12.C2 A small, 1-lb block is at rest at the top of a cylindrical surface. The block is given an initial velocity $\mathbf{v}_0$ to the right of magnitude 10 ft/s, which causes it to slide on the cylindrical surface. Using computational software calculate and plot the values of θ at which the block leaves the surface for values of μ_k, the coefficient of kinetic friction between the block and the surface, from 0 to 0.4.

12.C3 A block of mass m is attached to a spring of constant k. The block is released from rest when the spring is in a horizontal and undeformed position. Use computational software to determine, for various selected values of k/m and r_0, (a) the length of the spring and the magnitude and direction of the velocity of the block as the block passes directly under the point of suspension of the spring, (b) the value of k/m when $r_0 = 1$ m for which that velocity is horizontal.

Fig. P12.C2

Fig. P12.C3

12.C4 Use computational software to determine the ranges of values of θ for which the block E of Prob. 12.60 will not slide in the semicircular slot of the flat plate. Assuming a coefficient of static friction of 0.35, determine the ranges of values when the constant rate of rotation of the plate is (a) 14 rad/s, (b) 2 rad/s.

A golf ball will deform upon impact as shown by this high-speed photo. The maximum deformation will occur when the club head velocity and the ball velocity are the same. In this chapter impacts will be analyzed using the coefficient of restitution and conservation of linear momentum. The kinetics of particles using energy and momentum methods is the subject of this chapter.

Kinetics of Particles: Energy and Momentum Methods

Chapter 13 Kinetics of Particles: Energy and Momentum Methods

13.1 INTRODUCTION

In the preceding chapter, most problems dealing with the motion of particles were solved through the use of the fundamental equation of motion $\mathbf{F} = m\mathbf{a}$. Given a particle acted upon by a force $\mathbf{F}$, we could solve this equation for the acceleration $\mathbf{a}$; then, by applying the principles of kinematics, we could determine from $\mathbf{a}$ the velocity and position of the particle at any time.

Using the equation $\mathbf{F} = m\mathbf{a}$ together with the principles of kinematics allows us to obtain two additional methods of analysis, the *method of work and energy* and the *method of impulse and momentum*. The advantage of these methods lies in the fact that they make the determination of the acceleration unnecessary. Indeed, the method of work and energy directly relates force, mass, velocity, and displacement, while the method of impulse and momentum relates force, mass, velocity, and time.

The method of work and energy will be considered first. In Secs. 13.2 through 13.4, the *work of a force* and the *kinetic energy of a particle* are discussed and the principle of work and energy is applied to the solution of engineering problems. The concepts of *power* and *efficiency* of a machine are introduced in Sec. 13.5.

Sections 13.6 through 13.8 are devoted to the concept of *potential energy* of a conservative force and to the application of the principle of conservation of energy to various problems of practical interest. In Sec. 13.9, the principles of conservation of energy and of conservation of angular momentum are used jointly to solve problems of space mechanics.

The second part of the chapter is devoted to the *principle of impulse and momentum* and to its application to the study of the motion of a particle. As you will see in Sec. 13.11, this principle is particularly effective in the study of the *impulsive motion* of a particle, where very large forces are applied for a very short time interval.

In Secs. 13.12 through 13.14, the *central impact* of two bodies will be considered. It will be shown that a certain relation exists between the relative velocities of the two colliding bodies before and after impact. This relation, together with the fact that the total momentum of the two bodies is conserved, can be used to solve a number of problems of practical interest.

Finally, in Sec. 13.15, you will learn to select from the three fundamental methods presented in Chaps. 12 and 13 the method best suited for the solution of a given problem. You will also see how the principle of conservation of energy and the method of impulse and momentum can be combined to solve problems involving only conservative forces, except for a short impact phase during which impulsive forces must also be taken into consideration.

13.2 WORK OF A FORCE

We will first define the terms *displacement* and *work* as they are used in mechanics.† Consider a particle which moves from a point

---

†The definition of work was given in Sec. 10.2, and the basic properties of the work of a force were outlined in Secs. 10.2 and 10.6. For convenience, we repeat here the portions of this material which relate to the kinetics of particles.

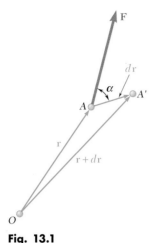

Fig. 13.1

A to a neighboring point A' (Fig. 13.1). If $\mathbf{r}$ denotes the position vector corresponding to point A, the small vector joining A and A' can be denoted by the differential $d\mathbf{r}$; the vector $d\mathbf{r}$ is called the *displacement* of the particle. Now, let us assume that a force $\mathbf{F}$ is acting on the particle. The *work of the force* $\mathbf{F}$ *corresponding to the displacement* $d\mathbf{r}$ is defined as the quantity

$$dU = \mathbf{F} \cdot d\mathbf{r} \qquad (13.1)$$

obtained by forming the scalar product of the force $\mathbf{F}$ and the displacement $d\mathbf{r}$. Denoting by F and ds, respectively, the magnitudes of the force and of the displacement, and by α the angle formed by $\mathbf{F}$ and $d\mathbf{r}$, and recalling the definition of the scalar product of two vectors (Sec. 3.9), we write

$$dU = F \, ds \cos \alpha \qquad (13.1')$$

Using formula (3.30), we can also express the work dU in terms of the rectangular components of the force and of the displacement:

$$dU = F_x \, dx + F_y \, dy + F_z \, dz \qquad (13.1'')$$

Being a *scalar quantity*, work has a magnitude and a sign but no direction. We also note that work should be expressed in units obtained by multiplying units of length by units of force. Thus, if U.S. customary units are used, work should be expressed in ft · lb or in · lb. If SI units are used, work should be expressed in N · m. The unit of work N · m is called a *joule* (J).† Recalling the conversion factors indicated in Sec. 12.4, we write

$$1 \text{ ft} \cdot \text{lb} = (1 \text{ ft})(1 \text{ lb}) = (0.3048 \text{ m})(4.448 \text{ N}) = 1.356 \text{ J}$$

It follows from (13.1') that the work dU is positive if the angle α is acute and negative if α is obtuse. Three particular cases are of special

---

†The joule (J) is the SI unit of *energy*, whether in mechanical form (work, potential energy, kinetic energy) or in chemical, electrical, or thermal form. We should note that even though N · m = J, the moment of a force must be expressed in N · m and not in joules, since the moment of a force is not a form of energy.

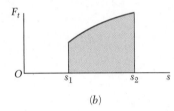

Fig. 13.2

interest. If the force $\mathbf{F}$ has the same direction as $d\mathbf{r}$, the work dU reduces to $F\,ds$. If $\mathbf{F}$ has a direction opposite to that of $d\mathbf{r}$, the work is $dU = -F\,ds$. Finally, if $\mathbf{F}$ is perpendicular to $d\mathbf{r}$, the work dU is zero.

The work of $\mathbf{F}$ during a *finite* displacement of the particle from A_1 to A_2 (Fig. 13.2a) is obtained by integrating Eq. (13.1) along the path described by the particle. This work, denoted by $U_{1\rightarrow2}$, is

$$U_{1\rightarrow2} = \int_{A_1}^{A_2} \mathbf{F} \cdot d\mathbf{r} \tag{13.2}$$

Using the alternative expression (13.1′) for the elementary work dU, and observing that $F\cos\alpha$ represents the tangential component F_t of the force, we can also express the work $U_{1\rightarrow2}$ as

$$U_{1\rightarrow2} = \int_{s_1}^{s_2} (F\cos\alpha)\,ds = \int_{s_1}^{s_2} F_t\,ds \tag{13.2′}$$

where the variable of integration s measures the distance traveled by the particle along the path. The work $U_{1\rightarrow2}$ is represented by the area under the curve obtained by plotting $F_t = F\cos\alpha$ against s (Fig. 13.2b).

When the force $\mathbf{F}$ is defined by its rectangular components, the expression (13.1″) can be used for the elementary work. We then write

$$U_{1\rightarrow2} = \int_{A_1}^{A_2} (F_x\,dx + F_y\,dy + F_z\,dz) \tag{13.2″}$$

where the integration is to be performed along the path described by the particle.

Work of a Constant Force in Rectilinear Motion. When a particle moving in a straight line is acted upon by a force $\mathbf{F}$ of constant magnitude and of constant direction (Fig. 13.3), formula (13.2′) yields

$$U_{1\rightarrow2} = (F\cos\alpha)\,\Delta x \tag{13.3}$$

where α = angle the force forms with direction of motion
Δx = displacement from A_1 to A_2

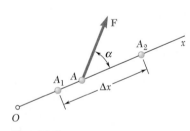

Fig. 13.3

Work of the Force of Gravity. The work of the weight $\mathbf{W}$ of a body, i.e., of the force of gravity exerted on that body, is obtained by substituting the components of $\mathbf{W}$ into (13.1″) and (13.2″). With the y axis chosen upward (Fig. 13.4), we have $F_x = 0$, $F_y = -W$, and $F_z = 0$, and we write

$$dU = -W\,dy$$
$$U_{1\rightarrow2} = -\int_{y_1}^{y_2} W\,dy = Wy_1 - Wy_2 \tag{13.4}$$

or

$$U_{1\rightarrow2} = -W(y_2 - y_1) = -W\,\Delta y \tag{13.4′}$$

where Δy is the vertical displacement from A_1 to A_2. The work of the weight $\mathbf{W}$ is thus equal to *the product of W and the vertical*

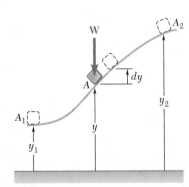

Fig. 13.4

displacement of the center of gravity of the body. The work is *positive* when $\Delta y < 0$, that is, *when the body moves down*.

Work of the Force Exerted by a Spring.

Consider a body A attached to a fixed point B by a spring; it is assumed that the spring is undeformed when the body is at A_0 (Fig. 13.5a). Experimental evidence shows that the magnitude of the force $\mathbf{F}$ exerted by the spring on body A is proportional to the deflection x of the spring measured from the position A_0. We have

$$F = kx \qquad (13.5)$$

where k is the *spring constant,* expressed in N/m or kN/m if SI units are used and in lb/ft or lb/in. if U.S. customary units are used.†

The work of the force $\mathbf{F}$ exerted by the spring during a finite displacement of the body from $A_1(x = x_1)$ to $A_2(x = x_2)$ is obtained by writing

$$dU = -F\,dx = -kx\,dx$$

$$U_{1\to2} = -\int_{x_1}^{x_2} kx\,dx = \tfrac{1}{2}kx_1^2 - \tfrac{1}{2}kx_2^2 \qquad (13.6)$$

Care should be taken to express k and x in consistent units. For example, if U.S. customary units are used, k should be expressed in lb/ft and x in feet, or k in lb/in. and x in inches; in the first case, the work is obtained in ft · lb, in the second case, in in · lb. We note that the work of the force $\mathbf{F}$ exerted by the spring on the body is *positive* when $x_2 < x_1$, that is, *when the spring is returning to its undeformed position*.

Since Eq. (13.5) is the equation of a straight line of slope k passing through the origin, the work $U_{1\to2}$ of $\mathbf{F}$ during the displacement from A_1 to A_2 can be obtained by evaluating the area of the trapezoid shown in Fig. 13.5b. This is done by computing F_1 and F_2 and multiplying the base Δx of the trapezoid by its mean height $\tfrac{1}{2}(F_1 + F_2)$. Since the work of the force $\mathbf{F}$ exerted by the spring is positive for a negative value of Δx, we write

$$U_{1\to2} = -\tfrac{1}{2}(F_1 + F_2)\,\Delta x \qquad (13.6')$$

Formula (13.6$'$) is usually more convenient to use than (13.6) and affords fewer chances of confusing the units involved.

Work of a Gravitational Force.

We saw in Sec. 12.10 that two particles of mass M and m at a distance r from each other attract each other with equal and opposite forces $\mathbf{F}$ and $-\mathbf{F}$, directed along the line joining the particles and of magnitude

$$F = G\frac{Mm}{r^2}$$

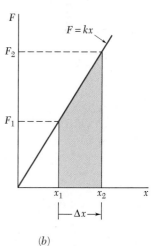

Fig. 13.5

---

†The relation $F = kx$ is correct under static conditions only. Under dynamic conditions, formula (13.5) should be modified to take the inertia of the spring into account. However, the error introduced by using the relation $F = kx$ in the solution of kinetics problems is small if the mass of the spring is small compared with the other masses in motion.

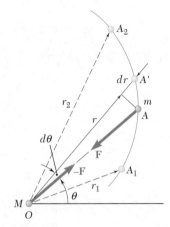

Fig. 13.6

Let us assume that the particle M occupies a fixed position O while the particle m moves along the path shown in Fig. 13.6. The work of the force $\mathbf{F}$ exerted on the particle m during an infinitesimal displacement of the particle from A to A' can be obtained by multiplying the magnitude F of the force by the radial component dr of the displacement. Since $\mathbf{F}$ is directed toward O, the work is negative and we write

$$dU = -F\,dr = -G\frac{Mm}{r^2}dr$$

The work of the gravitational force $\mathbf{F}$ during a finite displacement from $A_1(r = r_1)$ to $A_2(r = r_2)$ is therefore

$$U_{1\to2} = -\int_{r_1}^{r_2} \frac{GMm}{r^2}dr = \frac{GMm}{r_2} - \frac{GMm}{r_1} \qquad (13.7)$$

where M is the mass of the earth. This formula can be used to determine the work of the force exerted by the earth on a body of mass m at a distance r from the center of the earth, when r is larger than the radius R of the earth. Recalling the first of the relations (12.29), we can replace the product GMm in Eq. (13.7) by WR^2, where R is the radius of the earth ($R = 6.37 \times 10^6$ m or 3960 mi) and W is the weight of the body at the surface of the earth.

A number of forces frequently encountered in problems of kinetics *do no work*. They are forces applied to fixed points ($ds = 0$) or acting in a direction perpendicular to the displacement ($\cos \alpha = 0$). Among the forces which do no work are the following: the reaction at a frictionless pin when the body supported rotates about the pin, the reaction at a frictionless surface when the body in contact moves along the surface, the reaction at a roller moving along its track, and the weight of a body when its center of gravity moves horizontally.

13.3 KINETIC ENERGY OF A PARTICLE. PRINCIPLE OF WORK AND ENERGY

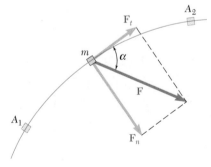

Fig. 13.7

Consider a particle of mass m acted upon by a force $\mathbf{F}$ and moving along a path which is either rectilinear or curved (Fig. 13.7). Expressing Newton's second law in terms of the tangential components of the force and of the acceleration (see Sec. 12.5), we write

$$F_t = ma_t \qquad \text{or} \qquad F_t = m\frac{dv}{dt}$$

where v is the speed of the particle. Recalling from Sec. 11.9 that $v = ds/dt$, we obtain

$$F_t = m\frac{dv}{ds}\frac{ds}{dt} = mv\frac{dv}{ds}$$
$$F_t\,ds = mv\,dv$$

Integrating from A_1, where $s = s_1$ and $v = v_1$, to A_2, where $s = s_2$ and $v = v_2$, we write

$$\int_{s_1}^{s_2} F_t\,ds = m\int_{v_1}^{v_2} v\,dv = \tfrac{1}{2}mv_2^2 - \tfrac{1}{2}mv_1^2 \qquad (13.8)$$

The left-hand member of Eq. (13.8) represents the work $U_{1\to2}$ of the force $\mathbf{F}$ exerted on the particle during the displacement from A_1 to

A_2; as indicated in Sec. 13.2, the work $U_{1\rightarrow 2}$ is a scalar quantity. The expression $\frac{1}{2}mv^2$ is also a scalar quantity; it is defined as the kinetic energy of the particle and is denoted by T. We write

$$T = \tfrac{1}{2}mv^2 \tag{13.9}$$

Substituting into (13.8), we have

$$U_{1\rightarrow 2} = T_2 - T_1 \tag{13.10}$$

which expresses that, when a particle moves from A_1 to A_2 under the action of a force **F**, *the work of the force* **F** *is equal to the change in kinetic energy of the particle*. This is known as the *principle of work and energy*. Rearranging the terms in (13.10), we write

$$T_1 + U_{1\rightarrow 2} = T_2 \tag{13.11}$$

Thus, *the kinetic energy of the particle at A_2 can be obtained by adding to its kinetic energy at A_1 the work done during the displacement from A_1 to A_2 by the force* **F** *exerted on the particle*. Like Newton's second law from which it is derived, the principle of work and energy applies only with respect to a newtonian frame of reference (Sec. 12.2). The speed v used to determine the kinetic energy T should therefore be measured with respect to a newtonian frame of reference.

Since both work and kinetic energy are scalar quantities, their sum can be computed as an ordinary algebraic sum, the work $U_{1\rightarrow 2}$ being considered as positive or negative according to the direction of **F**. When several forces act on the particle, the expression $U_{1\rightarrow 2}$ represents the total work of the forces acting on the particle; it is obtained by adding algebraically the work of the various forces.

As noted above, the kinetic energy of a particle is a scalar quantity. It further appears from the definition $T = \frac{1}{2}mv^2$ that regardless of the direction of motion of the particle the kinetic energy is always positive. Considering the particular case when $v_1 = 0$ and $v_2 = v$, and substituting $T_1 = 0$ and $T_2 = T$ into (13.10), we observe that the work done by the forces acting on the particle is equal to T. Thus, the kinetic energy of a particle moving with a speed v represents the work which must be done to bring the particle from rest to the speed v. Substituting $T_1 = T$ and $T_2 = 0$ into (13.10), we also note that when a particle moving with a speed v is brought to rest, the work done by the forces acting on the particle is $-T$. Assuming that no energy is dissipated into heat, we conclude that the work done by the forces exerted *by the particle* on the bodies which cause it to come to rest is equal to T. Thus, the kinetic energy of a particle also represents *the capacity to do work associated with the speed of the particle*.

The kinetic energy is measured in the same units as work, i.e., in joules if SI units are used and in ft · lb if U.S. customary units are used. We check that, in SI units,

$$T = \tfrac{1}{2}mv^2 = \text{kg}(\text{m/s})^2 = (\text{kg} \cdot \text{m/s}^2)\text{m} = \text{N} \cdot \text{m} = \text{J}$$

while, in customary units,

$$T = \tfrac{1}{2}mv^2 = (\text{lb} \cdot \text{s}^2/\text{ft})(\text{ft/s})^2 = \text{ft} \cdot \text{lb}$$

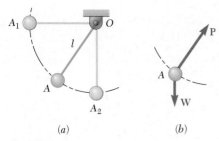

Fig. 13.8

13.4 APPLICATIONS OF THE PRINCIPLE OF WORK AND ENERGY

The application of the principle of work and energy greatly simplifies the solution of many problems involving forces, displacements, and velocities. Consider, for example, the pendulum OA consisting of a bob A of weight W attached to a cord of length l (Fig. 13.8a). The pendulum is released with no initial velocity from a horizontal position OA_1 and allowed to swing in a vertical plane. We wish to determine the speed of the bob as it passes through A_2, directly under O.

We first determine the work done during the displacement from A_1 to A_2 by the forces acting on the bob. We draw a free-body diagram of the bob, showing all the *actual* forces acting on it, i.e., the weight $\mathbf{W}$ and the force $\mathbf{P}$ exerted by the cord (Fig. 13.8b). (An inertia vector is not an actual force and *should not* be included in the free-body diagram.) We note that the force $\mathbf{P}$ does no work, since it is normal to the path; the only force which does work is thus the weight $\mathbf{W}$. The work of $\mathbf{W}$ is obtained by multiplying its magnitude W by the vertical displacement l (Sec. 13.2); since the displacement is downward, the work is positive. We therefore write $U_{1\rightarrow 2} = Wl$.

Now considering the kinetic energy of the bob, we find $T_1 = 0$ at A_1 and $T_2 = \frac{1}{2}(W/g)v_2^2$ at A_2. We can now apply the principle of work and energy; recalling formula (13.11), we write

$$T_1 + U_{1\rightarrow 2} = T_2 \qquad 0 + Wl = \frac{1}{2}\frac{W}{g}v_2^2$$

Solving for v_2, we find $v_2 = \sqrt{2gl}$. We note that the speed obtained is that of a body falling freely from a height l.

The example we have considered illustrates the following advantages of the method of work and energy:

1. In order to find the speed at A_2, there is no need to determine the acceleration in an intermediate position A and to integrate the expression obtained from A_1 to A_2.
2. All quantities involved are scalars and can be added directly, without using x and y components.
3. Forces which do no work are eliminated from the solution of the problem.

What is an advantage in one problem, however, may be a disadvantage in another. It is evident, for instance, that the method of work and energy cannot be used to directly determine an acceleration. It is also evident that in determining a force which is normal to the path of the particle, a force which does no work, the method of work and energy must be supplemented by the direct application of Newton's second law. Suppose, for example, that we wish to determine the tension in the cord of the pendulum of Fig. 13.8a as the bob passes through A_2. We draw a free-body diagram of the bob in that position (Fig. 13.9) and express Newton's second law in terms of tangential and normal components. The equations $\Sigma F_t = ma_t$ and $\Sigma F_n = ma_n$ yield, respectively, $a_t = 0$ and

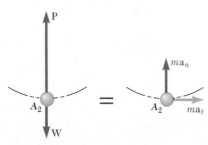

Fig. 13.9

$$P - W = ma_n = \frac{W}{g} \frac{v_2^2}{l}$$

But the speed at A_2 was determined earlier by the method of work and energy. Substituting $v_2^2 = 2gl$ and solving for P, we write

$$P = W + \frac{W}{g} \frac{2gl}{l} = 3W$$

When a problem involves two particles or more, the principle of work and energy can be applied to each particle separately. Adding the kinetic energies of the various particles, and considering the work of all the forces acting on them, we can also write a single equation of work and energy for all the particles involved. We have

$$T_1 + U_{1\rightarrow 2} = T_2 \qquad (13.11)$$

where T represents the arithmetic sum of the kinetic energies of the particles involved (all terms are positive) and $U_{1\rightarrow 2}$ is the work of all the forces acting on the particles, *including the forces of action and reaction exerted by the particles on each other*. In problems involving bodies connected by *inextensible cords or links*, however, the work of the forces exerted by a given cord or link on the two bodies it connects cancels out, since the points of application of these forces move through equal distances (see Sample Prob. 13.2).†

Since friction forces have a direction opposite of that of the displacement of the body on which they act, *the work of friction forces is always negative*. This work represents energy dissipated into heat and always results in a decrease in the kinetic energy of the body involved (see Sample Prob. 13.3).

13.5 POWER AND EFFICIENCY

Power is defined as the time rate at which work is done. In the selection of a motor or engine, power is a much more important criterion than is the actual amount of work to be performed. Either a small motor or a large power plant can be used to do a given amount of work; but the small motor may require a month to do the work done by the power plant in a matter of minutes. If ΔU is the work done during the time interval Δt, then the average power during that time interval is

$$\text{Average power} = \frac{\Delta U}{\Delta t}$$

Letting Δt approach zero, we obtain at the limit

$$\text{Power} = \frac{dU}{dt} \qquad (13.12)$$

---

†The application of the method of work and energy to a system of particles is discussed in detail in Chap. 14.

Substituting the scalar product $\mathbf{F} \cdot d\mathbf{r}$ for dU, we can also write

$$\text{Power} = \frac{dU}{dt} = \frac{\mathbf{F} \cdot d\mathbf{r}}{dt}$$

and, recalling that $d\mathbf{r}/dt$ represents the velocity $\mathbf{v}$ of the point of application of $\mathbf{F}$,

$$\text{Power} = \mathbf{F} \cdot \mathbf{v} \tag{13.13}$$

Since power was defined as the time rate at which work is done, it should be expressed in units obtained by dividing units of work by the unit of time. Thus, if SI units are used, power should be expressed in J/s; this unit is called a *watt* (W). We have

$$1 \text{ W} = 1 \text{ J/s} = 1 \text{ N} \cdot \text{m/s}$$

If U.S. customary units are used, power should be expressed in ft · lb/s or in *horsepower* (hp), with the latter defined as

$$1 \text{ hp} = 550 \text{ ft} \cdot \text{lb/s}$$

Recalling from Sec. 13.2 that 1 ft · lb = 1.356 J, we verify that

$$1 \text{ ft} \cdot \text{lb/s} = 1.356 \text{ J/s} = 1.356 \text{ W}$$
$$1 \text{ hp} = 550(1.356 \text{ W}) = 746 \text{ W} = 0.746 \text{ kW}$$

The *mechanical efficiency* of a machine was defined in Sec. 10.5 as the ratio of the output work to the input work:

$$\eta = \frac{\text{output work}}{\text{input work}} \tag{13.14}$$

This definition is based on the assumption that work is done at a constant rate. The ratio of the output to the input work is therefore equal to the ratio of the rates at which output and input work are done, and we have

$$\eta = \frac{\text{power output}}{\text{power input}} \tag{13.15}$$

Because of energy losses due to friction, the output work is always smaller than the input work, and consequently the power output is always smaller than the power input. The mechanical efficiency of a machine is therefore always less than 1.

When a machine is used to transform mechanical energy into electric energy, or thermal energy into mechanical energy, its *overall efficiency* can be obtained from formula (13.15). The overall efficiency of a machine is always less than 1; it provides a measure of all the various energy losses involved (losses of electric or thermal energy as well as frictional losses). Note that it is necessary to express the power output and the power input in the same units before using formula (13.15).

SAMPLE PROBLEM 13.1

An automobile weighing 4000 lb is driven down a 5° incline at a speed of 60 mi/h when the brakes are applied, causing a constant total braking force (applied by the road on the tires) of 1500 lb. Determine the distance traveled by the automobile as it comes to a stop.

SOLUTION

Kinetic Energy

Position 1: $v_1 = \left(60\,\dfrac{mi}{h}\right)\left(\dfrac{5280\,ft}{1\,mi}\right)\left(\dfrac{1\,h}{3600\,s}\right) = 88\,ft/s$

$T_1 = \tfrac{1}{2}mv_1^2 = \tfrac{1}{2}(4000/32.2)(88)^2 = 481{,}000\,ft\cdot lb$

Position 2: $v_2 = 0 \qquad T_2 = 0$

Work $U_{1\to2} = -1500x + (4000\sin 5°)x = -1151x$

Principle of Work and Energy

$$T_1 + U_{1\to2} = T_2$$
$$481{,}000 - 1151x = 0 \qquad\qquad x = 418\,ft \;\blacktriangleleft$$

SAMPLE PROBLEM 13.2

Two blocks are joined by an inextensible cable as shown. If the system is released from rest, determine the velocity of block A after it has moved 2 m. Assume that the coefficient of kinetic friction between block A and the plane is $\mu_k = 0.25$ and that the pulley is weightless and frictionless.

SOLUTION

Work and Energy for Block A. We denote the friction force by $\mathbf{F}_A$ and the force exerted by the cable by $\mathbf{F}_C$, and write

$$m_A = 200\,kg \qquad W_A = (200\,kg)(9.81\,m/s^2) = 1962\,N$$
$$F_A = \mu_k N_A = \mu_k W_A = 0.25(1962\,N) = 490\,N$$
$$T_1 + U_{1\to2} = T_2: \qquad 0 + F_C(2\,m) - F_A(2\,m) = \tfrac{1}{2}m_A v^2$$
$$F_C(2\,m) - (490\,N)(2\,m) = \tfrac{1}{2}(200\,kg)v^2 \qquad (1)$$

Work and Energy for Block B. We write

$$m_B = 300\,kg \qquad W_B = (300\,kg)(9.81\,m/s^2) = 2940\,N$$
$$T_1 + U_{1\to2} = T_2: \qquad 0 + W_B(2\,m) - F_C(2\,m) = \tfrac{1}{2}m_B v^2$$
$$(2940\,N)(2\,m) - F_C(2\,m) = \tfrac{1}{2}(300\,kg)v^2 \qquad (2)$$

Adding the left-hand and right-hand members of (1) and (2), we observe that the work of the forces exerted by the cable on A and B cancels out:

$$(2940\,N)(2\,m) - (490\,N)(2\,m) = \tfrac{1}{2}(200\,kg + 300\,kg)v^2$$
$$4900\,J = \tfrac{1}{2}(500\,kg)v^2 \qquad v = 4.43\,m/s \;\blacktriangleleft$$

SAMPLE PROBLEM 13.3

A spring is used to stop a 60-kg package which is sliding on a horizontal surface. The spring has a constant $k = 20$ kN/m and is held by cables so that it is initially compressed 120 mm. Knowing that the package has a velocity of 2.5 m/s in the position shown and that the maximum additional deflection of the spring is 40 mm, determine (a) the coefficient of kinetic friction between the package and the surface, (b) the velocity of the package as it passes again through the position shown.

SOLUTION

a. Motion from Position 1 to Position 2
Kinetic Energy   *Position 1:* $v_1 = 2.5$ m/s

$$T_1 = \tfrac{1}{2}mv_1^2 = \tfrac{1}{2}(60 \text{ kg})(2.5 \text{ m/s})^2 = 187.5 \text{ N} \cdot \text{m} = 187.5 \text{ J}$$

Position 2: (maximum spring deflection): $v_2 = 0$ $T_2 = 0$
Work
Friction Force **F**. We have

$$F = \mu_k N = \mu_k W = \mu_k mg = \mu_k(60 \text{ kg})(9.81 \text{ m/s}^2) = (588.6 \text{ N})\mu_k$$

The work of **F** is negative and equal to

$$(U_{1\to2})_f = -Fx = -(588.6 \text{ N})\mu_k(0.600 \text{ m} + 0.040 \text{ m}) = -(377 \text{ J})\mu_k$$

Spring Force **P**. The variable force **P** exerted by the spring does an amount of negative work equal to the area under the force-deflection curve of the spring force. We have

$$P_{min} = kx_0 = (20 \text{ kN/m})(120 \text{ mm}) = (20\,000 \text{ N/m})(0.120 \text{ m}) = 2400 \text{ N}$$
$$P_{max} = P_{min} + k\,\Delta x = 2400 \text{ N} + (20 \text{ kN/m})(40 \text{ mm}) = 3200 \text{ N}$$
$$(U_{1\to2})_e = -\tfrac{1}{2}(P_{min} + P_{max})\,\Delta x = -\tfrac{1}{2}(2400 \text{ N} + 3200 \text{ N})(0.040 \text{ m}) = -112.0 \text{ J}$$

The total work is thus

$$U_{1\to2} = (U_{1\to2})_f + (U_{1\to2})_e = -(377 \text{ J})\mu_k - 112.0 \text{ J}$$

Principle of Work and Energy

$$T_1 + U_{1\to2} = T_2: \qquad 187.5 \text{ J} - (377 \text{ J})\mu_k - 112.0 \text{ J} = 0 \quad \mu_k = 0.20 \blacktriangleleft$$

b. Motion from Position 2 to Position 3
Kinetic Energy.   *Position 2:* $v_2 = 0$ $T_2 = 0$
Position 3: $T_3 = \tfrac{1}{2}mv_3^2 = \tfrac{1}{2}(60 \text{ kg})v_3^2$

Work.   Since the distances involved are the same, the numerical values of the work of the friction force **F** and of the spring force **P** are the same as above. However, while the work of **F** is still negative, the work of **P** is now positive.

$$U_{2\to3} = -(377 \text{ J})\mu_k + 112.0 \text{ J} = -75.5 \text{ J} + 112.0 \text{ J} = +36.5 \text{ J}$$

Principle of Work and Energy

$$T_2 + U_{2\to3} = T_3: \qquad 0 + 36.5 \text{ J} = \tfrac{1}{2}(60 \text{ kg})v_3^2$$
$$v_3 = 1.103 \text{ m/s} \qquad \mathbf{v}_3 = 1.103 \text{ m/s} \leftarrow \blacktriangleleft$$

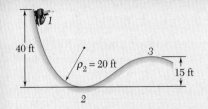

SAMPLE PROBLEM 13.4

A 2000-lb car starts from rest at point 1 and moves without friction down the track shown. (a) Determine the force exerted by the track on the car at point 2, where the radius of curvature of the track is 20 ft. (b) Determine the minimum safe value of the radius of curvature at point 3.

SOLUTION

a. Force Exerted by the Track at Point 2. The principle of work and energy is used to determine the velocity of the car as it passes through point 2.

Kinetic Energy. $\quad T_1 = 0 \qquad T_2 = \frac{1}{2}mv_2^2 = \frac{1}{2}\frac{W}{g}v_2^2$

Work. The only force which does work is the weight $\mathbf{W}$. Since the vertical displacement from point 1 to point 2 is 40 ft downward, the work of the weight is

$$U_{1\to2} = +W(40\text{ ft})$$

Principle of Work and Energy

$$T_1 + U_{1\to2} = T_2 \qquad 0 + W(40\text{ ft}) = \frac{1}{2}\frac{W}{g}v_2^2$$

$$v_2^2 = 80g = 80(32.2) \qquad v_2 = 50.8\text{ ft/s}$$

Newton's Second Law at Point 2. The acceleration $\mathbf{a}_n$ of the car at point 2 has a magnitude $a_n = v_2^2/\rho$ and is directed upward. Since the external forces acting on the car are $\mathbf{W}$ and $\mathbf{N}$, we write

$$+\uparrow \Sigma F_n = ma_n: \qquad -W + N = ma_n$$
$$= \frac{W}{g}\frac{v_2^2}{\rho}$$
$$= \frac{W}{g}\frac{80g}{20}$$

$$N = 5W \qquad N = 10{,}000\text{ lb} \uparrow \quad \blacktriangleleft$$

b. Minimum Value of ρ at Point 3. *Principle of Work and Energy.* Applying the principle of work and energy between point 1 and point 3, we obtain

$$T_1 + U_{1\to3} = T_3 \qquad 0 + W(25\text{ ft}) = \frac{1}{2}\frac{W}{g}v_3^2$$

$$v_3^2 = 50g = 50(32.2) \qquad v_3 = 40.1\text{ ft/s}$$

Newton's Second Law at Point 3. The minimum safe value of ρ occurs when $\mathbf{N} = 0$. In this case, the acceleration $\mathbf{a}_n$, of magnitude $a_n = v_3^2/\rho$, is directed downward, and we write

$$+\downarrow \Sigma F_n = ma_n: \qquad W = \frac{W}{g}\frac{v_3^2}{\rho}$$
$$= \frac{W}{g}\frac{50g}{\rho}$$

$$\rho = 50\text{ ft} \quad \blacktriangleleft$$

SAMPLE PROBLEM 13.5

The dumbwaiter D and its load have a combined weight of 600 lb, while the counterweight C weighs 800 lb. Determine the power delivered by the electric motor M when the dumbwaiter (a) is moving up at a constant speed of 8 ft/s, (b) has an instantaneous velocity of 8 ft/s and an acceleration of 2.5 ft/s², both directed upward.

SOLUTION

Since the force $\mathbf{F}$ exerted by the motor cable has the same direction as the velocity $\mathbf{v}_D$ of the dumbwaiter, the power is equal to Fv_D, where $v_D = 8$ ft/s. To obtain the power, we must first determine $\mathbf{F}$ in each of the two given situations.

a. Uniform Motion. We have $\mathbf{a}_C = \mathbf{a}_D = 0$; both bodies are in equilibrium.

Free Body C: $+\uparrow\Sigma F_y = 0$: $2T - 800\text{ lb} = 0$ $T = 400\text{ lb}$
Free Body D: $+\uparrow\Sigma F_y = 0$: $F + T - 600\text{ lb} = 0$

$$F = 600\text{ lb} - T = 600\text{ lb} - 400\text{ lb} = 200\text{ lb}$$
$$Fv_D = (200\text{ lb})(8\text{ ft/s}) = 1600\text{ ft}\cdot\text{lb/s}$$
$$\text{Power} = (1600\,\text{ft}\cdot\text{lb/s})\frac{1\,\text{hp}}{550\,\text{ft}\cdot\text{lb/s}} = 2.91\,\text{hp} \quad \blacktriangleleft$$

b. Accelerated Motion. We have

$$\mathbf{a}_D = 2.5\text{ ft/s}^2\uparrow \qquad \mathbf{a}_C = -\tfrac{1}{2}\mathbf{a}_D = 1.25\text{ ft/s}^2\downarrow$$

The equations of motion are

Free Body C: $+\downarrow\Sigma F_y = m_C a_C$: $800 - 2T = \dfrac{800}{32.2}(1.25)$ $T = 384.5\text{ lb}$

Free Body D: $+\uparrow\Sigma F_y = m_D a_D$: $F + T - 600 = \dfrac{600}{32.2}(2.5)$

$$F + 384.5 - 600 = 46.6 \qquad F = 262.1\text{ lb}$$
$$Fv_D = (262.1\text{ lb})(8\text{ ft/s}) = 2097\text{ ft}\cdot\text{lb/s}$$
$$\text{Power} = (2097\text{ ft}\cdot\text{lb/s})\frac{1\,\text{hp}}{550\,\text{ft}\cdot\text{lb/s}} = 3.81\,\text{hp} \quad \blacktriangleleft$$

SOLVING PROBLEMS
ON YOUR OWN

$\mathbf{I}$n the preceding chapter, you solved problems dealing with the motion of a particle by using the fundamental equation $\mathbf{F} = m\mathbf{a}$ to determine the acceleration $\mathbf{a}$. By applying the principles of kinematics you were then able to determine from $\mathbf{a}$ the velocity and displacement of the particle at any time. In this lesson we combined $\mathbf{F} = m\mathbf{a}$ and the principles of kinematics to obtain an additional method of analysis called the *method of work and energy*. This eliminates the need to calculate the acceleration and will enable you to relate the velocities of the particle at two points along its path of motion. To solve a problem by the method of work and energy you will follow these steps:

1. Computing the work of each of the forces. The work $U_{1\to2}$ of a given force $\mathbf{F}$ during the finite displacement of the particle from A_1 to A_2 is defined as

$$U_{1\to2} = \int \mathbf{F} \cdot d\mathbf{r} \qquad \text{or} \qquad U_{1\to2} = \int (F \cos \alpha)\, ds \quad \textbf{(13.2, 13.2')}$$

where α is the angle between $\mathbf{F}$ and the displacement $d\mathbf{r}$. The work $U_{1\to2}$ is a scalar quantity and is expressed in ft $\cdot$ lb or in $\cdot$ lb in the U.S. customary system of units and in N $\cdot$ m or joules (J) in the SI system of units. Note that the work done is zero for a force perpendicular to the displacement ($\alpha = 90°$). Negative work is done for $90° < \alpha < 180°$ and in particular for a friction force, which is always opposite in direction to the displacement ($\alpha = 180°$).

The work $U_{1\to2}$ can be easily evaluated in the following cases that you will encounter:

a. Work of a constant force in rectilinear motion

$$U_{1\to2} = (F \cos \alpha)\, \Delta x \qquad\qquad \textbf{(13.3)}$$

where α = angle the force forms with the direction of motion
Δx = displacement from A_1 to A_2 (Fig. 13.3)

b. Work of the force of gravity

$$U_{1\to2} = -W\, \Delta y \qquad\qquad \textbf{(13.4')}$$

where Δy is the vertical displacement of the center of gravity of the body of weight W. Note that the work is positive when Δy is negative, that is, when the body moves down (Fig. 13.4).

c. Work of the force exerted by a spring

$$U_{1\to2} = \tfrac{1}{2}kx_1^2 - \tfrac{1}{2}kx_2^2 \qquad\qquad \textbf{(13.6)}$$

where k is the spring constant and x_1 and x_2 are the elongations of the spring corresponding to the positions A_1 and A_2 (Fig. 13.5).

(continued)

d. Work of a gravitational force

$$U_{1\to2} = \frac{GMm}{r_2} - \frac{GMm}{r_1} \qquad (13.7)$$

for a displacement of the body from $A_1(r = r_1)$ to $A_2(r = r_2)$ (Fig. 13.6).

2. Calculate the kinetic energy at A_1 and A_2. The kinetic energy T is

$$T = \tfrac{1}{2}mv^2 \qquad (13.9)$$

where m is the mass of the particle and v is the magnitude of its velocity. The units of kinetic energy are the same as the units of work, that is, ft · lb or in · lb if U.S. customary units are used and N · m or joules (J) if SI units are used.

3. Substitute the values for the work done $U_{1\to2}$ and the kinetic energies T_1 and T_2 into the equation

$$T_1 + U_{1\to2} = T_2 \qquad (13.11)$$

You will now have *one equation* which you can solve for *one unknown*. Note that this equation does not yield the time of travel or the acceleration directly. However, if you know the radius of curvature ρ of the path of the particle at a point where you have obtained the velocity v, you can express the normal component of the acceleration as $a_n = v^2/\rho$ and obtain the normal component of the force exerted on the particle by writing $F_n = mv^2/\rho$.

4. Power was introduced in this lesson as the time rate at which work is done, $P = dU/dt$. Power is measured in ft · lb/s or *horsepower* (hp) in U.S. customary units and in J/s or *watts* (W) in the SI system of units. To calculate the power, you can use the equivalent formula,

$$P = \mathbf{F} \cdot \mathbf{v} \qquad (13.13)$$

where $\mathbf{F}$ and $\mathbf{v}$ denote the force and the velocity, respectively, at a given time [Sample Prob. 13.5]. In some problems [see, e.g., Prob. 13.47], you will be asked for the *average power*, which can be obtained by dividing the total work by the time interval during which the work is done.

PROBLEMS

CONCEPT QUESTION

13.CQ1 Block A is traveling with a speed v_0 on a smooth surface when the surface suddenly becomes rough with a coefficient of friction of μ causing the block to stop after a distance d. If block A were traveling twice as fast, that is, at a speed $2v_0$, how far will it travel on the rough surface before stopping?

    a. $d/2$

    b. d

    c. $\sqrt{2}d$

    d. $2d$

    e. $4d$

Fig. P13.CQ1

END-OF-SECTION PROBLEMS

13.1 A 400-kg satellite was placed in a circular orbit 1500 km above the surface of the earth. At this elevation the acceleration of gravity is 6.43 m/s². Determine the kinetic energy of the satellite, knowing that its orbital speed is 25.6×10^3 km/h.

13.2 A 1-lb stone is dropped down the "bottomless pit" at Carlsbad Caverns and strikes the ground with a speed of 95 ft/s. Neglecting air resistance, (*a*) determine the kinetic energy of the stone as it strikes the ground and the height h from which it was dropped. (*b*) Solve part *a* assuming that the same stone is dropped down a hole on the moon. (Acceleration of gravity on the moon = 5.31 ft/s².)

Fig. P13.2

13.3 A baseball player hits a 5.1-oz baseball with an initial velocity of 140 ft/s at an angle of 40° with the horizontal as shown. Determine (*a*) the kinetic energy of the ball immediately after it is hit, (*b*) the kinetic energy of the ball when it reaches its maximum height, (*c*) the maximum height above the ground reached by the ball.

Fig. P13.3

13.4 A 500-kg communications satellite is in a circular geosynchronous orbit and completes one revolution about the earth in 23 h and 56 min at an altitude of 35 800 km above the surface of the earth. Knowing that the radius of the earth is 6370 km, determine the kinetic energy of the satellite.

13.5 In an ore-mixing operation, a bucket full of ore is suspended from a traveling crane which moves along a stationary bridge. The bucket is to swing no more than 10 ft horizontally when the crane is brought to a sudden stop. Determine the maximum allowable speed v of the crane.

13.6 In an ore-mixing operation, a bucket full of ore is suspended from a traveling crane which moves along a stationary bridge. The crane is traveling at a speed of 10 ft/s when it is brought to a sudden stop. Determine the maximum horizontal distance through which the bucket will swing.

Fig. P13.5 and P13.6

Fig. P13.8

Fig. P13.9

Fig. P13.11 and P13.12

Fig. P13.13 and P13.14

13.7 Determine the maximum theoretical speed that may be achieved over a distance of 110 m by a car starting from rest assuming there is no slipping. The coefficient of static friction between the tires and pavement is 0.75, and 60 percent of the weight of the car is distributed over its front wheels and 40 percent over its rear wheels. Assume (*a*) front-wheel drive, (*b*) rear-wheel drive.

13.8 Skid marks on a drag racetrack indicate that the rear (drive) wheels of a car slip for the first 20 m of the 400-m track. (*a*) Knowing that the coefficient of kinetic friction is 0.60, determine the speed of the car at the end of the first 20-m portion of the track if it starts from rest and the front wheels are just off the ground. (*b*) What is the maximum theoretical speed of the car at the finish line if, after skidding for 20 m, it is driven without the wheels slipping for the remainder of the race? Assume that while the car is rolling without slipping, 60 percent of the weight of the car is on the rear wheels and the coefficient of static friction is 0.75. Ignore air resistance and rolling resistance.

13.9 A package is projected up a 15° incline at *A* with an initial velocity of 8 m/s. Knowing that the coefficient of kinetic friction between the package and the incline is 0.12, determine (*a*) the maximum distance *d* that the package will move up the incline, (*b*) the velocity of the package as it returns to its original position.

13.10 A 1.4-kg model rocket is launched vertically from rest with a constant thrust of 25 N until the rocket reaches an altitude of 15 m and the thrust ends. Neglecting air resistance, determine (*a*) the speed of the rocket when the thrust ends, (*b*) the maximum height reached by the rocket, (*c*) the speed of the rocket when it returns to the ground.

13.11 Packages are thrown down an incline at *A* with a velocity of 1 m/s. The packages slide along the surface *ABC* to a conveyor belt which moves with a velocity of 2 m/s. Knowing that $\mu_k = 0.25$ between the packages and the surface *ABC*, determine the distance *d* if the packages are to arrive at *C* with a velocity of 2 m/s.

13.12 Packages are thrown down an incline at *A* with a velocity of 1 m/s. The packages slide along the surface *ABC* to a conveyor belt which moves with a velocity of 2 m/s. Knowing that $d = 7.5$ m and $\mu_k = 0.25$ between the packages and all surfaces, determine (*a*) the speed of the package at *C*, (*b*) the distance a package will slide on the conveyor belt before it comes to rest relative to the belt.

13.13 Boxes are transported by a conveyor belt with a velocity $\mathbf{v}_0$ to a fixed incline at *A* where they slide and eventually fall off at *B*. Knowing that $\mu_k = 0.40$, determine the velocity of the conveyor belt if the boxes leave the incline at *B* with a velocity of 8 ft/s.

13.14 Boxes are transported by a conveyor belt with a velocity $\mathbf{v}_0$ to a fixed incline at *A* where they slide and eventually fall off at *B*. Knowing that $\mu_k = 0.40$, determine the velocity of the conveyor belt if the boxes are to have zero velocity at *B*.

13.15 A 1200-kg trailer is hitched to a 1400-kg car. The car and trailer are traveling at 72 km/h when the driver applies the brakes on both the car and the trailer. Knowing that the braking forces exerted on the car and the trailer are 5000 N and 4000 N, respectively, determine (*a*) the distance traveled by the car and trailer before they come to a stop, (*b*) the horizontal component of the force exerted by the trailer hitch on the car.

Fig. P13.15

13.16 A trailer truck enters a 2 percent uphill grade traveling at 72 km/h and reaches a speed of 108 km/h in 300 m. The cab has a mass of 1800 kg and the trailer 5400 kg. Determine (*a*) the average force at the wheels of the cab, (*b*) the average force in the coupling between the cab and the trailer.

Fig. P13.16

13.17 The subway train shown is traveling at a speed of 30 mi/h when the brakes are fully applied on the wheels of cars *B* and *C*, causing them to slide on the track, but are not applied on the wheels of car *A*. Knowing that the coefficient of kinetic friction is 0.35 between the wheels and the track, determine (*a*) the distance required to bring the train to a stop, (*b*) the force in each coupling.

Fig. P13.17 and P13.18

13.18 The subway train shown is traveling at a speed of 30 mi/h when the brakes are fully applied on the wheels of car *A*, causing it to slide on the track, but are not applied on the wheels of cars *B* or *C*. Knowing that the coefficient of kinetic friction is 0.35 between the wheels and the track, determine (*a*) the distance required to bring the train to a stop, (*b*) the force in each coupling.

13.19 Blocks *A* and *B* weigh 25 lb and 10 lb, respectively, and they are both at a height 6 ft above the ground when the system is released from rest. Just before hitting the ground block *A* is moving at a speed of 9 ft/s. Determine (*a*) the amount of energy dissipated in friction by the pulley, (*b*) the tension in each portion of the cord during the motion.

Fig. P13.19

13.20 The system shown is at rest when a constant 30-lb force is applied to collar B. (a) If the force acts through the entire motion, determine the speed of collar B as it strikes the support at C. (b) After what distance d should the 30-lb force be removed if the collar is to reach support C with zero velocity?

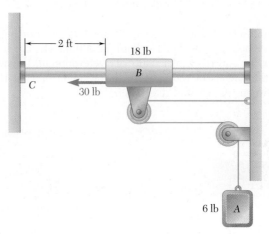

Fig. P13.20

13.21 Car B is towing car A at a constant speed of 10 m/s on an uphill grade when the brakes of car A are fully applied causing all four wheels to skid. The driver of car B does not change the throttle setting or change gears. The masses of the cars A and B are 1400 kg and 1200 kg, respectively, and the coefficient of kinetic friction is 0.8. Neglecting air resistance and rolling resistance, determine (a) the distance traveled by the cars before they come to a stop, (b) the tension in the cable.

Fig. P13.21

13.22 The system shown is at rest when a constant 250-N force is applied to block A. Neglecting the masses of the pulleys and the effect of friction in the pulleys and between block A and the horizontal surface, determine (a) the velocity of block B after block A has moved 2 m, (b) the tension in the cable.

13.23 The system shown is at rest when a constant 250-N force is applied to block A. Neglecting the masses of the pulleys and the effect of friction in the pulleys and assuming that the coefficients of friction between block A and the horizontal surface are $\mu_s = 0.25$ and $\mu_k = 0.20$, determine (a) the velocity of block B after block A has moved 2 m, (b) the tension in the cable.

Fig. P13.22 and P13.23

13.24 Two blocks A and B, of mass 4 kg and 5 kg, respectively, are connected by a cord which passes over pulleys as shown. A 3-kg collar C is placed on block A and the system is released from rest. After the blocks have moved 0.9 m, collar C is removed and blocks A and B continue to move. Determine the speed of block A just before it strikes the ground.

13.25 Four packages, each weighing 6 lb, are held in place by friction on a conveyor which is disengaged from its drive motor. When the system is released from rest, package *1* leaves the belt at A just as package *4* comes onto the inclined portion of the belt at B. Determine (*a*) the speed of package *2* as it leaves the belt at A, (*b*) the speed of package *3* as it leaves the belt at A. Neglect the mass of the belt and rollers.

Fig. P13.24

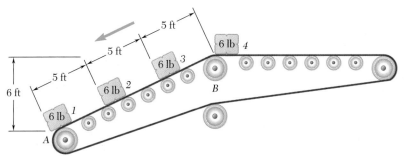

Fig. P13.25

13.26 A 3-kg block rests on top of a 2-kg block supported by but not attached to a spring of constant 40 N/m. The upper block is suddenly removed. Determine (*a*) the maximum speed reached by the 2-kg block, (*b*) the maximum height reached by the 2-kg block.

13.27 Solve Prob. 13.26, assuming that the 2-kg block is attached to the spring.

13.28 An 8-lb collar C slides on a horizontal rod between springs A and B. If the collar is pushed to the right until spring B is compressed 2 in. and released, determine the distance through which the collar will travel assuming (*a*) no friction between the collar and the rod, (*b*) a coefficient of friction $\mu_k = 0.35$.

Fig. P13.26

Fig. P13.28

13.29 A 6-lb block is attached to a cable and to a spring as shown. The constant of the spring is $k = 8$ lb/in. and the tension in the cable is 3 lb. If the cable is cut, determine (a) the maximum displacement of the block, (b) the maximum speed of the block.

13.30 A 10-kg block is attached to spring A and connected to spring B by a cord and pulley. The block is held in the position shown with both springs unstretched when the support is removed and the block is released with no initial velocity. Knowing that the constant of each spring is 2 kN/m, determine (a) the velocity of the block after it has moved down 50 mm, (b) the maximum velocity achieved by the block.

Fig. P13.29

Fig. P13.30

13.31 A 5-kg collar A is at rest on top of, but not attached to, a spring with stiffness $k_1 = 400$ N/m when a constant 150-N force is applied to the cable. Knowing A has a speed of 1 m/s when the upper spring is compressed 75 mm, determine the spring stiffness k_2. Ignore friction and the mass of the pulley.

Fig. P13.31

13.32 A piston of mass m and cross-sectional area A is in equilibrium under the pressure p at the center of a cylinder closed at both ends. Assuming that the piston is moved to the left a distance $a/2$ and released, and knowing that the pressure on each side of the piston varies inversely with the volume, determine the velocity of the piston as it again reaches the center of the cylinder. Neglect friction between the piston and the cylinder and express your answer in terms of m, a, p, and A.

Fig. P13.32

13.33 An uncontrolled automobile traveling at 65 mph strikes squarely a highway crash cushion of the type shown in which the automobile is brought to rest by successively crushing steel barrels. The magnitude F of the force required to crush the barrels is shown as a function of the distance x the automobile has moved into the cushion. Knowing that the weight of the automobile is 2250 lb and neglecting the effect of friction, determine (a) the distance the automobile will move into the cushion before it comes to rest, (b) the maximum deceleration of the automobile.

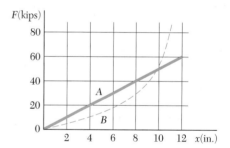

Fig. P13.33

13.34 Two types of energy-absorbing fenders designed to be used on a pier are statically loaded. The force-deflection curve for each type of fender is given in the graph. Determine the maximum deflection of each fender when a 90-ton ship moving at 1 mi/h strikes the fender and is brought to rest.

Fig. P13.34

13.35 Nonlinear springs are classified as hard or soft, depending upon the curvature of their force-deflection curve (see figure). If a delicate instrument having a mass of 5 kg is placed on a spring of length l so that its base is just touching the undeformed spring and then inadvertently released from that position, determine the maximum deflection x_m of the spring and the maximum force F_m exerted by the spring, assuming (a) a linear spring of constant $k = 3$ kN/m, (b) a hard, nonlinear spring, for which $F = (3 \text{ kN/m})(x + 160x^3)$.

Fig. P13.35

13.36 A rocket is fired vertically from the surface of the moon with a speed v_0. Derive a formula for the ratio h_n/h_u of heights reached with a speed v, if Newton's law of gravitation is used to calculate h_n and a uniform gravitational field is used to calculate h_u. Express your answer in terms of the acceleration of gravity g_m on the surface of the moon, the radius R_m of the moon, and the speeds v and v_0.

Fig. P13.38

Fig. P13.39 and P13.40

Fig. P13.41

13.37 Express the acceleration of gravity g_h at an altitude h above the surface of the earth in terms of the acceleration of gravity g_0 at the surface of the earth, the altitude h, and the radius R of the earth. Determine the percent error if the weight that an object has on the surface of earth is used as its weight at an altitude of (a) 1 km, (b) 1000 km.

13.38 A golf ball struck on earth rises to a maximum height of 60 m and hits the ground 230 m away. How high will the same golf ball travel on the moon if the magnitude and direction of its velocity are the same as they were on earth immediately after the ball was hit? Assume that the ball is hit and lands at the same elevation in both cases and that the effect of the atmosphere on the earth is neglected, so that the trajectory in both cases is a parabola. The acceleration of gravity on the moon is 0.165 times that on earth.

13.39 The sphere at A is given a downward velocity $\mathbf{v}_0$ of magnitude 5 m/s and swings in a vertical plane at the end of a rope of length $l = 2$ m attached to a support at O. Determine the angle θ at which the rope will break, knowing that it can withstand a maximum tension equal to twice the weight of the sphere.

13.40 The sphere at A is given a downward velocity $\mathbf{v}_0$ and swings in a vertical circle of radius l and center O. Determine the smallest velocity $\mathbf{v}_0$ for which the sphere will reach point B as it swings about point O (a) if AO is a rope, (b) if AO is a slender rod of negligible mass.

13.41 A small sphere B of weight W is released from rest in the position shown and swings freely in a vertical plane, first about O and then about the peg A after the cord comes in contact with the peg. Determine the tension in the cord (a) just before the sphere comes in contact with the peg, (b) just after it comes in contact with the peg.

13.42 A roller coaster starts from rest at A, rolls down the track to B, describes a circular loop of 40-ft diameter, and moves up and down past point E. Knowing that $h = 60$ ft and assuming no energy loss due to friction, determine (a) the force exerted by his seat on a 160-lb rider at B and D, (b) the minimum value of the radius of curvature at E if the roller coaster is not to leave the track at that point.

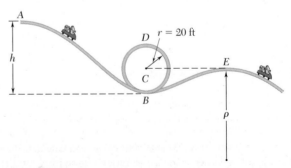

Fig. P13.42

13.43 In Prob. 13.42, determine the range of values of h for which the roller coaster will not leave the track at D or E, knowing that the radius of curvature at E is $\rho = 75$ ft. Assume no energy loss due to friction.

13.44 A small block slides at a speed v on a horizontal surface. Knowing that $h = 0.9$ m, determine the required speed of the block if it is to leave the cylindrical surface BCD when $\theta = 30°$.

13.45 A small block slides at a speed $v = 8$ ft/s on a horizontal surface at a height $h = 3$ ft above the ground. Determine (a) the angle θ at which it will leave the cylindrical surface BCD, (b) the distance x at which it will hit the ground. Neglect friction and air resistance.

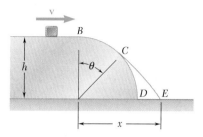

Fig. P13.44 and P13.45

13.46 A chair-lift is designed to transport 1000 skiers per hour from the base A to the summit B. The average mass of a skier is 70 kg and the average speed of the lift is 75 m/min. Determine (a) the average power required, (b) the required capacity of the motor if the mechanical efficiency is 85 percent and if a 300-percent overload is to be allowed.

Fig. P13.46

13.47 It takes 15 s to raise a 1200-kg car and the supporting 300-kg hydraulic car-lift platform to a height of 2.8 m. Determine (a) the average output power delivered by the hydraulic pump to lift the system, (b) the average electric power required, knowing that the overall conversion efficiency from electric to mechanical power for the system is 82 percent.

13.48 The velocity of the lift of Prob. 13.47 increases uniformly from zero to its maximum value at mid-height in 7.5 s and then decreases uniformly to zero in 7.5 s. Knowing that the peak power output of the hydraulic pump is 6 kW when the velocity is maximum, determine the maximum lift force provided by the pump.

Fig. P13.47

13.49 (a) A 120-lb woman rides a 15-lb bicycle up a 3-percent slope at a constant speed of 5 ft/s. How much power must be developed by the woman? (b) A 180-lb man on an 18-lb bicycle starts down the same slope and maintains a constant speed of 20 ft/s by braking. How much power is dissipated by the brakes? Ignore air resistance and rolling resistance.

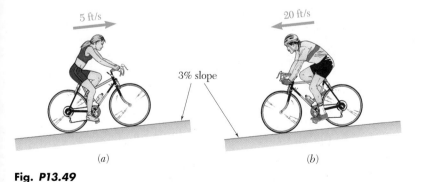

Fig. P13.49

13.50 A power specification formula is to be derived for electric motors which drive conveyor belts moving solid material at different rates to different heights and distances. Denoting the efficiency of a motor by η and neglecting the power needed to drive the belt itself, derive a formula (a) in the SI system of units for the power P in kW, in terms of the mass flow rate m in kg/h, the height b and horizontal distance l in meters and (b) in U.S. customary units, for the power in hp, in terms of the material flow rate w in tons/h, and the height b and horizontal distance l in feet.

Fig. P13.50

Fig. P13.51

13.51 In an automobile drag race, the rear (drive) wheels of a 1000-kg car skid for the first 20 m and roll with sliding impending during the remaining 380 m. The front wheels of the car are just off the ground for the first 20 m, and for the remainder of the race 80 percent of the weight is on the rear wheels. Knowing that the coefficients of friction are $\mu_s = 0.90$ and $\mu_k = 0.68$, determine the power developed by the car at the drive wheels (a) at the end of the 20-m portion of the race, (b) at the end of the race. Give your answer in kW and in hp. Ignore the effect of air resistance and rolling friction.

13.52 The frictional resistance of a ship is known to vary directly as the 1.75 power of the speed v of the ship. A single tugboat at full power can tow the ship at a constant speed of 4.5 km/h by exerting a constant force of 300 kN. Determine (a) the power developed by the tugboat, (b) the maximum speed at which two tugboats, capable of delivering the same power, can tow the ship.

13.53 A train of total mass equal to 500 Mg starts from rest and accelerates uniformly to a speed of 90 km/h in 50 s. After reaching this speed, the train travels with a constant velocity. The track is horizontal and axle friction and rolling resistance result in a total force of 15 kN in a direction opposite to the direction of motion. Determine the power required as a function of time.

Fig. P13.54

13.54 The elevator E has a weight of 6600 lb when fully loaded and is connected as shown to a counterweight W of weight of 2200 lb. Determine the power in hp delivered by the motor (a) when the elevator is moving down at a constant speed of 1 ft/s, (b) when it has an upward velocity of 1 ft/s and a deceleration of 0.18 ft/s^2.

13.6 POTENTIAL ENERGY†

Let us consider again a body of weight **W** which moves along a curved path from a point A_1 of elevation y_1 to a point A_2 of elevation y_2 (Fig. 13.4). We recall from Sec. 13.2 that the work of the force of gravity **W** during this displacement is

$$U_{1 \to 2} = Wy_1 - Wy_2 \qquad (13.4)$$

The work of **W** may thus be obtained by subtracting the value of the function Wy corresponding to the second position of the body from its value corresponding to the first position. The work of **W** is independent of the actual path followed; it depends only upon the initial and final values of the function Wy. This function is called the *potential energy* of the body with respect to the *force of gravity* **W** and is denoted by V_g. We write

$$U_{1 \to 2} = (V_g)_1 - (V_g)_2 \qquad \text{with } V_g = Wy \qquad (13.16)$$

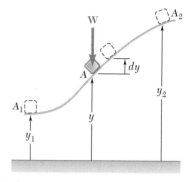

Fig. 13.4 (repeated)

We note that if $(V_g)_2 > (V_g)_1$, that is, *if the potential energy increases during the displacement* (as in the case considered here), *the work $U_{1 \to 2}$ is negative.* If, on the other hand, the work of **W** is positive, the potential energy decreases. Therefore, the potential energy V_g of the body provides a measure of the work which can be done by its weight **W**. Since only the *change* in potential energy, and not the actual value of V_g, is involved in formula (13.16), an arbitrary constant can be added to the expression obtained for V_g. In other words, the level, or datum, from which the elevation y is measured can be chosen arbitrarily. Note that potential energy is expressed in the same units as work, i.e., in joules if SI units are used and in ft · lb or in · lb if U.S. customary units are used.

It should be noted that the expression just obtained for the potential energy of a body with respect to gravity is valid only as long as the weight **W** of the body can be assumed to remain constant, i.e., as long as the displacements of the body are small compared with the radius of the earth. In the case of a space vehicle, however, we should take into consideration the variation of the force of gravity with the distance r from the center of the earth. Using the expression obtained in Sec. 13.2 for the work of a gravitational force, we write (Fig. 13.6)

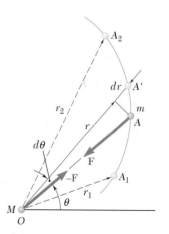

Fig. 13.6 (repeated)

$$U_{1 \to 2} = \frac{GMm}{r_2} - \frac{GMm}{r_1} \qquad (13.7)$$

The work of the force of gravity can therefore be obtained by subtracting the value of the function $-GMm/r$ corresponding to the second position of the body from its value corresponding to the first position. Thus, the expression which should be used for the potential energy V_g when the variation in the force of gravity cannot be neglected is

$$V_g = -\frac{GMm}{r} \qquad (13.17)$$

†Some of the material in this section has already been considered in Sec. 10.7.

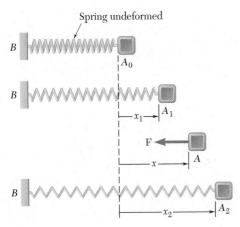

Fig. 13.5 *(repeated)*

Taking the first of the relations (12.29) into account, we write V_g in the alternative form

$$V_g = -\frac{WR^2}{r} \qquad (13.17')$$

where R is the radius of the earth and W is the value of the weight of the body at the surface of the earth. When either of the relations (13.17) or (13.17') is used to express V_g, the distance r should, of course, be measured from the center of the earth.† Note that V_g is always negative and that it approaches zero for very large values of r.

Consider now a body attached to a spring and moving from a position A_1, corresponding to a deflection x_1 of the spring, to a position A_2, corresponding to a deflection x_2 of the spring (Fig. 13.5). We recall from Sec. 13.2 that the work of the force $\mathbf{F}$ exerted by the spring on the body is

$$U_{1 \to 2} = \tfrac{1}{2}kx_1^2 - \tfrac{1}{2}kx_2^2 \qquad (13.6)$$

The work of the elastic force is thus obtained by subtracting the value of the function $\tfrac{1}{2}kx^2$ corresponding to the second position of the body from its value corresponding to the first position. This function is denoted by V_e and is called the *potential energy* of the body with respect to the *elastic force* $\mathbf{F}$. We write

$$U_{1 \to 2} = (V_e)_1 - (V_e)_2 \qquad \text{with } V_e = \tfrac{1}{2}kx^2 \qquad (13.18)$$

and observe that during the displacement considered, the work of the force $\mathbf{F}$ exerted by the spring on the body is negative and the potential energy V_e increases. You should note that the expression obtained for V_e is valid only if the deflection of the spring is measured from its undeformed position. On the other hand, formula (13.18) can be used even when the spring is rotated about its fixed end (Fig. 13.10a). The work of the elastic force depends only upon the initial and final deflections of the spring (Fig. 13.10b).

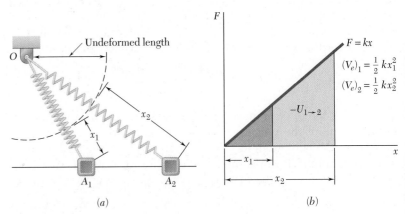

Fig. 13.10

† The expressions given for V_g in (13.17) and (13.17') are valid only when $r \geq R$, that is, when the body considered is above the surface of the earth.

The concept of potential energy can be used when forces other than gravity forces and elastic forces are involved. Indeed, it remains valid as long as the work of the force considered is independent of the path followed by its point of application as this point moves from a given position A_1 to a given position A_2. Such forces are said to be *conservative forces;* the general properties of conservative forces are studied in the following section.

*13.7 CONSERVATIVE FORCES

As indicated in the preceding section, a force $\mathbf{F}$ acting on a particle A is said to be conservative *if its work $U_{1\rightarrow2}$ is independent of the path followed by the particle A as it moves from A_1 to A_2* (Fig. 13.11a). We can then write

$$U_{1\rightarrow2} = V(x_1, y_1, z_1) - V(x_2, y_2, z_2) \qquad (13.19)$$

or, for short,

$$U_{1\rightarrow2} = V_1 - V_2 \qquad (13.19')$$

The function $V(x, y, z)$ is called the potential energy, or *potential function,* of $\mathbf{F}$.

We note that if A_2 is chosen to coincide with A_1, that is, if the particle describes a closed path (Fig. 13.11b), we have $V_1 = V_2$ and the work is zero. Thus for any conservative force $\mathbf{F}$ we can write

$$\oint \mathbf{F} \cdot d\mathbf{r} = 0 \qquad (13.20)$$

where the circle on the integral sign indicates that the path is closed.

Let us now apply (13.19) between two neighboring points $A(x, y, z)$ and $A'(x + dx, y + dy, z + dz)$. The elementary work dU corresponding to the displacement $d\mathbf{r}$ from A to A' is

$$dU = V(x, y, z) - V(x + dx, y + dy, z + dz)$$

or

$$dU = -dV(x, y, z) \qquad (13.21)$$

Thus, the elementary work of a conservative force is an *exact differential.*

Substituting for dU in (13.21) the expression obtained in (13.1″) and recalling the definition of the differential of a function of several variables, we write

$$F_x\,dx + F_y\,dy + F_z\,dz = -\left(\frac{\partial V}{\partial x}dx + \frac{\partial V}{\partial y}dy + \frac{\partial V}{\partial z}dz\right)$$

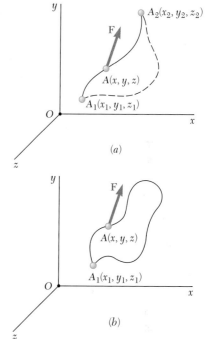

Fig. 13.11

from which it follows that

$$F_x = -\frac{\partial V}{\partial x} \qquad F_y = -\frac{\partial V}{\partial y} \qquad F_z = -\frac{\partial V}{\partial z} \qquad (13.22)$$

It is clear that the components of **F** must be functions of the coordinates x, y, and z. Thus, a *necessary* condition for a conservative force is that it depend only upon the position of its point of application. The relations (13.22) can be expressed more concisely if we write

$$\mathbf{F} = F_x\mathbf{i} + F_y\mathbf{j} + F_z\mathbf{k} = -\left(\frac{\partial V}{\partial x}\mathbf{i} + \frac{\partial V}{\partial y}\mathbf{j} + \frac{\partial V}{\partial z}\mathbf{k}\right)$$

The vector in parentheses is known as the *gradient of the scalar function* V and is denoted by **grad** V. We thus write for any conservative force

$$\mathbf{F} = -\mathbf{grad}\ V \qquad (13.23)$$

The relations (13.19) to (13.23) were shown to be satisfied by any conservative force. It can also be shown that if a force **F** satisfies one of these relations, **F** must be a conservative force.

13.8 CONSERVATION OF ENERGY

We saw in the preceding two sections that the work of a conservative force, such as the weight of a particle or the force exerted by a spring, can be expressed as a change in potential energy. When a particle moves under the action of conservative forces, the principle of work and energy stated in Sec. 13.3 can be expressed in a modified form. Substituting for $U_{1\rightarrow2}$ from (13.19′) into (13.10), we write

$$V_1 - V_2 = T_2 - T_1$$

$$T_1 + V_1 = T_2 + V_2 \qquad (13.24)$$

Formula (13.24) indicates that when a particle moves under the action of conservative forces, *the sum of the kinetic energy and of the potential energy of the particle remains constant.* The sum $T + V$ is called the *total mechanical energy* of the particle and is denoted by E.

Consider, for example, the pendulum analyzed in Sec. 13.4, which is released with no velocity from A_1 and allowed to swing in a vertical plane (Fig. 13.12). Measuring the potential energy from the level of A_2, we have, at A_1,

$$T_1 = 0 \qquad V_1 = Wl \qquad T_1 + V_1 = Wl$$

Recalling that at A_2 the speed of the pendulum is $v_2 = \sqrt{2gl}$, we have

$$T_2 = \tfrac{1}{2}mv_2^2 = \frac{1}{2}\frac{W}{g}(2gl) = Wl \qquad V_2 = 0$$

$$T_2 + V_2 = Wl$$

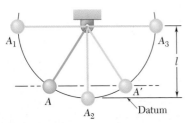

Fig. 13.12

We thus check that the total mechanical energy $E = T + V$ of the pendulum is the same at A_1 and A_2. Whereas the energy is entirely potential at A_1, it becomes entirely kinetic at A_2, and as the pendulum keeps swinging to the right, the kinetic energy is transformed back into potential energy. At A_3, $T_3 = 0$ and $V_3 = Wl$.

Since the total mechanical energy of the pendulum remains constant and since its potential energy depends only upon its elevation, the kinetic energy of the pendulum will have the same value at any two points located on the same level. Thus, the speed of the pendulum is the same at A and at A' (Fig. 13.12). This result can be extended to the case of a particle moving along any given path, regardless of the shape of the path, as long as the only forces acting on the particle are its weight and the normal reaction of the path. The particle of Fig. 13.13, for example, which slides in a vertical plane along a frictionless track, will have the same speed at A, A', and A''.

Fig. 13.13

While the weight of a particle and the force exerted by a spring are conservative forces, *friction forces are nonconservative forces*. In other words, *the work of a friction force cannot be expressed as a change in potential energy*. The work of a friction force depends upon the path followed by its point of application; and while the work $U_{1\to2}$ defined by (13.19) is positive or negative according to the sense of motion, *the work of a friction force*, as we noted in Sec. 13.4, *is always negative*. It follows that when a mechanical system involves friction, its total mechanical energy does not remain constant but decreases. The energy of the system, however, is not lost; it is transformed into heat, and the sum of the *mechanical energy* and of the *thermal energy* of the system remains constant.

Other forms of energy can also be involved in a system. For instance, a generator converts mechanical energy into *electric energy;* a gasoline engine converts *chemical energy* into mechanical energy; a nuclear reactor converts *mass* into thermal energy. If all forms of energy are considered, the energy of any system can be considered as constant and the principle of conservation of energy remains valid under all conditions.

13.9 MOTION UNDER A CONSERVATIVE CENTRAL FORCE. APPLICATION TO SPACE MECHANICS

We saw in Sec. 12.9 that when a particle P moves under a central force $\mathbf{F}$, the angular momentum $\mathbf{H}_O$ of the particle about the center of force O is constant. If the force $\mathbf{F}$ is also conservative, there exists a potential energy V associated with $\mathbf{F}$, and the total energy $E = T + V$ of the particle is constant (Sec. 13.8). Thus, when a particle moves under a conservative central force, both the principle of conservation of angular momentum and the principle of conservation of energy can be used to study its motion.

Consider, for example, a space vehicle of mass m moving under the earth's gravitational force. Let us assume that it begins its free flight at point P_0 at a distance r_0 from the center of the earth, with a velocity $\mathbf{v}_0$ forming an angle ϕ_0 with the radius vector OP_0

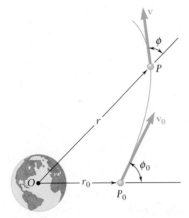

Fig. 13.14

(Fig. 13.14). Let P be a point of the trajectory described by the vehicle; we denote by r the distance from O to P, by $\mathbf{v}$ the velocity of the vehicle at P, and by ϕ the angle formed by $\mathbf{v}$ and the radius vector OP. Applying the principle of conservation of angular momentum about O between P_0 and P (Sec. 12.9), we write

$$r_0 m v_0 \sin \phi_0 = r m v \sin \phi \qquad (13.25)$$

Recalling the expression (13.17) obtained for the potential energy due to a gravitational force, we apply the principle of conservation of energy between P_0 and P and write

$$T_0 + V_0 = T + V$$
$$\tfrac{1}{2}mv_0^2 - \frac{GMm}{r_0} = \tfrac{1}{2}mv^2 - \frac{GMm}{r} \qquad (13.26)$$

where M is the mass of the earth.

Equation (13.26) can be solved for the magnitude v of the velocity of the vehicle at P when the distance r from O to P is known; Eq. (13.25) can then be used to determine the angle ϕ that the velocity forms with the radius vector OP.

Equations (13.25) and (13.26) can also be used to determine the maximum and minimum values of r in the case of a satellite launched from P_0 in a direction forming an angle ϕ_0 with the vertical OP_0 (Fig. 13.15). The desired values of r are obtained by making $\phi = 90°$ in (13.25) and eliminating v between Eqs. (13.25) and (13.26).

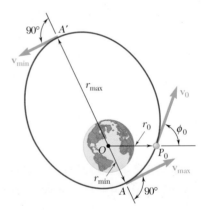

Fig. 13.15

It should be noted that the application of the principles of conservation of energy and of conservation of angular momentum leads to a more fundamental formulation of the problems of space mechanics than does the method indicated in Sec. 12.12. In all cases involving oblique launchings, it will also result in much simpler computations. And while the method of Sec. 12.12 must be used when the actual trajectory or the periodic time of a space vehicle is to be determined, the calculations will be simplified if the conservation principles are first used to compute the maximum and minimum values of the radius vector r.

SAMPLE PROBLEM 13.6

A 20-lb collar slides without friction along a vertical rod as shown. The spring attached to the collar has an undeformed length of 4 in. and a constant of 3 lb/in. If the collar is released from rest in position 1, determine its velocity after it has moved 6 in. to position 2.

SOLUTION

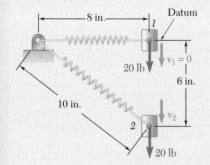

Position 1. *Potential Energy.* The elongation of the spring is

$$x_1 = 8 \text{ in.} - 4 \text{ in.} = 4 \text{ in.}$$

and we have

$$V_e = \tfrac{1}{2}kx_1^2 = \tfrac{1}{2}(3 \text{ lb/in.})(4 \text{ in.})^2 = 24 \text{ in} \cdot \text{lb}$$

Choosing the datum as shown, we have $V_g = 0$. Therefore,

$$V_1 = V_e + V_g = 24 \text{ in} \cdot \text{lb} = 2 \text{ ft} \cdot \text{lb}$$

Kinetic Energy. Since the velocity in position 1 is zero, $T_1 = 0$.

Position 2. *Potential Energy.* The elongation of the spring is

$$x_2 = 10 \text{ in.} - 4 \text{ in.} = 6 \text{ in.}$$

and we have

$$V_e = \tfrac{1}{2}kx_2^2 = \tfrac{1}{2}(3 \text{ lb/in.})(6 \text{ in.})^2 = 54 \text{ in} \cdot \text{lb}$$
$$V_g = Wy = (20 \text{ lb})(-6 \text{ in.}) = -120 \text{ in} \cdot \text{lb}$$

Therefore,

$$V_2 = V_e + V_g = 54 - 120 = -66 \text{ in} \cdot \text{lb}$$
$$= -5.5 \text{ ft} \cdot \text{lb}$$

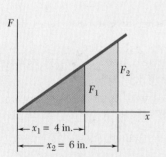

Kinetic Energy

$$T_2 = \tfrac{1}{2}mv_2^2 = \frac{1}{2}\frac{20}{32.2}v_2^2 = 0.311v_2^2$$

Conservation of Energy. Applying the principle of conservation of energy between positions 1 and 2, we write

$$T_1 + V_1 = T_2 + V_2$$
$$0 + 2 \text{ ft} \cdot \text{lb} = 0.311v_2^2 - 5.5 \text{ ft} \cdot \text{lb}$$
$$v_2 = \pm 4.91 \text{ ft/s}$$

$$\mathbf{v}_2 = 4.91 \text{ ft/s} \downarrow \ \blacktriangleleft$$

SAMPLE PROBLEM 13.7

The 0.5-lb pellet is pushed against the spring at A and released from rest. Neglecting friction, determine the smallest deflection of the spring for which the pellet will travel around the loop ABCDE and remain at all times in contact with the loop.

SOLUTION

Required Speed at Point D. As the pellet passes through the highest point D, its potential energy with respect to gravity is maximum and, thus, its kinetic energy and speed are minimum. Since the pellet must remain in contact with the loop, the force **N** exerted on the pellet by the loop must be equal to or greater than zero. Setting **N** = 0, we compute the smallest possible speed v_D.

$$+\downarrow \Sigma F_n = ma_n: \qquad W = ma_n \qquad mg = ma_n \qquad a_n = g$$

$$a_n = \frac{v_D^2}{r}: \qquad v_D^2 = ra_n = rg = (2 \text{ ft})(32.2 \text{ ft/s}^2) = 64.4 \text{ ft}^2/\text{s}^2$$

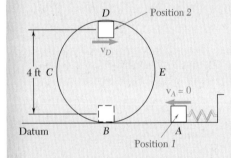

Position 1. Potential Energy. Denoting by x the deflection of the spring and noting that $k = 3$ lb/in. = 36 lb/ft, we write

$$V_e = \tfrac{1}{2}kx^2 = \tfrac{1}{2}(36 \text{ lb/ft})x^2 = 18x^2$$

Choosing the datum at A, we have $V_g = 0$; therefore

$$V_1 = V_e + V_g = 18x^2$$

Kinetic Energy. Since the pellet is released from rest, $v_A = 0$ and we have $T_1 = 0$.

Position 2. Potential Energy. The spring is now undeformed; thus $V_e = 0$. Since the pellet is 4 ft above the datum, we have

$$V_g = Wy = (0.5 \text{ lb})(4 \text{ ft}) = 2 \text{ ft} \cdot \text{lb}$$
$$V_2 = V_e + V_g = 2 \text{ ft} \cdot \text{lb}$$

Kinetic Energy. Using the value of v_D^2 obtained above, we write

$$T_2 = \tfrac{1}{2}mv_D^2 = \frac{1}{2}\frac{0.5 \text{ lb}}{32.2 \text{ ft/s}^2}(64.4 \text{ ft}^2/\text{s}^2) = 0.5 \text{ ft} \cdot \text{lb}$$

Conservation of Energy. Applying the principle of conservation of energy between positions 1 and 2, we write

$$T_1 + V_1 = T_2 + V_2$$
$$0 + 18x^2 = 0.5 \text{ ft} \cdot \text{lb} + 2 \text{ ft} \cdot \text{lb}$$
$$x = 0.3727 \text{ ft} \qquad\qquad x = 4.47 \text{ in.} \blacktriangleleft$$

SAMPLE PROBLEM 13.8

A sphere of mass $m = 0.6$ kg is attached to an elastic cord of constant $k = 100$ N/m, which is undeformed when the sphere is located at the origin O. Knowing that the sphere may slide without friction on the horizontal surface and that in the position shown its velocity $\mathbf{v}_A$ has a magnitude of 20 m/s, determine (a) the maximum and minimum distances from the sphere to the origin O, (b) the corresponding values of its speed.

SOLUTION

The force exerted by the cord on the sphere passes through the fixed point O, and its work can be expressed as a change in potential energy. It is therefore a conservative central force, and both the total energy of the sphere and its angular momentum about O are conserved.

Conservation of Angular Momentum About O. At point B, where the distance from O is maximum, the velocity of the sphere is perpendicular to OB and the angular momentum is $r_m m v_m$. A similar property holds at point C, where the distance from O is minimum. Expressing conservation of angular momentum between A and B, we write

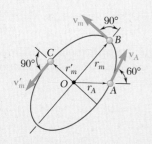

$$r_A m v_A \sin 60° = r_m m v_m$$
$$(0.5 \text{ m})(0.6 \text{ kg})(20 \text{ m/s}) \sin 60° = r_m (0.6 \text{ kg}) v_m$$
$$v_m = \frac{8.66}{r_m} \qquad (1)$$

Conservation of Energy

At point A: $\quad T_A = \frac{1}{2} m v_A^2 = \frac{1}{2}(0.6 \text{ kg})(20 \text{ m/s})^2 = 120 \text{ J}$
$\qquad\qquad V_A = \frac{1}{2} k r_A^2 \;=\; \frac{1}{2}(100 \text{ N/m})(0.5 \text{ m})^2 = 12.5 \text{ J}$

At point B: $\quad T_B = \frac{1}{2} m v_m^2 = \frac{1}{2}(0.6 \text{ kg}) v_m^2 = 0.3 v_m^2$
$\qquad\qquad V_B = \frac{1}{2} k r_m^2 \;=\; \frac{1}{2}(100 \text{ N/m}) r_m^2 = 50 r_m^2$

Applying the principle of conservation of energy between points A and B, we write

$$T_A + V_A = T_B + V_B$$
$$120 + 12.5 = 0.3 v_m^2 + 50 r_m^2 \qquad (2)$$

a. Maximum and Minimum Values of Distance. Substituting for v_m from Eq. (1) into Eq. (2) and solving for r_m^2, we obtain

$$r_m^2 = 2.468 \text{ or } 0.1824 \qquad r_m = 1.571 \text{ m}, \; r_m' = 0.427 \text{ m} \quad \blacktriangleleft$$

b. Corresponding Values of Speed. Substituting the values obtained for r_m and r_m' into Eq. (1), we have

$$v_m = \frac{8.66}{1.571} \qquad\qquad v_m = 5.51 \text{ m/s} \quad \blacktriangleleft$$

$$v_m' = \frac{8.66}{0.427} \qquad\qquad v_m' = 20.3 \text{ m/s} \quad \blacktriangleleft$$

Note. It can be shown that the path of the sphere is an ellipse of *center O*.

SAMPLE PROBLEM 13.9

A satellite is launched in a direction parallel to the surface of the earth with a velocity of 36 900 km/h from an altitude of 500 km. Determine (a) the maximum altitude reached by the satellite, (b) the maximum allowable error in the direction of launching if the satellite is to go into orbit and come no closer than 200 km to the surface of the earth.

SOLUTION

a. Maximum Altitude. We denote by A' the point of the orbit farthest from the earth and by r_1 the corresponding distance from the center of the earth. Since the satellite is in free flight between A and A', we apply the principle of conservation of energy:

$$T_A + V_A = T_{A'} + V_{A'}$$

$$\tfrac{1}{2}mv_0^2 - \frac{GMm}{r_0} = \tfrac{1}{2}mv_1^2 - \frac{GMm}{r_1} \qquad (1)$$

Since the only force acting on the satellite is the force of gravity, which is a central force, the angular momentum of the satellite about O is conserved. Considering points A and A', we write

$$r_0 m v_0 = r_1 m v_1 \qquad v_1 = v_0 \frac{r_0}{r_1} \qquad (2)$$

Substituting this expression for v_1 into Eq. (1), dividing each term by the mass m, and rearranging the terms, we obtain

$$\tfrac{1}{2}v_0^2\left(1 - \frac{r_0^2}{r_1^2}\right) = \frac{GM}{r_0}\left(1 - \frac{r_0}{r_1}\right) \qquad 1 + \frac{r_0}{r_1} = \frac{2GM}{r_0 v_0^2} \qquad (3)$$

Recalling that the radius of the earth is $R = 6370$ km, we compute

$$r_0 = 6370 \text{ km} + 500 \text{ km} = 6870 \text{ km} = 6.87 \times 10^6 \text{ m}$$
$$v_0 = 36\ 900 \text{ km/h} = (36.9 \times 10^6 \text{ m})/(3.6 \times 10^3 \text{ s}) = 10.25 \times 10^3 \text{ m/s}$$
$$GM = gR^2 = (9.81 \text{ m/s}^2)(6.37 \times 10^6 \text{ m})^2 = 398 \times 10^{12} \text{ m}^3/\text{s}^2$$

Substituting these values into (3), we obtain $r_1 = 66.8 \times 10^6$ m.

Maximum altitude $= 66.8 \times 10^6$ m $- 6.37 \times 10^6$ m $= 60.4 \times 10^6$ m $=$
$$60\ 400 \text{ km} \blacktriangleleft$$

b. Allowable Error in Direction of Launching. The satellite is launched from P_0 in a direction forming an angle ϕ_0 with the vertical OP_0. The value of ϕ_0 corresponding to $r_{min} = 6370$ km $+ 200$ km $= 6570$ km is obtained by applying the principles of conservation of energy and of conservation of angular momentum between P_0 and A:

$$\tfrac{1}{2}mv_0^2 - \frac{GMm}{r_0} = \tfrac{1}{2}mv_{max}^2 - \frac{GMm}{r_{min}} \qquad (4)$$

$$r_0 m v_0 \sin \phi_0 = r_{min} m v_{max} \qquad (5)$$

Solving (5) for v_{max} and then substituting for v_{max} into (4), we can solve (4) for $\sin \phi_0$. Using the values of v_0 and GM computed in part a and noting that $r_0/r_{min} = 6870/6570 = 1.0457$, we find

$$\sin \phi_0 = 0.9801 \qquad \phi_0 = 90° \pm 11.5° \qquad \text{Allowable error} = \pm\ 11.5° \blacktriangleleft$$

SOLVING PROBLEMS ON YOUR OWN

In this lesson you learned that when the work done by a force **F** acting on a particle A *is independent of the path followed by the particle* as it moves from a given position A_1 to a given position A_2 (Fig. 13.11*a*), then a function V, called *potential energy,* can be defined for the force **F**. Such forces are said to be *conservative forces,* and you can write

$$U_{1\to2} = V(x_1, y_1, z_1) - V(x_2, y_2, z_2) \tag{13.19}$$

or, for short,

$$U_{1\to2} = V_1 - V_2 \tag{13.19'}$$

Note that the work is negative when the change in the potential energy is positive, i.e., when $V_2 > V_1$.

Substituting the above expression into the equation for work and energy, you can write

$$T_1 + V_1 = T_2 + V_2 \tag{13.24}$$

which shows that when a particle moves under the action of a conservative force *the sum of the kinetic and potential energies of the particle remains constant.*

Your solution of problems using the above formula will consist of the following steps.

1. Determine whether all the forces involved are conservative. If some of the forces are not conservative, for example if friction is involved, you must use the method of work and energy from the previous lesson, since the work done by such forces depends upon the path followed by the particle and a potential function does not exist. If there is no friction and if all the forces are conservative, you can proceed as follows.

2. Determine the kinetic energy $T = \frac{1}{2}mv^2$ at each end of the path.

3. Compute the potential energy for all the forces involved at each end of the path. You will recall that the following expressions for the potential energy were derived in this lesson.

a. The potential energy of a weight W close to the surface of the earth and at a height y above a given datum,

$$V_g = Wy \qquad (13.16)$$

b. The potential energy of a mass m located at a distance r from the center of the earth, large enough so that the variation of the force of gravity must be taken into account,

$$V_g = -\frac{GMm}{r} \qquad (13.17)$$

where the distance r is measured from the center of the earth and V_g is equal to zero at $r = \infty$.

c. The potential energy of a body with respect to an elastic force F = kx,

$$V_e = \tfrac{1}{2}kx^2 \qquad (13.18)$$

where the distance x is the deflection of the elastic spring measured from its *undeformed* position and k is the spring constant. Note that V_e *depends only upon the deflection* x and not upon the path of the body attached to the spring. Also, V_e is always positive, whether the spring is compressed or elongated.

4. Substitute your expressions for the kinetic and potential energies into Eq. (13.24). You will be able to solve this equation for one unknown, for example, for a velocity [Sample Prob. 13.6]. If more than one unknown is involved, you will have to search for another condition or equation, such as the minimum speed [Sample Prob. 13.7] or the minimum potential energy of the particle. For problems involving a central force, a second equation can be obtained by using conservation of angular momentum [Sample Prob. 13.8]. This is especially useful in applications to space mechanics [Sec. 13.9].

PROBLEMS

CONCEPT QUESTIONS

13.CQ2 Two small balls A and B with masses $2m$ and m, respectively, are released from rest at a height h above the ground. Neglecting air resistance, which of the following statements is true when the two balls hit the ground?

    a. The kinetic energy of A is the same as the kinetic energy of B.

    b. The kinetic energy of A is half the kinetic energy of B.

    c. The kinetic energy of A is twice the kinetic energy of B.

    d. The kinetic energy of A is four times the kinetic energy of B.

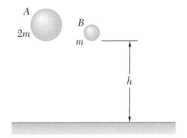

Fig. P13.CQ2

13.CQ3 A small block A is released from rest and slides down the frictionless ramp to the loop. The maximum height h of the loop is the same as the initial height of the block. Will A make it completely around the loop without losing contact with the track?

    a. Yes

    b. No

    c. Need more information

Fig. P13.CQ3

END-OF-SECTION PROBLEMS

13.55 A force **P** is slowly applied to a plate that is attached to two springs and causes a deflection x_0. In each of the two cases shown, derive an expression for the constant k_e, in terms of k_1 and k_2, of the single spring equivalent to the given system, that is, of the single spring which will undergo the same deflection x_0 when subjected to the same force **P**.

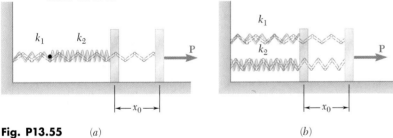

Fig. P13.55   (*a*)   (*b*)

13.56 A loaded railroad car of mass m is rolling at a constant velocity $\mathbf{v_0}$ when it couples with a massless bumper system. Determine the maximum deflection of the bumper assuming the two springs are (*a*) in series (as shown), (*b*) in parallel.

Fig. P13.56

Dimensions in mm

Fig. P13.57

Fig. P13.59

Fig. P13.60

13.57 A 600-g collar C may slide along a horizontal, semicircular rod ABD. The spring CE has an undeformed length of 250 mm and a spring constant of 135 N/m. Knowing that the collar is released from rest at A and neglecting friction, determine the speed of the collar (a) at B, (b) at D.

13.58 A 3-lb collar is attached to a spring and slides without friction along a circular rod in a *horizontal* plane. The spring has an undeformed length of 7 in. and a constant $k = 1.5$ lb/in. Knowing that the collar is in equilibrium at A and is given a slight push to get it moving, determine the velocity of the collar (a) as it passes through B, (b) as it passes through C.

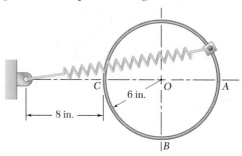

Fig. P13.58

13.59 A 3-lb collar C may slide without friction along a horizontal rod. It is attached to three springs, each of constant $k = 2$ lb/in. and 6-in. undeformed length. Knowing that the collar is released from rest in the position shown, determine the maximum speed it will reach in the ensuing motion.

13.60 A 500-g collar can slide without friction on the curved rod BC in a *horizontal* plane. Knowing that the undeformed length of the spring is 80 mm and that $k = 400$ kN/m, determine (a) the velocity that the collar should be given at A to reach B with zero velocity, (b) the velocity of the collar when it eventually reaches C.

13.61 An elastic cord is stretched between two points A and B, located 800 mm apart in the same horizontal plane. When stretched directly between A and B, the tension is 40 N. The cord is then stretched as shown until its midpoint C has moved through 300 mm to C'; a force of 240 N is required to hold the cord at C'. A 0.1-kg pellet is placed at C', and the cord is released. Determine the speed of the pellet as it passes through C.

Fig. P13.61

13.62 An elastic cable is to be designed for bungee jumping from a tower 130 ft high. The specifications call for the cable to be 85 ft long when unstretched, and to stretch to a total length of 100 ft when a 600-lb weight is attached to it and dropped from the tower. Determine (*a*) the required spring constant k of the cable, (*b*) how close to the ground a 186-lb man will come if he uses this cable to jump from the tower.

13.63 It is shown in mechanics of materials that the stiffness of an elastic cable is $k = AE/L$, where A is the cross-sectional area of the cable, E is the modulus of elasticity, and L is the length of the cable. A winch is lowering a 4000-lb piece of machinery using a constant speed of 3 ft/s when the winch suddenly stops. Knowing that the steel cable has a diameter of 0.4 in., $E = 29 \times 10^6$ lb/in^2, and when the winch stops $L = 30$ ft, determine the maximum downward displacement of the piece of machinery from the point it was when the winch stopped.

Fig. P13.62

Fig. P13.63

13.64 A 2-kg collar is attached to a spring and slides without friction in a vertical plane along the curved rod *ABC*. The spring is undeformed when the collar is at *C* and its constant is 600 N/m. If the collar is released at *A* with no initial velocity, determine its velocity (*a*) as it passes through *B*, (*b*) as it reaches *C*.

Fig. P13.64

13.65 A 1-kg collar can slide along the rod shown. It is attached to an elastic cord anchored at *F*, which has an undeformed length of 250 mm and spring constant of 75 N/m. Knowing that the collar is released from rest at *A* and neglecting friction, determine the speed of the collar (*a*) at *B*, (*b*) at *E*.

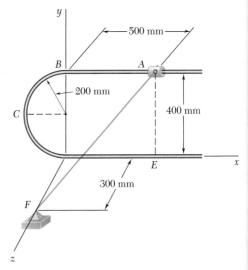

Fig. P13.65

13.66 A thin circular rod is supported in a *vertical plane* by a bracket at
A. Attached to the bracket and loosely wound around the rod is a
spring of constant $k = 3$ lb/ft and undeformed length equal to the
arc of circle AB. An 8-oz collar C, not attached to the spring, can
slide without friction along the rod. Knowing that the collar is
released from rest at an angle θ with the vertical, determine
(a) the smallest value of θ for which the collar will pass through
D and reach point A, (b) the velocity of the collar as it reaches
point A.

Fig. P13.66

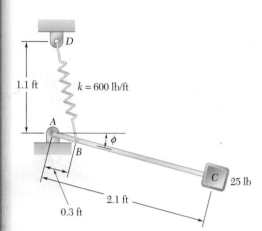

Fig. P13.67

13.67 The system shown is in equilibrium when $\phi = 0$. Knowing that
initially $\phi = 90°$ and that block C is given a slight nudge when the
system is in that position, determine the speed of the block as it
passes through the equilibrium position $\phi = 0$. Neglect the weight
of the rod.

13.68 A spring is used to stop a 50-kg package which is moving down a
20° incline. The spring has a constant $k = 30$ kN/m and is held by
cables so that it is initially compressed 50 mm. Knowing that the
velocity of the package is 2 m/s when it is 8 m from the spring and
neglecting friction, determine the maximum additional deformation
of the spring in bringing the package to rest.

Fig. P13.68

13.69 Solve Prob. 13.68 assuming the kinetic coefficient of friction
between the package and the incline is 0.2.

13.70 A section of track for a roller coaster consists of two circular arcs
AB and CD joined by a straight portion BC. The radius of AB is
27 m and the radius of CD is 72 m. The car and its occupants, of
total mass 250 kg, reach point A with practically no velocity and
then drop freely along the track. Determine the normal force
exerted by the track on the car as the car reaches point B. Ignore
air resistance and rolling resistance.

13.71 A section of track for a roller coaster consists of two circular arcs
AB and CD joined by a straight portion BC. The radius of AB is
27 m and the radius of CD is 72 m. The car and its occupants, of
total mass 250 kg, reach point A with practically no velocity and
then drop freely along the track. Determine the maximum and
minimum values of the normal force exerted by the track on the
car as the car travels from A to D. Ignore air resistance and rolling
resistance.

Fig. P13.70 and P13.71

13.72 A 1-lb collar is attached to a spring and slides without friction along a circular rod in a *vertical* plane. The spring has an undeformed length of 5 in. and a constant $k = 10$ lb/ft. Knowing that the collar is released from being held at A, determine the speed of the collar and the normal force between the collar and the rod as the collar passes through B.

13.73 A 10-lb collar is attached to a spring and slides without friction along a fixed rod in a vertical plane. The spring has an undeformed length of 14 in. and a constant $k = 4$ lb/in. Knowing that the collar is released from rest in the position shown, determine the force exerted by the rod on the collar at (*a*) point A, (*b*) point B. Both these points are on the curved portion of the rod.

Fig. P13.72

Fig. P13.73

13.74 An 8-oz package is projected upward with a velocity $\mathbf{v}_0$ by a spring at A; it moves around a frictionless loop and is deposited at C. For each of the two loops shown, determine (*a*) the smallest velocity $\mathbf{v}_0$ for which the package will reach C, (*b*) the corresponding force exerted by the package on the loop just before the package leaves the loop at C.

Fig. P13.74 and P13.75

13.75 If the package of Prob. 13.74 is not to hit the horizontal surface at C with a speed greater than 10 ft/s, (*a*) show that this requirement can be satisfied only by the second loop, (*b*) determine the largest allowable initial velocity $\mathbf{v}_0$ when the second loop is used.

13.76 A small package of weight W is projected into a vertical return loop at A with a velocity $\mathbf{v}_0$. The package travels without friction along a circle of radius r and is deposited on a horizontal surface at C. For each of the two loops shown, determine (a) the smallest velocity $\mathbf{v}_0$ for which the package will reach the horizontal surface at C, (b) the corresponding force exerted by the loop on the package as it passes point B.

Fig. P13.76

13.77 The 1-kg ball at A is suspended by an inextensible cord and given an initial horizontal velocity of 5 m/s. If $l = 0.6$ m and $x_B = 0$, determine y_B so that the ball will enter the basket.

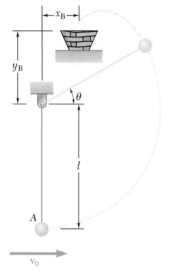

Fig. P13.77

*13.78 Packages are moved from point A on the upper floor of a warehouse to point B on the lower floor, 12 ft directly below A, by means of a chute, the centerline of which is in the shape of a helix of vertical axis y and radius $R = 8$ ft. The cross section of the chute is to be banked in such a way that each package, after being released at A with no velocity, will slide along the centerline of the chute without ever touching its edges. Neglecting friction, (a) express as a function of the elevation y of a given point P of the centerline the angle ϕ formed by the normal to the surface of the chute at P and the principal normal of the centerline at that point, (b) determine the magnitude and direction of the force exerted by the chute on a 20-lb package as it reaches point B. *Hint:* The principal normal to the helix at any point P is horizontal and directed toward the y axis, and the radius of curvature of the helix is $\rho = R[1 + (h/2\pi R)^2]$.

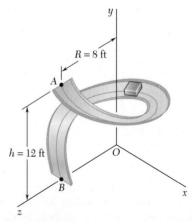

Fig. P13.78

*13.79 Prove that a force $F(x, y, z)$ is conservative if, and only if, the following relations are satisfied:

$$\frac{\partial F_x}{\partial y} = \frac{\partial F_y}{\partial x} \qquad \frac{\partial F_y}{\partial z} = \frac{\partial F_z}{\partial y} \qquad \frac{\partial F_z}{\partial x} = \frac{\partial F_x}{\partial z}$$

13.80 The force $\mathbf{F} = (yz\mathbf{i} + zx\mathbf{j} + xy\mathbf{k})/xyz$ acts on the particle $P(x, y, z)$ which moves in space. (*a*) Using the relation derived in Prob. 13.79, show that this force is a conservative force. (*b*) Determine the potential function associated with $\mathbf{F}$.

*13.81 A force F acts on a particle $P(x, y)$ which moves in the xy plane. Determine whether $\mathbf{F}$ is a conservative force and compute the work of $\mathbf{F}$ when P describes in a clockwise sense the path A, B, C, A including the quarter circle $x^2 + y^2 = a^2$, if (*a*) $\mathbf{F} = ky\mathbf{i}$, (*b*) $\mathbf{F} = k(y\mathbf{i} + x\mathbf{j})$.

Fig. P13.81

*13.82 The potential function associated with a force $\mathbf{P}$ in space is known to be $V(x, y, z) = -(x^2 + y^2 + z^2)^{1/2}$. (*a*) Determine the x, y, and z components of $\mathbf{P}$. (*b*) Calculate the work done by $\mathbf{P}$ from O to D by integrating along the path $OABD$, and show that it is equal to the negative of the change in potential from O to D.

*13.83 (*a*) Calculate the work done from D to O by the force $\mathbf{P}$ of Prob. 13.82 by integrating along the diagonal of the cube. (*b*) Using the result obtained and the answer to part *b* of Prob. 13.82, verify that the work done by a conservative force around the closed path $OABDO$ is zero.

*13.84 The force $\mathbf{F} = (x\mathbf{i} + y\mathbf{j} + z\mathbf{k})/(x^2 + y^2 + z^2)^{3/2}$ acts on the particle $P(x, y, z)$ which moves in space. (*a*) Using the relations derived in Prob. 13.79, prove that $\mathbf{F}$ is a conservative force. (*b*) Determine the potential function $V(x, y, z)$ associated with $\mathbf{F}$.

13.85 (*a*) Determine the kinetic energy per unit mass which a missile must have after being fired from the surface of the earth if it is to reach an infinite distance from the earth. (*b*) What is the initial velocity of the missile (called the *escape velocity*)? Give your answers in SI units and show that the answer to part *b* is independent of the firing angle.

13.86 A satellite describes an elliptic orbit of minimum altitude 606 km above the surface of the earth. The semimajor and semiminor axes are 17 440 km and 13 950 km, respectively. Knowing that the speed of the satellite at point C is 4.78 km/s, determine (*a*) the speed at point A, the perigee, (*b*) the speed at point B, the apogee.

Fig. P13.82

Fig. P13.86

13.87 While describing a circular orbit 200 mi above the earth a space vehicle launches a 6000-lb communications satellite. Determine (*a*) the additional energy required to place the satellite in a geosynchronous orbit at an altitude of 22,000 mi above the surface of the earth, (*b*) the energy required to place the satellite in the same orbit by launching it from the surface of the earth, excluding the energy needed to overcome air resistance. (A *geosynchronous orbit* is a circular orbit in which the satellite appears stationary with respect to the ground.)

13.88 A lunar excursion module (LEM) was used in the Apollo moon-landing missions to save fuel by making it unnecessary to launch the entire Apollo spacecraft from the moon's surface on its return trip to earth. Check the effectiveness of this approach by computing the energy per pound required for a spacecraft (as weighed on the earth) to escape the moon's gravitational field if the spacecraft starts from (*a*) the moon's surface, (*b*) a circular orbit 50 mi above the moon's surface. Neglect the effect of the earth's gravitational field. (The radius of the moon is 1081 mi and its mass is 0.0123 times the mass of the earth.)

13.89 Knowing that the velocity of an experimental space probe fired from the earth has a magnitude $v_A = 32.5$ Mm/h at point *A*, determine the speed of the probe as it passes through point *B*.

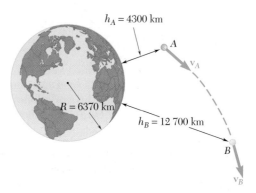

Fig. P13.89

13.90 A spacecraft is describing a circular orbit at an altitude of 1500 km above the surface of the earth. As it passes through point *A*, its speed is reduced by 40 percent and it enters an elliptic crash trajectory with the apogee at point *A*. Neglecting air resistance, determine the speed of the spacecraft when it reaches the earth's surface at point *B*.

13.91 Observations show that a celestial body traveling at 1.2×10^6 mi/h appears to be describing about point *B* a circle of radius equal to 60 light years. Point *B* is suspected of being a very dense concentration of mass called a black hole. Determine the ratio M_B/M_S of the mass at *B* to the mass of the sun. (The mass of the sun is 330,000 times the mass of the earth, and a light year is the distance traveled by light in 1 year at 186,300 mi/s.)

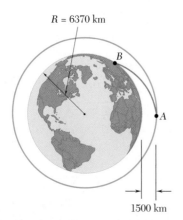

Fig. P13.90

13.92 (a) Show that, by setting $r = R + y$ in the right-hand member of Eq. (13.17′) and expanding that member in a power series in y/R, the expression in Eq. (13.16) for the potential energy V_g due to gravity is a first-order approximation for the expression given in Eq. (13.17′). (b) Using the same expansion, derive a second-order approximation for V_g.

13.93 Collar A has a mass of 3 kg and is attached to a spring of constant 1200 N/m and of undeformed length equal to 0.5 m. The system is set in motion with $r = 0.3$ m, $v_\theta = 2$ m/s, and $v_r = 0$. Neglecting the mass of the rod and the effect of friction, determine the radial and transverse components of the velocity of the collar when $r = 0.6$ m.

13.94 Collar A has a mass of 3 kg and is attached to a spring of constant 1200 N/m and of undeformed length equal to 0.5 m. The system is set in motion with $r = 0.3$ m, $v_\theta = 2$ m/s, and $v_r = 0$. Neglecting the mass of the rod and the effect of friction, determine (a) the maximum distance between the origin and the collar, (b) the corresponding speed. (*Hint:* Solve the equation obtained for r by trial and error.)

Fig. P13.93 and P13.94

13.95 A 4-lb collar A and a 1.5-lb collar B can slide without friction on a frame, consisting of the horizontal rod OE and the vertical rod CD, which is free to rotate about CD. The two collars are connected by a cord running over a pulley that is attached to the frame at O. At the instant shown, the velocity $\mathbf{v}_A$ of collar A has a magnitude of 6 ft/s and a stop prevents collar B from moving. If the stop is suddenly removed, determine (a) the velocity of collar A when it is 8 in. from O, (b) the velocity of collar A when collar B comes to rest. (Assume that collar B does not hit O, that collar A does not come off rod OE, and that the mass of the frame is negligible.)

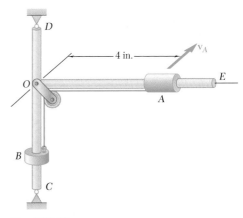

Fig. P13.95

13.96 A 1.5-lb ball that can slide on a *horizontal* frictionless surface is attached to a fixed point O by means of an elastic cord of constant $k = 1$ lb/in. and undeformed length 2 ft. The ball is placed at point A, 3 ft from O, and given an initial velocity $\mathbf{v}_0$ perpendicular to OA. Determine (a) the smallest allowable value of the initial speed v_0 if the cord is not to become slack, (b) the closest distance d that the ball will come to point O if it is given half the initial speed found in part a.

13.97 A 1.5-lb ball that can slide on a *horizontal* frictionless surface is attached to a fixed point O by means of an elastic cord of constant $k = 1$ lb/in. and undeformed length 2 ft. The ball is placed at point A, 3 ft from O, and given an initial velocity $\mathbf{v}_0$ perpendicular to OA, allowing the ball to come within a distance $d = 9$ in. of point O after the cord has become slack. Determine (a) the initial speed v_0 of the ball, (b) its maximum speed.

Fig. P13.96 and P13.97

13.98 Using the principles of conservation of energy and conservation of angular momentum, solve part a of Sample Prob. 12.9.

13.99 Solve Sample Prob. 13.8, assuming that the elastic cord is replaced by a central force $\mathbf{F}$ of magnitude $(80/r^2)$ N directed toward O.

Fig. P13.101

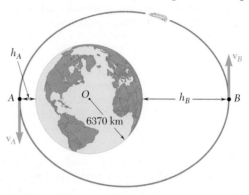

13.100 A spacecraft is describing an elliptic orbit of minimum altitude $h_A = 2400$ km and maximum altitude $h_B = 9600$ km above the surface of the earth. Determine the speed of the spacecraft at A.

Fig. P13.100

13.101 While describing a circular orbit, 185 mi above the surface of the earth, a space shuttle ejects at point A an inertial upper stage (IUS) carrying a communications satellite to be placed in a geosynchronous orbit (see Prob. 13.87) at an altitude of 22,230 mi above the surface of the earth. Determine (a) the velocity of the IUS relative to the shuttle after its engine has been fired at A, (b) the increase in velocity required at B to place the satellite in its final orbit.

13.102 A spacecraft approaching the planet Saturn reaches point A with a velocity $\mathbf{v}_A$ of magnitude 68.8×10^3 ft/s. It is to be placed in an elliptic orbit about Saturn so that it will be able to periodically examine Tethys, one of Saturn's moons. Tethys is in a circular orbit of radius 183×10^3 mi about the center of Saturn, traveling at a speed of 37.2×10^3 ft/s. Determine (a) the decrease in speed required by the spacecraft at A to achieve the desired orbit, (b) the speed of the spacecraft when it reaches the orbit of Tethys at B.

13.103 A spacecraft traveling along a parabolic path toward the planet Jupiter is expected to reach point A with a velocity $\mathbf{v}_A$ of magnitude 26.9 km/s. Its engines will then be fired to slow it down, placing it into an elliptic orbit which will bring it to within 100×10^3 km of Jupiter. Determine the decrease in speed Δv at point A which will place the spacecraft into the required orbit. The mass of Jupiter is 319 times the mass of the earth.

Fig. P13.102

Fig. P13.103

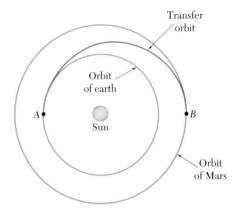

13.104 As a first approximation to the analysis of a space flight from the earth to Mars, it is assumed that the orbits of the earth and Mars are circular and coplanar. The mean distances from the sun to the earth and to Mars are 149.6×10^6 km and 227.8×10^6 km, respectively. To place the spacecraft into an elliptical transfer orbit at point A, its speed is increased over a short interval of time to v_A which is faster than the earth's orbital speed. When the spacecraft reaches point B on the elliptical transfer orbit, its speed v_B is increased to the orbital speed of Mars. Knowing that the mass of the sun is 332.8×10^3 times the mass of the earth, determine the increase in velocity required (*a*) at A, (*b*) at B.

13.105 The optimal way of transferring a space vehicle from an inner circular orbit to an outer coplanar circular orbit is to fire its engines as it passes through A to increase its speed and place it in an elliptic transfer orbit. Another increase in speed as it passes through B will place it in the desired circular orbit. For a vehicle in a circular orbit about the earth at an altitude $h_1 = 200$ mi, which is to be transferred to a circular orbit at an altitude $h_2 = 500$ mi, determine (*a*) the required increases in speed at A and at B, (*b*) the total energy per unit mass required to execute the transfer.

Fig. P13.104

Fig. P13.105

13.106 During a flyby of the earth, the velocity of a spacecraft is 10.4 km/s as it reaches its minimum altitude of 990 km above the surface at point A. At point B the spacecraft is observed to have an altitude of 8350 km. Determine (*a*) the magnitude of the velocity at point B, (*b*) the angle ϕ_B.

Fig. P13.106

13.107 A space platform is in a circular orbit about the earth at an altitude of 300 km. As the platform passes through A, a rocket carrying a communications satellite is launched from the platform with a relative velocity of magnitude 3.44 km/s in a direction tangent to the orbit of the platform. This was intended to place the rocket in an elliptic transfer orbit bringing it to point B, where the rocket would again be fired to place the satellite in a geosynchronous orbit of radius 42 140 km. After launching, it was discovered that the relative velocity imparted to the rocket was too large. Determine the angle γ at which the rocket will cross the intended orbit at point C.

Fig. P13.107

13.108 A satellite is projected into space with a velocity $\mathbf{v}_0$ at a distance r_0 from the center of the earth by the last stage of its launching rocket. The velocity $\mathbf{v}_0$ was designed to send the satellite into a circular orbit of radius r_0. However, owing to a malfunction of control, the satellite is not projected horizontally but at an angle α with the horizontal and, as a result, is propelled into an elliptic orbit. Determine the maximum and minimum values of the distance from the center of the earth to the satellite.

Fig. P13.108

Fig. P13.109

13.109 Upon the LEM's return to the command module, the Apollo spacecraft was turned around so that the LEM faced to the rear. The LEM was then cast adrift with a velocity of 200 m/s relative to the command module. Determine the magnitude and direction (angle ϕ formed with the vertical OC) of the velocity $\mathbf{v}_C$ of the LEM just before it crashed at C on the moon's surface.

13.110 A space vehicle is in a circular orbit at an altitude of 225 mi above the earth. To return to earth, it decreases its speed as it passes through A by firing its engine for a short interval of time in a direction opposite to the direction of its motion. Knowing that the velocity of the space vehicle should form an angle $\phi_B = 60°$ with the vertical as it reaches point B at an altitude of 40 mi, determine (*a*) the required speed of the vehicle as it leaves its circular orbit at A, (*b*) its speed at point B.

*13.111 In Prob. 13.110, the speed of the space vehicle was decreased as it passed through A by firing its engine in a direction opposite to the direction of motion. An alternative strategy for taking the space vehicle out of its circular orbit would be to turn it around so that its engine would point away from the earth and then give it an incremental velocity $\Delta\mathbf{v}_A$ toward the center O of the earth. This would likely require a smaller expenditure of energy when firing the engine at A, but might result in too fast a descent at B. Assuming this strategy is used with only 50 percent of the energy expenditure used in Prob. 13.110, determine the resulting values of ϕ_B and v_B.

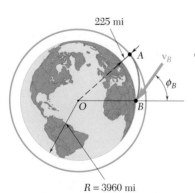

Fig. P13.110

13.112 Show that the values v_A and v_P of the speed of an earth satellite at the apogee A and the perigee P of an elliptic orbit are defined by the relations

$$v_A^2 = \frac{2GM}{r_A + r_P}\frac{r_P}{r_A} \qquad v_P^2 = \frac{2GM}{r_A + r_P}\frac{r_A}{r_P}$$

where M is the mass of the earth, and r_A and r_P represent, respectively, the maximum and minimum distances of the orbit to the center of the earth.

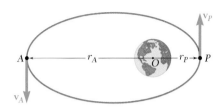

Fig. P13.112 and P13.113

13.113 Show that the total energy E of an earth satellite of mass m describing an elliptic orbit is $E = -GMm/(r_A + r_P)$, where M is the mass of the earth, and r_A and r_P represent, respectively, the maximum and minimum distances of the orbit to the center of the earth. (Recall that the gravitational potential energy of a satellite was defined as being zero at an infinite distance from the earth.)

*13.114 A space probe describes a circular orbit of radius nR with a velocity $\mathbf{v}_0$ about a planet of radius R and center O. Show that (a) in order for the probe to leave its orbit and hit the planet at an angle θ with the vertical, its velocity must be reduced to αv_0, where

$$\alpha = \sin\theta\sqrt{\frac{2(n-1)}{n^2 - \sin^2\theta}}$$

(b) the probe will not hit the planet if α is larger than $\sqrt{2/(1+n)}$.

13.115 A missile is fired from the ground with an initial velocity $\mathbf{v}_0$ forming an angle ϕ_0 with the vertical. If the missile is to reach a maximum altitude equal to αR, where R is the radius of the earth, (a) show that the required angle ϕ_0 is defined by the relation

$$\sin\phi_0 = (1 + \alpha)\sqrt{1 - \frac{\alpha}{1 + \alpha}\left(\frac{v_{esc}}{v_0}\right)^2}$$

where v_{esc} is the escape velocity, (b) determine the range of allowable values of v_0.

13.116 A spacecraft of mass m describes a circular orbit of radius r_1 around the earth. (a) Show that the additional energy ΔE which must be imparted to the spacecraft to transfer it to a circular orbit of larger radius r_2 is

$$\Delta E = \frac{GMm(r_2 - r_1)}{2r_1r_2}$$

where M is the mass of the earth. (b) Further show that if the transfer from one circular orbit to the other is executed by placing the spacecraft on a transitional semielliptic path AB, the amounts of energy ΔE_A and ΔE_B which must be imparted at A and B are, respectively, proportional to r_2 and r_1:

$$\Delta E_A = \frac{r_2}{r_1 + r_2}\Delta E \qquad \Delta E_B = \frac{r_1}{r_1 + r_2}\Delta E$$

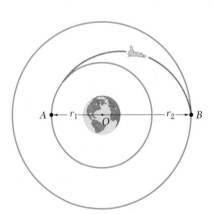

Fig. P13.116

*13.117 Using the answers obtained in Prob. 13.108, show that the intended circular orbit and the resulting elliptic orbit intersect at the ends of the minor axis of the elliptic orbit.

*13.118 (a) Express in terms of r_{min} and v_{max} the angular momentum per unit mass, h, and the total energy per unit mass, E/m, of a space vehicle moving under the gravitational attraction of a planet of mass M (Fig. 13.15). (b) Eliminating v_{max} between the equations obtained, derive the formula

$$\frac{1}{r_{min}} = \frac{GM}{h^2}\left[1 + \sqrt{1 + \frac{2E}{m}\left(\frac{h}{GM}\right)^2}\right]$$

(c) Show that the eccentricity £ of the trajectory of the vehicle can be expressed as

$$\varepsilon = \sqrt{1 + \frac{2E}{m}\left(\frac{h}{GM}\right)^2}$$

(d) Further show that the trajectory of the vehicle is a hyperbola, an ellipse, or a parabola, depending on whether E is positive, negative, or zero.

13.10 PRINCIPLE OF IMPULSE AND MOMENTUM

A third basic method for the solution of problems dealing with the motion of particles will be considered now. This method is based on the principle of impulse and momentum and can be used to solve problems involving force, mass, velocity, and time. It is of particular interest in the solution of problems involving impulsive motion and problems involving impact (Secs. 13.11 and 13.12).

Consider a particle of mass m acted upon by a force $\mathbf{F}$. As we saw in Sec. 12.3, Newton's second law can be expressed in the form

$$\mathbf{F} = \frac{d}{dt}(m\mathbf{v}) \tag{13.27}$$

where $m\mathbf{v}$ is the linear momentum of the particle. Multiplying both sides of Eq. (13.27) by dt and integrating from a time t_1 to a time t_2, we write

$$\mathbf{F}\,dt = d(m\mathbf{v})$$
$$\int_{t_1}^{t_2} \mathbf{F}\,dt = m\mathbf{v}_2 - m\mathbf{v}_1$$

or, transposing the last term,

$$m\mathbf{v}_1 + \int_{t_1}^{t_2} \mathbf{F}\,dt = m\mathbf{v}_2 \tag{13.28}$$

The integral in Eq. (13.28) is a vector known as the *linear impulse*, or simply the *impulse*, of the force $\mathbf{F}$ during the interval of time considered. Resolving $\mathbf{F}$ into rectangular components, we write

$$\mathbf{Imp}_{1\to2} = \int_{t_1}^{t_2} \mathbf{F}\,dt$$
$$= \mathbf{i}\int_{t_1}^{t_2} F_x\,dt + \mathbf{j}\int_{t_1}^{t_2} F_y\,dt + \mathbf{k}\int_{t_1}^{t_2} F_z\,dt \tag{13.29}$$

Photo 13.1

Photo 13.2 This impact test between an F-4 Phantom and a rigid reinforced target was to determine the impact force as a function of time.

and note that the components of the impulse of the force **F** are, respectively, equal to the areas under the curves obtained by plotting the components F_x, F_y, and F_z against t (Fig. 13.16). In the case of a force **F** of constant magnitude and direction, the impulse is represented by the vector $\mathbf{F}(t_2 - t_1)$, which has the same direction as **F**.

If SI units are used, the magnitude of the impulse of a force is expressed in N · s. But, recalling the definition of the newton, we have

$$\text{N} \cdot \text{s} = (\text{kg} \cdot \text{m/s}^2) \cdot \text{s} = \text{kg} \cdot \text{m/s}$$

which is the unit obtained in Sec. 12.4 for the linear momentum of a particle. We thus check that Eq. (13.28) is dimensionally correct. If U.S. customary units are used, the impulse of a force is expressed in lb · s, which is also the unit obtained in Sec. 12.4 for the linear momentum of a particle.

Equation (13.28) expresses that when a particle is acted upon by a force **F** during a given time interval, *the final momentum $m\mathbf{v}_2$ of the particle can be obtained by adding vectorially its initial momentum $m\mathbf{v}_1$ and the impulse of the force **F** during the time interval considered*

Fig. **13.16**

Fig. 13.17

(Fig. 13.17). We write

$$m\mathbf{v}_1 + \mathbf{Imp}_{1 \to 2} = m\mathbf{v}_2 \tag{13.30}$$

We note that while kinetic energy and work are scalar quantities, momentum and impulse are vector quantities. To obtain an analytic solution, it is thus necessary to replace Eq. (13.30) by the corresponding component equations

$$(mv_x)_1 + \int_{t_1}^{t_2} F_x \, dt = (mv_x)_2$$

$$(mv_y)_1 + \int_{t_1}^{t_2} F_y \, dt = (mv_y)_2 \tag{13.31}$$

$$(mv_z)_1 + \int_{t_1}^{t_2} F_z \, dt = (mv_z)_2$$

When several forces act on a particle, the impulse of each of the forces must be considered. We have

$$m\mathbf{v}_1 + \Sigma\,\mathbf{Imp}_{1 \to 2} = m\mathbf{v}_2 \tag{13.32}$$

Again, the equation obtained represents a relation between vector quantities; in the actual solution of a problem, it should be replaced by the corresponding component equations.

When a problem involves two particles or more, each particle can be considered separately and Eq. (13.32) can be written for each particle. We can also add vectorially the momenta of all the particles and the impulses of all the forces involved. We write then

$$\Sigma m\mathbf{v}_1 + \Sigma \,\mathbf{Imp}_{1\rightarrow2} = \Sigma m\mathbf{v}_2 \qquad (13.33)$$

Since the forces of action and reaction exerted by the particles on each other form pairs of equal and opposite forces, and since the time interval from t_1 to t_2 is common to all the forces involved, the impulses of the forces of action and reaction cancel out and only the impulses of the external forces need be considered.†

If no external force is exerted on the particles or, more generally, if the sum of the external forces is zero, the second term in Eq. (13.33) vanishes and Eq. (13.33) reduces to

$$\Sigma m\mathbf{v}_1 = \Sigma m\mathbf{v}_2 \qquad (13.34)$$

which expresses that *the total momentum of the particles is conserved*. Consider, for example, two boats, of mass m_A and m_B, initially at rest, which are being pulled together (Fig. 13.18). If the resistance

Fig. 13.18

of the water is neglected, the only external forces acting on the boats are their weights and the buoyant forces exerted on them. Since these forces are balanced, we write

$$\Sigma m\mathbf{v}_1 = \Sigma m\mathbf{v}_2$$
$$0 = m_A\mathbf{v}'_A + m_B\mathbf{v}'_B$$

where $\mathbf{v}'_A$ and $\mathbf{v}'_B$ represent the velocities of the boats after a finite interval of time. The equation obtained indicates that the boats move in opposite directions (toward each other) with velocities inversely proportional to their masses.‡

---

†We should note the difference between this statement and the corresponding statement made in Sec. 13.4 regarding the work of the forces of action and reaction between several particles. While the sum of the impulses of these forces is always zero, the sum of their work is zero only under special circumstances, e.g., when the various bodies involved are connected by inextensible cords or links and are thus constrained to move through equal distances.

‡Blue equals signs are used in Fig. 13.18 and throughout the remainder of this chapter to express that two systems of vectors are *equipollent*, i.e., that they have the same resultant and moment resultant (cf. Sec. 3.19). Red equals signs will continue to be used to indicate that two systems of vectors are *equivalent*, i.e., that they have the same effect. This and the concept of conservation of momentum for a system of particles will be discussed in greater detail in Chap. 14.

13.11 IMPULSIVE MOTION

A force acting on a particle during a very short time interval that is large enough to produce a definite change in momentum is called an *impulsive force* and the resulting motion is called an *impulsive motion*. For example, when a baseball is struck, the contact between bat and ball takes place during a very short time interval Δt. But the average value of the force **F** exerted by the bat on the ball is very large, and the resulting impulse **F** Δt is large enough to change the sense of motion of the ball (Fig. 13.19).

When impulsive forces act on a particle, Eq. (13.32) becomes

Fig. 13.19

$$m\mathbf{v}_1 + \Sigma\mathbf{F}\,\Delta t = m\mathbf{v}_2 \qquad (13.35)$$

Any force which is not an impulsive force may be neglected, since the corresponding impulse **F** Δt is very small. *Nonimpulsive forces* include the weight of the body, the force exerted by a spring, or any other force which is *known* to be small compared with an impulsive force. Unknown reactions may or may not be impulsive; their impulses should therefore be included in Eq. (13.35) as long as they have not been proved negligible. The impulse of the weight of the baseball considered above, for example, may be neglected. If the motion of the bat is analyzed, the impulse of the weight of the bat can also be neglected. The impulses of the reactions of the player's hands on the bat, however, should be included; these impulses will not be negligible if the ball is incorrectly hit.

We note that the method of impulse and momentum is particularly effective in the analysis of the impulsive motion of a particle, since it involves only the initial and final velocities of the particle and the impulses of the forces exerted on the particle. The direct application of Newton's second law, on the other hand, would require the determination of the forces as functions of the time and the integration of the equations of motion over the time interval Δt.

In the case of the impulsive motion of several particles, Eq. (13.33) can be used. It reduces to

$$\Sigma m\mathbf{v}_1 + \Sigma\mathbf{F}\,\Delta t = \Sigma m\mathbf{v}_2 \qquad (13.36)$$

where the second term involves only impulsive, external forces. If all the external forces acting on the various particles are nonimpulsive, the second term in Eq. (13.36) vanishes and this equation reduces to Eq. (13.34). We write

$$\Sigma m\mathbf{v}_1 = \Sigma m\mathbf{v}_2 \qquad (13.34)$$

which expresses that the total momentum of the particles is conserved. This situation occurs, for example, when two particles which are moving freely collide with one another. We should note, however, that while the total momentum of the particles is conserved, their total energy is generally *not* conserved. Problems involving the collision or *impact* of two particles will be discussed in detail in Secs. 13.12 through 13.14.

SAMPLE PROBLEM 13.10

An automobile weighing 4000 lb is driven down a 5° incline at a speed of 60 mi/h when the brakes are applied, causing a constant total braking force (applied by the road on the tires) of 1500 lb. Determine the time required for the automobile to come to a stop.

SOLUTION

We apply the principle of impulse and momentum. Since each force is constant in magnitude and direction, each corresponding impulse is equal to the product of the force and of the time interval t.

$$m\mathbf{v}_1 + \Sigma \, \mathbf{Imp}_{1\to2} = m\mathbf{v}_2$$

$+\searrow$ components: $\quad mv_1 + (W \sin 5°)t - Ft = 0$

$(4000/32.2)(88 \text{ ft/s}) + (4000 \sin 5°)t - 1500t = 0$ $\qquad t = 9.49 \text{ s}$ ◄

SAMPLE PROBLEM 13.11

A 4-oz baseball is pitched with a velocity of 80 ft/s toward a batter. After the ball is hit by the bat B, it has a velocity of 120 ft/s in the direction shown. If the bat and ball are in contact 0.015 s, determine the average impulsive force exerted on the ball during the impact.

SOLUTION

We apply the principle of impulse and momentum to the ball. Since the weight of the ball is a nonimpulsive force, it can be neglected.

$$m\mathbf{v}_1 + \Sigma \, \mathbf{Imp}_{1\to2} = m\mathbf{v}_2$$

$\xrightarrow{+} x$ components: $\qquad -mv_1 + F_x \Delta t = mv_2 \cos 40°$

$$-\frac{\frac{4}{16}}{32.2}(80 \text{ ft/s}) + F_x(0.015 \text{ s}) = \frac{\frac{4}{16}}{32.2}(120 \text{ ft/s})\cos 40°$$

$$F_x = +89.0 \text{ lb}$$

$+\uparrow y$ components: $\qquad 0 + F_y \Delta t = mv_2 \sin 40°$

$$F_y(0.015 \text{ s}) = \frac{\frac{4}{16}}{32.2}(120 \text{ ft/s})\sin 40°$$

$$F_y = +39.9 \text{ lb}$$

From its components F_x and F_y we determine the magnitude and direction of the force $\mathbf{F}$:

$$\mathbf{F} = 97.5 \text{ lb} \; \measuredangle \; 24.2° \; ◄$$

SAMPLE PROBLEM 13.12

A 10-kg package drops from a chute into a 25-kg cart with a velocity of 3 m/s. Knowing that the cart is initially at rest and can roll freely, determine (a) the final velocity of the cart, (b) the impulse exerted by the cart on the package, (c) the fraction of the initial energy lost in the impact.

SOLUTION

We first apply the principle of impulse and momentum to the package-cart system to determine the velocity $\mathbf{v}_2$ of the cart and package. We then apply the same principle to the package alone to determine the impulse $\mathbf{F}\,\Delta t$ exerted on it.

a. Impulse-Momentum Principle: Package and Cart

$$m_P\mathbf{v}_1 + \Sigma\,\mathbf{Imp}_{1\rightarrow 2} = (m_P + m_C)\mathbf{v}_2$$

$\xrightarrow{+}x$ components:
$$m_Pv_1\cos 30° + 0 = (m_P + m_C)v_2$$
$$(10\text{ kg})(3\text{ m/s})\cos 30° = (10\text{ kg} + 25\text{ kg})v_2$$
$$\mathbf{v}_2 = 0.742\text{ m/s}\rightarrow \blacktriangleleft$$

We note that the equation used expresses conservation of momentum in the x direction.

b. Impulse-Momentum Principle: Package

$$m_P\mathbf{v}_1 + \Sigma\,\mathbf{Imp}_{1\rightarrow 2} = m_P\mathbf{v}_2$$

$\xrightarrow{+}x$ components: $\quad (10\text{ kg})(3\text{ m/s})\cos 30° + F_x\,\Delta t = (10\text{ kg})(0.742\text{ m/s})$
$$F_x\,\Delta t = -18.56\text{ N}\cdot\text{s}$$

$+\uparrow y$ components: $\quad -m_Pv_1\sin 30° + F_y\,\Delta t = 0$
$$-(10\text{ kg})(3\text{ m/s})\sin 30° + F_y\,\Delta t = 0$$
$$F_y\,\Delta t = +15\text{ N}\cdot\text{s}$$

The impulse exerted on the package is $\quad \mathbf{F}\,\Delta t = 23.9\text{ N}\cdot\text{s}\,\measuredangle\,38.9° \quad\blacktriangleleft$

c. Fraction of Energy Lost.
The initial and final energies are
$$T_1 = \tfrac{1}{2}m_Pv_1^2 = \tfrac{1}{2}(10\text{ kg})(3\text{ m/s})^2 = 45\text{ J}$$
$$T_2 = \tfrac{1}{2}(m_P + m_C)v_2^2 = \tfrac{1}{2}(10\text{ kg} + 25\text{ kg})(0.742\text{ m/s})^2 = 9.63\text{ J}$$

The fraction of energy lost is
$$\frac{T_1 - T_2}{T_1} = \frac{45\text{J} - 9.63\text{J}}{45\text{J}} = 0.786 \quad\blacktriangleleft$$

SOLVING PROBLEMS
ON YOUR OWN

In this lesson we integrated Newton's second law to derive the *principle of impulse and momentum* for a particle. Recalling that the *linear momentum* of a particle was defined as the product of its mass m and its velocity $\mathbf{v}$ [Sec. 12.3], we wrote

$$m\mathbf{v}_1 + \Sigma\,\mathbf{Imp}_{1\to2} = m\mathbf{v}_2 \tag{13.32}$$

This equation expresses that the linear momentum $m\mathbf{v}_2$ of a particle at time t_2 can be obtained by adding to its linear momentum $m\mathbf{v}_1$ at time t_1 the *impulses* of the forces exerted on the particle during the time interval t_1 to t_2. For computing purposes, the momenta and impulses may be expressed in terms of their rectangular components, and Eq. (13.32) can be replaced by the equivalent scalar equations. The units of momentum and impulse are $\mathrm{N \cdot s}$ in the SI system of units and $\mathrm{lb \cdot s}$ in U.S. customary units. To solve problems using this equation you can follow these steps:

1. Draw a diagram showing the particle, its momentum at t_1 and at t_2, and the impulses of the forces exerted on the particle during the time interval t_1 to t_2.

2. Calculate the impulse of each force, expressing it in terms of its rectangular components if more than one direction is involved. You may encounter the following cases:

  a. The time interval is finite and the force is constant.

$$\mathbf{Imp}_{1\to2} = \mathbf{F}(t_2 - t_1)$$

  b. The time interval is finite and the force is a function of *t*.

$$\mathbf{Imp}_{1\to2} = \int_{t_1}^{t_2} \mathbf{F}(t)\,dt$$

  c. The time interval is very small and the force is very large. The force is called an *impulsive force* and its impulse over the time interval $t_2 - t_1 = \Delta t$ is

$$\mathbf{Imp}_{1\to2} = \mathbf{F}\,\Delta t$$

Note that this impulse is *zero for a nonimpulsive force* such as the *weight* of a body, the force exerted by a *spring*, or any other force which is known to be small by comparison with the impulsive forces. Unknown reactions, however, *cannot be assumed* to be nonimpulsive and their impulses should be taken into account.

3. Substitute the values obtained for the impulses into Eq. (13.32) or into the equivalent scalar equations. You will find that the forces and velocities in the problems of this lesson are contained in a plane. You will, therefore, write two scalar equations and solve these equations for *two unknowns*. These unknowns may be a *time* [Sample Prob. 13.10], a *velocity* and an *impulse* [Sample Prob. 13.12], or an *average impulsive force* [Sample Prob. 13.11].

4. When several particles are involved, a separate diagram should be drawn for each particle, showing the initial and final momentum of the particle, as well as the impulses of the forces exerted on the particle.

a. It is usually convenient, however, to first consider a diagram including all the particles. This diagram leads to the equation

$$\Sigma m\mathbf{v}_1 + \Sigma \ \mathbf{Imp}_{1\rightarrow2} = \Sigma m\mathbf{v}_2 \qquad (13.33)$$

where the impulses of *only the forces external to the system* need be considered. Therefore, the two equivalent scalar equations will not contain any of the impulses of the unknown internal forces.

b. If the sum of the impulses of the external forces is zero, Eq. (13.33) reduces to

$$\Sigma m\mathbf{v}_1 = \Sigma m\mathbf{v}_2 \qquad (13.34)$$

which expresses that *the total momentum of the particles is conserved*. This occurs either if the resultant of the external forces is zero or, when the time interval Δt is very short (impulsive motion), if all the external forces are nonimpulsive. Keep in mind, however, that the total momentum may be conserved *in one direction*, but not in another [Sample Prob. 13.12].

PROBLEMS

CONCEPT QUESTIONS

13.CQ4 A large insect impacts the front windshield of a sports car traveling down a road. Which of the following statements is true during the collision?

    a. The car exerts a greater force on the insect than the insect exerts on the car.

    b. The insect exerts a greater force on the car than the car exerts on the insect.

    c. The car exerts a force on the insect, but the insect does not exert a force on the car.

    d. The car exerts the same force on the insect as the insect exerts on the car.

    e. Neither exerts a force on the other; the insect gets smashed simply because it gets in the way of the car.

Case 1

13.CQ5 The expected damages associated with two types of perfectly plastic collisions are to be compared. In the first case, two identical cars traveling at the same speed impact each other head-on. In the second case, the car impacts a massive concrete wall. In which case would you expect the car to be more damaged?

    a. Case 1

    b. Case 2

    c. The same damage in each case

Case 2

Fig. P13.CQ5

IMPULSE-MOMENTUM PRACTICE PROBLEMS

13.F1 The initial velocity of the block in position A is 30 ft/s. The coefficient of kinetic friction between the block and the plane is $\mu_k = 0.30$. Draw the impulse-momentum diagram that can be used to determine the time it takes for the block to reach B with zero velocity, if $\theta = 20°$.

13.F2 A 4-lb collar which can slide on a frictionless vertical rod is acted upon by a force **P** which varies in magnitude as shown. Knowing that the collar is initially at rest, draw the impulse-momentum diagram that can be used to determine its velocity at $t = 3$ s.

Fig. P13.F1

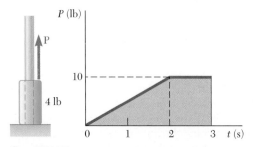

Fig. P13.F2

13.F3 The 15-kg suitcase A has been propped up against one end of a 40-kg luggage carrier B and is prevented from sliding down by other luggage. When the luggage is unloaded and the last heavy trunk is removed from the carrier, the suitcase is free to slide down, causing the 40-kg carrier to move to the left with a velocity v_B of magnitude 0.8 m/s. Neglecting friction, draw the impulse-momentum diagrams that can be used to determine (a) the velocity of A as it rolls on the carrier, (b) the velocity of the carrier after the suitcase hits the right side of the carrier without bouncing back.

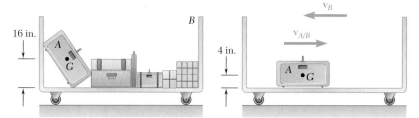

Fig. P13.F3

13.F4 Car A was traveling west at a speed of 15 m/s and car B was traveling north at an unknown speed when they slammed into each other at an intersection. Upon investigation it was found that after the crash the two cars got stuck and skidded off at an angle of 50° north of east. Knowing the masses of A and B are m_A and m_B, respectively, draw the impulse-momentum diagram that can be used to determine the velocity of B before impact.

Fig. P13.F4

13.F5 Two identical spheres A and B, each of mass m, are attached to an inextensible inelastic cord of length L and are resting at a distance a from each other on a frictionless horizontal surface. Sphere B is given a velocity $\mathbf{v}_0$ in a direction perpendicular to line AB and moves it without friction until it reaches B' where the cord becomes taut. Draw the impulse-momentum diagram that can be used to determine the magnitude of the velocity of each sphere immediately after the cord has become taut.

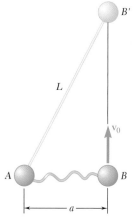

Fig. P13.F5

END-OF-SECTION PROBLEMS

13.119 A 35 000-Mg ocean liner has an initial velocity of 4 km/h. Neglecting the frictional resistance of the water, determine the time required to bring the liner to rest by using a single tugboat which exerts a constant force of 150 kN.

Fig. P13.121

13.120 A 2500-lb automobile is moving at a speed of 60 mi/h when the brakes are fully applied, causing all four wheels to skid. Determine the time required to stop the automobile (a) on dry pavement ($\mu_k = 0.75$), (b) on an icy road ($\mu_k = 0.10$).

13.121 A sailboat weighing 980 lb with its occupants is running down wind at 8 mi/h when its spinnaker is raised to increase its speed. Determine the net force provided by the spinnaker over the 10-s interval that it takes for the boat to reach a speed of 12 mi/h.

13.122 A truck is hauling a 300-kg log out of a ditch using a winch attached to the back of the truck. Knowing the winch applies a constant force of 2500 N and the coefficient of kinetic friction between the ground and the log is 0.45, determine the time for the log to reach a speed of 0.5 m/s.

Fig. *P13.122*

13.123 A truck is traveling down a road with a 3-percent grade at a speed of 55 mi/h when the brakes are applied. Knowing the coefficients of friction between the load and the flatbed trailer shown are $\mu_s = 0.40$ and $\mu_k = 0.35$, determine the shortest time in which the rig can be brought to a stop if the load is not to shift.

Fig. P13.123

Fig. P13.124

13.124 Steep safety ramps are built beside mountain highways to enable vehicles with defective brakes to stop. A 10-ton truck enters a 15° ramp at a high speed $v_0 = 108$ ft/s and travels for 6 s before its speed is reduced to 36 ft/s. Assuming constant deceleration, determine (a) the magnitude of the braking force, (b) the additional time required for the truck to stop. Neglect air resistance and rolling resistance.

13.125 Baggage on the floor of the baggage car of a high-speed train is not prevented from moving other than by friction. The train is traveling down a 5-percent grade when it decreases its speed at a constant rate from 120 mi/h to 60 mi/h in a time interval of 12 s. Determine the smallest allowable value of the coefficient of static friction between a trunk and the floor of the baggage car if the trunk is not to slide.

13.126 A 2-kg particle is acted upon by the force, expressed in newtons, $\mathbf{F} = (8 - 6t)\mathbf{i} + (4 - t^2)\mathbf{j} + (4 + t)\mathbf{k}$. Knowing that the velocity of the particle is $\mathbf{v} = (150 \text{ m/s})\mathbf{i} + (100 \text{ m/s})\mathbf{j} - (250 \text{ m/s})\mathbf{k}$ at $t = 0$, determine (a) the time at which the velocity of the particle is parallel to the yz plane, (b) the corresponding velocity of the particle.

13.127 A truck is traveling down a road with a 4-percent grade at a speed of 60 mi/h when its brakes are applied to slow it down to 20 mi/h. An antiskid braking system limits the braking force to a value at which the wheels of the truck are just about to slide. Knowing that the coefficient of static friction between the road and the wheels is 0.60, determine the shortest time needed for the truck to slow down.

13.128 Skid marks on a drag race track indicate that the rear (drive) wheels of a car slip for the first 20 m of the 400-m track. (*a*) Knowing that the coefficient of kinetic friction is 0.60, determine the shortest possible time for the car to travel the initial 20-m portion of the track if it starts from rest with its front wheels just off the ground. (*b*) Determine the minimum time for the car to run the whole race if, after skidding for 20 m, the wheels roll without sliding for the remainder of the race. Assume for the rolling portion of the race that 65 percent of the weight is on the rear wheels and that the coefficient of static friction is 0.85. Ignore air resistance and rolling resistance.

Fig. *P13.128*

13.129 The subway train shown is traveling at a speed of 30 mi/h when the brakes are fully applied on the wheels of cars *B* and *C*, causing them to slide on the track, but are not applied on the wheels of car *A*. Knowing that the coefficient of kinetic friction is 0.35 between the wheels and the track, determine (*a*) the time required to bring the train to a stop, (*b*) the force in each coupling.

Fig. P13.129

13.130 Solve Prob. 13.129, assuming that the brakes are applied only on the wheels of car *A*.

13.131 A trailer truck with a 2000-kg cab and an 8000-kg trailer is traveling on a level road at 90 km/h. The brakes on the trailer fail and the antiskid system of the cab provides the largest possible force which will not cause the wheels of the cab to slide. Knowing that the coefficient of static friction is 0.65, determine (*a*) the shortest time for the rig to come to a stop, (*b*) the force in the coupling during that time.

Fig. P13.131

13.132 The system shown is at rest when a constant 150-N force is applied to collar *B*. Neglecting the effect of friction, determine (*a*) the time at which the velocity of collar *B* will be 2.5 m/s to the left, (*b*) the corresponding tension in the cable.

Fig. P13.132

Fig. P13.133

13.133 An 8-kg cylinder C rests on a 4-kg platform A supported by a cord which passes over the pulleys D and E and is attached to a 4-kg block B. Knowing that the system is released from rest, determine (*a*) the velocity of block B after 0.8 s, (*b*) the force exerted by the cylinder on the platform.

13.134 An estimate of the expected load on over-the-shoulder seat belts is to be made before designing prototype belts that will be evaluated in automobile crash tests. Assuming that an automobile traveling at 45 mi/h is brought to a stop in 110 ms, determine (*a*) the average impulsive force exerted by a 200-lb man on the belt, (*b*) the maximum force F_m exerted on the belt if the force-time diagram has the shape shown.

Fig. P13.134

13.135 A 60-g model rocket is fired vertically. The engine applies a thrust **P** which varies in magnitude as shown. Neglecting air resistance and the change in mass of the rocket, determine (*a*) the maximum speed of the rocket as it goes up, (*b*) the time for the rocket to reach its maximum elevation.

Fig. P13.135

Fig. P13.136

13.136 A simplified model consisting of a single straight line is to be obtained for the variation of pressure inside the 10-mm-diameter barrel of a rifle as a 20-g bullet is fired. Knowing that it takes 1.6 ms for the bullet to travel the length of the barrel and that the velocity of the bullet upon exit is 700 m/s, determine the value of p_0.

13.137 A 125-lb block initially at rest is acted upon by a force **P** which varies as shown. Knowing that the coefficients of friction between the block and the horizontal surface are $\mu_s = 0.50$ and $\mu_k = 0.40$, determine (a) the time at which the block will start moving, (b) the maximum speed reached by the block, (c) the time at which the block will stop moving.

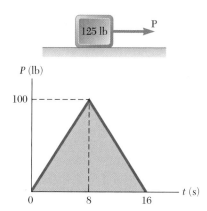

Fig. P13.137

13.138 Solve Prob. 13.137, assuming that the weight of the block is 175 lb.

13.139 A baseball player catching a ball can soften the impact by pulling his hand back. Assuming that a 5-oz ball reaches his glove at 90 mi/h and that the player pulls his hand back during the impact at an average speed of 30 ft/s over a distance of 6 in., bringing the ball to a stop, determine the average impulsive force exerted on the player's hand.

Fig. P13.139

13.140 A 1.62-oz golf ball is hit with a golf club and leaves it with a velocity of 100 mi/h. We assume that for $0 \leq t \leq t_0$, where t_0 is the duration of the impact, the magnitude F of the force exerted on the ball can be expressed as $F = F_m \sin(\pi t/t_0)$. Knowing that $t_0 = 0.5$ ms, determine the maximum value F_m of the force exerted on the ball.

13.141 The triple jump is a track-and-field event in which an athlete gets a running start and tries to leap as far as he can with a hop, step, and jump. Shown in the figure is the initial hop of the athlete. Assuming that he approaches the takeoff line from the left with a horizontal velocity of 10 m/s, remains in contact with the ground for 0.18 s, and takes off at a 50° angle with a velocity of 12 m/s, determine the vertical component of the average impulsive force exerted by the ground on his foot. Give your answer in terms of the weight W of the athlete.

Fig. P13.141

13.142 The last segment of the triple jump track-and-field event is the jump, in which the athlete makes a final leap, landing in a sand-filled pit. Assuming that the velocity of a 80-kg athlete just before landing is 9 m/s at an angle of 35° with the horizontal and that the athlete comes to a complete stop in 0.22 s after landing, determine the horizontal component of the average impulsive force exerted on his feet during landing.

Fig. P13.142

13.143 The design for a new cementless hip implant is to be studied using an instrumented implant and a fixed simulated femur. Assuming the punch applies an average force of 2 kN over a time of 2 ms to the 200-g implant, determine (a) the velocity of the implant immediately after impact, (b) the average resistance of the implant to penetration if the implant moves 1 mm before coming to rest.

Fig. P13.143

13.144 A 25-g steel-jacketed bullet is fired horizontally with a velocity of 600 m/s and ricochets off a steel plate along the path CD with a velocity of 400 m/s. Knowing that the bullet leaves a 10-mm scratch on the plate and assuming that its average speed is 500 m/s while it is in contact with the plate, determine the magnitude and direction of the average impulsive force exerted by the bullet on the plate.

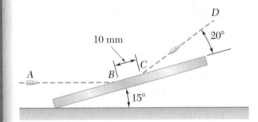

Fig. P13.144

13.145 A 25-ton railroad car moving at 2.5 mi/h is to be coupled to a 50-ton car which is at rest with locked wheels ($\mu_k = 0.30$). Determine (a) the velocity of both cars after the coupling is completed, (b) the time it takes for both cars to come to rest.

Fig. P13.145

13.146 At an intersection car B was traveling south and car A was traveling 30° north of east when they slammed into each other. Upon investigation it was found that after the crash the two cars got stuck and skidded off at an angle of 10° north of east. Each driver claimed that he was going at the speed limit of 50 km/h and that he tried to slow down but couldn't avoid the crash because the other driver was going a lot faster. Knowing that the masses of cars A and B were 1500 kg and 1200 kg, respectively, determine (a) which car was going faster, (b) the speed of the faster of the two cars if the slower car was traveling at the speed limit.

Fig. P13.146

13.147 The 650-kg hammer of a drop-hammer pile driver falls from a height of 1.2 m onto the top of a 140-kg pile, driving it 110 mm into the ground. Assuming perfectly plastic impact ($e = 0$), determine the average resistance of the ground to penetration.

Fig. P13.147

13.148 A small rivet connecting two pieces of sheet metal is being clinched by hammering. Determine the impulse exerted on the rivet and the energy absorbed by the rivet under each blow, knowing that the head of the hammer has a weight of 1.5 lb and that it strikes the rivet with a velocity of 20 ft/s. Assume that the hammer does not rebound and that the anvil is supported by springs and (a) has an infinite mass (rigid support), (b) has a weight of 9 lb.

Fig. P13.148

13.149 Bullet B weighs 0.5 oz and blocks A and C both weigh 3 lb. The coefficient of friction between the blocks and the plane is $\mu_k = 0.25$. Initially the bullet is moving at v_0 and blocks A and C are at rest (Fig. 1). After the bullet passes through A it becomes embedded in block C and all three objects come to stop in the positions shown (Fig. 2). Determine the initial speed of the bullet v_0.

Fig. P13.149

13.150 A 180-lb man and a 120-lb woman stand at opposite ends of a 300-lb boat, ready to dive, each with a 16-ft/s velocity relative to the boat. Determine the velocity of the boat after they have both dived, if (a) the woman dives first, (b) the man dives first.

Fig. P13.150

Fig. P13.151

13.151 A 75-g ball is projected from a height of 1.6 m with a horizontal velocity of 2 m/s and bounces from a 400-g smooth plate supported by springs. Knowing that the height of the rebound is 0.6 m, determine (a) the velocity of the plate immediately after the impact, (b) the energy lost due to the impact.

13.152 A 2-kg sphere A is connected to a fixed point O by an inextensible cord of length 1.2 m. The sphere is resting on a frictionless horizontal surface at a distance of 0.5 m from O when it is given a velocity v_0 in a direction perpendicular to line OA. It moves freely until it reaches position A', when the cord becomes taut. Determine the maximum allowable velocity v_0 if the impulse of the force exerted on the cord is not to exceed 3 N · s.

Fig. P13.152

13.153 A 1-oz bullet is traveling with a velocity of 1400 ft/s when it impacts and becomes embedded in a 5-lb wooden block. The block can move vertically without friction. Determine (a) the velocity of the bullet and block immediately after the impact, (b) the horizontal and vertical components of the impulse exerted by the block on the bullet.

13.154 In order to test the resistance of a chain to impact, the chain is suspended from a 240-lb rigid beam supported by two columns. A rod attached to the last link is then hit by a 60-lb block dropped from a 5-ft height. Determine the initial impulse exerted on the chain and the energy absorbed by the chain, assuming that the block does not rebound from the rod and that the columns supporting the beam are (a) perfectly rigid, (b) equivalent to two perfectly elastic springs.

Fig. P13.153

Fig. P13.154

13.12 IMPACT

A collision between two bodies which occurs in a very small interval of time and during which the two bodies exert relatively large forces on each other is called an *impact*. The common normal to the surfaces in contact during the impact is called the *line of impact*. If the mass centers on the two colliding bodies are located on this line, the impact is a *central impact*. Otherwise, the impact is said to be *eccentric*. Our present study will be limited to the central impact of two particles. The analysis of the eccentric impact of two rigid bodies will be considered later, in Sec. 17.12.

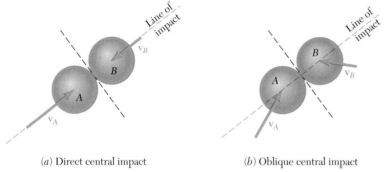

(a) Direct central impact   (b) Oblique central impact

Fig. 13.20

If the velocities of the two particles are directed along the line of impact, the impact is said to be a *direct impact* (Fig. 13.20a). If either or both particles move along a line other than the line of impact, the impact is said to be an *oblique impact* (Fig. 13.20b).

13.13 DIRECT CENTRAL IMPACT

Consider two particles A and B, of mass m_A and m_B, which are moving in the same straight line and to the right with known velocities $\mathbf{v}_A$ and $\mathbf{v}_B$ (Fig. 13.21a). If $\mathbf{v}_A$ is larger than $\mathbf{v}_B$, particle A will eventually strike particle B. Under the impact, the two particles will *deform* and, at the end of the period of deformation, they will have the same velocity $\mathbf{u}$ (Fig. 13.21b). A period of *restitution* will then take place, at the end of which, depending upon the magnitude of the impact forces and upon the materials involved, the two particles either will have regained their original shape or will stay permanently deformed. Our purpose here is to determine the velocities $\mathbf{v}'_A$ and $\mathbf{v}'_B$ of the particles at the end of the period of restitution (Fig. 13.21c).

Considering first the two particles as a single system, we note that there is no impulsive, external force. Thus, the total momentum of the two particles is conserved, and we write

$$m_A\mathbf{v}_A + m_B\mathbf{v}_B = m_A\mathbf{v}'_A + m_B\mathbf{v}'_B$$

Since all the velocities considered are directed along the same axis, we can replace the equation obtained by the following relation involving only scalar components:

$$m_A v_A + m_B v_B = m_A v'_A + m_B v'_B \tag{13.37}$$

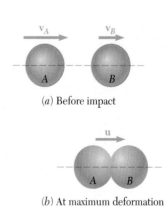

(a) Before impact

(b) At maximum deformation

(c) After impact

Fig. 13.21

A positive value for any of the scalar quantities v_A, v_B, v'_A, or v'_B means that the corresponding vector is directed to the right; a negative value indicates that the corresponding vector is directed to the left.

To obtain the velocities $\mathbf{v}'_A$ and $\mathbf{v}'_B$, it is necessary to establish a second relation between the scalars v'_A and v'_B. For this purpose, let us now consider the motion of particle A during the period of deformation and apply the principle of impulse and momentum. Since the only impulsive force acting on A during this period is the force $\mathbf{P}$ exerted by B (Fig. 13.22a), we write, using again scalar components,

$$m_A v_A - \int P\, dt = m_A u \qquad (13.38)$$

where the integral extends over the period of deformation. Considering now the motion of A during the period of restitution, and denoting by $\mathbf{R}$ the force exerted by B on A during this period (Fig. 13.22b), we write

$$m_A u - \int R\, dt = m_A v'_A \qquad (13.39)$$

where the integral extends over the period of restitution.

(a) Period of deformation

(b) Period of restitution

Fig. 13.22

In general, the force $\mathbf{R}$ exerted on A during the period of restitution differs from the force $\mathbf{P}$ exerted during the period of deformation, and the magnitude $\int R\, dt$ of its impulse is smaller than the magnitude $\int P\, dt$ of the impulse of $\mathbf{P}$. The ratio of the magnitudes of the impulses corresponding, respectively, to the period of restitution and to the period of deformation is called the *coefficient of restitution* and is denoted by e. We write

$$e = \frac{\int R\, dt}{\int P\, dt} \qquad (13.40)$$

The value of the coefficient e is always between 0 and 1. It depends to a large extent on the two materials involved, but it also varies considerably with the impact velocity and the shape and size of the two colliding bodies.

Solving Eqs. (13.38) and (13.39) for the two impulses and substituting into (13.40), we write

$$e = \frac{u - v'_A}{v_A - u} \qquad (13.41)$$

A similar analysis of particle B leads to the relation

$$e = \frac{v'_B - u}{u - v_B} \tag{13.42}$$

Since the quotients in (13.41) and (13.42) are equal, they are also equal to the quotient obtained by adding, respectively, their numerators and their denominators. We have, therefore,

$$e = \frac{(u - v'_A) + (v'_B - u)}{(v_A - u) + (u - v_B)} = \frac{v'_B - v'_A}{v_A - v_B}$$

and

$$v'_B - v'_A = e(v_A - v_B) \tag{13.43}$$

Photo 13.3 The height the tennis ball bounces decreases after each impact because it has a coefficient of restitution less than one and energy is lost with each bounce.

Since $v'_B - v'_A$ represents the relative velocity of the two particles after impact and $v_A - v_B$ represents their relative velocity before impact, formula (13.43) expresses that *the relative velocity of the two particles after impact can be obtained by multiplying their relative velocity before impact by the coefficient of restitution.* This property is used to determine experimentally the value of the coefficient of restitution of two given materials.

The velocities of the two particles after impact can now be obtained by solving Eqs. (13.37) and (13.43) simultaneously for v'_A and v'_B. It is recalled that the derivation of Eqs. (13.37) and (13.43) was based on the assumption that particle B is located to the right of A, and that both particles are initially moving to the right. If particle B is initially moving to the left, the scalar v_B should be considered negative. The same sign convention holds for the velocities after impact: A positive sign for v'_A will indicate that particle A moves to the right after impact, and a negative sign will indicate that it moves to the left.

Two particular cases of impact are of special interest:

1. $e = 0$, *Perfectly Plastic Impact.* When $e = 0$, Eq. (13.43) yields $v'_B = v'_A$. There is no period of restitution, and both particles stay together after impact. Substituting $v'_B = v'_A = v'$ into Eq. (13.37), which expresses that the total momentum of the particles is conserved, we write

$$m_A v_A + m_B v_B = (m_A + m_B)v' \tag{13.44}$$

This equation can be solved for the common velocity v' of the two particles after impact.

2. $e = 1$, *Perfectly Elastic Impact.* When $e = 1$, Eq. (13.43) reduces to

$$v'_B - v'_A = v_A - v_B \tag{13.45}$$

which expresses that the relative velocities before and after impact are equal. The impulses received by each particle during the period of deformation and during the period of restitution are equal. The particles move away from each other after impact with the same velocity with which they approached each

other before impact. The velocities v'_A and v'_B can be obtained
by solving Eqs. (13.37) and (13.45) simultaneously.

It is worth noting that *in the case of a perfectly elastic impact, the
total energy of the two particles,* as well as their total momentum, *is
conserved.* Equations (13.37) and (13.45) can be written as follows:

$$m_A(v_A - v'_A) = m_B(v'_B - v_B) \qquad (13.37')$$
$$v_A + v'_A = v_B + v'_B \qquad (13.45')$$

Multiplying (13.37′) and (13.45′) member by member, we have

$$m_A(v_A - v'_A)(v_A + v'_A) = m_B(v'_B - v_B)(v'_B + v_B)$$
$$m_A v_A^2 - m_A(v'_A)^2 = m_B(v'_B)^2 - m_B v_B^2$$

Rearranging the terms in the equation obtained and multiplying by $\frac{1}{2}$,
we write

$$\tfrac{1}{2}m_A v_A^2 + \tfrac{1}{2}m_B v_B^2 = \tfrac{1}{2}m_A(v'_A)^2 + \tfrac{1}{2}m_B(v'_B)^2 \qquad (13.46)$$

which expresses that the kinetic energy of the particles is conserved.
It should be noted, however, that *in the general case of impact,* i.e.,
when e is not equal to 1, *the total energy of the particles is not conserved.* This can be shown in any given case by comparing the kinetic
energies before and after impact. The lost kinetic energy is in part
transformed into heat and in part spent in generating elastic waves
within the two colliding bodies.

13.14 OBLIQUE CENTRAL IMPACT

Let us now consider the case when the velocities of the two colliding
particles are *not* directed along the line of impact (Fig. 13.23). As
indicated in Sec. 13.12, the impact is said to be *oblique.* Since the

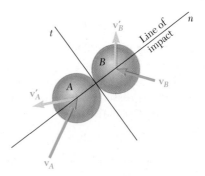

Fig. 13.23

velocities $\mathbf{v}'_A$ and $\mathbf{v}'_B$ of the particles after impact are unknown in
direction as well as in magnitude, their determination will require
the use of four independent equations.

We choose as coordinate axes the n axis along the line of impact,
i.e., along the common normal to the surfaces in contact, and the t
axis along their common tangent. Assuming that the particles are
perfectly *smooth and frictionless,* we observe that the only impulses

Photo 13.4 When pool balls strike each other
there is a transfer of momentum.

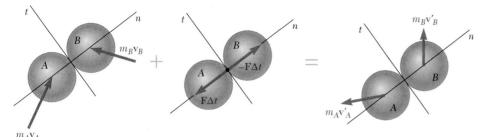

Fig. 13.24

exerted on the particles during the impact are due to internal forces directed along the line of impact, i.e., along the n axis (Fig. 13.24). It follows that

1. The component along the t axis of the momentum of each particle, considered separately, is conserved; hence the t component of the velocity of each particle remains unchanged. We write

$$(v_A)_t = (v'_A)_t \qquad (v_B)_t = (v'_B)_t \qquad (13.47)$$

2. The component along the n axis of the total momentum of the two particles is conserved. We write

$$m_A(v_A)_n + m_B(v_B)_n = m_A(v'_A)_n + m_B(v'_B)_n \qquad (13.48)$$

3. The component along the n axis of the relative velocity of the two particles after impact is obtained by multiplying the n component of their relative velocity before impact by the coefficient of restitution. Indeed, a derivation similar to that given in Sec. 13.13 for direct central impact yields

$$(v'_B)_n - (v'_A)_n = e[(v_A)_n - (v_B)_n] \qquad (13.49)$$

We have thus obtained four independent equations which can be solved for the components of the velocities of A and B after impact. This method of solution is illustrated in Sample Prob. 13.15.

Our analysis of the oblique central impact of two particles has been based so far on the assumption that both particles moved freely before and after the impact. Let us now examine the case when one or both of the colliding particles is constrained in its motion. Consider, for instance, the collision between block A, which is constrained to move on a horizontal surface, and ball B, which is free to move in the plane of the figure (Fig. 13.25). Assuming no friction between the block and the ball, or between the block and the horizontal surface, we note that the impulses exerted on the system consist of the impulses of the internal forces $\mathbf{F}$ and $-\mathbf{F}$ directed along the line of impact, i.e., along the n axis, and of the impulse of the external force $\mathbf{F}_{ext}$ exerted by the horizontal surface on block A and directed along the vertical (Fig. 13.26).

The velocities of block A and ball B immediately after the impact are represented by three unknowns: the magnitude of the velocity $\mathbf{v}'_A$ of block A, which is known to be horizontal, and the magnitude and

Fig. 13.25

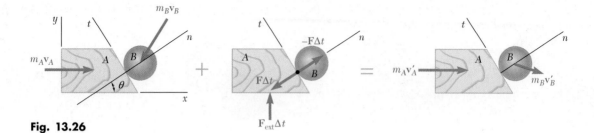

Fig. 13.26

direction of the velocity $\mathbf{v}'_B$ of ball B. We must therefore write three equations by expressing that

1. The component along the t axis of the momentum of ball B is conserved; hence the t component of the velocity of ball B remains unchanged. We write

$$(v_B)_t = (v'_B)_t \tag{13.50}$$

2. The component along the horizontal x axis of the total momentum of block A and ball B is conserved. We write

$$m_A v_A + m_B(v_B)_x = m_A v'_A + m_B(v'_B)_x \tag{13.51}$$

3. The component along the n axis of the relative velocity of block A and ball B after impact is obtained by multiplying the n component of their relative velocity before impact by the coefficient of restitution. We write again

$$(v'_B)_n - (v'_A)_n = e[(v_A)_n - (v_B)_n] \tag{13.49}$$

We should note, however, that in the case considered here, the validity of Eq. (13.49) cannot be established through a mere extension of the derivation given in Sec. 13.13 for the direct central impact of two particles moving in a straight line. Indeed, these particles were not subjected to any external impulse, while block A in the present analysis is subjected to the impulse exerted by the horizontal surface. To prove that Eq. (13.49) is still valid, we will first apply the principle of impulse and momentum to block A over the period of deformation (Fig. 13.27). Considering only the horizontal components, we write

$$m_A v_A - (\smallint P \, dt) \cos \theta = m_A u \tag{13.52}$$

where the integral extends over the period of deformation and where $\mathbf{u}$ represents the velocity of block A at the end of that period. Considering now the period of restitution, we write in a similar way

$$m_A u - (\smallint R \, dt) \cos \theta = m_A v'_A \tag{13.53}$$

where the integral extends over the period of restitution.

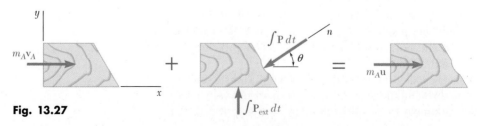

Fig. 13.27

Recalling from Sec. 13.13 the definition of the coefficient of restitution, we write

$$e = \frac{\int R\, dt}{\int P\, dt} \tag{13.40}$$

Solving Eqs. (13.52) and (13.53) for the integrals $\int P\, dt$ and $\int R\, dt$, and substituting into Eq. (13.40), we have, after reductions,

$$e = \frac{u - v'_A}{v_A - u}$$

or, multiplying all velocities by $\cos\theta$ to obtain their projections on the line of impact.

$$e = \frac{u_n - (v'_A)_n}{(v_A)_n - u_n} \tag{13.54}$$

We note that Eq. (13.54) is identical to Eq. (13.41) of Sec. 13.13, except for the subscripts n which are used here to indicate that we are considering velocity components along the line of impact. Since the motion of ball B is unconstrained, the proof of Eq. (13.49) can be completed in the same manner as the derivation of Eq. (13.43) of Sec. 13.13. Thus, we conclude that the relation (13.49) between the components along the line of impact of the relative velocities of two colliding particles remains valid when one of the particles is constrained in its motion. The validity of this relation is easily extended to the case when both particles are constrained in their motion.

13.15 PROBLEMS INVOLVING ENERGY AND MOMENTUM

You now have at your disposal three different methods for the solution of kinetics problems: the direct application of Newton's second law, $\Sigma\mathbf{F} = m\mathbf{a}$; the method of work and energy; and the method of impulse and momentum. To derive maximum benefit from these three methods, you should be able to choose the method best suited for the solution of a given problem. You should also be prepared to use different methods for solving the various parts of a problem when such a procedure seems advisable.

You have already seen that the method of work and energy is in many cases more expeditious than the direct application of Newton's second law. As indicated in Sec. 13.4, however, the method of work and energy has limitations, and it must sometimes be supplemented by the use of $\Sigma\mathbf{F} = m\mathbf{a}$. This is the case, for example, when you wish to determine an acceleration or a normal force.

For the solution of problems involving no impulsive forces, it will usually be found that the equation $\Sigma\mathbf{F} = m\mathbf{a}$ yields a solution just as fast as the method of impulse and momentum and that the method of work and energy, if it applies, is more rapid and more convenient. However, in problems of impact, the method of impulse and momentum is the only practicable method. A solution based on the direct application of $\Sigma\mathbf{F} = m\mathbf{a}$ would be unwieldy, and the method of work and energy cannot be used since impact (unless perfectly elastic) involves a loss of mechanical energy.

Many problems involve only conservative forces, except for a short impact phase during which impulsive forces act. The solution of such problems can be divided into several parts. The part corresponding to the impact phase calls for the use of the method of impulse and momentum and of the relation between relative velocities, and the other parts can usually be solved by the method of work and energy. If the problem involves the determination of a normal force, however, the use of $\Sigma\mathbf{F} = m\mathbf{a}$ is necessary.

Consider, for example, a pendulum A, of mass m_A and length l, which is released with no velocity from a position A_1 (Fig. 13.28a). The pendulum swings freely in a vertical plane and hits a second pendulum B, of mass m_B and same length l, which is initially at rest. After the impact (with coefficient of restitution e), pendulum B swings through an angle θ that we wish to determine.

The solution of the problem can be divided into three parts:

1. *Pendulum A Swings from A_1 to A_2.* The principle of conservation of energy can be used to determine the velocity $(\mathbf{v}_A)_2$ of the pendulum at A_2 (Fig. 13.28b).
2. *Pendulum A Hits Pendulum B.* Using the fact that the total momentum of the two pendulums is conserved and the relation between their relative velocities, we determine the velocities $(\mathbf{v}_A)_3$ and $(\mathbf{v}_B)_3$ of the two pendulums after impact (Fig. 13.28c).
3. *Pendulum B Swings from B_3 to B_4.* Applying the principle of conservation of energy to pendulum B, we determine the maximum elevation y_4 reached by that pendulum (Fig. 13.28d). The angle θ can then be determined by trigonometry.

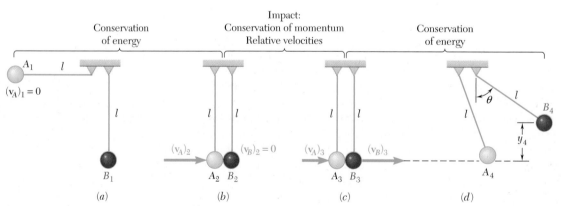

Fig. 13.28

We note that if the tensions in the cords holding the pendulums are to be determined, the method of solution just described should be supplemented by the use of $\Sigma\mathbf{F} = m\mathbf{a}$.

SAMPLE PROBLEM 13.13

A 20-Mg railroad car moving at a speed of 0.5 m/s to the right collides with a 35-Mg car which is at rest. If after the collision the 35-Mg car is observed to move to the right at a speed of 0.3 m/s, determine the coefficient of restitution between the two cars.

SOLUTION

We express that the total momentum of the two cars is conserved.

$$m_A\mathbf{v}_A + m_B\mathbf{v}_B = m_A\mathbf{v}'_A + m_B\mathbf{v}'_B$$
$$(20 \text{ Mg})(+0.5 \text{ m/s}) + (35 \text{ Mg})(0) = (20 \text{ Mg})v'_A + (35 \text{ Mg})(+0.3 \text{ m/s})$$
$$v'_A = -0.025 \text{ m/s} \qquad \mathbf{v}'_A = 0.025 \text{ m/s} \leftarrow$$

The coefficient of restitution is obtained by writing

$$e = \frac{v'_B - v'_A}{v_A - v_B} = \frac{+0.3 - (-0.025)}{+0.5 - 0} = \frac{0.325}{0.5} \qquad e = 0.65 \ \blacktriangleleft$$

SAMPLE PROBLEM 13.14

A ball is thrown against a frictionless, vertical wall. Immediately before the ball strikes the wall, its velocity has a magnitude v and forms an angle of 30° with the horizontal. Knowing that $e = 0.90$, determine the magnitude and direction of the velocity of the ball as it rebounds from the wall.

SOLUTION

We resolve the initial velocity of the ball into components respectively perpendicular and parallel to the wall:

$$v_n = v \cos 30° = 0.866v \qquad v_t = v \sin 30° = 0.500v$$

Motion Parallel to the Wall. Since the wall is frictionless, the impulse it exerts on the ball is perpendicular to the wall. Thus, the component parallel to the wall of the momentum of the ball is conserved and we have

$$\mathbf{v}'_t = \mathbf{v}_t = 0.500v \uparrow$$

Motion Perpendicular to the Wall. Since the mass of the wall (and earth) is essentially infinite, expressing that the total momentum of the ball and wall is conserved would yield no useful information. Using the relation (13.49) between relative velocities, we write

$$0 - v'_n = e(v_n - 0)$$
$$v'_n = -0.90(0.866v) = -0.779v \qquad \mathbf{v}'_n = 0.779v \leftarrow$$

Resultant Motion. Adding vectorially the components $\mathbf{v}'_n$ and $\mathbf{v}'_t$,

$$\mathbf{v}' = 0.926v \ \angle \ 32.7° \ \blacktriangleleft$$

SAMPLE PROBLEM 13.15

The magnitude and direction of the velocities of two identical frictionless balls before they strike each other are as shown. Assuming $e = 0.90$, determine the magnitude and direction of the velocity of each ball after the impact.

SOLUTION

The impulsive forces that the balls exert on each other during the impact are directed along a line joining the centers of the balls called the *line of impact*. Resolving the velocities into components directed, respectively, along the line of impact and along the common tangent to the surfaces in contact, we write

$$(v_A)_n = v_A \cos 30° = +26.0 \text{ ft/s}$$
$$(v_A)_t = v_A \sin 30° = +15.0 \text{ ft/s}$$
$$(v_B)_n = -v_B \cos 60° = -20.0 \text{ ft/s}$$
$$(v_B)_t = v_B \sin 60° = +34.6 \text{ ft/s}$$

Principle of Impulse and Momentum. In the adjoining sketches we show in turn the initial momenta, the impulses, and the final momenta.

Motion Along the Common Tangent. Considering only the t components, we apply the principle of impulse and momentum to each ball *separately*. Since the impulsive forces are directed along the line of impact, the t component of the momentum, and hence the t component of the velocity of each ball, is unchanged. We have

$$(\mathbf{v}'_A)_t = 15.0 \text{ ft/s} \uparrow \qquad (\mathbf{v}'_B)_t = 34.6 \text{ ft/s} \uparrow$$

Motion Along the Line of Impact. In the n direction, we consider the two balls as a single system and note that by Newton's third law, the internal impulses are, respectively, $\mathbf{F} \, \Delta t$ and $-\mathbf{F} \, \Delta t$ and cancel. We thus write that the total momentum of the balls is conserved:

$$m_A(v_A)_n + m_B(v_B)_n = m_A(v'_A)_n + m_B(v'_B)_n$$
$$m(26.0) + m(-20.0) = m(v'_A)_n + m(v'_B)_n$$
$$(v'_A)_n + (v'_B)_n = 6.0 \quad (1)$$

Using the relation (13.49) between relative velocities, we write

$$(v'_B)_n - (v'_A)_n = e[(v_A)_n - (v_B)_n]$$
$$(v'_B)_n - (v'_A)_n = (0.90)[26.0 - (-20.0)]$$
$$(v'_B)_n - (v'_A)_n = 41.4 \quad (2)$$

Solving Eqs. (1) and (2) simultaneously, we obtain

$$(v'_A)_n = -17.7 \qquad (v'_B)_n = +23.7$$
$$(\mathbf{v}'_A)_n = 17.7 \text{ ft/s} \leftarrow \qquad (\mathbf{v}'_B)_n = 23.7 \text{ ft/s} \rightarrow$$

Resultant Motion. Adding vectorially the velocity components of each ball, we obtain

$$\mathbf{v}'_A = 23.2 \text{ ft/s} \, \searrow \, 40.3° \qquad \mathbf{v}'_B = 41.9 \text{ ft/s} \, \nearrow \, 55.6° \quad \blacktriangleleft$$

840

SAMPLE PROBLEM 13.16

Ball B is hanging from an inextensible cord BC. An identical ball A is released from rest when it is just touching the cord and acquires a velocity $\mathbf{v}_0$ before striking ball B. Assuming perfectly elastic impact ($e = 1$) and no friction, determine the velocity of each ball immediately after impact.

SOLUTION

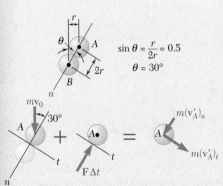

Since ball B is constrained to move in a circle of center C, its velocity $\mathbf{v}_B$ after impact must be horizontal. Thus the problem involves three unknowns: the magnitude v'_B of the velocity of B, and the magnitude and direction of the velocity $\mathbf{v}'_A$ of A after impact.

Impulse-Momentum Principle: Ball A

$$m\mathbf{v}_A + \mathbf{F}\,\Delta t = m\mathbf{v}'_A$$

$+\searrow t$ components:
$$mv_0 \sin 30° + 0 = m(v'_A)_t$$
$$(v'_A)_t = 0.5v_0 \tag{1}$$

We note that the equation used expresses conservation of the momentum of ball A along the common tangent to balls A and B.

Impulse-Momentum Principle: Balls A and B

$$m\mathbf{v}_A + \mathbf{T}\,\Delta t = m\mathbf{v}'_A + m\mathbf{v}'_B$$

$\xrightarrow{+} x$ components:
$$0 = m(v'_A)_t \cos 30° - m(v'_A)_n \sin 30° - mv'_B$$

We note that the equation obtained expresses conservation of the total momentum in the x direction. Substituting for $(v'_A)_t$ from Eq. (1) and rearranging terms, we write

$$0.5(v'_A)_n + v'_B = 0.433v_0 \tag{2}$$

Relative Velocities Along the Line of Impact.   Since $e = 1$, Eq. (13.49) yields

$$(v'_B)_n - (v'_A)_n = (v_A)_n - (v_B)_n$$
$$v'_B \sin 30° - (v'_A)_n = v_0 \cos 30° - 0$$
$$0.5v'_B - (v'_A)_n = 0.866v_0 \tag{3}$$

Solving Eqs. (2) and (3) simultaneously, we obtain

$$(v'_A)_n = -0.520v_0 \qquad v'_B = 0.693v_0$$
$$\mathbf{v}'_B = 0.693v_0 \leftarrow \quad \blacktriangleleft$$

Recalling Eq. (1) we draw the adjoining sketch and obtain by trigonometry

$$v'_A = 0.721v_0 \qquad \beta = 46.1° \qquad \alpha = 46.1° - 30° = 16.1°$$
$$\mathbf{v}'_A = 0.721v_0 \measuredangle\, 16.1° \quad \blacktriangleleft$$

SAMPLE PROBLEM 13.17

A 30-kg block is dropped from a height of 2 m onto the 10-kg pan of a spring scale. Assuming the impact to be perfectly plastic, determine the maximum deflection of the pan. The constant of the spring is $k = 20$ kN/m.

SOLUTION

The impact between the block and the pan *must* be treated separately; therefore we divide the solution into three parts.

Conservation of Energy. Block: $W_A = (30 \text{ kg})(9.81 \text{ m/s}^2) = 294$ N

$$T_1 = \tfrac{1}{2}m_A(v_A)_1^2 = 0 \qquad V_1 = W_A y = (294 \text{ N})(2 \text{ m}) = 588 \text{ J}$$
$$T_2 = \tfrac{1}{2}m_A(v_A)_2^2 = \tfrac{1}{2}(30 \text{ kg})(v_A)_2^2 \qquad V_2 = 0$$
$$T_1 + V_1 = T_2 + V_2: \qquad 0 + 588 \text{ J} = \tfrac{1}{2}(30 \text{ kg})(v_A)_2^2 + 0$$
$$(v_A)_2 = +6.26 \text{ m/s} \qquad (\mathbf{v}_A)_2 = 6.26 \text{ m/s} \downarrow$$

Impact: Conservation of Momentum. Since the impact is perfectly plastic, $e = 0$; the block and pan move together after the impact.

$$m_A(v_A)_2 + m_B(v_B)_2 = (m_A + m_B)v_3$$
$$(30 \text{ kg})(6.26 \text{ m/s}) + 0 = (30 \text{ kg} + 10 \text{ kg})v_3$$
$$v_3 = +4.70 \text{ m/s} \qquad \mathbf{v}_3 = 4.70 \text{ m/s} \downarrow$$

Conservation of Energy. Initially the spring supports the weight W_B of the pan; thus the initial deflection of the spring is

$$x_3 = \frac{W_B}{k} = \frac{(10 \text{ kg})(9.81 \text{ m/s}^2)}{20 \times 10^3 \text{ N/m}} = \frac{98.1 \text{ N}}{20 \times 10^3 \text{ N/m}} = 4.91 \times 10^{-3} \text{ m}$$

Denoting by x_4 the total maximum deflection of the spring, we write

$$T_3 = \tfrac{1}{2}(m_A + m_B)v_3^2 = \tfrac{1}{2}(30 \text{ kg} + 10 \text{ kg})(4.70 \text{ m/s})^2 = 442 \text{ J}$$
$$V_3 = V_g + V_e = 0 + \tfrac{1}{2}kx_3^2 = \tfrac{1}{2}(20 \times 10^3)(4.91 \times 10^{-3})^2 = 0.241 \text{ J}$$
$$T_4 = 0$$
$$V_4 = V_g + V_e = (W_A + W_B)(-h) + \tfrac{1}{2}kx_4^2 = -(392)h + \tfrac{1}{2}(20 \times 10^3)x_4^2$$

Noting that the displacement of the pan is $h = x_4 - x_3$, we write

$$T_3 + V_3 = T_4 + V_4:$$
$$442 + 0.241 = 0 - 392(x_4 - 4.91 \times 10^{-3}) + \tfrac{1}{2}(20 \times 10^3)x_4^2$$
$$x_4 = 0.230 \text{ m} \qquad h = x_4 - x_3 = 0.230 \text{ m} - 4.91 \times 10^{-3} \text{ m}$$
$$h = 0.225 \text{ m} \qquad\qquad h = 225 \text{ mm} \blacktriangleleft$$

SOLVING PROBLEMS
ON YOUR OWN

This lesson deals with the *impact of two bodies,* i.e., with a collision occurring in a very small interval of time. You will solve a number of impact problems by expressing that the total momentum of the two bodies is conserved and noting the relationship which exists between the relative velocities of the two bodies before and after impact.

1. As a first step in your solution you should select and draw the following coordinate axes: the t axis, which is tangent to the surfaces of contact of the two colliding bodies, and the n axis, which is normal to the surfaces of contact and defines the *line of impact.* In all the problems of this lesson the line of impact passes through the mass centers of the colliding bodies, and the impact is referred to as a *central impact.*

2. Next you will draw a diagram showing the momenta of the bodies before impact, the impulses exerted on the bodies during impact, and the final momenta of the bodies after impact (Fig. 13.24). You will then observe whether the impact is a *direct central impact* or an *oblique central impact.*

3. Direct central impact. This occurs when the velocities of bodies A and B before impact are *both directed along the line of impact* (Fig. 13.20a).

   a. Conservation of momentum. Since the impulsive forces are internal to the system, you can write that the *total momentum of A and B is conserved,*

$$m_A v_A + m_B v_B = m_A v_A' + m_B v_B' \qquad (13.37)$$

where v_A and v_B denote the velocities of bodies A and B before impact and v_A' and v_B' denote their velocities after impact.

   b. Coefficient of restitution. You can also write the following relation between the *relative velocities* of the two bodies before and after impact,

$$v_B' - v_A' = e(v_A - v_B) \qquad (13.43)$$

where e is the coefficient of restitution between the two bodies.

Note that Eqs. (13.37) and (13.43) are scalar equations which can be solved for two unknowns. Also, be careful to adopt a consistent sign convention for all velocities.

4. Oblique central impact. This occurs when *one or both* of the initial velocities of the two bodies is *not directed* along the line of impact (Fig. 13.20b). To solve problems of this type, you should *first resolve into components* along the t axis and the n axis the momenta and impulses shown in your diagram.

(continued)

843

a. Conservation of momentum. Since the impulsive forces act along the line of impact, i.e., along the n axis, the component along the t axis of the momentum *of each body* is conserved. Therefore, you can write for each body that the t components of its velocity before and after impact are equal,

$$(v_A)_t = (v_A')_t \qquad (v_B)_t = (v_B')_t \tag{13.47}$$

Also, the component along the n axis of the *total momentum* of the system is conserved,

$$m_A(v_A)_n + m_B(v_B)_n = m_A(v_A')_n + m_B(v_B')_n \tag{13.48}$$

b. Coefficient of restitution. The relation between the relative velocities of the two bodies before and after impact can be written in the n direction only,

$$(v_B')_n - (v_A')_n = e[(v_A)_n - (v_B)_n] \tag{13.49}$$

You now have four equations that you can solve for four unknowns. Note that after finding all the velocities, you can determine the impulse exerted by body A on body B by drawing an impulse-momentum diagram for B alone and equating components in the n direction.

c. When the motion of one of the colliding bodies is constrained, you must include the impulses of the external forces in your diagram. You will then observe that some of the above relations do not hold. However, in the example shown in Fig. 13.26 the total momentum of the system is conserved in a direction perpendicular to the external impulse. You should also note that when a body A bounces off a fixed surface B, the only conservation of momentum equation which can be used is the first of Eqs. (13.47) [Sample Prob. 13.14].

5. Remember that energy is lost during most impacts. The only exception is for *perfectly elastic* impacts ($e = 1$), where energy is conserved. Thus, in the general case of impact, where $e < 1$, the energy is not conserved. Therefore, be careful *not to apply* the principle of conservation of energy through an impact situation. Instead, apply this principle separately to the motions preceding and following the impact [Sample Prob. 13.17].

PROBLEMS

CONCEPT QUESTION

13.CQ6 A 5-kg ball A strikes a 1-kg ball B that is initially at rest. Is it possible that after the impact A is not moving and B has a speed of $5v$?

    a. Yes

    b. No

Explain your answer.

Fig. P13.CQ6

IMPULSE-MOMENTUM PRACTICE PROBLEMS

13.F6 A sphere with a speed v_0 rebounds after striking a frictionless inclined plane as shown. Draw the impulse-momentum diagram that can be used to find the velocity of the sphere after the impact.

Fig. P13.F6

13.F7 An 80-Mg railroad engine A coasting at 6.5 km/h strikes a 20-Mg flatcar C carrying a 30-Mg load B which can slide along the floor of the car ($\mu_k = 0.25$). The flatcar was at rest with its brakes released. Instead of A and C coupling as expected, it is observed that A rebounds with a speed of 2 km/h after the impact. Draw impulse-momentum diagrams that can be used to determine (a) the coefficient of restitution and the speed of the flatcar immediately after impact, (b) the time it takes the load to slide to a stop relative to the car.

Fig. P13.F7

13.F8 Two frictionless balls strike each other as shown. The coefficient of restitution between the balls is e. Draw the impulse-momentum diagram that could be used to find the velocities of A and B after the impact.

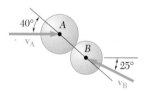

Fig. P13.F8

13.F9 A 10-kg ball A moving horizontally at 12 m/s strikes a 10-kg block B. The coefficient of restitution of the impact is 0.4 and the coefficient of kinetic friction between the block and the inclined surface is 0.5. Draw the impulse-momentum diagram that can be used to determine the speeds of A and B after the impact.

Fig. P13.F9

845

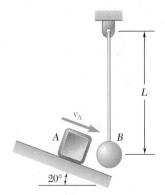

Fig. P13.F10

13.F10 Block A of mass m_A strikes ball B of mass m_B with a speed of v_A as shown. Draw the impulse-momentum diagram that can be used to determine the speeds of A and B after the impact and the impulse during the impact.

END-OF-SECTION PROBLEMS

13.155 The coefficient of restitution between the two collars is known to be 0.70. Determine (a) their velocities after impact, (b) the energy loss during impact.

Fig. P13.155

13.156 Collars A and B, of the same mass m, are moving toward each other with identical speeds as shown. Knowing that the coefficient of restitution between the collars is e, determine the energy lost in the impact as a function of m, e, and v.

Fig. P13.156

13.157 One of the requirements for tennis balls to be used in official competition is that, when dropped onto a rigid surface from a height of 100 in., the height of the first bounce of the ball must be in the range 53 in. $\leq h \leq 58$ in. Determine the range of the coefficients of restitution of the tennis balls satisfying this requirement.

13.158 Two disks sliding on a frictionless horizontal plane with opposite velocities of the same magnitude v_0 hit each other squarely. Disk A is known to have a weight of 6 lb and is observed to have zero velocity after impact. Determine (a) the weight of disk B, knowing that the coefficient of restitution between the two disks is 0.5, (b) the range of possible values of the weight of disk B if the coefficient of restitution between the two disks is unknown.

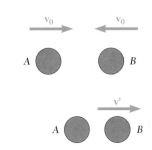

Fig. P13.158

13.159 To apply shock loading to an artillery shell, a 20-kg pendulum A is released from a known height and strikes impactor B at a known velocity $\mathbf{v}_0$. Impactor B then strikes the 1-kg artillery shell C. Knowing the coefficient of restitution between all objects is e, determine the mass of B to maximize the impulse applied to the artillery shell C.

13.160 Two identical cars A and B are at rest on a loading dock with brakes released. Car C, of a slightly different style but of the same weight, has been pushed by dockworkers and hits car B with a velocity of 1.5 m/s. Knowing that the coefficient of restitution is 0.8 between B and C and 0.5 between A and B, determine the velocity of each car after all collisions have taken place.

Fig. P13.159

Fig. P13.160

13.161 Three steel spheres of equal weight are suspended from the ceiling by cords of equal length which are spaced at a distance slightly greater than the diameter of the spheres. After being pulled back and released, sphere A hits sphere B, which then hits sphere C. Denoting by e the coefficient of restitution between the spheres and by $\mathbf{v}_0$ the velocity of A just before it hits B, determine (a) the velocities of A and B immediately after the first collision, (b) the velocities of B and C immediately after the second collision. (c) Assuming now that n spheres are suspended from the ceiling and that the first sphere is pulled back and released as described above, determine the velocity of the last sphere after it is hit for the first time. (d) Use the result of part c to obtain the velocity of the last sphere when n = 5 and e = 0.9.

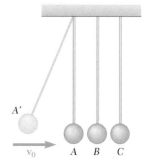

Fig. P13.161

13.162 At an amusement park there are 200-kg bumper cars A, B, and C that have riders with masses of 40 kg, 60 kg, and 35 kg, respectively. Car A is moving to the right with a velocity $\mathbf{v}_A$ = 2 m/s and car C has a velocity $\mathbf{v}_B$ = 1.5 m/s to the left, but car B is initially at rest. The coefficient of restitution between each car is 0.8. Determine the final velocity of each car, after all impacts, assuming (a) cars A and C hit car B at the same time, (b) car A hits car B before car C does.

Fig. *P13.162* and P13.163

13.163 At an amusement park there are 200-kg bumper cars A, B, and C that have riders with masses of 40 kg, 60 kg, and 35 kg, respectively. Car A is moving to the right with a velocity $\mathbf{v}_A$ = 2 m/s when it hits stationary car B. The coefficient of restitution between each car is 0.8. Determine the velocity of car C so that after car B collides with car C the velocity of car B is zero.

13.164 Two identical billiard balls can move freely on a horizontal table. Ball A has a velocity $\mathbf{v}_0$ as shown and hits ball B, which is at rest, at a point C defined by θ = 45°. Knowing that the coefficient of restitution between the two balls is e = 0.8 and assuming no friction, determine the velocity of each ball after impact.

Fig. P13.164

13.165 The coefficient of restitution is 0.9 between the two 2.37-in.-diameter billiard balls A and B. Ball A is moving in the direction shown with a velocity of 3 ft/s when it strikes ball B, which is at rest. Knowing that after impact B is moving in the x direction, determine (a) the angle θ, (b) the velocity of B after impact.

Fig. P13.165

Fig. P13.166

Fig. P13.167

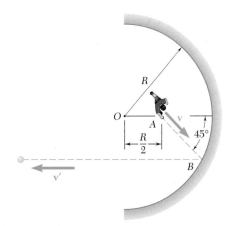

Fig. P13.169

13.166 A 600-g ball A is moving with a velocity of magnitude 6 m/s when it is hit as shown by a 1-kg ball B which has a velocity of magnitude 4 m/s. Knowing that the coefficient of restitution is 0.8 and assuming no friction, determine the velocity of each ball after impact.

13.167 Two identical hockey pucks are moving on a hockey rink at the same speed of 3 m/s and in perpendicular directions when they strike each other as shown. Assuming a coefficient of restitution $e = 0.9$, determine the magnitude and direction of the velocity of each puck after impact.

13.168 Two identical pool balls of 57.15 mm diameter may move freely on a pool table. Ball B is at rest and ball A has an initial velocity $\mathbf{v} = v_0\mathbf{i}$. (a) Knowing that $b = 50$ mm and $e = 0.7$, determine the velocity of each ball after impact. (b) Show that if $e = 1$, the final velocities of the balls form a right angle for all values of b.

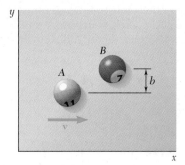

Fig. P13.168

13.169 A boy located at point A halfway between the center O of a semicircular wall and the wall itself throws a ball at the wall in a direction forming an angle of 45° with OA. Knowing that after hitting the wall the ball rebounds in a direction parallel to OA, determine the coefficient of restitution between the ball and the wall.

13.170 The Mars Pathfinder spacecraft used large airbags to cushion its impact with the planet's surface when landing. Assuming the spacecraft had an impact velocity of 18.5 m/s at an angle of 45° with respect to the horizontal, the coefficient of restitution is 0.85 and neglecting friction, determine (a) the height of the first bounce, (b) the length of the first bounce. (Acceleration of gravity on Mars = 3.73 m/s².)

Fig. P13.170

13.171 A girl throws a ball at an inclined wall from a height of 3 ft, hitting the wall at A with a horizontal velocity $\mathbf{v}_0$ of magnitude 25 ft/s. Knowing that the coefficient of restitution between the ball and the wall is 0.9 and neglecting friction, determine the distance d from the foot of the wall to the point B where the ball will hit the ground after bouncing off the wall.

Fig. P13.171

13.172 A sphere rebounds as shown after striking an inclined plane with a vertical velocity $\mathbf{v}_0$ of magnitude $v_0 = 5$ m/s. Knowing that $\alpha = 30°$ and $e = 0.8$ between the sphere and the plane, determine the height h reached by the sphere.

13.173 A sphere rebounds as shown after striking an inclined plane with a vertical velocity $\mathbf{v}_0$ of magnitude $v_0 = 6$ m/s. Determine the value of α that will maximize the horizontal distance the ball travels before reaching its maximum height h assuming the coefficient of restitution between the ball and the ground is (a) $e = 1$, (b) $e = 0.8$.

Fig. P13.172 and P13.173

13.174 Two cars of the same mass run head-on into each other at C. After the collision, the cars skid with their brakes locked and come to a stop in the positions shown in the lower part of the figure. Knowing that the speed of car A just before impact was 5 mi/h and that the coefficient of kinetic friction between the pavement and the tires of both cars is 0.30, determine (a) the speed of car B just before impact, (b) the effective coefficient of restitution between the two cars.

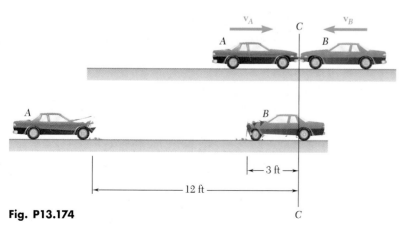

Fig. P13.174

13.175 A 1-kg block B is moving with a velocity $\mathbf{v}_0$ of magnitude $v_0 = 2$ m/s as it hits the 0.5-kg sphere A, which is at rest and hanging from a cord attached at O. Knowing that $\mu_k = 0.6$ between the block and the horizontal surface and $e = 0.8$ between the block and the sphere, determine after impact (a) the maximum height h reached by the sphere, (b) the distance x traveled by the block.

Fig. P13.175

13.176 A 0.25-lb ball thrown with a horizontal velocity $\mathbf{v}_0$ strikes a 1.5-lb plate attached to a vertical wall at a height of 36 in. above the ground. It is observed that after rebounding, the ball hits the ground at a distance of 24 in. from the wall when the plate is rigidly attached to the wall (Fig. 1) and at a distance of 10 in. when a foam-rubber mat is placed between the plate and the wall (Fig. 2). Determine (a) the coefficient of restitution e between the ball and the plate, (b) the initial velocity $\mathbf{v}_0$ of the ball.

(1) (2)

Fig. P13.176

13.177 After having been pushed by an airline employee, an empty 40-kg luggage carrier A hits with a velocity of 5 m/s an identical carrier B containing a 15-kg suitcase equipped with rollers. The impact causes the suitcase to roll into the left wall of carrier B. Knowing that the coefficient of restitution between the two carriers is 0.80 and that the coefficient of restitution between the suitcase and the wall of carrier B is 0.30, determine (a) the velocity of carrier B after the suitcase hits its wall for the first time, (b) the total energy lost in that impact.

Fig. P13.177

13.178 Blocks A and B each weigh 0.8 lb and block C weighs 2.4 lb. The coefficient of friction between the blocks and the plane is $\mu_k = 0.30$. Initially block A is moving at a speed $v_0 = 15$ ft/s and blocks B and C are at rest (Fig. 1). After A strikes B and B strikes C, all three blocks come to a stop in the positions shown (Fig. 2). Determine (a) the coefficients of restitution between A and B and between B and C, (b) the displacement x of block C.

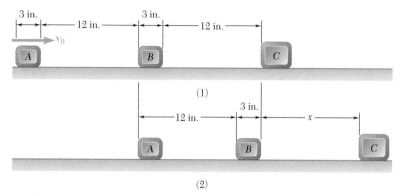

Fig. P13.178

13.179 A 0.5-kg sphere A is dropped from a height of 0.6 m onto a 1.0-kg plate B, which is supported by a nested set of springs and is initially at rest. Knowing that the coefficient of restitution between the sphere and the plate is $e = 0.8$, determine (a) the height h reached by the sphere after rebound, (b) the constant k of the single spring equivalent to the given set if the maximum deflection of the plate is observed to be equal to $3h$.

13.180 A 0.5-kg sphere A is dropped from a height of 0.6 m onto 1.0-kg plate B, which is supported by a nested set of springs and is initially at rest. Knowing that the set of springs is equivalent to a single spring of constant $k = 900$ N/m, determine (a) the value of the coefficient of restitution between the sphere and the plate for which the height h reached by the sphere after rebound is maximum, (b) the corresponding value of h, (c) the corresponding value of the maximum deflection of the plate.

Fig. P13.179 and P13.180

13.181 The three blocks shown are identical. Blocks B and C are at rest when block B is hit by block A, which is moving with a velocity $\mathbf{v}_A$ of 3 ft/s. After the impact, which is assumed to be perfectly plastic $(e = 0)$, the velocity of blocks A and B decreases due to friction, while block C picks up speed, until all three blocks are moving with the same velocity $\mathbf{v}$. Knowing that the coefficient of kinetic friction between all surfaces is $\mu_k = 0.20$, determine (a) the time required for the three blocks to reach the same velocity, (b) the total distance traveled by each block during that time.

Fig. P13.181

13.182 Block A is released from rest and slides down the frictionless surface of B until it hits a bumper on the right end of B. Block A has a mass of 10 kg and object B has a mass of 30 kg and B can roll freely on the ground. Determine the velocities of A and B immediately after impact when (a) $e = 0$, (b) $e = 0.7$.

Fig. P13.182

13.183 A 20-g bullet fired into a 4-kg wooden block suspended from cords AC and BD penetrates the block at point E, halfway between C and D, without hitting cord BD. Determine (a) the maximum height h to which the block and the embedded bullet will swing after impact, (b) the total impulse exerted on the block by the two cords during the impact.

Fig. P13.183

13.184 A 2-lb ball A is suspended from a spring of constant 10 lb/in. and is initially at rest when it is struck by 1-lb ball B as shown. Neglecting friction and knowing the coefficient of restitution between the balls is 0.6, determine (a) the velocities of A and B after the impact, (b) the maximum height reached by A.

Fig. P13.184

13.185 Ball B is hanging from an inextensible cord. An identical ball A is released from rest when it is just touching the cord and drops through the vertical distance $h_A = 8$ in. before striking ball B. Assuming $e = 0.9$ and no friction, determine the resulting maximum vertical displacement h_B of the ball B.

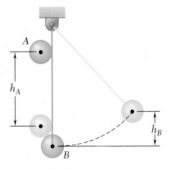

Fig. P13.185

13.186 A 70-g ball B dropped from a height $h_0 = 1.5$ m reaches a height $h_2 = 0.25$ m after bouncing twice from identical 210-g plates. Plate A rests directly on hard ground, while plate C rests on a foam-rubber mat. Determine (a) the coefficient of restitution between the ball and the plates, (b) the height h_1 of the ball's first bounce.

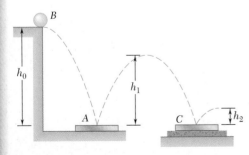

Fig. P13.186

13.187 A 700-g sphere A moving with a velocity $\mathbf{v}_0$ parallel to the ground strikes the inclined face of a 2.1-kg wedge B which can roll freely on the ground and is initially at rest. After impact the sphere is observed from the ground to be moving straight up. Knowing that the coefficient of restitution between the sphere and the wedge is $e = 0.6$, determine (a) the angle θ that the inclined face of the wedge makes with the horizontal, (b) the energy lost due to the impact.

Fig. P13.187

13.188 When the rope is at an angle of $\alpha = 30°$ the 1-lb sphere A has a speed $v_0 = 4$ ft/s. The coefficient of restitution between A and the 2-lb wedge B is 0.7 and the length of rope $l = 2.6$ ft. The spring constant has a value of 2 lb/in. and $\theta = 20°$. Determine (a) the velocities of A and B immediately after the impact, (b) the maximum deflection of the spring assuming A does not strike B again before this point.

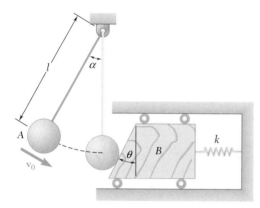

Fig. P13.188 and P13.189

13.189 When the rope is at an angle of $\alpha = 30°$ the 1-kg sphere A has a speed $v_0 = 0.6$ m/s. The coefficient of restitution between A and the 2-kg wedge B is 0.8 and the length of rope $l = 0.9$ m. The spring constant has a value of 1500 N/m and $\theta = 20°$. Determine (a) the velocities of A and B immediately after the impact, (b) the maximum deflection of the spring assuming A does not strike B again before this point.

This chapter was devoted to the method of work and energy and to the method of impulse and momentum. In the first half of the chapter we studied the method of work and energy and its application to the analysis of the motion of particles.

Work of a force

We first considered a force **F** acting on a particle A and defined the *work of* **F** *corresponding to the small displacement* $d\mathbf{r}$ [Sec. 13.2] as the quantity

$$dU = \mathbf{F} \cdot d\mathbf{r} \tag{13.1}$$

or, recalling the definition of the scalar product of two vectors,

$$dU = F \, ds \cos \alpha \tag{13.1'}$$

where α is the angle between **F** and $d\mathbf{r}$ (Fig. 13.29). The work of **F** during a finite displacement from A_1 to A_2, denoted by $U_{1\rightarrow2}$, was obtained by integrating Eq. (13.1) along the path described by the particle:

$$U_{1\rightarrow2} = \int_{A_1}^{A_2} \mathbf{F} \cdot d\mathbf{r} \tag{13.2}$$

For a force defined by its rectangular components, we wrote

$$U_{1\rightarrow2} = \int_{A_1}^{A_2} (F_x \, dx + F_y \, dy + F_z \, dz) \tag{13.2''}$$

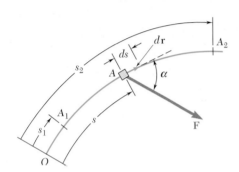

Fig. 13.29

Work of a weight

The work of the weight **W** of a body as its center of gravity moves from the elevation y_1 to y_2 (Fig. 13.30) was obtained by substituting $F_x = F_z = 0$ and $F_y = -W$ into Eq. (13.2'') and integrating. We found

$$U_{1\rightarrow2} = -\int_{y_1}^{y_2} W \, dy = Wy_1 - Wy_2 \tag{13.4}$$

Fig. 13.30

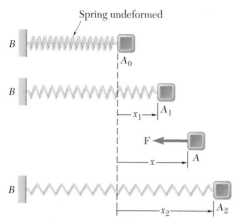

Fig. 13.31

The work of a force **F** exerted by a spring on a body A during a finite displacement of the body (Fig. 13.31) from $A_1(x = x_1)$ to $A_2(x = x_2)$ was obtained by writing

Work of the force exerted by a spring

$$dU = -F \, dx = -kx \, dx$$

$$U_{1\rightarrow 2} = -\int_{x_1}^{x_2} kx \, dx = \tfrac{1}{2}kx_1^2 - \tfrac{1}{2}kx_2^2 \qquad (13.6)$$

The work of **F** is therefore positive *when the spring is returning to its undeformed position.*

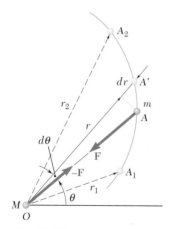

Fig. 13.32

The *work of the gravitational force* **F** exerted by a particle of mass M located at O on a particle of mass m as the latter moves from A_1 to A_2 (Fig. 13.32) was obtained by recalling from Sec. 12.10 the expression for the magnitude of **F** and writing

Work of the gravitational force

$$U_{1\rightarrow 2} = -\int_{r_1}^{r_2} \frac{GMm}{r^2} \, dr = \frac{GMm}{r_2} - \frac{GMm}{r_1} \qquad (13.7)$$

The *kinetic energy of a particle* of mass m moving with a velocity **v** [Sec. 13.3] was defined as the scalar quantity

Kinetic energy of a particle

$$T = \tfrac{1}{2}mv^2 \qquad (13.9)$$

Principle of work and energy

From Newton's second law we derived the *principle of work and energy*, which states that *the kinetic energy of a particle at A_2 can be obtained by adding to its kinetic energy at A_1 the work done during the displacement from A_1 to A_2 by the force $\mathbf{F}$ exerted on the particle:*

$$T_1 + U_{1 \to 2} = T_2 \tag{13.11}$$

Method of work and energy

The method of work and energy simplifies the solution of many problems dealing with forces, displacements, and velocities, since it does not require the determination of accelerations [Sec. 13.4]. We also note that it involves only scalar quantities and that forces which do no work need not be considered [Sample Probs. 13.1 and 13.3]. However, this method should be supplemented by the direct application of Newton's second law to determine a force normal to the path of the particle [Sample Prob. 13.4].

Power and mechanical efficiency

The power developed by a machine and its mechanical efficiency were discussed in Sec. 13.5. Power was defined as the time rate at which work is done:

$$\text{Power} = \frac{dU}{dt} = \mathbf{F} \cdot \mathbf{v} \tag{13.12, 13.13}$$

where $\mathbf{F}$ is the force exerted on the particle and $\mathbf{v}$ the velocity of the particle [Sample Prob. 13.5]. The *mechanical efficiency*, denoted by η, was expressed as

$$\eta = \frac{\text{power output}}{\text{power input}} \tag{13.15}$$

Conservative force. Potential energy

When the work of a force $\mathbf{F}$ is independent of the path followed [Secs. 13.6 and 13.7], the force $\mathbf{F}$ is said to be a *conservative force*, and its work is equal to *minus the change in the potential energy* V associated with $\mathbf{F}$:

$$U_{1 \to 2} = V_1 - V_2 \tag{13.19'}$$

The following expressions were obtained for the potential energy associated with each of the forces considered earlier:

Force of gravity (weight): $\qquad V_g = Wy \tag{13.16}$

Gravitational force: $\qquad V_g = -\frac{GMm}{r} \tag{13.17}$

Elastic force exerted by a spring: $\quad V_e = \frac{1}{2}kx^2 \tag{13.18}$

Substituting for $U_{1 \rightarrow 2}$ from Eq. (13.19') into Eq. (13.11) and rearranging the terms [Sec. 13.8], we obtained

$$T_1 + V_1 = T_2 + V_2 \qquad (13.24)$$

This is the *principle of conservation of energy,* which states that when a particle moves under the action of conservative forces, *the sum of its kinetic and potential energies remains constant.* The application of this principle facilitates the solution of problems involving only conservative forces [Sample Probs. 13.6 and 13.7].

Principle of conservation of energy

Recalling from Sec. 12.9 that, when a particle moves under a central force **F**, its angular momentum about the center of force O remains constant, we observed [Sec. 13.9] that, if the central force **F** is also conservative, the principles of conservation of angular momentum and of conservation of energy can be used jointly to analyze the motion of the particle [Sample Prob. 13.8]. Since the gravitational force exerted by the earth on a space vehicle is both central and conservative, this approach was used to study the motion of such vehicles [Sample Prob. 13.9] and was found particularly effective in the case of an *oblique launching.* Considering the initial position P_0 and an arbitrary position P of the vehicle (Fig. 13.33), we wrote

Motion under a gravitational force

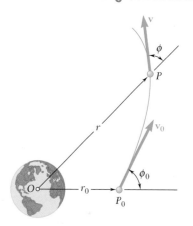

$$(H_O)_0 = H_O: \qquad r_0 m v_0 \sin \phi_0 = r m v \sin \phi \qquad (13.25)$$

$$T_0 + V_0 = T + V: \qquad \tfrac{1}{2} m v_0^2 - \frac{GMm}{r_0} = \tfrac{1}{2} m v^2 - \frac{GMm}{r} \qquad (13.26)$$

where m was the mass of the vehicle and M the mass of the earth.

Fig. 13.33

The second half of the chapter was devoted to the method of impulse and momentum and to its application to the solution of various types of problems involving the motion of particles.

Principle of impulse and momentum for a particle

The *linear momentum of a particle* was defined [Sec. 13.10] as the product $m\mathbf{v}$ of the mass m of the particle and its velocity **v**. From Newton's second law, $\mathbf{F} = m\mathbf{a}$, we derived the relation

$$m\mathbf{v}_1 + \int_{t_1}^{t_2} \mathbf{F} \, dt = m\mathbf{v}_2 \qquad (13.28)$$

where $m\mathbf{v}_1$ and $m\mathbf{v}_2$ represent the momentum of the particle at a time t_1 and a time t_2, respectively, and where the integral defines the *linear impulse of the force* **F** during the corresponding time interval. We wrote therefore

$$m\mathbf{v}_1 + \mathbf{Imp}_{1 \rightarrow 2} = m\mathbf{v}_2 \qquad (13.30)$$

which expresses the principle of impulse and momentum for a particle.

When the particle considered is subjected to several forces, the sum of the impulses of these forces should be used; we had

$$m\mathbf{v}_1 + \Sigma \, \mathbf{Imp}_{1\to2} = m\mathbf{v}_2 \tag{13.32}$$

Since Eqs. (13.30) and (13.32) involve *vector quantities*, it is necessary to consider their x and y components separately when applying them to the solution of a given problem [Sample Probs. 13.10 and 13.11].

Impulsive motion The method of impulse and momentum is particularly effective in the study of the *impulsive motion* of a particle, when very large forces, called *impulsive forces,* are applied for a very short interval of time Δt, since this method involves the impulses $\mathbf{F} \, \Delta t$ of the forces, rather than the forces themselves [Sec. 13.11]. Neglecting the impulse of any nonimpulsive force, we wrote

$$m\mathbf{v}_1 + \Sigma\mathbf{F} \, \Delta t = m\mathbf{v}_2 \tag{13.35}$$

In the case of the impulsive motion of several particles, we had

$$\Sigma m\mathbf{v}_1 + \Sigma\mathbf{F} \, \Delta t = \Sigma m\mathbf{v}_2 \tag{13.36}$$

where the second term involves only impulsive, external forces [Sample Prob. 13.12].

In the particular case *when the sum of the impulses of the external forces is zero*, Eq. (13.36) reduces to $\Sigma m\mathbf{v}_1 = \Sigma m\mathbf{v}_2$; that is, *the total momentum of the particles is conserved.*

Direct central impact In Secs. 13.12 through 13.14, we considered the *central impact* of two colliding bodies. In the case of a *direct central impact* [Sec. 13.13], the two colliding bodies A and B were moving along the *line of impact* with velocities $\mathbf{v}_A$ and $\mathbf{v}_B$, respectively (Fig. 13.34). Two equations could be used to determine their velocities $\mathbf{v}'_A$ and $\mathbf{v}'_B$ after the impact.

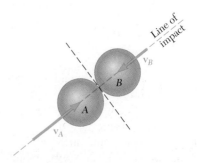

Fig. 13.34

The first expressed conservation of the total momentum of the two bodies,

$$m_A v_A + m_B v_B = m_A v_A' + m_B v_B' \qquad (13.37)$$

where a positive sign indicates that the corresponding velocity is directed to the right, while the second related the *relative velocities* of the two bodies before and after the impact,

$$v_B' - v_A' = e(v_A - v_B) \qquad (13.43)$$

The constant e is known as the *coefficient of restitution;* its value lies between 0 and 1 and depends in a large measure on the materials involved. When $e = 0$, the impact is said to be *perfectly plastic;* when $e = 1$, it is said to be *perfectly elastic* [Sample Prob. 13.13].

In the case of an *oblique central impact* [Sec. 13.14], the velocities of the two colliding bodies before and after the impact were resolved into n components along the line of impact and t components along the common tangent to the surfaces in contact (Fig. 13.35). We observed that the t component of the velocity of each body remained

Oblique central impact

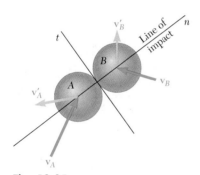

Fig. 13.35

unchanged, while the n components satisfied equations similar to Eqs. (13.37) and (13.43) [Sample Probs. 13.14 and 13.15]. It was shown that although this method was developed for bodies moving freely before and after the impact, it could be extended to the case when one or both of the colliding bodies is constrained in its motion [Sample Prob. 13.16].

In Sec. 13.15, we discussed the relative advantages of the three fundamental methods presented in this chapter and the preceding one, namely, Newton's second law, work and energy, and impulse and momentum. We noted that the method of work and energy and the method of impulse and momentum can be combined to solve problems involving a short impact phase during which impulsive forces must be taken into consideration [Sample Prob. 13.17].

Using the three fundamental methods of kinetic analysis

REVIEW PROBLEMS

13.190 A 32,000-lb airplane lands on an aircraft carrier and is caught by an arresting cable. The cable is inextensible and is paid out at A and B from mechanisms located below deck and consisting of pistons moving in long oil-filled cylinders. Knowing that the piston-cylinder system maintains a constant tension of 85 kips in the cable during the entire landing, determine the landing speed of the airplane if it travels a distance $d = 95$ ft after being caught by the cable.

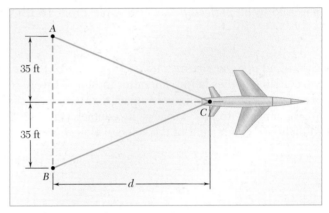

Fig. P13.190

13.191 A 2-oz pellet shot vertically from a spring-loaded pistol on the surface of the earth rises to a height of 300 ft. The same pellet shot from the same pistol on the surface of the moon rises to a height of 1900 ft. Determine the energy dissipated by aerodynamic drag when the pellet is shot on the surface of the earth. (The acceleration of gravity on the surface of the moon is 0.165 times that on the surface of the earth.)

13.192 A satellite describes an elliptic orbit about a planet of mass M. The minimum and maximum values of the distance r from the satellite to the center of the planet are, respectively, r_0 and r_1. Use the principles of conservation of energy and conservation of angular momentum to derive the relation

$$\frac{1}{r_0} + \frac{1}{r_1} = \frac{2GM}{h^2}$$

where h is the angular momentum per unit mass of the satellite and G is the constant of gravitation.

13.193 A 60-g steel sphere attached to a 200-mm cord can swing about point O in a vertical plane. It is subjected to its own weight and to a force F exerted by a small magnet embedded in the ground. The magnitude of that force expressed in newtons is $F = 3000/r^2$, where r is the distance from the magnet to the sphere expressed in millimeters. Knowing that the sphere is released from rest at A, determine its speed as it passes through point B.

Fig. P13.192

Fig. P13.193

13.194 A shuttle is to rendezvous with a space station which is in a circular orbit at an altitude of 250 mi above the surface of the earth. The shuttle has reached an altitude of 40 mi when its engine is turned off at point B. Knowing that at that time the velocity $\mathbf{v}_0$ of the shuttle forms an angle $\phi_0 = 55°$ with the vertical, determine the required magnitude of $\mathbf{v}_0$ if the trajectory of the shuttle is to be tangent at A to the orbit of the space station.

13.195 A 300-g block is released from rest after a spring of constant $k = 600$ N/m has been compressed 160 mm. Determine the force exerted by the loop $ABCD$ on the block as the block passes through (a) point A, (b) point B, (c) point C. Assume no friction.

Fig. **P13.194**

Fig. **P13.195**

13.196 A small sphere B of mass m is attached to an inextensible cord of length $2a$, which passes around the fixed peg A and is attached to a fixed support at O. The sphere is held close to the support at O and released with no initial velocity. It drops freely to point C, where the cord becomes taut, and swings in a vertical plane, first about A and then about O. Determine the vertical distance from line OD to the highest point C'' that the sphere will reach.

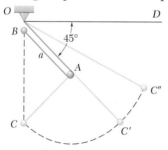

Fig. **P13.196**

13.197 A 300-g collar A is released from rest, slides down a frictionless rod, and strikes a 900-g collar B which is at rest and supported by a spring of constant 500 N/m. Knowing that the coefficient of restitution between the two collars is 0.9, determine (a) the maximum distance collar A moves up the rod after impact, (b) the maximum distance collar B moves down the rod after impact.

Fig. **P13.197**

Fig. P13.198

13.198 Blocks A and B are connected by a cord which passes over pulleys and through a collar C. The system is released from rest when $x = 1.7$ m. As block A rises, it strikes collar C with perfectly plastic impact ($e = 0$). After impact, the two blocks and the collar keep moving until they come to a stop and reverse their motion. As A and C move down, C hits the ledge and blocks A and B keep moving until they come to another stop. Determine (a) the velocity of the blocks and collar immediately after A hits C, (b) the distance the blocks and collar move after the impact before coming to a stop, (c) the value of x at the end of one complete cycle.

13.199 A 2-kg ball B is traveling horizontally at 10 m/s when it strikes 2-kg ball A. Ball A is initially at rest and is attached to a spring with constant 100 N/m and an unstretched length of 1.2 m. Knowing the coefficient of restitution between A and B is 0.8 and friction between all surfaces is negligible, determine the normal force between A and the ground when it is at the bottom of the hill.

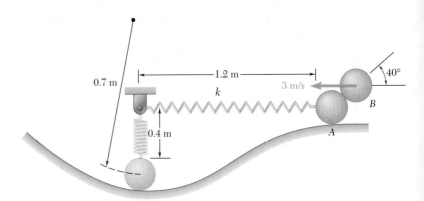

Fig. P13.199

13.200 A 2-kg block A is pushed up against a spring compressing it a distance $x = 0.1$ m. The block is then released from rest and slides down the 20° incline until it strikes a 1-kg sphere B which is suspended from a 1-m inextensible rope. The spring constant $k = 800$ N/m, the coefficient of friction between A and the ground is 0.2, the distance A slides from the unstretched length of the spring $d = 1.5$ m, and the coefficient of restitution between A and B is 0.8. When $\alpha = 40°$, determine (a) the speed of B, (b) the tension in the rope.

Fig. P13.200

*13.201 The 2-lb ball at A is suspended by an inextensible cord and given an initial horizontal velocity of v_0. If $l = 2$ ft, $x_B = 0.3$ ft, and $y_B = 0.4$ ft, determine the initial velocity v_0 so that the ball will enter the basket. (*Hint:* Use a computer to solve the resulting set of equations.)

Fig. P13.201

COMPUTER PROBLEMS

13.C1 A 12-lb collar is attached to a spring anchored at point C and can slide on a frictionless rod forming an angle of 30° with the vertical. The spring is of constant k and is unstretched when the collar is at A. Knowing that the collar is released from rest at A, use computational software to determine the velocity of the collar at point B for values of k from 0.1 to 2.0 lb/in.

Fig. P13.C1

13.C2 Skid marks on a drag race track indicate that the rear (drive) wheels of a 2000-lb car slip with the front wheels just off the ground for the first 60 ft of the 1320-ft track. The car is driven with slipping impending, with 60 percent of its weight on the rear wheels, for the remaining 1260 ft of the race. Knowing that the coefficients of kinetic and static friction are 0.60 and 0.85, respectively, and that the force due to the aerodynamic drag is $F_d = 0.0098v^2$, where the speed v is expressed in ft/s and the force F_d in lb, use computational software to determine the time elapsed and the speed of the car at various points along the track, (a) taking the force F_d into account, (b) ignoring the force F_d. If you write a computer program use increments of distance $\Delta x = 0.1$ ft in your calculations, and tabulate your results every 5 ft for the first 60 ft and every 90 ft for the remaining 1260 ft. [*Hint:* The time Δt_i required for the car to move through the increment of distance Δx_i can be obtained by dividing Δx_i by the average velocity $\frac{1}{2}(v_i + v_{i+1})$ of the car over Δx_i if the acceleration of the car is assumed to remain constant over Δx_i.]

13.C3 A 5-kg bag is gently pushed off the top of a wall and swings in a vertical plane at the end of a 2.4-m rope which can withstand a maximum tension F_m. For F_m from 40 to 140 N use computational software to determine (a) the difference in elevation h between point A and point B where the rope will break, (b) the distance d from the vertical wall to the point where the bag will strike the floor.

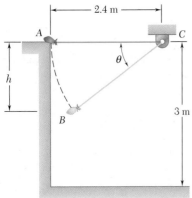

Fig. P13.C3

13.C4 Use computational software to determine (a) the time required for the system of Prob. 13.198 to complete 10 successive cycles of the motion described in that problem, starting with $x = 1.7$ m, (b) the value of x at the end of the tenth cycle.

13.C5 A 700-g ball B is hanging from an inextensible cord attached to a support at C. A 350-g ball A strikes B with a velocity $\mathbf{v}_0$ at an angle θ_0 with the vertical. Assuming no friction and denoting by e the coefficient of restitution, use computational software to determine the magnitudes v_A' and v_B' of the velocities of the balls immediately after impact and the percentage of energy lost in the collision for $v_0 = 6$ m/s and values of θ_0 from $20°$ to $150°$, assuming (a) $e = 1$, (b) $e = 0.75$, (c) $e = 0$.

Fig. P13.C5

13.C6 In Prob. 13.110, a space vehicle was in a circular orbit at an altitude of 225 mi above the surface of the earth. To return to earth it decreased its speed as it passed through A by firing its engine for a short interval of time in a direction opposite to the direction of its motion. Its resulting velocity as it reached point B at an altitude of 40 mi formed an angle $\phi_B = 60°$ with the vertical. An alternative strategy for taking the space vehicle out of its circular orbit would be to turn it around so that its engine pointed away from the earth and then give it an incremental velocity $\Delta\mathbf{v}_A$ toward the center O of the earth. This would likely require a smaller expenditure of energy when firing the engine at A, but might result in too fast a descent at B. Assuming that this strategy is used, use computational software to determine the values of ϕ_B and v_B for an energy expenditure ranging from 5 to 100 percent of that needed in Prob. 13.110.

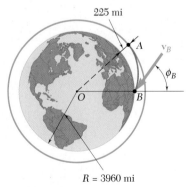

Fig. P13.C6

The thrust for this XR-5M15 prototype engine is produced by gas particles being ejected at a high velocity. The determination of the forces on the test stand is based on the analysis of the motion of a *variable system of particles,* i.e., the motion of a large number of air particles considered together rather than separately.

14.1 INTRODUCTION

In this chapter you will study the motion of *systems of particles*, i.e., the motion of a large number of particles considered together. The first part of the chapter is devoted to systems consisting of well-defined particles; the second part considers the motion of variable systems, i.e., systems which are continually gaining or losing particles, or doing both at the same time.

In Sec. 14.2, Newton's second law will first be applied to each particle of the system. Defining the *effective force* of a particle as the product $m_i\mathbf{a}_i$ of its mass m_i and its acceleration $\mathbf{a}_i$, we will show that the *external forces* acting on the various particles form a system equipollent to the system of the effective forces, i.e., both systems have the same resultant and the same moment resultant about any given point. In Sec. 14.3, it will be further shown that the resultant and moment resultant of the external forces are equal, respectively, to the rate of change of the total linear momentum and of the total angular momentum of the particles of the system.

In Sec. 14.4, the *mass center* of a system of particles is defined and the motion of that point is described, while in Sec. 14.5 the motion of the particles about their mass center is analyzed. The conditions under which the linear momentum and the angular momentum of a system of particles are conserved are discussed in Sec. 14.6, and the results obtained in that section are applied to the solution of various problems.

Sections 14.7 and 14.8 deal with the application of the work-energy principle to a system of particles, and Sec. 14.9 with the application of the impulse-momentum principle. These sections also contain a number of problems of practical interest.

It should be noted that while the derivations given in the first part of this chapter are carried out for a system of independent particles, they remain valid when the particles of the system are rigidly connected, i.e., when they form a rigid body. In fact, the results obtained here will form the foundation of our discussion of the kinetics of rigid bodies in Chaps. 16 through 18.

The second part of this chapter is devoted to the study of variable systems of particles. In Sec. 14.11 you will consider steady streams of particles, such as a stream of water diverted by a fixed vane, or the flow of air through a jet engine, and learn to determine the force exerted by the stream on the vane and the thrust developed by the engine. Finally, in Sec. 14.12, you will learn how to analyze systems which gain mass by continually absorbing particles or lose mass by continually expelling particles. Among the various practical applications of this analysis will be the determination of the thrust developed by a rocket engine.

14.2 APPLICATION OF NEWTON'S LAWS TO THE MOTION OF A SYSTEM OF PARTICLES. EFFECTIVE FORCES

In order to derive the equations of motion for a system of n particles, let us begin by writing Newton's second law for each individual particle of the system. Consider the particle P_i, where $1 \leq i \leq n$. Let

14.2 Application of Newton's Laws to the **869**
Motion of a System of Particles.
Effective Forces

m_i be the mass of P_i and $\mathbf{a}_i$ its acceleration with respect to the new-tonian frame of reference $Oxyz$. The force exerted on P_i by another particle P_j of the system (Fig. 14.1), called an *internal force*, will be denoted by $\mathbf{f}_{ij}$. The resultant of the internal forces exerted on P_i by all the other particles of the system is thus $\sum\limits_{j=1}^{n} \mathbf{f}_{ij}$ (where $\mathbf{f}_{ii}$ has no meaning and is assumed to be equal to zero). Denoting, on the other hand, by $\mathbf{F}_i$ the resultant of all the *external forces* acting on P_i, we write Newton's second law for the particle P_i as follows:

$$\mathbf{F}_i + \sum_{j=1}^{n} \mathbf{f}_{ij} = m_i\mathbf{a}_i \tag{14.1}$$

Denoting by $\mathbf{r}_i$ the position vector of P_i and taking the moments about O of the various terms in Eq. (14.1), we also write

$$\mathbf{r}_i \times \mathbf{F}_i + \sum_{j=1}^{n} (\mathbf{r}_i \times \mathbf{f}_{ij}) = \mathbf{r}_i \times m_i\mathbf{a}_i \tag{14.2}$$

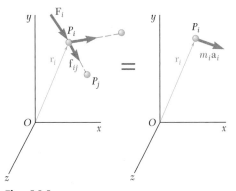

Fig. 14.1

Repeating this procedure for each particle P_i of the system, we obtain n equations of the type (14.1) and n equations of the type (14.2), where i takes successively the values $1, 2, \ldots, n$. The vectors $m_i\mathbf{a}_i$ are referred to as the *effective forces* of the particles.† Thus the equations obtained express the fact that the external forces $\mathbf{F}_i$ and the internal forces $\mathbf{f}_{ij}$ acting on the various particles form a system equivalent to the system of the effective forces $m_i\mathbf{a}_i$ (i.e., one system may be replaced by the other) (Fig. 14.2).

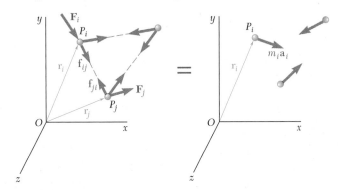

Fig. 14.2

Before proceeding further with our derivation, let us examine the internal forces $\mathbf{f}_{ij}$. We note that these forces occur in pairs $\mathbf{f}_{ij}$, $\mathbf{f}_{ji}$, where $\mathbf{f}_{ij}$ represents the force exerted by the particle P_j on the particle P_i and $\mathbf{f}_{ji}$ represents the force exerted by P_i on P_j (Fig. 14.2). Now, according to Newton's third law (Sec. 6.1), as extended by Newton's law of gravitation to particles acting at a distance (Sec. 12.10), the forces $\mathbf{f}_{ij}$ and $\mathbf{f}_{ji}$ are equal and opposite and have the same line of action. Their sum is therefore $\mathbf{f}_{ij} + \mathbf{f}_{ji} = 0$, and the sum of their moments about O is

$$\mathbf{r}_i \times \mathbf{f}_{ij} + \mathbf{r}_j \times \mathbf{f}_{ji} = \mathbf{r}_i \times (\mathbf{f}_{ij} + \mathbf{f}_{ji}) + (\mathbf{r}_j - \mathbf{r}_i) \times \mathbf{f}_{ji} = 0$$

†Since these vectors represent the resultants of the forces acting on the various particles of the system, they can truly be considered as forces.

since the vectors $\mathbf{r}_j - \mathbf{r}_i$ and $\mathbf{f}_{ji}$ in the last term are collinear. Adding all the internal forces of the system and summing their moments about O, we obtain the equations

$$\sum_{i=1}^{n} \sum_{j=1}^{n} \mathbf{f}_{ij} = 0 \qquad \sum_{i=1}^{n} \sum_{j=1}^{n} (\mathbf{r}_i \times \mathbf{f}_{ij}) = 0 \qquad (14.3)$$

which express the fact that the resultant and the moment resultant of the internal forces of the system are zero.

Returning now to the n equations (14.1), where $i = 1, 2, \ldots, n$, we sum their left-hand members and sum their right-hand members. Taking into account the first of Eqs. (14.3), we obtain

$$\sum_{i=1}^{n} \mathbf{F}_i = \sum_{i=1}^{n} m_i \mathbf{a}_i \qquad (14.4)$$

Proceeding similarly with Eq. (14.2) and taking into account the second of Eqs. (14.3), we have

$$\sum_{i=1}^{n} (\mathbf{r}_i \times \mathbf{F}_i) = \sum_{i=1}^{n} (\mathbf{r}_i \times m_i \mathbf{a}_i) \qquad (14.5)$$

Equations (14.4) and (14.5) express the fact that the system of the external forces $\mathbf{F}_i$ and the system of the effective forces $m_i \mathbf{a}_i$ have the same resultant and the same moment resultant. Referring to the definition given in Sec. 3.19 for two equipollent systems of vectors, we can therefore state that *the system of the external forces acting on the particles and the system of the effective forces of the particles are equipollent*† (Fig. 14.3).

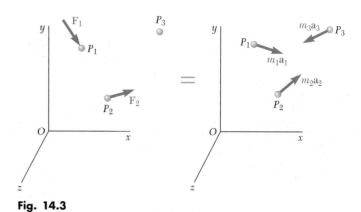

Fig. 14.3

†The result just obtained is often referred to as *d'Alembert's principle*, after the French mathematician Jean le Rond d'Alembert (1717–1783). However, d'Alembert's original statement refers to the motion of a system of connected bodies, with $\mathbf{f}_{ij}$ representing constraint forces which if applied by themselves will not cause the system to move. Since, as it will now be shown, this is in general not the case for the internal forces acting on a system of free particles, the consideration of d'Alembert's principle will be postponed until the motion of rigid bodies is considered (Chap. 16).

Equations (14.3) express the fact that the system of the internal forces $\mathbf{f}_{ij}$ is equipollent to zero. Note, however, that it does *not* follow that the internal forces have no effect on the particles under consideration. Indeed, the gravitational forces that the sun and the planets exert on one another are internal to the solar system and equipollent to zero. Yet these forces are alone responsible for the motion of the planets about the sun.

Similarly, it does not follow from Eqs. (14.4) and (14.5) that two systems of external forces which have the same resultant and the same moment resultant will have the same effect on a given system of particles. Clearly, the systems shown in Figs. 14.4a and 14.4b have

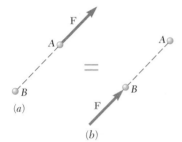

Fig. 14.4

the same resultant and the same moment resultant; yet the first system accelerates particle A and leaves particle B unaffected, while the second accelerates B and does not affect A. It is important to recall that when we stated in Sec. 3.19 that two equipollent systems of forces acting on a rigid body are also equivalent, we specifically noted that this property could *not* be extended to a system of forces acting on a set of independent particles such as those considered in this chapter.

In order to avoid any confusion, blue equals signs are used to connect equipollent systems of vectors, such as those shown in Figs. 14.3 and 14.4. These signs indicate that the two systems of vectors have the same resultant and the same moment resultant. Red equals signs will continue to be used to indicate that two systems of vectors are equivalent, i.e., that one system can actually be replaced by the other (Fig. 14.2).

14.3 LINEAR AND ANGULAR MOMENTUM OF A SYSTEM OF PARTICLES

Equations (14.4) and (14.5), obtained in the preceding section for the motion of a system of particles, can be expressed in a more condensed form if we introduce the linear and the angular momentum of the system of particles. Defining the linear momentum $\mathbf{L}$ of the system of particles as the sum of the linear momenta of the various particles of the system (Sec. 12.3), we write

$$\mathbf{L} = \sum_{i=1}^{n} m_i \mathbf{v}_i \qquad (14.6)$$

Defining the angular momentum $\mathbf{H}_O$ about O of the system of particles in a similar way (Sec. 12.7), we have

$$\mathbf{H}_O = \sum_{i=1}^{n} (\mathbf{r}_i \times m_i\mathbf{v}_i) \tag{14.7}$$

Differentiating both members of Eqs. (14.6) and (14.7) with respect to t, we write

$$\dot{\mathbf{L}} = \sum_{i=1}^{n} m_i\dot{\mathbf{v}}_i = \sum_{i=1}^{n} m_i\mathbf{a}_i \tag{14.8}$$

and

$$\begin{aligned}
\dot{\mathbf{H}}_O &= \sum_{i=1}^{n} (\dot{\mathbf{r}}_i \times m_i\mathbf{v}_i) + \sum_{i=1}^{n} (\mathbf{r}_i \times m_i\dot{\mathbf{v}}_i) \\
&= \sum_{i=1}^{n} (\mathbf{v}_i \times m_i\mathbf{v}_i) + \sum_{i=1}^{n} (\mathbf{r}_i \times m_i\mathbf{a}_i)
\end{aligned}$$

which reduces to

$$\dot{\mathbf{H}}_O = \sum_{i=1}^{n} (\mathbf{r}_i \times m_i\mathbf{a}_i) \tag{14.9}$$

since the vectors $\mathbf{v}_i$ and $m_i\mathbf{v}_i$ are collinear.

We observe that the right-hand members of Eqs. (14.8) and (14.9) are respectively identical with the right-hand members of Eqs. (14.4) and (14.5). It follows that the left-hand members of these equations are respectively equal. Recalling that the left-hand member of Eq. (14.5) represents the sum of the moments $\mathbf{M}_O$ about O of the external forces acting on the particles of the system, and omitting the subscript i from the sums, we write

$$\Sigma\mathbf{F} = \dot{\mathbf{L}} \tag{14.10}$$
$$\Sigma\mathbf{M}_O = \dot{\mathbf{H}}_O \tag{14.11}$$

These equations express that *the resultant and the moment resultant about the fixed point O of the external forces are respectively equal to the rates of change of the linear momentum and of the angular momentum about O of the system of particles.*

14.4 MOTION OF THE MASS CENTER OF A SYSTEM OF PARTICLES

Equation (14.10) may be written in an alternative form if the *mass center* of the system of particles is considered. The mass center of the system is the point G defined by the position vector $\bar{\mathbf{r}}$, which satisfies the relation

$$m\bar{\mathbf{r}} = \sum_{i=1}^{n} m_i \mathbf{r}_i \qquad (14.12)$$

where m represents the total mass $\sum_{i=1}^{n} m_i$ of the particles. Resolving the position vectors $\bar{\mathbf{r}}$ and $\mathbf{r}_i$ into rectangular components, we obtain the following three scalar equations, which can be used to determine the coordinates $\bar{x}, \bar{y}, \bar{z}$ of the mass center:

$$m\bar{x} = \sum_{i=1}^{n} m_i x_i \qquad m\bar{y} = \sum_{i=1}^{n} m_i y_i \qquad m\bar{z} = \sum_{i=1}^{n} m_i z_i \qquad (14.12')$$

Since $m_i g$ represents the weight of the particle P_i, and mg the total weight of the particles, G is also the center of gravity of the system of particles. However, in order to avoid any confusion, G will be referred to as the *mass center* of the system of particles when properties associated with the *mass* of the particles are being discussed, and as the *center of gravity* of the system when properties associated with the *weight* of the particles are being considered. Particles located outside the gravitational field of the earth, for example, have a mass but no weight. We can then properly refer to their mass center, but obviously not to their center of gravity.[†]

Differentiating both members of Eq. (14.12) with respect to t, we write

$$m\dot{\bar{\mathbf{r}}} = \sum_{i=1}^{n} m_i \dot{\mathbf{r}}_i$$

or

$$m\bar{\mathbf{v}} = \sum_{i=1}^{n} m_i \mathbf{v}_i \qquad (14.13)$$

where $\bar{\mathbf{v}}$ represents the velocity of the mass center G of the system of particles. But the right-hand member of Eq. (14.13) is, by definition, the linear momentum $\mathbf{L}$ of the system (Sec. 14.3). We therefore have

$$\mathbf{L} = m\bar{\mathbf{v}} \qquad (14.14)$$

and, differentiating both members with respect to t,

$$\dot{\mathbf{L}} = m\bar{\mathbf{a}} \qquad (14.15)$$

[†]It may also be pointed out that the mass center and the center of gravity of a system of particles do not exactly coincide, since the weights of the particles are directed toward the center of the earth and thus do not truly form a system of parallel forces.

where $\bar{\mathbf{a}}$ represents the acceleration of the mass center G. Substituting for $\dot{\mathbf{L}}$ from (14.15) into (14.10), we write the equation

$$\Sigma \mathbf{F} = m\bar{\mathbf{a}} \tag{14.16}$$

which defines the motion of the mass center G of the system of particles.

We note that Eq. (14.16) is identical with the equation we would obtain for a particle of mass m equal to the total mass of the particles of the system, acted upon by all the external forces. We therefore state that *the mass center of a system of particles moves as if the entire mass of the system and all the external forces were concentrated at that point.*

This principle is best illustrated by the motion of an exploding shell. We know that if air resistance is neglected, it can be assumed that a shell will travel along a parabolic path. After the shell has exploded, the mass center G of the fragments of shell will continue to travel along the same path. Indeed, point G must move as if the mass and the weight of all fragments were concentrated at G; it must, therefore, move as if the shell had not exploded.

It should be noted that the preceding derivation does not involve the moments of the external forces. Therefore, *it would be wrong to assume* that the external forces are equipollent to a vector $m\bar{\mathbf{a}}$ attached at the mass center G. This is not in general the case since, as you will see in the next section, the sum of the moments about G of the external forces is not in general equal to zero.

14.5 ANGULAR MOMENTUM OF A SYSTEM OF PARTICLES ABOUT ITS MASS CENTER

In some applications (for example, in the analysis of the motion of a rigid body) it is convenient to consider the motion of the particles of the system with respect to a centroidal frame of reference $Gx'y'z'$ which translates with respect to the newtonian frame of reference $Oxyz$ (Fig. 14.5). Although a centroidal frame is not, in general, a newtonian frame of reference, it will be seen that the fundamental relation (14.11) holds when the frame $Oxyz$ is replaced by $Gx'y'z'$.

Denoting, respectively, by $\mathbf{r}'_i$ and $\mathbf{v}'_i$ the position vector and the velocity of the particle P_i relative to the moving frame of reference $Gx'y'z'$, we define the *angular momentum* $\mathbf{H}'_G$ of the system of particles *about the mass center* G as follows:

$$\mathbf{H}'_G = \sum_{i=1}^{n} (\mathbf{r}'_i \times m_i \mathbf{v}'_i) \tag{14.17}$$

We now differentiate both members of Eq. (14.17) with respect to t. This operation is similar to that performed in Sec. 14.3 on Eq. (14.7), and so we write immediately

$$\dot{\mathbf{H}}'_G = \sum_{i=1}^{n} (\mathbf{r}'_i \times m_i \mathbf{a}'_i) \tag{14.18}$$

Fig. 14.5

where $\mathbf{a}'_i$ denotes the acceleration of P_i relative to the moving frame of reference. Referring to Sec. 11.12, we write

$$\mathbf{a}_i = \bar{\mathbf{a}} + \mathbf{a}'_i$$

where $\mathbf{a}_i$ and $\bar{\mathbf{a}}$ denote, respectively, the accelerations of P_i and G relative to the frame $Oxyz$. Solving for $\mathbf{a}'_i$ and substituting into (14.18), we have

$$\dot{\mathbf{H}}'_G = \sum_{i=1}^{n} (\mathbf{r}'_i \times m_i \mathbf{a}_i) - \left(\sum_{i=1}^{n} m_i \mathbf{r}'_i \right) \times \bar{\mathbf{a}} \qquad (14.19)$$

But, by (14.12), the second sum in Eq. (14.19) is equal to $m\bar{\mathbf{r}}'$ and thus to zero, since the position vector $\bar{\mathbf{r}}'$ of G relative to the frame $Gx'y'z'$ is clearly zero. On the other hand, since $\mathbf{a}_i$ represents the acceleration of P_i relative to a newtonian frame, we can use Eq. (14.1) and replace $m_i\mathbf{a}_i$ by the sum of the internal forces $\mathbf{f}_{ij}$ and of the resultant $\mathbf{F}_i$ of the external forces acting on P_i. But a reasoning similar to that used in Sec. 14.2 shows that the moment resultant about G of the internal forces $\mathbf{f}_{ij}$ of the entire system is zero. The first sum in Eq. (14.19) therefore reduces to the moment resultant about G of the external forces acting on the particles of the system, and we write

$$\Sigma \mathbf{M}_G = \dot{\mathbf{H}}'_G \qquad (14.20)$$

which expresses that *the moment resultant about G of the external forces is equal to the rate of change of the angular momentum about G of the system of particles.*

It should be noted that in Eq. (14.17) we defined the angular momentum $\mathbf{H}'_G$ as the sum of the moments about G of the momenta of the particles $m_i\mathbf{v}'_i$ in their motion relative to the centroidal frame of reference $Gx'y'z'$. We may sometimes want to compute the sum $\mathbf{H}_G$ of the moments about G of the momenta of the particles $m_i\mathbf{v}_i$ in their absolute motion, i.e., in their motion as observed from the newtonian frame of reference $Oxyz$ (Fig. 14.6):

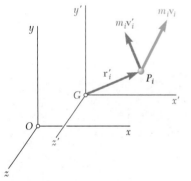

Fig. 14.6

$$\mathbf{H}_G = \sum_{i=1}^{n} (\mathbf{r}'_i \times m_i \mathbf{v}_i) \qquad (14.21)$$

Remarkably, the angular momenta $\mathbf{H}'_G$ and $\mathbf{H}_G$ are identically equal. This can be verified by referring to Sec. 11.12 and writing

$$\mathbf{v}_i = \bar{\mathbf{v}} + \mathbf{v}'_i \qquad (14.22)$$

Substituting for $\mathbf{v}_i$ from (14.22) into Eq. (14.21), we have

$$\mathbf{H}_G = \left(\sum_{i=1}^{n} m_i \mathbf{r}'_i \right) \times \bar{\mathbf{v}} + \sum_{i=1}^{n} (\mathbf{r}'_i \times m_i \mathbf{v}'_i)$$

But, as observed earlier, the first sum is equal to zero. Thus $\mathbf{H}_G$ reduces to the second sum, which, by definition, is equal to $\mathbf{H}'_G$.[†]

†Note that this property is peculiar to the centroidal frame $Gx'y'z'$ and does not, in general, hold for other frames of reference (see Prob. 14.29).

Taking advantage of the property we have just established, we simplify our notation by dropping the prime ($'$) from Eq. (14.20) and writing

$$\Sigma \mathbf{M}_G = \dot{\mathbf{H}}_G \qquad (14.23)$$

where it is understood that the angular momentum $\mathbf{H}_G$ can be computed by forming the moments about G of the momenta of the particles in their motion with respect to either the newtonian frame $Oxyz$ or the centroidal frame $Gx'y'z'$:

$$\mathbf{H}_G = \sum_{i=1}^{n} (\mathbf{r}'_i \times m_i \mathbf{v}_i) = \sum_{i=1}^{n} (\mathbf{r}'_i \times m_i \mathbf{v}'_i) \qquad (14.24)$$

14.6 CONSERVATION OF MOMENTUM FOR A SYSTEM OF PARTICLES

If no external force acts on the particles of a system, the left-hand members of Eqs. (14.10) and (14.11) are equal to zero and these equations reduce to $\dot{\mathbf{L}} = 0$ and $\dot{\mathbf{H}}_O = 0$. We conclude that

$$\mathbf{L} = \text{constant} \qquad \mathbf{H}_O = \text{constant} \qquad (14.25)$$

The equations obtained express that the linear momentum of the system of particles and its angular momentum about the fixed point O are conserved.

In some applications, such as problems involving central forces, the moment about a fixed point O of each of the external forces can be zero without any of the forces being zero. In such cases, the second of Eqs. (14.25) still holds; the angular momentum of the system of particles about O is conserved.

The concept of conservation of momentum can also be applied to the analysis of the motion of the mass center G of a system of particles and to the analysis of the motion of the system about G. For example, if the sum of the external forces is zero, the first of Eqs. (14.25) applies. Recalling Eq. (14.14), we write

$$\bar{\mathbf{v}} = \text{constant} \qquad (14.26)$$

which expresses that the mass center G of the system moves in a straight line and at a constant speed. On the other hand, if the sum of the moments about G of the external forces is zero, it follows from Eq. (14.23) that the angular momentum of the system about its mass center is conserved:

$$\mathbf{H}_G = \text{constant} \qquad (14.27)$$

Photo 14.1 If no external forces are acting on the two stages of this rocket, the linear and angular momentum of the system will be conserved.

SAMPLE PROBLEM 14.1

A 200-kg space vehicle is observed at $t = 0$ to pass through the origin of a newtonian reference frame $Oxyz$ with velocity $\mathbf{v}_0 = (150 \text{ m/s})\mathbf{i}$ relative to the frame. Following the detonation of explosive charges, the vehicle separates into three parts A, B, and C, of mass 100 kg, 60 kg, and 40 kg, respectively. Knowing that at $t = 2.5$ s the positions of parts A and B are observed to be $A(555, -180, 240)$ and $B(255, 0, -120)$, where the coordinates are expressed in meters, determine the position of part C at that time.

SOLUTION

Since there is no external force, the mass center G of the system moves with the constant velocity $\mathbf{v}_0 = (150 \text{ m/s})\mathbf{i}$. At $t = 2.5$ s, its position is

$$\bar{\mathbf{r}} = \mathbf{v}_0 t = (150 \text{m/s})\mathbf{i}(2.5 \text{s}) = (375 \text{m})\mathbf{i}$$

Recalling Eq. (14.12), we write

$$m\bar{\mathbf{r}} = m_A \mathbf{r}_A + m_B \mathbf{r}_B + m_C \mathbf{r}_C$$
$$(200 \text{ kg})(375 \text{ m})\mathbf{i} = (100 \text{ kg})[(555 \text{ m})\mathbf{i} - (180 \text{ m})\mathbf{j} + (240 \text{ m})\mathbf{k}]$$
$$+ (60 \text{ kg})[(255 \text{ m})\mathbf{i} - (120 \text{ m})\mathbf{k}] + (40 \text{ kg})\mathbf{r}_C$$

$$\mathbf{r}_C = (105 \text{ m})\mathbf{i} + (450 \text{ m})\mathbf{j} - (420 \text{ m})\mathbf{k} \quad \blacktriangleleft$$

SAMPLE PROBLEM 14.2

A 20-lb projectile is moving with a velocity of 100 ft/s when it explodes into two fragments A and B, weighing 5 lb and 15 lb, respectively. Knowing that immediately after the explosion, fragments A and B travel in directions defined respectively by $\theta_A = 45°$ and $\theta_B = 30°$, determine the velocity of each fragment.

SOLUTION

Since there is no external force, the linear momentum of the system is conserved, and we write

$$m_A \mathbf{v}_A + m_B \mathbf{v}_B = m\mathbf{v}_0$$
$$(5/g)\mathbf{v}_A + (15/g)\mathbf{v}_B = (20/g)\mathbf{v}_0$$

$\xrightarrow{+} x$ components: $\qquad 5v_A \cos 45° + 15v_B \cos 30° = 20(100)$

$+\uparrow y$ components: $\qquad 5v_A \sin 45° - 15v_B \sin 30° = 0$

Solving simultaneously the two equations for v_A and v_B, we have

$$v_A = 207 \text{ ft/s} \qquad v_B = 97.6 \text{ ft/s}$$

$$\mathbf{v}_A = 207 \text{ ft/s} \measuredangle 45° \qquad \mathbf{v}_B = 97.6 \text{ ft/s} \searrow 30° \quad \blacktriangleleft$$

SOLVING PROBLEMS
ON YOUR OWN

This chapter deals with the motion of *systems of particles,* that is, with the motion of a large number of particles considered together, rather than separately. In this first lesson you learned to compute the *linear momentum* and the *angular momentum* of a system of particles. We defined the linear momentum **L** of a system of particles as the sum of the linear momenta of the particles and we defined the angular momentum **H**$_O$ of the system as the sum of the angular momenta of the particles about O:

$$\mathbf{L} = \sum_{i=1}^{n} m_i \mathbf{v}_i \qquad \mathbf{H}_O = \sum_{i=1}^{n} (\mathbf{r}_i \times m_i \mathbf{v}_i) \qquad (14.6, \ 14.7)$$

In this lesson, you will solve a number of problems of practical interest, either by observing that the linear momentum of a system of particles is conserved or by considering the motion of the mass center of a system of particles.

1. Conservation of the linear momentum of a system of particles. This occurs *when the resultant of the external forces acting on the particles of the system is zero.* You may encounter such a situation in the following types of problems.

a. Problems involving the rectilinear motion of objects such as colliding automobiles and railroad cars. After you have checked that the resultant of the external forces is zero, equate the algebraic sums of the initial momenta and final momenta to obtain an equation which can be solved for one unknown.

b. Problems involving the two-dimensional or three-dimensional motion of objects such as exploding shells, or colliding aircraft, automobiles, or billiard balls. After you have checked that the resultant of the external forces is zero, add vectorially the initial momenta of the objects, add vectorially their final momenta, and equate the two sums to obtain a vector equation expressing that the linear momentum of the system is conserved.

In the case of a two-dimensional motion, this equation can be replaced by two scalar equations which can be solved for two unknowns, while in the case of a three-dimensional motion it can be replaced by three scalar equations which can be solved for three unknowns.

2. Motion of the mass center of a system of particles. You saw in Sec. 14.4 that *the mass center of a system of particles moves as if the entire mass of the system and all of the external forces were concentrated at that point.*

a. In the case of a body exploding while in motion, it follows that the mass center of the resulting fragments moves as the body itself would have moved if the explosion had not occurred. Problems of this type can be solved by writing the equation of motion of the mass center of the system in vectorial form and expressing the position vector of the mass center in terms of the position vectors of the various fragments [Eq. (14.12)]. You can then rewrite the vector equation as two or three scalar equations and solve the equations for an equivalent number of unknowns.

b. In the case of the collision of several moving bodies, it follows that the motion of the mass center of the various bodies is unaffected by the collision. Problems of this type can be solved by writing the equation of motion of the mass center of the system in vectorial form and expressing its position vector before and after the collision in terms of the position vectors of the relevant bodies [Eq. (14.12)]. You can then rewrite the vector equation as two or three scalar equations and solve these equations for an equivalent number of unknowns.

PROBLEMS

14.1 A 30-g bullet is fired with a horizontal velocity of 450 m/s and becomes embedded in block B which has a mass of 3 kg. After the impact, block B slides on 30-kg carrier C until it impacts the end of the carrier. Knowing the impact between B and C is perfectly plastic and the coefficient of kinetic friction between B and C is 0.2, determine (*a*) the velocity of the bullet and B after the first impact, (*b*) the final velocity of the carrier.

Fig. P14.1

14.2 A 30-g bullet is fired with a horizontal velocity of 450 m/s through 3-kg block B and becomes embedded in carrier C which has a mass of 30 kg. After the impact, block B slides 0.3 m on C before coming to rest relative to the carrier. Knowing the coefficient of kinetic friction between B and C is 0.2, determine (*a*) the velocity of the bullet immediately after passing through B, (*b*) the final velocity of the carrier.

14.3 Car A weighing 4000 lb and car B weighing 3700 lb are at rest on a 22-ton flatcar which is also at rest. Cars A and B then accelerate and quickly reach constant speeds relative to the flatcar of 7 ft/s and 3.5 ft/s, respectively, before decelerating to a stop at the opposite end of the flatcar. Neglecting friction and rolling resistance, determine the velocity of the flatcar when the cars are moving at constant speeds.

Fig. P14.3

14.4 A bullet is fired with a horizontal velocity of 1500 ft/s through a 6-lb block A and becomes embedded in a 4.95-lb block B. Knowing that blocks A and B start moving with velocities of 5 ft/s and 9 ft/s, respectively, determine (*a*) the weight of the bullet, (*b*) its velocity as it travels from block A to block B.

Fig. P14.4

14.5 Two swimmers A and B, of weight 190 lb and 125 lb, respectively, are at diagonally opposite corners of a floating raft when they realize that the raft has broken away from its anchor. Swimmer A immediately starts walking toward B at a speed of 2 ft/s relative to the raft. Knowing that the raft weighs 300 lb, determine (*a*) the speed of the raft if B does not move, (*b*) the speed with which B must walk toward A if the raft is not to move.

Fig. P14.5

879

14.6 A 180-lb man and a 120-lb woman stand side by side at the same end of a 300-lb boat, ready to dive, each with a 16-ft/s velocity relative to the boat. Determine the velocity of the boat after they have both dived, if (a) the woman dives first, (b) the man dives first.

Fig. P14.6

14.7 A 40-Mg boxcar A is moving in a railroad switchyard with a velocity of 9 km/h toward cars B and C, which are both at rest with their brakes off at a short distance from each other. Car B is a 25-Mg flatcar supporting a 30-Mg container, and car C is a 35-Mg boxcar. As the cars hit each other they get automatically and tightly coupled. Determine the velocity of car A immediately after each of the two couplings, assuming that the container (a) does not slide on the flatcar, (b) slides after the first coupling but hits a stop before the second coupling occurs, (c) slides and hits the stop only after the second coupling has occurred.

Fig. P14.7

14.8 Packages in an automobile parts supply house are transported to the loading dock by pushing them along on a roller track with very little friction. At the instant shown packages B and C are at rest and package A has a velocity of 2 m/s. Knowing that the coefficient of restitution between the packages is 0.3, determine (a) the velocity of package C after A hits B and B hits C, (b) the velocity of A after it hits B for the second time.

Fig. P14.8

14.9 A system consists of three particles A, B, and C. We know that $m_A = 3$ kg, $m_B = 2$ kg, and $m_C = 4$ kg and that the velocities of the particles expressed in m/s are, respectively, $\mathbf{v}_A = 4\mathbf{i} + 2\mathbf{j} + 2\mathbf{k}$, $\mathbf{v}_B = 4\mathbf{i} + 3\mathbf{j}$, and $\mathbf{v}_C = -2\mathbf{i} + 4\mathbf{j} + 2\mathbf{k}$. Determine the angular momentum $\mathbf{H}_O$ of the system about O.

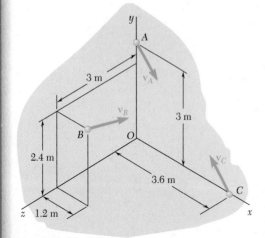

Fig. P14.9

14.10 For the system of particles of Prob. 14.9, determine (*a*) the position vector $\bar{\mathbf{r}}$ of the mass center *G* of the system, (*b*) the linear momentum $m\bar{\mathbf{v}}$ of the system, (*c*) the angular momentum $\mathbf{H}_G$ of the system about *G*. Also verify that the answers to this problem and to Prob. 14.9 satisfy the equation given in Prob. 14.27.

14.11 A system consists of three particles *A*, *B*, and *C*. We know that $W_A = 5$ lb, $W_B = 4$ lb, and $W_C = 3$ lb and that the velocities of the particles expressed in ft/s are, respectively, $\mathbf{v}_A = 2\mathbf{i} + 3\mathbf{j} - 2\mathbf{k}$, $\mathbf{v}_B = v_x\mathbf{i} + v_y\mathbf{j} + v_z\mathbf{k}$, and $\mathbf{v}_C = -3\mathbf{i} - 2\mathbf{j} + \mathbf{k}$. Determine (*a*) the components v_x and v_y of the velocity of particle *B* for which the angular momentum $\mathbf{H}_O$ of the system about *O* is parallel to the *x* axis, (*b*) the value of $\mathbf{H}_O$.

14.12 For the system of particles of Prob. 14.11, determine (*a*) the components v_x and v_z of the velocity of particle *B* for which the angular momentum $\mathbf{H}_O$ of the system about *O* is parallel to the *z* axis, (*b*) the value of $\mathbf{H}_O$.

Fig. P14.11

14.13 A system consists of three particles *A*, *B*, and *C*. We know that $m_A = 3$ kg, $m_B = 4$ kg, and $m_c = 5$ kg and that the velocities of the particles expressed in m/s are, respectively, $\mathbf{v}_A = -4\mathbf{i} + 4\mathbf{j} + 6\mathbf{k}$, $\mathbf{v}_B = -6\mathbf{i} + 8\mathbf{j} + 4\mathbf{k}$, and $\mathbf{v}_C = 2\mathbf{i} - 6\mathbf{j} - 4\mathbf{k}$. Determine the angular momentum $\mathbf{H}_O$ of the system about *O*.

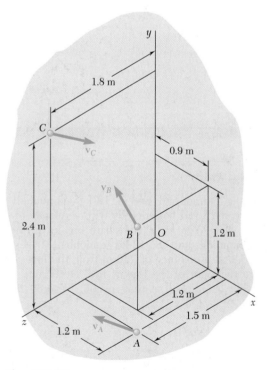

Fig. P14.13

14.14 For the system of particles of Prob. 14.13, determine (*a*) the position vector $\bar{\mathbf{r}}$ of the mass center *G* of the system, (*b*) the linear momentum $m\bar{\mathbf{v}}$ of the system, (*c*) the angular momentum $\mathbf{H}_G$ of the system about *G*. Also verify that the answers to this problem and to Prob. 14.13 satisfy the equation given in Prob. 14.27.

14.15 A 13-kg projectile is passing through the origin O with a velocity $\mathbf{v}_0 = (35$ m/s$)\mathbf{i}$ when it explodes into two fragments A and B, of mass 5 kg and 8 kg, respectively. Knowing that 3 s later the position of fragment A is (90 m, 7 m, -14 m), determine the position of fragment B at the same instant. Assume $a_y = -g = -9.81$ m/s^2 and neglect air resistance.

14.16 A 300-kg space vehicle traveling with a velocity $\mathbf{v}_0 = (360$ m/s$)\mathbf{i}$ passes through the origin O at $t = 0$. Explosive charges then separate the vehicle into three parts A, B, and C, with mass, respectively, 150 kg, 100 kg, and 50 kg. Knowing that at $t = 4$ s, the positions of parts A and B are observed to be A (1170 m, -290 m, -585 m) and B (1975 m, 365 m, 800 m), determine the corresponding position of part C. Neglect the effect of gravity.

14.17 A 2-kg model rocket is launched vertically and reaches an altitude of 70 m with a speed of 30 m/s at the end of powered flight, time $t = 0$. As the rocket approaches its maximum altitude it explodes into two parts of masses $m_A = 0.7$ kg and $m_B = 1.3$ kg. Part A is observed to strike the ground 80 m west of the launch point at $t = 6$ s. Determine the position of part B at that time.

Fig. P14.17

14.18 An 18-kg cannonball and a 12-kg cannonball are chained together and fired horizontally with a velocity of 165 m/s from the top of a 15-m wall. The chain breaks during the flight of the cannonballs and the 12-kg cannonball strikes the ground at $t = 1.5$ s, at a distance of 240 m from the foot of the wall, and 7 m to the right of the line of fire. Determine the position of the other cannonball at that instant. Neglect the resistance of the air.

Fig. P14.18

14.19 and 14.20 Car A was traveling east at high speed when it collided at point O with car B, which was traveling north at 45 mi/h. Car C, which was traveling west at 60 mi/h, was 32 ft east and 10 ft north of point O at the time of the collision. Because the pavement was wet, the driver of car C could not prevent his car from sliding into the other two cars, and the three cars, stuck together, kept sliding until they hit the utility pole P. Knowing that the weights of cars A, B, and C are, respectively, 3000 lb, 2600 lb, and 2400 lb, and neglecting the forces exerted on the cars by the wet pavement, solve the problems indicated.

14.19 Knowing that the speed of car A was 75 mi/h and that the time elapsed from the first collision to the stop at P was 2.4 s, determine the coordinates of the utility pole P.

14.20 Knowing that the coordinates of the utility pole are $x_p = 46$ ft and $y_p = 59$ ft, determine (a) the time elapsed from the first collision to the stop at P, (b) the speed of car A.

Fig. P14.19 and P14.20

14.21 An expert archer demonstrates his ability by hitting tennis balls thrown by an assistant. A 2-oz tennis ball has a velocity of $(32 \text{ ft/s})\mathbf{i} - (7 \text{ ft/s})\mathbf{j}$ and is 33 ft above the ground when it is hit by a 1.2-oz arrow traveling with a velocity of $(165 \text{ ft/s})\mathbf{j} + (230 \text{ ft/s})\mathbf{k}$ where $\mathbf{j}$ is directed upwards. Determine the position P where the ball and arrow will hit the ground, relative to point O located directly under the point of impact.

14.22 Two spheres, each of mass m, can slide freely on a frictionless, horizontal surface. Sphere A is moving at a speed $v_0 = 16$ ft/s when it strikes sphere B which is at rest and the impact causes sphere B to break into two pieces, each of mass $m/2$. Knowing that 0.7 s after the collision one piece reaches point C and 0.9 s after the collision the other piece reaches point D, determine (a) the velocity of sphere A after the collision, (b) the angle θ and the speeds of the two pieces after the collision.

Fig. P14.22

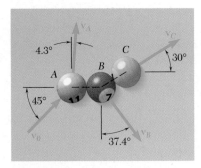

Fig. P14.23

14.23 In a game of pool, ball A is moving with a velocity $\mathbf{v}_0$ when it strikes balls B and C which are at rest and aligned as shown. Knowing that after the collision the three balls move in the directions indicated and that $v_0 = 12$ ft/s and $v_C = 6.29$ ft/s, determine the magnitude of the velocity of (a) ball A, (b) ball B.

14.24 A 6-kg shell moving with a velocity $\mathbf{v}_0 = (12 \text{ m/s})\mathbf{i} - (9 \text{ m/s})\mathbf{j} - (360 \text{ m/s})\mathbf{k}$ explodes at point D into three fragments A, B, and C of mass, respectively, 3 kg, 2 kg, and 1 kg. Knowing that the fragments hit the vertical wall at the points indicated, determine the speed of each fragment immediately after the explosion. Assume that elevation changes due to gravity may be neglected.

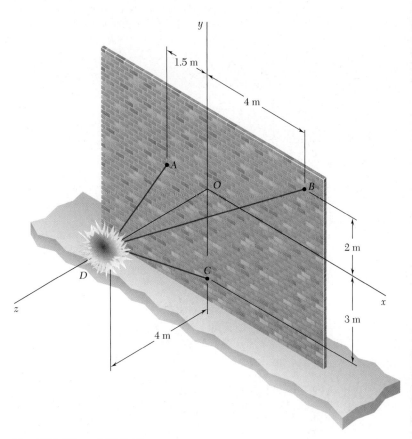

Fig. P14.24 and P14.25

14.25 A 6-kg shell moving with a velocity $\mathbf{v}_0 = (12 \text{ m/s})\mathbf{i} - (9 \text{ m/s})\mathbf{j} - (360 \text{ m/s})\mathbf{k}$ explodes at point D into three fragments A, B, and C of mass, respectively, 2 kg, 1 kg, and 3 kg. Knowing that the fragments hit the vertical wall at the points indicated, determine the speed of each fragment immediately after the explosion. Assume that elevation changes due to gravity may be neglected.

14.26 In a scattering experiment, an alpha particle A is projected with the velocity $\mathbf{u}_0 = -(600 \text{ m/s})\mathbf{i} + (750 \text{ m/s})\mathbf{j} - (800 \text{ m/s})\mathbf{k}$ into a stream of oxygen nuclei moving with a common velocity $\mathbf{v}_0 = (600 \text{ m/s})\mathbf{j}$. After colliding successively with the nuclei B and C, particle A is observed to move along the path defined by the points A_1 (280, 240, 120) and A_2 (360, 320, 160), while nuclei B and C are observed to move along paths defined, respectively, by B_1 (147, 220, 130) and B_2 (114, 290, 120), and by C_1 (240, 232, 90) and C_2 (240, 280, 75). All paths are along straight lines and all coordinates are expressed in millimeters. Knowing that the mass of an oxygen nucleus is four times that of an alpha particle, determine the speed of each of the three particles after the collisions.

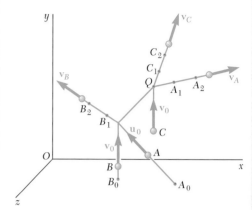

Fig. P14.26

14.27 Derive the relation

$$\mathbf{H}_O = \bar{\mathbf{r}} \times m\bar{\mathbf{v}} + H_G$$

between the angular momenta $\mathbf{H}_O$ and $\mathbf{H}_G$ defined in Eqs. (14.7) and (14.24), respectively. The vectors $\bar{\mathbf{r}}$ and $\bar{\mathbf{v}}$ define, respectively, the position and velocity of the mass center G of the system of particles relative to the newtonian frame of reference $Oxyz$, and m represents the total mass of the system.

14.28 Show that Eq. (14.23) may be derived directly from Eq. (14.11) by substituting for $\mathbf{H}_O$ the expression given in Prob. 14.27.

14.29 Consider the frame of reference $Ax'y'z'$ in translation with respect to the newtonian frame of reference $Oxyz$. We define the angular momentum $\mathbf{H}'_A$ of a system of n particles about A as the sum

$$\mathbf{H}'_A = \sum_{i=1}^{n} \mathbf{r}'_i \times m_i \mathbf{v}'_i \qquad (1)$$

of the moments about A of the momenta $m_i \mathbf{v}'_i$ of the particles in their motion relative to the frame $Ax'y'z'$. Denoting by $\mathbf{H}_A$ the sum

$$\mathbf{H}_A = \sum_{i=1}^{n} \mathbf{r}'_i \times m_i \mathbf{v}_i$$

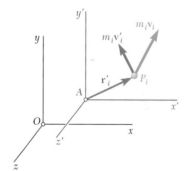

Fig. P14.29

of the moments about A of the momenta $m_i \mathbf{v}_i$ of the particles in their motion relative to the newtonian frame $Oxyz$, show that $\mathbf{H}_A = \mathbf{H}'_A$ at a given instant if, and only if, one of the following conditions is satisfied at that instant: (a) A has zero velocity with respect to the frame $Oxyz$, (b) A coincides with the mass center G of the system, (c) the velocity $\mathbf{v}_A$ relative to $Oxyz$ is directed along the line AG.

14.30 Show that the relation $\Sigma \mathbf{M}_A = \dot{\mathbf{H}}'_A$, where $\mathbf{H}'_A$ is defined by Eq. (1) of Prob. 14.29 and where $\Sigma \mathbf{M}_A$ represents the sum of the moments about A of the external forces acting on the system of particles, is valid if, and only if, one of the following conditions is satisfied: (a) the frame $Ax'y'z'$ is itself a newtonian frame of reference, (b) A coincides with the mass center G, (c) the acceleration $\mathbf{a}_A$ of A relative to $Oxyz$ is directed along the line AG.

14.7 KINETIC ENERGY OF A SYSTEM OF PARTICLES

The kinetic energy T of a system of particles is defined as the sum of the kinetic energies of the various particles of the system. Referring to Sec. 13.3, we therefore write

$$T = \frac{1}{2} \sum_{i=1}^{n} m_i v_i^2 \qquad (14.28)$$

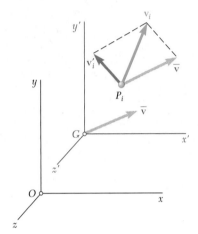

Fig. 14.7

Using a Centroidal Frame of Reference. It is often convenient when computing the kinetic energy of a system comprising a large number of particles (as in the case of a rigid body) to consider separately the motion of the mass center G of the system and the motion of the system relative to a moving frame attached to G.

Let P_i be a particle of the system, $\mathbf{v}_i$ its velocity relative to the newtonian frame of reference $Oxyz$, and $\mathbf{v}_i'$ its velocity relative to the moving frame $Gx'y'z'$ which is in translation with respect to $Oxyz$ (Fig. 14.7). We recall from the preceding section that

$$\mathbf{v}_i = \bar{\mathbf{v}} + \mathbf{v}_i' \qquad (14.22)$$

where $\bar{\mathbf{v}}$ denotes the velocity of the mass center G relative to the newtonian frame $Oxyz$. Observing that v_i^2 is equal to the scalar product $\mathbf{v}_i \cdot \mathbf{v}_i$, we express the kinetic energy T of the system relative to the newtonian frame $Oxyz$ as follows:

$$T = \frac{1}{2} \sum_{i=1}^{n} m_i v_i^2 = \frac{1}{2} \sum_{i=1}^{n} (m_i \mathbf{v}_i \cdot \mathbf{v}_i)$$

or, substituting for $\mathbf{v}_i$ from (14.22),

$$T = \frac{1}{2} \sum_{i=1}^{n} [m_i(\bar{\mathbf{v}} + \mathbf{v}_i') \cdot (\bar{\mathbf{v}} + \mathbf{v}_i')]$$

$$= \frac{1}{2} \left(\sum_{i=1}^{n} m_i \right) \bar{v}^2 + \bar{\mathbf{v}} \cdot \sum_{i=1}^{n} m_i \mathbf{v}_i' + \frac{1}{2} \sum_{i=1}^{n} m_i v_i'^2$$

The first sum represents the total mass m of the system. Recalling Eq. (14.13), we note that the second sum is equal to $m\bar{\mathbf{v}}'$ and thus to zero, since $\bar{\mathbf{v}}'$, which represents the velocity of G relative to the frame $Gx'y'z'$, is clearly zero. We therefore write

$$T = \tfrac{1}{2} m\bar{v}^2 + \frac{1}{2} \sum_{i=1}^{n} m_i v_i'^2 \qquad (14.29)$$

This equation shows that the kinetic energy T of a system of particles can be obtained *by adding the kinetic energy of the mass center G (assuming the entire mass concentrated at G) and the kinetic energy of the system in its motion relative to the frame $Gx'y'z'$.*

14.8 WORK-ENERGY PRINCIPLE. CONSERVATION OF ENERGY FOR A SYSTEM OF PARTICLES

The principle of work and energy can be applied to each particle P_i of a system of particles. We write

$$T_1 + U_{1 \rightarrow 2} = T_2 \qquad (14.30)$$

for each particle P_i, where $U_{1 \rightarrow 2}$ represents the work done by the internal forces $\mathbf{f}_{ij}$ and the resultant external force $\mathbf{F}_i$ acting on P_i. Adding the kinetic energies of the various particles of the system and considering the work of all the forces involved, we can apply Eq. (14.30) to the entire system. The quantities T_1 and T_2 now represent the kinetic energy of the entire system and can be computed from either Eq. (14.28) or Eq. (14.29). The quantity $U_{1 \rightarrow 2}$ represents the work of all the forces acting on the particles of the system. Note that while the internal forces $\mathbf{f}_{ij}$ and $\mathbf{f}_{ji}$ are equal and opposite, the work of these forces will not, in general, cancel out, since the particles P_i and P_j on which they act will, in general, undergo different displacements. Therefore, in computing $U_{1 \rightarrow 2}$, *we must consider the work of the internal forces $\mathbf{f}_{ij}$ as well as the work of the external forces $\mathbf{F}_i$.*

If all the forces acting on the particles of the system are conservative, Eq. (14.30) can be replaced by

$$T_1 + V_1 = T_2 + V_2 \qquad (14.31)$$

where V represents the potential energy associated with the internal and external forces acting on the particles of the system. Equation (14.31) expresses the principle of *conservation of energy* for the system of particles.

14.9 PRINCIPLE OF IMPULSE AND MOMENTUM FOR A SYSTEM OF PARTICLES

Integrating Eqs. (14.10) and (14.11) in t from t_1 to t_2, we write

$$\sum \int_{t_1}^{t_2} \mathbf{F}\, dt = \mathbf{L}_2 - \mathbf{L}_1 \qquad (14.32)$$

$$\sum \int_{t_1}^{t_2} \mathbf{M}_O\, dt = (\mathbf{H}_O)_2 - (\mathbf{H}_O)_1 \qquad (14.33)$$

Recalling the definition of the linear impulse of a force given in Sec. 13.10, we observe that the integrals in Eq. (14.32) represent the linear impulses of the external forces acting on the particles of the system. We shall refer in a similar way to the integrals in Eq. (14.33) as the *angular impulses* about O of the external forces. Thus, Eq. (14.32) expresses that the sum of the linear impulses of the external forces acting on the system is equal to the change in linear momentum of the system. Similarly, Eq. (14.33) expresses that the sum of the angular impulses about O of the external forces is equal to change in angular momentum about O of the system.

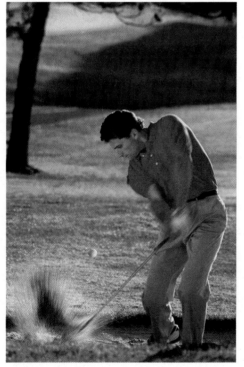

Photo 14.2 When a golf ball is hit out of a sand trap, some of the momentum of the club is transferred to the golf ball and any sand that is hit.

In order to make clear the physical significance of Eqs. (14.32) and (14.33), we will rearrange the terms in these equations and write

$$\mathbf{L}_1 + \sum \int_{t_1}^{t_2} \mathbf{F}\, dt = \mathbf{L}_2 \tag{14.34}$$

$$(\mathbf{H}_O)_1 + \sum \int_{t_1}^{t_2} \mathbf{M}_O\, dt = (\mathbf{H}_O)_2 \tag{14.35}$$

In parts a and c of Fig. 14.8 we have sketched the momenta of the particles of the system at times t_1 and t_2, respectively. In part b we have shown a vector equal to the sum of the linear impulses of the external forces and a couple of moment equal to the sum of the angular impulses about O of the external forces. For simplicity, the particles have been

Fig. 14.8

assumed to move in the plane of the figure, but the present discussion remains valid in the case of particles moving in space. Recalling from Eq. (14.6) that $\mathbf{L}$, by definition, is the resultant of the momenta $m_i\mathbf{v}_i$, we note that Eq. (14.34) expresses that the resultant of the vectors shown in parts a and b of Fig. 14.8 is equal to the resultant of the vectors shown in part c of the same figure. Recalling from Eq. (14.7) that $\mathbf{H}_O$ is the moment resultant of the momenta $m_i\mathbf{v}_i$, we note that Eq. (14.35) similarly expresses that the moment resultant of the vectors in parts a and b of Fig. 14.8 is equal to the moment resultant of the vectors in part c. Together, Eqs. (14.34) and (14.35) thus express that *the momenta of the particles at time t_1 and the impulses of the external forces from t_1 to t_2 form a system of vectors equipollent to the system of the momenta of the particles at time t_2.* This has been indicated in Fig. 14.8 by the use of blue plus and equals signs.

If no external force acts on the particles of the system, the integrals in Eqs. (14.34) and (14.35) are zero, and these equations yield

$$\mathbf{L}_1 = \mathbf{L}_2 \tag{14.36}$$
$$(\mathbf{H}_O)_1 = (\mathbf{H}_O)_2 \tag{14.37}$$

We thus check the result obtained in Sec. 14.6: If no external force acts on the particles of a system, the linear momentum and the angular momentum about O of the system of particles are conserved. The system of the initial momenta is equipollent to the system of the final momenta, and it follows that the angular momentum of the system of particles about *any* fixed point is conserved.

SAMPLE PROBLEM 14.3

For the 200-kg space vehicle of Sample Prob. 14.1, it is known that at $t = 2.5$ s, the velocity of part A is $\mathbf{v}_A = (270 \text{ m/s})\mathbf{i} - (120 \text{ m/s})\mathbf{j} + (160 \text{ m/s})\mathbf{k}$ and the velocity of part B is parallel to the xz plane. Determine the velocity of part C.

SOLUTION

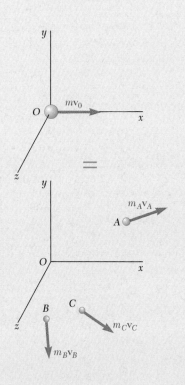

Since there is no external force, the initial momentum $m\mathbf{v}_0$ is equipollent to the system of the final momenta. Equating first the sums of the vectors in both parts of the adjoining sketch, and then the sums of their moments about O, we write

$$\mathbf{L}_1 = \mathbf{L}_2: \qquad m\mathbf{v}_0 = m_A\mathbf{v}_A + m_B\mathbf{v}_B + m_C\mathbf{v}_C \qquad (1)$$

$$(\mathbf{H}_O)_1 = (\mathbf{H}_O)_2: \qquad 0 = \mathbf{r}_A \times m_A\mathbf{v}_A + \mathbf{r}_B \times m_B\mathbf{v}_B + \mathbf{r}_C \times m_C\mathbf{v}_C \qquad (2)$$

Recalling from Sample Prob. 14.1 that $\mathbf{v}_0 = (150 \text{ m/s})\mathbf{i}$,

$$m_A = 100 \text{ kg} \qquad m_B = 60 \text{ kg} \qquad m_C = 40 \text{ kg}$$
$$\mathbf{r}_A = (555 \text{ m})\mathbf{i} - (180 \text{ m})\mathbf{j} + (240 \text{ m})\mathbf{k}$$
$$\mathbf{r}_B = (255 \text{ m})\mathbf{i} - (120 \text{ m})\mathbf{k}$$
$$\mathbf{r}_C = (105 \text{ m})\mathbf{i} + (450 \text{ m})\mathbf{j} - (420 \text{ m})\mathbf{k}$$

and using the information given in the statement of this problem, we rewrite Eqs. (1) and (2) as follows:

$$200(150\mathbf{i}) = 100(270\mathbf{i} - 120\mathbf{j} + 160\mathbf{k}) + 60[(v_B)_x\mathbf{i} + (v_B)_z\mathbf{k}] \\ + 40[(v_C)_x\mathbf{i} + (v_C)_y\mathbf{j} + (v_C)_z\mathbf{k}] \quad (1')$$

$$0 = 100 \begin{vmatrix} \mathbf{i} & \mathbf{j} & \mathbf{k} \\ 555 & -180 & 240 \\ 270 & -120 & 160 \end{vmatrix} + 60 \begin{vmatrix} \mathbf{i} & \mathbf{j} & \mathbf{k} \\ 255 & 0 & -120 \\ (v_B)_x & 0 & (v_B)_z \end{vmatrix}$$
$$+ 40 \begin{vmatrix} \mathbf{i} & \mathbf{j} & \mathbf{k} \\ 105 & 450 & -420 \\ (v_C)_x & (v_C)_y & (v_C)_z \end{vmatrix} \quad (2')$$

Equating to zero the coefficient of $\mathbf{j}$ in $(1')$ and the coefficients of $\mathbf{i}$ and $\mathbf{k}$ in $(2')$, we write, after reductions, the three scalar equations

$$(v_C)_y - 300 = 0$$
$$450(v_C)_z + 420(v_C)_y = 0$$
$$105(v_C)_y - 450(v_C)_x - 45\,000 = 0$$

which yield, respectively,

$$(v_C)_y = 300 \qquad (v_C)_z = -280 \qquad (v_C)_x = -30$$

The velocity of part C is thus

$$\mathbf{v}_C = -(30 \text{ m/s})\mathbf{i} + (300 \text{ m/s})\mathbf{j} - (280 \text{ m/s})\mathbf{k} \quad \blacktriangleleft$$

SAMPLE PROBLEM 14.4

Ball B, of mass m_B, is suspended from a cord of length l attached to cart A, of mass m_A, which can roll freely on a frictionless horizontal track. If the ball is given an initial horizontal velocity $\mathbf{v}_0$ while the cart is at rest, determine (a) the velocity of B as it reaches its maximum elevation, (b) the maximum vertical distance h through which B will rise. (It is assumed that $v_0^2 < 2gl$.)

SOLUTION

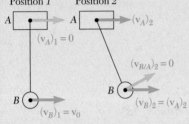

The impulse-momentum principle and the principle of conservation of energy will be applied to the cart-ball system between its initial position 1 and position 2, when B reaches its maximum elevation.

Velocities **Position 1:** $\quad (\mathbf{v}_A)_1 = 0 \qquad (\mathbf{v}_B)_1 = \mathbf{v}_0$ $\qquad\qquad$ (1)

Position 2: When ball B reaches its maximum elevation, its velocity $(\mathbf{v}_{B/A})_2$ relative to its support A is zero. Thus, at that instant, its absolute velocity is

$$(\mathbf{v}_B)_2 = (\mathbf{v}_A)_2 + (\mathbf{v}_{B/A})_2 = (\mathbf{v}_A)_2 \qquad\qquad (2)$$

Impulse-Momentum Principle. Noting that the external impulses consist of $\mathbf{W}_A t$, $\mathbf{W}_B t$, and $\mathbf{R}t$, where $\mathbf{R}$ is the reaction of the track on the cart, and recalling (1) and (2), we draw the impulse-momentum diagram and write

$$\Sigma m\mathbf{v}_1 + \Sigma \ \mathbf{Ext\ Imp}_{1\to 2} = \Sigma m\mathbf{v}_2$$

$\xrightarrow{+}x$ components: $\qquad m_B v_0 = (m_A + m_B)(v_A)_2$

which expresses that the linear momentum of the system is conserved in the horizontal direction. Solving for $(v_A)_2$:

$$(v_A)_2 = \frac{m_B}{m_A + m_B}v_0 \qquad (\mathbf{v}_B)_2 = (\mathbf{v}_A)_2 = \frac{m_B}{m_A + m_B}v_0 \rightarrow \ \blacktriangleleft$$

Conservation of Energy

Position 1. **Potential Energy:** $\quad V_1 = m_A gl$
$\qquad\qquad$ **Kinetic Energy:** $\quad T_1 = \frac{1}{2}m_B v_0^2$

Position 2. **Potential Energy:** $\quad V_2 = m_A gl + m_B gh$
$\qquad\qquad$ **Kinetic Energy:** $\quad T_2 = \frac{1}{2}(m_A + m_B)(v_A)_2^2$

$T_1 + V_1 = T_2 + V_2: \qquad \frac{1}{2}m_B v_0^2 + m_A gl = \frac{1}{2}(m_A + m_B)(v_A)_2^2 + m_A gl + m_B gh$

Solving for h, we have

$$h = \frac{v_0^2}{2g} - \frac{m_A + m_B}{m_B}\frac{(v_A)_2^2}{2g}$$

or, substituting for $(v_A)_2$ the expression found above,

$$h = \frac{v_0^2}{2g} - \frac{m_B}{m_A + m_B}\frac{v_0^2}{2g} \qquad h = \frac{m_A}{m_A + m_B}\frac{v_0^2}{2g} \ \blacktriangleleft$$

Remarks. (1) Recalling that $v_0^2 < 2gl$, it follows from the last equation that $h < l$; we thus check that B stays below A as assumed in our solution.

(2) For $m_A \gg m_B$, the answers obtained reduce to $(\mathbf{v}_B)_2 = (\mathbf{v}_A)_2 = 0$ and $h = v_0^2/2g$; B oscillates as a simple pendulum with A fixed. For $m_A \ll m_B$, they reduce to $(\mathbf{v}_B)_2 = (\mathbf{v}_A)_2 = \mathbf{v}_0$ and $h = 0$; A and B move with the same constant velocity $\mathbf{v}_0$.

SAMPLE PROBLEM 14.5

In a game of billiards, ball A is given an initial velocity $\mathbf{v}_0$ of magnitude $v_0 = 10$ ft/s along line DA parallel to the axis of the table. It hits ball B and then ball C, which are both at rest. Knowing that A and C hit the sides of the table squarely at points A' and C', respectively, that B hits the side obliquely at B', and assuming frictionless surfaces and perfectly elastic impacts, determine the velocities $\mathbf{v}_A$, $\mathbf{v}_B$, and $\mathbf{v}_C$ with which the balls hit the sides of the table. (*Remark:* In this sample problem and in several of the problems which follow, the billiard balls are assumed to be particles moving freely in a horizontal plane, rather than the rolling and sliding spheres they actually are.)

SOLUTION

Conservation of Momentum. Since there is no external force, the initial momentum $m\mathbf{v}_0$ is equipollent to the system of momenta after the two collisions (and before any of the balls hits the side of the table). Referring to the adjoining sketch, we write

$\xrightarrow{+} x$ components:
$$m(10 \text{ ft/s}) = m(v_B)_x + mv_C \qquad (1)$$

$+\uparrow y$ components:
$$0 = mv_A - m(v_B)_y \qquad (2)$$

$+\curvearrowleft$ moments about O:
$$-(2 \text{ ft})m(10 \text{ ft/s}) = (8 \text{ ft})mv_A$$
$$-(7 \text{ ft})m(v_B)_y - (3 \text{ ft})mv_C \qquad (3)$$

Solving the three equations for v_A, $(v_B)_x$, and $(v_B)_y$ in terms of v_C,

$$v_A = (v_B)_y = 3v_C - 20 \qquad (v_B)_x = 10 - v_C \qquad (4)$$

Conservation of Energy. Since the surfaces are frictionless and the impacts are perfectly elastic, the initial kinetic energy $\frac{1}{2}mv_0^2$ is equal to the final kinetic energy of the system:

$$\tfrac{1}{2}mv_0^2 = \tfrac{1}{2}m_Av_A^2 + \tfrac{1}{2}m_Bv_B^2 + \tfrac{1}{2}m_Cv_C^2$$
$$v_A^2 + (v_B)_x^2 + (v_B)_y^2 + v_C^2 = (10 \text{ ft/s})^2 \qquad (5)$$

Substituting for v_A, $(v_B)_x$, and $(v_B)_y$ from (4) into (5), we have

$$2(3v_C - 20)^2 + (10 - v_C)^2 + v_C^2 = 100$$
$$20v_C^2 - 260v_C + 800 = 0$$

Solving for v_C, we find $v_C = 5$ ft/s and $v_C = 8$ ft/s. Since only the second root yields a positive value for v_A after substitution into Eqs. (4), we conclude that $v_C = 8$ ft/s and

$$v_A = (v_B)_y = 3(8) - 20 = 4 \text{ ft/s} \qquad (v_B)_x = 10 - 8 = 2 \text{ ft/s}$$

$$\mathbf{v}_A = 4 \text{ ft/s} \uparrow \qquad \mathbf{v}_B = 4.47 \text{ ft/s} \searrow 63.4° \qquad \mathbf{v}_C = 8 \text{ ft/s} \rightarrow \quad \blacktriangleleft$$

SOLVING PROBLEMS ON YOUR OWN

In the preceding lesson we defined the linear momentum and the angular momentum of a system of particles. In this lesson we defined the *kinetic energy T* of a system of particles:

$$T = \frac{1}{2} \sum_{i=1}^{n} m_i v_i^2 \tag{14.28}$$

The solutions of the problems in the preceding lesson were based on the conservation of the linear momentum of a system of particles or on the observation of the motion of the mass center of a system of particles. In this lesson you will solve problems involving the following:

1. Computation of the kinetic energy lost in collisions. The kinetic energy T_1 of the system of particles before the collisions and its kinetic energy T_2 after the collisions are computed from Eq. (14.28) and are subtracted from each other. Keep in mind that, while linear momentum and angular momentum are vector quantities, kinetic energy is a *scalar* quantity.

2. Conservation of linear momentum and conservation of energy. As you saw in the preceding lesson, when the resultant of the external forces acting on a system of particles is zero, the linear momentum of the system is conserved. In problems involving two-dimensional motion, expressing that the initial linear momentum and the final linear momentum of the system are equipollent yields two algebraic equations. Equating the initial total energy of the system of particles (including potential energy as well as kinetic energy) to its final total energy yields an additional equation. Thus, you can write three equations which can be solved for three unknowns [Sample Prob. 14.5]. Note that if the resultant of the external forces is not zero but has a fixed direction, the component of the linear momentum in a direction perpendicular to the resultant is still conserved; the number of equations which can be used is then reduced to two [Sample Prob. 14.4].

3. Conservation of linear and angular momentum. When no external forces act on a system of particles, both the linear momentum of the system and its angular momentum about some arbitrary point are conserved. In the case of three-dimensional motion, this will enable you to write as many as six equations, although you may need to solve only some of them to obtain the desired answers [Sample Prob. 14.3]. In the case of two-dimensional motion, you will be able to write three equations which can be solved for three unknowns.

4. Conservation of linear and angular momentum and conservation of energy. In the case of the two-dimensional motion of a system of particles which are not subjected to any external forces, you will obtain two algebraic equations by expressing that the linear momentum of the system is conserved, one equation by writing that the angular momentum of the system about some arbitrary point is conserved, and a fourth equation by expressing that the total energy of the system is conserved. These equations can be solved for four unknowns.

PROBLEMS

14.31 Determine the energy lost due to friction and the impacts for Prob. 14.1.

14.32 In Prob. 14.4, determine the energy lost as the bullet (a) passes through block A, (b) becomes embedded in block B.

14.33 In Prob. 14.6, determine the work done by the woman and by the man as each dives from the boat, assuming that the woman dives first.

14.34 Determine the energy lost as a result of the series of collisions described in Prob. 14.8.

14.35 Two automobiles A and B, of mass m_A and m_B, respectively, are traveling in opposite directions when they collide head on. The impact is assumed perfectly plastic, and it is further assumed that the energy absorbed by each automobile is equal to its loss of kinetic energy with respect to a moving frame of reference attached to the mass center of the two-vehicle system. Denoting by E_A and E_B, respectively, the energy absorbed by automobile A and by automobile B, (a) show that $E_A/E_B = m_B/m_A$, that is, the amount of energy absorbed by each vehicle is inversely proportional to its mass, (b) compute E_A and E_B, knowing that $m_A = 1600$ kg and $m_B = 900$ kg and that the speeds of A and B are, respectively, 90 km/h and 60 km/h.

Fig. P14.35

14.36 It is assumed that each of the two automobiles involved in the collision described in Prob. 14.35 had been designed to safely withstand a test in which it crashed into a solid, immovable wall at the speed v_0. The severity of the collision of Prob. 14.35 may then be measured for each vehicle by the ratio of the energy it absorbed in the collision to the energy it absorbed in the test. On that basis, show that the collision described in Prob. 14.35 is $(m_A/m_B)^2$ times more severe for automobile B than for automobile A.

14.37 Solve Sample Prob. 14.4, assuming that cart A is given an initial horizontal velocity $\mathbf{v}_0$ while ball B is at rest.

14.38 Two hemispheres are held together by a cord which maintains a spring under compression (the spring is not attached to the hemispheres). The potential energy of the compressed spring is 120 J and the assembly has an initial velocity $\mathbf{v}_0$ of magnitude $v_0 = 8$ m/s. Knowing that the cord is severed when $\theta = 30°$, causing the hemispheres to fly apart, determine the resulting velocity of each hemisphere.

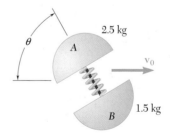

Fig. P14.38

14.39 A 15-lb block B starts from rest and slides on the 25-lb wedge A, which is supported by a horizontal surface. Neglecting friction, determine (a) the velocity of B relative to A after it has slid 3 ft down the inclined surface of the wedge, (b) the corresponding velocity of A.

Fig. P14.39

893

Fig. P14.40

14.40 A 40-lb block B is suspended from a 6-ft cord attached to a 60-lb cart A, which may roll freely on a frictionless, horizontal track. If the system is released from rest in the position shown, determine the velocities of A and B as B passes directly under A.

14.41 and 14.42 In a game of pool, ball A is moving with a velocity $\mathbf{v}_0$ of magnitude $v_0 = 15$ ft/s when it strikes balls B and C, which are at rest and aligned as shown. Knowing that after the collision the three balls move in the directions indicated and assuming frictionless surfaces and perfectly elastic impact (i.e., conservation of energy), determine the magnitudes of the velocities $\mathbf{v}_A$, $\mathbf{v}_B$, and $\mathbf{v}_C$.

Fig. P14.41   Fig. P14.42

Fig. P14.43

14.43 Three spheres, each of mass m, can slide freely on a frictionless, horizontal surface. Spheres A and B are attached to an inextensible, inelastic cord of length l and are at rest in the position shown when sphere B is struck squarely by sphere C which is moving to the right with a velocity $\mathbf{v}_0$. Knowing that the cord is slack when sphere B is struck by sphere C and assuming perfectly elastic impact between B and C, determine (a) the velocity of each sphere immediately after the cord becomes taut, (b) the fraction of the initial kinetic energy of the system which is dissipated when the cord becomes taut.

Fig. P14.44

14.44 In a game of pool, ball A is moving with the velocity $\mathbf{v}_0 = v_0\mathbf{i}$ when it strikes balls B and C, which are at rest side by side. Assuming frictionless surfaces and perfectly elastic impact (i.e., conservation of energy), determine the final velocity of each ball, assuming that the path of A is (a) perfectly centered and that A strikes B and C simultaneously, (b) not perfectly centered and that A strikes B slightly before it strikes C.

Fig. P14.45

14.45 Two small spheres A and B, of mass 2.5 kg and 1 kg, respectively, are connected by a rigid rod of negligible mass. The two spheres are resting on a horizontal, frictionless surface when A is suddenly given the velocity $\mathbf{v}_0 = (3.5 \text{ m/s})\mathbf{i}$. Determine (a) the linear momentum of the system and its angular momentum about its mass center G, (b) the velocities of A and B after the rod AB has rotated through 180°.

14.46 A 900-lb space vehicle traveling with a velocity $\mathbf{v}_0 = (1500 \text{ ft/s})\mathbf{k}$ passes through the origin O. Explosive charges then separate the vehicle into three parts A, B, and C, with masses of 150 lb, 300 lb, and 450 lb, respectively. Knowing that shortly thereafter the positions of the three parts are, respectively, A(250, 250, 2250), B(600, 1300, 3200), and C(−475, −950, 1900), where the coordinates are expressed in feet, that the velocity of B is $\mathbf{v}_B = (500 \text{ ft/s})\mathbf{i} + (1100 \text{ ft/s})\mathbf{j} + (2100 \text{ ft/s})\mathbf{k}$, and that the x component of the velocity of C is −400 ft/s, determine the velocity of part A.

14.47 Four small disks A, B, C, and D can slide freely on a frictionless horizontal surface. Disks B, C, and D are connected by light rods and are at rest in the position shown when disk B is struck squarely by disk A which is moving to the right with a velocity $\mathbf{v}_0 = (38.5 \text{ ft/s})\mathbf{i}$. The weights of the disks are $W_A = W_B = W_C = 15 \text{ lb}$, and $W_D = 30 \text{ lb}$. Knowing that the velocities of the disks immediately after the impact are $\mathbf{v}_A = \mathbf{v}_B = (8.25 \text{ ft/s})\mathbf{i}$, $\mathbf{v}_C = v_C\mathbf{i}$, and $\mathbf{v}_D = v_D\mathbf{i}$, determine (a) the speeds v_C and v_D, (b) the fraction of the initial kinetic energy of the system which is dissipated during the collision.

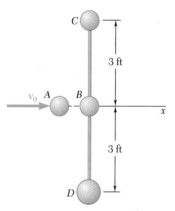

Fig. P14.47

14.48 In the scattering experiment of Prob. 14.26, it is known that the alpha particle is projected from A_0(300, 0, 300) and that it collides with the oxygen nucleus C at Q(240, 200, 100), where all coordinates are expressed in millimeters. Determine the coordinates of point B_0 where the original path of nucleus B intersects the zx plane. (*Hint.* Express that the angular momentum of the three particles about Q is conserved.)

14.49 Three identical small spheres, each of weight 2 lb, can slide freely on a horizontal frictionless surface. Spheres B and C are connected by a light rod and are at rest in the position shown when sphere B is struck squarely by sphere A which is moving to the right with a velocity $\mathbf{v}_0 = (8 \text{ ft/s})\mathbf{i}$. Knowing that $\theta = 45°$ and that the velocities of spheres A and B immediately after the impact are $\mathbf{v}_A = 0$ and $\mathbf{v}_B = (6 \text{ ft/s})\mathbf{i} + (v_B)_y\mathbf{j}$, determine $(v_B)_y$ and the velocity of C immediately after impact.

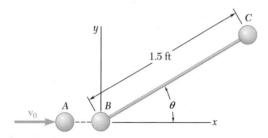

Fig. P14.49

14.50 Three small spheres A, B, and C, each of mass m, are connected to a small ring D of negligible mass by means of three inextensible, inelastic cords of length l. The spheres can slide freely on a frictionless horizontal surface and are rotating initially at a speed v_0 about ring D which is at rest. Suddenly the cord CD breaks. After the other two cords have again become taut, determine (a) the speed of ring D, (b) the relative speed at which spheres A and B rotate about D, (c) the fraction of the original energy of spheres A and B which is dissipated when cords AD and BD again became taut.

Fig. P14.50

Fig. P14.51

14.51 In a game of billiards, ball A is given an initial velocity $\mathbf{v}_0$ along the longitudinal axis of the table. It hits ball B and then ball C, which are both at rest. Balls A and C are observed to hit the sides of the table squarely at A' and C', respectively, and ball B is observed to hit the side obliquely at B'. Knowing that $v_0 = 4$ m/s, $v_A = 1.92$ m/s, and $a = 1.65$ m, determine (a) the velocities $\mathbf{v}_B$ and $\mathbf{v}_C$ of balls B and C, (b) the point C' where ball C hits the side of the table. Assume frictionless surfaces and perfectly elastic impacts (i.e., conservation of energy).

14.52 For the game of billiards of Prob. 14.51, it is now assumed that $v_0 = 5$ m/s, $v_C = 3.2$ m/s, and $c = 1.22$ m. Determine (a) the velocities $\mathbf{v}_A$ and $\mathbf{v}_B$ of balls A and B, (b) the point A' where ball A hits the side of the table.

14.53 Two small disks A and B, of mass 3 kg and 1.5 kg, respectively, may slide on a horizontal, frictionless surface. They are connected by a cord, 600 mm long, and spin counterclockwise about their mass center G at the rate of 10 rad/s. At $t = 0$, the coordinates of G are $\bar{x}_0 = 0$, $\bar{y}_0 = 2$ m, and its velocity $\bar{\mathbf{v}}_0 = (1.2 \text{ m/s})\mathbf{i} + (0.96 \text{ m/s})\mathbf{j}$. Shortly thereafter the cord breaks; disk A is then observed to move along a path parallel to the y axis and disk B along a path which intersects the x axis at a distance $b = 7.5$ m from O. Determine (a) the velocities of A and B after the cord breaks, (b) the distance a from the y axis to the path of A.

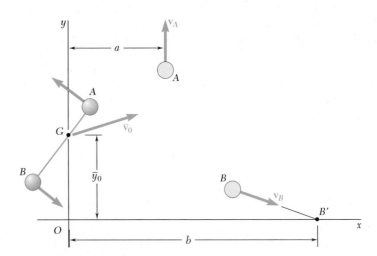

Fig. P14.53 and P14.54

14.54 Two small disks A and B, of mass 2 kg and 1 kg, respectively, may slide on a horizontal and frictionless surface. They are connected by a cord of negligible mass and spin about their mass center G. At $t = 0$, G is moving with the velocity $\bar{\mathbf{v}}_0$ and its coordinates are $\bar{x}_0 = 0$, $\bar{y}_0 = 1.89$ m. Shortly thereafter, the cord breaks and disk A is observed to move with a velocity $\mathbf{v}_A = (5 \text{ m/s})\mathbf{j}$ in a straight line and at a distance $a = 2.56$ m from the y axis, while B moves with a velocity $\mathbf{v}_B = (7.2 \text{ m/s})\mathbf{i} - (4.6 \text{ m/s})\mathbf{j}$ along a path intersecting the x axis at a distance $b = 7.48$ m from the origin O. Determine (a) the initial velocity $\bar{\mathbf{v}}_0$ of the mass center G of the two disks, (b) the length of the cord initially connecting the two disks, (c) the rate in rad/s at which the disks were spinning about G.

14.55 Three small identical spheres A, B, and C, which can slide on a horizontal, frictionless surface, are attached to three 9-in.-long strings, which are tied to a ring G. Initially the spheres rotate clockwise about the ring with a relative velocity of 2.6 ft/s and the ring moves along the x axis with a velocity $\mathbf{v}_0 = (1.3 \text{ ft/s})\mathbf{i}$. Suddenly the ring breaks and the three spheres move freely in the xy plane with A and B following paths parallel to the y axis at a distance $a = 1.0$ ft from each other and C following a path parallel to the x axis. Determine (a) the velocity of each sphere, (b) the distance d.

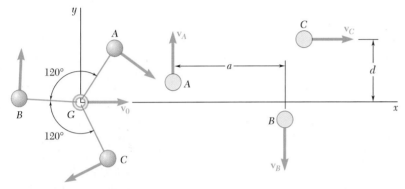

Fig. P14.55 and P14.56

14.56 Three small identical spheres A, B, and C, which can slide on a horizontal, frictionless surface, are attached to three strings of length l which are tied to a ring G. Initially the spheres rotate clockwise about the ring which moves along the x axis with a velocity $\mathbf{v}_0$. Suddenly the ring breaks and the three spheres move freely in the xy plane. Knowing that $\mathbf{v}_A = (3.5 \text{ ft/s})\mathbf{j}$, $\mathbf{v}_C = (6.0 \text{ ft/s})\mathbf{i}$, $a = 16$ in., and $d = 9$ in., determine (a) the initial velocity of the ring, (b) the length l of the strings, (c) the rate in rad/s at which the spheres were rotating about G.

*14.10 VARIABLE SYSTEMS OF PARTICLES

All the systems of particles considered so far consisted of well-defined particles. These systems did not gain or lose any particles during their motion. In a large number of engineering applications, however, it is necessary to consider *variable systems of particles,* i.e., systems which are continually gaining or losing particles, or doing both at the same time. Consider, for example, a hydraulic turbine. Its analysis involves the determination of the forces exerted by a stream of water on rotating blades, and we note that the particles of water in contact with the blades form an everchanging system which continually acquires and loses particles. Rockets furnish another example of variable systems, since their propulsion depends upon the continual ejection of fuel particles.

We recall that all the kinetics principles established so far were derived for constant systems of particles, which neither gain nor lose particles. We must therefore find a way to reduce the analysis of a variable system of particles to that of an auxiliary constant system. The procedure to follow is indicated in Secs. 14.11 and 14.12 for two broad categories of applications: a steady stream of particles and a system that is gaining or losing mass.

*14.11 STEADY STREAM OF PARTICLES

Fig. 14.9

Consider a steady stream of particles, such as a stream of water diverted by a fixed vane or a flow of air through a duct or through a blower. In order to determine the resultant of the forces exerted on the particles in contact with the vane, duct, or blower, we isolate these particles and denote by S the system thus defined (Fig. 14.9). We observe that S is a variable system of particles, since it continually gains particles flowing in and loses an equal number of particles flowing out. Therefore, the kinetics principles that have been established so far cannot be directly applied to S.

However, we can easily define an auxiliary system of particles which does remain constant for a short interval of time Δt. Consider at time t the system S *plus* the particles which will enter S during the interval at time Δt (Fig. 14.10a). Next, consider at time $t + \Delta t$ the system S *plus* the particles which have left S during the interval Δt (Fig. 14.10c). Clearly, *the same particles are involved in both cases,* and we can apply to those particles the principle of impulse and momentum. Since the total mass m of the system S remains constant, the particles entering the system and those leaving the system in the time Δt must have the same mass Δm. Denoting by $\mathbf{v}_A$ and $\mathbf{v}_B$, respectively, the velocities of the particles entering S at A and leaving S at B, we represent the momentum of the particles entering S by $(\Delta m)\mathbf{v}_A$ (Fig. 14.10a) and the momentum of the particles leaving S by $(\Delta m)\mathbf{v}_B$ (Fig. 14.10c). We also represent by appropriate vectors the momenta $m_i\mathbf{v}_i$ of the particles forming S and the impulses of the forces exerted on S and indicate by blue plus and equals signs that the system of the momenta and impulses in parts a and b of Fig. 14.10 is equipollent to the system of the momenta in part c of the same figure.

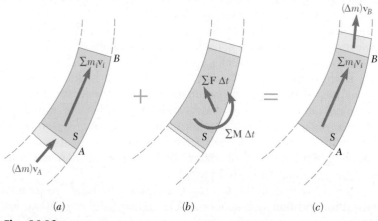

| (a) | (b) | (c) |

Fig. 14.10

The resultant $\Sigma m_i \mathbf{v}_i$ of the momenta of the particles of S is found on both sides of the equals sign and can thus be omitted. We conclude that *the system formed by the momentum* $(\Delta m)\mathbf{v}_A$ *of the particles entering* S *in the time* Δt *and the impulses of the forces exerted on* S *during that time is equipollent to the momentum* $(\Delta m)\mathbf{v}_B$ *of the particles leaving* S *in the same time* Δt. We can therefore write

$$(\Delta m)\mathbf{v}_A + \Sigma \mathbf{F} \, \Delta t = (\Delta m)\mathbf{v}_B \qquad (14.38)$$

A similar equation can be obtained by taking the moments of the vectors involved (see Sample Prob. 14.5). Dividing all terms of Eq. (14.38) by Δt and letting Δt approach zero, we obtain at the limit

$$\Sigma \mathbf{F} = \frac{dm}{dt}(\mathbf{v}_B - \mathbf{v}_A) \qquad (14.39)$$

where $\mathbf{v}_B - \mathbf{v}_A$ represents the difference between the *vector* $\mathbf{v}_B$ and the *vector* $\mathbf{v}_A$.

If SI units are used, dm/dt is expressed in kg/s and the velocities in m/s; we check that both members of Eq. (14.39) are expressed in the same units (newtons). If U.S. customary units are used, dm/dt must be expressed in slugs/s and the velocities in ft/s; we check again that both members of the equation are expressed in the same units (pounds).†

The principle we have established can be used to analyze a large number of engineering applications. Some of the more common of these applications will be considered next.

Fluid Stream Diverted by a Vane. If the vane is fixed, the method of analysis given above can be applied directly to find the force $\mathbf{F}$ exerted by the vane on the stream. We note that $\mathbf{F}$ is the only force which needs to be considered since the pressure in the stream is constant (atmospheric pressure). The force exerted by the stream on the vane will be equal and opposite to $\mathbf{F}$. If the vane moves with a constant velocity, the stream is not steady. However, it will appear steady to an observer moving with the vane. We should therefore choose a system of axes moving with the vane. Since this system of axes is not accelerated, Eq. (14.38) can still be used, but $\mathbf{v}_A$ and $\mathbf{v}_B$ must be replaced by the *relative velocities* of the stream with respect to the vane (see Sample Prob. 14.7).

Fluid Flowing Through a Pipe. The force exerted by the fluid on a pipe transition such as a bend or a contraction can be determined by considering the system of particles S in contact with the transition. Since, in general, the pressure in the flow will vary, the forces exerted on S by the adjoining portions of the fluid should also be considered.

---

†It is often convenient to express the mass rate of flow dm/dt as the product ρQ, where ρ is the density of the stream (mass per unit volume) and Q its volume rate of flow (volume per unit time). If SI units are used, ρ is expressed in kg/m³ (for instance, $\rho = 1000$ kg/m³ for water) and Q in m³/s. However, if U.S. customary units are used, ρ will generally have to be computed from the corresponding specific weight γ (weight per unit volume), $\rho = \gamma/g$. Since γ is expressed in lb/ft³ (for instance, $\gamma = 62.4$ lb/ft³ for water), ρ is obtained in slugs/ft³. The volume rate of flow Q is expressed in ft³/s.

Fig. 14.11

Fig. 14.12

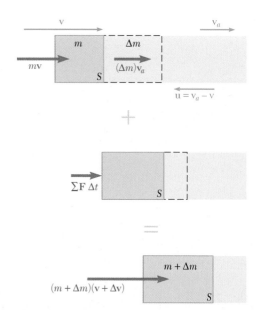

Fig. 14.13

Jet Engine. In a jet engine, air enters with no velocity through the front of the engine and leaves through the rear with a high velocity. The energy required to accelerate the air particles is obtained by burning fuel. The mass of the burned fuel in the exhaust gases will usually be small enough compared with the mass of the air flowing through the engine that it can be neglected. Thus, the analysis of a jet engine reduces to that of an airstream. This stream can be considered as a steady stream if all velocities are measured with respect to the airplane. It will be assumed, therefore, that the airstream enters the engine with a velocity $\mathbf{v}$ of magnitude equal to the speed of the airplane and leaves with a velocity $\mathbf{u}$ equal to the relative velocity of the exhaust gases (Fig. 14.11). Since the intake and exhaust pressures are nearly atmospheric, the only external force which needs to be considered is the force exerted by the engine on the airstream. This force is equal and opposite to the thrust.†

Fan. We consider the system of particles S shown in Fig. 14.12. The velocity $\mathbf{v}_A$ of the particles entering the system is assumed equal to zero, and the velocity $\mathbf{v}_B$ of the particles leaving the system is the velocity of the *slipstream*. The rate of flow can be obtained by multiplying v_B by the cross-sectional area of the slipstream. Since the pressure all around S is atmospheric, the only external force acting on S is the thrust of the fan.

Helicopter. The determination of the thrust created by the rotating blades of a hovering helicopter is similar to the determination of the thrust of a fan. The velocity $\mathbf{v}_A$ of the air particles as they approach the blades is assumed to be zero, and the rate of flow is obtained by multiplying the magnitude of the velocity $\mathbf{v}_B$ of the slipstream by its cross-sectional area.

*14.12 SYSTEMS GAINING OR LOSING MASS

Let us now analyze a different type of variable system of particles, namely, a system which gains mass by continually absorbing particles or loses mass by continually expelling particles. Consider the system S shown in Fig. 14.13. Its mass, equal to m at the instant t, increases by Δm in the interval of time Δt. In order to apply the principle of impulse and momentum to the analysis of this system, we must consider at time t the system S *plus* the particles of mass Δm which S absorbs during the time interval Δt. The velocity of S at time t is denoted by $\mathbf{v}$, the velocity of S at time $t + \Delta t$ is denoted by $\mathbf{v} + \Delta\mathbf{v}$, and the absolute velocity of the particles absorbed is denoted by $\mathbf{v}_a$. Applying the principle of impulse and momentum, we write

$$m\mathbf{v} + (\Delta m)\mathbf{v}_a + \Sigma\mathbf{F}\,\Delta t = (m + \Delta m)(\mathbf{v} + \Delta\mathbf{v}) \qquad (14.40)$$

†Note that if the airplane is accelerated, it cannot be used as a newtonian frame of reference. The same result will be obtained for the thrust, however, by using a reference frame at rest with respect to the atmosphere, since the air particles will then be observed to enter the engine with no velocity and to leave it with a velocity of magnitude $u - v$.

Solving for the sum $\Sigma\mathbf{F}\,\Delta t$ of the impulses of the external forces acting on S (excluding the forces exerted by the particles being absorbed), we have

$$\Sigma\mathbf{F}\,\Delta t = m\Delta\mathbf{v} + \Delta m(\mathbf{v} - \mathbf{v}_a) + (\Delta m)(\Delta\mathbf{v}) \qquad (14.41)$$

Introducing the *relative velocity* $\mathbf{u}$ with respect to S of the particles which are absorbed, we write $\mathbf{u} = \mathbf{v}_a - \mathbf{v}$ and note, since $v_a < v$, that the relative velocity $\mathbf{u}$ is directed to the left, as shown in Fig. 14.13. Neglecting the last term in Eq. (14.41), which is of the second order, we write

$$\Sigma\mathbf{F}\,\Delta t = m\,\Delta\mathbf{v} - (\Delta m)\mathbf{u}$$

Dividing through by Δt and letting Δt approach zero, we have at the limit†

$$\Sigma\mathbf{F} = m\frac{d\mathbf{v}}{dt} - \frac{dm}{dt}\mathbf{u} \qquad (14.42)$$

Rearranging the terms and recalling that $d\mathbf{v}/dt = \mathbf{a}$, where $\mathbf{a}$ is the acceleration of the system S, we write

$$\Sigma\mathbf{F} + \frac{dm}{dt}\mathbf{u} = m\mathbf{a} \qquad (14.43)$$

which shows that the action on S of the particles being absorbed is equivalent to a thrust

$$\mathbf{P} = \frac{dm}{dt}\mathbf{u} \qquad (14.44)$$

which tends to slow down the motion of S, since the relative velocity $\mathbf{u}$ of the particles is directed to the left. If SI units are used, dm/dt is expressed in kg/s, the relative velocity u in m/s, and the corresponding thrust in newtons. If U.S. customary units are used, dm/dt must be expressed in slugs/s, u in ft/s, and the corresponding thrust in pounds.‡

The equations obtained can also be used to determine the motion of a system S losing mass. In this case, the rate of change of mass is negative, and the action on S of the particles being expelled is equivalent to a thrust in the direction of $-\mathbf{u}$, that is, in the direction opposite to that in which the particles are being expelled. A *rocket* represents a typical case of a system continually losing mass (see Sample Prob. 14.8).

Photo 14.3 As the shuttle's booster rockets are fired, the gas particles they eject provide the thrust required for liftoff.

---

†When the absolute velocity $\mathbf{v}_a$ of the particles absorbed is zero, $\mathbf{u} = -\mathbf{v}$, and formula (14.42) becomes

$$\Sigma\mathbf{F} = \frac{d}{dt}(m\mathbf{v})$$

Comparing the formula obtained to Eq. (12.3) of Sec. 12.3, we observe that Newton's second law can be applied to a system gaining mass, *provided that the particles absorbed are initially at rest*. It may also be applied to a system losing mass, *provided that the velocity of the particles expelled is zero* with respect to the frame of reference selected.

‡See footnote on page 899.

SAMPLE PROBLEM 14.6

Grain falls from a hopper onto a chute CB at the rate of 240 lb/s. It hits the chute at A with a velocity of 20 ft/s and leaves at B with a velocity of 15 ft/s, forming an angle of 10° with the horizontal. Knowing that the combined weight of the chute and of the grain it supports is a force **W** of magnitude 600 lb applied at G, determine the reaction at the roller support B and the components of the reaction at the hinge C.

SOLUTION

We apply the principle of impulse and momentum for the time interval Δt to the system consisting of the chute, the grain it supports, and the amount of grain which hits the chute in the interval Δt. Since the chute does not move, it has no momentum. We also note that the sum $\Sigma m_i \mathbf{v}_i$ of the momenta of the particles supported by the chute is the same at t and $t + \Delta t$ and can thus be omitted.

Since the system formed by the momentum $(\Delta m)\mathbf{v}_A$ and the impulses is equipollent to the momentum $(\Delta m)\mathbf{v}_B$, we write

$\xrightarrow{+} x$ components: $\qquad C_x \, \Delta t = (\Delta m)v_B \cos 10°$ $\qquad$ (1)

$+\uparrow y$ components: $\qquad -(\Delta m)v_A + C_y \, \Delta t - W \, \Delta t + B \, \Delta t$
$$= -(\Delta m)v_B \sin 10° \quad (2)$$

$+\gamma$ moments about C: $\qquad -3(\Delta m)v_A - 7(W \, \Delta t) + 12(B \, \Delta t)$
$$= 6(\Delta m)v_B \cos 10° - 12(\Delta m)v_B \sin 10° \quad (3)$$

Using the given data, $W = 600$ lb, $v_A = 20$ ft/s, $v_B = 15$ ft/s, and $\Delta m/\Delta t = 240/32.2 = 7.45$ slugs/s, and solving Eq. (3) for B and Eq. (1) for C_x,

$$12B = 7(600) + 3(7.45)(20) + 6(7.45)(15)(\cos 10° - 2 \sin 10°)$$
$$12B = 5075 \qquad B = 423 \text{ lb} \qquad\qquad \mathbf{B} = 423 \text{ lb} \uparrow \quad \blacktriangleleft$$

$$C_x = (7.45)(15) \cos 10° = 110.1 \text{ lb} \qquad \mathbf{C}_x = 110.1 \text{ lb} \rightarrow \quad \blacktriangleleft$$

Substituting for B and solving Eq. (2) for C_y,

$$C_y = 600 - 423 + (7.45)(20 - 15 \sin 10°) = 307 \text{ lb}$$
$$\mathbf{C}_y = 307 \text{ lb} \uparrow \quad \blacktriangleleft$$

SAMPLE PROBLEM 14.7

A nozzle discharges a stream of water of cross-sectional area A with a velocity $\mathbf{v}_A$. The stream is deflected by a *single* blade which moves to the right with a constant velocity $\mathbf{V}$. Assuming that the water moves along the blade at constant speed, determine (a) the components of the force $\mathbf{F}$ exerted by the blade on the stream, (b) the velocity $\mathbf{V}$ for which maximum power is developed.

SOLUTION

a. Components of Force Exerted on Stream. We choose a coordinate system which moves with the blade at a constant velocity $\mathbf{V}$. The particles of water strike the blade with a relative velocity $\mathbf{u}_A = \mathbf{v}_A - \mathbf{V}$ and leave the blade with a relative velocity $\mathbf{u}_B$. Since the particles move along the blade at a constant speed, the relative velocities $\mathbf{u}_A$ and $\mathbf{u}_B$ have the same magnitude u. Denoting the density of water by ρ, the mass of the particles striking the blade during the time interval Δt is $\Delta m = A\rho(v_A - V)\,\Delta t$; an equal mass of particles leaves the blade during Δt. We apply the principle of impulse and momentum to the system formed by the particles in contact with the blade and the particles striking the blade in the time Δt.

Recalling that $\mathbf{u}_A$ and $\mathbf{u}_B$ have the same magnitude u, and omitting the momentum $\Sigma m_i \mathbf{v}_i$ which appears on both sides, we write

$\xrightarrow{+} x$ components: $(\Delta m)u - F_x\,\Delta t = (\Delta m)u \cos\theta$

$+\uparrow y$ components: $+F_y\,\Delta t = (\Delta m)u \sin\theta$

Substituting $\Delta m = A\rho(v_A - V)\,\Delta t$ and $u = v_A - V$, we obtain

$$\mathbf{F}_x = A\rho(v_A - V)^2(1 - \cos\theta)\leftarrow \qquad \mathbf{F}_y = A\rho(v_A - V)^2 \sin\theta\uparrow \ \blacktriangleleft$$

b. Velocity of Blade for Maximum Power. The power is obtained by multiplying the velocity V of the blade by the component F_x of the force exerted by the stream on the blade.

$$\text{Power} = F_x V = A\rho(v_A - V)^2(1 - \cos\theta)V$$

Differentiating the power with respect to V and setting the derivative equal to zero, we obtain

$$\frac{d(\text{power})}{dV} = A\rho(v_A^2 - 4v_A V + 3V^2)(1 - \cos\theta) = 0$$

$$V = v_A \qquad V = \tfrac{1}{3}v_A \qquad \text{For maximum power } \mathbf{V} = \tfrac{1}{3}v_A \rightarrow \ \blacktriangleleft$$

Note. These results are valid only when a *single* blade deflects the stream. Different results are obtained when a series of blades deflects the stream, as in a Pelton-wheel turbine. (See Prob. 14.81.)

SAMPLE PROBLEM 14.8

A rocket of initial mass m_0 (including shell and fuel) is fired vertically at time $t = 0$. The fuel is consumed at a constant rate $q = dm/dt$ and is expelled at a constant speed u relative to the rocket. Derive an expression for the magnitude of the velocity of the rocket at time t, neglecting the resistance of the air.

SOLUTION

At time t, the mass of the rocket shell and remaining fuel is $m = m_0 - qt$, and the velocity is $\mathbf{v}$. During the time interval Δt, a mass of fuel $\Delta m = q\,\Delta t$ is expelled with a speed u relative to the rocket. Denoting by $\mathbf{v}_e$ the absolute velocity of the expelled fuel, we apply the principle of impulse and momentum between time t and time $t + \Delta t$.

We write

$$(m_0 - qt)v - g(m_0 - qt)\,\Delta t = (m_0 - qt - q\,\Delta t)(v + \Delta v) - q\,\Delta t(u - v)$$

Dividing through by Δt and letting Δt approach zero, we obtain

$$-g(m_0 - qt) = (m_0 - qt)\frac{dv}{dt} - qu$$

Separating variables and integrating from $t = 0,\, v = 0$ to $t = t,\, v = v$,

$$dv = \left(\frac{qu}{m_0 - qt} - g\right) dt \qquad \int_0^v dv = \int_0^t \left(\frac{qu}{m_0 - qt} - g\right) dt$$

$$v = [-u \ln(m_0 - qt) - gt]_0^t \qquad v = u \ln\frac{m_0}{m_0 - qt} - gt \quad \blacktriangleleft$$

Remark. The mass remaining at time t_f, after all the fuel has been expended, is equal to the mass of the rocket shell $m_s = m_0 - qt_f$, and the maximum velocity attained by the rocket is $v_m = u \ln(m_0/m_s) - gt_f$. Assuming that the fuel is expelled in a relatively short period of time, the term gt_f is small and we have $v_m \approx u \ln(m_0/m_s)$. In order to escape the gravitational field of the earth, a rocket must reach a velocity of 11.18 km/s. Assuming $u = 2200$ m/s and $v_m = 11.18$ km/s, we obtain $m_0/m_s = 161$. Thus, to project each kilogram of the rocket shell into space, it is necessary to consume more than 161 kg of fuel if a propellant yielding $u = 2200$ m/s is used.

SOLVING PROBLEMS
ON YOUR OWN

This lesson is devoted to the study of the motion of *variable systems of particles*, i.e., systems which are continually *gaining or losing particles* or doing both at the same time. The problems you will be asked to solve will involve (1) *steady streams of particles* and (2) *systems gaining or losing mass*.

1. To solve problems involving a steady stream of particles, you will consider a portion S of the stream and express that the system formed by the momentum of the particles entering S at A in the time Δt and the impulses of the forces exerted on S during that time is equipollent to the momentum of the particles leaving S at B in the same time Δt (Fig. 14.10). Considering only the resultants of the vector systems involved, you can write the vector equation

$$(\Delta m)\mathbf{v}_A + \Sigma \mathbf{F} \, \Delta t = (\Delta m)\mathbf{v}_B \qquad (14.38)$$

You may want to consider as well the moments about a given point of the vector systems involved to obtain an additional equation [Sample Prob. 14.6], but many problems can be solved using Eq. (14.38) or the equation obtained by dividing all terms by Δt and letting Δt approach zero,

$$\Sigma \mathbf{F} = \frac{dm}{dt}(\mathbf{v}_B - \mathbf{v}_A) \qquad (14.39)$$

where $\mathbf{v}_B - \mathbf{v}_A$ represents a *vector subtraction* and where the mass rate of flow dm/dt can be expressed as the product ρQ of the density ρ of the stream (mass per unit volume) and the volume rate of flow Q (volume per unit time). If U.S. customary units are used, ρ is expressed as the ratio γ/g, where γ is the specific weight of the stream and g is the acceleration of gravity.

Typical problems involving a steady stream of particles have been described in Sec. 14.11. You may be asked to determine the following:

   a. Thrust caused by a diverted flow. Equation (14.39) is applicable, but you will get a better understanding of the problem if you use a solution based on Eq. (14.38).

   b. Reactions at supports of vanes or conveyor belts. First draw a diagram showing on one side of the equals sign the momentum $(\Delta m)\mathbf{v}_A$ of the particles impacting the vane or belt in the time Δt, as well as the impulses of the loads and reactions at the supports during that time, and showing on the other side the momentum $(\Delta m)\mathbf{v}_B$ of the particles leaving the vane or belt in the time Δt [Sample Prob. 14.6]. Equating the x components, y components, and moments of the quantities on both sides of the equals sign will yield three scalar equations which can be solved for three unknowns.

   c. Thrust developed by a jet engine, a propeller, or a fan. In most cases, a single unknown is involved, and that unknown can be obtained by solving the scalar equation derived from Eq. (14.38) or Eq. (14.39).

(continued)

2. To solve problems involving systems gaining mass, you will consider the system S, which has a mass m and is moving with a velocity $\mathbf{v}$ at time t, and the particles of mass Δm with velocity $\mathbf{v}_a$ that S will absorb in the time interval Δt (Fig. 14.13). You will then express that the total momentum of S and of the particles that will be absorbed, *plus* the impulse of the external forces exerted on S, are equipollent to the momentum of S at time $t + \Delta t$. Noting that the mass of S and its velocity at that time are, respectively, $m + \Delta m$ and $\mathbf{v} + \Delta \mathbf{v}$, you will write the vector equation

$$m\mathbf{v} + (\Delta m)\mathbf{v}_a + \Sigma \mathbf{F}\, \Delta t = (m + \Delta m)(\mathbf{v} + \Delta \mathbf{v}) \qquad (14.40)$$

As was shown in Sec. 14.12, if you introduce the relative velocity $\mathbf{u} = \mathbf{v}_a - \mathbf{v}$ of the particles being absorbed, you obtain the following expression for the resultant of the external forces applied to S:

$$\Sigma \mathbf{F} = m\frac{d\mathbf{v}}{dt} - \frac{dm}{dt}\mathbf{u} \qquad (14.42)$$

Furthermore, it was shown that the action on S of the particles being absorbed is equivalent to a thrust

$$\mathbf{P} = \frac{dm}{dt}\mathbf{u} \qquad (14.44)$$

exerted in the direction of the relative velocity of the particles being absorbed.

Examples of systems gaining mass are conveyor belts and moving railroad cars being loaded with gravel or sand, and chains being pulled out of a pile.

3. To solve problems involving systems losing mass, such as rockets and rocket engines, you can use Eqs. (14.40) through (14.44), provided that you give negative values to the increment of mass Δm and to the rate of change of mass dm/dt. It follows that the thrust defined by Eq. (14.44) will be exerted in a direction opposite to the direction of the relative velocity of the particles being ejected.

PROBLEMS

14.57 A stream of water of cross-section area A_1 and velocity $\mathbf{v}_1$ strikes a circular plate which is held motionless by a force $\mathbf{P}$. A hole in the circular plate of area A_2 results in a discharge jet having a velocity $\mathbf{v}_1$. Determine the magnitude of $\mathbf{P}$.

Fig. P14.57

14.58 A jet ski is placed in a channel and is tethered so that it is stationary. Water enters the jet ski with velocity $\mathbf{v}_1$ and exits with velocity $\mathbf{v}_2$. Knowing the inlet area is A_1 and the exit area is A_2, determine the tension in the tether.

Fig. P14.58

14.59 A stream of water of cross-section area A and velocity $\mathbf{v}_1$ strikes a plate which is held motionless by a force $\mathbf{P}$. Determine the magnitude of $\mathbf{P}$, knowing that $A = 0.75$ in^2, $v_1 = 80$ ft/s, and $V = 0$.

Fig. P14.59 and P14.60

14.60 A stream of water of cross-section area A and velocity $\mathbf{v}_1$ strikes a plate which moves to the right with a velocity $\mathbf{V}$. Determine the magnitude of $\mathbf{V}$, knowing that $A = 1$ in^2, $v_1 = 100$ ft/s, and $P = 90$ lb.

Fig. P14.61

14.61 A rotary power plow is used to remove snow from a level section of railroad track. The plow car is placed ahead of an engine which propels it at a constant speed of 20 km/h. The plow car clears 160 Mg of snow per minute, projecting it in the direction shown with a velocity of 12 m/s relative to the plow car. Neglecting friction, determine (*a*) the force exerted by the engine on the plow car, (*b*) the lateral force exerted by the track on the plow.

14.62 Tree limbs and branches are being fed at *A* at the rate of 5 kg/s into a shredder which spews the resulting wood chips at *C* with a velocity of 20 m/s. Determine the horizontal component of the force exerted by the shredder on the truck hitch at *D*.

Fig. P14.62

14.63 Sand falls from three hoppers onto a conveyor belt at a rate of 90 lb/s for each hopper. The sand hits the belt with a vertical velocity $v_1 = 10$ ft/s and is discharged at *A* with a horizontal velocity $v_2 = 13$ ft/s. Knowing that the combined mass of the beam, belt system, and the sand it supports is 1300 lb with a mass center at *G*, determine the reaction at *E*.

Fig. P14.63

14.64 The stream of water shown flows at a rate of 550 L/min and moves with a velocity of magnitude 18 m/s at both A and B. The vane is supported by a pin and bracket at C and by a load cell at D which can exert only a horizontal force. Neglecting the weight of the vane, determine the components of the reactions at C and D.

14.65 The nozzle discharges water at the rate of 340 gal/min. Knowing the velocity of the water at both A and B has a magnitude of 65 ft/s and neglecting the weight of the vane, determine the components of the reactions at C and D (1 ft³ = 7.48 gal).

Fig. P14.64

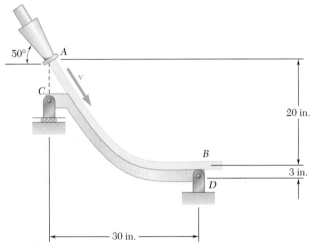

Fig. P14.65

14.66 A high-speed jet of air issues from nozzle A with a velocity of v_A and mass flow rate of 0.36 kg/s. The air impinges on a vane causing it to rotate to the position shown. The vane has a mass of 6 kg. Knowing that the magnitude of the air velocity is equal at A and B, determine (a) the magnitude of the velocity at A, (b) the components of the reactions at O.

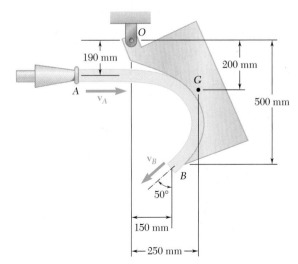

Fig. P14.66

14.67 Coal is being discharged from a first conveyor belt at the rate of 120 kg/s. It is received at A by a second belt which discharges it again at B. Knowing that $v_1 = 3$ m/s and $v_2 = 4.25$ m/s and that the second belt assembly and the coal it supports have a total mass of 472 kg, determine the components of the reactions at C and D.

Fig. P14.67

14.68 A mass q of sand is discharged per unit time from a conveyor belt moving with a velocity v_0. The sand is deflected by a plate at A so that it falls in a vertical stream. After falling a distance h the sand is again deflected by a curved plate at B. Neglecting the friction between the sand and the plates, determine the force required to hold in the position shown (a) plate A, (b) plate B.

14.69 The total drag due to air friction on a jet airplane traveling at 900 km/h is 35 kN. Knowing that the exhaust velocity is 600 m/s relative to the airplane, determine the mass of air which must pass through the engine per second to maintain the speed of 900 km/h in level flight.

14.70 While cruising in level flight at a speed of 600 mi/h, a jet plane scoops in air at the rate of 200 lb/s and discharges it with a velocity of 2100 ft/s relative to the airplane. Determine the total drag due to air friction on the airplane.

14.71 In order to shorten the distance required for landing, a jet airplane is equipped with movable vanes which partially reverse the direction of the air discharged by each of its engines. Each engine scoops in the air at a rate of 120 kg/s and discharges it with a velocity of 600 m/s relative to the engine. At an instant when the speed of the airplane is 270 km/h, determine the reverse thrust provided by each of the engines.

Fig. P14.68

Fig. P14.71

14.72 The helicopter shown can produce a maximum downward air speed of 80 ft/s in a 30-ft-diameter slipstream. Knowing that the weight of the helicopter and its crew is 3500 lb and assuming $\gamma = 0.076$ lb/ft^3 for air, determine the maximum load that the helicopter can lift while hovering in midair.

← 30 ft →

Fig. P14.72

14.73 A floor fan designed to deliver air at a maximum velocity of 6 m/s in a 400-mm-diameter slipstream is supported by a 200-mm-diameter circular base plate. Knowing that the total weight of the assembly is 60 N and that its center of gravity is located directly above the center of the base plate, determine the maximum height h at which the fan may be operated if it is not to tip over. Assume $\rho = 1.21$ kg/m^3 for air and neglect the approach velocity of the air.

400 mm

Fig. P14.73

14.74 The jet engine shown scoops in air at A at a rate of 200 lb/s and discharges it at B with a velocity of 2000 ft/s relative to the airplane. Determine the magnitude and line of action of the propulsive thrust developed by the engine when the speed of the airplane is (a) 300 mi/h, (b) 600 mi/h.

12 ft

Fig. P14.74

14.75 A jet airliner is cruising at a speed of 900 km/h with each of its three engines discharging air with a velocity of 800 m/s relative to the plane. Determine the speed of the airliner after it has lost the use of (a) one of its engines, (b) two of its engines. Assume that the drag due to air friction is proportional to the square of the speed and that the remaining engines keep operating at the same rate.

Fig. P14.75

Fig. P14.76

Fig. P14.78 and P14.79

82.5 m

Fig. P14.81

14.76 A 16-Mg jet airplane maintains a constant speed of 774 km/h while climbing at an angle $\alpha = 18°$. The airplane scoops in air at a rate of 300 kg/s and discharges it with a velocity of 665 m/s relative to the airplane. If the pilot changes to a horizontal flight while maintaining the same engine setting, determine (*a*) the initial acceleration of the plane, (*b*) the maximum horizontal speed that will be attained. Assume that the drag due to air friction is proportional to the square of the speed.

14.77 The propeller of a small airplane has a 2-m-diameter slipstream and produces a thrust of 3600 N when the airplane is at rest on the ground. Assuming $\rho = 1.225$ kg/m^3 for air, determine (*a*) the speed of the air in the slipstream, (*b*) the volume of air passing through the propeller per second, (*c*) the kinetic energy imparted per second to the air in the slipstream.

14.78 The wind turbine–generator shown has an output-power rating of 1.5 MW for a wind speed of 36 km/h. For the given wind speed, determine (*a*) the kinetic energy of the air particles entering the 82.5-m-diameter circle per second, (*b*) the efficiency of this energy conversion system. Assume $\rho = 1.21$ kg/m^3 for air.

14.79 A wind turbine-generator system having a diameter of 82.5 m produces 1.5 MW at a wind speed of 12 m/s. Determine the diameter of blade necessary to produce 10 MW of power assuming the efficiency is the same for both designs and $\rho = 1.21$ kg/m^3 for air.

14.80 While cruising in level flight at a speed of 570 mi/h, a jet airplane scoops in air at a rate of 240 lb/s and discharges it with a velocity of 2200 ft/s relative to the airplane. Determine (*a*) the power actually used to propel the airplane, (*b*) the total power developed by the engine, (*c*) the mechanical efficiency of the airplane.

14.81 In a Pelton-wheel turbine, a stream of water is deflected by a series of blades so that the rate at which water is deflected by the blades is equal to the rate at which water issues from the nozzle ($\Delta m/\Delta t = A\rho v_A$). Using the same notation as in Sample Prob. 14.7, (*a*) determine the velocity **V** of the blades for which maximum power is developed, (*b*) derive an expression for the maximum power, (*c*) derive an expression for the mechanical efficiency.

14.82 A circular reentrant orifice (also called Borda's mouthpiece) of diameter D is placed at a depth h below the surface of a tank. Knowing that the speed of the issuing stream is $v = \sqrt{2gh}$ and assuming that the speed of approach v_1 is zero, show that the diameter of the stream is $d = D/\sqrt{2}$. (*Hint:* Consider the section of water indicated, and note that P is equal to the pressure at a depth h multiplied by the area of the orifice.)

Fig. P14.82

14.83 Gravel falls with practically zero velocity onto a conveyor belt at the constant rate $q = dm/dt$. (a) Determine the magnitude of the force **P** required to maintain a constant belt speed v. (b) Show that the kinetic energy acquired by the gravel in a given time interval is equal to half the work done in that interval by the force **P**. Explain what happens to the other half of the work done by **P**.

Fig. P14.83

Fig. P14.84

*__**14.84**__ The depth of water flowing in a rectangular channel of width b at a speed v_1 and a depth d_1 increases to a depth d_2 at a *hydraulic jump*. Express the rate of flow Q in terms of b, d_1, and d_2.

*__**14.85**__ Determine the rate of flow in the channel of Prob. 14.84, knowing that $b = 12$ ft, $d_1 = 4$ ft, and $d_2 = 5$ ft.

14.86 A chain of length l and mass m lies in a pile on the floor. If its end A is raised vertically at a constant speed v, express in terms of the length y of chain which is off the floor at any given instant (a) the magnitude of the force **P** applied to A, (b) the reaction of the floor.

14.87 Solve Prob. 14.86, assuming that the chain is being *lowered* to the floor at a constant speed v.

14.88 The ends of a chain lie in piles at A and C. When released from rest at time $t = 0$, the chain moves over the pulley at B, which has a negligible mass. Denoting by L the length of chain connecting the two piles and neglecting friction, determine the speed v of the chain at time t.

Fig. P14.86

Fig. P14.88

14.89 A toy car is propelled by water that squirts from an internal tank at a constant 6 ft/s relative to the car. The weight of the empty car is 0.4 lb and it holds 2 lb of water. Neglecting other tangential forces, determine the top speed of the car.

Fig. P14.89 and P14.90

14.90 A toy car is propelled by water that squirts from an internal tank. The weight of the empty car is 0.4 lb and it holds 2 lb of water. Knowing the top speed of the car is 8 ft/s determine the relative velocity of the water that is being ejected.

14.91 The main propulsion system of a space shuttle consists of three identical rocket engines which provide a total thrust of 6 MN. Determine the rate at which the hydrogen-oxygen propellant is burned by each of the three engines, knowing that it is ejected with a relative velocity of 3750 m/s.

Fig. P14.91 and *P14.92*

14.92 The main propulsion system of a space shuttle consists of three identical rocket engines, each of which burns the hydrogen-oxygen propellant at the rate of 750 lb/s and ejects it with a relative velocity of 12,000 ft/s. Determine the total thrust provided by the three engines.

14.93 A rocket weighs 2600 lb, including 2200 lb of fuel, which is consumed at a rate of 25 lb/s and ejected with a relative velocity of 13,000 ft/s. Knowing that the rocket is fired vertically from the ground, determine its acceleration (*a*) as it is fired, (*b*) as the last particle of fuel is being consumed.

14.94 A space vehicle describing a circular orbit about the earth at a speed of 24×10^3 km/h releases at its front end a capsule which has a gross mass of 600 kg, including 400 kg of fuel. If the fuel is consumed at the rate of 18 kg/s and ejected with a relative velocity of 3000 m/s, determine (*a*) the tangential acceleration of the capsule as its engine is fired, (*b*) the maximum speed attained by the capsule.

Fig. P14.94

14.95 A 540-kg spacecraft is mounted on top of a rocket with a mass of 19 Mg, including 17.8 Mg of fuel. Knowing that the fuel is consumed at a rate of 225 kg/s and ejected with a relative velocity of 3600 m/s, determine the maximum speed imparted to the spacecraft if the rocket is fired vertically from the ground.

Fig. P14.95   **Fig. P14.96**

14.96 The rocket used to launch the 540-kg spacecraft of Prob. 14.95 is redesigned to include two stages A and B, each of mass 9.5 Mg, including 8.9 Mg of fuel. The fuel is again consumed at a rate of 225 kg/s and ejected with a relative velocity of 3600 m/s. Knowing that when stage A expels its last particle of fuel, its casing is released and jettisoned, determine (a) the speed of the rocket at that instant, (b) the maximum speed imparted to the spacecraft.

14.97 A communications satellite weighing 10,000 lb, including fuel, has been ejected from a space shuttle describing a low circular orbit around the earth. After the satellite has slowly drifted to a safe distance from the shuttle, its engine is fired to increase its velocity by 8000 ft/s as a first step to its transfer to a geosynchronous orbit. Knowing that the fuel is ejected with a relative velocity of 13,750 ft/s, determine the weight of fuel consumed in this maneuver.

Fig. P14.97

14.98 Determine the increase in velocity of the communications satellite of Prob. 14.97 after 2500 lb of fuel has been consumed.

14.99 Determine the distance separating the communications satellite of Prob. 14.97 from the space shuttle 60 s after its engine has been fired, knowing that the fuel is consumed at a rate of 37.5 lb/s.

14.100 For the rocket of Prob. 14.93, determine (a) the altitude at which all of the fuel has been consumed, (b) the velocity of the rocket at this time.

14.101 Determine the altitude reached by the spacecraft of Prob. 14.95 when all the fuel of its launching rocket has been consumed.

14.102 For the spacecraft and the two-stage launching rocket of Prob. 14.96, determine the altitude at which (*a*) stage *A* of the rocket is released, (*b*) the fuel of both stages has been consumed.

14.103 In a jet airplane, the kinetic energy imparted to the exhaust gases is wasted as far as propelling the airplane is concerned. The useful power is equal to the product of the force available to propel the airplane and the speed of the airplane. If v is the speed of the airplane and u is the relative speed of the expelled gases, show that the mechanical efficiency of the airplane is $\eta = 2v/(u + v)$. Explain why $\eta = 1$ when $u = v$.

14.104 In a rocket, the kinetic energy imparted to the consumed and ejected fuel is wasted as far as propelling the rocket is concerned. The useful power is equal to the product of the force available to propel the rocket and the speed of the rocket. If v is the speed of the rocket and u is the relative speed of the expelled fuel, show that the mechanical efficiency of the rocket is $\eta = 2uv/(u^2 + v^2)$. Explain why $\eta = 1$ when $u = v$.

REVIEW AND SUMMARY

In this chapter we analyzed the motion of *systems of particles,* i.e., the motion of a large number of particles considered together. In the first part of the chapter we considered systems consisting of well-defined particles, while in the second part we analyzed systems which are continually gaining or losing particles, or doing both at the same time.

We first defined the *effective force* of a particle P_i of a given system as the product $m_i\mathbf{a}_i$ of its mass m_i and its acceleration $\mathbf{a}_i$ with respect to a newtonian frame of reference centered at O [Sec. 14.2]. We then showed that *the system of the external forces acting on the particles and the system of the effective forces of the particles are equipollent;* i.e., both systems have the *same resultant* and the *same moment resultant* about O:

Effective forces

$$\sum_{i=1}^{n} \mathbf{F}_i = \sum_{i=1}^{n} m_i\mathbf{a}_i \qquad (14.4)$$

$$\sum_{i=1}^{n} (\mathbf{r}_i \times \mathbf{F}_i) = \sum_{i=1}^{n} (\mathbf{r}_i \times m_i\mathbf{a}_i) \qquad (14.5)$$

Defining the *linear momentum* $\mathbf{L}$ and the *angular momentum* $\mathbf{H}_O$ *about point O* of the system of particles [Sec. 14.3] as

Linear and angular momentum of a system of particles

$$\mathbf{L} = \sum_{i=1}^{n} m_i\mathbf{v}_i \qquad \mathbf{H}_O = \sum_{i=1}^{n} (\mathbf{r}_i \times m_i\mathbf{v}_i) \qquad (14.6,\ 14.7)$$

we showed that Eqs. (14.4) and (14.5) can be replaced by the equations

$$\Sigma\mathbf{F} = \dot{\mathbf{L}} \qquad \Sigma\mathbf{M}_O = \dot{\mathbf{H}}_O \qquad (14.10,\ 14.11)$$

which express that *the resultant and the moment resultant about O of the external forces are, respectively, equal to the rates of change of the linear momentum and of the angular momentum about O of the system of particles.*

In Sec. 14.4, we defined the mass center of a system of particles as the point G whose position vector $\bar{\mathbf{r}}$ satisfies the equation

Motion of the mass center of a system of particles

$$m\bar{\mathbf{r}} = \sum_{i=1}^{n} m_i\mathbf{r}_i \qquad (14.12)$$

where m represents the total mass $\sum_{i=1}^{n} m_i$ of the particles. Differentiating both members of Eq. (14.12) twice with respect to t, we obtained the relations

$$\mathbf{L} = m\overline{\mathbf{v}} \qquad \dot{\mathbf{L}} = m\overline{\mathbf{a}} \qquad (14.14, \ 14.15)$$

where $\overline{\mathbf{v}}$ and $\overline{\mathbf{a}}$ represent, respectively, the velocity and the acceleration of the mass center G. Substituting for $\dot{\mathbf{L}}$ from (14.15) into (14.10), we obtained the equation

$$\Sigma\mathbf{F} = m\overline{\mathbf{a}} \qquad (14.16)$$

from which we concluded that *the mass center of a system of particles moves as if the entire mass of the system and all the external forces were concentrated at that point* [Sample Prob. 14.1].

Angular momentum of a system of particles about its mass center

In Sec. 14.5 we considered the motion of the particles of a system with respect to a centroidal frame $Gx'y'z'$ attached to the mass center G of the system and in translation with respect to the newtonian frame $Oxyz$ (Fig. 14.14). We defined the *angular momentum* of the system *about its mass center* G as the sum of the moments about G of the momenta $m_i\mathbf{v}_i'$ of the particles in their motion relative to the frame $Gx'y'z'$. We also noted that the same result can be obtained by considering the moments about G of the momenta $m_i\mathbf{v}_i$ of the particles in their absolute motion. We therefore wrote

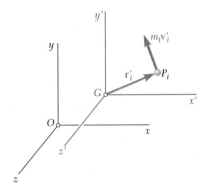

Fig. 14.14

$$\mathbf{H}_G = \sum_{i=1}^{n} (\mathbf{r}_i' \times m_i\mathbf{v}_i) = \sum_{i=1}^{n} (\mathbf{r}_i' \times m_i\mathbf{v}_i') \qquad (14.24)$$

and derived the relation

$$\Sigma\mathbf{M}_G = \dot{\mathbf{H}}_G \qquad (14.23)$$

which expresses that *the moment resultant about G of the external forces is equal to the rate of change of the angular momentum about G of the system of particles*. As will be seen later, this relation is fundamental to the study of the motion of rigid bodies.

Conservation of momentum

When no external force acts on a system of particles [Sec. 14.6], it follows from Eqs. (14.10) and (14.11) that the linear momentum $\mathbf{L}$ and the angular momentum $\mathbf{H}_O$ of the system are conserved [Sample Probs. 14.2 and 14.3]. In problems involving central forces, the angular momentum of the system about the center of force O will also be conserved.

Kinetic energy of a system of particles

The kinetic energy T of a system of particles was defined as the sum of the kinetic energies of the particles [Sec. 14.7]:

$$T = \frac{1}{2}\sum_{i=1}^{n} m_i v_i^2 \qquad (14.28)$$

Using the centroidal frame of reference $Gx'y'z'$ of Fig. 14.14, we noted that the kinetic energy of the system can also be obtained by adding the kinetic energy $\frac{1}{2}m\bar{v}^2$ associated with the motion of the mass center G and the kinetic energy of the system in its motion relative to the frame $Gx'y'z'$:

$$T = \tfrac{1}{2}m\bar{v}^2 + \frac{1}{2}\sum_{i=1}^{n} m_i v_i'^2 \qquad (14.29)$$

Principle of work and energy

The *principle of work and energy* can be applied to a system of particles as well as to individual particles [Sec. 14.8]. We wrote

$$T_1 + U_{1\to2} = T_2 \qquad (14.30)$$

and noted that $U_{1\to2}$ represents the work of *all* the forces acting on the particles of the system, internal as well as external.

Conservation of energy

If all the forces acting on the particles of the system are *conservative*, we can determine the potential energy V of the system and write

$$T_1 + V_1 = T_2 + V_2 \qquad (14.31)$$

which expresses the *principle of conservation of energy* for a system of particles.

Principle of impulse and momentum

We saw in Sec. 14.9 that the *principle of impulse and momentum* for a system of particles can be expressed graphically as shown in Fig. 14.15. It states that the momenta of the particles at time t_1 and the impulses of the external forces from t_1 to t_2 form a system of vectors equipollent to the system of the momenta of the particles at time t_2.

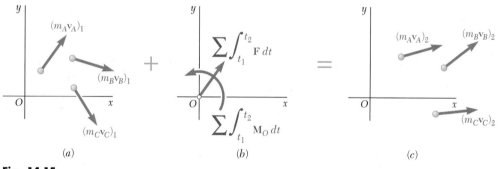

Fig. 14.15

If no external force acts on the particles of the system, the systems of momenta shown in parts a and c of Fig. 14.15 are equipollent and we have

$$\mathbf{L}_1 = \mathbf{L}_2 \qquad (\mathbf{H}_O)_1 = (\mathbf{H}_O)_2 \qquad (14.36, 14.37)$$

Use of conservation principles in the solution of problems involving systems of particles

Many problems involving the motion of systems of particles can be solved by applying simultaneously the principle of impulse and momentum and the principle of conservation of energy [Sample Prob. 14.4] or by expressing that the linear momentum, angular momentum, and energy of the system are conserved [Sample Prob. 14.5].

In the second part of the chapter, we considered *variable systems of particles*. First we considered a *steady stream of particles*, such as a stream of water diverted by a fixed vane or the flow of air through a jet engine [Sec. 14.11]. Applying the principle of impulse and momentum to a system S of particles during a time interval Δt, and including the particles which enter the system at A during that time interval and those (of the same mass Δm) which leave the system at B, we concluded that *the system formed by the momentum $(\Delta m)\mathbf{v}_A$ of the particles entering S in the time Δt and the impulses of the forces exerted on S during that time is equipollent to the momentum $(\Delta m)\mathbf{v}_B$ of the particles leaving S in the same time Δt* (Fig. 14.16). Equating

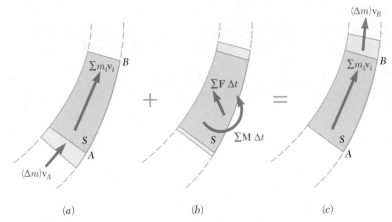

(a) (b) (c)

Fig. 14.16

the x components, y components, and moments about a fixed point of the vectors involved, we could obtain as many as three equations, which could be solved for the desired unknowns [Sample Probs. 14.6 and 14.7]. From this result, we could also derive the following expression for the resultant $\Sigma\mathbf{F}$ of the forces exerted on S,

$$\Sigma\mathbf{F} = \frac{dm}{dt}(\mathbf{v}_B - \mathbf{v}_A) \tag{14.39}$$

where $\mathbf{v}_B - \mathbf{v}_A$ represents the difference between the *vectors* $\mathbf{v}_B$ and $\mathbf{v}_A$ and where dm/dt is the mass rate of flow of the stream (see footnote, page 899).

Considering next a system of particles gaining mass by continually absorbing particles or losing mass by continually expelling particles [Sec. 14.12], as in the case of a rocket, we applied the principle of impulse and momentum to the system during a time interval Δt, being careful to include the particles gained or lost during that time interval [Sample Prob. 14.8]. We also noted that the action on a system S of the particles being *absorbed* by S was equivalent to a thrust

$$\mathbf{P} = \frac{dm}{dt}\mathbf{u} \tag{14.44}$$

where dm/dt is the rate at which mass is being absorbed, and $\mathbf{u}$ is the velocity of the particles *relative to* S. In the case of particles being *expelled* by S, the rate dm/dt is negative and the thrust $\mathbf{P}$ is exerted in a direction opposite to that in which the particles are being expelled.

REVIEW PROBLEMS

14.105 Three identical cars are being unloaded from an automobile carrier. Cars B and C have just been unloaded and are at rest with their brakes off when car A leaves the unloading ramp with a velocity of 5.76 ft/s and hits car B, which hits car C. Car A then again hits car B. Knowing that the velocity of car B is 5.04 ft/s after the first collision, 0.630 ft/s after the second collision, and 0.709 ft/s after the third collision, determine (a) the final velocities of cars A and C, (b) the coefficient of restitution for each of the collisions.

Fig. P14.105

14.106 A 30-g bullet is fired with a velocity of 480 m/s into block A, which has a mass of 5 kg. The coefficient of kinetic friction between block A and cart BC is 0.50. Knowing that the cart has a mass of 4 kg and can roll freely, determine (a) the final velocity of the cart and block, (b) the final position of the block on the cart.

Fig. P14.106

14.107 An 80-Mg railroad engine A coasting at 6.5 km/h strikes a 20-Mg flatcar C carrying a 30-Mg load B which can slide along the floor of the car ($\mu_k = 0.25$). Knowing that the car was at rest with its brakes released and that it automatically coupled with the engine upon impact, determine the velocity of the car (a) immediately after impact, (b) after the load has slid to a stop relative to the car.

Fig. P14.107

14.108 In a game of pool, ball A is moving with a velocity $\mathbf{v}_0$ when it strikes balls B and C which are at rest and aligned as shown. Knowing that after the collision the three balls move in the directions indicated and that $v_0 = 12$ ft/s and $v_C = 6.29$ ft/s, determine the magnitude of the velocity of (a) ball A, (b) ball B.

Fig. P14.108

14.109 Mass C, which has a mass of 4 kg, is suspended from a cord attached to cart A, which has a mass of 5 kg and can roll freely on a frictionless horizontal track. A 60-g bullet is fired with a speed $v_0 = 500$ m/s and gets lodged in block C. Determine (a) the velocity of C as it reaches its maximum elevation, (b) the maximum vertical distance h through which C will rise.

Fig. P14.109

14.110 A 15-lb block B is at rest and a spring of constant $k = 72$ lb/in is held compressed 3 in. by a cord. After 5-lb block A is placed against the end of the spring the cord is cut causing A and B to move. Neglecting friction, determine the velocities of blocks A and B immediately after A leaves B.

Fig. P14.110

14.111 Car A was at rest 9.28 m south of point O when it was struck in the rear by car B, which was traveling north at a speed v_B. Car C, which was traveling west at a speed v_C, was 40 m east of point O at the time of the collision. Cars A and B stuck together and, because the pavement was covered with ice, they slid into the intersection and were struck by car C which had not changed its speed. Measurements based on a photograph taken from a traffic helicopter shortly after the second collision indicated that the positions of the cars, expressed in meters, were $\mathbf{r}_A = -10.1\mathbf{i} + 16.9\mathbf{j}$, $\mathbf{r}_B = -10.1\mathbf{i} + 20.4\mathbf{j}$, and $\mathbf{r}_C = -19.8\mathbf{i} - 15.2\mathbf{j}$. Knowing that the masses of cars A, B, and C are, respectively, 1400 kg, 1800 kg, and 1600 kg, and that the time elapsed between the first collision and the time the photograph was taken was 3.4 s, determine the initial speeds of cars B and C.

Fig. P14.111

14.112 The nozzle shown discharges water at the rate of 200 gal/min. Knowing that at both B and C the stream of water moves with a velocity of magnitude 100 ft/s, and neglecting the weight of the vane, determine the force-couple system which must be applied at A to hold the vane in place (1 ft³ = 7.48 gal).

Fig. P14.112

14.113 Prior to takeoff, the pilot of a 6000-lb twin-engine airplane tests the reversible-pitch propellers with the brakes at point B locked. Knowing that the velocity of the air in the two 6.6-ft-diameter slipstreams is 60 ft/s and that point G is the center of gravity of the airplane, determine the reactions at points A and B. Assume $\gamma = 0.075$ lb/ft^3 and neglect the approach velocity of the air.

Fig. P14.113

14.114 A railroad car of length L and mass m_0 when empty is moving freely on a horizontal track while being loaded with sand from a stationary chute at a rate $dm/dt = q$. Knowing that the car was approaching the chute at a speed v_0, determine (a) the mass of the car and its load after the car has cleared the chute, (b) the speed of the car at that time.

Fig. P14.114

14.115 A garden sprinkler has four rotating arms, each of which consists of two horizontal straight sections of pipe forming an angle of 120° with each other. Each arm discharges water at a rate of 20 L/min with a velocity of 18 m/s relative to the arm. Knowing that the friction between the moving and stationary parts of the sprinkler is equivalent to a couple of magnitude $M = 0.375$ N · m, determine the constant rate at which the sprinkler rotates.

Fig. P14.115

14.116 A chain of length l and mass m falls through a small hole in a plate. Initially, when y is very small, the chain is at rest. In each case shown, determine (a) the acceleration of the first link A as a function of y, (b) the velocity of the chain as the last link passes through the hole. In case 1 assume that the individual links are at rest until they fall through the hole; in case 2 assume that at any instant all links have the same speed. Ignore the effect of friction.

Fig. P14.116

COMPUTER PROBLEMS

Fig. P14.C1

14.C1 A man and a woman, of weights 180 lb and 120 lb, respectively, stand at opposite ends of a stationary boat of weight 300 lb, ready to dive with velocities v_m and v_w, respectively, relative to the boat. Use computational software to determine the velocity of the boat after both swimmers have dived if (a) the woman dives first, (b) the man dives first. Solve that problem assuming that the velocities of the woman and the man relative to the boat are, respectively, (i) 14 ft/s and 18 ft/s, (ii) 18 ft/s and 14 ft/s.

14.C2 A system of particles consists of n particles A_i of mass m_i and coordinates x_i, y_i, and z_i, having velocities of components $(v_x)_i$, $(v_y)_i$, and $(v_z)_i$. Derive expressions for the components of the angular momentum of the system about the origin O of the coordinates. Use computational software to solve Probs. 14.11 and 14.13.

14.C3 A shell moving with a velocity of known components v_x, v_y, and v_z explodes into three fragments of weights W_1, W_2, and W_3 at point A_0 at a distance d from a vertical wall. Use computational software to determine the speed of each fragment immediately after the explosion, knowing the coordinates x_i and y_i of the points A_i ($i = 1, 2, 3$) where the fragments hit the wall. Use this program to solve (a) Prob. 14.24, (b) Prob. 14.25.

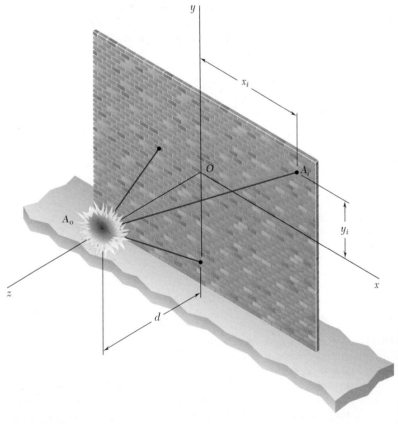

Fig. P14.C3

14.C4 As a 6000-kg training plane lands on an aircraft carrier at a speed of 180 km/h, its tail hooks into the end of an 80-m long chain which lies in a pile below deck. Knowing that the chain has a mass per unit length of 50 kg/m and assuming no other retarding force, use computational software to determine the distance traveled by the plane while the chain is being pulled out and the corresponding values of the time and of the velocity and deceleration of the plane.

Fig. P14.C4

14.C5 A 16-Mg jet airplane maintains a constant speed of 774 km/h while climbing at an angle $\alpha = 18°$. The airplane scoops in air at a rate of 300 kg/s and discharges it with a velocity of 665 m/s relative to the airplane. Knowing that the pilot changes the angle of climb α while maintaining the same engine setting, use computational software to calculate and plot values of α from 0 to 20° (a) the initial acceleration of the plane, (b) the maximum speed that will be attained. Assume that the drag due to air friction is proportional to the square of the speed.

14.C6 A rocket has a weight of 2400 lb, including 2000 lb of fuel, which is consumed at the rate of 25 lb/s and ejected with a relative velocity of 12,000 ft/s. Knowing that the rocket is fired vertically from the ground, assuming a constant value for the acceleration of gravity, and using 4-s time intervals, use computational software to determine and plot from the time of ignition to the time when the last particle of fuel is being consumed (a) the acceleration a of the rocket in ft/s², (b) its velocity v in ft/s, (c) its elevation h above the ground in miles. (*Hint:* Use for v the expression derived in Sample Prob. 14.8, and integrate this expression analytically to obtain h.)

Fig. P14.C5

This huge crank belongs to a Wartsila-Sulzer RTA96-C turbocharged two-stroke diesel engine. In this chapter you will learn to perform the *kinematic* analysis of rigid bodies that undergo *translation, fixed axis rotation,* and *general plane motion.*

Kinematics of Rigid Bodies

15.1 INTRODUCTION

In this chapter, the kinematics of *rigid bodies* will be considered. You will investigate the relations existing between the time, the positions, the velocities, and the accelerations of the various particles forming a rigid body. As you will see, the various types of rigid-body motion can be conveniently grouped as follows:

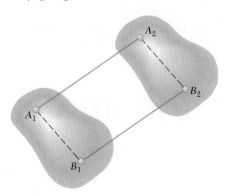

Fig. 15.1

1. *Translation.* A motion is said to be a translation if any straight line inside the body keeps the same direction during the motion. It can also be observed that in a translation all the particles forming the body move along parallel paths. If these paths are straight lines, the motion is said to be a *rectilinear translation* (Fig. 15.1); if the paths are curved lines, the motion is a *curvilinear translation* (Fig. 15.2).

2. *Rotation About a Fixed Axis.* In this motion, the particles forming the rigid body move in parallel planes along circles centered on the same fixed axis (Fig. 15.3). If this axis, called the *axis of rotation,* intersects the rigid body, the particles located on the axis have zero velocity and zero acceleration.

   Rotation should not be confused with certain types of curvilinear translation. For example, the plate shown in Fig. 15.4a is in curvilinear translation, with all its particles moving along *parallel* circles, while the plate shown in Fig. 15.4b is in rotation, with all its particles moving along *concentric* circles.

Fig. 15.2

Fig. 15.3

(*a*) Curvilinear translation

Fig. 15.4

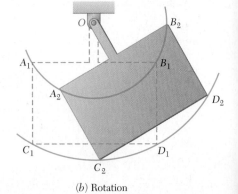

(*b*) Rotation

In the first case, any given straight line drawn on the plate will maintain the same direction, whereas in the second case, point O remains fixed.

Because each particle moves in a given plane, the rotation of a body about a fixed axis is said to be a *plane motion*.

3. *General Plane Motion.* There are many other types of plane motion, i.e., motions in which all the particles of the body move in parallel planes. Any plane motion that is neither a rotation nor a translation is referred to as a general plane motion. Two examples of general plane motion are given in Fig. 15.5.

(*a*) Rolling wheel   (*b*) Sliding rod

Fig. 15.5

4. *Motion About a Fixed Point.* The three-dimensional motion of a rigid body attached at a fixed point O, e.g., the motion of a top on a rough floor (Fig. 15.6), is known as motion about a fixed point.
5. *General Motion.* Any motion of a rigid body that does not fall in any of the categories above is referred to as a general motion.

Fig. 15.6

After a brief discussion in Sec. 15.2 of the motion of translation, the rotation of a rigid body about a fixed axis is considered in Sec. 15.3. The *angular velocity* and the *angular acceleration* of a rigid body about a fixed axis will be defined, and you will learn to express the velocity and the acceleration of a given point of the body in terms of its position vector and the angular velocity and angular acceleration of the body.

The following sections are devoted to the study of the general plane motion of a rigid body and to its application to the analysis of mechanisms such as gears, connecting rods, and pin-connected linkages. Resolving the plane motion of a slab into a translation and a rotation (Secs. 15.5 and 15.6), we will then express the velocity of a point B of the slab as the sum of the velocity of a reference point A and of the velocity of B relative to a frame of reference translating with A (i.e., moving with A but not rotating). The same approach is used later in Sec. 15.8 to express the acceleration of B in terms of the acceleration of A and of the acceleration of B relative to a frame translating with A.

An alternative method for the analysis of velocities in plane motion, based on the concept of *instantaneous center of rotation*, is given in Sec. 15.7; and still another method of analysis, based on the use of parametric expressions for the coordinates of a given point, is presented in Sec. 15.9.

The motion of a particle relative to a rotating frame of reference and the concept of *Coriolis acceleration* are discussed in Secs. 15.10 and 15.11, and the results obtained are applied to the analysis of the plane motion of mechanisms containing parts which slide on each other.

The remaining part of the chapter is devoted to the analysis of the three-dimensional motion of a rigid body, namely, the motion of a rigid body with a fixed point and the general motion of a rigid body. In Secs. 15.12 and 15.13, a fixed frame of reference or a frame of reference in translation will be used to carry out this analysis; in Secs. 15.14 and 15.15, the motion of the body relative to a rotating frame or to a frame in general motion will be considered, and the concept of Coriolis acceleration will again be used.

15.2 TRANSLATION

Photo 15.1 This replica of a battering ram at *Château des Baux, France* undergoes curvilinear translation.

Consider a rigid body in translation (either rectilinear or curvilinear translation), and let A and B be any two of its particles (Fig. 15.7a). Denoting, respectively, by $\mathbf{r}_A$ and $\mathbf{r}_B$ the position vectors of A and B with respect to a fixed frame of reference and by $\mathbf{r}_{B/A}$ the vector joining A and B, we write

$$\mathbf{r}_B = \mathbf{r}_A + \mathbf{r}_{B/A} \tag{15.1}$$

Let us differentiate this relation with respect to t. We note that from the very definition of a translation, the vector $\mathbf{r}_{B/A}$ must maintain a constant direction; its magnitude must also be constant, since A and B

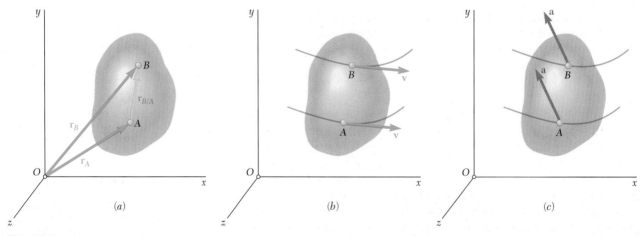

Fig. 15.7

belong to the same rigid body. Thus, the derivative of $\mathbf{r}_{B/A}$ is zero and we have

$$\mathbf{v}_B = \mathbf{v}_A \qquad (15.2)$$

Differentiating once more, we write

$$\mathbf{a}_B = \mathbf{a}_A \qquad (15.3)$$

Thus, *when a rigid body is in translation, all the points of the body have the same velocity and the same acceleration at any given instant* (Fig. 15.7*b* and *c*). In the case of curvilinear translation, the velocity and acceleration change in direction as well as in magnitude at every instant. In the case of rectilinear translation, all particles of the body move along parallel straight lines, and their velocity and acceleration keep the same direction during the entire motion.

15.3 ROTATION ABOUT A FIXED AXIS

Consider a rigid body which rotates about a fixed axis AA'. Let P be a point of the body and $\mathbf{r}$ its position vector with respect to a fixed frame of reference. For convenience, let us assume that the frame is centered at point O on AA' and that the z axis coincides with AA' (Fig. 15.8). Let B be the projection of P on AA'; since P must remain at a constant distance from B, it will describe a circle of center B and of radius $r \sin \phi$, where ϕ denotes the angle formed by $\mathbf{r}$ and AA'.

The position of P and of the entire body is completely defined by the angle θ the line BP forms with the zx plane. The angle θ is known as the *angular coordinate* of the body and is defined as positive when viewed as counterclockwise from A'. The angular coordinate will be expressed in radians (rad) or, occasionally, in degrees (°) or revolutions (rev). We recall that

$$1 \text{ rev} = 2\pi \text{ rad} = 360°$$

We recall from Sec. 11.9 that the velocity $\mathbf{v} = d\mathbf{r}/dt$ of a particle P is a vector tangent to the path of P and of magnitude $v = ds/dt$. Observing that the length Δs of the arc described by P when the body rotates through $\Delta \theta$ is

$$\Delta s = (BP)\,\Delta\theta = (r \sin \phi)\,\Delta\theta$$

and dividing both members by Δt, we obtain at the limit, as Δt approaches zero,

$$v = \frac{ds}{dt} = r\dot{\theta}\sin\phi \qquad (15.4)$$

where $\dot{\theta}$ denotes the time derivative of θ. (Note that the angle θ depends on the position of P within the body, but the rate of change $\dot{\theta}$ is itself independent of P.) We conclude that the velocity $\mathbf{v}$ of P is a vector perpendicular to the plane containing AA' and $\mathbf{r}$, and of

Fig. 15.8

Photo 15.2 For the central gear rotating about a fixed axis, the angular velocity and angular acceleration of that gear are vectors directed along the vertical axis of rotation.

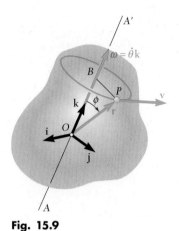

Fig. 15.9

magnitude v defined by (15.4). But this is precisely the result we would obtain if we drew along AA' a vector $\boldsymbol{\omega} = \dot\theta\mathbf{k}$ and formed the vector product $\boldsymbol{\omega} \times \mathbf{r}$ (Fig. 15.9). We thus write

$$\mathbf{v} = \frac{d\mathbf{r}}{dt} = \boldsymbol{\omega} \times \mathbf{r} \tag{15.5}$$

The vector

$$\boldsymbol{\omega} = \omega\mathbf{k} = \dot\theta\mathbf{k} \tag{15.6}$$

which is directed along the axis of rotation, is called the *angular velocity* of the body and is equal in magnitude to the rate of change $\dot\theta$ of the angular coordinate; its sense may be obtained by the right-hand rule (Sec. 3.6) from the sense of rotation of the body.[†]

The acceleration $\mathbf{a}$ of the particle P will now be determined. Differentiating (15.5) and recalling the rule for the differentiation of a vector product (Sec. 11.10), we write

$$\begin{aligned}
\mathbf{a} = \frac{d\mathbf{v}}{dt} &= \frac{d}{dt}(\boldsymbol{\omega} \times \mathbf{r}) \\
&= \frac{d\boldsymbol{\omega}}{dt} \times \mathbf{r} + \boldsymbol{\omega} \times \frac{d\mathbf{r}}{dt} \\
&= \frac{d\boldsymbol{\omega}}{dt} \times \mathbf{r} + \boldsymbol{\omega} \times \mathbf{v} \tag{15.7}
\end{aligned}$$

The vector $d\boldsymbol{\omega}/dt$ is denoted by $\boldsymbol{\alpha}$ and is called the *angular acceleration* of the body. Substituting also for $\mathbf{v}$ from (15.5), we have

$$\mathbf{a} = \boldsymbol{\alpha} \times \mathbf{r} + \boldsymbol{\omega} \times (\boldsymbol{\omega} \times \mathbf{r}) \tag{15.8}$$

Differentiating (15.6) and recalling that $\mathbf{k}$ is constant in magnitude and direction, we have

$$\boldsymbol{\alpha} = \alpha\mathbf{k} = \dot\omega\mathbf{k} = \ddot\theta\mathbf{k} \tag{15.9}$$

Thus, the angular acceleration of a body rotating about a fixed axis is a vector directed along the axis of rotation, and is equal in magnitude to the rate of change $\dot\omega$ of the angular velocity. Returning to (15.8), we note that the acceleration of P is the sum of two vectors. The first vector is equal to the vector product $\boldsymbol{\alpha} \times \mathbf{r}$; it is tangent to the circle described by P and therefore represents the tangential component of the acceleration. The second vector is equal to the *vector triple product* $\boldsymbol{\omega} \times (\boldsymbol{\omega} \times \mathbf{r})$ obtained by forming the vector product of $\boldsymbol{\omega}$ and $\boldsymbol{\omega} \times \mathbf{r}$; since $\boldsymbol{\omega} \times \mathbf{r}$ is tangent to the circle described by P, the vector triple product is directed toward the center B of the circle and therefore represents the normal component of the acceleration.

[†]It will be shown in Sec. 15.12 in the more general case of a rigid body rotating simultaneously about axes having different directions that angular velocities obey the parallelogram law of addition and thus are actually vector quantities.

Rotation of a Representative Slab. The rotation of a rigid body about a fixed axis can be defined by the motion of a representative slab in a reference plane perpendicular to the axis of rotation. Let us choose the xy plane as the reference plane and assume that it coincides with the plane of the figure, with the z axis pointing out of the paper (Fig. 15.10). Recalling from (15.6) that $\boldsymbol{\omega} = \omega\mathbf{k}$, we

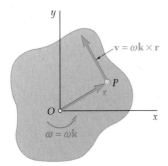

Fig. 15.10

note that a positive value of the scalar ω corresponds to a counterclockwise rotation of the representative slab, and a negative value to a clockwise rotation. Substituting $\omega\mathbf{k}$ for $\boldsymbol{\omega}$ into Eq. (15.5), we express the velocity of any given point P of the slab as

$$\mathbf{v} = \omega\mathbf{k} \times \mathbf{r} \tag{15.10}$$

Since the vectors $\mathbf{k}$ and $\mathbf{r}$ are mutually perpendicular, the magnitude of the velocity $\mathbf{v}$ is

$$v = r\omega \tag{15.10'}$$

and its direction can be obtained by rotating $\mathbf{r}$ through $90°$ in the sense of rotation of the slab.

Substituting $\boldsymbol{\omega} = \omega\mathbf{k}$ and $\boldsymbol{\alpha} = \alpha\mathbf{k}$ into Eq. (15.8), and observing that cross-multiplying $\mathbf{r}$ twice by $\mathbf{k}$ results in a $180°$ rotation of the vector $\mathbf{r}$, we express the acceleration of point P as

$$\mathbf{a} = \alpha\mathbf{k} \times \mathbf{r} - \omega^2\mathbf{r} \tag{15.11}$$

Resolving $\mathbf{a}$ into tangential and normal components (Fig. 15.11), we write

$$\mathbf{a}_t = \alpha\mathbf{k} \times \mathbf{r} \qquad a_t = r\alpha \tag{15.11'}$$
$$\mathbf{a}_n = -\omega^2\mathbf{r} \qquad a_n = r\omega^2$$

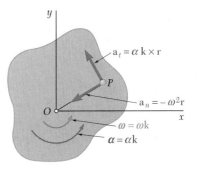

Fig. 15.11

The tangential component $\mathbf{a}_t$ points in the counterclockwise direction if the scalar α is positive, and in the clockwise direction if α is negative. The normal component $\mathbf{a}_n$ always points in the direction opposite to that of $\mathbf{r}$, that is, toward O.

15.4 EQUATIONS DEFINING THE ROTATION OF A RIGID BODY ABOUT A FIXED AXIS

The motion of a rigid body rotating about a fixed axis AA' is said to be *known* when its angular coordinate θ can be expressed as a known function of t. In practice, however, the rotation of a rigid body is seldom defined by a relation between θ and t. More often, the conditions of motion will be specified by the type of angular acceleration that the body possesses. For example, α may be given as a function of t, as a function of θ, or as a function of ω. Recalling the relations (15.6) and (15.9), we write

$$\omega = \frac{d\theta}{dt} \tag{15.12}$$

$$\alpha = \frac{d\omega}{dt} = \frac{d^2\theta}{dt^2} \tag{15.13}$$

or, solving (15.12) for dt and substituting into (15.13),

$$\alpha = \omega \frac{d\omega}{d\theta} \tag{15.14}$$

Photo 15.3 If the lower roll has a constant angular velocity, the speed of the paper being wound onto it increases as the radius of the roll increases.

Since these equations are similar to those obtained in Chap. 11 for the rectilinear motion of a particle, their integration can be performed by following the procedure outlined in Sec. 11.3.

Two particular cases of rotation are frequently encountered:

1. *Uniform Rotation.* This case is characterized by the fact that the angular acceleration is zero. The angular velocity is thus constant, and the angular coordinate is given by the formula

$$\theta = \theta_0 + \omega t \tag{15.15}$$

2. *Uniformly Accelerated Rotation.* In this case, the angular acceleration is constant. The following formulas relating angular velocity, angular coordinate, and time can then be derived in a manner similar to that described in Sec. 11.5. The similarity between the formulas derived here and those obtained for the rectilinear uniformly accelerated motion of a particle is apparent.

$$\omega = \omega_0 + \alpha t$$
$$\theta = \theta_0 + \omega_0 t + \tfrac{1}{2}\alpha t^2 \tag{15.16}$$
$$\omega^2 = \omega_0^2 + 2\alpha(\theta - \theta_0)$$

It should be emphasized that formula (15.15) can be used only when $\alpha = 0$, and formulas (15.16) can be used only when $\alpha =$ constant. In any other case, the general formulas (15.12) to (15.14) should be used.

SAMPLE PROBLEM 15.1

Load B is connected to a double pulley by one of the two inextensible cables shown. The motion of the pulley is controlled by cable C, which has a constant acceleration of 9 in./s² and an initial velocity of 12 in./s, both directed to the right. Determine (a) the number of revolutions executed by the pulley in 2 s, (b) the velocity and change in position of the load B after 2 s, and (c) the acceleration of point D on the rim of the inner pulley at $t = 0$.

SOLUTION

a. Motion of Pulley. Since the cable is inextensible, the velocity of point D is equal to the velocity of point C and the tangential component of the acceleration of D is equal to the acceleration of C.

$$(\mathbf{v}_D)_0 = (\mathbf{v}_C)_0 = 12 \text{ in./s} \rightarrow \qquad (\mathbf{a}_D)_t = \mathbf{a}_C = 9 \text{ in./s}^2 \rightarrow$$

Noting that the distance from D to the center of the pulley is 3 in., we write

$$\begin{aligned} (v_D)_0 = r\omega_0 \qquad & 12 \text{ in./s} = (3 \text{ in.})\omega_0 \qquad \omega_0 = 4 \text{ rad/s} \downarrow \\ (a_D)_t = r\alpha \qquad & 9 \text{ in./s}^2 = (3 \text{ in.})\alpha \qquad \alpha = 3 \text{ rad/s}^2 \downarrow \end{aligned}$$

Using the equations of uniformly accelerated motion, we obtain, for $t = 2$ s,

$$\omega = \omega_0 + \alpha t = 4 \text{ rad/s} + (3 \text{ rad/s}^2)(2 \text{ s}) = 10 \text{ rad/s}$$
$$\boldsymbol{\omega} = 10 \text{ rad/s} \downarrow$$
$$\theta = \omega_0 t + \tfrac{1}{2}\alpha t^2 = (4 \text{ rad/s})(2 \text{ s}) + \tfrac{1}{2}(3 \text{ rad/s}^2)(2 \text{ s})^2 = 14 \text{ rad}$$
$$\theta = 14 \text{ rad} \downarrow$$

$$\text{Number of revolutions} = (14 \text{ rad})\left(\frac{1 \text{ rev}}{2\pi \text{ rad}}\right) = 2.23 \text{ rev} \quad \blacktriangleleft$$

b. Motion of Load B. Using the following relations between linear and angular motion, with $r = 5$ in., we write

$$v_B = r\omega = (5 \text{ in.})(10 \text{ rad/s}) = 50 \text{ in./s} \qquad \mathbf{v}_B = 50 \text{ in./s} \uparrow \quad \blacktriangleleft$$
$$\Delta y_B = r\theta = (5 \text{ in.})(14 \text{ rad}) = 70 \text{ in.} \qquad \Delta y_B = 70 \text{ in. upward} \quad \blacktriangleleft$$

c. Acceleration of Point D at t = 0. The tangential component of the acceleration is

$$(\mathbf{a}_D)_t = \mathbf{a}_C = 9 \text{ in./s}^2 \rightarrow$$

Since, at $t = 0$, $\omega_0 = 4$ rad/s, the normal component of the acceleration is

$$(a_D)_n = r_D\omega_0^2 = (3 \text{ in.})(4 \text{ rad/s})^2 = 48 \text{ in./s}^2 \qquad (\mathbf{a}_D)_n = 48 \text{ in./s}^2 \downarrow$$

The magnitude and direction of the total acceleration can be obtained by writing

$$\tan \phi = (48 \text{ in./s}^2)/(9 \text{ in./s}^2) \qquad \phi = 79.4°$$
$$a_D \sin 79.4° = 48 \text{ in./s}^2 \qquad a_D = 48.8 \text{ in./s}^2$$
$$\mathbf{a}_D = 48.8 \text{ in./s}^2 \searrow 79.4° \quad \blacktriangleleft$$

SOLVING PROBLEMS
ON YOUR OWN

In this lesson we began the study of the motion of rigid bodies by considering two particular types of motion of rigid bodies: *translation* and *rotation* about a *fixed axis*.

1. Rigid body in translation. At any given instant, all the points of a rigid body in translation have the *same velocity* and the *same acceleration* (Fig. 15.7).

2. Rigid body rotating about a fixed axis. The position of a rigid body rotating about a fixed axis was defined at any given instant by the *angular coordinate* θ, which is usually measured in *radians*. Selecting the unit vector $\mathbf{k}$ along the fixed axis and in such a way that the rotation of the body appears counterclockwise as seen from the tip of $\mathbf{k}$, we defined the *angular velocity* $\boldsymbol{\omega}$ and the *angular acceleration* $\boldsymbol{\alpha}$ of the body:

$$\boldsymbol{\omega} = \dot{\theta}\mathbf{k} \qquad \boldsymbol{\alpha} = \ddot{\theta}\mathbf{k} \qquad\qquad (15.6, \ 15.9)$$

In solving problems, keep in mind that the vectors $\boldsymbol{\omega}$ and $\boldsymbol{\alpha}$ are both directed along the fixed axis of rotation and that their sense can be obtained by the right-hand rule.

    a. The velocity of a point P of a body rotating about a fixed axis was found to be

$$\mathbf{v} = \boldsymbol{\omega} \times \mathbf{r} \qquad\qquad (15.5)$$

where $\boldsymbol{\omega}$ is the angular velocity of the body and $\mathbf{r}$ is the position vector drawn from any point on the axis of rotation to point P (Fig. 15.9).

    b. The acceleration of point P was found to be

$$\mathbf{a} = \boldsymbol{\alpha} \times \mathbf{r} + \boldsymbol{\omega} \times (\boldsymbol{\omega} \times \mathbf{r}) \qquad\qquad (15.8)$$

Since vector products are not commutative, *be sure to write the vectors in the order shown* when using either of the above two equations.

3. Rotation of a representative slab. In many problems, you will be able to reduce the analysis of the rotation of a three-dimensional body about a fixed axis to the study of the rotation of a representative slab in a plane perpendicular to the fixed axis. The z axis should be directed along the axis of rotation and point out of the paper. Thus, the representative slab will be rotating in the xy plane about the origin O of the coordinate system (Fig. 15.10).

To solve problems of this type you should do the following:

    a. Draw a diagram of the representative slab, showing its dimensions, its angular velocity and angular acceleration, as well as the vectors representing the velocities and accelerations of the points of the slab for which you have or seek information.

b. Relate the rotation of the slab and the motion of points of the slab by writing the equations

$$v = r\omega \tag{15.10'}$$
$$a_t = r\alpha \qquad a_n = r\omega^2 \tag{15.11'}$$

Remember that the velocity $\mathbf{v}$ and the component $\mathbf{a}_t$ of the acceleration of a point P of the slab are tangent to the circular path described by P. The directions of $\mathbf{v}$ and $\mathbf{a}_t$ are found by rotating the position vector $\mathbf{r}$ through $90°$ in the sense indicated by $\boldsymbol{\omega}$ and $\boldsymbol{\alpha}$, respectively. The normal component $\mathbf{a}_n$ of the acceleration of P is always directed toward the axis of rotation.

4. Equations defining the rotation of a rigid body. You must have been pleased to note the similarity existing between the equations defining the rotation of a rigid body about a fixed axis [Eqs. (15.12) through (15.16)] and those in Chap. 11 defining the rectilinear motion of a particle [Eqs. (11.1) through (11.8)]. All you have to do to obtain the new set of equations is to substitute θ, ω, and α for x, v, and a in the equations of Chap. 11.

PROBLEMS

CONCEPT QUESTIONS

15.CQ1 A rectangular plate swings from arms of equal length as shown. What is the magnitude of the angular velocity of the plate?

    a. 0 rad/s
    b. 1 rad/s
    c. 2 rad/s
    d. 3 rad/s
    e. Need to know the location of the center of gravity.

Fig. P15.CQ1

Fig. P15.CQ2

15.CQ2 Knowing that wheel A rotates with a constant angular velocity and that no slipping occurs between ring C and wheel A and wheel B, which of the following statements concerning the angular speeds of the three objects is true?

    a. $\omega_a = \omega_b$
    b. $\omega_a > \omega_b$
    c. $\omega_a < \omega_b$
    d. $\omega_a = \omega_c$
    e. The contact points between A and C have the same acceleration.

END-OF-SECTION PROBLEMS

15.1 The brake drum is attached to a larger flywheel that is not shown. The motion of the brake drum is defined by the relation $\theta = 36t - 1.6t^2$, where θ is expressed in radians and t in seconds. Determine (*a*) the angular velocity at $t = 2$ s, (*b*) the number of revolutions executed by the brake drum before coming to rest.

Fig. P15.1

15.2 The motion of an oscillating crank is defined by the relation $\theta = \theta_0 \sin(\pi t/T) - (0.5\theta_0) \sin(2\pi t/T)$, where θ is expressed in radians and t in seconds. Knowing that $\theta_0 = 6$ rad and $T = 4$ s, determine the angular coordinate, the angular velocity, and the angular acceleration of the crank when (*a*) $t = 0$, (*b*) $t = 2$ s.

15.3 The motion of a disk rotating in an oil bath is defined by the relation $\theta = \theta_0(1 - e^{-t/4})$, where θ is expressed in radians and t in seconds. Knowing that $\theta_0 = 0.40$ rad, determine the angular coordinate, velocity, and acceleration of the disk when (*a*) $t = 0$, (*b*) $t = 3$ s, (*c*) $t = \infty$.

15.4 The rotor of a gas turbine is rotating at a speed of 6900 rpm when the turbine is shut down. It is observed that 4 min is required for the rotor to coast to rest. Assuming uniformly accelerated motion, determine (*a*) the angular acceleration, (*b*) the number of revolutions that the rotor executes before coming to rest.

938

15.5 A small grinding wheel is attached to the shaft of an electric motor which has a rated speed of 3600 rpm. When the power is turned on, the unit reaches its rated speed in 5 s, and when the power is turned off, the unit coasts to rest in 70 s. Assuming uniformly accelerated motion, determine the number of revolutions that the motor executes (*a*) in reaching its rated speed, (*b*) in coasting to rest.

Fig. P15.5

15.6 A connecting rod is supported by a knife-edge at point *A*. For small oscillations the angular acceleration of the connecting rod is governed by the relation $\alpha = -60\theta$ where α is expressed in rad/s^2 and θ in radians. Knowing that the connecting rod is released from rest when $\theta = 20°$, determine (*a*) the maximum angular velocity, (*b*) the angular position when $t = 2$ s.

Fig. P15.6

15.7 When studying whiplash resulting from rear-end collisions, the rotation of the head is of primary interest. An impact test was performed, and it was found that the angular acceleration of the head is defined by the relation $\alpha = 700 \cos \theta + 70 \sin \theta$, where α is expressed in rad/s^2 and θ in radians. Knowing that the head is initially at rest, determine the angular velocity of the head when $\theta = 30°$.

Fig. P15.7

15.8 The angular acceleration of an oscillating disk is defined by the relation $\alpha = -k\theta$. Determine (*a*) the value of *k* for which $\omega = 8$ rad/s when $\theta = 0$ and $\theta = 4$ rad when $\omega = 0$, (*b*) the angular velocity of the disk when $\theta = 3$ rad.

15.9 The angular acceleration of a shaft is defined by the relation $\alpha = -0.25\omega$, where α is expressed in rad/s^2 and ω in rad/s. Knowing that at $t = 0$ the angular velocity of the shaft is 20 rad/s, determine (*a*) the number of revolutions the shaft will execute before coming to rest, (*b*) the time required for the shaft to come to rest, (*c*) the time required for the angular velocity of the shaft to be reduced to 1 percent of its initial value.

15.10 The bent rod *ABCDE* rotates about a line joining points *A* and *E* with a constant angular velocity of 9 rad/s. Knowing that the rotation is clockwise as viewed from *E*, determine the velocity and acceleration of corner *C*.

15.11 In Prob. 15.10, determine the velocity and acceleration of corner *B*, assuming that the angular velocity is 9 rad/s and increases at the rate of 45 rad/s^2.

Fig. P15.10

Fig. P15.12

15.12 The assembly shown consists of the straight rod *ABC* which passes through and is welded to the rectangular plate *DEFH*. The assembly rotates about the axis *AC* with a constant angular velocity of 9 rad/s. Knowing that the motion when viewed from *C* is counterclockwise, determine the velocity and acceleration of corner *F*.

15.13 In Prob. 15.12, determine the acceleration of corner *H*, assuming that the angular velocity is 9 rad/s and decreases at a rate of 18 rad/s².

15.14 A circular plate of 120-mm radius is supported by two bearings *A* and *B* as shown. The plate rotates about the rod joining *A* and *B* with a constant angular velocity of 26 rad/s. Knowing that, at the instant considered, the velocity of point *C* is directed to the right, determine the velocity and acceleration of point *E*.

Fig. P15.14

15.15 In Prob. 15.14, determine the velocity and acceleration of point *E*, assuming that the angular velocity is 26 rad/s and increases at the rate of 65 rad/s².

15.16 The earth makes one complete revolution around the sun in 365.24 days. Assuming that the orbit of the earth is circular and has a radius of 93,000,000 mi, determine the velocity and acceleration of the earth.

15.17 The earth makes one complete revolution on its axis in 23 h 56 min. Knowing that the mean radius of the earth is 3960 mi, determine the linear velocity and acceleration of a point on the surface of the earth (*a*) at the equator, (*b*) at Philadelphia, latitude 40° north, (*c*) at the North Pole.

15.18 A series of small machine components being moved by a conveyor belt pass over a 120-mm-radius idler pulley. At the instant shown, the velocity of point *A* is 300 mm/s to the left and its acceleration is 180 mm/s² to the right. Determine (*a*) the angular velocity and angular acceleration of the idler pulley, (*b*) the total acceleration of the machine component at *B*.

15.19 A series of small machine components being moved by a conveyor belt pass over a 120-mm-radius idler pulley. At the instant shown, the angular velocity of the idler pulley is 4 rad/s clockwise. Determine the angular acceleration of the pulley for which the magnitude of the total acceleration of the machine component at *B* is 2400 mm/s².

Fig. P15.18 and P15.19

15.20 The belt sander shown is initially at rest. If the driving drum B has a constant angular acceleration of 120 rad/s² counterclockwise, determine the magnitude of the acceleration of the belt at point C when (*a*) $t = 0.5$ s, (*b*) $t = 2$ s.

Fig. *P15.20* and *P15.21*

15.21 The rated speed of drum B of the belt sander shown is 2400 rpm. When the power is turned off, it is observed that the sander coasts from its rated speed to rest in 10 s. Assuming uniformly decelerated motion, determine the velocity and acceleration of point C of the belt, (*a*) immediately before the power is turned off, (*b*) 9 s later.

15.22 The two pulleys shown may be operated with the V belt in any of three positions. If the angular acceleration of shaft A is 6 rad/s² and if the system is initially at rest, determine the time required for shaft B to reach a speed of 400 rpm with the belt in each of the three positions.

Fig. P15.22

15.23 Three belts move over two pulleys without slipping in the speed reduction system shown. At the instant shown, the velocity of point A on the input belt is 2 ft/s to the right, decreasing at the rate of 6 ft/s². Determine, at this instant, (*a*) the velocity and acceleration of point C on the output belt, (*b*) the acceleration of point B on the output pulley.

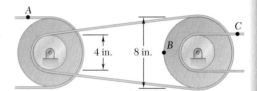

Fig. P15.23

15.24 A gear reduction system consists of three gears A, B, and C. Knowing that gear A rotates clockwise with a constant angular velocity $\omega_A = 600$ rpm, determine (*a*) the angular velocities of gears B and C, (*b*) the accelerations of the points on gears B and C which are in contact.

Fig. P15.24

15.25 A belt is pulled to the right between cylinders A and B. Knowing that the speed of the belt is a constant 5 ft/s and no slippage occurs, determine (*a*) the angular velocities of A and B, (*b*) the accelerations of the points which are in contact with the belt.

Fig. P15.25

15.26 Ring C has an inside radius of 55 mm and an outside radius of 60 mm and is positioned between two wheels A and B, each of 24-mm outside radius. Knowing that wheel A rotates with a constant angular velocity of 300 rpm and that no slipping occurs, determine (a) the angular velocity of ring C and of wheel B, (b) the acceleration of the points A and B which are in contact with C.

Fig. P15.26

Fig. P15.27

15.27 Ring B has an inside radius r_2 and hangs from the horizontal shaft A as shown. Shaft A rotates with a constant angular velocity of 25 rad/s and no slipping occurs. Knowing that $r_1 = 12$ mm, $r_2 = 30$ mm, and $r_3 = 40$ mm, determine (a) the angular velocity of ring B, (b) the accelerations of the points of shaft A and ring B which are in contact, (c) the magnitude of the acceleration of a point on the outside surface of ring B.

15.28 A plastic film moves over two drums. During a 4-s interval the speed of the tape is increased uniformly from $v_0 = 2$ ft/s to $v_1 = 4$ ft/s. Knowing that the tape does not slip on the drums, determine (a) the angular acceleration of drum B, (b) the number of revolutions executed by drum B during the 4-s interval.

Fig. P15.28

Fig. P15.29 and P15.30

15.29 A pulley and two loads are connected by inextensible cords as shown. Load A has a constant acceleration of 300 mm/s^2 and an initial velocity of 240 mm/s, both directed upward. Determine (a) the number of revolutions executed by the pulley in 3 s, (b) the velocity and position of load B after 3 s, (c) the acceleration of point D on the rim of the pulley at $t = 0$.

15.30 A pulley and two loads are connected by inextensible cords as shown. The pulley starts from rest at $t = 0$ and is accelerated at the uniform rate of 2.4 rad/s^2 clockwise. At $t = 4$ s, determine the velocity and position (a) of load A, (b) of load B.

15.31 A load is to be raised 20 ft by the hoisting system shown. Assuming gear A is initially at rest, accelerates uniformly to a speed of 120 rpm in 5 s, and then maintains a constant speed of 120 rpm, determine (a) the number of revolutions executed by gear A in raising the load, (b) the time required to raise the load.

Fig. P15.31

15.32 Disk B is at rest when it is brought into contact with disk A which is rotating freely at 450 rpm clockwise. After 6 s of slippage, during which each disk has a constant angular acceleration, disk A reaches a final angular velocity of 140 rpm clockwise. Determine the angular acceleration of each disk during the period of slippage.

Fig. P15.32 and P15.33

15.33 and 15.34 A simple friction drive consists of two disks A and B. Initially, disk A has a clockwise angular velocity of 500 rpm and disk B is at rest. It is known that disk A will coast to rest in 60 s. However, rather than waiting until both disks are at rest to bring them together, disk B is given a constant angular acceleration of 2.5 rad/s² counterclockwise. Determine (a) at what time the disks can be brought together if they are not to slip, (b) the angular velocity of each disk as contact is made.

Fig. P15.34 and P15.35

15.35 Two friction disks A and B are both rotating freely at 240 rpm counterclockwise when they are brought into contact. After 8 s of slippage, during which each disk has a constant angular acceleration, disk A reaches a final angular velocity of 60 rpm counterclockwise. Determine (a) the angular acceleration of each disk during the period of slippage, (b) the time at which the angular velocity of disk B is equal to zero.

Fig. P15.36

*15.36 Steel tape is being wound onto a spool which rotates with a constant angular velocity ω_0. Denoting by r the radius of the spool and tape at any given time and by b the thickness of the tape, derive an expression for the acceleration of the tape as it approaches the spool.

*15.37 In a continuous printing process, paper is drawn into the presses at a constant speed v. Denoting by r the radius of the paper roll at any given time and by b the thickness of the paper, derive an expression for the angular acceleration of the paper roll.

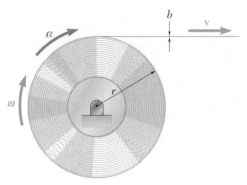

Fig. P15.37

15.5 GENERAL PLANE MOTION

As indicated in Sec. 15.1, we understand by general plane motion a plane motion which is neither a translation nor a rotation. As you will presently see, however, *a general plane motion can always be considered as the sum of a translation and a rotation.*

Consider, for example, a wheel rolling on a straight track (Fig. 15.12). Over a certain interval of time, two given points A and B will have moved, respectively, from A_1 to A_2 and from B_1 to B_2. The same result could be obtained through a translation which would bring A and B into A_2 and B'_1 (the line AB remaining vertical), followed by a rotation about A bringing B into B_2. Although the original rolling motion differs from the combination of translation and rotation when these motions are taken in succession, the original motion can be exactly duplicated by a combination of simultaneous translation and rotation.

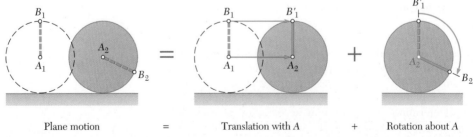

| Plane motion | = | Translation with A | + | Rotation about A |

Fig. 15.12

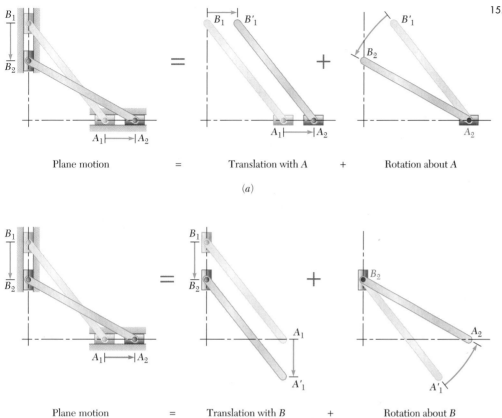

Fig. 15.13

Another example of plane motion is given in Fig. 15.13, which represents a rod whose extremities slide along a horizontal and a vertical track, respectively. This motion can be replaced by a translation in a horizontal direction and a rotation about A (Fig. 15.13a) or by a translation in a vertical direction and a rotation about B (Fig. 15.13b).

In the general case of plane motion, we will consider a small displacement which brings two particles A and B of a representative slab, respectively, from A_1 and B_1 into A_2 and B_2 (Fig. 15.14). This displacement can be divided into two parts: in one, the particles move into A_2 and B'_1 while the line AB maintains the same direction; in the other, B moves into B_2 while A remains fixed. The first part of the motion is clearly a translation and the second part a rotation about A.

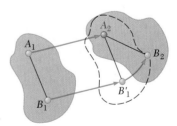

Fig. 15.14

Recalling from Sec. 11.12 the definition of the relative motion of a particle with respect to a moving frame of reference—as opposed to its absolute motion with respect to a fixed frame of reference—we can restate as follows the result obtained above: Given two particles A and B of a rigid slab in plane motion, the relative motion of B with respect to a frame attached to A and of fixed orientation is a rotation. To an observer moving with A but not rotating, particle B will appear to describe an arc of circle centered at A.

Photo 15.4 Planetary gear systems are used to high reduction ratios with minimum space and weight. The small gears undergo general plane motion.

15.6 ABSOLUTE AND RELATIVE VELOCITY IN PLANE MOTION

We saw in the preceding section that any plane motion of a slab can be replaced by a translation defined by the motion of an arbitrary reference point A and a simultaneous rotation about A. The absolute velocity $\mathbf{v}_B$ of a particle B of the slab is obtained from the relative-velocity formula derived in Sec. 11.12,

$$\mathbf{v}_B = \mathbf{v}_A + \mathbf{v}_{B/A} \tag{15.17}$$

where the right-hand member represents a vector sum. The velocity $\mathbf{v}_A$ corresponds to the translation of the slab with A, while the relative velocity $\mathbf{v}_{B/A}$ is associated with the rotation of the slab about A and is measured with respect to axes centered at A and of fixed orientation (Fig. 15.15). Denoting by $\mathbf{r}_{B/A}$ the position vector of B relative to A, and by $\omega\mathbf{k}$ the angular velocity of the slab with respect to axes of fixed orientation, we have from (15.10) and (15.10′)

$$\mathbf{v}_{B/A} = \omega\mathbf{k} \times \mathbf{r}_{B/A} \qquad v_{B/A} = r\omega \tag{15.18}$$

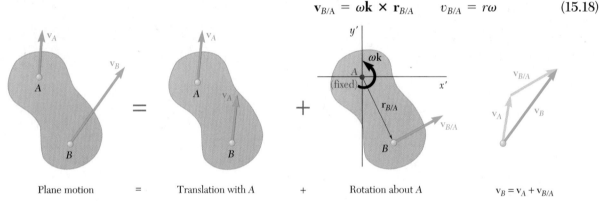

Plane motion = Translation with A + Rotation about A $\qquad v_B = v_A + v_{B/A}$

Fig. 15.15

where r is the distance from A to B. Substituting for $\mathbf{v}_{B/A}$ from (15.18) into (15.17), we can also write

$$\mathbf{v}_B = \mathbf{v}_A + \omega\mathbf{k} \times \mathbf{r}_{B/A} \tag{15.17′}$$

As an example, let us again consider the rod AB of Fig. 15.13. Assuming that the velocity $\mathbf{v}_A$ of end A is known, we propose to find the velocity $\mathbf{v}_B$ of end B and the angular velocity $\boldsymbol{\omega}$ of the rod, in terms of the velocity $\mathbf{v}_A$, the length l, and the angle θ. Choosing A as a reference point, we express that the given motion is equivalent to a translation with A and a simultaneous rotation about A (Fig. 15.16). The absolute velocity of B must therefore be equal to the vector sum

$$\mathbf{v}_B = \mathbf{v}_A + \mathbf{v}_{B/A} \tag{15.17}$$

We note that while the direction of $\mathbf{v}_{B/A}$ is known, its magnitude $l\omega$ is unknown. However, this is compensated for by the fact that the direction of $\mathbf{v}_B$ is known. We can therefore complete the diagram of Fig. 15.16. Solving for the magnitudes v_B and ω, we write

$$v_B = v_A \tan\theta \qquad \omega = \frac{v_{B/A}}{l} = \frac{v_A}{l\cos\theta} \tag{15.19}$$

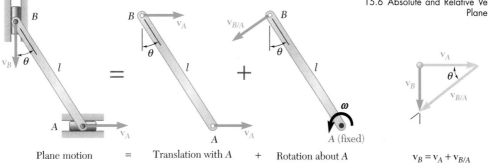

Plane motion = Translation with A + Rotation about A $\quad\quad v_B = v_A + v_{B/A}$

Fig. 15.16

The same result can be obtained by using B as a point of reference. Resolving the given motion into a translation with B and a simultaneous rotation about B (Fig. 15.17), we write the equation

$$\mathbf{v}_A = \mathbf{v}_B + \mathbf{v}_{A/B} \tag{15.20}$$

which is represented graphically in Fig. 15.17. We note that $\mathbf{v}_{A/B}$ and $\mathbf{v}_{B/A}$ have the same magnitude $l\omega$ but opposite sense. The sense of the relative velocity depends, therefore, upon the point of reference which has been selected and should be carefully ascertained from the appropriate diagram (Fig. 15.16 or 15.17).

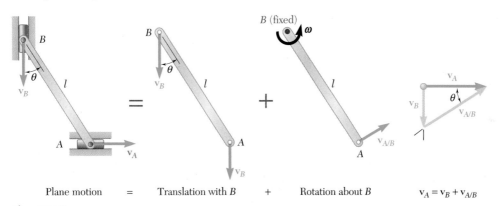

Plane motion = Translation with B + Rotation about B $\quad\quad v_A = v_B + v_{A/B}$

Fig. 15.17

Finally, we observe that the angular velocity $\boldsymbol{\omega}$ of the rod in its rotation about B is the same as in its rotation about A. It is measured in both cases by the rate of change of the angle θ. This result is quite general; we should therefore bear in mind that *the angular velocity $\boldsymbol{\omega}$ of a rigid body in plane motion is independent of the reference point.*

Most mechanisms consist not of one but of *several* moving parts. When the various parts of a mechanism are pin-connected, the analysis of the mechanism can be carried out by considering each part as a rigid body, keeping in mind that the points where two parts are connected must have the same absolute velocity (see Sample Prob. 15.3). A similar analysis can be used when gears are involved, since the teeth in contact must also have the same absolute velocity. However, when a mechanism contains parts which slide on each other, the relative velocity of the parts in contact must be taken into account (see Secs. 15.10 and 15.11).

SAMPLE PROBLEM 15.2

The double gear shown rolls on the stationary lower rack; the velocity of its center A is 1.2 m/s directed to the right. Determine (a) the angular velocity of the gear, (b) the velocities of the upper rack R and of point D of the gear.

SOLUTION

a. Angular Velocity of the Gear. Since the gear rolls on the lower rack, its center A moves through a distance equal to the outer circumference $2\pi r_1$ for each full revolution of the gear. Noting that 1 rev = 2π rad, and that when A moves to the right ($x_A > 0$) the gear rotates clockwise ($\theta < 0$), we write

$$\frac{x_A}{2\pi r_1} = -\frac{\theta}{2\pi} \qquad x_A = -r_1\theta$$

Differentiating with respect to the time t and substituting the known values $v_A = 1.2$ m/s and $r_1 = 150$ mm $= 0.150$ m, we obtain

$$v_A = -r_1\omega \qquad 1.2 \text{ m/s} = -(0.150 \text{ m})\omega \qquad \omega = -8 \text{ rad/s}$$
$$\omega = \omega\mathbf{k} = -(8 \text{ rad/s})\mathbf{k} \quad \blacktriangleleft$$

where $\mathbf{k}$ is a unit vector pointing out of the paper.

b. Velocities. The rolling motion is resolved into two component motions: a translation with the center A and a rotation about the center A. In the translation, all points of the gear move with the same velocity $\mathbf{v}_A$. In the rotation, each point P of the gear moves about A with a relative velocity $\mathbf{v}_{P/A} = \omega\mathbf{k} \times \mathbf{r}_{P/A}$, where $\mathbf{r}_{P/A}$ is the position vector of P relative to A.

| Translation | + | Rotation | = | Rolling Motion |

Velocity of Upper Rack. The velocity of the upper rack is equal to the velocity of point B; we write

$$\mathbf{v}_R = \mathbf{v}_B = \mathbf{v}_A + \mathbf{v}_{B/A} = \mathbf{v}_A + \omega\mathbf{k} \times \mathbf{r}_{B/A}$$
$$= (1.2 \text{ m/s})\mathbf{i} - (8 \text{ rad/s})\mathbf{k} \times (0.100 \text{ m})\mathbf{j}$$
$$= (1.2 \text{ m/s})\mathbf{i} + (0.8 \text{ m/s})\mathbf{i} = (2 \text{ m/s})\mathbf{i}$$
$$\mathbf{v}_R = 2 \text{ m/s} \rightarrow \quad \blacktriangleleft$$

Velocity of Point D

$$\mathbf{v}_D = \mathbf{v}_A + \mathbf{v}_{D/A} = \mathbf{v}_A + \omega\mathbf{k} \times \mathbf{r}_{D/A}$$
$$= (1.2 \text{ m/s})\mathbf{i} - (8 \text{ rad/s})\mathbf{k} \times (-0.150 \text{ m})\mathbf{i}$$
$$= (1.2 \text{ m/s})\mathbf{i} + (1.2 \text{ m/s})\mathbf{j}$$
$$\mathbf{v}_D = 1.697 \text{ m/s} \measuredangle 45° \quad \blacktriangleleft$$

SAMPLE PROBLEM 15.3

In the engine system shown, the crank AB has a constant clockwise angular velocity of 2000 rpm. For the crank position indicated, determine (a) the angular velocity of the connecting rod BD, (b) the velocity of the piston P.

SOLUTION

Motion of Crank AB. The crank AB rotates about point A. Expressing ω_{AB} in rad/s and writing $v_B = r\omega_{AB}$, we obtain

$$\omega_{AB} = \left(2000\frac{\text{rev}}{\text{min}}\right)\left(\frac{1 \text{ min}}{60 \text{ s}}\right)\left(\frac{2\pi \text{ rad}}{1 \text{ rev}}\right) = 209.4 \text{ rad/s}$$

$$v_B = (AB)\omega_{AB} = (3 \text{ in.})(209.4 \text{ rad/s}) = 628.3 \text{ in./s}$$
$$\mathbf{v}_B = 628.3 \text{ in./s} \searrow 50°$$

Motion of Connecting Rod BD. We consider this motion as a general plane motion. Using the law of sines, we compute the angle β between the connecting rod and the horizontal:

$$\frac{\sin 40°}{8 \text{ in.}} = \frac{\sin \beta}{3 \text{ in.}} \qquad \beta = 13.95°$$

The velocity $\mathbf{v}_D$ of the point D where the rod is attached to the piston must be horizontal, while the velocity of point B is equal to the velocity $\mathbf{v}_B$ obtained above. Resolving the motion of BD into a translation with B and a rotation about B, we obtain

Plane motion = Translation + Rotation

Expressing the relation between the velocities $\mathbf{v}_D$, $\mathbf{v}_B$, and $\mathbf{v}_{D/B}$, we write

$$\mathbf{v}_D = \mathbf{v}_B + \mathbf{v}_{D/B}$$

We draw the vector diagram corresponding to this equation. Recalling that $\beta = 13.95°$, we determine the angles of the triangle and write

$$\frac{v_D}{\sin 53.95°} = \frac{v_{D/B}}{\sin 50°} = \frac{628.3 \text{ in./s}}{\sin 76.05°}$$

$$v_{D/B} = 495.9 \text{ in./s} \qquad \mathbf{v}_{D/B} = 495.9 \text{ in./s} \measuredangle 76.05°$$
$$v_D = 523.4 \text{ in./s} = 43.6 \text{ ft/s} \qquad \mathbf{v}_D = 43.6 \text{ ft/s} \rightarrow$$
$$\mathbf{v}_P = \mathbf{v}_D = 43.6 \text{ ft/s} \rightarrow \quad \blacktriangleleft$$

Since $v_{D/B} = l\omega_{BD}$, we have

$$495.9 \text{ in./s} = (8 \text{ in.})\omega_{BD} \qquad \omega_{BD} = 62.0 \text{ rad/s} \upharpoonleft \quad \blacktriangleleft$$

SOLVING PROBLEMS
ON YOUR OWN

In this lesson you learned to analyze the velocity of bodies in *general plane motion*. You found that a general plane motion can always be considered as the sum of the two motions you studied in the last lesson, namely, *a translation and a rotation*.

To solve a problem involving the velocity of a body in plane motion you should take the following steps.

1. Whenever possible determine the velocity of the points of the body where the body is connected to another body whose motion is known. That other body may be an arm or crank rotating with a given angular velocity [Sample Prob. 15.3].

2. Next start drawing a "diagram equation" to use in your solution (Figs. 15.15 and 15.16). This "equation" will consist of the following diagrams.

   a. Plane motion diagram: Draw a diagram of the body including all dimensions and showing those points for which you know or seek the velocity.

   b. Translation diagram: Select a reference point A for which you know the direction and/or the magnitude of the velocity $\mathbf{v}_A$, and draw a second diagram showing the body in translation with all of its points having the same velocity $\mathbf{v}_A$.

   c. Rotation diagram: Consider point A as a fixed point and draw a diagram showing the body in rotation about A. Show the angular velocity $\boldsymbol{\omega} = \omega\mathbf{k}$ of the body and the relative velocities with respect to A of the other points, such as the velocity $\mathbf{v}_{B/A}$ of B relative to A.

3. Write the relative-velocity formula

$$\mathbf{v}_B = \mathbf{v}_A + \mathbf{v}_{B/A}$$

While you can solve this vector equation analytically by writing the corresponding scalar equations, you will usually find it easier to solve it by using a vector triangle (Fig. 15.16).

4. A different reference point can be used to obtain an equivalent solution. For example, if point B is selected as the reference point, the velocity of point A is expressed as

$$\mathbf{v}_A = \mathbf{v}_B + \mathbf{v}_{A/B}$$

Note that the relative velocities $\mathbf{v}_{B/A}$ and $\mathbf{v}_{A/B}$ have the same magnitude but opposite sense. Relative velocities, therefore, depend upon the reference point that has been selected. The angular velocity, however, is independent of the choice of reference point.

PROBLEMS

CONCEPT QUESTIONS

15.CQ3 The ball rolls without slipping on the fixed surface as shown. What is the direction of the velocity of point A?

   a. →   **b.** ↗   **c.** ↑   **d.** ↓   **e.** ↘

Fig. P15.CQ3

15.CQ4 Three uniform rods—ABC, DCE, and FGH—are connected as shown. Which of the following statements concerning the angular speed of the three objects is true?

   a. $\omega_{ABC} = \omega_{DCE} = \omega_{FGH}$

   b. $\omega_{DCE} > \omega_{ABC} > \omega_{FGH}$

   c. $\omega_{DCE} < \omega_{ABC} < \omega_{FGH}$

   d. $\omega_{ABC} > \omega_{DCE} > \omega_{FGH}$

   e. $\omega_{FGH} = \omega_{DCE} < \omega_{ABC}$

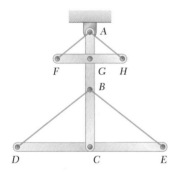

Fig. P15.CQ4

END-OF-SECTION PROBLEMS

15.38 An automobile travels to the right at a constant speed of 48 mi/h. If the diameter of a wheel is 22 in., determine the velocities of points B, C, D, and E on the rim of the wheel.

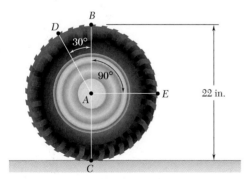

Fig. P15.38

15.39 The motion of rod AB is guided by pins attached at A and B which slide in the slots shown. At the instant shown, $\theta = 40°$ and the pin at B moves upward to the left with a constant velocity of 6 in./s. Determine (*a*) the angular velocity of the rod, (*b*) the velocity of the pin at end A.

Fig. P15.39

Fig. P15.43

15.40 Collar B moves upward with a constant velocity of 1.5 m/s. At the instant when $\theta = 50°$, determine (a) the angular velocity of rod AB, (b) the velocity of end A of the rod.

Fig. P15.40

15.41 Collar B moves downward to the left with a constant velocity of 1.6 m/s. At the instant shown when $\theta = 40°$, determine (a) the angular velocity of rod AB, (b) the velocity of collar A.

15.42 Collar A moves upward with a constant velocity of 1.2 m/s. At the instant shown when $\theta = 25°$, determine (a) the angular velocity of rod AB, (b) the velocity of collar B.

15.43 Rod AB moves over a small wheel at C while end A moves to the right with a constant velocity of 25 in./s. At the instant shown, determine (a) the angular velocity of the rod, (b) the velocity of end B of the rod.

15.44 The plate shown moves in the xy plane. Knowing that $(v_A)_x = 120$ mm/s, $(v_B)_y = 300$ mm/s, and $(v_C)_y = -60$ mm/s, determine (a) the angular velocity of the plate, (b) the velocity of point A.

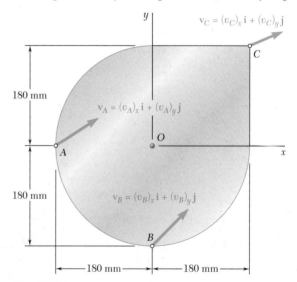

Fig. P15.44

15.45 In Prob. 15.44, determine (a) the velocity of point B, (b) the point of the plate with zero velocity.

15.46 The plate shown moves in the xy plane. Knowing that $(v_A)_x =$ 250 mm/s, $(v_B)_y = -450$ mm/s, and $(v_C)_x = -500$ mm/s, determine (a) the angular velocity of the plate, (b) the velocity of point A.

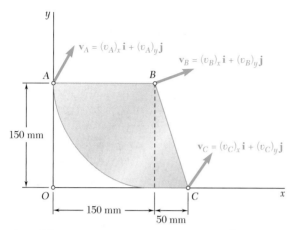

Fig. P15.46

15.47 The plate shown moves in the xy plane. Knowing that $(v_A)_x = 12$ in./s, $(v_B)_x = -4$ in./s, and $(v_C)_y = -24$ in./s, determine (a) the angular velocity of the plate, (b) the velocity of point B.

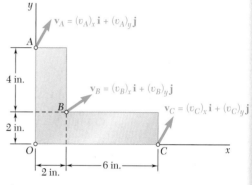

Fig. P15.47

15.48 In the planetary gear system shown, the radius of gears A, B, C, and D is a and the radius of the outer gear E is $3a$. Knowing that the angular velocity of gear A is ω_A clockwise and that the outer gear E is stationary, determine (a) the angular velocity of each planetary gear, (b) the angular velocity of the spider connecting the planetary gears.

Fig. P15.48 and P15.49

15.49 In the planetary gear system shown, the radius of gears A, B, C, and D is 30 mm and the radius of the outer gear E is 90 mm. Knowing that gear E has an angular velocity of 180 rpm clockwise and that the central gear A has an angular velocity of 240 rpm clockwise, determine (a) the angular velocity of each planetary gear, (b) the angular velocity of the spider connecting the planetary gears.

15.50 Arm AB rotates with an angular velocity of 20 rad/s counterclockwise. Knowing that the outer gear C is stationary, determine (a) the angular velocity of gear B, (b) the velocity of the gear tooth located at point D.

Fig. P15.50

Fig. P15.51

15.51 In the simplified sketch of a ball bearing shown, the diameter of the inner race A is 60 mm and the diameter of each ball is 12 mm. The outer race B is stationary while the inner race has an angular velocity of 3600 rpm. Determine (*a*) the speed of the center of each ball, (*b*) the angular velocity of each ball, (*c*) the number of times per minute each ball describes a complete circle.

15.52 A simplified gear system for a mechanical watch is shown. Knowing that gear A has a constant angular velocity of 1 rev/h and gear C has a constant angular velocity of 1 rpm, determine (*a*) the radius r, (*b*) the magnitudes of the accelerations of the points on gear B that are in contact with gears A and C.

Fig. *P15.52*

15.53 and 15.54 Arm ACB rotates about point C with an angular velocity of 40 rad/s counterclockwise. Two friction disks A and B are pinned at their centers to arm ACB as shown. Knowing that the disks roll without slipping at surfaces of contact, determine the angular velocity of (*a*) disk A, (*b*) disk B.

Fig. P15.53

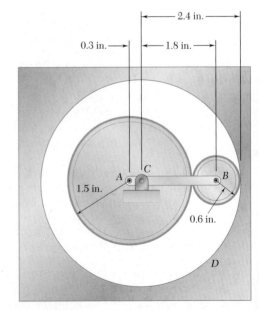

Fig. *P15.54*

15.55 Knowing that at the instant shown the velocity of collar A is 900 mm/s to the left, determine (*a*) the angular velocity of rod *ADB*, (*b*) the velocity of point B.

15.56 Knowing that at the instant shown the angular velocity of rod *DE* is 2.4 rad/s clockwise, determine (*a*) the velocity of collar A, (*b*) the velocity of point B.

15.57 A straight rack rests on a gear of radius r and is attached to a block B as shown. Denoting by ω_D the clockwise angular velocity of gear D and by θ the angle formed by the rack and the horizontal, derive expressions for the velocity of block B and the angular velocity of the rack in terms of r, θ, and ω_D.

Fig. P15.55 and P15.56

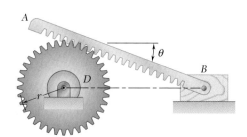

Fig. P15.57 and P15.58

15.58 A straight rack rests on a gear of radius $r = 2.5$ in. and is attached to a block B as shown. Knowing that at the instant shown the velocity of block B is 8 in./s to the right and $\theta = 25°$, determine (*a*) the angular velocity of gear D, (*b*) the angular velocity of the rack.

15.59 Knowing that at the instant shown the angular velocity of crank AB is 2.7 rad/s clockwise, determine (*a*) the angular velocity of link BD, (*b*) the velocity of collar D, (*c*) the velocity of the midpoint of link BD.

15.60 In the eccentric shown, a disk of 2-in. radius revolves about shaft O that is located 0.5 in. from the center A of the disk. The distance between the center A of the disk and the pin at B is 8 in. Knowing that the angular velocity of the disk is 900 rpm clockwise, determine the velocity of the block when $\theta = 30°$.

Fig. P15.59

Fig. P15.60

Fig. P15.61 and P15.62

15.61 In the engine system shown, $l = 160$ mm and $b = 60$ mm. Knowing that the crank AB rotates with a constant angular velocity of 1000 rpm clockwise, determine the velocity of the piston P and the angular velocity of the connecting rod when (*a*) $\theta = 0$, (*b*) $\theta = 90°$.

15.62 In the engine system shown, $l = 160$ mm and $b = 60$ mm. Knowing that crank AB rotates with a constant angular velocity of 1000 rpm clockwise, determine the velocity of the piston P and the angular velocity of the connecting rod when $\theta = 60°$.

15.63 Knowing that at the instant shown the angular velocity of rod AB is 15 rad/s clockwise, determine (*a*) the angular velocity of rod BD, (*b*) the velocity of the midpoint of rod BD.

Fig. P15.63

15.64 and 15.65 In the position shown, bar AB has an angular velocity of 4 rad/s clockwise. Determine the angular velocity of bars BD and DE.

Fig. P15.64

Fig. P15.65

15.66 Roberts linkage is named after Richard Roberts (1789–1864) and can be used to draw a close approximation to a straight line by locating a pen at point F. The distance AB is the same as BF, DF, and DE. Knowing that the angular velocity of bar AB is 5 rad/s clockwise in the position shown, determine (a) the angular velocity of bar DE, (b) the velocity of point F.

15.67 Roberts linkage is named after Richard Roberts (1789–1864) and can be used to draw a close approximation to a straight line by locating a pen at point F. The distance AB is the same as BF, DF, and DE. Knowing that the angular velocity of plate BDF is 2 rad/s counterclockwise when $\theta = 90°$, determine (a) the angular velocities of bars AB and DE, (b) the velocity of point F. When $\theta = 90°$, point F may be assumed to coincide with point E, with negligible error in the velocity analysis.

Fig. P15.66 and P15.67

15.68 In the position shown, bar DE has a constant angular velocity of 10 rad/s clockwise. Knowing that $h = 500$ mm, determine (a) the angular velocity of bar FBD, (b) the velocity of point F.

Fig. P15.68 and P15.69

15.69 In the position shown, bar DE has a constant angular velocity of 10 rad/s clockwise. Determine (a) the distance h for which the velocity of point F is vertical, (b) the corresponding velocity of point F.

15.70 Both 6-in.-radius wheels roll without slipping on the horizontal surface. Knowing that the distance AD is 5 in., the distance BE is 4 in., and D has a velocity of 6 in./s to the right, determine the velocity of point E.

Fig. P15.70

15.71 The 80-mm-radius wheel shown rolls to the left with a velocity of 900 mm/s. Knowing that the distance AD is 50 mm, determine the velocity of the collar and the angular velocity of rod AB when (a) $\beta = 0$, (b) $\beta = 90°$.

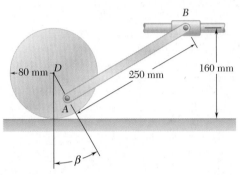

Fig. P15.71

*15.72 For the gearing shown, derive an expression for the angular velocity ω_C of gear C and show that ω_C is independent of the radius of gear B. Assume that point A is fixed and denote the angular velocities of rod ABC and gear A by ω_{ABC} and ω_A, respectively.

Fig. P15.72

15.7 INSTANTANEOUS CENTER OF ROTATION IN PLANE MOTION

Consider the general plane motion of a slab. We propose to show that at any given instant the velocities of the various particles of the slab are the same as if the slab were rotating about a certain axis perpendicular to the plane of the slab, called the *instantaneous axis of rotation*. This axis intersects the plane of the slab at a point C, called the *instantaneous center of rotation* of the slab.

We first recall that the plane motion of a slab can always be replaced by a translation defined by the motion of an arbitrary reference point A and by a rotation about A. As far as the velocities are concerned, the translation is characterized by the velocity $\mathbf{v}_A$ of the reference point A and the rotation is characterized by the angular velocity $\boldsymbol{\omega}$ of the slab (which is independent of the choice of A). Thus, the velocity $\mathbf{v}_A$ of point A and the angular velocity $\boldsymbol{\omega}$ of the slab define

Photo 15.5 If the tires of this car are rolling without sliding, the instantaneous center of rotation of a tire is the point of contact between the road and the tire.

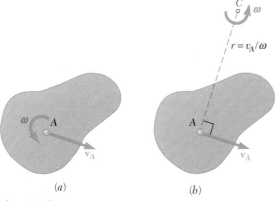

(a)   (b)

Fig. 15.18

completely the velocities of all the other particles of the slab (Fig. 15.18a). Now let us assume that $\mathbf{v}_A$ and $\boldsymbol{\omega}$ are known and that they are both different from zero. (If $\mathbf{v}_A = 0$, point A is itself the instantaneous center of rotation, and if $\boldsymbol{\omega} = 0$, all the particles have the same velocity $\mathbf{v}_A$.) These velocities could be obtained by letting the slab rotate with the angular velocity $\boldsymbol{\omega}$ about a point C located on the perpendicular to $\mathbf{v}_A$ at a distance $r = v_A/\omega$ from A as shown in Fig. 15.18b. We check that the velocity of A would be perpendicular to AC and that its magnitude would be $r\omega = (v_A/\omega)\omega = v_A$. Thus the velocities of all the other particles of the slab would be the same as originally defined. Therefore, *as far as the velocities are concerned, the slab seems to rotate about the instantaneous center C at the instant considered.*

The position of the instantaneous center can be defined in two other ways. If the directions of the velocities of two particles A and B of the slab are known and if they are different, the instantaneous center C is obtained by drawing the perpendicular to $\mathbf{v}_A$ through A and the perpendicular to $\mathbf{v}_B$ through B and determining the point in which these two lines intersect (Fig. 15.19a). If the velocities $\mathbf{v}_A$ and $\mathbf{v}_B$ of two particles A and B are perpendicular to the line AB and if their magnitudes are known, the instantaneous center can be found by intersecting the line AB with the line joining the extremities of the vectors $\mathbf{v}_A$ and $\mathbf{v}_B$ (Fig. 15.19b). Note that if $\mathbf{v}_A$ and $\mathbf{v}_B$ were parallel

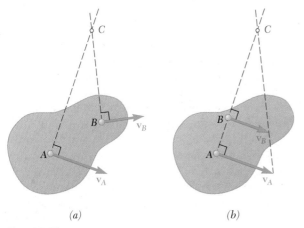

(a)   (b)

Fig. 15.19

in Fig. 15.19a or if $\mathbf{v}_A$ and $\mathbf{v}_B$ had the same magnitude in Fig. 15.19b, the instantaneous center C would be at an infinite distance and $\boldsymbol{\omega}$ would be zero; all points of the slab would have the same velocity.

To see how the concept of instantaneous center of rotation can be put to use, let us consider again the rod of Sec. 15.6. Drawing the perpendicular to $\mathbf{v}_A$ through A and the perpendicular to $\mathbf{v}_B$ through B (Fig. 15.20), we obtain the instantaneous center C. At the

Fig. 15.20

instant considered, the velocities of all the particles of the rod are thus the same as if the rod rotated about C. Now, if the magnitude v_A of the velocity of A is known, the magnitude ω of the angular velocity of the rod can be obtained by writing

$$\omega = \frac{v_A}{AC} = \frac{v_A}{l \cos \theta}$$

The magnitude of the velocity of B can then be obtained by writing

$$v_B = (BC)\omega = l \sin \theta \frac{v_A}{l \cos \theta} = v_A \tan \theta$$

Note that only *absolute* velocities are involved in the computation.

The instantaneous center of a slab in plane motion can be located either on the slab or outside the slab. If it is located on the slab, the particle C coinciding with the instantaneous center at a given instant t must have zero velocity at that instant. However, it should be noted that the instantaneous center of rotation is valid only at a given instant. Thus, the particle C of the slab which coincides with the instantaneous center at time t will generally not coincide with the instantaneous center at time $t + \Delta t$; while its velocity is zero at time t, it will probably be different from zero at time $t + \Delta t$. This means that, in general, the particle C *does not have zero acceleration* and, therefore, that the *accelerations* of the various particles of the slab *cannot* be determined as if the slab were rotating about C.

As the motion of the slab proceeds, the instantaneous center moves in space. But it was just pointed out that the position of the instantaneous center on the slab keeps changing. Thus, the instantaneous center describes one curve in space, called the *space centrode*, and another curve on the slab, called the *body centrode* (Fig. 15.21). It can be shown that at any instant, these two curves are tangent at C and that as the slab moves, the body centrode appears to *roll* on the space centrode.

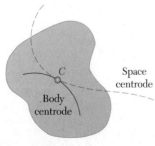

Fig. 15.21

SAMPLE PROBLEM 15.4

Solve Sample Prob. 15.2, using the method of the instantaneous center of rotation.

SOLUTION

a. Angular Velocity of the Gear. Since the gear rolls on the stationary lower rack, the point of contact C of the gear with the rack has no velocity; point C is therefore the instantaneous center of rotation. We write

$$v_A = r_A\omega \qquad 1.2 \text{ m/s} = (0.150 \text{ m})\omega$$
$$\omega = 8 \text{ rad/s} \downarrow \quad \blacktriangleleft$$

b. Velocities. As far as velocities are concerned, all points of the gear seem to rotate about the instantaneous center.

Velocity of Upper Rack. Recalling that $v_R = v_B$, we write

$$v_R = v_B = r_B\omega \qquad v_R = (0.250 \text{ m})(8 \text{ rad/s}) = 2 \text{ m/s}$$
$$\mathbf{v}_R = 2 \text{ m/s} \rightarrow \quad \blacktriangleleft$$

Velocity of Point D. Since $r_D = (0.150 \text{ m})\sqrt{2} = 0.2121$ m, we write

$$v_D = r_D\omega \qquad v_D = (0.2121 \text{ m})(8 \text{ rad/s}) = 1.697 \text{ m/s}$$
$$\mathbf{v}_D = 1.697 \text{ m/s} \angle 45° \quad \blacktriangleleft$$

SAMPLE PROBLEM 15.5

Solve Sample Prob. 15.3, using the method of the instantaneous center of rotation.

SOLUTION

Motion of Crank AB. Referring to Sample Prob. 15.3, we obtain the velocity of point B; $\mathbf{v}_B = 628.3$ in./s $\searrow 50°$.

Motion of the Connecting Rod BD. We first locate the instantaneous center C by drawing lines perpendicular to the absolute velocities $\mathbf{v}_B$ and $\mathbf{v}_D$. Recalling from Sample Prob. 15.3 that $\beta = 13.95°$ and that $BD = 8$ in., we solve the triangle BCD.

$$\gamma_B = 40° + \beta = 53.95° \qquad \gamma_D = 90° - \beta = 76.05°$$
$$\frac{BC}{\sin 76.05°} = \frac{CD}{\sin 53.95°} = \frac{8 \text{ in.}}{\sin 50°}$$
$$BC = 10.14 \text{ in.} \qquad CD = 8.44 \text{ in.}$$

Since the connecting rod BD seems to rotate about point C, we write

$$v_B = (BC)\omega_{BD}$$
$$628.3 \text{ in./s} = (10.14 \text{ in.})\omega_{BD}$$
$$\omega_{BD} = 62.0 \text{ rad/s} \uparrow \quad \blacktriangleleft$$
$$v_D = (CD)\omega_{BD} = (8.44 \text{ in.})(62.0 \text{ rad/s})$$
$$= 523 \text{ in./s} = 43.6 \text{ ft/s}$$
$$\mathbf{v}_P = \mathbf{v}_D = 43.6 \text{ ft/s} \rightarrow \quad \blacktriangleleft$$

SOLVING PROBLEMS
ON YOUR OWN

In this lesson we introduced the *instantaneous center of rotation* in plane motion. This provides us with an alternative way for solving problems involving the *velocities* of the various points of a body in plane motion.

As its name suggests, the *instantaneous center of rotation* is the point about which you can assume a body is rotating at a given instant, as you determine the velocities of the points of the body at that instant.

A. To determine the instantaneous center of rotation of a body in plane motion, you should use one of the following procedures.

1. If the velocity v_A of a point A and the angular velocity ω of the body are both known (Fig. 15.18):

 a. Draw a sketch of the body, showing point A, its velocity $\mathbf{v}_A$, and the angular velocity $\boldsymbol{\omega}$ of the body.

 b. From A draw a line perpendicular to v_A on the side of $\mathbf{v}_A$ from which this velocity is viewed as having *the same sense as $\boldsymbol{\omega}$*.

 c. Locate the instantaneous center C on this line, at a distance $r = v_A/\omega$ from point A.

2. If the directions of the velocities of two points A and B are known and are different (Fig. 15.19a):

 a. Draw a sketch of the body, showing points A and B and their velocities $\mathbf{v}_A$ and $\mathbf{v}_B$.

 b. From A and B draw lines perpendicular to v_A and v_B, respectively. The instantaneous center C is located at the point where the two lines intersect.

 c. If the velocity of one of the two points is known, you can determine the angular velocity of the body. For example, if you know $\boldsymbol{v}_A$, you can write $\omega = v_A/AC$, where AC is the distance from point A to the instantaneous center C.

3. If the velocities of two points A and B are known and are both perpendicular to the line AB (Fig. 15.19b):

 a. Draw a sketch of the body, showing points A and B with their velocities $\mathbf{v}_A$ and $\mathbf{v}_B$ *drawn to scale*.

 b. Draw a line through points A and B, and another line through the tips of the vectors $\mathbf{v}_A$ and $\mathbf{v}_B$. The instantaneous center C is located at the point where the two lines intersect.

c. The angular velocity of the body is obtained by either dividing $\mathbf{v}_A$ by AC or $\mathbf{v}_B$ by BC.

d. If the velocities v_A and v_B have the same magnitude, the two lines drawn in part b do not intersect; the instantaneous center C is at an infinite distance. The angular velocity $\boldsymbol{\omega}$ is zero and *the body is in translation.*

B. Once you have determined the instantaneous center and the angular velocity of a body, you can determine the velocity $\mathbf{v}_P$ of any point P of the body in the following way.

1. Draw a sketch of the body, showing point P, the instantaneous center of rotation C, and the angular velocity $\boldsymbol{\omega}$.

2. Draw a line from P to the instantaneous center C and measure or calculate the distance from P to C.

3. The velocity v_P is a vector perpendicular to the line PC, of the same sense as $\boldsymbol{\omega}$, and of magnitude $v_P = (PC)\omega$.

Finally, keep in mind that the instantaneous center of rotation can be used *only* to determine velocities. *It cannot be used to determine accelerations.*

PROBLEMS

CONCEPT QUESTIONS

15.CQ5 The disk rolls without sliding on the fixed horizontal surface. At the instant shown, the instantaneous center of zero velocity for rod *AB* would be located in which region?

    a. Region 1
    b. Region 2
    c. Region 3
    d. Region 4
    e. Region 5
    f. Region 6

Fig. P15.CQ5

Fig. P15.CQ6

15.CQ6 Bar *BDE* is pinned to two links, *AB* and *CD*. At the instant shown, the angular velocities of link *AB*, link *CD*, and bar *BDE* are ω_{AB}, ω_{CD}, and ω_{BDE}, respectively. Which of the following statements concerning the angular speeds of the three objects is true at this instant?

    a. $\omega_{AB} = \omega_{CD} = \omega_{BDE}$
    b. $\omega_{BDE} > \omega_{AB} > \omega_{CD}$
    c. $\omega_{AB} = \omega_{CD} > \omega_{BDE}$
    d. $\omega_{AB} > \omega_{CD} > \omega_{BDE}$
    e. $\omega_{CD} > \omega_{AB} > \omega_{BDE}$

END-OF-SECTION PROBLEMS

15.73 A juggling club is thrown vertically into the air. The center of gravity *G* of the 20-in. club is located 12 in. from the knob. Knowing that at the instant shown, *G* has a velocity of 4 ft/s upwards and the club has an angular velocity of 30 rad/s counterclockwise, determine (*a*) the speeds of points *A* and *B*, (*b*) the location of the instantaneous center of rotation.

Fig. P15.73

15.74 A 10-ft beam AE is being lowered by means of two overhead cranes. At the instant shown, it is known that the velocity of point D is 24 in./s downward and the velocity of point E is 36 in./s downward. Determine (*a*) the instantaneous center of rotation of the beam, (*b*) the velocity of point A.

Fig. P15.74

Fig. P15.75

15.75 A helicopter moves horizontally in the x direction at a speed of 120 mi/h. Knowing that the main blades rotate clockwise with an angular velocity of 180 rpm, determine the instantaneous axis of rotation of the main blades.

15.76 and 15.77 A 60-mm-radius drum is rigidly attached to a 100-mm-radius drum as shown. One of the drums rolls without sliding on the surface shown, and a cord is wound around the other drum. Knowing that end E of the cord is pulled to the left with a velocity of 120 mm/s, determine (*a*) the angular velocity of the drums, (*b*) the velocity of the center of the drums, (*c*) the length of cord wound or unwound per second.

Fig. P15.76

Fig. P15.77

Fig. P15.78 and P15.79

15.78 The spool of tape shown and its frame assembly are pulled upward at a speed $v_A = 750$ mm/s. Knowing that the 80-mm-radius spool has an angular velocity of 15 rad/s clockwise and that at the instant shown the total thickness of the tape on the spool is 20 mm, determine (*a*) the instantaneous center of rotation of the spool, (*b*) the velocities of points B and D.

15.79 The spool of tape shown and its frame assembly are pulled upward at a speed $v_A = 100$ mm/s. Knowing that end B of the tape is pulled downward with a velocity of 300 mm/s and that at the instant shown the total thickness of the tape on the spool is 20 mm, determine (*a*) the instantaneous center of rotation of the spool, (*b*) the velocity of point D of the spool.

Fig. P15.80

15.80 The arm *ABC* rotates with an angular velocity of 4 rad/s counterclockwise. Knowing that the angular velocity of the intermediate gear *B* is 8 rad/s counterclockwise, determine (*a*) the instantaneous centers of rotation of gears *A* and *C*, (*b*) the angular velocities of gears *A* and *C*.

15.81 The double gear rolls on the stationary left rack *R*. Knowing that the rack on the right has a constant velocity of 2 ft/s, determine (*a*) the angular velocity of the gear, (*b*) the velocities of points *A* and *D*.

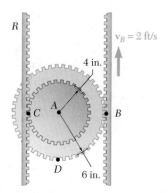

Fig. P15.81

15.82 An overhead door is guided by wheels at *A* and *B* that roll in horizontal and vertical tracks. Knowing that when $\theta = 40°$ the velocity of wheel *B* is 1.5 ft/s upward, determine (*a*) the angular velocity of the door, (*b*) the velocity of end *D* of the door.

Fig. P15.82

Fig. P15.83

15.83 Rod *ABD* is guided by wheels at *A* and *B* that roll in horizontal and vertical tracks. Knowing that at the instant $\beta = 60°$ and the velocity of wheel *B* is 40 in./s downward, determine (*a*) the angular velocity of the rod, (*b*) the velocity of point *D*.

15.84 Rod *BDE* is partially guided by a roller at *D* which moves in a vertical track. Knowing that at the instant shown the angular velocity of crank *AB* is 5 rad/s clockwise and that $\beta = 25°$, determine (*a*) the angular velocity of the rod, (*b*) the velocity of point *E*.

15.85 Rod *BDE* is partially guided by a roller at *D* which moves in a vertical track. Knowing that at the instant shown $\beta = 30°$, point *E* has a velocity of 2 m/s down and to the right, determine the angular velocities of rod *BDE* and crank *AB*.

15.86 Knowing that at the instant shown, the velocity of collar *D* is 1.6 m/s upward, determine (*a*) the angular velocity of rod *AD*, (*b*) the velocity of point *B*, (*c*) the velocity of point *A*.

Fig. P15.84 and P15.85

Fig. P15.86 and P15.87

15.87 Knowing that at the instant shown, the angular velocity of rod *BE* is 4 rad/s counterclockwise, determine (*a*) the angular velocity of rod *AD*, (*b*) the velocity of collar *D*, (*c*) the velocity of point *A*.

15.88 Rod *AB* can slide freely along the floor and the inclined plane. Denoting by $\mathbf{v}_A$ the velocity of point *A*, derive an expression for (*a*) the angular velocity of the rod, (*b*) the velocity of end *B*.

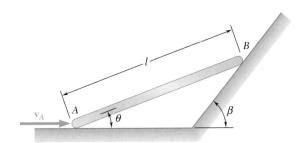

Fig. P15.88 and P15.89

15.89 Rod *AB* can slide freely along the floor and the inclined plane. Knowing that $\theta = 20°$, $\beta = 50°$, $l = 2$ ft, and $v_A = 8$ ft/s, determine (*a*) the angular velocity of the rod, (*b*) the velocity of end *B*.

15.90 Two slots have been cut in plate *FG* and the plate has been placed so that the slots fit two fixed pins *A* and *B*. Knowing that at the instant shown the angular velocity of crank *DE* is 6 rad/s clockwise, determine (*a*) the velocity of point *F*, (*b*) the velocity of point *G*.

Fig. P15.90

15.91 The disk is released from rest and rolls down the incline. Knowing that the speed of *A* is 1.2 m/s when $\theta = 0°$, determine at that instant (*a*) the angular velocity of the rod, (*b*) the velocity of *B*. (Only portions of the two tracks are shown.)

15.92 Arm *ABD* is connected by pins to a collar at *B* and to crank *DE*. Knowing that the velocity of collar *B* is 400 mm/s upward, determine (*a*) the angular velocity of arm *ABD*, (*b*) the velocity of point *A*.

Fig. P15.91

Fig. P15.92 and P15.93

15.93 Arm *ABD* is connected by pins to a collar at *B* and to crank *DE*. Knowing that the angular velocity of crank *DE* is 1.2 rad/s counterclockwise, determine (*a*) the angular velocity of arm *ABD*, (*b*) the velocity of point *A*.

15.94 Two links *AB* and *BD*, each 25 in. long, are connected at *B* and guided by hydraulic cylinders attached at *A* and *D*. Knowing that *D* is stationary and that the velocity of *A* is 30 in./s to the right, determine at the instant shown (*a*) the angular velocity of each link, (*b*) the velocity of *B*.

Fig. P15.94

15.95 Two 25-in. rods are pin-connected at D as shown. Knowing that B moves to the left with a constant velocity of 24 in./s, determine at the instant shown (*a*) the angular velocity of each rod, (*b*) the velocity of E.

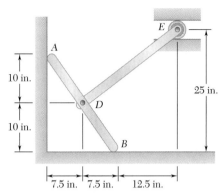

Fig. P15.95

15.96 Two rods ABD and DE are connected to three collars as shown. Knowing that the angular velocity of ABD is 5 rad/s clockwise, determine at the instant shown (*a*) the angular velocity of DE, (*b*) the velocity of collar E.

Fig. P15.96

15.97 Two collars C and D move along the vertical rod shown. Knowing that the velocity of collar C is 660 mm/s downward, determine (*a*) the velocity of collar D, (*b*) the angular velocity of member AB.

15.98 Two rods AB and DE are connected as shown. Knowing that point D moves to the left with a velocity of 40 in./s, determine (*a*) the angular velocity of each rod, (*b*) the velocity of point A.

Fig. P15.98

Fig. P15.97

15.99 Describe the space centrode and the body centrode of rod ABD of Prob. 15.83. (*Hint:* The body centrode need not lie on a physical portion of the rod.)

15.100 Describe the space centrode and the body centrode of the gear of Sample Prob. 15.2 as the gear rolls on the stationary horizontal rack.

15.101 Using the method of Sec. 15.7, solve Prob. 15.60.

15.102 Using the method of Sec. 15.7, solve Prob. 15.64.

15.103 Using the method of Sec. 15.7, solve Prob. 15.65.

15.104 Using the method of Sec. 15.7, solve Prob. 15.38.

15.8 ABSOLUTE AND RELATIVE ACCELERATION IN PLANE MOTION

We saw in Sec. 15.5 that any plane motion can be replaced by a translation defined by the motion of an arbitrary reference point A and a simultaneous rotation about A. This property was used in Sec. 15.6 to determine the velocity of the various points of a moving slab. The same property will now be used to determine the acceleration of the points of the slab.

We first recall that the absolute acceleration $\mathbf{a}_B$ of a particle of the slab can be obtained from the relative-acceleration formula derived in Sec. 11.12,

$$\mathbf{a}_B = \mathbf{a}_A + \mathbf{a}_{B/A} \tag{15.21}$$

where the right-hand member represents a vector sum. The acceleration $\mathbf{a}_A$ corresponds to the translation of the slab with A, while the relative acceleration $\mathbf{a}_{B/A}$ is associated with the rotation of the slab about A and is measured with respect to axes centered at A and of fixed orientation. We recall from Sec. 15.3 that the relative acceleration $\mathbf{a}_{B/A}$ can be resolved into two components, a *tangential component* $(\mathbf{a}_{B/A})_t$ perpendicular to the line AB, and a *normal component* $(\mathbf{a}_{B/A})_n$ directed toward A (Fig. 15.22). Denoting by $\mathbf{r}_{B/A}$ the position vector of B relative to A and, respectively, by $\omega\mathbf{k}$ and $\alpha\mathbf{k}$ the angular velocity and angular acceleration of the slab with respect to axes of fixed orientation, we have

$$\begin{matrix} (\mathbf{a}_{B/A})_t = \alpha\mathbf{k} \times \mathbf{r}_{B/A} & (a_{B/A})_t = r\alpha \\ (\mathbf{a}_{B/A})_n = -\omega^2\mathbf{r}_{B/A} & (a_{B/A})_n = r\omega^2 \end{matrix} \tag{15.22}$$

where r is the distance from A to B. Substituting into (15.21) the expressions obtained for the tangential and normal components of $\mathbf{a}_{B/A}$, we can also write

$$\mathbf{a}_B = \mathbf{a}_A + \alpha\mathbf{k} \times \mathbf{r}_{B/A} - \omega^2\mathbf{r}_{B/A} \tag{15.21'}$$

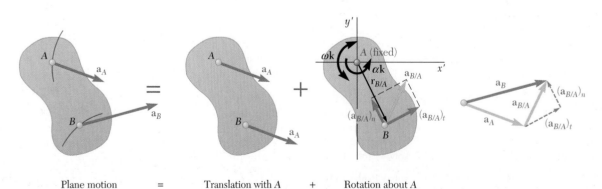

Plane motion = Translation with A + Rotation about A

Fig. 15.22

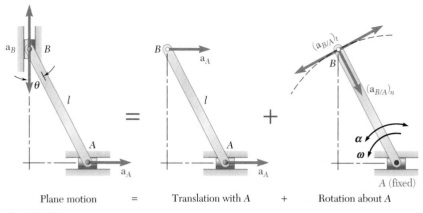

Plane motion   =   Translation with A +   Rotation about A

Fig. 15.23

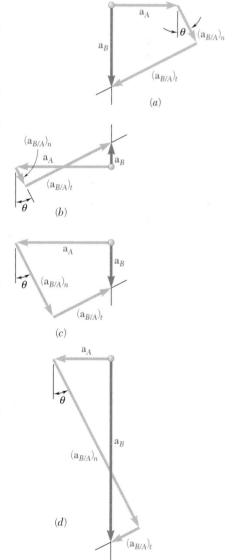

Fig. 15.24

As an example, let us again consider the rod AB whose extremities slide, respectively, along a horizontal and a vertical track (Fig. 15.23). Assuming that the velocity $\mathbf{v}_A$ and the acceleration $\mathbf{a}_A$ of A are known, we propose to determine the acceleration $\mathbf{a}_B$ of B and the angular acceleration $\boldsymbol{\alpha}$ of the rod. Choosing A as a reference point, we express that the given motion is equivalent to a translation with A and a rotation about A. The absolute acceleration of B must be equal to the sum

$$\mathbf{a}_B = \mathbf{a}_A + \mathbf{a}_{B/A}$$
$$= \mathbf{a}_A + (\mathbf{a}_{B/A})_n + (\mathbf{a}_{B/A})_t \qquad (15.23)$$

where $(\mathbf{a}_{B/A})_n$ has the magnitude $l\omega^2$ and is *directed toward A*, while $(\mathbf{a}_{B/A})_t$ has the magnitude $l\alpha$ and is perpendicular to AB. Students should note that there is no way to tell whether the tangential component $(\mathbf{a}_{B/A})_t$ is directed to the left or to the right, and therefore both possible directions for this component are indicated in Fig. 15.23. Similarly, both possible senses for $\mathbf{a}_B$ are indicated, since it is not known whether point B is accelerated upward or downward.

Equation (15.23) has been expressed geometrically in Fig. 15.24. Four different vector polygons can be obtained, depending upon the sense of $\mathbf{a}_A$ and the relative magnitude of a_A and $(a_{B/A})_n$. If we are to determine a_B and α from one of these diagrams, we must know not only a_A and θ but also ω. The angular velocity of the rod should therefore be separately determined by one of the methods indicated in Secs. 15.6 and 15.7. The values of a_B and α can then be obtained by considering successively the x and y components of the vectors shown in Fig. 15.24. In the case of polygon a, for example, we write

$\xrightarrow{+} x$ components: $0 = a_A + l\omega^2 \sin\theta - l\alpha \cos\theta$
$+\uparrow y$ components: $-a_B = -l\omega^2 \cos\theta - l\alpha \sin\theta$

and solve for a_B and α. The two unknowns can also be obtained by direct measurement on the vector polygon. In that case, care should be taken to draw first the known vectors $\mathbf{a}_A$ and $(\mathbf{a}_{B/A})_n$.

It is quite evident that the determination of accelerations is considerably more involved than the determination of velocities. Yet in the example considered here, the extremities A and B of the rod were moving along straight tracks, and the diagrams drawn were relatively simple. If A and B had moved along curved tracks, it would have been necessary to resolve the accelerations $\mathbf{a}_A$ and $\mathbf{a}_B$ into normal and tangential components and the solution of the problem would have involved six different vectors.

When a mechanism consists of several moving parts which are pin-connected, the analysis of the mechanism can be carried out by considering each part as a rigid body, keeping in mind that the points at which two parts are connected must have the same absolute acceleration (see Sample Prob. 15.7). In the case of meshed gears, the tangential components of the accelerations of the teeth in contact are equal, but their normal components are different.

*15.9 ANALYSIS OF PLANE MOTION IN TERMS OF A PARAMETER

In the case of certain mechanisms, it is possible to express the coordinates x and y of all the significant points of the mechanism by means of simple analytic expressions containing a single parameter. It is sometimes advantageous in such a case to determine the absolute velocity and the absolute acceleration of the various points of the mechanism directly, since the components of the velocity and of the acceleration of a given point can be obtained by differentiating the coordinates x and y of that point.

Let us consider again the rod AB whose extremities slide, respectively, in a horizontal and a vertical track (Fig. 15.25). The coordinates x_A and y_B of the extremities of the rod can be expressed in terms of the angle θ the rod forms with the vertical:

Fig. 15.25

$$x_A = l \sin \theta \qquad y_B = l \cos \theta \qquad (15.24)$$

Differentiating Eqs. (15.24) twice with respect to t, we write

$$v_A = \dot{x}_A = l\dot{\theta} \cos \theta$$
$$a_A = \ddot{x}_A = -l\dot{\theta}^2 \sin \theta + l\ddot{\theta} \cos \theta$$

$$v_B = \dot{y}_B = -l\dot{\theta} \sin \theta$$
$$a_B = \ddot{y}_B = -l\dot{\theta}^2 \cos \theta - l\ddot{\theta} \sin \theta$$

Recalling that $\dot{\theta} = \omega$ and $\ddot{\theta} = \alpha$, we obtain

$$v_A = l\omega \cos \theta \qquad\qquad v_B = -l\omega \sin \theta \qquad (15.25)$$

$$a_A = -l\omega^2 \sin \theta + l\alpha \cos \theta \qquad a_B = -l\omega^2 \cos \theta - l\alpha \sin \theta$$
$$(15.26)$$

We note that a positive sign for v_A or a_A indicates that the velocity $\mathbf{v}_A$ or the acceleration $\mathbf{a}_A$ is directed to the right; a positive sign for v_B or a_B indicates that $\mathbf{v}_B$ or $\mathbf{a}_B$ is directed upward. Equations (15.25) can be used, for example, to determine v_B and ω when v_A and θ are known. Substituting for ω in (15.26), we can then determine a_B and α if a_A is known.

SAMPLE PROBLEM 15.6

The center of the double gear of Sample Prob. 15.2 has a velocity of 1.2 m/s to the right and an acceleration of 3 m/s^2 to the right. Recalling that the lower rack is stationary, determine (a) the angular acceleration of the gear, (b) the acceleration of points B, C, and D of the gear.

SOLUTION

a. Angular Acceleration of the Gear. In Sample Prob. 15.2, we found that $x_A = -r_1\theta$ and $v_A = -r_1\omega$. Differentiating the latter with respect to time, we obtain $a_A = -r_1\alpha$.

$$v_A = -r_1\omega \qquad 1.2 \text{ m/s} = -(0.150 \text{ m})\omega \qquad \omega = -8 \text{ rad/s}$$
$$a_A = -r_1\alpha \qquad 3 \text{ m/s}^2 = -(0.150 \text{ m})\alpha \qquad \alpha = -20 \text{ rad/s}^2$$
$$\alpha = \alpha\mathbf{k} = -(20 \text{ rad/s}^2)\mathbf{k} \quad \blacktriangleleft$$

b. Accelerations. The rolling motion of the gear is resolved into a translation with A and a rotation about A.

Translation   +   Rotation   =   Rolling motion

Acceleration of Point B. Adding vectorially the accelerations corresponding to the translation and to the rotation, we obtain

$$\mathbf{a}_B = \mathbf{a}_A + \mathbf{a}_{B/A} = \mathbf{a}_A + (\mathbf{a}_{B/A})_t + (\mathbf{a}_{B/A})_n$$
$$= \mathbf{a}_A + \alpha\mathbf{k} \times \mathbf{r}_{B/A} - \omega^2\mathbf{r}_{B/A}$$
$$= (3 \text{ m/s}^2)\mathbf{i} - (20 \text{ rad/s}^2)\mathbf{k} \times (0.100 \text{ m})\mathbf{j} - (8 \text{ rad/s})^2(0.100 \text{ m})\mathbf{j}$$
$$= (3 \text{ m/s}^2)\mathbf{i} + (2 \text{ m/s}^2)\mathbf{i} - (6.40 \text{ m/s}^2)\mathbf{j}$$
$$\mathbf{a}_B = 8.12 \text{ m/s}^2 \searrow 52.0° \quad \blacktriangleleft$$

Acceleration of Point C

$$\mathbf{a}_C = \mathbf{a}_A + \mathbf{a}_{C/A} = \mathbf{a}_A + \alpha\mathbf{k} \times \mathbf{r}_{C/A} - \omega^2\mathbf{r}_{C/A}$$
$$= (3 \text{ m/s}^2)\mathbf{i} - (20 \text{ rad/s}^2)\mathbf{k} \times (-0.150 \text{ m})\mathbf{j} - (8 \text{ rad/s})^2(-0.150 \text{ m})\mathbf{j}$$
$$= (3 \text{ m/s}^2)\mathbf{i} - (3 \text{ m/s}^2)\mathbf{i} + (9.60 \text{ m/s}^2)\mathbf{j}$$
$$\mathbf{a}_C = 9.60 \text{ m/s}^2 \uparrow \quad \blacktriangleleft$$

Acceleration of Point D

$$\mathbf{a}_D = \mathbf{a}_A + \mathbf{a}_{D/A} = \mathbf{a}_A + \alpha\mathbf{k} \times \mathbf{r}_{D/A} - \omega^2\mathbf{r}_{D/A}$$
$$= (3 \text{ m/s}^2)\mathbf{i} - (20 \text{ rad/s}^2)\mathbf{k} \times (-0.150 \text{ m})\mathbf{i} - (8 \text{ rad/s})^2(-0.150 \text{ m})\mathbf{i}$$
$$= (3 \text{ m/s}^2)\mathbf{i} + (3 \text{ m/s}^2)\mathbf{j} + (9.60 \text{ m/s}^2)\mathbf{i}$$
$$\mathbf{a}_D = 12.95 \text{ m/s}^2 \measuredangle 13.4° \quad \blacktriangleleft$$

SAMPLE PROBLEM 15.7

Crank AB of the engine system of Sample Prob. 15.3 has a constant clockwise angular velocity of 2000 rpm. For the crank position shown, determine the angular acceleration of the connecting rod BD and the acceleration of point D.

SOLUTION

Motion of Crank AB. Since the crank rotates about A with constant $\omega_{AB} = 2000$ rpm $= 209.4$ rad/s, we have $\alpha_{AB} = 0$. The acceleration of B is therefore directed toward A and has a magnitude

$$a_B = r\omega_{AB}^2 = (\tfrac{3}{12} \text{ ft})(209.4 \text{ rad/s})^2 = 10{,}962 \text{ ft/s}^2$$
$$\mathbf{a}_B = 10{,}962 \text{ ft/s}^2 \; \nearrow \; 40°$$

Motion of the Connecting Rod BD. The angular velocity $\boldsymbol{\omega}_{BD}$ and the value of β were obtained in Sample Prob. 15.3:

$$\boldsymbol{\omega}_{BD} = 62.0 \text{ rad/s} \uparrow \qquad \beta = 13.95°$$

The motion of BD is resolved into a translation with B and a rotation about B. The relative acceleration $\mathbf{a}_{D/B}$ is resolved into normal and tangential components:

$$(a_{D/B})_n = (BD)\omega_{BD}^2 = (\tfrac{8}{12} \text{ ft})(62.0 \text{ rad/s})^2 = 2563 \text{ ft/s}^2$$
$$(\mathbf{a}_{D/B})_n = 2563 \text{ ft/s}^2 \; \searrow \; 13.95°$$
$$(a_{D/B})_t = (BD)\alpha_{BD} = (\tfrac{8}{12})\alpha_{BD} = 0.6667\alpha_{BD}$$
$$(\mathbf{a}_{D/B})_t = 0.6667\alpha_{BD} \; \swarrow \; 76.05°$$

While $(\mathbf{a}_{D/B})_t$ must be perpendicular to BD, its sense is not known.

Plane motion $\qquad = \qquad$ Translation $\qquad + \qquad$ Rotation

Noting that the acceleration $\mathbf{a}_D$ must be horizontal, we write

$$\mathbf{a}_D = \mathbf{a}_B + \mathbf{a}_{D/B} = \mathbf{a}_B + (\mathbf{a}_{D/B})_n + (\mathbf{a}_{D/B})_t$$
$$[a_D \leftrightarrow] = [10{,}962 \; \nearrow \; 40°] + [2563 \; \searrow \; 13.95°] + [0.6667\alpha_{BD} \; \swarrow \; 76.05°]$$

Equating x and y components, we obtain the following scalar equations:

$\xrightarrow{+} x$ components:
$$-a_D = -10{,}962 \cos 40° - 2563 \cos 13.95° + 0.6667\alpha_{BD} \sin 13.95°$$

$+\uparrow y$ components:
$$0 = -10{,}962 \sin 40° + 2563 \sin 13.95° + 0.6667\alpha_{BD} \cos 13.95°$$

Solving the equations simultaneously, we obtain $\alpha_{BD} = +9940$ rad/s^2 and $a_D = +9290$ ft/s^2. The positive signs indicate that the senses shown on the vector polygon are correct; we write

$$\alpha_{BD} = 9940 \text{ rad/s}^2 \uparrow \quad \blacktriangleleft$$
$$\mathbf{a}_D = 9290 \text{ ft/s}^2 \leftarrow \quad \blacktriangleleft$$

SAMPLE PROBLEM 15.8

The linkage $ABDE$ moves in the vertical plane. Knowing that in the position shown crank AB has a constant angular velocity $\boldsymbol{\omega}_1$ of 20 rad/s counterclockwise, determine the angular velocities and angular accelerations of the connecting rod BD and of the crank DE.

SOLUTION

$$r_B = 8i + 14j$$
$$r_D = -17i + 17j$$
$$r_{D/B} = 12i + 3j$$

This problem could be solved by the method used in Sample Prob. 15.7. In this case, however, the vector approach will be used. The position vectors $\mathbf{r}_B$, $\mathbf{r}_D$, and $\mathbf{r}_{D/B}$ are chosen as shown in the sketch.

Velocities. Since the motion of each element of the linkage is contained in the plane of the figure, we have

$$\boldsymbol{\omega}_{AB} = \omega_{AB}\mathbf{k} = (20 \text{ rad/s})\mathbf{k} \qquad \boldsymbol{\omega}_{BD} = \omega_{BD}\mathbf{k} \qquad \boldsymbol{\omega}_{DE} = \omega_{DE}\mathbf{k}$$

where $\mathbf{k}$ is a unit vector pointing out of the paper. We now write

$$\mathbf{v}_D = \mathbf{v}_B + \mathbf{v}_{D/B}$$
$$\omega_{DE}\mathbf{k} \times \mathbf{r}_D = \omega_{AB}\mathbf{k} \times \mathbf{r}_B + \omega_{BD}\mathbf{k} \times \mathbf{r}_{D/B}$$
$$\omega_{DE}\mathbf{k} \times (-17\mathbf{i} + 17\mathbf{j}) = 20\mathbf{k} \times (8\mathbf{i} + 14\mathbf{j}) + \omega_{BD}\mathbf{k} \times (12\mathbf{i} + 3\mathbf{j})$$
$$-17\omega_{DE}\mathbf{j} - 17\omega_{DE}\mathbf{i} = 160\mathbf{j} - 280\mathbf{i} + 12\omega_{BD}\mathbf{j} - 3\omega_{BD}\mathbf{i}$$

Equating the coefficients of the unit vectors $\mathbf{i}$ and $\mathbf{j}$, we obtain the following two scalar equations:

$$-17\omega_{DE} = -280 - 3\omega_{BD}$$
$$-17\omega_{DE} = +160 + 12\omega_{BD}$$
$$\boldsymbol{\omega}_{BD} = -(29.33 \text{ rad/s})\mathbf{k} \qquad \boldsymbol{\omega}_{DE} = (11.29 \text{ rad/s})\mathbf{k} \quad \blacktriangleleft$$

Accelerations. Noting that at the instant considered crank AB has a constant angular velocity, we write

$$\boldsymbol{\alpha}_{AB} = 0 \qquad \boldsymbol{\alpha}_{BD} = \alpha_{BD}\mathbf{k} \qquad \boldsymbol{\alpha}_{DE} = \alpha_{DE}\mathbf{k}$$
$$\mathbf{a}_D = \mathbf{a}_B + \mathbf{a}_{D/B} \tag{1}$$

Each term of Eq. (1) is evaluated separately:

$$\mathbf{a}_D = \alpha_{DE}\mathbf{k} \times \mathbf{r}_D - \omega_{DE}^2\mathbf{r}_D$$
$$= \alpha_{DE}\mathbf{k} \times (-17\mathbf{i} + 17\mathbf{j}) - (11.29)^2(-17\mathbf{i} + 17\mathbf{j})$$
$$= -17\alpha_{DE}\mathbf{j} - 17\alpha_{DE}\mathbf{i} + 2170\mathbf{i} - 2170\mathbf{j}$$
$$\mathbf{a}_B = \alpha_{AB}\mathbf{k} \times \mathbf{r}_B - \omega_{AB}^2\mathbf{r}_B = 0 - (20)^2(8\mathbf{i} + 14\mathbf{j})$$
$$= -3200\mathbf{i} - 5600\mathbf{j}$$
$$\mathbf{a}_{D/B} = \alpha_{BD}\mathbf{k} \times \mathbf{r}_{D/B} - \omega_{BD}^2\mathbf{r}_{D/B}$$
$$= \alpha_{BD}\mathbf{k} \times (12\mathbf{i} + 3\mathbf{j}) - (29.33)^2(12\mathbf{i} + 3\mathbf{j})$$
$$= 12\alpha_{BD}\mathbf{j} - 3\alpha_{BD}\mathbf{i} - 10{,}320\mathbf{i} - 2580\mathbf{j}$$

Substituting into Eq. (1) and equating the coefficients of $\mathbf{i}$ and $\mathbf{j}$, we obtain

$$-17\alpha_{DE} + 3\alpha_{BD} = -15{,}690$$
$$-17\alpha_{DE} - 12\alpha_{BD} = -6010$$
$$\boldsymbol{\alpha}_{BD} = -(645 \text{ rad/s}^2)\mathbf{k} \qquad \boldsymbol{\alpha}_{DE} = (809 \text{ rad/s}^2)\mathbf{k} \quad \blacktriangleleft$$

SOLVING PROBLEMS
ON YOUR OWN

This lesson was devoted to the determination of the *accelerations* of the points of a *rigid body in plane motion*. As you did previously for velocities, you will again consider the plane motion of a rigid body as the sum of two motions, namely, *a translation and a rotation*.

To solve a problem involving accelerations in plane motion you should use the following steps:

1. Determine the angular velocity of the body. To find $\boldsymbol{\omega}$ you can either

a. Consider the motion of the body as the sum of a translation and a rotation as you did in Sec. 15.6, or

b. Use the instantaneous center of rotation of the body as you did in Sec. 15.7. However, *keep in mind that you cannot use the instantaneous center to determine accelerations.*

2. Start drawing a "diagram equation" to use in your solution. This "equation" will involve the following diagrams (Fig. 15.22).

a. Plane motion diagram. Draw a sketch of the body, including all dimensions, as well as the angular velocity $\boldsymbol{\omega}$. Show the angular acceleration $\boldsymbol{\alpha}$ with its magnitude and sense if you know them. Also show those points for which you know or seek the accelerations, indicating all that you know about these accelerations.

b. Translation diagram. Select a reference point A for which you know the direction, the magnitude, or a component of the acceleration $\mathbf{a}_A$. Draw a second diagram showing the body in translation with each point having the same acceleration as point A.

c. Rotation diagram. Considering point A as a fixed reference point, draw a third diagram showing the body in rotation about A. Indicate the normal and tangential components of the relative accelerations of other points, such as the components $(\mathbf{a}_{B/A})_n$ and $(\mathbf{a}_{B/A})_t$ of the acceleration of point B with respect to point A.

3. Write the relative-acceleration formula

$$\mathbf{a}_B = \mathbf{a}_A + \mathbf{a}_{B/A} \qquad \text{or} \qquad \mathbf{a}_B = \mathbf{a}_A + (\mathbf{a}_{B/A})_n + (\mathbf{a}_{B/A})_t$$

The sample problems illustrate three different ways to use this vector equation:

a. If α is given or can easily be determined, you can use this equation to determine the accelerations of various points of the body [Sample Prob. 15.6].

b. If α cannot easily be determined, select for point B a point for which you know the direction, the magnitude, or a component of the acceleration $\mathbf{a}_B$ and draw a vector diagram of the equation. Starting at the same point, draw all known acceleration components in tip-to-tail fashion for each member of the equation. Complete the diagram by drawing the two remaining vectors in appropriate directions and in such a way that the two sums of vectors end at a common point.

The magnitudes of the two remaining vectors can be found either graphically or analytically. Usually an analytic solution will require the solution of two simultaneous equations [Sample Prob. 15.7]. However, by first considering the components of the various vectors in a direction perpendicular to one of the unknown vectors, you may be able to obtain an equation in a single unknown.

One of the two vectors obtained by the method just described will be $(\mathbf{a}_{B/A})_t$, from which you can compute α. Once α has been found, the vector equation can be used to determine the acceleration of any other point of the body.

c. A full vector approach can also be used to solve the vector equation. This is illustrated in Sample Prob. 15.8.

4. The analysis of plane motion in terms of a parameter completed this lesson. This method should be used *only if it is possible* to express the coordinates x and y of all significant points of the body in terms of a single parameter (Sec. 15.9). By differentiating twice with respect to t the coordinates x and y of a given point, you can determine the rectangular components of the absolute velocity and absolute acceleration of that point.

PROBLEMS

CONCEPT QUESTION

15.CQ7 A rear-wheel-drive car starts from rest and accelerates to the left so that the tires do not slip on the road. What is the direction of the acceleration of the point on the tire in contact with the road, that is, point A?

a. ← **b.** ↖ **c.** ↑ **d.** ↓ **e.** ↙

Fig. P15.CQ7

END-OF-SECTION PROBLEMS

15.105 A 3.5-m steel beam is lowered by means of two cables unwinding at the same speed from overhead cranes. As the beam approaches the ground, the crane operators apply brakes to slow down the unwinding motion. At the instant considered, the deceleration of the cable attached at A is 4 m/s², while that of the cable at B is 1.5 m/s². Determine (a) the angular acceleration of the beam, (b) the acceleration of point C.

15.106 The acceleration of point C is 0.3 m/s² downward and the angular acceleration of the beam is 0.8 rad/s² clockwise. Knowing that the angular velocity of the beam is zero at the instant considered, determine the acceleration of each cable.

15.107 A 900-mm rod rests on a horizontal table. A force P applied as shown produces the following accelerations: $\mathbf{a}_A = 3.6$ m/s² to the right, $\alpha = 6$ rad/s² counterclockwise as viewed from above. Determine the acceleration (a) of point G, (b) of point B.

15.108 In Prob. 15.107, determine the point of the rod that (a) has no acceleration, (b) has an acceleration of 2.4 m/s² to the right.

15.109 Knowing that at the instant shown crank BC has a constant angular velocity of 45 rpm clockwise, determine the acceleration (a) of point A, (b) of point D.

15.110 End A of rod AB moves to the right with a constant velocity of 6 ft/s. For the position shown, determine (a) the angular acceleration of rod AB, (b) the acceleration of the midpoint G of rod AB.

Fig. P15.105 and P15.106

Fig. P15.107 and P15.108

Fig. P15.109

Fig. P15.110

978

15.111 An automobile travels to the left at a constant speed of 72 km/h. Knowing that the diameter of the wheel is 560 mm, determine the acceleration (*a*) of point *B*, (*b*) of point *C*, (*c*) of point *D*.

15.112 The 18-in.-radius flywheel is rigidly attached to a 1.5-in.-radius shaft that can roll along parallel rails. Knowing that at the instant shown the center of the shaft has a velocity of 1.2 in./s and an acceleration of 0.5 in./s², both directed down to the left, determine the acceleration (*a*) of point *A*, (*b*) of point *B*.

Fig. P15.111

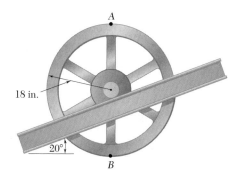

Fig. P15.112

15.113 and 15.114 A 3-in.-radius drum is rigidly attached to a 5-in.-radius drum as shown. One of the drums rolls without sliding on the surface shown, and a cord is wound around the other drum. Knowing that at the instant shown end *D* of the cord has a velocity of 8 in./s and an acceleration of 30 in./s², both directed to the left, determine the accelerations of points *A*, *B*, and *C* of the drums.

Fig. P15.113

Fig. P15.114

15.115 A carriage *C* is supported by a caster *A* and a cylinder *B*, each of 50-mm diameter. Knowing that at the instant shown the carriage has an acceleration of 2.4 m/s² and a velocity of 1.5 m/s, both directed to the left, determine (*a*) the angular accelerations of the caster and of the cylinder, (*b*) the accelerations of the centers of the caster and of the cylinder.

Fig. P15.115

Fig. P15.116

15.116 A wheel rolls without slipping on a fixed cylinder. Knowing that at the instant shown the angular velocity of the wheel is 10 rad/s clockwise and its angular acceleration is 30 rad/s² counterclockwise, determine the acceleration of (*a*) point *A*, (*b*) point *B*, (*c*) point *C*.

15.117 The 100-mm-radius drum rolls without slipping on a portion of a belt which moves downward to the left with a constant velocity of 120 mm/s. Knowing that at a given instant the velocity and acceleration of the center *A* of the drum are as shown, determine the acceleration of point *D*.

Fig. P15.117

Fig. P15.118

15.118 In the planetary gear system shown, the radius of gears *A*, *B*, *C*, and *D* is 3 in. and the radius of the outer gear *E* is 9 in. Knowing that gear *A* has a constant angular velocity of 150 rpm clockwise and that the outer gear *E* is stationary, determine the magnitude of the acceleration of the tooth of gear *D* that is in contact with (*a*) gear *A*, (*b*) gear *E*.

15.119 The 200-mm-radius disk rolls without sliding on the surface shown. Knowing that the distance *BG* is 160 mm and that at the instant shown the disk has an angular velocity of 8 rad/s counterclockwise and an angular acceleration of 2 rad/s² clockwise, determine the acceleration of *A*.

Fig. P15.119

Fig. P15.120 and P15.121

15.120 Knowing that crank *AB* rotates about point *A* with a constant angular velocity of 900 rpm clockwise, determine the acceleration of the piston *P* when $\theta = 60°$.

15.121 Knowing that crank *AB* rotates about point *A* with a constant angular velocity of 900 rpm clockwise, determine the acceleration of the piston *P* when $\theta = 120°$.

15.122 In the two-cylinder air compressor shown the connecting rods BD and BE are each 190 mm long and crank AB rotates about the fixed point A with a constant angular velocity of 1500 rpm clockwise. Determine the acceleration of each piston when $\theta = 0$.

15.123 The disk shown has a constant angular velocity of 500 rpm counterclockwise. Knowing that rod BD is 10 in. long, determine the acceleration of collar D when (a) $\theta = 90°$, (b) $\theta = 180°$.

Fig. P15.122

Fig. P15.123

15.124 Arm AB has a constant angular velocity of 16 rad/s counterclockwise. At the instant when $\theta = 90°$, determine the acceleration (a) of collar D, (b) of the midpoint G of bar BD.

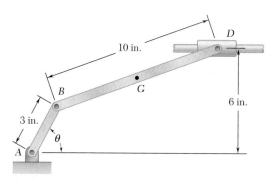

Fig. P15.124 and P15.125

15.125 Arm AB has a constant angular velocity of 16 rad/s counterclockwise. At the instant when $\theta = 60°$, determine the acceleration of collar D.

15.126 A straight rack rests on a gear of radius $r = 3$ in. and is attached to a block B as shown. Knowing that at the instant shown $\theta = 20°$, the angular velocity of gear D is 3 rad/s clockwise, and it is speeding up at a rate of 2 rad/s^2, determine (a) the angular acceleration of AB, (b) the acceleration of block B.

Fig. P15.126

Fig. P15.127 and P15.128

15.127 Knowing that at the instant shown rod AB has a constant angular velocity of 6 rad/s clockwise, determine the acceleration of point D.

15.128 Knowing that at the instant shown rod AB has a constant angular velocity of 6 rad/s clockwise, determine (a) the angular acceleration of member BDE, (b) the acceleration of point E.

15.129 Knowing that at the instant shown bar AB has a constant angular velocity of 19 rad/s clockwise, determine (a) the angular acceleration of bar BGD, (b) the angular acceleration of bar DE.

Fig. P15.129 and P15.130

15.130 Knowing that at the instant shown bar DE has a constant angular velocity of 18 rad/s clockwise, determine (a) the acceleration of point B, (b) the acceleration of point G.

15.131 and 15.132 Knowing that at the instant shown bar AB has a constant angular velocity of 4 rad/s clockwise, determine the angular acceleration (a) of bar BD, (b) of bar DE.

Fig. P15.131 and P15.133

Fig. P15.132 and P15.134

15.133 and 15.134 Knowing that at the instant shown bar AB has an angular velocity of 4 rad/s and an angular acceleration of 2 rad/s^2, both clockwise, determine the angular acceleration (a) of bar BD, (b) of bar DE by using the vector approach as is done in Sample Prob. 15.8.

15.135 Roberts linkage is named after Richard Roberts (1789–1864) and can be used to draw a close approximation to a straight line by locating a pen at point *F*. The distance *AB* is the same as *BF*, *DF*, and *DE*. Knowing that at the instant shown, bar *AB* has a constant angular velocity of 4 rad/s clockwise, determine (*a*) the angular acceleration of bar *DE*, (*b*) the acceleration of point *F*.

15.136 For the oil pump rig shown, link *AB* causes the beam *BCE* to oscillate as the crank *OA* revolves. Knowing that *OA* has a radius of 0.6 m and a constant clockwise angular velocity of 20 rpm, determine the velocity and acceleration of point *D* at the instant shown.

Fig. P15.135

Fig. P15.136

15.137 Denoting by $\mathbf{r}_A$ the position vector of a point *A* of a rigid slab that is in plane motion, show that (*a*) the position vector $\mathbf{r}_C$ of the instantaneous center of rotation is

$$\mathbf{r}_C = \mathbf{r}_A + \frac{\boldsymbol{\omega} \times \mathbf{v}_A}{\omega^2}$$

where $\boldsymbol{\omega}$ is the angular velocity of the slab and $\mathbf{v}_A$ is the velocity of point *A*, (*b*) the acceleration of the instantaneous center of rotation is zero if, and only if,

$$\mathbf{a}_A = \frac{\alpha}{\omega} \mathbf{v}_A + \boldsymbol{\omega} \times \mathbf{v}_A$$

where $\boldsymbol{\alpha} = \alpha \mathbf{k}$ is the angular acceleration of the slab.

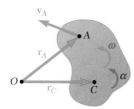

Fig. P15.137

*15.138** The drive disk of the Scotch crosshead mechanism shown has an angular velocity $\boldsymbol{\omega}$ and an angular acceleration $\boldsymbol{\alpha}$, both directed counterclockwise. Using the method of Sec. 15.9, derive expressions for the velocity and acceleration of point *B*.

Fig. P15.138

*15.139 The wheels attached to the ends of rod AB roll along the surfaces shown. Using the method of Sec. 15.9, derive an expression for the angular velocity of the rod in terms of v_B, θ, l, and β.

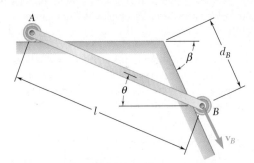

Fig. P15.139 and P15.140

*15.140 The wheels attached to the ends of rod AB roll along the surfaces shown. Using the method of Sec. 15.9 and knowing that the acceleration of wheel B is zero, derive an expression for the angular acceleration of the rod in terms of v_B, θ, l, and β.

*15.141 A disk of radius r rolls to the right with a constant velocity $\mathbf{v}$. Denoting by P the point of the rim in contact with the ground at $t = 0$, derive expressions for the horizontal and vertical components of the velocity of P at any time t.

*15.142 Rod AB moves over a small wheel at C while end A moves to the right with a constant velocity $\mathbf{v}_A$. Using the method of Sec. 15.9, derive expressions for the angular velocity and angular acceleration of the rod.

*15.143 Rod AB moves over a small wheel at C while end A moves to the right with a constant velocity $\mathbf{v}_A$. Using the method of Sec. 15.9, derive expressions for the horizontal and vertical components of the velocity of point B.

Fig. P15.142 and P15.143

15.144 Crank AB rotates with a constant clockwise angular velocity $\boldsymbol{\omega}$. Using the method of Sec. 15.9, derive expressions for the angular velocity of rod BD and the velocity of the point on the rod coinciding with point E in terms of θ, ω, b, and l.

Fig. P15.144 and P15.145

15.145 Crank AB rotates with a constant clockwise angular velocity $\boldsymbol{\omega}$. Using the method of Sec. 15.9, derive an expression for the angular acceleration of rod BD in terms of θ, ω, b, and l.

15.146 Pin C is attached to rod CD and slides in a slot cut in arm AB. Knowing that rod CD moves vertically upward with a constant velocity $\mathbf{v}_0$, derive an expression for (a) the angular velocity of arm AB, (b) the components of the velocity of point A, (c) an expression for the angular acceleration of arm AB.

*15.147 The position of rod AB is controlled by a disk of radius r which is attached to yoke CD. Knowing that the yoke moves vertically upward with a constant velocity $\mathbf{v}_0$, derive expressions for the angular velocity and angular acceleration of rod AB.

Fig. P15.146

Fig. P15.147

*15.148 A wheel of radius r rolls without slipping along the inside of a fixed cylinder of radius R with a constant angular velocity $\boldsymbol{\omega}$. Denoting by P the point of the wheel in contact with the cylinder at $t = 0$, derive expressions for the horizontal and vertical components of the velocity of P at any time t. (The curve described by point P is a *hypocycloid*.)

*15.149 In Prob. 15.148, show that the path of P is a vertical straight line when $r = R/2$. Derive expressions for the corresponding velocity and acceleration of P at any time t.

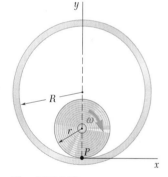

Fig. P15.148

15.10 RATE OF CHANGE OF A VECTOR WITH RESPECT TO A ROTATING FRAME

We saw in Sec. 11.10 that the rate of change of a vector is the same with respect to a fixed frame and with respect to a frame in translation. In this section, the rates of change of a vector $\mathbf{Q}$ with respect to a fixed frame and with respect to a rotating frame of reference will be considered.† You will learn to determine the rate of change of $\mathbf{Q}$ with respect to one frame of reference when $\mathbf{Q}$ is defined by its components in another frame.

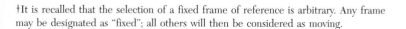

†It is recalled that the selection of a fixed frame of reference is arbitrary. Any frame may be designated as "fixed"; all others will then be considered as moving.

Photo 15.7 A Geneva mechanism is used to convert rotary motion into intermittent motion.

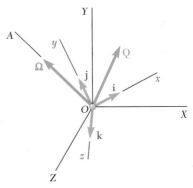

Fig. 15.26

Consider two frames of reference centered at O, a fixed frame $OXYZ$ and a frame $Oxyz$ which rotates about the fixed axis OA; let $\mathbf{\Omega}$ denote the angular velocity of the frame $Oxyz$ at a given instant (Fig. 15.26). Consider now a vector function $\mathbf{Q}(t)$ represented by the vector $\mathbf{Q}$ attached at O; as the time t varies, both the direction and the magnitude of $\mathbf{Q}$ change. Since the variation of $\mathbf{Q}$ is viewed differently by an observer using $OXYZ$ as a frame of reference and by an observer using $Oxyz$, we should expect the rate of change of $\mathbf{Q}$ to depend upon the frame of reference which has been selected. Therefore, the rate of change of $\mathbf{Q}$ with respect to the fixed frame $OXYZ$ will be denoted by $(\dot{\mathbf{Q}})_{OXYZ}$, and the rate of change of $\mathbf{Q}$ with respect to the rotating frame $Oxyz$ will be denoted by $(\dot{\mathbf{Q}})_{Oxyz}$. We propose to determine the relation existing between these two rates of change.

Let us first resolve the vector $\mathbf{Q}$ into components along the x, y, and z axes of the rotating frame. Denoting by $\mathbf{i}$, $\mathbf{j}$, and $\mathbf{k}$ the corresponding unit vectors, we write

$$\mathbf{Q} = Q_x\mathbf{i} + Q_y\mathbf{j} + Q_z\mathbf{k} \tag{15.27}$$

Differentiating (15.27) with respect to t and considering the unit vectors $\mathbf{i}$, $\mathbf{j}$, $\mathbf{k}$ as fixed, we obtain the rate of change of $\mathbf{Q}$ *with respect to the rotating frame Oxyz:*

$$(\dot{\mathbf{Q}})_{Oxyz} = \dot{Q}_x\mathbf{i} + \dot{Q}_y\mathbf{j} + \dot{Q}_z\mathbf{k} \tag{15.28}$$

To obtain the rate of change of $\mathbf{Q}$ *with respect to the fixed frame OXYZ*, we must consider the unit vectors $\mathbf{i}$, $\mathbf{j}$, $\mathbf{k}$ as variable when differentiating (15.27). We therefore write

$$(\dot{\mathbf{Q}})_{OXYZ} = \dot{Q}_x\mathbf{i} + \dot{Q}_y\mathbf{j} + \dot{Q}_z\mathbf{k} + Q_x\frac{d\mathbf{i}}{dt} + Q_y\frac{d\mathbf{j}}{dt} + Q_z\frac{d\mathbf{k}}{dt} \tag{15.29}$$

Recalling (15.28), we observe that the sum of the first three terms in the right-hand member of (15.29) represents the rate of change $(\dot{\mathbf{Q}})_{Oxyz}$. We note, on the other hand, that the rate of change $(\dot{\mathbf{Q}})_{OXYZ}$ would reduce to the last three terms in (15.29) if the vector $\mathbf{Q}$ were fixed within the frame $Oxyz$, since $(\dot{\mathbf{Q}})_{Oxyz}$ would then be zero. But in that case, $(\dot{\mathbf{Q}})_{OXYZ}$ would represent the velocity of a particle located at the tip of $\mathbf{Q}$ and belonging to a body rigidly attached to the frame $Oxyz$. Thus, the last three terms in (15.29) represent the velocity of that particle; since the frame $Oxyz$ has an angular velocity $\mathbf{\Omega}$ with respect to $OXYZ$ at the instant considered, we write, by (15.5),

$$Q_x\frac{d\mathbf{i}}{dt} + Q_y\frac{d\mathbf{j}}{dt} + Q_z\frac{d\mathbf{k}}{dt} = \mathbf{\Omega} \times \mathbf{Q} \tag{15.30}$$

Substituting from (15.28) and (15.30) into (15.29), we obtain the fundamental relation

$$(\dot{\mathbf{Q}})_{OXYZ} = (\dot{\mathbf{Q}})_{Oxyz} + \mathbf{\Omega} \times \mathbf{Q} \tag{15.31}$$

We conclude that the rate of change of the vector $\mathbf{Q}$ with respect to the fixed frame $OXYZ$ is made of two parts: The first part represents the rate of change of $\mathbf{Q}$ with respect to the rotating frame $Oxyz$; the second part, $\mathbf{\Omega} \times \mathbf{Q}$, is induced by the rotation of the frame $Oxyz$.

The use of relation (15.31) simplifies the determination of the rate of change of a vector $\mathbf{Q}$ with respect to a fixed frame of reference $OXYZ$ when the vector $\mathbf{Q}$ is defined by its components along the axes of a rotating frame $Oxyz$, since this relation does not require the separate computation of the derivatives of the unit vectors defining the orientation of the rotating frame.

15.11 PLANE MOTION OF A PARTICLE RELATIVE TO A ROTATING FRAME. CORIOLIS ACCELERATION

Consider two frames of reference, both centered at O and both in the plane of the figure, a fixed frame OXY and a rotating frame Oxy (Fig. 15.27). Let P be a particle moving in the plane of the figure. The position vector $\mathbf{r}$ of P is the same in both frames, but its rate of change depends upon the frame of reference which has been selected.

The absolute velocity $\mathbf{v}_P$ of the particle is defined as the velocity observed from the fixed frame OXY and is equal to the rate of change $(\dot{\mathbf{r}})_{OXY}$ of $\mathbf{r}$ with respect to that frame. We can, however, express $\mathbf{v}_P$ in terms of the rate of change $(\dot{\mathbf{r}})_{Oxy}$ observed from the rotating frame if we make use of Eq. (15.31). Denoting by $\mathbf{\Omega}$ the angular velocity of the frame Oxy with respect to OXY at the instant considered, we write

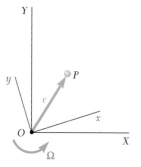

Fig. 15.27

$$\mathbf{v}_P = (\dot{\mathbf{r}})_{OXY} = \mathbf{\Omega} \times \mathbf{r} + (\dot{\mathbf{r}})_{Oxy} \qquad (15.32)$$

But $(\dot{\mathbf{r}})_{Oxy}$ defines the velocity of the particle P relative to the rotating frame Oxy. Denoting the rotating frame by $\mathcal{F}$ for short, we represent the velocity $(\dot{\mathbf{r}})_{Oxy}$ of P relative to the rotating frame by $\mathbf{v}_{P/\mathcal{F}}$. Let us imagine that a rigid slab has been attached to the rotating frame. Then $v_{P/\mathcal{F}}$ represents the velocity of P along the path that it describes on that slab (Fig. 15.28), and the term $\mathbf{\Omega} \times \mathbf{r}$ in (15.32) represents the velocity $\mathbf{v}_{P'}$ of the point P' of the slab—or rotating frame—which coincides with P at the instant considered. Thus, we have

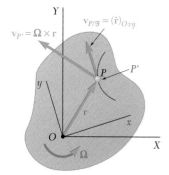

$$\mathbf{v}_P = \mathbf{v}_{P'} + \mathbf{v}_{P/\mathcal{F}} \qquad (15.33)$$

where $\mathbf{v}_P$ = absolute velocity of particle P
$\mathbf{v}_{P'}$ = velocity of point P' of moving frame $\mathcal{F}$ coinciding with P
$\mathbf{v}_{P/\mathcal{F}}$ = velocity of P relative to moving frame $\mathcal{F}$

Fig. 15.28

The absolute acceleration $\mathbf{a}_P$ of the particle is defined as the rate of change of $\mathbf{v}_P$ with respect to the fixed frame OXY. Computing the rates of change with respect to OXY of the terms in (15.32), we write

$$\mathbf{a}_P = \dot{\mathbf{v}}_P = \dot{\mathbf{\Omega}} \times \mathbf{r} + \mathbf{\Omega} \times \dot{\mathbf{r}} + \frac{d}{dt}[(\dot{\mathbf{r}})_{Oxy}] \qquad (15.34)$$

where all derivatives are defined with respect to OXY, except where indicated otherwise. Referring to Eq. (15.31), we note that the last term in (15.34) can be expressed as

$$\frac{d}{dt}[(\dot{\mathbf{r}})_{Oxy}] = (\ddot{\mathbf{r}})_{Oxy} + \mathbf{\Omega} \times (\dot{\mathbf{r}})_{Oxy}$$

On the other hand, $\dot{\mathbf{r}}$ represents the velocity $\mathbf{v}_P$ and can be replaced by the right-hand member of Eq. (15.32). After completing these two substitutions into (15.34), we write

$$\mathbf{a}_P = \dot{\boldsymbol{\Omega}} \times \mathbf{r} + \boldsymbol{\Omega} \times (\boldsymbol{\Omega} \times \mathbf{r}) + 2\boldsymbol{\Omega} \times (\dot{\mathbf{r}})_{Oxy} + (\ddot{\mathbf{r}})_{Oxy} \qquad (15.35)$$

Referring to the expression (15.8) obtained in Sec. 15.3 for the acceleration of a particle in a rigid body rotating about a fixed axis, we note that the sum of the first two terms represents the acceleration $\mathbf{a}_{P'}$ of the point P' of the rotating frame which coincides with P at the instant considered. On the other hand, the last term defines the acceleration $\mathbf{a}_{P/\mathcal{F}}$ of P relative to the rotating frame. If it were not for the third term, which has not been accounted for, a relation similar to (15.33) could be written for the accelerations, and $\mathbf{a}_P$ could be expressed as the sum of $\mathbf{a}_{P'}$ and $\mathbf{a}_{P/\mathcal{F}}$. However, it is clear that *such a relation would be incorrect* and that we must include the additional term. This term, which will be denoted by $\mathbf{a}_c$, is called the *complementary acceleration*, or *Coriolis acceleration*, after the French mathematician de Coriolis (1792–1843). We write

$$\mathbf{a}_P = \mathbf{a}_{P'} + \mathbf{a}_{P/\mathcal{F}} + \mathbf{a}_c \qquad (15.36)$$

where $\mathbf{a}_P$ = absolute acceleration of particle P
$\quad\mathbf{a}_{P'}$ = acceleration of point P' of moving frame $\mathcal{F}$ coinciding with P
$\quad\mathbf{a}_{P/\mathcal{F}}$ = acceleration of P relative to moving frame $\mathcal{F}$
$\quad\mathbf{a}_c = 2\boldsymbol{\Omega} \times (\dot{\mathbf{r}})_{Oxy} = 2\boldsymbol{\Omega} \times \mathbf{v}_{P/\mathcal{F}}$
$\quad\quad$ = complementary, or Coriolis, acceleration†

We note that since point P' moves in a circle about the origin O, its acceleration $\mathbf{a}_{P'}$ has, in general, two components: a component $(\mathbf{a}_{P'})_t$ tangent to the circle, and a component $(\mathbf{a}_{P'})_n$ directed toward O. Similarly, the acceleration $\mathbf{a}_{P/\mathcal{F}}$ generally has two components: a component $(\mathbf{a}_{P/\mathcal{F}})_t$ tangent to the path that P describes on the rotating slab, and a component $(\mathbf{a}_{P/\mathcal{F}})_n$ directed toward the center of curvature of that path. We further note that since the vector $\boldsymbol{\Omega}$ is perpendicular to the plane of motion, and thus to $\mathbf{v}_{P/\mathcal{F}}$, the magnitude of the Coriolis acceleration $\mathbf{a}_c = 2\boldsymbol{\Omega} \times \mathbf{v}_{P/\mathcal{F}}$ is equal to $2\Omega v_{P/\mathcal{F}}$, and its direction can be obtained by rotating the vector $\mathbf{v}_{P/\mathcal{F}}$ through 90° in the sense of rotation of the moving frame (Fig. 15.29). The Coriolis acceleration reduces to zero when either $\boldsymbol{\Omega}$ or $\mathbf{v}_{P/\mathcal{F}}$ is zero.

The following example will help in understanding the physical meaning of the Coriolis acceleration. Consider a collar P which is

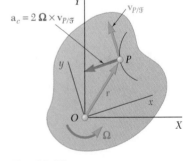

Fig. 15.29

---

†It is important to note the difference between Eq. (15.36) and Eq. (15.21) of Sec. 15.8. When we wrote

$$\mathbf{a}_B = \mathbf{a}_A + \mathbf{a}_{B/A} \qquad (15.21)$$

in Sec. 15.8, we were expressing the absolute acceleration of point B as the sum of its acceleration $\mathbf{a}_{B/A}$ relative to a *frame in translation* and of the acceleration $\mathbf{a}_A$ of a point of that frame. We are now trying to relate the absolute acceleration of point P to its acceleration $\mathbf{a}_{P/\mathcal{F}}$ relative to a *rotating frame* $\mathcal{F}$ and to the acceleration $\mathbf{a}_{P'}$ of the point P' of that frame which coincides with P; Eq. (15.36) shows that because the frame is rotating, it is necessary to include an additional term representing the Coriolis acceleration $\mathbf{a}_c$.

made to slide at a constant relative speed u along a rod OB rotating at a constant angular velocity $\boldsymbol{\omega}$ about O (Fig. 15.30a). According to formula (15.36), the absolute acceleration of P can be obtained by adding vectorially the acceleration $\mathbf{a}_A$ of the point A of the rod coinciding with P, the relative acceleration $\mathbf{a}_{P/OB}$ of P with respect to the rod, and the Coriolis acceleration $\mathbf{a}_c$. Since the angular velocity $\boldsymbol{\omega}$ of the rod is constant, $\mathbf{a}_A$ reduces to its normal component $(\mathbf{a}_A)_n$ of magnitude $r\omega^2$; and since u is constant, the relative acceleration $\mathbf{a}_{P/OB}$ is zero. According to the definition given above, the Coriolis acceleration is a vector perpendicular to OB, of magnitude $2\omega u$, and directed as shown in the figure. The acceleration of the collar P consists, therefore, of the two vectors shown in Fig. 15.30a. Note that the result obtained can be checked by applying the relation (11.44).

To understand better the significance of the Coriolis acceleration, let us consider the absolute velocity of P at time t and at time $t + \Delta t$ (Fig. 15.30b). The velocity at time t can be resolved into its components $\mathbf{u}$ and $\mathbf{v}_A$; the velocity at time $t + \Delta t$ can be resolved into its components $\mathbf{u}'$ and $\mathbf{v}_{A'}$. Drawing these components from the same origin (Fig. 15.30c), we note that the change in velocity during the time Δt can be represented by the sum of three vectors, $\overrightarrow{RR'}$, $\overrightarrow{TT''}$, and $\overrightarrow{T''T'}$. The vector $\overrightarrow{TT''}$ measures the change in direction of the velocity $\mathbf{v}_A$, and the quotient $\overrightarrow{TT''}/\Delta t$ represents the acceleration $\mathbf{a}_A$ when Δt approaches zero. We check that the direction of $\overrightarrow{TT''}$ is that of $\mathbf{a}_A$ when Δt approaches zero and that

$$\lim_{\Delta t \to 0} \frac{TT''}{\Delta t} = \lim_{\Delta t \to 0} v_A \frac{\Delta\theta}{\Delta t} = r\omega\omega = r\omega^2 = a_A$$

The vector $\overrightarrow{RR'}$ measures the change in direction of $\mathbf{u}$ due to the rotation of the rod; the vector $\overrightarrow{T''T'}$ measures the change in magnitude of $\mathbf{v}_A$ due to the motion of P on the rod. The vectors $\overrightarrow{RR'}$ and $\overrightarrow{T''T'}$ result from the *combined effect* of the relative motion of P and of the rotation of the rod; they would vanish if *either* of these two motions stopped. It is easily verified that the sum of these two vectors defines the Coriolis acceleration. Their direction is that of $\mathbf{a}_c$ when Δt approaches zero, and since $RR' = u\,\Delta\theta$ and $T''T' = v_{A'} - v_A = (r + \Delta r)\omega - r\omega = \omega\,\Delta r$, we check that a_c is equal to

$$\lim_{\Delta t \to 0}\left(\frac{RR'}{\Delta t} + \frac{T''T'}{\Delta t}\right) = \lim_{\Delta t \to 0}\left(u\frac{\Delta\theta}{\Delta t} + \omega\frac{\Delta r}{\Delta t}\right) = u\omega + \omega u = 2\omega u$$

Formulas (15.33) and (15.36) can be used to analyze the motion of mechanisms which contain parts sliding on each other. They make it possible, for example, to relate the absolute and relative motions of sliding pins and collars (see Sample Probs. 15.9 and 15.10). The concept of Coriolis acceleration is also very useful in the study of long-range projectiles and of other bodies whose motions are appreciably affected by the rotation of the earth. As was pointed out in Sec. 12.2, a system of axes attached to the earth does not truly constitute a newtonian frame of reference; such a system of axes should actually be considered as rotating. The formulas derived in this section will therefore facilitate the study of the motion of bodies with respect to axes attached to the earth.

(a)

(b)

(c)

Fig. 15.30

Disk S

$R = 50$ mm

P

R

$\phi = 135°$

O

B

$l = \sqrt{2}R$

Disk D

SAMPLE PROBLEM 15.9

The Geneva mechanism shown is used in many counting instruments and in other applications where an intermittent rotary motion is required. Disk D rotates with a constant counterclockwise angular velocity ω_D of 10 rad/s. A pin P is attached to disk D and slides along one of several slots cut in disk S. It is desirable that the angular velocity of disk S be zero as the pin enters and leaves each slot; in the case of four slots, this will occur if the distance between the centers of the disks is $l = \sqrt{2}\,R$.

At the instant when $\phi = 150°$, determine (a) the angular velocity of disk S, (b) the velocity of pin P relative to disk S.

SOLUTION

Disk S P Disk D

R

r

$\phi = 150°$

β

P'

O

B

$l = \sqrt{2}R$

We solve triangle OPB, which corresponds to the position $\phi = 150°$. Using the law of cosines, we write

$$r^2 = R^2 + l^2 - 2Rl \cos 30° = 0.551R^2 \qquad r = 0.742R = 37.1 \text{ mm}$$

From the law of sines,

$$\frac{\sin \beta}{R} = \frac{\sin 30°}{r} \qquad \sin \beta = \frac{\sin 30°}{0.742} \qquad \beta = 42.4°$$

Since pin P is attached to disk D, and since disk D rotates about point B, the magnitude of the absolute velocity of P is

$$v_P = R\omega_D = (50 \text{ mm})(10 \text{ rad/s}) = 500 \text{ mm/s}$$
$$\mathbf{v}_P = 500 \text{ mm/s} \ \angle\!\!\!\nearrow \ 60°$$

We consider now the motion of pin P along the slot in disk S. Denoting by P' the point of disk S which coincides with P at the instant considered and selecting a rotating frame $\mathscr{S}$ attached to disk S, we write

$$\mathbf{v}_P = \mathbf{v}_{P'} + \mathbf{v}_{P/\mathscr{S}}$$

Noting that $\mathbf{v}_{P'}$ is perpendicular to the radius OP and that $\mathbf{v}_{P/\mathscr{S}}$ is directed along the slot, we draw the velocity triangle corresponding to the equation above. From the triangle, we compute

$$\gamma = 90° - 42.4° - 30° = 17.6°$$
$$v_{P'} = v_P \sin \gamma = (500 \text{ mm/s}) \sin 17.6°$$
$$\mathbf{v}_{P'} = 151.2 \text{ mm/s} \ \nwarrow \ 42.4°$$
$$v_{P/\mathscr{S}} = v_P \cos \gamma = (500 \text{ mm/s}) \cos 17.6°$$
$$\mathbf{v}_{P/S} = \mathbf{v}_{P/\mathscr{S}} = 477 \text{ mm/s} \ \angle\!\!\!\nearrow \ 42.4° \quad \blacktriangleleft$$

Since $\mathbf{v}_{P'}$ is perpendicular to the radius OP, we write

$$v_{P'} = r\omega_{\mathscr{S}} \qquad 151.2 \text{ mm/s} = (37.1 \text{ mm})\omega_{\mathscr{S}}$$
$$\omega_S = \omega_{\mathscr{S}} = 4.08 \text{ rad/s} \ \downarrow \quad \blacktriangleleft$$

$v_{P'}$

v_P

γ

$v_{P/S}$

$30°$

$\beta = 42.4°$

SAMPLE PROBLEM 15.10

In the Geneva mechanism of Sample Prob. 15.9, disk D rotates with a constant counterclockwise angular velocity ω_D of 10 rad/s. At the instant when $\phi = 150°$, determine the angular acceleration of disk S.

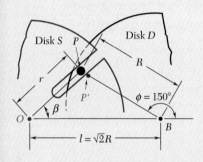

SOLUTION

Referring to Sample Prob. 15.9, we obtain the angular velocity of the frame $\mathscr{S}$ attached to disk S and the velocity of the pin relative to $\mathscr{S}$:

$$\omega_{\mathscr{S}} = 4.08 \text{ rad/s} \downarrow$$
$$\beta = 42.4° \qquad v_{P/\mathscr{S}} = 477 \text{ mm/s} \nearrow 42.4°$$

Since pin P moves with respect to the rotating frame $\mathscr{S}$, we write

$$\mathbf{a}_P = \mathbf{a}_{P'} + \mathbf{a}_{P/\mathscr{S}} + \mathbf{a}_c \qquad (1)$$

Each term of this vector equation is investigated separately.

Absolute Acceleration $\mathbf{a}_P$. Since disk D rotates with a constant angular velocity, the absolute acceleration $\mathbf{a}_P$ is directed toward B. We have

$$a_P = R\omega_D^2 = (500 \text{ mm})(10 \text{ rad/s})^2 = 5000 \text{ mm/s}^2$$
$$\mathbf{a}_P = 5000 \text{ mm/s}^2 \searrow 30°$$

Acceleration $\mathbf{a}_{P'}$ of the Coinciding Point P'. The acceleration $\mathbf{a}_{P'}$ of the point P' of the frame $\mathscr{S}$ which coincides with P at the instant considered is resolved into normal and tangential components. (We recall from Sample Prob. 15.9 that $r = 37.1$ mm.)

$$(a_{P'})_n = r\omega_{\mathscr{S}}^2 = (37.1 \text{ mm})(4.08 \text{ rad/s})^2 = 618 \text{ mm/s}^2$$
$$(\mathbf{a}_{P'})_n = 618 \text{ mm/s}^2 \nearrow 42.4°$$
$$(a_{P'})_t = r\alpha_{\mathscr{S}} = 37.1\alpha_{\mathscr{S}} \qquad (\mathbf{a}_{P'})_t = 37.1\alpha_{\mathscr{S}} \nwarrow 42.4°$$

Relative Acceleration $\mathbf{a}_{P/\mathscr{S}}$. Since the pin P moves in a straight slot cut in disk S, the relative acceleration $\mathbf{a}_{P/\mathscr{S}}$ must be parallel to the slot; i.e., its direction must be $\swarrow 42.4°$.

Coriolis Acceleration $\mathbf{a}_c$. Rotating the relative velocity $\mathbf{v}_{P/\mathscr{S}}$ through 90° in the sense of $\omega_{\mathscr{S}}$, we obtain the direction of the Coriolis component of the acceleration: $\searrow 42.4°$. We write

$$a_c = 2\omega_{\mathscr{S}}v_{P/\mathscr{S}} = 2(4.08 \text{ rad/s})(477 \text{ mm/s}) = 3890 \text{ mm/s}^2$$
$$\mathbf{a}_c = 3890 \text{ mm/s}^2 \searrow 42.4°$$

We rewrite Eq. (1) and substitute the accelerations found above:

$$\mathbf{a}_P = (\mathbf{a}_{P'})_n + (\mathbf{a}_{P'})_t + \mathbf{a}_{P/\mathscr{S}} + \mathbf{a}_c$$
$$[5000 \searrow 30°] = [618 \nearrow 42.4°] + [37.1\alpha_{\mathscr{S}} \nwarrow 42.4°]$$
$$+ [a_{P/\mathscr{S}} \swarrow 42.4°] + [3890 \searrow 42.4°]$$

Equating components in a direction perpendicular to the slot,

$$5000 \cos 17.6° = 37.1\alpha_{\mathscr{S}} - 3890$$
$$\alpha_S = \alpha_{\mathscr{S}} = 233 \text{ rad/s}^2 \downarrow \quad \blacktriangleleft$$

SOLVING PROBLEMS
ON YOUR OWN

In this lesson you studied the rate of change of a vector with respect to a rotating frame and then applied your knowledge to the analysis of the plane motion of a particle relative to a rotating frame.

1. Rate of change of a vector with respect to a fixed frame and with respect to a rotating frame. Denoting by $(\dot{\mathbf{Q}})_{OXYZ}$ the rate of change of a vector $\mathbf{Q}$ with respect to a fixed frame $OXYZ$ and by $(\dot{\mathbf{Q}})_{Oxyz}$ its rate of change with respect to a rotating frame $Oxyz$, we obtained the fundamental relation

$$(\dot{\mathbf{Q}})_{OXYZ} = (\dot{\mathbf{Q}})_{Oxyz} + \mathbf{\Omega} \times \mathbf{Q} \tag{15.31}$$

where $\mathbf{\Omega}$ is the angular velocity of the rotating frame.

This fundamental relation will now be applied to the solution of two-dimensional problems.

2. Plane motion of a particle relative to a rotating frame. Using the above fundamental relation and designating by $\mathscr{F}$ the rotating frame, we obtained the following expressions for the velocity and the acceleration of a particle P:

$$\mathbf{v}_P = \mathbf{v}_{P'} + \mathbf{v}_{P/\mathscr{F}} \tag{15.33}$$
$$\mathbf{a}_P = \mathbf{a}_{P'} + \mathbf{a}_{P/\mathscr{F}} + \mathbf{a}_c \tag{15.36}$$

In these equations:

   a. The subscript P refers to the absolute motion of the particle P, that is, to its motion with respect to a fixed frame of reference OXY.

   b. The subscript P' refers to the motion of the point P' of the rotating frame $\mathscr{F}$ which coincides with P at the instant considered.

   c. The subscript $P/\mathscr{F}$ refers to the motion of the particle P relative to the rotating frame $\mathscr{F}$.

   d. The term $\mathbf{a}_c$ represents the Coriolis acceleration of point P. Its magnitude is $2\Omega v_{P/\mathscr{F}}$, and its direction is found by rotating $\mathbf{v}_{P/\mathscr{F}}$ through $90°$ in the sense of rotation of the frame $\mathscr{F}$.

You should keep in mind that the Coriolis acceleration should be taken into account whenever a part of the mechanism you are analyzing is moving with respect to another part that is rotating. The problems you will encounter in this lesson involve collars that slide on rotating rods, booms that extend from cranes rotating in a vertical plane, etc.

When solving a problem involving a rotating frame, you will find it convenient to draw vector diagrams representing Eqs. (15.33) and (15.36), respectively, and use these diagrams to obtain either an analytical or a graphical solution.

PROBLEMS

CONCEPT QUESTION

15.CQ8 A person walks radially inward on a platform that is rotating counterclockwise about its center. Knowing that the platform has a constant angular velocity $\boldsymbol{\omega}$ and the person walks with a constant speed **u** relative to the platform, what is the direction of the acceleration of the person at the instant shown?

    a. Negative x
    b. Negative y
    c. Negative x and positive y
    d. Positive x and positive y
    e. Negative x and negative y

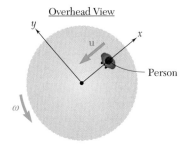

Fig. P15.CQ8

END-OF-SECTION PROBLEMS

15.150 and 15.151 Pin P is attached to the collar shown; the motion of the pin is guided by a slot cut in rod BD and by the collar that slides on rod AE. Knowing that at the instant considered the rods rotate clockwise with constant angular velocities, determine for the given data the velocity of pin P.

    15.150 $\omega_{AE} = 8$ rad/s, $\omega_{BD} = 3$ rad/s
    15.151 $\omega_{AE} = 7$ rad/s, $\omega_{BD} = 4.8$ rad/s

Fig. P15.150 and P15.151

15.152 and 15.153 Two rotating rods are connected by slider block P. The rod attached at A rotates with a constant clockwise angular velocity ω_A. For the given data, determine for the position shown (*a*) the angular velocity of the rod attached at B, (*b*) the relative velocity of slider block P with respect to the rod on which it slides.

    15.152 $b = 8$ in., $\omega_A = 6$ rad/s
    15.153 $b = 300$ mm, $\omega_A = 10$ rad/s

Fig. P15.152

Fig. P15.153

15.154 Pin P is attached to the wheel shown and slides in a slot cut in bar BD. The wheel rolls to the right without slipping with a constant angular velocity of 20 rad/s. Knowing that $x = 480$ mm when $\theta = 0$, determine the angular velocity of the bar and the relative velocity of pin P with respect to the rod when (a) $\theta = 0$, (b) $\theta = 90°$.

Fig. P15.154

15.155 Bar AB rotates clockwise with a constant angular velocity of 8 rad/s and rod EF rotates clockwise with a constant angular velocity of 6 rad/s. Determine at the instant shown (a) the angular velocity of bar BD, (b) the relative velocity of collar D with respect to rod EF.

15.156 Bar AB rotates clockwise with a constant angular velocity of 4 rad/s. Knowing that the magnitude of the velocity of collar D is 20 ft/s and that the angular velocity of bar BD is counterclockwise at the instant shown, determine (a) the angular velocity of bar EF, (b) the relative velocity of collar D with respect to rod EF.

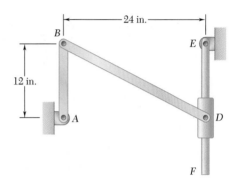

Fig. P15.155 and P15.156

15.157 The motion of pin P is guided by slots cut in rods AD and BE. Knowing that bar AD has a constant angular velocity of 4 rad/s clockwise and bar BE has an angular velocity of 5 rad/s counterclockwise and is slowing down at a rate of 2 rad/s², determine the velocity of P for the position shown.

15.158 Four pins slide in four separate slots cut in a circular plate as shown. When the plate is at rest, each pin has a velocity directed as shown and of the same constant magnitude u. If each pin maintains the same velocity relative to the plate when the plate rotates about O with a constant counterclockwise angular velocity $\boldsymbol{\omega}$, determine the acceleration of each pin.

Fig. P15.157

Fig. P15.158

15.159 Solve Prob. 15.158, assuming that the plate rotates about O with a constant clockwise angular velocity $\boldsymbol{\omega}$.

15.160 Pin P slides in the circular slot cut in the plate shown at a constant relative speed $u = 500$ mm/s. Assuming that at the instant shown the angular velocity of the plate is 6 rad/s and is increasing at the rate of 20 rad/s², determine the acceleration of pin P when $\theta = 90°$.

15.161 The cage of a mine elevator moves downward at a constant speed of 40 ft/s. Determine the magnitude and direction of the Coriolis acceleration of the cage if the elevator is located (*a*) at the equator, (*b*) at latitude 40° north, (*c*) at latitude 40° south.

15.162 A rocket sled is tested on a straight track that is built along a meridian. Knowing that the track is located at latitude 40° north, determine the Coriolis acceleration of the sled when it is moving north at a speed of 900 km/h.

15.163 The motion of blade D is controlled by the robot arm ABC. At the instant shown the arm is rotating clockwise at the constant rate $\omega = 1.8$ rad/s and the length of portion BC of the arm is being decreased at the constant rate of 250 mm/s. Determine (*a*) the velocity of D, (*b*) the acceleration of D.

15.164 At the instant shown the length of the boom AB is being *decreased* at the constant rate of 0.2 m/s and the boom is being lowered at the constant rate of 0.08 rad/s. Determine (*a*) the velocity of point B, (*b*) the acceleration of point B.

Fig. P15.160

Fig. P15.163

Fig. P15.164 and P15.165

15.165 At the instant shown the length of the boom AB is being *increased* at the constant rate of 0.2 m/s and the boom is being lowered at the constant rate of 0.08 rad/s. Determine (*a*) the velocity of point B, (*b*) the acceleration of point B.

15.166 and 15.167 The sleeve BC is welded to an arm that rotates about A with a constant angular velocity $\boldsymbol{\omega}$. In the position shown rod DF is being moved to the left at a constant speed $u = 400$ mm/s relative to the sleeve. For the given angular velocity $\boldsymbol{\omega}$, determine the acceleration (*a*) of point D, (*b*) of the point of rod DF that coincides with point E.

      15.166 $\boldsymbol{\omega} = (3$ rad/s$)$ **i**
      15.167 $\boldsymbol{\omega} = (3$ rad/s$)$ **j**

Fig. P15.166 and P15.167

Fig. P15.168 and P15.169

15.168 and 15.169 A chain is looped around two gears of radius 40 mm that can rotate freely with respect to the 320-mm arm *AB*. The chain moves about arm *AB* in a clockwise direction at the constant rate of 80 mm/s relative to the arm. Knowing that in the position shown arm *AB* rotates clockwise about *A* at the constant rate $\omega = 0.75$ rad/s, determine the acceleration of each of the chain links indicated.

  15.168 Links 1 and 2
  15.169 Links 3 and 4

15.170 A basketball player shoots a free throw in such a way that his shoulder can be considered a pin joint at the moment of release as shown. Knowing that at the instant shown the upper arm *SE* has a constant angular velocity of 2 rad/s counterclockwise and the forearm *EW* has a constant clockwise angular velocity of 4 rad/s with respect to *SE*, determine the velocity and acceleration of the wrist *W*.

Fig. P15.170

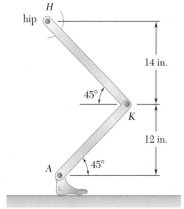

Fig. P15.171

15.171 The human leg can be crudely approximated as two rigid bars (the femur and the tibia) connected with a pin joint. At the instant shown, the velocity of the ankle *A* is zero, the tibia *AK* has an angular velocity of 1.5 rad/s counterclockwise and an angular acceleration of 1 rad/s² counterclockwise. Determine the relative angular velocity and relative angular acceleration of the femur *KH* with respect to *AK* so that the velocity and acceleration of *H* are both straight up at this instant.

15.172 The collar *P* slides outward at a constant relative speed *u* along rod *AB*, which rotates counterclockwise with a constant angular velocity of 20 rpm. Knowing that $r = 250$ mm when $\theta = 0$ and that the collar reaches *B* when $\theta = 90°$, determine the magnitude of the acceleration of the collar *P* just as it reaches *B*.

Fig. P15.172

15.173 Pin *P* slides in a circular slot cut in the plate shown at a constant relative speed *u* = 90 mm/s. Knowing that at the instant shown the plate rotates clockwise about *A* at the constant rate *ω* = 3 rad/s, determine the acceleration of the pin if it is located at (*a*) point *A*, (*b*) point *B*, (*c*) point *C*.

15.174 Pin *P* slides in a circular slot cut in the plate shown at a constant relative speed *u* = 90 mm/s. Knowing that at the instant shown the angular velocity *ω* of the plate is 3 rad/s clockwise and is decreasing at the rate of 5 rad/s², determine the acceleration of the pin if it is located at (*a*) point *A*, (*b*) point *B*, (*c*) point *C*.

15.175 Pin *P* is attached to the wheel shown and slides in a slot cut in bar *BD*. The wheel rolls to the right without slipping with a constant angular velocity of 20 rad/s. Knowing that *x* = 480 mm when *θ* = 0, determine (*a*) the angular acceleration of the bar, (*b*) the relative acceleration of pin *P* with respect to the bar when *θ* = 0.

15.176 Knowing that at the instant shown the rod attached at *A* has an angular velocity of 5 rad/s counterclockwise and an angular acceleration of 2 rad/s² clockwise, determine the angular velocity and the angular acceleration of the rod attached at *B*.

Fig. P15.173 and P15.174

Fig. P15.175

Fig. P15.176

15.177 The Geneva mechanism shown is used to provide an intermittent rotary motion of disk *S*. Disk *D* rotates with a constant counterclockwise angular velocity *ω*_D of 8 rad/s. A pin *P* is attached to disk *D* and can slide in one of the six equally spaced slots cut in disk *S*. It is desirable that the angular velocity of disk *S* be zero as the pin enters and leaves each of the six slots; this will occur if the distance between the centers of the disks and the radii of the disks are related as shown. Determine the angular velocity and angular acceleration of disk *S* at the instant when *ϕ* = 150°.

15.178 In Prob. 15.177, determine the angular velocity and angular acceleration of disk *S* at the instant when *ϕ* = 135°.

15.179 At the instant shown bar *BC* has an angular velocity of 3 rad/s and an angular acceleration of 2 rad/s², both counterclockwise; determine the angular acceleration of the plate.

15.180 At the instant shown bar *BC* has an angular velocity of 3 rad/s and an angular acceleration of 2 rad/s², both clockwise; determine the angular acceleration of the plate.

Fig. P15.177

Fig. P15.179 and P15.180

Fig. P15.181

*15.181 Rod AB passes through a collar which is welded to link DE. Knowing that at the instant shown block A moves to the right at a constant speed of 75 in./s, determine (a) the angular velocity of rod AB, (b) the velocity relative to the collar of the point of the rod in contact with the collar, (c) the acceleration of the point of the rod in contact with the collar. (*Hint:* Rod AB and link DE have the same $\boldsymbol{\omega}$ and the same $\boldsymbol{\alpha}$.)

*15.182 Solve Prob. 15.181 assuming block A moves to the left at a constant speed of 75 in./s.

*15.183 In Prob. 15.157, determine the acceleration of pin P.

*15.12 MOTION ABOUT A FIXED POINT

In Sec. 15.3 the motion of a rigid body constrained to rotate about a fixed axis was considered. The more general case of the motion of a rigid body which has a fixed point O will now be examined.

First, it will be proved that *the most general displacement of a rigid body with a fixed point O is equivalent to a rotation of the body about an axis through O.*† Instead of considering the rigid body itself, we can detach a sphere of center O from the body and analyze the motion of that sphere. Clearly, the motion of the sphere completely characterizes the motion of the given body. Since three points define the position of a solid in space, the center O and two points A and B on the surface of the sphere will define the position of the sphere and thus the position of the body. Let A_1 and B_1 characterize the position of the sphere at one instant, and let A_2 and B_2 characterize its position at a later instant (Fig. 15.31a). Since the sphere is rigid, the lengths of the arcs of great circle A_1B_1 and A_2B_2 must be equal, but except for this requirement, the positions of A_1, A_2, B_1, and B_2 are arbitrary. We propose to prove that the points A and B can be brought, respectively, from A_1 and B_1 into A_2 and B_2 by a single rotation of the sphere about an axis.

For convenience, and without loss of generality, we select point B so that its initial position coincides with the final position of A; thus, $B_1 = A_2$ (Fig. 15.31b). We draw the arcs of great circle A_1A_2, A_2B_2 and the arcs bisecting, respectively, A_1A_2 and A_2B_2. Let C be the point of intersection of these last two arcs; we complete the construction by drawing A_1C, A_2C, and B_2C. As pointed out above, because of the rigidity of the sphere, $A_1B_1 = A_2B_2$. Since C is by construction equidistant from A_1, A_2, and B_2, we also have $A_1C = A_2C = B_2C$. As a result, the spherical triangles A_1CA_2 and B_1CB_2 are congruent and the angles A_1CA_2 and B_1CB_2 are equal. Denoting by θ the common value of these angles, we conclude that the sphere can be brought from its initial position into its final position by a single rotation through θ about the axis OC.

(a)

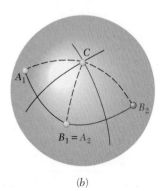

(b)

Fig. 15.31

†This is known as *Euler's theorem.*

It follows that the motion during a time interval Δt of a rigid body with a fixed point O can be considered as a rotation through $\Delta\theta$ about a certain axis. Drawing along that axis a vector of magnitude $\Delta\theta/\Delta t$ and letting Δt approach zero, we obtain at the limit the *instantaneous axis of rotation* and the angular velocity $\boldsymbol{\omega}$ of the body at the instant considered (Fig. 15.32). The velocity of a particle P of the body can then be obtained, as in Sec. 15.3, by forming the vector product of $\boldsymbol{\omega}$ and of the position vector $\mathbf{r}$ of the particle:

$$\mathbf{v} = \frac{d\mathbf{r}}{dt} = \boldsymbol{\omega} \times \mathbf{r} \tag{15.37}$$

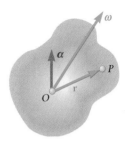

Fig. 15.32

The acceleration of the particle is obtained by differentiating (15.37) with respect to t. As in Sec. 15.3 we have

$$\mathbf{a} = \boldsymbol{\alpha} \times \mathbf{r} + \boldsymbol{\omega} \times (\boldsymbol{\omega} \times \mathbf{r}) \tag{15.38}$$

where the angular acceleration $\boldsymbol{\alpha}$ is defined as the derivative

$$\boldsymbol{\alpha} = \frac{d\boldsymbol{\omega}}{dt} \tag{15.39}$$

of the angular velocity $\boldsymbol{\omega}$.

In the case of the motion of a rigid body with a fixed point, the direction of $\boldsymbol{\omega}$ and of the instantaneous axis of rotation changes from one instant to the next. The angular acceleration $\boldsymbol{\alpha}$ therefore reflects the change in direction of $\boldsymbol{\omega}$ as well as its change in magnitude and, in general, *is not directed along the instantaneous axis of rotation.* While the particles of the body located on the instantaneous axis of rotation have zero velocity at the instant considered, they do not have zero acceleration. Also, the accelerations of the various particles of the body *cannot* be determined as if the body were rotating permanently about the instantaneous axis.

Recalling the definition of the velocity of a particle with position vector $\mathbf{r}$, we note that the angular acceleration $\boldsymbol{\alpha}$, as expressed in (15.39), represents the velocity of the tip of the vector $\boldsymbol{\omega}$. This property may be useful in the determination of the angular acceleration of a rigid body. For example, it follows that the vector $\boldsymbol{\alpha}$ is tangent to the curve described in space by the tip of the vector $\boldsymbol{\omega}$.

We should note that the vector $\boldsymbol{\omega}$ moves within the body, as well as in space. It thus generates two cones called, respectively, the *body cone* and the *space cone* (Fig. 15.33).† It can be shown that at any given instant, the two cones are tangent along the instantaneous axis of rotation and that as the body moves, the body cone appears to *roll* on the space cone.

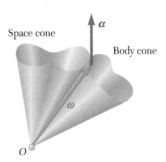

Fig. 15.33

†It is recalled that a *cone* is, by definition, a surface generated by a straight line passing through a fixed point. In general, the cones considered here *will not be circular cones.*

Photo 15.8 When the ladder rotates about its fixed base, its angular velocity can be obtained by adding the angular velocities which correspond to simultaneous rotations about two different axes.

Before concluding our analysis of the motion of a rigid body with a fixed point, we should prove that angular velocities are actually vectors. As indicated in Sec. 2.3, some quantities, such as the *finite rotations* of a rigid body, have magnitude and direction but do not obey the parallelogram law of addition; these quantities cannot be considered as vectors. In contrast, angular velocities (and also *infinitesimal rotations*), as will be demonstrated presently, *do obey* the parallelogram law and thus are truly vector quantities.

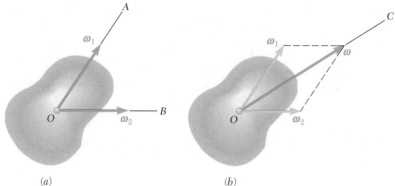

(a)   (b)

Fig. 15.34

Consider a rigid body with a fixed point O which at a given instant rotates simultaneously about the axes OA and OB with angular velocities $\boldsymbol{\omega}_1$ and $\boldsymbol{\omega}_2$ (Fig. 15.34a). We know that this motion must be equivalent at the instant considered to a single rotation of angular velocity $\boldsymbol{\omega}$. We propose to show that

$$\boldsymbol{\omega} = \boldsymbol{\omega}_1 + \boldsymbol{\omega}_2 \qquad (15.40)$$

i.e., that the resulting angular velocity can be obtained by adding $\boldsymbol{\omega}_1$ and $\boldsymbol{\omega}_2$ by the parallelogram law (Fig. 15.34b).

Consider a particle P of the body, defined by the position vector $\mathbf{r}$. Denoting, respectively, by $\mathbf{v}_1$, $\mathbf{v}_2$, and $\mathbf{v}$ the velocity of P when the body rotates about OA only, about OB only, and about both axes simultaneously, we write

$$\mathbf{v} = \boldsymbol{\omega} \times \mathbf{r} \qquad \mathbf{v}_1 = \boldsymbol{\omega}_1 \times \mathbf{r} \qquad \mathbf{v}_2 = \boldsymbol{\omega}_2 \times \mathbf{r} \qquad (15.41)$$

But the vectorial character of *linear* velocities is well established (since they represent the derivatives of position vectors). We therefore have

$$\mathbf{v} = \mathbf{v}_1 + \mathbf{v}_2$$

where the plus sign indicates vector addition. Substituting from (15.41), we write

$$\boldsymbol{\omega} \times \mathbf{r} = \boldsymbol{\omega}_1 \times \mathbf{r} + \boldsymbol{\omega}_2 \times \mathbf{r}$$
$$\boldsymbol{\omega} \times \mathbf{r} = (\boldsymbol{\omega}_1 + \boldsymbol{\omega}_2) \times \mathbf{r}$$

where the plus sign still indicates vector addition. Since the relation obtained holds for an arbitrary $\mathbf{r}$, we conclude that (15.40) must be true.

*15.13 GENERAL MOTION

The most general motion of a rigid body in space will now be considered. Let A and B be two particles of the body. We recall from Sec. 11.12 that the velocity of B with respect to the fixed frame of reference $OXYZ$ can be expressed as

$$\mathbf{v}_B = \mathbf{v}_A + \mathbf{v}_{B/A} \tag{15.42}$$

where $\mathbf{v}_{B/A}$ is the velocity of B relative to a frame $AX'Y'Z'$ attached to A and of fixed orientation (Fig. 15.35). Since A is fixed in this frame, the motion of the body relative to $AX'Y'Z'$ is the motion of a body with a fixed point. The relative velocity $\mathbf{v}_{B/A}$ can therefore be obtained from (15.37) after $\mathbf{r}$ has been replaced by the position vector $\mathbf{r}_{B/A}$ of B relative to A. Substituting for $\mathbf{v}_{B/A}$ into (15.42), we write

$$\mathbf{v}_B = \mathbf{v}_A + \boldsymbol{\omega} \times \mathbf{r}_{B/A} \tag{15.43}$$

where $\boldsymbol{\omega}$ is the angular velocity of the body at the instant considered.

The acceleration of B is obtained by a similar reasoning. We first write

$$\mathbf{a}_B = \mathbf{a}_A + \mathbf{a}_{B/A}$$

and, recalling Eq. (15.38),

$$\mathbf{a}_B = \mathbf{a}_A + \boldsymbol{\alpha} \times \mathbf{r}_{B/A} + \boldsymbol{\omega} \times (\boldsymbol{\omega} \times \mathbf{r}_{B/A}) \tag{15.44}$$

where $\boldsymbol{\alpha}$ is the angular acceleration of the body at the instant considered.

Equations (15.43) and (15.44) show that *the most general motion of a rigid body is equivalent, at any given instant, to the sum of a translation,* in which all the particles of the body have the same velocity and acceleration as a reference particle A, *and of a motion in which particle A is assumed to be fixed.*[†]

It is easily shown, by solving (15.43) and (15.44) for $\mathbf{v}_A$ and $\mathbf{a}_A$, that the motion of the body with respect to a frame attached to B would be characterized by the same vectors $\boldsymbol{\omega}$ and $\boldsymbol{\alpha}$ as its motion relative to $AX'Y'Z'$. The angular velocity and angular acceleration of a rigid body at a given instant are thus independent of the choice of reference point. On the other hand, one should keep in mind that whether the moving frame is attached to A or to B, it should maintain a fixed orientation; that is, it should remain parallel to the fixed reference frame $OXYZ$ throughout the motion of the rigid body. In many problems it will be more convenient to use a moving frame which is allowed to rotate as well as to translate. The use of such moving frames will be discussed in Secs. 15.14 and 15.15.

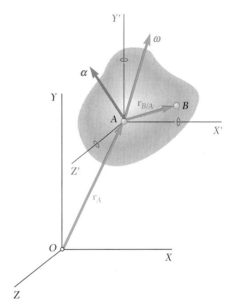

Fig. 15.35

[†]It is recalled from Sec. 15.12 that, in general, the vectors $\boldsymbol{\omega}$ and $\boldsymbol{\alpha}$ are not collinear, and that the accelerations of the particles of the body in their motion relative to the frame $AX'Y'Z'$ cannot be determined as if the body were rotating permanently about the instantaneous axis through A.

SAMPLE PROBLEM 15.11

The crane shown rotates with a constant angular velocity $\boldsymbol{\omega}_1$ of 0.30 rad/s. Simultaneously, the boom is being raised with a constant angular velocity $\boldsymbol{\omega}_2$ of 0.50 rad/s relative to the cab. Knowing that the length of the boom OP is $l = 12$ m, determine (a) the angular velocity $\boldsymbol{\omega}$ of the boom, (b) the angular acceleration $\boldsymbol{\alpha}$ of the boom, (c) the velocity $\mathbf{v}$ of the tip of the boom, (d) the acceleration $\mathbf{a}$ of the tip of the boom.

SOLUTION

a. Angular Velocity of Boom. Adding the angular velocity $\boldsymbol{\omega}_1$ of the cab and the angular velocity $\boldsymbol{\omega}_2$ of the boom relative to the cab, we obtain the angular velocity $\boldsymbol{\omega}$ of the boom at the instant considered:

$$\boldsymbol{\omega} = \boldsymbol{\omega}_1 + \boldsymbol{\omega}_2 \qquad \boldsymbol{\omega} = (0.30 \text{ rad/s})\mathbf{j} + (0.50 \text{ rad/s})\mathbf{k} \quad \blacktriangleleft$$

b. Angular Acceleration of Boom. The angular acceleration $\boldsymbol{\alpha}$ of the boom is obtained by differentiating $\boldsymbol{\omega}$. Since the vector $\boldsymbol{\omega}_1$ is constant in magnitude and direction, we have

$$\boldsymbol{\alpha} = \dot{\boldsymbol{\omega}} = \dot{\boldsymbol{\omega}}_1 + \dot{\boldsymbol{\omega}}_2 = 0 + \dot{\boldsymbol{\omega}}_2$$

where the rate of change $\dot{\boldsymbol{\omega}}_2$ is to be computed with respect to the fixed frame $OXYZ$. However, it is more convenient to use a frame $Oxyz$ attached to the cab and rotating with it, since the vector $\boldsymbol{\omega}_2$ also rotates with the cab and therefore has zero rate of change with respect to that frame. Using Eq. (15.31) with $\mathbf{Q} = \boldsymbol{\omega}_2$ and $\boldsymbol{\Omega} = \boldsymbol{\omega}_1$, we write

$$(\dot{\mathbf{Q}})_{OXYZ} = (\dot{\mathbf{Q}})_{Oxyz} + \boldsymbol{\Omega} \times \mathbf{Q}$$
$$(\dot{\boldsymbol{\omega}}_2)_{OXYZ} = (\dot{\boldsymbol{\omega}}_2)_{Oxyz} + \boldsymbol{\omega}_1 \times \boldsymbol{\omega}_2$$
$$\boldsymbol{\alpha} = (\dot{\boldsymbol{\omega}}_2)_{OXYZ} = 0 + (0.30 \text{ rad/s})\mathbf{j} \times (0.50 \text{ rad/s})\mathbf{k}$$

$$\boldsymbol{\alpha} = (0.15 \text{ rad/s}^2)\mathbf{i} \quad \blacktriangleleft$$

c. Velocity of Tip of Boom. Noting that the position vector of point P is $\mathbf{r} = (10.39 \text{ m})\mathbf{i} + (6 \text{ m})\mathbf{j}$ and using the expression found for $\boldsymbol{\omega}$ in part a, we write

$$\mathbf{v} = \boldsymbol{\omega} \times \mathbf{r} = \begin{vmatrix} \mathbf{i} & \mathbf{j} & \mathbf{k} \\ 0 & 0.30 \text{ rad/s} & 0.50 \text{ rad/s} \\ 10.39 \text{ m} & 6 \text{ m} & 0 \end{vmatrix}$$

$$\mathbf{v} = -(3 \text{ m/s})\mathbf{i} + (5.20 \text{ m/s})\mathbf{j} - (3.12 \text{ m/s})\mathbf{k} \quad \blacktriangleleft$$

d. Acceleration of Tip of Boom. Recalling that $\mathbf{v} = \boldsymbol{\omega} \times \mathbf{r}$, we write

$$\mathbf{a} = \boldsymbol{\alpha} \times \mathbf{r} + \boldsymbol{\omega} \times (\boldsymbol{\omega} \times \mathbf{r}) = \boldsymbol{\alpha} \times \mathbf{r} + \boldsymbol{\omega} \times \mathbf{v}$$

$$\mathbf{a} = \begin{vmatrix} \mathbf{i} & \mathbf{j} & \mathbf{k} \\ 0.15 & 0 & 0 \\ 10.39 & 6 & 0 \end{vmatrix} + \begin{vmatrix} \mathbf{i} & \mathbf{j} & \mathbf{k} \\ 0 & 0.30 & 0.50 \\ -3 & 5.20 & -3.12 \end{vmatrix}$$

$$= 0.90\mathbf{k} - 0.94\mathbf{i} - 2.60\mathbf{i} - 1.50\mathbf{j} + 0.90\mathbf{k}$$

$$\mathbf{a} = -(3.54 \text{ m/s}^2)\mathbf{i} - (1.50 \text{ m/s}^2)\mathbf{j} + (1.80 \text{ m/s}^2)\mathbf{k} \quad \blacktriangleleft$$

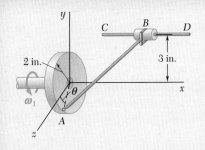

SAMPLE PROBLEM 15.12

The rod AB, of length 7 in., is attached to the disk by a ball-and-socket connection and to the collar B by a clevis. The disk rotates in the yz plane at a constant rate $\omega_1 = 12$ rad/s, while the collar is free to slide along the horizontal rod CD. For the position $\theta = 0$, determine (a) the velocity of the collar, (b) the angular velocity of the rod.

SOLUTION

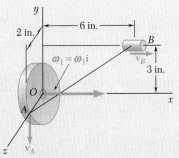

$$\omega_1 = 12\,\mathbf{i}$$
$$\mathbf{r}_A = 2\mathbf{k}$$
$$\mathbf{r}_B = 6\mathbf{i} + 3\mathbf{j}$$
$$\mathbf{r}_{B/A} = 6\mathbf{i} + 3\mathbf{j} - 2\mathbf{k}$$

a. Velocity of Collar. Since point A is attached to the disk and since collar B moves in a direction parallel to the x axis, we have

$$\mathbf{v}_A = \omega_1 \times \mathbf{r}_A = 12\mathbf{i} \times 2\mathbf{k} = -24\mathbf{j} \qquad \mathbf{v}_B = v_B\mathbf{i}$$

Denoting by ω the angular velocity of the rod, we write

$$\mathbf{v}_B = \mathbf{v}_A + \mathbf{v}_{B/A} = \mathbf{v}_A + \omega \times \mathbf{r}_{B/A}$$

$$v_B\mathbf{i} = -24\mathbf{j} + \begin{vmatrix} \mathbf{i} & \mathbf{j} & \mathbf{k} \\ \omega_x & \omega_y & \omega_z \\ 6 & 3 & -2 \end{vmatrix}$$

$$v_B\mathbf{i} = -24\mathbf{j} + (-2\omega_y - 3\omega_z)\mathbf{i} + (6\omega_z + 2\omega_x)\mathbf{j} + (3\omega_x - 6\omega_y)\mathbf{k}$$

Equating the coefficients of the unit vectors, we obtain

$$v_B = \qquad -2\omega_y \quad -3\omega_z \tag{1}$$
$$24 = 2\omega_x \qquad\qquad +6\omega_z \tag{2}$$
$$0 = 3\omega_x \quad -6\omega_y \tag{3}$$

Multiplying Eqs. (1), (2), (3), respectively, by 6, 3, −2 and adding, we write

$$6v_B + 72 = 0 \qquad v_B = -12 \qquad \mathbf{v}_B = -(12 \text{ in./s})\mathbf{i} \quad \blacktriangleleft$$

b. Angular Velocity of Rod AB. We note that the angular velocity cannot be determined from Eqs. (1), (2), and (3), since the determinant formed by the coefficients of ω_x, ω_y, and ω_z is zero. We must therefore obtain an additional equation by considering the constraint imposed by the clevis at B.

The collar-clevis connection at B permits rotation of AB about the rod CD and also about an axis perpendicular to the plane containing AB and CD. It prevents rotation of AB about the axis EB, which is perpendicular to CD and lies in the plane containing AB and CD. Thus the projection of ω on $\mathbf{r}_{E/B}$ must be zero and we write†

$$\omega \cdot \mathbf{r}_{E/B} = 0 \qquad (\omega_x\mathbf{i} + \omega_y\mathbf{j} + \omega_z\mathbf{k}) \cdot (-3\mathbf{j} + 2\mathbf{k}) = 0$$
$$-3\omega_y + 2\omega_z = 0 \tag{4}$$

Solving Eqs. (1) through (4) simultaneously, we obtain

$$v_B = -12 \qquad \omega_x = 3.69 \qquad \omega_y = 1.846 \qquad \omega_z = 2.77$$
$$\omega = (3.69 \text{ rad/s})\mathbf{i} + (1.846 \text{ rad/s})\mathbf{j} + (2.77 \text{ rad/s})\mathbf{k} \quad \blacktriangleleft$$

†We could also note that the direction of EB is that of the vector triple product $\mathbf{r}_{B/C} \times (\mathbf{r}_{B/C} \times \mathbf{r}_{B/A})$ and write $\omega \cdot [\mathbf{r}_{B/C} \times (\mathbf{r}_{B/C} \times \mathbf{r}_{B/A})] = 0$. This formulation would be particularly useful if the rod CD were skew.

SOLVING PROBLEMS ON YOUR OWN

In this lesson you started the study of the *kinematics of rigid bodies in three dimensions*. You first studied the *motion of a rigid body about a fixed point* and then the *general motion of a rigid body*.

A. Motion of a rigid body about a fixed point. To analyze the motion of a point B of a body rotating about a fixed point O you may have to take some or all of the following steps.

1. Determine the position vector r connecting the fixed point O to point B.

2. Determine the angular velocity ω of the body with respect to a fixed frame of reference. The angular velocity ω will often be obtained by adding two component angular velocities ω_1 and ω_2 [Sample Prob. 15.11].

3. Compute the velocity of B by using the equation

$$\mathbf{v} = \boldsymbol{\omega} \times \mathbf{r} \tag{15.37}$$

Your computation will usually be facilitated if you express the vector product as a determinant.

4. Determine the angular acceleration α of the body. The angular acceleration $\boldsymbol{\alpha}$ represents the rate of change $(\dot{\boldsymbol{\omega}})_{OXYZ}$ of the vector $\boldsymbol{\omega}$ *with respect to a fixed frame of reference OXYZ* and reflects both *a change in magnitude and a change in direction* of the angular velocity. However, when computing $\boldsymbol{\alpha}$ you may find it convenient to first compute the rate of change $(\dot{\boldsymbol{\omega}})_{Oxyz}$ of $\boldsymbol{\omega}$ with respect to a rotating frame of reference $Oxyz$ of your choice and use Eq. (15.31) of the preceding lesson to obtain $\boldsymbol{\alpha}$. You will write

$$\boldsymbol{\alpha} = (\dot{\boldsymbol{\omega}})_{OXYZ} = (\dot{\boldsymbol{\omega}})_{Oxyz} + \boldsymbol{\Omega} \times \boldsymbol{\omega}$$

where $\boldsymbol{\Omega}$ is the angular velocity of the rotating frame $Oxyz$ [Sample Prob. 15.11].

5. Compute the acceleration of B by using the equation

$$\mathbf{a} = \boldsymbol{\alpha} \times \mathbf{r} + \boldsymbol{\omega} \times (\boldsymbol{\omega} \times \mathbf{r}) \tag{15.38}$$

Note that the vector product $(\boldsymbol{\omega} \times \mathbf{r})$ represents the velocity of point B and was computed in step 3. Also, the computation of the first vector product in (15.38) will be facilitated if you express this product in determinant form. Remember that, as was the case with the plane motion of a rigid body, the instantaneous axis of rotation *cannot* be used to determine accelerations.

B. General motion of a rigid body. The general motion of a rigid body may be considered as *the sum of a translation and a rotation.* Keep the following in mind:

   a. In the translation part of the motion, all the points of the body have the *same velocity* $\mathbf{v}_A$ *and the same acceleration* $\mathbf{a}_A$ as the point A of the body that has been selected as the reference point.

   b. In the rotation part of the motion, the same reference point A is assumed to be a *fixed point.*

1. To determine the velocity of a point B of the rigid body when you know the velocity $\mathbf{v}_A$ of the reference point A and the angular velocity $\boldsymbol{\omega}$ of the body, you simply add $\mathbf{v}_A$ to the velocity $\mathbf{v}_{B/A} = \boldsymbol{\omega} \times \mathbf{r}_{B/A}$ of B in its rotation about A:

$$\mathbf{v}_B = \mathbf{v}_A + \boldsymbol{\omega} \times \mathbf{r}_{B/A} \tag{15.43}$$

As indicated earlier, the computation of the vector product will usually be facilitated if you express this product in determinant form.

Equation (15.43) can also be used to determine the magnitude of $\mathbf{v}_B$ when its direction is known, even if $\boldsymbol{\omega}$ is not known. While the corresponding three scalar equations are linearly dependent and the components of $\boldsymbol{\omega}$ are indeterminate, these components can be eliminated and $\mathbf{v}_A$ can be found by using an appropriate linear combination of the three equations [Sample Prob. 15.12, part a]. Alternatively, you can assign an arbitrary value to one of the components of $\boldsymbol{\omega}$ and solve the equations for $\mathbf{v}_A$. However, an additional equation must be sought in order to determine the true values of the components of $\boldsymbol{\omega}$ [Sample Prob. 15.12, part b].

2. To determine the acceleration of a point B of the rigid body when you know the acceleration $\mathbf{a}_A$ of the reference point A and the angular acceleration $\boldsymbol{\alpha}$ of the body, you simply add $\mathbf{a}_A$ to the acceleration of B in its rotation about A, as expressed by Eq. (15.38):

$$\mathbf{a}_B = \mathbf{a}_A + \boldsymbol{\alpha} \times \mathbf{r}_{B/A} + \boldsymbol{\omega} \times (\boldsymbol{\omega} \times \mathbf{r}_{B/A}) \tag{15.44}$$

Note that the vector product $(\boldsymbol{\omega} \times \mathbf{r}_{B/A})$ represents the velocity $\mathbf{v}_{B/A}$ of B relative to A and may already have been computed as part of your calculation of $\mathbf{v}_B$. We also remind you that the computation of the other two vector products will be facilitated if you express these products in determinant form.

The three scalar equations associated with Eq. (15.44) can also be used to determine the magnitude of $\mathbf{a}_B$ when its direction is known, even if $\boldsymbol{\omega}$ and $\boldsymbol{\alpha}$ are not known. While the components of $\boldsymbol{\omega}$ and $\boldsymbol{\alpha}$ are indeterminate, you can assign arbitrary values to one of the components of $\boldsymbol{\omega}$ and to one of the components of $\boldsymbol{\alpha}$ and solve the equations for $\mathbf{a}_B$.

PROBLEMS

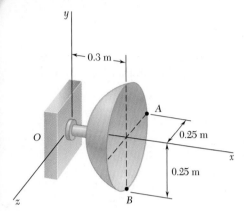

Fig. P15.184 and P15.185

15.184 At the instant considered the radar antenna shown rotates about the origin of coordinates with an angular velocity $\boldsymbol{\omega} = \omega_x\mathbf{i} + \omega_y\mathbf{j} + \omega_z\mathbf{k}$. Knowing that $(v_A)_y = 300$ mm/s, $(v_B)_y = 180$ mm/s, and $(v_B)_z = 360$ mm/s, determine (a) the angular velocity of the antenna, (b) the velocity of point A.

15.185 At the instant considered the radar antenna shown rotates about the origin of coordinates with an angular velocity $\boldsymbol{\omega} = \omega_x\mathbf{i} + \omega_y\mathbf{j} + \omega_z\mathbf{k}$. Knowing that $(v_A)_x = 100$ mm/s, $(v_A)_y = -90$ mm/s, and $(v_B)_z = 120$ mm/s, determine (a) the angular velocity of the antenna, (b) the velocity of point A.

15.186 Plate ABD and rod OB are rigidly connected and rotate about the ball-and-socket joint O with an angular velocity $\boldsymbol{\omega} = \omega_x\mathbf{i} + \omega_y\mathbf{j} + \omega_z\mathbf{k}$. Knowing that $\mathbf{v}_A = (80\text{ mm/s})\mathbf{i} + (360\text{ mm/s})\mathbf{j} + (v_A)_z\mathbf{k}$ and $\omega_x = 1.5$ rad/s, determine (a) the angular velocity of the assembly, (b) the velocity of point D.

Fig. P15.186

15.187 The bowling ball shown rolls without slipping on the horizontal xz plane with an angular velocity $\boldsymbol{\omega} = \omega_x\mathbf{i} + \omega_y\mathbf{j} + \omega_z\mathbf{k}$. Knowing that $\mathbf{v}_A = (14.4\text{ ft/s})\mathbf{i} - (14.4\text{ ft/s})\mathbf{j} + (10.8\text{ ft/s})\mathbf{k}$ and $\mathbf{v}_D = (28.8\text{ ft/s})\mathbf{i} + (21.6\text{ ft/s})\mathbf{k}$, determine (a) the angular velocity of the bowling ball, (b) the velocity of its center C.

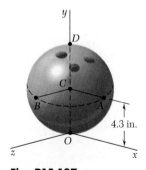

Fig. P15.187

15.188 The rotor of an electric motor rotates at the constant rate ω_1 = 1800 rpm. Determine the angular acceleration of the rotor as the motor is rotated about the y axis with a constant angular velocity $\boldsymbol{\omega}_2$ of 6 rpm counterclockwise when viewed from the positive y axis.

15.189 The disk of a portable sander rotates at the constant rate ω_1 = 4400 rpm as shown. Determine the angular acceleration of the disk as a worker rotates the sander about the z axis with an angular velocity of 0.5 rad/s and an angular acceleration of 2.5 rad/s^2, both clockwise when viewed from the positive z axis.

Fig. P15.188

Fig. P15.189

15.190 Knowing that the turbine rotor shown rotates at a constant rate ω_1 = 9000 rpm, determine the angular acceleration of the rotor if the turbine housing has a constant angular velocity of 2.4 rad/s clockwise as viewed from (a) the positive y axis, (b) the positive z axis.

Fig. P15.190

15.191 In the system shown, disk A is free to rotate about the horizontal rod OA. Assuming that disk B is stationary ($\omega_2 = 0$), and that shaft OC rotates with a constant angular velocity $\boldsymbol{\omega}_1$, determine (a) the angular velocity of disk A, (b) the angular acceleration of disk A.

15.192 In the system shown, disk A is free to rotate about the horizontal rod OA. Assuming that shaft OC and disk B rotate with constant angular velocities $\boldsymbol{\omega}_1$ and $\boldsymbol{\omega}_2$, respectively, both counterclockwise, determine (a) the angular velocity of disk A, (b) the angular acceleration of disk A.

Fig. P15.191 and P15.192

Fig. P15.193

15.193 The L-shaped arm BCD rotates about the z axis with a constant angular velocity $\omega_1 = 5$ rad/s. Knowing that the 150-mm-radius disk rotates about BC with a constant angular velocity $\omega_2 = 4$ rad/s, determine (a) the velocity of point A, (b) the acceleration of point A.

15.194 A gun barrel of length $OP = 4$ m is mounted on a turret as shown. To keep the gun aimed at a moving target the azimuth angle β is being increased at the rate $d\beta/dt = 30°/s$ and the elevation angle γ is being increased at the rate $d\gamma/dt = 10°/s$. For the position $\beta = 90°$ and $\gamma = 30°$, determine (a) the angular velocity of the barrel, (b) the angular acceleration of the barrel, (c) the velocity and acceleration of point P.

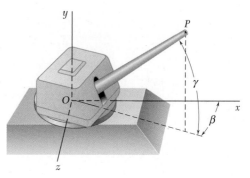

Fig. P15.194

15.195 A 3-in.-radius disk spins at the constant rate $\omega_2 = 4$ rad/s about an axis held by a housing attached to a horizontal rod that rotates at the constant rate $\omega_1 = 5$ rad/s. For the position shown, determine (a) the angular acceleration of the disk, (b) the acceleration of point P on the rim of the disk if $\theta = 0$, (c) the acceleration of point P on the rim of the disk if $\theta = 90°$.

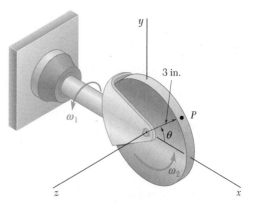

Fig. P15.195 and P15.196

15.196 A 3-in.-radius disk spins at the constant rate $\omega_2 = 4$ rad/s about an axis held by a housing attached to a horizontal rod that rotates at the constant rate $\omega_1 = 5$ rad/s. Knowing that $\theta = 30°$, determine the acceleration of point P on the rim of the disk.

15.197 A 30-mm-radius wheel is mounted on an axle *OB* of length 100 mm. The wheel rolls without sliding on the horizontal floor, and the axle is perpendicular to the plane of the wheel. Knowing that the system rotates about the *y* axis at a constant rate $\omega_1 = 2.4$ rad/s, determine (*a*) the angular velocity of the wheel, (*b*) the angular acceleration of the wheel, (*c*) the acceleration of point *C* located at the highest point on the rim of the wheel.

15.198 At the instant shown, the robotic arm *ABC* is being rotated simultaneously at the constant rate $\omega_1 = 0.15$ rad/s about the *y* axis, and at the constant rate $\omega_2 = 0.25$ rad/s about the *z* axis. Knowing that the length of arm *ABC* is 1 m, determine (*a*) the angular acceleration of the arm, (*b*) the velocity of point *C*, (*c*) the acceleration of point *C*.

Fig. P15.197

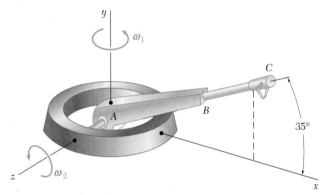

Fig. P15.198

15.199 In the planetary gear system shown, gears *A* and *B* are rigidly connected to each other and rotate as a unit about the inclined shaft. Gears *C* and *D* rotate with constant angular velocities of 30 rad/s and 20 rad/s, respectively (both counterclockwise when viewed from the right). Choosing the *x* axis to the right, the *y* axis upward, and the *z* axis pointing out of the plane of the figure, determine (*a*) the common angular velocity of gears *A* and *B*, (*b*) the angular velocity of shaft *FH*, which is rigidly attached to the inclined shaft.

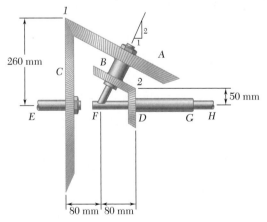

Fig. P15.199

15.200 In Prob. 15.199, determine (*a*) the common angular acceleration of gears *A* and *B*, (*b*) the acceleration of the tooth of gear *A* which is in contact with gear *C* at point *l*.

15.201 Several rods are brazed together to form the robotic guide arm shown which is attached to a ball-and-socket joint at O. Rod OA slides in a straight inclined slot while rod OB slides in a slot parallel to the z axis. Knowing that at the instant shown $\mathbf{v}_B = (9 \text{ in./s})\mathbf{k}$, determine (a) the angular velocity of the guide arm, (b) the velocity of point A, (c) the velocity of point C.

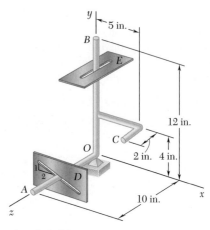

Fig. P15.201

15.202 In Prob. 15.201, the speed of point B is known to be constant. For the position shown, determine (a) the angular acceleration of the guide arm, (b) the acceleration of point C.

15.203 Rod AB of length 25 in. is connected by ball-and-socket joints to collars A and B, which slide along the two rods shown. Knowing that collar B moves toward point E at a constant speed of 20 in./s, determine the velocity of collar A as collar B passes through point D.

15.204 Rod AB, of length 11 in., is connected by ball-and-socket joints to collars A and B, which slide along the two rods shown. Knowing that collar B moves downward at a constant speed of 54 in./s, determine the velocity of collar A when $c = 2$ in.

Fig. P15.203

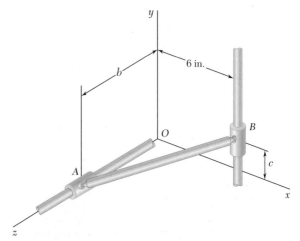

Fig. P15.204

15.205 Rod BC and BD are each 840 mm long and are connected by ball-and-socket joints to collars which may slide on the fixed rods shown. Knowing that collar B moves toward A at a constant speed of 390 mm/s, determine the velocity of collar C for the position shown.

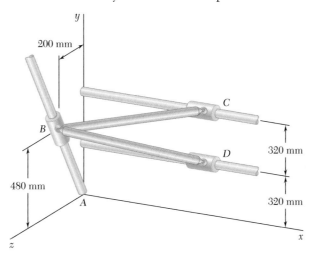

Fig. P15.205

15.206 Rod AB is connected by ball-and-socket joints to collar A and to the 16-in.-diameter disk C. Knowing that disk C rotates counterclockwise at the constant rate $\omega_0 = 3$ rad/s in the zx plane, determine the velocity of collar A for the position shown.

15.207 Rod AB of length 29 in. is connected by ball-and-socket joints to the rotating crank BC and to the collar A. Crank BC is of length 8 in. and rotates in the horizontal xy plane at the constant rate $\omega_0 = 10$ rad/s. At the instant shown, when crank BC is parallel to the z axis, determine the velocity of collar A.

Fig. P15.206

Fig. P15.207

Fig. P15.208 and P15.209

Fig. P15.212

Fig. P15.213

15.208 Rod *AB* of length 300 mm is connected by ball-and-socket joints to collars *A* and *B*, which slide along the two rods shown. Knowing that collar *B* moves toward point *D* at a constant speed of 50 mm/s, determine the velocity of collar *A* when $c = 80$ mm.

15.209 Rod *AB* of length 300 mm is connected by ball-and-socket joints to collars *A* and *B*, which slide along the two rods shown. Knowing that collar *B* moves toward point *D* at a constant speed of 50 mm/s, determine the velocity of collar *A* when $c = 120$ mm.

15.210 Two shafts *AC* and *EG*, which lie in the vertical *yz* plane, are connected by a universal joint at *D*. Shaft *AC* rotates with a constant angular velocity $\boldsymbol{\omega}_1$ as shown. At a time when the arm of the crosspiece attached to shaft *AC* is vertical, determine the angular velocity of shaft *EG*.

Fig. P15.210

15.211 Solve Prob. 15.210, assuming that the arm of the crosspiece attached to shaft *AC* is horizontal.

15.212 In Prob. 15.206, the ball-and-socket joint between the rod and collar *A* is replaced by the clevis shown. Determine (*a*) the angular velocity of the rod, (*b*) the velocity of collar *A*.

15.213 In Prob. 15.205, the ball-and-socket joint between the rod and collar *C* is replaced by the clevis connection shown. Determine (*a*) the angular velocity of the rod, (*b*) the velocity of collar *C*.

15.214 In Prob. 15.204, determine the acceleration of collar *A* when $c = 2$ in.

***15.215** In Prob. 15.205, determine the acceleration of collar *C*.

15.216 In Prob. 15.206, determine the acceleration of collar *A*.

15.217 In Prob. 15.207, determine the acceleration of collar *A*.

15.218 In Prob. 15.208, determine the acceleration of collar *A*.

15.219 In Prob. 15.209, determine the acceleration of collar *A*.

15.14   Three-Dimensional Motion of a Particle   **1013**
Relative to a Rotating Frame. Coriolis
Acceleration

*15.14   THREE-DIMENSIONAL MOTION OF A PARTICLE RELATIVE TO A ROTATING FRAME. CORIOLIS ACCELERATION

We saw in Sec. 15.10 that given a vector function $\mathbf{Q}(t)$ and two frames of reference centered at O—a fixed frame $OXYZ$ and a rotating frame $Oxyz$—the rates of change of $\mathbf{Q}$ with respect to the two frames satisfy the relation

$$(\dot{\mathbf{Q}})_{OXYZ} = (\dot{\mathbf{Q}})_{Oxyz} + \mathbf{\Omega} \times \mathbf{Q} \qquad (15.31)$$

We had assumed at the time that the frame $Oxyz$ was constrained to rotate about a fixed axis OA. However, the derivation given in Sec. 15.10 remains valid when the frame $Oxyz$ is constrained only to have a fixed point O. Under this more general assumption, the axis OA represents the *instantaneous* axis of rotation of the frame $Oxyz$ (Sec. 15.12) and the vector $\mathbf{\Omega}$, its angular velocity at the instant considered (Fig. 15.36).

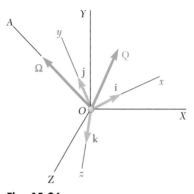

Fig. 15.36

Let us now consider the three-dimensional motion of a particle P relative to a rotating frame $Oxyz$ constrained to have a fixed origin O. Let $\mathbf{r}$ be the position vector of P at a given instant and $\mathbf{\Omega}$ be the angular velocity of the frame $Oxyz$ with respect to the fixed frame $OXYZ$ at the same instant (Fig. 15.37). The derivations given in Sec. 15.11 for the two-dimensional motion of a particle can be readily extended to the three-dimensional case, and the absolute velocity $\mathbf{v}_P$ of P (i.e., its velocity with respect to the fixed frame $OXYZ$) can be expressed as

$$\mathbf{v}_P = \mathbf{\Omega} \times \mathbf{r} + (\dot{\mathbf{r}})_{Oxyz} \qquad (15.45)$$

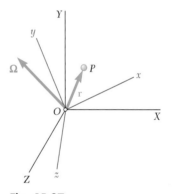

Fig. 15.37

Denoting by $\mathscr{F}$ the rotating frame $Oxyz$, we write this relation in the alternative form

$$\mathbf{v}_P = \mathbf{v}_{P'} + \mathbf{v}_{P/\mathscr{F}} \qquad (15.46)$$

where $\mathbf{v}_P$ = absolute velocity of particle P
$\mathbf{v}_{P'}$ = velocity of point P' of moving frame $\mathscr{F}$ coinciding with P
$\mathbf{v}_{P/\mathscr{F}}$ = velocity of P relative to moving frame $\mathscr{F}$

The absolute acceleration $\mathbf{a}_P$ of P can be expressed as

$$\mathbf{a}_P = \dot{\mathbf{\Omega}} \times \mathbf{r} + \mathbf{\Omega} \times (\mathbf{\Omega} \times \mathbf{r}) + 2\mathbf{\Omega} \times (\dot{\mathbf{r}})_{Oxyz} + (\ddot{\mathbf{r}})_{Oxyz} \qquad (15.47)$$

An alternative form is

$$\mathbf{a}_P = \mathbf{a}_{P'} + \mathbf{a}_{P/\mathscr{F}} + \mathbf{a}_c \qquad (15.48)$$

where $\mathbf{a}_P$ = absolute acceleration of particle P

$\mathbf{a}_{P'}$ = acceleration of point P' of moving frame $\mathcal{F}$ coinciding with P

$\mathbf{a}_{P/\mathcal{F}}$ = acceleration of P relative to moving frame $\mathcal{F}$

$\mathbf{a}_c = 2\mathbf{\Omega} \times (\dot{\mathbf{r}})_{Oxyz} = 2\mathbf{\Omega} \times \mathbf{v}_{P/\mathcal{F}}$
= complementary, or Coriolis, acceleration†

We note that the Coriolis acceleration is perpendicular to the vectors $\mathbf{\Omega}$ and $\mathbf{v}_{P/\mathcal{F}}$. However, since these vectors are usually not perpendicular to each other, the magnitude of $\mathbf{a}_c$ is in general *not* equal to $2\Omega v_{P/\mathcal{F}}$, as was the case for the plane motion of a particle. We further note that the Coriolis acceleration reduces to zero when the vectors $\mathbf{\Omega}$ and $\mathbf{v}_{P/\mathcal{F}}$ are parallel, or when either of them is zero.

Rotating frames of reference are particularly useful in the study of the three-dimensional motion of rigid bodies. If a rigid body has a fixed point O, as was the case for the crane of Sample Prob. 15.11, we can use a frame $Oxyz$ which is neither fixed nor rigidly attached to the rigid body. Denoting by $\mathbf{\Omega}$ the angular velocity of the frame $Oxyz$, we then resolve the angular velocity $\boldsymbol{\omega}$ of the body into the components $\mathbf{\Omega}$ and $\boldsymbol{\omega}_{B/\mathcal{F}}$, where the second component represents the angular velocity of the body relative to the frame $Oxyz$ (see Sample Prob. 15.14). An appropriate choice of the rotating frame often leads to a simpler analysis of the motion of the rigid body than would be possible with axes of fixed orientation. This is especially true in the case of the general three-dimensional motion of a rigid body, i.e., when the rigid body under consideration has no fixed point (see Sample Prob. 15.15).

*15.15 FRAME OF REFERENCE IN GENERAL MOTION

Consider a fixed frame of reference $OXYZ$ and a frame $Axyz$ which moves in a known, but arbitrary, fashion with respect to $OXYZ$ (Fig. 15.38). Let P be a particle moving in space. The position of P is defined at any instant by the vector $\mathbf{r}_P$ in the fixed frame, and by the vector $\mathbf{r}_{P/A}$ in the moving frame. Denoting by $\mathbf{r}_A$ the position vector of A in the fixed frame, we have

$$\mathbf{r}_P = \mathbf{r}_A + \mathbf{r}_{P/A} \tag{15.49}$$

The absolute velocity $\mathbf{v}_P$ of the particle is obtained by writing

$$\mathbf{v}_P = \dot{\mathbf{r}}_P = \dot{\mathbf{r}}_A + \dot{\mathbf{r}}_{P/A} \tag{15.50}$$

where the derivatives are defined with respect to the fixed frame $OXYZ$. The first term in the right-hand member of (15.50) thus represents the velocity $\mathbf{v}_A$ of the origin A of the moving axes. On the other hand, since the rate of change of a vector is the same with respect to a fixed frame and with respect to a frame in translation (Sec. 11.10), the second term can be regarded as the velocity $\mathbf{v}_{P/A}$ of

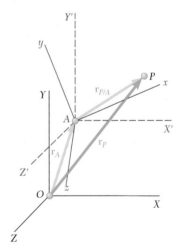

Fig. 15.38

---

†It is important to note the difference between Eq. (15.48) and Eq. (15.21) of Sec. 15.8. See the footnote on page 988.

P relative to the frame $AX'Y'Z'$ of the same orientation as $OXYZ$ and the same origin as $Axyz$. We therefore have

$$\mathbf{v}_P = \mathbf{v}_A + \mathbf{v}_{P/A} \tag{15.51}$$

But the velocity $\mathbf{v}_{P/A}$ of P relative to $AX'Y'Z'$ can be obtained from (15.45) by substituting $\mathbf{r}_{P/A}$ for $\mathbf{r}$ in that equation. We write

$$\mathbf{v}_P = \mathbf{v}_A + \mathbf{\Omega} \times \mathbf{r}_{P/A} + (\dot{\mathbf{r}}_{P/A})_{Axyz} \tag{15.52}$$

where $\mathbf{\Omega}$ is the angular velocity of the frame $Axyz$ at the instant considered.

The absolute acceleration $\mathbf{a}_P$ of the particle is obtained by differentiating (15.51) and writing

$$\mathbf{a}_P = \dot{\mathbf{v}}_P = \dot{\mathbf{v}}_A + \dot{\mathbf{v}}_{P/A} \tag{15.53}$$

where the derivatives are defined with respect to either of the frames $OXYZ$ or $AX'Y'Z'$. Thus, the first term in the right-hand member of (15.53) represents the acceleration $\mathbf{a}_A$ of the origin A of the moving axes and the second term represents the acceleration $\mathbf{a}_{P/A}$ of P relative to the frame $AX'Y'Z'$. This acceleration can be obtained from (15.47) by substituting $\mathbf{r}_{P/A}$ for $\mathbf{r}$. We therefore write

$$\mathbf{a}_P = \mathbf{a}_A + \dot{\mathbf{\Omega}} \times \mathbf{r}_{P/A} + \mathbf{\Omega} \times (\mathbf{\Omega} \times \mathbf{r}_{P/A}) \\ + 2\mathbf{\Omega} \times (\dot{\mathbf{r}}_{P/A})_{Axyz} + (\ddot{\mathbf{r}}_{P/A})_{Axyz} \tag{15.54}$$

Formulas (15.52) and (15.54) make it possible to determine the velocity and acceleration of a given particle with respect to a fixed frame of reference, when the motion of the particle is known with respect to a moving frame. These formulas become more significant, and considerably easier to remember, if we note that the sum of the first two terms in (15.52) represents the velocity of the point P' of the moving frame which coincides with P at the instant considered, and that the sum of the first three terms in (15.54) represents the acceleration of the same point. Thus, the relations (15.46) and (15.48) of the preceding section are still valid in the case of a reference frame in general motion, and we can write

$$\mathbf{v}_P = \mathbf{v}_{P'} + \mathbf{v}_{P/\mathcal{F}} \tag{15.46}$$
$$\mathbf{a}_P = \mathbf{a}_{P'} + \mathbf{a}_{P/\mathcal{F}} + \mathbf{a}_c \tag{15.48}$$

where the various vectors involved have been defined in Sec. 15.14.

It should be noted that if the moving reference frame $\mathcal{F}$ (or $Axyz$) is in translation, the velocity and acceleration of the point P' of the frame which coincides with P become, respectively, equal to the velocity and acceleration of the origin A of the frame. On the other hand, since the frame maintains a fixed orientation, $\mathbf{a}_c$ is zero, and the relations (15.46) and (15.48) reduce, respectively, to the relations (11.33) and (11.34) derived in Sec. 11.12.

Photo 15.9 The motion of air particles in a hurricane can be considered as motion relative to a frame of reference attached to the Earth and rotating with it.

SAMPLE PROBLEM 15.13

The bent rod OAB rotates about the vertical OB. At the instant considered, its angular velocity and angular acceleration are, respectively, 20 rad/s and 200 rad/s², both clockwise when viewed from the positive Y axis. The collar D moves along the rod, and at the instant considered, $OD = 8$ in. The velocity and acceleration of the collar relative to the rod are, respectively, 50 in./s and 600 in./s², both upward. Determine (a) the velocity of the collar, (b) the acceleration of the collar.

SOLUTION

Frames of Reference. The frame $OXYZ$ is fixed. We attach the rotating frame $Oxyz$ to the bent rod. Its angular velocity and angular acceleration relative to $OXYZ$ are therefore $\mathbf{\Omega} = (-20 \text{ rad/s})\mathbf{j}$ and $\dot{\mathbf{\Omega}} = (-200 \text{ rad/s}^2)\mathbf{j}$, respectively. The position vector of D is

$$\mathbf{r} = (8 \text{ in.})(\sin 30°\mathbf{i} + \cos 30°\mathbf{j}) = (4 \text{ in.})\mathbf{i} + (6.93 \text{ in.})\mathbf{j}$$

a. Velocity $\mathbf{v}_D$. Denoting by D' the point of the rod which coincides with D and by $\mathscr{F}$ the rotating frame $Oxyz$, we write from Eq. (15.46)

$$\mathbf{v}_D = \mathbf{v}_{D'} + \mathbf{v}_{D/\mathscr{F}} \qquad (1)$$

where

$$\mathbf{v}_{D'} = \mathbf{\Omega} \times \mathbf{r} = (-20 \text{ rad/s})\mathbf{j} \times [(4 \text{ in.})\mathbf{i} + (6.93 \text{ in.})\mathbf{j}] = (80 \text{ in./s})\mathbf{k}$$
$$\mathbf{v}_{D/\mathscr{F}} = (50 \text{ in./s})(\sin 30°\mathbf{i} + \cos 30°\mathbf{j}) = (25 \text{ in./s})\mathbf{i} + (43.3 \text{ in./s})\mathbf{j}$$

Substituting the values obtained for $\mathbf{v}_{D'}$ and $\mathbf{v}_{D/\mathscr{F}}$ into (1), we find

$$\mathbf{v}_D = (25 \text{ in./s})\mathbf{i} + (43.3 \text{ in./s})\mathbf{j} + (80 \text{ in./s})\mathbf{k} \qquad \blacktriangleleft$$

b. Acceleration $\mathbf{a}_D$. From Eq. (15.48) we write

$$\mathbf{a}_D = \mathbf{a}_{D'} + \mathbf{a}_{D/\mathscr{F}} + \mathbf{a}_c \qquad (2)$$

where

$$\begin{aligned}
\mathbf{a}_{D'} &= \dot{\mathbf{\Omega}} \times \mathbf{r} + \mathbf{\Omega} \times (\mathbf{\Omega} \times \mathbf{r}) \\
&= (-200 \text{ rad/s}^2)\mathbf{j} \times [(4 \text{ in.})\mathbf{i} + (6.93 \text{ in.})\mathbf{j}] - (20 \text{ rad/s})\mathbf{j} \times (80 \text{ in./s})\mathbf{k} \\
&= +(800 \text{ in./s}^2)\mathbf{k} - (1600 \text{ in./s}^2)\mathbf{i} \\
\mathbf{a}_{D/\mathscr{F}} &= (600 \text{ in./s}^2)(\sin 30°\mathbf{i} + \cos 30°\mathbf{j}) = (300 \text{ in./s}^2)\mathbf{i} + (520 \text{ in./s}^2)\mathbf{j} \\
\mathbf{a}_c &= 2\mathbf{\Omega} \times \mathbf{v}_{D/\mathscr{F}} \\
&= 2(-20 \text{ rad/s})\mathbf{j} \times [(25 \text{ in./s})\mathbf{i} + (43.3 \text{ in./s})\mathbf{j}] = (1000 \text{ in./s}^2)\mathbf{k}
\end{aligned}$$

Substituting the values obtained for $\mathbf{a}_{D'}$, $\mathbf{a}_{D/\mathscr{F}}$, and $\mathbf{a}_c$ into (2),

$$\mathbf{a}_D = -(1300 \text{ in./s}^2)\mathbf{i} + (520 \text{ in./s}^2)\mathbf{j} + (1800 \text{ in./s}^2)\mathbf{k} \qquad \blacktriangleleft$$

SAMPLE PROBLEM 15.14

The crane shown rotates with a constant angular velocity $\boldsymbol{\omega}_1$ of 0.30 rad/s. Simultaneously, the boom is being raised with a constant angular velocity $\boldsymbol{\omega}_2$ of 0.50 rad/s relative to the cab. Knowing that the length of the boom OP is $l = 12$ m, determine (a) the velocity of the tip of the boom, (b) the acceleration of the tip of the boom.

SOLUTION

Frames of Reference. The frame $OXYZ$ is fixed. We attach the rotating frame $Oxyz$ to the cab. Its angular velocity with respect to the frame $OXYZ$ is therefore $\boldsymbol{\Omega} = \boldsymbol{\omega}_1 = (0.30 \text{ rad/s})\mathbf{j}$. The angular velocity of the boom relative to the cab and the rotating frame $Oxyz$ (or $\mathscr{F}$, for short) is $\boldsymbol{\omega}_{B/\mathscr{F}} = \boldsymbol{\omega}_2 = (0.50 \text{ rad/s})\mathbf{k}$.

a. Velocity $\mathbf{v}_P$. From Eq. (15.46) we write

$$\mathbf{v}_P = \mathbf{v}_{P'} + \mathbf{v}_{P/\mathscr{F}} \tag{1}$$

where $\mathbf{v}_{P'}$ is the velocity of the point P' of the rotating frame which coincides with P:

$$\mathbf{v}_{P'} = \boldsymbol{\Omega} \times \mathbf{r} = (0.30 \text{ rad/s})\mathbf{j} \times [(10.39 \text{ m})\mathbf{i} + (6 \text{ m})\mathbf{j}] = -(3.12 \text{ m/s})\mathbf{k}$$

and where $\mathbf{v}_{P/\mathscr{F}}$ is the velocity of P relative to the rotating frame $Oxyz$. But the angular velocity of the boom relative to $Oxyz$ was found to be $\boldsymbol{\omega}_{B/\mathscr{F}} = (0.50 \text{ rad/s})\mathbf{k}$. The velocity of its tip P relative to $Oxyz$ is therefore

$$\mathbf{v}_{P/\mathscr{F}} = \boldsymbol{\omega}_{B/\mathscr{F}} \times \mathbf{r} = (0.50 \text{ rad/s})\mathbf{k} \times [(10.39 \text{ m})\mathbf{i} + (6 \text{ m})\mathbf{j}]$$
$$= -(3 \text{ m/s})\mathbf{i} + (5.20 \text{ m/s})\mathbf{j}$$

Substituting the values obtained for $\mathbf{v}_{P'}$ and $\mathbf{v}_{P/\mathscr{F}}$ into (1), we find

$$\mathbf{v}_P = -(3 \text{ m/s})\mathbf{i} + (5.20 \text{ m/s})\mathbf{j} - (3.12 \text{ m/s})\mathbf{k} \quad \blacktriangleleft$$

b. Acceleration $\mathbf{a}_P$. From Eq. (15.48) we write

$$\mathbf{a}_P = \mathbf{a}_{P'} + \mathbf{a}_{P/\mathscr{F}} + \mathbf{a}_c \tag{2}$$

Since $\boldsymbol{\Omega}$ and $\boldsymbol{\omega}_{B/\mathscr{F}}$ are both constant, we have

$$\mathbf{a}_{P'} = \boldsymbol{\Omega} \times (\boldsymbol{\Omega} \times \mathbf{r}) = (0.30 \text{ rad/s})\mathbf{j} \times (-3.12 \text{ m/s})\mathbf{k} = -(0.94 \text{ m/s}^2)\mathbf{i}$$
$$\mathbf{a}_{P/\mathscr{F}} = \boldsymbol{\omega}_{B/\mathscr{F}} \times (\boldsymbol{\omega}_{B/\mathscr{F}} \times \mathbf{r})$$
$$= (0.50 \text{ rad/s})\mathbf{k} \times [-(3 \text{ m/s})\mathbf{i} + (5.20 \text{ m/s})\mathbf{j}]$$
$$= -(1.50 \text{ m/s}^2)\mathbf{j} - (2.60 \text{ m/s}^2)\mathbf{i}$$
$$\mathbf{a}_c = 2\boldsymbol{\Omega} \times \mathbf{v}_{P/\mathscr{F}}$$
$$= 2(0.30 \text{ rad/s})\mathbf{j} \times [-(3 \text{ m/s})\mathbf{i} + (5.20 \text{ m/s})\mathbf{j}] = (1.80 \text{ m/s}^2)\mathbf{k}$$

Substituting for $\mathbf{a}_{P'}$, $\mathbf{a}_{P/\mathscr{F}}$, and $\mathbf{a}_c$ into (2), we find

$$\mathbf{a}_P = -(3.54 \text{ m/s}^2)\mathbf{i} - (1.50 \text{ m/s}^2)\mathbf{j} + (1.80 \text{ m/s}^2)\mathbf{k} \quad \blacktriangleleft$$

SAMPLE PROBLEM 15.15

Disk D, of radius R, is pinned to end A of the arm OA of length L located in the plane of the disk. The arm rotates about a vertical axis through O at the constant rate ω_1, and the disk rotates about A at the constant rate ω_2. Determine (a) the velocity of point P located directly above A, (b) the acceleration of P, (c) the angular velocity and angular acceleration of the disk.

SOLUTION

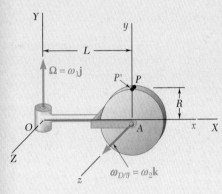

Frames of Reference. The frame $OXYZ$ is fixed. We attach the moving frame $Axyz$ to the arm OA. Its angular velocity with respect to the frame $OXYZ$ is therefore $\boldsymbol{\Omega} = \omega_1 \mathbf{j}$. The angular velocity of disk D relative to the moving frame $Axyz$ (or $\mathscr{F}$, for short) is $\boldsymbol{\omega}_{D/\mathscr{F}} = \omega_2 \mathbf{k}$. The position vector of P relative to O is $\mathbf{r} = L\mathbf{i} + R\mathbf{j}$, and its position vector relative to A is $\mathbf{r}_{P/A} = R\mathbf{j}$.

a. Velocity $\mathbf{v}_P$. Denoting by P' the point of the moving frame which coincides with P, we write from Eq. (15.46)

$$\mathbf{v}_P = \mathbf{v}_{P'} + \mathbf{v}_{P/\mathscr{F}} \qquad (1)$$

where $\mathbf{v}_{P'} = \boldsymbol{\Omega} \times \mathbf{r} = \omega_1 \mathbf{j} \times (L\mathbf{i} + R\mathbf{j}) = -\omega_1 L\mathbf{k}$

$$\mathbf{v}_{P/\mathscr{F}} = \boldsymbol{\omega}_{D/\mathscr{F}} \times \mathbf{r}_{P/A} = \omega_2 \mathbf{k} \times R\mathbf{j} = -\omega_2 R\mathbf{i}$$

Substituting the values obtained for $\mathbf{v}_{P'}$ and $\mathbf{v}_{P/\mathscr{F}}$ into (1), we find

$$\mathbf{v}_P = -\omega_2 R\mathbf{i} - \omega_1 L\mathbf{k} \qquad \blacktriangleleft$$

b. Acceleration $\mathbf{a}_P$. From Eq. (15.48) we write

$$\mathbf{a}_P = \mathbf{a}_{P'} + \mathbf{a}_{P/\mathscr{F}} + \mathbf{a}_c \qquad (2)$$

Since $\boldsymbol{\Omega}$ and $\boldsymbol{\omega}_{D/\mathscr{F}}$ are both constant, we have

$$\mathbf{a}_{P'} = \boldsymbol{\Omega} \times (\boldsymbol{\Omega} \times \mathbf{r}) = \omega_1 \mathbf{j} \times (-\omega_1 L\mathbf{k}) = -\omega_1^2 L\mathbf{i}$$
$$\mathbf{a}_{P/\mathscr{F}} = \boldsymbol{\omega}_{D/\mathscr{F}} \times (\boldsymbol{\omega}_{D/\mathscr{F}} \times \mathbf{r}_{P/A}) = \omega_2 \mathbf{k} \times (-\omega_2 R\mathbf{i}) = -\omega_2^2 R\mathbf{j}$$
$$\mathbf{a}_c = 2\boldsymbol{\Omega} \times \mathbf{v}_{P/\mathscr{F}} = 2\omega_1 \mathbf{j} \times (-\omega_2 R\mathbf{i}) = 2\omega_1 \omega_2 R\mathbf{k}$$

Substituting the values obtained into (2), we find

$$\mathbf{a}_P = -\omega_1^2 L\mathbf{i} - \omega_2^2 R\mathbf{j} + 2\omega_1 \omega_2 R\mathbf{k} \qquad \blacktriangleleft$$

c. Angular Velocity and Angular Acceleration of Disk.

$$\boldsymbol{\omega} = \boldsymbol{\Omega} + \boldsymbol{\omega}_{D/\mathscr{F}} \qquad\qquad \boldsymbol{\omega} = \omega_1 \mathbf{j} + \omega_2 \mathbf{k} \qquad \blacktriangleleft$$

Using Eq. (15.31) with $\mathbf{Q} = \boldsymbol{\omega}$, we write

$$\boldsymbol{\alpha} = (\dot{\boldsymbol{\omega}})_{OXYZ} = (\dot{\boldsymbol{\omega}})_{Axyz} + \boldsymbol{\Omega} \times \boldsymbol{\omega}$$
$$= 0 + \omega_1 \mathbf{j} \times (\omega_1 \mathbf{j} + \omega_2 \mathbf{k})$$
$$\boldsymbol{\alpha} = \omega_1 \omega_2 \mathbf{i} \qquad \blacktriangleleft$$

SOLVING PROBLEMS
ON YOUR OWN

In this lesson you concluded your study of the kinematics of rigid bodies by learning how to use an auxiliary frame of reference $\mathscr{F}$ to analyze the three-dimensional motion of a rigid body. This auxiliary frame may be a *rotating frame* with a fixed origin O, or it may be a *frame in general motion*.

A. Using a rotating frame of reference. As you approach a problem involving the use of a rotating frame $\mathscr{F}$ you should take the following steps.

1. Select the rotating frame $\mathscr{F}$ that you wish to use and draw the corresponding coordinate axes x, y, and z from the fixed point O.

2. Determine the angular velocity $\mathbf{\Omega}$ of the frame $\mathscr{F}$ with respect to a fixed frame $OXYZ$. In most cases, you will have selected a frame which is attached to some rotating element of the system; $\mathbf{\Omega}$ will then be the angular velocity of that element.

3. Designate as P' the point of the rotating frame $\mathscr{F}$ that coincides with the point P of interest at the instant you are considering. Determine the velocity $\mathbf{v}_{P'}$ and the acceleration $\mathbf{a}_{P'}$ of point P'. Since P' is part of $\mathscr{F}$ and has the same position vector $\mathbf{r}$ as P, you will find that

$$\mathbf{v}_{P'} = \mathbf{\Omega} \times \mathbf{r} \qquad \text{and} \qquad \mathbf{a}_{P'} = \boldsymbol{\alpha} \times \mathbf{r} + \mathbf{\Omega} \times (\mathbf{\Omega} \times \mathbf{r})$$

where $\boldsymbol{\alpha}$ is the angular acceleration of $\mathscr{F}$. However, in many of the problems that you will encounter, the angular velocity of $\mathscr{F}$ is constant in both magnitude and direction, and $\boldsymbol{\alpha} = 0$.

4. Determine the velocity and acceleration of point P with respect to the frame $\mathscr{F}$. As you are trying to determine $\mathbf{v}_{P/\mathscr{F}}$ and $\mathbf{a}_{P/\mathscr{F}}$ you will find it useful to visualize the motion of P on frame $\mathscr{F}$ when the frame is not rotating. If P is a point of a rigid body $\mathscr{B}$ which has an angular velocity $\boldsymbol{\omega}_{\mathscr{B}}$ and an angular acceleration $\boldsymbol{\alpha}_{\mathscr{B}}$ relative to $\mathscr{F}$ [Sample Prob. 15.14], you will find that

$$\mathbf{v}_{P/\mathscr{F}} = \boldsymbol{\omega}_{\mathscr{B}} \times \mathbf{r} \qquad \text{and} \qquad \mathbf{a}_{P/\mathscr{F}} = \boldsymbol{\alpha}_{\mathscr{B}} \times \mathbf{r} + \boldsymbol{\omega}_{\mathscr{B}} \times (\boldsymbol{\omega}_{\mathscr{B}} \times \mathbf{r})$$

In many of the problems that you will encounter, the angular velocity of body $\mathscr{B}$ relative to frame $\mathscr{F}$ is constant in both magnitude and direction, and $\boldsymbol{\alpha}_{\mathscr{B}} = 0$.

5. Determine the Coriolis acceleration. Considering the angular velocity $\mathbf{\Omega}$ of frame $\mathscr{F}$ and the velocity $\mathbf{v}_{P/\mathscr{F}}$ of point P relative to that frame, which was computed in the previous step, you write

$$\mathbf{a}_c = 2\mathbf{\Omega} \times \mathbf{v}_{P/\mathscr{F}}$$

(continued)

6. The velocity and the acceleration of P with respect to the fixed frame OXYZ can now be obtained by adding the expressions you have determined:

$$\mathbf{v}_P = \mathbf{v}_{P'} + \mathbf{v}_{P/\mathcal{F}} \tag{15.46}$$

$$\mathbf{a}_P = \mathbf{a}_{P'} + \mathbf{a}_{P/\mathcal{F}} + \mathbf{a}_c \tag{15.48}$$

B. Using a frame of reference in general motion. The steps that you will take differ only slightly from those listed under A. They consist of the following:

1. Select the frame $\mathcal{F}$ that you wish to use and a reference point A in that frame, from which you will draw the coordinate axes, x, y, and z defining that frame. You will consider the motion of the frame as the sum of a *translation with A and a rotation about A.*

2. Determine the velocity $\mathbf{v}_A$ of point A and the angular velocity Ω of the frame. In most cases, you will have selected a frame which is attached to some element of the system; Ω will then be the angular velocity of that element.

3. Designate as P' the point of frame $\mathcal{F}$ that coincides with the point P of interest at the instant you are considering, and determine the velocity $\mathbf{v}_{P'}$ and the acceleration $\mathbf{a}_{P'}$ of that point. In some cases, this can be done by visualizing the motion of P if that point were prevented from moving with respect to $\mathcal{F}$ [Sample Prob. 15.15]. A more general approach is to recall that the motion of P' is the sum of a translation with the reference point A and a rotation about A. The velocity $\mathbf{v}_{P'}$ and the acceleration $\mathbf{a}_{P'}$ of P', therefore, can be obtained by adding $\mathbf{v}_A$ and $\mathbf{a}_A$, respectively, to the expressions found in paragraph A3 and replacing the position vector $\mathbf{r}$ by the vector $\mathbf{r}_{P/A}$ drawn from A to P:

$$\mathbf{v}_{P'} = \mathbf{v}_A + \Omega \times \mathbf{r}_{P/A} \qquad \mathbf{a}_{P'} = \mathbf{a}_A + \boldsymbol{\alpha} \times \mathbf{r}_{P/A} + \Omega \times (\Omega \times \mathbf{r}_{P/A})$$

Steps 4, 5, and 6 are the same as in Part A, except that the vector $\mathbf{r}$ should again be replaced by $\mathbf{r}_{P/A}$. Thus, Eqs. (15.46) and (15.48) can still be used to obtain the velocity and the acceleration of P with respect to the fixed frame of reference $OXYZ$.

PROBLEMS

END-OF-SECTION PROBLEMS

15.220 A square plate of side 18 in. is hinged at A and B to a clevis. The plate rotates at the constant rate $\omega_2 = 4$ rad/s with respect to the clevis, which itself rotates at the constant rate $\omega_1 = 3$ rad/s about the Y axis. For the position shown, determine (a) the velocity of point C, (b) the acceleration of point C.

15.221 A square plate of side 18 in. is hinged at A and B to a clevis. The plate rotates at the constant rate $\omega_2 = 4$ rad/s with respect to the clevis, which itself rotates at the constant rate $\omega_1 = 3$ rad/s about the Y axis. For the position shown, determine (a) the velocity of corner D, (b) the acceleration of corner D.

15.222 and 15.223 The rectangular plate shown rotates at the constant rate $\omega_2 = 12$ rad/s with respect to arm AE, which itself rotates at the constant rate $\omega_1 = 9$ rad/s about the Z axis. For the position shown, determine the velocity and acceleration of the point of the plate indicated.

    15.222 Corner B
    15.223 Corner C

Fig. P15.220 and P15.221

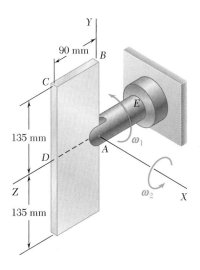

Fig. P15.222 and P15.223

15.224 Rod AB is welded to the 0.3-m-radius plate which rotates at the constant rate $\omega_1 = 6$ rad/s. Knowing that collar D moves toward end B of the rod at a constant speed $u = 1.3$ m/s, determine, for the position shown, (a) the velocity of D, (b) the acceleration of D.

Fig. P15.224

Fig. P15.225

Fig. P15.227

15.225 The bent rod ABC rotates at the constant rate $\omega_1 = 4$ rad/s. Knowing that collar D moves downward along the rod at a constant relative speed $u = 65$ in./s, determine, for the position shown, (a) the velocity of D, (b) the acceleration of D.

15.226 The bent pipe shown rotates at the constant rate $\omega_1 = 10$ rad/s. Knowing that a ball bearing D moves in portion BC of the pipe toward end C at a constant relative speed $u = 2$ ft/s, determine at the instant shown (a) the velocity of D, (b) the acceleration of D.

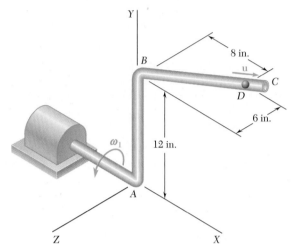

Fig. P15.226

15.227 The circular plate shown rotates about its vertical diameter at the constant rate $\omega_1 = 10$ rad/s. Knowing that in the position shown the disk lies in the XY plane and point D of strap CD moves upward at a constant relative speed $u = 1.5$ m/s, determine (a) the velocity of D, (b) the acceleration of D.

15.228 Manufactured items are spray-painted as they pass through the automated work station shown. Knowing that the bent pipe ACE rotates at the constant rate $\omega_1 = 0.4$ rad/s and that at point D the paint moves through the pipe at a constant relative speed $u = 150$ mm/s, determine, for the position shown, (a) the velocity of the paint at D, (b) the acceleration of the paint at D.

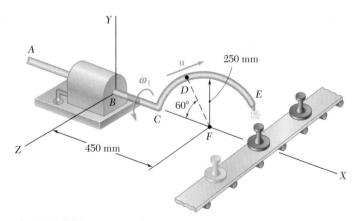

Fig. P15.228

15.229 Solve Prob. 15.227, assuming that at the instant shown the angular velocity ω_1 of the plate is 10 rad/s and is decreasing at the rate of 25 rad/s^2, while the relative speed u of point D of strap CD is 1.5 m/s and is decreasing at the rate of 3 m/s^2.

15.230 Solve Prob. 15.226 assuming that at the instant shown the angular velocity ω_1 of the pipe is 10 rad/s and is decreasing at the rate of 15 rad/s^2, while the relative speed u of the ball bearing is 2 ft/s and is increasing at the rate of 10 ft/s^2.

15.231 Using the method of Sec. 15.14, solve Prob. 15.192.

15.232 Using the method of Sec. 15.14, solve Prob. 15.196.

15.233 Using the method of Sec. 15.14, solve Prob. 15.198.

15.234 A disk of radius 120 mm rotates at the constant rate $\omega_2 = 5$ rad/s with respect to the arm AB, which itself rotates at the constant rate $\omega_1 = 3$ rad/s. For the position shown, determine the velocity and acceleration of point C.

15.235 A disk of radius 120 mm rotates at the constant rate $\omega_2 = 5$ rad/s with respect to the arm AB, which itself rotates at the constant rate $\omega_1 = 3$ rad/s. For the position shown, determine the velocity and acceleration of point D.

15.236 The arm AB of length 16 ft is used to provide an elevated platform for construction workers. In the position shown, arm AB is being raised at the constant rate $d\theta/dt = 0.25$ rad/s; simultaneously, the unit is being rotated about the Y axis at the constant rate $\omega_1 = 0.15$ rad/s. Knowing that $\theta = 20°$, determine the velocity and acceleration of point B.

15.237 The remote manipulator system (RMS) shown is used to deploy payloads from the cargo bay of space shuttles. At the instant shown, the whole RMS is rotating at the constant rate $\omega_1 = 0.03$ rad/s about the axis AB. At the same time, portion BCD rotates as a rigid body at the constant rate $\omega_2 = d\beta/dt = 0.04$ rad/s about an axis through B parallel to the X axis. Knowing that $\beta = 30°$, determine (a) the angular acceleration of BCD, (b) the velocity of D, (c) the acceleration of D.

Fig. P15.234 and P15.235

Fig. P15.236

Fig. P15.237

Fig. P15.238

Fig. P15.240 and P15.241

Fig. P15.242 and P15.243

15.238 The body AB and rod BC of the robotic component shown rotate at the constant rate $\omega_1 = 0.60$ rad/s about the Y axis. Simultaneously a wire-and-pulley control causes arm CD to rotate about C at the constant rate $\omega = d\beta/dt = 0.45$ rad/s. Knowing $\beta = 120°$, determine (*a*) the angular acceleration of arm CD, (*b*) the velocity of D, (*c*) the acceleration of D.

15.239 The crane shown rotates at the constant rate $\omega_1 = 0.25$ rad/s; simultaneously, the telescoping boom is being lowered at the constant rate $\omega_2 = 0.40$ rad/s. Knowing that at the instant shown the length of the boom is 20 ft and is increasing at the constant rate $u = 1.5$ ft/s, determine the velocity and acceleration of point B.

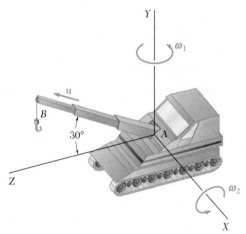

Fig. P15.239

15.240 The vertical plate shown is welded to arm EFG, and the entire unit rotates at the constant rate $\omega_1 = 1.6$ rad/s about the Y axis. At the same time, a continuous link belt moves around the perimeter of the plate at a constant speed $u = 4.5$ in./s. For the position shown, determine the acceleration of the link of the belt located (*a*) at point A, (*b*) at point B.

15.241 The vertical plate shown is welded to arm EFG, and the entire unit rotates at the constant rate $\omega_1 = 1.6$ rad/s about the Y axis. At the same time, a continuous link belt moves around the perimeter of the plate at a constant speed $u = 4.5$ in./s. For the position shown, determine the acceleration of the link of the belt located (*a*) at point C, (*b*) at point D.

15.242 A disk of 180-mm radius rotates at the constant rate $\omega_2 = 12$ rad/s with respect to arm CD, which itself rotates at the constant rate $\omega_1 = 8$ rad/s about the Y axis. Determine at the instant shown the velocity and acceleration of point A on the rim of the disk.

15.243 A disk of 180-mm radius rotates at the constant rate $\omega_2 = 12$ rad/s with respect to arm CD, which itself rotates at the constant rate $\omega_1 = 8$ rad/s about the Y axis. Determine at the instant shown the velocity and acceleration of point B on the rim of the disk.

15.244 A square plate of side $2r$ is welded to a vertical shaft which rotates with a constant angular velocity $\boldsymbol{\omega}_1$. At the same time, rod AB of length r rotates about the center of the plate with a constant angular velocity $\boldsymbol{\omega}_2$ with respect to the plate. For the position of the plate shown, determine the acceleration of end B of the rod if (a) $\theta = 0$, (b) $\theta = 90°$, (c) $\theta = 180°$.

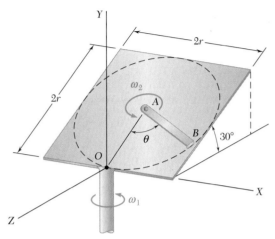

Fig. P15.244

15.245 Two disks, each of 130-mm radius, are welded to the 500-mm rod CD. The rod-and-disks unit rotates at the constant rate $\omega_2 = 3$ rad/s with respect to arm AB. Knowing that at the instant shown $\omega_1 = 4$ rad/s, determine the velocity and acceleration of (a) point E, (b) point F.

15.246 In Prob. 15.245, determine the velocity and acceleration of (a) point G, (b) point H.

15.247 The position of the stylus tip A is controlled by the robot shown. In the position shown, the stylus moves at a constant speed $u = 180$ mm/s relative to the solenoid BC. At the same time, arm CD rotates at the constant rate $\omega_2 = 1.6$ rad/s with respect to component DEG. Knowing that the entire robot rotates about the X axis at the constant rate $\omega_1 = 1.2$ rad/s, determine (a) the velocity of A, (b) the acceleration of A.

Fig. P15.245

Fig. P15.247

REVIEW AND SUMMARY

This chapter was devoted to the study of the kinematics of rigid bodies.

Rigid body in translation

We first considered the *translation* of a rigid body [Sec. 15.2] and observed that in such a motion, *all points of the body have the same velocity and the same acceleration at any given instant.*

Rigid body in rotation about a fixed axis

We next considered the *rotation* of a rigid body about a fixed axis [Sec. 15.3]. The position of the body is defined by the angle θ that the line BP, drawn from the axis of rotation to a point P of the body, forms with a fixed plane (Fig. 15.39). We found that the magnitude of the velocity of P is

Fig. 15.39

$$v = \frac{ds}{dt} = r\dot{\theta} \sin \phi \tag{15.4}$$

where $\dot{\theta}$ is the time derivative of θ. We then expressed the velocity of P as

$$\mathbf{v} = \frac{d\mathbf{r}}{dt} = \boldsymbol{\omega} \times \mathbf{r} \tag{15.5}$$

where the vector

$$\boldsymbol{\omega} = \omega\mathbf{k} = \dot{\theta}\mathbf{k} \tag{15.6}$$

is directed along the fixed axis of rotation and represents the *angular velocity* of the body.

Denoting by $\boldsymbol{\alpha}$ the derivative $d\boldsymbol{\omega}/dt$ of the angular velocity, we expressed the acceleration of P as

$$\mathbf{a} = \boldsymbol{\alpha} \times \mathbf{r} + \boldsymbol{\omega} \times (\boldsymbol{\omega} \times \mathbf{r}) \tag{15.8}$$

Differentiating (15.6), and recalling that $\mathbf{k}$ is constant in magnitude and direction, we found that

$$\boldsymbol{\alpha} = \alpha\mathbf{k} = \dot{\omega}\mathbf{k} = \ddot{\theta}\mathbf{k} \tag{15.9}$$

The vector $\boldsymbol{\alpha}$ represents the *angular acceleration* of the body and is directed along the fixed axis of rotation.

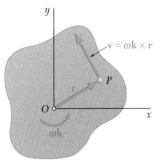

Fig. 15.40   **Fig. 15.41**

Next we considered the motion of a representative slab located in a plane perpendicular to the axis of rotation of the body (Fig. 15.40). Since the angular velocity is perpendicular to the slab, the velocity of a point P of the slab was expressed as

Rotation of a representative slab

$$\mathbf{v} = \omega\mathbf{k} \times \mathbf{r} \qquad (15.10)$$

where $\mathbf{v}$ is contained in the plane of the slab. Substituting $\boldsymbol{\omega} = \omega\mathbf{k}$ and $\boldsymbol{\alpha} = \alpha\mathbf{k}$ into (15.8), we found that the acceleration of P could be resolved into tangential and normal components (Fig. 15.41) respectively equal to

Tangential and normal components

$$\begin{aligned}\mathbf{a}_t &= \alpha\mathbf{k} \times \mathbf{r} & a_t &= r\alpha \\ \mathbf{a}_n &= -\omega^2\mathbf{r} & a_n &= r\omega^2\end{aligned} \qquad (15.11')$$

Recalling Eqs. (15.6) and (15.9), we obtained the following expressions for the *angular velocity* and the *angular acceleration* of the slab [Sec. 15.4]:

Angular velocity and angular acceleration of rotating slab

$$\omega = \frac{d\theta}{dt} \qquad (15.12)$$

$$\alpha = \frac{d\omega}{dt} = \frac{d^2\theta}{dt^2} \qquad (15.13)$$

or

$$\alpha = \omega\frac{d\omega}{d\theta} \qquad (15.14)$$

We noted that these expressions are similar to those obtained in Chap. 11 for the rectilinear motion of a particle.

Two particular cases of rotation are frequently encountered: *uniform rotation* and *uniformly accelerated rotation.* Problems involving either of these motions can be solved by using equations similar to those used in Secs. 11.4 and 11.5 for the uniform rectilinear motion and the uniformly accelerated rectilinear motion of a particle, but where x, v, and a are replaced by θ, ω, and α, respectively [Sample Prob. 15.1].

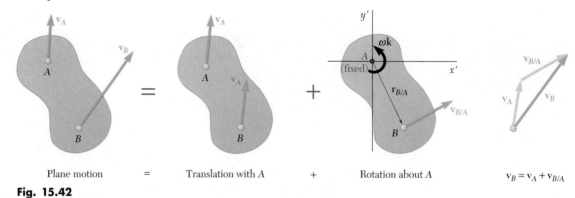

Plane motion   =   Translation with A +   Rotation about A $v_B = v_A + v_{B/A}$

Fig. 15.42

Velocities in plane motion

The *most general plane motion* of a rigid slab can be considered as the *sum of a translation and a rotation* [Sec. 15.5]. For example, the slab shown in Fig. 15.42 can be assumed to translate with point A, while simultaneously rotating about A. It follows [Sec. 15.6] that the velocity of any point B of the slab can be expressed as

$$\mathbf{v}_B = \mathbf{v}_A + \mathbf{v}_{B/A} \tag{15.17}$$

where $\mathbf{v}_A$ is the velocity of A and $\mathbf{v}_{B/A}$ the relative velocity of B with respect to A or, more precisely, with respect to axes $x'y'$ translating with A. Denoting by $\mathbf{r}_{B/A}$ the position vector of B relative to A, we found that

$$\mathbf{v}_{B/A} = \omega\mathbf{k} \times \mathbf{r}_{B/A} \qquad v_{B/A} = r\omega \tag{15.18}$$

The fundamental equation (15.17) relating the absolute velocities of points A and B and the relative velocity of B with respect to A was expressed in the form of a vector diagram and used to solve problems involving the motion of various types of mechanisms [Sample Probs. 15.2 and 15.3].

Instantaneous center of rotation

Another approach to the solution of problems involving the velocities of the points of a rigid slab in plane motion was presented in Sec. 15.7 and used in Sample Probs. 15.4 and 15.5. It is based on the determination of the *instantaneous center of rotation C* of the slab (Fig. 15.43).

(a)   (b)

Fig. 15.43

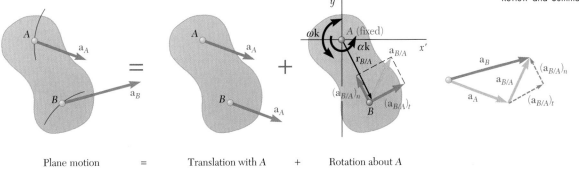

Plane motion   =   Translation with A +   Rotation about A

Fig. 15.44

Accelerations in plane motion

The fact that any plane motion of a rigid slab can be considered as the sum of a translation of the slab with a reference point A and a rotation about A was used in Sec. 15.8 to relate the absolute accelerations of any two points A and B of the slab and the relative acceleration of B with respect to A. We had

$$\mathbf{a}_B = \mathbf{a}_A + \mathbf{a}_{B/A} \qquad (15.21)$$

where $\mathbf{a}_{B/A}$ consisted of a *normal component* $(\mathbf{a}_{B/A})_n$ of magnitude $r\omega^2$ directed toward A, and a *tangential component* $(\mathbf{a}_{B/A})_t$ of magnitude $r\alpha$ perpendicular to the line AB (Fig. 15.44). The fundamental relation (15.21) was expressed in terms of vector diagrams or vector equations and used to determine the accelerations of given points of various mechanisms [Sample Probs. 15.6 through 15.8]. It should be noted that the instantaneous center of rotation C considered in Sec. 15.7 cannot be used for the determination of accelerations, since point C, in general, does *not* have zero acceleration.

Coordinates expressed in terms of a parameter

In the case of certain mechanisms, it is possible to express the coordinates x and y of all significant points of the mechanism by means of simple analytic expressions containing a *single parameter*. The components of the absolute velocity and acceleration of a given point are then obtained by differentiating twice with respect to the time t the coordinates x and y of that point [Sec. 15.9].

Rate of change of a vector with respect to a rotating frame

While the rate of change of a vector is the same with respect to a fixed frame of reference and with respect to a frame in translation, the rate of change of a vector with respect to a rotating frame is different. Therefore, in order to study the motion of a particle relative to a rotating frame we first had to compare the rates of change of a general vector $\mathbf{Q}$ with respect to a fixed frame $OXYZ$ and with respect to a frame $Oxyz$ rotating with an angular velocity $\mathbf{\Omega}$ [Sec. 15.10] (Fig. 15.45). We obtained the fundamental relation

$$(\dot{\mathbf{Q}})_{OXYZ} = (\dot{\mathbf{Q}})_{Oxyz} + \mathbf{\Omega} \times \mathbf{Q} \qquad (15.31)$$

and we concluded that the rate of change of the vector $\mathbf{Q}$ with respect to the fixed frame $OXYZ$ is made of two parts: The first part represents the rate of change of $\mathbf{Q}$ with respect to the rotating frame $Oxyz$; the second part, $\mathbf{\Omega} \times \mathbf{Q}$, is induced by the rotation of the frame $Oxyz$.

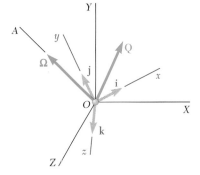

Fig. 15.45

Plane motion of a particle relative to a rotating frame

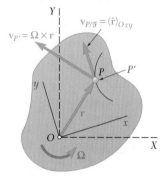

Fig. 15.46

The next part of the chapter [Sec. 15.11] was devoted to the two-dimensional kinematic analysis of a particle P moving with respect to a frame $\mathcal{F}$ rotating with an angular velocity $\mathbf{\Omega}$ about a fixed axis (Fig. 15.46). We found that the absolute velocity of P could be expressed as

$$\mathbf{v}_P = \mathbf{v}_{P'} + \mathbf{v}_{P/\mathcal{F}} \tag{15.33}$$

where $\mathbf{v}_P$ = absolute velocity of particle P
$\quad\mathbf{v}_{P'}$ = velocity of point P' of moving frame $\mathcal{F}$ coinciding with P
$\quad\mathbf{v}_{P/\mathcal{F}}$ = velocity of P relative to moving frame $\mathcal{F}$

We noted that the same expression for $\mathbf{v}_P$ is obtained if the frame is in translation rather than in rotation. However, when the frame is in rotation, the expression for the acceleration of P is found to contain an additional term $\mathbf{a}_c$ called the *complementary acceleration* or *Coriolis acceleration*. We wrote

$$\mathbf{a}_P = \mathbf{a}_{P'} + \mathbf{a}_{P/\mathcal{F}} + \mathbf{a}_c \tag{15.36}$$

where $\mathbf{a}_P$ = absolute acceleration of particle P
$\quad\mathbf{a}_{P'}$ = acceleration of point P' of moving frame $\mathcal{F}$ coinciding with P
$\quad\mathbf{a}_{P/\mathcal{F}}$ = acceleration of P relative to moving frame $\mathcal{F}$
$\quad\mathbf{a}_c = 2\mathbf{\Omega} \times (\dot{\mathbf{r}})_{Oxy} = 2\mathbf{\Omega} \times \mathbf{v}_{P/\mathcal{F}}$
$\quad\quad$ = complementary, or Coriolis, acceleration

Since $\mathbf{\Omega}$ and $\mathbf{v}_{P/\mathcal{F}}$ are perpendicular to each other in the case of plane motion, the Coriolis acceleration was found to have a magnitude $a_c = 2\Omega v_{P/\mathcal{F}}$ and to point in the direction obtained by rotating the vector $\mathbf{v}_{P/\mathcal{F}}$ through 90° in the sense of rotation of the moving frame. Formulas (15.33) and (15.36) can be used to analyze the motion of mechanisms which contain parts sliding on each other [Sample Probs. 15.9 and 15.10].

Motion of a rigid body with a fixed point

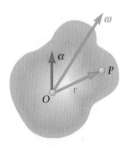

Fig. 15.47

The last part of the chapter was devoted to the study of the kinematics of rigid bodies in three dimensions. We first considered the motion of a rigid body with a fixed point [Sec. 15.12]. After proving that the most general displacement of a rigid body with a fixed point O is equivalent to a rotation of the body about an axis through O, we were able to define the angular velocity $\boldsymbol{\omega}$ and the *instantaneous axis of rotation* of the body at a given instant. The velocity of a point P of the body (Fig. 15.47) could again be expressed as

$$\mathbf{v} = \frac{d\mathbf{r}}{dt} = \boldsymbol{\omega} \times \mathbf{r} \tag{15.37}$$

Differentiating this expression, we also wrote

$$\mathbf{a} = \boldsymbol{\alpha} \times \mathbf{r} + \boldsymbol{\omega} \times (\boldsymbol{\omega} \times \mathbf{r}) \tag{15.38}$$

However, since the direction of $\boldsymbol{\omega}$ changes from one instant to the next, the angular acceleration $\boldsymbol{\alpha}$ is, in general, not directed along the instantaneous axis of rotation [Sample Prob. 15.11].

It was shown in Sec. 15.13 that *the most general motion of a rigid body in space is equivalent, at any given instant, to the sum of a translation and a rotation.* Considering two particles A and B of the body, we found that

$$\mathbf{v}_B = \mathbf{v}_A + \mathbf{v}_{B/A} \tag{15.42}$$

where $\mathbf{v}_{B/A}$ is the velocity of B relative to a frame $AX'Y'Z'$ attached to A and of fixed orientation (Fig. 15.48). Denoting by $\mathbf{r}_{B/A}$ the position vector of B relative to A, we wrote

$$\mathbf{v}_B = \mathbf{v}_A + \boldsymbol{\omega} \times \mathbf{r}_{B/A} \tag{15.43}$$

where $\boldsymbol{\omega}$ is the angular velocity of the body at the instant considered [Sample Prob. 15.12]. The acceleration of B was obtained by a similar reasoning. We first wrote

$$\mathbf{a}_B = \mathbf{a}_A + \mathbf{a}_{B/A}$$

and, recalling Eq. (15.38),

$$\mathbf{a}_B = \mathbf{a}_A + \boldsymbol{\alpha} \times \mathbf{r}_{B/A} + \boldsymbol{\omega} \times (\boldsymbol{\omega} \times \mathbf{r}_{B/A}) \tag{15.44}$$

In the final two sections of the chapter we considered the three-dimensional motion of a particle P relative to a frame $Oxyz$ rotating with an angular velocity $\boldsymbol{\Omega}$ with respect to a fixed frame $OXYZ$ (Fig. 15.49). In Sec. 15.14 we expressed the absolute velocity $\mathbf{v}_P$ of P as

$$\mathbf{v}_P = \mathbf{v}_{P'} + \mathbf{v}_{P/\mathscr{F}} \tag{15.46}$$

where $\mathbf{v}_P$ = absolute velocity of particle P
$\mathbf{v}_{P'}$ = velocity of point P' of moving frame $\mathscr{F}$ coinciding with P
$\mathbf{v}_{P/\mathscr{F}}$ = velocity of P relative to moving frame $\mathscr{F}$

General motion in space

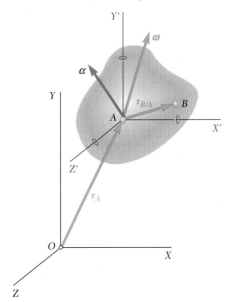

Fig. 15.48

Three-dimensional motion of a particle relative to a rotating frame

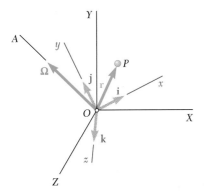

Fig. 15.49

The absolute acceleration $\mathbf{a}_P$ of P was then expressed as

$$\mathbf{a}_P = \mathbf{a}_{P'} + \mathbf{a}_{P/\mathcal{F}} + \mathbf{a}_c \qquad (15.48)$$

where $\mathbf{a}_P$ = absolute acceleration of particle P
  $\mathbf{a}_{P'}$ = acceleration of point P' of moving frame $\mathcal{F}$ coinciding with P
  $\mathbf{a}_{P/\mathcal{F}}$ = acceleration of P relative to moving frame $\mathcal{F}$
  $\mathbf{a}_c = 2\mathbf{\Omega} \times (\dot{\mathbf{r}})_{Oxyz} = 2\mathbf{\Omega} \times \mathbf{v}_{P/\mathcal{F}}$
    = complementary, or Coriolis, acceleration

It was noted that the magnitude a_c of the Coriolis acceleration is not equal to $2\Omega v_{P/\mathcal{F}}$ [Sample Prob. 15.13] except in the special case when $\mathbf{\Omega}$ and $\mathbf{v}_{P/\mathcal{F}}$ are perpendicular to each other.

Frame of reference in general motion

We also observed [Sec. 15.15] that Eqs. (15.46) and (15.48) remain valid when the frame $Axyz$ moves in a known, but arbitrary, fashion with respect to the fixed frame $OXYZ$ (Fig. 15.50), provided that the motion of A is included in the terms $\mathbf{v}_{P'}$ and $\mathbf{a}_{P'}$ representing the absolute velocity and acceleration of the coinciding point P'.

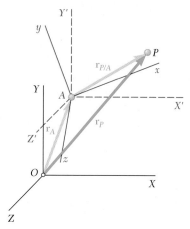

Fig. 15.50

Rotating frames of reference are particularly useful in the study of the three-dimensional motion of rigid bodies. Indeed, there are many cases where an appropriate choice of the rotating frame will lead to a simpler analysis of the motion of the rigid body than would be possible with axes of fixed orientation [Sample Probs. 15.14 and 15.15].

REVIEW PROBLEMS

15.248 The angular acceleration of the 600-mm-radius circular plate shown is defined by the relation $\alpha = \alpha_0 e^{-t}$. Knowing that the plate is at rest when $t = 0$ and that $\alpha_0 = 10$ rad/s^2, determine the magnitude of the total acceleration of point B when (a) $t = 0$, (b) $t = 0.5$ s, (c) $t = \infty$.

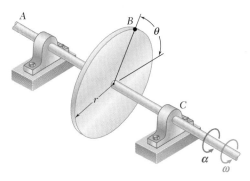

Fig. P15.248

15.249 Cylinder A is moving downward with a velocity of 9 ft/s when the brake is suddenly applied to the drum. Knowing that the cylinder moves 18 ft downward before coming to rest and assuming uniformly accelerated motion, determine (a) the angular acceleration of the drum, (b) the time required for the cylinder to come to rest.

15.250 A baseball pitching machine is designed to deliver a baseball with a ball speed of 70 mph and a ball rotation of 300 rpm clockwise. Knowing that there is no slipping between the wheels and the baseball during the ball launch, determine the angular velocities of wheels A and B.

Fig. P15.249

Fig. P15.250

1033

40 mm 80 mm

80 mm

A

B

C

Fig. P15.251

|← 4 in. →|← 5 in. →|← 5 in. →|← 4 in. →|

A

B *G* *D* *E*
3 in.

Fig. P15.253

15.251 Knowing that inner gear *A* is stationary and outer gear *C* starts from rest and has a constant angular acceleration of 4 rad/s² clockwise, determine at *t* = 5 s (*a*) the angular velocity of arm *AB*, (*b*) the angular velocity of gear *B*, (*c*) the acceleration of the point on gear *B* that is in contact with gear *A*.

15.252 Knowing that at the instant shown bar *AB* has an angular velocity of 10 rad/s clockwise and it is slowing down at a rate of 2 rad/s², determine the angular accelerations of bar *BD* and bar *DE*.

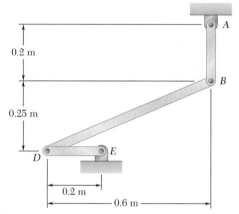

A

0.2 m

B

0.25 m

D *E*

0.2 m

0.6 m

Fig. P15.252

15.253 Knowing that at the instant shown rod *AB* has zero angular acceleration and an angular velocity of 15 rad/s counterclockwise, determine (*a*) the angular acceleration of arm *DE*, (*b*) the acceleration of point *D*.

15.254 Rod *AB* is attached to a collar at *A* and is fitted with a wheel at *B* that has a radius *r* = 15 mm. Knowing that when *θ* = 60° the collar has a velocity of 250 mm/s upward and it is slowing down at a rate of 150 mm/s², determine (*a*) the angular acceleration of rod *AB*, (*b*) the angular acceleration of the wheel.

C

300 mm

A

200 mm

θ

r

B

Fig. P15.254

15.255 Water flows through a curved pipe AB that rotates with a constant clockwise angular velocity of 90 rpm. If the velocity of the water relative to the pipe is 8 m/s, determine the total acceleration of a particle of water at point P.

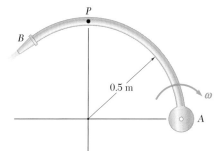

Fig. P15.255

15.256 A disk of 0.15-m radius rotates at the constant rate ω_2 with respect to plate BC, which itself rotates at the constant rate ω_1 about the y axis. Knowing that $\omega_1 = \omega_2 = 3$ rad/s, determine, for the position shown, the velocity and acceleration (a) of point D, (b) of point F.

Fig. P15.256

15.257 Two rods AE and BD pass through holes drilled into a hexagonal block. (The holes are drilled in different planes so that the rods will not touch each other.) Knowing that rod AE has an angular velocity of 20 rad/s clockwise and an angular acceleration of 4 rad/s² counterclockwise when $\theta = 90°$, determine (a) the relative velocity of the block with respect to each rod, (b) the relative acceleration of the block with respect to each rod.

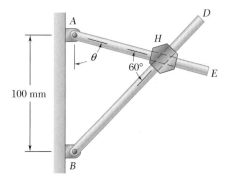

Fig. P15.257

15.258 Rod BC of length 24 in. is connected by ball-and-socket joints to a rotating arm AB and to a collar C that slides on the fixed rod DE. Knowing that the length of arm AB is 4 in. and that it rotates at the constant rate $\omega_1 = 10$ rad/s, determine the velocity of collar C when $\theta = 0$.

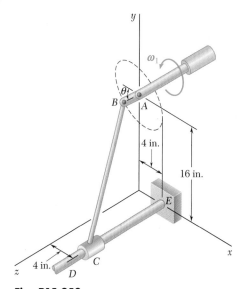

Fig. P15.258

15.259 In the position shown the thin rod moves at a constant speed $u = 3$ in./s out of the tube BC. At the same time tube BC rotates at the constant rate $\omega_2 = 1.5$ rad/s with respect to arm CD. Knowing that the entire assembly rotates about the X axis at the constant rate $\omega_1 = 1.2$ rad/s, determine the velocity and acceleration of end A of the rod.

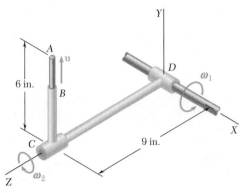

Fig. P15.259

COMPUTER PROBLEMS

15.C1 The disk shown has a constant angular velocity of 500 rpm counterclockwise. Knowing that rod *BD* is 250 mm long, use computational software to determine and plot for values of θ from 0 to 360° and using 30° increments, the velocity of collar *D* and the angular velocity of rod *BD*. Determine the two values of θ for which the speed of collar *D* is zero.

Fig. P15.C1

15.C2 Two rotating rods are connected by a slider block *P* as shown. Knowing that rod *BP* rotates with a constant angular velocity of 6 rad/s counterclockwise, use computational software to determine and plot for values of θ from 0 to 180° the angular velocity and angular acceleration of rod *AE*. Determine the value of θ for which the angular acceleration α_{AE} of rod *AE* is maximum and the corresponding value of α_{AE}.

Fig. P15.C2

Fig. P15.C3

15.C3 In the engine system shown, $l = 160$ mm and $b = 60$ mm. Knowing that crank AB rotates with a constant angular velocity of 1000 rpm clockwise, use computational software to determine and plot for values of θ from 0 to 180° and using 10° increments, (*a*) the angular velocity and angular acceleration of rod BD, (*b*) the velocity and acceleration of the piston P.

15.C4 Rod AB moves over a small wheel at C while end A moves to the right with a constant velocity of 180 mm/s. Use computational software to determine and plot for values of θ from 20° to 90° and using 5° increments, the velocity of point B and the angular acceleration of the rod. Determine the value of θ for which the angular acceleration α of the rod is maximum and the corresponding value of α.

Fig. P15.C4

15.C5 Rod BC of length 24 in. is connected by ball-and-socket joints to the rotating arm AB and to collar C that slides on the fixed rod DE. Arm AB of length 4 in. rotates in the XY plane with a constant angular velocity of 10 rad/s. Use computational software to determine and plot for values of θ from 0 to 360° the velocity of collar C. Determine the two values of θ for which the velocity of collar C is zero.

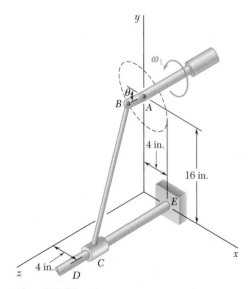

Fig. P15.C5

15.C6 Rod AB of length 25 in. is connected by ball-and-socket joints to collars A and B, which slide along the two rods shown. Collar B moves toward support E at a constant speed of 20 in./s. Denoting by d the distance from point C to collar B, use computational software to determine and plot the velocity of collar A for values of d from 0 to 15 in.

Fig. P15.C6

Three-bladed wind turbines, similar to the ones shown in this picture of a wind farm, are currently the most common design. In this chapter you will learn to analyze the motion of a rigid body by considering the motion of its mass center, the motion relative to its mass center, and the external forces acting on it.

CHAPTER 16

Plane Motion of Rigid Bodies: Forces and Accelerations

16.1 INTRODUCTION

In this chapter and in Chaps. 17 and 18, you will study the *kinetics of rigid bodies*, i.e., the relations existing between the forces acting on a rigid body, the shape and mass of the body, and the motion produced. In Chaps. 12 and 13, you studied similar relations, assuming then that the body could be considered as a particle, i.e., that its mass could be concentrated in one point and that all forces acted at that point. The shape of the body, as well as the exact location of the points of application of the forces, will now be taken into account. You will also be concerned not only with the motion of the body as a whole but also with the motion of the body about its mass center.

Our approach will be to consider rigid bodies as made of large numbers of particles and to use the results obtained in Chap. 14 for the motion of systems of particles. Specifically, two equations from Chap. 14 will be used: Eq. (14.16), $\Sigma\mathbf{F} = m\bar{\mathbf{a}}$, which relates the resultant of the external forces and the acceleration of the mass center G of the system of particles, and Eq. (14.23), $\Sigma\mathbf{M}_G = \dot{\mathbf{H}}_G$, which relates the moment resultant of the external forces and the angular momentum of the system of particles about G.

Except for Sec. 16.2, which applies to the most general case of the motion of a rigid body, the results derived in this chapter will be limited in two ways: (1) They will be restricted to the *plane motion* of rigid bodies, i.e., to a motion in which each particle of the body remains at a constant distance from a fixed reference plane. (2) The rigid bodies considered will consist only of plane slabs and of bodies which are symmetrical with respect to the reference plane.† The study of the plane motion of nonsymmetrical three-dimensional bodies and, more generally, the motion of rigid bodies in three-dimensional space will be postponed until Chap. 18.

In Sec. 16.3, we define the angular momentum of a rigid body in plane motion and show that the rate of change of the angular momentum $\dot{\mathbf{H}}_G$ about the mass center is equal to the product $\bar{I}\boldsymbol{\alpha}$ of the centroidal mass moment of inertia $\bar{I}$ and the angular acceleration $\boldsymbol{\alpha}$ of the body. D'Alembert's principle, introduced in Sec. 16.4, is used to prove that the external forces acting on a rigid body are equivalent to a vector $m\bar{\mathbf{a}}$ attached at the mass center and a couple of moment $\bar{I}\boldsymbol{\alpha}$.

In Sec. 16.5, we derive the principle of transmissibility using only the parallelogram law and Newton's laws of motion, allowing us to remove this principle from the list of axioms (Sec. 1.2) required for the study of the statics and dynamics of rigid bodies.

Free-body-diagram equations are introduced in Sec. 16.6 and will be used in the solution of all problems involving the plane motion of rigid bodies.

After considering the plane motion of connected rigid bodies in Sec. 16.7, you will be prepared to solve a variety of problems involving the translation, centroidal rotation, and unconstrained motion of rigid bodies. In Sec. 16.8 and in the remaining part of the chapter, the solution of problems involving noncentroidal rotation, rolling motion, and other partially constrained plane motions of rigid bodies will be considered.

†Or, more generally, bodies which have a principal centroidal axis of inertia perpendicular to the reference plane.

16.2 EQUATIONS OF MOTION FOR A RIGID BODY

Consider a rigid body acted upon by several external forces $\mathbf{F}_1$, $\mathbf{F}_2$, $\mathbf{F}_3$, . . . (Fig. 16.1). We can assume that the body is made of a large number n of particles of mass Δm_i ($i = 1, 2, \ldots, n$) and apply the results obtained in Chap. 14 for a system of particles (Fig. 16.2). Considering first the motion of the mass center G of the body with respect to the newtonian frame of reference $Oxyz$, we recall Eq. (14.16) and write

$$\Sigma\mathbf{F} = m\bar{\mathbf{a}} \qquad (16.1)$$

where m is the mass of the body and $\bar{\mathbf{a}}$ is the acceleration of the mass center G. Turning now to the motion of the body relative to the centroidal frame of reference $Gx'y'z'$, we recall Eq. (14.23) and write

$$\Sigma\mathbf{M}_G = \dot{\mathbf{H}}_G \qquad (16.2)$$

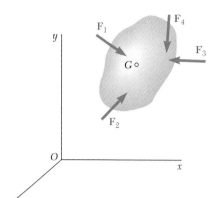

Fig. 16.1

where $\dot{\mathbf{H}}_G$ represents the rate of change of $\mathbf{H}_G$, the angular momentum about G of the system of particles forming the rigid body. In the following, $\mathbf{H}_G$ will simply be referred to as the *angular momentum of the rigid body about its mass center G*. Together Eqs. (16.1) and (16.2) express that *the system of the external forces is equipollent to the system consisting of the vector $m\bar{\mathbf{a}}$ attached at G and the couple of moment $\dot{\mathbf{H}}_G$* (Fig. 16.3).†

Fig. 16.3

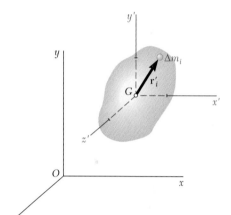

Fig. 16.2

Equations (16.1) and (16.2) apply in the most general case of the motion of a rigid body. In the rest of this chapter, however, our analysis will be limited to the *plane motion* of rigid bodies, i.e., to a motion in which each particle remains at a constant distance from a fixed reference plane, and it will be assumed that the rigid bodies considered consist only of plane slabs and of bodies which are symmetrical with respect to the reference plane. Further study of the plane motion of nonsymmetrical three-dimensional bodies and of the motion of rigid bodies in three-dimensional space will be postponed until Chap. 18.

†Since the systems involved act on a rigid body, we could conclude at this point, by referring to Sec. 3.19, that the two systems are *equivalent* as well as equipollent and use red rather than blue equals signs in Fig. 16.3. However, by postponing this conclusion, we will be able to arrive at it independently (Secs. 16.4 and 18.5), thereby eliminating the necessity of including the principle of transmissibility among the axioms of mechanics (Sec. 16.5).

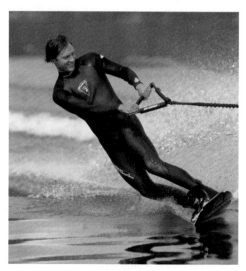

Photo 16.1 The system of external forces acting on the man and wakeboard includes the weights, the tension in the tow rope, and the forces exerted by the water and the air.

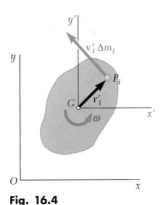

Fig. 16.4

16.3 ANGULAR MOMENTUM OF A RIGID BODY IN PLANE MOTION

Consider a rigid slab in plane motion. Assuming that the slab is made of a large number n of particles P_i of mass Δm_i and recalling Eq. (14.24) of Sec. 14.5, we note that the angular momentum $\mathbf{H}_G$ of the slab about its mass center G can be computed by taking the moments about G of the momenta of the particles of the slab in their motion with respect to either of the frames Oxy or $Gx'y'$ (Fig. 16.4). Choosing the latter course, we write

$$\mathbf{H}_G = \sum_{i=1}^{n} (\mathbf{r}'_i \times \mathbf{v}'_i \, \Delta m_i) \tag{16.3}$$

where $\mathbf{r}'_i$ and $\mathbf{v}'_i \, \Delta m_i$ denote, respectively, the position vector and the linear momentum of the particle P_i relative to the centroidal frame of reference $Gx'y'$. But since the particle belongs to the slab, we have $\mathbf{v}'_i = \boldsymbol{\omega} \times \mathbf{r}'_i$, where $\boldsymbol{\omega}$ is the angular velocity of the slab at the instant considered. We write

$$\mathbf{H}_G = \sum_{i=1}^{n} [\mathbf{r}'_i \times (\boldsymbol{\omega} \times \mathbf{r}'_i) \, \Delta m_i]$$

Referring to Fig. 16.4, we easily verify that the expression obtained represents a vector of the same direction as $\boldsymbol{\omega}$ (i.e., perpendicular to the slab) and of magnitude equal to $\omega \Sigma r'^2_i \, \Delta m_i$. Recalling that the sum $\Sigma r'^2_i \, \Delta m_i$ represents the moment of inertia $\bar{I}$ of the slab about a centroidal axis perpendicular to the slab, we conclude that the angular momentum $\mathbf{H}_G$ of the slab about its mass center is

$$\mathbf{H}_G = \bar{I}\boldsymbol{\omega} \tag{16.4}$$

Differentiating both members of Eq. (16.4) we obtain

$$\dot{\mathbf{H}}_G = \bar{I}\dot{\boldsymbol{\omega}} = \bar{I}\boldsymbol{\alpha} \tag{16.5}$$

Thus the rate of change of the angular momentum of the slab is represented by a vector of the same direction as $\boldsymbol{\alpha}$ (i.e., perpendicular to the slab) and of magnitude $\bar{I}\alpha$.

It should be kept in mind that the results obtained in this section have been derived for a rigid slab in plane motion. As you will see in Chap. 18, they remain valid in the case of the plane motion of rigid bodies which are symmetrical with respect to the reference plane.† However, they do not apply in the case of nonsymmetrical bodies or in the case of three-dimensional motion.

Photo 16.2 The hard disk and pick-up arms of a hard disk computer undergo centroidal rotation.

---

†Or, more generally, bodies which have a principal centroidal axis of inertia perpendicular to the reference plane.

16.4 PLANE MOTION OF A RIGID BODY. D'ALEMBERT'S PRINCIPLE

Consider a rigid slab of mass m moving under the action of several external forces $\mathbf{F}_1$, $\mathbf{F}_2$, $\mathbf{F}_3$, . . . , contained in the plane of the slab (Fig. 16.5). Substituting for $\dot{\mathbf{H}}_G$ from Eq. (16.5) into Eq. (16.2) and writing the fundamental equations of motion (16.1) and (16.2) in scalar form, we have

$$\Sigma F_x = m\bar{a}_x \qquad \Sigma F_y = m\bar{a}_y \qquad \Sigma M_G = \bar{I}\alpha \qquad (16.6)$$

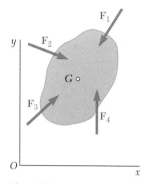

Fig. 16.5

Equations (16.6) show that the acceleration of the mass center G of the slab and its angular acceleration $\boldsymbol{\alpha}$ are easily obtained once the resultant of the external forces acting on the slab and their moment resultant about G have been determined. Given appropriate initial conditions, the coordinates $\bar{x}$ and $\bar{y}$ of the mass center and the angular coordinate θ of the slab can then be obtained by integration at any instant t. Thus *the motion of the slab is completely defined by the resultant and moment resultant about G of the external forces acting on it.*

This property, which will be extended in Chap. 18 to the case of the three-dimensional motion of a rigid body, is characteristic of the motion of a rigid body. Indeed, as we saw in Chap. 14, the motion of a system of particles which are not rigidly connected will in general depend upon the specific external forces acting on the various particles, as well as upon the internal forces.

Since the motion of a rigid body depends only upon the resultant and moment resultant of the external forces acting on it, it follows that *two systems of forces which are equipollent*, i.e., which have the same resultant and the same moment resultant, *are also equivalent;* that is, they have exactly the same effect on a given rigid body.[†]

Consider in particular the system of the external forces acting on a rigid body (Fig. 16.6a) and the system of the effective forces associated with the particles forming the rigid body (Fig. 16.6b). It was shown in Sec. 14.2 that the two systems thus defined are equipollent. But since the particles considered now form a rigid body, it follows from the discussion above that the two systems are also equivalent. We can thus state that *the external forces acting on a rigid body are equivalent to the effective forces of the various particles forming the body.* This statement is referred to as *d'Alembert's principle* after the French mathematician Jean le Rond d'Alembert (1717–1783), even though d'Alembert's original statement was written in a somewhat different form.

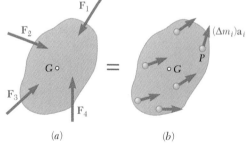

Fig. 16.6

The fact that the system of external forces is *equivalent* to the system of the effective forces has been emphasized by the use of a red equals sign in Fig. 16.6 and also in Fig. 16.7, where using results obtained earlier in this section, we have replaced the effective forces by a vector $m\bar{\mathbf{a}}$ attached at the mass center G of the slab and a couple of moment $\bar{I}\boldsymbol{\alpha}$.

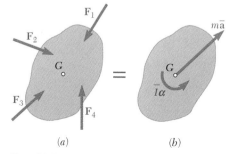

Fig. 16.7

[†]This result has already been derived in Sec. 3.19 from the principle of transmissibility (Sec. 3.3). The present derivation is independent of that principle, however, and will make possible its elimination from the axioms of mechanics (Sec. 16.5).

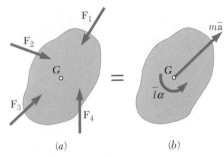

(a)　　　　　(b)

Fig. 16.7 (repeated)

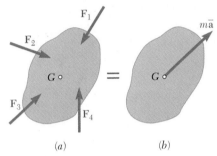

(a)　　　　　(b)

Fig. 16.8 Translation.

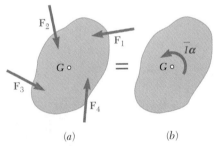

(a)　　　　　(b)

Fig. 16.9 Centroidal rotation.

Translation.　In the case of a body in translation, the angular acceleration of the body is identically equal to zero and its effective forces reduce to the vector $m\overline{\mathbf{a}}$ attached at G (Fig. 16.8). Thus, the resultant of the external forces acting on a rigid body in translation passes through the mass center of the body and is equal to $m\overline{\mathbf{a}}$.

Centroidal Rotation.　When a slab, or, more generally, a body symmetrical with respect to the reference plane, rotates about a fixed axis perpendicular to the reference plane and passing through its mass center G, we say that the body is in *centroidal rotation*. Since the acceleration $\overline{\mathbf{a}}$ is identically equal to zero, the effective forces of the body reduce to the couple $\overline{I}\boldsymbol{\alpha}$ (Fig. 16.9). Thus, the external forces acting on a body in centroidal rotation are equivalent to a couple of moment $\overline{I}\boldsymbol{\alpha}$.

General Plane Motion.　Comparing Fig. 16.7 with Figs. 16.8 and 16.9, we observe that from the point of view of *kinetics,* the most general plane motion of a rigid body symmetrical with respect to the reference plane can be replaced by the sum of a translation and a centroidal rotation. We should note that this statement is more restrictive than the similar statement made earlier from the point of view of *kinematics* (Sec. 15.5), since we now require that the mass center of the body be selected as the reference point.

Referring to Eqs. (16.6), we observe that the first two equations are identical with the equations of motion of a particle of mass m acted upon by the given forces $\mathbf{F}_1$, $\mathbf{F}_2$, $\mathbf{F}_3$, . . . We thus check that *the mass center G of a rigid body in plane motion moves as if the entire mass of the body were concentrated at that point, and as if all the external forces acted on it.* We recall that this result has already been obtained in Sec. 14.4 in the general case of a system of particles, the particles being not necessarily rigidly connected. We also note, as we did in Sec. 14.4, that the system of the external forces does not, in general, reduce to a single vector $m\overline{\mathbf{a}}$ attached at G. Therefore, in the general case of the plane motion of a rigid body, *the resultant of the external forces acting on the body does not pass through the mass center of the body.*

Finally, it should be observed that the last of Eqs. (16.6) would still be valid if the rigid body, while subjected to the same applied forces, were constrained to rotate about a fixed axis through G. Thus, *a rigid body in plane motion rotates about its mass center as if this point were fixed.*

*16.5　A REMARK ON THE AXIOMS OF THE MECHANICS OF RIGID BODIES

The fact that two equipollent systems of external forces acting on a rigid body are also equivalent, i.e., have the same effect on that rigid body, has already been established in Sec. 3.19. But there it was derived from the *principle of transmissibility,* one of the axioms used in our study of the statics of rigid bodies. It should be noted that this axiom has not been used in the present chapter because Newton's second and third laws of motion make its use unnecessary in the study of the dynamics of rigid bodies.

In fact, the principle of transmissibility may now be *derived* from the other axioms used in the study of mechanics. This principle stated, without proof (Sec. 3.3), that the conditions of equilibrium or motion of a rigid body remain unchanged if a force **F** acting at a given point of the rigid body is replaced by a force **F'** of the same magnitude and same direction, but acting at a different point, provided that the two forces have the same line of action. But since **F** and **F'** have the same moment about any given point, it is clear that they form two equipollent systems of external forces. Thus, we may now *prove*, as a result of what we established in the preceding section, that **F** and **F'** have the same effect on the rigid body (Fig. 3.3).

The principle of transmissibility can therefore be removed from the list of axioms required for the study of the mechanics of rigid bodies. These axioms are reduced to the parallelogram law of addition of vectors and to Newton's laws of motion.

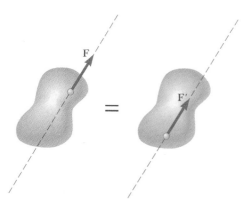

Fig. 3.3 (repeated)

16.6 SOLUTION OF PROBLEMS INVOLVING THE MOTION OF A RIGID BODY

We saw in Sec. 16.4 that when a rigid body is in plane motion, there exists a fundamental relation between the forces $\mathbf{F}_1$, $\mathbf{F}_2$, $\mathbf{F}_3$, . . . , acting on the body, the acceleration $\overline{\mathbf{a}}$ of its mass center, and the angular acceleration $\boldsymbol{\alpha}$ of the body. This relation, which is represented in Fig. 16.7 in the form of a *free-body-diagram equation*, can be used to determine the acceleration $\overline{\mathbf{a}}$ and the angular acceleration $\boldsymbol{\alpha}$ produced by a given system of forces acting on a rigid body or, conversely, to determine the forces which produce a given motion of the rigid body.

The three algebraic equations (16.6) can be used to solve problems of plane motion.† However, our experience in statics suggests that the solution of many problems involving rigid bodies could be simplified by an appropriate choice of the point about which the moments of the forces are computed. It is therefore preferable to remember the relation existing between the forces and the accelerations in the pictorial form shown in Fig. 16.7 and to derive from this fundamental relation the component or moment equations which fit best the solution of the problem under consideration.

The fundamental relation shown in Fig. 16.7 can be presented in an alternative form if we add to the external forces an inertia vector $-m\overline{\mathbf{a}}$ of sense opposite to that of $\overline{\mathbf{a}}$, attached at G, and an inertia couple $-\overline{I}\boldsymbol{\alpha}$ of moment equal in magnitude to $\overline{I}\alpha$ and of sense opposite to that of $\boldsymbol{\alpha}$ (Fig. 16.10). The system obtained is equivalent to zero, and the rigid body is said to be in *dynamic equilibrium.*

Whether the principle of equivalence of external and effective forces is directly applied, as in Fig. 16.7, or whether the concept of dynamic equilibrium is introduced, as in Fig. 16.10, the use of free-body-diagram equations showing vectorially the relationship existing between the forces applied on the rigid body and the resulting linear and angular accelerations presents considerable advantages over the blind application of formulas (16.6). These advantages can be summarized as follows:

1. The use of a pictorial representation provides a much clearer understanding of the effect of the forces on the motion of the body.

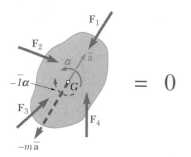

Fig. 16.10

†We recall that the last of Eqs. (16.6) is valid only in the case of the plane motion of a rigid body symmetrical with respect to the reference plane. In all other cases, the methods of Chap. 18 should be used.

Fig. 16.7 (repeated)

Fig. 16.10 (repeated)

Photo 16.3 The forklift and moving load can be analyzed as a system of two connected rigid bodies in plane motion.

2. This approach makes it possible to divide the solution of a dynamics problem into two parts: In the first part, the analysis of the kinematic and kinetic characteristics of the problem leads to the free-body diagrams of Fig. 16.7 or 16.10; in the second part, the diagram obtained is used to analyze the various forces and vectors involved by the methods of Chap. 3.

3. A unified approach is provided for the analysis of the plane motion of a rigid body, regardless of the particular type of motion involved. While the kinematics of the various motions considered may vary from one case to the other, the approach to the kinetics of the motion is consistently the same. In every case a diagram will be drawn showing the external forces, the vector $m\bar{\mathbf{a}}$ associated with the motion of G, and the couple $\bar{I}\boldsymbol{\alpha}$ associated with the rotation of the body about G.

4. The resolution of the plane motion of a rigid body into a translation and a centroidal rotation, which is used here, is a basic concept which can be applied effectively throughout the study of mechanics. It will be used again in Chap. 17 with the method of work and energy and the method of impulse and momentum.

5. As you will see in Chap. 18, this approach can be extended to the study of the general three-dimensional motion of a rigid body. The motion of the body will again be resolved into a translation and a rotation about the mass center, and free-body-diagram equations will be used to indicate the relationship existing between the external forces and the rates of change of the linear and angular momentum of the body.

16.7 SYSTEMS OF RIGID BODIES

The method described in the preceding section can also be used in problems involving the plane motion of several connected rigid bodies. For each part of the system, a diagram similar to Fig. 16.7 or Fig. 16.10 can be drawn. The equations of motion obtained from these diagrams are solved simultaneously.

In some cases, as in Sample Prob. 16.3, a single diagram can be drawn for the entire system. This diagram should include all the external forces, as well as the vectors $m\bar{\mathbf{a}}$ and the couples $\bar{I}\boldsymbol{\alpha}$ associated with the various parts of the system. However, internal forces such as the forces exerted by connecting cables, can be omitted since they occur in pairs of equal and opposite forces and are thus equipollent to zero. The equations obtained by expressing that the system of the external forces is equipollent to the system of the effective forces can be solved for the remaining unknowns.[†]

It is not possible to use this second approach in problems involving more than three unknowns, since only three equations of motion are available when a single diagram is used. We need not elaborate upon this point, since the discussion involved would be completely similar to that given in Sec. 6.11 in the case of the equilibrium of a system of rigid bodies.

---

†Note that we cannot speak of *equivalent* systems since we are not dealing with a single rigid body.

SAMPLE PROBLEM 16.1

When the forward speed of the truck shown was 30 ft/s, the brakes were suddenly applied, causing all four wheels to stop rotating. It was observed that the truck skidded to rest in 20 ft. Determine the magnitude of the normal reaction and of the friction force at each wheel as the truck skidded to rest.

SOLUTION

Kinematics of Motion. Choosing the positive sense to the right and using the equations of uniformly accelerated motion, we write

$$\bar{v}_0 = +30 \text{ ft/s} \qquad \bar{v}^2 = \bar{v}_0^2 + 2\bar{a}\bar{x} \qquad 0 = (30)^2 + 2\bar{a}(20)$$
$$\bar{a} = -22.5 \text{ ft/s}^2 \qquad \mathbf{a} = 22.5 \text{ ft/s}^2 \leftarrow$$

Equations of Motion. The external forces consist of the weight **W** of the truck and of the normal reactions and friction forces at the wheels. (The vectors $\mathbf{N}_A$ and $\mathbf{F}_A$ represent the sum of the reactions at the rear wheels, while $\mathbf{N}_B$ and $\mathbf{F}_B$ represent the sum of the reactions at the front wheels.) Since the truck is in translation, the effective forces reduce to the vector $m\bar{a}$ attached at G. Three equations of motion are obtained by expressing that the system of the external forces is equivalent to the system of the effective forces.

$$+\uparrow \Sigma F_y = \Sigma(F_y)_{\text{eff}}: \qquad N_A + N_B - W = 0$$

Since $F_A = \mu_k N_A$ and $F_B = \mu_k N_B$, where μ_k is the coefficient of kinetic friction, we find that

$$F_A + F_B = \mu_k(N_A + N_B) = \mu_k W$$

$$\xrightarrow{+} \Sigma F_x = \Sigma(F_x)_{\text{eff}}: \qquad -(F_A + F_B) = -m\bar{a}$$
$$-\mu_k W = -\frac{W}{32.2 \text{ ft/s}^2}(22.5 \text{ ft/s}^2)$$
$$\mu_k = 0.699$$
$$+\gamma \Sigma M_A = \Sigma(M_A)_{\text{eff}}: \qquad -W(5 \text{ ft}) + N_B(12 \text{ ft}) = m\bar{a}(4 \text{ ft})$$
$$-W(5 \text{ ft}) + N_B(12 \text{ ft}) = \frac{W}{32.2 \text{ ft/s}^2}(22.5 \text{ ft/s}^2)(4 \text{ ft})$$
$$N_B = 0.650W$$
$$F_B = \mu_k N_B = (0.699)(0.650W) \qquad F_B = 0.454W$$
$$+\uparrow \Sigma F_y = \Sigma(F_y)_{\text{eff}}: \qquad N_A + N_B - W = 0$$
$$N_A + 0.650W - W = 0$$
$$N_A = 0.350W$$
$$F_A = \mu_k N_A = (0.699)(0.350W) \qquad F_A = 0.245W$$

Reactions at Each Wheel. Recalling that the values computed above represent the sum of the reactions at the two front wheels or the two rear wheels, we obtain the magnitude of the reactions at each wheel by writing

$$N_{\text{front}} = \tfrac{1}{2}N_B = 0.325W \qquad N_{\text{rear}} = \tfrac{1}{2}N_A = 0.175W \blacktriangleleft$$
$$F_{\text{front}} = \tfrac{1}{2}F_B = 0.227W \qquad F_{\text{rear}} = \tfrac{1}{2}F_A = 0.122W \blacktriangleleft$$

SAMPLE PROBLEM 16.2

The thin plate $ABCD$ of mass 8 kg is held in the position shown by the wire BH and two links AE and DF. Neglecting the mass of the links, determine immediately after wire BH has been cut (a) the acceleration of the plate, (b) the force in each link.

SOLUTION

Kinematics of Motion. After wire BH has been cut, we observe that corners A and D move along parallel circles of radius 150 mm centered, respectively, at E and F. The motion of the plate is thus a curvilinear translation; the particles forming the plate move along parallel circles of radius 150 mm.

At the instant wire BH is cut, the velocity of the plate is zero. Thus the acceleration $\bar{\mathbf{a}}$ of the mass center G of the plate is tangent to the circular path which will be described by G.

Equations of Motion. The external forces consist of the weight $\mathbf{W}$ and the forces $\mathbf{F}_{AE}$ and $\mathbf{F}_{DF}$ exerted by the links. Since the plate is in translation, the effective forces reduce to the vector $m\bar{\mathbf{a}}$ attached at G and directed along the t axis. A free-body-diagram equation is drawn to show that the system of the external forces is equivalent to the system of the effective forces.

a. Acceleration of the Plate.

$$+\nearrow \Sigma F_t = \Sigma(F_t)_{\text{eff}}:$$

$$W \cos 30° = m\bar{a}$$
$$mg \cos 30° = m\bar{a}$$
$$\bar{a} = g \cos 30° = (9.81 \text{ m/s}^2) \cos 30° \tag{1}$$
$$\bar{\mathbf{a}} = 8.50 \text{ m/s}^2 \nearrow 60° \quad \blacktriangleleft$$

b. Forces in Links AE and DF.

$$+\nwarrow\Sigma F_n = \Sigma(F_n)_{\text{eff}}: \qquad F_{AE} + F_{DF} - W \sin 30° = 0 \tag{2}$$
$$+\downarrow\Sigma M_G = \Sigma(M_G)_{\text{eff}}:$$

$$(F_{AE} \sin 30°)(250 \text{ mm}) - (F_{AE} \cos 30°)(100 \text{ mm})$$
$$+ (F_{DF} \sin 30°)(250 \text{ mm}) + (F_{DF} \cos 30°)(100 \text{ mm}) = 0$$
$$38.4F_{AE} + 211.6F_{DF} = 0$$
$$F_{DF} = -0.1815F_{AE} \tag{3}$$

Substituting for F_{DF} from (3) into (2), we write

$$F_{AE} - 0.1815F_{AE} - W \sin 30° = 0$$
$$F_{AE} = 0.6109W$$
$$F_{DF} = -0.1815(0.6109W) = -0.1109W$$

Noting that $W = mg = (8 \text{ kg})(9.81 \text{ m/s}^2) = 78.48 \text{ N}$, we have

$$F_{AE} = 0.6109(78.48 \text{ N}) \qquad F_{AE} = 47.9 \text{ N } T \quad \blacktriangleleft$$
$$F_{DF} = -0.1109(78.48 \text{ N}) \qquad F_{DF} = 8.70 \text{ N } C \quad \blacktriangleleft$$

10 in.

6 in.

B

A
5 lb

10 lb

SAMPLE PROBLEM 16.3

A pulley weighing 12 lb and having a radius of gyration of 8 in. is connected to two blocks as shown. Assuming no axle friction, determine the angular acceleration of the pulley and the acceleration of each block.

SOLUTION

Sense of Motion. Although an arbitrary sense of motion can be assumed (since no friction forces are involved) and later checked by the sign of the answer, we may prefer to determine the actual sense of rotation of the pulley first. The weight of block B required to maintain the equilibrium of the pulley when it is acted upon by the 5-lb block A is first determined. We write

$$+\circlearrowleft \Sigma M_G = 0: \qquad W_B(6 \text{ in.}) - (5 \text{ lb})(10 \text{ in.}) = 0 \qquad W_B = 8.33 \text{ lb}$$

Since block B actually weighs 10 lb, the pulley will rotate counterclockwise.

Kinematics of Motion. Assuming $\boldsymbol{\alpha}$ counterclockwise and noting that $a_A = r_A \alpha$ and $a_B = r_B \alpha$, we obtain

$$\mathbf{a}_A = (\tfrac{10}{12} \text{ ft})\alpha \uparrow \qquad \mathbf{a}_B = (\tfrac{6}{12} \text{ ft})\alpha \downarrow$$

Equations of Motion. A single system consisting of the pulley and the two blocks is considered. Forces external to this system consist of the weights of the pulley and the two blocks and of the reaction at G. (The forces exerted by the cables on the pulley and on the blocks are internal to the system considered and cancel out.) Since the motion of the pulley is a centroidal rotation and the motion of each block is a translation, the effective forces reduce to the couple $\bar{I}\boldsymbol{\alpha}$ and the two vectors $m\mathbf{a}_A$ and $m\mathbf{a}_B$. The centroidal moment of inertia of the pulley is

$$\bar{I} = m\bar{k}^2 = \frac{W}{g}\bar{k}^2 = \frac{12 \text{ lb}}{32.2 \text{ ft/s}^2}(\tfrac{8}{12} \text{ ft})^2 = 0.1656 \text{ lb} \cdot \text{ft} \cdot \text{s}^2$$

Since the system of the external forces is equipollent to the system of the effective forces, we write

$$+\circlearrowleft \Sigma M_G = \Sigma (M_G)_{\text{eff}}:$$

$$(10 \text{ lb})(\tfrac{6}{12} \text{ ft}) - (5 \text{ lb})(\tfrac{10}{12} \text{ ft}) = +\bar{I}\alpha + m_B a_B(\tfrac{6}{12} \text{ ft}) + m_A a_A(\tfrac{10}{12} \text{ ft})$$

$$(10)(\tfrac{6}{12}) - (5)(\tfrac{10}{12}) = 0.1656\alpha + \tfrac{10}{32.2}(\tfrac{6}{12}\alpha)(\tfrac{6}{12}) + \tfrac{5}{32.2}(\tfrac{10}{12}\alpha)(\tfrac{10}{12})$$

$$\alpha = +2.374 \text{ rad/s}^2 \qquad\qquad \alpha = 2.37 \text{ rad/s}^2 \circlearrowleft \quad \blacktriangleleft$$

$$a_A = r_A\alpha = (\tfrac{10}{12} \text{ ft})(2.374 \text{ rad/s}^2) \qquad \mathbf{a}_A = 1.978 \text{ ft/s}^2 \uparrow \quad \blacktriangleleft$$

$$a_B = r_B\alpha = (\tfrac{6}{12} \text{ ft})(2.374 \text{ rad/s}^2) \qquad \mathbf{a}_B = 1.187 \text{ ft/s}^2 \downarrow \quad \blacktriangleleft$$

SAMPLE PROBLEM 16.4

A cord is wrapped around a homogeneous disk of radius $r = 0.5$ m and mass $m = 15$ kg. If the cord is pulled upward with a force $\mathbf{T}$ of magnitude 180 N, determine (a) the acceleration of the center of the disk, (b) the angular acceleration of the disk, (c) the acceleration of the cord.

SOLUTION

Equations of Motion. We assume that the components $\bar{\mathbf{a}}_x$ and $\bar{\mathbf{a}}_y$ of the acceleration of the center are directed, respectively, to the right and upward and that the angular acceleration of the disk is counterclockwise. The external forces acting on the disk consist of the weight $\mathbf{W}$ and the force $\mathbf{T}$ exerted by the cord. This system is equivalent to the system of the effective forces, which consists of a vector of components $m\bar{\mathbf{a}}_x$ and $m\bar{\mathbf{a}}_y$ attached at G and a couple $\bar{I}\boldsymbol{\alpha}$. We write

$$\xrightarrow{+}\Sigma F_x = \Sigma(F_x)_{\text{eff}}: \qquad\qquad 0 = m\bar{a}_x \qquad\qquad \bar{\mathbf{a}}_x = 0 \quad \blacktriangleleft$$

$$+\uparrow\Sigma F_y = \Sigma(F_y)_{\text{eff}}: \qquad\qquad T - W = m\bar{a}_y$$

$$\bar{a}_y = \frac{T - W}{m}$$

Since $T = 180$ N, $m = 15$ kg, and $W = (15 \text{ kg})(9.81 \text{ m/s}^2) = 147.1$ N, we have

$$\bar{a}_y = \frac{180 \text{ N} - 147.1 \text{ N}}{15 \text{ kg}} = +2.19 \text{ m/s}^2 \qquad \bar{\mathbf{a}}_y = 2.19 \text{ m/s}^2\uparrow \quad \blacktriangleleft$$

$$+\!\curvearrowleft\!\Sigma M_G = \Sigma(M_G)_{\text{eff}}: \qquad\qquad -Tr = \bar{I}\alpha$$

$$-Tr = (\tfrac{1}{2}mr^2)\alpha$$

$$\alpha = -\frac{2T}{mr} = -\frac{2(180 \text{ N})}{(15 \text{ kg})(0.5 \text{ m})} = -48.0 \text{ rad/s}^2$$

$$\boldsymbol{\alpha} = 48.0 \text{ rad/s}^2 \,\curvearrowright \quad \blacktriangleleft$$

Acceleration of Cord. Since the acceleration of the cord is equal to the tangential component of the acceleration of point A on the disk, we write

$$\mathbf{a}_{\text{cord}} = (\mathbf{a}_A)_t = \bar{\mathbf{a}} + (\mathbf{a}_{A/G})_t$$

$$= [2.19 \text{ m/s}^2\uparrow] + [(0.5 \text{ m})(48 \text{ rad/s}^2)\uparrow]$$

$$\mathbf{a}_{\text{cord}} = 26.2 \text{ m/s}^2\uparrow \quad \blacktriangleleft$$

A uniform sphere of mass m and radius r is projected along a rough horizontal surface with a linear velocity $\bar{\mathbf{v}}_0$ and no angular velocity. Denoting by μ_k the coefficient of kinetic friction between the sphere and the floor, determine (a) the time t_1 at which the sphere will start rolling without sliding, (b) the linear velocity and angular velocity of the sphere at time t_1.

SOLUTION

Equations of Motion. The positive sense is chosen to the right for $\bar{\mathbf{a}}$ and clockwise for $\boldsymbol{\alpha}$. The external forces acting on the sphere consist of the weight $\mathbf{W}$, the normal reaction $\mathbf{N}$, and the friction force $\mathbf{F}$. Since the point of the sphere in contact with the surface is sliding to the right, the friction force $\mathbf{F}$ is directed to the left. While the sphere is sliding, the magnitude of the friction force is $F = \mu_k N$. The effective forces consist of the vector $m\bar{\mathbf{a}}$ attached at G and the couple $\bar{I}\boldsymbol{\alpha}$. Expressing that the system of the external forces is equivalent to the system of the effective forces, we write

$+\uparrow \Sigma F_y = \Sigma(F_y)_{\text{eff}}$: $\qquad N - W = 0$

$\qquad\qquad\qquad N = W = mg \qquad F = \mu_k N = \mu_k mg$

$\xrightarrow{+} \Sigma F_x = \Sigma(F_x)_{\text{eff}}$: $\qquad -F = m\bar{a} \qquad -\mu_k mg = m\bar{a} \qquad \bar{a} = -\mu_k g$

$+\downarrow \Sigma M_G = \Sigma(M_G)_{\text{eff}}$: $\qquad Fr = \bar{I}\alpha$

Noting that $\bar{I} = \frac{2}{5}mr^2$ and substituting the value obtained for F, we write

$$(\mu_k mg)r = \tfrac{2}{5}mr^2\alpha \qquad \alpha = \frac{5}{2}\frac{\mu_k g}{r}$$

Kinematics of Motion. As long as the sphere both rotates and slides, its linear and angular motions are uniformly accelerated.

$$t = 0, \bar{v} = \bar{v}_0 \qquad \bar{v} = \bar{v}_0 + \bar{a}t = \bar{v}_0 - \mu_k g t \tag{1}$$

$$t = 0, \omega_0 = 0 \qquad \omega = \omega_0 + \alpha t = 0 + \left(\frac{5}{2}\frac{\mu_k g}{r}\right)t \tag{2}$$

The sphere will start rolling without sliding when the velocity $\mathbf{v}_C$ of the point of contact C is zero. At that time, $t = t_1$, point C becomes the instantaneous center of rotation, and we have

$$\bar{v}_1 = r\omega_1 \tag{3}$$

Substituting in (3) the values obtained for $\bar{v}_1$ and ω_1 by making $t = t_1$ in (1) and (2), respectively, we write

$$\bar{v}_0 - \mu_k g t_1 = r\left(\frac{5}{2}\frac{\mu_k g}{r}t_1\right) \qquad t_1 = \frac{2}{7}\frac{\bar{v}_0}{\mu_k g} \blacktriangleleft$$

Substituting for t_1 into (2), we have

$$\omega_1 = \frac{5}{2}\frac{\mu_k g}{r}t_1 = \frac{5}{2}\frac{\mu_k g}{r}\left(\frac{2}{7}\frac{\bar{v}_0}{\mu_k g}\right) \qquad \omega_1 = \frac{5}{7}\frac{\bar{v}_0}{r} \qquad \omega_1 = \frac{5}{7}\frac{\bar{v}_0}{r} \downarrow \blacktriangleleft$$

$$\bar{v}_1 = r\omega_1 = r\left(\frac{5}{7}\frac{\bar{v}_0}{r}\right) \qquad \bar{v}_1 = \tfrac{5}{7}\bar{v}_0 \qquad \mathbf{v}_1 = \tfrac{5}{7}\bar{v}_0 \rightarrow \blacktriangleleft$$

SOLVING PROBLEMS
ON YOUR OWN

This chapter deals with the *plane motion* of rigid bodies, and in this first lesson we considered rigid bodies that are free to move under the action of applied forces.

1. Effective forces. We first recalled that a rigid body consists of a large number of particles. The effective forces of the particles forming the body were found to be equivalent to a vector $m\mathbf{a}$ attached at the mass center G of the body and a couple of moment $\bar{I}\boldsymbol{\alpha}$ [Fig. 16.7]. Noting that the applied forces are equivalent to the effective forces, we wrote

$$\Sigma F_x = m\bar{a}_x \qquad \Sigma F_y = m\bar{a}_y \qquad \Sigma M_G = \bar{I}\alpha \qquad (16.5)$$

where $\bar{a}_x$ and $\bar{a}_y$ are the x and y components of the acceleration of the mass center G of the body and α is the angular acceleration of the body. It is important to note that when these equations are used, *the moments of the applied forces must be computed with respect to the mass center of the body*. However, you learned a more efficient method of solution based on the use of a free-body-diagram equation.

2. Free-body-diagram equation. Your first step in the solution of a problem should be to draw a *free-body-diagram equation*.

a. A free-body-diagram equation consists of two diagrams representing two equivalent systems of vectors. *In the first diagram* you should show *the forces exerted on the body*, including the applied forces, the reactions at the supports, and the weight of the body. *In the second diagram* you should show the vector $m\overline{\mathbf{a}}$ and the couple $\bar{I}\boldsymbol{\alpha}$ representing *the effective forces*.

b. Using a free-body-diagram equation allows you to *sum components in any direction and to sum moments about any point*. When writing the three equations of motion needed to solve a given problem, you can therefore select one or more equations involving a single unknown. Solving these equations first and substituting the values obtained for the unknowns into the remaining equation(s) will yield a simpler solution.

3. Plane motion of a rigid body. The problems that you will be asked to solve will fall into one of the following categories.

a. Rigid body in translation. For a body in translation, the angular acceleration is zero. The effective forces reduce to *the vector* $m\overline{\mathbf{a}}$ applied at the mass center [Sample Probs. 16.1 and 16.2].

b. Rigid body in centroidal rotation. For a body in centroidal rotation, the acceleration of the mass center is zero. The effective forces reduce to *the couple* $\overline{I}\boldsymbol{\alpha}$ [Sample Prob. 16.3].

c. Rigid body in general plane motion. You can consider the general plane motion of a rigid body as the sum of a translation and a centroidal rotation. The effective forces are equivalent to the vector $m\overline{\mathbf{a}}$ and the couple $\overline{I}\boldsymbol{\alpha}$ [Sample Probs. 16.4 and 16.5].

4. Plane motion of a system of rigid bodies. You first should draw a free-body-diagram equation that includes all the rigid bodies of the system. A vector $m\overline{\mathbf{a}}$ and a couple $\overline{I}\boldsymbol{\alpha}$ are attached to each body. However, the forces exerted on each other by the various bodies of the system can be omitted, since they occur in pairs of equal and opposite forces.

a. If no more than three unknowns are involved, you can use this free-body-diagram equation and sum components in any direction and sum moments about any point to obtain equations that can be solved for the desired unknowns [Sample Prob. 16.3].

b. If more than three unknowns are involved, you must draw a separate free-body-diagram equation for each of the rigid bodies of the system. Both internal forces and external forces should be included in each of the free-body-diagram equations, and care should be taken to represent with equal and opposite vectors the forces that two bodies exert on each other.

PROBLEMS

CONCEPT QUESTIONS

16.CQ1 Two pendulums, A and B, with the masses and lengths shown are released from rest. Which system has a larger mass moment of inertia about its pivot point?

a. A

b. B

c. They are the same.

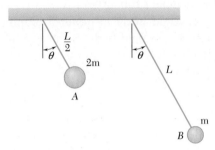

Fig. P16.CQ1 and P16.CQ2

16.CQ2 Two pendulums, A and B, with the masses and lengths shown are released from rest. Which system has a larger angular acceleration immediately after release?

a. A

b. B

c. They are the same.

16.CQ3 Two solid cylinders, A and B, have the same mass m and the radii $2r$ and r, respectively. Each is accelerated from rest with a force applied as shown. In order to impart identical angular accelerations to both cylinders, what is the relationship between F_1 and F_2?

a. $F_1 = 0.5F_2$

b. $F_1 = F_2$

c. $F_1 = 2F_2$

d. $F_1 = 4F_2$

e. $F_1 = 8F_2$

Fig. P16.CQ3

16.F1 A 6-ft board is placed in a truck with one end resting against a block secured to the floor and the other leaning against a vertical partition. Draw the FBD and KD necessary to determine the maximum allowable acceleration of the truck if the board is to remain in the position shown.

Fig. P16.F1

16.F2 A uniform circular plate of mass 3 kg is attached to two links *AC* and *BD* of the same length. Knowing that the plate is released from rest in the position shown, in which lines joining *G* to *A* and *B* are, respectively, horizontal and vertical, draw the FBD and KD for the plate.

Fig. P16.F2

16.F3 Two uniform disks and two cylinders are assembled as indicated. Disk *A* weighs 20 lb and disk *B* weighs 12 lb. Knowing that the system is released from rest, draw the FBD and KD for the whole system.

Fig. P16.F3

16.F4 The 400-lb crate shown is lowered by means of two overhead cranes. Knowing the tension in each cable, draw the FBD and KD that can be used to determine the angular acceleration of the crate and the acceleration of the center of gravity.

Fig. P16.F4

END-OF-SECTION PROBLEMS

16.1 A conveyor system is fitted with vertical panels, and a 15-in. rod *AB* weighing 5 lb is lodged between two panels as shown. If the rod is to remain in the position shown, determine the maximum allowable acceleration of the system.

Fig. P16.1 and P16.2

16.2 A conveyor system is fitted with vertical panels, and a 15-in. rod *AB* weighing 5 lb is lodged between two panels as shown. Knowing that the acceleration of the system is 3 ft/s² to the left, determine (*a*) the force exerted on the rod at *C*, (*b*) the reaction at *B*.

16.3 Knowing that the coefficient of static friction between the tires and the road is 0.80 for the automobile shown, determine the maximum possible acceleration on a level road, assuming (*a*) four-wheel drive, (*b*) rear-wheel drive, (*c*) front-wheel drive.

Fig. P16.3

16.4 The motion of the 2.5-kg rod *AB* is guided by two small wheels which roll freely in horizontal slots. If a force **P** of magnitude 8 N is applied at *B*, determine (*a*) the acceleration of the rod, (*b*) the reactions at *A* and *B*.

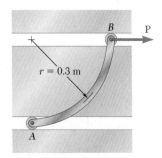

Fig. P16.4

16.5 A uniform rod *BC* of mass 4 kg is connected to a collar *A* by a 250-mm cord *AB*. Neglecting the mass of the collar and cord, determine (*a*) the smallest constant acceleration $\mathbf{a}_A$ for which the cord and the rod will lie in a straight line, (*b*) the corresponding tension in the cord.

Fig. P16.5

16.6 A 2000-kg truck is being used to lift a 400-kg boulder B that is on a 50-kg pallet A. Knowing the acceleration of the rear-wheel-drive truck is 1 m/s², determine (a) the reaction at each of the front wheels, (b) the force between the boulder and the pallet.

Fig. P16.6

16.7 The support bracket shown is used to transport a cylindrical can from one elevation to another. Knowing that $\mu_s = 0.25$ between the can and the bracket, determine (a) the magnitude of the upward acceleration **a** for which the can will slide on the bracket, (b) the smallest ratio h/d for which the can will tip before it slides.

Fig. P16.7

16.8 Solve Prob. 16.7, assuming that the acceleration **a** of the bracket is directed downward.

16.9 A 20-kg cabinet is mounted on casters that allow it to move freely ($\mu = 0$) on the floor. If a 100-N force is applied as shown, determine (a) the acceleration of the cabinet, (b) the range of values of h for which the cabinet will not tip.

Fig. P16.9

16.10 Solve Prob. 16.9, assuming that the casters are locked and slide on the rough floor ($\mu_k = 0.25$).

16.11 A completely filled barrel and its contents have a combined mass of 90 kg. A cylinder C is connected to the barrel at a height $h = 550$ mm as shown. Knowing $\mu_s = 0.40$ and $\mu_k = 0.35$, determine the maximum mass of C so the barrel will not tip.

Fig. P16.11

16.12 A 40-kg vase has a 200-mm-diameter base and is being moved using a 100-kg utility cart as shown. The cart moves freely ($\mu = 0$) on the ground. Knowing the coefficient of static friction between the vase and the cart is $\mu_s = 0.4$, determine the maximum force **F** that can be applied if the vase is not to slide or tip.

Fig. P16.12

16.13 The retractable shelf shown is supported by two identical linkage-and-spring systems; only one of the systems is shown. A 20-kg machine is placed on the shelf so that half of its weight is supported by the system shown. If the springs are removed and the system is released from rest, determine (a) the acceleration of the machine, (b) the tension in link AB. Neglect the weight of the shelf and links.

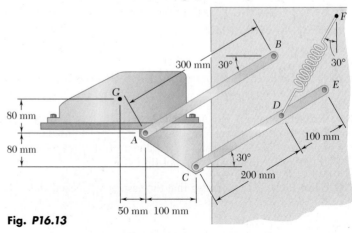

Fig. P16.13

16.14 A uniform rectangular plate has a mass of 5 kg and is held in position by three ropes as shown. Knowing that $\theta = 30°$, determine, immediately after rope CF has been cut, (*a*) the acceleration of the plate, (*b*) the tension in ropes AD and BE.

Fig. P16.14 and P16.15

16.15 A uniform rectangular plate has a mass of 5 kg and is held in position by three ropes as shown. Determine the largest value of θ for which both ropes AD and BE remain taut immediately after rope CF has been cut.

16.16 Three bars, each of mass 3 kg, are welded together and pin-connected to two links BE and CF. Neglecting the weight of the links, determine the force in each link immediately after the system is released from rest.

Fig. P16.16

16.17 Members ACE and DCB are each 600 mm long and are connected by a pin at C. The mass center of the 10-kg member AB is located at G. Determine (*a*) the acceleration of AB immediately after the system has been released from rest in the position shown, (*b*) the corresponding force exerted by roller A on member AB. Neglect the weight of members ACE and DCB.

16.18 The 15-lb rod BC connects a disk centered at A to crank CD. Knowing that the disk is made to rotate at the constant speed of 180 rpm, determine for the position shown the vertical components of the forces exerted on rod BC by pins at B and C.

Fig. P16.17

Fig. P16.18

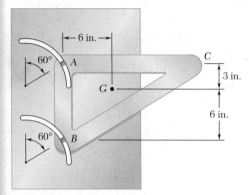

Fig. P16.19

16.19 The triangular weldment ABC is guided by two pins that slide freely in parallel curved slots of radius 6 in. cut in a vertical plate. The weldment weighs 16 lb and its mass center is located at point G. Knowing that at the instant shown the velocity of each pin is 30 in./s downward along the slots, determine (a) the acceleration of the weldment, (b) the reactions at A and B.

16.20 The coefficients of friction between the 30-lb block and the 5-lb platform BD are $\mu_s = 0.50$ and $\mu_k = 0.40$. Determine the accelerations of the block and of the platform immediately after wire AB has been cut.

Fig. P16.20

16.21 Draw the shear and bending-moment diagrams for the vertical rod AB of Prob. 16.16.

***16.22** Draw the shear and bending-moment diagrams for the connecting rod BC of Prob. 16.18.

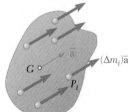

Fig. P16.23

16.23 For a rigid slab in translation, show that the system of the effective forces consists of vectors $(\Delta m_i)\bar{\mathbf{a}}$ attached to the various particles of the slab, where $\bar{\mathbf{a}}$ is the acceleration of the mass center G of the slab. Further show, by computing their sum and the sum of their moments about G, that the effective forces reduce to a single vector $m\bar{\mathbf{a}}$ attached at G.

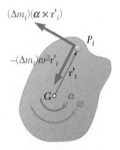

Fig. P16.24

16.24 For a rigid slab in centroidal rotation, show that the system of the effective forces consists of vectors $-(\Delta m_i)\omega^2\mathbf{r}'_i$ and $(\Delta m_i)(\boldsymbol{\alpha} \times \mathbf{r}'_i)$ attached to the various particles P_i of the slab, where $\boldsymbol{\omega}$ and $\boldsymbol{\alpha}$ are the angular velocity and angular acceleration of the slab, and where $\mathbf{r}'_i$ denotes the position vector of the particle P_i relative to the mass center G of the slab. Further show, by computing their sum and the sum of their moments about G, that the effective forces reduce to a couple $\bar{I}\boldsymbol{\alpha}$.

16.25 The rotor of an electric motor has an angular velocity of 3600 rpm when the load and power are cut off. The 50-kg rotor, which has a centroidal radius of gyration of 180 mm, then coasts to rest. Knowing that kinetic friction results in a couple of magnitude 3.5 N · m exerted on the rotor, determine the number of revolutions that the rotor executes before coming to rest.

16.26 It takes 10 min for a 6000-lb flywheel to coast to rest from an angular velocity of 300 rpm. Knowing that the radius of gyration of the flywheel is 36 in., determine the average magnitude of the couple due to kinetic friction in the bearings.

16.27 The 8-in.-radius brake drum is attached to a larger flywheel that is not shown. The total mass moment of inertia of the drum and the flywheel is 14 lb · ft · s^2 and the coefficient of kinetic friction between the drum and the brake shoe is 0.35. Knowing that the angular velocity of the flywheel is 360 rpm counterclockwise when a force **P** of magnitude 75 lb is applied to the pedal C, determine the number of revolutions executed by the flywheel before it comes to rest.

16.28 Solve Prob. 16.27, assuming that the initial angular velocity of the flywheel is 360 rpm clockwise.

16.29 The 100-mm-radius brake drum is attached to a flywheel which is not shown. The drum and flywheel together have a mass of 300 kg and a radius of gyration of 600 mm. The coefficient of kinetic friction between the brake band and the drum is 0.30. Knowing that a force **P** of magnitude 50 N is applied at A when the angular velocity is 180 rpm counterclockwise, determine the time required to stop the flywheel when $a = 200$ mm and $b = 160$ mm.

Fig. P16.27

Fig. P16.29

16.30 The 180-mm-radius disk is at rest when it is placed in contact with a belt moving at a constant speed. Neglecting the weight of the link AB and knowing that the coefficient of kinetic friction between the disk and the belt is 0.40, determine the angular acceleration of the disk while slipping occurs.

16.31 Solve Prob. 16.30, assuming that the direction of motion of the belt is reversed.

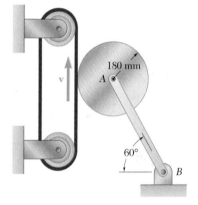

Fig. P16.30

16.32 In order to determine the mass moment of inertia of a flywheel of radius 600 mm, a 12-kg block is attached to a wire that is wrapped around the flywheel. The block is released and is observed to fall 3 m in 4.6 s. To eliminate bearing friction from the computation, a second block of mass 24 kg is used and is observed to fall 3 m in 3.1 s. Assuming that the moment of the couple due to friction remains constant, determine the mass moment of inertia of the flywheel.

16.33 The flywheel shown has a radius of 20 in., a weight of 250 lb, and a radius of gyration of 15 in. A 30-lb block A is attached to a wire that is wrapped around the flywheel, and the system is released from rest. Neglecting the effect of friction, determine (*a*) the acceleration of block A, (*b*) the speed of block A after it has moved 5 ft.

Fig. *P16.32* and P16.33

16.34 Each of the double pulleys shown has a mass moment of inertia of 15 lb · ft · s² and is initially at rest. The outside radius is 18 in., and the inner radius is 9 in. Determine (*a*) the angular acceleration of each pulley, (*b*) the angular velocity of each pulley after point *A* on the cord has moved 10 ft.

Fig. **P16.34**

16.35 Each of the gears *A* and *B* has a mass of 9 kg and has a radius of gyration of 200 mm; gear *C* has a mass of 3 kg and has a radius of gyration of 75 mm. If a couple **M** of constant magnitude 5 N-m is applied to gear *C*, determine (*a*) the angular acceleration of gear *A*, (*b*) the tangential force which gear *C* exerts on gear *A*.

Fig. **P16.35**

16.36 Solve Prob. 16.35, assuming that the couple **M** is applied to disk *A*.

16.37 Gear *A* weighs 1 lb and has a radius of gyration of 1.3 in; gear *B* weighs 6 lb and has a radius of gyration of 3 in.; gear *C* weighs 9 lb and has a radius of gyration of 4.3 in. Knowing a couple **M** of constant magnitude of 40 lb · in is applied to gear *A*, determine (*a*) the angular acceleration of gear *C*, (*b*) the tangential force which gear *B* exerts on gear *C*.

Fig. **P16.37**

16.38 Disks A and B are bolted together, and cylinders D and E are attached to separate cords wrapped on the disks. A single cord passes over disks B and C. Disk A weighs 20 lb and disks B and C each weigh 12 lb. Knowing that the system is released from rest and that no slipping occurs between the cords and the disks, determine the acceleration (a) of cylinder D, (b) of cylinder E.

Fig. P16.38

16.39 A belt of negligible mass passes between cylinders A and B and is pulled to the right with a force **P**. Cylinders A and B weigh, respectively, 5 and 20 lb. The shaft of cylinder A is free to slide in a vertical slot and the coefficients of friction between the belt and each of the cylinders are $\mu_s = 0.50$ and $\mu_k = 0.40$. For $P = 3.6$ lb, determine (a) whether slipping occurs between the belt and either cylinder, (b) the angular acceleration of each cylinder.

Fig. P16.39

16.40 Solve Prob. 16.39 for $P = 2.00$ lb.

16.41 Disk A has a mass of 6 kg and an initial angular velocity of 360 rpm clockwise; disk B has a mass of 3 kg and is initially at rest. The disks are brought together by applying a horizontal force of magnitude 20 N to the axle of disk A. Knowing that $\mu_k = 0.15$ between the disks and neglecting bearing friction, determine (a) the angular acceleration of each disk, (b) the final angular velocity of each disk.

Fig. P16.41

16.42 Solve Prob. 16.41, assuming that initially disk A is at rest and disk B has an angular velocity of 360 rpm clockwise.

Fig. P16.43 and P16.44

16.43 Disk A has a mass $m_A = 4$ kg, a radius $r_A = 300$ mm, and an initial angular velocity $\boldsymbol{\omega}_0 = 300$ rpm clockwise. Disk B has a mass $m_B = 1.6$ kg, a radius $r_B = 180$ mm, and is at rest when it is brought into contact with disk A. Knowing that $\mu_k = 0.35$ between the disks and neglecting bearing friction, determine (a) the angular acceleration of each disk, (b) the reaction at the support C.

16.44 Disk B is at rest when it is brought into contact with disk A, which has an initial angular velocity $\boldsymbol{\omega}_0$. (a) Show that the final angular velocities of the disks are independent of the coefficient of friction μ_k between the disks as long as $\mu_k \neq 0$. (b) Express the final angular velocity of disk A in terms of ω_0 and the ratio of the masses of the two disks m_A/m_B.

16.45 Cylinder A has an initial angular velocity of 720 rpm clockwise, and cylinders B and C are initially at rest. Disks A and B each weigh 5 lb and have radius $r = 4$ in. Disk C weighs 20 lb and has a radius of 8 in. The disks are brought together when C is placed gently onto A and B. Knowing that $\mu_k = 0.25$ between A and C and no slipping occurs between B and C, determine (a) the angular acceleration of each disk, (b) the final angular velocity of each disk.

Fig. P16.45

16.46 Show that the system of the effective forces for a rigid slab in plane motion reduces to a single vector, and express the distance from the mass center G of the slab to the line of action of this vector in terms of the centroidal radius of gyration $\bar{k}$ of the slab, the magnitude $\bar{a}$ of the acceleration of G, and the angular acceleration α.

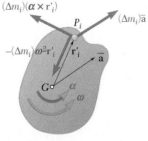

Fig. P16.47

16.47 For a rigid slab in plane motion, show that the system of the effective forces consists of vectors $(\Delta m_i)\bar{\mathbf{a}}$, $-(\Delta m_i)\omega^2\mathbf{r}'_i$, and $(\Delta m_i)(\boldsymbol{\alpha} \times \mathbf{r}'_i)$ attached to the various particles P_i of the slab, where $\bar{\mathbf{a}}$ is the acceleration of the mass center G of the slab, $\boldsymbol{\omega}$ is the angular velocity of the slab, $\boldsymbol{\alpha}$ is its angular acceleration, and $\mathbf{r}'_i$ denotes the position vector of the particle P_i, relative to G. Further show, by computing their sum and the sum of their moments about G, that the effective forces reduce to a vector $m\bar{\mathbf{a}}$ attached at G and a couple $\bar{I}\boldsymbol{\alpha}$.

16.48 A uniform slender rod AB rests on a frictionless horizontal surface, and a force **P** of magnitude 0.25 lb is applied at A in a direction perpendicular to the rod. Knowing that the rod weighs 1.75 lb, determine (*a*) the acceleration of point A, (*b*) the acceleration of point B, (*c*) the location of the point on the bar that has zero acceleration.

16.49 (*a*) In Prob. 16.48, determine the point of the rod AB at which the force **P** should be applied if the acceleration of point B is to be zero. (*b*) Knowing that $P = 0.25$ lb, determine the corresponding acceleration of point A.

16.50 A force **P** of magnitude 3 N is applied to a tape wrapped around a thin hoop of mass 2.4 kg. Knowing that the body rests on a frictionless horizontal surface, determine the acceleration of (*a*) point A, (*b*) point B.

Fig. P16.48

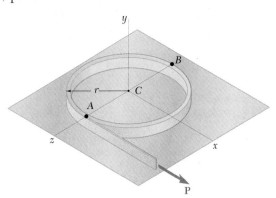

Fig. P16.50

16.51 A force **P** is applied to a tape wrapped around a uniform disk that rests on a frictionless horizontal surface. Show that for each 360° rotation of the disk the center of the disk will move a distance πr.

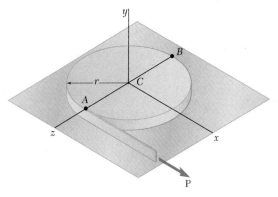

Fig. P16.51

16.52 A 250-lb satellite has a radius of gyration of 24 in. with respect to the y axis and is symmetrical with respect to the zx plane. Its orientation is changed by firing four small rockets A, B, C, and D, each of which produces a 4-lb thrust **T** directed as shown. Determine the angular acceleration of the satellite and the acceleration of its mass center G (*a*) when all four rockets are fired, (*b*) when all rockets except D are fired.

Fig. P16.52

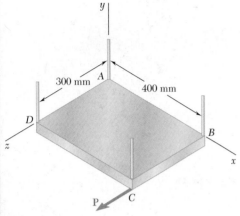

Fig. P16.53

16.53 A rectangular plate of mass 5 kg is suspended from four vertical wires, and a force **P** of magnitude 6 N is applied to corner C as shown. Immediately after **P** is applied, determine the acceleration of (a) the midpoint of edge BC, (b) corner B.

16.54 A uniform slender L-shaped bar ABC is at rest on a horizontal surface when a force **P** of magnitude 4 N is applied at point A. Neglecting friction between the bar and the surface and knowing that the mass of the bar is 2 kg, determine (a) the initial angular acceleration of the bar, (b) the initial acceleration of point B.

Fig. P16.54

16.55 By pulling on the string of a yo-yo, a person manages to make the yo-yo spin, while remaining at the same elevation above the floor. Denoting the mass of the yo-yo by m, the radius of the inner drum on which the string is wound by r, and the centroidal radius of gyration of the yo-yo by $\bar{k}$, determine the angular acceleration of the yo-yo.

Fig. P16.55 and P16.56

16.56 The 80-g yo-yo shown has a centroidal radius of gyration of 30 mm. The radius of the inner drum on which a string is wound is 6 mm. Knowing that at the instant shown the acceleration of the center of the yo-yo is 1 m/s² upward, determine (a) the required tension **T** in the string, (b) the corresponding angular acceleration of the yo-yo.

16.57 A 6-lb sprocket wheel has a centroidal radius of gyration of 2.75 in. and is suspended from a chain as shown. Determine the acceleration of points A and B of the chain, knowing that $T_A = 3$ lb and $T_B = 4$ lb.

16.58 The steel roll shown has a mass of 1200 kg, a centroidal radius of gyration of 150 mm, and is lifted by two cables looped around its shaft. Knowing that for each cable $T_A = 3100$ N and $T_B = 3300$ N, determine (*a*) the angular acceleration of the roll, (*b*) the acceleration of its mass center.

Fig. P16.57

Fig. P16.58 and P16.59

16.59 The steel roll shown has a mass of 1200 kg, has a centroidal radius of gyration of 150 mm, and is lifted by two cables looped around its shaft. Knowing that at the instant shown the acceleration of the roll is 150 mm/s² downward and that for each cable $T_A = 3000$ N, determine (*a*) the corresponding tension T_B, (*b*) the angular acceleration of the roll.

16.60 and 16.61 A 15-ft beam weighing 500 lb is lowered by means of two cables unwinding from overhead cranes. As the beam approaches the ground, the crane operators apply brakes to slow the unwinding motion. Knowing that the deceleration of cable A is 20 ft/s² and the deceleration of cable B is 2 ft/s², determine the tension in each cable.

Fig. P16.60

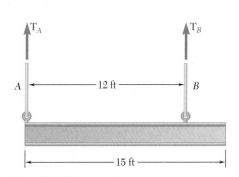

Fig. P16.61

16.62 Two uniform cylinders, each of weight $W = 14$ lb and radius $r = 5$ in., are connected by a belt as shown. If the system is released from rest, determine (*a*) the angular acceleration of each cylinder, (*b*) the tension in the portion of belt connecting the two cylinders, (*c*) the velocity of the center of the cylinder *A* after it has moved through 3 ft.

Fig. *P16.62*

Fig. P16.63

16.63 through 16.65   A beam *AB* of mass *m* and of uniform cross section is suspended from two springs as shown. If spring 2 breaks, determine at that instant (*a*) the angular acceleration of the bar, (*b*) the acceleration of point *A*, (*c*) the acceleration of point *B*.

Fig. P16.64

Fig. P16.65

16.66 through 16.68   A thin plate of the shape indicated and of mass *m* is suspended from two springs as shown. If spring 2 breaks, determine the acceleration at that instant (*a*) of point *A*, (*b*) of point *B*.

    16.66   A square plate of side *b*
    16.67   A circular plate of diameter *b*
    16.68   A rectangular plate of height *b* and width *a*

Fig. P16.66

Fig. P16.67

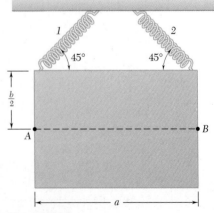

Fig. *P16.68*

16.69 A sphere of radius r and mass m is projected along a rough horizontal surface with the initial velocities indicated. If the final velocity of the sphere is to be zero, express, in terms of v_0, r, and μ_k, (a) the required magnitude of $\boldsymbol{\omega}_0$, (b) the time t_1 required for the sphere to come to rest, (c) the distance the sphere will move before coming to rest.

16.70 Solve Prob. 16.69, assuming that the sphere is replaced by a uniform thin hoop of radius r and mass m.

16.71 A bowler projects an 8-in.-diameter ball weighing 12 lb along an alley with a forward velocity $\mathbf{v}_0$ of 15 ft/s and a backspin $\boldsymbol{\omega}_0$ of 9 rad/s. Knowing that the coefficient of kinetic friction between the ball and the alley is 0.10, determine (a) the time t_1 at which the ball will start rolling without sliding, (b) the speed of the ball at time t_1, (c) the distance the ball will have traveled at time t_1.

Fig. P16.69

Fig. P16.71

16.72 Solve Prob. 16.71, assuming that the bowler projects the ball with the same forward velocity but with a backspin of 18 rad/s.

16.73 A uniform sphere of radius r and mass m is placed with no initial velocity on a belt that moves to the right with a constant velocity $\mathbf{v}_1$. Denoting by μ_k the coefficient of kinetic friction between the sphere and the belt, determine (a) the time t_1 at which the sphere will start rolling without sliding, (b) the linear and angular velocities of the sphere at time t_1.

Fig. P16.73

16.74 A sphere of radius r and mass m has a linear velocity $\mathbf{v}_0$ directed to the left and no angular velocity as it is placed on a belt moving to the right with a constant velocity $\mathbf{v}_1$. If after first sliding on the belt the sphere is to have no linear velocity relative to the ground as it starts rolling on the belt without sliding, determine in terms of v_1 and the coefficient of kinetic friction μ_k between the sphere and the belt (a) the required value of v_0, (b) the time t_1 at which the sphere will start rolling on the belt, (c) the distance the sphere will have moved relative to the ground at time t_1.

Fig. P16.74

16.8 CONSTRAINED PLANE MOTION

Most engineering applications deal with rigid bodies which are moving under given constraints. For example, cranks must rotate about a fixed axis, wheels must roll without sliding, and connecting rods must describe certain prescribed motions. In all such cases, definite relations exist between the components of the acceleration $\bar{\mathbf{a}}$ of the mass center G of the body considered and its angular acceleration $\boldsymbol{\alpha}$; the corresponding motion is said to be a *constrained motion*.

The solution of a problem involving a constrained plane motion calls first for a *kinematic analysis* of the problem. Consider, for example, a slender rod AB of length l and mass m whose extremities are connected to blocks of negligible mass which slide along horizontal and vertical frictionless tracks. The rod is pulled by a force $\mathbf{P}$ applied at A (Fig. 16.11). We know from Sec. 15.8 that the acceleration $\bar{\mathbf{a}}$ of the mass center G of the rod can be determined at any given instant from the position of the rod, its angular velocity, and its angular acceleration at that instant. Suppose, for example, that the values of θ, ω, and α are known at a given instant and that we wish to determine the corresponding value of the force $\mathbf{P}$, as well as the reactions at A and B. We should first *determine the components $\bar{a}_x$ and $\bar{a}_y$ of the acceleration of the mass center G* by the method of Sec. 15.8. We next apply d'Alembert's principle (Fig. 16.12), using the expressions obtained for $\bar{a}_x$ and $\bar{a}_y$. The unknown forces $\mathbf{P}$, $\mathbf{N}_A$, and $\mathbf{N}_B$ can then be determined by writing and solving the appropriate equations.

Fig. 16.11

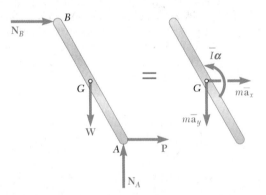

Fig. 16.12

Suppose now that the applied force $\mathbf{P}$, the angle θ, and the angular velocity ω of the rod are known at a given instant and that we wish to find the angular acceleration α of the rod and the components $\bar{a}_x$ and $\bar{a}_y$ of the acceleration of its mass center at that instant, as well as the reactions at A and B. The preliminary kinematic study of the problem will have for its object *to express the components $\bar{a}_x$ and $\bar{a}_y$ of the acceleration of G in terms of the angular acceleration α of the rod.* This will be done by first expressing the acceleration of a suitable reference point such as A in terms of the angular acceleration α. The components $\bar{a}_x$ and $\bar{a}_y$ of the acceleration of G can then be determined in terms of α, and the expressions obtained carried into Fig. 16.12. Three equations can then be derived in terms of α, N_A, and N_B and solved for the three unknowns (see Sample

Prob. 16.10). Note that the method of dynamic equilibrium can also be used to carry out the solution of the two types of problems we have considered (Fig. 16.13).

When a mechanism consists of *several moving parts*, the approach just described can be used with each part of the mechanism. The procedure required to determine the various unknowns is then similar to the procedure followed in the case of the equilibrium of a system of connected rigid bodies (Sec. 6.11).

Earlier, we analyzed two particular cases of constrained plane motion: the translation of a rigid body, in which the angular acceleration of the body is constrained to be zero, and the centroidal rotation, in which the acceleration $\bar{\mathbf{a}}$ of the mass center of the body is constrained to be zero. Two other particular cases of constrained plane motion are of special interest: the *noncentroidal rotation* of a rigid body and the *rolling motion* of a disk or wheel. These two cases can be analyzed by one of the general methods described above. However, in view of the range of their applications, they deserve a few special comments.

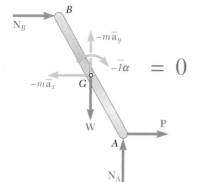

Fig. 16.13

Noncentroidal Rotation.

The motion of a rigid body constrained to rotate about a fixed axis which does not pass through its mass center is called *noncentroidal rotation*. The mass center G of the body moves along a circle of radius $\bar{r}$ centered at the point O, where the axis of rotation intersects the plane of reference (Fig. 16.14). Denoting, respectively, by $\boldsymbol{\omega}$ and $\boldsymbol{\alpha}$ the angular velocity and the angular acceleration of the line OG, we obtain the following expressions for the tangential and normal components of the acceleration of G:

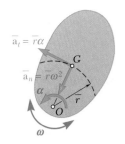

Fig. 16.14

$$\bar{a}_t = \bar{r}\alpha \qquad \bar{a}_n = \bar{r}\omega^2 \tag{16.7}$$

Since line OG belongs to the body, its angular velocity $\boldsymbol{\omega}$ and its angular acceleration $\boldsymbol{\alpha}$ also represent the angular velocity and the angular acceleration of the body in its motion relative to G. Equations (16.7) define, therefore, the kinematic relation existing between the motion of the mass center G and the motion of the body about G. They should be used to eliminate $\bar{a}_t$ and $\bar{a}_n$ from the equations obtained by applying d'Alembert's principle (Fig. 16.15) or the method of dynamic equilibrium (Fig. 16.16).

(a)   (b)

Fig. 16.15

Fig. 16.16

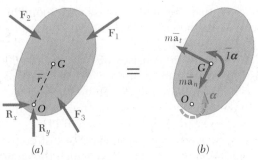

Fig. 16.15   (repeated)

An interesting relation is obtained by equating the moments about the fixed point O of the forces and vectors shown, respectively, in parts a and b of Fig. 16.15. We write

$$+\curvearrowleft\Sigma M_O = \bar{I}\alpha + (m\bar{r}\alpha)\bar{r} = (\bar{I} + m\bar{r}^2)\alpha$$

But according to the parallel-axis theorem, we have $\bar{I} + m\bar{r}^2 = I_O$, where I_O denotes the moment of inertia of the rigid body about the fixed axis. We therefore write

$$\Sigma M_O = I_O\alpha \qquad (16.8)$$

Although formula (16.8) expresses an important relation between the sum of the moments of the external forces about the fixed point O and the product $I_O\alpha$, it should be clearly understood that this formula does not mean that the system of the external forces is equivalent to a couple of moment $I_O\alpha$. The system of the effective forces, and thus the system of the external forces, reduces to a couple only when O coincides with G—that is, *only when the rotation is centroidal* (Sec. 16.4). In the more general case of noncentroidal rotation, the system of the external forces does not reduce to a couple.

A particular case of noncentroidal rotation is of special interest—the case of *uniform rotation,* in which the angular velocity $\boldsymbol{\omega}$ is constant. Since $\boldsymbol{\alpha}$ is zero, the inertia couple in Fig. 16.16 vanishes and the inertia vector reduces to its normal component. This component (also called *centrifugal force*) represents the tendency of the rigid body to break away from the axis of rotation.

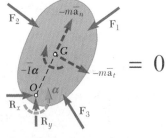

Fig. 16.16   (repeated)

Rolling Motion.   Another important case of plane motion is the motion of a disk or wheel rolling on a plane surface. If the disk is constrained to roll without sliding, the acceleration $\bar{\mathbf{a}}$ of its mass center G and its angular acceleration $\boldsymbol{\alpha}$ are not independent. Assuming that the disk is balanced, so that its mass center and its geometric center coincide, we first write that the distance $\bar{x}$ traveled by G during a rotation θ of the disk is $\bar{x} = r\theta$, where r is the radius of the disk. Differentiating this relation twice, we write

$$\bar{a} = r\alpha \qquad (16.9)$$

Recalling that the system of the effective forces in plane motion reduces to a vector $m\bar{\mathbf{a}}$ and a couple $\bar{I}\boldsymbol{\alpha}$, we find that in the particular case of the rolling motion of a balanced disk, the effective forces reduce to a vector of magnitude $mr\alpha$ attached at G and to a couple of magnitude $\bar{I}\alpha$. We may thus express that the external forces are equivalent to the vector and couple shown in Fig. 16.17.

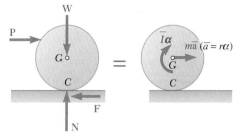

Fig. 16.17

When a disk *rolls without sliding*, there is no relative motion between the point of the disk in contact with the ground and the ground itself. Thus, as far as the computation of the friction force $\mathbf{F}$ is concerned, a rolling disk can be compared with a block at rest on a surface. The magnitude F of the friction force can have any value, as long as this value does not exceed the maximum value $F_m = \mu_s N$, where μ_s is the coefficient of static friction and N is the magnitude of the normal force. In the case of a rolling disk, the magnitude F of the friction force should therefore be determined independently of N by solving the equation obtained from Fig. 16.17.

When *sliding is impending*, the friction force reaches its maximum value $F_m = \mu_s N$ and can be obtained from N.

When the disk *rotates and slides* at the same time, a relative motion exists between the point of the disk which is in contact with the ground and the ground itself, and the force of friction has the magnitude $F_k = \mu_k N$, where μ_k is the coefficient of kinetic friction. In this case, however, the motion of the mass center G of the disk and the rotation of the disk about G are independent, and $\bar{a}$ is not equal to $r\alpha$.

These three different cases can be summarized as follows:

Rolling, no sliding: $F \le \mu_s N$ $\bar{a} = r\alpha$
Rolling, sliding impending: $F = \mu_s N$ $\bar{a} = r\alpha$
Rotating and sliding: $F = \mu_k N$ $\bar{a}$ *and* α *independent*

When it is not known whether or not a disk slides, it should first be assumed that the disk rolls without sliding. If F is found smaller than or equal to $\mu_s N$, the assumption is proved correct. If F is found larger than $\mu_s N$, the assumption is incorrect and the problem should be started again, assuming rotating and sliding.

When a disk is *unbalanced*, i.e., when its mass center G does not coincide with its geometric center O, the relation (16.9) does not hold between $\bar{a}$ and α. However, a similar relation holds between the magnitude a_O of the acceleration of the geometric center and the angular acceleration α of an unbalanced disk which rolls without sliding. We have

$$a_O = r\alpha \qquad (16.10)$$

To determine $\bar{a}$ in terms of the angular acceleration α and the angular velocity ω of the disk, we can use the relative-acceleration formula

$$\begin{aligned} \bar{\mathbf{a}} = \bar{\mathbf{a}}_G &= \mathbf{a}_O + \mathbf{a}_{G/O} \\ &= \mathbf{a}_O + (\mathbf{a}_{G/O})_t + (\mathbf{a}_{G/O})_n \end{aligned} \qquad (16.11)$$

where the three component accelerations obtained have the directions indicated in Fig. 16.18 and the magnitudes $a_O = r\alpha$, $(a_{G/O})_t = (OG)\alpha$, and $(a_{G/O})_n = (OG)\omega^2$.

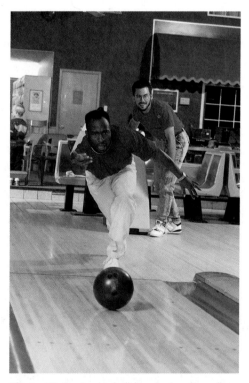

Photo 16.4 As the ball hits the bowling alley, it first spins and slides, then rolls without sliding.

Fig. 16.18

SAMPLE PROBLEM 16.6

The portion AOB of a mechanism consists of a 400-mm steel rod OB welded to a gear E of radius 120 mm which can rotate about a horizontal shaft O. It is actuated by a gear D and, at the instant shown, has a clockwise angular velocity of 8 rad/s and a counterclockwise angular acceleration of 40 rad/s². Knowing that rod OB has a mass of 3 kg and gear E a mass of 4 kg and a radius of gyration of 85 mm, determine (a) the tangential force exerted by gear D on gear E, (b) the components of the reaction at shaft O.

SOLUTION

In determining the effective forces of the rigid body AOB, gear E and rod OB will be considered separately. Therefore, the components of the acceleration of the mass center G_{OB} of the rod will be determined first:

$$(\bar{a}_{OB})_t = \bar{r}\alpha = (0.200 \text{ m})(40 \text{ rad/s}^2) = 8 \text{ m/s}^2$$
$$(\bar{a}_{OB})_n = \bar{r}\omega^2 = (0.200 \text{ m})(8 \text{ rad/s})^2 = 12.8 \text{ m/s}^2$$

Equations of Motion. Two sketches of the rigid body AOB have been drawn. The first shows the external forces consisting of the weight $\mathbf{W}_E$ of gear E, the weight $\mathbf{W}_{OB}$ of the rod OB, the force $\mathbf{F}$ exerted by gear D, and the components $\mathbf{R}_x$ and $\mathbf{R}_y$ of the reaction at O. The magnitudes of the weights are, respectively,

$$W_E = m_E g = (4 \text{ kg})(9.81 \text{ m/s}^2) = 39.2 \text{ N}$$
$$W_{OB} = m_{OB} g = (3 \text{ kg})(9.81 \text{ m/s}^2) = 29.4 \text{ N}$$

The second sketch shows the effective forces, which consist of a couple $\bar{I}_E\alpha$ (since gear E is in centroidal rotation) and of a couple and two vector components at the mass center of OB. Since the accelerations are known, we compute the magnitudes of these components and couples:

$$\bar{I}_E\alpha = m_E\bar{k}_E^2\alpha = (4 \text{ kg})(0.085 \text{ m})^2(40 \text{ rad/s}^2) = 1.156 \text{ N} \cdot \text{m}$$
$$m_{OB}(\bar{a}_{OB})_t = (3 \text{ kg})(8 \text{ m/s}^2) = 24.0 \text{ N}$$
$$m_{OB}(\bar{a}_{OB})_n = (3 \text{ kg})(12.8 \text{ m/s}^2) = 38.4 \text{ N}$$
$$\bar{I}_{OB}\alpha = (\tfrac{1}{12}m_{OB}L^2)\alpha = \tfrac{1}{12}(3 \text{ kg})(0.400 \text{ m})^2(40 \text{ rad/s}^2) = 1.600 \text{ N} \cdot \text{m}$$

Expressing that the system of the external forces is equivalent to the system of the effective forces, we write the following equations:

$+\circlearrowleft\Sigma M_O = \Sigma(M_O)_{\text{eff}}$:

$$F(0.120 \text{ m}) = \bar{I}_E\alpha + m_{OB}(\bar{a}_{OB})_t(0.200 \text{ m}) + \bar{I}_{OB}\alpha$$
$$F(0.120 \text{ m}) = 1.156 \text{ N} \cdot \text{m} + (24.0 \text{ N})(0.200 \text{ m}) + 1.600 \text{ N} \cdot \text{m}$$

$$F = 63.0 \text{ N} \qquad\qquad \mathbf{F} = 63.0 \text{ N} \downarrow \quad \blacktriangleleft$$

$\xrightarrow{+}\Sigma F_x = \Sigma(F_x)_{\text{eff}}$:
$$R_x = m_{OB}(\bar{a}_{OB})_t$$
$$R_x = 24.0 \text{ N} \qquad\qquad \mathbf{R}_x = 24.0 \text{ N} \rightarrow \quad \blacktriangleleft$$

$+\uparrow\Sigma F_y = \Sigma(F_y)_{\text{eff}}$:
$$R_y - F - W_E - W_{OB} = m_{OB}(\bar{a}_{OB})_n$$
$$R_y - 63.0 \text{ N} - 39.2 \text{ N} - 29.4 \text{ N} = 38.4 \text{ N}$$

$$R_y = 170.0 \text{ N} \qquad\qquad \mathbf{R}_y = 170.0 \text{ N} \uparrow \quad \blacktriangleleft$$

SAMPLE PROBLEM 16.7

A 6 × 8 in. rectangular plate weighing 60 lb is suspended from two pins A and B. If pin B is suddenly removed, determine (a) the angular acceleration of the plate, (b) the components of the reaction at pin A, immediately after pin B has been removed.

SOLUTION

a. Angular Acceleration. We observe that as the plate rotates about point A, its mass center G describes a circle of radius $\bar{r}$ with center at A.

Since the plate is released from rest ($\omega = 0$), the normal component of the acceleration of G is zero. The magnitude of the acceleration $\bar{\mathbf{a}}$ of the mass center G is thus $\bar{a} = \bar{r}\alpha$. We draw the diagram shown to express that the external forces are equivalent to the effective forces:

$$+\!\!\downarrow\!\Sigma M_A = \Sigma(M_A)_{\text{eff}}: \qquad W\bar{x} = (m\bar{a})\bar{r} + \bar{I}\alpha$$

Since $\bar{a} = \bar{r}\alpha$, we have

$$W\bar{x} = m(\bar{r}\alpha)\bar{r} + \bar{I}\alpha \qquad \alpha = \frac{W\bar{x}}{\dfrac{W}{g}\bar{r}^2 + \bar{I}} \qquad (1)$$

The centroidal moment of inertia of the plate is

$$\bar{I} = \frac{m}{12}(a^2 + b^2) = \frac{60 \text{ lb}}{12(32.2 \text{ ft/s}^2)}[(\tfrac{8}{12}\text{ ft})^2 + (\tfrac{6}{12}\text{ ft})^2]$$
$$= 0.1078 \text{ lb} \cdot \text{ft} \cdot \text{s}^2$$

Substituting this value of $\bar{I}$ together with $W = 60$ lb, $\bar{r} = \tfrac{5}{12}$ ft, and $\bar{x} = \tfrac{4}{12}$ ft into Eq. (1), we obtain

$$\alpha = +46.4 \text{ rad/s}^2 \qquad \alpha = 46.4 \text{ rad/s}^2 \downarrow \quad \blacktriangleleft$$

b. Reaction at A. Using the computed value of α, we determine the magnitude of the vector $m\bar{\mathbf{a}}$ attached at G.

$$m\bar{a} = m\bar{r}\alpha = \frac{60 \text{ lb}}{32.2 \text{ ft/s}^2}(\tfrac{5}{12}\text{ ft})(46.4 \text{ rad/s}^2) = 36.0 \text{ lb}$$

Showing this result on the diagram, we write the equations of motion

$$\xrightarrow{+}\Sigma F_x = \Sigma(F_x)_{\text{eff}}: \qquad A_x = -\tfrac{3}{5}(36 \text{ lb})$$
$$= -21.6 \text{ lb} \qquad\qquad \mathbf{A}_x = 21.6 \text{ lb} \leftarrow \quad \blacktriangleleft$$

$$+\!\!\uparrow\!\Sigma F_y = \Sigma(F_y)_{\text{eff}}: \qquad A_y - 60 \text{ lb} = -\tfrac{4}{5}(36 \text{ lb})$$
$$A_y = +31.2 \text{ lb} \qquad\qquad \mathbf{A}_y = 31.2 \text{ lb} \uparrow \quad \blacktriangleleft$$

The couple $\bar{I}\boldsymbol{\alpha}$ is not involved in the last two equations; nevertheless, it should be indicated on the diagram.

SAMPLE PROBLEM 16.8

A sphere of radius r and weight W is released with no initial velocity on the incline and rolls without slipping. Determine (a) the minimum value of the coefficient of static friction compatible with the rolling motion, (b) the velocity of the center G of the sphere after the sphere has rolled 10 ft, (c) the velocity of G if the sphere were to move 10 ft down a frictionless 30° incline.

SOLUTION

a. Minimum μ_s for Rolling Motion. The external forces $\mathbf{W}$, $\mathbf{N}$, and $\mathbf{F}$ form a system equivalent to the system of effective forces represented by the vector $m\bar{\mathbf{a}}$ and the couple $\bar{I}\boldsymbol{\alpha}$. Since the sphere rolls without sliding, we have $\bar{a} = r\alpha$.

$+\downarrow\Sigma M_C = \Sigma(M_C)_{\text{eff}}:$
$$(W \sin \theta)r = (m\bar{a})r + \bar{I}\alpha$$
$$(W \sin \theta)r = (mr\alpha)r + \bar{I}\alpha$$

Noting that $m = W/g$ and $\bar{I} = \frac{2}{5}mr^2$, we write

$$(W \sin \theta)r = \left(\frac{W}{g}r\alpha\right)r + \frac{2}{5}\frac{W}{g}r^2\alpha \qquad \alpha = +\frac{5g \sin \theta}{7r}$$

$$\bar{a} = r\alpha = \frac{5g \sin \theta}{7} = \frac{5(32.2 \text{ ft/s}^2) \sin 30°}{7} = 11.50 \text{ ft/s}^2$$

$+\searrow\Sigma F_x = \Sigma(F_x)_{\text{eff}}:$
$$W \sin \theta - F = m\bar{a}$$
$$W \sin \theta - F = \frac{W}{g}\frac{5g \sin \theta}{7}$$

$$F = +\tfrac{2}{7}W \sin \theta = \tfrac{2}{7}W \sin 30° \qquad \mathbf{F} = 0.143W \searrow 30°$$

$+\nearrow\Sigma F_y = \Sigma(F_y)_{\text{eff}}:$
$$N - W \cos \theta = 0$$
$$N = W \cos \theta = 0.866W \qquad \mathbf{N} = 0.866W \angle 60°$$

$$\mu_s = \frac{F}{N} = \frac{0.143W}{0.866W} \qquad \qquad \mu_s = 0.165 \blacktriangleleft$$

b. Velocity of Rolling Sphere. We have uniformly accelerated motion:

$$\bar{v}_0 = 0 \qquad \bar{a} = 11.50 \text{ ft/s}^2 \qquad \bar{x} = 10 \text{ ft} \qquad \bar{x}_0 = 0$$
$$\bar{v}^2 = \bar{v}_0^2 + 2\bar{a}(\bar{x} - \bar{x}_0) \qquad \bar{v}^2 = 0 + 2(11.50 \text{ ft/s}^2)(10 \text{ ft})$$
$$\bar{v} = 15.17 \text{ ft/s} \qquad \mathbf{v} = 15.17 \text{ ft/s} \searrow 30° \blacktriangleleft$$

c. Velocity of Sliding Sphere. Assuming now no friction, we have $F = 0$ and obtain

$+\downarrow\Sigma M_G = \Sigma(M_G)_{\text{eff}}:$
$$0 = \bar{I}\alpha \qquad \alpha = 0$$

$+\searrow\Sigma F_x = \Sigma(F_x)_{\text{eff}}:$
$$W \sin 30° = m\bar{a} \qquad 0.50W = \frac{W}{g}\bar{a}$$

$$\bar{a} = +16.1 \text{ ft/s}^2 \qquad \mathbf{a} = 16.1 \text{ ft/s}^2 \searrow 30°$$

Substituting $\bar{a} = 16.1 \text{ ft/s}^2$ into the equations for uniformly accelerated motion, we obtain

$$\bar{v}^2 = \bar{v}_0^2 + 2\bar{a}(\bar{x} - \bar{x}_0) \qquad \bar{v}^2 = 0 + 2(16.1 \text{ ft/s}^2)(10 \text{ ft})$$
$$\bar{v} = 17.94 \text{ ft/s} \qquad \mathbf{v} = 17.94 \text{ ft/s} \searrow 30° \blacktriangleleft$$

100 mm   60 mm

G

200 N

SAMPLE PROBLEM 16.9

A cord is wrapped around the inner drum of a wheel and pulled horizontally with a force of 200 N. The wheel has a mass of 50 kg and a radius of gyration of 70 mm. Knowing that $\mu_s = 0.20$ and $\mu_k = 0.15$, determine the acceleration of G and the angular acceleration of the wheel.

SOLUTION

α

G $\bar{a}$

$r = 0.100$ m

C

a. Assume Rolling without Sliding. In this case, we have

$$\bar{a} = r\alpha = (0.100 \text{ m})\alpha$$

We can determine whether this assumption is justified by comparing the friction force obtained with the maximum available friction force. The moment of inertia of the wheel is

$$\bar{I} = m\bar{k}^2 = (50 \text{ kg})(0.070 \text{ m})^2 = 0.245 \text{ kg} \cdot \text{m}^2$$

Equations of Motion

W

=

G $\bar{I}\alpha$ $m\bar{a}$

200 N   G

C   C   0.100 m

F   0.040 m

N

$+\downarrow \Sigma M_C = \Sigma(M_C)_{\text{eff}}$: $(200 \text{ N})(0.040 \text{ m}) = m\bar{a}(0.100 \text{ m}) + \bar{I}\alpha$
$8.00 \text{ N} \cdot \text{m} = (50 \text{ kg})(0.100 \text{ m})\alpha(0.100 \text{ m}) + (0.245 \text{ kg} \cdot \text{m}^2)\alpha$
$\alpha = +10.74 \text{ rad/s}^2$
$\bar{a} = r\alpha = (0.100 \text{ m})(10.74 \text{ rad/s}^2) = 1.074 \text{ m/s}^2$

$\xrightarrow{+} \Sigma F_x = \Sigma(F_x)_{\text{eff}}$: $F + 200 \text{ N} = m\bar{a}$
$F + 200 \text{ N} = (50 \text{ kg})(1.074 \text{ m/s}^2)$
$F = -146.3 \text{ N}$ $\mathbf{F} = 146.3 \text{ N} \leftarrow$

$+\uparrow \Sigma F_y = \Sigma(F_y)_{\text{eff}}$:
$N - W = 0$ $N - W = mg = (50 \text{ kg})(9.81 \text{ m/s}^2) = 490.5 \text{ N}$
$\mathbf{N} = 490.5 \text{ N} \uparrow$

Maximum Available Friction Force

$$F_{\text{max}} = \mu_s N = 0.20(490.5 \text{ N}) = 98.1 \text{ N}$$

Since $F > F_{\text{max}}$, the assumed motion is impossible.

b. Rotating and Sliding. Since the wheel must rotate and slide at the same time, we draw a new diagram, where $\bar{\mathbf{a}}$ and $\boldsymbol{\alpha}$ are independent and where

$$F = F_k = \mu_k N = 0.15(490.5 \text{ N}) = 73.6 \text{ N}$$

W

0.060 m

=

G $\bar{I}\alpha$ $m\bar{a}$

200 N   G

C   C   0.100 m

F = 73.6 N

N

From the computation of part *a*, it appears that $\mathbf{F}$ should be directed to the left. We write the following equations of motion:

$\xrightarrow{+} \Sigma F_x = \Sigma(F_x)_{\text{eff}}$: $200 \text{ N} - 73.6 \text{ N} = (50 \text{ kg})\bar{a}$
$\bar{a} = +2.53 \text{ m/s}^2$ $\bar{\mathbf{a}} = 2.53 \text{ m/s}^2 \rightarrow$ ◄

$+\downarrow \Sigma M_G = \Sigma(M_G)_{\text{eff}}$:
$(73.6 \text{ N})(0.100 \text{ m}) - (200 \text{ N})(0.060 \text{ m}) = (0.245 \text{ kg} \cdot \text{m}^2)\alpha$
$\alpha = -18.94 \text{ rad/s}^2$ $\boldsymbol{\alpha} = 18.94 \text{ rad/s}^2 \,\uparrow$ ◄

SAMPLE PROBLEM 16.10

The extremities of a 4-ft rod weighing 50 lb can move freely and with no friction along two straight tracks as shown. If the rod is released with no velocity from the position shown, determine (a) the angular acceleration of the rod, (b) the reactions at A and B.

SOLUTION

Kinematics of Motion. Since the motion is constrained, the acceleration of G must be related to the angular acceleration $\boldsymbol{\alpha}$. To obtain this relation, we first determine the magnitude of the acceleration $\mathbf{a}_A$ of point A in terms of α. Assuming that $\boldsymbol{\alpha}$ is directed counterclockwise and noting that $a_{B/A} = 4\alpha$, we write

$$\mathbf{a}_B = \mathbf{a}_A + \mathbf{a}_{B/A}$$
$$[a_B \searrow 45°] = [a_A \rightarrow] + [4\alpha \; \nearrow \; 60°]$$

Noting that $\phi = 75°$ and using the law of sines, we obtain

$$a_A = 5.46\alpha \qquad a_B = 4.90\alpha$$

The acceleration of G is now obtained by writing

$$\bar{\mathbf{a}} = \mathbf{a}_G = \mathbf{a}_A + \mathbf{a}_{G/A}$$
$$\bar{\mathbf{a}} = [5.46\alpha \rightarrow] + [2\alpha \; \nearrow \; 60°]$$

Resolving $\bar{\mathbf{a}}$ into x and y components, we obtain

$$\bar{a}_x = 5.46\alpha - 2\alpha \cos 60° = 4.46\alpha \qquad \bar{a}_x = 4.46\alpha \rightarrow$$
$$\bar{a}_y = -2\alpha \sin 60° = -1.732\alpha \qquad \bar{a}_y = 1.732\alpha \downarrow$$

Kinetics of Motion. We draw a free-body-diagram equation expressing that the system of the external forces is equivalent to the system of the effective forces represented by the vector of components $m\bar{a}_x$ and $m\bar{a}_y$ attached at G and the couple $\bar{I}\alpha$. We compute the following magnitudes:

$$\bar{I} = \tfrac{1}{12}ml^2 = \frac{1}{12}\frac{50 \text{ lb}}{32.2 \text{ ft/s}^2}(4 \text{ ft})^2 = 2.07 \text{ lb} \cdot \text{ft} \cdot \text{s}^2 \qquad \bar{I}\alpha = 2.07\alpha$$

$$m\bar{a}_x = \frac{50}{32.2}(4.46\alpha) = 6.93\alpha \qquad m\bar{a}_y = -\frac{50}{32.2}(1.732\alpha) = -2.69\alpha$$

Equations of Motion

$+\!\!\uparrow\!\Sigma M_E = \Sigma(M_E)_{\text{eff}}$:
$$(50)(1.732) = (6.93\alpha)(4.46) + (2.69\alpha)(1.732) + 2.07\alpha$$
$$\alpha = +2.30 \text{ rad/s}^2 \qquad \boldsymbol{\alpha = 2.30 \text{ rad/s}^2} \;\, \blacktriangleleft$$

$\xrightarrow{+} \Sigma F_x = \Sigma(F_x)_{\text{eff}}$: $\quad R_B \sin 45° = (6.93)(2.30) = 15.94$
$$R_B = 22.5 \text{ lb} \qquad \mathbf{R}_B = 22.5 \text{ lb} \;\measuredangle\; 45° \;\, \blacktriangleleft$$

$+\!\!\uparrow\!\Sigma F_y = \Sigma(F_y)_{\text{eff}}$: $\quad R_A + R_B \cos 45° - 50 = -(2.69)(2.30)$
$$R_A = -6.19 - 15.94 + 50 = 27.9 \text{ lb} \qquad \mathbf{R}_A = 27.9 \text{ lb} \uparrow \;\, \blacktriangleleft$$

SOLVING PROBLEMS
ON YOUR OWN

In this lesson we considered the *plane motion of rigid bodies under constraints.* We found that the types of constraints involved in engineering problems vary widely. For example, a rigid body may be constrained to rotate about a fixed axis or to roll on a given surface, or it may be pin-connected to collars or to other bodies.

1. Your solution of a problem involving the constrained motion of a rigid body, will, in general, consist of two steps. First, you will consider the *kinematics of the motion,* and then you will solve the *kinetics portion of the problem.*

2. The kinematic analysis of the motion is done by using the methods you learned in Chap. 15. Due to the constraints, linear and angular accelerations will be related. (They will *not* be independent, as they were in the last lesson.) You should establish *relationships among the accelerations* (angular as well as linear), and your goal should be to express all accelerations in terms of a *single unknown acceleration.* This is the first step taken in the solution of each of the sample problems in this lesson.

   a. For a body in noncentroidal rotation, the components of the acceleration of the mass center are $\bar{a}_t = \bar{r}\alpha$ and $\bar{a}_n = \bar{r}\omega^2$, where ω will generally be known [Sample Probs. 16.6 and 16.7].

   b. For a rolling disk or wheel, the acceleration of the mass center is $\bar{a} = r\alpha$ [Sample Prob. 16.8].

   c. For a body in general plane motion, your best course of action, if neither $\bar{a}$ nor α is known or readily obtainable, is to express $\bar{a}$ in terms of α [Sample Prob. 16.10].

3. The kinetic analysis of the motion is carried out as follows.

   a. Start by drawing a free-body-diagram equation. This was done in all the sample problems of this lesson. In each case the left-hand diagram shows the external forces, including the applied forces, the reactions, and the weight of the body. The right-hand diagram shows the vector $m\bar{\mathbf{a}}$ and the couple $\bar{I}\boldsymbol{\alpha}$.

   b. Next, reduce the number of unknowns in the free-body-diagram equation by using the relationships among the accelerations that you found in your kinematic analysis. You will then be ready to consider equations that can be written by summing components or moments. Choose first an equation that involves a single unknown. After solving for that unknown, substitute the value obtained into the other equations, which you will then solve for the remaining unknowns.

(continued)

4. When solving problems involving rolling disks or wheels, keep in mind the following.

   a. If sliding is impending, the friction force exerted on the rolling body has reached its maximum value, $F_m = \mu_s N$, where N is the normal force exerted on the body and μ_s is the coefficient of *static friction* between the surfaces of contact.

   b. If sliding is not impending, the friction force F can have *any value* smaller than F_m and should, therefore, be considered as an independent unknown. After you have determined F, be sure to check that it is smaller than F_m; if it is not, *the body does not roll*, but rotates and slides as described in the next paragraph.

   c. If the body rotates and slides at the same time, then the body is *not rolling* and the acceleration $\bar{a}$ of the mass center is *independent* of the angular acceleration α of the body: $\bar{a} \neq r\alpha$. On the other hand, the friction force has a well-defined value, $F = \mu_k N$, where μ_k is the coefficient of *kinetic friction* between the surfaces of contact.

   d. For an unbalanced rolling disk or wheel, the relation $\bar{a} = r\alpha$ between the acceleration $\bar{a}$ of the mass center G and the angular acceleration α of the disk or wheel *does not hold anymore*. However, a similar relation holds between the acceleration a_O of the *geometric center* O and the angular acceleration α of the disk or wheel: $a_O = r\alpha$. This relation can be used to express $\bar{a}$ in terms of α and ω (Fig. 16.18).

5. For a system of connected rigid bodies, the goal of your *kinematic analysis* should be to determine all the accelerations from the given data, or to express them all in terms of a single unknown. (For systems with several degrees of freedom, you will need to use as many unknowns as there are degrees of freedom.)

   Your *kinetic analysis* will generally be carried out by drawing a free-body-diagram equation for the entire system, as well as for one or several of the rigid bodies involved. In the latter case, both internal and external forces should be included, and care should be taken to represent with equal and opposite vectors the forces that two bodies exert on each other.

PROBLEMS

CONCEPT QUESTIONS

16.CQ4 A cord is attached to a spool when a force **P** is applied to the cord as shown. Assuming the spool rolls without slipping, what direction does the spool move for each case?

Case 1: **a.** left **b.** right **c.** It would not move.
Case 2: **a.** left **b.** right **c.** It would not move.
Case 3: **a.** left **b.** right **c.** It would not move.

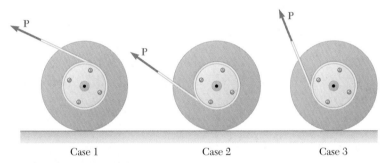

Case 1   Case 2   Case 3

Fig. P16.CQ4 and P16.CQ5

16.CQ5 A cord is attached to a spool when a force **P** is applied to the cord as shown. Assuming the spool rolls without slipping, in what direction does the friction force act for each case?

Case 1: **a.** left **b.** right **c.** The friction force would be zero.
Case 2: **a.** left **b.** right **c.** The friction force would be zero.
Case 3: **a.** left **b.** right **c.** The friction force would be zero.

16.CQ6 A front-wheel-drive car starts from rest and accelerates to the right. Knowing that the tires do not slip on the road, what is the direction of the friction force the road applies to the front tires?

a. left
b. right
c. The friction force is zero.

16.CQ7 A front-wheel-drive car starts from rest and accelerates to the right. Knowing that the tires do not slip on the road, what is the direction of the friction force the road applies to the rear tires?

a. left
b. right
c. The friction force is zero.

16.F5 A uniform 6 × 8-in. rectangular plate of mass m is pinned at A. Knowing the angular velocity of the plate at the instant shown is ω, draw the FBD and KD.

Fig. P16.F5

16.F6 Two identical 4-lb slender rods AB and BC are connected by a pin at B and by the cord AC. The assembly rotates in a vertical plane under the combined effect of gravity and a couple **M** applied to rod AB. Knowing that in the position shown the angular velocity of the assembly is ω, draw the FBD and KD that can be used to determine the angular acceleration of the assembly.

Fig. P16.F6

16.F7 The 4-lb uniform rod AB is attached to collars of negligible mass that slide without friction along the fixed rods shown. Rod AB is at rest in the position $\theta = 25°$ when a horizontal force **P** is applied to collar A causing it to start moving to the left. Draw the FBD and KD for the rod.

Fig. P16.F7

16.F8 A uniform disk of mass $m = 4$ kg and radius $r = 150$ mm is supported by a belt $ABCD$ that is bolted to the disk at B and C. If the belt suddenly breaks at a point located between A and B, draw the FBD and KD for the disk immediately after the break.

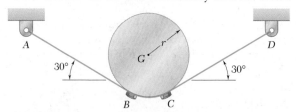

Fig. P16.F8

END-OF-SECTION PROBLEMS

16.75 Show that the couple $\bar{I}\alpha$ of Fig. 16.15 can be eliminated by attaching the vectors $m\bar{\mathbf{a}}_t$ and $m\bar{\mathbf{a}}_n$ at a point P called the *center of percussion,* located on line OG at a distance $GP = \bar{k}^2/\bar{r}$ from the mass center of the body.

Fig. P16.75

16.76 A uniform slender rod of length $L = 900$ mm and mass $m = 4$ kg is suspended from a hinge at C. A horizontal force $\mathbf{P}$ of magnitude 75 N is applied at end B. Knowing that $\bar{r} = 225$ mm, determine (*a*) the angular acceleration of the rod, (*b*) the components of the reaction at C.

Fig. P16.76

16.77 In Prob. 16.76, determine (*a*) the distance $\bar{r}$ for which the horizontal component of the reaction at C is zero, (*b*) the corresponding angular acceleration of the rod.

16.78 A uniform slender rod of length $L = 36$ in. and weight $W = 4$ lb hangs freely from a hinge at A. If a force $\mathbf{P}$ of magnitude 1.5 lb is applied at B horizontally to the left ($h = L$), determine (*a*) the angular acceleration of the rod, (*b*) the components of the reaction at A.

16.79 In Prob. 16.78, determine (*a*) the distance h for which the horizontal component of the reaction at A is zero, (*b*) the corresponding angular acceleration of the rod.

Fig. P16.78

16.80 The uniform slender rod AB is welded to the hub D, and the system rotates about the vertical axis DE with a constant angular velocity $\boldsymbol{\omega}$. (*a*) Denoting by w the mass per unit length of the rod, express the tension in the rod at a distance z from end A in terms of w, l, z, and $\boldsymbol{\omega}$, (*b*) Determine the tension in the rod for $w = 0.3$ kg/m, $l = 400$ mm, $z = 250$ mm, and $\omega = 150$ rpm.

Fig. P16.80

Fig. P16.81

Fig. P16.83

A $\qquad$ B

Fig. P16.84

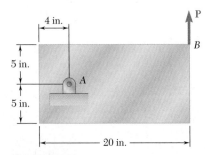

Fig. P16.86

16.81 The shutter shown was formed by removing one quarter of a disk of 0.75-in. radius and is used to interrupt a beam of light emanating from a lens at C. Knowing that the shutter weighs 0.125 lb and rotates at the constant rate of 24 cycles per second, determine the magnitude of the force exerted by the shutter on the shaft at A.

16.82 A 6-in.-diameter hole is cut as shown in a thin disk of 15-in. diameter. The disk rotates in a horizontal plane about its geometric center A at the constant rate of 480 rpm. Knowing that the disk has a mass of 60 lb after the hole has been cut, determine the horizontal component of the force exerted by the shaft on the disk at A.

Fig. P16.82

16.83 A turbine disk of mass 26 kg rotates at a constant rate of 9600 rpm. Knowing that the mass center of the disk coincides with the center of rotation O, determine the reaction at O immediately after a single blade at A, of mass 45 g, becomes loose and is thrown off.

16.84 and 16.85 A uniform rod of length L and mass m is supported as shown. If the cable attached at end B suddenly breaks, determine (a) the acceleration of end B, (b) the reaction at the pin support.

Fig. P16.85

16.86 A 12-lb uniform plate rotates about A in a vertical plane under the combined effect of gravity and of the vertical force $\mathbf{P}$. Knowing that at the instant shown the plate has an angular velocity of 20 rad/s and an angular acceleration of 30 rad/s² both counterclockwise, determine (a) the force $\mathbf{P}$, (b) the components of the reaction at A.

16.87 A 1.5-kg slender rod is welded to a 5-kg uniform disk as shown. The assembly swings freely about C in a vertical plane. Knowing that in the position shown the assembly has an angular velocity of 10 rad/s clockwise, determine (*a*) the angular acceleration of the assembly, (*b*) the components of the reaction at C.

Fig. P16.87

16.88 Two uniform rods, ABC of weight 6 lb and DCE of weight 8 lb, are connected by a pin at C and by two cords BD and BE. The T-shaped assembly rotates in a vertical plane under the combined effect of gravity and of a couple **M** which is applied to rod ABC. Knowing that at the instant shown the tension in cord BE is 2 lb and the tension in cord BD is 0.5 lb, determine (*a*) the angular acceleration of the assembly, (*b*) the couple **M**.

Fig. P16.88

16.89 The object ABC consists of two slender rods welded together at point B. Rod AB has a weight of 2 lb and bar BC has a weight of 4 lb. Knowing the magnitude of the angular velocity of ABC is 10 rad/s when $\theta = 0°$, determine the components of the reaction at point C when $\theta = 0°$.

Fig. P16.89

16.90 A 3.5-kg slender rod AB and a 2-kg slender rod BC are connected by a pin at B and by the cord AC. The assembly can rotate in a vertical plane under the combined effect of gravity and a couple **M** applied to rod BC. Knowing that in the position shown the angular velocity of the assembly is zero and the tension in cord AC is equal to 25 N, determine (*a*) the angular acceleration of the assembly, (*b*) the magnitude of the couple **M**.

Fig. P16.90

Fig. P16.91

16.91 A 9-kg uniform disk is attached to the 5-kg slender rod AB by means of frictionless pins at B and C. The assembly rotates in a vertical plane under the combined effect of gravity and of a couple **M** which is applied to rod AB. Knowing that at the instant shown the assembly has an angular velocity of 6 rad/s and an angular acceleration of 25 rad/s², both counterclockwise, determine (a) the couple **M**, (b) the force exerted by pin C on member AB.

16.92 Derive the equation $\Sigma M_C = I_C \alpha$ for the rolling disk of Fig. 16.17, where ΣM_C represents the sum of the moments of the external forces about the instantaneous center C, and I_C is the moment of inertia of the disk about C.

16.93 Show that in the case of an unbalanced disk, the equation derived in Prob. 16.92 is valid only when the mass center G, the geometric center O, and the instantaneous center C happen to lie in a straight line.

16.94 A wheel of radius r and centroidal radius of gyration $\bar{k}$ is released from rest on the incline and rolls without sliding. Derive an expression for the acceleration of the center of the wheel in terms of r, $\bar{k}$, β, and g.

Fig. P16.94

Fig. P16.95

16.95 A homogeneous sphere S, a uniform cylinder C, and a thin pipe P are in contact when they are released from rest on the incline shown. Knowing that all three objects roll without slipping, determine, after 4 s of motion, the clear distance between (a) the pipe and the cylinder, (b) the cylinder and the sphere.

16.96 A 40-kg flywheel of radius R = 0.5 m is rigidly attached to a shaft of radius r = 0.05 m that can roll along parallel rails. A cord is attached as shown and pulled with a force **P** of magnitude 150 N. Knowing the centroidal radius of gyration is $\bar{k}$ = 0.4 m, determine (a) the angular acceleration of the flywheel, (b) the velocity of the center of gravity after 5 s.

16.97 A 40-kg flywheel of radius R = 0.5 m is rigidly attached to a shaft of radius r = 0.05 m that can roll along parallel rails. A cord is attached as shown and pulled with a force **P**. Knowing the centroidal radius of gyration is $\bar{k}$ = 0.4 m and the coefficient of static friction is μ_s = 0.4, determine the largest magnitude of force **P** for which no slipping will occur.

Fig. P16.96 and P16.97

16.98 through 16.101 A drum of 60-mm radius is attached to a disk of 120-mm radius. The disk and drum have a total mass of 6 kg and a combined radius of gyration of 90 mm. A cord is attached as shown and pulled with a force **P** of magnitude 20 N. Knowing that the disk rolls without sliding, determine (*a*) the angular acceleration of the disk and the acceleration of *G*, (*b*) the minimum value of the coefficient of static friction compatible with this motion.

16.102 through 16.105 A drum of 4-in. radius is attached to a disk of 8-in. radius. The disk and drum have a combined weight of 10 lb and a combined radius of gyration of 6 in. A cord is attached as shown and pulled with a force **P** of magnitude 5 lb. Knowing that the coefficients of static and kinetic friction are $\mu_s = 0.25$ and $\mu_k = 0.20$, respectively, determine (*a*) whether or not the disk slides, (*b*) the angular acceleration of the disk and the acceleration of *G*.

Fig. P16.98 and P16.102

Fig. P16.99 and P16.103

Fig. *P16.100* and P16.104

Fig. *P16.101* and P16.105

16.106 and 16.107 A 12-in.-radius cylinder of weight 16 lb rests on a 6-lb carriage. The system is at rest when a force **P** of magnitude 4 lb is applied. Knowing that the cylinder rolls without sliding on the carriage and neglecting the mass of the wheels of the carriage, determine (*a*) the acceleration of the carriage, (*b*) the acceleration of point *A*, (*c*) the distance the cylinder has rolled with respect to the carriage after 0.5 s.

Fig. *P16.106*

Fig. P16.107

16.108 Gear *C* has a mass of 5 kg and a centroidal radius of gyration of 75 mm. The uniform bar *AB* has a mass of 3 kg and gear *D* is stationary. If the system is released from rest in the position shown, determine (*a*) the angular acceleration of gear *C*, (*b*) the acceleration of point *B*.

Fig. P16.108

Fig. P16.109

Fig. P16.110

16.109 Two uniform disks A and B, each of mass 2 kg, are connected by a 1.5-kg rod CD as shown. A counterclockwise couple **M** of moment 2.5 N-m is applied to disk A. Knowing that the disks roll without sliding, determine (a) the acceleration of the center of each disk, (b) the horizontal component of the force exerted on disk B by pin D.

16.110 A 10-lb cylinder of radius $r = 4$ in. is resting on a conveyor belt when the belt is suddenly turned on and it experiences an acceleration of magnitude $a = 6$ ft/s². The smooth vertical bar holds the cylinder in place when the belt is not moving. Knowing the cylinder rolls without slipping and the friction between the vertical bar and the cylinder is negligible, determine (a) the angular acceleration of the cylinder, (b) the components of the force the conveyor belt applies to the cylinder.

16.111 A hemisphere of weight W and radius r is released from rest in the position shown. Determine (a) the minimum value of μ_s for which the hemisphere starts to roll without sliding, (b) the corresponding acceleration of point B [Hint: Note that $OG = \frac{3}{8}r$ and that, by the parallel-axis theorem, $\bar{I} = \frac{2}{5}mr^2 - m(OG)^2$.]

Fig. P16.111

16.112 Solve Prob. 16.111, considering a half cylinder instead of a hemisphere. [Hint: Note that $OG = 4r/3\pi$ and that, by the parallel-axis theorem, $\bar{I} = \frac{1}{2}mr^2 - m(OG)^2$.]

16.113 The center of gravity G of a 1.5-kg unbalanced tracking wheel is located at a distance $r = 18$ mm from its geometric center B. The radius of the wheel is $R = 60$ mm and its centroidal radius of gyration is 44 mm. At the instant shown the center B of the wheel has a velocity of 0.35 m/s and an acceleration of 1.2 m/s², both directed to the left. Knowing that the wheel rolls without sliding and neglecting the mass of the driving yoke AB, determine the horizontal force **P** applied to the yoke.

Fig. P16.113

16.114 A small clamp of mass m_B is attached at B to a hoop of mass m_h. The system is released from rest when $\theta = 90°$ and rolls without sliding. Knowing that $m_h = 3m_B$, determine (a) the angular acceleration of the hoop, (b) the horizontal and vertical components of the acceleration of B.

16.115 A small clamp of mass m_B is attached at B to a hoop of mass m_h. Knowing that the system is released from rest and rolls without sliding, derive an expression for the angular acceleration of the hoop in terms of m_B, m_h, r, and θ.

Fig. P16.114 and P16.115

16.116 A 4-lb bar is attached to a 10-lb uniform cylinder by a square pin, P, as shown. Knowing that $r = 16$ in., $h = 8$ in., $\theta = 20°$, $L = 20$ in., and $\omega = 2$ rad/s at the instant shown, determine the reactions at P at this instant assuming that the cylinder rolls without sliding down the incline.

16.117 The ends of the 20-lb uniform rod AB are attached to collars of negligible mass that slide without friction along fixed rods. If the rod is released from rest when $\theta = 25°$, determine immediately after release (a) the angular acceleration of the rod, (b) the reaction at A, (c) the reaction at B.

Fig. P16.116

Fig. P16.117 and P16.118

16.118 The ends of the 20-lb uniform rod AB are attached to collars of negligible mass that slide without friction along fixed rods. A vertical force **P** is applied to collar B when $\theta = 25°$, causing the collar to start from rest with an upward acceleration of 40 ft/s². Determine (a) the force **P**, (b) the reaction at A.

16.119 The motion of the 3-kg uniform rod AB is guided by small wheels of negligible weight that roll along without friction in the slots shown. If the rod is released from rest in the position shown, determine immediately after release (a) the angular acceleration of the rod, (b) the reaction at B.

Fig. P16.119

16.120 A beam AB of length L and mass m is supported by two cables as shown. If cable BD breaks, determine at that instant the tension in the remaining cable as a function of its initial angular orientation θ.

Fig. P16.120

16.121 End A of a uniform 10-kg bar is attached to a horizontal rope and end B contacts a floor with negligible friction. Knowing that the bar is released from rest in the position shown, determine immediately after release (a) the angular acceleration of the bar, (b) the tension in the rope, (c) the reaction at B.

Fig. P16.121

Fig. P16.122

16.122 End A of the 8-kg uniform rod AB is attached to a collar that can slide without friction on a vertical rod. End B of the rod is attached to a vertical cable BC. If the rod is released from rest in the position shown, determine immediately after release (*a*) the angular acceleration of the rod, (*b*) the reaction at A.

16.123 A uniform thin plate $ABCD$ has a mass of 8 kg and is held in position by three inextensible cords AE, BF, and CG. If cord AE is cut, determine at that instant (*a*) if the plate is undergoing translation or general plane motion, (*b*) the tension in cords BF and CG.

Fig. P16.123

16.124 The 4-kg uniform rod ABD is attached to the crank BC and is fitted with a small wheel that can roll without friction along a vertical slot. Knowing that at the instant shown crank BC rotates with an angular velocity of 6 rad/s clockwise and an angular acceleration of 15 rad/s^2 counterclockwise, determine the reaction at A.

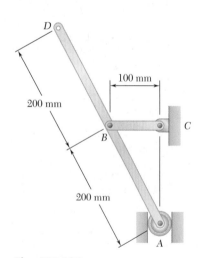

Fig. P16.124

16.125 The 7-lb uniform rod AB is connected to crank BD and to a collar of negligible weight, which can slide freely along rod EF. Knowing that in the position shown crank BD rotates with an angular velocity of 15 rad/s and an angular acceleration of 60 rad/s^2, both clockwise, determine the reaction at A.

Fig. P16.125

16.126 In Prob. 16.125, determine the reaction at A, knowing that in the position shown crank BD rotates with an angular velocity of 15 rad/s clockwise and an angular acceleration of 60 rad/s^2 counterclockwise.

16.127 The 250-mm uniform rod *BD*, of mass 5 kg, is connected as shown to disk *A* and to a collar of negligible mass, that may slide freely along a vertical rod. Knowing that disk *A* rotates counterclockwise at a constant rate of 500 rpm, determine the reactions at *D* when $\theta = 0$.

Fig. P16.127

16.128 Solve Prob. 16.127 when $\theta = 90°$.

16.129 The 4-kg uniform slender bar *BD* is attached to bar *AB* and a wheel of negligible mass that rolls on a circular surface. Knowing that at the instant shown bar *AB* has an angular velocity of 6 rad/s and no angular acceleration, determine the reaction at point *D*.

Fig. P16.129

16.130 The motion of the uniform slender rod of length $L = 0.5$ m and mass $m = 3$ kg is guided by pins at *A* and *B* that slide freely in frictionless slots, circular and horizontal, cut into a vertical plate as shown. Knowing that at the instant shown the rod has an angular velocity of 3 rad/s counterclockwise and $\theta = 30°$, determine the reactions at points *A* and *B*.

16.131 At the instant shown, the 20-ft-long, uniform 100-lb pole *ABC* has an angular velocity of 1 rad/s counterclockwise and point *C* is sliding to the right. A 120-lb horizontal force **P** acts at *B*. Knowing the coefficient of kinetic friction between the pole and the ground is 0.3, determine at this instant (*a*) the acceleration of the center of gravity, (*b*) the normal force between the pole and the ground.

Fig. P16.130

Fig. P16.131

Fig. P16.132

16.132 A driver starts his car with the door on the passenger's side wide open ($\theta = 0$). The 80-lb door has a centroidal radius of gyration $\bar{k} = 12.5$ in., and its mass center is located at a distance $r = 22$ in. from its vertical axis of rotation. Knowing that the driver maintains a constant acceleration of 6 ft/s^2, determine the angular velocity of the door as it slams shut ($\theta = 90°$).

16.133 For the car of Prob. 16.132, determine the smallest constant acceleration that the driver can maintain if the door is to close and latch, knowing that as the door hits the frame its angular velocity must be at least 2 rad/s for the latching mechanism to operate.

16.134 Two 8-lb uniform bars are connected to form the linkage shown. Neglecting the effect of friction, determine the reaction at D immediately after the linkage is released from rest in the position shown.

Fig. P16.134

*16.135 The 6-kg rod BC connects a 10-kg disk centered at A to a 5-kg rod CD. The motion of the system is controlled by the couple **M** applied to disk A. Knowing that at the instant shown disk A has an angular velocity of 36 rad/s clockwise and no angular acceleration, determine (*a*) the couple **M**, (*b*) the components of the force exerted at C on rod BC.

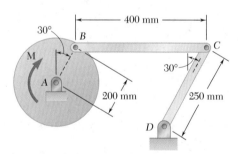

Fig. P16.135 and P16.136

*16.136 The 6-kg rod BC connects a 10-kg disk centered at A to a 5-kg rod CD. The motion of the system is controlled by the couple **M** applied to disk A. Knowing that at the instant shown disk A has an angular velocity of 36 rad/s clockwise and an angular acceleration of 150 rad/s^2 counterclockwise, determine (*a*) the couple **M**, (*b*) the components of the force exerted at C on rod BC.

16.137 In the engine system shown $l = 250$ mm and $b = 100$ mm. The connecting rod BD is assumed to be a 1.2-kg uniform slender rod and is attached to the 1.8-kg piston P. During a test of the system, crank AB is made to rotate with a constant angular velocity of 600 rpm clockwise with no force applied to the face of the piston. Determine the forces exerted on the connecting rod at B and D when $\theta = 180°$. (Neglect the effect of the weight of the rod.)

Fig. P16.137

16.138 Solve Prob. 16.137 when $\theta = 90°$.

16.139 The 4-lb rod *AB* and the 6-lb rod *BC* are connected as shown to a disk that is made to rotate in a vertical plane at a constant angular velocity of 6 rad/s clockwise. For the position shown, determine the forces exerted at *A* and *B* on rod *AB*.

16.140 The 4-lb rod *AB* and the 6-lb rod *BC* are connected as shown to a disk that is made to rotate in a vertical plane. Knowing that at the instant shown the disk has an angular acceleration of 18 rad/s² clockwise and no angular velocity, determine the components of the forces exerted at *A* and *B* on rod *AB*.

Fig. P16.139 and P16.140

16.141 Two rotating rods in the vertical plane are connected by a slider block *P* of negligible mass. The rod attached at *A* has a weight of 1.6 lb and a length of 8 in. Rod *BP* weighs 2 lb and is 10 in. long and the friction between block *P* and *AE* is negligible. The motion of the system is controlled by a couple **M** applied to rod *BP*. Knowing that rod *BP* has a constant angular velocity of 20 rad/s clockwise, determine (*a*) the couple **M**, (*b*) the components of the force exerted on *AE* by block *P*.

16.142 Two rotating rods in the vertical plane are connected by a slider block *P* of negligible mass. The rod attached at *A* has a mass of 0.8 kg and a length of 160 mm. Rod *BP* has a mass of 1 kg and is 200 mm long and the friction between block *P* and *AE* is negligible. The motion of the system is controlled by a couple **M** applied to bar *BP*. Knowing that at the instant shown rod *BP* has an angular velocity of 20 rad/s clockwise and an angular acceleration of 80 rad/s² clockwise, determine (*a*) the couple **M**, (*b*) the components of the force exerted on *AE* by block *P*.

Fig. P16.141 and P16.142

*16.143 Draw the shear and bending-moment diagrams for the rod of Prob. 16.77 immediately after the cable at *B* breaks.

*16.144 A uniform slender bar *AB* of mass *m* is suspended as shown from a uniform disk of the same mass *m*. Neglecting the effect of friction, determine the accelerations of points *A* and *B* immediately after a horizontal force **P** has been applied at *B*.

16.145 A uniform rod *AB*, of mass 15 kg and length 1 m, is attached to the 20-kg cart *C*. Neglecting friction, determine immediately after the system has been released from rest, (*a*) the acceleration of the cart, (*b*) the angular acceleration of the rod.

Fig. P16.144

Fig. P16.145

*16.146 The 5-kg slender rod AB is pin-connected to an 8-kg uniform disk as shown. Immediately after the system is released from rest, determine the acceleration of (*a*) point A, (*b*) point B.

Fig. P16.146

*16.147 and *16.148 The 6-lb cylinder B and the 4-lb wedge A are held at rest in the position shown by cord C. Assuming that the cylinder rolls without sliding on the wedge and neglecting friction between the wedge and the ground, determine, immediately after cord C has been cut, (*a*) the acceleration of the wedge, (*b*) the angular acceleration of the cylinder.

Fig. P16.147

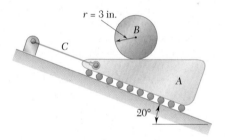

Fig. P16.148

*16.149 Each of the 3-kg bars AB and BC is of length $L = 500$ mm. A horizontal force **P** of magnitude 20 N is applied to bar BC as shown. Knowing that $b = L$ (**P** is applied at C), determine the angular acceleration of each bar.

*16.150 Each of the 3-kg bars AB and BC is of length $L = 500$ mm. A horizontal force **P** of magnitude 20 N is applied to bar BC. For the position shown, determine (*a*) the distance b for which the bars move as if they formed a single rigid body, (*b*) the corresponding angular acceleration of the bars.

Fig. P16.149 and P16.150

*16.151 (*a*) Determine the magnitude and the location of the maximum bending moment in the rod of Prob. 16.78. (*b*) Show that the answer to part *a* is independent of the weight of the rod.

*16.152 Draw the shear and bending-moment diagrams for the rod of Prob. 16.84 immediately after the cable at B breaks.

In this chapter, we studied the *kinetics of rigid bodies*, i.e., the relations existing between the forces acting on a rigid body, the shape and mass of the body, and the motion produced. Except for the first two sections, which apply to the most general case of the motion of a rigid body, our analysis was restricted to the *plane motion of rigid slabs* and rigid bodies symmetrical with respect to the reference plane. The study of the plane motion of nonsymmetrical rigid bodies and of the motion of rigid bodies in three-dimensional space will be considered in Chap. 18.

We first recalled [Sec. 16.2] the two fundamental equations derived in Chap. 14 for the motion of a system of particles and observed that they apply in the most general case of the motion of a rigid body. The first equation defines the motion of the mass center G of the body; we have

$$\Sigma \mathbf{F} = m\bar{\mathbf{a}} \qquad (16.1)$$

where m is the mass of the body and $\bar{\mathbf{a}}$ the acceleration of G. The second is related to the motion of the body relative to a centroidal frame of reference; we wrote

$$\Sigma \mathbf{M}_G = \dot{\mathbf{H}}_G \qquad (16.2)$$

where $\dot{\mathbf{H}}_G$ is the rate of change of the angular momentum $\mathbf{H}_G$ of the body about its mass center G. Together, Eqs. (16.1) and (16.2) express that *the system of the external forces is equipollent to the system consisting of the vector $m\bar{\mathbf{a}}$ attached at G and the couple of moment $\dot{\mathbf{H}}_G$* (Fig. 16.19).

Fundamental equations of motion for a rigid body

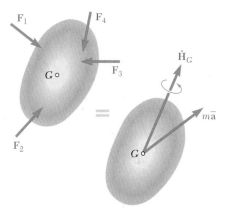

Fig. 16.19

Angular momentum in plane motion

Restricting our analysis at this point and for the rest of the chapter to the plane motion of rigid slabs and rigid bodies symmetrical with respect to the reference plane, we showed [Sec. 16.3] that the angular momentum of the body could be expressed as

$$\mathbf{H}_G = \bar{I}\boldsymbol{\omega} \qquad (16.4)$$

where $\bar{I}$ is the moment of inertia of the body about a centroidal axis perpendicular to the reference plane and $\boldsymbol{\omega}$ is the angular velocity of the body. Differentiating both members of Eq. (16.4), we obtained

$$\dot{\mathbf{H}}_G = \bar{I}\dot{\boldsymbol{\omega}} = \bar{I}\boldsymbol{\alpha} \qquad (16.5)$$

which shows that in the restricted case considered here, the rate of change of the angular momentum of the rigid body can be represented by a vector of the same direction as $\boldsymbol{\alpha}$ (i.e., perpendicular to the plane of reference) and of magnitude $\bar{I}\alpha$.

Equations for the plane motion of a rigid body

It follows from [Sec. 16.4] that the plane motion of a rigid slab or of a rigid body symmetrical with respect to the reference plane is defined by the three scalar equations

$$\Sigma F_x = m\bar{a}_x \qquad \Sigma F_y = m\bar{a}_y \qquad \Sigma M_G = \bar{I}\alpha \qquad (16.6)$$

D'Alembert's principle

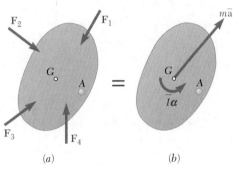

It further follows that *the external forces acting on the rigid body are actually **equivalent** to the effective forces of the various particles forming the body.* This statement, known as *d'Alembert's principle,* can be expressed in the form of the vector diagram shown in Fig. 16.20, where the effective forces have been represented by a vector $m\bar{\mathbf{a}}$ attached at G and a couple $\bar{I}\alpha$. In the particular case of a slab in *translation,* the effective forces shown in part b of this figure reduce to the single vector $m\bar{\mathbf{a}}$, while in the particular case of a slab in *centroidal rotation,* they reduce to the single couple $\bar{I}\alpha$; in any other case of plane motion, both the vector $m\bar{\mathbf{a}}$ and the couple $\bar{I}\alpha$ should be included.

(a) (b)

Fig. 16.20

Free-body-diagram equation

Any problem involving the plane motion of a rigid slab may be solved by drawing a *free-body-diagram equation* similar to that of Fig. 16.20 [Sec. 16.6]. Three equations of motion can then be obtained by equating the x components, y components, and moments about an arbitrary point A, of the forces and vectors involved [Sample Probs. 16.1, 16.2, 16.4, and 16.5]. An alternative solution can be obtained by adding to the external forces an *inertia vector* $-m\bar{\mathbf{a}}$ of sense opposite to that of $\bar{\mathbf{a}}$, attached at G, and an *inertia couple* $-\bar{I}\alpha$ of sense opposite to that of $\boldsymbol{\alpha}$. The system obtained in this way is equivalent to zero, and the slab is said to be in *dynamic equilibrium.*

Connected rigid bodies

The method described above can also be used to solve problems involving the plane motion of several connected rigid bodies [Sec. 16.7]. A free-body-diagram equation is drawn for each part of the system and the equations of motion obtained are solved simultaneously. In some cases, however, a single diagram can be drawn for the entire system, including all the external forces as well as the vectors $m\bar{\mathbf{a}}$ and the couples $\bar{I}\alpha$ associated with the various parts of the system [Sample Prob. 16.3].

Constrained plane motion

In the second part of the chapter, we were concerned with rigid bodies *moving under given constraints* [Sec. 16.8]. While the kinetic analysis of the constrained plane motion of a rigid slab is the same as above, it must be supplemented by a *kinematic analysis* which has for its object to express the components $\bar{a}_x$ and $\bar{a}_y$ of the acceleration of the mass center G of the slab in terms of its angular acceleration α. Problems solved in this way included the *noncentroidal rotation* of rods and plates [Sample Probs. 16.6 and 16.7], the *rolling motion* of spheres and wheels [Sample Probs. 16.8 and 16.9], and the plane motion of *various types of linkages* [Sample Prob. 16.10].

REVIEW PROBLEMS

16.153 A cyclist is riding a bicycle at a speed of 20 mph on a horizontal road. The distance between the axles is 42 in., and the mass center of the cyclist and the bicycle is located 26 in. behind the front axle and 40 in. above the ground. If the cyclist applies the brakes only on the front wheel, determine the shortest distance in which he can stop without being thrown over the front wheel.

16.154 The forklift truck shown weighs 2250 lb and is used to lift a crate of weight $W = 2500$ lb. The truck is moving to the left at a speed of 10 ft/s when the brakes are applied on all four wheels. Knowing that the coefficient of static friction between the crate and the fork lift is 0.30, determine the smallest distance in which the truck can be brought to a stop if the crate is not to slide and if the truck is not to tip forward.

Fig. P16.154

16.155 A 5-kg uniform disk is attached to the 3-kg uniform rod BC by means of a frictionless pin AB. An elastic cord is wound around the edge of the disk and is attached to a ring at E. Both ring E and rod BC can rotate freely about the vertical shaft. Knowing that the system is released from rest when the tension in the elastic cord is 15 N, determine (a) the angular acceleration of the disk, (b) the acceleration of the center of the disk.

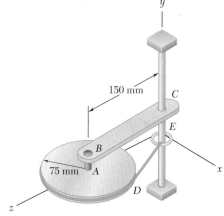

Fig. P16.155

16.156 Identical cylinders of mass m and radius r are pushed by a series of moving arms. Assuming the coefficient of friction between all surfaces to be $\mu < 1$ and denoting by a the magnitude of the acceleration of the arms, derive an expression for (a) the maximum allowable value of a if each cylinder is to roll without sliding, (b) the minimum allowable value of a if each cylinder is to move to the right without rotating.

Fig. P16.156

16.157 The uniform rod AB of weight W is released from rest when $\beta = 70°$. Assuming that the friction force between end A and the surface is large enough to prevent sliding, determine immediately after release (a) the angular acceleration of the rod, (b) the normal reaction at A, (c) the friction force at A.

Fig. P16.157 and P16.158

16.158 The uniform rod AB of weight W is released from rest when $\beta = 70°$. Assuming that the friction force is zero between end A and the surface, determine immediately after release (a) the angular acceleration of the rod, (b) the acceleration of the mass center of the rod, (c) the reaction at A.

16.159 A bar of mass $m = 5$ kg is held as shown between four disks, each of mass $m' = 2$ kg and radius $r = 75$ mm. Knowing that the normal forces on the disks are sufficient to prevent any slipping, for each of the cases shown determine the acceleration of the bar immediately after it has been released from rest.

Fig. P16.159

16.160 A uniform plate of mass m is suspended in each of the ways shown. For each case determine immediately after the connection B has been released (*a*) the angular acceleration of the plate, (*b*) the acceleration of its mass center.

(1)   (2)   (3)

Fig. P16.160

16.161 A cylinder with a circular hole is rolling without slipping on a fixed curved surface as shown. The cylinder would have a weight of 16 lb without the hole, but with the hole it has a weight of 15 lb. Knowing that at the instant shown the disk has an angular velocity of 5 rad/s clockwise, determine (*a*) the angular acceleration of the disk, (*b*) the components of the reaction force between the cylinder and the ground at this instant.

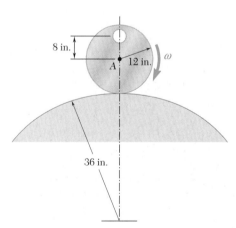

Fig. P16.161

16.162 The motion of a square plate of side 150 mm and mass 2.5 kg is guided by pins at corners A and B that slide in slots cut in a vertical wall. Immediately after the plate is released from rest in the position shown, determine (*a*) the angular acceleration of the plate, (*b*) the reaction at corner A.

Fig. P16.162

16.163 The motion of a square plate of side 150 mm and mass 2.5 kg is guided by a pin at corner A that slides in a horizontal slot cut in a vertical wall. Immediately after the plate is released from rest in the position shown, determine (*a*) the angular acceleration of the plate, (*b*) the reaction at corner A.

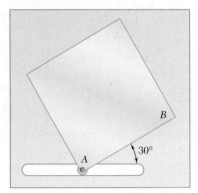

Fig. P16.163

16.164 The Geneva mechanism shown is used to provide an intermittent rotary motion of disk S. Disk D weighs 2 lb and has a radius of gyration of 0.9 in., and disk S weighs 6 lb and has a radius of gyration of 1.5 in. The motion of the system is controlled by a couple **M** applied to disk D. A pin P is attached to disk D and can slide in one of the six equally spaced slots cut in disk S. It is desirable that the angular velocity of disk S be zero as the pin enters and leaves each of the six slots; this will occur if the distance between the centers of the disks and the radii of the disks are related as shown. Knowing disk D rotates with a constant counterclockwise angular velocity of 8 rad/s and the friction between the slot and pin P is negligible, determine when $\phi = 150°$ (*a*) the couple **M**, (*b*) the magnitude of the force pin P applies to disk S.

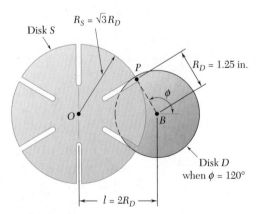

Fig. P16.164

COMPUTER PROBLEMS

16.C1 The 5-lb rod AB is released from rest in the position shown. (*a*) Assuming that the friction force between end A and the surface is large enough to prevent sliding, using software calculate the normal reaction and the friction force at A immediately after release for values of β from 0 to 85°. (*b*) Knowing that the coefficient of static friction between the rod and the floor is actually equal to 0.50, determine the range of values of β for which the rod will slip immediately after being released from rest.

16.C2 End A of the 5-kg rod AB is moved to the left at a constant speed $v_A = 1.5$ m/s. Using computational software calculate and plot the normal reactions at ends A and B of the rod for values of θ from 0 to 50°. Determine the value of θ at which end B of the rod loses contact with the wall.

16.C3 A 30-lb cylinder of diameter $b = 8$ in. and height $h = 6$ in. is placed on a 10-lb platform CD that is held in the position shown by three cables. It is desired to determine the minimum value of μ_s between the cylinder and the platform for which the cylinder does not slip on the platform, immediately after cable AB is cut. Using computational software calculate and plot the minimum allowable value of μ_s for values of θ from 0 to 30°. Knowing that the actual value of μ_s is 0.60, determine the value of θ at which slipping impends.

Fig. P16.C1

Fig. P16.C2

Fig. P16.C3

16.C4 For the engine system of Prob. 15.C3 of Chap. 15, the masses of piston P and the connecting rod BD are 2.5 kg and 3 kg, respectively. Knowing that during a test of the system no force is applied to the face of the piston, use computational software to calculate and plot the horizontal and vertical components of the dynamic reactions exerted on the connecting rod at B and D for values of θ from 0 to 180°.

16.C5 A uniform slender bar AB of mass m is suspended from springs AC and BD as shown. Using computational software calculate and plot the accelerations of ends A and B, immediately after spring AC has broken, for values of θ from 0 to 90°.

Fig. P16.C5

1103

In this chapter the energy and momentum methods will be added to the tools available for your study of the motion of rigid bodies. For example, by using the principle of conservation of energy and direct application of Newton's second law, the forces exerted on the hands of this gymnast can be determined as he swings from one stationary hold to another.

Plane Motion of Rigid Bodies: Energy and Momentum Methods

Photo 17.1 The work done by friction reduces the kinetic energy of the wheel.

17.1 INTRODUCTION

In this chapter the method of work and energy and the method of impulse and momentum will be used to analyze the plane motion of rigid bodies and of systems of rigid bodies.

The method of work and energy will be considered first. In Secs. 17.2 through 17.5, the work of a force and of a couple will be defined, and an expression for the kinetic energy of a rigid body in plane motion will be obtained. The principle of work and energy will then be used to solve problems involving displacements and velocities. In Sec. 17.6, the principle of conservation of energy will be applied to the solution of a variety of engineering problems.

In the second part of the chapter, the principle of impulse and momentum will be applied to the solution of problems involving velocities and time (Secs. 17.8 and 17.9) and the concept of conservation of angular momentum will be introduced and discussed (Sec. 17.10).

In the last part of the chapter (Secs. 17.11 and 17.12), problems involving the eccentric impact of rigid bodies will be considered. As was done in Chap. 13, where we analyzed the impact of particles, the coefficient of restitution between the colliding bodies will be used together with the principle of impulse and momentum in the solution of impact problems. It will also be shown that the method used is applicable not only when the colliding bodies move freely after the impact but also when the bodies are partially constrained in their motion.

17.2 PRINCIPLE OF WORK AND ENERGY FOR A RIGID BODY

The principle of work and energy will now be used to analyze the plane motion of rigid bodies. As was pointed out in Chap. 13, the method of work and energy is particularly well adapted to the solution of problems involving velocities and displacements. Its main advantage resides in the fact that the work of forces and the kinetic energy of particles are scalar quantities.

In order to apply the principle of work and energy to the analysis of the motion of a rigid body, it will again be assumed that the rigid body is made of a large number n of particles of mass Δm_i. Recalling Eq. (14.30) of Sec. 14.8, we write

$$T_1 + U_{1 \to 2} = T_2 \tag{17.1}$$

where T_1, T_2 = initial and final values of total kinetic energy of particles forming the rigid body

$U_{1 \to 2}$ = work of all forces acting on various particles of the body

The total kinetic energy

$$T = \frac{1}{2} \sum_{i=1}^{n} \Delta m_i v_i^2 \tag{17.2}$$

is obtained by adding positive scalar quantities and is itself a positive scalar quantity. You will see later how T can be determined for various types of motion of a rigid body.

The expression $U_{1\rightarrow2}$ in (17.1) represents the work of all the forces acting on the various particles of the body, whether these forces are internal or external. However, as you will see presently, the total work of the internal forces holding together the particles of a rigid body is zero. Consider two particles A and B of a rigid body and the two equal and opposite forces $\mathbf{F}$ and $-\mathbf{F}$ they exert on each other (Fig. 17.1). While, in general, small displacements $d\mathbf{r}$ and $d\mathbf{r}'$ of the two particles are different, the components of these displacements along AB must be equal; otherwise, the particles would not remain at the same distance from each other and the body would not be rigid. Therefore, the work of $\mathbf{F}$ is equal in magnitude and opposite in sign to the work of $-\mathbf{F}$, and their sum is zero. Thus, the total work of the internal forces acting on the particles of a rigid body is zero, and *the expression $U_{1\rightarrow2}$ in Eq. (17.1) reduces to the work of the external forces* acting on the body during the displacement considered.

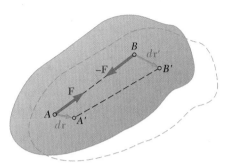

Fig. 17.1

17.3 WORK OF FORCES ACTING ON A RIGID BODY

We saw in Sec. 13.2 that the work of a force $\mathbf{F}$ during a displacement of its point of application from A_1 to A_2 is

$$U_{1\rightarrow2} = \int_{A_1}^{A_2} \mathbf{F} \cdot d\mathbf{r} \tag{17.3}$$

or

$$U_{1\rightarrow2} = \int_{s_1}^{s_2} (F \cos \alpha)\, ds \tag{17.3'}$$

where F is the magnitude of the force, α is the angle it forms with the direction of motion of its point of application A, and s is the variable of integration which measures the distance traveled by A along its path.

In computing the work of the external forces acting on a rigid body, it is often convenient to determine the work of a couple without considering separately the work of each of the two forces forming the couple. Consider the two forces $\mathbf{F}$ and $-\mathbf{F}$ forming a couple of moment $\mathbf{M}$ and acting on a rigid body (Fig. 17.2). Any small displacement of the rigid body bringing A and B, respectively, into A' and B'' can be divided into two parts: in one part points A and B undergo equal displacements $d\mathbf{r}_1$; in the other part A' remains fixed while B' moves into B'' through a displacement $d\mathbf{r}_2$ of magnitude $ds_2 = r\, d\theta$. In the first part of the motion, the work of $\mathbf{F}$ is equal in magnitude and opposite in sign to the work of $-\mathbf{F}$ and their sum is zero. In the second part of the motion, only force $\mathbf{F}$ works, and its work is $dU = F\, ds_2 = Fr\, d\theta$. But the product Fr is equal to the magnitude M of the moment of the couple. Thus, the work of a couple of moment $\mathbf{M}$ acting on a rigid body is

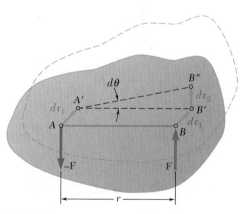

Fig. 17.2

$$dU = M\, d\theta \tag{17.4}$$

where $d\theta$ is the small angle expressed in radians through which the body rotates. We again note that work should be expressed in units obtained by multiplying units of force by units of length. The work of the couple during a finite rotation of the rigid body is obtained by integrating both members of (17.4) from the initial value θ_1 of the angle θ to its final value θ_2. We write

$$U_{1\rightarrow2} = \int_{\theta_1}^{\theta_2} M\,d\theta \tag{17.5}$$

When the moment **M** *of the couple is constant,* formula (17.5) reduces to

$$U_{1\rightarrow2} = M(\theta_2 - \theta_1) \tag{17.6}$$

It was pointed out in Sec. 13.2 that a number of forces encountered in problems of kinetics *do no work.* They are forces applied to fixed points or acting in a direction perpendicular to the displacement of their point of application. Among the forces which do no work the following have been listed: the reaction at a frictionless pin when the body supported rotates about the pin, the reaction at a frictionless surface when the body in contact moves along the surface, and the weight of a body when its center of gravity moves horizontally. We can add now that *when a rigid body rolls without sliding on a fixed surface, the friction force* **F** *at the point of contact C does no work.* The velocity $\mathbf{v}_C$ of the point of contact C is zero, and the work of the friction force **F** during a small displacement of the rigid body is

$$dU = F\,ds_C = F(v_C\,dt) = 0$$

17.4 KINETIC ENERGY OF A RIGID BODY IN PLANE MOTION

Consider a rigid body of mass m in plane motion. We recall from Sec. 14.7 that, if the absolute velocity $\mathbf{v}_i$ of each particle P_i of the body is expressed as the sum of the velocity $\overline{\mathbf{v}}$ of the mass center G of the body and of the velocity $\mathbf{v}_i'$ of the particle relative to a frame $Gx'y'$ attached to G and of fixed orientation (Fig. 17.3), the kinetic energy of the system of particles forming the rigid body can be written in the form

$$T = \tfrac{1}{2}m\overline{v}^2 + \frac{1}{2}\sum_{i=1}^{n}\Delta m_i v_i'^2 \tag{17.7}$$

But the magnitude v_i' of the relative velocity of P_i is equal to the product $r_i'\omega$ of the distance r_i' of P_i from the axis through G perpendicular to the plane of motion and of the magnitude ω of the angular velocity of the body at the instant considered. Substituting into (17.7), we have

$$T = \tfrac{1}{2}m\overline{v}^2 + \frac{1}{2}\left(\sum_{i=1}^{n} r_i'^2\,\Delta m_i\right)w^2 \tag{17.8}$$

or, since the sum represents the moment of inertia $\overline{I}$ of the body about the axis through G,

$$T = \tfrac{1}{2}m\overline{v}^2 + \tfrac{1}{2}\overline{I}\omega^2 \tag{17.9}$$

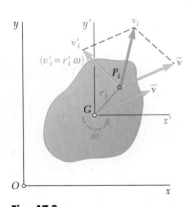

Fig. 17.3

We note that in the particular case of a body in translation ($\omega = 0$), the expression obtained reduces to $\frac{1}{2}m\bar{v}^2$, while in the case of a centroidal rotation ($\bar{v} = 0$), it reduces to $\frac{1}{2}\bar{I}\omega^2$. We conclude that the kinetic energy of a rigid body in plane motion can be separated into two parts: (1) the kinetic energy $\frac{1}{2}m\bar{v}^2$ associated with the motion of the mass center G of the body, and (2) the kinetic energy $\frac{1}{2}\bar{I}\omega^2$ associated with the rotation of the body about G.

Noncentroidal Rotation. The relation (17.9) is valid for any type of plane motion and can therefore be used to express the kinetic energy of a rigid body rotating with an angular velocity ω about a fixed axis through O (Fig. 17.4). In that case, however, the kinetic energy of the body can be expressed more directly by noting that the speed v_i of the particle P_i is equal to the product $r_i\omega$ of the distance r_i of P_i from the fixed axis and the magnitude ω of the angular velocity of the body at the instant considered. Substituting into (17.2), we write

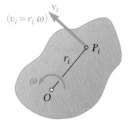

Fig. 17.4

$$T = \frac{1}{2}\sum_{i=1}^{n}\Delta m_i(r_i\omega)^2 = \frac{1}{2}\left(\sum_{i=1}^{n}r_i^2\,\Delta m_i\right)\omega^2$$

or, since the last sum represents the moment of inertia I_O of the body about the fixed axis through O,

$$T = \tfrac{1}{2}I_O\omega^2 \qquad (17.10)$$

We note that the results obtained are not limited to the motion of plane slabs or to the motion of bodies which are symmetrical with respect to the reference plane, and can be applied to the study of the plane motion of any rigid body, regardless of its shape. However, since Eq. (17.9) is applicable to any plane motion while Eq. (17.10) is applicable only in cases involving noncentroidal rotation, Eq. (17.9) will be used in the solution of all the sample problems.

17.5 SYSTEMS OF RIGID BODIES

When a problem involves several rigid bodies, each rigid body can be considered separately and the principle of work and energy can be applied to each body. Adding the kinetic energies of all the particles and considering the work of all the forces involved, we can also write the equation of work and energy for the entire system. We have

$$T_1 + U_{1\rightarrow 2} = T_2 \qquad (17.11)$$

where T represents the arithmetic sum of the kinetic energies of the rigid bodies forming the system (all terms are positive) and $U_{1\rightarrow 2}$ represents the work of all the forces acting on the various bodies, whether these forces are *internal* or *external* from the point of view of the system as a whole.

The method of work and energy is particularly useful in solving problems involving pin-connected members, blocks and pulleys connected by inextensible cords, and meshed gears. In all these cases, the internal forces occur by pairs of equal and opposite forces, and the points of application of the forces in each pair *move through equal distances* during a small displacement of the system. As a result, the work of the internal forces is zero and $U_{1 \to 2}$ reduces to the work of the *forces external to the system*.

17.6 CONSERVATION OF ENERGY

We saw in Sec. 13.6 that the work of conservative forces, such as the weight of a body or the force exerted by a spring, can be expressed as a change in potential energy. When a rigid body, or a system of rigid bodies, moves under the action of conservative forces, the principle of work and energy stated in Sec. 17.2 can be expressed in a modified form. Substituting for $U_{1 \to 2}$ from (13.19′) into (17.1), we write

$$T_1 + V_1 = T_2 + V_2 \qquad (17.12)$$

Formula (17.12) indicates that when a rigid body, or a system of rigid bodies, moves under the action of conservative forces, *the sum of the kinetic energy and of the potential energy of the system remains constant.* It should be noted that in the case of the plane motion of a rigid body, the kinetic energy of the body should include both the *translational* term $\frac{1}{2}m\bar{v}^2$ and the *rotational* term $\frac{1}{2}\bar{I}\omega^2$.

As an example of application of the principle of conservation of energy, let us consider a slender rod AB, of length l and mass m, whose extremities are connected to blocks of negligible mass sliding along horizontal and vertical tracks. We assume that the rod is released with no initial velocity from a horizontal position (Fig. 17.5a), and we wish to determine its angular velocity after it has rotated through an angle θ (Fig. 17.5b).

Since the initial velocity is zero, we have $T_1 = 0$. Measuring the potential energy from the level of the horizontal track, we write $V_1 = 0$. After the rod has rotated through θ, the center of gravity G of the rod is at a distance $\frac{1}{2}l \sin \theta$ below the reference level and we have

$$V_2 = -\tfrac{1}{2}Wl \sin \theta = -\tfrac{1}{2}mgl \sin \theta$$

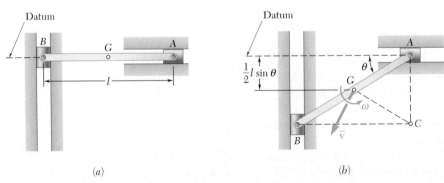

(a)   (b)

Fig. 17.5

Observing that in this position the instantaneous center of the rod is located at C and that $CG = \frac{1}{2}l$, we write $\bar{v}_2 = \frac{1}{2}l\omega$ and obtain

$$T_2 = \frac{1}{2}m\bar{v}_2^2 + \frac{1}{2}\bar{I}\omega_2^2 = \frac{1}{2}m(\tfrac{1}{2}l\omega)^2 + \frac{1}{2}(\tfrac{1}{12}ml^2)\omega^2$$
$$= \frac{1}{2}\frac{ml^2}{3}\omega^2$$

Applying the principle of conservation of energy, we write

$$T_1 + V_1 = T_2 + V_2$$
$$0 = \frac{1}{2}\frac{ml^2}{3}\omega^2 - \tfrac{1}{2}mgl\sin\theta$$
$$\omega = \left(\frac{3g}{l}\sin\theta\right)^{1/2}$$

The advantages of the method of work and energy, as well as its shortcomings, were indicated in Sec. 13.4. Here we should add that the method of work and energy must be supplemented by the application of d'Alembert's principle when reactions at fixed axles, rollers, or sliding blocks are to be determined. For example, in order to compute the reactions at the extremities A and B of the rod of Fig. 17.5b, a diagram should be drawn to express that the system of the external forces applied to the rod is equivalent to the vector $m\bar{a}$ and the couple $\bar{I}\alpha$. The angular velocity $\boldsymbol{\omega}$ of the rod, however, is determined by the method of work and energy before the equations of motion are solved for the reactions. The complete analysis of the motion of the rod and of the forces exerted on the rod requires, therefore, the combined use of the method of work and energy and of the principle of equivalence of the external and effective forces.

17.7 POWER

Power was defined in Sec. 13.5 as the time rate at which work is done. In the case of a body acted upon by a force $\mathbf{F}$, and moving with a velocity $\mathbf{v}$, the power was expressed as follows:

$$\text{Power} = \frac{dU}{dt} = \mathbf{F} \cdot \mathbf{v} \qquad (13.13)$$

In the case of a rigid body rotating with an angular velocity $\boldsymbol{\omega}$ and acted upon by a couple of moment $\mathbf{M}$ parallel to the axis of rotation, we have, by (17.4),

$$\text{Power} = \frac{dU}{dt} = \frac{M\,d\theta}{dt} = M\omega \qquad (17.13)$$

The various units used to measure power, such as the watt and the horsepower, were defined in Sec. 13.5.

1.25 ft

A

240 lb

SAMPLE PROBLEM 17.1

A 240-lb block is suspended from an inextensible cable which is wrapped around a drum of 1.25-ft radius rigidly attached to a flywheel. The drum and flywheel have a combined centroidal moment of inertia $\bar{I} = 10.5 \text{ lb} \cdot \text{ft} \cdot \text{s}^2$. At the instant shown, the velocity of the block is 6 ft/s directed downward. Knowing that the bearing at A is poorly lubricated and that the bearing friction is equivalent to a couple $\mathbf{M}$ of magnitude 60 lb · ft, determine the velocity of the block after it has moved 4 ft downward.

SOLUTION

We consider the system formed by the flywheel and the block. Since the cable is inextensible, the work done by the internal forces exerted by the cable cancels. The initial and final positions of the system and the external forces acting on the system are as shown.

ω_1 $M = 60 \text{ lb} \cdot \text{ft}$

A_y

A_x

$\bar{v}_1 = 6 \text{ ft/s}$ $s_1 = 0$

$W = 240 \text{ lb}$

Kinetic Energy. *Position 1.*

Block: $\qquad\qquad\qquad \bar{v}_1 = 6 \text{ ft/s}$

Flywheel: $\qquad\qquad w_1 = \dfrac{\bar{v}_1}{r} = \dfrac{6 \text{ ft/s}}{1.25 \text{ ft}} = 4.80 \text{ rad/s}$

$$
\begin{aligned}
T_1 &= \tfrac{1}{2}m\bar{v}_1^2 + \tfrac{1}{2}\bar{I}\omega_1^2 \\
&= \frac{1}{2}\frac{240 \text{ lb}}{32.2 \text{ ft/s}^2}(6 \text{ ft/s})^2 + \tfrac{1}{2}(10.5 \text{ lb} \cdot \text{ft} \cdot \text{s}^2)(4.80 \text{ rad/s})^2 \\
&= 255 \text{ ft} \cdot \text{lb}
\end{aligned}
$$

Position 2.   Noting that $\omega_2 = \bar{v}_2/1.25$, we write

$$
\begin{aligned}
T_2 &= \tfrac{1}{2}m\bar{v}_2^2 + \tfrac{1}{2}\bar{I}\omega_2^2 \\
&= \frac{1}{2}\frac{240}{32.2}(\bar{v}_2)^2 + (\tfrac{1}{2})(10.5)\left(\frac{\bar{v}_2}{1.25}\right)^2 = 7.09\bar{v}_2^2
\end{aligned}
$$

ω_2 $M = 60 \text{ lb} \cdot \text{ft}$

A_y

A_x

$s_1 = 0$

4 ft

$s_2 = 4 \text{ ft}$

$\bar{v}_2$

$W = 240 \text{ lb}$

Work.   During the motion, only the weight $\mathbf{W}$ of the block and the friction couple $\mathbf{M}$ do work. Noting that $\mathbf{W}$ does positive work and that the friction couple $\mathbf{M}$ does negative work, we write

$$
\begin{aligned}
s_1 &= 0 \qquad\qquad s_2 = 4 \text{ ft} \\
\theta_1 &= 0 \qquad\qquad \theta_2 = \frac{s_2}{r} = \frac{4 \text{ ft}}{1.25 \text{ ft}} = 3.20 \text{ rad}
\end{aligned}
$$

$$
\begin{aligned}
U_{1\to2} &= W(s_2 - s_1) - M(\theta_2 - \theta_1) \\
&= (240 \text{ lb})(4 \text{ ft}) - (60 \text{ lb} \cdot \text{ft})(3.20 \text{ rad}) \\
&= 768 \text{ ft} \cdot \text{lb}
\end{aligned}
$$

Principle of Work and Energy

$$
\begin{aligned}
T_1 + U_{1\to2} &= T_2 \\
255 \text{ ft} \cdot \text{lb} + 768 \text{ ft} \cdot \text{lb} &= 7.09\bar{v}_2^2 \\
\bar{v}_2 &= 12.01 \text{ ft/s} \qquad \mathbf{v}_2 = 12.01 \text{ ft/s} \downarrow \quad \blacktriangleleft
\end{aligned}
$$

$r_A = 250$ mm

$r_B = 100$ mm

SAMPLE PROBLEM 17.2

Gear A has a mass of 10 kg and a radius of gyration of 200 mm; gear B has a mass of 3 kg and a radius of gyration of 80 mm. The system is at rest when a couple **M** of magnitude 6 N · m is applied to gear B. Neglecting friction, determine (a) the number of revolutions executed by gear B before its angular velocity reaches 600 rpm, (b) the tangential force which gear B exerts on gear A.

SOLUTION

Motion of Entire System. Noting that the peripheral speeds of the gears are equal, we write

$$r_A\omega_A = r_B\omega_B \qquad \omega_A = \omega_B \frac{r_B}{r_A} = \omega_B \frac{100 \text{ mm}}{250 \text{ mm}} = 0.40\omega_B$$

For $\omega_B = 600$ rpm, we have

$$\omega_B = 62.8 \text{ rad/s} \qquad \omega_A = 0.40\omega_B = 25.1 \text{ rad/s}$$
$$\bar{I}_A = m_A\bar{k}_A^2 = (10 \text{ kg})(0.200 \text{ m})^2 = 0.400 \text{ kg} \cdot \text{m}^2$$
$$\bar{I}_B = m_B\bar{k}_B^2 = (3 \text{ kg})(0.080 \text{ m})^2 = 0.0192 \text{ kg} \cdot \text{m}^2$$

Kinetic Energy. Since the system is initially at rest, $T_1 = 0$. Adding the kinetic energies of the two gears when $\omega_B = 600$ rpm, we obtain

$$T_2 = \tfrac{1}{2}\bar{I}_A\omega_A^2 + \tfrac{1}{2}\bar{I}_B\omega_B^2$$
$$= \tfrac{1}{2}(0.400 \text{ kg} \cdot \text{m}^2)(25.1 \text{ rad/s})^2 + \tfrac{1}{2}(0.0192 \text{ kg} \cdot \text{m}^2)(62.8 \text{ rad/s})^2$$
$$= 163.9 \text{ J}$$

Work. Denoting by θ_B the angular displacement of gear B, we have

$$U_{1\to 2} = M\theta_B = (6\text{N} \cdot \text{m})(\theta_B \text{ rad}) = (6\theta_B) \text{ J}$$

Principle of Work and Energy

$$T_1 + U_{1\to 2} = T_2$$
$$0 + (6\theta_B) \text{ J} = 163.9 \text{ J}$$
$$\theta_B = 27.32 \text{ rad} \qquad \theta_B = 4.35 \text{ rev} \quad \blacktriangleleft$$

Motion of Gear A. *Kinetic Energy.* Initially, gear A is at rest, so $T_1 = 0$. When $\omega_B = 600$ rpm, the kinetic energy of gear A is

$$T_2 = \tfrac{1}{2}\bar{I}_A\omega_A^2 = \tfrac{1}{2}(0.400 \text{ kg} \cdot \text{m}^2)(25.1 \text{ rad/s})^2 = 126.0 \text{ J}$$

Work. The forces acting on gear A are as shown. The tangential force **F** does work equal to the product of its magnitude and of the length $\theta_A r_A$ of the arc described by the point of contact. Since $\theta_A r_A = \theta_B r_B$, we have

$$U_{1\to 2} = F(\theta_B r_B) = F(27.3 \text{ rad})(0.100 \text{ m}) = F(2.73 \text{ m})$$

Principle of Work and Energy

$$T_1 + U_{1\to 2} = T_2$$
$$0 + F(2.73 \text{ m}) = 126.0 \text{ J}$$
$$F = +46.2 \text{ N} \qquad F = 46.2 \text{ N} \quad \blacktriangleleft$$

SAMPLE PROBLEM 17.3

A sphere, a cylinder, and a hoop, each having the same mass and the same radius, are released from rest on an incline. Determine the velocity of each body after it has rolled through a distance corresponding to a change in elevation h.

SOLUTION

The problem will first be solved in general terms, and then results for each body will be found. We denote the mass by m, the centroidal moment of inertia by $\bar{I}$, the weight by W, and the radius by r.

Kinematics. Since each body rolls, the instantaneous center of rotation is located at C and we write

$$\omega = \frac{\bar{v}}{r}$$

Kinetic Energy

$$T_1 = 0$$
$$T_2 = \tfrac{1}{2}m\bar{v}^2 + \tfrac{1}{2}\bar{I}\omega^2$$
$$= \tfrac{1}{2}m\bar{v}^2 + \tfrac{1}{2}\bar{I}\left(\frac{\bar{v}}{r}\right)^2 = \tfrac{1}{2}\left(m + \frac{\bar{I}}{r^2}\right)\bar{v}^2$$

Work. Since the friction force $\mathbf{F}$ in rolling motion does no work,

$$U_{1\to2} = Wh$$

Principle of Work and Energy

$$T_1 + U_{1\to2} = T_2$$
$$0 + Wh = \tfrac{1}{2}\left(m + \frac{\bar{I}}{r^2}\right)\bar{v}^2 \qquad \bar{v}^2 = \frac{2Wh}{m + \bar{I}/r^2}$$

Noting that $W = mg$, we rearrange the result and obtain

$$\bar{v}^2 = \frac{2gh}{1 + \bar{I}/mr^2}$$

Velocities of Sphere, Cylinder, and Hoop. Introducing successively the particular expression for $\bar{I}$, we obtain

Sphere:	$\bar{I} = \tfrac{2}{5}mr^2$	$\bar{v} = 0.845\sqrt{2gh}$ ◄
Cylinder:	$\bar{I} = \tfrac{1}{2}mr^2$	$\bar{v} = 0.816\sqrt{2gh}$ ◄
Hoop:	$\bar{I} = mr^2$	$\bar{v} = 0.707\sqrt{2gh}$ ◄

Remark. Let us compare the results with the velocity attained by a frictionless block sliding through the same distance. The solution is identical to the above solution except that $\omega = 0$; we find $\bar{v} = \sqrt{2gh}$.

Comparing the results, we note that the velocity of the body is independent of both its mass and radius. However, the velocity does depend upon the quotient $\bar{I}/mr^2 = \bar{k}^2/r^2$, which measures the ratio of the rotational kinetic energy to the translational kinetic energy. Thus the hoop, which has the largest $\bar{k}$ for a given radius r, attains the smallest velocity, while the sliding block, which does not rotate, attains the largest velocity.

SAMPLE PROBLEM 17.4

A 30-lb slender rod AB is 5 ft long and is pivoted about a point O which is 1 ft from end B. The other end is pressed against a spring of constant $k = 1800$ lb/in. until the spring is compressed 1 in. The rod is then in a horizontal position. If the rod is released from this position, determine its angular velocity and the reaction at the pivot O as the rod passes through a vertical position.

SOLUTION

Position 1. *Potential Energy.* Since the spring is compressed 1 in., we have $x_1 = 1$ in.

$$V_e = \tfrac{1}{2} kx_1^2 = \tfrac{1}{2}(1800 \text{ lb/in.})(1 \text{ in.})^2 = 900 \text{ in} \cdot \text{lb}$$

Choosing the datum as shown, we have $V_g = 0$; therefore,

$$V_1 = V_e + V_g = 900 \text{ in} \cdot \text{lb} = 75 \text{ ft} \cdot \text{lb}$$

Kinetic Energy. Since the velocity in position *1* is zero, we have $T_1 = 0$.

Position 2. *Potential Energy.* The elongation of the spring is zero, and we have $V_e = 0$. Since the center of gravity of the rod is now 1.5 ft above the datum,

$$V_g = (30 \text{ lb})(+1.5 \text{ ft}) = 45 \text{ ft} \cdot \text{lb}$$
$$V_2 = V_e + V_g = 45 \text{ ft} \cdot \text{lb}$$

Kinetic Energy. Denoting by $\boldsymbol{\omega}_2$ the angular velocity of the rod in position 2, we note that the rod rotates about O and write $\bar{v}_2 = \bar{r}\omega_2 = 1.5\omega_2$.

$$\bar{I} = \tfrac{1}{12}ml^2 = \frac{1}{12}\frac{30 \text{ lb}}{32.2 \text{ ft/s}^2}(5 \text{ ft})^2 = 1.941 \text{ lb} \cdot \text{ft} \cdot \text{s}^2$$

$$T_2 = \tfrac{1}{2}m\bar{v}_2^2 + \tfrac{1}{2}\bar{I}\omega_2^2 = \frac{1}{2}\frac{30}{32.2}(1.5\omega_2)^2 + \tfrac{1}{2}(1.941)\omega_2^2 = 2.019\omega_2^2$$

Conservation of Energy

$$T_1 + V_1 = T_2 + V_2$$
$$0 + 75 \text{ ft} \cdot \text{lb} = 2.019\omega_2^2 + 45 \text{ ft} \cdot \text{lb}$$

$$\omega_2 = 3.86 \text{ rad/s} \quad \blacktriangleleft$$

Reaction in Position 2. Since $\omega_2 = 3.86$ rad/s, the components of the acceleration of G as the rod passes through position *2* are

$$\bar{a}_n = \bar{r}\omega_2^2 = (1.5 \text{ ft})(3.86 \text{ rad/s})^2 = 22.3 \text{ ft/s}^2 \qquad \bar{\mathbf{a}}_n = 22.3 \text{ ft/s}^2 \downarrow$$
$$\bar{a}_t = \bar{r}\alpha \qquad\qquad\qquad\qquad\qquad\qquad \bar{\mathbf{a}}_t = \bar{r}\alpha \rightarrow$$

We express that the system of external forces is equivalent to the system of effective forces represented by the vector of components $m\bar{\mathbf{a}}_t$ and $m\bar{\mathbf{a}}_n$ attached at G and the couple $\bar{I}\boldsymbol{\alpha}$.

$$+\downarrow\Sigma M_O = \Sigma(M_O)_{\text{eff}}: \qquad\qquad 0 = \bar{I}\alpha + m(\bar{r}\alpha)\bar{r} \qquad \alpha = 0$$
$$\xrightarrow{+}\Sigma F_x = \Sigma(F_x)_{\text{eff}}: \qquad\qquad R_x = m(\bar{r}\alpha) \qquad R_x = 0$$
$$+\uparrow\Sigma F_y = \Sigma(F_y)_{\text{eff}}: \qquad R_y - 30 \text{ lb} = -m\bar{a}_n$$
$$R_y - 30 \text{ lb} = -\frac{30 \text{ lb}}{32.2 \text{ ft/s}^2}(22.3 \text{ ft/s}^2)$$
$$R_y = +9.22 \text{ lb} \qquad \mathbf{R} = 9.22 \text{ lb} \uparrow \quad \blacktriangleleft$$

SAMPLE PROBLEM 17.5

Each of the two slender rods shown is 0.75 m long and has a mass of 6 kg. If the system is released from rest with $\beta = 60°$, determine (a) the angular velocity of rod AB when $\beta = 20°$, (b) the velocity of point D at the same instant.

SOLUTION

Kinematics of Motion When $\beta = 20°$. Since $\mathbf{v}_B$ is perpendicular to the rod AB and $\mathbf{v}_D$ is horizontal, the instantaneous center of rotation of rod BD is located at C. Considering the geometry of the figure, we obtain

$$BC = 0.75 \text{ m} \qquad CD = 2(0.75 \text{ m}) \sin 20° = 0.513 \text{ m}$$

Applying the law of cosines to triangle CDE, where E is located at the mass center of rod BD, we find $EC = 0.522$ m. Denoting by ω the angular velocity of rod AB, we have

$$\bar{v}_{AB} = (0.375 \text{ m})\omega \qquad \mathbf{\bar{v}}_{AB} = 0.375\omega \searrow$$
$$v_B = (0.75 \text{ m})\omega \qquad \mathbf{v}_B = 0.75\omega \searrow$$

Since rod BD seems to rotate about point C, we write

$$v_B = (BC)\omega_{BD} \qquad (0.75 \text{ m})\omega = (0.75 \text{ m})\omega_{BD} \qquad \boldsymbol{\omega}_{BD} = \omega \uparrow$$
$$\bar{v}_{BD} = (EC)\omega_{BD} = (0.522 \text{ m})\omega \qquad \mathbf{\bar{v}}_{BD} = 0.522\omega \searrow$$

Position 1. *Potential Energy.* Choosing the datum as shown, and observing that $W = (6 \text{ kg})(9.81 \text{ m/s}^2) = 58.86$ N, we have

$$V_1 = 2W\bar{y}_1 = 2(58.86 \text{ N})(0.325 \text{ m}) = 38.26 \text{ J}$$

Kinetic Energy. Since the system is at rest, $T_1 = 0$.

Position 2. *Potential Energy*

$$V_2 = 2W\bar{y}_2 = 2(58.86 \text{ N})(0.1283 \text{ m}) = 15.10 \text{ J}$$

Kinetic Energy

$$I_{AB} = \bar{I}_{BD} = \tfrac{1}{12}ml^2 = \tfrac{1}{12}(6 \text{ kg})(0.75 \text{ m})^2 = 0.281 \text{ kg} \cdot \text{m}^2$$
$$T_2 = \tfrac{1}{2}m\bar{v}_{AB}^2 + \tfrac{1}{2}\bar{I}_{AB}\omega_{AB}^2 + \tfrac{1}{2}m\bar{v}_{BD}^2 + \tfrac{1}{2}\bar{I}_{BD}\omega_{BD}^2$$
$$= \tfrac{1}{2}(6)(0.375\omega)^2 + \tfrac{1}{2}(0.281)\omega^2 + \tfrac{1}{2}(6)(0.522\omega)^2 + \tfrac{1}{2}(0.281)\omega^2$$
$$= 1.520\omega^2$$

Conservation of Energy

$$T_1 + V_1 = T_2 + V_2$$
$$0 + 38.26 \text{ J} = 1.520\omega^2 + 15.10 \text{ J}$$
$$\omega = 3.90 \text{ rad/s} \qquad \boldsymbol{\omega}_{AB} = 3.90 \text{ rad/s} \downarrow \quad \blacktriangleleft$$

Velocity of Point D

$$v_D = (CD)\omega = (0.513 \text{ m})(3.90 \text{ rad/s}) = 2.00 \text{ m/s}$$
$$\mathbf{v}_D = 2.00 \text{ m/s} \rightarrow \quad \blacktriangleleft$$

SOLVING PROBLEMS
ON YOUR OWN

In this lesson we introduced energy methods to determine the velocity of rigid bodies for various positions during their motion. As you found out previously in Chap. 13, energy methods should be considered for problems involving displacements and velocities.

1. The method of work and energy, when applied to all of the particles forming a rigid body, yields the equation

$$T_1 + U_{1 \to 2} = T_2 \tag{17.1}$$

where T_1 and T_2 are, respectively, the initial and final values of the total kinetic energy of the particles forming the body and $U_{1 \to 2}$ is the *work done by the external forces* exerted on the rigid body.

  a. Work of forces and couples. To the expression for the work of a force (Chap. 13), we added the expression for the work of a couple and wrote

$$U_{1 \to 2} = \int_{A_1}^{A_2} \mathbf{F} \cdot d\mathbf{r} \qquad U_{1 \to 2} = \int_{\theta_1}^{\theta_2} M \, d\theta \tag{17.3, 17.5}$$

When the moment of a couple is constant, the work of the couple is

$$U_{1 \to 2} = M(\theta_2 - \theta_1) \tag{17.6}$$

where θ_1 and θ_2 are expressed in radians [Sample Probs. 17.1 and 17.2].

  b. The kinetic energy of a rigid body in plane motion was found by considering the motion of the body as the sum of a translation with its mass center and a rotation about the mass center.

$$T = \tfrac{1}{2} m \bar{v}^2 + \tfrac{1}{2} \bar{I} \omega^2 \tag{17.9}$$

where $\bar{v}$ is the velocity of the mass center and ω is the angular velocity of the body [Sample Probs. 17.3 and 17.4].

2. For a system of rigid bodies we again used the equation

$$T_1 + U_{1 \to 2} = T_2 \tag{17.1}$$

where T is the sum of the kinetic energies of the bodies forming the system and U is the work done by *all the forces acting on the bodies,* internal as well as external. Your computations will be simplified if you keep the following in mind.

  a. The forces exerted on each other by pin-connected members or by meshed gears are equal and opposite, and, since they have the same point of application, they undergo equal small displacements. Therefore, *their total work is zero* and can be omitted from your calculations [Sample Prob. 17.2].

(continued)

b. The forces exerted by an inextensible cord on the two bodies it connects have the same magnitude and their points of application move through equal distances, but the work of one force is positive and the work of the other is negative. Therefore, *their total work is zero* and can again be omitted from your calculations [Sample Prob. 17.1].

c. The forces exerted by a spring on the two bodies it connects also have the same magnitude, but their points of application will generally move through different distances. Therefore, *their total work is usually not zero* and should be taken into account in your calculations.

3. The principle of conservation of energy can be expressed as

$$T_1 + V_1 = T_2 + V_2 \tag{17.12}$$

where V represents the potential energy of the system. This principle can be used when a body or a system of bodies is acted upon by conservative forces, such as the force exerted by a spring or the force of gravity [Sample Probs. 17.4 and 17.5].

4. The last section of this lesson was devoted to power, which is the time rate at which work is done. For a body acted upon by a couple of moment $\mathbf{M}$, the power can be expressed as

$$\text{Power} = M\omega \tag{17.13}$$

where ω is the angular velocity of the body expressed in rad/s. As you did in Chap. 13, you should express power either in watts or in horsepower (1 hp = 550 ft · lb/s).

PROBLEMS

CONCEPT QUESTIONS

17.CQ1 A round object of mass m and radius r is released from rest at the top of a curved surface and rolls without slipping until it leaves the surface with a horizontal velocity as shown. Will a solid sphere, a solid cylinder, or a hoop travel the greatest distance x?

  a. Solid sphere
  b. Solid cylinder
  c. Hoop
  d. They will all travel the same distance.

Fig. P17.CQ1

17.CQ2 A solid steel sphere A of radius r and mass m is released from rest and rolls without slipping down an incline as shown. After traveling a distance d, the sphere has a speed v. If a solid steel sphere of radius $2r$ is released from rest on the same incline, what will its speed be after rolling a distance d?

  a. $0.25\,v$
  b. $0.5\,v$
  c. v
  d. $2v$
  e. $4v$

Fig. P17.CQ2

17.CQ3 Slender bar A is rigidly connected to a massless rod BC in Case 1
and two massless cords in Case 2 as shown. The vertical thickness
of bar A is negligible compared to L. In both cases A is released
from rest at an angle $\theta = \theta_0$. When $\theta = 0°$, which system will have
the larger kinetic energy?

   a. Case 1

   b. Case 2

   c. The kinetic energy will be the same.

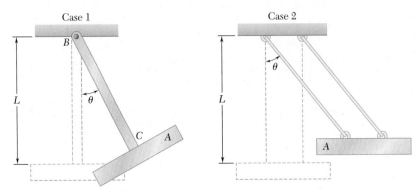

Fig. P17.CQ3 and P17.CQ5

17.CQ4 In Prob. 17.CQ3, how will the speeds of the centers of gravity
compare for the two cases when $\theta = 0°$?

   a. Case 1 will be larger.

   b. Case 2 will be larger.

   c. The speeds will be the same.

17.CQ5 Slender bar A is rigidly connected to a massless rod BC in Case 1
and two massless cords in Case 2 as shown. The vertical thickness of bar A is not negligible compared to L. In both cases A is
released from rest at an angle $\theta = \theta_0$. When $\theta = 0°$, which system
will have the largest kinetic energy?

   a. Case 1

   b. Case 2

   c. The kinetic energy will be the same.

END-OF-SECTION PROBLEMS

17.1 The rotor of an electric motor has an angular velocity of 3600 rpm
when the load and power are cut off. The 50-kg rotor then coasts
to rest after 5000 revolutions. Knowing that the kinetic friction of
the rotor produces a couple of magnitude 4 N · m, determine the
centroidal radius of gyration of the rotor.

17.2 It is known that 1500 revolutions are required for the 6000-lb
flywheel to coast to rest from an angular velocity of 300 rpm.
Knowing that the centroidal radius of gyration of the flywheel is
36 in., determine the average magnitude of the couple due to
kinetic friction in the bearings.

17.3 Two disks of the same material are attached to a shaft as shown. Disk A has a weight of 30 lb and a radius $r = 5$ in. Disk B is three times as thick as disk A. Knowing that a couple **M** of magnitude 15 lb $\cdot$ ft is to be applied to disk A when the system is at rest, determine the radius nr of disk B if the angular velocity of the system is to be 600 rpm after four revolutions.

17.4 Two disks of the same material are attached to a shaft as shown. Disk A is of radius r and has a thickness b, while disk B is of radius nr and thickness $3b$. A couple **M** of constant magnitude is applied when the system is at rest and is removed after the system has executed two revolutions. Determine the value of n which results in the largest final speed for a point on the rim of disk B.

Fig. P17.3 and P17.4

17.5 The flywheel of a small punch rotates at 300 rpm. It is known that 1800 ft $\cdot$ lb of work must be done each time a hole is punched. It is desired that the speed of the flywheel after one punching be not less than 90 percent of the original speed of 300 rpm. (*a*) Determine the required moment of inertia of the flywheel. (*b*) If a constant 25-lb $\cdot$ ft couple is applied to the shaft of the flywheel, determine the number of revolutions which must occur between each punching, knowing that the initial velocity is to be 300 rpm at the start of each punching.

17.6 The flywheel of a punching machine has a mass of 300 kg and a radius of gyration of 600 mm. Each punching operation requires 2500 J of work. (*a*) Knowing that the speed of the flywheel is 300 rpm just before a punching, determine the speed immediately after the punching. (*b*) If a constant 25-N $\cdot$ m couple is applied to the shaft of the flywheel, determine the number of revolutions executed before the speed is again 300 rpm.

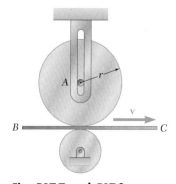

17.7 Disk A, of weight 10 lb and radius $r = 6$ in., is at rest when it is placed in contact with belt BC, which moves to the right with a constant speed $v = 40$ ft/s. Knowing that $\mu_k = 0.20$ between the disk and the belt, determine the number of revolutions executed by the disk before it attains a constant angular velocity.

Fig. P17.7 and P17.8

17.8 Disk A is of constant thickness and is at rest when it is placed in contact with belt BC, which moves with a constant velocity **v**. Denoting by μ_k the coefficient of kinetic friction between the disk and the belt, derive an expression for the number of revolutions executed by the disk before it attains a constant angular velocity.

17.9 The 10-in.-radius brake drum is attached to a larger flywheel which is not shown. The total mass moment of inertia of the flywheel and drum is 16 lb $\cdot$ ft $\cdot$ s^2 and the coefficient of kinetic friction between the drum and the brake shoe is 0.40. Knowing that the initial angular velocity is 240 rpm clockwise, determine the force which must be exerted by the hydraulic cylinder if the system is to stop in 75 revolutions.

17.10 Solve Prob. 17.9, assuming that the initial angular velocity of the flywheel is 240 rpm counterclockwise.

Fig. P17.9

80 mm — A   B — 80 mm

200 mm — C — **M**

Fig. P17.11

r nr

r nr

A   B   C

M_0

Fig. P17.13

17.11 Each of the gears A and B has a mass of 2.4 kg and a radius of gyration of 60 mm, while gear C has a mass of 12 kg and a radius of gyration of 150 mm. A couple **M** of constant magnitude 10 N · m is applied to gear C. Determine (a) the number of revolutions of gear C required for its angular velocity to increase from 100 to 450 rpm, (b) the corresponding tangential force acting on gear A.

17.12 Solve Prob. 17.11, assuming that the 10-N · m couple is applied to gear B.

17.13 The gear train shown consists of four gears of the same thickness and of the same material; two gears are of radius r, and the other two are of radius nr. The system is at rest when the couple M_0 is applied to shaft C. Denoting by I_0 the moment of inertia of a gear of radius r, determine the angular velocity of shaft A if the couple M_0 is applied for one revolution of shaft C.

17.14 The double pulley shown has a mass of 15 kg and a centroidal radius of gyration of 160 mm. Cylinder A and block B are attached to cords that are wrapped on the pulleys as shown. The coefficient of kinetic friction between block B and the surface is 0.2. Knowing that the system is at rest in the position shown when a constant force **P** = 200 N is applied to cylinder A, determine (a) the velocity of cylinder A as it strikes the ground, (b) the total distance that block B moves before coming to rest.

150 mm

C

P

250 mm

A
5 kg

1 m

30°   B
15 kg

Fig. P17.14

17.15 Gear A has a mass of 1 kg and a radius of gyration of 30 mm; gear B has a mass of 4 kg and a radius of gyration of 75 mm; gear C has a mass of 9 kg and a radius of gyration of 100 mm. The system is at rest when a couple M_0 of constant magnitude 4 N · m is applied to gear C. Assuming that no slipping occurs between the gears, determine the number of revolutions required for disk A to reach an angular velocity of 300 rpm.

50 mm
A   B   C

M_0

50 mm 100 mm   150 mm

Fig. P17.15

A   B

l

Fig. P17.16

17.16 A slender rod of length l and weight W is pivoted at one end as shown. It is released from rest in a horizontal position and swings freely. (a) Determine the angular velocity of the rod as it passes through a vertical position and determine the corresponding reaction at the pivot. (b) Solve part a for W = 1.8 lb and l = 3 ft.

17.17 A slender rod of length l is pivoted about a point C located at a distance b from its center G. It is released from rest in a horizontal position and swings freely. Determine (a) the distance b for which the angular velocity of the rod as it passes through a vertical position is maximum, (b) the corresponding values of its angular velocity and of the reaction at C.

Fig. P17.17

17.18 and 17.19 A slender 9-lb rod can rotate in a vertical plane about a pivot at B. A spring of constant $k = 30$ lb/ft and of unstretched length 6 in. is attached to the rod as shown. Knowing that the rod is released from rest in the position shown, determine its angular velocity after it has rotated through 90°.

Fig. P17.18

Fig. P17.19

17.20 A 160-lb gymnast is executing a series of full-circle swings on the horizontal bar. In the position shown he has a small and negligible clockwise angular velocity and will maintain his body straight and rigid as he swings downward. Assuming that during the swing the centroidal radius of gyration of his body is 1.5 ft, determine his angular velocity and the force exerted on his hands after he has rotated through (a) 90°, (b) 180°.

Fig. P17.20

17.21 A collar with a mass of 1 kg is rigidly attached at a distance $d = 300$ mm from the end of a uniform slender rod AB. The rod has a mass of 3 kg and is of length $L = 600$ mm. Knowing that the rod is released from rest in the position shown, determine the angular velocity of the rod after it has rotated through 90°.

Fig. P17.21 and P17.22

17.22 A collar with a mass of 1 kg is rigidly attached to a slender rod AB of mass 3 kg and length $L = 600$ mm. The rod is released from rest in the position shown. Determine the distance d for which the angular velocity of the rod is maximum after it has rotated through 90°.

17.23 Two identical slender rods AB and BC are welded together to form an L-shaped assembly. The assembly is pressed against a spring at D and released from the position shown. Knowing that the maximum angle of rotation of the assembly in its subsequent motion is 90° counterclockwise, determine the magnitude of the angular velocity of the assembly as it passes through the position where rod AB forms an angle of 30° with the horizontal.

Fig. P17.23

Fig. P17.24

Fig. P17.27

17.24 The 30-kg turbine disk has a centroidal radius of gyration of 175 mm and is rotating clockwise at a constant rate of 60 rpm when a small blade of weight 0.5 N at point A becomes loose and is thrown off. Neglecting friction, determine the change in the angular velocity of the turbine disk after it has rotated through (a) 90°, (b) 270°.

17.25 A rope is wrapped around a cylinder of radius r and mass m as shown. Knowing that the cylinder is released from rest, determine the velocity of the center of the cylinder after it has moved downward a distance s.

Fig. P17.25

17.26 Solve Prob. 17.25, assuming that the cylinder is replaced by a thin-walled pipe of radius r and mass m.

17.27 A 45-lb uniform cylindrical roller, initially at rest, is acted upon by a 20-lb force as shown. Knowing that the body rolls without slipping, determine (a) the velocity of its center G after it has moved 5 ft, (b) the friction force required to prevent slipping.

17.28 A small sphere of mass m and radius r is released from rest at A and rolls without sliding on the curved surface to point B where it leaves the surface with a horizontal velocity. Knowing that $a = 1.5$ m and $b = 1.2$ m, determine (a) the speed of the sphere as it strikes the ground at C, (b) the corresponding distance c.

Fig. P17.28

Fig. P17.29

17.29 The mass center G of a 3-kg wheel of radius $R = 180$ mm is located at a distance $r = 60$ mm from its geometric center C. The centroidal radius of gyration of the wheel is $\bar{k} = 90$ mm. As the wheel rolls without sliding, its angular velocity is observed to vary. Knowing that $\omega = 8$ rad/s in the position shown, determine (a) the angular velocity of the wheel when the mass center G is directly above the geometric center C, (b) the reaction at the horizontal surface at the same instant.

17.30 A half section of pipe of mass m and radius r is released from rest in the position shown. Knowing that the pipe rolls without sliding, determine (a) its angular velocity after it has rolled through 90°, (b) the reaction at the horizontal surface at the same instant. [*Hint:* Note that $GO = 2r/\pi$ and that, by the parallel-axis theorem, $\bar{I} = mr^2 - m(GO)^2$.]

17.31 A sphere of mass m and radius r rolls without slipping inside a curved surface of radius R. Knowing that the sphere is released from rest in the position shown, derive an expression for (a) the linear velocity of the sphere as it passes through B, (b) the magnitude of the vertical reaction at that instant.

Fig. P17.30

Fig. P17.31

17.32 Two uniform cylinders, each of weight $W = 14$ lb and radius $r = 5$ in., are connected by a belt as shown. Knowing that at the instant shown the angular velocity of cylinder B is 30 rad/s clockwise, determine (a) the distance through which cylinder A will rise before the angular velocity of cylinder B is reduced to 5 rad/s, (b) the tension in the portion of belt connecting the two cylinders.

17.33 Two uniform cylinders, each of weight $W = 14$ lb and radius $r = 5$ in., are connected by a belt as shown. If the system is released from rest, determine (a) the velocity of the center of cylinder A after it has moved through 3 ft, (b) the tension in the portion of belt connecting the two cylinders.

Fig. P17.32 and P17.33

17.34 A bar of mass $m = 5$ kg is held as shown between four disks each of mass $m' = 2$ kg and radius $r = 75$ mm. Knowing that the forces exerted on the disks are sufficient to prevent slipping and that the bar is released from rest, for each of the cases shown determine the velocity of the bar after it has moved through the distance h.

Fig. P17.34

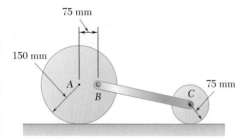

75 mm

150 mm

A

B

C

75 mm

Fig. P17.35

17.35 The 5-kg rod BC is attached by pins to two uniform disks as shown. The mass of the 150-mm-radius disk is 6 kg and that of the 75-mm-radius disk is 1.5 kg. Knowing that the system is released from rest in the position shown, determine the velocity of the rod after disk A has rotated through 90°.

17.36 The motion of the uniform rod AB is guided by small wheels of negligible mass that roll on the surface shown. If the rod is released from rest when $\theta = 0$, determine the velocities of A and B when $\theta = 30°$.

A

60°

θ

B

L

Fig. P17.36

17.37 A 5-m-long ladder has a mass of 15 kg and is placed against a house at an angle $\theta = 20°$. Knowing that the ladder is released from rest, determine the angular velocity of the ladder and the velocity of end A when $\theta = 45°$. Assume the ladder can slide freely on the horizontal ground and on the vertical wall.

17.38 A long ladder of length l, mass m, and centroidal mass moment of inertia $\bar{I}$ is placed against a house at an angle $\theta = \theta_0$. Knowing that the ladder is released from rest, determine the angular velocity of the ladder when $\theta = \theta_2$. Assume the ladder can slide freely on the horizontal ground and on the vertical wall.

B

θ

A

Fig. P17.37 and P17.38

17.39 The ends of a 9-lb rod AB are constrained to move along slots cut in a vertical plate as shown. A spring of constant $k = 3$ lb/in. is attached to end A in such a way that its tension is zero when $\theta = 0$. If the rod is released from rest when $\theta = 50°$, determine the angular velocity of the rod and the velocity of end B when $\theta = 0$.

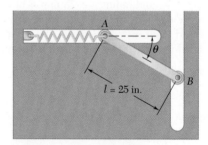

A

θ

$l = 25$ in.

B

Fig. P17.39 and P17.40

17.40 The ends of a 9-lb rod AB are constrained to move along slots cut in a vertical plate as shown. A spring of constant $k = 3$ lb/in. is attached to end A in such a way that its tension is zero when $\theta = 0$. If the rod is released from rest when $\theta = 0$, determine the angular velocity of the rod and the velocity of end B when $\theta = 30°$.

17.41 The motion of a slender rod of length R is guided by pins at A and B which slide freely in slots cut in a vertical plate as shown. If end B is moved slightly to the left and then released, determine the angular velocity of the rod and the velocity of its mass center (*a*) at the instant when the velocity of end B is zero, (*b*) as end B passes through point D.

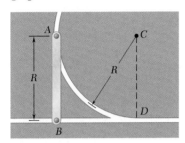

Fig. P17.41

17.42 Each of the two rods shown is of length $L = 1$ m and has a mass of 5 kg. Point D is connected to a spring of constant $k = 20$ N/m and is constrained to move along a vertical slot. Knowing that the system is released from rest when rod BD is horizontal and the spring connected to point D is initially unstretched, determine the velocity of point D when it is directly to the right of point A.

Fig. P17.42

17.43 The 4-kg rod AB is attached to a collar of negligible mass at A and to a flywheel at B. The flywheel has a mass of 16 kg and a radius of gyration of 180 mm. Knowing that in the position shown the angular velocity of the flywheel is 60 rpm clockwise, determine the velocity of the flywheel when point B is directly below C.

17.44 If in Prob. 17.43 the angular velocity of the flywheel is to be the same in the position shown and when point B is directly above C, determine the required value of its angular velocity in the position shown.

240 mm

Fig. P17.43 and P17.44

17.45 The uniform rods AB and BC weigh 2.4 kg and 4 kg, respectively, and the small wheel at C is of negligible weight. If the wheel is moved slightly to the right and then released, determine the velocity of pin B after rod AB has rotated through 90°.

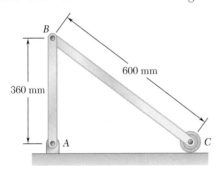

Fig. P17.45 and P17.46

17.46 The uniform rods AB and BC weigh 2.4 kg and 4 kg, respectively, and the small wheel at C is of negligible weight. Knowing that in the position shown the velocity of wheel C is 2 m/s to the right, determine the velocity of pin B after rod AB has rotated through 90°.

17.47 The 80-mm-radius gear shown has a mass of 5 kg and a centroidal radius of gyration of 60 mm. The 4-kg rod AB is attached to the center of the gear and to a pin at B that slides freely in a vertical slot. Knowing that the system is released from rest when $\theta = 60°$, determine the velocity of the center of the gear when $\theta = 20°$.

80 mm

320 mm

Fig. P17.47

17.48 Knowing that the maximum allowable couple that can be applied to a shaft is 15.5 kip · in., determine the maximum horsepower that can be transmitted by the shaft at (a) 180 rpm, (b) 480 rpm.

17.49 Three shafts and four gears are used to form a gear train which will transmit 7.5 kW from the motor at A to a machine tool at F. (Bearings for the shafts are omitted from the sketch.) Knowing that the frequency of the motor is 30 Hz, determine the magnitude of the couple which is applied to shaft (a) AB, (b) CD, (c) EF.

17.50 The shaft-disk-belt arrangement shown is used to transmit 2.4 kW from point A to point D. Knowing that the maximum allowable couples that can be applied to shafts AB and CD are 25 N · m and 80 N · m, respectively, determine the required minimum speed of shaft AB.

17.51 The experimental setup shown is used to measure the power output of a small turbine. When the turbine is operating at 200 rpm, the readings of the two spring scales are 10 and 22 lb, respectively. Determine the power being developed by the turbine.

F

E

180 mm

C

180 mm

75 mm

D

B

75 mm

A

Fig. P17.49

B

30 mm

A

C

D

120 mm

Fig. P17.50

A

B

18 in.

Fig. P17.51

17.8 PRINCIPLE OF IMPULSE AND MOMENTUM FOR THE PLANE MOTION OF A RIGID BODY

The principle of impulse and momentum will now be applied to the analysis of the plane motion of rigid bodies and of systems of rigid bodies. As was pointed out in Chap. 13, the method of impulse and momentum is particularly well adapted to the solution of problems involving time and velocities. Moreover, the principle of impulse and momentum provides the only practicable method for the solution of problems involving impulsive motion or impact (Secs. 17.11 and 17.12).

Considering again a rigid body as made of a large number of particles P_i, we recall from Sec. 14.9 that the system formed by the momenta of the particles at time t_1 and the system of the impulses of the external forces applied from t_1 to t_2 are together equipollent to the system formed by the momenta of the particles at time t_2. Since the vectors associated with a rigid body can be considered as sliding vectors, it follows (Sec. 3.19) that the systems of vectors shown in Fig. 17.6 are not only equipollent but truly *equivalent* in

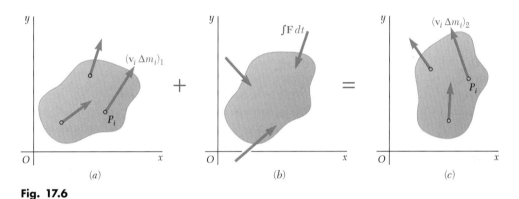

Fig. 17.6

the sense that the vectors on the left-hand side of the equals sign can be transformed into the vectors on the right-hand side through the use of the fundamental operations listed in Sec. 3.13. We therefore write

$$\textbf{Syst Momenta}_1 + \textbf{Syst Ext Imp}_{1\rightarrow2} = \textbf{Syst Momenta}_2 \quad (17.14)$$

But the momenta $\mathbf{v}_i \, \Delta m_i$ of the particles can be reduced to a vector attached at G, equal to their sum

$$\mathbf{L} = \sum_{i=1}^{n} \mathbf{v}_i \, \Delta m_i$$

and a couple of moment equal to the sum of their moments about G

$$\mathbf{H}_G = \sum_{i=1}^{n} \mathbf{r}_i' \times \mathbf{v}_i \, \Delta m_i$$

We recall from Sec. 14.3 that $\mathbf{L}$ and $\mathbf{H}_G$ define, respectively, the linear momentum and the angular momentum about G of the system

Photo 17.2 A Charpy impact test is used to determine the amount of energy absorbed by a material during impact by subtracting the final gravitation potential energy of the arm from its initial gravitational potential energy.

of particles forming the rigid body. We also note from Eq. (14.14) that $\mathbf{L} = m\bar{\mathbf{v}}$. On the other hand, restricting the present analysis to the plane motion of a rigid slab or of a rigid body symmetrical with respect to the reference plane, we recall from Eq. (16.4) that $\mathbf{H}_G = \bar{I}\boldsymbol{\omega}$. We thus conclude that the system of the momenta $\mathbf{v}_i\,\Delta m_i$ is equivalent to the *linear momentum vector* $m\bar{\mathbf{v}}$ attached at G and to the *angular momentum couple* $\bar{I}\boldsymbol{\omega}$ (Fig. 17.7). Observing that the

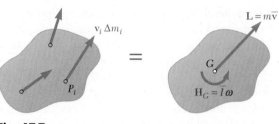

Fig. 17.7

system of momenta reduces to the vector $m\bar{\mathbf{v}}$ in the particular case of a translation ($\boldsymbol{\omega} = 0$) and to the couple $\bar{I}\boldsymbol{\omega}$ in the particular case of a centroidal rotation ($\bar{\mathbf{v}} = \mathbf{0}$), we verify once more that the plane motion of a rigid body symmetrical with respect to the reference plane can be resolved into a translation with the mass center G and a rotation about G.

Replacing the system of momenta in parts a and c of Fig. 17.6 by the equivalent linear momentum vector and angular momentum couple, we obtain the three diagrams shown in Fig. 17.8. This figure

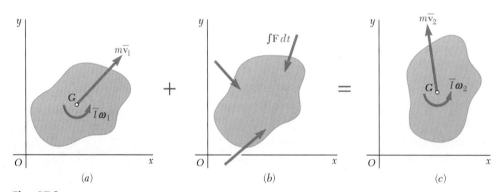

Fig. 17.8

expresses as a free-body-diagram equation the fundamental relation (17.14) in the case of the plane motion of a rigid slab or of a rigid body symmetrical with respect to the reference plane.

Three equations of motion can be derived from Fig. 17.8. Two equations are obtained by summing and equating the x and y *components* of the momenta and impulses, and the third equation is obtained by summing and equating the *moments* of these vectors *about any given point*. The coordinate axes can be chosen fixed in space, or allowed to move with the mass center of the body while maintaining a fixed direction. In either case, the point about which moments are taken should keep the same position relative to the coordinate axes during the interval of time considered.

In deriving the three equations of motion for a rigid body, care should be taken not to add linear and angular momenta indiscriminately. Confusion can be avoided by remembering that $m\bar{v}_x$ and $m\bar{v}_y$ represent the *components of a vector*, namely, the linear momentum vector $m\mathbf{v}$, while $\bar{I}\omega$ represents the *magnitude of a couple*, namely, the angular momentum couple $\bar{I}\omega$. Thus the quantity $\bar{I}\omega$ should be added only to the *moment* of the linear momentum $m\mathbf{v}$, never to this vector itself nor to its components. All quantities involved will then be expressed in the same units, namely $N \cdot m \cdot s$ or $lb \cdot ft \cdot s$.

Noncentroidal Rotation. In this particular case of plane motion, the magnitude of the velocity of the mass center of the body is $\bar{v} = \bar{r}\omega$, where $\bar{r}$ represents the distance from the mass center to the fixed axis of rotation and $\boldsymbol{\omega}$ represents the angular velocity of the body at the instant considered; the magnitude of the momentum vector attached at G is thus $m\bar{v} = m\bar{r}\omega$. Summing the moments about O of the momentum vector and momentum couple (Fig. 17.9)

Fig. 17.9

and using the parallel-axis theorem for moments of inertia, we find that the angular momentum $\mathbf{H}_O$ of the body about O has the magnitude†

$$\bar{I}\omega + (m\bar{r}\omega)\bar{r} = (\bar{I} + m\bar{r}^2)\omega = I_O\omega \qquad (17.15)$$

Equating the moments about O of the momenta and impulses in (17.14), we write

$$I_O\omega_1 + \sum \int_{t_1}^{t_2} M_O \, dt = I_O\omega_2 \qquad (17.16)$$

In the general case of plane motion of a rigid body symmetrical with respect to the reference plane, Eq. (17.16) can be used with respect to the instantaneous axis of rotation under certain conditions. It is recommended, however, that all problems of plane motion be solved by the general method described earlier in this section.

---

†Note that the sum $\mathbf{H}_A$ of the moments about an arbitrary point A of the momenta of the particles of a rigid slab is, in general, *not* equal to $I_A\boldsymbol{\omega}$. (See Prob. 17.67.)

17.9 SYSTEMS OF RIGID BODIES

The motion of several rigid bodies can be analyzed by applying the principle of impulse and momentum to each body separately (Sample Prob. 17.6). However, in solving problems involving no more than three unknowns (including the impulses of unknown reactions), it is often convenient to apply the principle of impulse and momentum to the system as a whole. The momentum and impulse diagrams are drawn for the entire system of bodies. For each moving part of the system, the diagrams of momenta should include a momentum vector, a momentum couple, or both. Impulses of forces internal to the system can be omitted from the impulse diagram, since they occur in pairs of equal and opposite vectors. Summing and equating successively the x components, y components, and moments of all vectors involved, one obtains three relations which express that the momenta at time t_1 and the impulses of the external forces form a system equipollent to the system of the momenta at time t_2.† Again, care should be taken not to add linear and angular momenta indiscriminately; each equation should be checked to make sure that consistent units have been used. This approach has been used in Sample Prob. 17.8 and, further on, in Sample Probs. 17.9 and 17.10.

17.10 CONSERVATION OF ANGULAR MOMENTUM

When no external force acts on a rigid body or a system of rigid bodies, the impulses of the external forces are zero and the system of the momenta at time t_1 is equipollent to the system of the momenta at time t_2. Summing and equating successively the x components, y components, and moments of the momenta at times t_1 and t_2, we conclude that the total linear momentum of the system is conserved in any direction and that its total angular momentum is conserved about any point.

There are many engineering applications, however, in which *the linear momentum is not conserved* yet *the angular momentum* $\mathbf{H}_O$ of the system about a given point O *is conserved* that is, in which

$$(\mathbf{H}_O)_1 = (\mathbf{H}_O)_2 \tag{17.17}$$

Such cases occur when the lines of action of all external forces pass through O or, more generally, when the sum of the angular impulses of the external forces about O is zero.

Problems involving *conservation of angular momentum* about a point O can be solved by the general method of impulse and momentum, i.e., by drawing momentum and impulse diagrams as described in Secs. 17.8 and 17.9. Equation (17.17) is then obtained by summing and equating moments about O (Sample Prob. 17.8). As you will see later in Sample Prob. 17.9, two additional equations can be written by summing and equating x and y components and these equations can be used to determine two unknown linear impulses, such as the impulses of the reaction components at a fixed point.

Photo 17.3 A figure skater at the beginning and at the end of a spin. By using the principle of conservation of angular momentum you will find that her angular velocity is much higher at the end of the spin.

---

†Note that as in Sec. 16.7, we cannot speak of *equivalent* systems since we are not dealing with a single rigid body.

$r_A = 250$ mm

$r_B = 100$ mm

SAMPLE PROBLEM 17.6

Gear A has a mass of 10 kg and a radius of gyration of 200 mm, and gear B has a mass of 3 kg and a radius of gyration of 80 mm. The system is at rest when a couple **M** of magnitude 6 N · m is applied to gear B. (These gears were considered in Sample Prob. 17.2.) Neglecting friction, determine (*a*) the time required for the angular velocity of gear B to reach 600 rpm, (*b*) the tangential force which gear B exerts on gear A.

SOLUTION

We apply the principle of impulse and momentum to each gear separately. Since all forces and the couple are constant, their impulses are obtained by multiplying them by the unknown time t. We recall from Sample Prob. 17.2 that the centroidal moments of inertia and the final angular velocities are

$$\bar{I}_A = 0.400 \text{ kg} \cdot \text{m}^2 \qquad \bar{I}_B = 0.0192 \text{ kg} \cdot \text{m}^2$$
$$(\omega_A)_2 = 25.1 \text{ rad/s} \qquad (\omega_B)_2 = 62.8 \text{ rad/s}$$

Principle of Impulse and Momentum for Gear A. The systems of initial momenta, impulses, and final momenta are shown in three separate sketches.

Syst Momenta$_1$ + Syst Ext Imp$_{1\to2}$ = Syst Momenta$_2$

+↺moments about A: $\qquad 0 - Ftr_A = -\bar{I}_A(\omega_A)_2$

$$Ft(0.250 \text{ m}) = (0.400 \text{ kg} \cdot \text{m}^2)(25.1 \text{ rad/s})$$
$$Ft = 40.2 \text{ N} \cdot \text{s}$$

Principle of Impulse and Momentum for Gear B.

Syst Momenta$_1$ + Syst Ext Imp$_{1\to2}$ = Syst Momenta$_2$

+↺moments about B: $\qquad 0 + Mt - Ftr_B = \bar{I}_B(\omega_B)_2$

$$+(6 \text{ N} \cdot \text{m})t - (40.2 \text{ N} \cdot \text{s})(0.100 \text{ m}) = (0.0192 \text{ kg} \cdot \text{m}^2)(62.8 \text{ rad/s})$$

$$t = 0.871 \text{ s} \quad \blacktriangleleft$$

Recalling that $Ft = 40.2$ N · s, we write

$$F(0.871 \text{ s}) = 40.2 \text{ N} \cdot \text{s} \qquad F = +46.2 \text{ N}$$

Thus, the force exerted by gear B on gear A is $\qquad \mathbf{F} = 46.2 \text{ N} \swarrow \quad \blacktriangleleft$

SAMPLE PROBLEM 17.7

A uniform sphere of mass m and radius r is projected along a rough horizontal surface with a linear velocity $\bar{\mathbf{v}}_1$ and no angular velocity. Denoting by μ_k the coefficient of kinetic friction between the sphere and the surface, determine (a) the time t_2 at which the sphere will start rolling without sliding, (b) the linear and angular velocities of the sphere at time t_2.

SOLUTION

While the sphere is sliding relative to the surface, it is acted upon by the normal force $\mathbf{N}$, the friction force $\mathbf{F}$, and its weight $\mathbf{W}$ of magnitude $W = mg$.

Principle of Impulse and Momentum. We apply the principle of impulse and momentum to the sphere from the time $t_1 = 0$ when it is placed on the surface until the time $t_2 = t$ when it starts rolling without sliding.

$$\text{Syst Momenta}_1 + \text{Syst Ext Imp}_{1\rightarrow2} = \text{Syst Momenta}_2$$

$+\uparrow y$ components:
$$Nt - Wt = 0 \qquad (1)$$

$\overset{+}{\rightarrow} x$ components:
$$m\bar{v}_1 - Ft = m\bar{v}_2 \qquad (2)$$

$+\downarrow$ moments about G:
$$Ftr = \bar{I}\omega_2 \qquad (3)$$

From (1) we obtain $N = W = mg$. During the entire time interval considered, sliding occurs at point C and we have $F = \mu_k N = \mu_k mg$. Substituting CS for F into (2), we write

$$m\bar{v}_1 - \mu_k mgt = m\bar{v}_2 \qquad \bar{v}_2 = \bar{v}_1 - \mu_k gt \qquad (4)$$

Substituting $F = \mu_k mg$ and $\bar{I} = \frac{2}{5}mr^2$ into (3),

$$\mu_k mgtr = \frac{2}{5}mr^2\omega_2 \qquad \omega_2 = \frac{5}{2}\frac{\mu_k g}{r}t \qquad (5)$$

The sphere will start rolling without sliding when the velocity $\mathbf{v}_C$ of the point of contact is zero. At that time, point C becomes the instantaneous center of rotation, and we have $\bar{v}_2 = r\omega_2$. Substituting from (4) and (5), we write

$$\bar{v}_2 = r\omega_2 \qquad \bar{v}_1 - \mu_k gt = r\left(\frac{5}{2}\frac{\mu_k g}{r}t\right) \qquad t = \frac{2}{7}\frac{\bar{v}_1}{\mu_k g} \quad \blacktriangleleft$$

Substituting this expression for t into (5),

$$\omega_2 = \frac{5}{2}\frac{\mu_k g}{r}\left(\frac{2}{7}\frac{\bar{v}_1}{\mu_k g}\right) \qquad \omega_2 = \frac{5}{7}\frac{\bar{v}_1}{r} \qquad \omega_2 = \frac{5}{7}\frac{\bar{v}_1}{r} \downarrow \quad \blacktriangleleft$$

$$\bar{v}_2 = r\omega_2 \qquad \bar{v}_2 = r\left(\frac{5}{7}\frac{v_1}{r}\right) \qquad \bar{\mathbf{v}}_2 = \frac{5}{7}\bar{v}_1 \rightarrow \quad \blacktriangleleft$$

SAMPLE PROBLEM 17.8

Two solid spheres of radius 3 in., weighing 2 lb each, are mounted at A and B on the horizontal rod $A'B'$, which rotates freely about the vertical with a counterclockwise angular velocity of 6 rad/s. The spheres are held in position by a cord which is suddenly cut. Knowing that the centroidal moment of inertia of the rod and pivot is $\bar{I}_R = 0.25$ lb · ft · s², determine (a) the angular velocity of the rod after the spheres have moved to positions A' and B', (b) the energy lost due to the plastic impact of the spheres and the stops at A' and B'.

SOLUTION

a. Principle of Impulse and Momentum. In order to determine the final angular velocity of the rod, we will express that the initial momenta of the various parts of the system and the impulses of the external forces are together equipollent to the final momenta of the system.

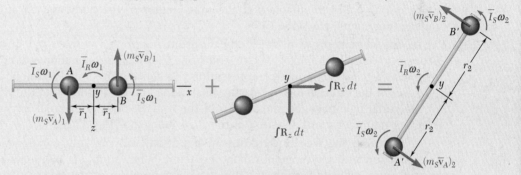

$$\text{Syst Momenta}_1 + \text{Syst Ext Imp}_{1\rightarrow2} = \text{Syst Momenta}_2$$

Observing that the external forces consist of the weights and the reaction at the pivot, which have no moment about the y axis, and noting that $\bar{v}_A = \bar{v}_B = \bar{r}\omega$, we equate moments about the y axis:

$$2(m_S\bar{r}_1\omega_1)\bar{r}_1 + 2\bar{I}_S\omega_1 + \bar{I}_R\omega_1 = 2(m_S\bar{r}_2\omega_2)\bar{r}_2 + 2\bar{I}_S\omega_2 + \bar{I}_R\omega_2$$
$$(2m_S\bar{r}_1^2 + 2\bar{I}_S + \bar{I}_R)\omega_1 = (2m_S\bar{r}_2^2 + 2\bar{I}_S + \bar{I}_R)\omega_2 \tag{1}$$

which expresses that *the angular momentum of the system about the y axis is conserved.* We now compute

$$\bar{I}_S = \tfrac{2}{5}m_Sa^2 = \tfrac{2}{5}(2 \text{ lb/32.2 ft/s}^2)(\tfrac{3}{12} \text{ ft})^2 = 0.00155 \text{ lb} \cdot \text{ft} \cdot \text{s}^2$$
$$m_S\bar{r}_1^2 = (2/32.2)(\tfrac{5}{12})^2 = 0.0108 \qquad m_S\bar{r}_2^2 = (2/32.2)(\tfrac{25}{12})^2 = 0.2696$$

Substituting these values, and $\bar{I}_R = 0.25$ and $\omega_1 = 6$ rad/s into (1):

$$0.275(6 \text{ rad/s}) = 0.792\omega_2 \qquad \omega_2 = 2.08 \text{ rad/s} \text{ ↰} \quad \blacktriangleleft$$

b. Energy Lost. The kinetic energy of the system at any instant is

$$T = 2(\tfrac{1}{2}m_S\bar{v}^2 + \tfrac{1}{2}\bar{I}_S\omega^2) + \tfrac{1}{2}\bar{I}_R\omega^2 = \tfrac{1}{2}(2m_S\bar{r}^2 + 2\bar{I}_S + \bar{I}_R)\omega^2$$

Recalling the numerical values found above, we have

$$T_1 = \tfrac{1}{2}(0.275)(6)^2 = 4.95 \text{ ft} \cdot \text{lb} \qquad T_2 = \tfrac{1}{2}(0.792)(2.08)^2 = 1.713 \text{ ft} \cdot \text{lb}$$
$$\Delta T = T_2 - T_1 = 1.71 - 4.95 \qquad \Delta T = -3.24 \text{ ft} \cdot \text{lb} \quad \blacktriangleleft$$

SOLVING PROBLEMS
ON YOUR OWN

In this lesson you learned to use the method of impulse and momentum to solve problems involving the plane motion of rigid bodies. As you found out previously in Chap. 13, this method is most effective when used in the solution of problems involving velocities and time.

1. The principle of impulse and momentum for the plane motion of a rigid body is expressed by the following vector equation:

$$\textbf{Syst Momenta}_1 + \textbf{Syst Ext Imp}_{1\rightarrow2} = \textbf{Syst Momenta}_2 \quad (17.14)$$

where **Syst Momenta** represents the system of the momenta of the particles forming the rigid body, and **Syst Ext Imp** represents the system of all the external impulses exerted during the motion.

a. The system of the momenta of a rigid body is equivalent to a linear momentum vector $m\bar{\textbf{v}}$ attached at the mass center of the body and an angular momentum couple $\bar{I}\boldsymbol{\omega}$ (Fig. 17.7).

b. You should draw a free-body-diagram equation for the rigid body to express graphically the above vector equation. Your diagram equation will consist of three sketches of the body, representing respectively the initial momenta, the impulses of the external forces, and the final momenta. It will show that the system of the initial momenta and the system of the impulses of the external forces are together equivalent to the system of the final momenta (Fig. 17.8).

c. By using the free-body-diagram equation, you can sum components in any direction and sum moments about any point. When summing moments about a point, remember to include the *angular momentum* $\bar{I}\boldsymbol{\omega}$ of the body, as well as the *moments* of the components of its *linear momentum*. In most cases you will be able to select and solve an equation that involves only one unknown. This was done in all the sample problems of this lesson.

2. In problems involving a system of rigid bodies, you can apply the principle of impulse and momentum to the system as a whole. Since internal forces occur in equal and opposite pairs, they will not be part of your solution [Sample Prob. 17.8].

3. Conservation of angular momentum about a given axis occurs when, for a system of rigid bodies, *the sum of the moments of the external impulses about that axis is zero.* You can indeed easily observe from the free-body-diagram equation that the initial and final angular momenta of the system about that axis are equal and, thus, that *the angular momentum of the system about the given axis is conserved.* You can then sum the angular momenta of the various bodies of the system and the moments of their linear momenta about that axis to obtain an equation which can be solved for one unknown [Sample Prob. 17.8].

PROBLEMS

CONCEPT QUESTIONS

17.CQ6 Slender bar A is rigidly connected to a massless rod BC in Case 1 and two massless cords in Case 2 as shown. The vertical thickness of bar A is negligible compared to L. If bullet D strikes A with a speed v_0 and becomes embedded in it, how will the speeds of the center of gravity of A immediately after the impact compare for the two cases?

   a. Case 1 will be larger.

   b. Case 2 will be larger.

   c. The speeds will be the same.

 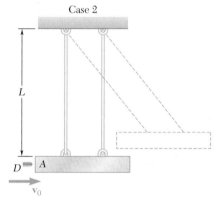

Fig. P17.CQ6

17.CQ7 A 1-m-long uniform slender bar AB has an angular velocity of 12 rad/s and its center of gravity has a velocity of 2 m/s as shown. About which point is the angular momentum of A smallest at this instant?

   a. P_1

   b. P_2

   c. P_3

   d. P_4

   e. It is the same about all the points.

Fig. P17.CQ7

IMPULSE-MOMENTUM PRACTICE PROBLEMS

17.F1 The 350-kg flywheel of a small hoisting engine has a radius of gyration of 600 mm. If the power is cut off when the angular velocity
of the flywheel is 100 rpm clockwise, draw an impulse-momentum
diagram that can be used to determine the time required for the
system to come to rest.

225 mm

120 kg

Fig. P17.F1

17.F2 A sphere of radius r and mass m is placed on a horizontal floor with
no linear velocity but with a clockwise angular velocity ω_0. Denoting by μ_k the coefficient of kinetic friction between the sphere
and the floor, draw the impulse-momentum diagram that can be
used to determine the time t_1 at which the sphere will start rolling
without sliding.

ω_0

Fig. P17.F2

17.F3 Two panels A and B are attached with hinges to a rectangular plate
and held by a wire as shown. The plate and the panels are made
of the same material and have the same thickness. The entire
assembly is rotating with an angular velocity ω_0 when the wire
breaks. Draw the impulse-momentum diagram that is needed to
determine the angular velocity of the assembly after the panels
have come to rest against the plate.

Fig. P17.F3

17.52 The rotor of an electric motor has a mass of 25 kg, and it is observed that 4.2 min is required for the rotor to coast to rest from an angular velocity of 3600 rpm. Knowing that kinetic friction produces a couple of magnitude 1.2 N · m, determine the centroidal radius of gyration for the rotor.

17.53 A small grinding wheel is attached to the shaft of an electric motor which has a rated speed of 3600 rpm. When the power is turned off, the unit coasts to rest in 70 s. The grinding wheel and rotor have a combined weight of 6 lb and a combined radius of gyration of 2 in. Determine the average magnitude of the couple due to kinetic friction in the bearings of the motor.

Fig. P17.53

17.54 A bolt located 50 mm from the center of an automobile wheel is tightened by applying the couple shown for 0.10 s. Assuming that the wheel is free to rotate and is initially at rest, determine the resulting angular velocity of the wheel. The wheel has a mass of 19 kg and has a radius of gyration of 250 mm.

100 N · 460 mm · 100 N

Fig. P17.54

17.55 Two disks of the same thickness and same material are attached to a shaft as shown. The 8-lb disk A has a radius $r_A = 3$ in., and disk B has a radius $r_B = 4.5$ in. Knowing that a couple **M** of magnitude 20 lb · in. is applied to disk A when the system is at rest, determine the time required for the angular velocity of the system to reach 960 rpm.

Fig. P17.55 and *P17.56*

17.56 Two disks of the same thickness and same material are attached to a shaft as shown. The 3-kg disk A has a radius $r_A = 100$ mm, and disk B has a radius $r_B = 125$ mm. Knowing that the angular velocity of the system is to be increased from 200 rpm to 800 rpm during a 3-s interval, determine the magnitude of the couple **M** that must be applied to disk A.

17.57 A disk of constant thickness, initially at rest, is placed in contact with a belt that moves with a constant velocity **v**. Denoting by μ_k the coefficient of kinetic friction between the disk and the belt, derive an expression for the time required for the disk to reach a constant angular velocity.

17.58 Disk A, of weight 5 lb and radius $r = 3$ in., is at rest when it is placed in contact with a belt which moves at a constant speed $v = 50$ ft/s. Knowing that $\mu_k = 0.20$ between the disk and the belt, determine the time required for the disk to reach a constant angular velocity.

Fig. *P17.57* and P17.58

17.59 A cylinder of radius r and weight W with an initial counterclockwise angular velocity $\boldsymbol{\omega}_0$ is placed in the corner formed by the floor and a vertical wall. Denoting by μ_k the coefficient of kinetic friction between the cylinder and the wall and the floor, derive an expression for the time required for the cylinder to come to rest.

Fig. P17.59

17.60 and 17.61 Two uniform disks and two cylinders are assembled as indicated. Disk A has a mass of 10 kg and disk B has a mass of 6 kg. Knowing that the system is released from rest, determine the time required for cylinder C to have a speed of 0.5 m/s.

17.60 Disks A and B are bolted together and the cylinders are attached to separate cords wrapped on the disks.

17.61 The cylinders are attached to a single cord that passes over the disks. Assume that no slipping occurs between the cord and the disks.

Fig. P17.60

Fig. P17.61

17.62 Disk B has an initial angular velocity $\boldsymbol{\omega}_0$ when it is brought into contact with disk A which is at rest. Show that the final angular velocity of disk B depends only on ω_0 and the ratio of the masses m_A and m_B of the two disks.

17.63 The 7.5-lb disk A has a radius $r_A = 6$ in. and is initially at rest. The 10-lb disk B has a radius $r_B = 8$ in. and an angular velocity ω_0 of 900 rpm when it is brought into contact with disk A. Neglecting friction in the bearings, determine (a) the final angular velocity of each disk, (b) the total impulse of the friction force exerted on disk A.

Fig. P17.62 and P17.63

17.64 A tape moves over the two drums shown. Drum A weighs 1.4 lb and has a radius of gyration of 0.75 in., while drum B weighs 3.5 lb and has a radius of gyration of 1.25 in. In the lower portion of the tape the tension is constant and equal to $T_A = 0.75$ lb. Knowing that the tape is initially at rest, determine (*a*) the required constant tension T_B if the velocity of the tape is to be $v = 10$ ft/s after 0.24 s, (*b*) the corresponding tension in the portion of the tape between the drums.

Fig. P17.64

17.65 Show that the system of momenta for a rigid slab in plane motion reduces to a single vector, and express the distance from the mass center G to the line of action of this vector in terms of the centroidal radius of gyration $\bar{k}$ of the slab, the magnitude $\bar{v}$ of the velocity of G, and the angular velocity $\boldsymbol{\omega}$.

17.66 Show that, when a rigid slab rotates about a fixed axis through O perpendicular to the slab, the system of the momenta of its particles is equivalent to a single vector of magnitude $m\bar{r}\omega$, perpendicular to the line OG, and applied to a point P on this line, called the *center of percussion*, at a distance $GP = \bar{k}^2/\bar{r}$ from the mass center of the slab.

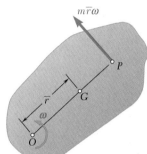

Fig. P17.66

17.67 Show that the sum $\mathbf{H}_A$ of the moments about a point A of the momenta of the particles of a rigid slab in plane motion is equal to $I_A\boldsymbol{\omega}$, where $\boldsymbol{\omega}$ is the angular velocity of the slab at the instant considered and I_A the moment of inertia of the slab about A, if and only if one of the following conditions is satisfied: (*a*) A is the mass center of the slab, (*b*) A is the instantaneous center of rotation, (*c*) the velocity of A is directed along a line joining point A and the mass center G.

17.68 Consider a rigid slab initially at rest and subjected to an impulsive force $\mathbf{F}$ contained in the plane of the slab. We define the *center of percussion* P as the point of intersection of the line of action of $\mathbf{F}$ with the perpendicular drawn from G. (*a*) Show that the instantaneous center of rotation C of the slab is located on line GP at a distance $GC = \bar{k}^2/GP$ on the opposite side of G. (*b*) Show that if the center of percussion were located at C the instantaneous center of rotation would be located at P.

Fig. P17.68

17.69 A flywheel is rigidly attached to a 1.5-in.-radius shaft that rolls without sliding along parallel rails. Knowing that after being released from rest the system attains a speed of 6 in./s in 30 s, determine the centroidal radius of gyration of the system.

Fig. P17.69

Fig. P17.71

Fig. P17.74 and P17.75

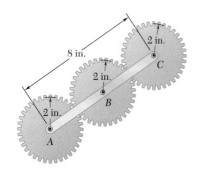

Fig. P17.76

17.70 A wheel of radius r and centroidal radius of gyration $\bar{k}$ is released from rest on the incline shown at time $t = 0$. Assuming that the wheel rolls without sliding, determine (a) the velocity of its center at time t, (b) the coefficient of static friction required to prevent slipping.

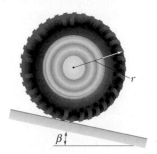

Fig. P17.70

17.71 The double pulley shown has a mass of 3 kg and a radius of gyration of 100 mm. Knowing that when the pulley is at rest, a force **P** of magnitude 24 N is applied to cord B, determine (a) the velocity of the center of the pulley after 1.5 s, (b) the tension in cord C.

17.72 and 17.73 A 9-in.-radius cylinder of weight 18 lb rests on a 6-lb carriage. The system is at rest when a force **P** of magnitude 2.5 lb is applied as shown for 1.2 s. Knowing that the cylinder rolls without sliding on the carriage and neglecting the mass of the wheels of the carriage, determine the resulting velocity of (a) the carriage, (b) the center of the cylinder.

Fig. P17.72

Fig. P17.73

17.74 Two uniform cylinders, each of mass $m = 6$ kg and radius $r = 125$ mm, are connected by a belt as shown. If the system is released from rest when $t = 0$, determine (a) the velocity of the center of cylinder B at $t = 3$ s, (b) the tension in the portion of belt connecting the two cylinders.

17.75 Two uniform cylinders, each of mass $m = 6$ kg and radius $r = 125$ mm, are connected by a belt as shown. Knowing that at the instant shown the angular velocity of cylinder A is 30 rad/s counterclockwise, determine (a) the time required for the angular velocity of cylinder A to be reduced to 5 rad/s, (b) the tension in the portion of belt connecting the two cylinders.

17.76 In the gear arrangement shown, gears A and C are attached to rod ABC, which is free to rotate about B, while the inner gear B is fixed. Knowing that the system is at rest, determine the magnitude of the couple **M** which must be applied to rod ABC, if 2.5 s later the angular velocity of the rod is to be 240 rpm clockwise. Gears A and C weigh 2.5 lb each and may be considered as disks of radius 2 in.; rod ABC weighs 4 lb.

17.77 A sphere of radius r and mass m is projected along a rough horizontal surface with the initial velocities shown. If the final velocity of the sphere is to be zero, express (a) the required magnitude of $\boldsymbol{\omega}_0$ in terms of v_0 and r, (b) the time required for the sphere to come to rest in terms of v_0 and the coefficient of kinetic friction μ_k.

Fig. P17.77

17.78 A bowler projects an 8.5-in.-diameter ball weighing 16 lb along an alley with a forward velocity $\mathbf{v}_0$ of 25 ft/s and a backspin ω_0 of 9 rad/s. Knowing that the coefficient of kinetic friction between the ball and the alley is 0.10, determine (a) the time t_1 at which the ball will start rolling without sliding, (b) the speed of the ball at time t_1.

Fig. P17.78

17.79 Four rectangular panels, each of length b and height $\frac{1}{2}b$, are attached with hinges to a circular plate of diameter $\sqrt{2}\,b$ and held by a wire loop in the position shown. The plate and the panels are made of the same material and have the same thickness. The entire assembly is rotating with an angular velocity ω_0 when the wire breaks. Determine the angular velocity of the assembly after the panels have come to rest in a horizontal position.

Fig. P17.79

17.80 A 2.5-lb disk of radius 4 in. is attached to the yoke BCD by means of short shafts fitted in bearings at B and D. The 1.5-lb yoke has a radius of gyration of 3 in. about the x axis. Initially the assembly is rotating at 120 rpm with the disk in the plane of the yoke ($\theta = 0$). If the disk is slightly disturbed and rotates with respect to the yoke until $\theta = 90°$, where it is stopped by a small bar at D, determine the final angular velocity of the assembly.

Fig. P17.80

17.81 Two 10-lb disks and a small motor are mounted on a 15-lb rectangular platform which is free to rotate about a central vertical spindle. The normal operating speed of the motor is 180 rpm. If the motor is started when the system is at rest, determine the angular velocity of all elements of the system after the motor has attained its normal operating speed. Neglect the mass of the motor and of the belt.

Fig. P17.81

Fig. P17.82

17.82 A 3-kg rod of length 800 mm can slide freely in the 240-mm cylinder *DE*, which in turn can rotate freely in a horizontal plane. In the position shown the assembly is rotating with an angular velocity of magnitude $\omega = 40$ rad/s and end *B* of the rod is moving toward the cylinder at a speed of 75 mm/s relative to the cylinder. Knowing that the centroidal mass moment of inertia of the cylinder about a vertical axis is 0.025 kg · m² and neglecting the effect of friction, determine the angular velocity of the assembly as end *B* of the rod strikes end *E* of the cylinder.

17.83 A 1.6-kg tube *AB* can slide freely on rod *DE* which in turn can rotate freely in a horizontal plane. Initially the assembly is rotating with an angular velocity $\omega = 5$ rad/s and the tube is held in position by a cord. The moment of inertia of the rod and bracket about the vertical axis of rotation is 0.30 kg · m² and the centroidal moment of inertia of the tube about a vertical axis is 0.0025 kg · m². If the cord suddenly breaks, determine (*a*) the angular velocity of the assembly after the tube has moved to end *E*, (*b*) the energy lost during the plastic impact at *E*.

Fig. P17.83

17.84 In the helicopter shown, a vertical tail propeller is used to prevent rotation of the cab as the speed of the main blades is changed. Assuming that the tail propeller is not operating, determine the final angular velocity of the cab after the speed of the main blades has been changed from 180 to 240 rpm. (The speed of the main blades is measured relative to the cab, and the cab has a centroidal moment of inertia of 650 lb · ft · s². Each of the four main blades is assumed to be a slender 14-ft rod weighing 55 lb.)

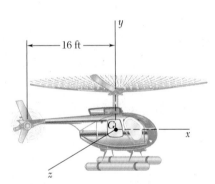

Fig. P17.84

17.85 Assuming that the tail propeller in Prob. 17.84 is operating and that the angular velocity of the cab remains zero, determine the final horizontal velocity of the cab when the speed of the main blades is changed from 180 to 240 rpm. The cab weighs 1250 lb and is initially at rest. Also determine the force exerted by the tail propeller if the change in speed takes place uniformly in 12 s.

17.86 The circular platform *A* is fitted with a rim of 200-mm inner radius and can rotate freely about the vertical shaft. It is known that the platform-rim unit has a mass of 5 kg and a radius of gyration of 175 mm with respect to the shaft. At a time when the platform is rotating with an angular velocity of 50 rpm, a 3-kg disk *B* of radius 80 mm is placed on the platform with no velocity. Knowing that disk *B* then slides until it comes to rest relative to the platform against the rim, determine the final angular velocity of the platform.

Fig. P17.86

17.87 Two 4-kg disks and a small motor are mounted on a 6-kg rectangular platform which is free to rotate about a central vertical spindle. The normal operating speed of the motor is 240 rpm. If the motor is started when the system is at rest, determine the angular velocity of all elements of the system after the motor has attained its normal operating speed. Neglect the mass of the motor and of the belt.

Fig. P17.87

17.88 The 4-kg rod *AB* can slide freely inside the 6-kg tube *CD*. The rod was entirely within the tube ($x = 0$) and released with no initial velocity relative to the tube when the angular velocity of the assembly was 5 rad/s. Neglecting the effect of friction, determine the speed of the rod relative to the tube when $x = 400$ mm.

Fig. P17.88

17.89 A 1.8-kg collar *A* and a 0.7-kg collar *B* can slide without friction on a frame, consisting of the horizontal rod *OE* and the vertical rod *CD*, which is free to rotate about its vertical axis of symmetry. The two collars are connected by a cord running over a pulley that is attached to the frame at *O*. At the instant shown, the velocity v_A of collar *A* has a magnitude of 2.1 m/s and a stop prevents collar *B* from moving. The stop is suddenly removed and collar *A* moves toward *E*. As it reaches a distance of 0.12 m from *O*, the magnitude of its velocity is observed to be 2.5 m/s. Determine at that instant the magnitude of the angular velocity of the frame and the moment of inertia of the frame and pulley system about *CD*.

Fig. P17.89

Fig. P17.90

17.90 A 6-lb collar C is attached to a spring and can slide on rod AB, which in turn can rotate in a horizontal plane. The mass moment of inertia of rod AB with respect to end A is 0.35 lb · ft · s^2. The spring has a constant $k = 15$ lb/in. and an undeformed length of 10 in. At the instant shown the velocity of the collar relative to the rod is zero and the assembly is rotating with an angular velocity of 12 rad/s. Neglecting the effect of friction, determine (a) the angular velocity of the assembly as the collar passes through a point located 7.5 in. from end A of the rod, (b) the corresponding velocity of the collar relative to the rod.

17.91 A small 4-lb collar C can slide freely on a thin ring of weight 6 lb and radius 10 in. The ring is welded to a short vertical shaft, which can rotate freely in a fixed bearing. Initially the ring has an angular velocity of 35 rad/s and the collar is at the top of the ring ($\theta = 0$) when it is given a slight nudge. Neglecting the effect of friction, determine (a) the angular velocity of the ring as the collar passes through the position $\theta = 90°$, (b) the corresponding velocity of the collar relative to the ring.

Fig. P17.91

Fig. P17.92

17.92 A uniform rod AB, of mass 7 kg and length 1.2 m, is attached to the 11-kg cart C. Knowing that the system is released from rest in the position shown and neglecting friction, determine (a) the velocity of point B as rod AB passes through a vertical position, (b) the corresponding velocity of cart C.

17.93 In Prob. 17.82, determine the velocity of rod AB relative to cylinder DE as end B of the rod strikes end E of the cylinder.

17.94 In Prob. 17.83, determine the velocity of the tube relative to the rod as the tube strikes end E of the assembly.

17.95 The 6-lb steel cylinder A and the 10-lb wooden cart B are at rest in the position shown when the cylinder is given a slight nudge, causing it to roll without sliding along the top surface of the cart. Neglecting friction between the cart and the ground, determine the velocity of the cart as the cylinder passes through the lowest point of the surface at C.

Fig. P17.95

17.11 IMPULSIVE MOTION

You saw in Chap. 13 that the method of impulse and momentum is the only practicable method for the solution of problems involving the impulsive motion of a particle. Now you will find that problems involving the impulsive motion of a rigid body are particularly well suited to a solution by the method of impulse and momentum. Since the time interval considered in the computation of linear impulses and angular impulses is very short, the bodies involved can be assumed to occupy the same position during that time interval, making the computation quite simple.

17.12 ECCENTRIC IMPACT

In Secs. 13.13 and 13.14, you learned to solve problems of *central impact*, i.e., problems in which the mass centers of the two colliding bodies are located on the line of impact. You will now analyze the *eccentric impact* of two rigid bodies. Consider two bodies which collide, and denote by $\mathbf{v}_A$ and $\mathbf{v}_B$ the velocities before impact *of the two points of contact A and B* (Fig. 17.10a). Under the impact, the two

(a)

(b)

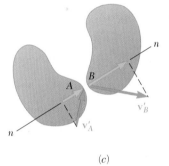
(c)

Fig. 17.10

bodies will *deform*, and at the end of the period of deformation, the velocities $\mathbf{u}_A$ and $\mathbf{u}_B$ of A and B will have equal components along the line of impact nn (Fig. 17.10b). A period of *restitution* will then take place, at the end of which A and B will have velocities $\mathbf{v}'_A$ and $\mathbf{v}'_B$ (Fig. 17.10c). Assuming that the bodies are frictionless, we find that the forces they exert on each other are directed along the line of impact. Denoting the magnitude of the impulse of one of these forces during the period of deformation by $\int P\, dt$ and the magnitude of its impulse during the period of restitution by $\int R\, dt$, we recall that the coefficient of restitution e is defined as the ratio

$$e = \frac{\int R\, dt}{\int P\, dt} \tag{17.18}$$

We propose to show that the relation established in Sec. 13.13 between the relative velocities of two particles before and after impact also holds between the components along the line of impact

Photo 17.4 When the rotating bat contacts the ball it applies an impulsive force to the ball requiring the method of impulse and momentum to be used to determine the final velocities of the ball and bat.

of the relative velocities of the two points of contact A and B. We propose to show, therefore, that

$$(v_B')_n - (v_A')_n = e[(v_A)_n - (v_B)_n] \tag{17.19}$$

It will first be assumed that the motion of each of the two colliding bodies of Fig. 17.10 is unconstrained. Thus the only impulsive forces exerted on the bodies during the impact are applied at A and B, respectively. Consider the body to which point A belongs and draw the three momentum and impulse diagrams corresponding to the period of deformation (Fig. 17.11). We denote by $\bar{\mathbf{v}}$ and $\bar{\mathbf{u}}$,

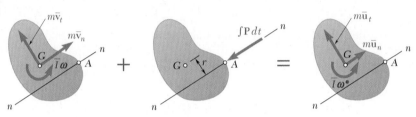

Fig. 17.11

respectively, the velocity of the mass center at the beginning and at the end of the period of deformation, and we denote by $\boldsymbol{\omega}$ and $\boldsymbol{\omega}^\circ$ the angular velocity of the body at the same instants. Summing and equating the components of the momenta and impulses along the line of impact nn, we write

$$m\bar{v}_n - \int P\,dt = m\bar{u}_n \tag{17.20}$$

Summing and equating the moments about G of the momenta and impulses, we also write

$$\bar{I}\omega - r\int P\,dt = \bar{I}\omega^\circ \tag{17.21}$$

where r represents the perpendicular distance from G to the line of impact. Considering now the period of restitution, we obtain in a similar way

$$m\bar{u}_n - \int R\,dt = m\bar{v}_n' \tag{17.22}$$
$$\bar{I}\omega^\circ - r\int R\,dt = \bar{I}\omega' \tag{17.23}$$

where $\bar{\mathbf{v}}'$ and $\boldsymbol{\omega}'$ represent, respectively, the velocity of the mass center and the angular velocity of the body after impact. Solving (17.20) and (17.22) for the two impulses and substituting into (17.18), and then solving (17.21) and (17.23) for the same two impulses and substituting again into (17.18), we obtain the following two alternative expressions for the coefficient of restitution:

$$e = \frac{\bar{u}_n - \bar{v}_n'}{\bar{v}_n - \bar{u}_n} \qquad e = \frac{\omega^\circ - \omega'}{\omega - \omega^\circ} \tag{17.24}$$

Multiplying by r the numerator and denominator of the second expression obtained for e, and adding respectively to the numerator and denominator of the first expression, we have

$$e = \frac{\bar{u}_n + r\omega^* - (\bar{v}'_n + r\omega')}{\bar{v}_n + r\omega - (\bar{u}_n + r\omega^*)} \tag{17.25}$$

Observing that $\bar{v}_n + r\omega$ represents the component $(v_A)_n$ along nn of the velocity of the point of contact A and that, similarly, $\bar{u}_n + r\omega^*$ and $\bar{v}'_n + r\omega'$ represent, respectively, the components $(u_A)_n$ and $(v'_A)_n$, we write

$$e = \frac{(u_A)_n - (v'_A)_n}{(v_A)_n - (u_A)_n} \tag{17.26}$$

The analysis of the motion of the second body leads to a similar expression for e in terms of the components along nn of the successive velocities of point B. Recalling that $(u_A)_n = (u_B)_n$, and eliminating these two velocity components by a manipulation similar to the one used in Sec. 13.13, we obtain relation (17.19).

If one or both of the colliding bodies is constrained to rotate about a fixed point O, as in the case of a compound pendulum (Fig. 17.12a), an impulsive reaction will be exerted at O (Fig. 17.12b).

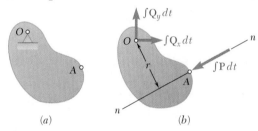

(a) (b)

Fig. 17.12

Let us verify that while their derivation must be modified, Eqs. (17.26) and (17.19) remain valid. Applying formula (17.16) to the period of deformation and to the period of restitution, we write

$$I_O\omega - r\int P \, dt = I_O\omega^* \tag{17.27}$$
$$I_O\omega^* - r\int R \, dt = I_O\omega' \tag{17.28}$$

where r represents the perpendicular distance from the fixed point O to the line of impact. Solving (17.27) and (17.28) for the two impulses and substituting into (17.18), and then observing that $r\omega$, $r\omega^*$, and $r\omega'$ represent the components along nn of the successive velocities of point A, we write

$$e = \frac{\omega^* - \omega'}{\omega - \omega^*} = \frac{r\omega^* - r\omega'}{r\omega - r\omega^*} = \frac{(u_A)_n - (v'_A)_n}{(v_A)_n - (u_A)_n}$$

and check that Eq. (17.26) still holds. Thus Eq. (17.19) remains valid when one or both of the colliding bodies is constrained to rotate about a fixed point O.

In order to determine the velocities of the two colliding bodies after impact, relation (17.19) should be used in conjunction with one or several other equations obtained by applying the principle of impulse and momentum (Sample Prob. 17.10).

SAMPLE PROBLEM 17.9

A 0.05-lb bullet B is fired with a horizontal velocity of 1500 ft/s into the side of a 20-lb square panel suspended from a hinge at A. Knowing that the panel is initially at rest, determine (a) the angular velocity of the panel immediately after the bullet becomes embedded, (b) the impulsive reaction at A, assuming that the bullet becomes embedded in 0.0006 s.

SOLUTION

Principle of Impulse and Momentum. We consider the bullet and the panel as a single system and express that the initial momenta of the bullet and panel and the impulses of the external forces are together equipollent to the final momenta of the system. Since the time interval $\Delta t = 0.0006$ s is very short, we neglect all nonimpulsive forces and consider only the external impulses $\mathbf{A}_x \, \Delta t$ and $\mathbf{A}_y \, \Delta t$.

$$\text{Syst Momenta}_1 + \text{Syst Ext Imp}_{1\to2} = \text{Syst Momenta}_2$$

$+\!\uparrow$moments about A: $\qquad m_B v_B(\tfrac{14}{12}\text{ ft}) + 0 = m_P \bar{v}_2(\tfrac{9}{12}\text{ ft}) + \bar{I}_P \omega_2 \qquad (1)$

$\xrightarrow{+} x$ components: $\qquad m_B v_B + A_x \, \Delta t = m_P \bar{v}_2 \qquad (2)$

$+\!\uparrow y$ components: $\qquad 0 + A_y \, \Delta t = 0 \qquad (3)$

The centroidal mass moment of inertia of the square panel is

$$\bar{I}_P = \tfrac{1}{6}m_P b^2 = \frac{1}{6}\left(\frac{20\text{ lb}}{32.2}\right)\left(\tfrac{18}{12}\text{ ft}\right)^2 = 0.2329\text{ lb}\cdot\text{ft}\cdot\text{s}^2$$

Substituting this value as well as the given data into (1) and noting that

$$\bar{v}_2 = (\tfrac{9}{12}\text{ ft})\omega_2$$

we write

$$\left(\frac{0.05}{32.2}\right)(1500)(\tfrac{14}{12}) = 0.2329\omega_2 + \left(\frac{20}{32.2}\right)(\tfrac{9}{12}\omega_2)(\tfrac{9}{12})$$

$$\omega_2 = 4.67\text{ rad/s} \qquad\qquad \mathbf{\omega_2 = 4.67\text{ rad/s}}\!\uparrow \blacktriangleleft$$

$$\bar{v}_2 = (\tfrac{9}{12}\text{ ft})\omega_2 = (\tfrac{9}{12}\text{ ft})(4.67\text{ rad/s}) = 3.50\text{ ft/s}$$

Substituting $\bar{v}_2 = 3.50$ ft/s, $\Delta t = 0.0006$ s, and the given data into Eq. (2), we have

$$\left(\frac{0.05}{32.2}\right)(1500) + A_x(0.0006) = \left(\frac{20}{32.2}\right)(3.50)$$

$$A_x = -259\text{ lb} \qquad\qquad \mathbf{A}_x = 259\text{ lb} \leftarrow \blacktriangleleft$$

From Eq. (3), we find $\qquad A_y = 0 \qquad\qquad\qquad \mathbf{A}_y = 0 \blacktriangleleft$

SAMPLE PROBLEM 17.10

A 2-kg sphere moving horizontally to the right with an initial velocity of 5 m/s strikes the lower end of an 8-kg rigid rod AB. The rod is suspended from a hinge at A and is initially at rest. Knowing that the coefficient of restitution between the rod and the sphere is 0.80, determine the angular velocity of the rod and the velocity of the sphere immediately after the impact.

SOLUTION

Principle of Impulse and Momentum. We consider the rod and sphere as a single system and express that the initial momenta of the rod and sphere and the impulses of the external forces are together equipollent to the final momenta of the system. We note that the only impulsive force external to the system is the impulsive reaction at A.

Syst Momenta₁ + Syst Ext Imp₁→₂ = Syst Momenta₂

$+\uparrow$ moments about A:

$$m_s v_s (1.2 \text{ m}) = m_s v_s' (1.2 \text{ m}) + m_R \bar{v}_R'(0.6 \text{ m}) + \bar{I}\omega' \tag{1}$$

Since the rod rotates about A, we have $\bar{v}_R' = \bar{r}\omega' = (0.6 \text{ m})\omega'$. Also,

$$\bar{I} = \tfrac{1}{12}mL^2 = \tfrac{1}{12}(8 \text{ kg})(1.2 \text{ m})^2 = 0.96 \text{ kg} \cdot \text{m}^2$$

Substituting these values and the given data into Eq. (1), we have

$$(2 \text{ kg})(5 \text{ m/s})(1.2 \text{ m}) = (2 \text{ kg})v_s'(1.2 \text{ m}) + (8 \text{ kg})(0.6 \text{ m})\omega'(0.6 \text{ m})$$
$$+ (0.96 \text{ kg} \cdot \text{m}^2)\omega'$$
$$12 = 2.4v_s' + 3.84\omega' \tag{2}$$

Relative Velocities. Choosing positive to the right, we write

$$v_B' - v_s' = e(v_s - v_B)$$

Substituting $v_s = 5$ m/s, $v_B = 0$, and $e = 0.80$, we obtain

$$v_B' - v_s' = 0.80(5 \text{ m/s}) \tag{3}$$

Again noting that the rod rotates about A, we write

$$v_B' = (1.2 \text{ m})\omega' \tag{4}$$

Solving Eqs. (2) to (4) simultaneously, we obtain

$$\omega' = 3.21 \text{ rad/s} \qquad \omega' = 3.21 \text{ rad/s} \; \uparrow \; \blacktriangleleft$$
$$v_s' = -0.143 \text{ m/s} \qquad \mathbf{v}_s' = -0.143 \text{ m/s} \leftarrow \; \blacktriangleleft$$

SAMPLE PROBLEM 17.11

A square package of side a and mass m moves down a conveyor belt A with a constant velocity $\bar{\mathbf{v}}_1$. At the end of the conveyor belt, the corner of the package strikes a rigid support at B. Assuming that the impact at B is perfectly plastic, derive an expression for the smallest magnitude of the velocity $\bar{\mathbf{v}}_1$ for which the package will rotate about B and reach conveyor belt C.

SOLUTION

Principle of Impulse and Momentum. Since the impact between the package and the support is perfectly plastic, the package rotates about B during the impact. We apply the principle of impulse and momentum to the package and note that the only impulsive force external to the package is the impulsive reaction at B.

$$\textbf{Syst Momenta}_1 + \textbf{Syst Ext Imp}_{1\to2} = \textbf{Syst Momenta}_2$$

$+\,\gamma$ moments about B: $\qquad (m\bar{v}_1)(\tfrac{1}{2}a) + 0 = (m\bar{v}_2)(\tfrac{1}{2}\sqrt{2}a) + \bar{I}\omega_2 \qquad (1)$

Since the package rotates about B, we have $\bar{v}_2 = (GB)\omega_2 = \tfrac{1}{2}\sqrt{2}a\omega_2$. We substitute this expression, together with $\bar{I} = \tfrac{1}{6}ma^2$, into Eq. (1):

$$(m\bar{v}_1)(\tfrac{1}{2}a) = m(\tfrac{1}{2}\sqrt{2}a\omega_2)(\tfrac{1}{2}\sqrt{2}a) + \tfrac{1}{6}ma^2\omega_2 \qquad \bar{v}_1 = \tfrac{4}{3}a\omega_2 \qquad (2)$$

Principle of Conservation of Energy. We apply the principle of conservation of energy between position 2 and position 3.

Position 2. $V_2 = Wh_2$. Recalling that $\bar{v}_2 = \tfrac{1}{2}\sqrt{2}a\omega_2$, we write

$$T_2 = \tfrac{1}{2}m\bar{v}_2^2 + \tfrac{1}{2}\bar{I}\omega_2^2 = \tfrac{1}{2}m(\tfrac{1}{2}\sqrt{2}a\omega_2)^2 + \tfrac{1}{2}(\tfrac{1}{6}ma^2)\omega_2^2 = \tfrac{1}{3}ma^2\omega_2^2$$

Position 3. Since the package must reach conveyor belt C, it must pass through position 3 where G is directly above B. Also, since we wish to determine the smallest velocity for which the package will reach this position, we choose $\bar{v}_3 = \omega_3 = 0$. Therefore $T_3 = 0$ and $V_3 = Wh_3$.

Conservation of Energy

$$T_2 + V_2 = T_3 + V_3$$
$$\tfrac{1}{3}ma^2\omega_2^2 + Wh_2 = 0 + Wh_3$$
$$\omega_2^2 = \frac{3W}{ma^2}(h_3 - h_2) = \frac{3g}{a^2}(h_3 - h_2) \qquad (3)$$

Substituting the computed values of h_2 and h_3 into Eq. (3), we obtain

$$\omega_2^2 = \frac{3g}{a^2}(0.707a - 0.612a) = \frac{3g}{a^2}(0.095a) \qquad \omega_2 = \sqrt{0.285g/a}$$

$$\bar{v}_1 = \tfrac{4}{3}a\omega_2 = \tfrac{4}{3}a\sqrt{0.285g/a} \qquad \bar{v}_1 = 0.712\sqrt{ga} \quad \blacktriangleleft$$

Position 2

$$GB = \tfrac{1}{2}\sqrt{2}a = 0.707a$$
$$h_2 = GB\sin(45° + 15°)$$
$$= 0.612a$$

Position 3

$$h_3 = GB = 0.707a$$

SOLVING PROBLEMS
ON YOUR OWN

This lesson was devoted to the *impulsive motion* and to the *eccentric impact of rigid bodies.*

1. Impulsive motion occurs when a rigid body is subjected to a very large force $\mathbf{F}$ for a very short interval of time Δt; the resulting impulse $\mathbf{F}\,\Delta t$ is both finite and different from zero. Such forces are referred to as *impulsive forces* and are encountered whenever there is an impact between two rigid bodies. Forces for which the impulse is zero are referred to as *nonimpulsive forces.* As you saw in Chap. 13, the following forces can be assumed to be nonimpulsive: the *weight* of a body, the force exerted by a *spring,* and any other force which is *known* to be small by comparison with the impulsive forces. Unknown reactions, however, *cannot be assumed* to be nonimpulsive.

2. Eccentric impact of rigid bodies. You saw that when two bodies collide, the velocity components along the line of impact of the *points of contact A and B* before and after impact satisfy the following equation:

$$(v'_B)_n - (v'_A)_n = e[(v_A)_n - (v_B)_n] \tag{17.19}$$

where the left-hand member is the *relative velocity after the impact,* and the right-hand member is the product of the coefficient of restitution and the *relative velocity before the impact.*

This equation expresses the same relation between the velocity components of the points of contact before and after an impact that you used for particles in Chap. 13.

3. To solve a problem involving an impact you should use the *method of impulse and momentum* and take the following steps.

    a. Draw a free-body-diagram equation of the body that will express that the system consisting of the momenta immediately before impact and of the impulses of the external forces is equivalent to the system of the momenta immediately after impact.

    b. The free-body-diagram equation will relate the velocities before and after impact and the impulsive forces and reactions. In some cases, you will be able to determine the unknown velocities and impulsive reactions by solving equations obtained by summing components and moments [Sample Prob. 17.9].

    c. In the case of an impact in which $e > 0$, the number of unknowns will be greater than the number of equations that you can write by summing components and moments, and you should supplement the equations obtained from the free-body-diagram equation with Eq. (17.19), which relates the relative velocities of the points of contact before and after impact [Sample Prob. 17.10].

    d. During an impact you must use the method of impulse and momentum. However, *before and after the impact* you can, if necessary, use some of the other methods of solution that you have learned, such as the method of work and energy [Sample Prob. 17.11].

PROBLEMS

IMPULSE-MOMENTUM PRACTICE PROBLEMS

17.F4 A uniform slender rod AB of mass m is at rest on a frictionless horizontal surface when hook C engages a small pin at A. Knowing that the hook is pulled upward with a constant velocity $\mathbf{v}_0$, draw the impulse-momentum diagram that is needed to determine the impulse exerted on the rod at A and B. Assume that the velocity of the hook is unchanged and that the impact is perfectly plastic.

Fig. P17.F4

17.F5 A uniform slender rod AB of length L is falling freely with a velocity $\mathbf{v}_0$ when cord AC suddenly becomes taut. Assuming that the impact is perfectly plastic, draw the impulse-momentum diagram that is needed to determine the angular velocity of the rod and the velocity of its mass center immediately after the cord becomes taut.

Fig. P17.F5

17.F6 A slender rod CDE of length L and mass m is attached to a pin support at its midpoint D. A second and identical rod AB is rotating about a pin support at A with an angular velocity $\boldsymbol{\omega}_1$ when its end B strikes end C of rod CDE. The coefficient of restitution between the rods is e. Draw the impulse-momentum diagrams that are needed to determine the angular velocity of each rod immediately after the impact.

Fig. P17.F6

17.96 At what height h above its center G should a billiard ball of radius r be struck horizontally by a cue if the ball is to start rolling without sliding?

Fig. P17.96

17.97 A bullet weighing 0.08 lb is fired with a horizontal velocity of 1800 ft/s into the lower end of a slender 15-lb bar of length $L = 30$ in. Knowing that $h = 12$ in. and that the bar is initially at rest, determine (*a*) the angular velocity of the bar immediately after the bullet becomes embedded, (*b*) the impulsive reaction at C, assuming that the bullet becomes embedded in 0.001 s.

Fig. P17.97

17.98 In Prob. 17.97, determine (*a*) the required distance h if the impulsive reaction at C is to be zero, (*b*) the corresponding angular velocity of the bar immediately after the bullet becomes embedded.

17.99 An 16-lb wooden panel is suspended from a pin support at A and is initially at rest. A 4-lb metal sphere is released from rest at B and falls into a hemispherical cup C attached to the panel at a point located on its top edge. Assuming that the impact is perfectly plastic, determine the velocity of the mass center G of the panel immediately after the impact.

17.100 A 16-lb wooden panel is suspended from a pin support at A and is initially at rest. A 4-lb metal sphere is released from rest at B' and falls into a hemispherical cup C' attached to the panel at the same level as the mass center G. Assuming that the impact is perfectly plastic, determine the velocity of the mass center G of the panel immediately after the impact.

Fig. P17.99 and P17.100

17.101 A 45-g bullet is fired with a velocity of 400 m/s at $\theta = 30°$ into a 9-kg square panel of side $b = 200$ mm. Knowing that $h = 150$ mm and that the panel is initially at rest, determine (*a*) the velocity of the center of the panel immediately after the bullet becomes embedded, (*b*) the impulsive reaction at A, assuming that the bullet becomes embedded in 2 ms.

17.102 A 45-g bullet is fired with a velocity of 400 m/s at $\theta = 5°$ into a 9-kg square panel of side $b = 200$ mm. Knowing that the panel is initially at rest, determine (*a*) the required distance h if the horizontal component of the impulsive reaction at A is to be zero, (*b*) the corresponding velocity of the center of the panel immediately after the bullet becomes embedded.

Fig. P17.101 and P17.102

Fig. P17.103

17.103 Two uniform rods, each of mass m, form the L-shaped rigid body ABC which is initially at rest on the frictionless horizontal surface when hook D of the carriage E engages a small pin at C. Knowing that the carriage is pulled to the right with a constant velocity $\mathbf{v}_0$, determine immediately after the impact (a) the angular velocity of the body, (b) the velocity of corner B. Assume that the velocity of the carriage is unchanged and that the impact is perfectly plastic.

17.104 The uniform slender rod AB of weight 5 lb and length 30 in. forms an angle $\beta = 30°$ with the vertical as it strikes the smooth corner shown with a vertical velocity $\mathbf{v}_1$ of magnitude 8 ft/s and no angular velocity. Assuming that the impact is perfectly plastic, determine the angular velocity of the rod immediately after the impact.

Fig. P17.104

Fig. P17.105

17.105 A bullet weighing 0.08 lb is fired with a horizontal velocity of 1800 ft/s into the 15-lb wooden rod AB of length $L = 30$ in. The rod, which is initially at rest, is suspended by a cord of length $L = 30$ in. Determine the distance h for which, immediately after the bullet becomes embedded, the instantaneous center of rotation of the rod is point C.

17.106 A uniform sphere of radius r rolls down the incline shown without slipping. It hits a horizontal surface and, after slipping for a while, it starts rolling again. Assuming that the sphere does not bounce as it hits the horizontal surface, determine its angular velocity and the velocity of its mass center after it has resumed rolling.

Fig. P17.106

17.107 A uniformly loaded rectangular crate is released from rest in the position shown. Assuming that the floor is sufficiently rough to prevent slipping and that the impact at B is perfectly plastic, determine the smallest value of the ratio a/b for which corner A will remain in contact with the floor.

Fig. P17.107

17.108 A bullet of mass m is fired with a horizontal velocity $\mathbf{v}_0$ and at a height $h = \frac{1}{2}R$ into a wooden disk of much larger mass M and radius R. The disk rests on a horizontal plane and the coefficient of friction between the disk and the plane is finite. (*a*) Determine the linear velocity $\bar{v}_1$ and the angular velocity ω_1 of the disk immediately after the bullet has penetrated the disk. (*b*) Describe the ensuing motion of the disk and determine its linear velocity after the motion has become uniform.

17.109 Determine the height h at which the bullet of Prob. 17.108 should be fired (*a*) if the disk is to roll without sliding immediately after impact, (*b*) if the disk is to slide without rolling immediately after impact.

Fig. P17.108 and P17.109

17.110 A uniform slender bar of length $L = 200$ mm and mass $m = 0.5$ kg is supported by a frictionless horizontal table. Initially the bar is spinning about its mass center G with a constant angular speed $\omega_1 = 6$ rad/s. Suddenly latch D is moved to the right and is struck by end A of the bar. Knowing that the coefficient of restitution between A and D is $e = 0.6$, determine the angular velocity of the bar and the velocity of its mass center immediately after the impact.

17.111 A uniform slender rod of length L is dropped onto rigid supports at A and B. Since support B is slightly lower than support A, the rod strikes A with a velocity $\bar{v}_1$ before it strikes B. Assuming perfectly elastic impact at both A and B, determine the angular velocity of the rod and the velocity of its mass center immediately after the rod (*a*) strikes support A, (*b*) strikes support B, (*c*) again strikes support A.

Fig. P17.110

Fig. P17.111

17.112 The slender rod AB of length L forms an angle β with the vertical as it strikes the frictionless surface shown with a vertical velocity $\bar{v}_1$ and no angular velocity. Assuming that the impact is perfectly plastic, derive an expression for the angular velocity of the rod immediately after the impact.

Fig. P17.112

Fig. P17.113

17.113 The slender rod AB of length $L = 1$ m forms an angle $\beta = 30°$ with the vertical as it strikes the frictionless surface shown with a vertical velocity $\bar{\mathbf{v}}_1 = 2$ m/s and no angular velocity. Knowing that the coefficient of restitution between the rod and the ground is $e = 0.8$, determine the angular velocity of the rod immediately after the impact.

17.114 The trapeze/lanyard air drop (t/LAD) launch is a proposed innovative method for airborne launch of a payload-carrying rocket. The release sequence involves several steps as shown in (1) where the payload rocket is shown at various instances during the launch. To investigate the first step of this process, where the rocket body drops freely from the carrier aircraft until the 2-m lanyard stops the vertical motion of B, a trial rocket is tested as shown in (2). The rocket can be considered a uniform 1×7-m rectangle with a mass of 4000 kg. Knowing that the rocket is released from rest and falls vertically 2 m before the lanyard becomes taut, determine the angular velocity of the rocket immediately after the lanyard is taut.

(1)

(2)

Fig. P17.114

17.115 The uniform rectangular block shown is moving along a frictionless surface with a velocity $\bar{\mathbf{v}}_1$ when it strikes a small obstruction at B. Assuming that the impact between corner A and obstruction B is perfectly plastic, determine the magnitude of the velocity $\bar{\mathbf{v}}_1$ for which the maximum angle θ through which the block will rotate will be 30°.

Fig. P17.115

17.116 A slender rod of length L and mass m is released from rest in the position shown. It is observed that after the rod strikes the vertical surface it rebounds to form an angle of 30° with the vertical. (*a*) Determine the coefficient of restitution between knob K and the surface. (*b*) Show that the same rebound can be expected for any position of knob K.

17.117 A slender rod of mass m and length L is released from rest in the position shown and hits edge D. Assuming perfectly plastic impact at D, determine for $b = 0.6L$, (*a*) the angular velocity of the rod immediately after the impact, (*b*) the maximum angle through which the rod will rotate after the impact.

Fig. P17.116

Fig. P17.117

17.118 A uniformly loaded square crate is released from rest with its corner D directly above A; it rotates about A until its corner B strikes the floor, and then rotates about B. The floor is sufficiently rough to prevent slipping and the impact at B is perfectly plastic. Denoting by $\boldsymbol{\omega}_0$ the angular velocity of the crate immediately before B strikes the floor, determine (*a*) the angular velocity of the crate immediately after B strikes the floor, (*b*) the fraction of the kinetic energy of the crate lost during the impact, (*c*) the angle θ through which the crate will rotate after B strikes the floor.

Fig. P17.118

17.119 A 1-oz bullet is fired with a horizontal velocity of 750 mi/h into the 18-lb wooden beam AB. The beam is suspended from a collar of negligible mass that can slide along a horizontal rod. Neglecting friction between the collar and the rod, determine the maximum angle of rotation of the beam during its subsequent motion.

17.120 For the beam of Prob. 17.119, determine the velocity of the 1-oz bullet for which the maximum angle of rotation of the beam will be 90°.

Fig. P17.119

17.121 The plank *CDE* has a mass of 15 kg and rests on a small pivot at *D*. The 55-kg gymnast *A* is standing on the plank at *C* when the 70-kg gymnast *B* jumps from a height of 2.5 m and strikes the plank at *E*. Assuming perfectly plastic impact and that gymnast *A* is standing absolutely straight, determine the height to which gymnast *A* will rise.

Fig. P17.123

Fig. P17.121

17.122 Solve Prob. 17.121, assuming that the gymnasts change places so that gymnast *A* jumps onto the plank while gymnast *B* stands at *C*.

17.123 A small plate *B* is attached to a cord that is wrapped around a uniform 8-lb disk of radius *R* = 9 in. A 3-lb collar *A* is released from rest and falls through a distance *h* = 15 in. before hitting plate *B*. Assuming that the impact is perfectly plastic and neglecting the weight of the plate, determine immediately after the impact (*a*) the velocity of the collar, (*b*) the angular velocity of the disk.

17.124 Solve Prob. 17.123, assuming that the coefficient of restitution between *A* and *B* is 0.8.

17.125 Two identical slender rods may swing freely from the pivots shown. Rod *A* is released from rest in a horizontal position and swings to a vertical position, at which time the small knob *K* strikes rod *B* which was at rest. If $h = \frac{1}{2}l$ and $e = \frac{1}{2}$, determine (*a*) the angle through which rod *B* will swing, (*b*) the angle through which rod *A* will rebound.

17.126 A 2-kg solid sphere of radius *r* = 40 mm is dropped from a height *h* = 200 mm and lands on a uniform slender plank *AB* of mass 4 kg and length *L* = 500 mm which is held by two inextensible cords. Knowing that the impact is perfectly plastic and that the sphere remains attached to the plank at a distance *a* = 40 mm from the left end, determine the velocity of the sphere immediately after impact. Neglect the thickness of the plank.

Fig. P17.125

Fig. P17.126

17.127 and 17.128 Member *ABC* has a mass of 2.4 kg and is attached to a pin support at *B*. An 800-g sphere *D* strikes the end of member *ABC* with a vertical velocity $\mathbf{v}_1$ of 3 m/s. Knowing that $L = 750$ mm and that the coefficient of restitution between the sphere and member *ABC* is 0.5, determine immediately after the impact (*a*) the angular velocity of member *ABC*, (*b*) the velocity of the sphere.

Fig. P17.127

Fig. P17.128

17.129 Sphere *A* of mass $m_A = 2$ kg and radius $r = 40$ mm rolls without slipping with a velocity $\bar{\mathbf{v}}_1 = 2$ m/s on a horizontal surface when it hits squarely a uniform slender bar *B* of mass is $m_B = 0.5$ kg and length $L = 100$ mm that is standing on end and is at rest. Denoting by μ_k the coefficient of kinetic friction between the sphere and the horizontal surface, neglecting friction between the sphere and the bar, and knowing the coefficient of restitution between *A* and *B* is 0.1, determine the angular velocities of the sphere and the bar immediately after the impact.

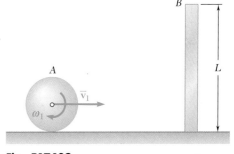

Fig. P17.129

17.130 A large 3-lb sphere with a radius $r = 3$ in. is thrown into a light basket at the end of a thin, uniform rod weighing 2 lb and length $L = 10$ in. as shown. Immediately before the impact the angular velocity of the rod is 3 rad/s counterclockwise and the velocity of the sphere is 2 ft/s down. Assume the sphere sticks in the basket. Determine after the impact (*a*) the angular velocity of the bar and sphere, (*b*) the components of the reactions at *A*.

17.131 A small rubber ball of radius *r* is thrown against a rough floor with a velocity $\bar{\mathbf{v}}_A$ of magnitude v_0 and a backspin $\boldsymbol{\omega}_A$ of magnitude ω_0. It is observed that the ball bounces from *A* to *B*, then from *B* to *A*, then from *A* to *B*, etc. Assuming perfectly elastic impact, determine the required magnitude ω_0 of the backspin in terms of $\bar{v}_0$ and *r*.

Fig. P17.130

Fig. P17.131

17.132 Sphere A of mass m and radius r rolls without slipping with a velocity $\bar{\mathbf{v}}_1$ on a horizontal surface when it hits squarely an identical sphere B that is at rest. Denoting by μ_k the coefficient of kinetic friction between the spheres and the surface, neglecting friction between the spheres, and assuming perfectly elastic impact, determine (*a*) the linear and angular velocities of each sphere immediately after the impact, (*b*) the velocity of each sphere after it has started rolling uniformly.

Fig. P17.132

17.133 In a game of pool, ball A is rolling without slipping with a velocity $\bar{\mathbf{v}}_0$ as it hits obliquely ball B, which is at rest. Denoting by r the radius of each ball and by μ_k the coefficient of kinetic friction between a ball and the table, and assuming perfectly elastic impact, determine (*a*) the linear and angular velocity of each ball immediately after the impact, (*b*) the velocity of ball B after it has started rolling uniformly.

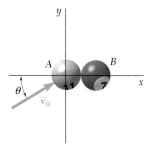

Fig. P17.133

17.134 Each of the bars AB and BC is of length $L = 400$ mm and mass $m = 1.2$ kg. Determine the angular velocity of each bar immediately after the impulse $\mathbf{Q}\Delta t = (1.5 \text{ N} \cdot \text{s})\mathbf{i}$ is applied at C.

Fig. P17.134

REVIEW AND SUMMARY

In this chapter we again considered the method of work and energy and the method of impulse and momentum. In the first part of the chapter we studied the method of work and energy and its application to the analysis of the motion of rigid bodies and systems of rigid bodies.

In Sec. 17.2, we first expressed the principle of work and energy for a rigid body in the form

Principle of work and energy for a rigid body

$$T_1 + U_{1\rightarrow2} = T_2 \tag{17.1}$$

where T_1 and T_2 represent the initial and final values of the kinetic energy of the rigid body and $U_{1\rightarrow2}$ represents the work of the *external forces* acting on the rigid body.

In Sec. 17.3, we recalled the expression found in Chap. 13 for the work of a force $\mathbf{F}$ applied at a point A, namely

Work of a force or a couple

$$U_{1\rightarrow2} = \int_{s_1}^{s_2} (F \cos \alpha)\,ds \tag{17.3'}$$

where F was the magnitude of the force, α the angle it formed with the direction of motion of A, and s the variable of integration measuring the distance traveled by A along its path. We also derived the expression for the *work of a couple of moment* $\mathbf{M}$ applied to a rigid body during a rotation in θ of the rigid body:

$$U_{1\rightarrow2} = \int_{\theta_1}^{\theta_2} M\,d\theta \tag{17.5}$$

We then derived an expression for the kinetic energy of a rigid body in plane motion [Sec. 17.4]. We wrote

Kinetic energy in plane motion

$$T = \tfrac{1}{2}m\bar{v}^2 + \tfrac{1}{2}\bar{I}\omega^2 \tag{17.9}$$

where $\bar{v}$ is the velocity of the mass center G of the body, ω is the angular velocity of the body, and $\bar{I}$ is its moment of inertia about an axis through G perpendicular to the plane of reference (Fig. 17.13) [Sample Prob. 17.3]. We noted that the kinetic energy of a rigid body in plane motion can be separated into two parts: (1) the kinetic energy $\tfrac{1}{2}m\bar{v}^2$ associated with the motion of the mass center G of the body, and (2) the kinetic energy $\tfrac{1}{2}\bar{I}\omega^2$ associated with the rotation of the body about G.

Fig. 17.13

1163

Kinetic energy in rotation

For a rigid body rotating about a fixed axis through O with an angular velocity $\boldsymbol{\omega}$, we had

$$T = \tfrac{1}{2}I_O\omega^2 \qquad (17.10)$$

where I_O was the moment of inertia of the body about the fixed axis. We noted that the result obtained is not limited to the rotation of plane slabs or of bodies symmetrical with respect to the reference plane, but is valid regardless of the shape of the body or of the location of the axis of rotation.

Systems of rigid bodies

Equation (17.1) can be applied to the motion of systems of rigid bodies [Sec. 17.5] as long as all the forces acting on the various bodies involved—internal as well as external to the system—are included in the computation of $U_{1\to2}$. However, in the case of systems consisting of pin-connected members, or blocks and pulleys connected by inextensible cords, or meshed gears, the points of application of the internal forces move through equal distances and the work of these forces cancels out [Sample Probs. 17.1 and 17.2].

Conservation of energy

When a rigid body, or a system of rigid bodies, moves under the action of conservative forces, the principle of work and energy can be expressed in the form

$$T_1 + V_1 = T_2 + V_2 \qquad (17.12)$$

which is referred to as the *principle of conservation of energy* [Sec. 17.6]. This principle can be used to solve problems involving conservative forces such as the force of gravity or the force exerted by a spring [Sample Probs. 17.4 and 17.5]. However, when a reaction is to be determined, the principle of conservation of energy must be supplemented by the application of d'Alembert's principle [Sample Prob. 17.4].

Power

In Sec. 17.7, we extended the concept of power to a rotating body subjected to a couple, writing

$$\text{Power} = \frac{dU}{dt} = \frac{M\,d\theta}{dt} = M\omega \qquad (17.13)$$

where M is the magnitude of the couple and ω the angular velocity of the body.

The middle part of the chapter was devoted to the method of impulse and momentum and its application to the solution of various types of problems involving the plane motion of rigid slabs and rigid bodies symmetrical with respect to the reference plane.

Principle of impulse and momentum for a rigid body

We first recalled the *principle of impulse and momentum* as it was derived in Sec. 14.9 for a system of particles and applied it to the *motion of a rigid body* [Sec. 17.8]. We wrote

$$\textbf{Syst Momenta}_1 + \textbf{Syst Ext Imp}_{1\to2} = \textbf{Syst Momenta}_2 \qquad (17.14)$$

Next we showed that for a rigid slab or a rigid body symmetrical with respect to the reference plane, the system of the momenta of the particles forming the body is equivalent to a vector $m\bar{\mathbf{v}}$ attached at the mass center G of the body and a couple $\bar{I}\boldsymbol{\omega}$ (Fig. 17.14). The vector

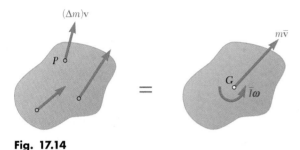

Fig. 17.14

$m\bar{\mathbf{v}}$ is associated with the translation of the body with G and represents the *linear momentum* of the body, while the couple $\bar{I}\boldsymbol{\omega}$ corresponds to the rotation of the body about G and represents the *angular momentum* of the body about an axis through G.

Equation (17.14) can be expressed graphically as shown in Fig. 17.15 by drawing three diagrams representing respectively the system of the initial momenta of the body, the impulses of the external forces acting on the body, and the system of the final momenta of the body.

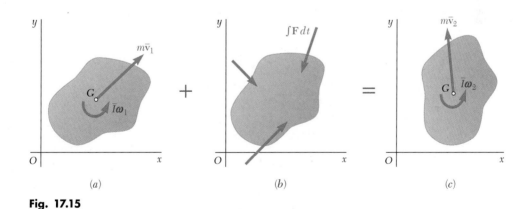

Fig. 17.15

Summing and equating respectively the *x components*, the *y components*, and the *moments about any given point* of the vectors shown in that figure, we obtain three equations of motion which can be solved for the desired unknowns [Sample Probs. 17.6 and 17.7].

In problems dealing with several connected rigid bodies [Sec. 17.9], each body can be considered separately [Sample Prob. 17.6], or, if no more than three unknowns are involved, the principle of impulse

and momentum can be applied to the entire system, considering the
impulses of the external forces only [Sample Prob. 17.8].

Conservation of angular momentum

When the lines of action of all the external forces acting on a system
of rigid bodies pass through a given point O, the angular momentum
of the system about O is conserved [Sec. 17.10]. It was suggested
that problems involving conservation of angular momentum be solved
by the general method described above [Sample Prob. 17.8].

Impulsive motion

The last part of the chapter was devoted to the *impulsive motion* and
the *eccentric impact* of rigid bodies. In Sec. 17.11, we recalled that
the method of impulse and momentum is the only practicable method
for the solution of problems involving impulsive motion and that the
computation of impulses in such problems is particularly simple
[Sample Prob. 17.9].

Eccentric impact

In Sec. 17.12, we recalled that the eccentric impact of two rigid
bodies is defined as an impact in which the mass centers of the colliding bodies are *not* located on the line of impact. It was shown that
in such a situation a relation similar to that derived in Chap. 13 for
the central impact of two particles and involving the coefficient of
restitution e still holds, but that *the velocities of points A and B where
contact occurs during the impact should be used.* We have

$$(v'_B)_n - (v'_A)_n = e[(v_A)_n - (v_B)_n] \qquad (17.19)$$

where $(v_A)_n$ and $(v_B)_n$ are the components along the line of impact
of the velocities of A and B before the impact, and $(v'_A)_n$ and $(v'_B)_n$
are their components after the impact (Fig. 17.16). Equation (17.19)

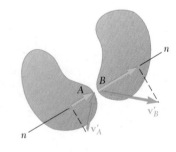

(a) Before impact   (b) After impact

Fig. 17.16

is applicable not only when the colliding bodies move freely after the
impact but also when the bodies are partially constrained in their
motion. It should be used in conjunction with one or several other
equations obtained by applying the principle of impulse and momentum [Sample Prob. 17.10]. We also considered problems where the
method of impulse and momentum and the method of work and
energy can be combined [Sample Prob. 17.11].

REVIEW PROBLEMS

17.135 A uniform disk of constant thickness and initially at rest is placed in contact with the belt shown, which moves at a constant speed $v = 80$ ft/s. Knowing that the coefficient of kinetic friction between the disk and the belt is 0.15, determine (a) the number of revolutions executed by the disk before it reaches a constant angular velocity, (b) the time required for the disk to reach that constant angular velocity.

Fig. P17.135

17.136 The 8-in.-radius brake drum is attached to a larger flywheel that is not shown. The total mass moment of inertia of the flywheel and drum is 14 lb · ft · s^2 and the coefficient of kinetic friction between the drum and the brake shoe is 0.35. Knowing that the initial angular velocity of the flywheel is 360 rpm counterclockwise, determine the vertical force **P** that must be applied to the pedal C if the system is to stop in 100 revolutions.

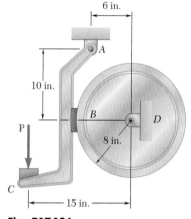

Fig. P17.136

17.137 A 6 × 8-in. rectangular plate is suspended by pins at A and B. The pin at B is removed and the plate swings freely about pin A. Determine (a) the angular velocity of the plate after it has rotated through 90°, (b) the maximum angular velocity attained by the plate as it swings freely.

Fig. P17.137

17.138 The gear shown has a radius $R = 150$ mm and a radius of gyration $\bar{k} = 125$ mm. The gear is rolling without sliding with a velocity $\bar{\mathbf{v}}_1$ of magnitude 3 m/s when it strikes a step of height $h = 75$ mm. Because the edge of the step engages the gear teeth, no slipping occurs between the gear and the step. Assuming perfectly plastic impact, determine (a) the angular velocity of the gear immediately after the impact, (b) the angular velocity of the gear after it has rotated to the top of the step.

Fig. P17.138

Fig. P17.139

17.139 A uniform slender rod is placed at corner B and is given a slight clockwise motion. Assuming that the corner is sharp and becomes slightly embedded in the end of the rod, so that the coefficient of static friction at B is very large, determine (a) the angle β through which the rod will have rotated when it loses contact with the corner, (b) the corresponding velocity of end A.

17.140 The motion of the slender 250-mm rod AB is guided by pins at A and B that slide freely in slots cut in a vertical plate as shown. Knowing that the rod has a mass of 2 kg and is released from rest when $\theta = 0$, determine the reactions at A and B when $\theta = 90°$.

Fig. P17.140

17.141 A 35-g bullet B is fired horizontally with a velocity of 400 m/s into the side of a 3-kg square panel suspended from a pin at A. Knowing that the panel is initially at rest, determine the components of the reaction at A after the panel has rotated 45°.

17.142 Two panels A and B are attached with hinges to a rectangular plate and held by a wire as shown. The plate and the panels are made of the same material and have the same thickness. The entire assembly is rotating with an angular velocity ω_0 when the wire breaks. Determine the angular velocity of the assembly after the panels have come to rest against the plate.

Fig. P17.141

Fig. P17.142

17.143 Disks A and B are made of the same material and are of the same thickness; they can rotate freely about the vertical shaft. Disk B is at rest when it is dropped onto disk A, which is rotating with an angular velocity of 500 rpm. Knowing that disk A has a mass of 8 kg, determine (a) the final angular velocity of the disks, (b) the change in kinetic energy of the system.

17.144 A square block of mass m is falling with a velocity $\bar{\mathbf{v}}_1$ when it strikes a small obstruction at B. Knowing that the coefficient of restitution for the impact between corner A and the obstruction B is $e = 0.5$, determine immediately after the impact (a) the angular velocity of the block, (b) the velocity of its mass center G.

Fig. P17.143

Fig. P17.144

17.145 A 3-kg bar AB is attached by a pin at D to a 4-kg square plate, which can rotate freely about a vertical axis. Knowing that the angular velocity of the plate is 120 rpm when the bar is vertical, determine (a) the angular velocity of the plate after the bar has swung into a horizontal position and has come to rest against pin C, (b) the energy lost during the plastic impact at C.

Fig. P17.145

17.146 A 1.8-lb javelin DE impacts a 10-lb slender rod ABC with a horizontal velocity of $\mathbf{v}_0 = 30$ ft/s as shown. Knowing that the javelin becomes embedded into the end of the rod at point C and does not penetrate very far into it, determine immediately after the impact (a) the angular velocity of the rod ABC, (b) the components of the reaction at B. Assume the javelin and the rod move as a single rigid body after the impact.

Fig. P17.146

COMPUTER PROBLEMS

Fig. P17.C1

17.C1 Rod AB has a mass of 3 kg and is attached at A to a 5-kg cart C. Knowing that the system is released from rest when $\theta = 30°$ and neglecting friction, use computational software to determine the velocity of the cart and the velocity of end B of the rod for values of θ from $+30°$ to $-90°$. Determine the value of θ for which the velocity of the cart to the left is maximum and the corresponding value of the velocity.

17.C2 The uniform slender rod AB of length $L = 800$ mm and mass 5 kg rests on a small wheel at D and is attached to a collar of negligible mass that can slide freely on the vertical rod EF. Knowing that $a = 200$ mm and that the rod is released from rest when $\theta = 0$, use computational software to calculate and plot the angular velocity of the rod and the velocity of end A for values of θ from 0 to 50°. Determine the maximum angular velocity of the rod and the corresponding value of θ.

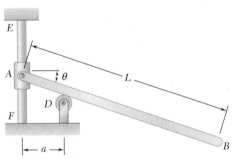

Fig. P17.C2

17.C3 A uniform 10-in.-radius sphere rolls over a series of parallel horizontal bars equally spaced at a distance d. As it rotates without slipping about a given bar, the sphere strikes the next bar and starts rotating about that bar without slipping, until it strikes the next bar, and so on. Assuming perfectly plastic impact and knowing that the sphere has an angular velocity $\boldsymbol{\omega}_0$ of 1.5 rad/s as its mass center G is directly above bar A, use computational software to calculate values of the spacing d from 1 to 6 in. (*a*) the angular velocity $\boldsymbol{\omega}_1$ of the sphere as G passes directly above bar B, (*b*) the number of bars over which the sphere will roll after leaving bar A.

Fig. P17.C3

17.C4 Collar C has a mass of 2.5 kg and can slide without friction on rod AB. A spring of constant 750 N/m and an unstretched length $r_0 = 500$ mm is attached as shown to the collar and to the hub B. The total mass moment of inertia of the rod, hub, and spring is known to be 0.3 kg $\cdot$ m^2 about B. Initially the collar is held at a distance of 500 mm from the axis of rotation by a small pin protruding from the rod. The pin is suddenly removed as the assembly is rotating in a horizontal plane with an angular velocity $\boldsymbol{\omega}_0$ of 10 rad/s. Denoting by r the distance of the collar from the axis of rotation, use computational software to calculate and plot the angular velocity of the assembly and the velocity of the collar relative to the rod for values of r from 500 to 700 mm. Determine the maximum value of r in the ensuing motion.

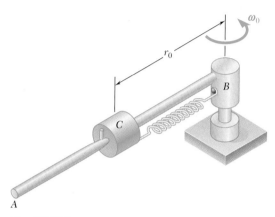

Fig. P17.C4

17.C5 Each of the two identical slender bars shown has a length $L = 30$ in. Knowing that the system is released from rest when the bars are horizontal, use computational software to calculate and plot the angular velocity of rod AB and the velocity of point D for values of θ from 0 to 90°.

Fig. P17.C5

While the general principles that you learned in earlier chapters can be used again to solve problems involving the three-dimensional motion of rigid bodies, the solution of these problems requires a new approach and is considerably more involved than the solution of two-dimensional problems. One example is the determination of the forces acting on the space shuttle's robotic arm.

CHAPTER

18

Kinetics of Rigid Bodies
in Three Dimensions

*18.1 INTRODUCTION

In Chaps. 16 and 17 we were concerned with the plane motion of rigid bodies and of systems of rigid bodies. In Chap. 16 and in the second half of Chap. 17 (momentum method), our study was further restricted to that of plane slabs and of bodies symmetrical with respect to the reference plane. However, many of the fundamental results obtained in these two chapters remain valid in the case of the motion of a rigid body in three dimensions.

For example, the two fundamental equations

$$\Sigma \mathbf{F} = m\bar{\mathbf{a}} \tag{18.1}$$
$$\Sigma \mathbf{M}_G = \dot{\mathbf{H}}_G \tag{18.2}$$

on which the analysis of the plane motion of a rigid body was based, remain valid in the most general case of motion of a rigid body. As indicated in Sec. 16.2, these equations express that the system of the external forces is equipollent to the system consisting of the vector $m\bar{\mathbf{a}}$ attached at G and the couple of moment $\dot{\mathbf{H}}_G$ (Fig. 18.1). However,

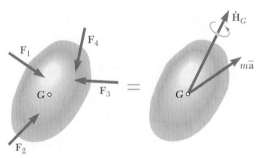

Fig. 18.1

the relation $\mathbf{H}_G = \bar{I}\boldsymbol{\omega}$, which enabled us to determine the angular momentum of a rigid slab and which played an important part in the solution of problems involving the plane motion of slabs and bodies symmetrical with respect to the reference plane, ceases to be valid in the case of nonsymmetrical bodies or three-dimensional motion. Thus in the first part of the chapter, in Sec. 18.2, a more general method for computing the angular momentum $\mathbf{H}_G$ of a rigid body in three dimensions will be developed.

Similarly, although the main feature of the impulse-momentum method discussed in Sec. 17.7, namely, the reduction of the momenta of the particles of a rigid body to a linear momentum vector $m\bar{\mathbf{v}}$ attached at the mass center G of the body and an angular momentum couple $\mathbf{H}_G$, remains valid, the relation $\mathbf{H}_G = \bar{I}\boldsymbol{\omega}$ must be discarded and replaced by the more general relation developed in Sec. 18.2 before this method can be applied to the three-dimensional motion of a rigid body (Sec. 18.3).

We also note that the work-energy principle (Sec. 17.2) and the principle of conservation of energy (Sec. 17.6) still apply in the case of the motion of a rigid body in three dimensions. However, the expression obtained in Sec. 17.4 for the kinetic energy of a rigid body in plane motion will be replaced by a new expression developed in Sec. 18.4 for a rigid body in three-dimensional motion.

In the second part of the chapter, you will first learn to determine the rate of change $\dot{\mathbf{H}}_G$ of the angular momentum $\mathbf{H}_G$ of a three-dimensional rigid body, using a rotating frame of reference with respect to which the moments and products of inertia of the body remain constant (Sec. 18.5). Equations (18.1) and (18.2) will then be expressed in the form of free-body-diagram equations, which can be used to solve various problems involving the three-dimensional motion of rigid bodies (Secs. 18.6 through 18.8).

The last part of the chapter (Secs. 18.9 through 18.11) is devoted to the study of the motion of the gyroscope or, more generally, of an axisymmetrical body with a fixed point located on its axis of symmetry. In Sec. 18.10, the particular case of the steady precession of a gyroscope will be considered, and, in Sec. 18.11, the motion of an axisymmetrical body subjected to no force, except its own weight, will be analyzed.

*18.2 ANGULAR MOMENTUM OF A RIGID BODY IN THREE DIMENSIONS

In this section you will see how the angular momentum $\mathbf{H}_G$ of a body about its mass center G can be determined from the angular velocity $\boldsymbol{\omega}$ of the body in the case of three-dimensional motion.

According to Eq. (14.24), the angular momentum of the body about G can be expressed as

$$\mathbf{H}_G = \sum_{i=1}^{n} (\mathbf{r}_i' \times \mathbf{v}_i' \, \Delta m_i) \tag{18.3}$$

where $\mathbf{r}_i'$ and $\mathbf{v}_i'$ denote, respectively, the position vector and the velocity of the particle P_i, of mass Δm_i, relative to the centroidal frame $Gxyz$ (Fig. 18.2). But $\mathbf{v}_i' = \boldsymbol{\omega} \times \mathbf{r}_i'$, where $\boldsymbol{\omega}$ is the angular

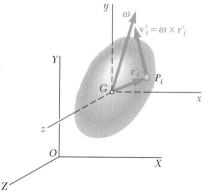

Fig. 18.2

velocity of the body at the instant considered. Substituting into (18.3), we have

$$\mathbf{H}_G = \sum_{i=1}^{n} [\mathbf{r}_i' \times (\boldsymbol{\omega} \times \mathbf{r}_i') \, \Delta m_i]$$

Recalling the rule for determining the rectangular components of a vector product (Sec. 3.5), we obtain the following expression for the x component of the angular momentum:

$$
\begin{aligned}
H_x &= \sum_{i=1}^{n} [y_i(\boldsymbol{\omega} \times \mathbf{r}_i')_z - z_i(\boldsymbol{\omega} \times \mathbf{r}_i')_y] \, \Delta m_i \\
&= \sum_{i=1}^{n} [y_i(\omega_x y_i - \omega_y x_i) - z_i(\omega_z x_i - \omega_x z_i)] \, \Delta m_i \\
&= \omega_x \sum_i (y_i^2 + z_i^2) \, \Delta m_i - \omega_y \sum_i x_i y_i \, \Delta m_i - \omega_z \sum_i z_i x_i \, \Delta m_i
\end{aligned}
$$

Replacing the sums by integrals in this expression and in the two similar expressions which are obtained for H_y and H_z, we have

$$
\begin{aligned}
H_x &= \omega_x \!\int (y^2 + z^2) \, dm - \omega_y \!\int xy \, dm - \omega_z \!\int zx \, dm \\
H_y &= -\omega_x \!\int xy \, dm + \omega_y \!\int (z^2 + x^2) \, dm - \omega_z \!\int yz \, dm \\
H_z &= -\omega_x \!\int zx \, dm - \omega_y \!\int yz \, dm + \omega_z \!\int (x^2 + y^2) \, dm
\end{aligned}
\tag{18.4}
$$

We note that the integrals containing squares represent the *centroidal mass moments of inertia* of the body about the x, y, and z axes, respectively (Sec. 9.11); we have

$$
\bar{I}_x = \int (y^2 + z^2) \, dm \qquad \bar{I}_y = \int (z^2 + x^2) \, dm
$$
$$
\bar{I}_z = \int (x^2 + y^2) \, dm
\tag{18.5}
$$

Similarly, the integrals containing products of coordinates represent the *centroidal mass products of inertia* of the body (Sec. 9.16); we have

$$
\bar{I}_{xy} = \int xy \, dm \qquad \bar{I}_{yz} = \int yz \, dm \qquad \bar{I}_{zx} = \int zx \, dm
\tag{18.6}
$$

Substituting from (18.5) and (18.6) into (18.4), we obtain the components of the angular momentum $\mathbf{H}_G$ of the body about its mass center:

$$
\begin{aligned}
H_x &= +\bar{I}_x \, \omega_x - \bar{I}_{xy} \omega_y - \bar{I}_{xz} \omega_z \\
H_y &= -\bar{I}_{yx} \omega_x + \bar{I}_y \, \omega_y - \bar{I}_{yz} \omega_z \\
H_z &= -\bar{I}_{zx} \omega_x - \bar{I}_{zy} \omega_y + \bar{I}_z \, \omega_z
\end{aligned}
\tag{18.7}
$$

The relations (18.7) show that the operation which transforms the vector $\boldsymbol{\omega}$ into the vector $\mathbf{H}_G$ (Fig. 18.3) is characterized by the array of moments and products of inertia

$$\begin{pmatrix} \bar{I}_x & -\bar{I}_{xy} & -\bar{I}_{xz} \\ -\bar{I}_{yx} & \bar{I}_y & -\bar{I}_{yz} \\ -\bar{I}_{zx} & -\bar{I}_{zy} & \bar{I}_z \end{pmatrix} \qquad (18.8)$$

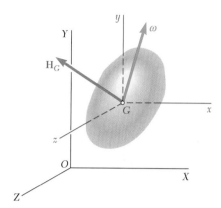

Fig. 18.3

The array (18.8) defines the *inertia tensor* of the body at its mass center G.† A new array of moments and products of inertia would be obtained if a different system of axes were used. The transformation characterized by this new array, however, would still be the same. Clearly, the angular momentum $\mathbf{H}_G$ corresponding to a given angular velocity $\boldsymbol{\omega}$ is independent of the choice of the coordinate axes. As was shown in Secs. 9.17 and 9.18, it is always possible to select a system of axes $Gx'y'z'$, called *principal axes of inertia*, with respect to which all the products of inertia of a given body are zero. The array (18.8) takes then the diagonalized form

$$\begin{pmatrix} \bar{I}_{x'} & 0 & 0 \\ 0 & \bar{I}_{y'} & 0 \\ 0 & 0 & \bar{I}_{z'} \end{pmatrix} \qquad (18.9)$$

where $\bar{I}_{x'}$, $\bar{I}_{y'}$, $\bar{I}_{z'}$ represent the *principal centroidal moments of inertia* of the body, and the relations (18.7) reduce to

$$H_{x'} = \bar{I}_{x'}\omega_{x'} \qquad H_{y'} = \bar{I}_{y'}\omega_{y'} \qquad H_{z'} = \bar{I}_{z'}\omega_{z'} \qquad (18.10)$$

We note that if the three principal centroidal moments of inertia $\bar{I}_{x'}$, $\bar{I}_{y'}$, $\bar{I}_{z'}$ are equal, the components $H_{x'}$, $H_{y'}$, $H_{z'}$ of the angular momentum about G are proportional to the components $\omega_{x'}$, $\omega_{y'}$, $\omega_{z'}$ of the angular velocity, and the vectors $\mathbf{H}_G$ and $\boldsymbol{\omega}$ are collinear. In general, however, the principal moments of inertia will be different, and the vectors $\mathbf{H}_G$ and $\boldsymbol{\omega}$ *will have different directions*, except when two of the three components of $\boldsymbol{\omega}$ happen to be zero, i.e., when $\boldsymbol{\omega}$ is directed along one of the coordinate axes. Thus, *the angular momentum $\mathbf{H}_G$ of a rigid body and its angular velocity $\boldsymbol{\omega}$ have the same direction if, and only if, $\boldsymbol{\omega}$ is directed along a principal axis of inertia.*‡

---

†Setting $\bar{I}_x = I_{11}$, $\bar{I}_y = I_{22}$, $\bar{I}_z = I_{33}$, and $-\bar{I}_{xy} = I_{12}$, $-\bar{I}_{xz} = I_{13}$, etc., we may write the inertia tensor (18.8) in the standard form

$$\begin{pmatrix} I_{11} & I_{12} & I_{13} \\ I_{21} & I_{22} & I_{23} \\ I_{31} & I_{32} & I_{33} \end{pmatrix}$$

Denoting by H_1, H_2, H_3 the components of the angular momentum $\mathbf{H}_G$ and by ω_1, ω_2, ω_3 the components of the angular velocity $\boldsymbol{\omega}$, we can write the relations (18.7) in the form

$$H_i = \sum_j I_{ij}\omega_j$$

where i and j take the values 1, 2, 3. The quantities I_{ij} are said to be the *components* of the inertia tensor. Since $I_{ij} = I_{ji}$, the inertia tensor is a *symmetric tensor of the second order*.

‡In the particular case when $\bar{I}_{x'} = \bar{I}_{y'} = \bar{I}_{z'}$, any line through G can be considered as a principal axis of inertia, and the vectors $\mathbf{H}_G$ and $\boldsymbol{\omega}$ are always collinear.

Photo 18.1 The design of a robotic welder for an automobile assembly line requires a three-dimensional study of both kinematics and kinetics.

Since this condition is satisfied in the case of the plane motion of a rigid body symmetrical with respect to the reference plane, we were able in Secs. 16.3 and 17.8 to represent the angular momentum $\mathbf{H}_G$ of such a body by the vector $\bar{I}\boldsymbol{\omega}$. We must realize, however, that this result cannot be extended to the case of the plane motion of a non-symmetrical body, or to the case of the three-dimensional motion of a rigid body. Except when $\boldsymbol{\omega}$ happens to be directed along a principal axis of inertia, the angular momentum and angular velocity of a rigid body have different directions, and the relation (18.7) or (18.10) must be used to determine $\mathbf{H}_G$ from $\boldsymbol{\omega}$.

Reduction of the Momenta of the Particles of a Rigid Body to a Momentum Vector and a Couple at G. We saw in Sec. 17.8 that the system formed by the momenta of the various particles of a rigid body can be reduced to a vector $\mathbf{L}$ attached at the mass center G of the body, representing the linear momentum of the body, and to a couple $\mathbf{H}_G$, representing the angular momentum of the body about G (Fig. 18.4). We are now in a position to determine the vector $\mathbf{L}$

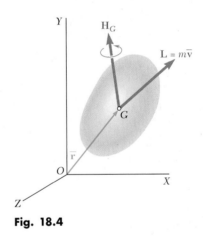

Fig. 18.4

and the couple $\mathbf{H}_G$ in the most general case of three-dimensional motion of a rigid body. As in the case of the two-dimensional motion considered in Sec. 17.8, the linear momentum $\mathbf{L}$ of the body is equal to the product $m\bar{\mathbf{v}}$ of its mass m and the velocity $\bar{\mathbf{v}}$ of its mass center G. The angular momentum $\mathbf{H}_G$, however, can no longer be obtained by simply multiplying the angular velocity $\boldsymbol{\omega}$ of the body by the scalar $\bar{I}$; it must now be obtained from the components of $\boldsymbol{\omega}$ and from the centroidal moments and products of inertia of the body through the use of Eq. (18.7) or (18.10).

We should also note that once the linear momentum $m\bar{\mathbf{v}}$ and the angular momentum $\mathbf{H}_G$ of a rigid body have been determined, its angular momentum $\mathbf{H}_O$ about any given point O can be obtained by adding the moments about O of the vector $m\bar{\mathbf{v}}$ and of the couple $\mathbf{H}_G$. We write

$$\mathbf{H}_O = \bar{\mathbf{r}} \times m\bar{\mathbf{v}} + \mathbf{H}_G \qquad (18.11)$$

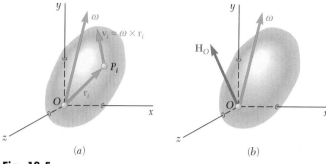

Fig. 18.5

Angular Momentum of a Rigid Body Constrained to Rotate about a Fixed Point. In the particular case of a rigid body constrained to rotate in three-dimensional space about a fixed point O (Fig. 18.5a), it is sometimes convenient to determine the angular momentum $\mathbf{H}_O$ of the body about the fixed point O. While $\mathbf{H}_O$ could be obtained by first computing $\mathbf{H}_G$ as indicated above and then using Eq. (18.11), it is often advantageous to determine $\mathbf{H}_O$ directly from the angular velocity $\boldsymbol{\omega}$ of the body and its moments and products of inertia with respect to a frame $Oxyz$ centered at the fixed point O. Recalling Eq. (14.7), we write

$$\mathbf{H}_O = \sum_{i=1}^{n} (\mathbf{r}_i \times \mathbf{v}_i \, \Delta m_i) \qquad (18.12)$$

where $\mathbf{r}_i$ and $\mathbf{v}_i$ denote, respectively, the position vector and the velocity of the particle P_i with respect to the fixed frame $Oxyz$. Substituting $\mathbf{v}_i = \boldsymbol{\omega} \times \mathbf{r}_i$, and after manipulations similar to those used in the earlier part of this section, we find that the components of the angular momentum $\mathbf{H}_O$ (Fig. 18.5b) are given by the relations

$$\begin{aligned}
H_x &= +I_x\,\omega_x - I_{xy}\omega_y - I_{xz}\omega_z \\
H_y &= -I_{yx}\omega_x + I_y\,\omega_y - I_{yz}\omega_z \\
H_z &= -I_{zx}\omega_x - I_{zy}\omega_y + I_z\,\omega_z
\end{aligned} \qquad (18.13)$$

where the moments of inertia I_x, I_y, I_z and the products of inertia I_{xy}, I_{yz}, I_{zx} are computed with respect to the frame $Oxyz$ centered at the fixed point O.

*18.3 APPLICATION OF THE PRINCIPLE OF IMPULSE AND MOMENTUM TO THE THREE-DIMENSIONAL MOTION OF A RIGID BODY

Before we can apply the fundamental equation (18.2) to the solution of problems involving the three-dimensional motion of a rigid body, we must learn to compute the derivative of the vector $\mathbf{H}_G$. This will be done in Sec. 18.5. The results obtained in the preceding section can, however, be used right away to solve problems by the impulse-momentum method.

Recalling that the system formed by the momenta of the particles of a rigid body reduces to a linear momentum vector $m\overline{\mathbf{v}}$

Photo 18.2 As a result of the impulsive force applied by the bowling ball, a pin acquires both linear momentum and angular momentum.

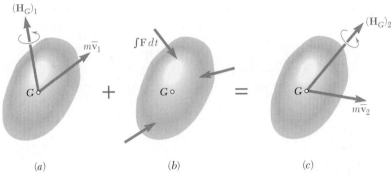

(a) (b) (c)

Fig. 18.6

attached at the mass center G of the body and an angular momentum couple $\mathbf{H}_G$, we represent graphically the fundamental relation

Syst Momenta$_1$ + Syst Ext Imp$_{1\to2}$ = Syst Momenta$_2$ (17.14)

by means of the three sketches shown in Fig. 18.6. To solve a given problem, we can use these sketches to write appropriate component and moment equations, keeping in mind that the components of the angular momentum $\mathbf{H}_G$ are related to the components of the angular velocity $\boldsymbol{\omega}$ by Eqs. (18.7) of the preceding section.

In solving problems dealing with the motion of a body rotating about a fixed point O, it will be convenient to eliminate the impulse of the reaction at O by writing an equation involving the moments of the momenta and impulses about O. We recall that the angular momentum $\mathbf{H}_O$ of the body about the fixed point O can be obtained either directly from Eqs. (18.13) or by first computing its linear momentum $m\bar{\mathbf{v}}$ and its angular momentum $\mathbf{H}_G$ and then using Eq. (18.11).

*18.4   KINETIC ENERGY OF A RIGID BODY IN THREE DIMENSIONS

Consider a rigid body of mass m in three-dimensional motion. We recall from Sec. 14.6 that if the absolute velocity $\mathbf{v}_i$ of each particle P_i of the body is expressed as the sum of the velocity $\bar{\mathbf{v}}$ of the mass center G of the body and of the velocity $\mathbf{v}_i'$ of the particle relative to a frame $Gxyz$ attached to G and of fixed orientation (Fig. 18.7), the kinetic energy of the system of particles forming the rigid body can be written in the form

$$T = \tfrac{1}{2}m\bar{v}^2 + \frac{1}{2}\sum_{i=1}^{n}\Delta m_i v_i'^2 \qquad (18.14)$$

where the last term represents the kinetic energy T' of the body relative to the centroidal frame $Gxyz$. Since $v_i' = |\mathbf{v}_i'| = |\boldsymbol{\omega} \times \mathbf{r}_i'|$, we write

$$T' = \frac{1}{2}\sum_{i=1}^{n}\Delta m_i v_i'^2 = \frac{1}{2}\sum_{i=1}^{n}|\boldsymbol{\omega} \times \mathbf{r}_i'|^2\,\Delta m_i$$

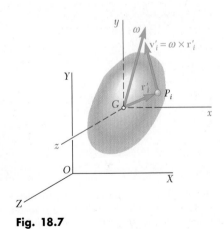

Fig. 18.7

Expressing the square in terms of the rectangular components of the vector product, and replacing the sums by integrals, we have

$$T' = \frac{1}{2}\int[(\omega_x y - \omega_y x)^2 + (\omega_y z - \omega_z y)^2 + (\omega_z x - \omega_x z)^2]\,dm$$
$$= \frac{1}{2}[\omega_x^2\int(y^2 + z^2)\,dm + \omega_y^2\int(z^2 + x^2)\,dm + \omega_z^2\int(x^2 + y^2)\,dm$$
$$- 2\omega_x\omega_y\int xy\,dm - 2\omega_y\omega_z\int yz\,dm - 2\omega_z\omega_x\int zx\,dm]$$

or, recalling the relations (18.5) and (18.6),

$$T' = \frac{1}{2}(\bar{I}_x\omega_x^2 + \bar{I}_y\omega_y^2 + \bar{I}_z\omega_z^2 - 2\bar{I}_{xy}\omega_x\omega_y - 2\bar{I}_{yz}\omega_y\omega_z - 2\bar{I}_{zx}\omega_z\omega_x)$$
(18.15)

Substituting into (18.14) the expression (18.15) we have just obtained for the kinetic energy of the body relative to centroidal axes, we write

$$T = \frac{1}{2}m\bar{v}^2 + \frac{1}{2}(\bar{I}_x\omega_x^2 + \bar{I}_y\omega_y^2 + \bar{I}_z\omega_z^2 - 2\bar{I}_{xy}\omega_x\omega_y$$
$$- 2\bar{I}_{yz}\omega_y\omega_z - 2\bar{I}_{zx}\omega_z\omega_x) \qquad (18.16)$$

If the axes of coordinates are chosen so that they coincide at the instant considered with the principal axes x', y', z' of the body, the relation obtained reduces to

$$T = \frac{1}{2}m\bar{v}^2 + \frac{1}{2}(\bar{I}_{x'}\omega_{x'}^2 + \bar{I}_{y'}\omega_{y'}^2 + \bar{I}_{z'}\omega_{z'}^2) \qquad (18.17)$$

where $\bar{\mathbf{v}}$ = velocity of mass center
$\boldsymbol{\omega}$ = angular velocity
m = mass of rigid body
$\bar{I}_{x'}, \bar{I}_{y'}, \bar{I}_{z'}$ = principal centroidal moments of inertia

The results we have obtained enable us to apply to the three-dimensional motion of a rigid body the principles of work and energy (Sec. 17.2) and conservation of energy (Sec. 17.6).

Kinetic Energy of a Rigid Body with a Fixed Point. In the particular case of a rigid body rotating in three-dimensional space about a fixed point O, the kinetic energy of the body can be expressed in terms of its moments and products of inertia with respect to axes attached at O (Fig. 18.8). Recalling the definition of the kinetic energy of a system of particles, and substituting $v_i = |\mathbf{v}_i| = |\boldsymbol{\omega} \times \mathbf{r}_i|$, we write

$$T = \frac{1}{2}\sum_{i=1}^{n}\Delta m_i v_i^2 = \frac{1}{2}\sum_{i=1}^{n}|\boldsymbol{\omega} \times \mathbf{r}_i|^2 \Delta m_i \qquad (18.18)$$

Manipulations similar to those used to derive Eq. (18.15) yield

$$T = \frac{1}{2}(I_x\omega_x^2 + I_y\omega_y^2 + I_z\omega_z^2 - 2I_{xy}\omega_x\omega_y - 2I_{yz}\omega_y\omega_z - 2I_{zx}\omega_z\omega_x)$$
(18.19)

or, if the principal axes x', y', z' of the body at the origin O are chosen as coordinate axes,

$$T = \frac{1}{2}(I_{x'}\omega_{x'}^2 + I_{y'}\omega_{y'}^2 + I_{z'}\omega_{z'}^2) \qquad (18.20)$$

Fig. 18.8

SAMPLE PROBLEM 18.1

A rectangular plate of mass m suspended from two wires at A and B is hit at D in a direction perpendicular to the plate. Denoting by $\mathbf{F}\,\Delta t$ the impulse applied at D, determine immediately after the impact (a) the velocity of the mass center G, (b) the angular velocity of the plate.

SOLUTION

Assuming that the wires remain taut and thus that the components $\bar{v}_y$ of $\bar{\mathbf{v}}$ and ω_z of $\boldsymbol{\omega}$ are zero after the impact, we have

$$\bar{\mathbf{v}} = \bar{v}_x\mathbf{i} + \bar{v}_z\mathbf{k} \qquad \boldsymbol{\omega} = \omega_x\mathbf{i} + \omega_y\mathbf{j}$$

and since the x, y, z axes are principal axes of inertia,

$$\mathbf{H}_G = \bar{I}_x\omega_x\mathbf{i} + \bar{I}_y\omega_y\mathbf{j} \qquad \mathbf{H}_G = \tfrac{1}{12}mb^2\omega_x\mathbf{i} + \tfrac{1}{12}ma^2\omega_y\mathbf{j} \qquad (1)$$

Principle of Impulse and Momentum. Since the initial momenta are zero, the system of the impulses must be equivalent to the system of the final momenta:

a. Velocity of Mass Center. Equating the components of the impulses and momenta in the x and z directions:

x components:	$0 = m\bar{v}_x$	$\bar{v}_x = 0$
z components:	$-F\,\Delta t = m\bar{v}_z$	$\bar{v}_z = -F\,\Delta t/m$

$$\bar{\mathbf{v}} = \bar{v}_x\mathbf{i} + \bar{v}_z\mathbf{k} \qquad \bar{\mathbf{v}} = -(F\,\Delta t/m)\mathbf{k} \quad \blacktriangleleft$$

b. Angular Velocity. Equating the moments of the impulses and momenta about the x and y axes:

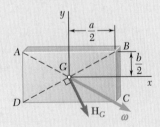

About x axis:	$\tfrac{1}{2}bF\,\Delta t = H_x$
About y axis:	$-\tfrac{1}{2}aF\,\Delta t = H_y$

$$\mathbf{H}_G = H_x\mathbf{i} + H_y\mathbf{j} \qquad \mathbf{H}_G = \tfrac{1}{2}bF\,\Delta t\,\mathbf{i} - \tfrac{1}{2}aF\,\Delta t\,\mathbf{j} \qquad (2)$$

Comparing Eqs. (1) and (2), we conclude that

$$\omega_x = 6F\,\Delta t/mb \qquad \omega_y = -6F\,\Delta t/ma$$
$$\boldsymbol{\omega} = \omega_x\mathbf{i} + \omega_y\mathbf{j} \qquad \boldsymbol{\omega} = (6F\,\Delta t/mab)(a\mathbf{i} - b\mathbf{j}) \quad \blacktriangleleft$$

We note that $\boldsymbol{\omega}$ is directed along the diagonal AC.

Remark: Equating the y components of the impulses and momenta, and their moments about the z axis, we obtain two additional equations which yield $T_A = T_B = \tfrac{1}{2}W$. We thus verify that the wires remain taut and that our assumption was correct.

SAMPLE PROBLEM 18.2

A homogeneous disk of radius r and mass m is mounted on an axle OG of length L and negligible mass. The axle is pivoted at the fixed point O, and the disk is constrained to roll on a horizontal floor. Knowing that the disk rotates counterclockwise at the rate ω_1 about the axle OG, determine (a) the angular velocity of the disk, (b) its angular momentum about O, (c) its kinetic energy, (d) the vector and couple at G equivalent to the momenta of the particles of the disk.

SOLUTION

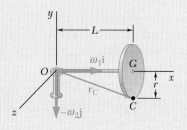

a. Angular Velocity. As the disk rotates about the axle OG it also rotates with the axle about the y axis at a rate ω_2 clockwise. The total angular velocity of the disk is therefore

$$\boldsymbol{\omega} = \omega_1\mathbf{i} - \omega_2\mathbf{j} \tag{1}$$

To determine ω_2 we write that the velocity of C is zero:

$$\mathbf{v}_C = \boldsymbol{\omega} \times \mathbf{r}_C = 0$$
$$(\omega_1\mathbf{i} - \omega_2\mathbf{j}) \times (L\mathbf{i} - r\mathbf{j}) = 0$$
$$(L\omega_2 - r\omega_1)\mathbf{k} = 0 \qquad \omega_2 = r\omega_1/L$$

Substituting into (1) for ω_2: $\qquad\qquad \boldsymbol{\omega} = \omega_1\mathbf{i} - (r\omega_1/L)\mathbf{j}$ ◀

b. Angular Momentum about O. Assuming the axle to be part of the disk, we can consider the disk to have a fixed point at O. Since the x, y, and z axes are principal axes of inertia for the disk,

$$H_x = I_x\omega_x = (\tfrac{1}{2}mr^2)\omega_1$$
$$H_y = I_y\omega_y = (mL^2 + \tfrac{1}{4}mr^2)(-r\omega_1/L)$$
$$H_z = I_z\omega_z = (mL^2 + \tfrac{1}{4}mr^2)0 = 0$$
$$\mathbf{H}_O = \tfrac{1}{2}mr^2\omega_1\mathbf{i} - m(L^2 + \tfrac{1}{4}r^2)(r\omega_1/L)\mathbf{j} \ ◀$$

c. Kinetic Energy. Using the values obtained for the moments of inertia and the components of $\boldsymbol{\omega}$, we have

$$T = \tfrac{1}{2}(I_x\omega_x^2 + I_y\omega_y^2 + I_z\omega_z^2) = \tfrac{1}{2}[\tfrac{1}{2}mr^2\omega_1^2 + m(L^2 + \tfrac{1}{4}r^2)(-r\omega_1/L)^2]$$
$$T = \tfrac{1}{8}mr^2\left(6 + \frac{r^2}{L^2}\right)\omega_1^2 \ ◀$$

d. Momentum Vector and Couple at G. The linear momentum vector $m\bar{\mathbf{v}}$ and the angular momentum couple $\mathbf{H}_G$ are

$$m\bar{\mathbf{v}} = mr\omega_1\mathbf{k} \ ◀$$

and

$$\mathbf{H}_G = \bar{I}_{x'}\omega_x\mathbf{i} + \bar{I}_{y'}\omega_y\mathbf{j} + \bar{I}_{z'}\omega_z\mathbf{k} = \tfrac{1}{2}mr^2\omega_1\mathbf{i} + \tfrac{1}{4}mr^2(-r\omega_1/L)\mathbf{j}$$
$$\mathbf{H}_G = \tfrac{1}{2}mr^2\omega_1\left(\mathbf{i} - \frac{r}{2L}\mathbf{j}\right) \ ◀$$

SOLVING PROBLEMS ON YOUR OWN

In this lesson you learned to compute the *angular momentum of a rigid body in three dimensions* and to apply the principle of impulse and momentum to the three-dimensional motion of a rigid body. You also learned to compute the *kinetic energy of a rigid body in three dimensions*. It is important for you to keep in mind that, except for very special situations, the angular momentum of a rigid body in three dimensions *cannot* be expressed as the product $\bar{I}\boldsymbol{\omega}$ and, therefore, *will not have the same direction as the angular velocity* $\boldsymbol{\omega}$ (Fig. 18.3).

1. To compute the angular momentum $\mathbf{H}_G$ of a rigid body about its mass center G, you must first determine the angular velocity $\boldsymbol{\omega}$ of the body with respect to a system of axes *centered at G and of fixed orientation*. Since you will be asked in this lesson to determine the angular momentum of the body *at a given instant only*, select the system of axes which will be most convenient for your computations.

a. If the principal axes of inertia of the body at G are known, use these axes as coordinate axes x', y', and z', since the corresponding products of inertia of the body will be equal to zero. Resolve $\boldsymbol{\omega}$ into components $\omega_{x'}$, $\omega_{y'}$, and $\omega_{z'}$ along these axes and compute the principal moments of inertia $\bar{I}_{x'}$, $\bar{I}_{y'}$, and $\bar{I}_{z'}$. The corresponding components of the angular momentum $\mathbf{H}_G$ are

$$H_{x'} = \bar{I}_{x'}\omega_{x'} \qquad H_{y'} = \bar{I}_{y'}\omega_{y'} \qquad H_{z'} = \bar{I}_{z'}\omega_{z'} \qquad \text{(18.10)}$$

b. If the principal axes of inertia of the body at G are not known, you must use Eqs. (18.7) to determine the components of the angular momentum $\mathbf{H}_G$. These equations require prior computation of the *products of inertia* of the body as well as prior computation of its moments of inertia with respect to the selected axes.

c. The magnitude and direction cosines of $\mathbf{H}_G$ are obtained from formulas similar to those used in Statics [Sec. 2.12]. We have

$$H_G = \sqrt{H_x^2 + H_y^2 + H_z^2}$$

$$\cos\theta_x = \frac{H_x}{H_G} \qquad \cos\theta_y = \frac{H_y}{H_G} \qquad \cos\theta_z = \frac{H_z}{H_G}$$

d. Once $\mathbf{H}_G$ has been determined, you can obtain the angular momentum of the body *about any given point O* by observing from Fig. (18.4) that

$$\mathbf{H}_O = \bar{\mathbf{r}} \times m\bar{\mathbf{v}} + \mathbf{H}_G \qquad \text{(18.11)}$$

where $\bar{\mathbf{r}}$ is the position vector of G relative to O, and $m\bar{\mathbf{v}}$ is the linear momentum of the body.

2. To compute the angular momentum $\mathbf{H}_O$ of a rigid body with a fixed point O, follow the procedure described in paragraph 1, except that you should now use axes centered at the fixed point O.

a. If the principal axes of inertia of the body at O are known, resolve $\boldsymbol{\omega}$ into components along these axes [Sample Prob. 18.2]. The corresponding components of the angular momentum $\mathbf{H}_G$ are obtained from equations similar to Eqs. (18.10).

b. If the principal axes of inertia of the body at O are not known, you must compute the products as well as the moments of inertia of the body with respect to the axes that you have selected and use Eqs. (18.13) to determine the components of the angular momentum $\mathbf{H}_O$.

3. To apply the principle of impulse and momentum to the solution of a problem involving the three-dimensional motion of a rigid body, you will use the same vector equation that you used for plane motion in Chap. 17,

$$\textbf{Syst Momenta}_1 + \textbf{Syst Ext Imp}_{1 \to 2} = \textbf{Syst Momenta}_2 \quad (17.14)$$

where the initial and final systems of momenta are each represented by a *linear-momentum vector* $m\overline{\mathbf{v}}$ and an *angular-momentum couple* $\mathbf{H}_G$. Now, however, these vector-and-couple systems should be represented in three dimensions as shown in Fig. 18.6, and $\mathbf{H}_G$ should be determined as explained in paragraph 1.

a. In problems involving the application of a known impulse to a rigid body, draw the free-body-diagram equation corresponding to Eq. (17.14). Equating the components of the vectors involved, you will determine the final linear momentum $m\overline{\mathbf{v}}$ of the body and, thus, the corresponding velocity $\overline{\mathbf{v}}$ of its mass center. Equating moments about G, you will determine the final angular momentum $\mathbf{H}_G$ of the body. You will then substitute the values obtained for the components of $\mathbf{H}_G$ into Eqs. (18.10) or (18.7) and solve these equations for the corresponding values of the components of the angular velocity $\boldsymbol{\omega}$ of the body [Sample Prob. 18.1].

b. In problems involving unknown impulses, draw the free-body-diagram equation corresponding to Eq. (17.4) and write equations which do not involve the unknown impulses. Such equations can be obtained by equating moments about the point or line of impact.

4. To compute the kinetic energy of a rigid body with a fixed point O, resolve the angular velocity $\boldsymbol{\omega}$ into components along axes of your choice and compute the moments and products of inertia of the body with respect to these axes. As was the case for the computation of the angular momentum, use the principal axes of inertia x', y', and z' if you can easily determine them. The products of inertia will then be zero [Sample Prob. 18.2], and the expression for the kinetic energy will reduce to

$$T = \tfrac{1}{2}(I_{x'}\omega_{x'}^2 + I_{y'}\omega_{y'}^2 + I_{z'}\omega_{x'}^2) \quad (18.20)$$

If you must use axes other than the principal axes of inertia, the kinetic energy of the body should be expressed as shown in Eq. (18.19).

5. To compute the kinetic energy of a rigid body in general motion, consider the motion as the sum of a *translation with the mass center G and a rotation about G*. The kinetic energy associated with the translation is $\tfrac{1}{2}m\overline{v}^2$. If principal axes of inertia can be used, the kinetic energy associated with the rotation about G can be expressed in the form used in Eq. (18.20). The total kinetic energy of the rigid body is then

$$T = \tfrac{1}{2}m\overline{v}^2 + \tfrac{1}{2}(\overline{I}_{x'}\omega_{x'}^2 + \overline{I}_{y'}\omega_{y'}^2 + \overline{I}_{z'}\omega_{z'}^2) \quad (18.17)$$

If you must use axes other than the principal axes of inertia to determine the kinetic energy associated with the rotation about G, the total kinetic energy of the body should be expressed as shown in Eq. (18.16).

PROBLEMS

Fig. P18.1

18.1 A thin, homogeneous disk of mass m and radius r spins at the constant rate ω_1 about an axle held by a fork-ended vertical rod which rotates at the constant rate ω_2. Determine the angular momentum $\mathbf{H}_G$ of the disk about its mass center G.

18.2 A thin rectangular plate of weight 15 lb rotates about its vertical diagonal AB with an angular velocity $\boldsymbol{\omega}$. Knowing that the z axis is perpendicular to the plate and that $\boldsymbol{\omega}$ is constant and equal to 5 rad/s, determine the angular momentum of the plate about its mass center G.

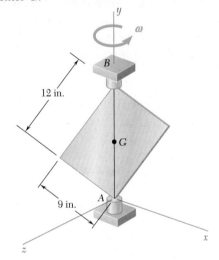

Fig. P18.2

18.3 Two uniform rods AB and CE, each of weight 3 lb and length 2 ft, are welded to each other at their midpoints. Knowing that this assembly has an angular velocity of constant magnitude $\omega = 12$ rad/s, determine the magnitude and direction of the angular momentum $\mathbf{H}_D$ of the assembly about D.

18.4 A homogeneous disk of weight $W = 6$ lb rotates at the constant rate $\omega_1 = 16$ rad/s with respect to arm ABC, which is welded to a shaft DCE rotating at the constant rate $\omega_2 = 8$ rad/s. Determine the angular momentum $\mathbf{H}_A$ of the disk about its center A.

Fig. P18.3

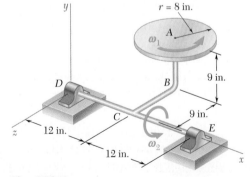

Fig. P18.4

18.5 A thin disk of mass $m = 4$ kg rotates at the constant rate $\omega_2 = 15$ rad/s with respect to arm ABC, which itself rotates at the constant rate $\omega_1 = 5$ rad/s about the y axis. Determine the angular momentum of the disk about its center C.

18.6 A solid rectangular parallelepiped of mass m has a square base of side a and a length $2a$. Knowing that it rotates at the constant rate ω about its diagonal AC' and that its rotation is observed from A as counterclockwise, determine (a) the magnitude of the angular momentum $\mathbf{H}_G$ of the parallelepiped about its mass center G, (b) the angle that $\mathbf{H}_G$ forms with the diagonal AC'.

Fig. P18.5

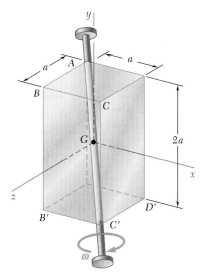

Fig. P18.6

18.7 Solve Prob. 18.6, assuming that the solid rectangular parallelepiped has been replaced by a hollow one consisting of six thin metal plates welded together.

18.8 A homogeneous disk of mass m and radius r is mounted on the vertical shaft AB. The normal to the disk at G forms an angle $\beta = 25°$ with the shaft. Knowing that the shaft has a constant angular velocity ω, determine the angle θ formed by the shaft AB and the angular momentum $\mathbf{H}_G$ of the disk about its mass center G.

18.9 Determine the angular momentum $\mathbf{H}_D$ of the disk of Prob. 18.4 about point D.

18.10 Determine the angular momentum of the disk of Prob. 18.5 about point A.

18.11 Determine the angular momentum $\mathbf{H}_O$ of the disk of Sample Prob. 18.2 from the expressions obtained for its linear momentum $m\bar{\mathbf{v}}$ and its angular momentum $\mathbf{H}_G$, using Eqs. (18.11). Verify that the result obtained is the same as that obtained by direct computation.

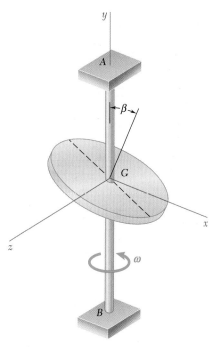

Fig. P18.8

18.12 The 100-kg projectile shown has a radius of gyration of 100 mm about its axis of symmetry Gx and a radius of gyration of 250 mm about the transverse axis Gy. Its angular velocity $\boldsymbol{\omega}$ can be resolved into two components; one component, directed along Gx, measures the *rate of spin* of the projectile, while the other component, directed along GD, measures its *rate of precession*. Knowing that $\theta = 6°$ and that the angular momentum of the projectile about its mass center G is $\mathbf{H}_G = (500 \text{ g} \cdot \text{m}^2/\text{s})\mathbf{i} - (10 \text{ g} \cdot \text{m}^2/\text{s})\mathbf{j}$, determine (*a*) the rate of spin, (*b*) the rate of precession.

Fig. P18.12

18.13 Determine the angular momentum $\mathbf{H}_A$ of the projectile of Prob. 18.12 about the center A of its base, knowing that its mass center G has a velocity $\bar{\mathbf{v}}$ of 750 m/s. Give your answer in terms of components respectively parallel to the x and y axes shown and to a third axis z pointing toward you.

18.14 (*a*) Show that the angular momentum $\mathbf{H}_B$ of a rigid body about point B can be obtained by adding to the angular momentum $\mathbf{H}_A$ of that body about point A the vector product of the vector $\mathbf{r}_{A/B}$ drawn from B to A and the linear momentum $m\bar{\mathbf{v}}$ of the body:

$$\mathbf{H}_B = \mathbf{H}_A + \mathbf{r}_{A/B} \times m\bar{\mathbf{v}}$$

(*b*) Further show that when a rigid body rotates about a fixed axis, its angular momentum is the same about any two points A and B located on the fixed axis ($\mathbf{H}_A = \mathbf{H}_B$) if, and only if, the mass center G of the body is located on the fixed axis.

Fig. P18.15

18.15 A 5-kg rod of uniform cross section is used to form the shaft shown. Knowing that the shaft rotates with a constant angular velocity $\boldsymbol{\omega}$ of magnitude 12 rad/s, determine (*a*) the angular momentum $\mathbf{H}_G$ of the shaft about its mass center G, (*b*) the angle formed by $\mathbf{H}_G$ and the axis AB.

18.16 Determine the angular momentum of the shaft of Prob. 18.15 about (*a*) point A, (*b*) point B.

18.17 Two L-shaped arms, each weighing 4 lb, are welded at the third points of the 2-ft shaft AB. Knowing that shaft AB rotates at the constant rate $\omega = 240$ rpm, determine (*a*) the angular momentum of the body about A, (*b*) the angle formed by the angular momentum and shaft AB.

18.18 For the body of Prob. 18.17, determine (*a*) the angular momentum about B, (*b*) the angle formed by the angular momentum about shaft BA.

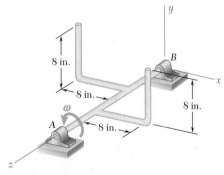

Fig. P18.17

18.19 The triangular plate shown has a mass of 7.5 kg and is welded to a vertical shaft AB. Knowing that the plate rotates at the constant rate $\omega = 12$ rad/s, determine its angular momentum about (*a*) point C, (*b*) point A. (*Hint:* To solve part *b*, find $\bar{\mathbf{v}}$ and use the property indicated in part *a* of Prob. 18.14.)

18.20 The triangular plate shown has a mass of 7.5 kg and is welded to a vertical shaft AB. Knowing that the plate rotates at the constant rate $\omega = 12$ rad/s, determine its angular momentum about (*a*) point C, (*b*) point B. (See hint of Prob. 18.19.)

18.21 One of the sculptures displayed on a university campus consists of a hollow cube made of six aluminum sheets, each 1.5×1.5 m, welded together and reinforced with internal braces of negligible weight. The cube is mounted on a fixed base at A and can rotate freely about its vertical diagonal AB. As she passes by this display on the way to a class in mechanics, an engineering student grabs corner C of the cube and pushes it for 1.2 s in a direction perpendicular to the plane ABC with an average force of 50 N. Having observed that it takes 5 s for the cube to complete one full revolution, she flips out her calculator and proceeds to determine the mass of the cube. What is the result of her calculation? (*Hint:* The perpendicular distance from the diagonal joining two vertices of a cube to any of its other six vertices can be obtained by multiplying the side of the cube by $\sqrt{2/3}$.)

Fig. P18.19 and P18.20

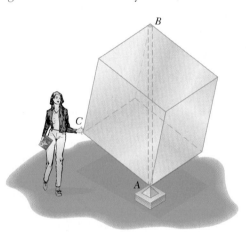

Fig. P18.21

18.22 If the aluminum cube of Prob. 18.21 were replaced by a cube of the same size, made of six plywood sheets with mass 8 kg each, how long would it take for that cube to complete one full revolution if the student pushed its corner C in the same way that she pushed the corner of the aluminum cube?

18.23 A uniform rod of total mass m is bent into the shape shown and is suspended by a wire attached at B. The bent rod is hit at D in a direction perpendicular to the plane containing the rod (in the negative z direction). Denoting the corresponding impulse by $\mathbf{F}\Delta t$, determine (*a*) the velocity of the mass center of the rod, (*b*) the angular velocity of the rod.

18.24 Solve Prob. 18.23, assuming that the bent rod is hit at C.

Fig. P18.23

18.25 Three slender rods, each of mass m and length $2a$, are welded together to form the assembly shown. The assembly is hit at A in a vertical downward direction. Denoting the corresponding impulse by $\mathbf{F}\,\Delta t$, determine immediately after the impact (a) the velocity of the mass center G, (b) the angular velocity of the rod.

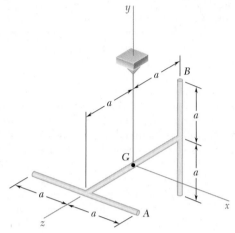

Fig. P18.25

18.26 Solve Prob. 18.25, assuming that the assembly is hit at B in a direction opposite to that of the x axis.

18.27 Two circular plates, each of mass 4 kg, are rigidly connected by a rod AB of negligible mass and are suspended from point A as shown. Knowing that an impulse $\mathbf{F}\,\Delta t = -(2.4 \text{ N} \cdot \text{s})\mathbf{k}$ is applied at point D, determine (a) the velocity of the mass center G of the assembly, (b) the angular velocity of the assembly.

18.28 Two circular plates, each of mass 4 kg, are rigidly connected by a rod AB of negligible mass and are suspended from point A as shown. Knowing that an impulse $\mathbf{F}\,\Delta t = (2.4 \text{ N} \cdot \text{s})\mathbf{j}$ is applied at point D, determine (a) the velocity of the mass center G of the assembly, (b) the angular velocity of the assembly.

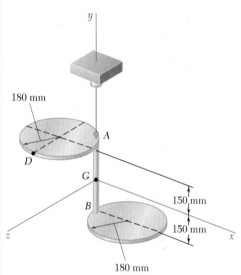

Fig. P18.27 and P18.28

18.29 A circular plate of mass m is falling with a velocity $\overline{\mathbf{v}}_0$ and no angular velocity when its edge C strikes an obstruction. A line passing the origin and parallel to the line CG makes a $45°$ angle with the x-axis. Assuming the impact to be perfectly plastic ($e = 0$), determine the angular velocity of the plate immediately after the impact.

Fig. P18.29

18.30 For the plate of Prob. 18.29, determine (a) the velocity of its mass center G immediately after the impact, (b) the impulse exerted on the plate by the obstruction during the impact.

18.31 A square plate of side a and mass m supported by a ball-and-socket joint at A is rotating about the y axis with a constant angular velocity $\boldsymbol{\omega} = \omega_0\mathbf{j}$ when an obstruction is suddenly introduced at B in the xy plane. Assuming the impact at B to be perfectly plastic ($e = 0$), determine immediately after the impact (a) the angular velocity of the plate, (b) the velocity of its mass center G.

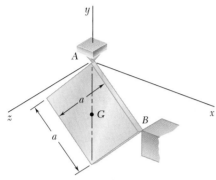

Fig. P18.31

18.32 Determine the impulse exerted on the plate of Prob. 18.31 during the impact by (a) the obstruction at B, (b) the support at A.

18.33 The coordinate axes shown represent the principal centroidal axes of inertia of a 3000-lb space probe whose radii of gyration are $k_x = 1.375$ ft, $k_y = 1.425$ ft, and $k_z = 1.250$ ft. The probe has no angular velocity when a 5-oz meteorite strikes one of its solar panels at point A with a velocity $\mathbf{v}_0 = (2400 \text{ ft/s})\mathbf{i} - (3000 \text{ ft/s})\mathbf{j} + (3200 \text{ ft/s})\mathbf{k}$ relative to the probe. Knowing that the meteorite emerges on the other side of the panel with no change in the direction of its velocity, but with a speed reduced by 20 percent, determine the final angular velocity of the probe.

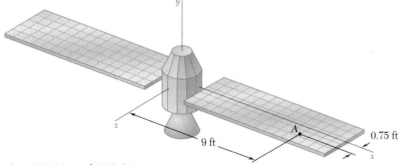

Fig. P18.33 and P18.34

18.34 The coordinate axes shown represent the principal centroidal axes of inertia of a 3000-lb space probe whose radii of gyration are $k_x = 1.375$ ft, $k_y = 1.425$ ft, and $k_z = 1.250$ ft. The probe has no angular velocity when a 5-oz meteorite strikes one of its solar panels at point A and emerges on the other side of the panel with no change in the direction of its velocity, but with a speed reduced by 25 percent. Knowing that the final angular velocity of the probe is $\boldsymbol{\omega} = (0.05 \text{ rad/s})\mathbf{i} - (0.12 \text{ rad/s})\mathbf{j} + \omega_z\mathbf{k}$ and that the x component of the resulting change in the velocity of the mass center of the probe is -0.675 in./s, determine (a) the component ω_z of the final angular velocity of the probe, (b) the relative velocity $\mathbf{v}_0$ with which the meteorite strikes the panel.

Fig. P18.35

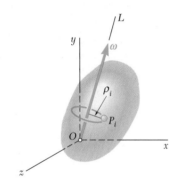

Fig. P18.38

18.35 A 2500-kg probe in orbit about the moon is 2.4 m high and has octagonal bases of sides 1.2 m. The coordinate axes shown are the principal centroidal axes of inertia of the probe, and its radii of gyration are $k_x = 0.98$ m, $k_y = 1.06$ m, and $k_z = 1.02$ m. The probe is equipped with a main 500-N thruster E and with four 20-N thrusters A, B, C, and D which can expel fuel in the positive y direction. The probe has an angular velocity $\boldsymbol{\omega} = (0.040 \text{ rad/s})\mathbf{i} + (0.060 \text{ rad/s})\mathbf{k}$ when two of the 20-N thrusters are used to reduce the angular velocity to zero. Determine (a) which of the thrusters should be used, (b) the operating time of each of these thrusters, (c) for how long the main thruster E should be activated if the velocity of the mass center of the probe is to remain unchanged.

18.36 Solve Prob. 18.35, assuming that the angular velocity of the probe is $\boldsymbol{\omega} = (0.060 \text{ rad/s})\mathbf{i} - (0.040 \text{ rad/s})\mathbf{k}$.

18.37 Denoting, respectively, by $\boldsymbol{\omega}$, $\mathbf{H}_O$, and T the angular velocity, the angular momentum, and the kinetic energy of a rigid body with a fixed point O, (a) prove that $\mathbf{H}_O \cdot \boldsymbol{\omega} = 2T$; (b) show that the angle θ between $\boldsymbol{\omega}$ and $\mathbf{H}_O$ will always be acute.

18.38 Show that the kinetic energy of a rigid body with a fixed point O can be expressed as $T = \frac{1}{2} I_{OL} \omega^2$, where $\boldsymbol{\omega}$ is the instantaneous angular velocity of the body and I_{OL} is its moment of inertia about the line of action OL of $\boldsymbol{\omega}$. Derive this expression (a) from Eqs. (9.46) and (18.19), (b) by considering T as the sum of the kinetic energies of particles P_i describing circles of radius ρ_i about line OL.

18.39 Determine the kinetic energy of the disk of Prob. 18.1.

18.40 Determine the kinetic energy of the plate of Prob. 18.2.

18.41 Determine the kinetic energy of the assembly of Prob. 18.3.

18.42 Determine the kinetic energy of the disk of Prob. 18.4.

18.43 Determine the kinetic energy of the disk of Prob. 18.5.

18.44 Determine the kinetic energy of the solid parallelepiped of Prob. 18.6.

18.45 Determine the kinetic energy of the hollow parallelepiped of Prob. 18.7.

18.46 Determine the kinetic energy of the disk of Prob. 18.8.

18.47 Determine the kinetic energy of the shaft of Prob. 18.15.

18.48 Determine the kinetic energy of the body of Prob. 18.17.

18.49 Determine the kinetic energy of the triangular plate of Prob. 18.19.

18.50 Determine the kinetic energy imparted to the cube of Prob. 18.21.

18.51 Determine the kinetic energy lost when edge C of the plate of Prob. 18.29 hits the obstruction.

18.52 Determine the kinetic energy lost when the plate of Prob. 18.31 hits the obstruction at B.

18.53 Determine the kinetic energy of the space probe of Prob. 18.33 in its motion about its mass center after its collision with the meteorite.

18.54 Determine the kinetic energy of the space probe of Prob. 18.34 in its motion about its mass center after its collision with the meteorite.

*18.5 MOTION OF A RIGID BODY IN THREE DIMENSIONS

As indicated in Sec. 18.2, the fundamental equations

$$\Sigma \mathbf{F} = m\bar{\mathbf{a}} \tag{18.1}$$

$$\Sigma \mathbf{M}_G = \dot{\mathbf{H}}_G \tag{18.2}$$

remain valid in the most general case of the motion of a rigid body. Before Eq. (18.2) could be applied to the three-dimensional motion of a rigid body, however, it was necessary to derive Eqs. (18.7), which relate the components of the angular momentum $\mathbf{H}_G$ and those of the angular velocity $\boldsymbol{\omega}$. It still remains for us to find an effective and convenient way for computing the components of the derivative $\dot{\mathbf{H}}_G$ of the angular momentum.

Since $\mathbf{H}_G$ represents the angular momentum of the body in its motion relative to centroidal axes $GX'Y'Z'$ of fixed orientation (Fig. 18.9), and since $\dot{\mathbf{H}}_G$ represents the rate of change of $\mathbf{H}_G$ with respect to the same axes, it would seem natural to use components of $\boldsymbol{\omega}$ and $\mathbf{H}_G$ along the axes X', Y', Z' in writing the relations (18.7). But since the body rotates, its moments and products of inertia would change continually, and it would be necessary to determine their values as functions of the time. It is therefore more convenient to use axes x, y, z attached to the body, ensuring that its moments and products of inertia will maintain the same values during the motion. This is permissible since, as indicated earlier, the transformation of $\boldsymbol{\omega}$ into $\mathbf{H}_G$ is independent of the system of coordinate axes selected. The angular velocity $\boldsymbol{\omega}$, however, should still be *defined* with respect to the frame $GX'Y'Z'$ of fixed orientation. The vector $\boldsymbol{\omega}$ may then be *resolved* into components along the rotating x, y, and z axes. Applying the relations (18.7), we obtain the *components* of the vector $\mathbf{H}_G$ along the rotating axes. The vector $\mathbf{H}_G$, however, represents the angular momentum about G of the body *in its motion relative to the frame $GX'Y'Z'$.*

Differentiating with respect to t the components of the angular momentum in (18.7), we define the rate of change of the vector $\mathbf{H}_G$ with respect to the rotating frame $Gxyz$:

$$(\dot{\mathbf{H}}_G)_{Gxyz} = \dot{H}_x \mathbf{i} + \dot{H}_y \mathbf{j} + \dot{H}_z \mathbf{k} \tag{18.21}$$

where $\mathbf{i}$, $\mathbf{j}$, $\mathbf{k}$ are the unit vectors along the rotating axes. Recalling from Sec. 15.10 that the rate of change $\dot{\mathbf{H}}_G$ of the vector $\mathbf{H}_G$ with respect to the frame $GX'Y'Z'$ is found by adding to $(\dot{\mathbf{H}}_G)_{Gxyz}$ the vector product $\boldsymbol{\Omega} \times \mathbf{H}_G$, where $\boldsymbol{\Omega}$ denotes the angular velocity of the rotating frame, we write

$$\dot{\mathbf{H}}_G = (\dot{\mathbf{H}}_G)_{Gxyz} + \boldsymbol{\Omega} \times \mathrm{H}_G \tag{18.22}$$

where $\mathbf{H}_G$ = angular momentum of body with respect to frame $GX'Y'Z'$ of fixed orientation

$(\dot{\mathbf{H}}_G)_{Gxyz}$ = rate of change of $\mathbf{H}_G$ with respect to rotating frame $Gxyz$, to be computed from the relations (18.7) and (18.21)

$\boldsymbol{\Omega}$ = angular velocity of rotating frame $Gxyz$

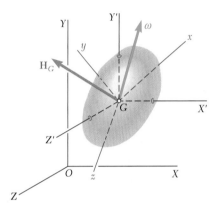

Fig. 18.9

Substituting for $\dot{\mathbf{H}}_G$ from (18.22) into (18.2), we have

$$\Sigma \mathbf{M}_G = (\dot{\mathbf{H}}_G)_{Gxyz} + \mathbf{\Omega} \times \mathbf{H}_G \qquad (18.23)$$

If the rotating frame is attached to the body, as has been assumed in this discussion, its angular velocity $\mathbf{\Omega}$ is identically equal to the angular velocity $\boldsymbol{\omega}$ of the body. There are many applications, however, where it is advantageous to use a frame of reference which is not actually attached to the body but rotates in an independent manner. For example, if the body considered is axisymmetrical, as in Sample Prob. 18.5 or Sec. 18.9, it is possible to select a frame of reference with respect to which the moments and products of inertia of the body remain constant, but which rotates less than the body itself.† As a result, it is possible to obtain simpler expressions for the angular velocity $\boldsymbol{\omega}$ and the angular momentum $\mathbf{H}_G$ of the body than could have been obtained if the frame of reference had actually been attached to the body. It is clear that in such cases the angular velocity $\mathbf{\Omega}$ of the rotating frame and the angular velocity $\boldsymbol{\omega}$ of the body are different.

*18.6 EULER'S EQUATIONS OF MOTION. EXTENSION OF D'ALEMBERT'S PRINCIPLE TO THE MOTION OF A RIGID BODY IN THREE DIMENSIONS

If the x, y, and z axes are chosen to coincide with the principal axes of inertia of the body, the simplified relations (18.10) can be used to determine the components of the angular momentum $\mathbf{H}_G$. Omitting the primes from the subscripts, we write

$$\mathbf{H}_G = \bar{I}_x \omega_x \mathbf{i} + \bar{I}_y \omega_y \mathbf{j} + \bar{I}_z \omega_z \mathbf{k} \qquad (18.24)$$

where $\bar{I}_x$, $\bar{I}_y$, and $\bar{I}_z$ denote the principal centroidal moments of inertia of the body. Substituting for $\mathbf{H}_G$ from (18.24) into (18.23) and setting $\mathbf{\Omega} = \boldsymbol{\omega}$, we obtain the three scalar equations

$$\begin{aligned}
\Sigma M_x &= \bar{I}_x \dot{\omega}_x - (\bar{I}_y - \bar{I}_z)\omega_y \omega_z \\
\Sigma M_y &= \bar{I}_y \dot{\omega}_y - (\bar{I}_z - \bar{I}_x)\omega_z \omega_x \\
\Sigma M_z &= \bar{I}_z \dot{\omega}_z - (\bar{I}_x - \bar{I}_y)\omega_x \omega_y
\end{aligned} \qquad (18.25)$$

These equations, called *Euler's equations of motion* after the Swiss mathematician Leonhard Euler (1707–1783), can be used to analyze the motion of a rigid body about its mass center. In the following sections, however, Eq. (18.23) will be used in preference to Eqs. (18.25), since the former is more general and the compact vectorial form in which it is expressed is easier to remember.

Writing Eq. (18.1) in scalar form, we obtain the three additional equations

$$\Sigma F_x = m\bar{a}_x \qquad \Sigma F_y = m\bar{a}_y \qquad \Sigma F_z = m\bar{a}_z \qquad (18.26)$$

which, together with Euler's equations, form a system of six differential equations. Given appropriate initial conditions, these differential

---

†More specifically, the frame of reference will have no spin (see Sec. 18.9).

equations have a unique solution. Thus, the motion of a rigid body in three dimensions is completely defined by the resultant and the moment resultant of the external forces acting on it. This result will be recognized as a generalization of a similar result obtained in Sec. 16.4 in the case of the plane motion of a rigid slab. It follows that in three as well as two dimensions, two systems of forces which are equipollent are also equivalent; that is, they have the same effect on a given rigid body.

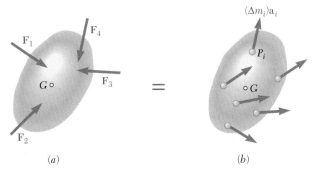

Fig. 18.10

Considering in particular the system of the external forces acting on a rigid body (Fig. 18.10a) and the system of the effective forces associated with the particles forming the rigid body (Fig. 18.10b), we can state that the two systems—which were shown in Sec. 14.2 to be equipollent—are also equivalent. This is the extension of d'Alembert's principle to the three-dimensional motion of a rigid body. Replacing the effective forces in Fig. 18.10b by an equivalent force-couple system, we verify that the system of the external forces acting on a rigid body in three-dimensional motion is equivalent to the system consisting of the vector $m\bar{\mathbf{a}}$ attached at the mass center G of the body and the couple of moment $\dot{\mathbf{H}}_G$ (Fig. 18.11), where $\dot{\mathbf{H}}_G$ is obtained from the relations (18.7) and (18.22). Note that the equivalence of the systems of vectors shown in Fig. 18.10 and in Fig. 18.11 has been indicated by *red* equals signs. Problems involving the three-dimensional motion of a rigid body can be solved by considering the free-body-diagram equation represented in Fig. 18.11 and writing appropriate scalar equations relating the components or moments of the external and effective forces (see Sample Prob. 18.3).

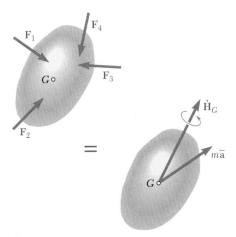

Fig. 18.11

*18.7 MOTION OF A RIGID BODY ABOUT A FIXED POINT

When a rigid body is constrained to rotate about a fixed point O, it is desirable to write an equation involving the moments about O of the external and effective forces, since this equation will not contain the unknown reaction at O. While such an equation can be obtained from Fig. 18.11, it may be more convenient to write it by considering the rate of change of the angular momentum $\mathbf{H}_O$ of the body about the fixed point O (Fig. 18.12). Recalling Eq. (14.11), we write

$$\Sigma \mathbf{M}_O = \dot{\mathbf{H}}_O \qquad (18.27)$$

where $\dot{\mathbf{H}}_O$ denotes the rate of change of the vector $\mathbf{H}_O$ with respect to the fixed frame $OXYZ$. A derivation similar to that used in Sec. 18.5

Fig. 18.12

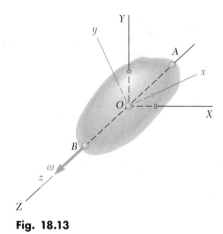

Fig. 18.13

enables us to relate $\dot{\mathbf{H}}_O$ to the rate of change $(\dot{\mathbf{H}}_O)_{Oxyz}$ of $\mathbf{H}_O$ with respect to the rotating frame $Oxyz$. Substitution into (18.27) leads to the equation

$$\Sigma\mathbf{M}_O = (\dot{\mathbf{H}}_O)_{Oxyz} + \boldsymbol{\Omega} \times \mathbf{H}_O \qquad (18.28)$$

where $\Sigma\mathbf{M}_O$ = sum of moments about O of forces applied to rigid body

$\mathbf{H}_O$ = angular momentum of body with respect to fixed frame $OXYZ$

$(\dot{\mathbf{H}}_O)_{Oxyz}$ = rate of change of $\mathbf{H}_O$ with respect to rotating frame $Oxyz$, to be computed from relations (18.13)

$\boldsymbol{\Omega}$ = angular velocity of rotating frame $Oxyz$

If the rotating frame is attached to the body, its angular velocity $\boldsymbol{\Omega}$ is identically equal to the angular velocity $\boldsymbol{\omega}$ of the body. However, as indicated in the last paragraph of Sec. 18.5, there are many applications where it is advantageous to use a frame of reference which is not actually attached to the body but rotates in an independent manner.

*18.8 ROTATION OF A RIGID BODY ABOUT A FIXED AXIS

Equation (18.28), which was derived in the preceding section, will be used to analyze the motion of a rigid body constrained to rotate about a fixed axis AB (Fig. 18.13). First, we note that the angular velocity of the body with respect to the fixed frame $OXYZ$ is represented by the vector $\boldsymbol{\omega}$ directed along the axis of rotation. Attaching the moving frame of reference $Oxyz$ to the body, with the z axis along AB, we have $\boldsymbol{\omega} = \omega\mathbf{k}$. Substituting $\omega_x = 0$, $\omega_y = 0$, $\omega_z = \omega$ into the relations (18.13), we obtain the components along the rotating axes of the angular momentum $\mathbf{H}_O$ of the body about O:

$$H_x = -I_{xz}\omega \qquad H_y = -I_{yz}\omega \qquad H_z = I_z\omega$$

Since the frame $Oxyz$ is attached to the body, we have $\boldsymbol{\Omega} = \boldsymbol{\omega}$ and Eq. (18.28) yields

$$\begin{aligned}
\Sigma\mathbf{M}_O &= (\dot{\mathbf{H}}_O)_{Oxyz} + \boldsymbol{\omega} \times \mathbf{H}_O \\
&= (-I_{xz}\mathbf{i} - I_{yz}\mathbf{j} + I_z\mathbf{k})\dot{\omega} + \omega\mathbf{k} \times (-I_{xz}\mathbf{i} - I_{yz}\mathbf{j} + I_z\mathbf{k})\omega \\
&= (-I_{xz}\mathbf{i} - I_{yz}\mathbf{j} + I_z\mathbf{k})\alpha + (-I_{xz}\mathbf{j} + I_{yz}\mathbf{i})\omega^2
\end{aligned}$$

The result obtained can be expressed by the three scalar equations

$$\begin{aligned}
\Sigma M_x &= -I_{xz}\alpha + I_{yz}\omega^2 \\
\Sigma M_y &= -I_{yz}\alpha - I_{xz}\omega^2 \qquad (18.29) \\
\Sigma M_z &= I_z\alpha
\end{aligned}$$

When the forces applied to the body are known, the angular acceleration α can be obtained from the last of Eqs. (18.29). The angular velocity ω is then determined by integration and the values obtained for α and ω substituted into the first two equations (18.29). These equations plus the three equations (18.26) which define the motion of the mass center of the body can then be used to determine the reactions at the bearings A and B.

It is possible to select axes other than the ones shown in Fig. 18.13 to analyze the rotation of a rigid body about a fixed axis. In many cases, the principal axes of inertia of the body will be found more advantageous. It is therefore wise to revert to Eq. (18.28) and to select the system of axes which best fits the problem under consideration.

If the rotating body is symmetrical with respect to the xy plane, the products of inertia I_{xz} and I_{yz} are equal to zero and Eqs. (18.29) reduce to

$$\Sigma M_x = 0 \qquad \Sigma M_y = 0 \qquad \Sigma M_z = I_z\alpha \qquad (18.30)$$

which is in accord with the results obtained in Chap. 16. If, on the other hand, the products of inertia I_{xz} and I_{yz} are different from zero, the sum of the moments of the external forces about the x and y axes will also be different from zero, even when the body rotates at a constant rate ω. Indeed, in the latter case, Eqs. (18.29) yield

$$\Sigma M_x = I_{yz}\omega^2 \qquad \Sigma M_y = -I_{xz}\omega^2 \qquad \Sigma M_z = 0 \qquad (18.31)$$

This last observation leads us to discuss the *balancing of rotating shafts.* Consider, for instance, the crankshaft shown in Fig. 18.14a, which is symmetrical about its mass center G. We first observe that when the crankshaft is at rest, it exerts no lateral thrust on its supports, since its center of gravity G is located directly above A. The shaft is said to be *statically balanced.* The reaction at A, often referred to as a *static reaction,* is vertical and its magnitude is equal to the weight W of the shaft. Let us now assume that the shaft rotates with a constant angular velocity $\boldsymbol{\omega}$. Attaching our frame of reference to the shaft, with its origin at G, the z axis along AB, and the y axis in the plane of symmetry of the shaft (Fig. 18.14b), we note that I_{xz} is zero and that I_{yz} is positive. According to Eqs. (18.31), the external forces include a couple of moment $I_{yz}\omega^2\mathbf{i}$. Since this couple is formed by the reaction at B and the horizontal component of the reaction at A, we have

$$\mathbf{A}_y = \frac{I_{yz}\omega^2}{l}\mathbf{j} \qquad \mathbf{B} = -\frac{I_{yz}\omega^2}{l}\mathbf{j} \qquad (18.32)$$

Since the bearing reactions are proportional to ω^2, the shaft will have a tendency to tear away from its bearings when rotating at high speeds. Moreover, since the bearing reactions $\mathbf{A}_y$ and $\mathbf{B}$, called *dynamic reactions,* are contained in the yz plane, they rotate with the shaft and cause the structure supporting it to vibrate. These undesirable effects will be avoided if, by rearranging the distribution of mass around the shaft or by adding corrective masses, we let I_{yz} become equal to zero. The dynamic reactions $\mathbf{A}_y$ and $\mathbf{B}$ will vanish and the reactions at the bearings will reduce to the static reaction $\mathbf{A}_z$, the direction of which is fixed. The shaft will then be *dynamically as well as statically balanced.*

Photo 18.4 The forces exerted by a rotating automobile crankshaft on its bearings are the static and dynamic reactions. The crankshaft can be designed to be dynamically as well as statically balanced.

(a)

(b)

Fig. 18.14

SAMPLE PROBLEM 18.3

A slender rod AB of length $L = 8$ ft and weight $W = 40$ lb is pinned at A to a vertical axle DE which rotates with a constant angular velocity $\boldsymbol{\omega}$ of 15 rad/s. The rod is maintained in position by means of a horizontal wire BC attached to the axle and to the end B of the rod. Determine the tension in the wire and the reaction at A.

SOLUTION

The effective forces reduce to the vector $m\bar{\mathbf{a}}$ attached at G and the couple $\dot{\mathbf{H}}_G$. Since G describes a horizontal circle of radius $\bar{r} = \frac{1}{2}L\cos\beta$ at the constant rate ω, we have

$$\bar{\mathbf{a}} = \mathbf{a}_n = -\bar{r}\omega^2\mathbf{I} = -(\tfrac{1}{2}L\cos\beta)\omega^2\mathbf{I} = -(450 \text{ ft/s}^2)\mathbf{I}$$
$$m\bar{\mathbf{a}} = \frac{40}{g}(-450\mathbf{I}) = -(559 \text{ lb})\mathbf{I}$$

Determination of $\dot{\mathbf{H}}_G$. We first compute the angular momentum $\mathbf{H}_G$. Using the principal centroidal axes of inertia x, y, z, we write

$$\bar{I}_x = \tfrac{1}{12}mL^2 \qquad \bar{I}_y = 0 \qquad \bar{I}_z = \tfrac{1}{12}mL^2$$
$$\omega_x = -\omega\cos\beta \qquad \omega_y = \omega\sin\beta \qquad \omega_z = 0$$
$$\mathbf{H}_G = \bar{I}_x\omega_x\mathbf{i} + \bar{I}_y\omega_y\mathbf{j} + \bar{I}_z\omega_z\mathbf{k}$$
$$\mathbf{H}_G = -\tfrac{1}{12}mL^2\omega\cos\beta\,\mathbf{i}$$

The rate of change $\dot{\mathbf{H}}_G$ of $\mathbf{H}_G$ with respect to axes of fixed orientation is obtained from Eq. (18.22). Observing that the rate of change $(\dot{\mathbf{H}}_G)_{Gxyz}$ of $\mathbf{H}_G$ with respect to the rotating frame $Gxyz$ is zero, and that the angular velocity $\boldsymbol{\Omega}$ of that frame is equal to the angular velocity $\boldsymbol{\omega}$ of the rod, we have

$$\dot{\mathbf{H}}_G = (\dot{\mathbf{H}}_G)_{Gxyz} + \boldsymbol{\omega} \times \mathbf{H}_G$$
$$\dot{\mathbf{H}}_G = 0 + (-\omega\cos\beta\,\mathbf{i} + \omega\sin\beta\,\mathbf{j}) \times (-\tfrac{1}{12}mL^2\omega\cos\beta\,\mathbf{i})$$
$$\dot{\mathbf{H}}_G = \tfrac{1}{12}mL^2\omega^2\sin\beta\cos\beta\,\mathbf{k} = (645 \text{ lb} \cdot \text{ft})\,\mathbf{k}$$

Equations of Motion. Expressing that the system of the external forces is equivalent to the system of the effective forces, we write

$$\Sigma\mathbf{M}_A = \Sigma(\mathbf{M}_A)_{\text{eff}}:$$
$$6.93\mathbf{J} \times (-T\mathbf{I}) + 2\mathbf{I} \times (-40\mathbf{J}) = 3.46\mathbf{J} \times (-559\mathbf{I}) + 645\mathbf{K}$$
$$(6.93T - 80)\mathbf{K} = (1934 + 645)\mathbf{K} \qquad T = 384 \text{ lb} \quad \blacktriangleleft$$
$$\Sigma\mathbf{F} = \Sigma\mathbf{F}_{\text{eff}}: \qquad A_X\mathbf{I} + A_Y\mathbf{J} + A_Z\mathbf{K} - 384\mathbf{I} - 40\mathbf{J} = -559\mathbf{I}$$
$$\mathbf{A} = -(175 \text{ lb})\mathbf{I} + (40 \text{ lb})\mathbf{J} \quad \blacktriangleleft$$

Remark. The value of T could have been obtained from $\mathbf{H}_A$ and Eq. (18.28). However, the method used here also yields the reaction at A. Moreover, it draws attention to the effect of the asymmetry of the rod on the solution of the problem by clearly showing that both the vector $m\bar{\mathbf{a}}$ and the couple $\dot{\mathbf{H}}_G$ must be used to represent the effective forces.

SAMPLE PROBLEM 18.4

Two 100-mm rods A and B, each of mass 300 g, are welded to shaft CD which is supported by bearings at C and D. If a couple $\mathbf{M}$ of magnitude equal to 6 N · m is applied to the shaft, determine the components of the dynamic reactions at C and D at the instant when the shaft has reached an angular velocity at 1200 rpm. Neglect the moment of inertia of the shaft itself.

SOLUTION

Angular Momentum About O. We attach to the body the frame of reference $Oxyz$ and note that the axes chosen are not principal axes of inertia for the body. Since the body rotates about the x axis, we have $\omega_x = \omega$ and $\omega_y = \omega_z = 0$. Substituting into Eqs. (18.13),

$$H_x = I_x\omega \qquad H_y = -I_{xy}\omega \qquad H_z = -I_{xz}\omega$$
$$\mathbf{H}_O = (I_x\mathbf{i} - I_{xy}\mathbf{j} - I_{xz}\mathbf{k})\omega$$

Moments of the External Forces About O. Since the frame of reference rotates with the angular velocity $\boldsymbol{\omega}$, Eq. (18.28) yields

$$\begin{aligned}
\Sigma\mathbf{M}_O &= (\dot{\mathbf{H}}_O)_{Oxyz} + \boldsymbol{\omega} \times \mathbf{H}_O \\
&= (I_x\mathbf{i} - I_{xy}\mathbf{j} - I_{xz}\mathbf{k})\alpha + \omega\mathbf{i} \times (I_x\mathbf{i} - I_{xy}\mathbf{j} - I_{xz}\mathbf{k})\omega \\
&= I_x\alpha\mathbf{i} - (I_{xy}\alpha - I_{xz}\omega^2)\mathbf{j} - (I_{xz}\alpha + I_{xy}\omega^2)\mathbf{k} \quad (1)
\end{aligned}$$

Dynamic Reaction at D. The external forces consist of the weights of the shaft and rods, the couple $\mathbf{M}$, the static reactions at C and D, and the dynamic reactions at C and D. Since the weights and static reactions are balanced, the external forces reduce to the couple $\mathbf{M}$ and the dynamic reactions $\mathbf{C}$ and $\mathbf{D}$ as shown in the figure. Taking moments about O, we have

$$\Sigma\mathbf{M}_O = L\mathbf{i} \times (D_y\mathbf{j} + D_z\mathbf{k}) + M\mathbf{i} = M\mathbf{i} - D_zL\mathbf{j} + D_yL\mathbf{k} \quad (2)$$

Equating the coefficients of the unit vector $\mathbf{i}$ in (1) and (2),

$$M = I_x\alpha \qquad M = 2(\tfrac{1}{3}mc^2)\alpha \qquad \alpha = 3M/2mc^2$$

Equating the coefficients of $\mathbf{k}$ and $\mathbf{j}$ in (1) and (2):

$$D_y = -(I_{xz}\alpha + I_{xy}\omega^2)/L \qquad D_z = (I_{xy}\alpha - I_{xz}\omega^2)/L \quad (3)$$

Using the parallel-axis theorem, and noting that the product of inertia of each rod is zero with respect to centroidal axes, we have

$$I_{xy} = \Sigma m\bar{x}\bar{y} = m(\tfrac{1}{2}L)(\tfrac{1}{2}c) = \tfrac{1}{4}mLc$$
$$I_{xz} = \Sigma m\bar{x}\bar{z} = m(\tfrac{1}{4}L)(\tfrac{1}{2}c) = \tfrac{1}{8}mLc$$

Substituting into (3) the values found for I_{xy}, I_{xz}, and α:

$$D_y = -\tfrac{3}{16}(M/c) - \tfrac{1}{4}mc\omega^2 \qquad D_z = \tfrac{3}{8}(M/c) - \tfrac{1}{8}mc\omega^2$$

Substituting $\omega = 1200$ rpm $= 125.7$ rad/s, $c = 0.100$ m, $M = 6$ N · m, and $m = 0.300$ kg, we have

$$D_y = -129.8 \text{ N} \qquad D_z = -36.8 \text{ N} \quad \blacktriangleleft$$

Dynamic Reaction at C. Using a frame of reference attached at D, we obtain equations similar to Eqs. (3), which yield

$$C_y = -152.2 \text{ N} \qquad C_z = -155.2 \text{ N} \quad \blacktriangleleft$$

SAMPLE PROBLEM 18.5

A homogeneous disk of radius r and mass m is mounted on an axle OG of length L and negligible mass. The axle is pivoted at the fixed point O and the disk is constrained to roll on a horizontal floor. Knowing that the disk rotates counterclockwise at the constant rate ω_1 about the axle, determine (a) the force (assumed vertical) exerted by the floor on the disk, (b) the reaction at the pivot O.

SOLUTION

The effective forces reduce to the vector $m\bar{\mathbf{a}}$ attached at G and the couple $\dot{\mathbf{H}}_G$. Recalling from Sample Prob. 18.2 that the axle rotates about the y axis at the rate $\omega_2 = r\omega_1/L$, we write

$$m\bar{\mathbf{a}} = -mL\omega_2^2\mathbf{i} = -mL(r\omega_1/L)^2\mathbf{i} = -(mr^2\omega_1^2/L)\mathbf{i} \quad (1)$$

Determination of $\dot{\mathbf{H}}_G$. We recall from Sample Prob. 18.2 that the angular momentum of the disk about G is

$$\mathbf{H}_G = \tfrac{1}{2}mr^2\omega_1\left(\mathbf{i} - \frac{r}{2L}\mathbf{j}\right)$$

where $\mathbf{H}_G$ is resolved into components along the rotating axes x', y', z', with x' along OG and y' vertical. The rate of change $\dot{\mathbf{H}}_G$ of $\mathbf{H}_G$ with respect to axes of fixed orientation is obtained from Eq. (18.22). Noting that the rate of change $(\dot{\mathbf{H}}_G)_{Gx'y'z'}$ of $\mathbf{H}_G$ with respect to the rotating frame is zero, and that the angular velocity $\boldsymbol{\Omega}$ of that frame is

$$\boldsymbol{\Omega} = -\omega_2\mathbf{j} = -\frac{r\omega_1}{L}\mathbf{j}$$

we have

$$\begin{aligned}
\dot{\mathbf{H}}_G &= (\dot{\mathbf{H}}_G)_{Gx'y'z'} + \boldsymbol{\Omega} \times \mathbf{H}_G \\
&= 0 - \frac{r\omega_1}{L}\mathbf{j} \times \tfrac{1}{2}mr^2\omega_1\left(\mathbf{i} - \frac{r}{2L}\mathbf{j}\right) \\
&= \tfrac{1}{2}mr^2(r/L)\omega_1^2\mathbf{k} \quad (2)
\end{aligned}$$

Equations of Motion. Expressing that the system of the external forces is equivalent to the system of the effective forces, we write

$$\Sigma\mathbf{M}_O = \Sigma(\mathbf{M}_O)_{\text{eff}}: \qquad L\mathbf{i} \times (N\mathbf{j} - W\mathbf{j}) = \dot{\mathbf{H}}_G$$
$$(N - W)L\mathbf{k} = \tfrac{1}{2}mr^2(r/L)\omega_1^2\mathbf{k}$$
$$N = W + \tfrac{1}{2}mr(r/L)^2\omega_1^2 \qquad \mathbf{N} = [W + \tfrac{1}{2}mr(r/L)^2\omega_1^2]\mathbf{j} \quad (3) \blacktriangleleft$$
$$\Sigma\mathbf{F} = \Sigma\mathbf{F}_{\text{eff}}: \qquad \mathbf{R} + N\mathbf{j} - W\mathbf{j} = m\bar{\mathbf{a}}$$

Substituting for N from (3), for $m\bar{\mathbf{a}}$ from (1), and solving for $\mathbf{R}$, we have

$$\mathbf{R} = -(mr^2\omega_1^2/L)\mathbf{i} - \tfrac{1}{2}mr(r/L)^2\omega_1^2\mathbf{j}$$

$$\mathbf{R} = -\frac{mr^2\omega_1^2}{L}\left(\mathbf{i} + \frac{r}{2L}\mathbf{j}\right) \blacktriangleleft$$

SOLVING PROBLEMS
ON YOUR OWN

In this lesson you will be asked to solve problems involving the *three-dimensional motion of rigid bodies*. The method you will use is basically the same that you used in Chap. 16 in your study of the plane motion of rigid bodies. You will draw a free-body-diagram equation showing that the system of the external forces is equivalent to the system of the effective forces, and you will equate sums of components and sums of moments on both sides of this equation. Now, however, the system of the effective forces will be represented by the vector $m\bar{\mathbf{a}}$ and a couple vector $\dot{\mathbf{H}}_G$, the determination of which will be explained in paragraphs 1 and 2 below.

To solve a problem involving the three-dimensional motion of a rigid body, you should take the following steps:

1. Determine the angular momentum H_G of the body about its mass center G from its angular velocity $\boldsymbol{\omega}$ with respect to a frame of reference $GX'Y'Z'$ of fixed orientation. This is an operation you learned to perform in the preceding lesson. However, since the configuration of the body will be changing with time, it will now be necessary for you to use an auxiliary system of axes $Gx'y'z'$ (Fig. 18.9) to compute the components of $\boldsymbol{\omega}$ and the moments and products of inertia of the body. These axes may be rigidly attached to the body, in which case their angular velocity is equal to $\boldsymbol{\omega}$ [Sample Probs. 18.3 and 18.4], or they may have an angular velocity $\boldsymbol{\Omega}$ of their own [Sample Prob. 18.5].

Recall the following from the preceding lesson:

    a. If the principal axes of inertia of the body at G are known, use these axes as coordinate axes x', y', and z', since the corresponding products of inertia of the body will be equal to zero. (Note that if the body is axisymmetric, these axes do not need to be rigidly attached to the body.) Resolve $\boldsymbol{\omega}$ into components $\omega_{x'}$, $\omega_{y'}$, and $\omega_{z'}$ along these axes and compute the principal moments of inertia $\bar{I}_{x'}$, $\bar{I}_{y'}$, and $\bar{I}_{z'}$. The corresponding components of the angular momentum $\mathbf{H}_G$ are

$$H_{x'} = \bar{I}_{x'}\omega_{x'} \qquad H_{y'} = \bar{I}_{y'}\omega_{y'} \qquad H_{z'} = \bar{I}_{z'}\omega_{z'} \qquad (18.10)$$

    b. If the principal axes of inertia of the body at G are not known, you must use Eqs. (18.7) to determine the components of the angular momentum $\mathbf{H}_G$. These equations require your prior computation of the *products of inertia* of the body, as well as of its moments of inertia, with respect to the selected axes.

(continued)

2. Compute the rate of change $\dot{\mathbf{H}}_G$ of the angular momentum $\mathbf{H}_G$ with respect to the frame $GX'Y'Z'$. Note that this frame has a *fixed orientation*, while the frame $Gx'y'z'$ you used when you calculated the components of the vector $\boldsymbol{\omega}$ was a *rotating frame*. We refer you to our discussion in Sec. 15.10 of the rate of change of a vector with respect to a rotating frame. Recalling Eq. (15.31), you will express the rate of change $\dot{\mathbf{H}}_G$ as follows:

$$\dot{\mathbf{H}}_G = (\dot{\mathbf{H}}_G)_{Gx'y'z'} + \boldsymbol{\Omega} \times \mathbf{H}_G \qquad (18.22)$$

The first term in the right-hand member of Eq. (18.22) represents the rate of change of $\mathbf{H}_G$ with respect to the rotating frame $Gx'y'z'$. This term will drop out if $\boldsymbol{\omega}$—and, thus, $\mathbf{H}_G$—remain constant in both magnitude and direction when viewed from that frame. On the other hand, if any of the time derivatives $\dot{\omega}_{x'}$, $\dot{\omega}_{y'}$, and $\dot{\omega}_{z'}$ is different from zero, $(\dot{\mathbf{H}}_G)_{Gx'y'z'}$ will also be different from zero, and its components should be determined by differentiating Eqs. (18.10) with respect to t. Finally, we remind you that if the rotating frame is rigidly attached to the body, its angular velocity will be the same as that of the body, and $\boldsymbol{\Omega}$ can be replaced by $\boldsymbol{\omega}$.

3. Draw the free-body-diagram equation for the rigid body, showing that the system of the external forces exerted on the body is equivalent to the vector $m\bar{\mathbf{a}}$ applied at G and the couple vector $\dot{\mathbf{H}}_G$ (Fig. 18.11). By equating components in any direction and moments about any point, you can write as many as six independent scalar equations of motion [Sample Probs. 18.3 and 18.5].

4. When solving problems involving the motion of a rigid body about a fixed point O, you may find it convenient to use the following equation, derived in Sec. 18.7, which eliminates the components of the reaction at the support O,

$$\Sigma \mathbf{M}_O = (\dot{\mathbf{H}}_O)_{Oxyz} + \boldsymbol{\Omega} \times \mathbf{H}_O \qquad (18.28)$$

where the first term in the right-hand member represents the rate of change of $\mathbf{H}_O$ with respect to the rotating frame $Oxyz$, and where $\boldsymbol{\Omega}$ is the angular velocity of that frame.

5. When determining the reactions at the bearings of a rotating shaft, use Eq. (18.28) and take the following steps:

    a. Place the fixed point O at one of the two bearings supporting the shaft and attach the rotating frame $Oxyz$ to the shaft, with one of the axes directed along it. Assuming, for instance, that the x axis has been aligned with the shaft, you will have $\boldsymbol{\Omega} = \boldsymbol{\omega} = \omega\mathbf{i}$ [Sample Prob. 18.4].

b. Since the selected axes, usually, will not be the principal axes of inertia at O, you must compute the *products of inertia* of the shaft, as well as its moments of inertia, with respect to these axes, and use Eqs. (18.13) to determine $\mathbf{H}_O$. Assuming again that the x axis has been aligned with the shaft, Eqs. (18.13) reduce to

$$H_x = I_x\omega \qquad H_y = -I_{yx}\omega \qquad H_z = -I_{zx}\omega \qquad (18.13')$$

which shows that $\mathbf{H}_O$ *will not be directed along the shaft.*

c. To obtain $\dot{\mathbf{H}}_O$, substitute the expressions obtained into Eq. (18.28), and let $\mathbf{\Omega} = \boldsymbol{\omega} = \omega\mathbf{i}$. If the angular velocity of the shaft is constant, the first term in the right-hand member of the equation will drop out. However, if the shaft has an angular acceleration $\boldsymbol{\alpha} = \alpha\mathbf{i}$, the first term will not be zero and must be determined by differentiating with respect to t the expressions in (18.13'). The result will be equations similar to Eqs. (18.13'), with ω replaced by α.

d. Since point O coincides with one of the bearings, the three scalar equations corresponding to Eq. (18.28) can be solved for the components of the dynamic reaction at the other bearing. If the mass center G of the shaft is located on the line joining the two bearings, the effective force $m\bar{\mathbf{a}}$ will be zero. Drawing the free-body-diagram equation of the shaft, you will then observe that the components of the dynamic reaction at the first bearing must be equal and opposite to those you have just determined. If G is not located on the line joining the two bearings, you can determine the reaction at the first bearing by placing the fixed point O at the second bearing and repeating the earlier procedure [Sample Prob. 18.4]; or you can obtain additional equations of motion from the free-body-diagram equation of the shaft, making sure to first determine and include the effective force $m\bar{\mathbf{a}}$ applied at G.

e. Most problems call for the determination of the "dynamic reactions" at the bearings, that is, for the *additional forces* exerted by the bearings on the shaft when the shaft is rotating. When determining dynamic reactions, ignore the effect of static loads, such as the weight of the shaft.

PROBLEMS

18.55 Determine the rate of change $\dot{\mathbf{H}}_G$ of the angular momentum $\mathbf{H}_G$ of the disk of Prob. 18.1.

18.56 Determine the rate of change $\dot{\mathbf{H}}_G$ of the angular momentum $\mathbf{H}_G$ of the plate of Prob. 18.2.

18.57 Determine the rate of change $\dot{\mathbf{H}}_D$ of the angular momentum $\mathbf{H}_D$ of the assembly of Prob. 18.3.

18.58 Determine the rate of change $\dot{\mathbf{H}}_A$ of the angular momentum $\mathbf{H}_A$ of the disk of Prob. 18.4.

18.59 Determine the rate of change $\dot{\mathbf{H}}_C$ of the angular momentum $\mathbf{H}_C$ of the disk of Prob. 18.5.

18.60 Determine the rate of change $\dot{\mathbf{H}}_G$ of the angular momentum $\mathbf{H}_G$ of the disk of Prob. 18.8.

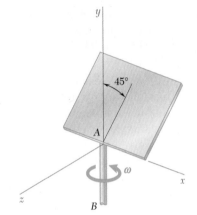

Fig. P18.63

18.61 Determine the rate of change $\dot{\mathbf{H}}_D$ of the angular momentum $\mathbf{H}_D$ of the assembly of Prob. 18.3, assuming that at the instant considered the assembly has an angular velocity $\boldsymbol{\omega} = (12 \text{ rad/s})\mathbf{i}$ and an angular acceleration $\boldsymbol{\alpha} = -(96 \text{ rad/s}^2)\mathbf{i}$.

18.62 Determine the rate of change $\dot{\mathbf{H}}_D$ of the angular momentum $\mathbf{H}_D$ of the assembly of Prob. 18.3, assuming that at the instant considered the assembly has an angular velocity $\boldsymbol{\omega} = (12 \text{ rad/s})\mathbf{i}$ and an angular acceleration $\boldsymbol{\alpha} = (96 \text{ rad/s}^2)\mathbf{i}$.

18.63 A thin, homogeneous square of mass m and side a is welded to a vertical shaft AB with which it forms an angle of 45°. Knowing that the shaft rotates with an angular velocity $\boldsymbol{\omega} = \omega\mathbf{j}$ and an angular acceleration $\boldsymbol{\alpha} = \alpha\mathbf{j}$, determine the rate of change $\dot{\mathbf{H}}_A$ of the angular momentum $\mathbf{H}_A$ of the plate assembly.

18.64 Determine the rate of change $\dot{\mathbf{H}}_G$ of the angular momentum $\mathbf{H}_G$ of the disk of Prob. 18.8, assuming that at the instant considered the assembly has an angular velocity $\boldsymbol{\omega} = \omega\mathbf{j}$ and an angular acceleration $\boldsymbol{\alpha} = \alpha\mathbf{j}$.

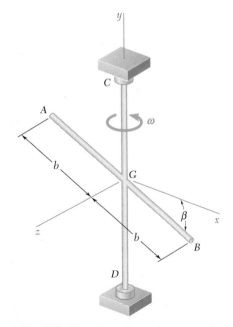

Fig. P18.65

18.65 A slender, uniform rod AB of mass m and a vertical shaft CD, each of length $2b$, are welded together at their midpoints G. Knowing that the shaft rotates at the constant rate ω, determine the dynamic reactions at C and D.

18.66 A thin, homogeneous triangular plate of weight 10 lb is welded to a light, vertical axle supported by bearings at A and B. Knowing that the plate rotates at the constant rate $\omega = 8$ rad/s, determine the dynamic reactions at A and B.

18.67 The assembly shown consists of pieces of sheet aluminum of uniform thickness and of total weight 2.7 lb welded to a light axle supported by bearings at A and B. Knowing that the assembly rotates at the constant rate $\omega = 240$ rpm, determine the dynamic reactions at A and B.

Fig. P18.66

Fig. P18.67

18.68 The 8-kg shaft shown has a uniform cross section. Knowing that the shaft rotates at the constant rate $\omega = 12$ rad/s, determine the dynamic reactions at A and B.

Fig. P18.68

18.69 After attaching the 18-kg wheel shown to a balancing machine and making it spin at the rate of 15 rev/s, a mechanic has found that to balance the wheel both statically and dynamically, he should use two corrective masses, a 170-g mass placed at B and a 56-g mass placed at D. Using a right-handed frame of reference rotating with the wheel (with the z axis perpendicular to the plane of the figure), determine before the corrective masses have been attached (*a*) the distance from the axis of rotation to the mass center of the wheel and the products of inertia I_{xy} and I_{zx}, (*b*) the force-couple system at C equivalent to the forces exerted by the wheel on the machine.

18.70 When the 18-kg wheel shown is attached to a balancing machine and made to spin at a rate of 12.5 rev/s, it is found that the forces exerted by the wheel on the machine are equivalent to a force-couple system consisting of a force $\mathbf{F} = (160\text{ N})\mathbf{j}$ applied at C and a couple $\mathbf{M}_C = (14.7\text{ N} \cdot \text{m})\mathbf{k}$, where the unit vectors form a triad which rotates with the wheel. (*a*) Determine the distance from the axis of rotation to the mass center of the wheel and the products of inertia I_{xy} and I_{zx}. (*b*) If only two corrective masses are to be used to balance the wheel statically and dynamically, what should these masses be and at which of the points A, B, D, or E should they be placed?

Fig. P18.69 and P18.70

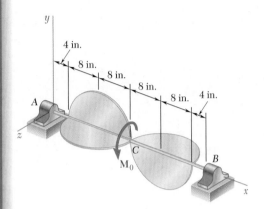

Fig. P18.75

18.71 Knowing that the assembly of Prob. 18.65 is initially at rest ($\omega = 0$) when a couple of moment $\mathbf{M}_0 = M_0\mathbf{j}$ is applied to shaft CD, determine (a) the resulting angular acceleration of the assembly, (b) the dynamic reactions at C and D immediately after the couple is applied.

18.72 Knowing that the plate of Prob. 18.66 is initially at rest ($\omega = 0$) when a couple of moment $\mathbf{M}_0 = (0.75 \text{ ft} \cdot \text{lb})\mathbf{j}$ is applied to it, determine (a) the resulting angular acceleration of the plate, (b) the dynamic reactions A and B immediately after the couple has been applied.

18.73 The assembly of Prob. 18.67 is initially at rest ($\omega = 0$) when a couple $\mathbf{M}_0$ is applied to axle AB. Knowing that the resulting angular acceleration of the assembly is $\boldsymbol{\alpha} = (150 \text{ rad/s}^2)\mathbf{i}$, determine (a) the couple $\mathbf{M}_0$, (b) the dynamic reactions at A and B immediately after the couple is applied.

18.74 The shaft of Prob. 18.68 is initially at rest ($\omega = 0$) when a couple $\mathbf{M}_0$ is applied to it. Knowing that the resulting angular acceleration of the shaft is $\boldsymbol{\alpha} = (20 \text{ rad/s}^2)\mathbf{i}$, determine (a) the couple $\mathbf{M}_0$, (b) the dynamic reactions at A and B immediately after the couple is applied.

18.75 The assembly shown weighs 12 lb and consists of 4 thin 16-in.-diameter semicircular aluminum plates welded to a light 40-in.-long shaft AB. The assembly is at rest ($\omega = 0$) at time $t = 0$ when a couple $\mathbf{M}_0$ is applied to it as shown, causing the assembly to complete one full revolution in 2 s. Determine (a) the couple $\mathbf{M}_0$, (b) the dynamic reactions at A and B at $t = 0$.

18.76 For the assembly of Prob. 18.75, determine the dynamic reactions at A and B at $t = 2$ s.

18.77 The sheet-metal component shown is of uniform thickness and has a mass of 600 g. It is attached to a light axle supported by bearings at A and B located 150 mm apart. The component is at rest when it is subjected to a couple $\mathbf{M}_0$ as shown. If the resulting angular acceleration is $\boldsymbol{\alpha} = (12 \text{ rad/s}^2)\mathbf{k}$, determine (a) the couple $\mathbf{M}_0$, (b) the dynamic reactions A and B immediately after the couple has been applied.

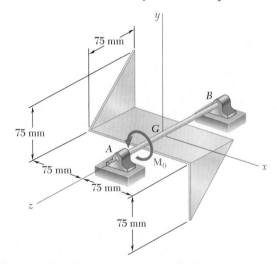

Fig. P18.77

18.78 For the sheet-metal component of Prob. 18.77, determine (a) the angular velocity of the component 0.6 s after the couple $\mathbf{M}_0$ has been applied to it, (b) the magnitude of the dynamic reactions at A and B at that time.

18.79 The blade of an oscillating fan and the rotor of its motor have a total mass of 300 g and a combined radius of gyration of 75 mm. They are supported by bearings at A and B, 125 mm apart, and rotate at the rate $\omega_1 = 1800$ rpm. Determine the dynamic reactions at A and B when the motor casing has an angular velocity $\omega_2 = (0.6 \text{ rad/s})\mathbf{j}$.

18.80 The blade of a portable saw and the rotor of its motor have a total weight of 2.5 lb and a combined radius of gyration of 1.5 in. Knowing that the blade rotates as shown at the rate $\omega_1 = 1500$ rpm, determine the magnitude and direction of the couple $\mathbf{M}$ that a worker must exert on the handle of the saw to rotate it with a constant angular velocity $\omega_2 = -(2.4 \text{ rad/s})\mathbf{j}$.

Fig. P18.79

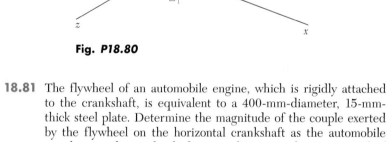

Fig. P18.80

18.81 The flywheel of an automobile engine, which is rigidly attached to the crankshaft, is equivalent to a 400-mm-diameter, 15-mm-thick steel plate. Determine the magnitude of the couple exerted by the flywheel on the horizontal crankshaft as the automobile travels around an unbanked curve of 200-m radius at a speed of 90 km/h, with the flywheel rotating at 2700 rpm. Assume the automobile to have (*a*) a rear-wheel drive with the engine mounted longitudinally, (*b*) a front-wheel drive with the engine mounted transversely. (Density of steel = 7860 kg/m³.)

18.82 Each wheel of an automobile has a mass of 22 kg, a diameter of 575 mm, and a radius of gyration of 225 mm. The automobile travels around an unbanked curve of radius 150 m at a speed of 95 km/h. Knowing that the transverse distance between the wheels is 1.5 m, determine the additional normal force exerted by the ground on each outside wheel due to the motion of the car.

18.83 The uniform, thin 5-lb disk spins at a constant rate $\omega_2 = 6$ rad/s about an axis held by a housing attached to a horizontal rod that rotates at the constant rate $\omega_1 = 3$ rad/s. Determine the couple which represents the dynamic reaction at the support A.

18.84 The essential structure of a certain type of aircraft turn indicator is shown. Each spring has a constant of 500 N/m, and the 200-g uniform disk of 40-mm radius spins at the rate of 10 000 rpm. The springs are stretched and exert equal vertical forces on yoke AB when the airplane is traveling in a straight path. Determine the angle through which the yoke will rotate when the pilot executes a horizontal turn of 750-m radius to the right at a speed of 800 km/h. Indicate whether point A will move up or down.

Fig. P18.83

Fig. P18.84

Fig. P18.85

18.85 A slender rod is bent to form a square frame of side 6 in. The frame is attached by a collar at A to a vertical shaft which rotates with a constant angular velocity $\boldsymbol{\omega}$. Determine the value of ω for which line AB forms an angle $\beta = 48°$ with the horizontal x axis.

18.86 A uniform semicircular plate of radius 120 mm is hinged at A and B to a clevis which rotates with a constant angular velocity $\boldsymbol{\omega}$ about a vertical axis. Determine (a) the angle β that the plate forms with the horizontal x axis when $\omega = 15$ rad/s, (b) the largest value of ω for which the plate remains vertical ($\beta = 90°$).

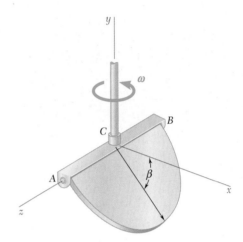

Fig. P18.86 and P18.87

18.87 A uniform semicircular plate of radius 120 mm is hinged at A and B to a clevis which rotates with a constant angular velocity $\boldsymbol{\omega}$ about a vertical axis. Determine the value of ω for which the plate forms an angle $\beta = 50°$ with the horizontal x axis.

18.88 and 18.89 The slender rod AB is attached by a clevis to arm BCD which rotates with a constant angular velocity $\boldsymbol{\omega}$ about the centerline of its vertical portion CD. Determine the magnitude of the angular velocity $\boldsymbol{\omega}$.

Fig. P18.88

Fig. P18.89

18.90 The 950-g gear A is constrained to roll on the fixed gear B, but is free to rotate about axle AD. Axle AD, of length 400 mm and negligible mass, is connected by a clevis to the vertical shaft DE which rotates as shown with a constant angular velocity ω_1. Assuming that gear A can be approximated by a thin disk of radius 80 mm, determine the largest allowable value of ω_1 if gear A is not to lose contact with gear B.

18.91 Determine the force $\mathbf{F}$ exerted by gear B on gear A of Prob. 18.90 when shaft DE rotates with the constant angular velocity $\omega_1 = 4$ rad/s. (*Hint:* The force $\mathbf{F}$ must be perpendicular to the line drawn from D to C.)

18.92 The essential structure of a certain type of aircraft turn indicator is shown. Springs AC and BD are initially stretched and exert equal vertical forces at A and B when the airplane is traveling in a straight path. Each spring has a constant of 600 N/m and the uniform disk has a mass of 250 g and spins at the rate of 12 000 rpm. Determine the angle through which the yoke will rotate when the pilot executes a horizontal turn of 800-m radius to the right at a speed of 720 km/h. Indicate whether point A will move up or down.

18.93 The 10-oz disk shown spins at the rate $\omega_1 = 750$ rpm, while axle AB rotates as shown with an angular velocity ω_2 of 6 rad/s. Determine the dynamic reactions at A and B.

Fig. P18.90

Fig. P18.92

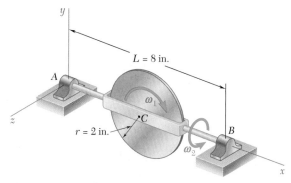

Fig. P18.93 and P18.94

18.94 The 10-oz disk shown spins at the rate $\omega_1 = 750$ rpm, while axle AB rotates as shown with an angular velocity ω_2. Determine the maximum allowable magnitude of ω_2 if the dynamic reactions at A and B are not to exceed 0.25 lb each.

18.95 Two disks, each of mass 5 kg and radius 100 mm, spin as shown at the rate $\omega_1 = 1500$ rpm about a rod AB of negligible mass which rotates about a vertical axis at the rate $\omega_2 = 45$ rpm. (*a*) Determine the dynamic reactions at C and D. (*b*) Solve part *a* assuming that the direction of spin of disk B is reversed.

18.96 Two disks, each of mass 5 kg and radius 100 mm, spin as shown at the rate $\omega_1 = 1500$ rpm about a rod AB of negligible mass which rotates about a vertical axis at a rate ω_2. Determine the maximum allowable value of ω_2 if the dynamic reactions at C and D are not to exceed 250 N each.

Fig. P18.95 and P18.96

Fig. P18.97

Fig. P18.99

18.97 A stationary horizontal plate is attached to the ceiling by means of a fixed vertical tube. A wheel of radius a and mass m is mounted on a light axle AC which is attached by means of a clevis at A to a rod AB fitted inside the vertical tube. The rod AB is made to rotate with a constant angular velocity Ω causing the wheel to roll on the lower face of the stationary plate. Determine the minimum angular velocity Ω for which contact is maintained between the wheel and the plate. Consider the particular cases (a) when the mass of the wheel is concentrated in the rim, (b) when the wheel is equivalent to a thin disk of radius a.

18.98 Assuming that the wheel of Prob. 18.97 weighs 8 lb, has a radius $a = 4$ in., and a radius of gyration of 3 in., and that $R = 20$ in., determine the force exerted by the plate on the wheel when $\Omega = 25$ rad/s.

18.99 A thin disk of mass $m = 4$ kg rotates with an angular velocity $\boldsymbol{\omega}_2$ with respect to arm ABC, which itself rotates with an angular velocity $\boldsymbol{\omega}_1$ about the y axis. Knowing that $\omega_1 = 5$ rad/s and $\omega_2 = 15$ rad/s and that both are constant, determine the force-couple system representing the dynamic reaction at the support at A.

18.100 An experimental Fresnel-lens solar-energy concentrator can rotate about the horizontal axis AB which passes through its mass center G. It is supported at A and B by a steel framework which can rotate about the vertical y axis. The concentrator has a mass of 30 Mg, a radius of gyration of 12 m about its axis of symmetry CD, and a radius of gyration of 10 m about any transverse axis through G. Knowing that the angular velocities $\boldsymbol{\omega}_1$ and $\boldsymbol{\omega}_2$ have constant magnitudes equal to 0.20 rad/s and 0.25 rad/s, respectively, determine for the position $\theta = 60°$ (a) the forces exerted on the concentrator at A and B, (b) the couple $M_2\mathbf{k}$ applied to the concentrator at that instant.

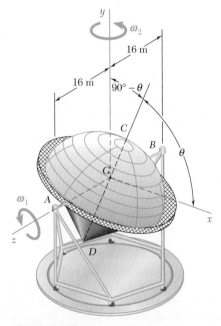

Fig. P18.100

18.101 A 6-lb homogeneous disk of radius 3 in. spins as shown at the constant rate $\omega_1 = 60$ rad/s. The disk is supported by the fork-ended rod AB, which is welded to the vertical shaft CBD. The system is at rest when a couple $\mathbf{M}_0 = (0.25$ ft $\cdot$ lb)$\mathbf{j}$ is applied to the shaft for 2 s and then removed. Determine the dynamic reactions at C and D after the couple has been removed.

18.102 A 6-lb homogeneous disk of radius 3 in. spins as shown at the constant rate $\omega_1 = 60$ rad/s. The disk is supported by the fork-ended rod AB, which is welded to the vertical shaft CBD. The system is at rest when a couple $\mathbf{M}_0$ is applied as shown to the shaft for 3 s and then removed. Knowing that the maximum angular velocity reached by the shaft is 18 rad/s, determine (a) the couple $\mathbf{M}_0$, (b) the dynamic reactions at C and D after the couple has been removed.

18.103 A 2.5-kg homogeneous disk of radius 80 mm rotates with an angular velocity $\boldsymbol{\omega}_1$ with respect to arm ABC, which is welded to a shaft DCE rotating as shown at the constant rate $\omega_2 = 12$ rad/s. Friction in the bearing at A causes ω_1 to decrease at the rate of 15 rad/s². Determine the dynamic reactions at D and E at a time when ω_1 has decreased to 50 rad/s.

Fig. P18.101 and P18.102

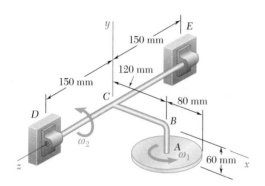

Fig. P18.103 and P18.104

18.104 A 2.5-kg homogeneous disk of radius 80 mm rotates at the constant rate $\omega_1 = 50$ rad/s with respect to arm ABC, which is welded to a shaft DCE. Knowing that at the instant shown, shaft DCE has an angular velocity $\boldsymbol{\omega}_2 = (12$ rad/s)$\mathbf{k}$ and an angular acceleration $\boldsymbol{\alpha}_2 = (8$ rad/s²)$\mathbf{k}$, determine (a) the couple which must be applied to shaft DCE to produce that acceleration, (b) the corresponding dynamic reactions at D and E.

18.105 For the disk of Prob. 18.99, determine (a) the couple $M_1\mathbf{j}$ which should be applied to arm ABC to give it an angular acceleration $\boldsymbol{\alpha}_1 = -(7.5$ rad/s²)$\mathbf{j}$ when $\omega_1 = 5$ rad/s, knowing that the disk rotates at the constant rate $\omega_2 = 15$ rad/s, (b) the force-couple system representing the dynamic reaction at A at that instant. Assume that ABC has a negligible mass.

*18.106 A slender homogeneous rod AB of mass m and length L is made to rotate at a constant rate ω_2 about the horizontal z axis, while frame CD is made to rotate at the constant rate ω_1 about the y axis. Express as a function of the angle θ (a) the couple $\mathbf{M}_1$ required to maintain the rotation of the frame, (b) the couple $\mathbf{M}_2$ required to maintain the rotation of the rod, (c) the dynamic reactions at the supports C and D.

Fig. P18.106

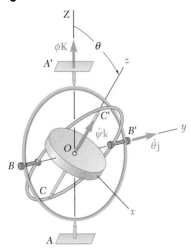

(a)

(b)

Fig. 18.15

Fig. 18.16

*18.9 MOTION OF A GYROSCOPE. EULERIAN ANGLES

A *gyroscope* consists essentially of a rotor which can spin freely about its geometric axis. When mounted in a Cardan's suspension (Fig. 18.15), a gyroscope can assume any orientation, but its mass center must remain fixed in space. In order to define the position of a gyroscope at a given instant, let us select a fixed frame of reference $OXYZ$, with the origin O located at the mass center of the gyroscope and the Z axis directed along the line defined by the bearings A and A' of the outer gimbal. We will consider a reference position of the gyroscope in which the two gimbals and a given diameter DD' of the rotor are located in the fixed YZ plane (Fig. 18.15a). The gyroscope can be brought from this reference position into any arbitrary position (Fig. 18.15b) by means of the following steps: (1) a rotation of the outer gimbal through an angle ϕ about the axis AA', (2) a rotation of the inner gimbal through θ about BB', and (3) a rotation of the rotor through ψ about CC'. The angles ϕ, θ, and ψ are called the *Eulerian angles;* they completely characterize the position of the gyroscope at any given instant. Their derivatives $\dot{\phi}$, $\dot{\theta}$, and $\dot{\psi}$ define, respectively, the rate of *precession,* the rate of *nutation,* and the rate of *spin* of the gyroscope at the instant considered.

In order to compute the components of the angular velocity and of the angular momentum of the gyroscope, we will use a rotating system of axes $Oxyz$ *attached to the inner gimbal,* with the y axis along BB' and the z axis along CC' (Fig. 18.16). These axes are principal axes of inertia for the gyroscope. While they follow it in its precession and nutation, however, they do not spin; for that reason, they are more convenient to use than axes actually attached to the gyroscope. The angular velocity $\boldsymbol{\omega}$ of the gyroscope with respect to the fixed frame of reference $OXYZ$ will now be expressed as the sum of three partial angular velocities corresponding, respectively, to the precession, the nutation, and the spin of the gyroscope. Denoting by $\mathbf{i}$, $\mathbf{j}$, and $\mathbf{k}$ the unit vectors along the rotating axes, and by $\mathbf{K}$ the unit vector along the fixed Z axis, we have

$$\boldsymbol{\omega} = \dot{\phi}\mathbf{K} + \dot{\theta}\mathbf{j} + \dot{\psi}\mathbf{k} \qquad (18.33)$$

Since the vector components obtained for $\boldsymbol{\omega}$ in (18.33) are not orthogonal (Fig. 18.16), the unit vector $\mathbf{K}$ will be resolved into components along the x and z axes; we write

$$\mathbf{K} = -\sin\theta\,\mathbf{i} + \cos\theta\,\mathbf{k} \qquad (18.34)$$

and, substituting for $\mathbf{K}$ into (18.33),

$$\boldsymbol{\omega} = -\dot{\phi}\,\sin\theta\,\mathbf{i} + \dot{\theta}\mathbf{j} + (\dot{\psi} + \dot{\phi}\,\cos\theta)\mathbf{k} \qquad (18.35)$$

Since the coordinate axes are principal axes of inertia, the components of the angular momentum $\mathbf{H}_O$ can be obtained by multiplying the components of $\boldsymbol{\omega}$ by the moments of inertia of the rotor about the x, y, and z axes, respectively. Denoting by I the moment of inertia of the rotor about its spin axis, by I' its moment of inertia about a transverse axis through O, and neglecting the mass of the gimbals, we write

$$\mathbf{H}_O = -I'\dot{\phi}\sin\theta\,\mathbf{i} + I'\dot{\theta}\mathbf{j} + I(\dot{\psi} + \dot{\phi}\cos\theta)\mathbf{k} \qquad (18.36)$$

Recalling that the rotating axes are attached to the inner gimbal and thus do not spin, we express their angular velocity as the sum

$$\boldsymbol{\Omega} = \dot{\phi}\mathbf{K} + \dot{\theta}\mathbf{j} \qquad (18.37)$$

or, substituting for $\mathbf{K}$ from (18.34),

$$\boldsymbol{\Omega} = -\dot{\phi}\sin\theta\,\mathbf{i} + \dot{\theta}\mathbf{j} + \dot{\phi}\cos\theta\,\mathbf{k} \qquad (18.38)$$

Substituting for $\mathbf{H}_O$ and $\boldsymbol{\Omega}$ from (18.36) and (18.38) into the equation

$$\Sigma\mathbf{M}_O = (\dot{\mathbf{H}}_O)_{Oxyz} + \boldsymbol{\Omega} \times \mathbf{H}_O \qquad (18.28)$$

we obtain the three differential equations

$$\begin{aligned}
\Sigma M_x &= -I'(\ddot{\phi}\sin\theta + 2\dot{\theta}\dot{\phi}\cos\theta) + I\dot{\theta}(\dot{\psi} + \dot{\phi}\cos\theta) \\
\Sigma M_y &= I'(\ddot{\theta} - \dot{\phi}^2\sin\theta\cos\theta) + I\dot{\phi}\sin\theta(\dot{\psi} + \dot{\phi}\cos\theta) \\
\Sigma M_z &= I\frac{d}{dt}(\dot{\psi} + \dot{\phi}\cos\theta)
\end{aligned} \qquad (18.39)$$

The equations (18.39) define the motion of a gyroscope subjected to a given system of forces when the mass of its gimbals is neglected. They can also be used to define the motion of an *axisymmetrical body* (or body of revolution) attached at a point on its axis of symmetry, and to define the motion of an axisymmetrical body about its mass center. While the gimbals of the gyroscope helped us visualize the Eulerian angles, it is clear that these angles can be used to define the position of any rigid body with respect to axes centered at a point of the body, regardless of the way in which the body is actually supported.

Since the equations (18.39) are nonlinear, it will not be possible, in general, to express the Eulerian angles ϕ, θ, and ψ as analytical functions of the time t, and numerical methods of solution may have to be used. However, as you will see in the following sections, there are several particular cases of interest which can be analyzed easily.

Photo 18.5 A gyroscope can be used for measuring orientation and is capable of maintaining the same absolute direction in space.

Fig. 18.17

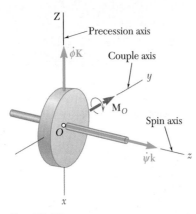

Fig. 18.18

*18.10 STEADY PRECESSION OF A GYROSCOPE

Let us now investigate the particular case of gyroscopic motion in which the angle θ, the rate of precession $\dot{\phi}$, and the rate of spin $\dot{\psi}$ remain constant. We propose to determine the forces which must be applied to the gyroscope to maintain this motion, known as the *steady precession* of a gyroscope.

Instead of applying the general equations (18.39), we will determine the sum of the moments of the required forces by computing the rate of change of the angular momentum of the gyroscope in the particular case considered. We first note that the angular velocity $\boldsymbol{\omega}$ of the gyroscope, its angular momentum $\mathbf{H}_O$, and the angular velocity $\boldsymbol{\Omega}$ of the rotating frame of reference (Fig. 18.17) reduce, respectively, to

$$\boldsymbol{\omega} = -\dot{\phi}\,\sin\theta\,\mathbf{i} + \omega_z\mathbf{k} \qquad (18.40)$$
$$\mathbf{H}_O = -I'\dot{\phi}\,\sin\theta\,\mathbf{i} + I\omega_z\mathbf{k} \qquad (18.41)$$
$$\boldsymbol{\Omega} = -\dot{\phi}\,\sin\theta\,\mathbf{i} + \dot{\phi}\,\cos\theta\,\mathbf{k} \qquad (18.42)$$

where $\omega_z = \dot{\psi} + \dot{\phi}\,\cos\theta$ = rectangular component along spin axis of total angular velocity of gyroscope

Since θ, $\dot{\phi}$, and $\dot{\psi}$ are constant, the vector $\mathbf{H}_O$ is constant in magnitude and direction with respect to the rotating frame of reference and its rate of change $(\dot{\mathbf{H}}_O)_{Oxyz}$ with respect to that frame is zero. Thus Eq. (18.28) reduces to

$$\Sigma\mathbf{M}_O = \boldsymbol{\Omega} \times \mathbf{H}_O \qquad (18.43)$$

which yields, after substitutions from (18.41) and (18.42),

$$\Sigma\mathbf{M}_O = (I\omega_z - I'\dot{\phi}\,\cos\theta)\dot{\phi}\,\sin\theta\,\mathbf{j} \qquad (18.44)$$

Since the mass center of the gyroscope is fixed in space, we have, by (18.1), $\Sigma\mathbf{F} = 0$; thus, the forces which must be applied to the gyroscope to maintain its steady precession reduce to a couple of moment equal to the right-hand member of Eq. (18.44). We note that *this couple should be applied about an axis perpendicular to the precession axis and to the spin axis of the gyroscope* (Fig. 18.18).

In the particular case when the precession axis and the spin axis are at a right angle to each other, we have $\theta = 90°$ and Eq. (18.44) reduces to

$$\Sigma\mathbf{M}_O = I\dot{\psi}\dot{\phi}\,\mathbf{j} \qquad (18.45)$$

Thus, if we apply to the gyroscope a couple $\mathbf{M}_O$ about an axis perpendicular to its axis of spin, the gyroscope will precess about an axis perpendicular to both the spin axis and the couple axis, in a sense such that the vectors representing the spin, the couple, and the precession, respectively, form a right-handed triad (Fig. 18.19).

Because of the relatively large couples required to change the orientation of their axles, gyroscopes are used as stabilizers in torpedoes

Fig. 18.19

and ships. Spinning bullets and shells remain tangent to their trajectory because of gyroscopic action. And a bicycle is easier to keep balanced at high speeds because of the stabilizing effect of its spinning wheels. However, gyroscopic action is not always welcome and must be taken into account in the design of bearings supporting rotating shafts subjected to forced precession. The reactions exerted by its propellers on an airplane which changes its direction of flight must also be taken into consideration and compensated for whenever possible.

*18.11 MOTION OF AN AXISYMMETRICAL BODY UNDER NO FORCE

In this section you will analyze the motion about its mass center of an axisymmetrical body under no force except its own weight. Examples of such a motion are furnished by projectiles, if air resistance is neglected, and by artificial satellites and space vehicles after burnout of their launching rockets.

Since the sum of the moments of the external forces about the mass center G of the body is zero, Eq. (18.2) yields $\dot{\mathbf{H}}_G = 0$. It follows that the angular momentum $\mathbf{H}_G$ of the body about G is constant. Thus, the direction of $\mathbf{H}_G$ is fixed in space and can be used to define the Z axis, or axis of precession (Fig. 18.20). Selecting a rotating system of axes $Gxyz$ with the z axis along the axis of symmetry of the body, the x axis in the plane defined by the Z and z axes, and the y axis pointing away from you, we have

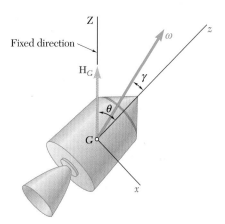

Fig. 18.20

$$H_x = -H_G \sin \theta \qquad H_y = 0 \qquad H_z = H_G \cos \theta \qquad (18.46)$$

where θ represents the angle formed by the Z and z axes, and H_G denotes the constant magnitude of the angular momentum of the body about G. Since the x, y, and z axes are principal axes of inertia for the body considered, we can write

$$H_x = I'\omega_x \qquad H_y = I'\omega_y \qquad H_z = I\omega_z \qquad (18.47)$$

where I denotes the moment of inertia of the body about its axis of symmetry and I' denotes its moment of inertia about a transverse axis through G. It follows from Eqs. (18.46) and (18.47) that

$$\omega_x = -\frac{H_G \sin \theta}{I'} \qquad \omega_y = 0 \qquad \omega_2 = \frac{H_G \cos \theta}{I} \qquad (18.48)$$

The second of the relations obtained shows that the angular velocity $\boldsymbol{\omega}$ has no component along the y axis, i.e., along an axis perpendicular to the Zz plane. Thus, the angle θ formed by the Z and z axes remains constant and *the body is in steady precession about the Z axis.*

Dividing the first and third of the relations (18.48) member by member, and observing from Fig. 18.21 that $-\omega_x/\omega_z = \tan \gamma$, we obtain the following relation between the angles γ and θ that the

Fig. 18.21

(a)   *(b)*

Fig. 18.22

Fig. 18.23

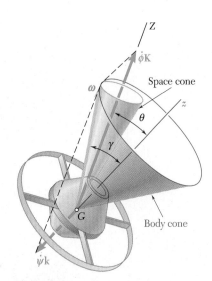

Fig. 18.24

vectors $\boldsymbol{\omega}$ and $\mathbf{H}_G$, respectively, form with the axis of symmetry of the body:

$$\tan \gamma = \frac{I}{I'} \tan \theta \qquad (18.49)$$

There are two particular cases of motion of an axisymmetrical body under no force which involve no precession: (1) If the body is set to spin about its axis of symmetry, we have $\omega_x = 0$ and, by (18.47), $H_x = 0$; the vectors $\boldsymbol{\omega}$ and $\mathbf{H}_G$ have the same orientation and the body keeps spinning about its axis of symmetry (Fig. 18.22a). (2) If the body is set to spin about a transverse axis, we have $\omega_z = 0$ and, by (18.47), $H_z = 0$; again $\boldsymbol{\omega}$ and $\mathbf{H}_G$ have the same orientation and the body keeps spinning about the given transverse axis (Fig. 18.22b).

Considering now the general case represented in Fig. 18.21, we recall from Sec. 15.12 that the motion of a body about a fixed point—or about its mass center—can be represented by the motion of a body cone rolling on a space cone. In the case of steady precession, the two cones are circular, since the angles γ and $\theta - \gamma$ that the angular velocity $\boldsymbol{\omega}$ forms, respectively, with the axis of symmetry of the body and with the precession axis are constant. Two cases should be distinguished:

1. $I < I'$. This is the case of an elongated body, such as the space vehicle of Fig. 18.23. By (18.49) we have $\gamma < \theta$; the vector $\boldsymbol{\omega}$ lies inside the angle ZGz; the space cone and the body cone are tangent externally; the spin and the precession are both observed as counterclockwise from the positive z axis. The precession is said to be *direct*.
2. $I > I'$. This is the case of a flattened body, such as the satellite of Fig. 18.24. By (18.49) we have $\gamma > \theta$; since the vector $\boldsymbol{\omega}$ must lie outside the angle ZGz, the vector $\dot{\psi}\mathbf{k}$ has a sense opposite to that of the z axis; the space cone is inside the body cone; the precession and the spin have opposite senses. The precession is said to be *retrograde*.

SAMPLE PROBLEM 18.6

A space satellite of mass m is known to be dynamically equivalent to two thin disks of equal mass. The disks are of radius $a = 800$ mm and are rigidly connected by a light rod of length $2a$. Initially the satellite is spinning freely about its axis of symmetry at the rate $\omega_0 = 60$ rpm. A meteorite, of mass $m_0 = m/1000$ and traveling with a velocity $\mathbf{v}_0$ of 2000 m/s relative to the satellite, strikes the satellite and becomes embedded at C. Determine (a) the angular velocity of the satellite immediately after impact, (b) the precession axis of the ensuing motion, (c) the rates of precession and spin of the ensuing motion.

SOLUTION

Moments of Inertia. We note that the axes shown are principal axes of inertia for the satellite and write

$$I = I_z = \tfrac{1}{2}ma^2 \qquad I' = I_x = I_y = 2[\tfrac{1}{4}(\tfrac{1}{2}m)a^2 + (\tfrac{1}{2}m)a^2] = \tfrac{5}{4}ma^2$$

Principle of Impulse and Momentum. We consider the satellite and the meteorite as a single system. Since no external force acts on this system, the momenta before and after impact are equipollent. Taking moments about G, we write

$$-a\mathbf{j} \times m_0 v_0 \mathbf{k} + I\omega_0 \mathbf{k} = \mathbf{H}_G$$
$$\mathbf{H}_G = -m_0 v_0 a\mathbf{i} + I\omega_0 \mathbf{k} \qquad (1)$$

Angular Velocity After Impact. Substituting the values obtained for the components of $\mathbf{H}_G$ and for the moments of inertia into

$$H_x = I_x \omega_x \qquad H_y = I_y \omega_y \qquad H_z = I_z \omega_z$$

we write

$$-m_0 v_0 a = I' \omega_x = \tfrac{5}{4}ma^2 \omega_x \qquad 0 = I' \omega_y \qquad I\omega_0 = I\omega_z$$
$$\omega_x = -\frac{4}{5}\frac{m_0 v_0}{ma} \qquad \omega_y = 0 \qquad \omega_z = \omega_0 \qquad (2)$$

For the satellite considered we have $\omega_0 = 60$ rpm $= 6.283$ rad/s, $m_0/m = \frac{1}{1000}$, $a = 0.800$ m, and $v_0 = 2000$ m/s; we find

$$\omega_x = -2 \text{ rad/s} \qquad \omega_y = 0 \qquad \omega_z = 6.283 \text{ rad/s}$$
$$\omega = \sqrt{\omega_x^2 + \omega_z^2} = 6.594 \text{ rad/s} \qquad \tan \gamma = \frac{-\omega_x}{\omega_z} = +0.3183$$
$$\omega = 63.0 \text{ rpm} \qquad \gamma = 17.7° \quad \blacktriangleleft$$

Precession Axis. Since in free motion the direction of the angular momentum $\mathbf{H}_G$ is fixed in space, the satellite will precess about this direction. The angle θ formed by the precession axis and the z axis is

$$\tan \theta = \frac{-H_x}{H_z} = \frac{m_0 v_0 a}{I\omega_0} = \frac{2m_0 v_0}{ma\omega_0} = 0.796 \qquad \theta = 38.5° \quad \blacktriangleleft$$

Rates of Precession and Spin. We sketch the space and body cones for the free motion of the satellite. Using the law of sines, we compute the rates of precession and spin.

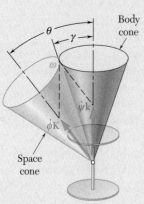

Body cone

Space cone

$$\frac{\omega}{\sin \theta} = \frac{\dot{\phi}}{\sin \gamma} = \frac{\dot{\psi}}{\sin (\theta - \gamma)}$$
$$\dot{\phi} = 30.8 \text{ rpm} \qquad \dot{\psi} = 35.9 \text{ rpm} \quad \blacktriangleleft$$

SOLVING PROBLEMS
ON YOUR OWN

In this lesson we analyzed the motion of *gyroscopes* and of other *axisymmetrical bodies* with a fixed point O. In order to define the position of these bodies at any given instant, we introduced the three *Eulerian angles* ϕ, θ, and ψ (Fig. 18.15), and noted that their time derivatives define, respectively, the rate of *precession,* the rate of *nutation,* and the rate of *spin* (Fig. 18.16). The problems you will encounter fall into one of the following categories.

1. Steady precession. This is the motion of a gyroscope or other axisymmetrical body with a fixed point located on its axis of symmetry, in which the angle θ, the rate of precession $\dot{\phi}$, and the rate of spin $\dot{\psi}$ all remain constant.

a. Using the rotating frame of reference *Oxyz* shown in Fig. 18.17, which *precesses* with the body, *but does not spin* with it, we obtained the following expressions for the angular velocity $\boldsymbol{\omega}$ of the body, its angular momentum $\mathbf{H}_O$, and the angular velocity $\boldsymbol{\Omega}$ of the frame *Oxyz*:

$$\boldsymbol{\omega} = -\dot{\phi}\sin\theta\,\mathbf{i} + \omega_z\mathbf{k} \tag{18.40}$$

$$\mathbf{H}_O = -I'\dot{\phi}\sin\theta\,\mathbf{i} + I\,\omega_z\mathbf{k} \tag{18.41}$$

$$\boldsymbol{\Omega} = -\dot{\phi}\sin\theta\,\mathbf{i} + \dot{\phi}\cos\theta\,\mathbf{k} \tag{18.42}$$

where I = moment of inertia of body about its axis of symmetry
  I' = moment of inertia of body about a transverse axis through O
  ω_z = *rectangular* component of $\boldsymbol{\omega}$ along z axis = $\dot{\psi} + \dot{\phi}\cos\theta$

b. The sum of the moments about O of the forces applied to the body is equal to the rate of change of its angular momentum, as expressed by Eq. (18.28). But, since θ and the rates of change $\dot{\phi}$ and $\dot{\psi}$ are constant, it follows from Eq. (18.41) that $\mathbf{H}_O$ remains constant in magnitude and direction when viewed from the frame *Oxyz*. Thus, its rate of change is zero with respect to that frame and you can write

$$\Sigma\mathbf{M}_O = \boldsymbol{\Omega} \times \mathbf{H}_O \tag{18.43}$$

where $\boldsymbol{\Omega}$ and $\mathbf{H}_O$ are defined, respectively, by Eq. (18.42) and Eq. (18.41). The equation obtained shows that the moment resultant at O of the forces applied to the body is perpendicular to both the axis of precession and the axis of spin (Fig. 18.18).

c. Keep in mind that the method described applies, not only to gyroscopes, where the fixed point O coincides with the mass center G, but also *to any axisymmetrical body with a fixed point O located on its axis of symmetry.* This method, therefore, can be used to analyze the *steady precession of a top* on a rough floor.

d. When an axisymmetrical body has no fixed point, but is in steady precession about its mass center G, you should draw a *free-body-diagram equation* showing that the system of the external forces exerted on the body (including the body's weight) is equivalent to the vector $m\overline{\mathbf{a}}$ applied at G and the couple vector

$\dot{\mathbf{H}}_G$. You can use Eqs. (18.40) through (18.42), replacing $\mathbf{H}_O$ with $\mathbf{H}_G$, and express the moment of the couple as

$$\dot{\mathbf{H}}_G = \mathbf{\Omega} \times \mathbf{H}_G$$

You can then use the free-body-diagram equation to write as many as six independent scalar equations.

2. Motion of an axisymmetrical body under no force, except its own weight. We have $\Sigma \mathbf{M}_G = 0$ and, thus, $\dot{\mathbf{H}}_G = 0$; it follows that *the angular momentum* $\mathbf{H}_G$ is constant in magnitude and direction (Sec. 18.11). The body is in *steady precession* with the precession axis GZ directed along $\mathbf{H}_G$ (Fig. 18.20). Using the rotating frame $Gxyz$ and denoting by γ the angle that $\boldsymbol{\omega}$ forms with the spin axis Gz (Fig. 18.21), we obtained the following relation between γ and the angle θ formed by the precession and spin axes:

$$\tan \gamma = \frac{I}{I'} \tan \theta \tag{18.49}$$

The precession is said to be *direct* if $I < I'$ (Fig. 18.23) and *retrograde* if $I > I'$ (Fig. 18.24).

   a. In many of the problems dealing with the motion of an axisymmetrical body under no force, you will be asked to determine the *precession axis* and the *rates of precession and spin* of the body, knowing the magnitude of its *angular velocity* $\boldsymbol{\omega}$ and the angle γ that it forms with the axis of symmetry Gz (Fig. 18.21). From Eq. (18.49) you will determine the angle θ that the precession axis GZ forms with Gz and resolve $\boldsymbol{\omega}$ into its two *oblique components* $\dot{\phi}\mathbf{K}$ and $\dot{\psi}\mathbf{k}$. Using the law of sines, you will then determine the rate of precession $\dot{\phi}$ and the rate of spin $\dot{\psi}$.

   b. In other problems, the body will be subjected to *a given impulse* and you will first determine the resulting *angular momentum* $\mathbf{H}_G$. Using Eqs. (18.10), you will calculate the rectangular components of the angular velocity $\boldsymbol{\omega}$, its magnitude ω, and the angle γ that it forms with the axis of symmetry. You will then determine the *precession axis* and the *rates of precession and spin* as described above [Sample Prob. 18.6].

3. General motion of an axisymmetric body with a fixed point O located on its axis of symmetry, and subjected only to its own weight. This is a motion in which the angle θ is allowed to vary. At any given instant you should take into account the rate of precession $\dot{\phi}$, the rate of spin $\dot{\psi}$, *and the rate of nutation $\dot{\theta}$,* none of which will remain constant. An example of such a motion is the motion of a top, which is discussed in Probs. 18.137 and 18.138. The rotating frame of reference $Oxyz$ that you will use is still the one shown in Fig. 18.18, but this frame

(continued)

will now rotate about the y axis at the rate $\dot{\theta}$. Equations (18.40), (18.41), and (18.42), therefore, should be replaced by the following equations:

$$\boldsymbol{\omega} = -\dot{\phi} \sin \theta \, \mathbf{i} + \dot{\theta} \mathbf{j} + (\dot{\psi} + \dot{\phi} \cos \theta) \, \mathbf{k} \qquad (18.40')$$

$$\mathbf{H}_O = -I'\dot{\phi} \sin \theta \, \mathbf{i} + I'\dot{\theta} \mathbf{j} + I(\dot{\psi} + \dot{\phi} \cos \theta) \, \mathbf{k} \qquad (18.41')$$

$$\boldsymbol{\Omega} = -\dot{\phi} \sin \theta \, \mathbf{i} + \dot{\theta} \mathbf{j} + \dot{\phi} \cos \theta \, \mathbf{k} \qquad (18.42')$$

Since substituting these expressions into Eq. (18.44) would lead to nonlinear differential equations, it is preferable, whenever feasible, to apply the following conservation principles.

a. Conservation of energy. Denoting by c the distance between the fixed point O and the mass center G of the body, and by E the total energy, you will write

$$T + V = E: \qquad \tfrac{1}{2}(I'\omega_x^2 + I'\omega_y^2 + I\omega_z^2) + mgc \cos \theta = E$$

and substitute for the components of $\boldsymbol{\omega}$ the expressions obtained in Eq. (18.40'). Note that c will be positive or negative, depending upon the position of G relative to O. Also, $c = 0$ if G coincides with O; the *kinetic energy* is then conserved.

b. Conservation of the angular momentum about the axis of precession. Since the support at O is located on the Z axis, and since the weight of the body and the Z axis are both vertical and, thus, parallel to each other, it follows that $\Sigma M_Z = 0$ and, thus, that H_Z remains constant. This can be expressed by writing that the scalar product $\mathbf{K} \cdot \mathbf{H}_O$ is constant, where $\mathbf{K}$ is the unit vector along the Z axis.

c. Conservation of the angular momentum about the axis of spin. Since the support at O and the center of gravity G are both located on the z axis, it follows that $\Sigma M_z = 0$ and, thus, that H_z remains constant. This is expressed by writing that the coefficient of the unit vector $\mathbf{k}$ in Eq. (18.41') is constant. Note that this last conservation principle cannot be applied when the body is restrained from spinning about its axis of symmetry, but in that case the only variables are θ and ϕ.

PROBLEMS

18.107 A solid cone of height 9 in. with a circular base of radius 3 in. is supported by a ball-and-socket joint at A. Knowing that the cone is observed to precess about the vertical axis AC at the constant rate of 40 rpm in the sense indicated and that its axis of symmetry AB forms an angle $\beta = 40°$ with AC, determine the rate at which the cone spins about the axis AB.

18.108 A solid cone of height 9 in. with a circular base of radius 3 in. is supported by a ball-and-socket joint at A. Knowing that the cone is spinning about its axis of symmetry AB at the rate of 3000 rpm and that AB forms an angle $\beta = 60°$ with the vertical axis AC, determine the two possible rates of steady precession of the cone about the axis AC.

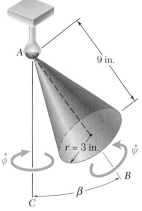

Fig. P18.107 and P18.108

18.109 The 85-g top shown is supported at the fixed point O. The radii of gyration of the top with respect to its axis of symmetry and with respect to a transverse axis through O are 21 mm and 45 mm, respectively. Knowing that $c = 37.5$ mm and that the rate of spin of the top about its axis of symmetry is 1800 rpm, determine the two possible rates of steady precession corresponding to $\theta = 30°$.

18.110 The top shown is supported at the fixed point O and its moments of inertia about its axis of symmetry and about a transverse axis through O are denoted, respectively, by I and I'. (a) Show that the condition for steady precession of the top is

$$(I\omega_z - I'\dot\phi \cos\theta)\dot\phi = Wc$$

where $\dot\phi$ is the rate of precession and ω_z is the rectangular component of the angular velocity along the axis of symmetry of the top. (b) Show that if the rate of spin $\dot\psi$ of the top is very large compared with its rate of precession $\dot\phi$, the condition for steady precession is $I\dot\psi\dot\phi \approx Wc$. (c) Determine the percentage error introduced when this last relation is used to approximate the slower of the two rates of precession obtained for the top of Prob. 18.109.

Fig. P18.109 and P18.110

18.111 A solid aluminum sphere of radius 4 in. is welded to the end of a 10-in.-long rod AB of negligible mass which is supported by a ball-and-socket joint at A. Knowing that the sphere is observed to precess about a vertical axis at the constant rate of 60 rpm in the sense indicated and that rod AB forms an angle $\beta = 20°$ with the vertical, determine the rate of spin of the sphere about line AB.

18.112 A solid aluminum sphere of radius 4 in. is welded to the end of a 10-in.-long rod AB of negligible mass which is supported by a ball-and-socket joint at A. Knowing that the sphere spins as shown about line AB at the rate of 600 rpm, determine the angle β for which the sphere will precess about a vertical axis at the constant rate of 60 rpm in the sense indicated.

Fig. P18.111 and P18.112

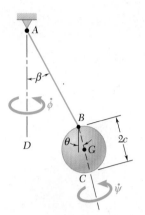

Fig. P18.113 and P18.114

18.113 A solid sphere of radius $c = 3$ in. is attached as shown to cord AB. The sphere is observed to precess at the constant rate $\dot{\phi} = 6$ rad/s about the vertical axis AD. Knowing that $\beta = 40°$, determine the angle θ that the diameter BC forms with the vertical when the sphere (a) has no spin, (b) spins about its diameter BC at the rate $\dot{\psi} = 50$ rad/s, (c) spins about BC at the rate $\dot{\psi} = -50$ rad/s.

18.114 A solid sphere of radius $c = 3$ in. is attached as shown to a cord AB of length 15 in. The sphere spins about its diameter BC and precesses about the vertical axis AD. Knowing that $\theta = 20°$ and $\beta = 35°$, determine (a) the rate of spin of the sphere, (b) its rate of precession.

18.115 A solid cube of side $c = 80$ mm is attached as shown to cord AB. It is observed to spin at the rate $\dot{\psi} = 40$ rad/s about its diagonal BC and to precess at the constant rate $\dot{\phi} = 5$ rad/s about the vertical axis AD. Knowing that $\beta = 30°$, determine the angle θ that the diagonal BC forms with the vertical. (*Hint:* The moment of inertia of a cube about an axis through its center is independent of the orientation of that axis.)

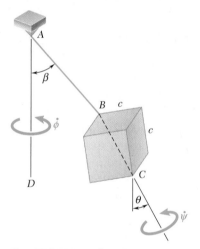

Fig. P18.115 and P18.116

18.116 A solid cube of side $c = 120$ mm is attached as shown to a cord AB of length 240 mm. The cube spins about its diagonal BC and precesses about the vertical axis AD. Knowing that $\theta = 25°$ and $\beta = 40°$, determine (a) the rate of spin of the cube, (b) its rate of precession. (See hint of Prob. 18.115.)

18.117 A high-speed photographic record shows that a certain projectile was fired with a horizontal velocity $\bar{\mathbf{v}}$ of 2000 ft/s and with its axis of symmetry forming an angle $\beta = 3°$ with the horizontal. The rate of spin $\dot{\psi}$ of the projectile was 6000 rpm, and the atmospheric drag was equivalent to a force $\mathbf{D}$ of 25 lb acting at the center of pressure C_P located at a distance $c = 6$ in. from G. (a) Knowing that the projectile has a weight of 45 lb and a radius of gyration of 2 in. with respect to its axis of symmetry, determine its approximate rate of steady precession. (b) If it is further known that the radius of gyration of the projectile with respect to a transverse axis through G is 8 in., determine the exact values of the two possible rates of precession.

Fig. P18.117

18.118 If the earth were a sphere, the gravitational attraction of the sun, moon, and planets would at all times be equivalent to a single force **R** acting at the mass center of the earth. However, the earth is actually an oblate spheroid and the gravitational system acting on the earth is equivalent to a force **R** and a couple **M**. Knowing that the effect of the couple **M** is to cause the axis of the earth to precess about the axis *GA* at the rate of one revolution in 25 800 years, determine the average magnitude of the couple **M** applied to the earth. Assume that the average density of the earth is 5.51 g/cm³, that the average radius of the earth is 6370 km, and that $\bar{I} = \frac{2}{5}mR^2$. (*Note:* This forced precession is known as the precession of the equinoxes and is not to be confused with the free precession discussed in Prob. 18.123.)

Fig. P18.118

18.119 Show that for an axisymmetrical body under no force, the rates of precession and spin can be expressed, respectively, as

$$\dot{\phi} = \frac{H_G}{I'}$$

and

$$\dot{\psi} = \frac{H_G \cos \theta (I' - I)}{II'}$$

where H_G is the constant value of the angular momentum of the body.

18.120 (*a*) Show that for an axisymmetrical body under no force, the rate of precession can be expressed as

$$\dot{\phi} = \frac{I\omega_z}{I' \cos \theta}$$

where ω_2 is the rectangular component of **ω** along the axis of symmetry of the body. (*b*) Use this result to check that the condition (18.44) for steady precession is satisfied by an axisymmetrical body under no force.

18.121 Show that the angular velocity vector **ω** of an axisymmetrical body under no force observed from the body itself to rotate about the axis of symmetry at the constant rate

$$n = \frac{I' - I}{I'} \omega_z$$

where ω_2 is the rectangular component of **ω** along the axis of symmetry of the body.

18.122 For an axisymmetrical body under no force, prove (*a*) that the rate of retrograde precession can never be less than twice the rate of spin of the body about its axis of symmetry, (*b*) that in Fig. 18.24 the axis of symmetry of the body can never lie within the space cone.

18.123 Using the relation given in Prob. 18.121, determine the period of precession of the north pole of the earth about the axis of symmetry of the earth. The earth may be approximated by an oblate spheroid of axial moment of inertia I and of transverse moment of inertia $I' = 0.9967I$. (*Note:* Actual observations show a period of precession of the north pole of about 432.5 mean solar days; the difference between the observed and computed periods is due to the fact that the earth is not a perfectly rigid body. The free precession considered here should not be confused with the much slower precession of the equinoxes, which is a forced precession. See Prob. 18.118.)

Fig. P18.124

18.124 A coin is tossed into the air. It is observed to spin at the rate of 600 rpm about an axis GC perpendicular to the coin and to precess about the vertical direction GD. Knowing that GC forms an angle of 15° with GD, determine (a) the angle that the angular velocity $\boldsymbol{\omega}$ of the coin forms with GD, (b) the rate of precession of the coin about GD.

18.125 The angular velocity vector of a football which has just been kicked is horizontal, and its axis of symmetry OC is oriented as shown. Knowing that the magnitude of the angular velocity is 200 rpm and that the ratio of the axis and transverse moments of inertia is $I/I' = \frac{1}{3}$, determine (a) the orientation of the axis of precession OA, (b) the rates of precession and spin.

Fig. P18.125

18.126 The space capsule has no angular velocity when the jet at A is activated for 1 s in a direction parallel to the x axis. Knowing that the capsule has a mass of 1000 kg, that its radii of gyration are $\bar{k}_x = \bar{k}_y = 1.00$ m and $\bar{k}_z = 1.25$ m, and that the jet at A produces a thrust of 50 N, determine the axis of precession and the rates of precession and spin after the jet has stopped.

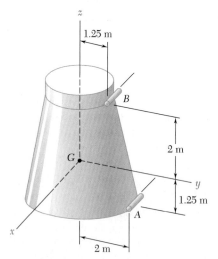

Fig. P18.126 and P18.127

18.127 The space capsule has an angular velocity $\boldsymbol{\omega} = (0.02 \text{ rad/s})\mathbf{j} + (0.10 \text{ rad/s})\mathbf{k}$ when the jet at B is activated for 1 s in a direction parallel to the x axis. Knowing that the capsule has a mass of 1000 kg, that its radii of gyration are $\bar{k}_x = \bar{k}_y = 1.00$ m and $\bar{k}_z = 1.25$ m, and that the jet at B produces a thrust of 50 N, determine the axis of precession and the rates of precession and spin after the jet has stopped.

18.128 Solve Sample Prob. 18.6, assuming that the meteorite strikes the satellite at C with a velocity $\mathbf{v}_0 = (2000 \text{ m/s})\mathbf{i}$.

18.129 An 800-lb geostationary satellite is spinning with an angular velocity $\boldsymbol{\omega}_0 = (1.5 \text{ rad/s})\mathbf{j}$ when it is hit at B by a 6-oz meteorite traveling with a velocity $\mathbf{v}_0 = -(1600 \text{ ft/s})\mathbf{i} + (1300 \text{ ft/s})\mathbf{j} + (4000 \text{ ft/s})\mathbf{k}$ relative to the satellite. Knowing that $b = 20$ in. and that the radii of gyration of the satellite are $\bar{k}_x = \bar{k}_z = 28.8$ in. and $\bar{k}_y = 32.4$ in., determine the precession axis and the rates of precession and spin of the satellite after the impact.

18.130 Solve Prob. 18.129, assuming that the meteorite hits the satellite at A instead of B.

18.131 A homogeneous rectangular plate of mass m and sides c and $2c$ is held at A and B by a fork-ended shaft of negligible mass which is supported by a bearing at C. The plate is free to rotate about AB, and the shaft is free to rotate about a horizontal axis through C. Knowing that, initially, $\theta_0 = 40°$, $\dot{\theta}_0 = 0$, and $\dot{\phi}_0 = 10$ rad/s, determine for the ensuing motion (a) the range of values of θ, (b) the minimum value of $\dot{\phi}$, (c) the maximum value of $\dot{\theta}$.

Fig. P18.129

Fig. P18.131 and P18.132

18.132 A homogeneous rectangular plate of mass m and sides c and $2c$ is held at A and B by a fork-ended shaft of negligible mass which is supported by a bearing at C. The plate is free to rotate about AB, and the shaft is free to rotate about a horizontal axis through C. Initially the plate lies in the plane of the fork ($\theta_0 = 0$) and the shaft has an angular velocity $\dot{\phi}_0 = 10$ rad/s. If the plate is slightly disturbed, determine for the ensuing motion (a) the minimum value of $\dot{\phi}$, (b) the maximum value of $\dot{\theta}$.

18.133 A homogeneous disk of radius 180 mm is welded to a rod AG of length 360 mm and of negligible mass which is connected by a clevis to a vertical shaft AB. The rod and disk can rotate freely about a horizontal axis AC, and shaft AB can rotate freely about a vertical axis. Initially rod AG is horizontal ($\theta_0 = 90°$) and has no angular velocity about AC. Knowing that the maximum value $\dot{\phi}_m$ of the angular velocity of shaft AB in the ensuing motion is twice its initial value $\dot{\phi}_0$, determine (a) the minimum value of θ, (b) the initial angular velocity $\dot{\phi}_0$ of shaft AB.

18.134 A homogeneous disk of radius 180 mm is welded to a rod AG of length 360 mm and of negligible mass which is connected by a clevis to a vertical shaft AB. The rod and disk can rotate freely about a horizontal axis AC, and shaft AB can rotate freely about a vertical axis. Initially rod AG is horizontal ($\theta_0 = 90°$) and has no angular velocity about AC. Knowing that the smallest value of θ in the ensuing motion is 30°, determine (a) the initial angular velocity of shaft AB, (b) its maximum angular velocity.

Fig. P18.133 and P18.134

Fig. P18.135

Fig. P18.136

Fig. P18.137 and P18.138

18.135 The slender homogeneous rod AB of mass m and length L is free to rotate about a horizontal axle through its mass center G. The axle is supported by a frame of negligible mass which is free to rotate about the vertical CD. Knowing that, initially, $\theta = \theta_0$, $\dot{\theta} = 0$, and $\dot{\phi} = \dot{\phi}_0$, show that the rod will oscillate about the horizontal axle and determine (a) the range of values of angle θ during this motion, (b) the maximum value of $\dot{\theta}$, (c) the minimum value of $\dot{\phi}$.

18.136 The gimbal $ABA'B'$ is of negligible mass and may rotate freely about the vertical AA'. The uniform disk of radius a and mass m may rotate freely about its diameter BB', which is also the horizontal diameter of the gimbal. (a) Applying the principle of conservation of energy, and observing that, since $\Sigma M_{AA'} = 0$, the component of the angular momentum of the disk along the fixed axis AA' must be constant, write two first-order differential equations defining the motion of the disk. (b) Given the initial conditions $\theta_0 \neq 0$, $\dot{\phi}_0 \neq 0$, and $\dot{\theta}_0 = 0$, express the rate of nutation $\dot{\theta}$ as a function of θ. (c) Show that the angle θ will never be larger than θ_0 during the ensuing motion.

*18.137 The top shown is supported at the fixed point O. Denoting by ϕ, θ, and ψ the Eulerian angles defining the position of the top with respect to a fixed frame of reference, consider the general motion of the top in which all Eulerian angles vary.

(a) Observing that $\Sigma M_Z = 0$ and $\Sigma M_z = 0$, and denoting by I and I', respectively, the moments of inertia of the top about its axis of symmetry and about a transverse axis through O, derive the two first-order differential equations of motion

$$I'\dot{\phi}\sin^2\theta + I(\dot{\psi} + \dot{\phi}\cos\theta)\cos\theta = \alpha \tag{1}$$

$$I(\dot{\psi} + \dot{\phi}\cos\theta) = \beta \tag{2}$$

where α and β are constants depending upon the initial conditions. These equations express that the angular momentum of the top is conserved about both the Z and z axes, i.e., that the rectangular component of $\mathbf{H}_O$ along each of these axes is constant.

(b) Use Eqs. (1) and (2) to show that the rectangular component ω_z of the angular velocity of the top is constant and that the rate of precession $\dot{\phi}$ depends upon the value of the angle of nutation θ.

*18.138 (a) Applying the principle of conservation of energy, derive a third differential equation for the general motion of the top of Prob. 18.137.

(b) Eliminating the derivatives $\dot{\phi}$ and $\dot{\psi}$ from the equation obtained and from the two equations of Prob. 18.137, show that the rate of nutation $\dot{\theta}$ is defined by the differential equation $\dot{\theta}^2 = f(\theta)$, where

$$f(\theta) = \frac{1}{I'}\left(2E - \frac{\beta^2}{I} - 2mgc\cos\theta\right) - \left(\frac{\alpha - \beta\cos\theta}{I'\sin\theta}\right)^2 \tag{1}$$

(c) Further show, by introducing the auxiliary variable $x = \cos\theta$, that the maximum and minimum values of θ can be obtained by solving for x the cubic equation

$$\left(2E - \frac{\beta^2}{I} - 2mgcx\right)(1 - x^2) - \frac{1}{I'}(\alpha - \beta x)^2 = 0 \tag{2}$$

*18.139 A solid cone of height 180 mm with a circular base of radius 60 mm is supported by a ball and socket at A. The cone is released from the position $\theta_0 = 30°$ with a rate of spin $\dot{\psi}_0 = 300$ rad/s, a rate of precession $\dot{\phi}_0 = 20$ rad/s, and a zero rate of nutation. Determine (a) the maximum value of θ in the ensuing motion, (b) the corresponding values of the rates of spin and precession. [*Hint:* Use Eq. (2) of Prob. 18.138; you can either solve this equation numerically or reduce it to a quadratic equation, since one of its roots is known.]

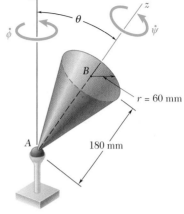

*18.140 A solid cone of height 180 mm with a circular base of radius 60 mm is supported by a ball and socket at A. The cone is released from the position $\theta_0 = 30°$ with a rate of spin $\dot{\psi}_0 = 300$ rad/s, a rate of precession $\dot{\phi}_0 = -4$ rad/s, and a zero rate of nutation. Determine (a) the maximum value of θ in the ensuing motion, (b) the corresponding values of the rates of spin and precession, (c) the value of θ for which the sense of the precession is reversed. (See hint of Prob. 18.139.)

Fig. P18.139 and P18.140

*18.141 A homogeneous sphere of mass m and radius a is welded to a rod AB of negligible mass, which is held by a ball-and-socket support at A. The sphere is released in the position $\beta = 0$ with a rate of precession $\dot{\phi}_0 = \sqrt{17g/11a}$ with no spin or nutation. Determine the largest value of β in the ensuing motion.

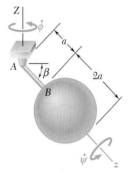

*18.142 A homogeneous sphere of mass m and radius a is welded to a rod AB of negligible mass, which is held by a ball-and-socket support at A. The sphere is released in the position $\beta = 0$ with a rate of precession $\dot{\phi} = \dot{\phi}_0$ with no spin or nutation. Knowing that the largest value of β in the ensuing motion is $30°$, determine (a) the rate of precession $\dot{\phi}_0$ of the sphere in its initial position, (b) the rates of precession and spin when $\beta = 30°$.

Fig. P18.141 and P18.142

*18.143 Consider a rigid body of arbitrary shape which is attached at its mass center O and subjected to no force other than its weight and the reaction of the support at O.

(a) Prove that the angular momentum $\mathbf{H}_O$ of the body about the fixed point O is constant in magnitude and direction, that the kinetic energy T of the body is constant, and that the projection along $\mathbf{H}_O$ of the angular velocity $\boldsymbol{\omega}$ of the body is constant.

(b) Show that the tip of the vector $\boldsymbol{\omega}$ describes a curve on a fixed plane in space (called the *invariable plane*), which is perpendicular to $\mathbf{H}_O$ and at a distance $2T/H_O$ from O.

(c) Show that with respect to a frame of reference attached to the body and coinciding with its principal axes of inertia, the tip of the vector $\boldsymbol{\omega}$ appears to describe a curve on an ellipsoid of equation

$$I_x\omega_x^2 + I_y\omega_y^2 + I_z\omega_z^2 = 2T = \text{constant}$$

The ellipsoid (called the *Poinsot ellipsoid*) is rigidly attached to the body and is of the same shape as the ellipsoid of inertia, but of a different size.

Fig. P18.143

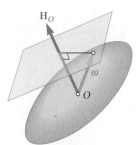

Fig. P18.144

*18.144 Referring to Prob. 18.143, (a) prove that the Poinsot ellipsoid is tangent to the invariable plane, (b) show that the motion of the rigid body must be such that the Poinsot ellipsoid appears to roll on the invariable plane. [*Hint:* In part a, show that the normal to the Poinsot ellipsoid at the tip of $\boldsymbol{\omega}$ is parallel to $\mathbf{H}_O$. It is recalled that the direction of the normal to a surface of equation $F(x, y, z) =$ constant at a point P is the same as that of **grad** F at point P.]

*18.145 Using the results obtained in Probs. 18.143 and 18.144, show that for an axisymmetrical body attached at its mass center O and under no force other than its weight and the reaction at O, the Poinsot ellipsoid is an ellipsoid of revolution and the space and body cones are both circular and are tangent to each other. Further show that (a) the two cones are tangent externally, and the precession is direct, when $I < I'$, where I and I' denote, respectively, the axial and transverse moment of inertia of the body, (b) the space cone is inside the body cone, and the precession is retrograde, when $I > I'$.

*18.146 Refer to Probs. 18.143 and 18.144.

(a) Show that the curve (called *polhode*) described by the tip of the vector $\boldsymbol{\omega}$ with respect to a frame of reference coinciding with the principal axes of inertia of the rigid body is defined by the equations

$$I_x\omega_x^2 + I_y\omega_y^2 + I_z\omega_z^2 = 2T = \text{constant} \qquad (1)$$
$$I_x^2\omega_x^2 + I_y^2\omega_y^2 + I_z^2\omega_z^2 = H_O^2 = \text{constant} \qquad (2)$$

and that this curve can, therefore, be obtained by intersecting the Poinsot ellipsoid with the ellipsoid defined by Eq. (2).

(b) Further show, assuming $I_x > I_y > I_z$, that the polhodes obtained for various values of H_O have the shapes indicated in the figure.

(c) Using the result obtained in part b, show that a rigid body under no force can rotate about a fixed centroidal axis if, and only if, that axis coincides with one of the principal axes of inertia of the body, and that the motion will be stable if the axis of rotation coincides with the major or minor axis of the Poinsot ellipsoid (z or x axis in the figure) and unstable if it coincides with the intermediate axis (y axis).

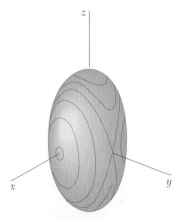

Fig. P18.146

REVIEW AND SUMMARY

This chapter was devoted to the kinetic analysis of the motion of rigid bodies in three dimensions.

Fundamental equations of motion for a rigid body

We first noted [Sec. 18.1] that the two fundamental equations derived in Chap. 14 for the motion of a system of particles

$$\Sigma \mathbf{F} = m\bar{\mathbf{a}} \qquad (18.1)$$
$$\Sigma \mathbf{M}_G = \dot{\mathbf{H}}_G \qquad (18.2)$$

provide the foundation of our analysis, just as they did in Chap. 16 in the case of the plane motion of rigid bodies. The computation of the angular momentum $\mathbf{H}_G$ of the body and of its derivative $\dot{\mathbf{H}}_G$, however, are now considerably more involved.

Angular momentum of a rigid body in three dimensions

In Sec. 18.2, we saw that the rectangular components of the angular momentum $\mathbf{H}_G$ of a rigid body can be expressed as follows in terms of the components of its angular velocity $\boldsymbol{\omega}$ and of its centroidal moments and products of inertia:

$$
\begin{aligned}
H_x &= +\bar{I}_x\,\omega_x - \bar{I}_{xy}\omega_y - \bar{I}_{xz}\omega_z \\
H_y &= -\bar{I}_{yx}\omega_x + \bar{I}_y\,\omega_y - \bar{I}_{yz}\omega_z \\
H_z &= -\bar{I}_{zx}\omega_x - \bar{I}_{zy}\omega_y + \bar{I}_z\,\omega_z
\end{aligned}
\qquad (18.7)
$$

If *principal axes of inertia* $Gx'y'z'$ are used, these relations reduce to

$$H_{x'} = \bar{I}_{x'}\omega_{x'} \qquad H_{y'} = \bar{I}_{y'}\omega_{y'} \qquad H_{z'} = \bar{I}_{z'}\omega_{z'} \qquad (18.10)$$

We observed that, in general, *the angular momentum $\mathbf{H}_G$ and the angular velocity $\boldsymbol{\omega}$ do not have the same direction* (Fig. 18.25). They will, however, have the same direction if $\boldsymbol{\omega}$ is directed along one of the principal axes of inertia of the body.

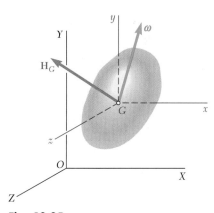

Fig. 18.25

Recalling that the system of the momenta of the particles forming a rigid body can be reduced to the vector $m\bar{\mathbf{v}}$ attached at G and the couple $\mathbf{H}_G$ (Fig. 18.26), we noted that, once the linear momentum $m\bar{\mathbf{v}}$ and the angular momentum $\mathbf{H}_G$ of a rigid body have been determined, the angular momentum $\mathbf{H}_O$ of the body about any given point O can be obtained by writing

$$\mathbf{H}_O = \bar{\mathbf{r}} \times m\bar{\mathbf{v}} + \mathbf{H}_G \qquad (18.11)$$

In the particular case of a rigid body *constrained to rotate about a fixed point* O, the components of the angular momentum $\mathbf{H}_O$ of the body about O can be obtained directly from the components of its angular velocity and from its moments and products of inertia with respect to axes through O. We wrote

$$\begin{aligned}
H_x &= +I_x\,\omega_x - I_{xy}\omega_y - I_{xz}\omega_z \\
H_y &= -I_{yx}\omega_x + I_y\,\omega_y - I_{yz}\omega_z \\
H_z &= -I_{zx}\omega_x - I_{zy}\omega_y + I_z\,\omega_z
\end{aligned} \qquad (18.13)$$

The *principle of impulse and momentum* for a rigid body in three-dimensional motion [Sec. 18.3] is expressed by the same fundamental formula that was used in Chap. 17 for a rigid body in plane motion,

Syst Momenta$_1$ + Syst Ext Imp$_{1\rightarrow 2}$ = Syst Momenta$_2$ (17.4)

but the systems of the initial and final momenta should now be represented as shown in Fig. 18.26, and $\mathbf{H}_G$ should be computed from the relations (18.7) or (18.10) [Sample Probs. 18.1 and 18.2].

The *kinetic energy* of a rigid body in three-dimensional motion can be divided into two parts [Sec. 18.4], one associated with the motion of its mass center G and the other with its motion about G. Using principal centroidal axes x', y', z', we wrote

$$T = \tfrac{1}{2}m\bar{v}^2 + \tfrac{1}{2}(\bar{I}_{x'}\omega_{x'}^2 + \bar{I}_{y'}\omega_{y'}^2 + \bar{I}_{z'}\omega_{z'}^2) \qquad (18.17)$$

where $\bar{\mathbf{v}}$ = velocity of mass center
$\boldsymbol{\omega}$ = angular velocity
m = mass of rigid body
$\bar{I}_{x'}, \bar{I}_{y'}, \bar{I}_{z'}$ = principal centroidal moments of inertia

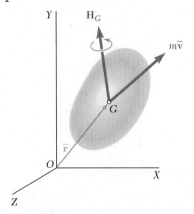

Fig. 18.26

We also noted that, in the case of a rigid body *constrained to rotate about a fixed point O*, the kinetic energy of the body can be expressed as

$$T = \tfrac{1}{2}(I_{x'}\omega_{x'}^2 + I_{y'}\omega_{y'}^2 + I_{z'}\omega_{z'}^2) \qquad (18.20)$$

where the x', y', and z' axes are the principal axes of inertia of the body at O. The results obtained in Sec. 18.4 make it possible to extend to the three-dimensional motion of a rigid body the application of the *principle of work and energy* and of the *principle of conservation of energy*.

The second part of the chapter was devoted to the application of the fundamental equations

Using a rotating frame to write the equations of motion of a rigid body in space

$$\Sigma \mathbf{F} = m\bar{\mathbf{a}} \qquad (18.1)$$

$$\Sigma \mathbf{M}_G = \dot{\mathbf{H}}_G \qquad (18.2)$$

to the motion of a rigid body in three dimensions. We first recalled [Sec. 18.5] that $\mathbf{H}_G$ represents the angular momentum of the body relative to a centroidal frame $GX'Y'Z'$ of fixed orientation (Fig. 18.27)

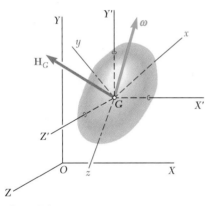

Fig. 18.27

and that $\dot{\mathbf{H}}_G$ in Eq. (18.2) represents the rate of change of $\mathbf{H}_G$ with respect to that frame. We noted that, as the body rotates, its moments and products of inertia with respect to the frame $GX'Y'Z'$ change continually. Therefore, it is more convenient to use a rotating frame $Gxyz$ when resolving $\boldsymbol{\omega}$ into components and computing the moments and products of inertia that will be used to determine $\mathbf{H}_G$ from Eqs. (18.7) or (18.10). However, since $\dot{\mathbf{H}}_G$ in Eq. (18.2) represents the rate of change of $\mathbf{H}_G$ with respect to the frame $GX'Y'Z'$ of fixed orientation, we must use the method of Sec. 15.10 to determine its value. Recalling Eq. (15.31), we wrote

$$\dot{\mathbf{H}}_G = (\dot{\mathbf{H}}_G)_{Gxyz} + \boldsymbol{\Omega} \times \mathbf{H}_G \qquad (18.22)$$

where $\mathbf{H}_G$ = angular momentum of body with respect to frame $GX'Y'Z'$ of fixed orientation

$(\dot{\mathbf{H}}_G)_{Gxyz}$ = rate of change of $\mathbf{H}_G$ with respect to rotating frame $Gxyz$, to be computed from relations (18.7)

$\boldsymbol{\Omega}$ = angular velocity of the rotating frame $Gxyz$

Substituting for $\dot{\mathbf{H}}_G$ from (18.22) into (18.2), we obtained

$$\Sigma\mathbf{M}_G = (\dot{\mathbf{H}}_G)_{Gxyz} + \mathbf{\Omega} \times \mathbf{H}_G \qquad (18.23)$$

If the rotating frame is actually attached to the body, its angular velocity $\mathbf{\Omega}$ is identically equal to the angular velocity $\boldsymbol{\omega}$ of the body. There are many applications, however, where it is advantageous to use a frame of reference which is not attached to the body but rotates in an independent manner [Sample Prob. 18.5].

Euler's equations of motion. D'Alembert's principle

Setting $\mathbf{\Omega} = \boldsymbol{\omega}$ in Eq. (18.23), using principal axes, and writing this equation in scalar form, we obtained *Euler's equations of motion* [Sec. 18.6]. A discussion of the solution of these equations and of the scalar equations corresponding to Eq. (18.1) led us to extend d'Alembert's principle to the three-dimensional motion of a rigid body and to conclude that the system of the external forces acting on the rigid body is not only equipollent, but actually *equivalent* to the effective forces of the body represented by the vector $m\bar{\mathbf{a}}$ and the couple $\dot{\mathbf{H}}_G$ (Fig. 18.28). Problems involving the three-dimensional motion of a rigid body can be solved by considering the free-body-diagram equation represented in Fig. 18.28 and writing appropriate scalar equations relating the components or moments of the external and effective forces [Sample Probs. 18.3 and 18.5].

Free-body-diagram equation

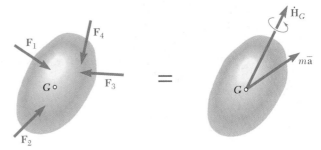

Fig. 18.28

Rigid body with a fixed point

In the case of a rigid body *constrained to rotate about a fixed point O*, an alternative method of solution, involving the moments of the forces and the rate of change of the angular momentum about point O, can be used. We wrote [Sec. 18.7]:

$$\Sigma\mathbf{M}_O = (\dot{\mathbf{H}}_O)_{Oxyz} + \mathbf{\Omega} \times \mathbf{H}_O \qquad (18.28)$$

where $\Sigma\mathbf{M}_O$ = sum of moments about O of forces applied to rigid body

$\mathbf{H}_O$ = angular momentum of body with respect to fixed frame $OXYZ$

$(\dot{\mathbf{H}}_O)_{Oxyz}$ = rate of change of $\mathbf{H}_O$ with respect to rotating frame $Oxyz$, to be computed from relations (18.13)

$\mathbf{\Omega}$ = angular velocity of rotating frame $Oxyz$

This approach can be used to solve certain problems involving the rotation of a rigid body about a fixed axis [Sec. 18.8], for example, an unbalanced rotating shaft [Sample Prob. 18.4].

In the last part of the chapter, we considered the motion of *gyroscopes* and other *axisymmetrical bodies*. Introducing the *Eulerian angles* ϕ, θ, and ψ to define the position of a gyroscope (Fig. 18.29), we observed that their derivatives $\dot{\phi}$, $\dot{\theta}$, and $\dot{\psi}$ represent, respectively, the rates of *precession*, *nutation*, and *spin* of the gyroscope [Sec. 18.9]. Expressing the angular velocity $\boldsymbol{\omega}$ in terms of these derivatives, we wrote

$$\boldsymbol{\omega} = -\dot{\phi} \sin\theta \, \mathbf{i} + \dot{\theta}\mathbf{j} + (\dot{\psi} + \dot{\phi} \cos\theta)\mathbf{k} \quad (18.35)$$

Motion of a gyroscope

Fig. 18.29

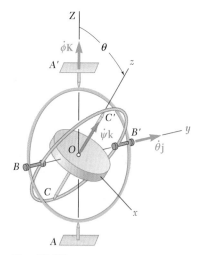

Fig. 18.30

where the unit vectors are associated with a frame $Oxyz$ attached to the inner gimbal of the gyroscope (Fig. 18.30) and rotate, therefore, with the angular velocity

$$\boldsymbol{\Omega} = -\dot{\phi} \sin\theta \, \mathbf{i} + \dot{\theta}\mathbf{j} + \dot{\phi} \cos\theta \, \mathbf{k} \quad (18.38)$$

Denoting by I the moment of inertia of the gyroscope with respect to its spin axis z and by I' its moment of inertia with respect to a transverse axis through O, we wrote

$$\mathbf{H}_O = -I' \dot{\phi} \sin\theta \, \mathbf{i} + I' \dot{\theta}\mathbf{j} + I(\dot{\psi} + \dot{\phi} \cos\theta)\mathbf{k} \quad (18.36)$$

Substituting for $\mathbf{H}_O$ and $\boldsymbol{\Omega}$ into Eq. (18.28) led us to the differential equations defining the motion of the gyroscope.

In the particular case of the *steady precession* of a gyroscope [Sec. 18.10], the angle θ, the rate of precession $\dot{\phi}$, and the rate of spin $\dot{\psi}$ remain constant. We saw that such a motion is possible only if the moments of the external forces about O satisfy the relation

$$\Sigma\mathbf{M}_O = (I\omega_z - I' \dot{\phi} \cos\theta)\dot{\phi} \sin\theta\mathbf{j} \quad (18.44)$$

i.e., if the external forces reduce to a couple of moment equal to the right-hand member of Eq. (18.44) and applied *about an axis perpendicular to the precession axis and to the spin axis* (Fig. 18.31). The chapter ended with a discussion of the motion of an axisymmetrical body spinning and precessing *under no force* [Sec. 18.11; Sample Prob. 18.6].

Steady precession

Fig. 18.31

18.147 Three 25-lb rotor disks are attached to a shaft which rotates at 720 rpm. Disk A is attached eccentrically so that its mass center is $\frac{1}{4}$ in. from the axis of rotation, while disks B and C are attached so that their mass centers coincide with the axis of rotation. Where should 2-lb weights be bolted to disks B and C to balance the system dynamically?

Fig. P18.147

18.148 A homogeneous disk of mass $m = 5$ kg rotates at the constant rate $\omega_1 = 8$ rad/s with respect to the bent axle ABC, which itself rotates at the constant rate $\omega_2 = 3$ rad/s about the y axis. Determine the angular momentum $\mathbf{H}_C$ of the disk about its center C.

18.149 A rod of uniform cross section is used to form the shaft shown. Denoting by m the total mass of the shaft and knowing that the shaft rotates with a constant angular velocity ω, determine (a) the angular momentum $\mathbf{H}_G$ of the shaft about its mass center G, (b) the angle formed by $\mathbf{H}_G$ and the axis AB, (c) the angular momentum of the shaft about point A.

Fig. P18.149

18.150 A uniform rod of mass m and length $5a$ is bent into the shape shown and is suspended from a wire attached at point B. Knowing that the rod is hit at point A in the negative y direction and denoting the corresponding impulse by $-(F \,\Delta t)\mathbf{j}$, determine immediately after the impact (a) the velocity of the mass center G, (b) the angular velocity of the rod.

Fig. P18.148

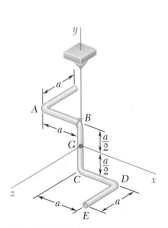

Fig. P18.150

18.151 A four-bladed airplane propeller has a mass of 160 kg and a radius of gyration of 800 mm. Knowing that the propeller rotates at 1600 rpm as the airplane is traveling in a circular path of 600-m radius at 540 km/h, determine the magnitude of the couple exerted by the propeller on its shaft due to the rotation of the airplane.

Fig. P18.151

18.152 A 2.4-kg piece of sheet steel with dimensions 160 × 640 mm was bent to form the component shown. The component is at rest ($\omega = 0$) when a couple $\mathbf{M}_0 = (0.8 \text{ N} \cdot \text{m})\mathbf{k}$ is applied to it. Determine (a) the angular acceleration of the component, (b) the dynamic reactions at A and B immediately after the couple is applied.

18.153 A homogeneous disk of weight $W = 6$ lb rotates at the constant rate $\omega_1 = 16$ rad/s with respect to arm ABC, which is welded to a shaft DCE rotating at the constant rate $\omega_2 = 8$ rad/s. Determine the dynamic reactions at D and E.

18.154 A 48-kg advertising panel of length $2a = 2.4$ m and width $2b = 1.6$ m is kept rotating at a constant rate ω_1 about its horizontal axis by a small electric motor attached at A to frame ACB. This frame itself is kept rotating at a constant rate ω_2 about a vertical axis by a second motor attached at C to the column CD. Knowing that the panel and the frame complete a full revolution in 6 s and 12 s, respectively, express, as a function of the angle θ, the dynamic reaction exerted on column CD by its support at D.

Fig. P18.152

Fig. P18.153

Fig. P18.154

18.155 A 2500-kg satellite is 2.4 m high and has octagonal bases of sides 1.2 m. The coordinate axes shown are the principal centroidal axes of inertia of the satellite, and its radii of gyration are $k_x = k_z = 0.90$ m and $k_y = 0.98$ m. The satellite is equipped with a main 500-N thruster E and four 20-N thrusters A, B, C, and D which can expel fuel in the positive y direction. The satellite is spinning at the rate of 36 rev/h about its axis of symmetry Gy, which maintains a fixed direction in space, when thrusters A and B are activated for 2 s. Determine (a) the precession axis of the satellite, (b) its rate of precession, (c) its rate of spin.

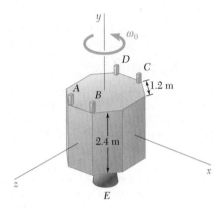

Fig. P18.155

18.156 A thin disk of weight $W = 8$ lb rotates with an angular velocity ω_2 with respect to arm OA, which itself rotates with an angular velocity ω_1 about the y axis. Determine (a) the couple $M_1\mathbf{j}$ which should be applied to arm OA to give it an angular acceleration $\boldsymbol{\alpha}_1 = (6 \text{ rad/s}^2)\mathbf{j}$ with $\omega_1 = 4$ rad/s, knowing that the disk rotates at the constant rate $\omega_2 = 12$ rad/s, (b) the force-couple system representing the dynamic reaction at O at that instant. Assume that arm OA has negligible mass.

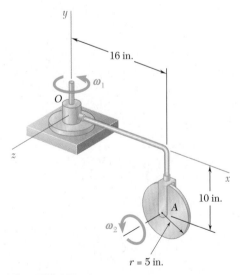

Fig. P18.156

18.157 A homogeneous disk of mass m is connected at A and B to a fork-ended shaft of negligible mass which is supported by a bearing at C. The disk is free to rotate about its horizontal diameter AB and the shaft is free to rotate about a vertical axis through C. Initially the disk lies in a vertical plane ($\theta_0 = 90°$) and the shaft has an angular velocity $\dot{\phi}_0 = 8$ rad/s. If the disk is slightly disturbed, determine for the ensuing motion (a) the minimum value of $\dot{\phi}$, (b) the maximum value of θ.

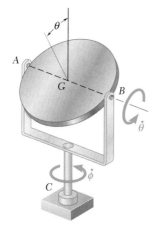

Fig. P18.157

18.158 The essential features of the gyrocompass are shown. The rotor spins at the rate $\dot{\psi}$ about an axis mounted in a single gimbal, which may rotate freely about the vertical axis AB. The angle formed by the axis of the rotor and the plane of the meridian is denoted by θ, and the latitude of the position on the earth is denoted by λ. We note that line OC is parallel to the axis of the earth, and we denote by $\boldsymbol{\omega}_e$ the angular velocity of the earth about its axis.

(a) Show that the equations of motion of the gyrocompass are

$$I'\ddot{\theta} + I\omega_z\omega_e \cos\lambda \sin\theta - I'\omega_e^2 \cos^2\lambda \sin\theta \cos\theta = 0$$

$$I\dot{\omega}_z = 0$$

where ω_z is the rectangular component of the total angular velocity $\boldsymbol{\omega}$ along the axis of the rotor, and I and I' are the moments of inertia of the rotor with respect to its axis of symmetry and a transverse axis through O, respectively.

(b) Neglecting the term containing ω_e^2, show that for small values of θ, we have

$$\ddot{\theta} + \frac{I\omega_z\omega_e \cos\lambda}{I'}\theta = 0$$

and that the axis of the gyrocompass oscillates about the north-south direction.

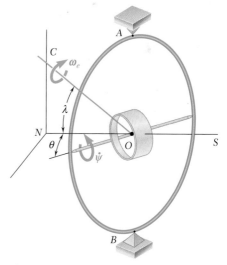

Fig. P18.158

COMPUTER PROBLEMS

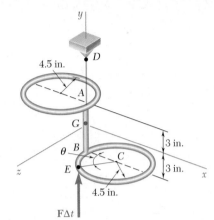

Fig. P18.C1

18.C1 A wire of uniform cross section weighing $\frac{5}{8}$ oz/ft is used to form the wire figure shown, which is suspended from cord AD. An impulse $\mathbf{F}\,\Delta t = (0.5\ \text{lb} \cdot \text{s})\mathbf{j}$ is applied to the wire figure at point E. Use computational software to calculate and plot immediately after the impact, for values of θ from 0 to 180°, (*a*) the velocity of the mass center of the wire figure, (*b*) the angular velocity of the figure.

18.C2 A 2500-kg probe in orbit about the moon is 2.4 m high and has octagonal bases of sides 1.2 m. The coordinate axes shown are the principal centroidal axes of inertia of the probe, and its radii of gyration are $k_x = 0.98$ m, $k_y = 1.06$ m, and $k_z = 1.02$ m. The probe is equipped with a main 500-N thruster E and four 20-N thrusters A, B, C, and D that can expel fuel in the positive y direction. The probe has an angular velocity $\boldsymbol{\omega} = \omega_x\,\mathbf{i} + \omega_z\,\mathbf{k}$ when two of the 20-N thrusters are used to reduce the angular velocity to zero. Use computational software to determine for any pair of values of ω_x and ω_z less than or equal to 0.06 rad/s, which of the thrusters should be used and for how long each of them should be activated. Apply this program assuming $\boldsymbol{\omega}$ to be (*a*) $\boldsymbol{\omega} = (0.040\ \text{rad/s})\mathbf{i} + (0.060\ \text{rad/s})\mathbf{k}$, (*b*) $\boldsymbol{\omega} = (0.060\ \text{rad/s})\mathbf{i} - (0.040\ \text{rad/s})\mathbf{k}$, (*c*) $\boldsymbol{\omega} = (0.06\ \text{rad/s})\mathbf{i} + (0.02\ \text{rad/s})\mathbf{k}$, (*d*) $\boldsymbol{\omega} = -(0.06\ \text{rad/s})\mathbf{i} - (0.02\ \text{rad/s})\mathbf{k}$.

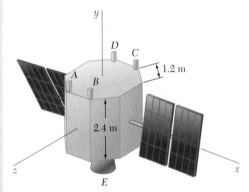

Fig. P18.C2

18.C3 A couple $\mathbf{M}_0 = (0.03\ \text{lb} \cdot \text{ft})\mathbf{i}$ is applied to an assembly consisting of pieces of sheet aluminum of uniform thickness and of total weight 2.7 lb, which are welded to a light axle supported by bearings at A and B. Use computational software to determine the dynamic reactions exerted by the bearings on the axle at any time t after the couple has been applied. Resolve these reactions into components directed along y and z axes rotating with the assembly. (*a*) Calculate and plot the components of the reactions from $t = 0$ to $t = 2$ s at 0.1-s intervals. (*b*) Determine the time at which the z components of the reactions at A and B are equal to zero.

Fig. P18.C3

18.C4 A 2.5-kg homogeneous disk of radius 80 mm can rotate with respect to arm ABC, which is welded to a shaft DCE supported by bearings at D and E. Both the arm and the shaft are of negligible mass. At time $t = 0$ a couple $\mathbf{M}_0 = (0.5\ \text{N} \cdot \text{m})\mathbf{k}$ is applied to shaft DCE. Knowing that at $t = 0$ the angular velocity of the disk is $\boldsymbol{\omega}_1 = (60\ \text{rad/s})\mathbf{j}$ and that friction in the bearing at A causes the magnitude of $\boldsymbol{\omega}_1$ to decrease at the rate of 15 rad/s², determine the dynamic reactions exerted on the shaft by the bearings at D and E at any time t. Resolve these reactions into components directed along x and y axes rotating with the shaft. Use computational software (*a*) to calculate the components of the reactions from $t = 0$ to $t = 4$ s, (*b*) to determine the times t_1 and t_2 at which the x and y components of the reaction at E are respectively equal to zero.

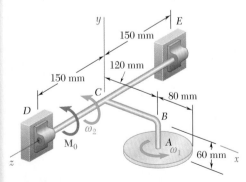

Fig. P18.C4

18.C5 A homogeneous disk of radius 180 mm is welded to a rod AG of length 360 mm and of negligible mass which is connected by a clevis to a vertical shaft AB. The rod and disk can rotate freely about a horizontal axis AC, and shaft AB can rotate freely about a vertical axis. Initially rod AG forms a given angle θ_0 with the downward vertical and its angular velocity $\dot{\theta}_0$ about AC is zero. Shaft AB is then given an angular velocity $\dot{\phi}_0$ about the vertical. Use computational software (a) to calculate the minimum value θ_m of the angle θ in the ensuing motion and the period of oscillation in θ, that is, the time required for θ to regain its initial value θ_0, (b) to compute the angular velocity $\dot{\phi}$ of shaft AB for values of θ from θ_0 to θ_m. Apply this program with the initial conditions (i) $\theta_0 = 90°$, $\dot{\phi}_0 = 5$ rad/s, (ii) $\theta_0 = 90°$, $\dot{\phi}_0 = 10$ rad/s, (iii) $\theta_0 = 60°$, $\dot{\phi}_0 = 5$ rad/s. [Hint: Use the principle of conservation of energy and the fact that the angular momentum of the body about the vertical through A is conserved to obtain an equation of the form $\dot{\theta}^2 = f(\theta)$. This equation can be integrated by a numerical method.]

18.C6 A homogeneous disk of radius 180 mm is welded to a rod AG of length 360 mm and of negligible mass which is supported by a ball-and-socket joint at A. The disk is released in the position $\theta = \theta_0$ with a rate of spin $\dot{\psi}_0$, a rate of precession $\dot{\phi}_0$, and a zero rate of nutation. Use computational software (a) to calculate the minimum value θ_m of the angle θ in the ensuing motion and the period of oscillation in θ, that is, the time required for θ to regain its initial value θ_0, (b) to compute the rate of spin $\dot{\psi}$ and the rate of precession $\dot{\phi}$ for values of θ from θ_0 to θ_m, using 2° decrements. Apply this program with the initial conditions (i) $\theta_0 = 90°$, $\dot{\psi}_0 = 50$ rad/s, $\dot{\phi}_0 = 0$, (ii) $\theta_0 = 90°$, $\dot{\psi}_0 = 0$, $\dot{\phi}_0 = 5$ rad/s, (iii) $\theta_0 = 90°$, $\dot{\psi}_0 = 50$ rad/s, $\dot{\phi}_0 = 5$ rad/s, (iv) $\theta_0 = 90°$, $\dot{\psi}_0 = 10$ rad/s, $\dot{\phi}_0 = 5$ rad/s, (v) $\theta_0 = 60°$, $\dot{\psi}_0 = 50$ rad/s, $\dot{\phi}_0 = 5$ rad/s. [Hint: Use the principle of conservation of energy and the fact that the angular momentum of the body is conserved about both the Z and z axes to obtain an equation of the form $\dot{\theta}^2 = f(\theta)$. This equation can be integrated by a numerical method.]

Fig. P18.C5

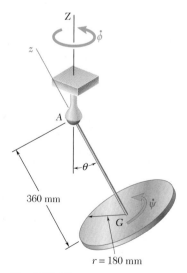

Fig. P18.C6

The Wind Damper inside of Taipei 101 helps protect against typhoons and earthquakes by reducing the effects of wind and vibrations on the building. Mechanical systems may undergo *free vibrations* or they may be subject to *forced vibrations*. The vibrations are *damped* when there is energy dissipation and *undamped* otherwise. This chapter is an introduction to many fundamental concepts in vibration analysis.

CHAPTER 19

Mechanical Vibrations

19.1 INTRODUCTION

A *mechanical vibration* is the motion of a particle or a body which oscillates about a position of equilibrium. Most vibrations in machines and structures are undesirable because of the increased stresses and energy losses which accompany them. They should therefore be eliminated or reduced as much as possible by appropriate design. The analysis of vibrations has become increasingly important in recent years owing to the current trend toward higher-speed machines and lighter structures. There is every reason to expect that this trend will continue and that an even greater need for vibration analysis will develop in the future.

The analysis of vibrations is a very extensive subject to which entire texts have been devoted. Our present study will therefore be limited to the simpler types of vibrations, namely, the vibrations of a body or a system of bodies with one degree of freedom.

A mechanical vibration generally results when a system is displaced from a position of stable equilibrium. The system tends to return to this position under the action of restoring forces (either elastic forces, as in the case of a mass attached to a spring, or gravitational forces, as in the case of a pendulum). But the system generally reaches its original position with a certain acquired velocity which carries it beyond that position. Since the process can be repeated indefinitely, the system keeps moving back and forth across its position of equilibrium. The time interval required for the system to complete a full cycle of motion is called the *period* of the vibration. The number of cycles per unit time defines the *frequency*, and the maximum displacement of the system from its position of equilibrium is called the *amplitude* of the vibration.

When the motion is maintained by the restoring forces only, the vibration is said to be a *free vibration* (Secs. 19.2 to 19.6). When a periodic force is applied to the system, the resulting motion is described as a *forced vibration* (Sec. 19.7). When the effects of friction can be neglected, the vibrations are said to be *undamped*. However, all vibrations are actually *damped* to some degree. If a free vibration is only slightly damped, its amplitude slowly decreases until, after a certain time, the motion comes to a stop. But if damping is large enough to prevent any true vibration, the system then slowly regains its original position (Sec. 19.8). A damped forced vibration is maintained as long as the periodic force which produces the vibration is applied. The amplitude of the vibration, however, is affected by the magnitude of the damping forces (Sec. 19.9).

VIBRATIONS WITHOUT DAMPING

19.2 FREE VIBRATIONS OF PARTICLES. SIMPLE HARMONIC MOTION

Consider a body of mass m attached to a spring of constant k (Fig. 19.1a). Since at the present time we are concerned only with the motion of its mass center, we will refer to this body as a particle. When the particle is in static equilibrium, the forces acting on it are its weight $\mathbf{W}$ and the force $\mathbf{T}$ exerted by the spring, of magnitude

$T = k\delta_{st}$, where δ_{st} denotes the elongation of the spring. We have, therefore,

$$W = k\delta_{st}$$

Suppose now that the particle is displaced through a distance x_m from its equilibrium position and released with no initial velocity. If x_m has been chosen smaller than δ_{st}, the particle will move back and forth through its equilibrium position; a vibration of amplitude x_m has been generated. Note that the vibration can also be produced by imparting a certain initial velocity to the particle when it is in its equilibrium position $x = 0$ or, more generally, by starting the particle from any given position $x = x_0$ with a given initial velocity $\mathbf{v}_0$.

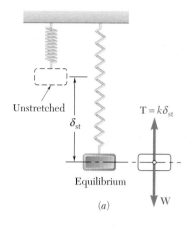

To analyze the vibration, let us consider the particle in a position P at some arbitrary time t (Fig. 19.1b). Denoting by x the displacement OP measured from the equilibrium position O (positive downward), we note that the forces acting on the particle are its weight $\mathbf{W}$ and the force $\mathbf{T}$ exerted by the spring which, in this position, has a magnitude $T = k(\delta_{st} + x)$. Recalling that $W = k\delta_{st}$, we find that the magnitude of the resultant $\mathbf{F}$ of the two forces (positive downward) is

$$F = W - k(\delta_{st} + x) = -kx \qquad (19.1)$$

Thus the *resultant* of the forces exerted on the particle is proportional to the displacement *OP measured from the equilibrium position*. Recalling the sign convention, we note that $\mathbf{F}$ is always directed *toward* the equilibrium position O. Substituting for F into the fundamental equation $F = ma$ and recalling that a is the second derivative $\ddot{x}$ of x with respect to t, we write

$$m\ddot{x} + kx = 0 \qquad (19.2)$$

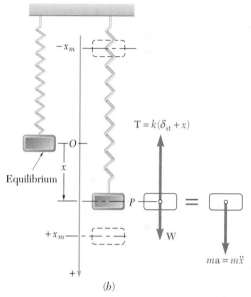

Note that the same sign convention should be used for the acceleration $\ddot{x}$ and for the displacement x, namely, positive downward.

The motion defined by Eq. (19.2) is called a *simple harmonic motion*. It is characterized by the fact that *the acceleration is proportional to the displacement and of opposite direction*. We can verify that each of the functions $x_1 = \sin(\sqrt{k/m}\,t)$ and $x_2 = \cos(\sqrt{k/m}\,t)$ satisfies Eq. (19.2). These functions, therefore, constitute two *particular solutions* of the differential equation (19.2). The *general solution* of Eq. (19.2) is obtained by multiplying each of the particular solutions by an arbitrary constant and adding. Thus, the general solution is expressed as

$$x = C_1 x_1 + C_2 x_2 = C_1 \sin\left(\sqrt{\frac{k}{m}}\,t\right) + C_2 \cos\left(\sqrt{\frac{k}{m}}\,t\right) \qquad (19.3)$$

We note that x is a *periodic function* of the time t and does, therefore, represent a vibration of the particle P. The coefficient of t in the expression we have obtained is referred to as the *natural circular frequency* of the vibration and is denoted by ω_n. We have

$$\text{Natural circular frequency} = \omega_n = \sqrt{\frac{k}{m}} \qquad (19.4)$$

Fig. 19.1

Substituting for $\sqrt{k/m}$ into Eq. (19.3), we write

$$x = C_1 \sin \omega_n t + C_2 \cos \omega_n t \qquad (19.5)$$

This is the general solution of the differential equation

$$\ddot{x} + \omega_n^2 x = 0 \qquad (19.6)$$

which can be obtained from Eq. (19.2) by dividing both terms by m and observing that $k/m = \omega_n^2$. Differentiating twice both members of Eq. (19.5) with respect to t, we obtain the following expressions for the velocity and the acceleration at time t:

$$v = \dot{x} = C_1 \omega_n \cos \omega_n t - C_2 \omega_n \sin \omega_n t \qquad (19.7)$$
$$a = \ddot{x} = -C_1 \omega_n^2 \sin \omega_n t - C_2 \omega_n^2 \cos \omega_n t \qquad (19.8)$$

The values of the constants C_1 and C_2 depend upon the *initial conditions* of the motion. For example, we have $C_1 = 0$ if the particle is displaced from its equilibrium position and released at $t = 0$ with no initial velocity, and we have $C_2 = 0$ if the particle is started from O at $t = 0$ with a certain initial velocity. In general, substituting $t = 0$ and the initial values x_0 and v_0 of the displacement and the velocity into Eqs. (19.5) and (19.7), we find that $C_1 = v_0/\omega_n$ and $C_2 = x_0$.

The expressions obtained for the displacement, velocity, and acceleration of a particle can be written in a more compact form if we observe that Eq. (19.5) expresses that the displacement $x = OP$ is the sum of the x components of two vectors $\mathbf{C}_1$ and $\mathbf{C}_2$, respectively, of magnitude C_1 and C_2, directed as shown in Fig. 19.2a. As t varies, both vectors rotate clockwise; we also note that the magnitude of their resultant $\overrightarrow{OQ}$ is equal to the maximum displacement x_m. The simple harmonic motion of P along the x axis can thus be obtained by projecting on this axis the motion of a point Q describing an *auxiliary circle* of radius x_m *with a constant angular velocity* ω_n (which explains the name of natural *circular* frequency given to ω_n). Denoting by ϕ the angle formed by the vectors $\overrightarrow{OQ}$ and $\mathbf{C}_1$, we write

$$OP = OQ \sin (\omega_n t + \phi) \qquad (19.9)$$

which leads to new expressions for the displacement, velocity, and acceleration of P:

$$x = x_m \sin (\omega_n t + \phi) \qquad (19.10)$$

$$v = \dot{x} = x_m \omega_n \cos (\omega_n t + \phi) \qquad (19.11)$$
$$a = \ddot{x} = -x_m \omega_n^2 \sin (\omega_n t + \phi) \qquad (19.12)$$

The displacement-time curve is represented by a sine curve (Fig. 19.2b); the maximum value x_m of the displacement is called the *amplitude* of the vibration, and the angle ϕ which defines the initial position of Q on the circle is called the *phase angle*. We note from Fig. 19.2 that a full *cycle* is described as the angle $\omega_n t$ increases by 2π rad. The corresponding value of t, denoted by τ_n, is called the *period* of the free vibration and is measured in seconds. We have

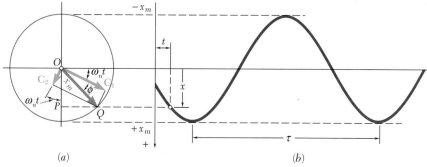

(a)   (b)

Fig. 19.2

$$\text{period} = \tau_n = \frac{2\pi}{\omega_n} \qquad (19.13)$$

The number of cycles described per unit of time is denoted by f_n and is known as the *natural frequency* of the vibration. We write

$$\text{Natural frequency} = f_n = \frac{1}{\tau_n} = \frac{\omega_n}{2\pi} \qquad (19.14)$$

The unit of frequency is a frequency of 1 cycle per second, corresponding to a period of 1 s. In terms of base units the unit of frequency is thus 1/s or s^{-1}. It is called a *hertz* (Hz) in the SI system of units. It also follows from Eq. (19.14) that a frequency of 1 s^{-1} or 1 Hz corresponds to a circular frequency of 2π rad/s. In problems involving angular velocities expressed in revolutions per minute (rpm), we have 1 rpm = $\frac{1}{60}$ s^{-1} = $\frac{1}{60}$ Hz, or 1 rpm = $(2\pi/60)$ rad/s.

Recalling that ω_n was defined in (19.4) in terms of the constant k of the spring and the mass m of the particle, we observe that the period and the frequency are independent of the initial conditions and of the amplitude of the vibration. Note that τ_n and f_n depend on the *mass* rather than on the *weight* of the particle and thus are independent of the value of g.

The velocity-time and acceleration-time curves can be represented by sine curves of the same period as the displacement-time curve, but with different phase angles. From Eqs. (19.11) and (19.12), we note that the maximum values of the magnitudes of the velocity and acceleration are

$$v_m = x_m\omega_n \qquad a_m = x_m\omega_n^2 \qquad (19.15)$$

Since the point Q describes the auxiliary circle, of radius x_m, at the constant angular velocity ω_n, its velocity and acceleration are equal, respectively, to the expressions (19.15). Recalling Eqs. (19.11) and (19.12), we find, therefore, that the velocity and acceleration of P can be obtained at any instant by projecting on the x axis vectors of magnitudes $v_m = x_m\omega_n$ and $a_m = x_m\omega_n^2$ representing, respectively, the velocity and acceleration of Q at the same instant (Fig. 19.3).

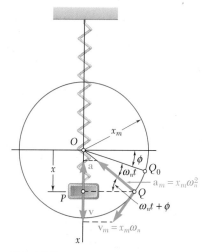

Fig. 19.3

The results obtained are not limited to the solution of the problem of a mass attached to a spring. They can be used to analyze the rectilinear motion of a particle *whenever the resultant* **F** *of the forces acting on the particle is proportional to the displacement x and directed toward O.* The fundamental equation of motion $F = ma$ can then be written in the form of Eq. (19.6), which is characteristic of a simple harmonic motion. Observing that the coefficient of x must be equal to ω_n^2, we can easily determine the natural circular frequency ω_n of the motion. Substituting the value obtained for ω_n into Eqs. (19.13) and (19.14), we then obtain the period τ_n and the natural frequency f_n of the motion.

(a)

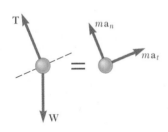

(b)

Fig. 19.4

19.3 SIMPLE PENDULUM (APPROXIMATE SOLUTION)

Most of the vibrations encountered in engineering applications can be represented by a simple harmonic motion. Many others, although of a different type, can be *approximated* by a simple harmonic motion, provided that their amplitude remains small. Consider, for example, a *simple pendulum,* consisting of a bob of mass m attached to a cord of length l, which can oscillate in a vertical plane (Fig. 19.4a). At a given time t, the cord forms an angle θ with the vertical. The forces acting on the bob are its weight **W** and the force **T** exerted by the cord (Fig. 19.4b). Resolving the vector $m\mathbf{a}$ into tangential and normal components, with $m\mathbf{a}_t$ directed to the right, i.e., in the direction corresponding to increasing values of θ, and observing that $a_t = l\alpha = l\ddot{\theta}$, we write

$$\Sigma F_t = ma_t: \qquad -W \sin \theta = ml\ddot{\theta}$$

Noting that $W = mg$ and dividing through by ml, we obtain

$$\ddot{\theta} + \frac{g}{l} \sin \theta = 0 \tag{19.16}$$

For oscillations of small amplitude, we can replace $\sin \theta$ by θ, expressed in radians, and write

$$\ddot{\theta} + \frac{g}{l} \theta = 0 \tag{19.17}$$

Comparison with Eq. (19.6) shows that the differential equation (19.17) is that of a simple harmonic motion with a natural circular frequency ω_n equal to $(g/l)^{1/2}$. The general solution of Eq. (19.17) can, therefore, be expressed as

$$\theta = \theta_m \sin (\omega_n t + \phi)$$

where θ_m is the amplitude of the oscillations and ϕ is a phase angle. Substituting into Eq. (19.13) the value obtained for ω_n, we get the following expression for the period of the small oscillations of a pendulum of length l:

$$\tau_n = \frac{2\pi}{\omega_n} = 2\pi \sqrt{\frac{l}{g}} \tag{19.18}$$

*19.4 SIMPLE PENDULUM (EXACT SOLUTION)

Formula (19.18) is only approximate. To obtain an exact expression for the period of the oscillations of a simple pendulum, we must return to Eq. (19.16). Multiplying both terms by $2\dot{\theta}$ and integrating from an initial position corresponding to the maximum deflection, that is, $\theta = \theta_m$ and $\dot{\theta} = 0$, we write

$$\left(\frac{d\theta}{dt}\right)^2 = \frac{2g}{l}(\cos \theta - \cos \theta_m)$$

Replacing $\cos \theta$ by $1 - 2 \sin^2 (\theta/2)$ and $\cos \theta_m$ by a similar expression, solving for dt, and integrating over a quarter period from $t = 0$, $\theta = 0$ to $t = \tau_n/4$, $\theta = \theta_m$, we have

$$\tau_n = 2\sqrt{\frac{l}{g}} \int_0^{\theta_m} \frac{d\theta}{\sqrt{\sin^2(\theta_m/2) - \sin^2(\theta/2)}}$$

The integral in the right-hand member is known as an *elliptic integral*; it cannot be expressed in terms of the usual algebraic or trigonometric functions. However, setting

$$\sin (\theta/2) = \sin (\theta_m/2) \sin \phi$$

we can write

$$\tau_n = 4\sqrt{\frac{l}{g}} \int_0^{\pi/2} \frac{d\phi}{\sqrt{1 - \sin^2(\theta_m/2) \sin^2 \phi}} \qquad (19.19)$$

where the integral obtained, commonly denoted by K, can be calculated by using a numerical method of integration. It can also be found in *tables of elliptic integrals* for various values of $\theta_m/2$.† In order to compare the result just obtained with that of the preceding section, we write Eq. (19.19) in the form

$$\tau_n = \frac{2K}{\pi}\left(2\pi \sqrt{\frac{l}{g}}\right) \qquad (19.20)$$

Formula (19.20) shows that the actual value of the period of a simple pendulum can be obtained by multiplying the approximate value given in Eq. (19.18) by the correction factor $2K/\pi$. Values of the correction factor are given in Table 19.1 for various values of the amplitude θ_m. We note that for ordinary engineering computations the correction factor can be omitted as long as the amplitude does not exceed 10°.

TABLE 19.1 **Correction Factor for the Period of a Simple Pendulum**

θ_m	0°	10°	20°	30°	60°	90°	120°	150°	180°
K	1.571	1.574	1.583	1.598	1.686	1.854	2.157	2.768	∞
$2K/\pi$	1.000	1.002	1.008	1.017	1.073	1.180	1.373	1.762	∞

†See, for example, *Standard Mathematical Tables*, Chemical Rubber Publishing Company, Cleveland, Ohio.

(a)

(b)

SAMPLE PROBLEM 19.1

A 50-kg block moves between vertical guides as shown. The block is pulled 40 mm down from its equilibrium position and released. For each spring arrangement, determine the period of the vibration, the maximum velocity of the block, and the maximum acceleration of the block.

SOLUTION

a. Springs Attached in Parallel. We first determine the constant k of a single spring equivalent to the two springs *by finding the magnitude of the force* **P** required to cause a given deflection δ. Since for a deflection δ the magnitudes of the forces exerted by the springs are, respectively, $k_1\delta$ and $k_2\delta$, we have

$$P = k_1\delta + k_2\delta = (k_1 + k_2)\delta$$

The constant k of the single equivalent spring is

$$k = \frac{P}{\delta} = k_1 + k_2 = 4 \text{ kN/m} + 6 \text{ kN/m} = 10 \text{ kN/m} = 10^4 \text{ N/m}$$

Period of Vibration: Since $m = 50$ kg, Eq. (19.4) yields

$$\omega_n^2 = \frac{k}{m} = \frac{10^4 \text{ N/m}}{50 \text{ kg}} \qquad \omega_n = 14.14 \text{ rad/s}$$

$$\tau_n = 2\pi/\omega_n \qquad \tau_n = 0.444 \text{ s} \quad \blacktriangleleft$$

Maximum Velocity: $\quad v_m = x_m\omega_n = (0.040 \text{ m})(14.14 \text{ rad/s})$

$$v_m = 0.566 \text{ m/s} \qquad \mathbf{v}_m = 0.566 \text{ m/s} \updownarrow \quad \blacktriangleleft$$

Maximum Acceleration: $\quad a_m = x_m\omega_n^2 = (0.040 \text{ m})(14.14 \text{ rad/s})^2$

$$a_m = 8.00 \text{ m/s}^2 \qquad \mathbf{a}_m = 8.00 \text{ m/s}^2 \updownarrow \quad \blacktriangleleft$$

b. Springs Attached in Series. We first determine the constant k of a single spring equivalent to the two springs *by finding the total elongation δ* of the springs under a given static load **P**. To facilitate the computation, a static load of magnitude $P = 12$ kN is used.

$$\delta = \delta_1 + \delta_2 = \frac{P}{k_1} + \frac{P}{k_2} = \frac{12 \text{ kN}}{4 \text{ kN/m}} + \frac{12 \text{ kN}}{6 \text{ kN/m}} = 5 \text{ m}$$

$$k = \frac{P}{\delta} = \frac{12 \text{ kN}}{5 \text{ m}} = 2.4 \text{ kN/m} = 2400 \text{ N/m}$$

Period of Vibration: $\qquad \omega_n^2 = \frac{k}{m} = \frac{2400 \text{ N/m}}{50 \text{ kg}} \qquad \omega_n = 6.93 \text{ rad/s}$

$$\tau_n = \frac{2\pi}{\omega_n} \qquad \tau_n = 0.907 \text{ s} \quad \blacktriangleleft$$

Maximum Velocity: $\quad v_m = x_m\omega_n = (0.040 \text{ m})(6.93 \text{ rad/s})$

$$v_m = 0.277 \text{ m/s} \qquad \mathbf{v}_m = 0.277 \text{ m/s} \updownarrow \quad \blacktriangleleft$$

Maximum Acceleration: $\quad a_m = x_m\omega_n^2 = (0.040 \text{ m})(6.93 \text{ rad/s})^2$

$$a_m = 1.920 \text{ m/s}^2 \qquad \mathbf{a}_m = 1.920 \text{ m/s}^2 \updownarrow \quad \blacktriangleleft$$

SOLVING PROBLEMS
ON YOUR OWN

This chapter deals with *mechanical vibrations*, i.e., with the motion of a particle or body oscillating about a position of equilibrium.

In this first lesson, we saw that a *free vibration* of a particle occurs when the particle is subjected to a force proportional to its displacement and of opposite direction, such as the force exerted by a spring (Fig. 19.1). The resulting motion, called a *simple harmonic motion*, is characterized by the differential equation

$$m\ddot{x} + kx = 0 \tag{19.2}$$

where x is the displacement of the particle, $\ddot{x}$ is its acceleration, m is its mass, and k is the constant of the spring. The solution of this differential equation was found to be

$$x = x_m \sin(\omega_n t + \phi) \tag{19.10}$$

where x_m = amplitude of the vibration
$\omega_n = \sqrt{k/m}$ = natural circular frequency (rad/s)
ϕ = phase angle (rad)

We also defined the *period* of the vibration as the time $\tau_n = 2\pi/\omega_n$ needed for the particle to complete one cycle, and the *natural frequency* as the number of cycles per second, $f_n = 1/\tau_n = \omega_n/2\pi$, expressed in Hz or s^{-1}. Differentiating Eq. (19.10) twice yields the velocity and the acceleration of the particle at any time. The maximum values of the velocity and acceleration were found to be

$$v_m = x_m \omega_n \qquad a_m = x_m \omega_n^2 \tag{19.15}$$

To determine the parameters in Eq. (19.10) you can follow these steps.

1. Draw a free-body diagram showing the forces exerted on the particle when the particle is at a distance x from its position of equilibrium. The resultant of these forces will be proportional to x and its direction will be opposite to the positive direction of x [Eq. (19.1)].

2. Write the differential equation of motion by equating to $m\ddot{x}$ the resultant of the forces found in step 1. Note that once a positive direction for x has been chosen, the same sign convention should be used for the acceleration $\ddot{x}$. After transposition, you will obtain an equation of the form of Eq. (19.2).

(continued)

3. Determine the natural circular frequency ω_n by dividing the coefficient of x by the coefficient of $\ddot{x}$ in this equation and taking the square root of the result obtained. Make sure that ω_n is expressed in rad/s.

4. Determine the amplitude x_m **and the phase angle** ϕ by substituting the value obtained for ω_n and the initial values of x and $\dot{x}$ into Eq. (19.10) and the equation obtained by differentiating Eq. (19.10) with respect to t.

Equation (19.10) and the two equations obtained by differentiating Eq. (19.10) twice with respect to t can now be used to find the displacement, velocity, and acceleration of the particle at any time. Equations (19.15) yield the maximum velocity v_m and the maximum acceleration a_m.

5. You also saw that for the small oscillations of a simple pendulum, the angle θ that the cord of the pendulum forms with the vertical satisfies the differential equation

$$\ddot{\theta} + \frac{g}{l}\theta = 0 \tag{19.17}$$

where l is the length of the cord and where θ is expressed in radians [Sec. 19.3]. This equation defines again a *simple harmonic motion*, and its solution is of the same form as Eq. (19.10),

$$\theta = \theta_m \sin(\omega_n t + \phi)$$

where the natural circular frequency $\omega_n = \sqrt{g/l}$ is expressed in rad/s. The determination of the various constants in this expression is carried out in a manner similar to that described above. Remember that the velocity of the bob is tangent to the path and that its magnitude is $v = l\dot{\theta}$, while the acceleration of the bob has a tangential component $\mathbf{a}_t$, of magnitude $a_t = l\ddot{\theta}$, and a normal component $\mathbf{a}_n$ directed toward the center of the path and of magnitude $a_n = l\dot{\theta}^2$.

PROBLEMS

19.1 Determine the maximum velocity and maximum acceleration of a particle which moves in simple harmonic motion with an amplitude of 3 mm and a frequency of 20 Hz.

19.2 A particle moves in simple harmonic motion. Knowing that the amplitude is 15 in. and the maximum acceleration is 15 ft/s², determine the maximum velocity of the particle and the frequency of its motion.

19.3 Determine the amplitude and maximum velocity of a particle which moves in simple harmonic motion with a maximum acceleration of 15 ft/s² and a frequency of 8 Hz.

19.4 A 32-kg block is attached to a spring and can move without friction in a slot as shown. The block is in its equilibrium position when it is struck by a hammer which imparts to the block an initial velocity of 250 mm/s. Determine (a) the period and frequency of the resulting motion, (b) the amplitude of the motion and the maximum acceleration of the block.

Fig. P19.4

19.5 A 13-kg block is supported by the spring shown. If the block is moved vertically downward from its equilibrium position and released, determine (a) the period and frequency of the resulting motion, (b) the maximum velocity and acceleration of the block if the amplitude of its motion is 50 mm.

Fig. P19.5

19.6 An instrument package A is bolted to a shaker table as shown. The table moves vertically in simple harmonic motion at the same frequency as the variable-speed motor which drives it. The package is to be tested at a peak acceleration of 150 ft/s². Knowing that the amplitude of the shaker table is 2.3 in., determine (a) the required speed of the motor in rpm, (b) the maximum velocity of the table.

Fig. P19.6

19.7 A simple pendulum consisting of a bob attached to a cord oscillates in a vertical plane with a period of 1.3 s. Assuming simple harmonic motion and knowing that the maximum velocity of the bob is 0.4 m/s, determine (a) the amplitude of the motion in degrees, (b) the maximum tangential acceleration of the bob.

19.8 A simple pendulum consisting of a bob attached to a cord of length $l = 800$ mm oscillates in a vertical plane. Assuming simple harmonic motion and knowing that the bob is released from rest when $\theta = 6°$, determine (a) the frequency of oscillation, (b) the maximum velocity of the bob.

Fig. P19.7 and P19.8

1251

Fig. P19.9

19.9 An instrument package B is placed on the shaking table C as shown. The table is made to move horizontally in simple harmonic motion with a frequency of 3 Hz. Knowing that the coefficient of static friction is $\mu_s = 0.40$ between the package and the table, determine the largest allowable amplitude of the motion if the package is not to slip on the table. Give the answers in both SI and U.S. customary units.

19.10 A 5-kg fragile glass vase is surrounded by packing material in a cardboard box of negligible weight. The packing material has negligible damping and a force-deflection relationship as shown. Knowing that the box is dropped from a height of 1 m and the impact with the ground is perfectly plastic, determine (a) the amplitude of vibration for the vase, (b) the maximum acceleration the vase experiences in g's.

Fig. P19.10

Fig. P19.11

19.11 A 3-lb block is supported as shown by a spring of constant $k = 2$ lb/in. which can act in tension or compression. The block is in its equilibrium position when it is struck from below by a hammer which imparts to the block an upward velocity of 90 in./s. Determine (a) the time required for the block to move 3 in. upward, (b) the corresponding velocity and acceleration of the block.

19.12 In Prob. 19.11, determine the position, velocity, and acceleration of the block 0.90 s after it has been struck by the hammer.

19.13 The bob of a simple pendulum of length $l = 40$ in. is released from rest when $\theta = +5°$. Assuming simple harmonic motion, determine 1.6 s after release (a) the angle θ, (b) the magnitudes of the velocity and acceleration of the bob.

Fig. P19.13

19.14 A 150-kg electromagnet is at rest and is holding 100 kg of scrap steel when the current is turned off and the steel is dropped. Knowing that the cable and the supporting crane have a total stiffness equivalent to a spring of constant 200 kN/m, determine (*a*) the frequency, the amplitude, and the maximum velocity of the resulting motion, (*b*) the minimum tension which will occur in the cable during the motion, (*c*) the velocity of the magnet 0.03 s after the current is turned off.

Fig. P19.14

19.15 A variable-speed motor is rigidly attached to beam *BC*. The rotor is slightly unbalanced and causes the beam to vibrate with a frequency equal to the motor speed. When the speed of the motor is less than 600 rpm or more than 1200 rpm, a small object placed at *A* is observed to remain in contact with the beam. For speeds between 600 rpm and 1200 rpm the object is observed to "dance" and actually to lose contact with the beam. Determine the amplitude of the motion of *A* when the speed of the motor is (*a*) 600 rpm, (*b*) 1200 rpm. Give answers in both SI and U.S. customary units.

Fig. P19.15

19.16 A small bob is attached to a cord of length 1.2 m and is released from rest when $\theta_A = 5°$. Knowing that $d = 0.6$ m, determine (*a*) the time required for the bob to return to point *A*, (*b*) the amplitude θ_C.

19.17 A 5-kg block, attached to the lower end of a spring whose upper end is fixed, vibrates with a period of 6.8 s. Knowing that the constant *k* of a spring is inversely proportional to its length, determine the period of a 3-kg block which is attached to the center of the same spring if the upper and lower ends of the spring are fixed.

Fig. P19.16

19.18 and 19.19 A 75-lb block is supported by the spring arrangement shown. The block is moved vertically downward from its equilibrium position and released. Knowing that the amplitude of the resulting motion is 2 in., determine (*a*) the period and frequency of the motion, (*b*) the maximum velocity and maximum acceleration of the block.

Fig. P19.18   **Fig. P19.19**

3.5 kN/m

2.8 kN/m

2.1 kN/m

Fig. P19.20

19.20 A 13.6-kg block is supported by the spring arrangement shown. If the block is moved from its equilibrium position 44 mm vertically downward and released, determine (*a*) the period and frequency of the resulting motion, (*b*) the maximum velocity and acceleration of the block.

19.21 An 11-lb block, attached to the lower end of a spring whose upper end is fixed, vibrates with a period of 7.2 s. Knowing that the constant *k* of a spring is inversely proportional to its length (e.g., if you cut a 10-lb/in. spring in half, the remaining two springs each have a spring constant of 20 lb/in.), determine the period of a 7-lb block which is attached to the center of the same spring if the upper and lower ends of the spring are fixed.

19.22 Block *A* of mass *m* is supported by the spring arrangement as shown. Knowing that the mass of the pulley is negligible and that the block is moved vertically downward from its equilibrium position and released, determine the frequency of the motion.

2*k*

k   *m*   *A*

Fig. P19.22

k_1

$k_2 = 20$ lb/in.

A

Fig. P19.23

19.23 The period of vibration of the system shown is observed to be 0.2 s. After the spring of constant $k_2 = 20$ lb/in. is removed and block *A* is connected to the spring of constant k_1, the period is observed to be 0.12 s. Determine (*a*) the constant k_1 of the remaining spring, (*b*) the weight of block *A*.

19.24 The period of vibration of the system shown is observed to be 0.8 s. If block *A* is removed, the period is observed to be 0.7 s. Determine (*a*) the mass of block *C*, (*b*) the period of vibration when both blocks *A* and *B* have been removed.

3 kg   *A*

3 kg   *B*

C

Fig. P19.24

19.25 The 100-lb platform *A* is attached to springs *B* and *D*, each of which has a constant *k* = 120 lb/ft. Knowing that the frequency of vibration of the platform is to remain unchanged when an 80-lb block is placed on it and a third spring *C* is added between springs *B* and *D*, determine the required constant of spring *C*.

A

B   *C*   *D*

Fig. P19.25

19.26 The period of vibration for a barrel floating in salt water is found to be 0.58 s when the barrel is empty and 1.8 s when it is filled with 55 gallons of crude oil. Knowing that the density of the oil is 900 kg/m³, determine (a) the mass of the empty barrel, (b) the density of the salt water, ρ_{sw}. [*Hint:* The force of the water on the bottom of the barrel can be modeled as a spring with constant $k = \rho_{sw}gA$.]

Fig. P19.26

19.27 From mechanics of materials it is known that for a cantilever beam of constant cross section a static load **P** applied at end B will cause a deflection $\delta_B = PL^3/3EI$, where L is the length of the beam, E is the modulus of elasticity, and I is the moment of inertia of the cross-sectional area of the beam. Knowing that $L = 10$ ft, $E = 29 \times 10^6$ lb/in², and $I = 12.4$ in⁴, determine (a) the equivalent spring constant of the beam, (b) the frequency of vibration of a 520-lb block attached to end B of the same beam.

Fig. P19.27

19.28 From mechanics of materials it is known that when a static load **P** is applied at the end B of a uniform metal rod fixed at end A, the length of the rod will increase by an amount $\delta = PL/AE$, where L is the length of the undeformed rod, A is its cross-sectional area, and E is the modulus of elasticity of the metal. Knowing that $L = 450$ mm and $E = 200$ GPa and that the diameter of the rod is 8 mm, and neglecting the mass of the rod, determine (a) the equivalent spring constant of the rod, (b) the frequency of the vertical vibrations of a block of mass $m = 8$ kg attached to end B of the same rod.

Fig. P19.28

19.29 Denoting by δ_{st} the static deflection of a beam under a given load, show that the frequency of vibration of the load is

$$f = \frac{1}{2\pi}\sqrt{\frac{g}{\delta_{st}}}$$

Neglect the mass of the beam, and assume that the load remains in contact with the beam.

19.30 A 40-mm deflection of the second floor of a building is measured directly under a newly installed 3500-kg piece of rotating machinery which has a slightly unbalanced rotor. Assuming that the deflection of the floor is proportional to the load it supports, determine (a) the equivalent spring constant of the floor system, (b) the speed in rpm of the rotating machinery that should be avoided if it is not to coincide with the natural frequency of the floor-machinery system.

Fig. P19.31

19.31 If $h = 700$ mm and $d = 500$ mm and each spring has a constant $k = 600$ N/m, determine the mass m for which the period of small oscillations is (a) 0.50 s, (b) infinite. Neglect the mass of the rod and assume that each spring can act in either tension or compression.

19.32 The force-deflection equation for a nonlinear spring fixed at one end is $F = 1.5x^{1/2}$ where F is the force, expressed in newtons, applied at the other end and x is the deflection expressed in meters. (a) Determine the deflection x_0 if a 4-oz block is suspended from the spring and is at rest. (b) Assuming that the slope of the force-deflection curve at the point corresponding to this loading can be used as an equivalent spring constant, determine the frequency of vibration of the block if it is given a very small downward displacement from its equilibrium position and released.

*19.33 Expanding the integrand in Eq. (19.19) of Sec. 19.4 into a series of even powers of $\sin \varphi$ and integrating, show that the period of a simple pendulum of length l may be approximated by the formula

$$\tau = 2\pi\sqrt{\frac{l}{g}}\left(1 + \tfrac{1}{4}\sin^2\frac{\theta_m}{2}\right)$$

where θ_m is the amplitude of the oscillations.

*19.34 Using the formula given in Prob. 19.33, determine the amplitude θ_m for which the period of a simple pendulum is $\tfrac{1}{2}$ percent longer than the period of the same pendulum for small oscillations.

*19.35 Using the data of Table 19.1, determine the period of a simple pendulum of length $l = 750$ mm (a) for small oscillations, (b) for oscillations of amplitude $\theta_m = 60°$, (c) for oscillations of amplitude $\theta_m = 90°$.

*19.36 Using the data of Table 19.1, determine the length in inches of a simple pendulum which oscillates with a period of 2 s and an amplitude of 90°.

19.5 FREE VIBRATIONS OF RIGID BODIES

The analysis of the vibrations of a rigid body or of a system of rigid bodies possessing a single degree of freedom is similar to the analysis of the vibrations of a particle. An appropriate variable, such as a distance x or an angle θ, is chosen to define the position of the body or system of bodies, and an equation relating this variable and its second derivative with respect to t is written. If the equation obtained is of the same form as (19.6), i.e., if we have

$$\ddot{x} + \omega_n^2 x = 0 \qquad \text{or} \qquad \ddot{\theta} + \omega_n^2\theta = 0 \qquad (19.21)$$

the vibration considered is a simple harmonic motion. The period and natural frequency of the vibration can then be obtained by identifying ω_n and substituting its value into Eqs. (19.13) and (19.14).

In general, a simple way to obtain one of Eqs. (19.21) is to express that the system of the external forces is equivalent to the system of the effective forces by drawing a free-body-diagram equation for an arbitrary value of the variable and writing the appropriate equation of motion. We recall that our goal should be *the determination* of the coefficient of the variable x or θ, *not* the determination of the variable itself or of the derivative $\ddot{x}$ or $\ddot{\theta}$. Setting this coefficient equal to ω_n^2, we obtain the natural circular frequency ω_n, from which τ_n and f_n can be determined.

The method we have outlined can be used to analyze vibrations which are truly represented by a simple harmonic motion, or vibrations of small amplitude which can be *approximated* by a simple harmonic motion. As an example, let us determine the period of the small oscillations of a square plate of side $2b$ which is suspended from the midpoint O of one of its sides (Fig. 19.5a). We consider the plate in an arbitrary position defined by the angle θ that the line OG forms with the vertical and draw a free-body-diagram equation to express that the weight $\mathbf{W}$ of the plate and the components $\mathbf{R}_x$ and $\mathbf{R}_y$ of the reaction at O are equivalent to the vectors $m\mathbf{a}_t$ and $m\mathbf{a}_n$ and to the couple $\bar{I}\boldsymbol{\alpha}$ (Fig. 19.5b). Since the angular velocity and angular acceleration of the plate are equal, respectively, to $\dot{\theta}$ and $\ddot{\theta}$, the magnitudes of the two vectors are, respectively, $mb\ddot{\theta}$ and $mb\dot{\theta}^2$, while the moment of the couple is $\bar{I}\ddot{\theta}$. In previous applications of this method (Chap. 16), we tried whenever possible to assume the correct sense for the acceleration. Here, however, we must assume the same positive sense for θ and $\ddot{\theta}$ in order to obtain an equation of the form (19.21). Consequently, the angular acceleration $\ddot{\theta}$ will be assumed positive counterclockwise, even though this assumption is obviously unrealistic. Equating moments about O, we write

$$+\uparrow \qquad -W(b \sin \theta) = (mb\ddot{\theta})b + \bar{I}\ddot{\theta}$$

Noting that $\bar{I} = \frac{1}{12}m[(2b)^2 + (2b)^2] = \frac{2}{3}mb^2$ and $W = mg$, we obtain

$$\ddot{\theta} + \frac{3}{5}\frac{g}{b}\sin\theta = 0 \qquad (19.22)$$

For oscillations of small amplitude, we can replace $\sin\theta$ by θ, expressed in radians, and write

$$\ddot{\theta} + \frac{3}{5}\frac{g}{b}\theta = 0 \qquad (19.23)$$

Comparison with (19.21) shows that the equation obtained is that of a simple harmonic motion and that the natural circular frequency ω_n of the oscillations is equal to $(3g/5b)^{1/2}$. Substituting into (19.13), we find that the period of the oscillations is

$$\tau_n = \frac{2\pi}{\omega_n} = 2\pi\sqrt{\frac{5b}{3g}} \qquad (19.24)$$

The result obtained is valid only for oscillations of small amplitude. A more accurate description of the motion of the plate is obtained by comparing Eqs. (19.16) and (19.22). We note that the two equations are identical if we choose l equal to $5b/3$. This means that the plate will oscillate as a simple pendulum of length $l = 5b/3$ and the results of Sec. 19.4 can be used to correct the value of the period given in (19.24). The point A of the plate located on line OG at a distance $l = 5b/3$ from O is defined as the *center of oscillation* corresponding to O (Fig. 19.5a).

(a)

(b)

Fig. 19.5

SAMPLE PROBLEM 19.2

A cylinder of weight W and radius r is suspended from a looped cord as shown. One end of the cord is attached directly to a rigid support, while the other end is attached to a spring of constant k. Determine the period and natural frequency of the vibrations of the cylinder.

SOLUTION

Kinematics of Motion. We express the linear displacement and the acceleration of the cylinder in terms of the angular displacement θ. Choosing the positive sense clockwise and measuring the displacements from the equilibrium position, we write

$$\bar{x} = r\theta \qquad \delta = 2\bar{x} = 2r\theta$$

$$\alpha = \ddot{\theta}\downarrow \qquad \bar{a} = r\alpha = r\ddot{\theta} \qquad \mathbf{a} = r\ddot{\theta}\downarrow \qquad (1)$$

Equations of Motion. The system of external forces acting on the cylinder consists of the weight $\mathbf{W}$ and of the forces $\mathbf{T}_1$ and $\mathbf{T}_2$ exerted by the cord. We express that this system is equivalent to the system of effective forces represented by the vector $m\bar{\mathbf{a}}$ attached at G and the couple $\bar{I}\alpha$.

$$+\!\downarrow\Sigma M_A = \Sigma(M_A)_{\text{eff}}: \qquad Wr - T_2(2r) = m\bar{a}r + \bar{I}\alpha \qquad (2)$$

When the cylinder is in its position of equilibrium, the tension in the cord is $T_0 = \frac{1}{2}W$. We note that for an angular displacement θ, the magnitude of $\mathbf{T}_2$ is

$$T_2 = T_0 + k\delta = \tfrac{1}{2}W + k\delta = \tfrac{1}{2}W + k(2r\theta) \qquad (3)$$

Substituting from (1) and (3) into (2), and recalling that $\bar{I} = \frac{1}{2}mr^2$, we write

$$Wr - (\tfrac{1}{2}W + 2kr\theta)(2r) = m(r\ddot{\theta})r + \tfrac{1}{2}mr^2\ddot{\theta}$$

$$\ddot{\theta} + \frac{8}{3}\frac{k}{m}\theta = 0$$

The motion is seen to be simple harmonic, and we have

$$\omega_n^2 = \frac{8}{3}\frac{k}{m} \qquad \omega_n = \sqrt{\frac{8}{3}\frac{k}{m}}$$

$$\tau_n = \frac{2\pi}{\omega_n} \qquad \tau_n = 2\pi\sqrt{\frac{3}{8}\frac{m}{k}} \quad \blacktriangleleft$$

$$f_n = \frac{\omega_n}{2\pi} \qquad f_n = \frac{1}{2\pi}\sqrt{\frac{8}{3}\frac{k}{m}} \quad \blacktriangleleft$$

SAMPLE PROBLEM 19.3

A circular disk, weighing 20 lb and of radius 8 in., is suspended from a wire as shown. The disk is rotated (thus twisting the wire) and then released; the period of the torsional vibration is observed to be 1.13 s. A gear is then suspended from the same wire, and the period of torsional vibration for the gear is observed to be 1.93 s. Assuming that the moment of the couple exerted by the wire is proportional to the angle of twist, determine (a) the torsional spring constant of the wire, (b) the centroidal moment of inertia of the gear, (c) the maximum angular velocity reached by the gear if it is rotated through 90° and released.

SOLUTION

a. Vibration of Disk. Denoting by θ the angular displacement of the disk, we express that the magnitude of the couple exerted by the wire is $M = K\theta$, where K is the torsional spring constant of the wire. Since this couple must be equivalent to the couple $\bar{I}\boldsymbol{\alpha}$ representing the effective forces of the disk, we write

$$+\curvearrowright \Sigma M_O = \Sigma(M_O)_{\text{eff}}: \qquad +K\theta = -\bar{I}\ddot{\theta}$$

$$\ddot{\theta} + \frac{K}{\bar{I}}\theta = 0$$

The motion is seen to be simple harmonic, and we have

$$\omega_n^2 = \frac{K}{\bar{I}} \qquad \tau_n = \frac{2\pi}{\omega_n} \qquad \tau_n = 2\pi\sqrt{\frac{\bar{I}}{K}} \qquad (1)$$

For the disk, we have

$$\tau_n = 1.13 \text{ s} \qquad \bar{I} = \tfrac{1}{2}mr^2 = \frac{1}{2}\left(\frac{20 \text{ lb}}{32.2 \text{ ft/s}^2}\right)\left(\frac{8}{12}\text{ ft}\right)^2 = 0.138 \text{ lb} \cdot \text{ft} \cdot \text{s}^2$$

Substituting into (1), we obtain

$$1.13 = 2\pi\sqrt{\frac{0.138}{K}} \qquad K = 4.27 \text{ lb} \cdot \text{ft/rad} \ \blacktriangleleft$$

b. Vibration of Gear. Since the period of vibration of the gear is 1.93 s and $K = 4.27$ lb $\cdot$ ft/rad, Eq. (1) yields

$$1.93 = 2\pi\sqrt{\frac{\bar{I}}{4.27}} \qquad \bar{I}_{\text{gear}} = 0.403 \text{ lb} \cdot \text{ft} \cdot \text{s}^2 \ \blacktriangleleft$$

c. Maximum Angular Velocity of Gear. Since the motion is simple harmonic, we have

$$\theta = \theta_m \sin \omega_n t \qquad \omega = \theta_m \omega_n \cos \omega_n t \qquad \omega_m = \theta_m \omega_n$$

Recalling that $\theta_m = 90° = 1.571$ rad and $\tau = 1.93$ s, we write

$$\omega_m = \theta_m \omega_n = \theta_m\left(\frac{2\pi}{\tau}\right) = (1.571 \text{ rad})\left(\frac{2\pi}{1.93 \text{ s}}\right)$$

$$\omega_m = 5.11 \text{ rad/s} \ \blacktriangleleft$$

SOLVING PROBLEMS
ON YOUR OWN

In this lesson you saw that a rigid body, or a system of rigid bodies, whose position can be defined by a single coordinate x or θ, will execute a simple harmonic motion if the differential equation obtained by applying Newton's second law is of the form

$$\ddot{x} + \omega_n^2 x = 0 \qquad \text{or} \qquad \ddot{\theta} + \omega_n^2 \theta = 0 \tag{19.21}$$

Your goal should be to determine ω_n, from which you can obtain the period τ_n and the natural frequency f_n. Taking into account the initial conditions, you can then write an equation of the form

$$x = x_m \sin(\omega_n t + \phi) \tag{19.10}$$

where x should be replaced by θ if a rotation is involved. To solve the problems in this lesson, you will follow these steps:

1. Choose a coordinate which will measure the displacement of the body from its equilibrium position. You will find that many of the problems in this lesson involve the rotation of a body about a fixed axis and that the angle measuring the rotation of the body from its equilibrium position is the most convenient coordinate to use. In problems involving the general plane motion of a body, where a coordinate x (and possibly a coordinate y) is used to define the position of the mass center G of the body, and a coordinate θ is used to measure its rotation about G, find kinematic relations which will allow you to express x (and y) in terms of θ [Sample Prob. 19.2].

2. Draw a free-body-diagram equation to express that the system of the external forces is equivalent to the system of the effective forces, which consists of the vector $m\bar{\mathbf{a}}$ and the couple $\bar{I}\boldsymbol{\alpha}$, where $\bar{a} = \ddot{x}$ and $\alpha = \ddot{\theta}$. Be sure that each applied force or couple is drawn in a direction consistent with the assumed displacement and that the senses of $\bar{\mathbf{a}}$ and $\boldsymbol{\alpha}$ are, respectively, those in which the coordinates x and θ are increasing.

3. Write the differential equations of motion by equating the sums of the components of the external and effective forces in the x and y directions and the sums of their moments about a given point. If necessary, use the kinematic relations developed in step 1 to obtain equations involving only the coordinate θ. If θ is a small angle, replace $\sin \theta$ by θ and $\cos \theta$ by 1, if these functions appear in your equations. Eliminating any unknown reactions, you will obtain an equation of the type of Eqs. (19.21). Note that in problems involving a body rotating about a fixed axis, you can immediately obtain such an equation by equating the moments of the external and effective forces about the fixed axis.

4. Comparing the equation you have obtained with one of Eqs. (19.21), you can identify ω_n^2 and, thus, determine the natural circular frequency ω_n. Remember that the object of your analysis is *not to solve* the differential equation you have obtained, *but to identify* ω_n^2.

5. Determine the amplitude and the phase angle ϕ by substituting the value obtained for ω_n and the initial values of the coordinate and its first derivative into Eq. (19.10) and the equation obtained by differentiating (19.10) with respect to t. From Eq. (19.10) and the two equations obtained by differentiating (19.10) twice with respect to t, and using the kinematic relations developed in step 1, you will be able to determine the position, velocity, and acceleration of any point of the body at any given time.

6. In problems involving torsional vibrations, the torsional spring constant K is expressed in N · m/rad or lb · ft/rad. The product of K and the angle of twist θ, expressed in radians, yields the moment of the restoring couple, which should be equated to the sum of the moments of the effective forces or couples about the axis of rotation [Sample Prob. 19.3].

PROBLEMS

19.37 The uniform rod shown has mass 6 kg and is attached to a spring of constant $k = 700$ N/m. If end B of the rod is depressed 10 mm and released, determine (a) the period of vibration, (b) the maximum velocity of end B.

Fig. P19.37

19.38 A belt is placed around the rim of a 500-lb flywheel and attached as shown to two springs, each of constant $k = 85$ lb/in. If end C of the belt is pulled 1.5 in. down and released, the period of vibration of the flywheel is observed to be 0.5 s. Knowing that the initial tension in the belt is sufficient to prevent slipping, determine (a) the maximum angular velocity of the flywheel, (b) the centroidal radius of gyration of the flywheel.

19.39 An 8-kg uniform rod AB is hinged to a fixed support at A and is attached by means of pins B and C to a 12-kg disk of radius 400 mm. A spring attached at D holds the rod at rest in the position shown. If point B is moved down 25 mm and released, determine (a) the period of vibration, (b) the maximum velocity of point B.

19.40 Solve Prob. 19.39, assuming that pin C is removed and that the disk can rotate freely about pin B.

19.41 A 15-lb slender rod AB is riveted to a 12-lb uniform disk as shown. A belt is attached to the rim of the disk and to a spring which holds the rod at rest in the position shown. If end A of the rod is moved 0.75 in. down and released, determine (a) the period of vibration, (b) the maximum velocity of end A.

18 in.

Fig. P19.38

Fig. P19.39

Fig. P19.41

19.42 A 30-lb uniform cylinder can roll without sliding on a 15° incline. A belt is attached to the rim of the cylinder, and a spring holds the cylinder at rest in the position shown. If the center of the cylinder is moved 2 in. down the incline and released, determine (*a*) the period of vibration, (*b*) the maximum acceleration of the center of the cylinder.

19.43 A square plate of mass m is held by eight springs, each of constant k. Knowing that each spring can act in either tension or compression, determine the frequency of the resulting vibration if (*a*) the plate is given a small vertical displacement and released, (*b*) the plate is rotated through a small angle about G and released.

Fig. P19.42

Fig. P19.43

19.44 Two small weights w are attached at A and B to the rim of a uniform disk of radius r and weight W. Denoting by τ_0 the period of small oscillations when $\beta = 0$, determine the angle β for which the period of small oscillations is $2\tau_0$.

Fig. P19.44 and P19.45

19.45 Two 40-g weights are attached at A and B to the rim of a 1.5-kg uniform disk of radius $r = 100$ mm. Determine the frequency of small oscillations when $\beta = 60°$.

19.46 A three-blade wind turbine used for research is supported on a shaft so that it is free to rotate about O. One technique to determine the centroidal mass moment of inertia of an object is to place a known weight at a known distance from the axis of rotation and to measure the frequency of oscillations after releasing it from rest with a small initial angle. In this case, a weight of $W_{add} = 50$ lb is attached to one of the blades at a distance $R = 20$ ft from the axis of rotation. Knowing that when the blade with the added weight is displaced slightly from the vertical axis, and the system is found to have a period of 7.6 s, determine the centroidal mass moment of inertia of the three-blade rotor.

Fig. P19.46

Fig. P19.47

Fig. P19.49

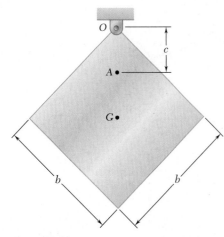

Fig. P19.51

19.47 A connecting rod is supported by a knife-edge at point A; the period of its small oscillations is observed to be 0.87 s. The rod is then inverted and supported by a knife edge at point B and the period of its small oscillations is observed to be 0.78 s. Knowing that $r_a + r_b = 10$ in., determine (a) the location of the mass center G, (b) the centroidal radius of gyration $\bar{k}$.

19.48 A 75-mm-radius hole is cut in a 200-mm-radius uniform disk which is attached to a frictionless pin at its geometric center O. Determine (a) the period of small oscillations of the disk, (b) the length of a simple pendulum which has the same period.

Fig. P19.48

19.49 A uniform disk of radius $r = 250$ mm is attached at A to a 650-mm rod AB of negligible mass which can rotate freely in a vertical plane about B. Determine the period of small oscillations (a) if the disk is free to rotate in a bearing at A, (b) if the rod is riveted to the disk at A.

19.50 A small collar of mass 1 kg is rigidly attached to a 3-kg uniform rod of length $L = 750$ mm. Determine (a) the distance d to maximize the frequency of oscillation when the rod is given a small initial displacement, (b) the corresponding period of oscillation.

Fig. P19.50

19.51 For the uniform square plate of side $b = 12$ in., determine (a) the period of small oscillations if the plate is suspended as shown, (b) the distance c from O to a point A from which the plate should be suspended for the period to be a minimum.

19.52 A *compound pendulum* is defined as a rigid slab which oscillates about a fixed point O, called the center of suspension. Show that the period of oscillation of a compound pendulum is equal to the period of a simple pendulum of length OA, where the distance from A to the mass center G is $GA = \bar{k}^2/\bar{r}$. Point A is defined as the center of oscillation and coincides with the center of percussion defined in Prob. 17.66.

Fig. P19.52 and P19.53

19.53 A rigid slab oscillates about a fixed point O. Show that the smallest period of oscillation occurs when the distance $\bar{r}$ from point O to the mass center G is equal to $\bar{k}$.

19.54 Show that if the compound pendulum of Prob. 19.52 is suspended from A instead of O, the period of oscillation is the same as before and the new center of oscillation is located at O.

19.55 The 8-kg uniform bar AB is hinged at C and is attached at A to a spring of constant $k = 500$ N/m. If end A is given a small displacement and released, determine (*a*) the frequency of small oscillations, (*b*) the smallest value of the spring constant k for which oscillations will occur.

19.56 Two uniform rods, each of mass $m = 12$ kg and length $L = 800$ mm, are welded together to form the assembly shown. Knowing that the constant of each spring is $k = 500$ N/m and that end A is given a small displacement and released, determine the frequency of the resulting motion.

19.57 A 45-lb uniform square plate is suspended from a pin located at the midpoint A of one of its 1.2-ft edges and is attached to springs, each of constant $k = 8$ lb/in. If corner B is given a small displacement and released, determine the frequency of the resulting vibration. Assume that each spring can act in either tension or compression.

Fig. P19.55

Fig. P19.56

Fig. P19.57

Fig. P19.58

19.58 A 1300-kg sports car has a center of gravity G located a distance h above a line connecting the front and rear axles. The car is suspended from cables that are attached to the front and rear axles as shown. Knowing that the periods of oscillation are 4.04 s when $L = 4$ m and 3.54 s when $L = 3$ m, determine h and the centroidal radius of gyration.

19.59 A 6-lb slender rod is suspended from a steel wire which is known to have a torsional spring constant $K = 1.5$ ft · lb/rad. If the rod is rotated through 180° about the vertical and released, determine (a) the period of oscillation, (b) the maximum velocity of end A of the rod.

Fig. P19.59

Fig. P19.60

19.60 A uniform disk of radius $r = 250$ mm is attached at A to a 650-mm rod AB of negligible mass which can rotate freely in a vertical plane about B. If the rod is displaced 2° from the position shown and released, determine the magnitude of the maximum velocity of point A, assuming that the disk is (a) free to rotate in a bearing at A, (b) riveted to the rod at A.

19.61 Two uniform rods, each of mass m and length l, are welded together to form the T-shaped assembly shown. Determine the frequency of small oscillations of the assembly.

Fig. P19.61

Fig. P19.62

19.62 A homogeneous wire bent to form the figure shown is attached to a pin support at A. Knowing that $r = 220$ mm and that point B is pushed down 20 mm and released, determine the magnitude of the velocity of B, 8 s later.

19.63 A horizontal platform P is held by several rigid bars which are connected to a vertical wire. The period of oscillation of the platform is found to be 2.2 s when the platform is empty and 3.8 s when an object A of unknown moment of inertia is placed on the platform with its mass center directly above the center of the plate. Knowing that the wire has a torsional constant $K = 27$ N $\cdot$ m/rad, determine the centroidal moment of inertia of object A.

19.64 A uniform disk of radius $r = 120$ mm is welded at its center to two elastic rods of equal length with fixed ends at A and B. Knowing that the disk rotates through an 8° angle when a 500-mN $\cdot$ m couple is applied to the disk and that it oscillates with a period of 1.3 s when the couple is removed, determine (a) the mass of the disk, (b) the period of vibration if one of the rods is removed.

Fig. P19.63

Fig. P19.64

19.65 A 5-kg uniform rod CD of length $l = 0.7$ m is welded at C to two elastic rods, which have fixed ends at A and B and are known to have a combined torsional spring constant $K = 24$ N $\cdot$ m/rad. Determine the period of small oscillations, if the equilibrium position of CD is (a) vertical as shown, (b) horizontal.

19.66 A 1.8-kg uniform plate in the shape of an equilateral triangle is suspended at its center of gravity from a steel wire which is known to have a torsional constant $K = 35$ mN $\cdot$ m/rad. If the plate is rotated 360° about the vertical and then released, determine (a) the period of oscillation, (b) the maximum velocity of one of the vertices of the triangle.

Fig. P19.65

Fig. P19.66

19.67 A period of 6.00 s is observed for the angular oscillations of a 4-oz gyroscope rotor suspended from a wire as shown. Knowing that a period of 3.80 s is obtained when a 1.25-in.-diameter steel sphere is suspended in the same fashion, determine the centroidal radius of gyration of the rotor. (Specific weight of steel = 490 lb/ft³.)

Fig. P19.67

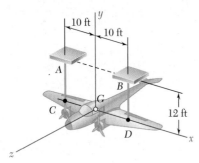

Fig. P19.68

19.68 The centroidal radius of gyration $\bar{k}_y$ of an airplane is determined by suspending the airplane by two 12-ft-long cables as shown. The airplane is rotated through a small angle about the vertical through G and then released. Knowing that the observed period of oscillation is 3.3 s, determine the centroidal radius of gyration $\bar{k}_y$.

19.6 APPLICATION OF THE PRINCIPLE OF CONSERVATION OF ENERGY

We saw in Sec. 19.2 that when a particle of mass m is in simple harmonic motion, the resultant **F** of the forces exerted on the particle has a magnitude proportional to the displacement x measured from the position of equilibrium O and is directed toward O; we write $F = -kx$. Referring to Sec. 13.6, we note that **F** is a *conservative force* and that the corresponding potential energy is $V = \frac{1}{2}kx^2$, where V is assumed equal to zero in the equilibrium position $x = 0$. Since the velocity of the particle is equal to $\dot{x}$, its kinetic energy is $T = \frac{1}{2}m\dot{x}^2$ and we can express that the total energy of the particle is conserved by writing

$$T + V = \text{constant} \qquad \tfrac{1}{2}m\dot{x}^2 + \tfrac{1}{2}kx^2 = \text{constant}$$

Dividing through by $m/2$ and recalling from Sec. 19.2 that $k/m = \omega_n^2$, where ω_n is the natural circular frequency of the vibration, we have

$$\dot{x}^2 + \omega_n^2 x^2 = \text{constant} \qquad (19.25)$$

Equation (19.25) is characteristic of a simple harmonic motion, since it can be obtained from Eq. (19.6) by multiplying both terms by $2\dot{x}$ and integrating.

The principle of conservation of energy provides a convenient way for determining the period of vibration of a rigid body or of a system of rigid bodies possessing a single degree of freedom, once it has been established that the motion of the system is a simple harmonic motion or that it can be approximated by a simple harmonic motion. Choosing an appropriate variable, such as a distance x or an angle θ, we consider two particular positions of the system:

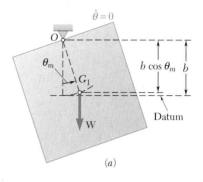

(a)

1. *The displacement of the system is maximum;* we have $T_1 = 0$, and V_1 can be expressed in terms of the amplitude x_m or θ_m (choosing $V = 0$ in the equilibrium position).
2. *The system passes through its equilibrium position;* we have $V_2 = 0$, and T_2 can be expressed in terms of the maximum velocity $\dot{x}_m$ or the maximum angular velocity $\dot{\theta}_m$.

We then express that the total energy of the system is conserved and write $T_1 + V_1 = T_2 + V_2$. Recalling from (19.15) that for simple harmonic motion the maximum velocity is equal to the product of the amplitude and of the natural circular frequency ω_n, we find that the equation obtained can be solved for ω_n.

As an example, let us consider again the square plate of Sec. 19.5. In the position of maximum displacement (Fig. 19.6a), we have

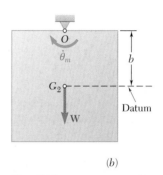

(b)

Fig. 19.6

$$T_1 = 0 \qquad V_1 = W(b - b \cos \theta_m) = Wb(1 - \cos \theta_m)$$

or, since $1 - \cos \theta_m = 2 \sin^2(\theta_m/2) \approx 2(\theta_m/2)^2 = \theta_m^2/2$ for oscillations of small amplitude,

$$T_1 = 0 \qquad V_1 = \tfrac{1}{2}Wb\theta_m^2 \qquad (19.26)$$

As the plate passes through its position of equilibrium (Fig. 19.6b), its velocity is maximum and we have

$$T_2 = \tfrac{1}{2}m\bar{v}_m^2 + \tfrac{1}{2}\bar{I}\omega_m^2 = \tfrac{1}{2}mb^2\dot{\theta}_m^2 + \tfrac{1}{2}\bar{I}\dot{\theta}_m^2 \qquad V_2 = 0$$

or, recalling from Sec. 19.5 that $\bar{I} = \tfrac{2}{3}mb^2$,

$$T_2 = \tfrac{1}{2}(\tfrac{5}{3}mb^2)\dot{\theta}_m^2 \qquad V_2 = 0 \qquad (19.27)$$

Substituting from (19.26) and (19.27) into $T_1 + V_1 = T_2 + V_2$, and noting that the maximum velocity $\dot{\theta}_m$ is equal to the product $\theta_m\omega_n$, we write

$$\tfrac{1}{2}Wb\theta_m^2 = \tfrac{1}{2}(\tfrac{5}{3}mb^2)\theta_m^2\omega_n^2 \qquad (19.28)$$

which yields $\omega_n^2 = 3g/5b$ and

$$\tau_n = \frac{2\pi}{\omega_n} = 2\pi\sqrt{\frac{5b}{3g}} \qquad (19.29)$$

as previously obtained.

SAMPLE PROBLEM 19.4

Determine the period of small oscillations of a cylinder of radius r which rolls without slipping inside a curved surface of radius R.

SOLUTION

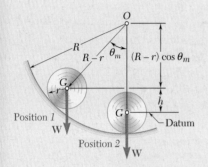

Position 1

Position 2

Datum

Position 2

We denote by θ the angle which line OG forms with the vertical. Since the cylinder rolls without slipping, we may apply the principle of conservation of energy between position 1, where $\theta = \theta_m$, and position 2, where $\theta = 0$.

Position 1
Kinetic Energy.   Since the velocity of the cylinder is zero, $T_1 = 0$.
Potential Energy.   Choosing a datum as shown and denoting by W the weight of the cylinder, we have

$$V_1 = Wh = W(R - r)(1 - \cos\theta)$$

Noting that for small oscillations $(1 - \cos\theta) = 2\sin^2(\theta/2) \approx \theta^2/2$, we have

$$V_1 = W(R - r)\frac{\theta_m^2}{2}$$

Position 2.
Denoting by $\dot{\theta}_m$ the angular velocity of line OG as the cylinder passes through position 2, and observing that point C is the instantaneous center of rotation of the cylinder, we write

$$\bar{v}_m = (R - r)\dot{\theta}_m \qquad \omega_m = \frac{\bar{v}_m}{r} = \frac{R - r}{r}\dot{\theta}_m$$

Kinetic Energy

$$
\begin{aligned}
T_2 &= \tfrac{1}{2}m\bar{v}_m^2 + \tfrac{1}{2}\bar{I}\omega_m^2 \\
&= \tfrac{1}{2}m(R - r)^2\dot{\theta}_m^2 + \tfrac{1}{2}(\tfrac{1}{2}mr^2)\left(\frac{R - r}{r}\right)^2\dot{\theta}_m^2 \\
&= \tfrac{3}{4}m(R - r)^2\dot{\theta}_m^2
\end{aligned}
$$

Potential Energy

$$V_2 = 0$$

Conservation of Energy

$$T_1 + V_1 = T_2 + V_2$$

$$0 + W(R - r)\frac{\theta_m^2}{2} = \tfrac{3}{4}m(R - r)^2\dot{\theta}_m^2 + 0$$

Since $\dot{\theta}_m = \omega_n\theta_m$ and $W = mg$, we write

$$mg(R - r)\frac{\theta_m^2}{2} = \tfrac{3}{4}m(R - r)^2(\omega_n\theta_m)^2 \qquad \omega_n^2 = \frac{2}{3}\frac{g}{R - r}$$

$$\tau_n = \frac{2\pi}{\omega_n} \qquad \tau_n = 2\pi\sqrt{\frac{3}{2}\frac{R - r}{g}} \quad \blacktriangleleft$$

SOLVING PROBLEMS
ON YOUR OWN

In the problems which follow you will be asked to use the *principle of conservation of energy* to determine the period or natural frequency of the simple harmonic motion of a particle or rigid body. Assuming that you choose an angle θ to define the position of the system (with $\theta = 0$ in the equilibrium position), as you will in most of the problems in this lesson, you will express that the total energy of the system is conserved, $T_1 + V_1 = T_2 + V_2$, between the position 1 of maximum displacement ($\theta_1 = \theta_m$, $\dot{\theta}_1 = 0$) and the position 2 of maximum velocity ($\dot{\theta}_2 = \dot{\theta}_m$, $\theta_2 = 0$). It follows that T_1 and V_2 will both be zero, and the energy equation will reduce to $V_1 = T_2$, where V_1 and T_2 are homogeneous quadratic expressions in θ_m and $\dot{\theta}_m$, respectively. Recalling that, for a simple harmonic motion, $\dot{\theta}_m = \theta_m \omega_n$ and substituting this product into the energy equation, you will obtain, after reduction, an equation that you can solve for ω_n^2. Once you have determined the natural circular frequency ω_n, you can obtain the period τ_n and the natural frequency f_n of the vibration.

The steps that you should take are as follows:

1. Calculate the potential energy V_1 of the system in its position of maximum displacement. Draw a sketch of the system in its position of maximum displacement and express the potential energy of all the forces involved (internal as well as external) in terms of the maximum displacement x_m or θ_m.

a. The potential energy associated with the weight W of a body is $V_g = Wy$, where y is the elevation of the center of gravity G of the body above its equilibrium position. If the problem you are solving involves the oscillation of a rigid body about a horizontal axis through a point O located at a distance b from G (Fig. 19.6), express y in terms of the angle θ that the line OG forms with the vertical: $y = b(1 - \cos \theta)$. But, for small values of θ, you can replace this expression with $y = \frac{1}{2}b\theta^2$ [Sample Prob. 19.4]. Therefore, when θ reaches its maximum value θ_m, and for oscillations of small amplitude, you can express V_g as

$$V_g = \tfrac{1}{2}Wb\theta_m^2$$

Note that *if G is located above O* in its equilibrium position (instead of below O, as we have assumed), the vertical displacement y will be negative and should be approximated as $y = -\frac{1}{2}b\theta^2$, which will result in a negative value for V_g. In the absence of other forces, the equilibrium position will be unstable, and the system will not oscillate. (See, for instance, Prob. 19.89.)

b. The potential energy associated with the elastic force exerted by a spring is $V_e = \frac{1}{2}kx^2$, where k is the constant of the spring and x its deflection. In problems involving the rotation of a body about an axis, you will generally have $x = a\theta$, where a is the distance from the axis of rotation to the point of the body

(continued)

where the spring is attached, and where θ is the angle of rotation. Therefore, when x reaches its maximum value x_m and θ reaches its maximum value θ_m, you can express V_e as

$$V_e = \tfrac{1}{2}kx_m^2 = \tfrac{1}{2}ka^2\theta_m^2$$

c. The potential energy V_1 of the system in its position of maximum displacement is obtained by adding the various potential energies that you have computed. It will be equal to the product of a constant and θ_m^2.

2. Calculate the kinetic energy T_2 of the system in its position of maximum velocity. Note that this position is also the equilibrium position of the system.

a. If the system consists of a single rigid body, the kinetic energy T_2 of the system will be the sum of the kinetic energy associated with the motion of the mass center G of the body and the kinetic energy associated with the rotation of the body about G. You will write, therefore,

$$T_2 = \tfrac{1}{2}m\bar{v}_m^2 + \tfrac{1}{2}\bar{I}\omega_m^2$$

Assuming that the position of the body has been defined by an angle θ, express $\bar{v}_m$ and ω_m in terms of the rate of change $\dot{\theta}_m$ of θ as the body passes through its equilibrium position. The kinetic energy of the body will thus be expressed as the product of a constant and $\dot{\theta}_m^2$. Note that if θ measures the rotation of the body about its mass center, as was the case for the plate of Fig. 19.6, then $\omega_m = \dot{\theta}_m$. In other cases, however, the kinematics of the motion should be used to derive a relation between ω_m and $\dot{\theta}_m$ [Sample Prob. 19.4].

b. If the system consists of several rigid bodies, repeat the above computation for each of the bodies, using the same coordinate θ, and add the results obtained.

3. Equate the potential energy V_1 of the system to its kinetic energy T_2,

$$V_1 = T_2$$

and, recalling the first of Eqs. (19.15), replace $\dot{\theta}_m$ in the right-hand term by the product of the amplitude θ_m and the circular frequency ω_n. Since both terms now contain the factor θ_m^2, this factor can be canceled and the resulting equation can be solved for the circular frequency ω_n.

PROBLEMS

19.69 A 1.8-kg collar A is attached to a spring of constant 800 N/m and can slide without friction on a horizontal rod. If the collar is moved 70 mm to the left from its equilibrium position and released, determine the maximum velocity and maximum acceleration of the collar during the resulting motion.

Fig. P19.69

19.70 Two blocks, each of weight 3 lb, are attached to links which are pin-connected to bar BC as shown. The weights of the links and bar are negligible, and the blocks can slide without friction. Block D is attached to a spring of constant $k = 4$ lb/in. Knowing that block A is moved 0.5 in. from its equilibrium position and released, determine the magnitude of the maximum velocity of block D during the resulting motion.

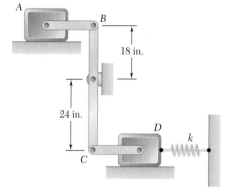

Fig. P19.70

19.71 A 14-oz sphere A and a 10-oz sphere C are attached to the ends of a rod AC of negligible weight which can rotate in a vertical plane about an axis at B. Determine the period of small oscillations of the rod.

19.72 Determine the period of small oscillations of a small particle which moves without friction inside a cylindrical surface of radius R.

Fig. P19.71

Fig. P19.72

19.73 The inner rim of an 85-lb flywheel is placed on a knife edge, and the period of its small oscillations is found to be 1.26 s. Determine the centroidal moment of inertia of the flywheel.

Fig. P19.73

Fig. P19.74

19.74 A connecting rod is supported by a knife edge at point A; the period of its small oscillations is observed to be 1.03 s. Knowing that the distance r_a is 6 in., determine the centroidal radius of gyration of the connecting rod.

19.75 A uniform rod AB can rotate in a vertical plane about a horizontal axis at C located at a distance c above the mass center G of the rod. For small oscillations determine the value of c for which the frequency of the motion will be maximum.

Fig. P19.75

19.76 A homogeneous wire of length $2l$ is bent as shown and allowed to oscillate about a frictionless pin at B. Denoting by τ_0 the period of small oscillations when $\beta = 0$, determine the angle β for which the period of small oscillations is $2\,\tau_0$.

19.77 A uniform disk of radius r and mass m can roll without slipping on a cylindrical surface and is attached to bar ABC of length L and negligible mass. The bar is attached to a spring of constant k and can rotate freely in the vertical plane about point B. Knowing that end A is given a small displacement and released, determine the frequency of the resulting oscillations in terms of m, L, k, and g.

Fig. P19.76

Fig. P19.77

19.78 Two uniform rods, each of weight $W = 1.2$ lb and length $l = 8$ in., are welded together to form the assembly shown. Knowing that the constant of each spring is $k = 0.6$ lb/in. and that end A is given a small displacement and released, determine the frequency of the resulting motion.

19.79 A 15-lb uniform cylinder can roll without sliding on an incline and is attached to a spring AB as shown. If the center of the cylinder is moved 0.4 in. down the incline and released, determine (*a*) the period of vibration, (*b*) the maximum velocity of the center of the cylinder.

Fig. P19.79

Fig. P19.78

19.80 A 3-kg slender rod AB is bolted to a 5-kg uniform disk. A spring of constant 280 N/m is attached to the disk and is unstretched in the position shown. If end B of the rod is given a small displacement and released, determine the period of vibration of the system.

19.81 A slender rod AB of mass m and length l is connected to two collars of negligible mass in a horizontal plane as shown. Collar A is attached to a spring of constant k. Knowing that the collars can slide freely on their respective rods and the system is in equilibrium in the position shown, determine the period of vibration if collar A is given a small displacement and released.

Fig. P19.80

Fig. *P19.81* and *P19.82*

19.82 A slender rod AB of mass m and length l is connected to two collars of mass m_c in a horizontal plane as shown. Collar A is attached to a spring of constant k. Knowing that the collars can slide freely on their respective rods and the system is in equilibrium in the position shown, determine the period of vibration if collar A is given a small displacement and released.

19.83 An 800-g rod AB is bolted to a 1.2-kg disk. A spring of constant $k = 12$ N/m is attached to the center of the disk at A and to the wall at C. Knowing that the disk rolls without sliding, determine the period of small oscillations of the system.

Fig. P19.83

19.84 Three identical rods are connected as shown. If $b = \frac{3}{4}l$, determine the frequency of small oscillations of the system.

Fig. P19.84

Fig. P19.85

19.85 A 14-oz sphere A and a 10-oz sphere C are attached to the ends of a 20-oz rod AC which can rotate in a vertical plane about an axis at B. Determine the period of small oscillations of the rod.

19.86 A 10-lb uniform rod CD is welded at C to a shaft of negligible mass which is welded to the centers of two 20-lb uniform disks A and B. Knowing that the disks roll without sliding, determine the period of small oscillations of the system.

19.87 and 19.88 Two uniform rods AB and CD, each of length l and mass m, are attached to gears as shown. Knowing that the mass of gear C is m and that the mass of gear A is $4m$, determine the period of small oscillations of the system.

Fig. P19.86

Fig. P19.87

Fig. P19.88

Fig. P19.89

19.89 An inverted pendulum consisting of a rigid bar ABC of length l and mass m is supported by a pin and bracket at C. A spring of constant k is attached to the bar at B and is undeformed when the bar is in the vertical position shown. Determine (a) the frequency of small oscillations, (b) the smallest value of a for which these oscillations will occur.

19.90 Two 12-lb uniform disks are attached to the 20-lb rod *AB* as shown. Knowing that the constant of the spring is 30 lb/in. and that the disks roll without sliding, determine the frequency of vibration of the system.

Fig. P19.90

19.91 The 20-lb rod *AB* is attached to two 8-lb disks as shown. Knowing that the disks roll without sliding, determine the frequency of small oscillations of the system.

Fig. P19.91

19.92 A half section of a uniform cylinder of radius *r* and mass *m* rests on two casters *A* and *B*, each of which is a uniform cylinder of radius *r*/4 and mass *m*/8. Knowing that the half cylinder is rotated through a small angle and released and that no slipping occurs, determine the frequency of small oscillations.

Fig. P19.92

19.93 The motion of the uniform rod *AB* is guided by the cord *BC* and by the small roller at *A*. Determine the frequency of oscillation when the end *B* of the rod is given a small horizontal displacement and released.

Fig. P19.93

19.94 A uniform rod of length *L* is supported by a ball-and-socket joint at *A* and by a vertical wire *CD*. Derive an expression for the period of oscillation of the rod if end *B* is given a small horizontal displacement and then released.

Fig. *P19.94*

19.95 A section of uniform pipe is suspended from two vertical cables attached at *A* and *B*. Determine the frequency of oscillation when the pipe is given a small rotation about the centroidal axis *OO'* and released.

Fig. P19.95

$k = 350$ N/m

A

200 mm

B

C

M

D

$k = 260$ N/m

300 mm

Fig. P19.96

Fig. P19.98

Fig. P19.7

$T = k(\delta_{st} + x)$

Equilibrium

x

W

P

$P = P_m \sin \omega_f t$

$ma = m\ddot{x}$

19.96 A 0.6-kg uniform arm ABC is supported by a pin at B and is attached to a spring at A. It is connected at C to a 1.4-kg mass M which is attached to a spring. Knowing that each spring can act in tension or compression, determine the frequency of small oscillations of the system when the weight is given a small vertical displacement and released.

*19.97 A thin plate of length l rests on a half cylinder of radius r. Derive an expression for the period of small oscillations of the plate.

l

r

Fig. P19.97

*19.98 As a submerged body moves through a fluid, the particles of the fluid flow around the body and thus acquire kinetic energy. In the case of a sphere moving in an ideal fluid, the total kinetic energy acquired by the fluid is $\frac{1}{4}\rho V v^2$, where ρ is the mass density of the fluid, V is the volume of the sphere, and v is the velocity of the sphere. Consider a 500-g hollow spherical shell of radius 80 mm which is held submerged in a tank of water by a spring of constant 500 N/m. (*a*) Neglecting fluid friction, determine the period of vibration of the shell when it is displaced vertically and then released. (*b*) Solve part *a*, assuming that the tank is accelerated upward at the constant rate of 8 m/s^2.

19.7 FORCED VIBRATIONS

The most important vibrations from the point of view of engineering applications are the *forced vibrations* of a system. These vibrations occur when a system is subjected to a periodic force or when it is elastically connected to a support which has an alternating motion.

Consider first the case of a body of mass m suspended from a spring and subjected to a periodic force $\mathbf{P}$ of magnitude $P = P_m \sin \omega_f t$, where ω_f is the circular frequency of $\mathbf{P}$ and is referred to as the *forced circular frequency* of the motion (Fig. 19.7). This force may be an actual external force applied to the body, or it may be a centrifugal force produced by the rotation of some unbalanced part of the body (see Sample Prob. 19.5). Denoting by x the displacement of the body measured from its equilibrium position, we write the equation of motion,

$$+\downarrow \Sigma F = ma: \quad P_m \sin \omega_f t + W - k(\delta_{st} + x) = m\ddot{x}$$

Recalling that $W = k\delta_{st}$, we have

$$m\ddot{x} + kx = P_m \sin \omega_f t \qquad (19.30)$$

Next we consider the case of a body of mass m suspended from a spring attached to a moving support whose displacement δ is equal to $\delta_m \sin \omega_f t$ (Fig. 19.8). Measuring the displacement x of the body from the position of static equilibrium corresponding to $\omega_f t = 0$, we find that the total elongation of the spring at time t is $\delta_{st} + x - \delta_m \sin \omega_f t$. The equation of motion is thus

$$+\downarrow \Sigma F = ma: \qquad W - k(\delta_{st} + x - \delta_m \sin \omega_f t) = m\ddot{x}$$

Recalling that $W = k\delta_{st}$, we have

$$m\ddot{x} + kx = k\delta_m \sin \omega_f t \qquad (19.31)$$

We note that Eqs. (19.30) and (19.31) are of the same form and that a solution of the first equation will satisfy the second if we set $P_m = k\delta_m$.

A differential equation such as (19.30) or (19.31), possessing a right-hand member different from zero, is said to be *nonhomogeneous*. Its general solution is obtained by adding a particular solution of the given equation to the general solution of the corresponding *homogeneous* equation (with right-hand member equal to zero). A *particular solution* of (19.30) or (19.31) can be obtained by trying a solution of the form

$$x_{part} = x_m \sin \omega_f t \qquad (19.32)$$

Substituting x_{part} for x into (19.30), we find

$$-m\omega_f^2 x_m \sin \omega_f t + kx_m \sin \omega_f t = P_m \sin \omega_f t$$

which can be solved for the amplitude,

$$x_m = \frac{P_m}{k - m\omega_f^2}$$

Recalling from (19.4) that $k/m = \omega_n^2$, where ω_n is the natural circular frequency of the system, we write

$$x_m = \frac{P_m/k}{1 - (\omega_f/\omega_n)^2} \qquad (19.33)$$

Substituting from (19.32) into (19.31), we obtain in a similar way

$$x_m = \frac{\delta_m}{1 - (\omega_f/\omega_n)^2} \qquad (19.33')$$

The homogeneous equation corresponding to (19.30) or (19.31) is Eq. (19.2), which defines the free vibration of the body. Its general solution, called the *complementary function*, was found in Sec. 19.2:

$$x_{comp} = C_1 \sin \omega_n t + C_2 \cos \omega_n t \qquad (19.34)$$

Fig. 19.8

Photo 19.1 A seismometer operates by measuring the amount of electrical energy needed to keep a mass centered in the housing in the presence of strong ground shaking.

Adding the particular solution (19.32) to the complementary function (19.34), we obtain the *general solution* of Eqs. (19.30) and (19.31):

$$x = C_1 \sin \omega_n t + C_2 \cos \omega_n t + x_m \sin \omega_f t \qquad (19.35)$$

We note that the vibration obtained consists of two superposed vibrations. The first two terms in Eq. (19.35) represent a *free vibration* of the system. The frequency of this vibration is the *natural frequency* of the system, which depends only upon the constant k of the spring and the mass m of the body, and the constants C_1 and C_2 can be determined from the initial conditions. This free vibration is also called a *transient* vibration, since in actual practice it will soon be damped out by friction forces (Sec. 19.9).

The last term in (19.35) represents the *steady-state* vibration produced and maintained by the impressed force or impressed support movement. Its frequency is the *forced frequency* imposed by this force or movement, and its amplitude x_m, defined by (19.33) or (19.33'), depends upon the *frequency ratio* ω_f/ω_n. The ratio of the amplitude x_m of the steady-state vibration to the static deflection P_m/k caused by a force P_m, or to the amplitude δ_m of the support movement, is called the *magnification factor*. From (19.33) and (19.33'), we obtain

$$\text{Magnification factor} = \frac{x_m}{P_m/k} = \frac{x_m}{\delta_m} = \frac{1}{1 - (\omega_f/\omega_n)^2} \qquad (19.36)$$

The magnification factor has been plotted in Fig. 19.9 against the frequency ratio ω_f/ω_n. We note that when $\omega_f = \omega_n$, the amplitude of the forced vibration becomes infinite. The impressed force or impressed support movement is said to be in *resonance* with the given system. Actually, the amplitude of the vibration remains finite because of damping forces (Sec. 19.9); nevertheless, such a situation should be avoided, and the forced frequency should not be chosen too close to the natural frequency of the system. We also note that for $\omega_f < \omega_n$ the coefficient of $\sin \omega_f t$ in (19.35) is positive, while for $\omega_f > \omega_n$ this coefficient is negative. In the first case the forced vibration is *in phase* with the impressed force or impressed support movement, while in the second case it is 180° *out of phase*.

Finally, let us observe that the velocity and the acceleration in the steady-state vibration can be obtained by differentiating twice with respect to t the last term of Eq. (19.35). Their maximum values are given by expressions similar to those of Eqs. (19.15) of Sec. 19.2, except that these expressions now involve the amplitude and the circular frequency of the forced vibration:

$$v_m = x_m \omega_f \qquad a_m = x_m \omega_f^2 \qquad (19.37)$$

Fig. 19.9

SAMPLE PROBLEM 19.5

A motor weighing 350 lb is supported by four springs, each having a constant of 750 lb/in. The unbalance of the rotor is equivalent to a weight of 1 oz located 6 in. from the axis of rotation. Knowing that the motor is constrained to move vertically, determine (a) the speed in rpm at which resonance will occur, (b) the amplitude of the vibration of the motor at a speed of 1200 rpm.

SOLUTION

a. Resonance Speed. The resonance speed is equal to the natural circular frequency ω_n (in rpm) of the free vibration of the motor. The mass of the motor and the equivalent constant of the supporting springs are

$$m = \frac{350 \text{ lb}}{32.2 \text{ ft/s}^2} = 10.87 \text{ lb} \cdot \text{s}^2/\text{ft}$$

$$k = 4(750 \text{ lb/in.}) = 3000 \text{ lb/in.} = 36{,}000 \text{ lb/ft}$$

$$\omega_n = \sqrt{\frac{k}{m}} = \sqrt{\frac{36{,}000}{10.87}} = 57.5 \text{ rad/s} = 549 \text{ rpm}$$

Resonance speed = 549 rpm ◄

b. Amplitude of Vibration at 1200 rpm. The angular velocity of the motor and the mass of the equivalent 1-oz weight are

$$\omega = 1200 \text{ rpm} = 125.7 \text{ rad/s}$$

$$m = (1 \text{ oz})\frac{1 \text{ lb}}{16 \text{ oz}}\frac{1}{32.2 \text{ ft/s}^2} = 0.001941 \text{ lb} \cdot \text{s}^2/\text{ft}$$

The magnitude of the centrifugal force due to the unbalance of the rotor is

$$P_m = ma_n = mr\omega^2 = (0.001941 \text{ lb} \cdot \text{s}^2/\text{ft})(\tfrac{6}{12} \text{ ft})(125.7 \text{ rad/s})^2 = 15.33 \text{ lb}$$

The static deflection that would be caused by a constant load P_m is

$$\frac{P_m}{k} = \frac{15.33 \text{ lb}}{3000 \text{ lb/in.}} = 0.00511 \text{ in.}$$

The forced circular frequency ω_f of the motion is the angular velocity of the motor,

$$\omega_f = \omega = 125.7 \text{ rad/s}$$

Substituting the values of P_m/k, ω_f, and ω_n into Eq. (19.33), we obtain

$$x_m = \frac{P_m/k}{1 - (\omega_f/\omega_n)^2} = \frac{0.00511 \text{ in.}}{1 - (125.7/57.5)^2} = -0.001352 \text{ in.}$$

$x_m = 0.001352$ in. (out of phase) ◄

Note. Since $\omega_f > \omega_n$, the vibration is 180° out of phase with the centrifugal force due to the unbalance of the rotor. For example, when the unbalanced mass is directly below the axis of rotation, the position of the motor is $x_m = 0.001352$ in. above the position of equilibrium.

SOLVING PROBLEMS
ON YOUR OWN

This lesson was devoted to the analysis of the *forced vibrations* of a mechanical system. These vibrations occur either when the system is subjected to a periodic force **P** (Fig. 19.7), or when it is elastically connected to a support which has an alternating motion (Fig. 19.8). In the first case, the motion of the system is defined by the differential equation

$$m\ddot{x} + kx = P_m \sin \omega_f t \tag{19.30}$$

where the right-hand member represents the magnitude of the force **P** at a given instant. In the second case, the motion is defined by the differential equation

$$m\ddot{x} + kx = k\delta_m \sin \omega_f t \tag{19.31}$$

where the right-hand member is the product of the spring constant k and the displacement of the support at a given instant. You will be concerned only with the *steady-state* motion of the system, which is defined by a *particular solution* of these equations, of the form

$$x_{\text{part}} = x_m \sin \omega_f t \tag{19.32}$$

1. If the forced vibration is caused by a periodic force P, of amplitude P_m and circular frequency ω_f, the amplitude of the vibration is

$$x_m = \frac{P_m/k}{1 - (\omega_f/\omega_n)^2} \tag{19.33}$$

where ω_n is the *natural circular frequency* of the system, $\omega_n = \sqrt{k/m}$, and k is the spring constant. Note that the circular frequency of the vibration is ω_f and that the amplitude x_m does not depend upon the initial conditions. For $\omega_f = \omega_n$, the denominator in Eq. (19.33) is zero and x_m is infinite (Fig. 19.9); the impressed force **P** is said to be in *resonance* with the system. Also, for $\omega_f < \omega_n$, x_m is positive and the vibration is *in phase* with **P**, while, for $\omega_f > \omega_n$, x_m is negative and the vibration is *out of phase*.

    a. In the problems which follow, you may be asked to determine one of the parameters in Eq. (19.33) when the others are known. We suggest that you keep Fig. 19.9 in front of you when solving these problems. For example, if you are asked to find the frequency at which the amplitude of a forced vibration has a given value, but you do not know whether the vibration is in or out of phase with respect to the impressed force, you should note from Fig. 19.9 that there can be two frequencies satisfying this requirement, one corresponding to a positive value of x_m and to a vibration in phase with the impressed force, and the other corresponding to a negative value of x_m and to a vibration out of phase with the impressed force.

b. Once you have obtained the amplitude x_m of the motion of a component of the system from Eq. (19.33), you can use Eqs. (19.37) to determine the maximum values of the velocity and acceleration of that component:

$$v_m = x_m \omega_f \qquad a_m = x_m \omega_f^2 \qquad (19.37)$$

c. When the impressed force P is due to the unbalance of the rotor of a motor, its maximum value is $P_m = mr\omega_f^2$, where m is the mass of the rotor, r is the distance between its mass center and the axis of rotation, and ω_f is equal to the angular velocity ω of the rotor expressed in rad/s [Sample Prob. 19.5].

2. If the forced vibration is caused by the simple harmonic motion of a support, of amplitude δ_m and circular frequency ω_f, the amplitude of the vibration is

$$x_m = \frac{\delta_m}{1 - (\omega_f/\omega_n)^2} \qquad (19.33')$$

where ω_n is the *natural circular frequency* of the system, $\omega_n = \sqrt{k/m}$. Again, note that the circular frequency of the vibration is ω_f and that the amplitude x_m does not depend upon the initial conditions.

a. Be sure to read our comments in paragraphs 1, 1a, and 1b, since they apply equally well to a vibration caused by the motion of a support.

b. If the maximum acceleration a_m **of the support is specified,** rather than its maximum displacement δ_m, remember that, since the motion of the support is a simple harmonic motion, you can use the relation $a_m = \delta_m \omega_f^2$ to determine δ_m; the value obtained is then substituted into Eq. (19.33').

PROBLEMS

$P = P_m \sin \omega_f t$

$k = 8$ kN/m

20 kg

Fig. P19.99 and P19.100

$P = P_m \sin \omega_f t$

Fig. P19.101 and P19.102

19.99 A 20-kg block is attached to a spring of constant $k = 8$ kN/m and can move without friction in a vertical slot as shown. The block is acted upon by a periodic force of magnitude $P = P_m \sin \omega_f t$, where $P_m = 100$ N. Determine the amplitude of the motion of the block if (a) $\omega_f = 10$ rad/s, (b) $\omega_f = 19$ rad/s, (c) $\omega_f = 30$ rad/s.

19.100 A 20-kg block is attached to a spring of constant $k = 8$ kN/m and can move without friction in a vertical slot as shown. The block is acted upon by a periodic force of magnitude $P = P_m \sin \omega_f t$, where $P_m = 10$ N. Knowing that the amplitude of the motion is 3 mm, determine the value of ω_f.

19.101 A 9-lb collar can slide on a frictionless horizontal rod and is attached to a spring of constant k. It is acted upon by a periodic force of magnitude $P = P_m \sin \omega_f t$, where $P_m = 2$ lb and $\omega_f = 5$ rad/s. Determine the value of the spring constant k knowing that the motion of the collar has an amplitude of 6 in. and is (a) in phase with the applied force, (b) out of phase with the applied force.

19.102 A collar of mass m which slides on a frictionless horizontal rod is attached to a spring of constant k and is acted upon by a periodic force of magnitude $P = P_m \sin \omega_f t$. Determine the range of values of ω_f for which the amplitude of the vibration exceeds two times the static deflection caused by a constant force of magnitude P_m.

19.103 A small 20-kg block A is attached to the rod BC of negligible mass which is supported at B by a pin and bracket and at C by a spring of constant $k = 2$ kN/m. The system can move in a vertical plane and is in equilibrium when the rod is horizontal. The rod is acted upon at C by a periodic force $\mathbf{P}$ of magnitude $P = P_m \sin \omega_f t$, where $P_m = 6$ N. Knowing that $b = 200$ mm, determine the range of values of ω_f for which the amplitude of vibration of block A exceeds 3.5 mm.

$P = P_m \sin \omega_f t$

Fig. P19.103

$T = T_m \sin \omega_f t$

19.104 An 8-kg uniform disk of radius 200 mm is welded to a vertical shaft with a fixed end at B. The disk rotates through an angle of $3°$ when a static couple of magnitude 50 N · m is applied to it. If the disk is acted upon by a periodic torsional couple of magnitude $T = T_m \sin \omega_f t$, where $T_m = 60$ N · m, determine the range of values of ω_f for which the amplitude of the vibration is less than the angle of rotation caused by a static couple of magnitude T_m.

Fig. P19.104

19.105 An 18-lb block A slides in a vertical frictionless slot and is connected to a moving support B by means of a spring AB of constant $k = 10$ lb/in. Knowing that the displacement of the support is $\delta = \delta_m \sin \omega_f t$, where $\delta_m = 6$ in., determine the range of values of ω_f for which the amplitude of the fluctuating force exerted by the spring on the block is less than 30 lb.

Fig. P19.105

19.106 A cantilever beam AB supports a block which causes a static deflection of 8 mm at B. Assuming that the support at A undergoes a vertical periodic displacement $\delta = \delta_m \sin \omega_f t$, where $\delta_m = 2$ mm, determine the range of values of ω_f for which the amplitude of the motion of the block will be less than 4 mm. Neglect the weight of the beam and assume that the block does not leave the beam.

Fig. P19.106

19.107 Rod AB is rigidly attached to the frame of a motor running at a constant speed. When a collar of mass m is placed on the spring, it is observed to vibrate with an amplitude of 15 mm. When two collars, each of mass m, are placed on the spring, the amplitude is observed to be 18 mm. What amplitude of vibration should be expected when three collars, each of mass m, are placed on the spring? (Obtain two answers.)

Fig. P19.107

19.108 The crude-oil-pumping rig shown is driven at 20 rpm. The inside diameter of the well pipe is 2 in., and the diameter of the pump rod is 0.75 in. The length of the pump rod and the length of the column of oil lifted during the stroke are essentially the same, and equal to 6000 ft. During the downward stroke, a valve at the lower end of the pump rod opens to let a quantity of oil into the well pipe, and the column of oil is then lifted to obtain a discharge into the connecting pipeline. Thus, the amount of oil pumped in a given time depends upon the stroke of the lower end of the pump rod. Knowing that the upper end of the rod at D is essentially sinusoidal with a stroke of 45 in. and the specific weight of crude oil is 56.2 lb/ft³, determine (a) the output of the well in ft³/min if the shaft is rigid, (b) the output of the well in ft³/min if the stiffness of the rod is 2210 N/m, the equivalent mass of the oil and shaft is 290 kg, and damping is negligible.

Fig. P19.108

Fig. P19.109 and P19.110

19.109 A simple pendulum of length l is suspended from collar C which is forced to move horizontally according to the relation $x_C = \delta_m \sin \omega_f t$. Determine the range of values of ω_f for which the amplitude of the motion of the bob is less than δ_m. (Assume that δ_m is small compared with the length l of the pendulum.)

19.110 The 2.75-lb bob of a simple pendulum of length $l = 24$ in. is suspended from a 3-lb collar C. The collar is forced to move according to the relation $x_C = \delta_m \sin \omega_f t$, with an amplitude $\delta_m = 0.4$ in. and a frequency $f_f = 0.5$ Hz. Determine (a) the amplitude of the motion of the bob, (b) the force that must be applied to collar C to maintain the motion.

19.111 An 18-lb block A slides in a vertical frictionless slot and is connected to a moving support B by means of a spring AB of constant $k = 8$ lb/ft. Knowing that the acceleration of the support is $a = a_m \sin \omega_f t$, where $a_m = 5$ ft/s^2 and $\omega_f = 6$ rad/s, determine (a) the maximum displacement of block A, (b) the amplitude of the fluctuating force exerted by the spring on the block.

Fig. P19.111

19.112 A variable-speed motor is rigidly attached to a beam BC. When the speed of the motor is less than 600 rpm or more than 1200 rpm, a small object placed at A is observed to remain in contact with the beam. For speeds between 600 and 1200 rpm the object is observed to "dance" and actually to lose contact with the beam. Determine the speed at which resonance will occur.

Fig. P19.112

19.113 A motor of mass M is supported by springs with an equivalent spring constant k. The unbalance of its rotor is equivalent to a mass m located at a distance r from the axis of rotation. Show that when the angular velocity of the motor is ω_f, the amplitude x_m of the motion of the motor is

$$x_m = \frac{r(m/M)(\omega_f/\omega_n)^2}{1 - (\omega_f/\omega_n)^2}$$

where $\omega_n = \sqrt{k/M}$.

19.114 As the rotational speed of a spring-supported 100-kg motor is increased, the amplitude of the vibration due to the unbalance of its 15-kg rotor first increases and then decreases. It is observed that as very high speeds are reached, the amplitude of the vibration approaches 3.3 mm. Determine the distance between the mass center of the rotor and its axis of rotation. (*Hint:* Use the formula derived in Prob. 19.113.)

19.115 A motor of weight 40 lb is supported by four springs, each of constant 225 lb/in. The motor is constrained to move vertically, and the amplitude of its motion is observed to be 0.05 in. at a speed of 1200 rpm. Knowing that the weight of the rotor is 9 lb, determine the distance between the mass center of the rotor and the axis of the shaft.

Fig. P19.115

19.116 A motor weighing 400 lb is supported by springs having a total constant of 1200 lb/in. The unbalance of the rotor is equivalent to a 1-oz weight located 8 in. from the axis of rotation. Determine the range of allowable values of the motor speed if the amplitude of the vibration is not to exceed 0.06 in.

Fig. P19.117

19.117 A 180-kg motor is bolted to a light horizontal beam. The unbalance of its rotor is equivalent to a 28-g mass located 150 mm from the axis of rotation, and the static deflection of the beam due to the weight of the motor is 12 mm. The amplitude of the vibration due to the unbalance can be decreased by adding a plate to the base of the motor. If the amplitude of vibration is to be less than 60 μm for motor speeds above 300 rpm, determine the required mass of the plate.

19.118 The unbalance of the rotor of a 400-lb motor is equivalent to a 3-oz weight located 6 in. from the axis of rotation. In order to limit to 0.2 lb the amplitude of the fluctuating force exerted on the foundation when the motor is run at speeds of 100 rpm and above, a pad is to be placed between the motor and the foundation. Determine (a) the maximum allowable spring constant k of the pad, (b) the corresponding amplitude of the fluctuating force exerted on the foundation when the motor is run at 200 rpm.

Fig. P19.118

19.119 A counter-rotating eccentric mass exciter consisting of two rotating 100-g masses describing circles of radius r at the same speed but in opposite senses is placed on a machine element to induce a steady-state vibration of the element. The total mass of the system is 300 kg, the constant of each spring is $k = 600$ kN/m, and the rotational speed of the exciter is 1200 rpm. Knowing that the amplitude of the total fluctuating force exerted on the foundation is 160 N, determine the radius r.

Fig. P19.119

19.120 A 360-lb motor is supported by springs of total constant 12.5 kips/ft. The unbalance of the rotor is equivalent to a 0.9-oz weight located 7.5 in. from the axis of rotation. Determine the range of speeds of the motor for which the amplitude of the fluctuating force exerted on the foundation is less than 5 lb.

19.121 Figures (1) and (2) show how springs can be used to support a block in two different situations. In Fig. (1) they help decrease the amplitude of the fluctuating force transmitted by the block to the foundation. In Fig. (2) they help decrease the amplitude of the fluctuating displacement transmitted by the foundation to the block. The ratio of the transmitted force to the impressed force or the ratio of the transmitted displacement to the impressed displacement is called the *transmissibility*. Derive an equation for the transmissibility for each situation. Give your answer in terms of the ratio ω_f/ω_n of the frequency ω_f of the impressed force or impressed displacement to the natural frequency ω_n of the spring-mass system. Show that in order to cause any reduction in transmissibility, the ratio ω_f/ω_n must be greater than $\sqrt{2}$.

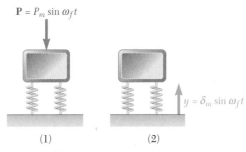

$\mathbf{P} = P_m \sin \omega_f t$

$y = \delta_m \sin \omega_f t$

(1)   (2)

Fig. P19.121

19.122 A vibrometer used to measure the amplitude of vibrations consists essentially of a box containing a mass-spring system with a known natural frequency of 120 Hz. The box is rigidly attached to a surface which is moving according to the equation $y = \delta_m \sin \omega_f t$. If the amplitude z_m of the motion of the mass relative to the box is used as a measure of the amplitude δ_m of the vibration of the surface, determine (*a*) the percent error when the frequency of the vibration is 600 Hz, (*b*) the frequency at which the error is zero.

$y = \delta_m \sin \omega_f t$

Fig. P19.122 and P19.123

19.123 A certain accelerometer consists essentially of a box containing a mass-spring system with a known natural frequency of 2200 Hz. The box is rigidly attached to a surface which is moving according to the equation $y = \delta_m \sin \omega_f t$. If the amplitude z_m of the motion of the mass relative to the box times a scale factor ω_n^2 is used as a measure of the maximum acceleration $\alpha_m = \delta_m \omega_f^2$ of the vibrating surface, determine the percent error when the frequency of the vibration is 600 Hz.

Fig. P19.124

19.124 Block A can move without friction in the slot as shown and is acted upon by a vertical periodic force of magnitude $P = P_m \sin \omega_f t$, where $\omega_f = 2$ rad/s and $P_m = 20$ N. A spring of constant k is attached to the bottom of block A and to a 22-kg block B. Determine (a) the value of the constant k which will prevent a steady-state vibration of block A, (b) the corresponding amplitude of the vibration of block B.

19.125 A 60-lb disk is attached with an eccentricity $e = 0.006$ in. to the midpoint of a vertical shaft AB which revolves at a constant angular velocity ω_f. Knowing that the spring constant k for horizontal movement of the disk is 40,000 lb/ft, determine (a) the angular velocity ω_f at which resonance will occur, (b) the deflection r of the shaft when $\omega_f = 1200$ rpm.

Fig. P19.125

Fig. P19.126

19.126 A small trailer and its load have a total mass of 250 kg. The trailer is supported by two springs, each of constant 10 kN/m, and is pulled over a road, the surface of which can be approximated by a sine curve with an amplitude of 40 mm and a wavelength of 5 m (i.e., the distance between successive crests is 5 m and the vertical distance from crest to trough is 80 mm). Determine (a) the speed at which resonance will occur, (b) the amplitude of the vibration of the trailer at a speed of 50 km/h.

DAMPED VIBRATIONS

*19.8 DAMPED FREE VIBRATIONS

The vibrating systems considered in the first part of this chapter were assumed free of damping. Actually all vibrations are damped to some degree by friction forces. These forces can be caused by *dry friction*, or *Coulomb friction*, between rigid bodies, by *fluid friction* when a rigid body moves in a fluid, or by *internal friction* between the molecules of a seemingly elastic body.

A type of damping of special interest is the *viscous damping* caused by fluid friction at low and moderate speeds. Viscous damping is characterized by the fact that the friction force is *directly proportional and opposite to the velocity* of the moving body. As an example, let us again consider a body of mass m suspended from a spring of constant k, assuming that the body is attached to the plunger of a dashpot (Fig. 19.10). The magnitude of the friction force exerted on the plunger by the surrounding fluid is equal to $c\dot{x}$, where the constant c, expressed in N · s/m or lb · s/ft and known as the *coefficient of viscous damping*, depends upon the physical properties of the fluid and the construction of the dashpot. The equation of motion is

$$+\downarrow \Sigma F = ma: \qquad W - k(\delta_{st} + x) - c\dot{x} = m\ddot{x}$$

Recalling that $W = k\delta_{st}$, we write

$$m\ddot{x} + c\dot{x} + kx = 0 \qquad (19.38)$$

Substituting $x = e^{\lambda t}$ into (19.38) and dividing through by $e^{\lambda t}$, we write the *characteristic equation*

$$m\lambda^2 + c\lambda + k = 0 \qquad (19.39)$$

and obtain the roots

$$\lambda = -\frac{c}{2m} \pm \sqrt{\left(\frac{c}{2m}\right)^2 - \frac{k}{m}} \qquad (19.40)$$

Defining the *critical damping coefficient* c_c as the value of c which makes the radical in Eq. (19.40) equal to zero, we write

$$\left(\frac{c_c}{2m}\right)^2 - \frac{k}{m} = 0 \qquad c_c = 2m\sqrt{\frac{k}{m}} = 2m\omega_n \qquad (19.41)$$

where ω_n is the natural circular frequency of the system in the absence of damping. We can distinguish three different cases of damping, depending upon the value of the coefficient c.

1. *Heavy damping: $c > c_c$.* The roots λ_1 and λ_2 of the characteristic equation (19.39) are real and distinct, and the general solution of the differential equation (19.38) is

$$x = C_1 e^{\lambda_1 t} + C_2 e^{\lambda_2 t} \qquad (19.42)$$

This solution corresponds to a nonvibratory motion. Since λ_1 and λ_2 are both negative, x approaches zero as t increases indefinitely. However, the system actually regains its equilibrium position after a finite time.

2. *Critical damping: $c = c_c$.* The characteristic equation has a double root $\lambda = -c_c/2m = -\omega_n$, and the general solution of (19.38) is

$$x = (C_1 + C_2 t)e^{-\omega_n t} \qquad (19.43)$$

The motion obtained is again nonvibratory. Critically damped systems are of special interest in engineering applications since

$T = k(\delta_{st} + x)$

Equilibrium

$ma = m\ddot{x}$

$c\dot{x}$

Fig. 19.10

they regain their equilibrium position in the shortest possible time without oscillation.

3. *Light damping: $c < c_c$.* The roots of Eq. (19.39) are complex and conjugate, and the general solution of (19.38) is of the form

$$x = e^{-(c/2m)t}(C_1 \sin \omega_d t + C_2 \cos \omega_d t) \qquad (19.44)$$

where ω_d is defined by the relation

$$\omega_d^2 = \frac{k}{m} - \left(\frac{c}{2m}\right)^2$$

Substituting $k/m = \omega_n^2$ and recalling (19.41), we write

$$\omega_d = \omega_n\sqrt{1 - \left(\frac{c}{c_c}\right)^2} \qquad (19.45)$$

where the constant c/c_c is known as the *damping factor*. Even though the motion does not actually repeat itself, the constant ω_d is commonly referred to as the *circular frequency* of the damped vibration. A substitution similar to the one used in Sec. 19.2 enables us to write the general solution of Eq. (19.38) in the form

$$x = x_0 e^{-(c/2m)t} \sin (\omega_d t + \phi) \qquad (19.46)$$

The motion defined by Eq. (19.46) is vibratory with diminishing amplitude (Fig. 19.11), and the time interval $\tau_d = 2\pi/\omega_d$ separating two successive points where the curve defined by Eq. (19.46) touches one of the limiting curves shown in Fig. 19.11 is commonly referred to as the *period of the damped vibration*. Recalling Eq. (19.45), we observe that $\omega_d < \omega_n$ and, thus, that τ_d is larger than the period of vibration τ_n of the corresponding undamped system.

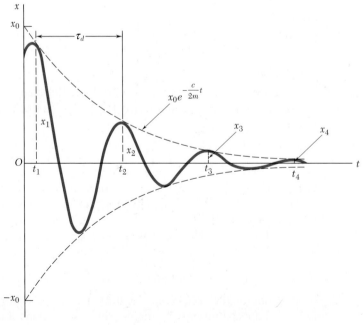

Fig. 19.11

*19.9 DAMPED FORCED VIBRATIONS

If the system considered in the preceding section is subjected to a periodic force **P** of magnitude $P = P_m \sin \omega_f t$, the equation of motion becomes

$$m\ddot{x} + c\dot{x} + kx = P_m \sin \omega_f t \qquad (19.47)$$

The general solution of (19.47) is obtained by adding a particular solution of (19.47) to the complementary function or general solution of the homogeneous equation (19.38). The complementary function is given by (19.42), (19.43), or (19.44), depending upon the type of damping considered. It represents a *transient* motion which is eventually damped out.

Our interest in this section is centered on the steady-state vibration represented by a particular solution of (19.47) of the form

$$x_{\text{part}} = x_m \sin (\omega_f t - \varphi) \qquad (19.48)$$

Substituting x_{part} for x into (19.47), we obtain

$$-m\omega_f^2 x_m \sin (\omega_f t - \varphi) + c\omega_f x_m \cos (\omega_f t - \varphi) + kx_m \sin (\omega_f t - \varphi)$$
$$= P_m \sin \omega_f t$$

Making $\omega_f t - \varphi$ successively equal to 0 and to $\pi/2$, we write

$$c\omega_f x_m = P_m \sin \varphi \qquad (19.49)$$
$$(k - m\omega_f^2) x_m = P_m \cos \varphi \qquad (19.50)$$

Squaring both members of (19.49) and (19.50) and adding, we have

$$[(k - m\omega_f^2)^2 + (c\omega_f)^2] x_m^2 = P_m^2 \qquad (19.51)$$

Solving (19.51) for x_m and dividing (19.49) and (19.50) member by member, we obtain, respectively,

$$x_m = \frac{P_m}{\sqrt{(k - m\omega_f^2)^2 + (c\omega_f)^2}} \qquad \tan \varphi = \frac{c\omega_f}{k - m\omega_f^2} \qquad (19.52)$$

Recalling from (19.4) that $k/m = \omega_n^2$, where ω_n is the circular frequency of the undamped free vibration, and from (19.41) that $2m\omega_n = c_c$, where c_c is the critical damping coefficient of the system, we write

$$\frac{x_m}{P_m/k} = \frac{x_m}{\delta_m} = \frac{1}{\sqrt{[1 - (\omega_f/\omega_n)^2]^2 + [2(c/c_c)(\omega_f/\omega_n)]^2}} \qquad (19.53)$$

$$\tan \varphi = \frac{2(c/c_c)(\omega_f/\omega_n)}{1 - (\omega_f/\omega_n)^2} \qquad (19.54)$$

Photo 19.2 The automobile suspension shown consists essentially of a spring and a shock absorber, which will cause the body of the car to undergo *damped forced vibrations* when the car is driven over an uneven road.

Photo 19.3 This truck is experiencing damped forced vibration in the vehicle dynamics test shown.

Formula (19.53) expresses the magnification factor in terms of the frequency ratio ω_f/ω_n and damping factor c/c_c. It can be used to determine the amplitude of the steady-state vibration produced by an impressed force of magnitude $P = P_m \sin \omega_f t$ or by an impressed support movement $\delta = \delta_m \sin \omega_f t$. Formula (19.54) defines in terms of the same parameters the *phase difference* φ between the impressed force or impressed support movement and the resulting steady-state vibration of the damped system. The magnification factor has been plotted against the frequency ratio in Fig. 19.12 for various values of the damping factor. We observe that the amplitude of a forced vibration can be kept small by choosing a large coefficient of viscous damping c or by keeping the natural and forced frequencies far apart.

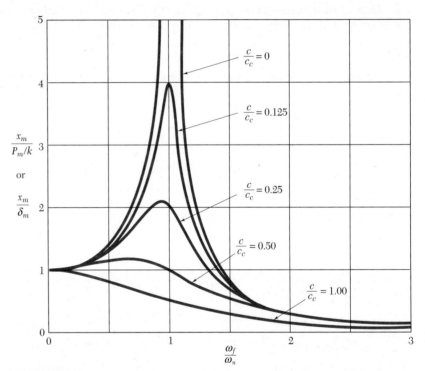

Fig. 19.12

*19.10 ELECTRICAL ANALOGUES

Oscillating electrical circuits are characterized by differential equations of the same type as those obtained in the preceding sections. Their analysis is therefore similar to that of a mechanical system, and the results obtained for a given vibrating system can be readily extended to the equivalent circuit. Conversely, any result obtained for an electrical circuit will also apply to the corresponding mechanical system.

Consider an electrical circuit consisting of an inductor of inductance L, a resistor of resistance R, and a capacitor of capacitance C, connected in series with a source of alternating voltage $E = E_m \sin \omega_f t$ (Fig. 19.13). It is recalled from elementary circuit theory† that if i denotes the current in the circuit and q denotes the electric charge on the capacitor, the drop in potential is $L(di/dt)$ across the inductor, Ri across the resistor, and q/C across the capacitor. Expressing that the algebraic sum of the applied voltage and of the drops in potential around the circuit loop is zero, we write

$$E_m \sin \omega_f t - L\frac{di}{dt} - Ri - \frac{q}{C} = 0 \qquad (19.55)$$

$E = E_m \sin \omega_f t$

Fig. 19.13

Rearranging the terms and recalling that at any instant the current i is equal to the rate of change $\dot{q}$ of the charge q, we have

$$L\ddot{q} + R\dot{q} + \frac{1}{C}q = E_m \sin \omega_f t \qquad (19.56)$$

We verify that Eq. (19.56), which defines the oscillations of the electrical circuit of Fig. 19.13, is of the same type as Eq. (19.47), which characterizes the damped forced vibrations of the mechanical system of Fig. 19.10. By comparing the two equations, we can construct a table of the analogous mechanical and electrical expressions.

Table 19.2 can be used to extend the results obtained in the preceding sections for various mechanical systems to their electrical analogues. For instance, the amplitude i_m of the current in the circuit of Fig. 19.13 can be obtained by noting that it corresponds to the

TABLE 19.2 Characteristics of a Mechanical System and of Its Electrical Analogue

Mechanical System		Electrical Circuit	
m	Mass	L	Inductance
c	Coefficient of viscous damping	R	Resistance
k	Spring constant	$1/C$	Reciprocal of capacitance
x	Displacement	q	Charge
v	Velocity	i	Current
P	Applied force	E	Applied voltage

†See C. R. Paul, S. A. Nasar, and L. E. Unnewehr, *Introduction to Electrical Engineering*, 2nd ed., McGraw-Hill, New York, 1992.

maximum value v_m of the velocity in the analogous mechanical system. Recalling from the first of Eqs. (19.37) that $v_m = x_m\omega_f$, substituting for x_m from Eq. (19.52), and replacing the constants of the mechanical system by the corresponding electrical expressions, we have

$$i_m = \frac{\omega_f E_m}{\sqrt{\left(\dfrac{1}{C} - L\omega_f^2\right)^2 + (R\omega_f)^2}}$$

$$i_m = \frac{E_m}{\sqrt{R^2 + \left(L\omega_f - \dfrac{1}{C\omega_f}\right)^2}} \qquad (19.57)$$

The radical in the expression obtained is known as the *impedance* of the electrical circuit.

Fig. 19.14

The analogy between mechanical systems and electrical circuits holds for transient as well as steady-state oscillations. The oscillations of the circuit shown in Fig. 19.14, for instance, are analogous to the damped free vibrations of the system of Fig. 19.10. As far as the initial conditions are concerned, we should note that closing the switch S when the charge on the capacitor is $q = q_0$ is equivalent to releasing the mass of the mechanical system with no initial velocity from the position $x = x_0$. We should also observe that if a battery of constant voltage E is introduced in the electrical circuit of Fig. 19.14, closing the switch S will be equivalent to suddenly applying a force of constant magnitude P to the mass of the mechanical system of Fig. 19.10.

The discussion above would be of questionable value if its only result were to make it possible for mechanics students to analyze electrical circuits without learning the elements of circuit theory. It is hoped that this discussion will instead encourage students to apply to the solution of problems in mechanical vibrations the mathematical techniques they may learn in later courses in circuit theory. The chief value of the concept of electrical analogue, however, resides in its application to *experimental methods* for the determination of the characteristics of a given mechanical system. Indeed, an electrical circuit is much more easily constructed than is a mechanical model, and the fact that its characteristics can be modified by varying the inductance, resistance, or capacitance of its various components makes the use of the electrical analogue particularly convenient.

To determine the electrical analogue of a given mechanical system, we focus our attention on each moving mass in the system and observe which springs, dashpots, or external forces are applied directly to it. An equivalent electrical loop can then be constructed to match each of the mechanical units thus defined; the various loops obtained in that way will together form the desired circuit. Consider, for instance, the mechanical system of Fig. 19.15. We observe that the mass m_1 is acted upon by two springs of constants k_1 and k_2 and by two dashpots characterized by the coefficients of viscous damping c_1 and c_2. The electrical circuit should therefore include a loop consisting of an inductor of inductance L_1 proportional to m_1, of two capacitors of capacitance C_1 and C_2 inversely proportional to k_1 and k_2, respectively, and of two resistors of resistance R_1 and R_2, proportional to c_1 and c_2, respectively. Since the mass m_2 is acted upon by the spring k_2 and the dashpot c_2, as well as the force $P = P_m \sin \omega_f t$, the circuit should also include a loop containing the capacitor C_2, the resistor R_2, the new inductor L_2, and the voltage source $E = E_m \sin \omega_f t$ (Fig. 19.16).

$P = P_m \sin \omega_f t$

Fig. 19.15

To check that the mechanical system of Fig. 19.15 and the electrical circuit of Fig. 19.16 actually satisfy the same differential equations, the equations of motion for m_1 and m_2 will first be derived. Denoting, respectively, by x_1 and x_2 the displacements of m_1 and m_2 from their equilibrium positions, we observe that the elongation of the spring k_1 (measured from the equilibrium position) is equal to x_1, while the elongation of the spring k_2 is equal to the relative displacement $x_2 - x_1$ of m_2 with respect to m_1. The equations of motion for m_1 and m_2 are therefore

$$m_1 \ddot{x}_1 + c_1 \dot{x}_1 + c_2(\dot{x}_1 - \dot{x}_2) + k_1 x_1 + k_2(x_1 - x_2) = 0 \quad \text{(19.58)}$$

$$m_2 \ddot{x}_2 + c_2(\dot{x}_2 - \dot{x}_1) + k_2(x_2 - x_1) = P_m \sin \omega_f t \quad \text{(19.59)}$$

$E = E_m \sin \omega_f t$

Fig. 19.16

Consider now the electrical circuit of Fig. 19.16; we denote, respectively, by i_1 and i_2 the current in the first and second loops, and by q_1 and q_2 the integrals $\int i_1 \, dt$ and $\int i_2 \, dt$. Noting that the charge on the capacitor C_1 is q_1, while the charge on C_2 is $q_1 - q_2$, we express that the sum of the potential differences in each loop is zero and obtain the following equations

$$L_1 \ddot{q}_1 + R_1 \dot{q}_1 + R_2(\dot{q}_1 - \dot{q}_2) + \frac{q_1}{C_1} + \frac{q_1 - q_2}{C_2} = 0 \quad \text{(19.60)}$$

$$L_2 \ddot{q}_2 + R_2(\dot{q}_2 - \dot{q}_1) + \frac{q_2 - q_1}{C_2} = E_m \sin \omega_f t \quad \text{(19.61)}$$

We easily check that Eqs. (19.60) and (19.61) reduce to (19.58) and (19.59), respectively, when the substitutions indicated in Table 19.2 are performed.

SOLVING PROBLEMS
ON YOUR OWN

In this lesson a more realistic model of a vibrating system was developed by including the effect of the *viscous damping* caused by fluid friction. Viscous damping was represented in Fig. 19.10 by the force exerted on the moving body by a plunger moving in a dashpot. This force is equal in magnitude to $c\dot{x}$, where the constant c, expressed in $N \cdot s/m$ or $lb \cdot s/ft$, is known as the *coefficient of viscous damping*. Keep in mind that the same sign convention should be used for x, $\dot{x}$, and $\ddot{x}$.

1. Damped free vibrations. The differential equation defining this motion was found to be

$$m\ddot{x} + c\dot{x} + kx = 0 \qquad (19.38)$$

To obtain the solution of this equation, calculate the *critical damping coefficient* c_c, using the formula

$$c_c = 2m\sqrt{k/m} = 2m\omega_n \qquad (19.41)$$

where ω_n is the natural circular frequency of the *undamped* system.

a. If $c > c_c$ (heavy damping), the solution of Eq. (19.38) is

$$x = C_1 e^{\lambda_1 t} + C_2 e^{\lambda_2 t} \qquad (19.42)$$

where

$$\lambda_{1,2} = -\frac{c}{2m} \pm \sqrt{\left(\frac{c}{2m}\right)^2 - \frac{k}{m}} \qquad (19.40)$$

and where the constants C_1 and C_2 can be determined from the initial conditions $x(0)$ and $\dot{x}(0)$. This solution corresponds to a nonvibratory motion.

b. If $c = c_c$ (critical damping), the solution of Eq. (19.38) is

$$x = (C_1 + C_2 t)e^{-\omega_n t} \qquad (19.43)$$

which also corresponds to a nonvibratory motion.

c. If $c < c_c$ (light damping), the solution of Eq. (19.38) is

$$x = x_0 e^{-(c/2m)t} \sin(\omega_d t + \phi) \qquad (19.46)$$

where

$$\omega_d = \omega_n \sqrt{1 - \left(\frac{c}{c_c}\right)^2} \qquad (19.45)$$

and where x_0 and ϕ can be determined from the initial conditions $x(0)$ *and* $\dot{x}(0)$. This solution corresponds to oscillations of decreasing amplitude and of period $\tau_d = 2\pi/\omega_d$ (Fig. 19.11).

2. Damped forced vibrations. These vibrations occur when a system with viscous damping is subjected to a periodic force **P** of magnitude $P = P_m \sin \omega_f t$ or when it is elastically connected to a support with an alternating motion $\delta = \delta_m \sin \omega_f t$. In the first case the motion is defined by the differential equation

$$m\ddot{x} + c\dot{x} + kx = P_m \sin \omega_f t \qquad (19.47)$$

and in the second case by a similar equation obtained by replacing P_m with $k\delta_m$. You will be concerned only with the *steady-state* motion of the system, which is defined by a *particular solution* of these equations, of the form

$$x_{\text{part}} = x_m \sin (\omega_f t - \varphi) \qquad (19.48)$$

where

$$\frac{x_m}{P_m/k} = \frac{x_m}{\delta_m} = \frac{1}{\sqrt{[1 - (\omega_f/\omega_n)^2]^2 + [2(c/c_c)(\omega_f/\omega_n)]^2}} \qquad (19.53)$$

and

$$\tan \varphi = \frac{2(c/c_c)(\omega_f/\omega_n)}{1 - (\omega_f/\omega_n)^2} \qquad (19.54)$$

The expression given in Eq. (19.53) is referred to as the *magnification factor* and has been plotted against the frequency ratio ω_f/ω_n in Fig. 19.12 for various values of the damping factor c/c_c. In the problems which follow, you may be asked to determine one of the parameters in Eqs. (19.53) and (19.54) when the others are known.

PROBLEMS

19.127 Show that in the case of heavy damping $(c > c_c)$, a body never passes through its position of equilibrium O if it is (a) released with no initial velocity from an arbitrary position, (b) started from O with an arbitrary initial velocity.

19.128 Show that in the case of heavy damping $(c > c_c)$, a body released from an arbitrary position with an arbitrary initial velocity cannot pass more than once through its equilibrium position.

19.129 In the case of light damping, the displacements x_1, x_2, x_3, shown in Fig. 19.11 may be assumed equal to the maximum displacements. Show that the ratio of any two successive maximum displacements x_n and x_{n+1} is a constant and that the natural logarithm of this ratio, called the *logarithmic decrement*, is

$$\ln \frac{x_n}{x_{n+1}} = \frac{2\pi(c/c_c)}{\sqrt{1 - (c/c_c)^2}}$$

19.130 In practice, it is often difficult to determine the logarithmic decrement of a system with light damping defined in Prob. 19.129 by measuring two successive maximum displacements. Show that the logarithmic decrement can also be expressed as $(1/k) \ln(x_n/x_{n+k})$, where k is the number of cycles between readings of the maximum displacement.

19.131 In a system with light damping $(c < c_c)$, the period of vibration is commonly defined as the time interval $\tau_d = 2\pi/\omega_d$ corresponding to two successive points where the displacement-time curve touches one of the limiting curves shown in Fig. 19.11. Show that the interval of time (a) between a maximum positive displacement and the following maximum negative displacement is $\frac{1}{2}\tau_d$, (b) between two successive zero displacements is $\frac{1}{2}\tau_d$, (c) between a maximum positive displacement and the following zero displacement is greater than $\frac{1}{4}\tau_d$.

19.132 A loaded railroad car weighing 30,000 lb is rolling at a constant velocity $\mathbf{v}_0$ when it couples with a spring and dashpot bumper system (Fig. 1). The recorded displacement-time curve of the loaded railroad car after coupling is as shown (Fig. 2). Determine (a) the damping constant, (b) the spring constant. (*Hint:* Use the definition of logarithmic decrement given in 19.129.)

Fig. P19.132   (1)

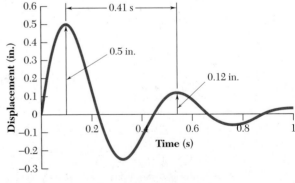

(2)

19.133 A torsional pendulum has a centroidal mass moment of inertia of $0.3 \text{ kg} \cdot \text{m}^2$ and when given an initial twist and released is found to have a frequency of oscillation of 200 rpm. Knowing that when this pendulum is immersed in oil and when given the same initial condition it is found to have a frequency of oscillation of 180 rpm, determine the damping constant for the oil.

19.134 The barrel of a field gun weighs 1500 lb and is returned into firing position after recoil by a recuperator of constant $c = 1100 \text{ lb} \cdot \text{s/ft}$. Determine (a) the constant k which should be used for the recuperator to return the barrel into firing position in the shortest possible time without any oscillation, (b) the time needed for the barrel to move back two-thirds of the way from its maximum-recoil position to its firing position.

19.135 A platform of weight 200 lb, supported by two springs each of constant $k = 250$ lb/in., is subjected to a periodic force of maximum magnitude equal to 125 lb. Knowing that the coefficient of damping is $12 \text{ lb} \cdot \text{s/in.}$, determine (a) the natural frequency in rpm of the platform if there were no damping, (b) the frequency in rpm of the periodic force corresponding to the maximum value of the magnification factor, assuming damping, (c) the amplitude of the actual motion of the platform for each of the frequencies found in parts a and b.

$\mathbf{P} = P_m \sin \omega_f t$

Fig. P19.135

19.136 A 4-kg block A is dropped from a height of 800 mm onto a 9-kg block B which is at rest. Block B is supported by a spring of constant $k = 1500$ N/m and is attached to a dashpot of damping coefficient $c = 230 \text{ N} \cdot \text{s/m}$. Knowing that there is no rebound, determine the maximum distance the blocks will move after the impact.

Fig. P19.136

19.137 A 3-kg slender rod AB is bolted to a 5-kg uniform disk. A dashpot of damping coefficient $c = 9 \text{ N} \cdot \text{s/m}$ is attached to the disk as shown. Determine (a) the differential equation of motion for small oscillations, (b) the damping factor c/c_c.

Fig. P19.137

19.138 A uniform rod of mass m is supported by a pin at A and a spring of constant k at B and is connected at D to a dashpot of damping coefficient c. Determine in terms of m, k, and c, for small oscillations, (a) the differential equation of motion, (b) the critical damping coefficient c_c.

Fig. P19.138

Fig. P19.139

19.139 A machine element weighing 800 lb is supported by two springs, each having a constant of 200 lb/in. A periodic force of maximum value 30 lb is applied to the element with a frequency of 2.5 cycles per second. Knowing that the coefficient of damping is 8 lb · s/in., determine the amplitude of the steady-state vibration of the element.

19.140 In Prob. 19.139, determine the required value of the coefficient of damping if the amplitude of the steady-state vibration of the element is to be 0.15 in.

19.141 In the case of the forced vibration of a system, determine the range of values of the damping factor c/c_c for which the magnification factor will always decrease as the frequency ratio ω_f/ω_n increases.

19.142 Show that for a small value of the damping factor c/c_c, the maximum amplitude of a forced vibration occurs when $\omega_f \approx \omega_n$ and that the corresponding value of the magnification factor is $\frac{1}{2}(c/c_c)$.

19.143 A counter-rotating eccentric mass exciter consisting of two rotating 14-oz weights describing circles of 6-in. radius at the same speed but in opposite senses is placed on a machine element to induce a steady-state vibration of the element and to determine some of the dynamic characteristics of the element. At a speed of 1200 rpm a stroboscope shows the eccentric masses to be exactly under their respective axes of rotation and the element to be passing through its position of static equilibrium. Knowing that the amplitude of the motion of the element at that speed is 0.6 in. and that the total mass of the system is 300 lb, determine (a) the combined spring constant k, (b) the damping factor c/c_c.

Fig. P19.143

19.144 A 15-kg motor is supported by four springs, each of constant 40 kN/m. The unbalance of the motor is equivalent to a mass of 20 g located 125 mm from the axis of rotation. Knowing that the motor is constrained to move vertically and that the damping factor c/c_c is equal to 0.4, determine the range of frequencies for which the amplitude of the steady-state vibration of the motor is less than 0.2 mm.

19.145 A 220-lb motor is supported by four springs, each of constant 500 lb/in., and is connected to the ground by a dashpot having a coefficient of damping $c = 35$ lb · s/in. The motor is constrained to move vertically, and the amplitude of its motion is observed to be 0.08 in. at a speed of 1200 rpm. Knowing that the weight of the rotor is 30 lb, determine the distance between the mass center of the rotor and the axis of the shaft.

Fig. P19.144 and P19.145

19.146 A 100-lb motor is directly supported by a light horizontal beam which has a static deflection of 0.2 in. due to the weight of the motor. The unbalance of the rotor is equivalent to a weight of 3.5 oz located 3 in. from the axis of rotation. Knowing that the amplitude of the vibration of the motor is 0.03 in. at a speed of 400 rpm, determine (a) the damping factor c/c_c, (b) the coefficient of damping c.

Fig. P19.146

19.147 A machine element is supported by springs and is connected to a dashpot as shown. Show that if a periodic force of magnitude $P = P_m \sin \omega_f t$ is applied to the element, the amplitude of the fluctuating force transmitted to the foundation is

$$F_m = P_m \sqrt{\frac{1 + [2(c/c_c)(\omega_f/\omega_n)]^2}{[1 - (\omega_f/\omega_n)^2]^2 + [2(c/c_c)(\omega_f/\omega_n)]^2}}$$

$\mathbf{P} = P_m \sin \omega_f t$

19.148 A 91-kg machine element supported by four springs, each of constant $k = 175$ N/m, is subjected to a periodic force of frequency 0.8 Hz and amplitude 89 N. Determine the amplitude of the fluctuating force transmitted to the foundation if (a) a dashpot with a coefficient of damping $c = 365$ N · s/m is connected to the machine element and to the ground, (b) the dashpot is removed.

Fig. P19.147 and P19.148

19.149 A simplified model of a washing machine is shown. A bundle of wet clothes forms a weight w_b of 20 lb in the machine and causes a rotating unbalance. The rotating weight is 40 lb (including w_b) and the radius of the washer basket e is 9 in. Knowing the washer has an equivalent spring constant $k = 70$ lb/ft and damping ratio $\zeta = c/c_c = 0.05$ and during the spin cycle the drum rotates at 250 rpm, determine the amplitude of the motion and the magnitude of the force transmitted to the sides of the washing machine.

Fig. P19.149

*19.150 For a steady-state vibration with damping under a harmonic force, show that the mechanical energy dissipated per cycle by the dashpot is $E = \pi c x_m^2 \omega_f$, where c is the coefficient of damping, x_m is the amplitude of the motion, and ω_f is the circular frequency of the harmonic force.

*19.151 The suspension of an automobile can be approximated by the simplified spring-and-dashpot system shown. (a) Write the differential equation defining the vertical displacement of the mass m when the system moves at a speed v over a road with a sinusoidal cross section of amplitude δ_m and wave length L. (b) Derive an expression for the amplitude of the vertical displacement of the mass m.

Fig. P19.151

$\mathbf{P} = P_m \sin \omega_f t$

Fig. P19.152

*19.152 Two blocks A and B, each of mass m, are supported as shown by three springs of the same constant k. Blocks A and B are connected by a dashpot and block B is connected to the ground by two dashpots, each dashpot having the same coefficient of damping c. Block A is subjected to a force of magnitude $P = P_m \sin \omega_f t$. Write the differential equations defining the displacements x_A and x_B of the two blocks from their equilibrium positions.

19.153 Express in terms of L, C, and E the range of values of the resistance R for which oscillations will take place in the circuit shown when switch S is closed.

Fig. P19.153

19.154 Consider the circuit of Prob. 19.153 when the capacitor C is removed. If switch S is closed at time $t = 0$, determine (a) the final value of the current in the circuit, (b) the time t at which the current will have reached $(1 - 1/e)$ times its final value. (The desired value of t is known as the *time constant* of the circuit.)

19.155 and 19.156 Draw the electrical analogue of the mechanical system shown. (*Hint:* Draw the loops corresponding to the free bodies m and A.)

Fig. P19.157 and P19.158

Fig. P19.155 and P19.156

19.157 and 19.158 Write the differential equations defining (a) the displacements of the mass m and of the point A, (b) the charges on the capacitors of the electrical analogue.

REVIEW AND SUMMARY

This chapter was devoted to the study of *mechanical vibrations*, i.e., to the analysis of the motion of particles and rigid bodies oscillating about a position of equilibrium. In the first part of the chapter [Secs. 19.2 through 19.7], we considered *vibrations without damping*, while the second part was devoted to *damped vibrations* [Secs. 19.8 through 19.10].

In Sec. 19.2, we considered the *free vibrations of a particle*, i.e., the motion of a particle P subjected to a restoring force proportional to the displacement of the particle—such as the force exerted by a spring. If the displacement x of the particle P is measured from its equilibrium position O (Fig. 19.17), the resultant **F** of the forces acting on P (including its weight) has a magnitude kx and is directed toward O. Applying Newton's second law $F = ma$ and recalling that $a = \ddot{x}$, we wrote the differential equation

$$m\ddot{x} + kx = 0 \tag{19.2}$$

or, setting $\omega_n^2 = k/m$,

$$\ddot{x} + \omega_n^2 x = 0 \tag{19.6}$$

Free vibrations of a particle

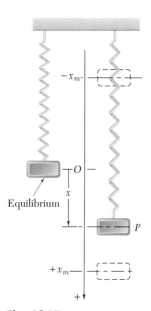

Fig. 19.17

1305

The motion defined by this equation is called a *simple harmonic motion.*

The solution of Eq. (19.6), which represents the displacement of the particle P, was expressed as

$$x = x_m \sin(\omega_n t + \phi) \qquad (19.10)$$

where x_m = amplitude of the vibration
   $\omega_n = \sqrt{k/m}$ = natural circular frequency
   ϕ = phase angle

The *period of the vibration* (i.e., the time required for a full cycle) and its *natural frequency* (i.e., the number of cycles per second) were expressed as

$$\text{Period} = \tau_n = \frac{2\pi}{\omega_n} \qquad (19.13)$$

$$\text{Natural frequency} = f_n = \frac{1}{\tau_n} = \frac{\omega_n}{2\pi} \qquad (19.14)$$

The velocity and acceleration of the particle were obtained by differentiating Eq. (19.10), and their maximum values were found to be

$$v_m = x_m \omega_n \qquad a_m = x_m \omega_n^2 \qquad (19.15)$$

Since all the above parameters depend directly upon the natural circular frequency ω_n and thus upon the ratio k/m, it is essential in any given problem to calculate the value of the constant k; this can be done by determining the relation between the restoring force and the corresponding displacement of the particle [Sample Prob. 19.1].

It was also shown that the oscillatory motion of the particle P can be represented by the projection on the x axis of the motion of a point Q describing an auxiliary circle of radius x_m with the constant angular velocity ω_n (Fig. 19.18). The instantaneous values of the velocity and acceleration of P can then be obtained by projecting on the x axis the vectors $\mathbf{v}_m$ and $\mathbf{a}_m$ representing, respectively, the velocity and acceleration of Q.

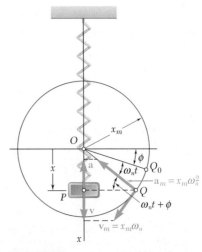

Fig. 19.18

While the motion of a *simple pendulum* is not truly a simple harmonic motion, the formulas given above can be used with $\omega_n^2 = g/l$ to calculate the period and natural frequency of the *small oscillations* of a simple pendulum [Sec. 19.3]. Large-amplitude oscillations of a simple pendulum were discussed in Sec. 19.4.

The *free vibrations of a rigid body* can be analyzed by choosing an appropriate variable, such as a distance x or an angle θ, to define the position of the body, drawing a free-body-diagram equation to express the equivalence of the external and effective forces, and writing an equation relating the selected variable and its second derivative [Sec. 19.5]. If the equation obtained is of the form

$$\ddot{x} + \omega_n^2 x = 0 \qquad \text{or} \qquad \ddot{\theta} + \omega_n^2 \theta = 0 \qquad (19.21)$$

the vibration considered is a simple harmonic motion and its period and natural frequency can be obtained *by identifying* ω_n and substituting its value into Eqs. (19.13) and (19.14) [Sample Probs. 19.2 and 19.3].

The *principle of conservation of energy* can be used as an alternative method for the determination of the period and natural frequency of the simple harmonic motion of a particle or rigid body [Sec. 19.6]. Choosing again an appropriate variable, such as θ, to define the position of the system, we express that the total energy of the system is conserved, $T_1 + V_1 = T_2 + V_2$, between the position of maximum displacement ($\theta_1 = \theta_m$) and the position of maximum velocity ($\dot{\theta}_2 = \dot{\theta}_m$). If the motion considered is simple harmonic, the two members of the equation obtained consist of homogeneous quadratic expressions in θ_m and $\dot{\theta}_m$, respectively.† Substituting $\dot{\theta}_m = \theta_m \omega_n$ in this equation, we can factor out θ_m^2 and solve for the circular frequency ω_n [Sample Prob. 19.4].

In Sec. 19.7, we considered the *forced vibrations* of a mechanical system. These vibrations occur when the system is subjected to a periodic force (Fig. 19.19) or when it is elastically connected to a support which has an alternating motion (Fig. 19.20). Denoting by ω_f the forced circular frequency, we found that in the first case, the motion of the system was defined by the differential equation

$$m\ddot{x} + kx = P_m \sin \omega_f t \qquad (19.30)$$

and that in the second case it was defined by the differential equation

$$m\ddot{x} + kx = k\delta_m \sin \omega_f t \qquad (19.31)$$

The general solution of these equations is obtained by adding a particular solution of the form

$$x_{\text{part}} = x_m \sin \omega_f t \qquad (19.32)$$

---

†If the motion considered can only be *approximated* by a simple harmonic motion, such as for the small oscillations of a body under gravity, the potential energy must be approximated by a quadratic expression in θ_m.

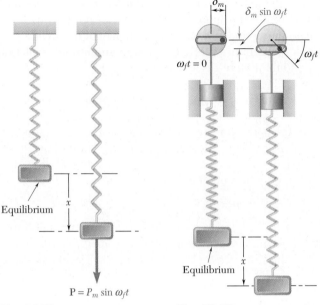

Fig. 19.19

Fig. 19.20

to the general solution of the corresponding homogeneous equation. The particular solution (19.32) represents a *steady-state vibration* of the system, while the solution of the homogeneous equation represents a *transient free vibration* which can generally be neglected.

Dividing the amplitude x_m of the steady-state vibration by P_m/k in the case of a periodic force, or by δ_m in the case of an oscillating support, we defined the *magnification factor* of the vibration and found that

$$\text{Magnification factor} = \frac{x_m}{P_m/k} = \frac{x_m}{\delta_m} = \frac{1}{1 - (\omega_f/\omega_n)^2} \quad (19.36)$$

According to Eq. (19.36), the amplitude x_m of the forced vibration *becomes infinite when* $\omega_f = \omega_n$, i.e., *when the forced frequency is equal to the natural frequency of the system.* The impressed force or impressed support movement is then said to be in *resonance* with the system [Sample Prob. 19.5]. Actually the amplitude of the vibration remains finite, due to damping forces.

Damped free vibrations   In the last part of the chapter, we considered the *damped vibrations* of a mechanical system. First, we analyzed the *damped free vibrations* of a system with *viscous damping* [Sec. 19.8]. We found that the motion of such a system was defined by the differential equation

$$m\ddot{x} + c\dot{x} + kx = 0 \quad (19.38)$$

where c is a constant called the *coefficient of viscous damping*. Defining the *critical damping coefficient* c_c as

$$c_c = 2m\sqrt{\frac{k}{m}} = 2m\omega_n \qquad (19.41)$$

where ω_n is the natural circular frequency of the system in the absence of damping, we distinguished three different cases of damping, namely, (1) *heavy damping*, when $c > c_c$; (2) *critical damping*, when $c = c_c$; and (3) *light damping*, when $c < c_c$. In the first two cases, the system when disturbed tends to regain its equilibrium position without any oscillation. In the third case, the motion is vibratory with diminishing amplitude.

In Sec. 19.9, we considered the *damped forced vibrations* of a mechanical system. These vibrations occur when a system with viscous damping is subjected to a periodic force **P** of magnitude $P = P_m \sin \omega_f t$ or when it is elastically connected to a support with an alternating motion $\delta = \delta_m \sin \omega_f t$. In the first case, the motion of the system was defined by the differential equation

Damped forced vibrations

$$m\ddot{x} + c\dot{x} + kx = P_m \sin \omega_f t \qquad (19.47)$$

and in the second case by a similar equation obtained by replacing P_m by $k\delta_m$ in (19.47).

The *steady-state vibration* of the system is represented by a particular solution of Eq. (19.47) of the form

$$x_{\text{part}} = x_m \sin (\omega_f t - \varphi) \qquad (19.48)$$

Dividing the amplitude x_m of the steady-state vibration by P_m/k in the case of a periodic force, or by δ_m in the case of an oscillating support, we obtained the following expression for the magnification factor:

$$\frac{x_m}{P_m/k} = \frac{x_m}{\delta_m} = \frac{1}{\sqrt{[1 - (\omega_f/\omega_n)^2]^2 + [2(c/c_c)(\omega_f/\omega_n)]^2}} \qquad (19.53)$$

where $\omega_n = \sqrt{k/m}$ = natural circular frequency of undamped system
$\qquad c_c = 2m\omega_n$ = critical damping coefficient
$\qquad c/c_c$ = damping factor

We also found that the *phase difference* φ between the impressed force or support movement and the resulting steady-state vibration of the damped system was defined by the relation

$$\tan \varphi = \frac{2(c/c_c)(\omega_f/\omega_n)}{1 - (\omega_f/\omega_n)^2} \qquad (19.54)$$

The chapter ended with a discussion of *electrical analogues* [Sec. 19.10], in which it was shown that the vibrations of mechnical systems and the oscillations of electrical circuits are defined by the same differential equations. Electrical analogues of mechanical systems can therefore be used to study or predict the behavior of these systems.

Electrical analogues

REVIEW PROBLEMS

19.159 An automobile wheel-and-tire assembly of total weight 47 lb is attached to a mounting plate of negligible weight which is suspended from a steel wire. The torsional spring constant of the wire is known to be $K = 0.40$ lb · in./rad. The wheel is rotated through 90° about the vertical and then released. Knowing that the period of oscillation is observed to be 30 s, determine the centroidal mass moment of inertia and the centroidal radius of gyration of the wheel-and-tire assembly.

Fig. P19.159

19.160 The period of vibration of the system shown is observed to be 0.6 s. After cylinder B has been removed, the period is observed to be 0.5 s. Determine (a) the weight of cylinder A, (b) the constant of the spring.

Fig. P19.160

19.161 Disks A and B weigh 30 lb and 12 lb, respectively, and a small 5-lb block C is attached to the rim of disk B. Assuming that no slipping occurs between the disks, determine the period of small oscillations of the system.

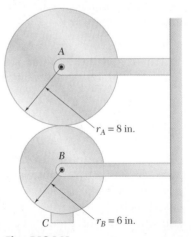

Fig. P19.161

19.162 For the uniform equilateral triangular plate of side $l = 300$ mm, determine the period of small oscillations if the plate is suspended from (a) one of its vertices, (b) the midpoint of one of its sides.

19.163 An 0.8-lb ball is connected to a paddle by means of an elastic cord AB of constant $k = 5$ lb/ft. Knowing that the paddle is moved vertically according to the relation $\delta = \delta_m \sin \omega_f t$, where $\delta_m = 8$ in., determine the maximum allowable circular frequency ω_f if the cord is not to become slack.

Fig. P19.162

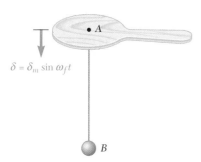

$\delta = \delta_m \sin \omega_f t$

Fig. P19.163

19.164 The block shown is depressed 1.2 in. from its equilibrium position and released. Knowing that after 10 cycles the maximum displacement of the block is 0.5 in., determine (a) the damping factor c/c_c, (b) the value of the coefficient of viscous damping. (*Hint:* See Probs. 19.129 and 19.130.)

$k = 8$ lb/ft

9 lb

c

Fig. P19.164

19.165 A 4-lb uniform rod is supported by a pin at O and a spring at A and is connected to a dashpot at B. Determine (a) the differential equation of motion for small oscillations, (b) the angle that the rod will form with the horizontal 5 s after end B has been pushed 0.9 in. down and released.

← 6 in. → ← 18 in. →

A O B

$k = 5$ lb/ft $c = 0.5$ lb·s/ft

Fig. P19.165

Fig. P19.166

19.166 A 400-kg motor supported by four springs, each of constant 150 kN/m, and a dashpot of constant $c = 6500$ N · s/m is constrained to move vertically. Knowing that the unbalance of the rotor is equivalent to a 23-g mass located at a distance of 100 mm from the axis of rotation, determine for a speed of 800 rpm (*a*) the amplitude of the fluctuating force transmitted to the foundation, (*b*) the amplitude of the vertical motion of the motor.

19.167 The compressor shown has a mass of 250 kg and operates at 2000 rpm. At this operating condition, the force transmitted to the ground is excessively high and is found to be $mr\omega_f^2$, where mr is the unbalance and ω_f is the forcing frequency. To fix this problem, it is proposed to isolate the compressor by mounting it on a square concrete block separated from the rest of the floor as shown. The density of concrete is 2400 kg/m^3 and the spring constant for the soil is found to be 80×10^6 N/m. The geometry of the compressor leads to choosing a block that is 1.5 m by 1.5 m. Determine the depth h that will reduce the force transmitted to the ground by 75 percent.

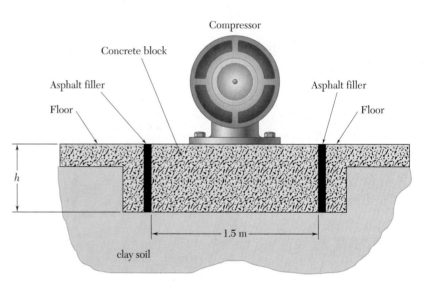

Fig. P19.167

19.168 A small ball of mass m attached at the midpoint of a tightly stretched elastic cord of length l can slide on a horizontal plane. The ball is given a small displacement in a direction perpendicular to the cord and released. Assuming the tension T in the cord to remain constant, (*a*) write the differential equation of motion of the ball, (*b*) determine the period of vibration.

Fig. P19.168

19.169 A certain vibrometer used to measure vibration amplitudes consists essentially of a box containing a slender rod to which a mass m is attached; the natural frequency of the mass-rod system is known to be 5 Hz. When the box is rigidly attached to the casing of a motor rotating at 600 rpm, the mass is observed to vibrate with an amplitude of 0.06 in. relative to the box. Determine the amplitude of the vertical motion of the motor.

Fig. P19.169

19.170 If either a simple or a compound pendulum is used to determine experimentally the acceleration of gravity g, difficulties are encountered. In the case of the simple pendulum, the string is not truly weightless, while in the case of the compound pendulum, the exact location of the mass center is difficult to establish. In the case of a compound pendulum, the difficulty can be eliminated by using a reversible, or Kater, pendulum. Two knife edges A and B are placed so that they are obviously not at the same distance from the mass center G, and the distance l is measured with great precision. The position of a counterweight D is then adjusted so that the period of oscillation τ is the same when either knife edge is used. Show that the period τ obtained is equal to that of a true simple pendulum of length l and that $g = 4\pi^2 l/\tau^2$.

Fig. P19.170

COMPUTER PROBLEMS

19.C1 By expanding the integrand in Eq. (19.19) into a series of even powers of $\sin \phi$ and integrating, it can be shown that the period of a simple pendulum of length l can be approximated by the expression

$$\tau_n = 2\pi \sqrt{\frac{l}{g}} \left[1 + \left(\frac{1}{2}\right)^2 c^2 + \left(\frac{1 \times 3}{2 \times 4}\right)^2 c^4 + \left(\frac{1 \times 3 \times 5}{2 \times 4 \times 6}\right)^2 c^6 + \cdots \right]$$

where $c = \sin \frac{1}{2}\theta_m$ and θ_m is the amplitude of the oscillations. Use computational software to calculate the sum of the series in brackets, using successively 1, 2, 4, 8, and 16 terms, for values of θ_m from 30 to 120° using 30° increments.

19.C2 The force-deflection equation for a class of nonlinear springs fixed at one end is $F = 5x^{1/n}$, where F is the magnitude, expressed in newtons, of the force applied at the other end of the spring and x is the deflection expressed in meters. Knowing that a block of mass m is suspended from the spring and is given a small downward displacement from its equilibrium position, use computational software to calculate and plot the frequency of vibration of the block for values of m equal to 0.2, 0.6, and 1.0 kg and values of n from 1 to 2. Assume that the slope of the force-deflection curve at the point corresponding to $F = mg$ can be used as an equivalent spring constant.

19.C3 A machine element supported by springs and connected to a dashpot is subjected to a periodic force of magnitude $P = P_m \sin \omega_f t$. The *transmissibility* T_m of the system is defined as the ratio F_m/P_m of the maximum value F_m of the fluctuating periodic force transmitted to the foundation to the maximum value P_m of the periodic force applied to the machine element. Use computational software to calculate and plot the value of T_m for frequency ratios ω_f/ω_n equal to 0.8, 1.4, and 2.0 and for damping factors c/c_c equal to 0, 1, and 2. (*Hint:* Use the formula given in Prob. 19.147.)

$$\mathbf{P} = P_m \sin \omega_f t$$

Fig. P19.C3

19.C4 A 15-kg motor is supported by four springs, each of constant 60 kN/m. The unbalance of the motor is equivalent to a mass of 20 g located 125 mm from the axis of rotation. Knowing that the motor is constrained to move vertically, use computational software to calculate and plot the amplitude of the vibration and the maximum acceleration of the motor for motor speeds of 1000 to 2500 rpm.

19.C5 Solve Prob. 19.C4, assuming that a dashpot having a coefficient of damping $c = 2.5$ kN · s/m has been connected to the motor base and to the ground.

19.C6 A small trailer and its load have a total mass of 250 kg. The trailer is supported by two springs, each of constant 10 kN/m, and is pulled over a road, the surface of which can be approximated by a sine curve with an amplitude of 40 mm and a wave length of 5 m (i.e., the distance between successive crests is 5 m and the vertical distance from crest to trough is 80 mm). (*a*) Neglecting the mass of the wheels and assuming that the wheels stay in contact with the ground, use computational software to calculate and plot the amplitude of the vibration and the maximum vertical acceleration of the trailer for speeds of 10 to 80 km/h. (*b*) Determine the range of values of the speed of the trailer for which the wheels will lose contact with the ground.

Fig. P19.C6

Fundamentals of Engineering Examination

Engineers are required to be licensed when their work directly affects the public health, safety, and welfare. The intent is to ensure that engineers have met minimum qualifications involving competence, ability, experience, and character. The licensing process involves an initial exam, called the *Fundamentals of Engineering Examination;* professional experience; and a second exam, called the *Principles and Practice of Engineering.* Those who successfully complete these requirements are licensed as a *Professional Engineer.* The exams are developed under the auspices of the *National Council of Examiners for Engineering and Surveying.*

The first exam, the *Fundamentals of Engineering Examination,* can be taken just before or after graduation from a four-year accredited engineering program. The exam stresses subject material in a typical undergraduate engineering program, including statics. The topics included in the exam cover much of the material in this book. The following is a list of the main topic areas, with references to the appropriate sections in this book. Also included are problems that can be solved to review this material.

Concurrent Force Systems (2.2–2.9; 2.12–2.14)
   Problems: 2.33, 2.35, 2.36, 2.37, 2.75, 2.84, 2.92, 2.93, 2.97

Vector Forces (3.4–3.11)
   Problems: 3.17, 3.18, 3.24, 3.33, 3.37, 3.39

Equilibrium in Two Dimensions (2.11; 4.1–4.7)
   Problems: 4.1, 4.9, 4.10, 4.17, 4.29, 4.32, 4.67, 4.81

Equilibrium in Three Dimensions (2.15; 4.8–4.9)
   Problems: 4.97, 4.99, 4.100, 4.105, 4.113, 4.115, 4.128, 4.129, 4.140

Centroid of an Area (5.2–5.7)
   Problems: 5.9, 5.17, 5.28, 5.36, 5.41, 5.56, 5.62, 5.97, 5.100, 5.103, 5.126

Analysis of Trusses (6.2–6.7)
   Problems; 6.2, 6.5, 6.33, 6.45, 6.46, 6.50

Equilibrium of Two-Dimensional Frames (6.9–6.11)
   Problems: 6.76, 6.82, 6.89, 6.93, 6.94

Shear and Bending Moment (7.3–7.6)
Problems: 7.22, 7.25, 7.31, 7.36, 7.45, 7.49, 7.70, 7.83

Friction (8.2–8.5; 8.10)
Problems: 8.11, 8.15, 8.21, 8.30, 8.50, 8.53, 8.101, 8.104, 8.105

Moments of Inertia (9.2–9.10)
Problems: 9.5, 9.31, 9.32, 9.33, 9.77, 9.78, 9.84, 9.89, 9.101, 9.103

Kinematics (11.1–11.6; 11.9–11.14; 15.2–15.8)
Problems: 11.3, 11.5, 11.35, 11.61, 11.67, 11.97, 15.8, 15.29, 15.39, 15.61, 15.65, 15.87, 15.112, 15.142

Force, Mass, and Acceleration (12.1–12.6; 16.2–16.8)
Problems: 12.6, 12.8, 12.30, 12.32, 12.36, 12.45, 12.52, 12.55, 16.1, 16.3, 16.7, 16.26, 16.27, 16.50, 16.61, 16.63, 16.78, 16.85, 16.137

Work and Energy (13.1–13.6; 13.8; 17.1–17.7)
Problems: 13.5, 13.7, 13.17, 13.22, 13.40, 13.41, 13.47, 13.64, 13.66, 13.68, 17.1, 17.2, 17.16, 17.22

Impulse and Momentum (13.10–13.15; 17.8–17.12)
Problems: 13.119, 13.121, 13.129, 13.134, 13.146, 13.155, 13.160, 13.171, 17.53, 17.58, 17.70, 17.72, 17.96, 17.97, 17.111

Vibration (19.1–19.3; 19.5–19.7)
Problems: 19.1, 19.2, 19.11, 19.18, 19.23, 19.28, 19.49, 19.55, 19.63, 19.74, 19.83, 19.85, 19.101, 19.105, 19.116

Friction (Problems involving friction occur in each of the above subjects)

Photo Credits

CHAPTER 1
Opener (crane): © Tom Paiva/Getty Images; (Manhattan Bridge): © Getty Images RF; (x-lift): © Tom Brakefield/SuperStock; (tugboats pushing tanker): © Daniel Sheehan; (communication tower): © Alfredo Maiquez/Getty Images; (construction worker): © Lester Lefkowitz/Getty Images; (Newton): © Lebrecht Music and Arts Photo Library/Alamy; **Photo 1.1:** © PhotoLink/Getty Images RF.

CHAPTER 2
Opener: © Tom Paiva/Getty Images; **Photo 2.1:** © H. David Seawell/Corbis; **Photo 2.2:** © WIN-Initiative/Getty Images.

CHAPTER 3
Opener: © Daniel Sheehan; **Photo 3.1:** © McGraw-Hill/Photo by Lucinda Dowell; **Photo 3.2:** © Steve Hix; **Photo 3.3:** © Jose Luis Pelaez/Getty Images; **Photo 3.4:** © Images-USA/Alamy RF; **Photo 3.5:** © Dana White/Photo Edit.

CHAPTER 4
Opener: © Alfredo Maiquez/Getty Images; **Photo 4.1, Photo 4.2, Photo 4.3:** © The McGraw-Hill Companies, Inc./Photo by Lucinda Dowell; **Photo 4.4, Photo 4.5:** Courtesy National Information Service for Earthquake Engineering, University of California, Berkeley; **Photo 4.6:** © McGraw-Hill/Photo by Lucinda Dowell; **Photo 4.7:** Courtesy of SKF, Limited.

CHAPTER 5
Opener: © Christopher Morris/Corbis; **Photo 5.1:** © Christies Images/SuperStock; **Photo 5.2:** Value RF/© Kevin Burke/Corbis; **Photo 5.3:** © Ghislain & Marie David de Lossy/Getty Images; **Photo 5.4:** NASA.

CHAPTER 6
Opener: © Alan Schein Photography/Corbis; **Photo 6.1:** Courtesy National Information Service for Earthquake Engineering, University of California, Berkeley; **Photo 6.2:** Courtesy of Ferdinand Beer; **Photo 6.3:** © The McGraw-Hill Companies, Inc./Photo by Sabina Dowell; **Photo 6.4:** © EyeWire Collection/Getty Images RF; **Photo 6.5:** © Mouse in the House/Alamy.

CHAPTER 7
Opener: © Getty Images RF; **Photo 7.1:** © The McGraw-Hill Companies, Inc./Photo by Sabina Dowell; **Photo 7.2:** © Alan Thornton/Getty Images; **Photo 7.3:** © Michael S. Yamashita/Corbis; **Photo 7.4:** © Steve Kahn.

CHAPTER 8
Opener: © Bicot (Marc Caya); **Photo 8.1:** © Chuck Savage/Corbis; **Photo 8.2:** © Ted Spiegel/Corbis; **Photo 8.3:** © Adam Woolfitt/Corbis.

CHAPTER 9
Opener: © Lester Lefkowitz/Getty Images; **Photo 9.1:** © Ed Eckstein/Corbis; **Photo 9.2:** Reprinted Courtesy of Caterpillar Inc.

CHAPTER 10
Opener: © Tom Brakefield/SuperStock; **Photo 10.1:** Courtesy of Altec Industries; **Photo 10.2:** Courtesy of DE-STA-CO.

CHAPTER 11
Opener: © NASA/Getty Images RF; **Photo 11.1:** U.S. Department of Energy; **Photo 11.2:** © Digital Vision/Getty RF; **Photo 11.3:** © Brand X Pictures/Jupiter Images RF; **Photo 11.4:** © Digital Vision/Getty RF; **Photo 11.5, Photo 11.6:** © Royalty-Free/Corbis.

CHAPTER 12
Opener: © Lester Lefkowitz/Corbis; **Photo 12.1:** © Royalty-Free/Corbis; **Photo 12.2:** © Brand X Pictures/PunchStock RF; **Photo 12.3:** © Royalty-Free/Corbis; **Photo 12.4:** © Russell Illig/Getty RF; **Photo 12.5:** © Royalty-Free/Corbis.

CHAPTER 13
Opener: © Tom Miles; **Photo 13.1, Photo 13.2:** © Scandia National Laboratories/Getty RF; **Photo 13.3:** © Andrew Davidhazy/RIT; **Photo 13.4:** © Tom McCarthy/Photolibrary.

CHAPTER 14
Opener: © XCOR; **Photo 14.1:** NASA; **Photo 14.2:** © Royalty-Free/Corbis; **Photo 14.3:** © Brand X Pictures/PunchStock RF.

CHAPTER 15
Opener: Courtesy A.P. Moller-Maersk; **(inset):** Courtesy of Wärtsilä Corporation; **Photo 15.1:** © Chris Hellier/Corbis; **Photo 15.2:** © Royalty-Free/Corbis; **Photo 15.3:** © Joseph Nettis/Stock Boston Inc.; **Photo 15.4:** © AGE Fotostock/Photolibrary; **Photo 15.5:** © George Tiedemann/NewSport/Corbis; **Photo 15.6:** © Royalty-Free/Corbis; **Photo 15.7:** Purdue University/Physics/PRIME Lab; **Photo 15.8:** © Northrop Grumman/Index Stock Imagery/Jupiter Images/Photolibrary; **Photo 15.9:** © Royalty-Free/Corbis.

CHAPTER 16
Opener, Photo 16.1: © Getty RF; **Photo 16.2:** Courtesy of Samsung Semiconductor, Inc.; **Photo 16.3:** © Tony Arruza/Corbis; **Photo 16.4:** © Robert E. Daemmrich.

CHAPTER 17
Opener: © AP Photo/Matt Dunham; **Photo 17.1:** © Richard McDowell/Alamy RF; **Photo 17.2:** Phillip Cornwell; **Photo 17.3:** © Photography by Leah; **17.4:** © Chuck Savage/Corbis.

CHAPTER 18
Opener: © Royalty-Free/Corbis; **Photo 18.1:** © age fotostock/SuperStock; **Photo 18.2:** © Matthias Kulka/Corbis; **Photo 18.3:** © Roger Ressmeyer/Corbis; **Photo 18.4:** Reprinted Courtesy of Caterpillar Inc.; **Photo 18.5:** © Lawrence Manning/Corbis RF.

CHAPTER 19
Opener: © Peter Tsai Photography; **Photo 19.1:** © Tony Freeman/Index Stock/Photolibrary; **Photo 19.2:** © The McGraw-Hill Companies, Inc., photo by Sabina Dowell; **Photo 19.3:** Courtesy of MTS Systems Corporation.

Index

A

Absolute and relative acceleration in plane motion, 708, 970–972, 1029

Absolute and relative velocity in plane motion, 946–958, 960, 1028

Absolute motion, 649

Abstract science, 2

Accelerated motion, 776

Acceleration-time curve, 636

Accelerations, 975–976, 1017–1020. *See also* Coriolis acceleration; Forces and accelerations

  angular, 929, 932, 973, 1002, 1004, 1018–1019, 1027, 1072

  of the coinciding point, 991

  determining, 612, 636, 960, 1004–1005

  in the *Fundamentals of Engineering Examination,* A45

  of gravity, 701

  instantaneous, 604–605

  in plane motion, absolute and relative, 970–972, 1029

  relationships among, 1081

  relative, 687, 991

  tangential component of, 668

Action, lines of, 16, 21, 77, 84, 307

Addition

  of couples, 113

  of forces

    parallelogram law for, 3

    in space, concurrent, 50–58, 68

    by summing X and Y components, 30–35

  of vectors, 18–20, A1

Allowable error, 798

Analysis

  of a frame, 314–315, 346

  of structures, 282–351

    computer problems, 350–351

    frames and machines, 314–343, 346

    introduction, 284–285

    review problems, 347–349

    summary, 344–346

  of trusses, 285–313, 344–345

    by the method of joints, 288–290, 344–345

    by the method of sections, 302–303, 345

Analytic solutions, 30, 40

Angles

  formed by two vectors, A5

  of friction, 415–416, 460

  lead, 432, 461

  of repose, 415

Angular acceleration, 929, 932, 1002, 1004, 1072

  in rotation about a fixed axis, 973, 1026–1027

Angular coordinates, 931

Angular impulses, 886

Angular momentum, 1165

  components of, 1201

  conservation of, 729–730, 733, 755–756, 794, 878, 892, 1132, 1136, 1220

  of a particle, 727–728, 755

  rate of change of, 755

  of a rigid body in three dimensions, 1175–1179, 1229

    constrained to rotate about a fixed point, 1179, 1230

    reduction of the momenta of the particles of a rigid body to a momentum vector and a couple, 1178–1179

  of a system of particles, 871–872, 917

    about its mass center, 874–876, 918

Angular momentum couple, 1130, 1183, 1185

Angular velocity, 929, 932, 1002–1004, 1018, 1182–1185

  constant, 1244

  in rotation about a fixed axis, 1026–1027

Answers, checking carefully, 13

Apogee, 744

Applied loads, 321

Applied science, 2

Approximate solutions, 635, 1257

  to the simple pendulum, 1246

Arbitrary axes, mass products of inertia, moments of inertia of a body with respect to, A28–A29, A43

Arbitrary shapes, determination of the principal axes and principal moments of inertia of a body of, 533–544

Archimedes, 2

Areal velocity, 729–730, 755–756

Areas

  symmetric and unsymmetric, 224

  units for (SI units), 5–8

Areas and lines, 220–257. *See also* Moments, of inertia, of areas

  center of gravity of a two-dimensional body, 220–221, 274

  centroids of, 222–223, 274

  composite plates and wires, 226–236

  determination of centroids by integration, 236–238, 275

  distributed loads on beams, 248, 276

  first moments of, 223–226, 274

  forces on submerged surfaces, 249–257

  theorems of Pappus-Guldinus, 238–247, 276

Aristotle, 2

Associative addition, 20

  of vectors, A1

Associative property, 96

Assumptions, 363–364

Average impulsive force, 820

Average power, 778

Average velocity, 603–604

Axes

  centroidal, 481, 488, 496, 512, A9

  of inertia, 1177, 1184–1185, 1201, 1229

    principal, 496–504, 531–532, 547, 550, A30–A32, A34–A36

  projection of a vector on, 97, 150

  of rotation, instantaneous, 958, 999, 1013, 1030

  of a wrench, 136, 138

Axial forces, 354–355, 357, 403

Axisymmetrical bodies, 1218–1219, 1233

Axle friction, 439–441, 445

B

Balance, 1197

Beams

  cantilever, 404

  forces in, 361–383

  relations among load, shear, and bending moment, 373–383, 404

  shear and bending-moment diagrams, 365–373

  shear and bending-moment in, 363–365, 404

  simply supported, 404

  various types of loading and support, 361–363

Bearings, 1202–1203

  journal, 439–441, 445

  thrust, 441–442, 445

Belt friction, 449–459

Bending, 354–355

  pure, 471

Bending moment, 355, 357

  in a beam, 363–365, 403–404

  diagrams, 365–373, 379

Binormal, 669, 674, 688

Body centrode, 960

Body cone, 999

Body of revolution, 238–239

Bound vectors, 17

Bowling ball, 1072, 1179

C

Cables

  catenary, 395–402, 405

  with concentrated loads, 383–384

  with distributed loads, 384–385, 405

  flexibility of, 383, 404

  forces in, 383–402

  parabolic, 385–395, 405

Answers to Problems

Answers to problems with a number set in straight type are given on this and the following pages. Answers to problems set in italic are not listed.

CHAPTER 2

2.1 3.30 kN ⤡ 66.6°.
2.2 139.1 lb ⤢ 67.0°.
2.4 8.03 kips ⤢ 3.8°.
2.5 (a) 101.4 N.
(b) 196.6 N.
2.6 (a) 3660 N.
(b) 3730 N.
2.7 2600 N ⤡ 53.5°.
2.8 (a) 853 lb.
(b) 567 lb.
2.10 (a) 37.1°.
(b) 73.2 N.
2.11 (a) 392 lb. (b) 346 lb.
2.13 (a) 368 lb →. (b) 213 lb.
2.14 (a) 21.1 N ↓. (b) 45.3 N.
2.15 139.1 lb ⤢ 67.0°.
2.16 8.03 kips ⤢ 3.8°.
2.18 100.3 N ⤢ 21.2°.
2.19 104.4 N ⤣ 86.7°.
2.21 (80 N) 61.3 N, 51.4 N; (120 N) 41.0 N, 112.8 N; (150 N) −122.9 N, 86.0 N.
2.22 (40 lb) 20.0 lb, −34.6 lb; (50 lb) −38.3 lb, −32.1 lb; (60 lb) 54.4 lb, 25.4 lb.
2.24 (102 lb) −48.0 lb, 90.0 lb; (106 lb) 56.0 lb, 90.0 lb; (200 lb) −160.0 lb, −120.0 lb.
2.25 (a) 2190 N. (b) 2060 N.
2.26 (a) 610 lb. (b) 500 lb.
2.28 (a) 373 lb. (b) 286 lb.
2.30 (a) 621 N. (b) 160.8 N.
2.31 654 N ⤡ 21.5°.
2.32 251 N ⤣ 85.3°.
2.33 54.9 lb ⤡ 48.9°.
2.35 309 N ⤢ 86.6°.
2.36 474 N ⤡ 32.5°.
2.37 203 lb ⤣ 8.46°.
2.39 (a) 21.7°. (b) 229 N.
2.40 (a) 26.5 N. (b) 623 N.
2.42 (a) 29.4°. (b) 371 lb.
2.43 (a) 6.37 kN. (b) 12.47 kN.
2.45 (a) 1244 lb. (b) 115.4 lb.
2.46 (a) 172.7 lb. (b) 231 lb.
2.48 (a) 305 N. (b) 514 N.
2.49 T_c = 5.87 kips; T_D = 9.14 kips.
2.51 (a) 312 N. (b) 144 N.
2.52 0 < P < 514 N.
2.53 (a) 1213 N. (b) 166.3 N
2.54 (a) 863 N. (b) 1216 N
2.55 F_A = 1303 lb; F_B = 420 lb.
2.57 (a) 1081 N. (b) 82.5°.
2.58 (a) 1294 N. (b) 62.5°.

2.59 (a) 5.00° ⤢. (b) 104.6 lb.
2.61 (a) 784 N. (b) 71.0°.
2.62 1.250 m.
2.63 (a) 10.98 lb. (b) 30.0 lb.
2.65 27.4° < α < 222.6°.
2.67 (a) 300 lb. (b) 300 lb. (c) 200 lb. (d) 200 lb. (e) 150.0 lb.
2.68 (b) 200 lb. (d) 150.0 lb.
2.69 (a) 1293 N. (b) 2220 N.
2.71 (a) −130.1 N; +816 N; +357 N. (b) 98.3°; 25.0°; 66.6°.
2.72 (a) +390 N; +614 N; +181.8 N. (b) 58.7°; 35.0°; 76.0°.
2.73 (a) −175.8 N; −257 N; +251 N. (b) 116.1°; 130.0°; 51.1°.
2.74 (a) +350 N; −169.0 N; +93.8 N. (b) 28.9°; 115.0°; 76.4°.
2.75 (a) −1861 lb; +3360 lb; +677 lb. (b) 118.5°; 30.5°; 80.0°.
2.77 (a) +56.4 lb; −103.9 lb; −20.5 lb. (b) 62.0°; 150.0°; 99.8°.
2.79 950 N; 43.4°; 71.6°; 127.6°.
2.81 (a) 43.9°. (b) F_x = +107.7 lb; F_z = 267 lb; F = 416 lb.
2.82 (a) 114.4°. (b) F_y = +694 N; F_z = +855 N; F = 1209.1 N.
2.83 (a) 194.0 N; 108.0 N. (b) 105.1°; 62.0°.
2.85 −1.260 kips; +1.213 kips; +0.970 kips.
2.86 −0.820 kips; +0.978 kips; −0.789 kips.
2.87 −1125 N; 750 N; 450 N.
2.89 240 N; −255 N; 160.0 N.
2.91 940 N; 65.7°; 28.2°; 76.4°.
2.92 940 N; 63.4°; 27.2°; 84.5°.
2.94 913 lb; 50.6°; 117.6°; 51.8°.
2.95 748 N; 120.1°; 52.5°; 128.0°.
2.96 3120 N; 37.4°; 122.0°; 72.6°.
2.97 130.0 lb.
2.99 13.98 kN.
2.101 926 N ↑.
2.103 2100 lb.
2.104 1868 lb.
2.106 T_{AB} = 571 lb; T_{AC} = 830 lb; T_{AD} = 528 lb.
2.107 960 N.
2.108 0 ≤ Q < 300 N.
2.109 845 N.
2.110 768 N.
2.112 3090 lb.
2.113 T_{AB} = 31.7 lb; T_{AC} = 64.3 lb.
2.115 T_{AB} = 510 N; T_{AC} = 56.2 N; T_{AD} = 536 N.
2.116 T_{AB} = 1340 N; T_{AC} = 1025 N; T_{AD} = 915 N.
2.117 T_{AB} = 1431 N; T_{AC} = 1560 N; T_{AD} = 183.0 N.
2.118 T_{AB} = 1249 N; T_{AC} = 490 N; T_{AD} = 1647 N.
2.119 T_{AB} = 842 lb; T_{AC} = 624 lb; T_{AD} = 1088 lb.
2.121 T_{BAC} = 76.7 lb; T_{AD} = 26.9 lb; T_{AE} = 49.2 lb.
2.122 P = 305 lb; T_{BAC} = 117.0 lb; T_{AD} = 40.9 lb.
2.123 P = 131.2 N; Q = 29.6 N.
2.125 (a) 1155 N. (b) 1012 N.
2.127 21.8 kN ⤡ 73.4°.
2.128 (a) 523 lb. (b) 428 lb.
2.129 (a) 95.1 lb. (b) 95.0 lb.

2.131 $F_C = 6.40$ kN; $F_D = 4.80$ kN.
2.133 (a) 288 N. (b) 67.5°, 30.0°, 108.7°.
2.134 (a) 114.4°. (b) $F_y = 694$ lb, $F_z = 855$ lb, $F = 1209$ lb.
2.135 515 N; $\theta_x = 70.2°$; $\theta_y = 27.6°$; $\theta_z = 71.5°$.
2.137 (a) 125.0 lb. (b) 45.0 lb.
2.C2 (1) (b) 20°; (c) 244 lb. (2) (b) −10°; (c) 467 lb. (3) (b) 10°; (c) 163.2 lb.
2.C3 (a) 1.001 m. (b) 4.01 kN. (c) 1.426 kN; 1.194 kN.

CHAPTER 3

3.1 115.7 lb-in. ↻.
3.2 115.7 lb-in. ↻.
3.4 (a) 196.2 N · m ↻. (b) 199.0 N ↘ 59.5°
3.5 (a) 196.2 N · m ↻. (b) 321 N ↗ 35.0°. (c) 231 N ↑ at point D.
3.6 (a) 20.5 N · m ↺. (b) 68.4 mm.
3.7 (a) 27.4 N · m ↺. (b) 228 N ↗ 42°.
3.9 1.120 kip · in. ↺.
3.10 493 lb · in. ↺.
3.11 (a) 760 N · m ↺. (b) 760 N · m ↺.
3.12 1224 N.
3.17 (a) $(-3\mathbf{i} - \mathbf{j} - \mathbf{k})/\sqrt{11}$. (b) $(2\mathbf{j} + 3\mathbf{k})/\sqrt{13}$.
3.18 2.21 m.
3.20 (a) $9\mathbf{i} + 22\mathbf{j} + 21\mathbf{k}$. (b) $22\mathbf{i} + 11\mathbf{k}$, (c) 0.
3.22 $(492 \text{ lb} \cdot \text{ft})\mathbf{i} + (144.0 \text{ lb} \cdot \text{ft})\mathbf{j} - (372 \text{ lb} \cdot \text{ft})\mathbf{k}$.
3.23 $-(25.4 \text{ lb} \cdot \text{ft})\mathbf{i} - (12.60 \text{ lb} \cdot \text{ft})\mathbf{j} - (12.60 \text{ lb} \cdot \text{ft})\mathbf{k}$.
3.24 $(1200 \text{ N} \cdot \text{m})\mathbf{i} - (1500 \text{ N} \cdot \text{m})\mathbf{j} - (900 \text{ N} \cdot \text{m})\mathbf{k}$.
3.25 $(7.50 \text{ N} \cdot \text{m})\mathbf{i} - (6.00 \text{ N} \cdot \text{m})\mathbf{j} - 10.39 \text{ N} \cdot \text{m})\mathbf{k}$.
3.27 100.8 mm.
3.28 144.8 mm.
3.29 4.86 ft.
3.30 5.17 ft.
3.32 2.36 m.
3.33 1.491 m.
3.35 $\mathbf{P} \cdot \mathbf{Q} = +1$; $\mathbf{P} \cdot \mathbf{S} = -11$; $\mathbf{Q} \cdot \mathbf{S} = +10$.
3.37 43.6°.
3.38 38.9°.
3.39 77.9°.
3.41 26.8°.
3.42 33.3.
3.43 (a) 52.9°. (b) 326 N.
3.45 7.
3.46 (a) 67. (b) 111.
3.47 $M_x = 0$, $M_y = -162.0 \text{ N} \cdot \text{m}$, $M_z = 270 \text{ N} \cdot \text{m}$.
3.48 $M_x = -576 \text{ N} \cdot \text{m}$, $M_y = -243 \text{ N} \cdot \text{m}$, $M_z = 405 \text{ N} \cdot \text{m}$.
3.49 61.5 lb.
3.51 283 lb.
3.53 $P = 125.0$ N; $\phi = 73.7°$, $\theta = 53.1°$.
3.54 23.0 N · m.
3.55 2.28 N · m.
3.56 −9.50 N · m.
3.57 +207 lb · ft.
3.59 −90 N · m.
3.60 −111.0 N · m.
3.61 $aP/\sqrt{2}$.
3.64 0.1198 m.
3.66 13.06 in.
3.67 12.69 in.
3.69 0.249 m.
3.70 (a) 336 lb · in. ↺. (b) 28.0 in. (c) 54.0°.
3.72 1.250 in.

3.73 (a) 26.7 N. (b) 50.0 N. (c) 23.5 N.
3.74 (a) 6.19 N · m. ↓. (b) 6.19 N · m ↓. (c) 6.19 N · m ↓.
3.76 $M = 3.22 \text{ N} \cdot \text{m}$; $\theta_x = 90.0°$; $\theta_y = 53.1°$, $\theta_z = 36.9°$.
3.77 $M = 2.72 \text{ N} \cdot \text{m}$; $\theta_x = 134.9°$, $\theta_y = 58.0°$, $\theta_z = 61.9°$.
3.78 $M = 604 \text{ lb} \cdot \text{in.}$; $\theta_x = 72.8°$, $\theta_y = 27.3°$, $\theta_z = 110.5°$.
3.79 $M = 1170 \text{ lb} \cdot \text{in.}$; $\theta_x = 81.2°$, $\theta_y = 13.70°$, $\theta_z = 100.4°$.
3.80 $M = 4.50 \text{ N} \cdot \text{m}$; $\theta_x = 90.0°$, $\theta_y = 177.1°$, $\theta_z = 87.1°$.
3.81 $\mathbf{F} = 260 \text{ lb} \nearrow 67.4°$; $\mathbf{M}_C = 200 \text{ lb} \cdot \text{in.}$ ↓.
3.82 (a) $\mathbf{F} = 30.0 \text{ lb} \downarrow$; $\mathbf{M} = 150.0 \text{ lb} \cdot \text{in.}$ ↖.
(b) $\mathbf{B} = 50.0 \text{ lb} \leftarrow$; $\mathbf{C} = 50.0 \text{ lb} \rightarrow$.
3.83 (a) $\mathbf{F}_B = 250 \text{ N} \searrow 60.0°$; $\mathbf{M}_B = 75.0 \text{ N} \cdot \text{m}$ ↓.
(b) $\mathbf{F}_A = 375 \text{ N} \searrow 60.0°$; $\mathbf{F}_B = 625 \text{ N} \searrow 60.0°$.
3.86 $\mathbf{F}_A = 389 \text{ N} \searrow 60.0°$; $\mathbf{F}_C = 651 \text{ N} \searrow 60.0°$.
3.87 (a) $\mathbf{F} = 216 \text{ N} \measuredangle 65.0°$; $\mathbf{M} = 33.0 \text{ N} \cdot \text{m}$ ↓.
(b) $\mathbf{F} = 216 \text{ N} \measuredangle 65.0°$ applied to the lever 267 mm to the left of B.
3.89 (a) $\mathbf{P} = 60.0 \text{ lb} \measuredangle 50.0°$; 3.24 in. from A.
(b) $\mathbf{P} = 60.0 \text{ lb} \measuredangle 50.0°$ 3.87 in. below A.
3.90 (a) 30.0°. (b) 65.7°.
3.91 $\mathbf{F} = 900 \text{ N} \downarrow$; $x = 50.0$ mm.
3.93 $\mathbf{F} = -(128.0 \text{ lb})\mathbf{i} - (256 \text{ lb})\mathbf{j} + (32.0 \text{ lb})\mathbf{k}$;
$\mathbf{M} = (4.10 \text{ kip} \cdot \text{ft})\mathbf{i} + (16.38 \text{ kip} \cdot \text{ft})\mathbf{k}$.
3.95 $\mathbf{F} = -(28.5 \text{ N})\mathbf{j} + (106.3 \text{ N})\mathbf{k}$; $\mathbf{M} = (12.35 \text{ N} \cdot \text{m})\mathbf{i} - (19.16 \text{ N} \cdot \text{m})\mathbf{j} - (5.13 \text{ N} \cdot \text{m})\mathbf{k}$.
3.96 $\mathbf{F} = -(1220 \text{ N})\mathbf{i}$; $\mathbf{M} = (73.2 \text{ N} \cdot \text{m})\mathbf{j} - (122 \text{ N} \cdot \text{m})\mathbf{k}$.
3.97 $\mathbf{F}_C = (5.00 \text{ N})\mathbf{i} + (150.0 \text{ N})\mathbf{j} - (90.0 \text{ N})\mathbf{k}$;
$\mathbf{M}_C = (77.4 \text{ N} \cdot \text{m})\mathbf{i} + (61.5 \text{ N} \cdot \text{m})\mathbf{j} + (106.8 \text{ N} \cdot \text{m})\mathbf{k}$.
3.98 $\mathbf{F} = (36.0 \text{ lb})\mathbf{i} - (28.0 \text{ lb})\mathbf{j} - (6.00 \text{ lb})\mathbf{k}$;
$\mathbf{M} = -(157.0 \text{ lb} \cdot \text{ft})\mathbf{i} + (22.5 \text{ lb} \cdot \text{ft})\mathbf{j} - (240 \text{ lb} \cdot \text{ft})\mathbf{k}$.
3.99 (a) 135.0 mm. (b) $\mathbf{F}_z = (42.0 \text{ N})\mathbf{i} + (42.0 \text{ N})\mathbf{j} - (49.0 \text{ N})\mathbf{k}$;
$\mathbf{M}_z = -(25.9 \text{ N} \cdot \text{m})\mathbf{i}. + (21.2 \text{ N} \cdot \text{m})\mathbf{j}$.
3.101 (a) Loading a 500 N ↓; 1000 N · m ↓.
Loading b 500 N ↑; 500 N · m ↺.
Loading c 500 N ↓; 500 N · m ↓.
Loading d 500 N ↓; 1100 N · m ↓.
Loading e 500 N ↓; 1000 N · m ↓.
Loading f 500 N ↓; 200 N · m ↓.
Loading g 500 N ↓; 2300 N · m ↺.
Loading h 500 N ↓; 650 N · m ↺.
(b) Loadings a and e are equivalent.
3.102 Equivalent to case f of problem 3.101.
3.104 Equivalent force-couple system at D.
3.106 (a) 39.6 in. to the right of D. (b) 33.1 in.
3.107 (a) 2.00 ft to the right of C. (b) 2.31 ft to the right of C.
3.108 (a) 34.0 ↘ 28.0°. (b) AB: 11.64 in. to the left of B; BC: 6.20 in. below B.
3.109 (a) 48.2 lb · in. ↺. (b) 240 lb · in. ↺. (c) 0.
3.111 (a) 0.365 m above G. (b) 0.227 m to the right of G.
3.112 (a) 0.299 m above G. (b) 0.259 m to the right of G.
3.113 773 lb ↗ 79.0°; 9.54 ft to the right of A.
3.114 (a) 665 lb ∡ 79.6°; 64.9 in. to the right of A.
(b) 22.9°.
3.116 (a) 1562 N ↘ 50.2°. (b) 250 mm to the right of C and 300 mm above C.
3.117 (a) 1308. N ∡ 66.6°. (b) 412 mm to the right of A and 250 mm to the right of C.
3.118 (a) $\mathbf{R} = F \nearrow \tan^{-1}(a^2/2bx)$;
$\mathbf{M} = 2Fb^2(x - x^3/a^2)/\sqrt{a^4 + 4b^2x^2}$ ↺. (b) 0.369 m.
3.119 $\mathbf{R} = -(21.0 \text{ N})\mathbf{i} - (29.0 \text{ N})\mathbf{j} + (16.00 \text{ N})\mathbf{k}$;
$\mathbf{M} = -(0.870 \text{ N} \cdot \text{m})\mathbf{i} + (0.630 \text{ N} \cdot \text{m})\mathbf{j} + (0.390 \text{ N} \cdot \text{m})\mathbf{k}$.
3.120 $\mathbf{R} = (420 \text{ N})\mathbf{j} - (339 \text{ N})\mathbf{k}$; $\mathbf{M} = (1.125 \text{ N} \cdot \text{m})\mathbf{i} + (163.9 \text{ N} \cdot \text{m})\mathbf{j} - (109.9 \text{ N} \cdot \text{m})\mathbf{k}$.

3.121 $\mathbf{R} = -(420 \text{ N})\mathbf{j} - (50.0 \text{ N})\mathbf{j} - (250 \text{ N})\mathbf{k}$;
$\mathbf{M} = (30.8 \text{ N} \cdot \text{m})\mathbf{j} - (22.0 \text{ N} \cdot \text{m})\mathbf{k}$.

3.122 (a) $\mathbf{B} = (2.50 \text{ lb})\mathbf{i}$, $\mathbf{C} = (0.1000 \text{ lb})\mathbf{i} - (2.47 \text{ lb})\mathbf{j} - (0.700 \text{ lb})\mathbf{k}$.
(b) $R_y = -2.47 \text{ lb}$; $M_x = 1.360 \text{ lb} \cdot \text{ft}$.

3.123 (a) $\mathbf{F}_B = -(80.0 \text{ N})\mathbf{k}$, $\mathbf{F}_C = -(30.0 \text{ N})\mathbf{i} + (40.0 \text{ N})\mathbf{k}$.
(b) $R_y = 0$, $R_z = 40.0 \text{ N}$. (c) When the slot in vertical

3.124 (a) $60.0°$. (b) $(20.0 \text{ lb})\mathbf{i} - (34.6 \text{ lb})\mathbf{j}$; $(520 \text{ lb} \cdot \text{in.})\mathbf{i}$.

3.127 1035 N; 2.57 m from OG and 3.05 m from OE.

3.128 2.32 m from OG and 1.165 m from OE.

3.129 405 lb; 12.60 ft to the right of AB and 2.94 ft. below BC.

3.130 $a = 0.722$ ft; $b = 20.6$ ft.

3.133 (a) P; $\theta_x = 90.0°$, $\theta_y = 90.0°$, $\theta_z = 0$. (b) $5a/2$. (c) Axis of the wrench is parallel to the z axis at $x = a$, $y = -a$.

3.134 (a) $P\sqrt{3}$; $\theta_x = \theta_y = \theta_z = 54.7°$. (b) $-a$.
(c) Axis of the wrench is diagonal OA.

3.135 (a) $-(21.0 \text{ lb})\mathbf{j}$. (b) 0.571 in. (c) Axis of wrench is parallel to the y axis at $x = 0$; $z = 1.667$ in.

3.137 (a) $-(84.0 \text{ N})\mathbf{j} - (80.0 \text{ N})\mathbf{k}$. (b) 0.477 m.
(c) $x = 0.526$ m, $y = 0$, $z = -0.1857$ m.

3.139 (a) $3P(2\mathbf{i} - 20\mathbf{j} - \mathbf{k})/25$. (b) $-0.0988\,a$.
(c) $x = 2.00\,a$, $z = -1.990\,a$.

3.142 $\mathbf{R} = (20.0 \text{ N})\mathbf{i} + (30.0 \text{ N})\mathbf{j} - (10.00 \text{ N})\mathbf{k}$; $y = -0.540$ m, $z = -0.420$ m.

3.143 $\mathbf{F}_A = (M/b)\mathbf{i} + R[1+(a/b)]\mathbf{k}$; $\mathbf{F}_B = -(M/b)\mathbf{i} - (aR/b)\mathbf{k}$.

3.147 41.7 N · m ↰. (b) 147.4 N ∠ 45.0°.

3.148 116.2 lb · ft ↰.

3.150 27.4°.

3.151 1.252 m.

3.153 (a) $\mathbf{F} = 560$ lb ↘ 20.0°; $\mathbf{M} = 7720$ lb · ft ↓.
(b) $\mathbf{F} = 560$ lb ↘ 20.0°; $\mathbf{M} = 4290$ lb · ft ↓.

3.154 $(0.227 \text{ lb})\mathbf{i} + (0.1057 \text{ lb})\mathbf{k}$; 63.6 in. to the right of B.

3.156 (a) 6.91 m. (b) 458 N; 3.16 m to the right of A.

3.158 $\mathbf{F}_B = 35.0$ kips ↓; $\mathbf{F}_F = 25.0$ kips ↓.

3.C3 4 sides: $\beta = 10°$, $\alpha = 44.1°$;
$\beta = 20°$, $\alpha = 41.6°$;
$\beta = 30°$, $\alpha = 37.8°$.

3.C4 $\theta = 0$ rev: $M = 97.0$ N · m;
$\theta = 6$ rev: $M = 63.3$ N · m;
$\theta = 12$ rev: $M = 9.17$ N · m.

3.C6 $d_{AB} = 36.0$ in.; $d_{CD} = 9.00$ in.; $d_{\min} = 58.3$ in.

CHAPTER 4

4.1 (a) 6.07 kN ↑. (b) 4.23 kN ↑.

4.2 (a) 4.89 kN ↑. (b) 3.69 kN ↑.

4.3 (a) $\mathbf{A} = 20.0$ lb ↓; $\mathbf{B} = 150.0$ lb ↑.
(b) $\mathbf{A} = 10.00$ lb ↓; $\mathbf{B} = 140.0$ lb ↑.

4.5 (a) 37.9 N ↑. (b) 373 N ↑.

4.6 (a) 2.76 N ↑. (b) 391 N ↑.

4.8 (a) 245 lb ↑. (b) 140.0 lb ↑.

4.9 2.00 in. ≤ a ≤ 10.00 in.

4.10 150.0 mm ≤ d ≤ 400 mm.

4.11 0.500 kN ≤ Q ≤ 11.00 kN.

4.14 6.00 kips ≤ P ≤ 42.0 kips.

4.15 (a) 2.00 kN. (b) 2.32 kN ∠ 46.4°.

4.17 (a) 150.0 lb. (b) 225 lb ↗ 32.3°.

4.18 111.2 lb →.

4.19 (a) 600 N. (b) 1253 N ∠ 69.8°.

4.21 (a) $\mathbf{A} = 225$ N ↑; $\mathbf{C} = 641$ N ↗ 20.6°.
(b) $\mathbf{A} = 365$ N ∠ 60.0°; $\mathbf{C} = 844$ N ↗ 22.0°.

4.23 (a) $\mathbf{A} = 150.0$ N ∠ 30.0°; $\mathbf{B} = 150.0$ N ↘ 30.0°.
(b) $\mathbf{A} = 433$ N ↘ 12.55°; $\mathbf{B} = 488$ N ↘ 30.0°.

4.25 (a) $\mathbf{A} = 44.7$ ↘ 26.6°; $\mathbf{B} = 30.0$ lb ↑.
(b) $\mathbf{A} = 30.2$ lb ↘ 41.4°; $\mathbf{B} = 34.6$ lb ↘ 60.0°.

4.26 (a) $\mathbf{A} = 20.0$ lb ↑; $\mathbf{B} = 50.0$ lb ↘ 36.9°.
(b) $\mathbf{A} = 23.1$ lb ∠ 60.0°; $\mathbf{B} = 59.6$ lb ↘ 30.2°.

4.27 (a) 190.9 N. (b) 142.3 N ∠ 18.43°.

4.28 (a) 324 N. (b) 270 N →.

4.29 (a) 117.0 lb. (b) 129.8 lb ↘ 56.3°.

4.30 (a) 195.0 lb. (b) 255 lb ↖ 45.0°.

4.31 $T = 80.0$ N; $\mathbf{C} = 89.4$ N ∠ 26.6°.

4.32 (a) 130.0 N. (b) 224 N ↗ 2.05°.

4.35 (a) 600 N. (b) $\mathbf{A} = 4.00$ kN ←; $\mathbf{B} = 4.00$ kN →.

4.36 (a) 105.1 N. (b) $\mathbf{A} = 147.2$ N ↑; $\mathbf{B} = 105.1$ N ←.

4.37 $T_{BE} = 50.0$ N; $\mathbf{A} = 18.75$ lb →; $\mathbf{D} = 18.75$ lb ←.

4.38 $\mathbf{A} = 69.3$ lb →; $\mathbf{B} = 34.6$ lb ↖ 60.0°; $\mathbf{C} = 173.2$ lb ↘ 60.0°.

4.39 $T = 80.0$ N; $\mathbf{A} = 160.0$ N ↖ 30.0°; $\mathbf{C} = 160.0$ N ↘ 30.0°.

4.40 $T = 69.3$ N; $\mathbf{A} = 140.0$ N ↖ 30.0°; $\mathbf{C} = 180.0$ N ↘ 30.0°.

4.43 (a) $\mathbf{D} = 20.0$ lb ↓; $\mathbf{M}_D = 20.0$ lb · ft ↰.
(b) $\mathbf{D} = 10.00$ lb ↓; $\mathbf{M}_D = 30.0$ lb · ft ↓.

4.45 (a) $\mathbf{A} = 78.5$ N ↑; $\mathbf{M}_A = 125.6$ N · m ↰.
(b) $\mathbf{A} = 111.0$ N ∠ 45.0°; $\mathbf{M}_A = 125.6$ N · m ↰.
(c) $\mathbf{A} = 157.0$ N ↑; $\mathbf{M}_A = 251$ N · m.

4.46 $\mathbf{C} = 28.3$ N ↘ 45.0°; $\mathbf{M}_C = 4.30$ N · m ↓.

4.47 $\mathbf{C} = 28.3$ N ↘ 45.0°; $\mathbf{M}_C = 4.50$ N · m ↓.

4.48 (a) $\mathbf{E} = 8.80$ kips ↑; $\mathbf{M}_E = 36.0$ kip · ft ↓.
(b) $\mathbf{E} = 4.80$ kips ↑; $\mathbf{M}_E = 51.0$ kip · ft ↓.

4.50 $\mathbf{A} = 1848$ N ∠ 82.6°; $\mathbf{M}_A = 1431$ N · m ↓.

4.51 (a) $\theta = 2 \cos^{-1}\left[\frac{1}{4}\left(\frac{W}{P} \pm \sqrt{\frac{W^2}{P^2} + 8}\right)\right]$. (b) $\theta = 65.1°$.

4.52 (a) $\theta = 2 \sin^{-1}(W/2P)$. (b) $\theta = 29.0°$.

4.53 (a) $T = \frac{1}{2}W(1 - \tan\theta)$. (b) $\theta = 39.8°$.

4.54 (a) $\sin\theta + \cos\theta = M/pl$. (b) 17.11° and 72.9°.

4.55 141.1°.

4.56 (a) $(1 - \cos\theta)\tan\theta = W/2kl$. (b) 49.7°.

4.59 (1) completely constrained; determinate; $\mathbf{A} = \mathbf{C} = 196.2$ N ↑.
(2) completely constrained; determinate; $\mathbf{B} = 0$, $\mathbf{C} = \mathbf{D} = 196.2$ N ↑.
(3) completely constrained; indeterminate; $\mathbf{A}_x = 294$ N →; $\mathbf{D}_x = 294$ N ←.
(4) improperly constrained; indeterminate; no equilibrium.
(5) partially constrained; determinate; equilibrium; $\mathbf{C} = \mathbf{D} = 196.2$ N ↑.
(6) completely constrained; determinate; $\mathbf{B} = 294$ N →, $\mathbf{D} = 491$ N ↘ 53.1°.
(7) partially constrained; no equilibrium.
(8) completely constrained; indeterminate; $\mathbf{B} = 196.2$ N ↑, $\mathbf{D}_y = 196.2$ N ↑.

4.61 $\mathbf{A} = 680$ N ∠ 28.1°; $\mathbf{B} = 600$ N ←.

4.62 200 mm.

4.64 $T = 289$ lb; $\mathbf{A} = 577$ lb ∠ 60.0°.

4.65 $\mathbf{A} = 63.6$ lb ↖ 45.0°; $\mathbf{C} = 87.5$ lb ↘ 59.0°.

4.67 $\mathbf{B} = 888$ N ↖ 41.3°; $\mathbf{D} = 943$ N ↘ 45.0°.

4.68 $\mathbf{B} = 1001$ N ↘ 48.2°; $\mathbf{D} = 943$ N ↖ 45.0°.

4.69 $\mathbf{A} = 778$ N ↓; $\mathbf{C} = 1012$ N ↘ 77.9°.

4.71 (a) 24.9 lb ↗ 30.0°. (b) 15.34 lb ∠ 30.0°.

4.72 $\mathbf{A} = 37.1$ lb ∠ 62.4°; $T = 18.57$ lb.

4.73 (a) 499 N. (b) 457 N ↘ 26.6°.

4.75 $\mathbf{A} = 163.1$ N ↖ 74.1°; $\mathbf{B} = 258$ N ↘ 65.0°.

4.77 (a) $2P$ ↘ 60.0°. (b) $1.239P$ ↘ 36.2°.

4.78 (a) $1.155P$ ↘ 30.0°. (b) $1.086P$ ∠ 22.9°.

4.79 (a) $\mathbf{A} = 150.0$ N ∠ 30.0°; $\mathbf{B} = 150.0$ N ↘ 30.0°.
(b) $\mathbf{A} = 433$ N ↖ 12.55°; $\mathbf{B} = 488$ N ↘ 30.0°.

4.80 (a) 119.3 lb. (b) 178.7 lb ↘ 60.5°.

4.81 $T = 100.0$ lb; $\mathbf{B} = 111.1$ lb $\searrow 30.3°$.
4.83 (a) 225 mm. (b) 23.1 N. (c) 12.21 N →.
4.84 32.5°.
4.87 (a) 59.4°. (b) $\mathbf{A} = 8.45$ lb →; 13.09 lb $\searrow 49.8°$.
4.88 60.0 mm.
4.89 $\tan\theta = 2\tan\beta$.
4.90 (a) 49.1°. (b) $\mathbf{A} = 45.3$ N ←; $\mathbf{B} = 90.6$ N $\measuredangle 60.0°$.
4.91 (a) 1200 N. (b) $\mathbf{C} = (400\text{ N})\mathbf{i} + (1200\text{ N})\mathbf{j}$;
$\mathbf{D} = -(1600\text{ N})\mathbf{i} - (480\text{ N})\mathbf{j}$.
4.93 $\mathbf{A} = (24.0\text{ lb})\mathbf{j} - (2.31\text{ lb})\mathbf{k}$; $\mathbf{B} = (16.00\text{ lb})\mathbf{j} - (9.24\text{ lb})\mathbf{k}$;
$\mathbf{C} = (11.55\text{ lb})\mathbf{k}$.
4.94 $\mathbf{A} = (14.00\text{ lb})\mathbf{j} + (4.50\text{ lb})\mathbf{k}$; $\mathbf{D} = (6.00\text{ lb})\mathbf{j} + (10.50\text{ lb})\mathbf{k}$.
4.95 $\mathbf{A} = (120.0\text{ N})\mathbf{j} + (133.3\text{ N})\mathbf{k}$; $\mathbf{D} = (60.0\text{ N})\mathbf{j} + (166.7\text{ N})\mathbf{k}$.
4.97 $T_A = 23.5$ N; $T_C = 11.77$ N; $T_D = 105.9$ N.
4.98 (a) 0.480 m. (b) $T_A = 23.5$ N; $T_C = 0$; $T_D = 117.7$ N.
4.99 $T_A = 5.63$ lb; $T_B = 16.88$ lb; $T_C = 22.5$ lb.
4.101 (a) 121.9 N. (b) −46.2 N. (c) 100.9 N.
4.102 (a) 95.6 N. (b) −7.36 N. (c) 88.3 N.
4.103 $T_A = 30.0$ lb; $T_B = 10.00$ lb; $T_C = 40.0$ lb.
4.105 $T_{AD} = 2.60$ kN; $T_{AE} = 2.80$ kN; $\mathbf{C} = (1.800\text{ kN})\mathbf{j} + (4.80\text{ kN})\mathbf{k}$.
4.106 $T_{AD} = 5.20$ kN; $T_{AE} = 5.60$ kN; $\mathbf{C} = (9.60\text{ kN})\mathbf{k}$.
4.107 $T_{BD} = T_{BE} = 1100$ lb; $\mathbf{A} = (1200\text{ lb})\mathbf{i} - (560\text{ lb})\mathbf{j}$.
4.108 $T_{BE} = T_{BF} = 17.50$ kN; $\mathbf{A} = -(7.00\text{ kN})\mathbf{i} + (22.4\text{ kN})\mathbf{j}$.
4.109 $T_{BE} = 6.62$ kN; $T_{BF} = 25.1$ kN; $\mathbf{A} = -(6.34\text{ kN})\mathbf{i} + (20.3\text{ kN})\mathbf{j} + (2.96\text{ kN})\mathbf{k}$.
4.112 $T_{DE} = T_{DF} = 262$ lb; $\mathbf{A} = -(801\text{ lb})\mathbf{i} + (1544\text{ lb})\mathbf{j}$.
4.113 (a) 345 N. (b) $\mathbf{A} = (114.5\text{ N})\mathbf{i} + (377\text{ N})\mathbf{j} + (144.5\text{ N})\mathbf{k}$;
$\mathbf{B} = (113.2\text{ N})\mathbf{j} + (185.5\text{ N})\mathbf{k}$.
4.115 (a) 49.5 lb. (b) $\mathbf{A} = -(12.00)\mathbf{i} + (22.5\text{ lb})\mathbf{j} - (4.00\text{ lb})\mathbf{k}$;
$\mathbf{B} = (15.00\text{ lb})\mathbf{j} + (34.0\text{ lb})\mathbf{k}$.
4.116 (a) 118.8 lb. (b) $\mathbf{A} = (93.8\text{ lb})\mathbf{i} + (22.5\text{ lb})\mathbf{j} + (70.8\text{ lb})\mathbf{k}$;
$\mathbf{B} = (15.00\text{ lb})\mathbf{j} - (8.33\text{ lb})\mathbf{k}$.
4.117 (a) 101.6 N. (b) $\mathbf{A} = -(26.3\text{ N})\mathbf{i}$; $\mathbf{B} = (98.1\text{ N})\mathbf{j}$.
4.119 (a) 49.5 lb. (b) $\mathbf{A} = -(12.00\text{ lb})\mathbf{i} + (37.5\text{ lb})\mathbf{j} + (30.0\text{ lb})\mathbf{k}$;
$\mathbf{M}_A = -(1020\text{ lb}\cdot\text{in.})\mathbf{j} + (450\text{ lb}\cdot\text{in.})\mathbf{k}$.
4.120 (a) 462 N. (b) $\mathbf{C} = (169.1\text{ N})\mathbf{j} + (400\text{ N})\mathbf{k}$;
$\mathbf{M}_C = (20.0\text{ N}\cdot\text{m})\mathbf{j} + (151.5\text{ N}\cdot\text{m})\mathbf{k}$.
4.121 $T_{CF} = 200$ N; $T_{DE} = 450$ N; $\mathbf{A} = (160.0\text{ N})\mathbf{i} + (270\text{ N})\mathbf{k}$;
$\mathbf{M}_A = -(16.20\text{ N}\cdot\text{m})\mathbf{i}$.
4.122 (a) 5.00 lb. (b) $\mathbf{C} = -(5.00\text{ lb})\mathbf{i} + (6.00\text{ lb})\mathbf{j} - (5.00\text{ lb})\mathbf{k}$;
$\mathbf{M}_C = (8.00\text{ lb}\cdot\text{in.})\mathbf{j} - (12.00\text{ lb}\cdot\text{in.})\mathbf{k}$.
4.125 $T_{BD} = 2.18$ kN; $T_{BE} = 3.96$ kN; $T_{CD} = 1.500$ kN.
4.126 $T_{BD} = 0$; $T_{BE} = 3.96$ kN; $T_{CO} = 3.00$ kN.
4.127 (a) $T_B = -0.366P$; $T_C = 1.219P$; $T_D = -0.853P$.
(b) $\mathbf{F} = -0.345P\mathbf{i} + P\mathbf{j} - 0.862\,P\mathbf{k}$.
4.129 $\mathbf{A} = (120.0\text{ lb})\mathbf{j} - (150.0\text{ lb})\mathbf{k}$; $\mathbf{B} = (180.0\text{ lb})\mathbf{i} + (150.0\text{ lb})\mathbf{k}$;
$\mathbf{C} = -(180.0\text{ lb})\mathbf{i} + (120.0\text{ lb})\mathbf{j}$.
4.130 $\mathbf{A} = (20.0\text{ lb})\mathbf{j} + (25.0\text{ lb})\mathbf{k}$; $\mathbf{B} = (30.0\text{ lb})\mathbf{i} - (25.0\text{ lb})\mathbf{k}$;
$\mathbf{C} = -(30.0\text{ lb})\mathbf{i} - (20.0\text{ lb})\mathbf{j}$.
4.131 $\mathbf{B} = (60.0\text{ N})\mathbf{k}$; $\mathbf{C} = (30.0\text{ N})\mathbf{j} - (16.00\text{ N})\mathbf{k}$;
$\mathbf{D} = -(30.0\text{ N})\mathbf{j} + (4.00\text{ N})\mathbf{k}$.
4.133 373 N.
4.135 $(45.0\text{ lb})\mathbf{j}$.
4.136 (a) $x = 4.00$ ft; $y = 8.00$ ft; (b) 10.73 lb.
4.137 (a) $x = 0$ ft; $y = 16.00$ ft. (b) 11.31 lb.
4.138 360 N.
4.140 85.3 lb.
4.141 181.7 lb.
4.142 42.0 N ↑.
4.143 (a) 80.8 lb ↓. (b) 216 lb $\measuredangle 22.0°$.
4.145 (a) 875 lb. (b) 1584 lb $\searrow 45.0°$.

4.147 $\mathbf{C} = 1951$ N $\searrow 88.5°$; $\mathbf{M}_C = 75.0$ N · m $\downarrow$.
4.149 $\mathbf{A} = 170.0$ N $\searrow 33.9°$. $\mathbf{C} = 160.0$ N $\measuredangle 28.1°$.
4.150 (a) $T_A = 6.00$ lb; $T_B = T_C = 9.00$ lb. (b) 15.00 in.
4.151 $T_{BE} = 975$ N; $T_{CF} = 600$ N; $T_{DG} = 625$ N; $\mathbf{A} = (2100\text{ N})\mathbf{i} + (175.0\text{ N})\mathbf{j} - (375\text{ N})\mathbf{k}$.
4.153 (a) $\mathbf{A} = 0.745P \measuredangle 63.4°$; $\mathbf{C} = 0.471P \searrow 45.0°$.
(b) $\mathbf{A} = 0.812P \measuredangle 60.0°$; $\mathbf{C} = 0.503P \nearrow 36.2°$.
(c) $\mathbf{A} = 0.448P \searrow 60.0°$; $\mathbf{C} = 0.652P \measuredangle 69.9°$.
(d) improperly constrained: no equilibrium.
4.C1 $\theta = 20°$; $T = 114.8$ lb; $\theta = 70°$; $T = 127.7$ lb, $T_{max} = 132.2$ lb at $\theta = 50.4°$.
4.C2 $x = 600$ mm: $P = 31.4$ N; $x = 150$ mm: $P = 37.7$ N; $P_{max} = 47.2$ N at $x = 283$ mm.
4.C3 $\theta = 30°$: $W = 9.66$ lb; $\theta = 60°$: $W = 36.6$ lb; $W = 5$ lb at $\theta = 22.9°$ [Also at $\theta = 175.7°$].
4.C4 $\theta = 30°$: $W = 0.80$ lb; $\theta = 60°$: $W = 4.57$ lb at $\theta = 62.6°$ [Also at $\theta = 159.6°$].
4.C5 $\theta = 30°$: $m = 7.09$ kg; $\theta = 60°$: $m = 11.02$ kg. When $m = 10$ kg, $\theta = 51.0°$.
4.C6 $\theta = 15°$: $T_{BD} = 10.30$ kN, $T_{BE} = 21.7$ kN; $\theta = 30°$: $T_{BD} = 5.69$ kN, $T_{BE} = 24.4$ kN; $T_{max} = 26.5$ kN at $\theta = 36.9°$.

CHAPTER 5

5.1 $\overline{X} = 1.045$ in., $\overline{Y} = 3.59$ in.
5.2 $\overline{X} = 36.0$ mm, $\overline{Y} = 48.0$ mm.
5.3 $\overline{X} = 19.27$ mm, $\overline{Y} = 26.6$ mm.
5.4 $\overline{X} = 5.67$ in., $\overline{Y} = 5.17$ in.
5.6 $\overline{X} = 7.22$ in., $\overline{Y} = 9.56$ in.
5.8 $\overline{X} = -10.00$ mm, $\overline{Y} = 87.5$ mm.
5.9 $\overline{X} = 92.0$ mm, $\overline{Y} = 23.3$ mm.
5.10 $\overline{X} = 0$, $\overline{Y} = 6.45$ in.
5.11 $\overline{X} = 0$, $\overline{Y} = 1.372$ in.
5.13 $\overline{X} = 50.5$ mm, $\overline{Y} = 19.34$ mm.
5.14 $\overline{X} = 3.20$ in. $\overline{Y} = 2.00$ in.
5.17 $a/r = 0.508$.
5.18 $\overline{Y} = \dfrac{2}{3}\left(\dfrac{r_2^3 - r_1^3}{r_2^2 - r_1^2}\right)\left(\dfrac{2\cos\alpha}{\pi - 2\alpha}\right)$.
5.19 $\overline{Y} = \left(\dfrac{r_1 + r_2}{\pi - 2\alpha}\right)\cos\alpha$.
5.20 0.235 in^3 for A, -0.235 in^3 for A_2.
5.22 459 N.
5.23 (a) $b(c^2 - y^2)/2$. (b) $y = 0$, $Q_x = bc^2/2$.
5.24 $\overline{X} = 36.6$ mm, $\overline{Y} = 47.6$ mm.
5.26 $\overline{X} = 5.52$ in., $\overline{Y} = 5.16$ in.
5.28 120.0 mm.
5.29 99.5 mm.
5.31 (a) 5.09 lb. (b) 9.48 lb $\searrow 57.5°$.
5.32 (a) $0.513\,a$. (b) $0.691\,a$.
5.34 $\bar{x} = a/3$, $\bar{y} = 2h/3$.
5.36 $\bar{x} = a/2$, $\bar{y} = 2h/5$.
5.37 $\bar{y} = \bar{x} = 9a/20$.
5.38 $\bar{x} = 2a/3(4 - \pi)$, $\bar{y} = 2b/3(4 - \pi)$.
5.40 $\bar{x} = 5a/8$, $\bar{y} = b/3$.
5.41 $\bar{x} = 3a/8$, $\bar{y} = b$.
5.43 $\bar{x} = a.$, $\bar{y} = 17b/35$.
5.44 $\bar{x} = 17a/130$, $\bar{y} = 11b/26$.
5.45 $-2\sqrt{2}r/3\pi$.
5.46 $2a/5$.
5.48 $\bar{x} = L/\pi$, $\bar{y} = \pi a/8$.
5.49 $\bar{x} = -9.27a$, $\bar{y} = 3.09a$.

5.50 $\bar{x} = 1.629$ in., $\bar{y} = 0.1853$ in.
5.51 $a = 1.901$ in. or 3.74 in.
5.52 (a) Volume $= 248$ in^3, Area $= 547$ in^2.
(b) Volume $= 72.3$ in^3, Area $= 169.6$ in^2.
5.53 (a) Volume $= 423 \times 10^3$ mm^3, Area $= 37.2 \times 10^3$ mm^2.
(b) Volume $= 847 \times 10^3$ mm^3, Area $= 72.5 \times 10^3$ mm^2.
5.54 (a) Volume $= 2.26 \times 10^6$ mm^3, Area $= 116.3 \times 10^3$ mm^2.
(b) Volume $= 1.471 \times 10^6$ mm^3, Area $= 116.3 \times 10^3$ mm^2.
5.56 $V = 3470$ mm^3; $A = 2320$ mm^2.
5.58 $V = 3.96$ in^3; $W = 1.211$ lb.
5.60 0.0305 kg.
5.62 0.0900 in^3.
5.63 308 in^2.
5.64 31.9 liters.
5.66 (a) $\mathbf{R} = 7.60$ kN $\downarrow$, $\bar{x} = 2.57$ m, (b) $\mathbf{A} = 4.35$ kN $\uparrow$; $\mathbf{B} = 3.25$ kN $\uparrow$.
5.67 (a) $\mathbf{R} = 1215$ lb $\downarrow$, $\bar{x} = 4.33$ ft. (b) $\mathbf{A} = 630$ lb $\uparrow$; $\mathbf{B} = 585$ lb $\uparrow$.
5.68 $\mathbf{A} = 105.0$ N $\uparrow$; $\mathbf{B} = 270$ N $\uparrow$.
5.69 $\mathbf{A} = 2860$ lb $\uparrow$; $\mathbf{B} = 740$ lb $\uparrow$.
5.71 $\mathbf{A} = 32.0$ kN $\uparrow$; $M_A = 124.0$ kN $\cdot$ m $\curvearrowright$.
5.72 $\mathbf{A} = 3.00$ kN $\uparrow$; $M_A = 12.60$ kN $\cdot$ m $\curvearrowright$.
5.74 $\mathbf{B} = 3770$ lb $\uparrow$; $\mathbf{C} = 429$ lb $\uparrow$.
5.75 (a) 900 lb/ft. (b) 7200 lb $\uparrow$.
5.76 (a) 0.536 m. (b) $\mathbf{A} = \mathbf{B} = 761$ N $\uparrow$.
5.78 $w_{BC} = 2810$ N/m; $w_{DE} = 3150$ N/m.
5.80 (a) $\mathbf{H} = 44.1$ kN $\rightarrow$; $\mathbf{V} = 228$ kN $\uparrow$. (b) 1.159 m to the right of A. (c) $\mathbf{R} = 59.1$ kN $\nearrow$ 41.6°.
5.81 (a) $\mathbf{H} = 254$ kN $\rightarrow$; $\mathbf{V} = 831$ kN $\uparrow$. (b) 3.25 m to the right of A (c) $\mathbf{R} = 268$ kN $\nearrow$ 18.43°.
5.82 12.00 in.
5.83 4.00 in.
5.84 $\mathbf{T} = 67.2$ kN $\leftarrow$; $\mathbf{A} = 141.2$ kN $\leftarrow$.
5.86 3.70 kips $\uparrow$.
5.87 $t = 35.7$ $\curvearrowright$; gate rotates clockwise.
5.88 0.683 m.
5.89 0.0711 m.
5.91 208 lb.
5.92 $\mathbf{A} = 1197$ N $\searrow$ 53.1°; $\mathbf{B} = 1511$ N $\searrow$ 53.1°.
5.93 3570 N.
5.94 6.00 ft.
5.96 (a) $0.0536\,a$ below base of cone.
(b) $0.0625\,a$ above base of cone.
5.97 (a) $0.548\,L$. (b) $2\sqrt{3}$.
5.98 $-(2h^2 - 3b^2)/2(4h - 3b)$.
5.99 $-a(4h - 2b)/\pi(4h - 3b)$.
5.100 -0.1403 in.
5.101 19.13 mm.
5.103 3.47 in.
5.104 18.28 mm.
5.106 $\overline{X} = 45.0$ mm; $\overline{Z} = -20.2$ mm.
5.107 $\overline{X} = 0.1402$ m; $\overline{Y} = 0.0944$ m; $\overline{Z} = 0.0959$ m.
5.108 $\overline{X} = 17.00$ in.; $\overline{Y} = 15.68$ in.; $\overline{Z} = 14.16$ in.
5.109 $\overline{X} = 46.5$ mm; $\overline{Y} = 27.2$ mm; $\overline{Z} = 30.0$ mm.
5.110 $\overline{X} = \overline{Z} = 4.21$ in.; $\overline{Y} = 7.03$ in.
5.113 $\overline{X} = 180.2$ mm; $\overline{Y} = 38.0$ mm; $\overline{Z} = 193.5$ mm.
5.114 $\overline{X} = 0$; $\overline{Y} = 10.05$ in.; $\overline{Z} = 5.15$ in.
5.115 $\overline{X} = 0.410$ m; $\overline{Y} = 0.510$ m; $\overline{Z} = 0.1500$ m.
5.116 $\overline{X} = 0.909$ m; $\overline{Y} = 0.1842$ m; $\overline{Z} = 0.884$ m.
5.118 $\overline{X} = \overline{Z} = 0$; $\overline{Y} = 83.3$ mm above the base.
5.120 $\overline{Y} = 0.526$ in. above the base.

5.121 $\overline{X} = 61.1$ mm from the end of the handle.
5.122 $(\bar{x}_1) = 21a/88$; $(\bar{x}_2) = 27a/40$.
5.123 $(\bar{x}_1) = 21h/88$; $(\bar{x}_2) = 27h/40$.
5.124 $(\bar{x}_1) = 2h/9$; $(\bar{x}_2) = 2h/3$.
5.126 $\bar{x} = 2.34$ m; $\bar{y} = \bar{z} = 0$.
5.128 $\bar{x} = 1.297\,a$; $\bar{y} = \bar{z} = 0$.
5.129 $\bar{x} = \bar{z} = 0$; $\bar{y} = 0.374\,b$.
5.132 (a) $\bar{x} = \bar{z} = 0$; $\bar{y} = -121.9$ mm. (b) $\bar{x} = \bar{z} = 0$; $\bar{y} = -90.2$ mm.
5.134 $\bar{x} = 0$; $\bar{y} = 5h/16$; $\bar{z} = -b/4$.
5.135 $V = 688$ ft^3; $\bar{x} = 15.91$ ft.
5.136 $\bar{x} = a/2$; $y = 8h/25$; $\bar{z} = b/2$.
5.137 $\bar{x} = 1.643$ in.; $\bar{y} = 17.46$ in.
5.138 $\bar{x} = 30.0$ mm; $\bar{y} = 64.8$ mm.
5.140 $\bar{x} = 2a/5$; $\bar{y} = 4h/7$.
5.141 $\bar{x} = 5L/4$; $\bar{y} = 33a/40$.
5.143 $\mathbf{B} = 1360$ lb $\uparrow$; $\mathbf{C} = 2360$ lb $\uparrow$.
5.144 $w_A = 10.00$ kN/m; $w_B = 50.0$ kN/m.
5.146 (a) $b/10$ to the left of base of cone
(b) $0.01136b$ to the right of base of cone.
5.147 $\overline{X} = 0.295$ m; $\overline{Y} = 0.423$ m; $\overline{Z} = 1.703$ m.
5.C1 (b) $\mathbf{A} = 1220$ lb $\uparrow$; $\mathbf{B} = 1830$ lb $\uparrow$.
(c) $\mathbf{A} = 1265$ lb $\uparrow$; $\mathbf{B} = 1601$ lb $\uparrow$.
5.C2 (a) $\overline{X} = 0$, $\overline{Y} = 0.278$ m. $\overline{Z} = 0.0878$ m.
(b) $\overline{X} = 0.0487$ mm, $\overline{Y} = 0.1265$ mm, $\overline{Z} = 0.0997$ mm.
(c) $\overline{X} = -0.0372$ m, $\overline{Y} = 0.1659$ m, $\overline{Z} = 0.1043$ m.
5.C3 $d = 1.00$ m: $\mathbf{F} = 5.66$ kN $\searrow$ 30°;
$d = 3.00$ m: $\mathbf{F} = 49.9$ kN $\searrow$ 27.7°.
5.C4 (a) $\overline{X} = 5.80$ in., $\overline{Y} = 1.492$ in. (b) $\overline{X} = 9.11$ in., $\overline{Y} = 2.78$ in.
(c) $\overline{X} = 8.49$ in., $\overline{Y} = 0.375$ in.
5.C5 With $n = 40$: (a) $\overline{X} = 60.2$ mm, $\overline{Y} = 23.4$ mm.
(b) $\overline{X} = 60.2$ mm, $\overline{Y} = 146.2$ mm.
(c) $\overline{X} = 68.7$ mm. $\overline{Y} = 20.4$ mm.
(d) $\overline{X} = 68.7$ mm. $\overline{Y} = 127.8$ mm.
5.C6 With $n = 40$: (a) $\overline{X} = 60.0$ mm, $\overline{Y} = 24.0$ mm.
(b) $\overline{X} = 60.0$ mm, $\overline{Y} = 150.0$ mm.
(c) $\overline{X} = 68.6$ mm, $\overline{Y} = 21.8$ mm.
(d) $\overline{X} = 68.6$ mm, $\overline{Y} = 136.1$ mm.
5.C7 (a) $V = 628$ ft^3.
(b) $\overline{X} = 8.65$ ft, $\overline{Y} = -4.53$ ft, $\overline{Z} = 9.27$ ft.

CHAPTER 6

6.1 $F_{AB} = 180.0$ kN T; $F_{AC} = 156.0$ kN C; $F_{BC} = 144.0$ kN T.
6.2 $F_{AB} = 720$ lb T; $F_{AC} = 1200$ lb C; $F_{BC} = 780$ lb C.
6.3 $F_{AB} = 52.0$ kN T; $F_{AC} = 64.0$ kN T; $F_{BC} = 80.0$ kN C.
6.5 $F_{AB} = F_{BC} = 0$; $F_{AD} = F_{CF} = 1.000$ kip C; $F_{BD} = F_{CF} = 6.80$ kips C; $F_{BE} = 2.40$ kips C; $F_{DE} = F_{EF} = 6.00$ kips T.
6.7 $F_{AB} = 20.0$ kN T; $F_{AD} = 20.6$ kN C; $F_{BC} = 30.0$ kN T; $F_{BD} = 30.0$ kN C; $F_{CD} = 10.00$ kN T.
6.8 $F_{AB} = 4.00$ kN T; $F_{AD} = 15.00$ kN C; $F_{BD} = 9.00$ kN C; $F_{BE} = 5.00$ kN T; $F_{CD} = 16.00$ kN C; $F_{DE} = 4.00$ kN C.
6.9 $F_{AB} = F_{FH} = 1500$ lb C; $F_{AC} = F_{CE} = F_{EG} = F_{GH} = 1200$ lb T; $F_{BC} = F_{FG} = 0$; $F_{BD} = F_{DF} = 1200$ lb C; $F_{BE} = F_{EF} = 60.0$ lb C; $F_{DE} = 72.0$ lb T.
6.10 $F_{AB} = F_{FH} = 1500$ lb C; $F_{AC} = F_{CE} = F_{EG} = F_{GH} = 1200$ lb T; $F_{BC} = F_{FG} = 0$; $F_{BD} = F_{DF} = 1000$ lb C; $F_{BE} = F_{EF} = 500$ lb C; $F_{DE} = 600$ lb T.
6.11 $F_{AB} = 47.2$ kN C; $F_{AC} = 44.6$ kN T; $F_{BC} = 10.50$ kN C; $F_{BD} = 47.2$ kN C; $F_{CD} = 17.50$ kN T; $F_{CE} = 30.6$ kN T; $F_{DE} = 0$.

6.13 $F_{AB} = 7.83$ kN C; $F_{AC} = 7.00$ kN T; $F_{BC} = 1.886$ kN C; $F_{BD} = 6.34$ kN C; $F_{CD} = 1.491$ kN T; $F_{CE} = 5.00$ kN T; $F_{DE} = 2.83$ kN C; $F_{DF} = 3.35$ kN C; $F_{EF} = 2.75$ kN T; $F_{EG} = 1.061$ kN C; $F_{EH} = 3.75$ kN T; $F_{FG} = 4.24$ kN C; $F_{GH} = 5.30$ kN C.

6.15 $F_{AB} = 2240$ lb C; $F_{AC} = F_{CE} = 2000$ lb T; $F_{BC} = F_{EH} = 0$; $F_{BD} = 1789$ lb C; $F_{BE} = 447$ lb C; $F_{DE} = 600$ lb C; $F_{DF} = 2010$ lb C; $F_{DG} = 224$ lb T; $F_{EG} = 1789$ lb T.

6.17 $F_{AB} = 9.90$ kN C; $F_{AC} = 7.83$ kN T; $F_{BC} = 0$; $F_{BD} = 7.07$ kN C; $F_{BE} = 2.00$ kN C; $F_{CE} = 7.83$ kN T; $F_{DE} = 1.000$ kN T; $F_{DF} = 5.03$ kN C; $F_{DG} = 0.559$ kN C; $F_{EG} = 5.59$ kN T.

6.18 $F_{FG} = 3.50$ kN T; $F_{FH} = 5.03$ kN C; $F_{GH} = 1.677$ kN T; $F_{GI} = F_{KL} = F_{KL} = 3.35$ kN T; $F_{HI} = F_{IJ} = F_{JK} = 0$; $F_{HJ} = F_{JL} = 4.24$ kN C.

6.19 $F_{AB} = F_{FG} = 7.50$ kips C; $F_{AC} = F_{EG} = 4.50$ kips T; $F_{BC} = F_{EF} = 7.50$ kips T; $F_{BD} = F_{DF} = 9.00$ kips C; $F_{CD} = F_{DE} = 0$; $F_{CE} = 9.00$ kips T.

6.21 $F_{AB} = F_{FH} = 7.50$ kips C; $F_{AC} = F_{GH} = 4.50$ kips T; $F_{BC} = F_{FG} = 4.00$ kips T; $F_{BD} = F_{DF} = 6.00$ kips C; $F_{BE} = F_{EF} = 2.50$ kips T; $F_{CE} = F_{EG} = 4.50$ kips T; $F_{DE} = 0$.

6.22 $F_{AB} = 6.25$ kips C; $F_{AC} = 3.75$ kips T; $F_{BC} = 4.00$ kips T; $F_{BD} = F_{DF} = 4.50$ kips C; $F_{BE} = 1.250$ kips C; $F_{CE} = 3.75$ kips T; $F_{DE} = F_{FG} = 0$; $F_{EF} = 3.75$ kips T; $F_{EG} = 2.25$ kips T; $F_{FH} = 3.75$ kips C.

6.23 $F_{AB} = F_{DF} = 2.29$ kN T; $F_{AC} = F_{EF} = 2.29$ kN C; $F_{BC} = F_{DE} = 0.600$ kN C; $F_{BD} = 2.21$ kN T; $F_{BE} = F_{EH} = 0$; $F_{CE} = 2.21$ kN C; $F_{CH} = F_{EJ} = 1.200$ kN C.

6.26 $F_{AB} = 9.39$ kN C; $F_{AC} = 8.40$ kN T; $F_{BC} = 2.26$ kN C; $F_{BD} = 7.60$ kN C; $F_{CD} = 0.128$ kN C; $F_{CE} = 7.07$ kN T; $F_{DE} = 2.14$ kN C; $F_{DF} = 6.10$ kN C; $F_{EF} = 2.23$ kN T.

6.27 $F_{AB} = 31.0$ kips C; $F_{AC} = 28.3$ kips C; $F_{AD} = 15.09$ kips T; $F_{AE} = 9.50$ kips T; $F_{BD} = 21.5$ kips T; $F_{BF} = 28.0$ kips C; $F_{CE} = 41.0$ kips T; $F_{CG} = 42.0$ kips T; $F_{DE} = 22.0$ kips T; $F_{DF} = 33.5$ kips T; $F_{EG} = 0$.

6.28 $F_{AB} = F_{BC} = F_{CD} = 36.0$ kips T; $F_{AE} = 57.6$ kips T; $F_{AF} = 45.0$ kips C; $F_{BF} = F_{BG} = F_{CG} = F_{CH} = 0$; $F_{DH} = F_{FG} = F_{GH} = 39.0$ kips C; $F_{EF} = 36.0$ kips C.

6.29 Truss of prob. 6.33a is the only simple truss.

6.30 Trusses of prob 6.31b and prob 6.33b are simple trusses.

6.31 (a) AI, BJ, CK, DI, EI, FK, GK. (b) FK, IO.

6.34 (a) GH, GJ, IJ. (b) BF, BG, CG, CH.

6.35 $F_{AB} = F_{AD} = 244$ lb C; $F_{AC} = 1040$ lb T; $F_{BC} = F_{CD} = 500$ lb C; $F_{BD} = 280$ lb T.

6.36 $F_{AB} = F_{AD} = 861$ N C; $F_{AC} = 676$ N C; $F_{BC} = F_{CD} = 162.5$ N T; $F_{BD} = 244$ N T.

6.37 $F_{AB} = F_{AD} = 2810$ N T; $F_{AC} = 5510$ N C; $F_{BC} = F_{CD} = 1325$ N T; $F_{BD} = 1908$ N C.

6.38 $F_{AB} = F_{AC} = 1061$ lb C; $F_{AD} = 2500$ lb T; $F_{BC} = 2100$ lb T; $F_{BD} = F_{CD} = 1250$ lb C; $F_{BE} = F_{CE} = 1250$ lb C; $F_{DE} = 1500$ lb T.

6.39 $F_{AB} = 840$ N C; $F_{AC} = 110.6$ N C; $F_{AD} = 394$ N C; $F_{AE} = 0$; $F_{BC} = 160.0$ N T; $F_{BE} = 200$ N T; $F_{CD} = 225$ N T; $F_{CE} = 233$ N C; $F_{DE} = 120.0$ N T.

6.40 $F_{AB} = F_{AE} = F_{BC} = 0$; $F_{AC} = 995$ N T; $F_{AD} = 1181$ N C; $F_{BE} = 600$ N T; $F_{CD} = 375$ V T; $F_{CE} = 700$ N C; $F_{DE} = 360$ N T.

6.43 $F_{CD} = 9.00$ kN C; $F_{DF} = 12.00$ kN C.

6.44 $F_{FG} = 5.00$ kN T; $F_{FH} = 20.0$ kN T.

6.45 $F_{CE} = 8000$ lb T; $F_{DE} = 2600$ lb T; $F_{DF} = 9000$ lb C.

6.46 $F_{EG} = 7500$ lb T; $F_{FG} = 3900$ lb C; $F_{FH} = 6000$ lb C.

6.49 $F_{AD} = 13.50$ kN C; $F_{CD} = 0$; $F_{CE} = 56.1$ kN T.

6.50 $F_{DG} = 75.0$ kN C; $F_{FG} = 56.1$ kN T; $F_{FH} = 69.7$ kN T.

6.51 $F_{AB} = 8.20$ kips T; $F_{AG} = 4.50$ kips T; F_{FG} 11.60 kips C.

6.52 $F_{AE} = 17.46$ kips T; $F_{EF} = 11.60$ kips C; $F_{FJ} = 18.45$ kips C.

6.53 $F_{CD} = 20.0$ kN C; $F_{DF} = 52.0$ kN C.

6.54 $F_{CE} = 36.0$ kN T; $F_{EF} = 15.00$ kN C.

6.55 $F_{FG} = 5.23$ kN C; $F_{EG} = 0.1476$ kN C; $F_{EH} = 5.08$ kN T.

6.56 $F_{KM} = 5.02$ kN T; $F_{LM} = 1.963$ kN C; $F_{LN} = 3.95$ kN C.

6.59 $F_{DE} = 25.0$ kips T; $F_{DF} = 13.00$ kips C.

6.60 $F_{EG} = 16.00$ kips T; $F_{EF} = 6.40$ kips C.

6.63 $F_{DG} = 3.75$ kN T; $F_{FI} = 3.75$ kN C.

6.64 $F_{GJ} = 11.25$ kN T; $F_{TK} = 11.25$ kN C.

6.65 $F_{BG} = 5.48$ kips T; $F_{DG} = 1.825$ kips T.

6.66 $F_{CF} = 3.65$ kips T; $F_{CH} = 7.30$ kips T.

6.67 (a) CJ. (b) 1.026 kN T.

6.68 (a) IO. (b) 2.05 kN T.

6.69 (a) improperly constrained. (b) completely constrained, determinate. (c) completely constrained, indeterminate.

6.70 (a) completely constrained, determine. (b) partially constrained. (c) improperly constrained.

6.71 (a) completely constrained, determinate. (b) completely constrained, indeterminate. (c) improperly constrained.

6.72 (a) partially constrained. (b) completely constrained, determinate. (c) completely constrained, indeterminate.

6.75 $F_{BD} = 780$ lb T; $\mathbf{C}_x = 720$ lb $\leftarrow$, $\mathbf{C}_y = 140.0$ lb $\downarrow$.

6.76 $F_{BD} = 255$ N C; $\mathbf{C}_x = 120.0$ N $\rightarrow$, $\mathbf{C}_y = 625$ N $\uparrow$.

6.77 $\mathbf{A}_x = 480$ N $\rightarrow$, $\mathbf{A}_y = 120.0$ N $\uparrow$; $\mathbf{B}_x = 480$ N $\leftarrow$, $\mathbf{B}_y = 320$ N $\downarrow$; $\mathbf{C} = 120.0$ N $\downarrow$; $\mathbf{D} = 320$ N $\uparrow$.

6.79 $\mathbf{A}_x = 25.0$ kips $\leftarrow$, $\mathbf{A}_y = 20.0$ kips $\uparrow$; $\mathbf{B}_x = 25.0$ kips, $\mathbf{B}_y = 10.00$ kips $\downarrow$; $\mathbf{C}_x = 50.0$ kips $\rightarrow$, $\mathbf{C}_y = 10.00$ kips $\downarrow$.

6.81 $\mathbf{A} = 375$ N; $\mathbf{B}_x = 375$ N $\leftarrow$, $\mathbf{B}_y = 150.0$ N $\uparrow$; $\mathbf{C} = 50.0$ N $\uparrow$; $\mathbf{D} = 200$ N $\downarrow$.

6.82 $\mathbf{A} = 150.0$ N $\rightarrow$; $\mathbf{B} = 0$; $\mathbf{C}_x = 150.0$ N $\leftarrow$, $\mathbf{C}_y = 100.0$ N $\uparrow$; $\mathbf{D} = 100.0$ N $\downarrow$.

6.83 (a) $\mathbf{A}_x = 300$ N $\leftarrow$, $\mathbf{A}_y = 660$ N $\uparrow$; $\mathbf{E}_x = 300$ N $\rightarrow$, $\mathbf{E}_y = 90.0$ N $\uparrow$. (b) $\mathbf{A}_x = 300$ N $\leftarrow$, $\mathbf{A}_y = 150.0$ N $\uparrow$; $\mathbf{E}_x = 300$ N $\rightarrow$, $\mathbf{E}_y = 600$ N $\uparrow$.

6.84 (a) $\mathbf{A}_x = 450$ N $\leftarrow$, $\mathbf{A}_y = 525$ N $\uparrow$; $\mathbf{E}_x = 450$ N $\rightarrow$, $\mathbf{E}_y = 225$ N $\uparrow$; (b) $\mathbf{A}_x = 450$ N $\leftarrow$, $\mathbf{A}_y = 150.0$ N $\uparrow$; $\mathbf{E}_x = 450$ N $\rightarrow$, $\mathbf{E}_y = 600$ N $\uparrow$.

6.87 (a) $\mathbf{A} = 65.0$ lb $\nearrow$ 22.6°; $\mathbf{C} = 120.0$ lb $\rightarrow$; $\mathbf{G} = 60.0$ lb $\leftarrow$; $\mathbf{I} = 25.0$ lb $\uparrow$. (b) $\mathbf{A} = 65.0$ lb $\nearrow$ 22.6°; $\mathbf{C} = 60.0$ lb $\rightarrow$; $\mathbf{G} = 0$; $\mathbf{I} = 25.0$ lb $\uparrow$.

6.88 (a) $\mathbf{A} = 65.0$ lb $\nearrow$ 22.6°; $\mathbf{C} = 120.0$ lb $\rightarrow$; $\mathbf{G} = 60.0$ lb $\leftarrow$; $\mathbf{I} = 25.0$ lb $\uparrow$.
(b) $\mathbf{A} = 65.0$ lb $\nearrow$ 22.6°; $\mathbf{C} = 100.0$ lb $\rightarrow$; $\mathbf{G} = 80.0$ lb $\leftarrow$; $\mathbf{I} = 25.0$ lb $\uparrow$.

6.89 (a) $\mathbf{A}_x = 80.0$ lb $\leftarrow$, $\mathbf{A}_y = 40.0$ lb $\uparrow$; $\mathbf{B}_x = 80.0$ lb $\rightarrow$, $\mathbf{B}_y = 60.0$ lb $\uparrow$. (b) $\mathbf{A}_x = 0$, $\mathbf{A}_y = 40.0$ lb $\uparrow$; $\mathbf{B}_x = 0$, $\mathbf{B}_y = 60.0$ lb $\uparrow$.

6.91 (a) $\mathbf{E}_x = 2.00$ kips $\leftarrow$, $\mathbf{E}_y = 2.25$ kips $\uparrow$.
(b) $\mathbf{C}_x = 4.00$ kips $\leftarrow$, $\mathbf{C}_y = 5.75$ kips $\uparrow$.

6.92 (a) $\mathbf{E}_x = 3.00$ kips $\leftarrow$, $\mathbf{E}_y = 1.500$ kips $\uparrow$.
(b) $\mathbf{C}_x = 3.00$ kips $\leftarrow$, $\mathbf{C}_y = 6.50$ kips $\uparrow$.

6.93 $\mathbf{A}_x = 150.0$ N $\leftarrow$, $\mathbf{A}_y = 250$ N $\uparrow$; $\mathbf{E}_x = 150.0$ N $\rightarrow$, $\mathbf{E}_y = 450$ N $\uparrow$.

6.94 $\mathbf{B}_x = 700$ N $\leftarrow$, $\mathbf{B}_y = 200$ N $\downarrow$; $\mathbf{E}_x = 700$ N $\rightarrow$, $\mathbf{E}_y = 500$ N $\uparrow$.

6.95 (a) $\mathbf{A} = 982$ lb $\uparrow$; $\mathbf{B} = 935$ lb $\uparrow$; $\mathbf{C} = 733$ lb $\uparrow$.
(b) $\Delta B = +291$ lb; $\Delta C = -72.7$ lb.

6.96 (a) 572 lb. (b) $\mathbf{A} = 1070$ lb $\uparrow$; $\mathbf{B} = 709$ lb $\uparrow$; $\mathbf{C} = 870$ lb $\uparrow$.

6.99 $\mathbf{A}_x = 13.00$ kN $\leftarrow$, $\mathbf{A}_y = 4.00$ kN $\downarrow$; $\mathbf{B}_x = 36.0$ kN $\rightarrow$, $\mathbf{B}_y = 6.00$ kN $\uparrow$; $\mathbf{E}_x = 23.0$ kN $\leftarrow$, $\mathbf{E}_y = 2.00$ kN $\downarrow$.

6.100 $\mathbf{A}_x = 2025$ N $\leftarrow$, $\mathbf{A}_y = 1800$ N $\downarrow$; $\mathbf{B}_x = 4050$ N $\rightarrow$, $\mathbf{B}_y = 1200$ N $\uparrow$; $\mathbf{E}_x = 2025$ N $\leftarrow$, $\mathbf{E}_y = 600$ N $\uparrow$.

6.101 $\mathbf{A}_x = 1110$ lb $\leftarrow$, $\mathbf{A}_y = 600$ lb $\uparrow$; $\mathbf{B}_x = 1110$ lb $\leftarrow$, $\mathbf{B}_y = 800$ lb $\downarrow$; $\mathbf{D}_x = 2220$ lb $\rightarrow$, $\mathbf{D}_y = 200$ lb $\uparrow$.

6.102 $\mathbf{A}_x = 660$ lb $\leftarrow$, $\mathbf{A}_y = 240$ lb $\uparrow$; $\mathbf{B}_x = 660$ lb $\leftarrow$, $\mathbf{B}_y = 320$ lb $\downarrow$; $\mathbf{D}_x = 1320$ lb $\rightarrow$, $\mathbf{D}_y = 80.0$ lb $\uparrow$.

6.103 $\mathbf{C}_x = 21.7$ lb $\rightarrow$, $\mathbf{C}_y = 37.5$ lb $\downarrow$; $\mathbf{D}_x = 21.7$ lb $\leftarrow$, $\mathbf{D}_y = 62.5$ lb $\uparrow$.

6.104 $\mathbf{C}_x = 78.0$ lb $\rightarrow$, $\mathbf{C}_j = 28.0$ lb $\uparrow$; $\mathbf{F}_x = 78.0$ lb $\leftarrow$, $\mathbf{F}_y = 12.00$ lb $\uparrow$.

6.107 $\mathbf{F}_x = 300$ lb $\leftarrow$, $\mathbf{F}_y = 1200$ lb $\uparrow$; $\mathbf{F}_{AE} = 1000$ lb C; $\mathbf{F}_{BD} = 500$ lb T.

6.108 $\mathbf{A} = 327$ lb $\rightarrow$; $\mathbf{B} = 827$ lb $\leftarrow$; $\mathbf{D} = 621$ lb $\uparrow$; $\mathbf{E} = 246$ lb $\uparrow$.

6.109 (a) $\mathbf{A}_x = 200$ kN $\rightarrow$, $\mathbf{A}_y = 122.0$ kN $\uparrow$. (b) $\mathbf{B}_x = 200$ kN $\leftarrow$, $\mathbf{B}_y = 10.00$ kN $\downarrow$.

6.110 (a) $\mathbf{A}_x = 205$ kN $\rightarrow$, $\mathbf{A}_y = 134.5$ kN $\uparrow$. (b) $\mathbf{B}_x = 205$ kN $\leftarrow$, $\mathbf{B}_y = 5.50$ kN $\uparrow$.

6.112 $\mathbf{F}_{AF} = P/4\ C$; $\mathbf{F}_{BG} = \mathbf{F}_{DG} = P/\sqrt{2}\ C$; $\mathbf{F}_{EH} = P/4\ T$.

6.113 $\mathbf{F}_{AG} = \sqrt{2}P/6\ C$; $\mathbf{F}_{BF} = 2\sqrt{2}P/3\ C$; $\mathbf{F}_{DI} = \sqrt{2}P/3\ C$; $\mathbf{F}_{EH} = \sqrt{2}P/6\ T$.

6.115 $\mathbf{F}_{AF} = \mathbf{M}_0/4a\ C$; $\mathbf{F}_{BG} = \mathbf{F}_{DG} = \mathbf{M}_0/\sqrt{2}a\ T$; $\mathbf{F}_{EH} = 3\mathbf{M}_0/4a\ C$.

6.116 $\mathbf{F}_{AF} = \mathbf{M}_0/6a\ T$; $\mathbf{F}_{BG} = \sqrt{2}\mathbf{M}_0/6a\ T$; $\mathbf{F}_{DG} = \sqrt{2}\mathbf{M}_0/3a\ T$; $\mathbf{F}_{EH} = \mathbf{M}_0/6a\ C$.

6.117 $\mathbf{E} = P/5\ \downarrow$; $\mathbf{F} = 8P/5\ \uparrow$; $\mathbf{G} = 4P/5\ \downarrow$; $\mathbf{H} = 2P/5\ \uparrow$.

6.118 $\mathbf{A} = P/15\ \uparrow$; $\mathbf{D} = 2P/15\ \uparrow$; $\mathbf{E} = 8P/15\ \uparrow$; $\mathbf{H} = 4P/15\ \uparrow$.

6.121 (a) $\mathbf{A} = 2.06P \measuredangle 14.04°$; $\mathbf{B} = 2.06 \measuredangle 14.04°$; frame is rigid. (b) Frame is not rigid. (c) $\mathbf{A} = 1.25P \measuredangle 36.9°$; $\mathbf{B} = 1.031P \measuredangle 14.04°$; frame is rigid.

6.122 (a) 2860 N $\downarrow$. (b) 2700 N $\measuredangle 68.5°$.

6.123 (a) 746 N $\downarrow$. (b) 565 N $\measuredangle 61.3°$.

6.126 (a) $(\mathbf{F}_{BD})_y = 96.0$ lb $\downarrow$. (b) $\mathbf{F}_{BD} = 100.0$ lb $\measuredangle 73.7°$.

6.127 (a) $(\mathbf{F}_{BD})_y = 240$ lb $\downarrow$. (b) $\mathbf{F}_{BD} = 250$ lb $\measuredangle 73.7°$.

6.128 (a) $\mathbf{P} = 109.8$ N $\rightarrow$. (b) 126.8 N T. (c) 139.8 N $\measuredangle 38.3°$.

6.129 (a) 160.8 N $\cdot$ m $\uparrow$. (b) 155.9 N $\cdot$ m $\uparrow$.

6.130 (a) 117.8 N $\cdot$ m $\uparrow$. (b) 47.9 N $\cdot$ m $\uparrow$.

6.131 (a) 21.0 kN $\leftarrow$. (b) 52.5 kN $\leftarrow$.

6.132 (a) 1143 N $\cdot$ m $\downarrow$. (b) 457 N $\cdot$ m $\downarrow$.

6.133 832 lb $\cdot$ in. $\uparrow$.

6.134 360 lb $\cdot$ in. $\uparrow$.

6.135 18.43 N $\cdot$ m $\downarrow$.

6.136 208 N $\cdot$ m $\downarrow$.

6.139 $\mathbf{F}_{AE} = 800$ N T; $\mathbf{F}_{DG} = 100.0$ N C.

6.140 $\mathbf{P} = 120.0$ N $\downarrow$; $\mathbf{Q} = 110.0$ N $\leftarrow$.

6.141 $\mathbf{D} = 30.0$ kN $\leftarrow$; $\mathbf{F} = 37.5$ kN $\measuredangle 36.9°$.

6.142 $\mathbf{D} = 150.0$ kN; $\mathbf{F} = 96.4$ kN $\measuredangle 13.50°$.

6.144 $\mathbf{F} = 3290$ lb $\measuredangle 15.12°$; $\mathbf{D} = 4450$ lb $\leftarrow$.

6.145 8.45 kN.

6.147 (a) 475 lb. (b) 528 lb $\measuredangle 63.3°$.

6.148 44.8 kN.

6.149 (a) 312 lb. (b) 135.0 lb $\cdot$ in. $\downarrow$.

6.150 140.0 N.

6.152 21.3 lb $\searrow$.

6.153 (a) 9.29 kN $\measuredangle 44.4°$. (b) 8.04 kN $\measuredangle 34.4°$.

6.155 (a) 2.86 kips C. (b) 9.43 kips C.

6.156 (a) 4.91 kips C. (b) 10.69 kips C.

6.159 (a) 27.0 mm. (b) 40.0 N $\cdot$ m $\downarrow$.

6.160 (a) $(90.0$ N $\cdot$ m$)\mathbf{i}$. (b) $\mathbf{A} = 0$; $\mathbf{M}_A = -(48.0$ N $\cdot$ m$)\mathbf{i}$; $\mathbf{B} = 0$; $\mathbf{M}_B = -(72.0$ N $\cdot$ m$)\mathbf{i}$.

6.163 $\mathbf{E}_x = 100.0$ kN $\rightarrow$, $\mathbf{E}_y = 154.9$ kN $\uparrow$; $\mathbf{F}_x = 26.5$ kN $\rightarrow$, $\mathbf{F}_y = 118.1$ kN $\downarrow$; $\mathbf{H}_x = 126.5$ kN $\leftarrow$, $\mathbf{H}_y = 36.8$ kN $\downarrow$.

6.164 $\mathbf{F}_{AB} = \mathbf{F}_{BD} = 0$; $\mathbf{F}_{AC} = 675$ N T; $\mathbf{F}_{AD} = 1125$ N C; $\mathbf{F}_{CD} = 900$ N T; $\mathbf{F}_{CE} = 2025$ N T; $\mathbf{F}_{CF} = 2250$ N C; $\mathbf{F}_{DF} = 675$ N C; $\mathbf{F}_{EF} = 1800$ N T.

6.165 $\mathbf{F}_{AB} = 6.24$ kN C; $\mathbf{F}_{AC} = 2.76$ kN T; $\mathbf{F}_{BC} = 2.50$ kN C; $\mathbf{F}_{BD} = 4.16$ kN C; $\mathbf{F}_{CD} = 1.867$ kN T; $\mathbf{F}_{CE} = 2.88$ kN T; $\mathbf{F}_D = 3.75$ kN C; $\mathbf{F}_{DF} = 0$; $\mathbf{F}_{EF} = 1.200$ kN C.

6.166 $\mathbf{F}_{DF} = 10.48$ kips C; $\mathbf{F}_{DG} = 3.35$ kips C; $\mathbf{F}_{EG} = 13.02$ kips T.

6.168 (a) 80.0 lb T. (b) 72.1 lb $\measuredangle 16.10°$.

6.170 $\mathbf{D}_x = 13.60$ kN $\rightarrow$, $\mathbf{D}_y = 7.50$ kN $\uparrow$; $\mathbf{E}_x = 13.60$ kN $\leftarrow$, $\mathbf{E}_y = 2.70$ kN $\downarrow$.

6.172 (a) 301 lb $\measuredangle 48.4°$. (b) 375 lb T.

6.173 764 N $\leftarrow$.

6.175 25.0 lb $\downarrow$.

6.C1 (a) $\theta = 30°$: $W = 472$ lb, $A_{AB} = 1.500$ in^2, $A_{AC} = A_{CE} = 1.299$ in^2, $A_{BC} = A_{BE} = 0.500$ in^2, $A_{BD} = 1.732$ in^2. (b) $\theta_{\text{opt}} = 56.8°$: $W = 312$ lb, $A_{AB} = 0.896$ in^2, $A_{AC} = A_{CE} = 0.491$ in^2, $A_{BC} = 0.500$ in^2, $A_{BE} = 0.299$ in^2, $A_{BD} = 0.655$ in^2.

6.C2 (a) Fox $x = 9.75$ m, $F_{BH} = 3.19$ kN T. (b) For $x = 3.75$ m, $F_{BH} = 1.313$ kN C. (c) Fox $x = 6$ m, $F_{GH} = 3.04$ kN T.

6.C3 $\theta = 30°$: $\mathbf{M} = 5860$ lb $\cdot$ ft $\uparrow$; $\mathbf{A} = 670$ lb $\measuredangle 75.5°$. (a) $M_{\text{max}} = 8680$ lb $\cdot$ ft when $\theta = 65.9°$. (b) $A_{\text{max}} = 1436$ lb when $\theta = 68.5°$.

6.C4 $\theta = 30°$: $\mathbf{M}_A = 1.669$ N $\cdot$ m $\uparrow$, $F = 11.79$ N, $\theta = 80°$; $\mathbf{M}_A = 3.21$ N $\cdot$ m $\uparrow$, $F = 11.98$ N.

6.C5 $d = 0.40$ in.: 634 lb C; $d = 0.55$ in.: 286 lb C; $d = 0.473$ in.: $F_{AB} = 500$ lb C.

6.C6 $\theta = 20°$: $M = 31.8$ N $\cdot$ m; $\theta = 75°$: $M = 12.75$ N $\cdot$ m; $\theta = 60.0°$: $M_{min} = 12.00$ N $\cdot$ m.

CHAPTER 7

7.1 $\mathbf{F} = 720$ lb $\rightarrow$; $\mathbf{V} = 140.0$ lb $\uparrow$; $\mathbf{M} = 1120$ lb $\cdot$ in. $\uparrow$ (On JC).

7.2 $\mathbf{F} = 1106$ lb $\measuredangle 20.6°$; $\mathbf{V} = 386$ lb $\measuredangle 69.4°$; $\mathbf{M} = 1650$ lb $\cdot$ ft $\downarrow$ (On JD).

7.3 $\mathbf{F} = 125.0$ N $\measuredangle 67.4°$; $\mathbf{V} = 300$ N $\measuredangle 22.6°$; $\mathbf{M} = 156.0$ N $\cdot$ m $\downarrow$ (On BJ).

7.4 $\mathbf{F} = 2330$ N $\measuredangle 67.4°$; $\mathbf{V} = 720$ N $\measuredangle 22.6°$; $\mathbf{M} = 374$ N $\cdot$ m $\downarrow$ (On BJ).

7.7 $\mathbf{F} = 12.50$ lb $\measuredangle 30.0°$; $\mathbf{V} = 21.7$ lb $\measuredangle 60.0°$; $\mathbf{M} = 75.0$ lb $\cdot$ in. $\downarrow$ (On BJ).

7.8 $\mathbf{F} = 108.3$ lb $\measuredangle 60.0°$; $\mathbf{V} = 62.5$ lb $\measuredangle 30.0°$; $\mathbf{M} = 100.5$ lb $\cdot$ in. $\uparrow$ (On DK).

7.9 $\mathbf{F} = 103.9$ N $\measuredangle 60.0°$; $\mathbf{V} = 60.0$ N $\measuredangle 30.0°$; $\mathbf{M} = 18.71$ N $\cdot$ m $\downarrow$ (On AJ).

7.10 $\mathbf{F} = 60.0$ N $\measuredangle 30.0°$; $\mathbf{V} = 103.9$ N $\measuredangle 60.0°$; $\mathbf{M} = 10.80$ N $\cdot$ m $\uparrow$ (On BK).

7.11 $\mathbf{F} = 194.6$ N $\measuredangle 60.0°$; $\mathbf{V} = 257$ N $\measuredangle 30.0°$; $\mathbf{M} = 24.7$ N $\cdot$ m $\downarrow$ (On AJ).

7.12 45.2 N $\cdot$ m for $\theta = 82.9°$.

7.15 $\mathbf{F} = 250$ N $\measuredangle 36.9°$; $\mathbf{V} = 120.0$ N $\measuredangle 53.1$; $\mathbf{M} = 120.0$ N $\cdot$ m $\uparrow$ (On BJ).

7.16 $\mathbf{F} = 560$ N $\leftarrow$; $\mathbf{V} = 90.0$ N $\downarrow$; $\mathbf{M} = 72.0$ N $\cdot$ m $\downarrow$ (On AK).

7.17 150.0 lb $\cdot$ in. at D.

7.18 105.0 lb $\cdot$ in. at E.

7.19 (a) $\mathbf{F} = 500$ N $\leftarrow$; $\mathbf{V} = 500$ N $\uparrow$; $\mathbf{M} = 300$ N $\cdot$ m $\downarrow$ (On AJ). (b) $\mathbf{F} = 970$ N $\uparrow$; $\mathbf{V} = 171.0$ N $\leftarrow$; $\mathbf{M} = 446$ N $\cdot$ m $\downarrow$ (On AK).

7.20 (a) $\mathbf{F} = 500$ N $\leftarrow$; $\mathbf{V} = 500$ N $\uparrow$; $\mathbf{M} = 300$ N $\cdot$ m $\downarrow$ (On AJ). (b) $\mathbf{F} = 933$ N $\uparrow$; $\mathbf{V} = 250$ N $\leftarrow$; $\mathbf{M} = 375$ N $\cdot$ m $\downarrow$ (On AK).

7.23 0.0557 Wr $\uparrow$ (On AJ).

7.24 0.289 Wr $\downarrow$ (On BJ).

7.25 0.1009 Wr for $\theta = 57.3°$.

7.26 $0.357\ Wr$ for $\theta = 49.3°$.
7.29 (b) $|V|_{\max} = wL$; $|M|_{\max} = wL^2/2$.
7.30 (b) $|V|_{\max} = wL/2$; $|M|_{\max} = w_0L^2/6$.
7.31 (b) $|V|_{\max} = 2P/3$; $|M|_{\max} = PL/9$.
7.32 (b) $|V|_{\max} = P$; $|M|_{\max} = PL/2$.
7.35 (b) $|V|_{\max} = 35.0$ kN; $|M|_{\max} = 12.50$ kN · m.
7.36 (b) $|V|_{\max} = 50.5$ kN; $|M|_{\max} = 39.8$ kN · m.
7.39 (b) $|V|_{\max} = 64.0$ kN; $|M|_{\max} = 92.0$ kN · m.
7.40 (b) $|V|_{\max} = 30.0$ kN; $|M|_{\max} = 72.0$ kN · m.
7.41 (b) $|V|_{\max} = 18.00$ kips; $|M|_{\max} = 48.5$ kip · ft.
7.42 (b) $|V|_{\max} = 15.30$ kips; $|M|_{\max} = 46.8$ kip · ft.
7.45 (b) $|V|_{\max} = 1.800$ kN; $|M|_{\max} = 0.225$ kN · m.
7.46 (b) $|V|_{\max} = 2.00$ kN; $|M|_{\max} = 0.500$ kN · m.
7.47 (a) $M \geq 0$ everywhere.
(b) $|V|_{\max} = 4.50$ kips; $|M|_{\max} = 13.50$ kip · ft.
7.48 (a) $M \leq 0$ everywhere.
(b) $|V|_{\max} = 4.50$ kips; $|M|_{\max} = 13.50$ kip · ft.
7.49 $|V|_{\max} = 180.0$ N; $|M|_{\max} = 36.0$ N · m.
7.50 $|V|_{\max} = 800$ N; $|M|_{\max} = 180.0$ N · m.
7.51 $|V|_{\max} = 90.0$ lb; $|M|_{\max} = 1400$ lb · in.
7.52 $|V|_{\max} = 165.0$ lb; $|M|_{\max} = 1625$ lb · in.
7.55 (a) $54.5°$. (b) 675 N · m.
7.56 (a) 1.236. (b) $0.1180\ wa^2$.
7.57 (a) 0.311 m. (b) 193.0 N · m.
7.58 (a) 0.840 m. (b) 1.680 N · m.
7.59 $0.207\ L$.
7.62 (a) $0.414\ wL$; $0.0858\ wL^2$. (b) $0.250\ wL$; $0.250\ wL^2$.
7.69 (b) $|V|_{\max} = 6.40$ kN; $|M|_{\max} = 4.00$ kN · m.
7.70 (b) $|V|_{\max} = 9.00$ kN; $|M|_{\max} = 14.00$ kN · m.
7.77 (b) 75.0 kN · m, 4.00m from A.
7.78 (b) 1.378 kN · m, 1.050m from A.
7.79 (b) 40.5 kN · m, 1.800m from A.
7.80 (b) 60.5 kN · m, 2.20m from A.
7.81 (a) 18.00 kip · ft, 3.00 ft from A.
(b) 34.1 kip · ft, 2.25 ft from A.
7.82 (a) 12.00 kip · ft at C. (b) 6.25 kip · ft, 2.50 ft from A.
7.86 (a) $V = (w_0/6L)(L^2 - 3x^2)$; $M = (w_0/6L)(L^2x - x^3)$.
(b) $0.0642\ w_0\ L^2$A $\measuredangle = 0.577L$.
7.87 (a) $V = (w_0/3L)(2x^2 - 3Lx + L^2)$; $M = (w_0/8L)(4x^3 - 9Lx + 6L^2\ x - L^3)$.
(b) $w_0/L^2/72$, at $x = L/2$.
7.89 (a) $\mathbf{P} = 4.00$ kN $\downarrow$; $\mathbf{Q} = 6.00$ kN $\downarrow$. (b) $M_C = -900$ N · m.
7.90 (a) $\mathbf{P} = 2.50$ kN $\downarrow$; $\mathbf{Q} = 7.50$ kN $\downarrow$. (b) $M_C = -900$ N · m.
7.91 (a) $\mathbf{P} = 1.350$ kips $\downarrow$; $\mathbf{Q} = 0.450$ kips $\downarrow$. (b) $V_{\max} = 2.70$ kips at A; $M_{\max} = 6.345$ kip · ft, 5.40 ft from A.
7.92 (a) $\mathbf{P} = 0.540$ kips $\downarrow$; $\mathbf{Q} = 1.860$ kips $\downarrow$.
(b) $V_{\max} = 3.14$ kips at B; $M_{\max} = 7.00$ kip · ft, 6.88 ft from A.
7.93 (a) $\mathbf{E}_x = 8.00$ kN $\rightarrow$; $\mathbf{E}_y = 5.00$ kN $\uparrow$. (b) 9.43 kN.
7.94 2.00 m.
7.95 (a) 838 lb $\measuredangle\ 17.35°$. (b) 971 lb $\measuredangle\ 34.5°$.
7.96 (a) 2670 lb $\measuredangle\ 2.10°$. (b) 2810 lb $\measuredangle\ 18.65$.
7.97 (a) $d_B = 1.733$ m; $d_D = 4.20$ m. (b) 21.5 kN $\measuredangle\ 3.81°$.
7.98 (a) 2.80 m. (b) $\mathbf{A} = 32.0$ kN $\measuredangle\ 38.7°$; $\mathbf{E} = 25.0$ kN $\rightarrow$.
7.101 196.2 N.
7.102 157.0 N.
7.103 (a) 240 lb. (b) 9.00 ft.
7.104 $a = 7.50$ ft; $b = 17.50$ ft.
7.107 (a) 2770 N. (b) 75.14 m.
7.109 (a) $66,900$ kips. (b) 4353 ft.
7.110 3.75 ft.
7.111 (a) $56,400$ kips. (b) 4284 ft.
7.112 (a) 6.75 m. (b) $T_{AB} = 615$ N; $T_{BC} = 600$ N.

7.114 (a) $\sqrt{3L\Delta/8}$. (b) 12.25 ft.
7.115 $h = 27.6$ mm; $\theta_A = 25.5°$; $\theta_C = 27.6°$.
7.116 (a) 4.05 m. (b) 6.41 m. (c) $A_x = 5890$ N $\leftarrow$, $A_y = 5300$ N $\uparrow$.
7.117 (a) $58,900$ kips, (b) $29.2°$.
7.118 (a) 16.00 ft to the left of B. (b) 2000 lb.
7.125 $Y = h[1 - \cos(\pi x/L)]$; $T_{\min} = w_0L^2/h\pi^2$;
$T_{\max} = (w_0L/\pi)\sqrt{(L^2/h^2\pi^2)} + 1$
7.127 (a) 9.89 m. (b) 60.3 N.
7.128 (a) 495 ft. (b) 1125 lb.
7.129 (a) 35.6 m. (b) 49.2 kg.
7.130 199.5 ft.
7.133 (a) 5.89 m. (b) 10.89 N $\rightarrow$.
7.134 10.05 ft
7.135 (a) 4.22 ft. (b) $80.3°$.
7.136 (a) 30.2 m. (b) 56.6 kg.
7.139 31.8 N.
7.140 29.8 N.
7.143 (a) $a = 79.0$ ft; $b = 60.0$ ft. (b) 103.9 ft.
7.144 (a) $a = 65.8$ ft; $b = 50.0$ ft. (b) 86.6 ft.
7.145 119.1 N $\rightarrow$.
7.146 177.6 N $\rightarrow$.
7.147 3.50 ft.
7.148 5.71 ft.
7.151 0.394 m and 10.97 m
7.152 0.1408.
7.153 (a) 0.338. (b) $56.5°$; $0.755\ wL$.
7.154 $\mathbf{F} = 625$ N $\downarrow$; $\mathbf{V} = 120.0$ N $\leftarrow$; $\mathbf{M} = 27.0$ N · m $\uparrow$ (on CJ).
7.156 $\mathbf{F} = 23.6$ lb $\measuredangle\ 76.0°$; $\mathbf{V} = 29.1$ lb $\measuredangle\ 14.04°$;
$\mathbf{M} = 540$ lb · in. $\downarrow$ (on BJ).
7.157 $\mathbf{F} = 200$ N $\measuredangle\ 36.9°$; $\mathbf{V} = 120.0$ N $\measuredangle\ 53.1°$; $\mathbf{M} = 120$ N · m$\uparrow$ (on BJ).
7.158 (a) 40.0 kips. (b) 40.0 kip · ft.
7.161 (b) 12.00 kip · ft, 6.00 ft from A.
7.163 (a) 2.28 m. (b) $\mathbf{D}_x = 13.67$ kN $\rightarrow$; $\mathbf{D}_y = 7.80$ kN $\uparrow$.
(c) 15.94 kN.
7.164 (a) 138.1 m. (b) 602 N.
7.165 (a) 56.3 ft. (b) 2.36 lb/ft.
7.C1 (a) $M_D = +39.8$ kN · m. (b) $M_D = +14.00$ kip · ft.
(c) $M_D = +1800$ lb · in.
7.C3 $a = 1.923$ m; $M_{\max} = 37.0$ kN · m at 4.64 from A.
7.C4 (b) $M_{\max} = 5.42$ kip · ft when $x = 8.5$ ft and 11.5 ft.
7.C8 $c/L = 0.300$; $h/L = 0.5225$; $s_{AB}/L = 1.532$; $T_0/wL = 0.300$;
$T_{\max}/wL = 0.823$.

CHAPTER 8

8.1 Equilibrium; $\mathbf{F} = 172.6$ N $\searrow$.
8.2 Block moves; $\mathbf{F} = 279$ N $\nwarrow$.
8.3 Block moves; $\mathbf{F} = 31.0$ lb. $\searrow$.
8.4 Equilibrium; $\mathbf{F} = 23.5$ lb $\searrow$.
8.5 (a) 74.8 lb. (b) 59.7 lb. (c) 6.76 lb.
8.6 (a) 170.5 N. (b) $14.04°$.
8.8 (a) 403 N. (b) 229 N.
8.10 143.0 N $\leq P \leq 483$ N.
8.11 $31.0°$.
8.12 $53.5°$.
8.13 (a) 353 N $\leftarrow$. (b) 196.2 N $\leftarrow$.
8.14 (a) 275 N $\leftarrow$. (b) 196.2 N $\leftarrow$.
8.17 (a) 36.0 lb $\rightarrow$. (b) 30.0 lb $\rightarrow$. (c) 12.86 lb $\rightarrow$.
8.18 (a) 36.0 lb $\rightarrow$. (b) 40.0 in.
8.19 8.34 lb.
8.20 7.50 lb.

8.21 151.5 N · m.

8.22 1.473 kN.

8.23 $6.35 \le L/a \le 10.81$.

8.25 0.208.

8.27 664 N ↓.

8.29 (*a*) Plate in equilibrium. (*b*) Plate moves downward.

8.30 10.00 lb < P < 36.7 lb.

8.32 0.860.

8.34 0.0533.

8.35 (*a*) 1.333. (*b*) 1.192. (*c*) 0.839.

8.36 (*b*) 2.69 lb.

8.37 0.225.

8.39 $168.4 \text{ N} \le P \le 308 \text{ N}$.

8.40 $9.38 \text{ N} \cdot \text{m} \le M \le 15.01 \text{ N} \cdot \text{m}$

8.41 135.0 lb.

8.43 (*a*) System slides; $P = 62.8$ N. (*b*) System rotates about B; $P = 73.2$ N.

8.44 35.8°.

8.45 20.5°.

8.46 1.225 W.

8.47 $46.4° \le \theta \le 52.4°$ and $67.6° \le \theta \le 79.4°$.

8.48 (*a*) 620 N ←. (*b*) $\mathbf{B}_x = 1390$ N ←; $\mathbf{B}_y = 1050$ N ↓.

8.49 (*a*) 234 N →. (*b*) $\mathbf{B}_x = 1824$ N ←; $\mathbf{B}_y = 1050$ N ↓.

8.52 313 lb →.

8.53 297 lb →.

8.54 9.86 kN ←.

8.55 9.13 N ←.

8.56 (*a*) 28.1°. (*b*) 728 N ∡ 14.04°.

8.57 (*a*) 62.7 lb. (*b*) 62.7 lb.

8.59 67.4 N.

8.60 1.400 lb.

8.62 (*a*) 197.0 lb →. (*b*) Base will not move.

8.63 (*a*) 280 lb ←. (*b*) Base moves.

8.64 (*b*) 283 N ←.

8.65 0.442.

8.66 0.1103.

8.67 0.1013.

8.71 1068 N · m.

8.72 153.1 lb · in.

8.73 41.4 lb · in.

8.75 4.18 N · m.

8.77 (*a*) 0.238. (*b*) 218 N ↓.

8.78 4.70 kips.

8.79 450 N.

8.80 412 N.

8.81 344 N.

8.82 376 N.

8.84 $T_{AB} = 77.5$ lb; $T_{CD} = 72.5$ lb. $T_{EF} = 67.8$ lb.

8.86 (*a*) 4.80 kN. (*b*) 1.375°.

8.88 22.0 lb ←.

8.89 1.948 lb ↓.

8.90 18.01 lb ←.

8.92 0.1670.

8.93 3.75 lb.

8.98 10.87 lb.

8.99 0.0600 in.

8.100 154.4 N.

8.101 300 mm.

8.102 (*a*) 1.288 kN. (*b*) 1.058 kN.

8.103 $73.0 \text{ lb} \le P \le 1233 \text{ lb}$.

8.104 (*a*) 0.329. (*b*) 2.67 turns.

8.105 (*a*) 22.8 kg. (*b*) 291 N.

8.106 (*a*) 109.7 kg. (*b*) 828 N.

8.109 44.9 N · m ↰.

8.110 (*a*) $T_A = 8.40$ lb; $T_B = 19.60$ lb. (*b*) 0.270.

8.111 (*a*) $T_A = 11.13$ lb; $T_B = 20.9$ lb. (*b*) 91.3 lb · in. ↲.

8.112 35.1 N · m.

8.113 (*a*) 27.0 N · m. (*b*) 675 N.

8.114 (*a*) 39.0 N · m. (*b*) 844 N.

8.117 4.49 in.

8.118 (*a*) 11.66 kg. (*b*) 38.6 kg. (*c*) 34.4 kg.

8.119 (*a*) 9.46 kg. (*b*) 167.2 kg. (*c*) 121.0 kg.

8.120 (*a*) 10.39 lb. (*b*) 58.5 lb.

8.121 (*a*) 28.9 lb. (*b*) 28.9 lb.

8.124 5.97 N.

8.125 9.56 N.

8.126 0.350.

8.128 (*a*) 30.3 lb · in. ↰. (*b*) 3.78 lb ↓.

8.129 (*a*) 17.23 lb · in. ↲. (*b*) 2.15 lb ↑.

8.133 (*a*) 51.0 N · m. (*b*) 875 N.

8.134 Block moves; $\mathbf{F} = 103.5$ N ↖.

8.136 (*a*) 0.300 Wr. (*b*) 0.349 Wr.

8.137 (*a*) 136.4°. (*b*) 0.928 W.

8.139 0.750.

8.140 $-46.8 \text{ N} \le P \le 34.3 \text{ N}$.

8.141 (*a*) $\mathbf{P} = 56.6$ lb ←. (*b*) $\mathbf{B}_x = 82.6$ lb ←; $\mathbf{B}_y = 96.0$ lb ↓.

8.143 169.7 lb · in.

8.144 0.226.

8.C1 *x = 500 mm:* 63.3 N; $P_{max} = 67.8$ N at $x = 355$ mm.

8.C2 $W_B = 10 \text{ lb}: \theta = 46.4°; W_B = 70 \text{ lb}: \theta = 21.3°$.

8.C3 $\mu_A = 0.25: M = 0.0603$ N · m.

8.C4 $\theta = 30°: 1.336 \text{ N} \cdot \text{m} \le M_A \le 2.23 \text{ N} \cdot \text{m}$.

8.C5 $\theta = 60°: \mathbf{P} = 16.40$ lb ↓; $R = 5.14$ lb.

8.C6 $\theta = 20°: 10.39$ N · m.

8.C7 $\theta = 20°: 30.3$ lb; 13.25 lb.

8.C8 (*a*) $x_0 = 0.600L; x_m = 0.604L; \theta_1 = 5.06°$. (*b*) $\theta_2 = 55.4°$.

CHAPTER 9

9.1 $a^3(h_1 + 3h_2)/12$.

9.2 $3a^3b/10$.

9.3 $ha^3/5$.

9.4 $4a^3b/21$.

9.6 $ab^3/6$.

9.8 $4ab^3/13$.

9.9 $ab^3/28$.

9.10 $(ab^3/3)/(3n + 1)$.

9.11 $0.1056 \, ab^3$.

9.12 $a^3b/20$.

9.15 $3ab^3/35; b\sqrt{9/35}$.

9.16 $\pi ab^3/8; b/2$.

9.17 $3a^3b/35; a\sqrt{9/35}$.

9.18 $\pi a^3b/8; a/2$.

9.21 $43a^4/48; 0.773a$.

9.22 $4ab(a^2 + 4b^2)/3; \sqrt{(a^2 + 4b^2)/3}$.

9.23 $64 \, a^4/15; 1.265 \, a$.

9.25 (*a*) $\pi(R_2^4 - R_1^4)/4$. (*b*) $I_x = I_y = \pi(R_2^4 - R_1^4)/8$.

9.26 (*b*) $-10.56\%; -2.99\%; -0.1248\%$.

9.28 $bh (12h^2 + b^2)/48; \sqrt{(12h^2 + b^2)/24}$.

9.31 $390 \times 10^3 \text{ mm}^4; 21.9$ mm.

9.32 $46.0 \text{ in}^4; 1.599$ in.

9.33 $64.3 \times 10^3 \text{ mm}^4; 8.87$ mm.

9.34 $46.5 \text{ in}^4; 1.607$ in.

9.37 $\bar{I}_x = 150.0 \text{ in}^4; \bar{I}_y = 300 \text{ in}^4$.

9.39 $A = 4000 \text{ mm}^2; \bar{I} = 500 \times 10^3 \text{ mm}^4$.

9.40 $46.2 \times 10^6 \text{ mm}^4$.

9.41 $\bar{I}_x = 1.874 \times 10^6 \text{ mm}^4; \bar{I}_y = 5.82 \times 10^6 \text{ mm}^4$.

9.42 $\bar{I}_x = 479 \times 10^3 \text{ mm}^4; \bar{I}_y = 149.7 \times 10^3 \text{ mm}^4$.

9.43 $\bar{I}_x = 191.3 \text{ in}^4; \bar{I}_y = 75.2 \text{ in}^4$.

9.44 $\bar{I}_x = 18.13 \text{ in}^4; \bar{I}_y = 4.51 \text{ in}^4$.

9.47 (a) $11.57 \times 10^6 \text{ mm}^4$. (b) $7.81 \times 10^6 \text{ mm}^4$.

9.48 (a) $3.13 \times 10^6 \text{ mm}^4$. (b) $2.41 \times 10^6 \text{ mm}^4$.

9.49 $\bar{I}_x = 186.7 \times 10^6 \text{ mm}^4; \bar{k}_x = 118.6 \text{ mm}; \bar{I}_y = 167.7 \times 10^6 \text{ mm}^4;$
$\bar{k}_y = 112.4 \text{ mm}$.

9.50 $\bar{I}_x = 44.5 \text{ in}^4; \bar{k}_x = 2.16 \text{ in.}; \bar{I}_y = 27.7 \text{ in}^4; \bar{k}_y = 1.709 \text{ in.}$

9.51 $\bar{I}_x = 254 \text{ in}^4; \bar{k}_x = 4.00 \text{ in.}; \bar{I}_y = 102.1 \text{ in}^4; \bar{k}_y = 2.54 \text{ in.}$

9.52 $\bar{I}_x = 260 \times 10^6 \text{ mm}^4; \bar{k}_x = 144.6 \text{ mm}; \bar{I}_y = 17.53 \text{ mm}^4;$
$\bar{k}_y = 37.6 \text{ mm}$.

9.54 $\bar{I}_x = 745 \times 10^6 \text{ mm}^4; \bar{I}_y = 91.3 \times 10^6 \text{ mm}^4$.

9.55 $\bar{I}_x = 3.55 \times 10^6 \text{ mm}^4; \bar{I}_y = 49.8 \times 10^6 \text{ mm}^4$.

9.57 $h/2$.

9.58 $(a + 3b) h/(2a + 4b)$.

9.59 $3\pi b/16$.

9.60 $4h/7$.

9.63 $5a/8$.

9.64 80.0 mm.

9.67 $a^4/2$.

9.68 $b^2 h^2/8$.

9.69 $a^2 b^2/16$.

9.71 $-1.760 \times 10^6 \text{ mm}^4$.

9.72 $2.40 \times 10^6 \text{ mm}^4$.

9.74 -0.380 in^4.

9.75 $471 \times 10^3 \text{ mm}^4$.

9.76 -9010 in^4.

9.78 $1.165 \times 10^6 \text{ mm}^4$.

9.79 (a) $\bar{I}_{x'} = 0.482a^4; \bar{I}_{y'} = 1.482a^4; \bar{I}_{x'y'} = -0.589a^4$.
(b) $\bar{I}_{x'} = 1.120a^4; \bar{I}_{y'} = 0.843a^4; \bar{I}_{x'y'} = 0.760a^4$.

9.80 $\bar{I}_{x'} = 2.12 \times 10^6 \text{ mm}^4; \bar{I}_{y'} = 8.28 \times 10^6 \text{ mm}^4;$
$\bar{I}_{x'y'} = -0.532 \times 10^6 \text{ mm}^4$.

9.81 $\bar{I}_{x'} = 1033 \text{ in}^4; \bar{I}_{y'} = 2020 \text{ in}^4; \bar{I}_{x'y'} = -873 \text{ in}^4$.

9.83 $\bar{I}_{x'} = 0.236 \text{ in}^4; \bar{I}_{y'} = 1.244 \text{ in}^4; \bar{I}_{x'y'} = 0.1132 \text{ in}^4$.

9.85 $20.2°; 1.754a^4; 0.209a^4$.

9.86 $25.1°; \bar{I}_{max} = 8.32 \times 10^6 \text{ mm}^4; \bar{I}_{min} = 2.08 \times 10^6 \text{ mm}^4$.

9.87 $29.7°; 2530 \text{ in}^4; 524 \text{ in}^4$.

9.89 $-23.7°$ and $66.3°; 1.257 \text{ in}^4; 0.224 \text{ in}^4$.

9.91 (a) $\bar{I}_{x'} = 0.482a^4; \bar{I}_{y'} = 1.482a^4; \bar{I}_{x'y'} = -0.589a^4$.
(b) $\bar{I}_{x'} = 1.120a^4; \bar{I}_{y'} = 0.843a^4; 0.760a^4$.

9.92 $\bar{I}_{x'} = 2.12 \times 10^6 \text{ mm}^4; \bar{I}_{y'} = 8.28 \times 10^6 \text{ mm}^4;$
$\bar{I}_{x'y'} = -0.532 \times 10^6 \text{ mm}^4$.

9.93 $\bar{I}_{x'} = 1033 \text{ in}^4; \bar{I}_{y'} = 2020 \text{ in}^4; \bar{I}_{x'y'} = -873 \text{ in}^4$.

9.95 $\bar{I}_{x'} = 0.236 \text{ in}^4; \bar{I}_{y'} = 1.244 \text{ in}^4; \bar{I}_{x'y'} = 0.1132 \text{ in}^4$.

9.97 $20.2°; 1.754a^4; 0.209a^4$.

9.98 $25.1°$ counterclockwise at $C; I_{max} = 8.32 \times 10^6 \text{ mm}^4;$
$\bar{I}_{min} = 2.08 \times 10^6 \text{ mm}^4$.

9.99 $-33.4°; 22.1 \times 10^3 \text{ in}^4; 2490 \text{ in}^4$.

9.100 $29.7°; 2530 \text{ in}^4; 524 \text{ in}^4$.

9.103 (a) -1.146 in^4. (b) $29.1°$ clockwise. (c) 3.39 in^4.

9.104 $23.8°$ clockwise; $0.524 \times 10^6 \text{ mm}^4; 0.0917 \times 10^6 \text{ mm}^4$.

9.105 $19.54°$ counterclockwise; $4.34 \times 10^6 \text{ mm}^4; 0.647 \times 10^6 \text{ mm}^4$.

9.106 (a) $25.3°$. (b) $1459 \text{ in}^4; 40.5 \text{ in}^4$.

9.107 (a) $88.0 \times 10^6 \text{ mm}^4$. (b) $96.3 \times 10^6 \text{ mm}^4; 39.7 \times 10^6 \text{ mm}^4$.

9.111 (a) $\bar{I}_{AA'} = \bar{I}_{BB'} = ma^2/24$. (b) $ma^2/12$.

9.112 (a) $5mb^2/4$. (b) $5m (a^2 + b^2)/4$.

9.113 (a) $0.0699 \, ma^2$. (b) $0.320 \, ma^2$.

9.114 (a) $25 \, mr_2^2/64$. (b) $0.1522 \, mr_2^2$.

9.117 (a) $ma^2/3$. (b) $3ma^2/2$.

9.118 (a) $7ma^2/6$. (b) $ma^2/2$.

9.119 $m(3a^2 + 4L^2)/12$.

9.120 $1.329 \, mh^2$.

9.121 (a) $0.241 \, mh^2$. (b) $m(3a^2 + 0.1204 \, h^2)$.

9.122 $m(b^2 + 3h^2)/5$.

9.124 $m(a^2 + 3h^2)/6$.

9.126 $I_x = I_y = ma^2/4; I_z = ma^2/2$.

9.127 $1.160 \times 10^{-6} \text{ lb} \cdot \text{ft} \cdot \text{s}^2; 0.341 \text{ in.}$

9.128 $837 \times 10^{-9} \text{ kg} \cdot \text{m}^2; 6.92 \text{ mm}$.

9.130 $ma^2/2; a/\sqrt{2}$.

9.131 (a) 2.30 in. (b) $20.6 \times 10^{-3} \text{ lb} \cdot \text{ft} \cdot \text{s}^2; 2.27 \text{ in.}$

9.132 (a) $\pi p l^2 \, [6a^2 t(5a^2/3l^2 + 2a/l + 1) + d^2 l/4]$. (b) 0.1851.

9.133 (a) 27.5 mm to the right of A. (b) 32.0 mm.

9.135 $I_x = 0.877 \text{ kg} \cdot \text{m}^2; I_y = 1.982 \text{ kg} \cdot \text{m}^2; I_z = 1.652 \text{ kg} \cdot \text{m}^2$.

9.136 $I_x = 175.5 \times 10^{-3} \text{ kg} \cdot \text{m}^2; I_y = 309 \times 10^{-3} \text{ kg} \cdot \text{m}^2;$
$I_z = 154.4 \times 10^{-3} \text{ kg} \cdot \text{m}^2$.

9.138 $I_x = 745 \times 10^{-6} \text{ lb} \cdot \text{ft} \cdot \text{s}^2; I_y = 896 \times 10^{-6} \text{ lb} \cdot \text{ft} \cdot \text{s}^2;$
$I_z = 304 \text{ lb} \cdot \text{ft} \cdot \text{s}^2$.

9.139 $I_x = 344 \times 10^{-6} \text{ lb} \cdot \text{ft} \cdot \text{s}^2; I_y = 132.1 \times 10^{-6} \text{ lb} \cdot \text{ft} \cdot \text{s}^2;$
$I_z = 453 \times 10^{-6} \text{ lb} \cdot \text{ft} \cdot \text{s}^2$.

9.141 (a) $13.99 \times 10^{-3} \text{ kg} \cdot \text{m}^2$. (b) $20.6 \times 10^{-3} \text{ kg} \cdot \text{m}^2$.
(c) $14.30 \times 10^{-3} \text{ kg} \cdot \text{m}^2$.

9.142 $I_x = 28.3 \times 10^{-3} \text{ kg} \cdot \text{m}^2; I_y = 183.8 \times 10^{-3} \text{ kg} \cdot \text{m}^2;$
$k_x = 42.9 \text{ mm}; k_y = 109.3 \text{ mm}$.

9.143 $0.1785 \text{ lb} \cdot \text{ft} \cdot \text{s}^2$.

9.145 (a) $26.4 \times 10^{-3} \text{ kg} \cdot \text{m}^2$. (b) $31.2 \times 10^{-3} \text{ kg} \cdot \text{m}^2$.
(c) $8.58 \times 10^{-3} \text{ kg} \cdot \text{m}^2$.

9.147 $I_x = 0.0392 \text{ lb} \cdot \text{ft} \cdot \text{s}^2; I_y = 0.0363 \text{ lb} \cdot \text{ft} \cdot \text{s}^2;$
$I_z = 0.0304 \text{ lb} \cdot \text{ft} \cdot \text{s}^2$.

9.148 $I_x = 0.323 \text{ kg} \cdot \text{m}^2; I_y = I_z = 0.419 \text{ kg} \cdot \text{m}^2$.

9.149 $I_{xy} = 2.50 \times 10^{-3} \text{ kg} \cdot \text{m}^2; I_{yz} = 4.06 \times 10^{-3} \text{ kg} \cdot \text{m}^2;$
$I_{zx} = 8.81 \times 10^{-3} \text{ kg} \cdot \text{m}^2$.

9.150 $I_{xy} = 286 \times 10^{-6} \text{ kg} \cdot \text{m}^2; I_{yz} = I_{zx} = 0$.

9.151 $I_{xy} = -1.726 \times 10^{-3} \text{ lb} \cdot \text{ft} \cdot \text{s}^2; I_{yz} = 0.507 \times 10^{-3} \text{ lb} \cdot \text{ft} \cdot \text{s}^2;$
$I_{zx} = -2.12 \times 10^{-3} \text{ lb} \cdot \text{ft} \cdot \text{s}^2$.

9.152 $I_{xy} = -538 \times 10^{-6} \text{ lb} \cdot \text{ft} \cdot \text{s}^2; I_{yz} = -171.4 \times 10^{-6} \text{ lb} \cdot \text{ft} \cdot \text{s}^2;$
$I_{zx} = 1120 \times 10^{-6} \text{ lb} \cdot \text{ft} \cdot \text{s}^2$.

9.155 $I_{xy} = -8.04 \times 10^{-3} \text{ kg} \cdot \text{m}^2; I_{yz} = 12.90 \times 10^{-3} \text{ kg} \cdot \text{m}^2;$
$I_{zx} = 94.0 \times 10^{-3} \text{ kg} \cdot \text{m}^2$.

9.156 $I_{xy} = 0; I_{yz} = 48.3 \times 10^{-6} \text{ kg} \cdot \text{m}^2;$
$I_{zx} = -4.43 \times 10^{-3} \text{ kg} \cdot \text{m}^2$.

9.157 $I_{xy} = 47.9 \times 10^{-6} \text{ kg} \cdot \text{m}^2; I_{yz} = 102.1 \times 10^{-6} \text{ kg} \cdot \text{m}^2;$
$I_{zx} = 64.1 \times 10^{-6} \text{ kg} \cdot \text{m}^2$.

9.158 $I_{xy} = -m' R_1^3/2; I_{yz} = m' R_1^3/2; I_{zx} = -m' R_2^3/2$.

9.159 $I_{xy} = wa^3 (1 - 5\pi)/g; I_{yz} = -11\pi \, wa^3/g;$
$I_{zx} = 4wa^3 (1 + 2\pi)/g$.

9.160 $I_{xy} = -11wa^3/g; I_{yz} = wa^3(\pi + 6)/2g; I_{zx} = -wa^3/4g$.

9.162 (a) $mac/20$. (b) $I_{xy} = mab/20; I_{yz} = mbc/20$.

9.165 $18.17 \times 10^{-3} \text{ kg} \cdot \text{m}^2$.

9.166 $11.81 \times 10^{-3} \text{ kg} \cdot \text{m}^2$.

9.167 $5Wa^2/18g$.

9.168 $4.41 \, rta^4/g$.

9.169 $281 \times 10^{-3} \text{ kg} \cdot \text{m}^2$.

9.170 $0.354 \text{ kg} \cdot \text{m}^2$.

9.173 (a) $1/\sqrt{3}$. (b) $\sqrt{7/12}$.

9.174 (a) $b/a = 2; c/a = 2$. (b) $b/a = 1; c/a = 0.5$.

9.175 (a) 2. (b) $\sqrt{2/3}$.

9.179 (a) $K_1 = 0.363ma^2; K_2 = 1.583ma^2; K_3 = 1.720ma^2$.
(b) $(\theta_x)_1 = (\theta_z)_1 = 49.7°, (\theta_y)_1 = 113.7°; (\theta_x)_2 = 45.0°,$
$(\theta_y)_2 = 90.0°, (\theta_z)_2 = 135.0°; (\theta_x)_3 = (\theta_z)_3 = 73.5°, (\theta_y)_3 = 23.7°$.

9.180 (a) $K_1 = 14.30 \times 10^{-3}$ kg · m^2; $K_2 = 13.96 \times 10^{-3}$ kg · m^2; $K_3 = 20.6 \times 10^{-3}$ kg · m^2.
(b) $(\theta_x)_1 = (\theta_y)_1 = 90.0°$, $(\theta_z)_1 = 0$; $(\theta_x)_2 = 3.42°$, $(\theta_y)_2 = 86.6°$, $(\theta_z)_2 = 90.0°$; $(\theta_x)_3 = 93.4°$, $(\theta_y)_3 = 3.43°$, $(\theta_z)_3 = 90.0°$.

9.182 (a) $K_1 = 0.1639 Wa^2/g$; $K_2 = 1.054 Wa^2/g$; $K_3 = 1.115 Wa^2/g$.
(b) $(\theta_x)_1 = 36.7°$, $(\theta_y)_1 = 71.6°$, $(\theta_z)_1 = 59.5°$; $(\theta_x)_2 = 74.9°$, $(\theta_y)_2 = 54.5°$, $(\theta_z)_2 = 140.5°$; $(\theta_x)_3 = 57.5°$, $(\theta_y)_3 = 138.8°$, $(\theta_z)_3 = 112.4°$.

9.183 (a) $K_1 = 2.26\gamma t a^4/g$; $K_2 = 17.27\gamma t a^4/g$; $K_3 = 19.08\gamma t a^4/g$.
(b) $(\theta_x)_1 = 85.0°$, $(\theta_y)_1 = 36.8°$, $(\theta_z)_1 = 53.7°$; $(\theta_x)_2 = 81.7°$, $(\theta_y)_2 = 54.7°$; $(\theta_z)_2 = 143.4°$; $(\theta_x)_3 = 9.70°$, $(\theta_y)_3 = 99.0°$, $(\theta_z)_3 = 86.3°$.

9.185 $I_x = a^4/8$; $I_y = 3a^4/2$.
9.186 $a^3b/6$; $a/\sqrt{3}$.
9.188 $I_x = 48.9 \times 10^3$ mm^4; $I_y = 8.35 \times 10^3$ mm^4.
9.189 (a) 80.9×10^6 mm^4. (b) 57.4×10^6 mm^4.
9.191 -2.81 in^4.
9.193 (a) $5ma^2/18$. (b) $3.61ma^2$.
9.195 $I_x = 26.0 \times 10^{-3}$ kg · m^2; $I_y = 38.2 \times 10^{-3}$ kg · m^2; $I_z = 17.55 \times 10^{-3}$ kg · m^2.
9.196 $I_x = 38.1 \times 10^{-3}$ kg · m^2; $k_x = 110.7$ mm.
9.C1 $\theta = 20°$; $I_{x'} = 14.20$ in^4, $I_{y'} = 3.15$ in^4, $I_{x'y'} = -3.93$ in^4.
9.C3 (a) $\bar{I}_{x'} = 371 \times 10^3$ mm^4, $\bar{I}_{y'} = 64.3 \times 10^3$ mm^4; $\bar{k}_{x'} = 21.3$ mm, $\bar{k}_{y'} = 8.87$ mm. (b) $\bar{I}_{x'} = 40.4$ in^4, $\bar{I}_{y'} = 46.5$ in^4; $\bar{k}_{x'} = 1.499$ in., $\bar{k}_{y'} = 1.607$ in. (c) $\bar{k}_x = 2.53$ in., $\bar{k}_y = 1.583$ in. (d) $\bar{k}_x = 1.904$ in., $\bar{k}_y = 0.950$ in.
9.C5 (a) 5.99×10^{-3} kg · m^2. (b) 77.4×10^{-3} kg · m^2.
9.C6 (a) 74.0×10^{-6} lb · ft · s^2. (b) 645×10^{-6} lb · ft · s^2. (c) 208×10^{-6} lb · ft · s^2.

CHAPTER 10

10.1 270 N ↑.
10.2 60.0 lb ↓.
10.3 32.4 N · m ↙.
10.4 600 lb · in. ↙.
10.5 500 N ↑.
10.6 750 N ↑.
10.9 $Q = 3P \tan \theta$.
10.10 $Q = P [(l/a)] \cos^3 \theta - 1]$.
10.12 $Q = 2P \sin \theta/\cos (\theta/2)$.
10.14 $Q = (3P/2) \tan \theta$.
10.15 $M = Pl/2 \tan \theta$.
10.16 $M = Pl(\sin \theta + \cos \theta)$.
10.17 $M = \frac{1}{2} Wl \tan \alpha \sin \theta$.
10.18 $M = PR \csc^2 \theta$.
10.19 85.2 lb · ft ↙.
10.20 22.8 lb ⤢ 70.0°.
10.23 39.2°.
10.26 19.81° and 51.9°.
10.27 36.4°.
10.28 67.1°.
10.29 40.2°.
10.31 15.60 in.
10.32 13.20 in.
10.34 57.2°.
10.35 38.7°.
10.36 60.4°.
10.37 22.6°.
10.38 51.1°.
10.39 59.0°.

10.40 78.7°, 324°, 379°.
10.43 12.03 kN ↘.
10.44 20.4°.
10.45 2370 lb ↖.
10.46 2550 lb ↖.
10.48 300 N · m, 81.8 N · m.
10.49 $\eta = 1/(1 + \mu \cot \alpha)$.
10.50 $\eta = \tan \theta/\tan (\theta + \phi_s)$.
10.52 37.6 N, 31.6 N.
10.53 7.75 kN ↑.
10.54 $H = 1.361$ kN ↑; $M_H = 550$ N · m ↑.
10.57 0.833 in. ↓.
10.58 0.625 in. →.
10.69 $\theta = -45.0°$, unstable; $\theta = 135.0°$, stable.
10.70 $\theta = -63.4°$, unstable; $\theta = 116.6°$, stable.
10.71 (a) 0, unstable. (b) 137.8°, stable.
10.72 $\theta = 0$ and $\theta = 180.0°$, unstable; $\theta = 75.5°$ and $\theta = 284°$, stable.
10.73 59.0°, stable.
10.74 78.7°, stable; 324°, unstable; 379°, stable.
10.77 357 mm.
10.78 252 mm.
10.80 9.39° and 90.0°, stable; 34.2°, unstable.
10.81 17.11°, stable; 72.9°, unstable.
10.83 49.1°.
10.86 16.88 m.
10.87 54.8°.
10.88 37.4°.
10.89 $P < kl/2$.
10.91 $k > 6.94$ lb/in.
10.92 15.00 in.
10.93 $P < 2kL/9$.
10.94 $P < kL/18$.
10.96 $P < 160.0$ N.
10.98 $P < 764$ N.
10.100 (a) P < 10.00 lb. (b) P < 20.0 lb.
10.101 120.0 lb →.
10.102 1200 lb · in. ↑.
10.103 (a) 60.0 N C, 8.00 mm ↓. (b) 300 N C, 40.0 mm ↓.
10.105 $M = 7Pa \cos \theta$.
10.107 19.40°.
10.108 7.13 in.
10.110 $P < k (l - a)^2/2l$.
10.112 (a) 22.0°. (b) 30.6°.
10.C1 $\theta = 60°$: 2.42 in.; $\theta = 120°$: 1.732 in.; $(M/P)_{max} = 2.25$ in. at $\theta = 73.7°$.
10.C2 $\theta = 60°$: 171.1 N C. For $32.5° \le \theta \le 134.3°$, $|F| \le 400$ N.
10.C3 $\theta = 60°$; 296 N T. For $\theta \le 125.7°$, $|F| \le 400$ N.
10.C4 (b) $\theta = 60°$, datum at C: V = −294 in · lb. (c) 34.2°, stable; 90°, unstable; 145.8°, stable.
10.C5 (b) $\theta = 50°$, datum at E: V = 100.5 J. $dV/d\theta = 22.9$ J. (c) $\theta = 0$, unstable; 30.4°.
10.C6 (b) $\theta = 60°$, datum at B: 30.0 J. (c) $\theta = 0$, unstable; 41.4°, stable.
10.C7 (b) $\theta = 60°$, datum at $\theta = 0$: −37.0 J. (c) 52.2°, stable.

CHAPTER 11

11.1 11.00 in., −8.00 in/s, −8.00 in/s^2.
11.2 1.000s, 15.00 ft, −6.00 ft/s^2; 2.00 s, 14.00 ft, 6.00 ft/s^2.
11.3 (a) 102.9 mm, −35.6 mm/s, −11.40 mm/s^2. (b) −36.1 mm/s, 72.1 mm/s^2.

11.4 (a) 0, 960 mm/s² →, 9220 mm/s² or 9.22 m/s² ←.
(b) 14.16 mm ←, 87.9 mm/s →, 3110 mm/s² or 3.11 m/s² →.

11.5 0.667 s, 0.259m, −8.56 m/s.

11.7 (a) 1.000 s and 4.00 s. (b) 1.500 m, 24.5 m.

11.9 (a) 77.5 ft/s. (b) 7.75 s.

11.10 −33.0 in/s, 2.00 in., 87.7 in.

11.11 $x(t) = t^4/108 + 10t + 24$ m.
$v(t) = t^3/27 + 10$ m/s.

11.12 (a) 6.00 m/s⁴. (b) $a = 6t^2$, $v = 2t^3 - 8$, $x = t^4/2 - 8t + 8$.

11.15 800 m/s² ↑.

11.16 (a) -2.43×10^6 ft/s². (b) 1.366×10^{-3} s.

11.17 (a) 5.89 ft/s. (b) 1.772 ft.

11.18 167.1 mm/s² ↑, 15.19 m/s² ↑.

11.21 (a) 1.250 m. (b) 0.866 s.

11.22 (a) 10.00 ft. (b) 1.833 ft/s, 0.440 ft/s².

11.23 (a) 42.0 ft. (b) 12.86 ft/s.

11.24 (a) 29.3 m/s. (b) 0.947 s.

11.25 (a) 3.33 m. (b) 2.22 s. (c) 1.667 s.

11.26 (a) 0.1457 s/m. (b) 145.2 m. (c) 6.86 m/s.

11.27 (a) −0.0525 m/s². (b) 6.17 s.

11.28 (a) 7.15 mi. (b) -275×10^{-6} ft/s². (c) 49.9 min.

11.31 (a) $2.36\,v_0 T$, $\pi\,v_0/T$. (b) $0.363\,v_0$.

11.33 (a) 9.62 m/s ↑. (b) 29.6. m/s ↓.

11.34 (a) −0.417 m/s². (b) 18.00 km/h.

11.35 (a) 5.28 ft/s². (b) 8.33 s.

11.36 (a) 252 ft/s. (b) 1076 ft.

11.39 (a) $a_A = -2.10$ m/s², $a_B = 2.06$ m/s².
(b) 2.59 s before A reaches the exchange zone.

11.40 (a) 1.563 m/s². (b) 3.13 m/s².

11.41 (a) 1760 ft. (b) 28.6 mi/h.

11.42 (a) 15.05 s, 734 ft from the ititial point of A.
(b) A: 42.5 mi/h. B: 23.7 mi/h.

11.43 (a) $\mathbf{a}_A = 0.767$ ft/s² ←, $\mathbf{a}_B = 0.834$ ft/s² →.
(b) 20.7 s. (c) 51.8 mi/h.

11.44 (a) 1.330 s. (b) 4.68 m below the man.

11.47 (a) 8.00 m/s ↑. (b) 4.00 m/s ↑. (c) 12.00 m/s ↑. (d) 8.00 m/s ↑.

11.48 (a) $\mathbf{a}_E = 2.40$ ft/s² ↑, $\mathbf{a}_C = 4.80$ ft/s² ↓.
(b) 12.00 ft/s ↑.

11.49 (a) 2.00 m/s ↑. (b) 2.00 m/s ↓. (c) 8.00 m/s ↑.

11.50 (a) $\mathbf{a}_A = 24.0$ ft/s² →, $\mathbf{a}_B = 8.00$ ft/s² ↓.
(b) 16.00 ft/s ↓, 16.00 ft.

11.51 (a) 200 mm/s →. (b) 600 mm/s →.
(c) 200 mm/s ←. (d) 400 mm/s →.

11.52 (a) $\mathbf{a}_A = 13.33$ mm/s² ←, $\mathbf{a}_B = 20.0$ mm/s² ←.
(b) 13.33 mm/s² →. (c) 70.0 mm/s →. 440 mm →.

11.55 (a) 1.500 s. (b) 3.00 s. (c) 10.00 in. ↑.

11.56 (a) 1.000 s. (b) 3.00 in. ↓.

11.57 (a) $\mathbf{a}_A = 240$ mm/s² ↓, $\mathbf{a}_B = 345$ mm/s² ↑.
(b) $(\mathbf{v}_A)_0 = 43.3$ mm/s ↑, $(v_C)_0 = 130.0$ mm/s →.
(c) 728 mm →.

11.58 (a) 10.00 mm/s →. (b) $\mathbf{a}_A = 2.00$ mm/s² ↑,
$\mathbf{a}_C = 6.00$ mm/s² →. (c) 175.0 mm ↑.

11.63 (a) 10 s to 26 s, $a = -5.00$ m/s²;
41 s to 46 s, $a = 3.00$ m/s²; otherwise $a = 0$.
(b) 1383 m. (c) 9.00 s, 49.5 s.

11.64 (a) Same as Prob. 11.63. (b) 420 m. (c) 10.69 s, 40.0 s.

11.65 10.50 s.

11.66 (a) 44.8 s. (b) 103.3 m/s².

11.69 (a) 0.600 s. (b) 0.200 m/s, 2.84 m.

11.70 (a) 60.0 m/s, 1194 m. (b) 59.3 m/s.

11.71 (a) A: 52.2 s, B: 52.0 s. (b) 1.879 m.

11.72 9.39 s.

11.73 8.54 s, 58.3 mi/h.

11.74 77.5 ft.

11.75 5.67 s.

11.78 (a) 18.00 s. (b) 178.8 m. (c) 34.7 km/h.

11.79 (a) 5.01 min. (b) 19.18 mi/h.

11.80 (a) 2.00 s. (b) 1.200 ft/s, 0.600 ft/s.

11.83 (a) 2.96 s. (b) 224 ft.

11.84 (a) 163.0 in/s². (b) 114.3 in/s².

11.89 (a) 6.28 m/s ⦨ 37.2°. (b) 7.49 m.

11.90 (a) 67.1 mm/s ∡ 63.4°, 256 mm/s² ⦩ 69.4°.
(b) 8.29 mm/s ∡ 36.2°, 336 mm/s² ⦩ 86.6°.

11.91 (a) $(-112.57$ in/s$)\mathbf{i}$, $(-39.5$ in/s²$)\mathbf{j}$. (b) $y = x^2/8 - 1$.

11.92 (a) max: 15.00 ft/s, min: 5.00 ft/s
(b) min: $t = 2\pi N$ s, $x = 20\pi N$ ft, $y = 5$ ft, $v_x = 5$ ft/s,
$v_y = 0$, $\theta = 0$.
max: $t = (2N + 1)\,\pi$ s, $x = 20\pi(N + 1)$ ft, $y = 15$ ft,
$v_x = 15$ ft/s, $v_y = 0$, $\theta = 0$.

11.95 $\sqrt{R^2(1 + w_n^2 t^2)} + c^2$, $Rw_n\sqrt{4 + w_n^2 t^2}$.

11.96 (a) 3.00 ft/s, 3.61 ft/s². (b) 3.82 s.

11.97 1140 ft.

11.98 (a) 330 m. (b) 149.9 m.

11.99 (a) 115.3 km/h $\le v_0 \le$ 148.0 km/h.
(b) $h = 0.788$ m, $\alpha = 6.66°$; $h = 1.068$ m, $\alpha = 4.05°$.

11.100 15.38 ft/s $< v_0 <$ 35.0 ft/s.

11.102 0.244 m $< h <$ 0.386 m.

11.103 (a) Ball clears the net. (b) 7.01 m from the net.

11.105 22.9 ft/s.

11.106 16.20 m/s $< v_0 <$ 21.0 m/s.

11.107 (a) 29.8 ft/s. (b) 29.6 ft/s.

11.108 37.7 m/s $< v_0 <$ 44.3 m/s.

11.111 (a) 10.38°. (b) 9.74°.

11.112 (a) 4.17°. (b) 285 m. (c) 15.89 s.

11.113 (a) 14.66°. (b) 0.1074 s.

11.114 (a) 4.98 m. (b) 23.8°.

11.117 17.80 ft/s ⦦ 50.9°.

11.118 $\mathbf{v}_A = 125$ mm/s ↑, $\mathbf{v}_B = 75$ mm/s ↓, $\mathbf{v}_C = 175$ mm/s ↓.

11.119 (a) 91.0 ft/s ⦩ 47.0°. (b) 364 ft ⦩ 47.0°. (c) 293 ft.

11.120 3.20 km/h ⦦ 17.8°.

11.123 (a) 7.01 in/s ⦩ 60.6°. (b) 11.69 in/s² ⦩ 60.6°.

11.124 (a) 8.53 in/s ⦦ 54.1°. (b) 6.40 in/s ⦦ 54.1°.

11.125 (a) 0.979 m. (b) 12.55 m/s ⦨ 86.5°.

11.126 (a) 0.835 mm/s² ⦦ 75°. (b) 8.35 mm/s ⦦ 75°.

11.127 (a) 5.18 ft/s ⦦ 15°. (b) 1.232 ft/s ⦦ 15°.

11.128 10.54 ft/s ⦩ 81.3°.

11.129 5.96 m/s ⦨ 82.8°.

11.131 15.79 km/h ⦨ 26.0°.

11.132 1.024 ft/s ⦦ 2.07°.

11.133 500 m.

11.134 97.6 km/h.

11.135 444 m/s².

11.136 (a) 792 ft. (b) 51.9 mi/h.

11.137 2.53 ft/s².

11.138 (a) 10.20 mm/s². (b) 25.2 s.

11.139 (a) 178.9 m. (b) 1.118 m/s².

11.141 (a) 189.5 km/h ⦨ 54.0°. (b) 21.8 m/s² ⦨ 5.3°.

11.143 (a) 27.2 ft/s ∡ 40°. (b) 13.48 ft.

11.144 (a) 2.46 m. (b) 42.7 mm.

11.145 (a) 281 m. (b) 209 m.

11.146 (a) 27.6 m. (b) 34.0 m.

11.147 (a) 0.634 m. (b) 9.07 m.

11.149 18.17 m/s ∡ 4.04° and 18.17 m/s ⦨ 4.04°.

11.151 $(R^2 + C^2)/2w_n R$.

11.152 2.50 ft.
11.153 149.8 Gm.
11.154 1425 Gm.
11.155 16 200 mi/h.
11.156 7740 mi/h.
11.159 1.606 h.
11.161 (a) $(1.624 \text{ in/s})\mathbf{e}_r - (15.56 \text{ in/s})\mathbf{e}_\theta$
(b) $(-49.9 \text{ in/s}^2)\mathbf{e}_r + (-9.74 \text{ in/s}^2)\mathbf{e}_\theta$
(c) $-3.25 \text{ in/s}^2)\mathbf{e}_r$.
11.162 (a) $(-2.50 \text{ in/s})\mathbf{e}_r + (2.50 \text{ in/s})\mathbf{e}_\theta$
(b) $(7.50 \text{ in/s}^2)\mathbf{e}_r + (-10.00 \text{ in/s}^2)\mathbf{e}_\theta$
(c) 7.07 in.
11.163 (a) $(2\pi \text{ m/s})\mathbf{e}_\theta, - (4\pi^2 \text{ m/s}^2)\mathbf{e}_r,$
(b) $-(\pi/2 \text{ m/s})\mathbf{e}_r + (\pi \text{ m/s})\mathbf{e}_\theta, -(\pi^2/2 \text{ m/s}^2)\mathbf{e}_r - (\pi^2 \text{ m/s}^2)\mathbf{e}_\theta.$
11.164 (a) $v = 2abt, a = 2ab\sqrt{1 + 4b^2t^4}.$
(b) $p = a$; The path is a circle.
11.165 (a) $\mathbf{v} = bk\mathbf{e}_\theta, \mathbf{a} = -(bk^2/2)\mathbf{e}_r.$
(b) $\mathbf{v} = 2bk\mathbf{e}_r + 2bk\mathbf{e}_\theta, \mathbf{a} = 2bk^2\mathbf{e}_r + 4bk^2\mathbf{e}_\theta.$
11.166 (a) $a = 4b\dot{\theta}^2$. (b) directed toward point A.
11.169 $\dot{r} = 370 \text{ ft/s}, \ddot{r} = 57.9 \text{ ft/s}^2, \dot{\theta} = -0.0924 \text{ rad/s},$
$\ddot{\theta} = 0.0315 \text{ rad/s}^2.$
11.170 (a) $\dot{r} = -dw/2, \dot{\theta} = w/2.$ (b) $\ddot{r} = -\sqrt{3}\, dw^2/4, \ddot{\theta} = 0.$
11.171 185.7 km/h.
11.172 61.8 mi/h, 49.7°.
11.175 $bw^2\sqrt{4 + \theta^4}/\theta^3.$
11.176 $(1 + b^2)w^2e^{b\theta}.$
11.177 $v = 2\pi\sqrt{A^2 + n^2B^2\cos^2 2\pi nt}$
$a = 4\pi^2\sqrt{A^2 + n^4B^2\sin^2 2\pi nt}$
11.179 (a) $v = \sqrt{A^2 + B^2}, a = \sqrt{(1 + 16\pi^2)A^2 + B^2}.$
(b) $v = 2\pi A, a = 4\pi^2 A.$
11.180 $\tan^{-1}[R(2 + w_n^2t^2)/c\sqrt{4 + w_n^2t^2}].$
11.181 (a) $\theta_x = 90°, \theta_y = 123.7°, \theta_z = 33.7°.$
(b) $\theta_x = 103.4°, \theta_y = 134.3°, \theta_z = 47.4°.$
11.182 (a) 1.00 s and 4.00 s. (b) 1.500 m, 24.5 m.
11.183 $A = -36.8 \text{ m}^2, k = 1.832 \text{ s}^{-2}.$
11.185 (a) 111.4 km/h $\measuredangle$ 10.50°. (b) 2.96 km.
11.187 (a) $\mathbf{a}_B = 2.00 \text{ in/s}^2 \uparrow, \mathbf{a}_C = 3.00 \text{ in/s}^2 \downarrow.$ (b) 0.667 s.
(c) 0.667 in. $\uparrow.$
11.188 (a) 38.1 m/s, 20.4 m. (b) 41.1 m/s, 29.6 m.
11.189 (a) 3.21 ft/s² $\searrow$ 22.4°. (b) 6.43 ft/s² $\searrow$ 22.4°.
11.190 15.95 ft/s².
11.191 (a) 23.4 ft/s. (b) 103.2 ft.

CHAPTER 12

12.1 (a) 844 lb. (b) 26.2 slugs.
12.2 (a) 0°: 4.987 lb, 45°: 5.000 lb, 90°: 5.013 lb.
(b) 5.000 lb. (c) 0.1554 lb · s²/ft.
12.3 2.84×10^6 kg · m/s.
12.4 (a) 66.8 N. (b) Load indicated = 73.6 N, m = 6.81 kg.
12.5 0.242 mi.
12.6 (a) 20.0 ft/s. (b) 0.0621.
12.7 0.414 m/s² $\searrow$ 15°.
12.8 (a) 110.5 km/h. (b) 85.6 km/h. (c) 69.9 km/h.
12.9 (a) 40.1 m. (b) 47.0 m.
12.10 (a) 2.22 s. (b) 3.32 m.
12.11 51.0 m.
12.12 (a) 234 m. (b) 3.33 kN (tension).
12.15 (a) (1): 10.73 ft/s² $\downarrow$, (2): 16.10 ft/s² $\downarrow$, (3): 0.749 ft/s² $\downarrow$.
(b) (1): 14.65 ft/s $\downarrow$, (2): 17.94 ft/s $\downarrow$, (3): 3.87 ft/s $\downarrow$.
(c) (1): 1.864 s, (2): 1.242 s, (3): 26.7 s.
12.16 $\mathbf{a}_A = 0.997 \text{ ft/s}^2 \measuredangle 15°, \mathbf{a}_B = 1.619 \text{ ft/s}^2 \measuredangle 15°.$

12.17 (a) 765 lb. (b) 1016 lb.
12.18 (a) 0.986 m/s² $\searrow$ 25°. (b) 51.7 N.
12.19 (a) 1.794 m/s² $\searrow$ 25°. (b) 58.2 N.
12.20 0.321 m $\rightarrow.$
12.23 $\mathbf{a}_1 = 19.53 \text{ m/s}^2 \measuredangle 65°, \mathbf{a}_2 = 4.24 \text{ m/s}^2 \nearrow 65°.$
12.24 1.598 km.
12.25 $0.347\, m_0 v_0^2/F_0.$
12.26 $x = Pt/K - kv/m.$
12.27 $\sqrt{k/m}\,(\sqrt{l^2 + x_0^2} - l)$
12.28 (a) 10.00 N. (b) 103.1 N.
12.29 (a) 8.94 ft/s² $\leftarrow$, 18.06 lb.
(b) 12.38 ft/s² $\leftarrow$, 15.38 lb. (c) Same as (b).
12.30 (a) 33.6 N. (b) $\mathbf{a}_A$ = 4.76 m/s² $\rightarrow$,
$\mathbf{a}_B$ = 3.08 m/s² $\downarrow$, $\mathbf{a}_C$ = 1.401 m/s² $\leftarrow$.
12.31 (a) 2.43 lb. (b) $\mathbf{a}_A$ = 3.14 ft/s² $\rightarrow$,
$\mathbf{a}_B$ = 0.881 m/s² $\rightarrow$, $\mathbf{a}_C$ = 5.41 m/s² $\downarrow$.
12.34 (a) 2.80 m/s² $\leftarrow$. (b) 8.32 m/s² $\searrow$ 25°.
12.35 (a) 5.94 m/s² $\searrow$ 75.6°. (b) 3.74 m/s $\searrow$ 20°.
12.36 (a) 49.9°. (b) 6.85 N.
12.37 (a) 80.4 N. (b) 2.30 m/s.
12.38 3.47 m/s.
12.39 3.01 m/s $\leq v \leq$ 3.96 m/s.
12.40 3.01 m/s $\leq v \leq$ 3.85 m/s.
12.42 9.00 ft/s $< v_C <$ 12.31 ft/s.
12.43 2.42 ft/s $< v <$ 13.85 ft/s.
12.44 (a) 122.2 lb. (b) 145.6 lb.
12.45 (a) 668 ft. (b) 120.0 lb $\uparrow.$
12.46 (a) 131.7 N. (b) 88.4 N.
12.47 (a) 4.63 m/s². (b) 1.962 m/s². (c) 0.1842 m.s².
12.48 24.1° $< \theta <$ 155.9°.
12.49 (a) 2.91 N. (b) 13.09°.
12.50 1126 N $\searrow$ 25.6°.
12.51 (a) 12.19 m/s. (b) 2290 N.
12.53 (a) 0.1858 W. (b) 10.28°.
12.55 7.67 m/s.
12.56 (a) 12.00 m/s. (b) 2.05×10^{-3} N.
12.57 0.236.
12.58 468 mm.
12.61 0.400.
12.62 (a) 0.1834. (b) left: 10.39°, right 169.6°.
12.63 (a) 2.98 ft/s. (b) left: 19.29°, right 160.7°.
12.64 $\delta = eVlL/mdv_0^2.$
12.65 $d/l > 1.054 \sqrt{eV/mv_0^2}$
12.66 (a) $F_r = -13.15$ lb, $F_\theta = 0.520$ lb.; $\theta = 0.$
(b) $F_r = -2.04$ lb, $F_\theta = 0.938$ lb; $\theta = 180°.$
12.67 (a) $F_r = -0.258$ lb, $F_\theta = -0.0504$ lb.; $\theta = 0.$
(b) $F_r = -0.00618$ lb, $F_\theta = 0.278$ lb; $\theta = -54.7°.$
12.68 2.00 s.
12.69 (a) 72.0 m/s² radically outward. (b) 1.250 N.
12.71 (a) 126.6 N. (b) 5.48 m/s² $\rightarrow$. (c) 4.75 m/s² $\downarrow$.
12.72 (a) 142.7 N. (b) 6.18 m/s² $\rightarrow$. (c) 4.10 m/s² $\downarrow$.
12.74 $v_r = v_0 \sin 2\theta/\sqrt{\cos 2\theta}, v_\theta = v_0\sqrt{\cos 2\theta}.$
12.77 (a) 0. (b) $8m\, v_0^2/r_0.$
12.78 413×10^{21} lb · s²/ft.
12.79 383×10^3 km, 238×10^3 mi.
12.80 (a) 35 800 km, 22 200 mi. (b) 3.07 km/s, 10.09×10^3 ft/s.
12.81 (b) 24.8 m/s².
12.82 (a) 1.998×10^{30} kg. (b) 276 m/s².
12.85 (a) 1684 N. (b) 2510 km. (c) 1.620 m/s².
12.86 (a) 1551 m/s. (b) −15.8 m/s.
12.87 5000 m/s.
12.88 (a) 5280 ft/s. (b) 8000 ft/s.

12.89 53 ft/s.

12.90 (a) $(a_A)_r = (a_A)_\theta = 0$. (b) 38.4 m/s². (c) 0.800 m/s.

12.91 (a) $(a_B)_r = (a_B)_\theta = 0$. (b) 61.4 ft/s². (c) 2.98 ft/s.

12.100 (a) 10.13 km/s. (b) 2.97 km/s.

12.101 1.147.

12.103 $\sqrt{2/(2 + \alpha)}$.

12.104 (a) 8.00×10^3 m/s. (b) 127 m/s.

12.107 (a) 52.4×10^3 ft/s. (b) A: 1318 ft/s, B': 3900 ft/s.

12.108 5.31×10^9 km.

12.109 91.8×10^3 yr.

12.112 4.95 h.

12.113 50 min 55 s.

12.114 $\cos^{-1}[(1 - n\beta^2)/(1 - \beta^2)]$.

12.115 (a) 4.00 km/s. (b) 0.684.

12.124 (a) 20.5 ft/s² $\nearrow$ 30°. (b) 17.75 ft/s² →.

12.125 (a) 1.088 ft/s² ←. (b) 233 lb.

12.126 (a) 5.79 m/s². (b) 2.45 m/s². (c) 0.230 m/s².

12.127 (a) $F_r = (5.76 \text{ N}) \tan^2 \theta \sec \theta$, $F_\theta = (5.76 \text{ N}) \tan \theta \sec \theta$.
(b) $\mathbf{P} = (5.76 \text{ N}) \tan \theta \sec^2 \theta \searrow \theta$, $\mathbf{Q} = (5.76 \text{ N}) \tan^2 \theta \sec^2 \theta$ →.

12.128 (a) 0.454, down. (b) 0.1796 down. (c) 0.218, up.

12.129 (a) 539 N. (b) 47.1 m.

12.132 106.1.

12.133 (a) 0.500 m, 0. (b) 0.270 m, −84.1 N.

CHAPTER 13

13.1 10.11 GJ.

13.2 (a) 140.1 ft · lb, 140.1 ft. (b) 140.1 ft · lb, 850 ft.

13.5 10.51 ft/s.

13.6 9.53 ft.

13.7 (a) 112.2 km/h. (b) 91.6 km/h.

13.8 (a) 15.34 m/s. (b) 59.9 m/s.

13.9 (a) 8.70 m. (b) 4.94 m/s $\nearrow$ 15°.

13.11 6.71 m.

13.12 (a) 2.90 m/s. (b) 0.893 m.

13.15 (a) 57.8 m. (b) 154 N →.

13.16 (a) 7.41 kN. (b) 5.56 kN (tension)

13.17 (a) 124.1 ft. (b) A to B: 19.38 kips (tension);
B to C: 8.62 kips (tension).

13.18 (a) 279 ft. (b) A to B: 19.38 kips (compression);
B to C: 8.62 kips (compression).

13.19 (a) 46.0 ft · lb.
(b) A: 19.76 lb; B: 12.10 lb.

13.20 (a) 7.43 ft/s. (b) 0.800 ft.

13.23 (a) 1.218 m/s ←. (b) 91.0 N.

13.24 1.190 m/s.

13.25 (a) 11.35 ft/s $\nearrow$ 23.6°.
(b) 16.05 ft/s $\nearrow$ 23.6°.

13.26 (a) 3.29 m/s. (b) 1.533 m.

13.27 (a) 3.29 m/s. (b) 1.472 m.

13.28 (a) 13.63 in. (b) 8.57 in.

13.29 (a) 0.750 in. ↓. (b) 8.51 in/s ↕.

13.30 (a) 0.597 m/s. (b) 0.617 m/s.

13.32 $0.759\sqrt{pAa/m}$.

13.33 (a) 13.43 ft. (b) 386 ft/s².

13.34 A: 5.37 in.; B: 7.21 in.

13.36 $1/[1 - (v_0^2 - v^2)/2g_m R_m]$.

13.37 (a) 0.0314%. (b) 25.3%.

13.38 364 m.

13.39 14.00°.

13.40 (a) $\sqrt{3gl}$. (b) $\sqrt{2gl}$.

13.41 (a) 1.500 W. (b) 2.50 W.

13.44 2.30 m/s.

13.45 (a) 27.4°. (b) 3.81 ft.

13.46 (a) 57.2 kW. (b) 269 kW.

13.47 (a) 2.75 kW. (b) 3.35 kW.

13.48 14.80 kN.

13.51 (a) 109.0 kW, 146.2 hp.
(b) 530 kW, 711 hp.

13.52 (a) 375 kW. (b) 5.79 km/h.

13.54 (a) 8.00 hp. (b) 7.91 hp.

13.55 (a) $k_1 k_2/(k_1 + k_2)$. (b) $k_1 + k_2$.

13.56 (a) $v_0\sqrt{(k_1 + k_2)/k_1 k_2}$. (b) $v_0\sqrt{m/(k_1 + k_2)}$.

13.57 (a) 4.22 m/s. (b) 4.42 m/s.

13.58 (a) 11.66 ft/s. (b) 15.01 ft/s.

13.59 9.35 ft/s (left and right).

13.62 (a) 533 lb/ft. (b) 37.0 ft.

13.64 (a) 2.48 m/s ←. (b) 1.732 m/s ↑.

13.65 (a) 3.31 m/s. (b) 3.90 m/s.

13.66 (a) 43.5°. (b) 8.02 ft/s ↓.

13.67 6.20 ft/s.

13.68 0.269 m.

13.69 0.1744 m.

13.70 731 N.

13.71 (max) 5520 N at D; (min) 731 N just above B.

13.72 14.34 ft/s, 13.77 lb ↑.

13.74 Loop 1: (a) 25.1 ft/s. (b) 1.500 lb ←.
Loop 2: (a) 24.1 ft/s. (b) 1.000 lb

13.76 Loop 1: (a) $\sqrt{5gr}$. (b) 3 W →.
Loop 2: (a) $\sqrt{4gr}$. (b) 2 W →.

13.77 0.488 m.

13.78 (a) $\cot \varphi = 0.243 (12 - y)$
(b) 60.0 lb; $\theta_x = 85.7°$, $\theta_y = 71.6°$, $\theta_z = 161.1°$.

13.80 $V = -\ln xyz$.

13.81 (a) $\pi ka^2/4$. (b) 0.

13.82 (a) $P_x = x/R$, $P_y = y/R$, $P_z = z/R$, where $R = (x^2 + y^2 + z^2)^{1/2}$
(b) $U_{\text{OABD}} = -\Delta V_{\text{OD}} = a\sqrt{3}$.

13.85 (a) 62.5 MJ/kg. (b) 11.18 km/s.

13.86 (a) 9.56 km/s. (b) 2.39 km/s.

13.87 (a) 50.1×10^9 ft · lb. (b) 115.9×10^9 ft · lb.

13.88 (a) 942×10^3 ft · lb/lb. (b) 450×10^3 ft · lb/lb.

13.89 25.1 Mm/h.

13.90 6.48 km/s.

13.93 $v_r = \pm 3.87$ m/s, $v_\theta = 1.000$ m/s.

13.94 (a) 0.720 m. (b) 0.834 m/s.

13.95 (a) $v_r = 3.71$ ft/s, $v_\theta = 3.00$ ft/s, $v = 4.77$ ft/s.
(b) $v_r = 0$, $v_\theta = 1.129$ ft/s, $v = 1.129$ ft/s.

13.96 (a) 14.36 ft/s. (b) 1.225 ft.

13.97 (a) 4.14 ft/s. (b) 16.58 ft/s.

13.100 27.6×10^3 km/h.

13.101 (a) 7960 ft/s. (b) 4820 ft/s.

13.102 (a) 16 800 ft/s. (b) 32 700 ft/s.

13.103 14.20 km/s.

13.106 (a) 7.35 km/s. (b) 45.0°.

13.107 68.9°.

13.108 $r_{\max} = r_0(1 + \sin \alpha)$, $r_{\min} = (1 - \sin \alpha)r_0$.

13.109 1555 m/s, 79.3°.

13.110 (a) 11.32×10^3 ft/s. (b) 13.68×10^3 ft/s.

13.111 30.9×10^3 ft/s, 58.9°.

13.115 (b) $v_{\text{esc}}\sqrt{\alpha/(1 + \alpha)} < v_0 < v_{\text{esc}}\sqrt{(1 + \alpha)/(2 + \alpha)}$.

13.119 4 min 19 s.

13.120 (a) 3.64 s. (b) 27.3 s.

13.121 17.86 lb.

13.123 6.77 s.

13.124 (a) 2280 lb. (b) 3.00 s.
13.125 0.278.
13.126 (a) 11.42 s. (b) $-(125.5$ m/s$)\mathbf{j} - (194.5$ m/s$)\mathbf{k}$.
13.129 (a) 5.64 s.
    (b) AB: 19 390 lb (tension); BC: 8620 lb (tension).
13.130 (a) 12.69 s. (b) AB: 19 390 lb (compression);
    BC: 8620 lb (compression).
13.131 (a) 19.60 s. (b) 10.20 kN (compression)
13.132 (a) 0.549 s. (b) 56.8 N.
13.134 (a) 3730 lb. (b) 7450 lb.
13.136 223 Mpa.
13.138 (a) 7.00 s. (b) 10.99 ft/s. (c) 13.49 s.
13.139 76.9 lb.
13.140 1.449 kips.
13.141 6.21 W.
13.142 2.68 kN.
13.145 (a) 0.833 km/h ←. (b) 0.190 s.
13.146 (a) car A. (b) 115. 2 km/h.
13.147 65.0 kN.
13.148 (a) 9.32 ft · lb, 0.932 lb · s.
    (b) 7.99 ft · lb, 0.799 lb · s.
13.149 497 ft/s.
13.150 (a) 2.80 ft/s ←. (b) 0.229 ft/s ←.
13.151 (a) 1.694 m/s ↓. (b) 0.1619 J.
13.152 1.650 m/s.
13.155 (a) $v_A = 0.594$ m/s ←, $v_B = 1.156$ m/s →. (b) 2.99 J.
13.156 $(1 - e^2)mv^2$.
13.157 $0.728 \le e \le 0.762$.
13.158 (a) 3.00 lb. (b) 2.00 lb $\le W_B \le 6.00$ lb.
13.161 (a) $v'_A = v_0(1 - e)/2$, $v'_B = v_0(1 + e)/2$.
    (b) $v'_C = v_0(1 + e)^2/4$, $v''_B = v_0(1 - e^2)/4$.
    (c) $v'_n = v_0(1 + e)^{(n-1)}/2^{(n-1)}$. (d) $0.815\,v_0$.
13.163 0.294 m/s ←.
13.164 $\mathbf{v}'_A = 0.711\,v_0 \measuredangle 39.3°$, $\mathbf{v}'_B = 0.636\,v_0 \measuredangle 45°$.
13.165 (a) 70.0°. (b) 0.972 ft/s →.
13.166 $\mathbf{v}'_A = 6.37$ m/s $\measuredangle 77.2°$, $\mathbf{v}'_B = 1.802$ m/s $\measuredangle 40°$.
13.167 $\mathbf{v}'_A = 1.322$ m/s $\measuredangle 70.9°$, $\mathbf{v}'_B = 3.85$ m/s $\measuredangle 27.0°$.
13.168 (a) $\mathbf{v}_A = 0.878\,v_0 \measuredangle 24.2°$, $v_B = 0.412\,v_0 \measuredangle 61.0°$.
13.169 0.837.
13.172 0.156 m.
13.173 (a) 22.5°. (b) 21.3°.
13.174 (a) 20.6 mi/h. (b) 0.203.
13.175 (a) 0.294 m. (b) 54.4 mm.
13.176 (a) 0.324. (b) 14.30 ft/s.
13.177 (a) 2.90 m/s. (b) 100.5 J.
13.179 (a) 0.0240 m. (b) 817 N/m.
13.180 (a) 1.000. (b) 0.0667 m. (c) 0.0762 m.
13.182 (a) $\mathbf{v}'_A = 0$, $\mathbf{v}'_B = 0$.
    (b) $\mathbf{v}'_A = 1.201$ m/s ←, $\mathbf{v}'_B = 0.400$ m/s →.
13.183 (a) 401 mm. (b) 4.10 N · s.
13.184 (a) $\mathbf{v}'_A = 1.002$ ft/s ↑, $\mathbf{v}'_B = 0.695$ ft/s $\measuredangle 10.4°$.
    (b) 0.274 in.
13.185 3.47 in.
13.186 (a) 0.923. (b) 1.278 m.
13.188 (a) $\mathbf{v}'_A = 2.36$ ft/s $\measuredangle 83.8°$, $\mathbf{v}'_B = 3.23$ ft/s →. (b) 1.97 in.
13.190 102.6 mi/h.
13.191 1.688 ft · lb.
13.194 12 990 ft/s.
13.195 (a) 13.31 N →. (b) 4.49 N ↓. (c) 13.31 N ←.
13.197 (a) 217 mm. (b) 69.1 mm.
13.198 (a) $v'_A = v'_B = v'_C = 1.368$ m/s. (b) 0.668 m. (c) 1.049 m.
13.200 (a) 2.94 m/s. (b) 16.14 N.

CHAPTER 14

14.1 (a) 4.46 m/s ←. (b) 0.409 m/s ←.
14.2 (a) 306 m/s ←. (b) 0.409 m/s ←.
14.3 0.792 ft/s →.
14.4 (a) 0.800 oz. (b) 900 ft/s →.
14.7 (a) 3.79 km/h →, 2.77 km/h →.
    (b) 5.54 km/h →, 2.77 km/h →.
    (c) 5.54 km/h →, 3.60 km/h →.
14.8 (a) 0.901 m/s →. (b) 0.807 m/s →.
14.9 $-(4.80$ kg · m^2/s$)\mathbf{j} + (9.60$ kg · m^2/s$)\mathbf{k}$.
14.10 (a) $(1.867$ m$)\mathbf{i} + (1.533$ m$)\mathbf{j} + (0.667$ m$)\mathbf{k}$.
    (b) $(12.00$ kg · m/s$)\mathbf{i} + (28.0$ kg · m/s$)\mathbf{j} + (14.00$ kg · m/s$)\mathbf{k}$.
    (c) $-(2.80$ kg · m^2/s$)\mathbf{i} + (13.33$ kg · m^2/s$)\mathbf{j} - (24.3$ kg · m^2/s$)\mathbf{k}$.
14.11 (a) $v_x = -0.750$ ft/s, $v_z = 0.438$ ft/s. (b) $-(3.39$ ft · lb · s$)\mathbf{k}$.
14.12 (a) $v_z = 7.25$ ft/s, $v_x = 8.33$ ft/s. (b) $-(4.51$ ft · lb · s$)\mathbf{k}$
    (b) $-(26.0$ kg · m/s$)\mathbf{i} + (14.00$ kg · m/s$)\mathbf{j} + (14.00$ kg · m/s$)\mathbf{k}$.
    (c) $-(29.5$ kg · m^2/s$)\mathbf{i} - (16.75$ kg · m^2/s$)\mathbf{j} + (3.20$ kg · m^2/s$)\mathbf{k}$.
14.15 $(114.4$ m$)\mathbf{i} - (76.1$ m$)\mathbf{j} + (8.75$ m$)\mathbf{k}$.
14.16 $(1180$ m$)\mathbf{i} + (140$ m$)\mathbf{j} + (155$ m$)\mathbf{k}$.
14.19 $x = 45.2$ ft, $y = 54.5$ ft.
14.20 (a) 2.00 s. (b) 92.8 mi/h.
14.21 $(81.5$ ft$)\mathbf{i} + (351$ ft$)\mathbf{k}$.
14.22 (a) 8.00 ft/s →. (b) 36.6°, $v_C = 10.39$ ft/s, $v_D = 8.72$ ft/s.
14.24 $v_A = 431$ m/s, $v_B = 395$ m/s, $v_C = 528$ m/s.
14.25 $v_A = 646$ m/s, $v_B = 789$ m/s, $v_C = 176$ m/s.
14.26 $v_A = 919$ m/s, $v_B = 717$ m/s, $v_C = 619$ m/s.
14.31 friction: 2.97 J, first impact: 3007 J, second impact: 24.3 J.
14.32 (a) 1116 ft · lb. (b) 623 ft · lb.
14.33 (woman) 382 ft · lb, (man) 447 ft · lb.
14.34 (A hits B) 4.86 J, (B hits C) 3.28 J, (A hits B again) 0.688 J.
14.35 (b) $E_A = 180.0$ kJ, $E_B = 320$ kJ.
14.37 (a) $\mathbf{v}_B = \dfrac{m_A v_0}{m_A + m_B} \to$. (b) $h = \dfrac{m_A}{m_A + m_B} \dfrac{v_0^2}{2g}$.
14.38 $\mathbf{v}_A = 4.11$ m/s $\measuredangle 46.9°$, $\mathbf{v}_B = 17.39$ m/s $\measuredangle 16.7°$.
14.39 (a) $\mathbf{v}_{B/A} = 11.59$ ft/s $\measuredangle 30°$. (b) $\mathbf{v}_A = 3.76$ ft/s →.
14.40 $\mathbf{v}_A = 3.11$ ft/s ←, $\mathbf{v}_B = 4.66$ ft/s →.
14.41 $v_A = 10.61$ ft/s, $v_B = 5.30$ ft/s, $v_C = 9.19$ ft/s.
14.42 $v_A = 7.50$ ft/s, $v_B = 9.19$ ft/s, $v_C = 9.19$ ft/s.
14.45 (a) $\mathbf{L} = (8.75$ kg · m/s$)\mathbf{i}$, $\mathbf{H}_G = -(0.500$ kg · m^2/s$)\mathbf{k}$.
    (b) $\mathbf{v}'_A = (1.500$ m/s$)\mathbf{i}$, $\mathbf{v}'_B = (5.00$ m/s$)\mathbf{i}$.
14.46 $(200$ ft/s$)\mathbf{i} + (172$ ft/s$)\mathbf{j} + (1560$ ft/s$)\mathbf{k}$.
14.47 (a) $v_C = 11.00$ ft/s, $v_D = 5.50$ ft/s. (b) 0.786.
14.48 $x = 181.7$ mm, $y = 0$, $z = 139.4$ mm.
14.51 (a) $\mathbf{v}_B = 2.40$ m/s $\measuredangle 53.1°$, $\mathbf{v}_C = 2.56$ m/s →, (b) $c = 1.059$ m
14.52 (a) $\mathbf{v}_A = 2.40$ m/s ↓, $\mathbf{v}_B = 3.00$ m/s $\measuredangle 53.1°$, (b) $a = 1.864$ m
14.55 (a) $\mathbf{v}_A = 2.25$ ft ↑, $\mathbf{v}_B = 2.25$ ft/s ↓, $\mathbf{v}_c = 3.90$ ft/s →. (b) 11.1 in.
14.56 (a) 2.00 ft/s →. (b) 0.760 ft. (c) 5.29 rad/s ↓.
14.57 $\rho v_1^2 (A_1 - A_2)$.
14.58 $\rho A_2 v_2^2 - \rho A_1 v_1^2 \cos \theta$.
14.59 64.6 lb.
14.60 18.2 ft/s.
14.61 (a) 14.8 kN. (b) 27.7 kN.
14.62 90.6 N ←.
14.64 $D_x = 329$ N, $D_y = 0$, $C_x = -203$ N, $C_y = 271$ N.
14.66 (a) 61.1 m/s. (b) $R_x = -38.8$ N, $R_y = 44.7$ N.
14.67 $C_x = 90.0$ N, $C_y = 2360$ N, $D_x = 0$, $D_y = 2900$ N.
14.68 $\mathbf{A} = qv_0$ ←, $\mathbf{B} = \sqrt{2gh} \measuredangle 30°$.
14.69 100 kg/s.
14.70 7580 lb.
14.71 33.6 kN ←.
14.72 7180 lb.

14.74 (a) 9690 lb, 3.38 ft. (b) 6960 lb, 9.43 ft.
14.76 (a) 3.03 m/s² ⦫ 18°. (b) 922 km/h.
14.77 (a) 30.6 m/s. (b) 96.1 m³/s. (c) 55 100 N · m/s.
14.78 (a) 3.23 MW. (b) 0.464.
14.79 213 m.
14.80 (a) 15 450 hp. (b) 28 060 hp. (c) 0.551.
14.83 (b) friction from gravel slipping on belt.
14.86 (a) $m(v^2 + gy)/l$. (b) $\mathbf{R} = mg(1 - y/l)$.
14.87 (a) mgy/l. (b) $m[g(l - y) + v^2]/l \uparrow$.
14.88 $\sqrt{gh}\,\tan \mathrm{h}(\sqrt{gh}\,t/L)$.
14.89 10.10 ft/s.
14.90 4.75 ft/s.
14.91 533 kg/s.
14.94 (a) 90.0 m/s². (b) 35.9×10^3 km/h.
14.95 7930 m/s.
14.96 (a) 1800 m/s. (b) 9240 m/s.
14.99 19.07 mi.
14.100 (a) 119.3 mi. (b) 14 660 mi/h.
14.101 186.8 km/h.
14.102 (a) 31.2 km. (b) 197.5 km.
14.106 (a) 1.595 m/s. (b) 0.370 m.
14.107 (a) 5.20 km/h →. (b) 4.00 km/h →.
14.108 (a) 6.05 ft/s. (b) 6.81 ft/s.
14.109 (a) 3.11 m/s →. (b) 1.356 m.
14.110 $\mathbf{v}_A = 15.38$ ft/s →, $\mathbf{v}_B = 5.13$ ft/s ←.
14.112 $\mathbf{A}_x = 55.5$ lb →, $\mathbf{A}_y = 20.2$ lb ↓, $\mathbf{m}_A = 41.4$ lb · ft ↓.
14.114 (a) $m_0 + qt_L = m_0\,e^{ql/mv_0}$. (b) $v_0 e^{-ql/mv_0}$.
14.115 414 rpm.
14.116 Case 1: (a) 0.333 g ↓. (b) $0.817\sqrt{gl}$.
  Case 2: (a) gy/l ↓. (b) $\sqrt{gl}$.

CHAPTER 15

15.1 (a) 29.6 rad/s. (b) 32.2 rev.
15.2 (a) 0, 0, 0. (b) 6.00 rad, 4.71 rad/s, −3.70 rad/s².
15.3 (a) 0, 0.1000 rad/s, −0.250 rad/s².
  (b) 0.211 rad, 0.0472 rad/s, −0.01181 rad/s².
  (c) 0.400 rad, 0, 0.
15.4 (a) −3.01 rad/s². (b) 13 800 rev.
15.5 (a) 150 rev. (b) 2100 rev.
15.6 (a) 0.855 rad/s. (b) 3.71°.
15.9 (a) 12.73 rev. (b) ∞, (c) 18.42 s.
15.10 −(0.450 m/s)**i** − (1.200 m/s)**j** + (1.500 m/s)**k**, (12.60 m/s²)**i** + (7.65 m/s²)**j** + (9.90 m/s²)**k**.
15.11 (0.750 m/s)**i** + (1.500 m/s)**k**, (12.75 m/s²)**i** + (11.25 m/s²)**j** + (3.00 m/s²)**k**.
15.12 −(1.333 ft/s)**i** − (4.67 ft/s)**j** − (2.33 ft/s)**k**, (28.0 ft/s²)**i** + (11.00 ft/s²)**j** −(38.0 ft/s²)**k**.
15.13 −(1.333 ft/s)**i** + (2.33 ft/s)**k**.
  −(6.67 ft/s²)**i** − (21.7 ft/s²)**j** −(10.00 ft/s²)**k**.
15.16 66 700 mi/h, 19.47×10^{-3} ft/s².
15.17 (a) 1525 ft/s, 0.1112 ft/s². (b) 1168 ft/s, 0.0852 ft/s². (c) 0, 0.
15.18 (a) 2.50 rad/s ↱, 1.500 rad/s² ↲. (b) 771 mm/s² ⦨ 76.5°.
15.19 12.00 rad/s² ↱ or 12.00 rad/s² ↲.
15.22 left: 3.49 s; middle: 6.98 s; right: 13.96 s.
15.23 (a) 0.500 ft/s →, 1.500 ft/s² ←. (b) 4.24 ft/s² ⦨ 45°.
15.24 (a) 300 rpm ↱, 100 rpm ↲. (b) $\mathbf{a}_B = 1974$ in/s² ←, $\mathbf{a}_C = 658$ in/s²→.
15.25 (a) A: 15.00 rad/s ↱; B: 7.50 rad/s ↲.
  (b) A: 75.0 ft/s² ↑; B: 37.5 ft/s² ↓.
15.26 (a) C: 120 rpm; B: 275 rpm.
  (b) A: 23.7 m/s² ↑; B: 19.90 m/s² ↓.

15.27 (a) 10.00 rad/s. (b) A: 7.50 m/s²; B: 3.00 m/s² ↓. (c) 4.00 m/s² ↓.
15.28 (a) 0.400 rad/s² ↲. (b) 1.528 rev.
15.29 (a) 2.75 rev. (b) 1.710 m/s ↓, 3.11 m ↓. (c) 849 m/s² ⦫ 32.0°.
15.30 (a) 1.152 m/s ↑, 2.30 m ↑. (b) 1.728 m/s ↓, 3.46 m ↓.
15.31 (a) 15.28 rev. (b) 10.14 s.
15.32 $\boldsymbol{\alpha}_A = 5.41$ rad/s² ↱, $\boldsymbol{\alpha}_B = 1.466$ rad/s² ↱.
15.33 (a) 10.39 s. (b) $\boldsymbol{\omega}_A = 413$ rpm ↲, $\boldsymbol{\omega}_B = 248$ rpm ↱.
15.36 $b\omega_0^2/2\pi$ →.
15.37 $bv^2/2\pi r^3$ ↲.
15.38 $\mathbf{v}_B = 140.8$ ft/s →, $\mathbf{v}_C = 0$, $\mathbf{v}_0 = 136.0$ ft/s ⦫ 15°, $\mathbf{v}_E = 99.6$ ft/s ⦩ 45°.
15.39 (a) 0.378 rad/s ↲. (b) 6.42 in/s ↑.
15.40 (a) 1.173 rad/s ↱. (b) 0.998 m/s ⦫ 25°.
15.41 (a) 3.62 rad/s ↱. (b) 1.963 m/s ↓.
15.44 (a) 2.00 rad/s ↲. (b) (120 mm/s)**i** + (660 mm/s)**j**.
15.45 (a) −(240 mm/s)**i** + (300 mm/s)**j**.
  (b) $x = 150$ mm, $y = -60$ mm.
15.47 (a) −(4.00 rad/s)**k** or 4.00 rad/s ↲.
  (b) −(4.00 in/s)**i**.
15.48 (a) $\boldsymbol{\omega}_B = \boldsymbol{\omega}_C = \boldsymbol{\omega}_D = \frac{1}{2}\boldsymbol{\omega}_A$ ↱. (b) $\boldsymbol{\omega}_S = 0.25\,\boldsymbol{\omega}_A$ ↲.
15.49 (a) $\boldsymbol{\omega}_B = \boldsymbol{\omega}_C = \boldsymbol{\omega}_D = 150$ rpm ↲. (b) $\boldsymbol{\omega}_S = 195$ rpm ↲.
15.50 (a) 48.0 rad/s ↲. (b) 3.39 m/s ⦫ 45°.
15.51 (a) 5.65 m/s ↑. (b) 9000 rpm, (c) 1500.
15.53 (a) 200 rad/s ↱. (b) 24.0 rad/s ↲.
15.55 (a) (6.00 rad/s)**k** or 6.00 rad/s ↱.
  (b) (360 mm/s)**i** −(672 mm/s)**j** or 762 mm/s ⦩ 61.8°.
15.56 (a) 540 mm/s →. (b) 457 mm/s ⦪ 61.8°.
15.57 (a) $r\omega_D/\cos\theta$ →. (b) $\omega_D \tan^2\theta$ ↱.
15.58 (a) 2.90 rad/s ↲. (b) 0.631 rad/s ↱.
15.59 (a) 1.500 rad/s ↱. (b) 18.00 in/s ↑. (c) 11.25 in/s ⦪ 53.1°.
15.61 (a) $\mathbf{v}_P = 0$, $\omega_{BD} = 39.3$ rad/s ↱.
  (b) $\mathbf{v}_P = 6.28$ m/s ↓, $\omega_{BD} = 0$.
15.62 $\mathbf{v}_P = 6.52$ m/s ↓, $\omega_{BD} = 20.8$ rad/s ↱.
15.63 (a) 12.00 rad/s ↱. (b) 3.90 m/s ⦨ 67.4°.
15.64 $\boldsymbol{\omega}_{DE} = 2.55$ rad/s ↲, $\boldsymbol{\omega}_{BD} = 0.955$ rad/s ↱.
15.65 $\boldsymbol{\omega}_{DE} = 6.40$ rad/s ↲, $\boldsymbol{\omega}_{BD} = 5.20$ rad/s ↲.
15.68 (a) 3.33 rad/s ↱. (b) 2.00 m/s ⦩ 56.3°.
15.69 (a) 1.500 m. (b) 5.00 m/s ↓.
15.70 14.76 in/s →.
15.71 (a) 338 mm/s ←, 0. (b) 710 mm/s ←, 2.37 rad/s ↲.
15.72 $(1 - r_A/r_C)\omega_{ABC}$.
15.74 (a) 1.00 ft to the right of A. (b) 4.00 in/s ↑.
15.75 $x = 0$, $z = 9.34$ ft.
15.76 (a) 3.00 rad/s ↱. (b) 300 mm/s ←. (c) 180.0 mm/s (wound).
15.77 (a) 3.00 rad/s ↲. (b) 180 mm/s →. (c) 300 mm/s (unwound).
15.78 (a) 50 mm to the right of the axle.
  (b) $\mathbf{v}_B = 750$ mm/s ↓, $\mathbf{v}_D = 1.950$ m/s ↑.
15.79 (a) 25 mm to the right of 0. (b) 420 mm/s ↑.
15.80 (a) A: 300 mm to the left of A.
    C: 600 mm to the left of C.
  (b) $\boldsymbol{\omega}_A = 4.00$ rad/s ↲, $\boldsymbol{\omega}_C = 2.00$ rad/s ↱.
15.82 (a) 0.467 rad/s ↱. (b) 3.49 ft/s ⦫ 59.2°.
15.83 (a) 3.08 rad/s ↱. (b) 83.3 m/s ⦨ 73.9°.
15.86 (a) 5.13 rad/s ↱. (b) 0.924 ft/s ←. (c) 1.870 m/s ⦨ 34.7°.
15.87 (a) 4.27 rad/s ↲. (b) 1.330 m/s ↓. (c) 1.557 m/s ⦫ 34.7°.
15.88 (a) $(v_A/l)\sin\beta/\cos(\beta - \theta)$. (b) $v_A \cos\theta/\cos(\beta - \theta)$.
15.89 (a) 3.54 rad/s ↱. (b) 8.68 ft/s ⦫ 50°.
15.90 (a) 0.900 rad/s ↲. (b) 411 mm/s ⦩ 20.5°.
15.94 (a) $\boldsymbol{\omega}_{AB} = 1.920$ rad/s ↲, $\boldsymbol{\omega}_{BD} = 1.200$ rad/s ↲.
  (b) 30.0 in/s ⦨ 73.7°.
15.95 (a) $\boldsymbol{\omega}_{AB} = 1.200$ rad/s ↲, $\boldsymbol{\omega}_{DE} = 0.450$ rad/s ↲.
  (b) 5.25 in/s ←.

15.96 (a) 5.00 rad/s. (b) 3.00 m/s $\downarrow$.

15.97 (a) 1260 mm/s $\uparrow$. (b) 1.250 rad/s $\uparrow$.

15.98 (a) $\boldsymbol{\omega}_{AB} = 1.177$ rad/s $\downarrow$, $\boldsymbol{\omega}_{DE} = 2.50$ rad/s $\downarrow$. (b) 29.4 in/s $\leftarrow$.

15.99 Space centrode: quarter circle, $r = 15$ in, centered at O. Body centrode: semicircle, $r = 7.5$ in., centered midway between A and B.

15.100 Space centrode: lower rack. Body centrode: circumference of gear.

15.102 $\omega_{BD} = 0.955$ rad/s $\downarrow$, $\omega_{DE} = 2.55$ rad/s $\uparrow$.

15.103 $\omega_{BD} = 5.20$ rad/s $\downarrow$, $\omega_{DE} = 6.40$ rad/s $\downarrow$.

15.105 (a) 0.833 rad/s^2 $\downarrow$, (b) 1.083 m/s^2 $\uparrow$.

15.106 $\mathbf{a}_A = 2.50$ m/s^2 $\uparrow$, $\mathbf{a}_B = 0.100$ m/s^2 $\uparrow$.

15.107 (a) 0.900 m/s^2 $\rightarrow$. (b) 1.800 m/s^2 $\leftarrow$.

15.108 (a) 0.600 m from A. (b) 0.200 m from A.

15.109 (a) 51.3 in/s^2 $\downarrow$. (b) 184.9 in/s^2 $\measuredangle$ 16.1°.

15.110 (a) 1.039 rad/s^2 $\downarrow$. (b) (2.60 ft/s^2)$\mathbf{i}$ + (4.50 ft/s^2)$\mathbf{j}$ or 5.20 ft/s^2 $\measuredangle$ 60°.

15.111 (a) 1430 m/s^2 $\downarrow$. (b) 1430 m/s^2 $\uparrow$, (c) 1430 m/s^2 $\sear{}$ 60°.

15.112 (a) 13.35 in/s^2 $\nearrow$ 61.0°. (b) 12.62 in/s^2 $\measuredangle$ 64.0°.

15.113 $\mathbf{a}_A = 56.6$ in/s^2 $\searrow$ 58.0°, $\mathbf{a}_B = 80.0$ in/s^2 $\uparrow$, $\mathbf{a}_C = 172.2$ in/s^2 $\searrow$ 25.8°.

15.114 $\mathbf{a}_A = 48.0$ in/s^2 $\uparrow$, $\mathbf{a}_B = 85.4$ in/s^2 $\searrow$ 69.4°. $\mathbf{a}_C = 82.8$ in/s^2 $\nearrow$ 65.0°.

15.115 $\alpha_A = 96.0$ rad/s^2 $\uparrow$, $\mathbf{a}_A = 2.40$ m/s^2 $\leftarrow$. $\alpha_B = 48.0$ rad/s^2 $\uparrow$, $\mathbf{a}_B = 1.200$ m/s^2 $\leftarrow$.

15.118 (a) 92.5 in/s^2. (b) 278 in/s^2.

15.120 148.3 m/s^2 $\downarrow$.

15.121 296 m/s^2 $\uparrow$.

15.122 $\mathbf{a}_D = 1558$ m/s^2 $\searrow$ 45°. $\mathbf{a}_E = 337$ m/s^2 $\measuredangle$ 45°.

15.124 (a) 242 in/s^2 $\leftarrow$. (b) 403 in/s^2 $\nearrow$ 72.5°.

15.125 694 in/s^2 $\leftarrow$.

15.127 1.745 m/s^2 $\nearrow$ 68.2°.

15.128 1.296 m/s^2 $\leftarrow$.

15.129 (a) 228 rad/s^2 $\uparrow$. (b) 92.0 rad/s^2 $\downarrow$.

15.130 (a) 138.1 ft/s^2 $\searrow$ 78.6°. (b) 203 ft/s^2 $\measuredangle$ 19.5°.

15.131 (a) 10.75 rad/s^2 $\uparrow$. (b) 2.30 rad/s^2 $\uparrow$.

15.132 (a) 4.18 rad/s^2 $\downarrow$. (b) 2.43 rad/s^2 $\downarrow$.

15.133 (a) 8.15 rad/s^2 $\uparrow$. (b) 0.896 rad/s^2 $\downarrow$.

15.134 (a) 3.70 rad/s^2 $\downarrow$. (b) 3.70 rad/s^2 $\downarrow$.

15.135 (a) 16.53 rad/s^2 $\uparrow$. (b) 193.6 in/s^2 $\nearrow$ 7.36°.

15.136 $\mathbf{v}_D = 1.382$ m/s $\downarrow$. $\mathbf{a}_D = 0.695$ m/s^2 $\downarrow$.

15.138 $\mathbf{v}_B = b\omega \cos \theta$, $a_B = b\alpha \cos \theta - b\omega^2 \sin \theta$.

15.139 $\mathbf{v}_B \sin \beta/l \cos \theta$.

15.140 $(v_B \sin \beta/l)^2 (\sin \theta/\cos^3 \theta)$.

15.141 $v_x = v[1 - \cos (vt/r)]$. $v_y = v \sin (vt/r)$.

15.142 $\boldsymbol{\omega} = bv_A(b^2 + x_A^2) \uparrow$, $\boldsymbol{\alpha} = 2bx_A v_A^2/(b^2 + x_A^2)^2 \uparrow$.

15.143 $v_{B_s} = v_A - \mathrm{lb}^2(b^2 + x_A^2)^{3/2} \rightarrow$, $(v_B)_y = \mathrm{lb} \, x_A v_A/(b^2 + x_A^2)^{3/2} \uparrow$.

15.144 $\omega_{BD} = b\omega(b + l \cos \theta)/(l^2 + b^2 + 2bl \cos \theta) \downarrow$, $\mathbf{v}_E = bl\omega \sin \theta/(l^2 + b^2 + 2bl \cos \theta) \searrow$ $\tan^{-1}[(b \sin \theta/(l + b \cos \theta)]$

15.145 $bl\omega^2(l^2 - b^2) \sin \theta/(l^2 + b^2 + 2bl \cos \theta) \uparrow$.

15.146 (a) $(v_0/b) \sin^2 \theta \downarrow$. (b) $(v_0 l/b) \sin^2 \theta \cos \theta \rightarrow$, $(v_0 l/b) \sin^3 \theta \uparrow$. (c) $2(v_0/b)^2 \sin^3 \theta \cos \theta \uparrow$.

15.147 $\boldsymbol{\omega} = v_0 \sin^2 \theta/r \cos \theta \uparrow$, $\boldsymbol{\alpha} = (v_0/r)^2 (1 + \cos^2 \theta) \tan^3 \theta \uparrow$.

15.148 $(v_\rho)_x = r\omega \left[\cos \dfrac{r\omega t}{R - r} - \cos \omega t \right]$,
$(v_\rho)_y = r\omega \left[\sin \dfrac{r\omega t}{R - r} + \sin \omega t \right]$

15.149 Path is the y axis. $\mathbf{v} = (R\omega \sin \omega t)\mathbf{j}$, $\mathbf{a} = (R\omega^2 \cos \omega t)\mathbf{j}$.

15.150 2.40 m/s $\searrow$ 73.9°.

15.151 2.87 m/s $\searrow$ 44.8°.

15.152 (a) 1.815 rad/s $\downarrow$. (b) 16.42 in/s $\searrow$ 20°.

15.153 (a) 5.16 rad/s $\downarrow$. (b) 1.399 m/s $\searrow$ 60°.

15.154 (a) 3.81 rad/s $\downarrow$, 6.53 m/s $\measuredangle$ 16.26°. (b) 3.00 rad/s $\downarrow$, 4.00 m/s $\rightarrow$.

15.155 (a) 14.00 rad/s $\downarrow$. (b) 28.0 ft/s $\downarrow$.

15.160 15.47 m/s^2 $\nearrow$ 77.3°.

15.161 (a) 5.88×10^{-3} ft/s^2 west. (b) 4.47×10^{-3} ft/s^2 west. (c) 4.47×10^{-3} ft/s^2 west.

15.162 0.0234 m/s^2 west.

15.163 (a) 0.809 m/s $\searrow$ 35.5°. (b) 1.723 m/s^2 $\searrow$ 67.6°.

15.164 (a) 0.520 m/s $\searrow$ 82.6°. 50.0 mm/s^2 $\searrow$ 9.8°.

15.165 (a) 0.520 m/s $\searrow$ 37.4°. 50.0 mm/s^2 $\nearrow$ 69.8°.

15.166 (a) $-(1.32$ m/s$)\mathbf{j}$ $- (2.70$ m/s$)\mathbf{k}$. (b) $-(1.32$ m/s$^2)\mathbf{j}$.

15.167 (a) $(2.4$ m/s$)\mathbf{i}$ $- (2.70$ m/s$)\mathbf{k}$. (b) $(2.4$ m/s$^2)\mathbf{i}$.

15.168 (1) 303 mm/s^2 $\rightarrow$; (2) 168.5 mm/s^2 $\nearrow$ 57.7°.

15.169 (3) 483 mm/s^2 $\leftarrow$; (4) 168.5 mm/s^2 $\searrow$ 57.7°.

15.170 0.750 m/s $\measuredangle$ 71.3°, 2.13 m/s^2 $\nearrow$ 61.9°.

15.171 2.79 rad/s $\downarrow$, 2.13 rad/s^2 $\downarrow$.

15.174 (a) 0.621 m/s^2 $\uparrow$. (b) 1.733 m/s^2 $\searrow$ 53.6°. (c) 2.62 m/s^2 $\nearrow$ 67.6°.

15.175 (a) 8.09 rad/s^2 $\downarrow$. (b) 8.43 m/s^2 $\nearrow$ 16.26°.

15.176 7.86 rad/s $\uparrow$, 81.1 rad/s^2 $\uparrow$.

15.177 3.81 rad/s $\downarrow$, 81.4 rad/s^2 $\downarrow$.

15.178 1.526 rad/s $\downarrow$, 57.6 rad/s^2 $\downarrow$.

15.181 (a) 3.61 rad/s $\uparrow$. (b) 86.6 in/s $\measuredangle$ 30°. (c) 563 in/s^2 $\nearrow$ 46.1°.

15.182 (a) 3.61 rad/s $\downarrow$. (b) 86.6 in/s $\nearrow$ 30°. (c) 563 in/s^2 $\nearrow$ 46.1°.

15.183 51.5 m/s^2 $\searrow$ 44.4°.

15.184 (a) (0.480 rad/s)$\mathbf{i}$ − (1.600 rad/s)$\mathbf{j}$ + (0.600 m/s)$\mathbf{k}$. (b) (400 mm/s)$\mathbf{i}$ + (300 mm/s)$\mathbf{j}$ + (480 mm/s)$\mathbf{k}$.

15.185 (a) −(0.400 rad/s)$\mathbf{i}$ − (0.360 rad/s)$\mathbf{k}$. (b) (100 mm/s)$\mathbf{i}$ − (90 mm/s)$\mathbf{j}$ + (120 mm/s)$\mathbf{k}$.

15.186 (a) (1.5 rad/s)$\mathbf{i}$ − (3.5 rad/s)$\mathbf{j}$ − (3.0 rad/s)$\mathbf{k}$. (b) (640 mm/s)$\mathbf{i}$ − (360 mm/s)$\mathbf{j}$ + (740 mm/s)$\mathbf{k}$.

15.187 (a) (30.1 rad/s)$\mathbf{i}$ − (40.2 rad/s)$\mathbf{k}$. (b) (14.4 ft/s)$\mathbf{i}$ + (10.8 ft/s)$\mathbf{k}$.

15.188 (118.4 rad/s^2)$\mathbf{i}$.

15.189 (230 rad/s^2)$\mathbf{i}$ − (2.5 rad/s^2)$\mathbf{k}$.

15.190 (a) −(2260 rad/s^2)$\mathbf{k}$. (b) (2260 rad/s^2)$\mathbf{j}$.

15.193 (a) −(0.600 m/s)$\mathbf{i}$ + (0.750 m/s)$\mathbf{j}$ − (0.600 m/s)$\mathbf{k}$. (b) −(6.15 m/s^2)$\mathbf{i}$ − (3.00 m/s^2)$\mathbf{j}$.

15.195 (a) −(20.0 rad/s^2)$\mathbf{j}$. (b) −(4.00 ft/s^2)$\mathbf{i}$ + (10.00 ft/s^2)$\mathbf{k}$. (c) −(10.25 ft/s^2)$\mathbf{j}$.

15.196 −(3.46 ft/s^2)$\mathbf{i}$ − (5.13 ft/s^2)$\mathbf{j}$ + (8.66 ft/s^2)$\mathbf{k}$.

15.197 (a) (8.00 rad/s^2)$\mathbf{i}$. (b) −(19.20 rad/s^2)$\mathbf{k}$. (c) −(1.103 m/s^2)$\mathbf{i}$ − (2.005 m/s^2)$\mathbf{j}$.

15.198 (a) (0.0375 rad/s^2)$\mathbf{i}$. (b) −(0.1434 m/s)$\mathbf{i}$ + (0.204 m/s)$\mathbf{j}$ − (0.1228 m/s)$\mathbf{k}$. (c) −(0.696 m/s^2)$\mathbf{i}$ − (0.0358 m/s^2)$\mathbf{j}$ + (0.0430 m/s^2)$\mathbf{k}$.

15.199 (a) (28.4 rad/s)$\mathbf{i}$ + (5.24 rad/s)$\mathbf{j}$. (b) (25.8 rad/s)$\mathbf{i}$.

15.200 (a) (135.1 rad/s^2)$\mathbf{k}$. (b) (5.77 m/s^2)$\mathbf{i}$. − (232 m/s^2)$\mathbf{j}$.

15.203 −(33.3 in/s)$\mathbf{j}$.

15.204 (12.00 in/s)$\mathbf{k}$.

15.205 −(34.5 mm/s)$\mathbf{i}$.

15.206 −(30.0 in/s)$\mathbf{j}$.

15.207 (45.7 in/s)$\mathbf{j}$.

15.210 ($\omega_2/\cos 25°$) ($-\sin 25°\mathbf{i}$ + $\cos 25°$)$\mathbf{k}$.

15.211 ($\omega_1 \cos 25°$) ($-\sin 25°\mathbf{i}$ + $\cos 25°\mathbf{k}$).

15.212 (a) −(0.348 rad/s)$\mathbf{i}$ + (0.279 rad/s)$\mathbf{j}$ + (1.089 rad/s)$\mathbf{k}$. (b) −(30.0 in/s)$\mathbf{j}$.

15.213 (a) (1.463 rad/s)$\mathbf{i}$ + (0.1052 rad/s)$\mathbf{j}$ + (0.0841 rad/s)$\mathbf{k}$. (b) −(34.5 mm/s)$\mathbf{j}$.

15.216 $-(45.0 \text{ in/s}^2)\mathbf{j}$.
15.217 $(205 \text{ in/s}^2)\mathbf{j}$.
15.218 $-(9.51 \text{ mm/s}^2)\mathbf{j}$.
15.219 $-(8.76 \text{ mm/s}^2)\mathbf{j}$.
15.220 (a) $(24.6 \text{ in/s})\mathbf{i} + (67.7 \text{ in/s})\mathbf{j} - (50.7 \text{ in/s})\mathbf{k}$.
(b) $-(423 \text{ in/s}^2)\mathbf{i} + (98.5 \text{ in/s}^2)\mathbf{j} - (147.8 \text{ in/s}^2)\mathbf{k}$.
15.221 (a) $(51.6 \text{ in/s})\mathbf{i} + (67.7 \text{ in/s})\mathbf{j} - (50.7 \text{ in/s})\mathbf{k}$.
(b) $-(423 \text{ in/s}^2)\mathbf{i} + (98.5 \text{ in/s}^2)\mathbf{j} - (229 \text{ in/s}^2)\mathbf{k}$.
15.222 (a) $-(1.215 \text{ m/s})\mathbf{i} + (1.620 \text{ m/s})\mathbf{k}$. ($b$) $-(30.4 \text{ m/s}^2)\mathbf{j}$.
15.223 (a) $-(1.215 \text{ m/s})\mathbf{i} - (1.080 \text{ m/s})\mathbf{j} + (1.620 \text{ m/s})\mathbf{k}$.
(b) $(19.44 \text{ m/s}^2)\mathbf{i} - (30.4 \text{ m/s}^2)\mathbf{j} - (12.96 \text{ m/s}^2)\mathbf{k}$.
15.224 (a) $(1.200 \text{ m/s})\mathbf{i} + (0.500 \text{ m/s})\mathbf{j} - (1.200 \text{ m/s})\mathbf{k}$.
(b) $-(7.20 \text{ m/s}^2)\mathbf{i} - (14.40 \text{ m/s}^2)\mathbf{k}$.
15.227 (a) $(0.750 \text{ m/s})\mathbf{i} + (1.299 \text{ m/s})\mathbf{j} - (1.732 \text{ m/s})\mathbf{k}$.
(b) $(27.1 \text{ m/s}^2)\mathbf{i} + (5.63 \text{ m/s}^2)\mathbf{j} - (15.00 \text{ m/s}^2)\mathbf{k}$.
15.228 (a) $(129.9 \text{ mm/s})\mathbf{i} + (75.0 \text{ mm/s})\mathbf{j} + (86.6 \text{ mm/s})\mathbf{k}$.
(b) $(45.0 \text{ mm/s}^2)\mathbf{i} - (112.6 \text{ mm/s}^2)\mathbf{j} + (60.0 \text{ mm/s}^2)\mathbf{k}$.
15.230 (a) $(1.600 \text{ ft/s})\mathbf{i} + (5.00 \text{ ft/s})\mathbf{j} + (8.80 \text{ ft/s})\mathbf{k}$.
(b) $(8.00 \text{ ft/s}^2)\mathbf{i} - (83.5 \text{ ft/s}^2)\mathbf{j} + (20.0 \text{ ft/s}^2)\mathbf{k}$.
15.231 (a) $\boldsymbol{\omega}_1 + (R/r)(\omega_1 - \omega_2)\mathbf{k}$. ($b$) $\omega_1(\omega_1 - \omega_2)(R/r)\mathbf{j}$.
15.232 $-(41.6 \text{ in/s}^2)\mathbf{i} - (61.5 \text{ in/s}^2)\mathbf{j} + (103.9 \text{ in/s}^2)\mathbf{k}$.
15.233 (a) $(0.0375 \text{ rad/s}^2)\mathbf{i}$.
(b) $-(0.143 \text{ m/s})\mathbf{i} + (0.205 \text{ m/s})\mathbf{j} - (0.123 \text{ m/s})\mathbf{k}$.
(c) $-(0.0696 \text{ m/s}^2)\mathbf{i} - (0.0358 \text{ m/s}^2)\mathbf{j} + (0.0430 \text{ m/s}^2)\mathbf{k}$.
15.234 (a) $(0.600 \text{ m/s})\mathbf{j} - (0.585 \text{ m/s})\mathbf{k}$. ($b$) $-(4.76 \text{ m/s}^2)\mathbf{j}$.
15.235 (a) $(0.600 \text{ m/s})\mathbf{i} - (0.225 \text{ m/s})\mathbf{k}$.
(b) $-(0.675 \text{ m/s}^2)\mathbf{i} + (3.00 \text{ m/s}^2)\mathbf{j} - (3.60 \text{ m/s}^2)\mathbf{k}$.
15.236. (a) $-(1.37 \text{ ft/s})\mathbf{i} + (3.76 \text{ ft/s})\mathbf{j} + (1.88 \text{ ft/s})\mathbf{k}$.
(b) $(1.22 \text{ ft/s}^2)\mathbf{i} - (3.42 \text{ ft/s}^2)\mathbf{j} - (0.410 \text{ ft/s}^2)\mathbf{k}$.
15.239 (a) $(4.33 \text{ ft/s})\mathbf{i} - (6.18 \text{ ft/s})\mathbf{j} + (5.30 \text{ ft/s})\mathbf{k}$.
(b) $(2.65 \text{ ft/s}^2)\mathbf{i} - (2.64 \text{ ft/s}^2)\mathbf{j} - (3.25 \text{ ft/s}^2)\mathbf{k}$.
15.240 (a) $(27.2 \text{ in/s}^2)\mathbf{i} - (6.75 \text{ in/s}^2)\mathbf{j}$.
(b) $(12.80 \text{ in/s}^2)\mathbf{i} - (7.68 \text{ in/s}^2)\mathbf{k}$.
15.241 (a) $-(1.600 \text{ in/s}^2)\mathbf{j} + (6.75 \text{ in/s}^2)\mathbf{j}$.
(b) $(12.80 \text{ in/s}^2)\mathbf{i} + (7.68 \text{ in/s}^2)\mathbf{k}$.
15.242 $-(5.04 \text{ m/s})\mathbf{i} - (1.200 \text{ m/s})\mathbf{k}$.
$-(9.60 \text{ m/s}^2)\mathbf{i} - (25.9 \text{ m/s}^2)\mathbf{j} + (57.6 \text{ m/s}^2)\mathbf{k}$.
15.243 $-(0.720 \text{ m/s})\mathbf{i} - (1.200 \text{ m/s})\mathbf{k}$,
$-(9.60 \text{ m/s}^2)\mathbf{i} + (25.9 \text{ m/s}^2)\mathbf{j} - (11.52 \text{ m/s}^2)\mathbf{k}$.
15.244 (a) $r\omega_2^2 \sin 30°\mathbf{j} - (r\omega_2^2 \cos 30° + 2r\omega_1\omega_2)\mathbf{k}$.
(b) $-r(\omega_1^2 + \omega_2^2 + 2\omega_1\omega_2 \cos 30°)\mathbf{i} + r\omega_1^2 \cos 30°\mathbf{k}$.
(c) $-r\omega_2^2 \sin 30°\mathbf{j} + r(2\omega_1^2 \cos 30° + \omega_2^2 \cos 30° + 2\omega_1\omega_2)\mathbf{k}$.
15.245 (a) $(0.610 \text{ m/s})\mathbf{k}, -(0.880 \text{ m/s}^2)\mathbf{i} + (1.170 \text{ m/s}^2)\mathbf{j}$.
(b) $(5.20 \text{ m/s})\mathbf{i} - (0.390 \text{ m/s})\mathbf{j} - (1.000 \text{ m/s})\mathbf{k}$,
$-(4.00 \text{ m/s}^2)\mathbf{i} - (3.25 \text{ m/s}^2)\mathbf{k}$.
15.248 (a) 6.00 m/s^2. (b) 9.98 m/s^2. (c) 60.0 m/s^2.
15.249 (a) $3.00 \text{ rad/s}^2 \downarrow$. ($b$) 4.00 s.
15.252 $\boldsymbol{\alpha}_{BD} = 306 \text{ rad/s}^2 \uparrow$, $\boldsymbol{\alpha}_{DE} = 737 \text{ rad/s}^2 \uparrow$.
15.253 (a) $1080 \text{ rad/s}^2 \downarrow$. ($b$) $460 \text{ ft/s}^2 \searrow 64.9°$.
15.255 $49.4 \text{ m/s}^2 \searrow 26.0°$.
15.256 (a) $(0.450 \text{ m/s})\mathbf{k}, (4.05 \text{ m/s}^2)\mathbf{i}$. ($b$) $-(1.350 \text{ m/s})\mathbf{k}, -(6.75 \text{ m/s}^2)\mathbf{i}$.
(b) AE: $23.5 \text{ m/s}^2 \leftarrow$, BD: $46.2 \text{ m/s}^2 \nearrow 60°$.
15.258 $(40.0 \text{ in/s})\mathbf{k}$.
15.259 $(9.00 \text{ in/s})\mathbf{i} - (7.80 \text{ in/s})\mathbf{j} + (7.20 \text{ in/s})\mathbf{k}$,
$(9.00 \text{ in/s}^2)\mathbf{i} - (22.1 \text{ in/s}^2)\mathbf{j} - (5.76 \text{ in/s}^2)\mathbf{k}$.

CHAPTER 16

16.1 11.72 ft/s^2
16.2 (a) $0.897 \text{ lb} \measuredangle 20°$. ($b$) $4.87 \text{ lb} \searrow 74.4°$.
16.3 (a) 25.8 ft/s^2. (b) 12.27 ft/s^2. (c) 13.32 ft/s^2.
16.4 (a) 3.20 m/s^2. (b) $\mathbf{A} = 3.82 \text{ N} \uparrow$, $\mathbf{B} = 20.7 \text{ N} \uparrow$.

16.5 (a) 4.09 m/s^2. (b) 42.5 N.
16.6 (a) $5270 \text{ N} \uparrow$. (b) 4120 N.
16.9 (a) $5.00 \text{ m/s}^2 \rightarrow$. ($b$) $0.311 \text{ m} \leq h \leq 1.489 \text{ m}$.
16.10 (a) $2.55 \text{ m/s}^2 \rightarrow$. ($b$) $h \leq 1.047 \text{ m}$.
16.11 195.9 kg.
16.14 (a) $4.91 \text{ m/s}^2 \nearrow 30°$. ($b$) $T_{AD} = 31.0 \text{ N}$, $T_{BE} = 11.43 \text{ N}$.
16.15 $51.3°$.
16.18 $\mathbf{B}_y = 40.3 \text{ lb} \downarrow$, $\mathbf{C}_y = 40.3 \text{ lb} \downarrow$.
16.19 (a) $30.6 \text{ ft/s}^2 \searrow 84.1°$. ($b$) $\mathbf{A} = 0.505 \text{ lb} \measuredangle 30°$,
$\mathbf{B} = 1.285 \text{ lb} \measuredangle 30°$.
16.20 Block: $17.01 \text{ ft/s}^2 \searrow 58.5°$; platform: $31.3 \text{ ft/s}^2 \searrow 30°$.
16.25 5230 rev.
16.26 $87.8 \text{ lb} \cdot \text{ft}$.
16.27 93.5 rev.
16.28 107.6 rev.
16.29 74.5 s.
16.30 $20.4 \text{ rad/s}^2 \downarrow$.
16.31 $32.7 \text{ rad/s}^2 \uparrow$.
16.33 (a) $5.66 \text{ ft/s}^2 \downarrow$. ($b$) $7.52 \text{ ft/s} \downarrow$.
16.34 (1): (a) $8.00 \text{ rad/s}^2 \uparrow$. ($b$) $14.61 \text{ rad/s} \uparrow$.
(2): (a) $6.74 \text{ rad/s}^2 \uparrow$. ($b$) $13.41 \text{ rad/s} \uparrow$.
(3): (a) $4.24 \text{ rad/s}^2 \uparrow$. ($b$) $10.64 \text{ rad/s} \uparrow$.
(4): (a) $5.83 \text{ rad/s}^2 \uparrow$. ($b$) $8.82 \text{ rad/s} \uparrow$.
16.36 (a) $6.06 \text{ rad/s}^2 \downarrow$. ($b$) $11.28 \text{ N} \nearrow$.
16.39 (a) No slipping on A; slipping on B.
(b) $\boldsymbol{\alpha}_A = 61.8 \text{ rad/s}^2 \uparrow$; $\boldsymbol{\alpha}_B = 9.66 \text{ rad/s}^2 \downarrow$.
16.40 (a) No slipping at either cylinder.
(b) $\boldsymbol{\alpha}_A = 15.46 \text{ rad/s}^2 \uparrow$; $\boldsymbol{\alpha}_B = 7.73 \text{ rad/s}^2 \downarrow$.
16.41 (a) $\boldsymbol{\alpha}_A = 12.50 \text{ rad/s}^2 \uparrow$, $\boldsymbol{\alpha}_B = 33.3 \text{ rad/s}^2 \uparrow$.
(b) $\boldsymbol{\omega}_A = 320 \text{ rpm} \downarrow$, $\boldsymbol{\omega}_B = 320 \text{ rpm} \uparrow$.
16.42 (a) $\boldsymbol{\alpha}_A = 12.50 \text{ rad/s}^2 \uparrow$, $\boldsymbol{\alpha}_B = 33.3 \text{ rpm} \uparrow$.
(b) $\boldsymbol{\omega}_A = 90.0 \text{ rpm} \uparrow$, $\boldsymbol{\omega}_B = 120.0 \text{ rpm} \downarrow$.
16.43 (a) $\boldsymbol{\alpha}_A = 9.16 \text{ rad/s}^2 \uparrow$, $\boldsymbol{\alpha}_B = 38.2 \text{ rad/s}^2 \uparrow$.
(b) $\mathbf{C} = 54.9 \text{ N} \uparrow$, $\mathbf{M}_C = 2.64 \text{ N} \cdot \text{m} \uparrow$.
16.44 (b) $\omega_0/(1 + m_B/m_A) \downarrow$.
16.48 (a) $18.40 \text{ ft/s}^2 \rightarrow$. ($b$) $9.20 \text{ ft/s}^2 \leftarrow$. ($c$) $z = 24.0 \text{ in}$.
16.49 (a) $12.0 \text{ in. from end } A$. ($b$) $9.20 \text{ ft/s}^2 \rightarrow$.
16.50 (a) $2.50 \text{ m/s}^2 \rightarrow$. ($b$) 0.
16.52 (a) $0, -1.374 \text{ rad/s}^2 \mathbf{j}$. ($b$) $-(0.515 \text{ ft/s}^2)\mathbf{i}, -1.030 \text{ rad/s}^2 \mathbf{j}$.
16.55 $rg/\bar{k}^2$.
16.56 (a) 0.865 N. (b) $72.1 \text{ rad/s}^2 \downarrow$.
16.57 $\mathbf{a}_A = 3.33 \text{ ft/s}^2 \downarrow$, $\mathbf{a}_B = 14.06 \text{ ft/s}^2 \uparrow$.
16.58 (a) $0.741 \text{ rad/s}^2 \uparrow$. ($b$) 0.857 m/s^2.
16.59 (a) 2800 N. (b) $15.11 \text{ rad/s}^2 \downarrow$.
16.60 $T_A = 359 \text{ lb}$, $T_B = 312 \text{ lb}$.
16.63 (a) $3 g/L \downarrow$. (b) $g \uparrow$. (c) $2 g \downarrow$.
16.64 (a) $2 g/L \downarrow$. (b) $g/3 \uparrow$. (c) $5 g/3 \downarrow$.
16.65 (a) $g/L \downarrow$. (b) $0.866 g \leftarrow$. (c) $0.5 g \downarrow$.
16.66 (a) $0.25 g \uparrow$. (b) $5 g/4 \downarrow$.
16.67 (a) $\frac{1}{2} g \uparrow$. (b) $\frac{3}{2} g \downarrow$.
16.68 (a) $g(\mathbf{i} - \mathbf{j})/2 + 1.5 ga (a + b)/(a^2 + b^2)\mathbf{j}$.
(b) $g(\mathbf{i} - \mathbf{j})/2 - 1.5 ga (a + b)/(a^2 + b^2)\mathbf{j}$.
16.69 (a) $5v_0/2r \uparrow$. (b) $v_0/\mu_k g$. (c) $v_0^2/2\mu_k g$.
16.70 (a) $v_0/r \uparrow$. (b) $v_0/\mu_k g$. (c) $v_0^2/2\mu_k g$.
16.71 (a) 1.597 s. (b) 9.86 ft/s. (c) 19.85 ft.
16.72 (a) 1.863 s. (b) 9.00 ft/s. (c) 22.4 ft.
16.76 (a) $107.1 \text{ rad/s}^2 \downarrow$. ($b$) $21.4 \text{ N} \leftarrow$, $39.2 \text{ N} \uparrow$.
16.77 (a) 150 mm. (b) $125 \text{ rad/s}^2 \downarrow$.
16.78 (a) $12.08 \text{ rad/s}^2 \downarrow$. ($b$) $0.750 \text{ lb} \leftarrow$, $4.00 \text{ lb} \uparrow$.
16.79 (a) $8.05 \text{ rad/s}^2 \downarrow$. ($b$) 24.0 in.
16.80 (a) $\omega z(l - z/2)\omega^2$. ($b$) 5.09 N.
16.83 $13.64 \text{ kN} \rightarrow$.

16.84 (a) 1.5 g ↓. (b) 0.25 mg ↑.

16.85 (a) 9g/7. (b) 4mg/7 ↑.

16.86 (a) 9.02 lb ↑. (b) 74.5 lb ←, 8.57 lb ↑.

16.87 (a) 43.6 rad/s². (b) 21.0 N ←, 54.6 N ↑.

16.88 (a) 10.87 rad/s². (b) 8.49 ft · lb ↘.

16.94 $r^2g \sin \beta/(r^2 + \bar{k}^2)$.

16.95 (a) 2.27 m (7.46 ft). (b) 0.649 m (2.13 ft).

16.98 (a) 17.78 rad/s² ↓, 2.13 m/s² →. (b) 0.122.

16.99 (a) 26.7 rad/s² ↓, 3.20 m/s² →. (b) 0.0136.

16.102 (a) no sliding. (b) 15.46 rad/s² ↓, 10.30 ft/s².

16.103 (a) no sliding. (b) 23.2 rad/s² ↓, 15.46 ft/s².

16.104 (a) slides. (b) 4.29 rad/s² ↘, 9.66 ft/s² →.

16.105 (a) slides. (b) 12.88 rad/s² ↘, 3.22 ft/s² ←.

16.107 (a) 6.63 ft/s² →. (b) 3.79 ft/s² →. (c) 0.355 ft →.

16.108 72.4 rad/s² ↘. (b) 7.24 m/s² ↓.

16.109 (a) $\mathbf{a}_A$ = 1.923 m/s² ←, $\mathbf{a}_B$ = 1.923 m/s² ←. (b) 4.33 N ←.

16.111 (a) 0.298. (b) 0.536 g →

16.112 (a) 0.322. (b) 0.566 g →

16.113 8.26 N ←.

16.114 (a) 0.125 g/r ↓. (b) 0.125 g →, 0.125 g ↓.

16.115 $m_Bg \sin \theta/[2r \{m_h + m_B (1 + \cos \theta)\}]$

16.116 3.43 lb ∠ 70.5°, 0.1550 ft · lb ↓.

16.117 (a) 10.94 rad/s² ↓. (b) 7.68 lb ↑. (c) 5.75 lb →.

16.118 (a) 20.1 lb ↑. (b) 12.42 lb ↑.

16.119 (a) 10.62 rad/s² ↘. (b) 4.25 N ←.

16.120 $mg \sin \theta/(1 + 3 \sin^2 \theta)$.

16.121 (a) 10.41 rad/s² ↓. (b) 36.8 N. (c) 61.3 N.

16.124 6.40 N ←.

16.125 7.95 lb ∠ 60°.

16.126 6.02 lb ∠ 60°.

16.127 171.7 N →.

16.128 60.0 N →.

16.129 25.9 N ↘ 60°.

16.131 (a) 37.8 ft/s² ↘ 26.1°. (b) 48.4 lb ↑.

16.134 0.330 lb ←.

16.135 (a) 36.3 N · m ↘. (b) 231 N ←, 524 N ↑.

16.136 (a) 82.3 N · m ↘. (b) 147.2 N ←, 479 N ↑.

16.137 **B** = 805 N ←, **D** = 426 N →.

16.138 **B** = 525 N ↗ 38.1°, **D** = 322 N ↘ 15.7°.

16.139 **A** = 1.565 lb ↑, **B** = 1.689 lb ↑.

16.140 **A** = 0.839 lb → + 2.00 lb ↑, **B** = 0.280 lb ← + 2.00 lb ↑.

16.143 (a) A: 0.400 g/r ↘, B: 0.400 g/r ↓. (b) 0.200 mg. (c) 0.800 g ↓.

16.144 A: 2P/7m →, B: 22P/7m ←.

16.145 (a) 5.63 m/s² ↘ 25°. (b) 7.66 rad/s² ↓.

16.146 (a) 13.55 m/s² ↓. (b) 2.34 m/s² ↓.

16.147 (a) 6.40 ft/s² →. (b) 45.4 rad/s² ↘.

*16.148 (a) 17.03 ft/s² ↘ 20°. (b) 42.7 rad/s² ↘.

16.151 M_{max} = 10.39 lb · in. located 20.8 in. below A.

16.153 20.6 ft.

16.154 12.34 ft.

16.156 (a) 2μg/(1 + 3μ). (b) 1.000 g.

16.157 (a) 0.513 g/L ↓. (b) 0.912 mg ↑. (c) 0.241 mg →.

16.158 (a) 1.519 g/L ↓. (b) 0.260 g ↓. (c) 0.740 mg ↑.

16.160 (1): (a) 1.200 g/c ↓. (b) 0.671 g ↗ 63.4°. (2): (a) 24 g/17c ↓. (b) 12 g/17 ↓. (3): (a) 2.40 g/c ↓. (b) 0.500 g ↓.

16.162 (a) 51.2 rad/s² ↓. (b) 21.0 N ↑.

16.163. (a) 59.8 rad/s² ↓. (b) 20.4 N ↑.

CHAPTER 17

17.1 188.1 mm.

17.2 58.7 lb · ft.

17.3 4.10 in.

17.4 0.760.

17.5 11.46 rev.

17.6 (a) 293 rpm. (b) 15.92 rev.

17.7 19.77 rev.

17.10 109.4 lb →.

17.11 (a) 6.35 rev. (b) 7.14 N.

17.12 (a) 2.54 rev. (b) 17.86 N.

17.13 80.7 lb ↓.

17.16 (a) $\sqrt{3g/L}$, 2.50 W ↑. (b) 5.67 rad/s, 4.50 lb ↑.

17.17 (a) 0.289 l. (b) 1.861$\sqrt{g/l}$, 2.00 mg ↑.

17.18 11.52 rad/s ↘.

17.19 4.61 rad/s ↓.

17.20 (a) 3.94 rad/s ↓, 271 lb ↘ 5.25°. (b) 5.58 rad/s ↓, 701 lb ↑.

17.23 7.09 rad/s.

17.24 (a) −0.250 rpm. (b) 0.249 rpm.

17.25 $\sqrt{4gs/3}$.

17.26 $\sqrt{gs}$.

17.27 (a) 9.77 ft/s. (b) 6.67 lb ←.

17.29 (a) 5.00 rad/s. (b) 24.9 N ↑.

17.30 (a) 1.324$\sqrt{g/l}$. (b) 2.12 mg ↑.

17.31 (a) $[10g (R - n) (1 - \cos \beta)/7]^{1/2}$. (b) mg(17 − 10 cos β)/7.

17.32 (a) 2.06 ft. (b) 4.00 lb.

17.33 (a) 7.43 ft/s ↓. (b) 4.00 lb.

17.35 292 mm/s →.

17.36 $\mathbf{v}_A$ = 0.775$\sqrt{gl}$ ←, $\mathbf{v}_B$ = 0.775$\sqrt{gl}$ ↗ 60°.

17.37 1.170 rad/s ↓, 5.07 m/s ←.

17.38 $[3g (\cos \theta_0 - \cos \theta_2)/L]^{1/2}$ ↓.

17.39 3.71 rad/s ↑, 7.74 ft/s ↑.

17.40 4.03 rad/s ↓, 7.27 ft/s ↓.

17.42 2.69 m/s ↓.

17.43 84.7 rpm ↓.

17.44 110.8 rpm ↓.

17.45 3.25 m/s ↓.

17.46 4.43 m/s ↓.

17.47 0.770 m/s ←.

17.48 (a) 44.3 hp. (b) 118.1 hp.

17.49 (a) 39.8 N · m. (b) 95.5 N · m. (c) 229 N · m.

17.50 1146 rpm.

17.51 0.343 hp.

17.52 179.1 mm.

17.53 0.335 lb · in.

17.54 3.87 rad/s.

17.55 2.84 s.

17.58 3.88 s.

17.59 $(1 + \mu_k^2) r\omega_0/2\mu_k(1 + \mu_k)g$.

17.62 $\omega_0/(1 + m_A/m_B)$.

17.63 (a) $\boldsymbol{\omega}_A$ = 686 rpm ↑, $\boldsymbol{\omega}_B$ = 514 rpm ↓. (b) 4.18 lb · s ↑.

17.64 (a) 5.15 lb. (b) 2.01 lb.

17.65 **X** = m**v**, $d = \bar{k}^2\omega/\bar{v}$.

17.69 2.79 ft.

17.70 (a) $r^2gt \sin \beta/(r^2 + \bar{k}^2)$ ↘ β. (b) $u_s \geq \bar{k}^2 \tan \beta/(r^2 + \bar{k}^2)$.

17.71 (a) 2.55 m/s ↑. (b) 10.53 N

17.72 (a) 8.05 ft/s →. (b) 2.68 ft/s →.

17.74 (a) 8.41 m/s ↓. (b) 16.82 N.

17.75 (a) 0.557 s. (b) 16.82 N.
17.77 (a) $2.50\,\bar{v}_0/r$. (b) $\bar{v}_0/\mu_k g$.
17.78 (a) 2.50 s. (b) 16.95 ft/s.
17.79 $0.614\,\omega_0$.
17.80 84.2 rpm.
17.81 A and B: 159.1 rpm ↓; platform 20.9 rpm ↑.
17.82 18.07 rad/s.
17.83 (a) 2.54 rad/s. (b) 1.902 J.
17.86 37.2 rpm.
17.87 A and B: 212 rpm ↓; platform 27.9 rpm ↑.
17.88 2.51 m/s.
17.89 18.83 rad/s, 0.0508 kg · m².
17.90 (a) 31.1 rad/s. (b) 18.13 ft/s.
17.91 (a) 15.00 rad/s. (b) 20.5 ft/s.
17.93 7.45 m/s.
17.94 1.542 m/s.
17.95 2.01 ft/s ←.
17.96 $0.400\,r$.
17.97 (a) 24.4 rad/s ↓. (b) 1545 lb →.
17.98 (a) 10.00 in. (b) 22.6 rad/s ↓.
17.101 (a) 2.16 m/s →. (b) 4.87 kN ∡ 66.9°.
17.102 (a) 79.2 mm. (b) 1.992 m/s →.
17.103 (a) $0.429\,v_0/L$ ↓. (b) $0.571\,v_0$ →.
17.104 2.40 rad/s ↓.
17.105 1.667 in.
17.106 $(2 + 5\cos\beta)\,\omega_1/7$ ↑, $(2 + 5\cos\beta)\bar{v}_1/7$.
17.107 1.414.
17.108 (a) mv_0/M →. (b) mv_0/MR ↑.
17.109 (a) 1.500 R. (b) 1.000 R.
17.112 $6\,(v_1/L)\sin\beta/(1 + 3\sin^2\beta)$.
17.115 2.38 m/s.
17.116 (a) 0.366.
17.117 (a) $0.437\sqrt{g/L}$. (b) 5.12°.
17.118 (a) $0.250\,\omega_0$ ↓. (b) 0.9375. (c) 1.50°.
17.119 48.7°.
17.120 1887 ft/s.
17.121 725 mm.
17.122 447 mm.
17.123 (a) 3.85 ft/s ↓. (b) 5.13 rad/s ↓.
17.124 (a) 0.256 ft/s ↑. (b) 9.23 rad/s ↓.
17.127 (a) 3.00 rad/s ↑. (b) 0.938 m/s ↑.
17.128 (a) 2.60 rad/s ↓. (b) 1.635 m/s ⭦ 53.4°.
17.131 $1.250\,v_0/r$.
17.132 (a) $\mathbf{v}_A = 0$, $\boldsymbol{\omega}_A = v_1/r$ ↓, $\mathbf{v}_B = v_1$ →, $\omega_B = 0$.
(b) $\mathbf{v}'_A = 0.286\,v_1$ →, $v'_B = 0.514\,v_1$ →.
17.133 (a) $\mathbf{v}_A = (v_0\sin\theta)\mathbf{j}$, $\mathbf{v}_B = (v_0\cos\theta)\mathbf{i}$, $\boldsymbol{\omega}_A = (v_0/r)\,(-\sin\theta\mathbf{i} + \cos\theta\mathbf{j})$, $\boldsymbol{\omega}_B = 0$.
(b) $0.514\,(v_0/r)\cos\theta\mathbf{i}$.
17.134 $\boldsymbol{\omega}_{AB} = 2.68$ rad/s ↓, $\boldsymbol{\omega}_{BC} = 13.39$ rad/s ↓.
17.135 (a) 106.7 rev. (b) 6.98 s.
17.136 70.1 lb ↓.
17.137 4.81 rad/s.
17.139 (a) 53.1°. (b) $1.095\sqrt{gL}$ ⭦ 53.1°.
17.140 $\mathbf{A} = 100.1$ N ↑, $\mathbf{B} = 43.9$ N →.
17.142 $0.778\,\omega_0$.
17.143 (a) 418 rpm. (b) −20.4 J.
17.145 (a) 68.6 rpm. (b) 2.82 J.

CHAPTER 18

18.1 $0.250\,mr^2\,\omega_2\,\mathbf{j} + 0.500\,mr^2\,\omega_1\mathbf{k}$.
18.2 $-(0.408\text{ slug} \cdot \text{ft}^2/\text{s})\mathbf{i} + (0.1398\text{ slug} \cdot \text{ft}^2/\text{s})\mathbf{j}$.

18.3 0.247 slug · ft²/s, $\theta_x = 48.6°$, $\theta_y = 41.4°$, $\theta_z = 90°$.
18.5 $(0.1125\text{ kg} \cdot \text{m}^2/\text{s})\mathbf{j} + (0.675\text{ kg} \cdot \text{m}^2/\text{s})\mathbf{k}$.
18.7 $0.432\,ma^2\omega$, 20.2°.
18.8 11.88°.
18.9 $(1.843\text{ lb} \cdot \text{ft} \cdot \text{s})\mathbf{i} - (0.455\text{ lb} \cdot \text{ft} \cdot \text{s})\mathbf{j} + (1.118\text{ lb} \cdot \text{ft} \cdot \text{s})\mathbf{k}$.
18.10 $-(1.747\text{ lb} \cdot \text{s} \cdot \text{ft})\mathbf{i} + (3.59\text{ lb} \cdot \text{s} \cdot \text{ft})\mathbf{j} + (0.0582\text{ lb} \cdot \text{s} \cdot \text{ft})\mathbf{k}$.
18.11 $0.500\,mr^2\omega_1\mathbf{i} - m(L^2 + 0.250\,r^2)\,(r\omega_1/L)\mathbf{j}$.
18.12 (a) 0.485 rad/s. (b) 0.01531 rad/s.
18.15 (a) $(1.563\text{ kg} \cdot \text{m}^2/\text{s})\mathbf{i} - (0.938\text{ kg} \cdot \text{m}^2/\text{s})\mathbf{k}$. (b) 31.0°.
18.16 (a) $(1.563\text{ kg} \cdot \text{m}^2/\text{s})\mathbf{i} - (0.938\text{ kg} \cdot \text{m}^2/\text{s})\mathbf{k}$.
(b) $(1.563\text{ kg} \cdot \text{m}^2/\text{s})\mathbf{i} - (0.938\text{ kg} \cdot \text{m}^2/\text{s})\mathbf{k}$.
18.17 (a) $-(1.041\text{ lb} \cdot \text{ft} \cdot \text{s})\mathbf{i} + (1.041\text{ lb} \cdot \text{ft} \cdot \text{s})\mathbf{j} + (2.31\text{ lb} \cdot \text{ft} \cdot \text{s})\mathbf{k}$.
(b) 147.5°.
18.18 (a) $-(1.041\text{ lb} \cdot \text{ft} \cdot \text{s})\mathbf{i} - (1.041\text{ lb} \cdot \text{ft} \cdot \text{s})\mathbf{j} + (2.31\text{ lb} \cdot \text{ft} \cdot \text{s})\mathbf{k}$.
(b) 32.5°.
18.21 93.6 kg.
18.22 2.57 s.
18.25 (a) 0. (b) $(3F\Delta t/ma)\,(\mathbf{i} - 4\mathbf{k})$.
18.26 (a) $-(F\Delta t/m)\mathbf{i}$. (b) $(3\,F\Delta t/8ma)\,(\mathbf{j} + 4\mathbf{k})$.
18.27 (a) $-(0.300\text{ m/s})\mathbf{i}$. (b) $-(0.962\text{ rad/s})\mathbf{i} - (0.577\text{ rad/s})\mathbf{j}$.
18.28 (a) $(0.300\text{ m/s})\mathbf{j}$.
(b) $-(3.46\text{ rad/s})\mathbf{i} + (1.923\text{ rad/s})\mathbf{j} - (0.857\text{ rad/s})\mathbf{k}$.
18.31 (a) $0.1250\,\omega_0\,(-\mathbf{i} + \mathbf{j})$. (b) $0.0884\,a\omega_0\mathbf{k}$.
18.32 (a) $0.1031\,ma\omega_0\mathbf{k}$. (b) $-0.01473\,ma\omega_0\mathbf{k}$.
18.33 $(0.0248\text{ rad/s})\mathbf{i} - (0.277\text{ rad/s})\mathbf{j} - (0.360\text{ rad/s})\mathbf{k}$.
18.34 (a) -0.726 rad/s.
(b) $-(2160\text{ ft/s})\mathbf{i} - (4860\text{ ft/s})\mathbf{j} + (860\text{ ft/s})\mathbf{k}$.
18.35 (a) C and B. (b) B: 4.84 s, C: 8.16 s. (c) 0.520 s.
18.36 (a) D and A. (b) A: 1.848 s, D: 6.82 s. (c) 0.347 s.
18.39 $0.1250\,mr^2\,(\omega_2^2 + 2\omega_1^2)$.
18.40 0.349 ft · lb.
18.41 0.978 ft · lb.
18.42 12.67 ft · lb.
18.43 15.47 J.
18.44 $0.1250\,ma^2\omega^2$.
18.45 $0.203\,ma^2\omega^2$.
18.47 9.38 J.
18.49 1.296 J.
18.50 46.2 J.
18.51 $0.1000\,m\bar{v}_0^2$.
18.53 16.75 ft · lb.
18.54 39.9 ft · lb.
18.55 $0.500\,mr^2\,\omega_1\omega_2\mathbf{i}$.
18.56 $(0.204\text{ ft} \cdot \text{lb})\mathbf{k}$.
18.57 $(2.22\text{ ft} \cdot \text{lb})\mathbf{k}$.
18.58 $(5.30\text{ lb} \cdot \text{ft})\mathbf{k}$.
18.59 $(3.38\text{ N} \cdot \text{m})\mathbf{i}$.
18.60 $0.0958\,mr^2\omega^2\mathbf{k}$.
18.64 $mr^2\,(0.0958\,\alpha\mathbf{i} + 0.455\,\alpha\mathbf{j} - 0.0958\,\omega^2\mathbf{k})$.
18.65 $\mathbf{C} = 0.1667\,mb\omega^2\sin\beta\cos\beta\mathbf{i}$.
$\mathbf{D} = -0.1667\,mb\omega^2\sin\beta\cos\beta\mathbf{i}$.
18.66 $\mathbf{A} = -(4.97\text{ lb})\mathbf{i}$, $\mathbf{B} = -(1.656\text{ lb})\mathbf{i}$.
18.67 $\mathbf{A} = -(1.103\text{ lb})\mathbf{j} - (0.920\text{ lb})\mathbf{k}$.
$\mathbf{B} = (1.103\text{ lb})\mathbf{j} + (0.920\text{ lb})\mathbf{k}$.
18.68 $\mathbf{A} = (14.4\text{ N})\mathbf{k}$, $\mathbf{B} = -(14.4\text{ N})\mathbf{k}$.
18.71 (a) $3M_0/mb^2\cos^2\beta$. (b) $\mathbf{C} = -\mathbf{D} = (M_0\tan\beta/2b)\mathbf{k}$.
18.72 (a) $(14.49\text{ rad/s}^2)\mathbf{j}$. (b) $\mathbf{A} = -(1.125\text{ lb})\mathbf{k}$, $\mathbf{B} = -(0.375\text{ lb})\mathbf{k}$.
18.73 (a) $(0.873\text{ lb} \cdot \text{ft})\mathbf{i}$ (b) $\mathbf{A} = -\mathbf{B} = -(0.218\text{ lb})\mathbf{j} + (0.262\text{ lb})\mathbf{k}$.
18.74 (a) $(2.67\text{ N} \cdot \text{m})\mathbf{i}$. (b) $\mathbf{A} = -\mathbf{B} = (2.00\text{ N})\mathbf{j}$.
18.75 (a) $(0.1301\text{ lb} \cdot \text{ft})\mathbf{i}$. (b) $\mathbf{A} = -\mathbf{B} = -(0.0331\text{ lb})\mathbf{i} + (0.0331\text{ lb})\mathbf{j}$.
18.76 $\mathbf{A} = -\mathbf{B} = -(0.449\text{ lb})\mathbf{j} - (0.383\text{ lb})\mathbf{k}$.
18.79 $\mathbf{A} = -\mathbf{B} = (1.527\text{ N})\mathbf{j}$.

18.81 (a) 10.47 N · m. (b) 10.47 N · m.

18.82 24.0 N ↑.

18.84 1.138°; up.

18.85 10.20 rad/s.

18.86 (a) 38.1°. (b) 11.78 rad/s.

18.87 13.46 rad/s.

18.88 7.00 rad/s.

18.90 5.45 rad/s.

18.91 2.11 N ∡ 18.7°.

18.93 $\mathbf{A} = -\mathbf{B} = (0.1906 \text{ lb})\mathbf{k}$.

18.94 7.87 rad/s.

18.95 (a) $\mathbf{C} = -\mathbf{D} = -(123.4 \text{ N})\mathbf{i}$. (b) $\mathbf{C} = \mathbf{D} = 0$.

18.96 91.2 rpm.

18.99 $-(45.0 \text{ N})\mathbf{i}$, $(3.38 \text{ N} \cdot \text{m})\mathbf{i} + (10.13 \text{ N} \cdot \text{m})\mathbf{k}$.

18.100 (a) $\mathbf{A} = (1.786 \text{ kN})\mathbf{i} + (143.5 \text{ kN})\mathbf{j}$,
$\mathbf{B} = -(1.786 \text{ kN})\mathbf{i} + (150.8 \text{ kN})\mathbf{j}$, (b) $-(35.7 \text{ kN} \cdot \text{m})\mathbf{k}$.

18.101 $\mathbf{C} = -(7.81 \text{ lb})\mathbf{i} + (7.43 \text{ lb})\mathbf{k}$,
$\mathbf{D} = -(7.81 \text{ lb})\mathbf{i} - (7.43 \text{ lb})\mathbf{k}$.

18.102 $\mathbf{C} = -(12.58 \text{ lb})\mathbf{i} + (9.43 \text{ lb})\mathbf{k}$,
$\mathbf{D} = -(12.58 \text{ lb})\mathbf{i} - (9.43 \text{ lb})\mathbf{k}$.

18.103 $\mathbf{D} = -(22.0 \text{ N})\mathbf{i} + (26.8 \text{ N})\mathbf{j}$, $\mathbf{E} = -(21.2 \text{ N})\mathbf{i} - (5.20 \text{ N})\mathbf{j}$.

18.104 (a) $(0.392 \text{ N} \cdot \text{m})\mathbf{k}$. (b) $\mathbf{D} = -(21.0 \text{ N})\mathbf{i} + (28.0 \text{ N})\mathbf{j}$,
$\mathbf{E} = -(21.0 \text{ N})\mathbf{i} - (4.00 \text{ N})\mathbf{j}$.

18.107 1666 rpm.

18.109 45.9 rpm, 533 rpm.

18.111 442 rpm.

18.112 68.1°.

18.113 (a) 40.0°. (b) 23.5°. (c) 85.3°.

18.114 (a) 56.1 rad/s. (b) 5.30 rad/s.

18.115 23.7°.

18.116 (a) 52.7 rad/s. (b) 6.44 rad/s.

18.117 (a) 4.89 rpm. (b) 4.96 rpm, 396 rpm.

18.124 (a) 13.19°. (b) 1242 rpm (retrograde)

18.125 (a) 23.8°. (b) precession: 82.6 rpm; spin: 128.8 rpm.

18.126 (a) 32.0°. (b) precession: 1.126 rpm; spin: 0.344 rpm.

18.127 (a) 52.0°. (b) precession: 0.1523 rad/s (retrograde);
spin: 0.0338 rad/s.

18.128 (a) 109.4 rpm, $r_x = 90°$, $r_y = 100.05°$, $r_z = 10.05°$.
(b) $\theta_x = 90°$, $\theta_y = 113.9°$, $\theta_z = 23.9°$.
(c) precession: 47.1 rpm; spin: 64.6 rpm.

18.130 (a) $\theta_x = 90.0°$, $\theta_y = 26.0°$, $\theta_z = 64.0°$.
(b) precession, 0.847 rad/s (retrograde); spin: 0.1593 rad/s.

18.131 (a) $40.0° < \theta < 140.0°$. (b) 5.31 rad/s. (c) 5.58 rad/s.

18.132 (a) 2.00 rad/s. (b) 8.94 rad/s.

18.135 (a) $\theta_0 \le \theta \le 180° - \theta_0$. (b) $\dot{\phi}_0 \sin \theta_0 \cos \theta_0$.
(c) $\dot{\phi}_0 \sin^2 \theta_0$.

18.136 (a) $(1 + \cos^2 \theta) \dot{\phi}^2 + \dot{\theta}^2 = \text{constant}$; $\dot{\phi}(1 + \cos^2 \theta) = \text{constant}$.
(b) $\dot{\theta} = \dot{\phi}_0[(1 + \cos^2 \theta_0)(\cos^2 \theta - \cos^2 \theta_0)/(1 + \cos^2 \theta)]^{1/2}$.
(c) $\theta < \theta_0$.

18.139 (a) 47.0°. (b) precession: 15.25 rad/s; spin: 307 rad/s.

18.140 (a) 76.3°. (b) precession: 9.62 rad/s; spin: 294 rad/s. (c) 36.5°.

18.148 $(0.234 \text{ kg} \cdot \text{m}^2/\text{s})\mathbf{j} + (1.250 \text{ kg} \cdot \text{m}^2/\text{s})\mathbf{k}$.

18.150 (a) 0. (b) $(F\Delta t/ma) (2.50\mathbf{i} - 1.454\mathbf{j} + 2.19\mathbf{k})$.

18.151 4.29 kN · m.

18.153 $\mathbf{D} = -(7.12 \text{ lb})\mathbf{j} + (4.47 \text{ lb})\mathbf{k}$, $\mathbf{E} = -(1.822 \text{ lb})\mathbf{j} + (4.47 \text{ lb})\mathbf{k}$.

18.154 $\mathbf{D} = 0$; $\mathbf{M}_D = (11.23 \text{ N} \cdot \text{m}) \cos^2 \theta \mathbf{i} +$
$(11.23 \text{ N} \cdot \text{m}) \sin \theta \cos \theta \mathbf{j} - (2.81 \text{ N} \cdot \text{m}) \sin \theta \cos \theta \mathbf{k}$.

18.155 (a) $\theta_x = 52.5°$, $\theta_y = 37.5°$, $\theta_z = 90°$.
(b) 53.8 rev/h. (c) 6.68 rev/h.

18.156 (a) $(2.71 \text{ lb} \cdot \text{ft})\mathbf{j}$. (b) $\mathbf{R} = -(5.30 \text{ lb})\mathbf{i} - (1.988 \text{ lb})\mathbf{k}$;
$\mathbf{M}_O = (2.69 \text{ lb} \cdot \text{ft})\mathbf{i} - (4.42 \text{ lb} \cdot \text{ft})\mathbf{k}$

18.157 (a) 4.00 rad/s. (b) 5.66 rad/s.

CHAPTER 19

19.1 0.377 m/s, 47.3 m/s^2.

19.2 4.33 ft/s, 0.551 Hz.

19.3 0.1026 in., 5.16 in/s.

19.4 (a) 0.324 s, 3.08 Hz. (b) 12.91 mm, 4.84 m/s^2.

19.5 (a) 0.308 s, 3.25 Hz. (b) 1.021 m/s, 20.8 m/s^2.

19.6 (a) 267 rpm. (b) 5.36 ft/s.

19.7 (a) 11.29°. (b) 1.933 m/s^2.

19.8 (a) 0.557 Hz. (b) 293 mm/s.

19.11 (a) 0.0352 s. (b) 6.34 ft/s ↑, 64.4 ft/s^2 ↓.

19.12 0.445 ft ↑, 2.27 ft/s ↓, 114.7 ft/s^2 ↓.

19.13 (a) 1.288°. (b) 0.874 ft/s, 0.759 ft/s^2.

19.14 (a) 4.91 mm, 5.81 Hz, 0.1791 m/s.
(b) 491 N, (c) 0.1592 m/s ↑.

19.17 2.63 s.

19.18 (a) 0.206 s, 4.85 Hz. (b) 5.08 ft/s, 154.6 ft/s^2.

19.19 (a) 0.413 s, 2.42 Hz. (b) 2.54 ft/s, 38.6 ft/s^2.

19.20 (a) 0.361 s, 2.77 Hz. (b) 0.765 m/s, 13.30 m/s^2.

19.23 (a) 35.6 lb/in. (b) 5.01 lb.

19.24 (a) 6.80 kg. (b) 0.583 s.

19.25 192 lb/ft.

19.26 (a) 21.7 kg. (b) 1011 kg/m^3.

19.28 (a) 22.3 MN/m. (b) 266 Hz.

19.30 (a) 858 N/mm. (b) 149.5 rpm.

19.31 (a) 3.56 kg. (b) 43.7 kg.

19.34 16.26°.

19.35 (a) 1.737 s. (b) 1.864 s. (c) 2.05 s.

19.36 28.1 in.

19.37 (a) 0.293 s. (b) 0.215 m/s.

19.38 (a) 1.047 rad/s. (b) 16.42 in.

19.39 (a) 1.740 s. (b) 90.3 mm/s.

19.41 (a) 0.491 s. (b) 9.60 in/s.

19.42 (a) 0.1947 s. (b) 171.7 ft/s^2.

19.44 75.5°.

19.45 0.346 Hz.

19.48 (a) 2.28 s. (b) 1.294 m.

19.49 (a) 1.617 s. (b) 1.676 s.

19.50 (a) 227 mm. (b) 1.352 s.

19.51 (a) 1.075 s. (b) 3.59 in.

19.55 (a) 2.21 Hz. (b) 115. 3 N/m.

19.56 0.945 Hz.

19.57 3.03 Hz.

19.59 (a) 0.426 s. (b) 15.44 ft/s.

19.60 (a) 88.1 mm. (b) 85.1 mm.

19.62 82.2 mm/s ↑.

19.63 6.57 kg · m^2.

19.64 (a) 21.3 kg. (b) 1.836 s.

19.67 0.672 in.

19.68 8.60 ft.

19.69 1.476 m/s, 31.1 m/s^2.

19.70 12.11 m/s.

19.71 3.18 s.

19.72 $6.28\sqrt{R/g}$.

19.75 $l/\sqrt{12}$

19.76 75.5°.

19.77 $0.363\sqrt{(2k/3m) + (4g/3L)}$.

19.78 2.10 Hz.

19.79 (a) 0.715 s. (b) 0.293 ft/s.

19.80 0.821 s.

19.83 1.327 s.

19.85 1.834 s.

19.86 2.39 s.

19.87 $2\pi\sqrt{(12r^2 + 2l^2)/3gl}$.

19.89 (a) $\sqrt{(6ka^2 - 3mgl)}/(2\pi)$. (b) $\sqrt{mgl/2k}$.

19.90 2.29 Hz.

19.91 0.911 Hz.

19.92 $0.1312\sqrt{g/r}$.

19.93 $0.1125\sqrt{g/l}$.

19.95 $0.276\sqrt{g/l}$.

19.96 2.59 Hz.

19.97 $1.814l/\sqrt{gr}$.

19.98 0.352 s.

19.99 (a) 16.67 mm (in-phase). (b) 128.2 mm (in-phase).
(c) 10.00 mm (out-of-phase).

19.100 15.28 rad/s, 23.8 rad/s.

19.101 (a) 10.99 lb/ft. (b) 2.99 lb/ft.

19.102 $\sqrt{k/2m} < \omega_f < \sqrt{3k/2m}$.

19.105 $\omega_f < 8.46$ rad/s, (no out-of-phase solution).

19.106 $\omega_f < 24.8$ rad/s (in-phase), $\omega_f > 42.9$ rad/s (out-of-phase).

19.107 22.5 mm, −5.63 mm.

19.108 (a) 1.406 ft³/min. (b) 3.31 ft³/min.

19.109 $\omega_f > \sqrt{2g/l}$.

19.110 (a) 1.034 in. (b) − 0.1033 sin πt (lb).

19.112 651 rpm.

19.114 22.0 mm.

19.115 0.0999 in.

19.116 $\omega_f < 322$ rpm.

19.117 (a) −0.00303 in. (b) 0.000758 in. (c) −0.01990 in.

19.119 149.3 mm.

19.120 (a) $\omega_f \leq 286$ rpm. (b) $\omega_f \geq 367$ rpm.

19.121 Force transmissibility: $1/(1 - \omega_f^2/\omega_n^2)$
Displacement transmissiblity: $1/(1 - \omega_f^2/\omega_n^2)$

19.122 (a) 4.17%. (b) 84.9 Hz.

19.123 8.04%.

19.125 (a) 1399 rpm. (b) 0.01670 in.

19.126 (a) 25.6 km/h. (b) −14.25 mm.

19.132 (a) 6.49 kip · s/ft. (b) 230 kips/ft.

19.133 5.48 N · m · s.

19.134 (a) 6490 lb/ft. (b) 0.1939 s.

19.135 (a) 297 rpm. (b) 252 rpm. (c) 0.335 in., 0.361 in.

19.136 56.9 mm.

19.137 0.0431.

19.138 (a) $\ddot{\theta} + (3c/m)\dot{\theta} + (3k/4m)\theta = 0$
(b) $\sqrt{km/3}$.

19.139 0.1791 in.

19.140 10.61 lb · s/in.

19.141 ≥ 0.707.

19.143 (a) 147 kip/ft. (b) 0.0292.

19.144 > 30.8 Hz and < 15.85 Hz.

19.145 0.539 in.

19.146 (a) 0.0924. (b) 25.2 lb · s/ft.

19.148 (a) 71.8 N. (b) 39.0 N.

19.149 (a) 4.90 in. (b) 30.3 lb.

19.151 (a) $m\ddot{x} + c\dot{x} + kx = (k \sin \omega_f t + c\omega_f \cos \omega_f t)\delta_m$.
(b) $x = x_m \sin (\omega_f t - \varphi + \psi)$, where
$x_m = \delta_m\sqrt{k^2 + (c\omega_f)^2}/\sqrt{(k - m\omega_f^2)^2 + (c\omega_f)^2}$,
$\tan \varphi = c\omega_f/(k - m\omega_f^2)$, $\tan \psi = c\omega_f/k$.

19.153 $R < 2\sqrt{L/C}$.

19.154 (a) E/R. (b) L/R.

19.157 (a) $C(\dot{x}_A - \dot{x}_m) + kx_A = 0$
$m\ddot{x}_m + c(\dot{x}_m - \dot{x}_A) = P_m \sin \omega_f t$
(b) $R(\dot{q}_A - \dot{q}_m) + (1/C)q_A = 0$
$L\ddot{q}_m + R(\dot{q}_m - \dot{q}_A) = E_m \sin \omega_f t$

19.159 0.760 lb · s² · ft, 8.66 in.

19.160 (a) 6.82 lb. (b) 33.4 lb/ft.

19.161 1.785 s.

19.162 (a) 2.28 s. (b) 1.294 m.

19.165 (a) $0.07246\ddot{\theta} + 0.3375\dot{\theta} + 1.25\theta = 0$.
(b) -19.05×10^{-6} degrees.

19.168 (a) $m\ddot{x} + 2T(2x/l) = 0$. (b) $\pi\sqrt{ml/T}$.

19.169 0.045 in.